ELSEVIER'S

DICTIONARY OF NUCLEAR SCIENCE AND TECHNOLOGY

ELSEVIER PUBLISHING COMPANY
335 JAN VAN GALENSTRAAT
P.O. BOX 211, AMSTERDAM, THE NETHERLANDS

ELSEVIER PUBLISHING CO. LTD.
BARKING, ESSEX, ENGLAND

AMERICAN ELSEVIER PUBLISHING COMPANY, INC.
52 VANDERBILT AVENUE
NEW YORK, NEW YORK 10017

Library of Congress Card Number: 72-103357

Standard Book Number: 444-40810-X

Copyright © 1970 by Elsevier Publishing Company, Amsterdam

Printed in The Netherlands

ELSEVIER'S

DICTIONARY OF NUCLEAR SCIENCE AND TECHNOLOGY

IN SIX LANGUAGES

ENGLISH/AMERICAN-FRENCH-SPANISH-ITALIAN
DUTCH AND GERMAN

Second, revised edition

Compiled and arranged on
an English alphabetical basis by

W.E.CLASON

Formerly Head of the Translation Department
N.V. Philips' Gloeilampenfabrieken
Eindhoven (The Netherlands)

ELSEVIER PUBLISHING COMPANY
AMSTERDAM / LONDON / NEW YORK
1970

FROM THE SAME PUBLISHER:

DICTIONARY OF GENERAL PHYSICS
English/American, French, Spanish, Italian, Dutch, German

DICTIONARY OF PURE AND APPLIED PHYSICS
English, German

DICTIONARY OF PURE AND APPLIED PHYSICS
German, English

NUCLEAR PHYSICS AND ATOMIC ENERGY
English, French, German, Russian

DICTIONARY OF AUTOMATION, COMPUTERS, CONTROL AND MEASURING
English/American, French, Spanish, Italian, Dutch, German

DICTIONARY OF AUTOMATIC CONTROL
English/American, French, German, Russian

LEXICON OF INTERNATIONAL AND NATIONAL UNITS
English/American, German, Spanish, French, Italian, Japanese,
Dutch, Portuguese, Polish, Swedish, Russian

Complete catalogue is available on request.

PUBLISHER'S NOTE

By the most conservative estimates, there are at least two thousand basic languages in current use today. By some definitions, this figure could reach more than five thousand languages, not including a vast number of dialects. Added to this, the incredible current advances in engineering and science have rendered most existing technical dictionaries obsolete. The accumulated result has been a fast growing need for specialized multilingual dictionaries, particularly in the face of international scientific cooperation and exchange of knowledge.

To meet this need, Elsevier publishes a number of multilingual technical dictionaries relating to special fields of science and industry. Edited under authoritative auspices, they draw upon rich sources of knowledge.

In planning this dictionary, the author and publishers have been guided by the principles proposed by the United Nations Educational, Scientific and Cultural Organization (UNESCO). The aim is to ensure that each dictionary will fit into place in a pattern which may progressively extend over all interrelated fields of science and technology and cover all necessary languages.

For each language, there is an alphabetical list of words referring to corresponding numbers in the basic table. The system of thumb-indexing enables one to find any language at once. The binding, smooth paper and convenient size result in an enjoyable and valuable reference book.

PREFACE

Nuclear science and technology, like electronics and space science and technology, is one of the areas of science that have experienced an unprecedented growth in recent years. Inevitably, this growth has been reflected in an expansion of the terminology: the number of terms has greatly increased and definitions have been reformulated.

The Editor has tried to take into account this expansion in preparing the second edition. In so far as he has been successful, this is due largely to the excellent cooperation he has received both in the Netherlands and abroad. Once again, he is most grateful to his wife for her careful preparation of the manuscript.

W. E. Clason

ABBREVIATIONS

ab	absorption	md	medicine
an	analysis	me	measuring
ar	alpha rays	mg	metallography and metallurgy
bi	biology	mi	mineralogy
br	beta rays	ms	miscellaneous applications
cd	constructional details	mt	materials
cg	cleaning	np	nuclear physics
ch	chemistry	nw	nuclear weapons
cl	cooling	pa	particle accelerators
cm	cosmic radiation	ph	photography
co	control	pp	plasma physics
cr	crystallography	qm	quantum mechanics and physics
cs	cross sections	ra	radiation
ct	counters	rt	reactor technology
ec	electrotechnics and electronics	rw	radioactive waste
fu	fuel	sa	safety appliances
ge	general	sp	spectra
gp	general physics	te	testing
gr	gamma rays	tr	transport
ic	ionization chambers	un	units
is	isotopes	vt	vacuum technique
ma	mathematics	xr	X-rays

---

adj	adjective	v	verb
f	feminine	f	François
GB	English, British usage	e	Español
m	masculine	i	Italiano
n	neuter	n	Nederlands
pl	plural	d	Deutsch
US	English, American usage		

BASIC TABLE

A

1 A BOMB,   nw
   ATOM BOMB, FISSION BOMB,
   NUCLEAR BOMB
   A nuclear weapon in which the energy is
   mainly derived from fission.
f bombe *f* atomique, bombe *f* nucléaire
e bomba *f* atómica, bomba *f* nuclear
i bomba *f* atomica, bomba *f* nucleare
n atoombom, kernbom
d Atombombe *f*, Kernbombe *f*,
   Spaltungsbombe *f*

2 ABERNATHYITE   mi
   A rare secondary mineral, containing
   52.8 % of uranium.
f abernathyite *f*
e abernatita *f*
i abernatite *f*
n abernathyiet *n*
d Abernathyit *m*

3 ABRAHAM ELECTRON MODEL   np
   A theoretical model of the electron
   proposed by Abraham in 1903.
f modèle *m* de l'électron d'Abraham
e modelo *m* del electrón de Abraham
i modello *m* dell'elettrone d'Abraham
n elektronmodel *n* volgens Abraham
d Elektronenmodell *n* nach Abraham

4 ABSOLUTE AGE   mi, np
   The age of a mineral determined by means
   of natural radioactivity.
f âge *m* absolu
e edad *f* absoluta
i età *f* assoluta
n absolute leeftijd
d absolutes Alter *n*

5 ABSOLUTE CALIBRATION   co, me, rt
   A calibration carried out to find the
   reactivity change caused by a control rod
   movement of one effective inch.
f étalonnage *m* absolu,
   étalonnement *m* absolu
e calibración *f* absoluta
i taratura *f* assoluta
n absolute ijking
d absolute Eichung *f*

6 ABSOLUTE DISINTEGRATION RATE   np
   Represented by n = $F_a F_b F_c F_d$ N, where
   N is the observed counting rate obtained
   experimentally and F_a, F_b, F_c and F_d
   are the degradation factors due to
   efficiency of the extraction process,
   self-absorption in the source, the geometry
   of the measuring arrangement and the
   detector efficiency respectively.
f taux *m* absolu de désintégration
e velocidad *f* absoluta de desintegración
i velocità *f* assoluta di disintegrazione

n absoluut desintegratietempo *n*
d absolute Zerfallsrate *f*

7 ABSORBED DOSE   ab, np, ra
   At a point of interest, the energy imparted
   by ionizing radiation to unit mass of
   irradiated material.
f dose *f* absorbée
e dosis *f* absorbida
i dose *f* assorbita
n dosis, geabsorbeerde dosis
d Energiedosis *f*

8 ABSORBED DOSE RATE   ra
   The absorbed dose per unit time.
f débit *m* de dose absorbée
e dosis *f* absorbida por unidad de tiempo,
   intensidad *f* de dosis absorbida
i intensità *f* di dose assorbita,
   tasso *m* di dose assorbita
n absorptiedoseringssnelheid,
   geabsorbeerde-dosistempo *n*
d Energiedosisleistung *f*

9 ABSORBENT,   abs
   ABSORBER
   Any material that absorbs or stops
   ionizing radiation.
f absorbant *m*, absorbeur *m*
e absorbedor *m*, absorbente *m*
i assorbente *m*, assorbitore *m*
n absorbens *n*, absorptiemiddel *n*
d Absorbens *n*, Absorptionsmittel *n*

10 ABSORBING ROD,   eo, rt
   CONTROL ROD
   A control member in the form of a rod.
f barre *f* de commande
e barra *f* de regulación
i barra *f* di regolazione
n regelstaaf
d Steuerstab *m*

11 ABSORPTION   ab
   A phenomenon in which incident radiation
   transfers some or all of its energy to the
   matter which it traverses.
f absorption *f*
e absorción *f*
i assorbimento *m*
n absorptie
d Absorption *f*

12 ABSORPTION ANALYSIS   ab, an
   Analysis method of an element in a
   substance by measuring the flux density
   of monoenergetic neutrons, before and
   after their passing through the substance.
f analyse *f* par absorption
e análisis *f* por absorción
i analisi *f* per assorbimento
n absorptieanalyse
d Absorptionsanalyse *f*

13 ABSORPTION BAND OF A   ab, sp
SCINTILLATING MATERIAL
The energy (or wavelength) band of the
photons which are most likely to be
absorbed by the scintillating material
(a scintillator).
f bande *f* d'absorption d'une matière
scintillante
e banda *f* de absorción de un material
de centelleo
i banda *f* d'assorbimento d'un materiale
scintillante
n absorptieband van een scintillerend
materiaal
d Absorptionsbande *f* eines szintillierendes
Materials

14 ABSORPTION COEFFICIENT,   ab
ENERGY ABSORPTION COEFFICIENT,
REAL ABSORPTION COEFFICIENT
Of a substance, for a parallel beam of
specified radiation, the quantity μ_{abs}
in the expression $\mu_{abs}\Delta_x$ for the
fraction absorbed in passing through
a thin layer of thickness Δ_x of that
substance.
f coefficient *m* d'absorption
e coeficiente *m* de absorción
i coefficiente *m* d'assorbimento
n absorptiecoëfficiënt
d Absorptionskoeffizient *m*

15 ABSORPTION CONTROL   co, rt
Control of a nuclear reactor by adjustment of the properties, position or
quantity of neutron absorbing material,
other than fuel, moderator and reflector
material, in such a way as to change the
reactivity.
f commande *f* par absorption
e regulacion *f* por absorción
i regolazione *f* per assorbimento
n regeling door absorptie
d Steuerung *f* durch Absorption

16 ABSORPTION CROSS SECTION   np
The sum of the cross sections for all
neutron reactions with an atom except
elastic and inelastic collisions.
f section *f* efficace d'absorption
e sección *f* eficaz de absorción
i sezione *f* d'urto d'assorbimento
n absorptiedoorsnede
d Absorptionsquerschnitt *m*

17 ABSORPTION CURVE,   ab
ATTENUATION CURVE,
TRANSMISSION CURVE
A graph of the intensity of the transmitted
radiation plotted against the absorber
thickness.
f courbe *f* d'absorption,
courbe *f* d'atténuation,
courbe *f* de transmission
e curva *f* de absorción,
curva *f* de atenuación,
curva *f* de transmisión
i curva *f* d'assorbimento,

curva *f* d'attenuazione,
curva *f* di trasmissione
n absorptiekromme
d Absorptionskurve *f*, Schwächungskurve *f*

18 ABSORPTION DISCONTINUITY,   ab
ABSORPTION EDGE,
ABSORPTION LIMIT
The X-energy or wavelength characteristic
of a given electron energy level in an atom
of a specified element at which an absorption discontinuity occurs.
f discontinuité *f* d'absorption
e discontinuidad *f* de absorción
i limite *m* d'assorbimento
n absorptiekant
d Absorptionskante *f*

19 ABSORPTION EXTRACTION   ab, ch
The operation in chemical engineering
whereby extraction is achieved by means
of absorption.
f extraction *f* par absorption
e extracción *f* por absorción
i estrazione *f* per assorbimento
n absorptie-extractie
d Absorptionsextraktion *f*

20 ABSORPTION FACTOR   ab
In any absorbing system the ratio of the
total unabsorbed radiation to the total
incident radiation, or to the total radiation
transmitted in the absence of the absorbing
substance.
f facteur *m* d'absorption
e factor *m* de absorción
i fattore *m* d'assorbimento
n absorptiefactor
d Absorptionsfaktor *m*

21 ABSORPTION LENGTH   np
A length, taken parallelly to the average
propagation direction of a given type of
particle in a given medium, at the end of
which the volume number of the particles
decreases in average of a factor e.
f longueur *f* d'absorption
e longitud *f* de absorción
i lunghezza *f* d'assorbimento
n absorptielengte
d Absorptionslänge *f*

22 ABSORPTION SPECTRUM   sp
The spectrum of which the lines or bands
correspond to the photon absorption by
the substance under examination.
f spectre *m* d'absorption
e espectro *m* de absorción
i spettro *m* d'assorbimento
n absorptiespectrum *n*
d Absorptionsspektrum *n*

23 ABSORPTIVE POWER,   ab
ABSORPTIVITY
A mathematical expression of the capacity
of a substance to absorb another substance
or form of energy.
f pouvoir *m* absorbant

e   poder *m* absorbente
i   potere *m* assorbente
n   absorptievermogen *n*
d   Absorptionsvermögen *n*

24 ABUKUMALITE   mi
    A mineral, related to apatite and containing
    a minor amount, about 0.8 %, of thorium.
f   abukumalite *f*
e   abukumalita *f*
i   abukumalite *f*
n   abukumaliet *n*
d   Abukumalit *m*

25 ABUNDANCE,   is
      CONCENTRATION
    The number of atoms of a particular
    isotope in a mixture of the isotopes of an
    element, expressed as a fraction of all
    the atoms of the element.
f   abondance *f* ·
e   abundancia *f*
i   abbondanza *f*
n   abondantie
d   Häufigkeit *f*

26 ABUNDANCE RATIO,   is
      ISOTOPIC RATIO,
      RELATIVE ABUNDANCE
    The ratio of the number of atoms of one
    isotope to the number of atoms of another
    isotope of the same element in a given
    sample.
f   rapport *m* des teneurs isotopiques
e   abundancia *f* isotópica relativa
i   abbondanza *f* isotopica relativa,
      ricchezza *f* isotopica
n   isotopenverhouding
d   Isotopenhäufigkeitsverhältnis *n*

27 ACCELERATING CHAMBER   pa
    An evacuated glass, metal or ceramic
    envelope in which charged particles
    are accelerated.
f   chambre *f* d'accélération
e   cámara *f* de aceleración
i   camera *f* d'accelerazione,
      spazio *m* d'accelerazione
n   versnellingskamer, versnellingsvat *n*
d   Beschleunigungskammer *f*,
      Beschleunigungsraum *m*

28 ACCELERATING PERIOD   gen, pa
    The period of time in which particle
    acceleration takes place.
f   durée *f* d'accélération
e   duración *f* de aceleración
i   durata *f* d'accelerazione
n   versnellingsduur
d   Beschleunigungszeitraum *m*

29 ACCELERATING TUBE   pa
    A tubular accelerating chamber.
f   tube *m* accélérateur
e   tubo *m* acelerador
i   tubo *m* acceleratore
n   versnelbuis
d   Beschleunigungsrohr *n*

30 ACCELERATION   gen, pa
    The time rate of change of velocity.
f   accélération *f*
e   aceleración *f*
i   accelerazione *f*
n   versnelling
d   Beschleunigung *f*

31 ACCELERATION GAP,   pa
      ACCELERATION SLIT
    The space between two field-free sections
    of a particle accelerator.
f   fente *f* d'accélération
e   ranura *f* de aceleración
i   fessura *j* d'accelerazione
n   versnellingsspleet
d   Beschleunigungsspalt *m*,
      Beschleunigungsstrecke *f*

32 ACCELERATOR,   pa
      ATOM SMASHER
    A device for imparting large amounts of
    kinetic energy to charged particles such as
    electrons, deuterons, etc.
f   accélérateur *m*
e   acelerador *m*
i   acceleratore *m*
n   accelerator, versneller
d   Beschleuniger *m*,
      Beschleunigungsvorrichtung *f*

33 ACCELERATOR APERTURE,   pa
      APERTURE OF AN ACCELERATOR
    In an accelerator, the vertical or horizontal
    distance available to the passing of
    particles.
f   ouverture *f* d'accélérateur
e   abertura *f* de acelerador
i   apertura *f* d'acceleratore
n   versnellerapertuur
d   Beschleunigerapertur *f* ·

34 ACCELERATOR FOCUSING   pa
    The action of electric and/or magnetic
    fields in directing accelerated particles
    along desired paths.
f   focalisation *f* d'accélérateur
e   enfoque *m* de acelerador
i   focalizzazione *f* d'acceleratore
n   focussering van een versneller
d   Fokussierung *f* eines Beschleunigers

35 ACCELERATOR RELATIVE   pa
      APERTURE
    The ratio of the minimum vertical or
    horizontal clearance for particle passage
    to the particle orbit radius.
f   ouverture *f* relative d'accélérateur
e   abertura *f* relativa de acelerador
i   apertura *f* relativa d'acceleratore
n   relatieve versnellerapertuur
d   relative Beschleunigerapertur *f*

36 ACCEPTOR   ec
    In a semiconductor, an imperfection which
    permits hole conduction by the acceptance
    of electrons.
f   accepteur *m*

e   aceptor *m*
i   accettore *m*
n   acceptor
d   Akzeptor *m*

37 ACCEPTOR ENERGY LEVEL,   ec
    ACCEPTOR LEVEL
    In the energy diagram of an extrinsic
    semiconductor, an intermediate level
    close to the normal band.
f   niveau *m* accepteur
e   nivel *m* aceptor
i   livello *m* accettore
n   acceptorniveau *n*
d   Akzeptorniveau *n*

38 ACCEPTOR IMPURITY,   ec
    ACCEPTOR TYPE IMPURITY
    In a semiconductor an impurity which
    may induce hole induction.
f   impureté *f* type accepteur,
    impureté *f* type P
e   impureza *f* aceptador
i   impurità *f* d'accettore
n   acceptorverontreiniging
d   Akzeptorverunreinigung *f*

39 ACCESSORY   ge
    A part, sub-assembly or assembly that
    contributes to the effectiveness of a
    piece of equipment without changing its
    basic function.
f   accessoire *m*
e   accesorio *m*
i   accessorio *m*
n   onderdeel *n*
d   Zubehörstück *n*, Zubehörteil *m*

40 ACCIDENTAL COINCIDENCE,   ct
    CHANCE COINCIDENCE,
    RANDOM COINCIDENCE
    A coincidence due to the fortuitous
    occurrence of unrelated counts in the
    separate detectors.
f   coïncidence *f* accidentelle,
    coïncidence *f* fortuite
e   coincidencia *f* accidental
i   coincidenza *f* accidentale
n   toevallige coïncidentie
d   Zufallskoinzidenz *f*

41 ACCIDENTAL COINCIDENCE   ct
    CORRECTION
    The correction made in coincidence
    counting to take into account the chance
    occurrence of unrelated signals within
    the resolving time of the apparatus.
f   correction *f* de la coïncidence acciden-
    telle
e   corrección *f* de la coincidencia accidental
i   correzione *f* della coincidenza
    accidentale
n   correctie voor toevallige coïncidentie
d   Korrektion *f* für Zufallskoinzidenz

42 ACCIDENTAL ERROR,   me
    RANDOM ERROR
    In repeated observations of a quantity which

is in principle constant, it is in general found
that slightly different values are obtained.
f   erreur *f* aléatoire
e   error *m* aleatorio, error *m* fortuito
i   errore *m* aleatorio, errore *m* casuale
n   toevallige fout, toevalsfout
d   Zufallsfehler *m*

43 ACCIDENTAL EXPOSURE   ra
    An exposure in excess of the maximum
    admissible values laid down in health
    physics regulations.
f   exposition *f* accidentelle
e   exposiciòn *f* accidental
i   esposizione *f* accidentale
n   onvoorziene exposie
d   nichtvorgesehene Exposition *f*

44 ACCIDENTAL HIGH EXPOSURE,   ra
    HIGH-LEVEL SINGLE EXPOSURE
    An exposure which leads to a non-admissible
    dose so high that it can be borne only once
    by the exposed person.
f   exposition *f* élevée accidentelle
e   exposición *f* elevada accidental
i   esposizione *f* elevata accidentale
n   onvoorziene ontoelaatbare exposie
d   nichtvorgesehene nichtzugelassene
    Exposition *f*

45 ACCIDENTAL HIGH EXPOSURE   ra
    TO EXTERNAL RADIATION
    Accidental exposure in excess of the maxi-
    mum admissible values laid down in
    health physics regulations.
f   exposition *f* externe élevée exceptionnelle
    non-concertée
e   exposición *f* externa elevada excepcional
    no planeada
i   esposizione *f* esterna elevata eccezionale
    non concordata
n   onvoorziene uitwendige ontoelaatbare exposie
d   nichtvorgesehene äussere nichtzugelassene
    Exposition *f*

46 ACCIDENTAL HIGH EXPOSURE TO   ra
    RADIOACTIVE MATERIALS
    Accidental exposure resulting in contami-
    nation of some inner organs.
f   contamination *f* interne exceptionnelle
    non-concertée
e   contaminación *f* interna no planeada
i   contaminazione *f* interna non concordata
n   onvoorziene interne besmetting
d   nichtvorgesehene innere Kontamination *f*

47 ACCUMULATOR TANK,   rw
    STORAGE TANK
    Tank of stainless steel in the interior of
    which a neoprene rubber bag is arranged
    to collect fission gases.
f   cuve *f* collectrice,
    réservoir *m* de gaz de fission
e   tanque *m* colector
i   serbatoio *m* collettore
n   verzameltank
d   Auffanggefäss *n*,
    Sammelbehälter *m*

48 ACCURACY IN MEASUREMENT   me
A number indicating the maximum error
of an instrument.
f précision f de mesure
e precisión f de medida
i precisione f di misura
n meetnauwkeurigheid
d Messgenauigkeit f

49 ACETYLENE IONIZATION   ic
     CHAMBER
An ionization chamber of which the
gas-filling consists of acetylene (C_2H_2).
f chambre f d'ionisation à acétylène
e cámara f de ionización con acetileno
i camera f d'ionizzazione ad acetilene
n met acetyleen gevuld ionisatievat n
d azetylengefüllte Ionisationskammer f

50 ACTINIDES   ch
The name by analogy with lanthanides, for
the elements of atomic numbers 89 to 103.
f actinides pl
e actínidos pl
i attinidi pl
n actiniden pl
d Aktiniden pl

51 ACTINIUM   ch
A radioactive element, symbol Ac,
atomic number 89.
f actinium m
e actinio m
i attinio m
n actinium n
d Aktinium n

52 ACTINIUM FAMILY,   ch, np
     ACTINO-URANIUM SERIES,
       4 n + 3 SERIES
The series of nuclides resulting from
the decay of U^{235}.
f famille f de l'actinium,
   famille f de l'actino-uranium
e familia f del actinio
i famiglia f attinica
n actiniumreeks
d Aktiniumzerfallsreihe f,
   Uran-Aktiniumzerfallsreihe f

53 ACTINODERMATITIS,   md, xr
     RADIODERMATITIS,
     X-RAY DERMATITIS
An inflammation of the skin produced by
excessive exposure to X-rays or to rays
emitted by radioactive substances.
f actinodermatose f, dermite f des rayons
   X, radiodermite f
e actinodermatitis f,
   dermatitis f radiográfica,
   radiodermatitis f
i attinodermatite f, dermatosi f attinica,
   radiodermatite f
n huidontsteking tengevolge van bestraling,
   radiodermatitis
d Radiodermatitis f,
   Röntgenhautentzündung f

54 ACTINON   ch
The isotope of the element radon having
a mass number 219.
f actinon m
e actinón m
i actinon m, attinon m
n actinon n
d Aktinon n

55 ACTINO-URANIUM,   np
     URANIUM-235
A common name for $8.810^8 yU^{235}$, the
natural parent of the actinium series.
f actino-uranium m, uranium-235 m
e actinouranio m, uranio-235 m
i attinouranio m, uranio-235 m
n actino-uranium n, uranium-235 n
d Aktinouran n, Uran-235 n

56 ACTION AT A DISTANCE   np
Theory of the interaction between charged
particles in which the electromagnetic
fields do not appear explicitly
f interaction f à distance
e interacción f de distancia
i interazione f a distanza
n afstandswisselwerking
d Abstandswechselwirkung f

57 ACTION INTEGRAL   ma
For a generalized co-ordinate q_i and its
conjugate momentum p_i, the phase integral
$J_i = \int p_i dq_i$, is called the action integral.
f intégrale f d'activation
e integral f de activación
i integrale m d'attivazione
n werkingsintegraal
d Wirkungsintegral n

58 ACTION VARIABLE   ma
If one of the momenta of a classical
dynamical system yields a closed curve
when plotted against the conjugate
co-ordinate, the area contained within this
curve is the action variable corresponding
to this degree of freedom.
f variable f d'activation
e variable f de activación
i variabile f d'attivazione
n werkingsvariabele
d Wirkungsvariable f

59 ACTIVATED CARBON,   ab, mt
     ACTIVATED CHARCOAL
Used as an absorbing material in
krypton-85 sources.
f charbon m actif
e carbón m activado
i carbone m attivo
n actieve kool, geactiveerde houtskool
d Aktivkohle f

60 ACTIVATED MOLECULE   ch, np
A molecule with one or more of its atoms
ionized.
f molécule f activée
e molécula f activada

i molecola *f* attivata
n geactiveerde molecule
d angeregtes Molekül *n*

61 ACTIVATED WATER ra
The passage of ionizing radiation through
water produces, temporarily, ions, atoms,
radicals or molecules in a chemically
reactive state. The combined effect of
all such entities is said to be due to
activated water.
f eau *f* activée
e agua *f* activada
i acqua *f* attivata
n geactiveerd water *n*
d aktiviertes Wasser *n*

62 ACTIVATION ra
The process of inducing radioactivity
by irradiation.
f activation *f*
e activación *f*
i attivazione *f*
n activering
d Aktivierung *f*

63 ACTIVATION ANALYSIS, ch, ra
RADIOACTIVATION ANALYSIS
A method of chemical analysis based on
the identification and measurement of
characteristic radiation of nuclides
formed by irradiation.
f analyse *f* par activation
e análisis *f* por activación
i analisi *f* per attivazione
n activeringsanalyse
d Aktivierungsanalyse *f*

64 ACTIVATION CROSS SECTION cs
The cross section for the formation of a
radionuclide by a specified interaction.
f section *f* efficace d'activation
e sección *f* eficaz de activación
i sezione *f* d'urto d'attivazione
n activeringsdoorsnede
d Aktivierungsquerschnitt *m*

65 ACTIVATION CURVE is
A curve which in function of time gives the
activity of an element contained in a
substance subject to suitable irradiation
or bombardment.
f courbe *f* d'activation
e curva *f* de activación
i curva *f* d'attivazione
n activeringskromme
d Aktivierungskurve *f*

66 ACTIVATION DETECTOR ra
Radiation detector enabling the measurement of the neutron flux density in connection with the activity induced by these
neutrons in a substance.
f détecteur *m* par activation
e detector *m* por activación
i rivelatore *m* per attivazione
n activeringsdetector
d Aktivierungsdetektor *m*

67 ACTIVATION ENERGY, np
INTENSITY OF ACTIVATION
The excess energy over the ground state
which must be acquired by an atomic system
in order that a particular process may
occur.
f énergie *f* d'activation
e energía *f* de activación
i energia *f* d'attivazione
n activeringsenergie
d Aktivierungsenergie *f*

68 ACTIVATION METHOD OF CROSS np
SECTION DETERMINATION
The determination of the thermal neutron
absorption cross section of a substance
by means of the measurement of the radioactivity of the product formed by the
neutron absorption.
f détermination *f* de la section efficace par
la méthode d'activation
e determinación *f* de la sección eficaz por
el método de activación
i determinazione *f* della sezione d'urto
per il metodo d'attivazione
n werkzame-doorsnedebepaling met behulp
van de activeringsmethode
d Wirkungsquerschnittbestimmung *f* mittels
Aktivierungsverfahren

69 ACTIVATOR ch
A chemical component added in very small
quantities to certain scintillators for the
essential purpose of increasing the intensity of the light emitted, and sometimes
also to change its wavelength.
f activateur *m*
e activador *m*
i attivatore *m*
n activator
d Aktivator *m*

70 ACTIVE np, rt
1. Fissile (fissionable).
2. Radioactive.
f actif adj
e activo adj
i attivo adj
n actief adj
d aktiv adj

71 ACTIVE AREA, sa
CONTROLLED AREA,
RESTRICTED AREA
The area or region where activity actually
exists owing to radiation or contamination.
f zone *f* contrôlée
e zona *f* controlada
i zona *f* controllata
n gecontroleerd gebied *n*
d überwachtes Gebiet *n*

72 ACTIVE BY-PRODUCT, np
RADIOACTIVE BY-PRODUCT
A by-product created during a nuclear
reaction process.
f sous-produit *m* radioactif
e subproducto *m* radiactivo

i   sottoprodotto *m* radioattivo
n   radioactief bijprodukt *n*
d   radioaktives Nebenprodukt *n*

73 ACTIVE CARBON,   ch
    CARBON-14, RADIOACTIVE CARBON
    An isotope of carbon.
f   carbone-14 *m*, carbone *m* radioactif
e   carbono-14 *m*, carbono *m* radiactivo
i   carbonio-14 *m*, carbonio *m* radioattivo
n   koolstof-14, radioactieve koolstof
d   Kohlenstoff-14 *m*,
    radioaktiver Kohlenstoff *m*

74 ACTIVE CORE   rt
    The space in which chain fissions can be
    produced.
f   espace *m* multiplicateur,
    milieu *m* multiplicateur
e   espacio *m* multiplicador
i   spazio *m* moltiplicatore
n   vermenigvuldigingsruimte
d   Vermehrungsraum *m*

75 ACTIVE DEPOSIT,   np
    RADIOACTIVE DEPOSIT
    1. The radioactive decay products deposited
    on a surface exposed to radon, actinon
    or thoron gas.
    2. Any radioactive decay products depo-
    sited on a surface exposed to a radio-
    active gas.
    3. Any radioactive material deposited on
    a surface.
f   dépôt *m* actif, dépôt *m* radioactif
e   depósito *m* activo, depósito *m* radiactivo
i   deposito *m* attivo, deposito *m* radioattivo
n   actieve neerslag, radioactieve neerslag
d   aktiver Niederschlag *m*,
    radioaktiver Niederschlag *m*

76 ACTIVE DUST,   np, nw
    RADIOACTIVE DUST
    Dust deposited from the air or any other
    source upon the earth.
f   poussière *f* radioactive
e   polvo *m* radiactivo
i   polvere *f* radioattiva
n   radioactief stof *n*
d   radioaktiver Staub *m*

77 ACTIVE EFFLUENT,   rw
    RADIOACTIVE EFFLUENT
    Any solid, liquid or gaseous radioactive
    waste material discharged from a system.
f   effluent *m* radioactif
e   efluente *m* radiactivo
i   scarico *m* attivo
n   radioactieve afvoer,
    radioactieve afvoerstroom
d   aktive Ausströmung *f*,
    aktiver Ausfluss *m*

78 ACTIVE EFFLUENT DISPOSAL,   rw
    RADIOACTIVE EFFLUENT DISPOSAL
    In nuclear power plants, the disposal of
    contaminated radioactive waste liquid.

f   élimination *f* des déchets radioactifs
    liquides
e   eliminación *f* de los desechos radiactivos
    líquidos
i   disposizione *f* dei rifiuti radioattivi
    liquidi
n   verwijdering van vloeibaar radioactief
    afval
d   Beseitigung *f* von flüssigem radioaktivem
    Abfall

79 ACTIVE EFFLUENT DRAIN PIPE,   rw
    RADIOACTIVE EFFLUENT DRAIN PIPE
f   tuyau *m* d'écoulement d'eau usée radio-
    active
e   alcantarilla *f* de agua efluente radiactiva
i   condotto *m* scaricatore d'acqua radioattiva
n   afvoerleiding voor radioactief afvalwater
d   Abflussleitung *f* für radioaktives Abwasser

80 ACTIVE EFFLUENT PLANT AREA,   rw
    RADIOACTIVE EFFLUENT PLANT
    AREA
    That part of a nuclear reactor installation
    where the first treatment is given to the
    active effluent.
f   station *f* de traitement d'effluents radio-
    actifs
e   instalación *f* de tratamiento de efluentes
    radiactivos
i   impianto *m* di trattamento di scarichi
    radioattivi
n   installatie voor de verwerking van de
    radioactieve afvoer
d   Anlage *f* zur Verarbeitung des aktiven
    Ausflusses

81 ACTIVE ELEMENT,   ch, np
    RADIOACTIVE ELEMENT,
    RADIOELEMENT
    An element which possesses one or more
    radioactive isotopes.
f   élément *m* actif, élément *m* radioactif
e   elemento *m* activo, elemento *m* radiactivo
i   elemento *m* attivo, elemento *m* radioattivo
n   actief element *n*, radioactief element *n*
d   aktives Element *n*, radioaktives Element *n*

82 ACTIVE FALL-OUT,   nw
    FALL-OUT, RADIOACTIVE FALL-OUT
    The deposition on the ground of radioactive
    substances resulting from the explosion
    of a nuclear weapon.
f   retombées *pl* radioactives
e   depósito *m* radiactivo, poso *m* radiactivo
i   ricaduta *f*, ricaduta *f* radioattiva
n   radioactieve neerslag
d   radioaktiver Ausfall *m*,
    radioaktiver Niederschlag *m*

83 ACTIVE FISSION PRODUCT,   np
    RADIOACTIVE FISSION PRODUCT
    A radioactive product produced during a
    nuclear fission process.
f   produit *m* de fission radioactif
e   producto *m* de fisión radiactivo
i   prodotto *m* di fissione radioattivo

n radioactief splijtprodukt *n*
d radioaktives Spaltprodukt *n*

84 ACTIVE LATTICE   rt
The core of a lattice reactor, in which
fissile (fissionable) and non fissile (non
fissionable) materials are arranged in a
regular pattern.
f réseau *m* multiplicateur
e celosía *f* espacial activa
i reticolo *m* attivo
n reactorrooster *n*
d Spaltzone *f*

85 ACTIVE LAUNDRY   cg, sa
An installation for cleaning and washing
contaminated clothing, etc.
f blanchisserie *f* de décontamination
e lavadero *m* de decontaminación
i lavanderia *f* di decontaminazione
n ontsmettingswasserij
d Dekontaminationswäscherei *f*

86 ACTIVE LAUNDRY   cg, sa
The clothes and other textile materials
worn by workers exposed to radiation.
f linge *m* radioactif
e ropa *f* radiactiva
i bucato *m* radioattivo
n besmet wasgoed *n*, radioactief wasgoed *n*
d radioaktive Wäsche *f*

87 ACTIVE LOOP,   rt
HOT LOOP
In a reactor, a piping system through
which a fluid may flow for experimental
purposes, and which also contains fissile
(fissionable) material.
f boucle *f* active
e lazo *m* activo
i cappio *m* attivo
n actieve kringloop, actieve lus
d aktiver Kreislauf *m*

88 ACTIVE MATERIAL,   np
RADIOACTIVE MATERIAL
A material of which one or more constituents exhibit radioactivity.
f matière *f* radioactive
e material *m* radiactivo
i materiale *m* radioattivo
n radioactief materiaal *n*
d radioaktiver Stoff *m*

89 ACTIVE POISON,   rt
RADIOACTIVE POISON
Parasitic, radioactive absorber of thermal
neutrons and acting to poison the chain
reaction.
f poison *m* radioactif
e veneno *m* radiactivo
i veleno *m* radioattivo
n radioactief gif *n*
d radioaktives Gift *n*

90 ACTIVE PRODUCT,   np
RADIOACTIVE PRODUCT
A radioactive decay product of a radio-

nuclide.
f produit *m* de décroissance radioactif
e producto *m* de desintegración radiactivo
i prodotto *m* di decadimento radioattivo
n radioactief vervalprodukt *n*
d radioaktives Zerfallsprodukt *n*

91 ACTIVE SAMPLING EQUIPMENT,   an, ch
RADIOACTIVE SAMPLING
EQUIPMENT
Equipment for collecting active samples
inside an active area and transporting it
outside towards an analyzing laboratory.
f installation *f* d'échantillonnage actif
e equipo *m* de muestreo activo
i impianto *m* di prelievo di campioni
attivi
n inrichting voor het nemen van radioactieve monsters
d Einrichtung *f* zur aktiven Probenahme

92 ACTIVE SECTION,   rt
ACTIVE ZONE
The part of a nuclear reactor in which the
reproduction factor, aside from leakage,
is unity or greater, so that the chain
reaction is potentially self-sustaining.
f zone *f* active
e zona *f* activa
i zona *f* attiva
n actieve ruimte
d aktive Zone *f*

93 ACTIVE SOURCE   ra, sp
Any device which supplies some extensive
active entity such as electric charge,
particles, matter, etc.
f source *f* active
e fuente *f* activa
i sorgente *f* attiva
n actieve bron
d aktive Quelle *f*

94 ACTIVE SOURCE AREA   ra, sp
The surface covered by an active source.
f aire *f* de source active
e área *f* de fuente activa
i area *f* di sorgente attiva
n actief bronoppervlak *n*
d aktive Quellenoberfläche *f*

95 ACTIVE WASTE,   ge
RADIOACTIVE WASTE
Useless radioactive material obtained in
the processing or handling of radioactive
materials.
f déchets *pl* radioactifs
e desechos *pl* radiactivos
i rifiuti *pl* radioattivi
n radioactief afval *n*
d radioaktive Abfälle *pl*

96 ACTIVE WASTE HANDLING BAY,   rw
RADIOACTIVE WASTE HANDLING
BAY
f salle *f* de traitement de déchets radioactifs
e sala *f* de tratamiento de desechos
radiactivos

i   sala *f* di trattamento di rifiuti radioattivi
n   verwerkingsruimte van radioactief afval
d   Verarbeitungsraum *m* für radioaktive
    Abfälle

97 ACTIVE WASTE INCINERATOR,   nw
    RADIOACTIVE WASTE INCINERATOR
    Incinerator used to burn active waste in
    order to reduce its volume.
f   incinérateur *m* de déchets radioactifs
e   incinerador *m* de desechos radiactivos
i   inceneratore *m* di rifiuti radioattivi
n   radioactief-afvalverbrander
d   Verbrennungsofen *m* für radioaktive
    Abfälle

98 ACTIVE WATER,   rw
    RADIOACTIVE WATER
    Water that has been activated by the waste
    products of a nuclear process.
f   eau *f* radioactive
e   agua *f* radiactiva
i   acqua *f* radioattiva
n   radioactief water *n*
d   radioaktives Wasser *n*

99 ACTIVE WATER HOMING,   ms
    RADIOACTIVE WATER HOMING
    Homing on the trail of radioactive sea
    water left by a nuclear-powered submarine.
f   ralliement *m* par eau radioactive
e   recalada *f* por agua radiactiva
i   allineamento *m* per acqua radioattiva
n   basisaansturing met behulp van radio-
    actief water
d   Zielansteuerung *f* mittels radioaktiven
    Wassers

100 ACTIVITY,   np
    NUCLEAR ACTIVITY
    The number of nuclear disintegrations
    per unit of time.
f   activité *f*, activité *f* nucléaire
e   actividad *f*, actividad *f* nuclear
i   attività *f*, attività nucleare
n   activiteit, kernactiviteit
d   Aktivität *f*, Kernaktivität *f*

101 ACTIVITY BUILD-UP   rt
    The phenomenon in which the level of
    radioactivity in a material or in an
    ambiance is enhanced by any reason.
f   augmentation *f* d'activité
e   aumento *m* de actividad
i   accumulo *m* d'attività
n   activiteitsverhoging
d   Aktivitätszunahme *f*

102 ACTIVITY COEFFICIENT   np
    The ratio of the thermodynamic activity
    to the true concentration of the substance.
f   coefficient *m* d'activité
e   coeficiente *m* de actividad
i   coefficiente *m* d'attività
n   activiteitscoëfficiënt
d   Aktivitätskoeffizient *m*

103 ACTIVITY CONCENTRATION,   is, np
    NUCLEAR ACTIVITY CONCEN-
    TRATION
    The activity of a material divided by its
    volume.
f   activité *f* nucléaire volumique
e   actividad *f* nuclear de volumen,
    concentración *f* de actividad
i   attività *f* nucleare di volume,
    concentrazione *f* d'attività
n   activiteitsconcentratie,
    volumieke kernactiviteit
d   Aktivitätskonzentration *f*,
    räumliche Kernaktivität *f*

104 ACTIVITY CURVE,   np
    DECAY CURVE,
    DISINTEGRATION CURVE
    A curve, usually derived from experimental
    data, showing the activity of a radioactive
    source as a function of time.
f   courbe *f* d'activité,
    courbe *f* de décroissance
e   curva *f* de actividad,
    curva *f* de decaimiento
i   curva *f* d'attività,
    curva *f* di decadimento
n   desintegratiekromme, vervalskromme
d   Aktivitätskurve *f*, Zerfallskurve *f*

105 ACTIVITY-MASS FORMULA   ma
    A formula for computing the activity
    of a given mass of a radioisotope.
f   formule *f* activité-masse
e   fórmula *f* actividad-masa
i   formula *f* attività-massa
n   activiteit-massa-formule
d   Aktivität-Masse-Formel *f*

106 ACTIVITY METER,   me
    RADIOACTIVITY METER
    An assembly designed to measure the
    activity of a radiation emitter and
    equipped with an indicating and/or recor-
    ding instrument.
f   activimètre *m*, radioactivimètre *m*
e   activímetro *m* de radiaciones
i   attivimetro *m*
n   activiteitsmeter, radioactiviteitsmeter
d   Gerät *n* zur Bestimmung der Aktivität,
    Aktivitätsmesser *m*

107 ACTIVITY METER WITH   me
    AUTOMATIC CHANGER
    An activity meter including an automatic
    sample changer in front of its detector,
    the présentation and the measure being
    carried out according to a determined
    program(me).
f   activimètre *m* à passeur automatique
e   activímetro *m* de cambiador automático,
    activímetro *m* de cambio automático
i   attivimetro *m* a cambia-campioni
    automatico,
    attivimetro *m* a scambiatore automatico
n   activiteitsmeter met automatische wisse-
    laar,

telopstelling met automatische preparaat-
wisselaar
d Aktivitätsmesser *m* mit selbsttätigem
Probenwechsler

108 ACTIVITY OF A SOURCE, ra
SOURCE ACTIVITY
The total amount of radioactivity contained in a source, expressed in curies
or millicuries.
f activité *f* de source
e actividad *f* de fuente
i attività *f* di sorgente
n bronactiviteit
d Quellenaktivität *f*

109 ACTIVITY RANGE is
The range over which a labelled (tagged)
compound is active.
f plage *f* d'activité
e campo *m* de actividad
i campo *m* d'attività
n activiteitsgebied *n*
d Aktivitätsbereich *m*

110 ACTUAL VALUE, co
INSTANTANEOUS VALUE
The value of a sinusoidal or otherwise
varying quantity at a particular instant.
f valeur *f* instantanée, valeur *f* momentanée
e valor *m* efectivo, valor *m* instantáneo
i valore *m* effettivo, valore *m* istantaneo
n werkelijke waarde
d Istwert *m*

111 ACUTE EXPOSURE ra
Irradiation over a short period of time,
so that any long term recovery process
does not reduce the effect of the radiation.
f exposition *f* aiguë
e exposición *f* aguda
i esposizione *f* acuta
n korte bestraling
d kurzzeitige Bestrahlung *f*

112 ADAMITE mi
A zinc arsenate, reported to contain
small amounts of uranium.
f adamite *f*
e adamita *f*
i adamite *f*
n adamiet *n*
d Adamit *m*

113 ADDED FILTER xr
Filter placed outside the röntgen tube
enclosure.
f filtre *m* additionnel
e filtro *m* adicional
i filtro *m* addizionale
n extra filter *n*
d Zusatzfilter *m*

114 ADDITIVITY ra
Refers to doses of different types of
radiation or radiation from different
sources which are received by the same
volume or tissue.

f additivité *f*
e aditividad *f*
i additività *f*
n additiviteit
d Additivität *f*

115 ADIABATIC APPROXIMATION ma
The assumption that the electronic wave
functions in a molecule or solid are distorted by the motion of the nuclei, but in
such a way that their energy is a function
only of the nuclear configuration at a
given moment and does not depend on the
rate at which the nuclei are moving.
f approximation *f* adiabatique
e aproximación *f* adiabática
i approssimazione *f* adiabatica
n adiabatische benadering
d adiabatische Näherung *f*

116 ADIABATIC COMPRESSION pp
For a plasma in a magnetic field a
compression sufficiently slow so that the
magnetic moment of the plasma particles
may be considered constant.
f compression *f* adiabatique
e compresión *f* adiabática
i compressione *f* adiabatica
n adiabatische compressie
d adiabatische Kompression *f*

117 ADIABATIC CONTAINMENT pp
The containment by an externally generated
magnetic field, usually cylindrical, arranged
so that at its ends there are regions of
high field strength.
f confinement *m* adiabatique
e confinamiento *m* adiabático
i confinamento *m* adiabatico
n adiabatische insluiting
d adiabatische Einschliessung *f*

118 ADIABATIC HEATING pp
The heating of the ions and electrons in
plasma physics to a multiple of their
original temperature resulting from compression of the plasma by a gradual
increase of field strength.
f chauffage *m* adiabatique
e calentamiento *m* adiabático
i riscaldamento *m* adiabatico
n adiabatische verhitting
d adiabatische Erhitzung *f*

119 ADIABATIC INVARIANT ma, pp
A physical quantity which remains constant
during the spatial or temporal variation
of a system.
f invariant *m* adiabatique
e invariante *m* adiabático
i invariante *m* adiabatico
n adiabatische invariant
d adiabatische Invariante *f*

120 ADIABATIC TRAP,   pp
    MAGNETIC BOTTLE,
    MIRROR MACHINE
Colloquial term for any configuration of
magnetic fields which will contain a
plasma.
f bouteille *f* magnétique,
   enceinte *f* magnétique,
   machine *f* à miroirs,
   piège *m* adiabatique
e botella *f* magnética, recinto *m* magnético
i bottiglia *f* magnetica,
   chiusura *f* magnetica
n magnetische fles, magnetische omhulling
d magnetische Flasche *f*
   magnetische Hülle *f*

121 ADION   ab
An ion absorbed on a surface that is held
so that it is free to move on the surface
but not away from it.
f adion *m*
e adión *m*
i adione *m*
n adion *n*
d Adion *n*

122 ADJOINT FUNCTION   ma
A mathematical concept which, in reactor
theory, includes the importance function
as a special case.
f fonction *f* adjointe
e función *f* adjunta
i funzione *f* aggiunta
n geadjugeerde functie
d adjungierte Funktion *f*

123 ADMINISTRATION   mc
The application of medicine by authorized
people to patients.
f administration *f*
e administración *f*
i somministrazione *f*
n toedienen *n*
d Verabreichung *f*

124 ADSORBATE   ab, ch
A substance which is adsorbed.
f substance *f* adsorbée
e adsorbato *m*
i sostanza *f* adsorbita
n adsorptief *n*
d Adsorptiv *n*

125 ADSORBENT   ab, ch
A substance or material which adsorbs.
f adsorbant *m*
e adsorbante *m*
i adsorbente *m*
n adsorbens *n*
d Adsorbens *n*

126 ADSORPTION   ab, ch
A process in which a substance or entity
concentrates or holds another substance
on its surface.
f adsorption *f*

e adsorción *f*
i adsorbimento *m*
n adsorptie
d Adsorption *f*

127 ADSORPTION STAGE   ch, is
The apparatus in which an adsorption
process takes place.
f étage *m* d'adsorption
e etapa *f* de adsorción
i stadio *m* d'adsorbimento
n adsorptietrap
d Adsorptionsstufe *f*

128 ADVANCE PULSE GENERATOR,   cd
    CLOCK GENERATOR,
    CLOCK PULSE GENERATOR
Component part of the 400 channel pulse
height analyzer.
f générateur *m* horloge
e generador *m* reloj
i generatore *m* orologio
n klokpulsgenerator
d Uhrimpulsgenerator *m*

129 ADVANCED GAS-COOLED   rt
    REACTOR
An improved type of gas-cooled reactor.
f réacteur *m* à gaz poussé,
   réacteur *m* gaz-graphite avancé
e reactor *m* de enfriamiento con gas
   avanzado
i reattore *m* a gas di tipo perfezionato,
   reattore *m* raffreddato a gas di tipo avanzato
n geavanceerde met gas gekoelde reactor
d fortgeschrittener gasgekühlter Reaktor *m*,
   AGR *m*

130 ADVANTAGE FACTOR   rt
In reactor engineering, the ratio of the
value of a specified radiation quantity at
a position where an enhanced effect is
produced to the value of the same radiation
quantity at some reference position.
f facteur *m* d'avantage,
   facteur *m* d'irradiation optimale
e factor *m* de ventaja
i fattore *m* di vantaggio
n gunstfactor, voordeelfactor
d Überhöhungsfaktor *m*

131 AERIAL SURVEY,   sa
    AIR SURVEY
Control of the contamination of the air in
a certain area by aircraft.
f contrôle *m* aérien de l'air,
   surveillance *f* aérienne de l'air
e control *m* aéreo del aire,
   vigilancia *f* aérea del aire
i controllo *m* aereo dell'aria,
   sorveglianza *f* aerea dell'aria
n luchtbesmettingsmeting door vliegtuigen,
   luchtcontrole door vliegtuigen
d Luftkontaminationskontrolle *f* durch
   Flugzeuge

132 AEROSOL ch
A dispersion of fine particles as dust or droplets in the air.
f aérosol *m*
e aerosol *m*
i aerosole *m*
n aërosol *n*
d Aerosol *n*

133 AFTERCOOLING rt
The cooling of a reactor after it has been shut down.
f refroidissement *m* postérieur
e enfriamiento *m* posterior, postrefrigeración *f*
i raffreddamento *m* posteriore
n nakoeling
d Nachkühlung *f*

134 AFTERGLOW, ec
DECAY, PERSISTENCE
Of a phosphor, the gradual decline of brightness after excitation.
f persistance *f*
e persistencia *f*
i persistenza *f*
n nalichten *n*
d Nachleuchten *n*

135 AFTERHEAT, fe
HEAT RELEASE
Heat resulting from residual radioactivity in reactor fuel or components after a reactor has been shut down.
f chaleur *f* résiduelle
e calor *m* residuo
i calore *m* residuo
n nawarmte
d Nachwärme *f*

136 AFTERLOADING SOURCE br, md, ra
A source intended to be inserted into a body cavity through a previously inserted non-radioactive guide device.
f source *f* à introduction guidée
e fuente *f* de introducción guiada
i sorgente *f* ad introduzione guidata
n bron met geleide inbrenging
d Quelle *f* mit gelenkter Einführung

137 AFTERPOWER, rt
SHUT-DOWN POWER
For a shut-down reactor, the power corresponding to the afterheat.
f puissance *f* résiduelle
e potencia *f* remanente
i potenza *f* susseguente
n restvermogen *n*
d Nachleistung *f*

138 AFTERPULSE ec
A spurious pulse induced in a multiplier phototube by a previous pulse.
f impulsion *f* de traînage
e impulso *m* posterior
i impulso *m* posteriore
n napuls, volgpuls
d Nachimpuls *m*

139 AGE, np
FERMI AGE (GB), NEUTRON AGE (US)
In nuclear reactor theory, the value calculated for the slowing-down area used in the Fermi age model.
f âge *m*, âge *m* de Fermi
e edad *f*, edad *f* de Fermi
i età *f*, età *f* Fermi
n fermileeftijd, leeftijd
d Alter *n*, Fermi-Alter *n*

140 AGE EQUATION, ma, np
FERMI AGE EQUATION
An equation, due to Fermi, by which the slowing down density may be calculated.
f équation *f* de Fermi, équation *f* de l'âge
e ecuación *f* de Fermi, ecuación *f* de la edad
i equazione *f* dell'età, equazione *f* di Fermi
n fermivergelijking, leeftijdsvergelijking
d Altersgleichung *f*, Fermi-Alter-Gleichung *f*

141 AGE EQUATION WITHOUT ma
CAPTURE
The equation $\nabla^2 q = \frac{\delta q}{\delta r}$, which describes the slowing down of neutrons in a scattering medium in which neutrons are not captured by the nuclei of the moderator.
f équation *f* de l'âge sans capture
e ecuación *f* de la edad sin captura
i equazione *f* dell'età senza cattura
n leeftijdsvergelijking zonder vangst
d Altersgleichung *f* ohne Einfang

142 AGE HARDENING, mg
PRECIPITATION HARDENING
A process of hardening by ag(e)ing at room temperature after an annealing process.
f durcissement *m* par précipitation, durcissement *m* par viellissement
e endurecimiento *m* por envejecimiento
i indurimento *m* per precipitazione
n precipitatieharding
d Ausscheidungshärtung *f*

143 AGEING (GB), ge
AGING (US)
To undergo change with age or lapse of time.
f vieillissement *m*
e envejecimiento *m*
i invecchiamento *m*
n veroudering
d Alterung *f*

144 AGGREGATE RECOIL np
The ejection, from the surface of a radioactive sample, of atoms additional to those which recoil on disintegrating.
f amas *m* de recul
e retroceso *m* en grupo
i rimbalzo *m* in gruppo
n aggregaatterugslag
d Aggregatrückstoss *m*

145 AGGREGATION gp
The gathering of particles, especially in the sense of their formation into larger

   aggregates or entities.
f agrégation *f*, amassement *m*
e acumulación *f*, agregación *f*
i aggregazione *f*, ammassamento *m*
n opeenhoping, samenballing
d Anhäufung *f*, Ansammlung *f*,
   Zusammenballung *f*

146 AGITATION,   ch
     STIRRING
   The creation and maintaining of motion
   in fluids.
f agitation *f*
e agitación *f*
i agitazione *f*
n roeren *n*
d Rühren *n*

147 AIR BURST,   nw
     ATOMIC AIR BURST
   The explosion of a nuclear weapon at such
   a height that the expanding ball of fire
   does not touch the earth's surface even
   when the luminosity is a maximum.
f explosion *f* aérienne,
   explosion *f* dans l'atmosphère
e explosión *f* en el aire
i esplosione *f* nell'atmosfera
n explosie in de atmosfeer
d Explosion *f* in der Atmosphäre,
   Hochexplosion *f*

148 AIR CONTAMINATION   sa
     INDICATOR
   An indicator designed to detect the
   presence of contamination by aerosols,
   vapo(u)rs or gas in air.
f signaleur *m* atmosphérique
e indicador *m* atmosférico
i segnalatore *m* atmosferico
n luchtbesmettingsindicator
d Luftkontaminationsanzeiger *m*

149 AIR CONTAMINATION METER   me
   An assembly for measuring the activity
   of aerosols, vapo(u)rs or gas per unit
   volume of air.
f contaminamètre *m* atmosphérique
e medidor *m* de contaminaciones
   atmosféricas
i contaminametro *m* atmosferico
n luchtbesmettingsmeter
d Gerät *n* zur Bestimmung der Luft-
   kontamination

150 AIR CONTAMINATION MONITOR,   sa
     AIR MONITOR
   A monitor designed to measure the
   activity of aerosols, vapo(u)rs or gas per
   unit volume of air and to give a warning
   when it exceeds a predetermined value.
f moniteur *m* atmosphérique
e monitor *m* atmosférico
i monitore *m* atmosferico
n luchtbesmettingsmonitor
d Luftkontaminationswarngerät *n*

151 AIR COOLANT,   cl
     COOLING AIR
   Air used for cooling purposes.
f air *m* de refroidissement
e aire *m* de enfriamiento
i aria *f* di raffreddamento
n koellucht
d Kühlluft *f*

152 AIR-COOLED GRAPHITE   rt
     MODERATED REACTOR
f réacteur *m* à refroidissement par air et
   à modérateur en graphite
e reactor *m* de enfriamiento por aire y
   moderado por grafito
i reattore *m* a raffreddamento ad aria e
   moderato per grafite
n met lucht gekoelde en met grafiet gemodereerde reactor
d luftgekühlter, graphitmoderierter
   Reaktor *m*

153 AIR-COOLED OIL VAPOR   vt
     DIFFUSION PUMP (US),
     AIR-COOLED OIL VAPOUR
     DIFFUSION PUMP (GB)
   A special type of diffusion pump for
   obtaining very high vacua.
f pompe *f* à diffusion d'huile à refroidissement par air
e bomba *f* de difusión de aceite enfriada
   por aire
i pompa *f* a diffusione d'olio con
   raffreddamento ad aria
n met lucht gekoelde oliediffusiepomp
d luftgekühlte Öldiffusionspumpe *f*

154 AIR COOLING   cl
   In general the cooling of parts of
   apparatus by increasing its radiating·
   surface by means of ribs, etc. so that it
   is exposed to a current of air.
f refroidissement *m* par air
e enfriamiento *m* por aire
i raffreddamento *m* ad aria
n luchtkoeling
d Luftkühlung *f*

155 AIR COUNT   ct
   A measurement in which the radioactivity
   in a standard volume of air is measured.
f comptage *m* dans un volume étalon d'air
e recuento *m* en un volumen patrón de aire
i conteggio *m* in un volume campione d'aria
n telling in een standaardhoeveelheid lucht
d Zählung *f* in Normalvolumen Luft

156 AIR DOSE,   ra
     FREE-AIR DOSE, IN-AIR DOSE
   A dose of radiation, measured in air,
   from which secondary radiation, apart from
   that arising from the air, or associated
   with the source, is excluded.
f dose *f* dans l'air
e dosis *f* atmosférica
i dose *f* in aria, dose *f*. misurata in aria
n dosis vrij in lucht
d Luftdosis *f*

157 AIR EQUIVALENT   ab
  Of a given absorber, the thickness of a
  layer of air at standard temperature and
  pressure which causes the same absorption
  or energy loss.
f équivalent-air *m*, équivalent *m* en air
e equivalente *m* en aire
i equivalente *m* in aria
n luchtequivalent *n*
d Luftäquivalent *n*

158 AIR-EQUIVALENT IONIZATION   ic
  CHAMBER,
      AIR-WALL IONIZATION CHAMBER
  An air-filled cavity ionization chamber
  whose walls consist of a substance such
  that the ionization produced inside the
  chamber is essentially the same as that
  which would be produced in air at the
  same point in the absence of the chamber.
f chambre *f* d'ionisation à paroi
  équivalente à l'air
e cámara *f* de ionización de pared equivalente al aire
i camera *f* d'ionizzazione con parete
  equivalente all'aria
n ionisatievat *n* met aan lucht equivalente
  wanden
d Ionisationskammer *f* mit luftäquivalenten
  Wänden, Luftwändekammer *f*

159 AIR-EQUIVALENT MATERIAL,   ic
      AIR-WALL MATERIAL
  A material, suitable for the walls of
  ionization chambers and having substantially the same effective atomic number as
  air.
f matière *f* équivalente à l'air
e materia *f* equivalente al aire
i materia *f* equivalente all'aria
n aan lucht equivalente stof
d luftäquivalenter Stoff *m*

160 AIR IONIZING ELECTRODE   ar, md, ra
  A platinum wire on the tip of which
  polonium is electrodeposited, the whole
  being encased in a brass tube.
f électrode *f* ionisante de l'air
e electrodo *m* ionizante del aire
i elettrodo *m* ionizzante dell'aria
n lucht ioniserende elektrode
d luftionisierende Elektrode *f*

161 AIR LOCK   rt
  A labyrinth passage through which access
  is obtained to e.g. a reactor sphere.
f labyrinthe *m* d'entrée
e laberinto *m* de entrada
i labirinto *m* d'ingresso
n toegangssluis
d Eintrittsschleuse *f*

162 AIR MONITORING   sa
  Periodic or continuous determination of
  the amount of ionizing radiation or radioactive contamination present in the air.
f supervision *f* d'air

e vigilancia *f* de aire
i controllo *m* di radiazione ionizzante in aria,
  controllo *m* di radioattività in aria
n atmosfeerbewaking, luchtcontrole
d Luftüberwachung *f*

163 AIR POLLUTION,   is
      ATMOSPHERIC POLLUTION
  One of the problems to be studied in using
  isotopic tracers.
f pollution *f* atmosphérique
e contaminación *f* atmosférica
i inquinamento *m* atmosferico
n luchtverontreiniging
d Luftverunreinigung *f*

164 AIR SAMPLING EQUIPMENT   co, sa
  An equipment to continuously take air
  samples to estimate the radioactivity of
  the air.
f appareil *m* de prélèvement d'aérosols
  atmosphériques,
  installation *f* de prise d'échantillons d'air
e equipo *m* muestreador de aire
i impianto *m* da disporre campioni d'aria
n installatie voor het nemen van lucht-
  monsters
d Luftprobenahmeanlage *f*

165 AIR SCATTER (GB),   ra
      SKY SHINE (US)
  Radiation from a source reflected by the
  atmosphere.
f rayonnement *m* diffusé,
  rayonnement *m* indirect
e radiación *f* indirecta
i radiazione *f* indiretta
n gereflecteerde straling
d reflektierte Strahlung *f*

166 AIR SCATTER SHIELD (GB),   ra
      SKY SHINE SHIELD (US)
f écran *m* contre le rayonnement indirect
e pantalla *f* contra la radiación indirecta
i schermo *m* contro la radiazione indiretta
n scherm *n* tegen gereflecteerde straling
d Schirm *m* gegen reflektierte Strahlung

167 AIR SURVEY, see 131

168 AIRBORNE CONTAMINANTS   nw
  Contaminated particles in solid or liquid
  state, in suspension in the atmosphere.
f contaminants *pl* dans l'air
e contaminantes *pl* aerotransportados
i contaminanti *pl* aerei
n in de lucht aanwezige actieve deeltjes *pl*
d in der Luft vorhandene aktive Teilchen *pl*

169 AIRBORNE PARTICULATES   np
  Radioactive particles of a large size
  transported by air currents.
f macroparticules *pl* en suspension dans
  l'air
e macropartículas *pl* en suspensión en el
  aire
i macroparticelle *pl* in sospensione nell'aria

n   door de lucht meegevoerde macrodeeltjes *pl*
d   mit der Luft mitgeführte Makroteilchen *pl*

170 AIRBORNE RADIOACTIVITY   np
Radioactivity spread over a certain area
by atmospheric agents.
f   radioactivité *f* dans l'air
e   radiactividad *f* en el aire
i   radioattività *f* hell'aria
n   in de lucht aanwezige radioactiviteit
d   in der Luft vorhandene Radioaktivität *f*,
Luftaktivität *f*

171 AIRLIFT   ch
The use of air or some other gas to
transfer liquid or solid material from one
part of a plant to another part.
f   engin *m* de levage pneumatique
e   elevador *m* al aire comprimido
i   pompa *f* all'aria compressa
n   pneumatische hefinrichting
d   Drucklufttheber *m*

172 AIRLIFT MIXER-SETTLER   ch, is
A solvent extraction unit in which mixing
is achieved by bubbling air through the
two phases, in a narrow chamber.
f   échangeur *m* de matière mélangeur-préci-
pitateur à levage pneumatique
e   cambiador *m* de materia mezclador-pre-
cipitador con elevador al aire comprimido
i   scambiatore *m* di materia mescolatore-
precipitatore all'aria compressa
n   meng-bezink-stofuitwisselaar met
pneumatische hefinrichting
d   Misch-Setz-Stoffaustauscher *m* mit
Drucklufttheber

173 AKERLUND GRID   ·xr
A modified form of anti-diffusion grid.
f   grille *f* d'Akerlund
e   rejilla *f* de Akerlund
i   griglia *f* d'Akerlund
n   akerlundraster *n*
d   Akerlund-Blende *f*

174 ALAMOGORDO BOMB,   nw
TRINITY BOMB
The first nuclear bomb, detonated July 16,
1945 at Alamogordo, New Mexico.
f   bombe *f* d'Alamogordo
e   bomba *f* de Alamogordo
i   bomba *f* d'Alamogordo
n   Alamogordobom
d   Alamogordo-Bombe *f*

175 ALARM DOSEMETER   gr, ra, sa
A dosemeter designed to give an aural
and/or visual alarm when the gamma dose
to the wearer has reached a certain level.
f   dosimètre *m* avertisseur
e   dosímetro *m* de alarma
i   dosimetro *m* d'allarme
n   waarschuwende dosismeter
d   Warnungsdosismesser *n*

176 ALBEDO,   np
NEUTRON ALBEDO
The probability under specified conditions
that a neutron entering into a region
through a surface will return through that
surface.
f   albédo *m* pour neutrons
e   albedo *m* para neutrones
i   albedo *m* per neutroni
n   albedo voor neutronen
d   Albedo *f* für Neutronen

177 ALDANITE   mi
A variety of thorianite, containing 14.0 to
29.0 % UO_2.
f   aldanite *f*
e   aldanita *f*
i   aldanite *f*
n   aldaniet *n*
d   Aldanit *m*

178 ALFVEN VELOCITY   pp
The phase velocity of an Alfven wave.
f   vitesse *f* d'Alfven
e   velocidad *f* de Alfven
i   velocità *f* d'Alfven
n   alfvensnelheid
d   Alfven-Geschwindigkeit *f*

179 ALFVEN WAVE   pp
A transverse hydromagnetic wave in an
electrically conducting fluid containing
a magnetic field.
f   onde *f* d'Alfven
e   onda *f* de Alfven
i   onda *f* d'Alfven
n   alfvengolf
d   Alfven-Welle *f*

180 ALIMENTARY TRACT,   md
DIGESTIVE CANAL,
GASTROINTESTINAL TRACT,
G.I. TRACT
The critical organ for swallowed radioactive materials during the early stages
of intake.
f   tube *m* digestif
e   conducto *m* digestivo, tubo *m* digestivo
i   canale *m* gastrointestinale,
tubo *m* digestivo
n   spijsverteringskanaal *n*
d   Verdauungstraktus *m*

181 ALIQUOT   ra
A small sample of radioactive material
assayed in order to determine the radioactivity of the whole.
f   échantillon *m* petit
e   muestra *f* pequeña
i   prova *f* piccola
n   klein monster *n*
d   kleine Probe *f*

182 ALLANITE   mi
A mineral containing about 0.02 % U and
up to 3.2 % Th.
f   allanite *f*

e alanita *f*
i allanite *f*
n allaniet *n*
d Allanit *m*

183 ALLOBARS   is
Forms of an element having different
atomic weights and therefore different
isotopic compositions.
f allobares *pl*
e alobaras *pl*
i allobari *pl*
n allobaren *pl*
d Allobare *pl*

184 ALLOWED BAND   ec, gp
In solid-state physics, those ranges of
energy levels or bands that electrons
occupy in a material.
f bande *f* permise
e banda *f* permitida
i banda *f* permessa
n toegestane band
d erlaubtes Band *n*

185 ALLOWED TRANSITION   np
The most probable type of transition
between two states of a quantum-mechanical system.
f transition *f* permise
e transición *f* permitida
i transizione *f* permessa
n toegestane overgang
d erlaubter Übergang *m*

186 ALPHA   ar
Pertaining to alpha particles.
f alpha
e alfa
i alfa
n alfa
d alpha

187 ALPHA CHANGE,   ar
    ALPHA EMISSION,
    ALPHA PARTICLE EMISSION
A nuclear change consisting of the
emission of an ∝-particle.
f émission *f* alpha
e emisión *f* alfa
i emissione *f* alfa
n alfa-emissie
d Alphaemission *f*

188 ALPHA CONTAMINATION   sa
    INDICATOR
An indicator designed to detect alpha
surface contamination, in which the output
pulses from the detector control a warning
signal.
f signaleur *m* de contamination alpha
e indicador *m* de contaminación alfa
i segnalatore *m* di contaminazione alfa
n indicator voor alfabesmetting
d Alphakontaminationsanzeiger *m*

189 ALPHA COUNTER   ct
A system for counting alpha particles

including a counter tube, amplifier, pulse
height discriminator, etc.
f compteur *m* alpha
e contador *m* alfa
i contatore *m* alfa
n alfateller
d Alphazähler *m*

190 ALPHA COUNTER TUBE   ct
A counter tube for the detection of alpha
particles.
f tube *m* compteur alpha
e tubo *m* contador alfa
i tubo *m* contatore alfa
n alfatelbuis
d Alphazählrohr *n*

191 ALPHA DECAY,   ar
    ALPHA DISINTEGRATION,
    ALPHA TRANSFORMATION
Radioactive decay in which an alpha particle is emitted.
f désintégration *f* alpha
e desintegración *f* alfa
i disintegrazione *f* alfa
n alfadesintegratie, alfaverval *n*
d Alphazerfall *m*

192 ALPHA DECAY ENERGY,   ar
    ALPHA DISINTEGRATION ENERGY
The sum of the energy of the alpha
·particle produced in the disintegration
process and the recoil energy of the
product atom.
f énergie *f* de désintégration alpha
e energía *f* de desintegración alfa
i energia *f* di disintegrazione alfa
n alfadesintegratie-energie
d Alphazerfallsenergie *f*

193 ALPHA EMITTER,   ar
    ALPHA RADIATOR
A radionuclide that undergoes a transformation by alpha particle emission.
f émetteur *m* alpha
e emisor *m* alfa
i emettitore *m* alfa
n alfastraler
d Alphastrahler *m*

194 ALPHA EMITTING FOIL   ar, ra
An active source used e.g. for static
elimination in industry.
f feuille *f* métallique à émission alpha
e hoja *f* delgada de emisión alfa
i foglia *f* ad emissione alfa
n alfastralen emitterende foelie
d alphastrahlenemittierende Folie *f*

195 ALPHA HAND CONTAMINATION   sa
    MONITOR
An assembly designed to measure the
hand contamination by alpha emitters and
including a device for indicating contamination exceeding a predetermined value.
f moniteur *m* de contamination alpha pour
   les mains
e vigía *f* de contaminación alfa en las manos

i monitore *m* di contaminazione alfa per le mani
n alfamonitor voor handbesmetting
d Alphahandmonitor *m*

196 ALPHA-NEUTRON REACTION np
The capture of a bombarding alpha particle by the target nucleus; the compound nucleus thus formed ejects a neutron and thereby changes into a new nucleus.
f réaction *f* alpha-neutron
e reacción *f* alfa-neutrón
i reazione *f* alfa-neutrone
n alfa-neutron-reactie
d Alpha-Neutron-Reaktion *f*

197 ALPHA PARTICLE, ar
ALPHA RAY
A helium-4 nucleus emitted during a nuclear transformation.
f particule *f* alpha
e partícula *f* alfa
i particella *f* alfa
n alfadeeltje *n*
d Alphateilchen *n*

198 ALPHA PARTICLE BINDING ar
ENERGY
The energy required to remove an alpha particle from a nucleus.
f énergie *f* de liaison d'une particule alpha
e energía *f* de enlace de una partícula alfa
i energia *f* di legame d'una particella alfa
n bindingsenergie van een alfadeeltje
d Alphabindungsenergie *f*,
Bindungsenergie *f* des Alphateilchens

199 ALPHA PARTICLE MODEL np
A nuclear model in which it is assumed that as many as possible of the nucleons are grouped to form alpha particles.
f modèle *m* de la structure en particules alpha,
modèle *m* nucléaire alpha
e modelo *m* de la estructura in partículas alfa
i modello *m* della struttura in particelle alfa
n alfakernmodel *n*
d Alphateilchenmodell *n*

200 ALPHA PARTICLE SPECTRUM, ar, sp
ALPHA SPECTRUM
The distribution in energy or momentum of the α-particles emitted by a pure radionuclide, or, less commonly, by a mixture of radionuclides.
f spectre *m* alpha
e espectro *m* alfa
i spettro *m* alfa
n alfaspectrum *n*
d Alphaspektrum *n*

201 ALPHA-PROTON REACTION np
The capture of a bombarding alpha particle by the target nucleus; the compound nucleus

thus formed ejects a proton and thereby changes into a new nucleus.
f réaction *f* alpha-proton
e reacción *f* alfa-protón
i reazione *f* alfa-protone
n alfa-proton-reactie
d Alpha-Proton-Reaktion *f*

202 ALPHA PULSE COUNTING me
ASSEMBLY
A pulse counting assembly which includes an alpha radiation detector whose output pulses are applied to a counting sub-assembly and/or to a counting rate measuring sub-assembly.
f ensemble *m* de mesure à impulsions pour particules alpha
e conjunto *m* contador por impulsos para partículas alfa
i complesso *m* di misura ad impulsi per particelle alfa
n telopstelling voor alfadeeltjes
d Alphazählanordnung *f*

203 ALPHA RADIATION ar
Alpha particles emerging from radioactive atoms.
f rayonnement *m* alpha
e radiación *f* alfa
i radiazione *f* alfa
n alfastraling
d Alphastrahlung *f*

204 ALPHA RADIOACTIVITY ar
The emission of an alpha radiation.
f radioactivité *f* alpha
e actividad *f* alfa
i attività *f* alfa
n alfa-activiteit
d Alphaaktivität *f*

205 ALPHA RATIO ar
As applied to fissionable nuclei, the ratio of the radiative capture cross section to the fission cross section.
f facteur *m* alpha
e factor *m* alfa
i fattore *m* alfa
n alfafactor
d Alphafaktor *m*,
Verhältniszahl *f* alpha bei spaltbaren Kernen

206 ALPHA RAY ar, me, sp
SPECTROMETER
A measuring assembly for determining the energy spectrum of alpha rays.
f spectromètre *m* à rayons alpha
e espectrómetro *m* alfa
i spettrometro *m* alfa
n alfaspectrometer
d Alphaspektrometer *n*

207 ALPHA RAYS ar
A stream of alpha particles.
f rayons *pl* alpha
e rayos *pl* alfa

i   raggi *pl* alfa
n   alfastralen *pl*
d   Alphastrahlen *pl*

208 ALPHA URANIUM   mg
That allotropic form of uranium metal
which is stable below approximately
660° C.
f   uranium *m* alpha
e   uranio *m* alfa
i   uranio *m* alfa
n   alfa-uranium *n*
d   Alpha-Uran *n*

209 ALPHATOPIC   ar
Pertaining to a relationship wherein the
masses of composition of two nuclei differ
by an alpha particle.
f   alphatopique adj
e   alfatópico adj
i   alfatopico adj
n   alfatoop adj
d   alphatopisch adj

210 ALPHATRON,   ar, me
     ALPHATRON GAGE (US),
     ALPHATRON GAUGE (GB)
An ionization ga(u)ge in which the
ionization is produced by alpha particles,
obtained from a radioactive source.
f   alphatron *m*
e   alfatrón *m*
i   alfatrone *m*
n   alfatron *n*
d   Alphatron *n*

211 ALTERNATING GRADIENT   pa
     ACCELERATOR,
        STRONG FOCUSING ACCELERATOR
A high-energy accelerator using alternating
gradient focusing.
f   accélérateur *m* à focalisation à gradient
    alternant
e   acelerador *m* de enfoque de gradiente
    alterno.
i   acceleratore *m* a focalizzazione a gradiente
    alternato
n   versneller met focussering met wisselende
    gradiënt,
    versneller met sterke focussering
d   Beschleuniger *m* mit AG-Fokussierung,
    Beschleuniger *m* mit starker Fokussierung

212 ALTERNATING GRADIENT   pa
     FOCUSING,
        STRONG FOCUSING
A principle of focusing used in high-energy
accelerators to prevent spreading of the
beam.
f   focalisation *f* à gradient alternant
e   enfoque *m* de gradiente alterno
i   focalizzazione *f* a gradiente alternato
n   focussering met wisselende gradiënt,
    sterke focussering
d   AG-Fokussierung *f*,
    starke Fokussierung *f*

213 ALTERNATION OF MULTI-   np
     PLICITIES
Since the multiplicity for an atom is 2S-1,
where S is the resultant spin of the electrons of the atom, and since S is integral
or half integral for an even or odd number
of electrons, respectively, it follows that
the multiplicity of an atomic term is odd
for an even number of electrons and even
for an odd number of electrons.
f   alternance *f* de multiplicités
e   alternación *f* de multiplicidades
i   alternazione *f* di moltiplicità
n   meervoudigheidsafwisseling
d   Vielfältigkeitsabwechslung *f*

214 ALTITUDE CURVE   cm
A curve showing the intensity of cosmic
rays at different altitudes.
f   courbe *f* d'altitude
e   curva *f* de altitud
i   curva *f* d'altezza
n   hoogtekromme
d   Höhenkurve *f*

215 ALTITUDE EFFECT   cm
The intensity of cosmic rays as a function
of altitude shows a maximum at a distance
below the top of the atmosphere corresponding to a pressure of about 45 mm Hg.
f   effet *m* d'altitude
e   efecto *m* de altitud
i   effetto *m* d'altezza
n   hoogte-effect *n*
d   Höheneffekt *m*

216 ALUMINA   mt
The oxide of alumin(i)um Al_2O_3.
f   alumine *f*
e   alúmina *f*
i   allumina *f*
n   aluinaarde, alumina *n*, aluminiumoxyde *n*
d   Alaunerde *f*, Aluminiumoxyd *n*, Tonerde *f*

217 ALUMINIUM (GB),   ch
     ALUMINUM (US)
Metallic element, symbol Al, atomic
number 13.
f   aluminium *m*
e   aluminio *m*
i   alluminio *m*
n   aluminium *n*
d   Aluminium *n*

218 ALUMINIUM-SILICON ALLOY   fu, rt
     (GB),
        ALUMINUM-SILICON ALLOY (US)
An Al-Si alloy used as an interlayer
between an alumin(i)um can and a uranium
fuel bar.
f   alliage *m* aluminium-silicium
e   aleación *f* aluminio-silicio
i   lega *f* alluminio-silicio
n   aluminium-siliciumlegering
d   Aluminium-Siliziumlegierung *f*

219 ALWAYS SAFE GEOMETRY,   rt
   INFINITELY SAFE GEOMETRY
   Said of a reactor when cylinders below a
   certain diameter or slabs less than a
   certain thickness are used, which cannot
   go critical regardless of other dimensions.
f géométrie *f* toujours sûre
e geometría *f* siempre segura
i geometria *f* sempre sicura
n absoluut veilige geometrie
d absolut sichere Geometrie *f*

220 AMBIPOLAR DIFFUSION   pp
   Diffusion occurring in a plasma in which
   there are ions of both signs; the faster
   ions tend to diffuse out of the plasma and
   leave an excess of positive charge, the
   space charge field of which then retards the
   electrons and accelerates the slower
   ions.
f diffusion *f* ambipolaire
e difusión *f* ambipolar
i diffusione *f* ambipolare
n ambipolaire diffusie
d ambipolare Diffusion *f*

221 AMERICIUM   ch
   Transuranic radioactive element, symbol
   Am, atomic number 95.
f américium *m*
e americio *m*
i americio *m*
n americium *n*
d Americium *n*

222 AMMONIUM DIURANATE   ch, mg, mi
   A yellow crystalline compound formed
   by treatment of uranyl solutions with
   ammonia, used for the purification of
   uranium ore concentrates.
f diuranate *m* d'ammonium
e diuranato *m* de amonio
i diuranato *m* d'ammonio
n ammoniumdiuranaat *n*
d Ammoniumdiuranat *n*

223 AMNIOGRAPHY   xr
   The radiological examination of the uterus
   following the injection of a contrast medium
   into the amniotic sac during late pregnancy.
f amniographie *f*
e amniografía *f*
i amniografia *f*
n amniografie
d Amniographie *f*

224 AMORPHOUS MATERIAL   mt
   Material in which no repeated orderly
   arrangements of atoms exist over a region
   large compared with atomic or molecular
   dimensions.
f matière *f* amorphe
e materia *f* amorfa
i materia *f* amorfa
n amorfe stof
d amorphes Material *n*

225 AMPANGABEITE,   mi
   HYDROEUXENITE
   Niobate and titanate of iron, uranium and
   rare earths.
f ampangabéite *f*, hydroeuxénite *f*
e ampangabeita *f*. hidroeuxenita *f*
i ampangabeite *f*, idroeusenite *f*
n ampangabeïet *n*, hydroeuxeniet *n*
d Ampangabeit *m*, Hydroeuxenit *m*

226 AMPHOTERIC ION,   ch
   ZWITTERION
   An ion which carries both a positive and
   a negative charge.
f ion *m* amphotérique
e ión *m* anfotérico
i ione *m* amfoterico
n amfoteer ion *n*
d Zwitterion *n*

227 AMPLIFIER   ec
f amplificateur *m*
e amplificador *m*
i amplificatore *m*
n versterker
d Verstärker *m*

228 AMPLITUDE ANALYZER,   ec
   KICK SORTER (GB),
   PULSE HEIGHT ANALYZER (US)
   A sub-assembly for determining the
   distribution function of a set of pulses in
   terms of their amplitude.
f analyseur *m* d'amplitude
e analizador *m* de amplitud
i analizzatore *m* d'ampiezza,
   classificatore *m* d'impulsi
n pulshoogteanalysator
d Amplitudenanalysator *m*,
   Impulshöhenanalysator *m*

229 AMPLITUDE ANALYZING   me
   ASSEMBLY
   A measuring assembly designed to analyze
   the output from its detector(s) as a
   function of the energy of the radiation.
f ensemble *m* d'analyse d'amplitude
e conjunto *m* analizador de amplitud
i complesso *m* d'analisi d'ampiezza
n opstelling voor pulshoogteanalyse,
   pulshoogteanalysator
d Anordnung *f* zur Impulshöhenanalyse

230 AMPLITUDE ANALYZING   me
   ASSEMBLY WITH STORAGE FUNCTION
   An amplitude analyzing assembly which
   includes a multichannel amplitude
   analyzer with storage function.
f ensemble *m* d'analyse d'amplitude à
   mémoire
e conjunto *m* analizador de amplitud con
   memoria
i complesso *m* d'analisi d'ampiezza a
   memoria
n pulshoogteanalysator met geheugen
d Anordnung *f* zur Impulshöhenanalyse mit
   Speicherung

231 AMPLITUDE DISCRIMINATOR,   ec
    PULSE AMPLITUDE DISCRIMINATOR,
    PULSE HEIGHT DISCRIMINATOR
    A device with an output pulse for each
    input pulse whose amplitude lies above
    a given threshold value.
f   discriminateur *m* d'amplitude
e   discriminador *m* de amplitud de impulsos
i   discriminatore *m* d'ampiezza
n   pulshoogtediscriminator
d   Amplitudendiskriminator *m*,
    Impulshöhendiskriminator *m*

232 AMPOULE,   md, ra
    AMPULLA
    In safety regulations with respect to radio-
    active materials, a plastic or glass recep-
    tacle immediately surrounding the radio-
    active material and used for unsealed
    sources.
f   ampoule *f*
e   ampolla *f*
i   ampolla *f*
n   ampul, ampulla
d   Ampulle *f*

233 AMU,   un
    ATOMIC MASS UNIT,
    PHYSICAL MASS UNIT
    One-sixteenth of the mass of a neutral atom
    of ^{16}O.
f   unité *f* de masse atomique
e   unidad *f* de masa atómica
i   unità *f* di massa atomica
n   atomaire massaeenheid
d   atomare Masseneinheit *f*

234 ANABATIC WIND FALL-OUT,   mo
    UPWIND FALL-OUT
f   retombées *pl* radioactives de vent
    anabatique
e   depósito *m* radiactivo de viento anabático
i   ricaduta *f* radioattiva di vento anabatico
n   radioactieve stijgwindneerslag
d   radioaktiver Aufwindausfall *m*

235 ANAEROBIC   md
    Occurring without the presence of or able
    to live without free oxygen.
f   anaérobique adj
e   anaeróbico adj
i   anaerobico adj
n   anaëroob adj
d   anaerob adj

236 ANALOG COMPUTER (US),   ma
    ANALOGUE COMPUTER (GB)
    A computer that solves problems by setting
    up equivalent electric circuits and making
    measurements as the variables are changed
    in accordance with the corresponding
    physical phenomena.
f   calculatrice *f* analogique
e   computadora *f* análoga
i   calcolatore *m* analogo
n   analogonrekentuig *n*
d   Analogrechner *m*

237 ANALOG SIGNAL (US),   ma
    ANALOGUE SIGNAL (GB)
    A signal whose value varies progressively
    1. with the value of a physical quantity of
    the same or a different nature.
    2. with the scope of the desired action, in
    the case of a demand.
f   signal *m* analogique
e   señal *f* análoga
i   segnale *m* analogo
n   analoog signaal *n*
d   analoges Signal *n*

238 ANALOG-TO-DIGITAL   ma
    CONVERTER (US),
    ANALOGUE-TO-DIGITAL CONVERTER
    (GB)
    A sub-assembly designed to provide an
    output signal which is a digital represen-
    tation of the analog(ue) input signal.
f   convertisseur *m* analogique-numérique,
    convertisseur *m* autonome
e   convertidor *m* análogo-numérico
i   convertitore *m* analogo-numerico
n   analoog-digitaalomzetter
d   Analog-Digitalwandler *m*

239 ANALYZING MAGNET,   an
    MAGNETIC ANALYZER
    An apparatus with a magnetic field for
    measuring the magnetic rigidity of particles
    and hence their energy.
f   analyseur *m* magnétique
e   analizador *m* magnético
i   analizzatore *m* magnetico
n   magnetische analysator
d   Analysemagnet *m*,
    magnetischer Analysator *m*

240 ANAPHASE   md
    A stage in nuclear division, either mitotic,
    or meiotic, in which the paired chromo-
    somes are separated towards opposite
    poles of the spindle.
f   anaphase *f*
e   anafase *f*
i   anafase *f*
n   anafaze
d   Anaphase *f*

241 ANAPHORESIS   ch
    The migration of particles through a fluid
    under the influence of an electric field.
f   anaphorèse *f*
e   anaforesis *f*
i   anaforesi *f*
n   anaforese
d   Anaphorese *f*

242 ANCILLARY EQUIPMENT,   ge
    AUXILIARY EQUIPMENT
    All equipment not belonging to the
    fundamental equipment.
f   appareillage *m* auxiliaire
e   equipo *m* auxiliar
i   apparecchiatura *f* ausiliaria
n   hulpapparatuur
d   Hilfsgeräte *pl*, Zusatzeinrichtung *f*

243 ANDERSONITE   mi
A rare secondary mineral containing
about 39.2 % U.
f andersonite *f*
e andersonita *f*
i andersonite *f*
n andersoniet *n*
d Andersonit *m*

244 ANDROGENESIS   md
Development of an egg with paternal
chromosomes only, in which the paired
chromosomes are separated towards
opposite poles of the spindle.
f androgénie *f*
e androgénesis *f*
i androgenesi *f*
n androgenesis
d Androgenese *f*

245 ANEMIA   md
A condition in which the blood is deficient
either in quantity or quality.
f anémie *f*
e anemia *f*
i anemia *f*
n anemie, bloedarmoede
d Anämie *f*, Blutarmut *f*

246 ANGIOCARDIOGRAPHY   xr
The radiological examination of the heart
and large vessels following injection of
a contrast medium.
f angiocardiographie *f*
e angiocardiografía *f*
i angiocardiografia *f*
n angiocardiografie
d Angiocardiographie *f*

247 ANGIOCARDIOPNEUMOGRAPHY   xr
Angiocardiography carried out in a
pressurized environment.
f angiopneumocardiographie *f*
e angiocardioneumografía *f*
i angiopneumocardiografia *f*
n angiopneumocardiografie
d Angiopneumocardiographie *f*

248 ANGIOGRAPHY   xr
The radiological examination of the blood
vessels following direct injection of a
contrast medium.
f angiographie *f*
e angiografía *f*
i angiografia *f*
n angiografie
d Angiographie *f*

249 ANGLE OF DEFLECTION   ec, ra
The angle between a deflected beam and
its original direction before its deflection.
f angle *m* de déviation
e ángulo *m* de desviación
i angolo *m* di deviazione
n afbuigingshoek
d Ablenkungswinkel *m*

250 ANGLE OF INCIDENCE   ra
The angle at which radiation strikes a
surface, measured from the line of
direction of the moving entry to a line
perpendicular to the surface at the point
of impact.
f angle *m* d'incidence
e ángulo *m* de incidencia
i angolo *m* d'incidenza
n hoek van inval, invalshoek
d Einfallswinkel *m*

251 ANGLE OF REFLECTION   gp
The angle between a wave or beam leaving
a surface and the perpendicular to the
surface.
f angle *m* de réflexion
e ángulo *m* de reflexión
i angolo *m* di reflessione
n hoek van uitval, uitvalshoek
d Ausfallwinkel *m*, Reflexionswinkel *m*

252 ANGLE STRAGGLING   ra
The variation in the direction of motion
of particles after passing through a
certain thickness of matter, the paths of
the particles initially being parallel.
f déviation *f* de direction
e desviación *f* de dirección
i deviazione *f* di direzione
n richtingsspreiding
d Richtungsstreuung *f*, Winkelstreuung *f*

253 ANGULAR ACCELERATION   gp
The time rate of change of angular velocity
ω, either in angular speed or in the
direction of the axis of rotation.
f accélération *f* angulaire
e aceleración *f* angular
i accelerazione *f* angolare
n hoekversnelling
d Winkelbeschleunigung *f*

254 ANGULAR CORRELATION   ra
For collision processes that result in the
emission of two successive radiations, the
correlation between the directions in which
individual radiations of each pair are
emitted.
f corrélation *f* angulaire
e correlación *f* angular
i correlazione *f* angolare
n hoekcorrelatie
d Winkelkorrelation *f*

255 ANGULAR DEPENDENCE OF   np
     SCATTERING
The change in wavelength of a scattered
photon is independent of the initial wavelength and depends only on the scattering
angle.
f dépendance *f* angulaire de la diffusion
e dependencia *f* angular de la dispersión
i dipendenza *f* angolare della deviazione
n verstrooiingshoekafhankelijkheid
d Streuwinkelabhängigkeit *f*

256 ANGULAR DISPLACEMENT ma
When a vector is displaced from one
position to another, the angular displacement is the angle between the original
direction of the vector and the displaced
vector.
f déplacement *m* angulaire
e desplazamiento *m* angular
i spostamento *m* angolare
n hoekverschuiving
d Winkelverschiebung *f*

257 ANGULAR DISTRIBUTION, np, ra
PHASE SPACE DISTRIBUTION
For collision processes, the variation
of the differential cross section for the
emission of a given radiation in a nuclear
reaction, with the angle of emission.
f densité *f* en phase,
distribution *f* angulaire,
répartition *f* angulaire
e distribución *f* angular
i distribuzione *f* angolare
n hoekverdeling
d Winkelverteilung *f*

258 ANGULAR FLUX, gp
VECTOR FLUX
$\phi(r,w,E)w = vQ(r,w,E)w$ where v is the
speed of the neutrons corresponding to
the energy E.
f flux *m* vectoriel
e flujo *m* vectorial
i flusso *m* vettoriale
n vectorflux
d Vektorfluss *m*

259 ANGULAR IMPULSE, ma
IMPULSIVE MOMENT
The time integral of a torque, especially
when applied for a short time.
f impulsion *f* angulaire,
moment *m* d'inertie géométrique
e impulso *m* angular,
momento *m* de inercia geométrica
i impulso *m* angolare,
momento *m* d'inerzia geometrica
n geometrisch traagheidsmoment *n*
d Flächenträgheitsmoment *n*

260 ANGULAR KINETIC ENERGY gp
The kinetic energy of a body rotating
with an angular velocity of a specified
value about an axis.
f énergie *f* cinétique de rotation
e energía *f* cinética de rotación
i energia *f* cinetica di rotazione
n draaiingsenergie
d Rotationsenergie *f*

261 ANGULAR MOMENTUM np
The moment of momentum of a particle
or system.
f moment *m* angulaire
e momento *m* angular
i momento *m* angolare
n impulsmoment *n*
d Drehimpuls *m*

262 ANGULAR-MOMENTUM ma, np
QUANTUM NUMBER
A quantum number that determines the
total angular momentum of a molecule
exclusive of nuclear spin.
f nombre *m* quantique du moment angulaire
e número *m* cuántico del momento angular
i numero *m* quantico del momento angolare
n quantumgetal *n* van het impulsmoment
d Drehimpulsquantenzahl *f*,
Impulsmomentquantenzahl *f*

263 ANHYDROUS HYDROCHLORI- ch, fu
NATION OF ZIRCONIUM BEARING
FUELS,
ZIRCEX PROCESS
f hydrochloruration *f* anhydrique de
combustibles zirconifères
e hidrocloruración *f* anhídrico de combustibles zirconíferos
i idroclorurazione *f* anidrico di combustibili
zirconiferi
n watervrije hydrochlorering van zirkonium-
houdende splijtstoffen
d wasserfreie Hydrochlorierung *f* zirkon-
haltiger Brennstoffe

264 ANION, ch
NEGATIVE ION
An ion having a negative charge.
f anion *m*, ion *m* négatif
e anión *m*, ión *m* negativo
i anione *m*, ione *m* negativo
n anion *n*, negatief ion *n*
d Anion *n*, negatives Ion *n*

265 ANION EXCHANGE ch, mi
An important process in the separation
of uranium from poor ores.
f échange *m* d'anions
e cambio *m* de aniones
i scambio *m* d'anioni
n anionenuitwisseling
d Anionenaustausch *m*

266 ANIONIC CURRENT ec
The portion of the electric current carried
by the anions.
f courant *m* anionique
e corriente *f* aniónica
i corrente *f* anionica
n negatieve ionenstroom
d negativer Ionenstrom *m*

267 ANNEALING mg
A process involving heating and cooling,
usually applied to induce softening.
f recuit *m*
e recocido *m*
i ricottura *f*
n uitgloeien *n*
d Ausglühen *n*

268 ANNERODITE mi
Samarskite mineral containing uranium
and thorium.
f annerodite *f*
e anerodita *f*

i annerodite *f*
n annerodiet *n*
d Annerodit *m*

269 ANNIHILATION   np
A collision between a particle and its
antiparticle in which they both disappear,
their energy being converted into
annihilation radiation.
f annihilation *f*, dématérialisation *f*
e aniquilación *f*
i annichilazione *f*
n annihilatie, verstraling
d Vernichtung *f*, Zerstrahlung *f*

270 ANNIHILATION FORCE   np
Contribution to the force between an
electron and a positron arising from the
fact that these párticles may virtually
annihilate each other, thereby making the
charges less effective and the attraction
less.
f force *f* d'annihilation
e fuerza *f* de aniquilación
i forza *f* d'annichilazione
n annihilatiekracht, verstralingskracht
d Vernichtungskraft *f*, Zerstrahlungskraft *f*

271 ANNIHILATION GAMMA   np
QUANTUM
The gamma quantum emitted upon
annihilation of an electron pair.
f quantum *m* gamma d'annihilation
e cuanto *m* gamma de aniquilación
i quanto *m* gamma d'annichilazione
n annihilatiegammaquantum *n*,
verstralingsgammaquantum *n*
d Vernichtungsgammaquant *m*,
Zerstrahlungsgammaquant *m*

272 ANNIHILATION PHOTON   np
A photon produced by the annihilation of
a positron and an electron.
f photon *m* d'annihilation
e fotón *m* de aniquilación
i fotone *m* d'annichilazione
n annihilatiefoton *n*, verstralingsfoton *n*
d Vernichtungsphoton *n*,
Zerstrahlungsphoton *n*

273 ANNIHILATION RADIATION   np
Electromagnetic radiation produced by
the union, and consequent annihilation,
of a positron and an electron.
f rayonnement *m* d'annihilation
e radiación *f* de aniquilación
i radiazione *f* d'annichilazione
n annihilatiestraling
d Vernichtungsstrahlung *f*

274 ANNULAR IMPACTOR   sa
A device for sorting the heavier particles
of plutonium and uranium from lighter
particles carrying natural activity.
f appareil *m* d'impact annulaire
e impactor *m* anular
i apparecchio *m* d'impatto anulare

n ringvormige opvanger
d ringförmiger Auffänger *m*

275 ANNULAR LINEAR INDUCTION   cl
PUMP
An induction pump with a polyphase
winding arranged as flat coils around an
annulus.
f pompe *f* à induction linéaire à espace
annulaire
e bomba *f* de inducción lineal con espacio
anular
i pompa *f* ad induzione lineare a spazio
anulare
n lineaire inductiepomp met ringvormige
ruimte
d lineare Induktionspumpe *f* mit Kreisring

276 ANNULUS   cl, rt
The space between the inner and outer
surfaces of two concentric cylinders.
f espace *m* annulaire
e espacio *m* anular
i spazio *m* anulare
n ringvormige ruimte
d Kreisring *m*

277 ANODE   ec
The electrode by which the current in a
system enters, i.e. by which the electrons
leave the medium.
f anode *f*
e ánodo *m*
i anodo *m*
n anode
d Anode *f*

278 ANOMALOUS ATOMIC   an, xr
SCATTERING METHOD
A proposed method of X-ray analysis in
which the variation with the X-ray wavelength of the scattering power of certain
atoms is to be exploited.
f méthode *f* analytique par diffusion
atomique anomale
e método *m* analítico por dispersión atómica
anómala
i metodo *m* analitico per deviazione atomica
anomala
n röntgenanalyse door middel van anomale
atomaire verstrooiing
d Röntgenanalyse *f* mittels anomaler
atomarer Streuung

279 ANOMALOUS ELECTRON-   np
MOMENT CORRECTION
A physical constant, equal to the magnetic
moment of an electron, μ_e, divided by the
Bohr magneton.
f correction *f* du moment magnétique
anomal
e corrección *f* del momento magnético
anómalo
i correzione *f* del momento magnetico
anomalo
n correctie van het anomaal magnetisch
moment

d Korrektion f des anomalen magnetischen
Momentes

280 ANOMALOUS MAGNETIC ma
MOMENT
Contribution $e^3 4\pi mc^2$ to the magnetic
moment of the electron arising from
radiative corrections to the value
$eh/2mc$ derived from the unquantized
Dirac electron theory.
f moment m magnétique anomal
e momento m magnético anómalo
i momento m magnetico anomalo
n anomaal magnetisch moment n
d anomales magnetisches Moment n

281 ANOMALOUS VALENCE ch
An exceptional valence that an element
has in certain compounds.
f valence f anomale
e valencia f anómalo
i valenza f anomalo
n anomale valentie
d anomale Valenz f

282 ANOXIC CELL md
A living cell, or organism, to which
oxygen is not freely accessible.
f cellule f anoxique
e célula f anóxica
i cella f anossica
n anoxiecel
d Anoxiezelle f

283 ANTERIOR POSTERIOR VIEW, xr
DORSAL PROJECTION,
POSTERIOR PROJECTION
A radiograph for which the röntgen ray
beam traverses the body from front to
back.
f projection f dorsale,
projection f postérieure,
vue f antéropostérieure
e proyección f dorsal,
proyección f posterior,
vista f anteroposterior
i proiezione f dorsale,
proiezione f posteriore,
vista f anteroposteriore
n ventro-dorsale projectie
d Projektion f von vorn nach hinten

284 ANTERIOR PROJECTION, xr
FRONTAL PROJECTION,
POSTERIOR-ANTERIOR VIEW
A radiograph for which the röntgen ray
beam traverses a body from back to front.
f projection f antérieure,
projection f frontale,
vue f postéro-antérieure
e proyección f anterior,
proyección f frontal,
vista f posteroanterior
i proiezione f anteriore,
proiezione f frontale,
vista f posteroanteriore
n dorso-ventrale projectie
d Projektion f von hinten nach vorn

285 ANTHRACENE mt
A crystalline organic compound, $C_{14}H_{10}$,
of a condensed benzene ring
structure, used frequently as a scintillation crystal.
f anthracène m
e antraceno m
i antracene m
n anthraceen n
d Anthrazen n

286 ANTIBARYON np
The antiparticle of a baryon.
f antibaryon m
e antibarión m
i antibarione m
n antibaryon n
d Antibaryon n

287 ANTIBONDING ORBITAL ch
A molecular orbital whose energy increases
as the two atoms are brought together
corresponding to a net repulsion.
f orbite f électronique non-commune
e órbita f electrónica no común
i orbita f elettronica non comune
n niet-gemeenschappelijke elektronenbaan
d nichtgemeinsame Elektronenbahn f

288 ANTICATHODE xr
In an X-ray tube, the target on which the
electron beam is focused and from which the
X-rays are radiated.
f anode f, anticathode f
e ánodo m, anticátodo m
i anodo m, anticatodo m
n anode, antikatode
d Anode f, Antikatode f

289 ANTICOINCIDENCE ct
The occurrence of a count in a specified
detector unaccompanied simultaneously
or within an assignable time interval
by a count in one or more specified
detectors.
f antocoïncidence f
e anticoincidencia f
i anticoincidenza f
n anticoïncidentie
d Antikoinzidenz f

290 ANTICOINCIDENCE CIRCUIT ct
An electronic circuit with two input
terminals which delivers an output pulse
if an input pulse is received at one specified terminal only, but can deliver no
output pulse within a given time interval
after the occurrence of a pulse at the
other input terminal.
f circuit m d'anticoïncidences,
circuit m de sélection des anticoïncidences
e circuito m de anticoincidencias
i circuito m d'anticoïncidenze
n anticoïncidentieschakeling
d Antikoinzidenzschaltung f

291 ANTICOINCIDENCE COUNTER   ct
An arrangement of counter tubes and
associated circuits which will record
a count if and only if an ionizing particle
passes through certain of the counter
tubes but not through the others.
f compteur *m* d'anticoïncidences
e contador *m* de anticoincidencias
i contatore *m* d'anticoincidenze
n anticoïncidentieteller
d Antikoinzidenzzähler *m*

292 ANTICOINCIDENCE SELECTOR   ct
A selector having an anticoincidence circuit
and pulse shaping circuits which are
placed between each radiation detector and
the corresponding input of the anticoincidence circuit.
f sélecteur *m* d'anticoïncidences
e selector *m* de anticoincidencias
i selettore *m* d'anticoincidenze
n anticoïncidentiekiezer
d Antikoinzidenzwähler *m*

293 ANTICOINCIDENCE SELECTOR   ct
UNIT
A basic function unit containing an anticoincidence selector.
f élément *m* sélecteur d'anticoïncidences
e unidad *f* selectora por anticoincidencias
i elemento *m* selettore per anticoincidenze
n anticoïncidentieschakel
d Antikoinzidenzeinheit *f*

294 ANTI-COMPTON GAMMA-RAY   gr, sp
SPECTROMETER
A gamma-ray spectrometer in which the
effect of Compton scattering is compensated.
f spectromètre *m* à rayons gamma
   anti-Compton
e espectrómetro *m* de rayos gamma
   anti-Compton
i spettrometro *m* a raggi gamma anti-Compton
n anticomptonspectrometer,
   gammaspectrometer met compensatie
   van het comptoneffect
d Anticomptonspektrometer *n*,
   Gammaspektrometer *n* mit Compton-
   Effektkompensation

295 ANTICONTAMINATION   sa
CLOTHING
Clothing provided for all persons in the
exclusion area who work in the immediate
reactor vicinity or enter any of the
laboratories designated as hot.
f vêtements *pl* de protection contre la
   contamination
e ropaje *m* de protección contra la contaminación
i vestiti *pl* di protezione contro la
   contaminazione
n tegen besmetting beschermende kleding
d Strahlenschutzkleidung *f*

296 ANTIDIFFUSION GRID   xr
A barrier consisting of alternating strips
of radiolucent and radio-opaque materials
allowing primary radiation (useful beam)
to pass, and absorb some oblique secondary
radiation.
f grille *f* antidiffusante
e rejilla *f* antidifusora
i griglia *f* antidiffondente
n strooistralenraster *n*
d Sekundärstrahlenblende *f*

297 ANTI-EXPLOSION VALVE,   sa
EXPLOSION TRAP
A pressure-operated valve which isolates
units of plant in the event of an explosion.
f clapet *m* anti-explosion
e válvula *f* de alivio contra explosiones
i valvola *f* antiesplosione
n explosieveilige klep
d explosionssicheres Ventil *n*

298 ANTILEPTON   np
The antiparticle of a lepton.
f antilepton *m*
e antileptón *m*
i antileptone *m*
n antilepton *n*
d Antilepton *n*

299 ANTIMATTER   np
A form of matter in which protons, electrons
and other particles have charges opposite
those with which they are normally
associated.
f antimatière *f*
e antimateria *f*
i antimateria *f*
n antimaterie
d Antimaterie *f*

300 ANTIMONY   ch
Metallic element, symbol Sb, atomic
number 51.
f antimoine *m*
e antimonio *m*
i antimonio *m*
n antimoon *n*
d Antimon *n*

301 ANTINEUTRINO   np
The kind of neutrino presumed to be
emitted during radioactive decay by β
emission or electron capture.
f antineutrino *m*
e antineutrino *m*
i antineutrino *m*
n antineutrino *n*
d Antineutrino *n*

302 ANTINEUTRON   np
A postulated fundamental particle of zero
charge, and mass equal to that of the
neutron, capable of combining with a
neutron and annihilating both particles
with the production of mesons.
f antineutron *m*

e   antineutrón *m*
i   antineutrone *m*
n   antineutron *n*
d   Antineutron *n*

303 ANTIPARTICLE   np
A charge carrier in nuclear physics that
is the counterpart of the particle.
f   antiparticule *f*
e   antipartícula *f*
i   antiparticella *f*
n   antideeltje *n*
d   Antipartikel *n*, Antiteilchen *n*

304 ANTIPROTON,   np
     NEGATIVE PROTON
The antiparticle of a proton.
f   antiproton *m*, proton *m* négatif
e   antiprotón *m*, protón *m* negativo
i   antiprotone *m*, protone *m* negativo
n   antiproton *n*, negatief proton *n*
d   Antiproton *n*, negatives Proton *n*

305 ANTIRATCHETTING GROOVE   fu
A groove preventing the ratchetting of the
envelope of the fuel element.
f   rainure *f* antirochetage,
    rainure *f* empêchant le mouvement de
    long en large
e   ranura *f* eliminadora del movimiento
    vaivén
i   scanalatura *f* anticorrugamenta
n   groef tegen ongewenste verschuiving
d   Rille *f* gegen unerwünschte Verschiebung

306 ANTISHADOWING   co
Shadow effect between control rods whereby the efficiency of one control rod is
enhanced when inserting another rod.
f   interaction *f* positive entre barres de
    commande
e   interacción *f* positiva entre barras de
    regulación
i   interazione *f* positiva fra barre di
    regolazione
n   gunstig schaduweffect *n*
d   günstige Schattenwirkung *f*,
    günstiger Schatteneffekt *m*

307 ANTISTOKES LINES   sp
Spectral lines that result from excitation
by radiation of atoms or molecules in
abnormal energy states.
f   raies *pl* anti-Stokes
e   líneas *pl* anti-Stokes, rayos *pl* anti-Stokes
i   righe *pl* anti-Stokes
n   antistokeslijnen *pl*
d   antistokessche Linien *pl*

308 AORTOGRAPHY   xr
Radiological examination of the aorta after
injection of a sodium iodide solution as a
contrast medium.
f   aortographie *f*
e   aortografía *f*
i   aortografia *f*
n   aortografie
d   Aortographie *f*

309 APERTURE OF AN ACCELERATOR,
     see 33

310 APPARATUS   ge
A general term used for designating
assemblies, sub-assemblies, basic
function units, detectors, etc., in a title
or text of general scope, when it is not
practical to specify them more precisely.
f   appareil *m*
e   aparato *m*
i   apparecchio *m*
n   apparaat *n*
d   Gerät *n*

311 APPARENT ABSORPTION   ab
That part of the total absorption which
arises from the deflection of radiation out
of the primary beam.
f   absorption *f* apparente
e   absorción *f* aparente
i   assorbimento *m* apparente
n   schijnbare absorptie
d   scheinbare Absorption *f*

312 APPEARANCE POTENTIAL,   ma
     THRESHOLD POTENTIAL
The minimum energy that the electron beam
in the ion source must have to produce
ions of particular species when a molecule
is ionized.
f   potentiel *m* d'apparition
e   potencial *m* de aparición
i   potenziale *m* d'apparenza
n   verschijningspotentiaal
d   Erscheinungspotential *n*

313 APPLICATOR   ra
A device containing radium or other
radioactive material, designed to be placed
in a known relationship to the part to be
treated.
f   applicateur *m*
e   aplicador *m*
i   applicatore *m*
n   capsule
d   Kapsel *f*

314 APPLICATOR,   xr
     LOCALIZER, TREATMENT CONE,
     X-RAY THERAPY LOCALIZER
An attachment to an X-ray therapy-tube
head or telecurie-therapy unit, designed
so that it defines the cross section of the
radiation beam.
f   cône *m*, localisateur *m*
e   localizador *m*
i   localizzatore *m*
n   bestralingsconus
d   Bestrahlungstubus *m*, Lokalisator *m*

315 APPROACH TO CRITICALITY   rt
The total of tests carried out in order to
determine experimentally the critical size
of a specific material composition and
geometric disposition.
f   approche *f* sous-critique
e   acercamiento *m* subcrítico

i avvicinamento *m* sottocritico
n criticiteitsnadering
d Kritizitätsnäherung *f*

316 AQUARIUM REACTOR rt
A type of swimming pool reactor in which emphasis is placed on being able to see the core.
f réacteur *m* piscine à coeur visible
e reactor *m* de piscina con núcleo visible
i reattore *m* a piscina con nocciolo visibile
n basisreactor met zichtbare kern
d Schwimmbadreaktor *m* mit offener Spaltzone

317 AQUEOUS HOMOGENEOUS rt
REACTOR,
  SOLUTION TYPE REACTOR
A homogeneous nuclear reactor of small dimensions in which the fuel is dissolved in water.
f réacteur *m* homogène aqueux
e reactor *m* homogéneo acuoso
i reattore *m* omogeneo acquoso
n homogene oplossingsreactor
d Wasserlösungsreaktor *m*

318 AQUEOUS REPROCESSING, fu
WET REPROCESSING
The reprocessing of irradiated fuel elements by solvent extraction and similar wet methods.
f traitement *m* du combustible irradié par voie humide
e método *m* húmedo de reprocesamiento
i trattamento *m* del combustibile irradiato per via umida
n natte opwerking van splijtstof
d nasse Brennstoffaufarbeitung *f*

319 ARC SPECTRUM sp
The spectrum of an element produced in an electric arc.
f spectre *m* d'arc
e espectro *m* del arco
i spettro *m* dell'arco
n boogspectrum *n*
d Bogenspektrum *n*

320 AREA DENSITY, ab
SUPERFICIAL DENSITY
The method of expressing the thickness of a layer of absorbing material as mass per unit surface area.
f densité *f* superficielle
e densidad *f* superficial
i densità *f* superficiale
n oppervlaktedichtheid
d Oberflächendichte *f*

321 AREA MONITOR me, sa
Any device used for detecting and /or measuring nuclear radiation levels at a given location for control purposes.
f moniteur *m* de terrain
e monitor *m* de área
i avvisatore *m* d'area
n terreinmonitor
d Gebietsmonitor *m*, Raummonitor *m*

322 AREA MONITORING sa
Periodic or continuous determination of the amount of ionizing radiation or radioactive contamination present in a specified area.
f supervision *f* de terrain
e vigilancia *f* de área
i controllo *m* di radiazione ionizzante d'area, controllo *m* di radioattività d'area
n gebiedsbewaking, terreinbewaking
d Gebietsüberwachung *f*, Raumüberwachung *f*

323 ARGON ch
Gaseous element, symbol A, atomic number 18.
f argon *m*
e argón *m*
i argon *m*
n argon *n*
d Argon *n*

324 ARGON CHAMBER ic
An ionization chamber filled with argon.
f chambre *f* d'ionisation à remplissage d'argon
e cámara *f* de ionización rellena de argón
i camera *f* d'ionizzazione a riempimento d'argon
n ionisatievat *n* met argonvulling
d Ionisationskammer *f* mit Argonfüllung

325 ARSENIC ch
Metallic element, symbol As, atomic number 33.
f arsenic *m*
e arsénico *m*
i arsenico *m*
n arseen *n*, arsenicum *n*
d Arsen *n*

326 ARTERIOGRAPHY xr
The radiological examination of arteries following direct injection of a contrast material.
f artériographie *f*
e arteriografía *f*
i arteriografia *f*
n arteriografie
d Arteriographie *f*

327 ARTHROGRAPHY xr
The radiological examination of a joint cavity following direct injection of air or other contrast medium.
f arthrographie *f*
e artrografía *f*
i artrografia *f*
n arthrografie
d Arthrographie *f*

328 ARTIFICIAL NUCLEAR np
DISINTEGRATION,
  ARTIFICIAL NUCLEAR TRANSFORMATION
Artificially induced transformation or change, involving nuclei.

f désintégration f nucléaire artificielle
e desintegración f nuclear artificial
i disintegrazione f nucleare artificiale
n kunstmatige kerndesintegratie
d künstlicher Kernzerfall m

329 ARTIFICIAL RADIOACTIVE   ch, np
 ELEMENT,
  ARTIFICIAL RADIONUCLIDE
 A radioactive element produced artificially
 in contrast with radioactive elements which
 occur in nature.
f élément m radioactif artificiel
e elemento m radiactivo artificial
i elemento m radioattivo artificiale
n kunstmatig radioactief element n
d künstlich-radioaktives Element n

330 ARTIFICIAL RADIOACTIVITY,   ra
 INDUCED RADIOACTIVITY
 Radioactivity induced by irradiation.
f radioactivité f artificielle,
 radioactivité f induite
e radiactividad f artificial,
 radiactividad f inducida
i radioattività f artificiale,
 radioattività f indotta
n geïnduceerde radioactiviteit,
 kunstmatige radioactiviteit
d induzierte Radioaktivität f

331 ARTIFICIAL TRANSMUTATION   ch, np
 · OF ELEMENTS
 The transmutation of one element into
 another by artificial means.
f transmutation f artificielle d'éléments
e transmutación f artificial de elementos
i trasmutazione f artificiale d'elementi
n kunstmatige transmutatie van elementen
d künstliche Transmutation f von
 Elementen

332 ARTIFICIALLY ACCELERATED   rt, sa
 SAFETY MECHANISM
 A safety mechanism in which an extra
 force is provided in order to activate
 the particular safety member into a
 nuclear reactor.
f mécanisme m de sécurité à accélération
 complémentaire
e mecanismo m de seguridad acelerado
i meccanismo m di sicurezza ad
 accelerazione complementare
n kunstmatig versneld veiligheidsmechanisme n
d künstlich beschleunigte Sicherheitsvorrichtung f

333 ASBESTOS WOOL,   mt
 COTTON ASBESTOS
 Material used e.g. in air filters.
f laine f d'amiante
e lana f de amianto
i lana f d'amianto
n asbestwol
d Asbestwolle f

334 ASSAY   mi
 Examination and determination of the
 constants and properties of ores.
f essai m
e ensayo m
i assaggio m
n analyse, essaaieren n
d Probe f, Untersuchung f

335 ASSAY METER   me
 An instrument used for the assay of radioactive material.
f instrument m d'essai
e equipo m de ensayo
i strumento m d'assaggio
n essaaieertoestel n
d Probiergerät n

336 ASSEMBLY   ge
 A well-defined set of members necessary
 and sufficient to achieve a specified total
 function.
f ensemble m
e conjunto m
i complesso m
n opstelling, samenstel n
d Anordnung f

337 ASSEMBLY FOR FAILED   me, rt
 ELEMENT LOCALIZATION
 An assembly designed for localization of
 failed elements by scanning the different
 channels of the nuclear reactor.
f ensemble m de localisation de rupture de
 gaine
e conjunto m de localización de averías
i complesso m per localizzazione di
 rottura di guaine
n meetopstelling voor de localisatie van
 beschadigde splijtstofelementen
d Anordnung f zur Lokalisierung schadhafter
 Brennelemente

338 ASSEMBLY FOR THE MEASURE-   ma
 MENT OF RADIOACTIVITY IN
 AN ORGANISM
 An assembly designed to detect the
 presence of radionuclides in an organism
 and to measure their activity, possibly
 with identification of the nuclides concerned.
f ensemble m de mesure de la radioactivité
 présente dans un organisme
e conjunto m de medida de la radiactividad
 en un organismo
i complesso m di misura della radioattività
 in un organismo
n opstelling voor de meting van de in een
 organisme aanwezige radioactiviteit
d Anordnung f zur Bestimmung der Radioaktivität in Organismen

339 ASSOCIATED CORPUSCULAR   xr
 EMISSION
 The full complement of secondary charged
 particles associated with the X-ray or
 γ-ray beam in its passage through air.
f émission f corpusculaire associée

e emisión *f* corpuscular asociada
i emissione *f* corpuscolare associata
n er mee gepaard gaande corpusculaire
straling,
secondaire corpusculaire straling
d sekundäre Korpuskularstrahlung *f*

340 ASSOCIATED WAVE np
A wave supposed to be associated with
material particles.
f onde *f* associée
e onda *f* asociada
i onda *f* associata
n begeleidende golf
d Begleitwelle *f*

341 ASSUMPTION OF CONGRUENCE ma
Equation used in.defining ambipolar
diffusion.
f supposition *f* de congruence
e suposición *f* de congruencia
i supposizione *f* di congruenza
n congruentieveronderstelling
d Kongruenzannahme *f*

342 ASTATINE ch
Radioactive element, symbol At, atomic
number 85.
f astate *m*, astatine *m*
e astatino *m*
i astatino *m*
n astaat *n*, astatium *n*
d Astatin *n*, Astatium *n*

343 ASTIGMATIC FOCUSING, sp
LINE FOCUSING
In particle spectrometers, the lens-like
action of a suitably designed field whereby
particles which diverge at first will come
together again in a line.
f focalisation *f* astigmatique
e enfoque *m* astigmático
i focalizzazione *f* astigmatica
n astigmatische focussering
d astigmatische Fokussierung *f*

344 ASTON WHOLE NUMBER RULE is, ma
The atomic weights of isotopes are whole
numbers when expressed in atomic weight
units, and the deviations from the whole
numbers of the atomic weights of the
elements are due chiefly to the presence
of isotopes with different weights.
f règle *f* des nombres entiers d'Aston
e regla *f* de los números interos de Aston
i regola *f* dei numeri interi d'Aston
n astonregel, regel van Aston
d Aston-Regel *f*

345 ASTRON pp
A system involving use of a cylindrical
sheath of rotating electrons, moving at
speeds approaching that of light, to produce
plasma pinch.
f système *m* astron
e sistema *m* astrón
i sistema *m* astron

n astronsysteem *n*
d Astron-System *n*

346 ASYMMETRY ENERGY, np
SYMMETRY ENERGY
The energy which arises from the nucleus
having unequal numbers of neutrons and
protons (sometimes called symmetry
energy).
f énergie *f* d'asymétrie
e energía *f* de asimetría
i energia *f* d'asimmetria
n asymmetrie-energie
d Asymmetrieenergie *f*

347 ASYMPTOTE ma
A line such that a point, tracing a given
curve and simultaneously receding to an
infinite distance from the origin,
approaches indefinitely near to the line.
f asymptote *f*
e asintota *f*
i asintoto *m*
n asymptoot
d Asymptote *f*

348 ASYMPTOTIC ma
Pertaining to, or of the nature of, an
asymptote.
f asymptotique adj
e asintótico adj
i asintotico adj
n asymptotisch adj
d asymptotisch adj

349 ASYMPTOTIC NEUTRON np
FLUX DENSITY
The flux density far from boundaries,
localized sources and localized absorbers.
f densité *f* asymptotique de flux de
neutrons
e densidad *f* asintótica de flujo de
neutrones
i densità *f* asintotica di flusso neutronico
n asymptotische neutronenfluxdichtheid
d asymptotische Neutronenflussdichte *f*

350 ATMOSPHERE, ge
EARTH'S ATMOSPHERE
The spheroidal gaseous envelope surrounding the earth.
f atmosphère *f*
e atmósfera *f*
i atmosfera *f*
n atmosfeer
d Atmosphäre *f*

351 ATMOSPHERIC POLLUTION,
see 163

352 ATMOSPHERIC RADIATION ra
ATTENUATION
Attenuation of radiation by the atmosphere.
f atténuation *f* atmosphérique du rayonnement
e atenuación *f* atmosférica de la radiación
i attenuazione *f* atmosferica della radiazione

n   atmosferische stralingsverzwakking
d   atmosphärische Strahlungsschwächung *f*

353 ATMOSPHERIC RADIOACTIVITY   cr, np
Radioactivity detectable in the atmosphere,
apparently emanating from minute floating
radioactive particles.
f   radioactivité *f* atmosphérique
e   radiactividad *f* atmosférica
i   radioattività *f* atmosferica
n   atmosferische radioactiviteit
d   atmosphärische Radioaktivität *f*

354 ATMOSPHERIC STABILITY   rw
A measure of the vertical acceleration
of small volumes of air by buoyancy forces
which is determined by the vertical air
density gradient.
f   stabilité *f* atmosphérique
e   estabilidad *f* atmosférica
i   stabilità *f* atmosferica
n   atmosferische stabiliteit
d   atmosphärische Stabilität *f*

355 ATMOSPHERIC TRANSMITTANCE   gp
The fraction or percentage of the thermal
energy received at a given location after
passage through the atmosphere relative to
that which would have been received at the
same location if the atmosphere were
present.
f   transmittance *f* atmosphérique
e   transmitancia *f* atmosférica
i   trasmettenza *f* atmosferica
n   warmtedoorlatend vermogen *n* van de
    atmosfeer
d   atmosphärische Wärmedurchlässigkeit *f*

356 ATOM   ch, np
A unit of matter consisting of a single
nucleus surrounded by one or more orbital
electrons.
f   atome *m*
e   átomo *m*
i   atomo *m*
n   atoom *n*
d   Atom *n*

357 ATOM BOMB, see 1

358 ATOM PERCENT   is, np
The number of atoms of an element or
isotope in a mixture expressed as percentage of the total number of atoms present.
f   pourcentage *m* atomique
e   porcentaje *m* atómico
i   percentuale *m* atomico
n   atoompercentage *n*
d   Atomprozentsatz *m*

359 ATOM PHYSICS   gp
The physics of the electron shells.
f   physique *f* atomique
e   física *f* atómica
i   fisica *f* atomica
n   atoomfysica
d   Atomphysik *f*

360 ATOM SMASHER, see 32

361 ATOMIC ABSORPTION   ab, np
     COEFFICIENT
The fractional decrease in intensity of a
beam of radiation per number of atoms
per unit area.
f   coefficient *m* d'absorption atomique
e   coeficiente *m* de absorción atómica
i   coefficiente *m* d'assorbimento atomico
n   atomaire absorptiecoëfficiënt
d   atomarer Absorptionskoeffizient *m*

362 ATOMIC ACCELERATOR   pa
An accelerator of atomic particles.
f   accélérateur *m* de particules atomiques
e   acelerador *m* de partículas atómicas
i   acceleratore *m* di particelle atomiche
n   versneller van atomaire deeltjes
d   Beschleuniger *m* von atomaren Teilchen

363 ATOMIC AGE,   ge
     NUCLEAR ERA
The historical period which commenced
with the first utilization of atomic
(nuclear) energy.
f   âge *m* atomique, ère *f* nucléaire
e   era *f* atómica, era *f* nuclear
i   età *f* atomica, età *f* nucleare
n   atoomtijdperk *n*, kernenergietijdperk n
d   Atomzeitalter *n*, Kernenergiezeitalter *n*

364 ATOMIC AIR BURST, see 147

365 ATOMIC ARRANGEMENT,   ch
     ATOMIC CONFIGURATION
Of a molecule, the spatial distribution
of its atoms.
f   disposition *f* atomique
e   configuración *f* atómica
i   disposizione *f* atomica
n   atoomrangschikking, configuratie
d   räumliche Anordnung *f* der Atome

366 ATOMIC ATTENUATION   ma
     COEFFICIENT
$\mu a = \mu/n$, where n is the number density
of atoms in the substance.
f   coefficient *m* d'atténuation atomique
e   coeficiente *m* de atenuación atómica
i   coefficiente *m* d'attenuazione atomica
n   atomaire verzwakkingscoëfficiënt,
    verzwakkingscoëfficiënt per atoom
d   atomarer Schwächungskoeffizient *m*

367 ATOMIC BATTERY,   ms
     NUCLEAR BATTERY,
     RADIOACTIVE BATTERY,
     RADIOISOTOPIC GENERATOR
A small power generator that converts the
heat released during radioactive decay
into electricity directly.
f   batterie *f* atomique
e   pila *f* atómica
i   batteria *f* atomica
n   atoombatterij, radioactieve spanningsbron
d   Atombatterie *f*,
    radioaktive Spannungsquelle *f*

368 ATOMIC BEAM   np
   Gas atoms emerging from a small aperture
   into a high vacuum and collimated by one
   or more additional apertures so as to form
   a narrow beam.
f faisceau *m* atomique,
   faisceau *m* d'atomes
e haz *m* atómico, haz *m* de átomos
i fascio *m* atomico, fascio *m* d'atomi
n atoombundel
d Atomstrahl *m*

369 ATOMIC-BEAM FREQUENCY   np
      STANDARD
   A frequency standard that provides one or
   more precise frequencies derived from
   an element such as cesium.
f fréquence *f* standard de faisceau atomique
e frecuencia *f* patrón de haz atómico
i frequenza *f* campione di fascio atomico
n standaardfrequentie voor atoomstralen
d Standardfrequenz *f* für Atomstrahlen.

370 ATOMIC BOND   np
   A valence linkage between atoms consisting
   of a pair of electrons, one of which has
   been contributed by each of the atoms
   bonded.
f liaison *f* atomique
e enlace *m* atómico
i legame *m* atomico
n atoombinding
d atomare Bindung *f*, Atombindung *f*

371 ATOMIC CHARGE   np
   The product of the number of electrons
   an atom has gained or lost in ionization and
   the charge of one electron.
f charge *f* atomique, nombre *m* de charge
e carga *f* atómica
i carica *f* atomica
n atoomlading
d Atomladung *f*

372 ATOMIC CLOCK   ms
   A device that uses the vibrations of atomic
   nuclei or molecules to measure time
   intervals.
f chronomètre *m* atomique
e cronómetro *m* atómico
i cronometro *m* atomico
n atoomklok
d Atomuhr *f*

373 ATOMIC CLOUD   nw
   The cloud of hot gases, smoke, dust and
   other matter that is carried aloft after the
   explosion of a nuclear weapon.
f nuage *m* de bombe atomique
e nube *f* de bomba atómica
i nube *f* da bomba atomica
n atoombomwolk
d Atombombenwolke *f*

374 ATOMIC CORE,   np
      ATOMIC KERNEL, ATOMIC TRUNK
   An atomic nucleus and its complement of
   closed electron shells without any

additional or valence electrons.
f tronc *m* de l'atome
e tronco *m* del átomo
i tronco *m* dell'atomo
n atoomromp
d Atomrumpf *m*

375 ATOMIC CROSS SECTION   np
   The cross section of an atom for a
   particular process.
f section *f* atomique
e sección *f* atómica
i sezione *f* atomica
n atoomdoorsnede
d Atomquerschnitt *m*

376 ATOMIC DENSITY   np
   The number of atoms in a unit volume
   of a substance.
f densité *f* atomique
e densidad *f* atómica
i densità *f* atomica
n atoomdichtheid
d Atomdichte *f*

377 ATOMIC DIAMETER   np
   The diameter of an individual atom.
f diamètre *m* atomique
e diámetro *m* atómico
i diametro *m* atomico
n atoomdoorsnede
d Atomquerschnitt *m*

378 ATOMIC DIFFUSION,   is
      ATOMIC SOLUTION DIFFUSION,
      AUTODIFFUSION
   One cause of internal friction in solid
   solutions is the diffusion of the constituent
   atoms.
f diffusion *f* interatomique
e difusión *f* interatómica
i diffusione *f* interatomica
n interatomaire diffusie
d interatomare Diffusion *f*

379 ATOMIC DISTANCE   ch
   The average separation between atoms.
f distance *f* des atomes dans la molécule
e distancia *f* de los átomos en la molécula
i distanza *f* degli atomi nella molecola
n atoomafstand in de molecule
d Atomabstand *m* im Molekül

380 ATOMIC ELECTRON   np
   One of the bound electrons that surround
   the nucleus of an atom.
f électron *m* de l'atome
e electrón *m* del átomo
i elettrone *m* dell'atomo
n atoomelektron *n*
d Atomelektron *n*

381 ATOMIC ENERGY,   np
      NUCLEAR ENERGY
   Energy released in nuclear reactions or
   transitions.
f énergie *f* nucléaire
e energía *f* nuclear

i   energia *f* nucleare
n   kernenergie
d   Kernenergie *f*

382 ATOMIC ENERGY LEVELS   np
  1. The values of the energy corresponding
  to the stationary states of an isolated atom.
  2. The set of stationary states in which
  an atom of a particular species may be
  found, including the ground state, or
  normal state, and the excited states.
f   niveaux *pl* d'énergie atomique
e   niveles *pl* de energía atómica
i   livelli *pl* d'energia atomica
n   atoomenergieniveaus *pl*
d   Atomenergieniveaus *pl*

383 ATOMIC EXCITATION FUNCTION   np
  The cross section for the excitation of an
  atom to a particular excited state expressed as a function of the energy of the
  incident electrons.
f   fonction *f* d'excitation atomique
e   función *f* de excitación atómica
i   funzione *f* d'eccitazione atomica
n   atomaire aanslagfunctie
d   atomare Anregungsfunktion *f*

384 ATOMIC FIELD   np
  The region surrounding an atom in which
  the repulsion forces for other particles
  are considerable.
f   champ *f* atomique
e   campo *m* atómico
i   campo *m* atomico
n   atoomveld *n*
d   Atomfeld *n*

385 ATOMIC FORM FACTOR,   xr
     ATOMIC SCATTERING FACTOR
  In the determination of the scattering
  efficiency of X-radiation by a given crystal
  atom, a factor dependent on the charge
  density, the wavelength, the duration and
  angle of incidence of the beam.
f   facteur *m* de diffusion atomique
e   factor *m* de dispersión atómica
i   fattore *m* di deviazione atomica
n   atoomvormfactor
d   atomarer Streufaktor *m*, Atomformfaktor *m*

386 ATOMIC FREQUENCY   np
  The vibrational frequency of an atom, used
  particularly with respect to the solid state.
f   fréquence *f* atomique
e   frecuencia *f* atómica
i   frequenza *f* atomica
n   atoomfrequentie
d   Atomfrequenz *f*

387 ATOMIC HEAT   ch
  Of an element, the thermal capacity of one
  gram(me)-atom, which is equivalent to
  the product of the atomic mass and the
  specific heat.
f   chaleur *f* atomique
e   calor *m* atómico

i   calore *m* atomico
n   atoomwarmte
d   Atomwärme *f*

388 ATOMIC INTERACTION   np
  Any force that occurs between atoms.
f   interaction *f* atomique
e   interacción *f* atómica
i   interazione *f* atomica
n   atomaire wisselwerking
d   atomare Wechselwirkung *f*

389 ATOMIC JUMP   np
  Displacement of an atom from a lattice
  position to a void position.
f   saut *m* atomique
e   salto *m* atómico
i   salto *m* atomico
n   atoomsprong
d   Atomsprung *m*

390 ATOMIC MASS   np
  The mass of a neutral atom of a nuclide.
f   masse *f* atomique
e   masa *f* atómica
i   massa *f* atomica
n   atoommassa
d   Atommasse *f*

391 ATOMIC MASS CONSTANT,   np
     UNIFIED ATOMIC MASS CONSTANT
  1/12 of the rest mass of an atom or
  nuclide C-12.
f   constante *f.* de masse atomique,
    constante *f* unifiée de masse atomique
e   constante *f* de masa atómica,
    constante *f* unificada de masa atómica
i   costante *f* di massa atomica,
    costante *f* unificata di massa atomica
n   geünificeerde atoommassaconstante
d   kollektive Atommassenkonstante *f*

392 ATOMIC MASS CONVERSION   np, un
     FACTOR,
     MASS CONVERSION FACTOR,
     MASS-ENERGY CONVERSION FORMULA
  The experimentally determined ratio of the
  atomic weight unit to the atomic mass unit.
f   facteur *m* de conversion awu en amu,
    facteur *m* de conversion de la masse
    atomique
e   factor *m* de conversión awu en amu,
    factor *m* de conversión de la masa atómica
i   fattore *m* di conversione awu in amu,
    fattore *m* di conversione della massa
    atomica
n   omrekeningsfactor voor de atoommassa
d   Atommassenumrechnungsfaktor *m*,
    Smythe-Faktor *m*,
    Umrechnungsfaktor *m* zwischen Atomge-
    wicht und ME

393 ATOMIC MASS UNIT, see 233

394 ATOMIC MIGRATION   ch, gp
  The progressive transfer of the valence
  electrons from atom to atom within a
  molecule.

f migration *f* atomique
e migración *f* atómica
i migrazione *f* atomica
n atoomzwerving
d Atomwanderung *f*

f paramètres *pl* atomiques
e parámetros *pl* atómicos
i parametri *pl* atomici
n atoomparameters *pl*
d Atomparameter *pl*

395 ATOMIC MODERATION RATIO,   rt
      ATOMIC RATIO OF MODERATOR
      TO FUEL,
      DEGREE OF MODERATION
The ratio of the number of atoms of a
moderator to the number of atoms of a
fissile (fissionable) nuclide, contained in
a specified volume.
f rapport *m* de modération atomique
e relación *f* de moderación atómica
i rapporto *m* di moderazione atomica
n atomaire moderatieverhouding
d atomares Bremsverhältnis *n*

396 ATOMIC NUCLEUS,   np
      NUCLEUS
The positively charged central portion of
an atom, with which is associated almost
the whole mass of the atom, but only a
minute part of its volume.
f noyau *m*, noyau *m* atomique
e núcleo *m*, núcleo *m* atómico
i nucleo *m*, nucleo *m* atomico
n atoomkern, kern
d Atomkern *m*, Kern *m*

397 ATOMIC NUMBER,   np
      PROTON NUMBER
Number of protons in a nucleus; number of
electrons in a neutral atom.
f numéro *m* atomique, nombre *m* de protons
e número *m* atómico, número *m* de protones
i numero *m* atomico, numero *m* di protoni
n atoomgetal *n*, protongetal *n*
d Atomzahl *f*, Protonzahl *f*

398 ATOMIC ORBIT   np
The closed trajectory followed by an atom
under the influence of a central force.
f orbite *f* de l'atome
e órbita *f* del átomo
i orbita *f* dell' atomo
n atoombaan
d Atombahn *f*

399 ATOMIC ORBITAL,   np
      ORBITAL
A particular energy state, or a particular
wave function in quantum mechanics, from
which the probability of finding an electron
at a given point can be calculated.
f fonction *f* de l'orbite électronique dans
   l'atome
e función *f* de la órbita electrónica en el
   átomo
i funzione *f* dell'orbita elettronica nell'atomo
n baanfunctie van de elektronen in het atoom
d Bahnfunktion *f* der Elektronen im Atom

400 ATOMIC PARAMETERS   cr
The inter-atomic distances in a crystal
lattice.

401 ATOMIC PERCENTAGE,   np
      ATOMIC RATIO
The ratio of quantities of different constituents in a given sample in terms of the
numbers of atoms present, as opposed to
the ratio of masses or volumes.
f rapport *m* atomique
e relación *f* atómica
i rapportc *m* atomico
n atoomverhouding
d Atomverhältnis *n*

402 ATOMIC PHOTOELECTRIC   np
      EFFECT,
      PHOTOIONIZATION
Ejection of a bound electron from an inner
orbit of a gas atom by a photon of light
or ultraviolet radiation.
f photoionisation *f*
e efecto *m* fotoeléctrico atómico,
   fotoionización *f*
i fotoionizzazione *f*
n foto-ionisatie
d Photoionisation *f*

403 ATOMIC PLANE,   cr
      LATTICE PLANE
A plane passed through the atoms of a
crystal space lattice, in accordance with
certain rules relating its position to the
crystallographic axes.
f plan *m* réticulair
e plano *m* de la red
i piano *m* del reticolo
n roostervlak *n*
d Gitterebene *f*

404 ATOMIC POLARIZABILITY   np
The susceptibility per atom.
f polarisabilité *f* atomique
e polarizabilidad *f* atómica
i polarizzabilità *f* atomica
n polarisatievermogen *n* van een atoom
d Polarisierbarkeit *f* eines Atoms

405 ATOMIC POWER,   np
      NUCLEAR POWER
Power obtained from nuclear reactions,
usually electric power for industrial
purposes or motive power for propulsion.
f énergie *f* nucléaire
e energía *f* nuclear
i energia *f* nucleare
n kernenergie
d Kernenergie *f*

406 ATOMIC PROPERTIES   gp
Properties of substances which are due to
the nature of the constituent atoms.
f propriétés *f* atomiques
e propiedades *pl* atómicas
i proprietà *pl* atomiche

n   atoomeigenschappen *pl*
d   Atomeigenschaften *pl*

407 ATOMIC RADIUS   gp
Can be assigned to each atom as if it were
a sphere of a definite size by a comparison
of the crystal structures of covalent
compounds.
f   rayon *m* atomique
e   radio *m* atómico
i   raggio *m* atomico
n   atoomradius
d   Atomradius *m*

408 ATOMIC RATIO, see 401

409 ATOMIC REFRACTION   gp
Of an element, the product of its atomic
weight and its specific refractive power.
f   réfraction *f* atomique
e   refracción *f* atómica
i   rifrazione *f* atomica
n   atoomrefractie
d   Atomrefraktion *f*

410 ATOMIC SCALE   ch
f   échelle *f* des poids atomiques
e   escala *f* de los pesos atómicos
i   scala *f* dei pesi atomici
n   atoomgewichtenschaal
d   Atomgewichtsskale *f*

411 ATOMIC SCATTERING   np
        COEFFICIENT
A measure of the power of a given material
to scatter an electron beam passing
through it.
f   coefficient *m* de diffusion atomique
e   coeficiente *m* de dispersión atómica
i   coefficiente *m* di deviazione atomica
n   atomaire verstrooiingscoëfficiënt
d   atomarer Streukoeffizient *m*

412 ATOMIC SCATTERING FACTOR   np
A factor representing the efficiency with
which X-rays of a given frequency are
scattered into a given direction by a given
atom usually measured in terms of the
corresponding factor for a point electron.
f   facteur *m* de diffusion atomique
e   factor *m* de dispersión atómica
i   fattore *m* di deviazione atomica
n   atomaire verstrooiingsfactor
d   atomarer Streufaktor *m*

413 ATOMIC SHELL STRUCTURE,   np
        SHELL STRUCTURE OF THE ATOM
A formerly accepted theory as to the
structure of the atom postulated electrons
moving about a nucleus somewhat as the
planets move about the sun.
f   structure *f* de l'atome en couches,
    structure *f* quantique de l'atome
e   estructura *f* de capas del átomo
i   struttura *f* di strato dell'atomo
n   schilstructuur van het atoom
d   Schalenaufbau *m* des Atomkerns,
    Schalenstruktur *f*

414 ATOMIC SPECTRUM   np
The spectrum of radiations from an atom
arising from internal changes.
f   spectre *m* atomique
e   espectro *m* atómico
i   spettro *m* atomico
n   atoomspectrum *n*
d   Atomspektrum *n*

415 ATOMIC STATE   np
The energy state of an atom.
f   état *m* de l'atome
e   estado *m* del átomo
i   stato *m* dell'atomo
n   atoomtoestand
d   Atomzustand *m*

416 ATOMIC STOPPING POWER,   np
        STOPPING CROSS SECTION
$S_a = S/n$, where n is the number density
of atoms in the substance.
f   pouvoir *m* d'arrêt atomique
e   poder *m* de frenado atómico
i   potere *m* di rallentamento atomico
n   atomair stoppend vermogen *n*,
    stoppend vermogen *n* per atoom
d   atomares Bremsvermögen *n*

417 ATOMIC STRUCTURE   np
The internal structure of the atom, including
its nucleus and the electrons that surround
the nucleus.
f   constitution *f* de l'atome,
    structure *f* atomique
e   estructura *f* atómica
i   costituzione *f* dell'atomo
n   atoomstructuur
d   Atombau *m*

418 ATOMIC SURFACE BURST,   nw
        SURFACE BURST
The explosion of a nuclear weapon at an
elevation such that the fireball touches
ground.
f   explosion *f* nucléaire superficielle
e   explosión *f* nuclear superficial
i   esplosione *f* nucleare superficiale
n   kernexplosie aan de oppervlakte
d   Oberflächenkernexplosion *f*

419 ATOMIC SUSCEPTIBILITY   gp
The change in magnetic moment of one
gram(me)-atom of a substance produced
by the application of a magnetic field
of unit strength.
f   susceptibilité *f* atomique
e   susceptibilidad *f* atómica
i   suscettibilità *f* atómica
n   atoomsusceptibiliteit
d   Atomsuszeptibilität *f*

420 ATOMIC THEORY   np
The assumption that matter is not infinitely
divisible but is composed of ultimate
particles or atoms.
f   théorie *f* atomique
e   teoría *f* atómica

i teoria f atomica
n atoomtheorie
d Atomtheorie f

421 ATOMIC TRANSMUTATION   np
The changing of an atom into an atom of
different atomic number, or in other words,
into an atom of a different element.
f transmutation f atomique
e transmutación f atómica
i trasmutazione f atomica
n atoomtransmutatie
d Atomtransmutation f

422 ATOMIC UNDERGROUND BURST,   nw
NUCLEAR UNDERGROUND BURST,
UNDERGROUND BURST
The explosion of an atomic weapon with
its centre(er) beneath the surface of the
ground.
f explosion f nucléaire souterraine
e explosión f nuclear subterránea
i esplosione f nucleare sotterranea
n ondergrondse kernexplosie
d unterirdische Kernexplosion f

423 ATOMIC UNDERWATER BURST,   nw
NUCLEAR UNDERWATER BURST,
UNDERWATER BURST
The explosion of a nuclear weapon with
its centre(er) beneath the surface of water.
f explosion f nucléaire sous-marine
e explosión f nuclear submarina
i esplosione f nucleare subacquea
n onderwaterkernexplosie
d Unterwasserkernexplosion f

424 ATOMIC UNITS,   un
HARTREE UNITS
n = the quantum of angular momentum,
m = the mass of the electron,
e = the charge on the electron.
f unités pl de Hartree
e unidades pl de Hartree
i unità pl di Hartree
n hartree-eenheden pl
d Hartree-Einheiten pl

425 ATOMIC VOLUME   ma
A numerical result obtained by dividing the
atomic weight of an element by its density.
f volume m atomique
e volumen m atómico
i volume m atomico
n atoomvolume n
d Atomvolumen n

426 ATOMIC WEIGHT   ch
For a given specimen of an element, the
mean weight of its atoms, expressed
either in amu or awu.
f poids m atomique
e peso m atómico
i peso m atomico
n atoomgewicht n
d Atomgewicht n

427 ATOMIC WEIGHT   ch, gp, ma
CONVERSION FACTOR
The factor for converting the atomic
weight values between the chemical and
physical scales.
f facteur m de conversion du poids atomique
e factor m de conversión del peso atómico
i fattore m di conversione del peso atomico
n atoomgewichtomrekeningsfactor
d Atomgewichtsumrechnungsfaktor m

428 ATOMIC WEIGHT UNIT,   ch
AWU
One-sixteenth of the mean mass of the
neutral atoms of naturally occurring
oxygen.
f unité f de poids atomique
e unidad f de peso atómico
i unità f di peso atomico
n eenheid van atoomgewicht
d Einheit f des Atomgewichts

429 ATTACHMENT COEFFICIENT   np
Defined as the number of negative ions
which are formed by one electron moving
along 1 cm path in field direction.
f coefficient m d'attachement
e coeficiente m de adhesión
i coefficiente m d'aderenza
n aanhechtingscoëfficiënt
d Anlagerungskoeffizient m

430 ATTACK RATE   mt
The rate at which the radiolytic attack in
reactor materials takes place.
f vitesse f d'attaque
e velocidad f de ataque
i velocità f d'attacco
n aantastingstempo n
d Angriffrate f

431 ATTENUATION   ra
The reduction of a radiation quantity upon
passage of radiation through matter
resulting from all types of interaction
with that matter.
f atténuation f
e atenuación f
i attenuazione f
n verzwakking
d Schwächung f

432 ATTENUATION COEFFICIENT   ra
Of a substance, for a parallel beam of
specified radiation, the quantity μ in the
expression $\mu \Delta x$ for the fraction removed
by attenuation in passing through a thin
layer of thickness Δx of that substance.
f coefficient m d'atténuation
e coeficiente m de atenuación
i coefficiente m d'attenuazione
n verzwakkingscoëfficiënt
d Schwächungskoeffizient m

433 ATTENUATION CURVE, see 17

434 ATTENUATION FACTOR np
For a given attenuating body in a given
configuration, the factor by which a
radiation quantity at some point of interest
is reducing owing to the interposition of
the body between the source of radiation
and the point of interest.
f facteur *m* d'atténuation
e factor *m* de atenuación
i fattore *m* d'attenuazione
n verzwakkingsfactor
d Schwächungsfaktor *m*

435 ATTENUATOR np
Material placed so as to reduce the intensity of radiation passing through it.
f atténuateur *m*
e atenuador *m*
i attenuatore *m*
n verzwakkend materiaal *n*
d Schwächungsstoff *m*

436 ATTRACTIVE FORCE, gp
FORCE OF ATTRACTION
A force acting on a particle such that the
acceleration of the particle is in the
direction of the agency responsible for the
force.
f force *f* d'attraction
e fuerza *f* de atracción
i forza *f* d'attrazione
n aantrekkingskracht
d Anziehungskraft *f*

437 AUGER COEFFICIENT np
The ratio of the Auger yield to fluorescence
yield.
f coefficient *m* Auger
e coeficiente *m* de Auger
i coefficiente *m* d'Auger
n augercoëfficiënt
d Auger-Koeffizient *m*

438 AUGER EFFECT np
A change in an atom from an excited to
a lower energy state, accompanied by the
emission of an electron but without
radiation.
f effet *m* Auger
e efecto *m* Auger
i effetto *m* Auger
n augereffect *n*
d Auger-Effekt *m*

439 AUGER ELECTRON np
An electron ejected from an excited atom
in the Auger effect.
f électron *m* Auger
e electrón *m* Auger
i elettrone *m* Auger
n augerelektron *n*
d Auger-Elektron *n*

440 AUGER SHOWER, np
EXTENSIVE SHOWER,
GIANT AIR SHOWER
A shower of Auger electrons or in a more
general sense any extensive shower.

f gerbe *f* d'Auger, gerbe *f* extensive
e chaparrón *m* de Auger, chaparrón *m* extenso
i sciame *m* d'Auger, sciame *m* estensivo
n augerbui, uitgebreide bui
d Auger-Schauer *m*, ausgedehnter Schauer *m*

441 AUGER TRANSITION np
Secondary electron emission process of
the alternative single state type in which
two electrons are involved at the same
time.
f transition *f* d'Auger
e transición *f* de Auger
i transizione *f* d'Auger
n augerovergang
d Auger-Übergang *m*

442 AUGER YIELD np
The ratio a/b where a is the number of
Auger electrons emitted, and b is the
number of events resulting in an inner-shell
electron vacancy.
f rendement *m* Auger
e rendimiento *m* Auger
i rendimento *m* Auger
n augerrendement *n*
d Auger-Ausbeute *f*

443 AUGMENTATION DISTANCE, np
EXTRAPOLATION DISTANCE
In the one-group model of neutron transport, the distance beyond the boundary of
a medium to a point at which the asymptotic
neutron flux density would go to zero if
it were represented by the same function
as within the boundary.
f longueur *f* d'extrapolation
e distancia *f* de extrapolación
i lunghezza *f* d'estrapolazione
n extrapolatieafstand
d Extrapolationslänge *f*,
Extrapolationsstrecke *f*

444 AUTOCLAVE ch, rt
An apparatus for conducting reactions
under pressure and usually at high
temperatures.
f autoclave *m*
e autoclave *f*
i autoclave *f*
n autoclaaf
d Autoklav *m*

445 AUTODIFFUSION, see 378

446 AUTO-ELECTRONIC EMISSION, ec
COLD EMISSION, FIELD EMISSION
The emission of electron from unheated
metal surfaces, produced by sufficiently
strong electric fields.
f émission *f* de champ, émission *f* froide
e emisión *f* de campo, emisión *f* fría
i emissione *f* di campo, emissione *f* fredda
n koude emissie, veldemissie
d Feldemission *f*, Kaltemission *f*

447 AUTO-IONIZATION,   ec, np
    PRE-IONIZATION
    An atomic process in which an atom under-
    goes self-ionization when originally at an
    energy state above the ionization level.
f auto-ionisation *f*
e autoionización *f*
i autoionizzazione *f*
n zelfionisatie
d Selbstionisation *f*

448 AUTOMATIC CONTROL   co
    ASSEMBLY
    An assembly designed to perform the
    automatic regulation of a quantity which is
    characteristic of the power of a reactor,
    in certain conditions, to change automa-
    tically the value of this quantity.
f ensemble *m* de réglage automatique
e conjunto *m* regulador automático
i complesso *m* di regolazione automatica
n automatische regelopstelling,
    opstelling voor automatische regeling
d automatische Regelanordnung *f*

449 AUTOMATIC CONTROL SYSTEM   co
    Any operable combination of one or more
    automatic controls connected in closed
    loops with one or more processes.
f pilote *m* automatique,
    système *m* de commande automatique
e sistema *m* de regulación automática
i sistema *m* di regolazione automatica
n automatisch regelsysteem *n*
d selbsttätiges Steuersystem *n*

450 AUTOMATIC CONTROLLER   co
    A device which measures the value of a
    variable quantity or condition and operates
    to correct or limit deviation of this
    measured value from a selected reference.
f régulateur *m*
e mando *m*, regulador *m*
i regolatore *m*, regolatore automatico
n regelaar
d Regler *m*

451 AUTOMATIC COUNTING AND   ct
    PRINTING ASSEMBLY
    Assembly for counting, scaling and
    printing of detector output pulses.
f ensemble *m* de comptage et d'impression
    automatique
e conjunto *m* de recuento y de impresión
    automático
i complesso *m* di conteggio e di
    stampaggio automatico
n automatische tel- en drukopstelling
d selbsttätige Zähl- und Druckanordnung *f*

452 AUTOMATIC DOUBLE   sp
    BEAM SPECTROMETER
    A spectrometer which can determine oxygen
    and nitrogen simultaneously in a metal
    sample.
f spectromètre *m* automatique à deux
    faisceaux

e espectrómetro *m* automático de dos haces
i spettrometro *m* automatico a due fasci
n automatische spectrometer met twee
    bundels
d selbsttätiges Doppelstrahlspektrometer *n*

453 AUTOMATIC FILTER   me, sa
    ACTIVITY METER
    Used for the automatic or manually
    controlled measurement of alpha or beta
    activity of filter dust.
f activimètre *m* automatique des filtres
e activímetro *m* automático de los filtros
i attivimetro *m* automatico dei filtri
n automatische filteractiviteitsmeter
d selbsttätiger Filteraktivitätsmesser *m*

454 AUTOMATIC PERIOD CONTROL   co
f réglage *m* automatique de la constante
    de temps
e regulación *f* automática del período
i regolazione *f* automatica del tempo di
    divergenza
n automatische regeling van de tijdconstante
d automatische Regelung *f* der Zeitkonstante

455 AUTOMATIC POWER REDUCTION rt, sa
    A power reduction in a nuclear reactor in
    the event of an accident or mistake in
    operation not necessitating a shut-down.
f réduction *f* automatique de puissance
e reducción *f* automática de potencia
i riduzione *f* automatica di potenza
n automatische vermogensreductie
d selbsttätige Leistungsherabsetzung *f*

456 AUTOMATIC RADIOCHROMATO-   an
    GRAM ANALYZER
    Device used for the automatic analysis of
    radiochromatographic strips.
f dérouleur *m* automatique de radio-
    chromatogrammes
e analizador *m* automático de radio-
    cromatogramas
i svolgitore *m* automatico di radio-
    cromatogrammi
n automatische radiochromatogram-
    analysator
d selbsttätiger Radiochromatogramm-
    analysator *m*

457 AUTOMATIC RANGE-SWITCHING   ec
    LINEAR AMPLIFIER
    An amplifier generally associated with an
    ionization chamber for health monitoring
    and reactor control as an accurate neutron
    flux monitor.
f amplificateur *m* linéaire à commutation
    automatique de sensibilité
e amplificador *m* lineal de conmutación
    automática de sensibilidad
i amplificatore *m* lineare di commutazione
    automatica di sensibilità
n versterker met automatische gebieds-
    schakelaar
d Linearverstärker *m* mit selbsttätigem
    Bereichschalter

458 AUTOMATIC SAMPLE   me
   CHANGER
An electromechanical assembly automatically presenting radioactive samples
before a detector, according to a
pre-established program(me).
f passeur *m* automatique d'échantillons
e cambiador *m* automático de muestras
i scambiatore *m* automatico di campioni
n automatische bronnenwisselaar
d selbsttätiger Probenwechsler *m*

459 AUTOMATIC SCALER   ct
A scaler which includes an automatic
stop-start circuit.
f échelle *f* automatique de comptage
e escala *f* automática
i unità *f* automatica per conteggio
n automatische pulsteller
d Zähler *m* mit Vorwahl

460 AUTORADIOGRAPH,   ra
   RADIOAUTOGRAPH
A photographic record of the radiation
from radioactive material in an object,
made by placing the object with its
surface close to a photographic emulsion.
f autoradiogramme *m*
e autorradiograma *m*
i autoradiogramma *m*
n autoröntgenfoto
d Autoradiogramm *n*

461 AUTORADIOGRAPHY   ra, xr
Process for the production of auto-
radiographs.
f autoradiographie *f*
e autorradiografía *f*
i autoradiografia *f*
n autoradiografie
d Autoradiographie *f*

462 AUTORADIOLYSIS   ch
Decomposition of a radioactive substance
in solution by the effect of its self-radiation.
f autoradiolyse *f*
e autorradiólisis *f*
i autoradiolisi *f*
n autoradiolyse
d Autoradiolyse *f*

463 AUTOSTEREOGRAM   xr
Any type of stereogram which can be seen
three-dimensionally without the need for
individual viewing devices.
f autostéréogramme *m*
e autoestereograma *m*
i autostereogramma *m*
n autostereogram *n*
d Autostereogramm *n*

464 AUTUNITE   ch, mi
A lemon-yellow phosphate of uranium and
calcium, containing from 45.4 to 48.2 % U.
f autunite *f*
e autunita *f*
i autunite *f*
n autuniet *n*
d Autunit *m*

465 AUXILIARY EQUIPMENT, see 242

466 AUXILIARY LINE   ec
Transmission line which, in the high
frequency system of e.g. a cyclotron,
facilitates the connection to a control
system.
f ligne *f* auxiliaire
e línea *f* auxiliar
i linea *f* ausiliaria
n hulpleiding
d Hilfsleitung *f*

467 AUXILIARY SAFETY DEVICES,   rt, sa
   BACKUP SAFETY DEVICES
Safety devices which only operate in case
of failure of the normal safety devices.
f appareils *pl* de sécurité auxiliaires
e aparatos *pl* de seguridad auxiliares
i apparecchi *pl* di sicurezza ausiliari
n noodveiligheidsapparaten *pl*
d Notschutzgeräte *pl*

468 AVALANCHE   np
All the ions produced from a single
primary ion through the process of cumulative ionization.
f avalanche *f*
e avalancha *f*
i valanga *f*
n lawine
d Lawine *f*

469 AVERAGE ACCELERATION   gen, pa
If the instantaneous velocity of a particle
is V_1 at a given instant and V_2 at a time
Δt later, the average acceleration during
the time Δt is defined as:
$$a_{av} = \frac{V_2 - V_1}{\Delta t}$$
f accélération *f* moyenne
e aceleración *f* media
i accelerazione *f* media
n gemiddelde versnelling
d mittlere Beschleunigung *f*

470 AVERAGE BEHAVIOR (US),   rt
   AVERAGE BEHAVIOUR (GB)
The behavio(u)r of a reactor over a certain
period.
f régime *m* global
e régimen *m* medio
i regime *m* medio
n gemiddeld gedrag *n*
d Mittelzeitverhalten *n*

471 AVERAGE BINDING ENERGY   np
   PER NUCLEON
The quotient of the total nuclear binding
energy and the mass number.
f énergie *f* de liaison moyenne par nucléon
e energía *f* de enlace media por nucleón
i energia *f* di legame media per nucleone

n gemiddelde bindingsenergie per kern
d mittlere Bindungsenergie f je Kern

472 AVERAGE CHARGE   np
The mean electrical charge of the particles constituting a charge multiplet.
f charge f moyenne
e carga f media
i carica f media
n gemiddelde lading
d mittlere Ladung f

473 AVERAGE CROSS SECTION   cs
f section f efficace moyenne
e sección f eficaz media
i sezione f d'urto media
n gemiddelde werkzame doorsnede
d mittlerer Wirkungsquerschnitt m

474 AVERAGE DIFFUSION TIME   np
The average time spent by a neutron
or other particle from the time it is
released to the time it is absorbed.
f temps m moyen de diffusion
e tiempo m medio de difusión
i tempo m medio di diffusione
n gemiddelde diffusietijd
d mittlere Diffusionszeit f

475 AVERAGE ENERGY PER ION   np
      PAIR FORMED
The average energy needed to produce
one ion pair.
f énergie f moyenne par paire d'ions
formée
e energía f media por par de iones formado
i energia f media per coppia d'ioni
prodotta
n gemiddelde energie per gevormd ionenpaar
d mittlerer Energieverlust m eines geladenen
Teilchens je erzeugtes Ionenpaar

476 AVERAGE EXCITATION ENERGY   np
An energy which can be calculated when
the total probability of all individual
levels are known.
f énergie f moyenne d'excitation
e energía f media de excitación
i energia f media d'eccitazione
n gemiddelde aanslagenergie
d mittlere Anregungsenergie f

477 AVERAGE EXCITATION   np
      POTENTIAL
Of an atom, the average energy absorption
necessary to cause excitations.
f potentiel m moyen d'excitation
e potencial m medio de excitación
i potenziale m medio d'eccitazione
n gemiddelde aanslagpotentiaal
d mittlere Anregungsspannung f

478 AVERAGE LIFE,   np
      MEAN LIFE, RADIOACTIVE PERIOD
The average lifetime for an atomic or
nuclear system in a specified state.

f vie f moyenne
e vida f media
i vita f media
n gemiddelde levensduur
d mittlere Lebensdauer f

479 AVERAGE LIFE OF AN   np
      ATOMIC STATE,
      MEAN LIFE OF AN ATOMIC STATE
The time r after which the number of
atoms left in the given atomic state is
$1/e$ of the original number.
f vie f moyenne d'un état atomique
e vida f media de un estado atómico
i vita f media d'uno stato atomico
n gemiddelde levensduur van een atomaire
toestand
d mittlere Lebensdauer f eines atomaren
Zustands

480 AVERAGE LOGARITHMIC   np
      ENERGY DECREMENT
The mean value of the increase in lethargy
per neutron collision.
f décrément m logarithmique moyen de
l'énergie,
paramètre m de ralentissement
e decremento m logarítmico medio de la
energía
i decremento m logaritmico medio
dell'energia
n gemiddelde logarithmische energievermindering
d mittleres logarithmisches Dekrement n
der Energie

481 AVOGADRO'S NUMBER,   ch, gp
      LOSCHMIDT NUMBER
The number of atoms:
$= 6.024.10^{23}$ contained in 16 gram(me)s
of oxygen under normal conditions.
f nombre m d'Avogadro
e número m de Avogadro
i numero m d'Avogadro
n getal n van Avogadro
d Avogadrosche Zahl f, Loschmidt-Zahl f

482 AWU, see 428

483 AXIAL DIFFUSION COEFFICIENT   np
Diffusion coefficient for the neutron flux
density along the axis of a multiplying
anisotropic lattice.
f coefficient m de diffusion axial
e coeficiente m de difusión axial
i coefficiente m di diffusione assiale
n axiale diffusiecoëfficiënt
d axialer Diffusionskoeffizient m

484 AXIAL PROJECTION,   xr
      AXIAL VIEW
A radiograph for which the röntgen rays
traverse a body in the same direction as
its longitudinal axis.
f projection f axiale, vue f axiale
e proyección f axial, yista f axial
i proiezione f assiale, vista f assiale

n axiale projectie
d Projektion f in Richtung der Längsachse des Körpers oder Gliedes

485 AXIAL RATIOS cr
The ratios of the length of pairs of edges of a unit cell.
f rapports pl des axes
e relaciones pl de los ejes
i rapporti pl degli assi
n assenverhoudingen pl
d Achsenverhältnisse pl

486 AXIAL REARRANGEMENT OF fu, rt
FUEL,
AXIAL SHUFFLING OF FUEL.
f altération f axiale de l'arrangement du combustible
e rearreglo m radial del combustible
i alterazione f radiale della disposizione del combustibile
n axiale wijziging van de splijtstofopstelling
d axiale Abänderung f der Brennstoffanordnung

487 AXIAL VECTOR, ma
PSEUDOVECTOR
A vector quantity which changes sign in the transition from a right-handed to a left-handed co-ordinate system.
f vecteur m axial
e vector m axial
i vettore m assiale
n axiale vector
d axialer Vektor m

488 AZEOTROPE ch
A liquid mixture which, on boiling at

constant pressure, yields a vapo(u)r the same in composition as itself.
f azéotrope m
e azeótropo m
i azeotropo m
n azeotroop n
d Azeotrop n

489 AZEOTROPIC DISTILLATION ch
Separation by distillation of a relatively volatile azeotrope formed from an added compound and one of the constituents of the original mixture.
f distillation f azéotropique
e destilación f azeotrópica
i distillazione f azeotropica
n azeotrope destillatie
d azeotrope Destillation f

490 AZIMUTHAL QUANTUM np, qm
NUMBER,
ORBITAL QUANTUM NUMBER,
SECONDARY QUANTUM NUMBER
The number which gives the angular momentum of the electron in its orbital motion round the nucleus.
f nombre m quantique azimutal
nombre m quantique secondaire
e número m cuántico acimutal,
número m cuántico secundario
i numero m quantico azimutale,
numero m quantico secondario
n azimutaal quantumgetal n,
nevenquantumgetal n
d azimutale Quantenzahl f,
Nebenquantenzahl f

B

491 BABYLINE DOSE RATEMETER,   ic
      CUTIE PIE
      A gamma dose ratemeter with a pistol grip
      and a large-diameter plastics barrel
      which contains an ionization chamber.
f débitmètre m d'exposition gamma à
      réponse linéaire
e debitómetro m de exposición gamma de
      respuesta lineal
i rateometro m d'esposizione gamma a
      risposta lineare
n gammadosistempometer met lineaire
      responsie
d C.P. Meter n

492 BACK-DIFFUSION OF ELECTRONS   np
      An effect contributed to the divergence
      between the maximum current as calcu-
      lated by means of the ionization coefficient
      and the observed value.
f diffusion f inverse des électrons
e difusión f inversa de los electrones
i diffusione f inversa degli elettroni
n terugdiffusie van elektronen
d Rückdiffusion f von Elektronen

493 BACK POINTER   xr
      A device used to localize the point of exit
      and to indicate the direction of the central
      ray.
f rétrocentreur m
e retroindicador m
i centratore m posteriore
n centrale-straalaanwijzer
d Zentralstrahlindex m

494 BACKGROUND   me
      The total of parasitic phenomena due to
      natural ionizing radiation contaminating
      measuring apparatus, electronic circuits,
      etc.
f bruit m de fond
e ruido m de fondo
i rumore m di fondo
n achtergrondruis
d Hintergrundrauschen n

495 BACKGROUND   sa
      In health physics, ambient radiation exist-
      ing in a given point and due to natural
      ionizing radiation and of the permanent
      sources in the neighbo(u)rhood of this
      source.
f fond m de rayonnement
e fondo m de radiación
i fondo m di radiazione
n stralingsachtergrond
d Strahlungshintergrund m

496 BACKGROUND COUNTING RATE   np
      The counting rate arising from background
      radiation.
f taux m de comptage de fond naturel

e velocidad f de recuento de fondo natural
i velocità f di conteggio di fondo naturale
n achtergrondteltempo n
d Hintergrundzählrate f

497 BACKGROUND COUNTS   ct
      Counts caused by radiation coming from
      sources other than that to be measured.
f impulsions pl de fond
e impulsos pl de fondo
i impulsi pl di fondo
n achtergrondtelimpulsen pl,
      nuleffectimpulsen pl
d Dunkelstösse pl, Hintergrundstösse pl,
      Nulleffektimpulse pl

498 BACKGROUND ERADICATION   ic
      The removal, by physical or chemical
      methods, of tracks accumulated in a
      photographic emulsion before the desired
      exposure.
f effacement m des traces de fond
e desarraigo m de las trazas de fondo,
      erradicación f de las trazas de fondo
i scartamento m delle traccie di fondo
n verwijdering van de achtergrondsporen,
      verwijdering van de nuleffectsporen
d Beseitigung f der Hintergrundspuren,
      Beseitigung f der Nulleffektspuren

499 BACKGROUND EXPOSURE   ra
      The exposure to natural background
      ionizing radiation.
f exposition f naturelle
e exposición f natural
i esposizione f naturale
n natuurlijke achtergrondbestraling
d natürliche Hintergrundbestrahlung f

500 BACKGROUND MONITOR   ra
      A monitor used to give indication of the
      prevailing level of background radiation.
f moniteur m du fond
e monitor m del fondo
i avvisatore m del fondo
n achtergrondmonitor, nuleffectmonitor
d Hintergrundmonitor m, Nulleffektmonitor m

501 BACKGROUND OF A DEVICE   ct, me
      The value indicated by a radiation measur-
      ing device in the absence of the source
      whose radiation is to be measured, when
      the device is placed under its normal
      conditions of operation.
f mouvement m propre d'un appareil
e fondo m de un aparato
i fondo m d'un apparecchio
n nuleffect n van een instrument
d Nulleffekt m eines Apparats

502 BACKGROUND RADIATION,   ra
      NATURAL BACKGROUND RADIATION
      Cosmic radiation and radiation arising
      from natural activity.

f fond *m* naturel de rayonnement,
rayonnement *m* ionisant naturel
e radiación *f* del fondo natural
i radiazione *f* del fondo naturale
n natuurlijke achtergrondstraling
d natürliche Hintergrundstrahlung *f*

503 BACKGROUND RESPONSE ra
The response caused by ionizing radiation
coming from sources other than that to
be measured by a radiation detector.
f réponse *f* de fond
e respuesta *f* de fondo
i risposta *f* di fondo
n achtergrondaandeel *n*, nuleffectaandeel *n*
d Hintergrundanteil *m*, Nulleffektanteil *m*

504 BACKING FOIL mt, ra
Component part of e.g. a Mössbauer source.
f feuille *f* métallique de support
e hoja *f* delgada de soporte
i foglia *f* di sopporto
n dragerfoelie
d Trägerfolie *f*

505 BACKING OFF me
An electrical or mechanical method of
suppressing all or part of a signal to
enable small variations of the signal to be
more easily observed.
f limitation *f* du signal
e limitación *f* de la señal
i limitazione *f* del segnale
n signaalbeperking
d Signaleinschränkung *f*

506 BACKSCATTER, np
BACKSCATTERING
The emergence of radiation from that
surface of a material through which it
entered.
f rétrodiffusion *f*
e retrodispersión *f*
i retrodeviazione *f*,
sparpagliamento *m* all'indietro
n terugverstrooiing
d Rückstreuung *f*

507 BACKSCATTER FACTOR ct, ra
1. In dosimetry, the ratio of the dose rate
at a point on the surface of an externally
irradiated body to the dose rate at the
same point in the absence of the body.
2. In counting, the factor by which the
counting rate is increased by the additional
radiation which is scattered back by the
source material and its support.
f facteur *m* de rétrodiffusion
e factor *m* de retrodispersión
i fattore *m* di sparpagliamento
all'indietro
n terugverstrooiingsfactor
d Rückstreuungsfaktor *m*

508 BACKSCATTER GEOMETRY is, ra
Most radioisotopes are mounted in a
robust holder of some kind. The rate at
which radiation emerges from the front

of the source is affected by the radius,
thickness, nature, etc., of the material on
which the radioactive source is mounted.
These properties of the backing material
are what are described as backscatter
geometry.
f influence *f* du matérial du support
e influencia *f* del material del soporte
i influenza *f* del materiale del sopporto
n invloed van het houdermateriaal
d Einfluss *m* des Haltermaterials

509 BACKSCATTERED ELECTRONS np
Occurring when the electrons start with
a small initial velocity and are accelerated
by the electric field but collide soon with
gas molecules thereby acquiring a random
velocity.
f électrons *pl* rétrodiffusés
e electrones *pl* retrodispersos
i elettroni *pl* retrodeviati
n terugverstrooide elektronen *pl*
d rückwärtsgestreute Elektronen *pl*

510 BACKSCATTERED RADIATION ra
Radiation the direction of which has been
changed due to its diffusion through
matter, along angles larger than 90° in
relation to its initial direction.
f rayonnement *m* rétrodiffusé
e radiación *f* retrodispersa
i radiazione *f* retrodeviata
n terugverstrooide straling
d rückwärtsgestreute Strahlung *f*

511 BACKSCATTERING GAGE (US), me
BACKSCATTERING GAUGE (GB)
An instrument used to measure thickness
of coatings and materials as e.g. concrete
and composition of certain heterogeneous
materials.
f épaisseurmètre *m* par rétrodiffusion
e calibrador *m* por retrodispersión
i spessimetro *m* per sparpagliamento
all'indietro
n diktemeter berustend op terugverstrooiing
d Gerät *n* zur Bestimmung der Dicke
mittels Rückwärtsstreuung

512 BACKUP SAFETY DEVICES, see 467

513 BACKWARD SCATTERING np
The deflection of particles or radiation
by scattering processes through angles
greater than 90° to the original direction
of motion.
f diffusion *f* en arrière,
dispersion *f* en arrière
e difusión *f* hacia atrás,
dispersión *f* hacia atrás
i diffusione *f* all'indietro,
dispersione *f* all'indietro
n achterwaartse verstrooiing
d Rückwärtsstreuung *f*

514 BACKWASH CASCADE, ch
STRIPPING CASCADE
The group of solvent extraction stages

used to wash the desired component back
into the aqueous phase.
f   cascade *f* de réextraction
e   cascada *f* de reextracción
i   cascata *f* di riestrazione
n   herextractiecascade
d   Rückwaschkaskade *f*

515 BACKWASHING (GB),   ch
     STRIPPING (US)
In the solvent extraction process, the
washing of the desired component back
into the aqueous phase.
f   réextraction *f*
e   reextracción *f*
i   riestrazione *f*
n   herextractie
d   Rückwaschung *f*

516 BAD GEOMETRY,   rt
     POOR GEOMETRY
A geometry which entails large corrections.
f   géométrie *f* fausse
e   mala geometría *f*
i   mala geometria *f*
n   slechte geometrie
d   schlechte Geometrie *f*,
    ungünstige Geometrie *f*

517 BAFFLE PLATE   ch, is
A tray or partition placed in a tower, heat
exchanger or other process equipment
to direct, or to change the direction of, the
flow of liquids.
f   chicane *f*, plaque *f* de déviation
e   chicana *f*, placa *f* deflectora
i   deflettore *m*
n   keerschot *n*
d   Prallplatte *f*

518 BALANCED HETEROGENEOUS   mt
     CONCRETE
Concrete mixture for a biological shield
with specific additions in order to obtain
an equilibrated screening action.
f   béton *m* hétérogène équilibré
e   hormigón *m* heterogéneo equilibrato
i   calcestruzzo *m* eterogeneo ad azione
    equilibrata
n   uitgebalanceerd heterogeen beton *n*
d   ausgeglichenes heterogenes Beton *n*

519 BALL-JOINT MANIPULATOR   rt
Manipulator of special construction used
in remote handling of radioactive material.
f   manipulateur *m* à articulation à rotule
e   manipulador *m* con articulación de rótula
i   manipolatore *m* a giunto sferico
n   kogelgewrichtmanipulator
d   Kugelgelenkmanipulator *m*

520 BALL OF FIRE,   nw
     FIREBALL
The luminous sphere of hot gases which
forms a few millionths of a second after
a nuclear explosion and immediately
starts to expand and cool.
f   boule *f* de feu

e   bola *f* de fuego
i   pallone *m* di fuoco
n   vuurbal
d   Feuerkugel *f*

521 BALL-SHAPED PLASMOID   pp
A discrete piece of plasma shaped as a ball.
f   plasmoïde *m* sphérique
e   plasmoide *m* esférico
i   plasmoide *m* sferico
n   bolvormig plasmoïde *n*
d   kugelförmiges Plasmoid *n*

522 BALMER SERIES   sp
A group of lines in the visible spectrum
of atomic hydrogen.
f   série *f* de Balmer
e   serie *f* de Balmer
i   serie *f* di Balmer
n   balmerserie
d   Balmer-Serie *f*

523 BAND   sp
The set of closely spaced spectral lines
produced by molecules of one kind when
there is a transition between two electronic-
vibrational states possessing rotational
fine structure.
f   bande *f*
e   banda *f*
i   banda *f*
n   band
d   Band *n*

524 BAND EDGE ENERGY   ec, np
The energy of the edge of the conduction
or valence band in a solid.
f   énergie *f* de la limite de bande
e   energía *f* del borde de banda
i   energia *f* del limite di banda
n   bandgrensenergie
d   Bandkantenenergie *f*

525 BAND HEAD   sp
The wavelength of the sharpest edge of a
spectral band.
f   tête *f* de bande
e   cabeza *f* de banda
i   testa *f* di banda
n   bandenkop
d   Bandkopf *m*

526 BAND SCHEME   ec, np
The classification of the electronic
states according to the energy bands to
which they belong.
f   table *f* des bandes d'énergie
e   esquema *m* de las bandas de energía
i   schema *m* delle bande d'energia
n   energiebandenschema *n*
d   Energiebänderübersicht *f*

527 BAND SPECTRUM   sp
A spectrum in which the radiation is
contained in diffuse bands.
f   spectre *m* de bandes ·
e   espectro *m* de bandas
i   spettro *m* di bande

n bandspectrum n
d Bandenspektrum n

528 BAND SPECTRUM CONSTANT   sp
A constant, symbol BI, used in atomic
spectroscopy.
f constante f de spectre de bandes
e constante f de espectro de bandas
i costante f di spettro di bande
n bandspectrumconstante
d Bandenspektrumkonstante f

529 BARE REACTOR   rt
A nuclear reactor without a reflector.
f réacteur m nu, réacteur m sans réflecteur
e reactor m sin reflector
i reattore m nudo,
   reattore m senza riflettore
n naakte reflector, reflectorloze reactor
d Reaktor m ohne Reflektor

530 BARIUM   ch
Metallic element, symbol Ba, atomic
number 56.
f baryum m
e bario m
i bario m
n barium n
d Barium n

531 BARIUM CONCRETE   mt
Concrete containing a high percentage of
barium compounds.
f béton m au baryum, béton m baryté
e hormigón m al bario
i calcestruzzo m al bario
n bariumbeton n
d Bariumbeton n

532 BARIUM ENEMA EXAMINATION md, xr
The radiological examination of the large
intestine following the reactor injection of
a contrast medium containing a barium
compound.
f examen m à clystère de contraste
e examen m con enema opaca
i esame m a clistere di contrasto
n onderzoek n na bariumclysma
d Untersuchung f mit Kontrasteinlauf

533 BARIUM MEAL EXAMINATION   md, xr
The radiological examination of the
alimentary tract following the ingestion
of a contrast medium containing barium.
f examen m à bouillie opaque
e examen m con papilla de bario
i esame m dopo pasto opaco
n onderzoek n na toediening van bariumpap
d Untersuchung f nach Verabfolgung eines
   Bariumbreis.

534 BARIUM MIXER   ch, md
An apparatus for preparing a homogeneous
mixture of barium containing contrast
media.
f mélangeur m de boisson d'épreuve du
   baryum,
   mélangeur m de bouillie opaque

e mezclador m de papilla barítica
i mescolatore m di pasto opaco
n bariumpapmenger
d Bariumbreimischer m

535 BARIUM PLASTER   mt
Builder's plaster containing a high
percentage of barium compounds.
f plâtre m au baryum, plâtre m baryté
e yeso m al bario
i gesso m al bario
n bariumgips n
d Bariumgipsmörtel m

536 BARIUM SWALLOW   md, xr
     EXAMINATION
The radiological examination of the
oesophagus and cardia during the passage
of a mouthful of a contrast medium containing barium.
f examen m à boisson d'épreuve au baryum
e examen m después de deglución del bario
i esame m dopo l'inghiottimento del bario
n onderzoek n na slikken van bariumpap
d Untersuchung f nach Speiseröhrenbrei-
   schluck

537 BARLOW RULE   ch
The volumes of space occupied by the
various atoms in a given molecule are
approximately proportional to the valences
of the atoms.
f loi f de Barlow
e ley f de Barlow
i legge f di Barlow
n wet van Barlow
d Barlowsches Gesetz n

538 BARN   un
A unit of area used in expressing a nuclear
cross section.
$1 \text{ barn} = 10^{-24} \text{cm}^2 = 10^{-28} \text{m}^2$.
f barn m
e barn m
i barn m
n barn
d Barn n

539 BARODIFFUSION   ch
Diffusion caused by non-uniformity of the
total pressure in a mixture.
f barodiffusion f
e barodifusión f
i barodiffusione f
n barodiffusie
d Barodiffusion f

540 BARODIFFUSION COEFFICIENT, ch, gp
     DIFFUSION COEFFICIENT,
     DIFFUSIVITY
The constant of proportionality in Fick's
law, which states that the current is
proportional to the negative gradient of
the flux or the density.
f coefficient m de diffusion
e coeficiente m de difusión
i coefficiente m di diffusione

n diffusiecoëfficiënt
d Diffusionskoeffizient *m*

541 BARRIER,   np
   GAMOW BARRIER, NUCLEAR BARRIER,
   POTENTIAL BARRIER
   The region of high potential energy through
   which a charged particle must pass on
   leaving or entering a nucleus.
f barrière *f* de potentiel
e barrera *f* de potencial
i barriera *f* di potenziale
n potentiaalstoep
d Potentialberg *m*, Potentialschwelle *f*,
   Potentialwall *m*

542 BARRIER (US),   is
   MEMBRANE (GB)
   In isotope separation by gaseous diffusion,
   a porous partition through which material
   transfer takes place, principally by
   molecular diffusion.
f barrière *f*, membrane *f*
e membrana *f*
i membrana *f*
n membraan *n*
d Membran *f*

543 BARRIER DIFFUSION METHOD   is
   A method of separating isotopes in which
   a gas that is an isotopic mixture is
   allowed to diffuse through a porous wall
   or barrier.
f procédé *m* de diffusion à barrière
e procedimiento *m* de difusión en barrera
i processo *m* di diffusione a barriera
n diffusiemethode met scheidingswand,
   scheidingswanddiffusieproces *n*
d Trennwanddiffusionsverfahren *n*

544 BARRIER HEIGHT   np
   The maximum energy of a Coulomb
   barrier.
f hauteur *f* de barrière
e altitud *f* de barrera
i altezza *f* di barriera
n stoephoogte
d Höhe *f* des Potentialwalls

545 BARTLETT FORCE   np
   Phenomenologically postulated force
   between two nucleons derivable from a
   potential in which there appears an
   operator which exchanges the spins of
   the two particles but not their positions.
f force *f* de Bartlett
e fuerza *f* de Bartlett
i forza *f* di Bartlett
n bartlettkracht
d Bartlett-Kraft *f*

546 BARYON   np
   A collective name for nucleons and
   hyperons.
f baryon *m*
e barión *m*
i barione *m*
n baryon *n*
d Baryon *n*

547 BARYON NUMBER   np
   The number of baryons minus the number
   of antibaryons in a system.
f nombre *m* de baryons
e número *m* de bariones
i numero *m* di barioni
n baryongetal *n*
d Baryonzahl *f*

548 BASE SURGE   nw
   A cloud which rolls outward from the bottom
   of the column produced by a sub-surface
   explosion.
f nuage *m* de base
e nube *f* de base
i nuvelone *m* di base
n basiswolk
d Basiswolke *f*

549 BASIC ELEMENT   ge, me
   A measurement component or group of
   components that performs one necessary
   and distinct function in a sequence of
   measurement operations.
f élément *m* fonctionnel
e elemento *m* funcional
i elemento *m* funzionale
n functioneel orgaan *n*
d funktionelles Organ *n*

550 BASIC FUNCTION UNIT   ge
   A well-defined set of components which
   effects one or possibly more than one
   elementary function in an assembly or
   sub-assembly, and which will appear as
   a physical unit when being removed.
f élément *m* fonctionnel
e unidad *f* funcional básica
i elemento *m* funzionale
n functionele schakel
d Funktionsgruppe *f*

551 BASIC OCCUPATIONAL   ra, xr
   EXPOSURE RATE
   The maximum permissible value of the
   specific absorbed dose in a medium during
   a predetermined period of an individual
   subject to irradiation for professional
   reasons.
f dose *f* étalon professionnelle
e dosis *f* normal profesional
i intensità *f* di dose professionale di norma
n professionele standaarddosis
d berufsmässige Normaldosis *f*

552 BASIC SUBSTANCE,   ch
   KEY SUBSTANCE
   The initial substance in a chemical or
   physical process.
f substance *f* de base
e substancia *f* clave
i sostanza *f* di base
n uitgangsstof
d Ausgangsstoff *m*

553 BASOPHIL GRANULE   ch
   A granule accepting, or being stained by,
   basic dyestuffs.

f granule *m* basophile
e gránulo *m* basófilo
i granulo *m* basofilo
n basofiel korreltje *n*
d basophiles Körnchen *n*

554 BASSETITE   mi
Rare secondary mineral containing up to
51 % of U.
f bassetite *f*
e basetita *f*
i bassetite *f*
n bassetiet *n*
d Bassetit *m*

555 BATCH   ch
A discrete quantity of material in a
chemical process.
f charge *f*, lot *m*
e lote *m*, partida *f*, tanda *f*
i carica *f* di materiale, lotto *m*
n lading, portie
d Absatz *m*, Posten *m*

556 BATCH DISTILLATION   ch
A distillation process in which a given
quantity of material is charged into a
still, and the distillation is continued
without additional charge to the still.
f distillation *f* discontinue
e destilación *f* discontinua
i distillazione *f* discontinua
n discontinue destillatie
d absatzweise Destillation *f*,
Blasendestillation *f*

557 BATCH PROCESS   ch, ge
A process in which the feed is introduced
as discrete charges, each of which is
processed to completion separately.
f procédé *m* à charges, procédé *m* discontinu,
procédé *m* en lots
e procedimiento *m* de lotes,
procedimiento *m* de partidas,
procedimiento *m* de tandas
i processo *m* a lotti,
processo *m* discontinuo
n discontinu proces *n*
d Postenverfahren *n*, Satzverfahren *n*

558 BAYLEYITE   mi
A rare secondary mineral containing
about 29 % of U.
f bayléyite *f*
e baileita *f*
i baileite *f*
n bayleyiet *n*
d Bayleyit *m*

559 BEAM   ec, pa
A unidirectional or approximately unidirectional flow of electromagnetic
radiation or of particles.
f faisceau *m*
e haz *m*
i fascio *m*
n bundel
d Bündel *n*, Strahlen *pl*

560 BEAM DIRECTION INDICATOR   xr
A device which indicates the direction of
the beam of radiation in the body.
f indicateur *m* de la direction du rayonnement
e indicador *m* de la dirección de la
radiación
i indicatore *m* della direzione della
radiazione
n stralingsrichtingaanwijzer
d Strahlungsrichtungsanzeiger *m*

561 BEAM HOLE (GB),   rt
GLORY HOLE (US)
A hole through a reactor shield into the
interior of the reactor for the passage
of a beam of radiation for experiments
outside the reactor.
f canal *m* expérimental
e agujero *m* de haz,
conducto *m* interior de un reactor
i canale *m* d'irradiazione,
canale *m* passante
n bundelgat *n*, stralingskanaal *n*
d Bestrahlungskanal *m*, Strahlenkanal *m*,
Strahlrohr *n*

562 BEAM HOLE SHIELDING   rt
The thickness of the shielding of a beam
tube.
f épaisseur *f* de blindage
e espesor *m* de blindaje
i spessore *m* di schermatura
n afschermingsdikte, schermdikte
d Abschirmungsdicke *f*, Schilddicke *f*

563 BEAM REACTOR   rt
A nuclear reactor specially designed to
produce external beams of neutrons for
research.
f réacteur *m* à faisceaux sortis
e reactor *m* de haces
i reattore *m* a fasci
n bundelreactor
d Bündelreaktor *m*

564 BEAM TRAP   ra
A system for absorbing a beam of radiation
so that the scattered radiation is reduced
to an acceptable level.
f piège *m* de faisceau
e trampa *f* de haz
i trappola *f* di fascio
n bundelval
d Bündelfalle *f*, Strahlenfanger *m*

565 BECQUEREL RAYS   ra
Penetrating radiation emitted by uranium
salts.
f rayons *pl* Becquerel
e rayos *pl* Becquerel
i raggi *pl* Becquerel
n becquerelstralen *pl*
d Becquerel-Strahlen *pl*

566 BECQUERELITE   mi
Uncommon, secondary mineral, iso-
structural with billietite, and containing
about 75.7 % of U.

f becquérélite *f*
e becquerelita *f*
i becquerelite *f*
n becquereliet *n*
d Becquerelit *m*

567 BEFORE-AND-AFTER TEST, rt
IRRADIATION TEST
In reactor engineering an irradiation test
in which the properties of a specimen are
measured before insertion and after its
removal from the reactor.
f essai *m* avant et après l'irradiation
e prueba *f* antes de y después de la
irradiación
i prova *f* prima e dopo l'irradiazione
n voor- en naproef
d Vor- und Nachbestrahlungsmessung *f*

568 BELL COUNTER TUBE ct
A type of end-window counter tube.
f tube *m* compteur à cloche
e tubo *m* contador acampanado,
tubo *m* contador de campana
i tubo *m* contatore a campana
n kloktelbuis
d Glockenzählrohr *n*

569 BEND LOSS cl
The loss in energy of a liquid flowing
through a bend in a pipe.
f perte *f* de coude
e pérdida *f* de codo
i perdita *f* di gomito
n bochtverlies *n*
d Krümmerverlust *m*

570 BENIGN GROWTH, md, xr
BENIGN NEOPLASM,
INNOCENT GROWTH
A new growth of cells, some degree of
growth restraint being present in the
absence of spread to distant parts.
f croissance *f* bénigne, néoplasma *m* bénin
e crecimiento *m* benigno,
neoplasma *m* benigno
i neoformazione *f* benigna,
neoplasma *m* benigno
n goedaardig neoplasma *n*,
goedaardige nieuwvorming
d gutartige Neubildung *f*,
gutartiges Neoplasma *n*

571 BENIGN TUMOR, md, xr
INNOCENT TUMOR
Tumor which cannot metastatize.
f tumeur *f* bénigne
e tumor *m* benigno
i tumore *m* benigno
n goedaardig gezwel *n*
d gutartige Geschwulst *f*

572 BENSON BOILER, ge, rt
FLASH BOILER,
ONCE-THROUGH STEAM BOILER
A steam boiler consisting of a long coil
of steel tube, usually heated by oil

burners, in which water is evaporated
as it is pumped through by the feed pump.
f chaudière *f* à passage unique
e caldera *f* de paso único
i caldaia *f* a rapida vaporizzazione
n stoomketel met enkele doorloop
d Dampfkessel *m* mit einmaligem Durchlauf

573 BERKELIUM ch
Radioactive element, symbol Bk, atomic
number 95.
f berkélium *m*
e berquelio *m*
i berchelio *m*
n berkelium *n*
d Berkelium *n*

574 BERNOULLI BINOMIAL DISTRI- ct, ma
BUTION,
BERNOULLI EQUATION
An equation expressing the probability P_n
that exactly n counts will be obtained
in a single period when there are N atoms
present, each with the probability P of
disintegrating and being counted in that
period.
f distribution *f* binomiale de Bernoulli
e distribución *f* binomial de Bernoulli
i distribuzione *f* binomiale di Bernoulli
n binomiale verdeling volgens Bernoulli
d binomiale Verteilung *f* nach Bernoulli

575 BERYL mt
A mineral, silicate of beryllium and
alumin(i)um, an important ore of
beryllium.
f béril *m*, béryl *m*
e berilo *m*
i berillo *m*
n beril, beryl
d Beryll *m*

576 BERYLLIA mt, rt
Oxide of beryllium, useful as a moderator
or reflector.
f glucine *f*, oxyde *m* de béryllium
e óxido *m* de berilio
i ossido *m* di berillio
n berilaarde, berylaarde
d Beryllerde *f*

577 BERYLLIOSIS md, sa
An industrial disease to which beryllium
process workers are liable, caused by the
inhalation of beryllium compounds as
smoke or aerosols.
f bérylliose *f*
e beriliosis *f*
i berilliosi *f*
n berylliose
d Berylliosis *f*

578 BERYLLIUM ch
Metallic element, symbol Be, atomic
number 4.
f béryllium *m*
e berilio *m*

i   berillio *m*
n   beryllium *n*
d   Beryllium *n*

579 BERYLLIUM CONTENT METER   me
A content meter to determine the
beryllium content of an ore sample by
means of the nuclear reaction.
f   teneurmètre *m* en béryllium
e   berilímetro *m*
i   berilliotenorimetro *m*
n   gehaltemeter voor beryllium
d   Gerät *n* zur Bestimmung des
Berylliumgehalts

580 BERYLLIUM MODERATED   rt
REACTOR,
BERYLLIUM REACTOR
A nuclear reactor in which the moderator
consists of beryllium.
f   réacteur *m* au béryllium,
réacteur *m* modéré au béryllium
e   reactor *m* de berilio,
reactor *m* moderado de berilio
i   reattore *m* a berillio,
reattore *m* moderato a berillio
n   met beryllium gemodereerde reactor
d   berylliummoderierter Reaktor *m*,
Berylliumreaktor *m*

581 BERYLLIUM-OXIDE MODERATED   rt
REACTOR
f   réacteur *m* modéré par oxyde de
béryllium
e   reactor *m* moderado por óxido de
berilio
i   reattore *m* moderato per ossido di
berillio
n   reactor met berylliumoxydemoderator
d   Reaktor *m* mit Berylliumoxydmoderator

582 BERYLLIUM PROSPECTING   me
METER
A measuring assembly to be used in
prospecting beryllium ores.
f   ensemble *m* de prospection de béryllium
e   radiámetro *m* para berilio
i   berilliotenorimetro *m* per prospezione
n   opstelling voor prospectie van beryllium
d   Gerät *n* für die Prospektion von
Beryllium

583 BETA   bp
Pertaining to beta particles.
f   bêta
e   beta
i   beta
n   bêta
d   beta

584 BETA ABSORPTION GAGE (US),   me
BETA ABSORPTION GAUGE (GB)
An instrument which measures the thickness or density of a sample by measuring
the absorption of beta-rays in the sample.
f   jauge *f* bêta à absorption
e   calibre *m* beta de absorción

i   calibro *m* beta ad assorbimento
n   bêta-absorptiemeter
d   Betaabsorptionsmesser *m*

585 BETA BACKSCATTER   me
THICKNESS METER
A thickness meter including a beta
radiation source and designed to determine
material thickness by measurement of the
radiation backscattered by this material.
f   épaisseurmètre *m* à rétrodiffusion bêta
e   calibrador *m* por retrodispersión beta
i   spessimetro *m* a sparpagliamento beta
all'indietro
n   diktemeter berustend op bêtaterug-
verstrooiing
d   Dickenmessgerät *n* nach der Betarück-
streumethode

586 BETA CONTAMINATION   sa
INDICATOR
An indicator designed to detect beta
contamination, in which the output pulses
control a warning signal.
f   signaleur *m* de contamination bêta
e   indicador *m* de contaminación beta
i   segnalatore *m* di contaminazione beta
n   indicator voor bêtabesmetting
d   Betakontaminationsanzeiger *m*

587 BETA DECAY,   br
BETA DISINTEGRATION,
BETA TRANSFORMATION
Radioactive decay in which a beta particle
is emitted or in which orbital electron
capture occurs.
f   désintégration *f* bêta
e   desintegración *f* beta
i   disintegrazione *f* beta
n   bêtaverval *n*
d   Betazerfall *m*

588 BETA$^+$DECAY,   np
POSITON DECAY,
POSITON DISINTEGRATION,
POSITRON DECAY,
POSITRON DISINTEGRATION
The emission of posit(r)ons by radioactive
nuclei which contain one or more protons
too many.
f   désintégration *f* avec émission de positons,
désintégration *f* positogène
e   desintegración *f* positónica
i   disintegrazione *f* positonica
n   bêta$^+$-verval *n*, positondesintegratie
d   Beta$^+$-Zerfall *m*, Positronenzerfall *m*

589 BETA DECAY ELECTRON   br
An electron emitted in a beta-disintegration
process.
f   électron *m* de désintégration bêta
e   electrón *m* de desintegración beta
i   elettrone *m* di disintegrazione beta
n   bêtadesintegratie-elektron *n*,
bêtavervalelektron *n*
d   Betazerfallselektron *n*

590 BETA DECAY ENERGY,   br
      BETA DISINTEGRATION ENERGY
   1. The disintegration energy of a beta
   decay process: symbol Q.
   2. Often the ground state beta disintegration energy.
f énergie f de désintégration bêta
e energía f de desintegración beta
i energia f di disintegrazione beta
n bêtadesintegratie-energie,
   bêtavervalenergie
d Betazerfallsenergie f

591 BETA DECAY-THEORY,   br
      BETA DISINTEGRATION THEORY
   The quantitative theory of beta decay
   based on the neutrino hypothesis of Pauli.
f théorie f de désintégration bêta
e teoría f de desintegración beta
i teoria f di disintegrazione beta
n bêtavervaltheorie
d Betazerfallstheorie f

592 BETA DUST AND GAS MONITOR   co, rt
   Used for the continuous and simultaneous
   monitoring of beta dust and gas at several
   points in a reactor building.
f détecteur m de poussière et de gaz
   radioactif
e monitor m de polvo y gas radiactivo
i monitore m di polvere e gas radioattivo
n radioactieve stof- en gasmonitor
d radioaktiver Staub- und Gasmonitor m

593 BETA EMITTER,   br
      BETA RADIATOR
   A radionuclide that disintegrates by beta
   particle emission.
f émetteur m bêta
e emisor m beta
i emettitore m beta
n bêtastraler
d Betastrahler m

594 BETA-EMITTING SILVER FOIL   md, ra
   Component part of a nasopharyngeal
   applicator containing up to 100mc
   strontium-90.
f feuille f en argent à émission bêta
e hoja f delgada en plata de emisión beta
i foglia f in argento ad emissione beta
n bêtastralen emitterende zilverfoelie
d betastrahlenemittierende Silberfolie f

595 BETA-GAMMA ANGULAR   np
      CORRELATION
   The angular correlation between a beta
   and a gamma particle.
f corrélation f angulaire bêta-gamma
e correlación f angular beta-gamma
i correlazione f angolare beta-gamma
n bêta-gamma-hoekcorrelatie
d Beta-Gamma-Winkelkorrelation f

596 BETA-GAMMA DOORWAY   sa
      MONITOR
   A monitor with detectors arranged about
   a doorway to give a measure of beta

and/or gamma contamination carried by
a person or things passing through the
doorway and to give a warning when it
exceeds a predetermined value.
f moniteur m portique bêta-gamma
e monitor m beta-gamma de acceso a puerta
i monitore m beta-gamma a portale
n bêta-gamma-poortmonitor
d Beta-Gamma-Türrahmenmonitor m

597 BETA-GAMMA EMITTER   br, gr
   An emitter whereby the beta disintegration
   is accompanied by the emission of gamma
   rays.
f émetteur m bêta-gamma
e emisor m beta-gamma
i emettitore m beta-gamma
n bêta-gamma-straler
d Beta-Gamma-Strahler m

598 BETA-GAMMA EXPOSURE   ic
      DOSE RATEMETER BBHF,
      FISH-POLE PROBE
   A high-level radiation meter similar to
   the cutie pie and capable of monitoring
   levels up to 100 röntgens per hour with
   about 10 % accuracy.
f débitmètre m d'exposition bêta-gamma
   à chambre d'ionisation BBHF
e debitómetro m de exposición beta-gamma
   de cámara de ionización BBHF
i rateometro m d'esposizione beta-gamma
   a camera d'ionizzazione BBHF
n dosistempometer voor bêta- en gammastralen met ingebouwde ionisatiekamer
d Dosisleistungsmesser m für Beta- und
   Gammastrahlen mit eingebauter
   Ionisationskammer

599 BETA HAND CONTAMINATION   br, sp
      MONITOR
   An assembly designed to measure the hand
   contamination by beta emitters and including a device for indicating contamination exceeding a predetermined value.
f moniteur m de contamination bêta pour
   les mains
e vigía f de contaminación beta en las manos
i monitore m di contaminazione beta per le
   mani
n bêtamonitor voor handbesmetting
d Betahandmonitor m

600 BETA MAXIMUM ENERGY   br, sp
   The maximum energy of the energy
   spectrum in a beta disintegration process.
f énergie f bêta maximale
e energía f beta máxima
i energia f beta massima
n maximale bêta-energie
d maximale Betaenergie f

601 BETA-NEUTRINO   np
      ANGULAR CORRELATION
   The angular correlation between a beta
   particle and a neutrino.
f corrélation f angulaire bêta-neutrino
e correlación f angular beta-neutrino

i   correlazione *f* angolare beta-neutrino
n   bêta-neutrino-hoekcorrelatie
d   Beta-Neutrino-Winkelkorrelation *f*

602 BETA PARTICLE,									br
	BETA RAY
	A negative electron of a positive electron
	(positron) emitted from a nucleus during
	beta decay.
f   particule *f* bêta
e   partícula *f* beta
i   particella *f* beta
n   bêtadeeltje *n*
d   Betateilchen *n*

603 BETA POINT SOURCE				br, md, ra
	A source in which the radioactive material
	is in the form of a glass bead and
	contained in a metal cylinder with a thin
	window at one end where necessary.
f   source *f* bêta ponctuelle
e   fuente *f* beta puntual
i   sorgente *f* beta a punta
n   puntvormige bêtabron
d   punktförmige Betaquelle *f*

604 BETA PROCESS,									br
	BETA TRANSFORMATION
	The transformation of a nucleus into its
	neighbo(u)ring isobar, with the emission
	of a positive or negative electron, or with
	the capture of an orbital electron.
f   transmutation *f* bêta
e   transmutación *f* beta
i   trasmutazione *f* beta
n   bêtatransmutatie
d   Betaumwandlung *f*

605 BETA PULSE COUNTING					me
	ASSEMBLY
	A pulse counting assembly which includes
	a beta radiation detector, whose output
	pulses are applied to a counting sub-
	assembly and/or to a counting rate
	measuring sub-assembly.
f   ensemble *m* de mesure à impulsions
	pour particules bêta
e   conjunto *m* contador por impulsos para
	partículas beta
i   complesso *m* di misura ad impulsi per
	particelle beta
n   telopstelling voor bêtadeeltjes
d   Betazählanordnung *f*

606 BETA QUENCH										mg
	The rapid cooling of uranium from a
	temperature in the beta region to random-
	ize the structure or to refine the grain
	size.
f   refroidissement *m* rapide dans la région
	de température bêta
e   enfriamiento *m* rápido en la región de
	temperatura beta
i   raffreddamento *m* brusco nella regione
	di temperatura beta
n   afschrikken *n* in het bêtatemperatuurgebied
d   Abschrecken *n* im Betatemperaturgebiet

607 BETA RADIATION								br, ra
	A radiation composed of beta rays.
f   rayonnement *m* bêta
e   radiación *f* beta
i   radiazione *f* beta
n   bêtastraling
d   Betastrahlung *f*

608 BETA RADIOACTIVITY						br, ra
	Emission of beta radiation or capture
	of a peripheral electron.
f   radioactivité *f* bêta
e   radiactividad *f* beta
i   radioattività *f* beta
n   bêta-activiteit
d   Betaaktivität *f*

609 BETA-RADIOGRAPHY DATING,				me
	CONTACT BETA-RADIOGRAPHY
	DATING
	A method used for historic document dating.
f   détermination *f* de l'âge par rayonnement
	bêta
e   datación *f* por radiación beta
i   datazione *f* per radiazione beta
n   leeftijdsbepaling met behulp van
	bêtastraling
d   Altersbestimmung *f* mittels Betastrahlung

610 BETA-RAY ELECTROSCOPE				br, me
	An electroscope used for detecting the
	presence of beta rays.
f   électroscope *m* pour détecter les rayons
	bêta
e   electroscopio *m* para detección de rayos
	beta
i   elettroscopio *m* per rivelazione di
	raggi beta
n   elektroscoop voor het opsporen van bêta-
	stralen
d   Elektroskop *n* zum Nachweis von Beta-
	strahlen

611 BETA-RAY EMISSION								br
	The emission of a beta particle.
f   émission *f* de rayons bêta
e   emisión *f* de rayos beta
i   emissione *f* di raggi beta
n   bêtastraling
d   Betastrahlung *f*

612 BETA-RAY EXTRA-							br, md, ra
	CORPOREAL IRRADIATOR,
	EXTRACORPOREAL BLOOD
	IRRADIATOR
	An irradiator for blood from outside
	containing a strontium-90 source in a
	suitable stainless steel holder.
f   appareil *m* d'irradiation bêta extra-
	corporéal du sang
e   irradiador *m* beta extracorporeo del
	sangre
i   irradiatore *m* beta estracorporeo del
	sangue
n   uitwendige bêtabloedbestraler
d   extrakorporaler Betablutbestrahler *m*

613 BETA-RAY PLAQUE   br, ra
A carrier for the radioisotope compound
consisting of an alumin(i)um alloy case
with a guard ring.
f plaque *f* bêta
e placa *f* beta
i placca *f* beta
n bêtaplaket
d Betaplakete *f*

614 BETA-RAY PLATE   br, ra
A carrier for the radioisotope compound
with an inactive border 3 mm wide
surrounding the active area, and without
guard ring.
f lame *f* bêta
e lámina *f* beta
i lamina *f* beta
n bêtaplaatje *n*
d Betaplatte *f*

615 BETA-RAY SPECTROMETER   br, sp
A measuring assembly for determining
the energy spectrum of beta rays.
f spectromètre *m* à rayons bêta
e espectrómetro *m* beta
i spettrometro *m* beta
n bêtaspectrometer
d Betaspektrometer *n*

616 BETA-RAY SPECTRUM   np
The distribution in energy or momentum
of the beta particles, not including
conversion electrons, emitted in a
beta-decay process.
f spectre *m* bêta
e espectro *m* beta
i spettro *m* beta
n bêtaspectrum *n*
d Betaspektrum *n*

617 BETA-RAY TELETHERAPY   br, md, ra
f télébêtathérapie *f*
e telebetaterapia *f*
i telebetaterapia *f*
n telebêtatherapie
d Betateletherapie *f*

618 BETA-RAY THERAPY   br, md, ra
f bêtathérapie *f*
e betaterapia *f*
i betaterapia *f*
n bêtatherapie
d Betatherapie *f*

619 BETA RAYS   br
A stream of beta particles.
f rayons *pl* bêta
e rayos *pl* beta
i raggi *pl* beta
n bêtastralen *pl*
d Betastrahlen *pl*

620 BETA-RECOIL NUCLEUS   np
       ANGULAR CORRELATION
The angular correlation between a beta
particle and a recoil nucleus.

f corrélation *f* angulaire bêta-noyau de
   recul
e correlación *f* angular beta-núcleo de
   rechazo
i correlazione *f* angolare beta-nucleo di
   rinculo
n bêta-terugslagkerncorrelatie
d Beta-Rückstosskernkorrelation *f*

621 BETA SCREEN   ic
A metal shield covering the ionization
chamber of a survey meter; when removed
an estimation of beta radiation can be
made.
f écran *m* bêta
e pantalla *f* beta
i schermo *m* beta
n bêtascherm *n*
d Betaschirm *m*

622 BETA-URANIUM   fu, mt
An allotropic modification of uranium that
is stable between approximately 660ºC
and 770ºC.
f uranium *m* bêta
e uranio *m* beta
i uranio *m* beta
n bêta-uranium *n*
d Beta-Uran *n*

623 BETA-URANOPHANE,   mi
       BETA-URANOTILE
A secondary uranium mineral containing
about 55.6 % of U.
f bêta-uranophane *m*
e beta-uranófano *m*
i beta-uranofano *m*
n bêta-uranofaan *n*
d Beta-Uranophan *n*

624 BETA VALUE   br, pp
In the controlled thermonuclear
program(me) the ratio of the outward
pressure exerted by a plasma to the inward
pressure that the confining magnetic field
is capable of exerting.
f valeur *f* bêta
e valor *m* beta
i valore *m* beta
n bêtawaarde
d Betawert *m*

625 BETAFITE,   mi
       BLOMSTRANDITE
A uranium and thorium containing ore of
uncertain composition.
f bêtafite *f*, blomstrandite *f*
e betafita *f*, blomstrandita *f*
i betafite *f*, blomstrandite *f*
n bêtafiet *n*, blomstrandiet *n*
d Betafit *m*, Blomstrandit *m*

626 BETATOPIC   np
Differing by or pertaining to a difference
by unit atomic number.
f bêtatopique adj
e betatópico adj

i   betatopico adj
n   bêtatoop adj
d   betatopisch adj

627 BETATOPIC NUCLIDES   np
   Nuclides differing by unit atomic number.
f   nucléides *pl* bêtatopiques
e   núclidos *pl* betatópicos
i   nuclidi *pl* betatopici
n   bêtatope nucliden *pl*
d   betatope Nuklide *pl*

628 BETATRON,   pa
   INDUCTION ACCELERATOR
   A device for accelerating electrons by
   means of magnetic induction.
f   bêtatron *m*
e   betatrón *m*
i   betatrone *m*
n   bêtatron *n*
d   Betatron *n*, Elektronenschleuder *f*

629 BETATRON OSCILLATIONS   pa
   Free particle oscillations around the
   equilibrium orbit in an accelerator using
   a guide field.
f   oscillations *pl* bêtatron,
   oscillations *pl* bêtatroniques
e   oscilaciones *pl* de betatrón
i   oscillazioni *pl* di betatrone
n   bêtatrontrillingen *pl*
d   Betatronschwingungen *pl*

630 BETHE-SALPETER   np
   EQUATION
   Equation in quantized field theory
   describing the bound state of two interacting particles, formulated in a
   completely relativistic manner and
   employing a relative time variable.
f   équation *f* de Bethe-Salpeter
e   ecuación *f* de Bethe-Salpeter
i   equazione *f* di Bethe-Salpeter
n   vergelijking van Bethe en Salpeter
d   Bethe-Salpetersche Gleichung *f*

631 BETHE-WEIZSÄCKER CYCLE,   ch, gp
   CARBON-NITROGEN CYCLE,
   C-N CYCLE
   A reaction cycle of protons with carbon,
   nitrogen and oxygen nuclei, taking place
   probably in stars.
f   cycle *m* carbone-azote,
   cycle *m* solaire de Bethe-Weizsäcker
e   ciclo *m* carbono-nitrógeno,
   ciclo *m* de Bethe-Weizsäcker
i   ciclo *m* carbonio-azoto,
   ciclo *m* di Bethe-Weizsäcker
n   cyclus van Bethe en Weizsäcker,
   koolstof-stikstof-kringloop
d   Bethe-Weizsäcker-Zyklus *m*,
   Kohlenstoff-Stickstoff-Kreislauf *m*

632 BeV (US),   un
   BILLION ELECTRON VOLTS (US),
   GeV, GIGA-ELECTRON VOLT
   A unit of energy, equal to 10^9 eV.

f   GeV *m*, giga-électron-volt *m*
e   GeV *m*, giga-electrón-voltio *m*
i   GeV *m*, giga-elettronevolt *m*
n   GeV, giga-elektronvolt
d   GeV *n*, Giga-Elektronenvolt *n*

633 BEVATRON   pa
   A six-thousand million volt accelerator
   of protons and other atomic particles.
f   bévatron *m*
e   bevatrón *m*
i   bevatrone *m*
n   bevatron *n*
d   Bevatron *n*

634 BIASED LINEAR AMPLIFIER   ec
   A pulse amplifier which, within the limits
   of its normal operating characteristics,
   has constant gain for that portion of an
   input pulse that exceeds the threshold
   value and that produces no output for
   pulses whose amplitude is below the
   threshold.
f   amplificateur *m* linéaire à seuil
e   amplificador *m* lineal de umbral
i   amplificatore *m* lineare a soglia
n   lineaire drempelversterker
d   Fensterverstärker *m*

635 BILLIETITE   mi
   A rare secondary mineral containing about
   70 % of U.
f   billiétite *f*
e   billietita *f*
i   billietite *f*
n   billietiet *n*
d   Billietit *m*

636 BINARY SCALER   ct
   A scaler using the binary system of
   presentation.
f   échelle *f* de comptage binaire
e   escalador *m* binario,
   escalímetro *m* binario
i   circuito *m* numeratore d'impulsi binario
n   binaire pulsteller
d   binärer Impulszähler *m*,
   Zweifachuntersetzer *m*

637 BINARY SIGNAL   ma
   Logical signal using binary symbols.
f   signal *m* binaire
e   señal *f* binaria
i   segnale *m* binario
n   binair signaal *n*
d   binäres Signal *n*

638 BINDING ENERGY,   np
   SEPARATION ENERGY
   For a nucleus, the energy required to
   remove a nucleon.
f   énergie *f* de liaison,
   énergie *f* de séparation
e   energía *f* de enlace,
   energía *f* de separación
i   energia *f* di legame,
   energia *f* di separazione

n bindingsenergie
d Bindungsenergie f

639 BINDING FRACTION   np
   $b = B_r/A$, where B_r is the relative mass
   effect and A the nucleon number.
f fraction f de liaison
e fracción f de enlace
i frazione f di legame
n bindingsfractie
d Bindungsanteil m

640 BINEUTRON   np
   A supposed neutral particle with a mass
   near 2; the existence has not yet been
   proved.
f bineutron m
e bineutrón m
i bineutrone m
n bineutron n
d Bineutron n

641 BIOFILTER,   rw
       PERCOLATING FILTER,
       SPRINKLING FILTER,
       TRICKLING FILTER
   Filter sometimes used to describe sand
   or other filters used in biological processes for the treatment of low activity
   radioactive effluent solutions.
f lit m bactérien, lit m percolateur
e filtro m de escurrimiento,
   filtro m percolador
i letto m percolatore, percolatore m
n biologisch filter n, druppelfilter n,
   percolator
d Perkolator m, Tropfkörper m

642 BIOLOGICAL CONCENTRATION   is
   The concentration of a radioactive element
   present in an organism or in an organ or
   tissue.
f concentration f biologique
e concentración f biológica
i concentrazione f biologica
n biologische concentratie
d biologische Konzentration f

643 BIOLOGICAL CONCENTRATION   is
       FACTOR
   Of a radioactive isotope present in the
   surroundings of an organism, the ratio
   of the concentration of the isotope in that
   organism to its concentration in the
   surroundings.
f facteur m de concentration biologique
e factor m de concentración biológica
i fattore m di concentrazione biologica
n biologische concentratiefactor
d biologischer Konzentrationsfaktor m

644 BIOLOGICAL EFFECT OF   md, ra
       RADIATION
   The effect of radioactive radiation on
   living tissues.
f effet m biologique du rayonnement
e efecto m biológico de la radiación

i effetto m biologico della radiazione
n biologisch stralingseffect n
d biologischer Strahlungseffekt m

645 BIOLOGICAL HALF-LIFE   bi
   The time required for the amount of a
   particular substance in a biological
   system to be reduced to one half of its
   value by biological processes when the
   rate of removal is approximately
   exponential.
f demi-vie f biologique,
   période f biologique
e perfodo m biológico
i tempo m biologico di dimezzamento
n biologische halveringstijd
d biologische Halbwertzeit f

646 BIOLOGICAL HOLE   bi, rt
   A cavity in a reactor to permit the
   placing of biological material in or near
   the core.
f cavité f pour matières biologiques
e cavidad f por materiales biológicos
i cavitá f per materiali biologici
n holte voor biologisch materiaal
d Hohlraum m für biologisches Material

647 BIOLOGICAL PROTECTION   sa
   Protection of living beings against the
   effect of radiation.
f protection f biologique
e protección f biológica
i protezione f biologica
n biologische bescherming
d biologischer Schutz m

648 BIOLOGICAL SHIELD   ra, sa
   A shield whose prime purpose is to
   reduce ionizing radiation to biologically
   permissible levels.
f bouclier m biologique
e blindaje m biológico
i schermo m biologico
n biologisch scherm n
d biologische Abschirmung f,
   biologischer Schild m

649 BIOMEDICAL REACTOR   rt
   A nuclear reactor used primarily for
   biomedical purposes.
f réacteur m biomédical
e reactor m biomedical
i reattore m biomedicale
n biomedische reactor
d biomedizinischer Reaktor m

650 BIOPHYSICS   bi, md
   The study of phenomena of living organisms
   by physical methods.
f biophysique f
e biofísica f
i biofisica f
n biofysica
d Biophysik f

651 BIPHENYL,   cl
    DIPHENYL
An organic material with useful heat
transfer properties and resistant to
corrosion.
f diphényle *m*
e difenilo *m*
i difenile *m*
n difenyl *n*
d Diphenyl *n*

652 BIRD CAGE   is
A facility to transport radioactive
material.
f cage *f* de transport
e jaula *f* de transporte
i gabbia *f* di trasporto
n transportkooi
d Transportkäfig *m*

653 BISCUIT,   mg
    SPONGE METAL
The solid metallic product which results
from reduction of a metallic compound
by an active metal in a bomb or furnace.
f métal *m* spongieux
e metal *m* esponjoso
i metallo *m* spongioso
n sponsmetaal *n*
d Metallschwamm *m*

654 BISMUTH   ch
Metallic element, symbol Bi, atomic
number 83.
f bismuth *m*
e bismuto *m*
i bismuto *m*
n bismut *n*
d Wismut *n*

655 BISMUTH PHOSPHATE   mg
    SEPARATION PROCESS
An early precipitation process for the
extraction and purification of plutonium
from irradiated uranium.
f procédé *m* de séparation à phosphate de
   bismuth
e procedimiento *m* de separación con
   fosfato de bismuto
i processo *m* di separazione a fosfato di
   bismuto
n bismutfosfaatscheidingsproces *n*
d Trennungsverfahren *n* mit Wismutphosphat

656 BLACK   rt
In reactor technology of a body or medium,
effectively absorbing all the neutrons of
some specified energy incident on it.
f noir adj
e negro adj
i nero adj
n zwart adj
d schwarz adj

657 BLACK BODY   rt
A body absorbing practically all incident
neutrons of a specified energy.
f corps *m* noir

e cuerpo *m* negro
i corpo *m* nero
n zwart lichaam *n*
d schwarzer Körper *m*

658 BLACK URANIUM OXIDE   ch
The uranium oxide with formula U_3O_8.
f oxyde *m* noir d'uranium
e óxido *m* negro de uranio
i ossido *m* nero d'uranio
n zwart uraniumoxyde *n*
d schwarzes Uranoxyd *n*

659 BLANKET   rt
A layer of materials in contact with the
reactor core surrounding the latter
completely and containing sometimes part
of the primary cooling circuit.
f couverture *f*
e envoltura *f*
i copertura *f*
n deklaag
d Decke *f*

660 BLANKET   rt
A region of fertile material placed around
or within the core of a reactor for the
purpose of conversion.
f couche *f* fertile
e capa *f* fértil, envoltura *f* fértil
i mantello *m*
n kweekzone, mantel
d Brutzone *f*

661 BLANKET CONVERSION RATIO,   rt
    EXTERNAL CONVERSION RATIO
Conversion ratio relating to the only
fissile (fissionable) nuclei produced in the
blanket surrounding the core of the
conversion factor.
f rapport *m* de conversion externe
e relación *f* de conversión externa
i rapporto *m* di conversione esterna
n uitwendige conversieverhouding
d äusseres Konversionsverhältnis *n*,
   äusseres Umwandlungsverhältnis *n*

662 BLANKET MATERIAL   rt
The material of the blanket.
f matière *f* fertile
e material *m* fértil
i materiale *m* del mantello
n mantelmateriaal *n*
d Brutzonematerial *n*

663 BLANKET POWER   rt
The power generated in the blanket of
a breeder reactor.
f puissance *f* de la couche fertile
e potencia *f* de la capa fértil
i potenza *f* del mantello
n mantelvermogen *n*
d Brutzoneleistung *f*

664 BLANKET VESSEL   rt
The container for the fertile material
which surrounds the core of a breeder
reactor.

f récipient *m* de la couche fertile
e recipiente *m* de la capa fértil
i serbatoio *m* del mantello
n mantelhouder
d Brutzonebehälter *m*

665 BLAST   nw
The mechanical effect in which the sudden
thermal expansion of the air around the
bomb builds up an enormous pressure,
which spreads out as a shock-wave
front at velocities greater than that of
sound.
f expansion *f*
e expansión *f*
i espansione *f*
n expansie
d Expansion *f*

666 BLAST LOADING   nw
The loading or force on an object caused
by the air blast from an explosion striking
and flowing around the object.
f force *f* de l'expansion
e fuerza *f* de la expansión
i forza *f* dell'espansione
n expansiekracht
d Expansionskraft *f*

667 BLAST SCALING LAWS   ma, nw
Formulas which permit the calculation of
the properties of a blast wave at any
distance from an explosion of specified
energy from the known variation with
distance of these properties for a
reference explosion of known energy.
f lois *pl* de calcul de l'onde expansive
e leyes *pl* de cálculo de la onda expansiva
i leggi *pl* di calcolo dell'onda espansiva
n mathematische formules *pl* van de
   expansiegolf
d mathematische Formeln *pl* der
   Expansionswelle

668 BLAST WAVE   nw
A pressure pulse of air, accompanied
by winds, propagated continuously from
an explosion.
f onde *f* expansive
e onda *f* expansiva
i onda *f* espansiva
n expansiegolf
d Expansionswelle *f*

669 BLENDING   ch, mg
A term used with reference to the mixing
of material enriched in fissile (fissionable)
uranium 235 with depleted fuel from
nuclear reactors.
f mélange *m*
e mezclado *m*
i mescolanza *f*
n menging
d Mischung *f*

670 BLISTER   mg
A defect on the surface of a metal in the

form of a raised spot or nodule produced
during processing.
f poquette *f*, pustule *f*
e verruga *f*
i rilievo *m*, ruvidezza *f*,
   soffiatura *f* superficiale
n gietgal
d Wanze *f*

671 BLOB COUNTING,   ic
    CLUSTER COUNTING
In the photographic emulsion technique,
a method of measuring the grain density
along the track by counting the clusters
of grains.
f comptage *m* d'amas
e recuento *m* de enjambres
i conteggio *m* di grappoli,
   conteggio *m* di sciami
n klontentelling, telling van opeenhopingen
d Cluster-Zählung *f*, Traubenzählung *f*

672 BLOCH BAND,   ec
    ENERGY BAND
The sets of the discrete but closely adjacent
energy levels, equal in number to the
number of atoms, which arise from each
of the quantum states of the atoms of a
substance when they condense to a solid
state from a non-degenerate gaseous
conduction.
f bande *f* d'énergie
e banda *f* de energía
i banda *f* d'energia
n energieband
d Energieband *n*

673 BLOCKAGE   cl
The blocking of the flow of the coolant in
a reactor which could lead to overheating
of the fuel.
f blocage *m*
e bloqueo *m*
i bloccaggio *m*
n blokkering
d Blockierung *f*

674 BLOCKING MEDIUM   ab
Material of appropriate radiation opacity
for applying to an object, either round the
edges or as a filling for holes, to reduce
the effect of scattered radiation.
f moyen *m* de blocage
e medio *m* de bloqueo
i mezzo *m* di bloccaggio
n spermedium *n*
d Sperrmedium *n*

675 BLOMSTRANDITE, see 625

676 BLOOD-CELL DIFFERENTIAL   ct, md
    COUNTER
Counter used for indicating numerically
the percentage of leucocytes identified in a
differential leucocyte count.
f compteur *m* différentiel de la numération
   leucocytaire

e   contador *m* diferencial de la numeración
    leucocitaria
i   contatore *m* differenziale della
    numerazione leucocitaria
n   differentiële teller van witte bloed-
    lichaampjes
d   differentieller Zähler *m* der weissen
    Blutkörperchen

677 BLOOD COUNT,   ct, md
     HEMACYTOMETRY,
     HEMATIMETRY
     The number of corpuscles per cubic
     millimetre(er) of blood.
f   nombre *m* de globules sanguins
e   número *m* de glóbulos sanguineos
i   numero *m* di globuli sanguigni
n   aantal *n* bloedlichaampjes
d   Blutkörperchenzahl *f*

678 BLOOD COUNTING   ct, md
     The counting of the number of corpuscles
     per cubic millimetre(er) of blood.
f   comptage *m* de corpuscules sanguins,
    numération *f* de corpuscules sanguins
e   numeración *f* de glóbulos sanguineos,
    recuento *m* de glóbulos sanguineos
i   conteggio *m* di globuli sanguigni,
    numerazione *f* di globuli sanguigni
n   telling van het aantal bloedlichaampjes
d   Blutkörperchenzählung *f*

679 BLOWDOWN   mt
     Violent rupture of a vessel due to internal
     pressure.
f   crevaison *f*
e   reventazón *m*
i   scoppio *m*
n   openbarsten *n*
d   Zerplatzen *n*

680 BLOWDOWN LINE   cl
     The tube venting waste coolant to
     atmosphere.
f   tuyau *m* d'échappement
e   tubo *m* de escape
i   tubo *m* di scappamento
n   uitlaatpijp
d   Ablassrohr *n*

681 BLUR,   ph, xr
     UNSHARPNESS
     Unsharpness of a picture or radiograph.
f   flou *m*
e   borrosidad *f*
i   sfocatura *f*, soffuso
n   onscherpte
d   Unschärfe *f*

682 BODENBENDERITE   mi
     A titanium silicate of ytterbium,
     alumin(i)um and manganese.
f   bodenbendérite *f*
e   bodenbenderita *f*
i   bodenbenderite *f*
n   bodenbenderiet *n*
d   Bodenbenderit *m*

683 BODILY ROTATION   pp
     Rotation of the plasma body as a whole in
     the containment space.
f   rotation *f* du corps de plasma
e   rotación *f* del cuerpo de plasma
i   rotazione *f* del corpo di plasma
n   rotatie van het plasmalichaam
d   Rotation *f* des Plasmakörpers

684 BODY APRON,   sa, xr
     LEAD IMPREGNATED APRON,
     LEAD-RUBBER APRON
'   A lead rubber apron worn by X-ray
     operators.
f   tablier *m* en caoutchouc au plomb,
    tablier *m* opaque, tablier *m* plombé
e   delantal *m* plomado,
    mandil *m* de caucho al plomo
i   grembiale *m* di gomma al piombo
n   loodrubberschort
d   Bleigummischürze *f*, Bleischürze *f*

685 BODY BURDEN   is, md
     Activity of a radioelement present in an
     organism or in an organ.
f   charge *f* corporelle
e   carga *f* corporal
i   carico *m* corporale
n   lichaamsbelasting
d   Körperbelastung *f*

686 BODY CENTERED   cr
     CUBIC LATTICE,
     BODY CENTERED STRUCTURE
     A crystal structure characterized by a
     cubic cell with an atom in each corner
     and an atom in the centre(er) of the cell.
f   réseau *m* cubique centré
e   red *f* cúbica centrada en el cuerpo
i   reticolo *m* cubico a corpo centrato
n   midbloks kubisch rooster *n*,
    ruimtelijk gecentreerd kubisch rooster *n*
d   raumzentriertes kubisches Gitter *n*

687 BODY RADIOCARTOGRAPH,   me
     RADIOCARTOGRAPH
     An assembly designed to establish the
     radioactivity chart of a part of the human
     body, after absorption of a suitable
     radionuclide.
f   anthroporadiocartographe *m*,
    radiocartographe *m*
e   radiocartógrafo *m* corporal
i   radiocartografo *m*
n   radiocartograaf voor delen van het lichaam
d   Körperradiokartograph *m*

688 BODY SECTION DEVICE,   xr
     LAMINAGRAPH, PLANIGRAPH,
     STRATIGRAPH, TOMOGRAPH
     An instrument that produces a relatively
     sharp image of a thin layer of the object,
     all other layers being blurred by pre-
     determined relative motion of the röntgen
     tube, film and subject.
f   dispositif *m* pour radiographie en coupes,
    planigraphe *m*, stratigraphe *m*,
    tomographe *m*

e dispositivo *m* para radiografía por secciones, estratógrafo *m*, planígrafo *m*, tomógrafo *m*
i dispositivo *m* per radiografia a sezione, planigrafo *m*, stratigrafo *m*, tomografo *m*
n apparaat *n* voor snedeopname, stratigraaf, tomograaf
d Planigraph *m*, Stratigraph *m*, Tomograph *m*

689 BODY SECTION xr
RÖNTGENOGRAPHY,
LAMINAGRAPHY, STRATIGRAPHY,
TOMOGRAPHY
A röntgen-diagnostic technique of making röntgenograms of thin layers of objects.
f radiographie *f* en coupe, stratigraphie *f*, tomographie *f*
e estratografía *f*, radiografía por secciones, tomografía *f*
i radiografia *f* per sezioni, stratigrafia *f*, tomografia *f*
n tomografie
d Körperschichtaufnahme *f*

690 BOHR ATOM np
A model of the hydrogen atom based on the nuclear atom of Rutherford and on classical mechanics except for the superposition of non-classical, quantum, conditions.
f atome *m* de Bohr
e átomo *m* de Bohr
i atomo *m* di Bohr
n atoommodel *n* volgens Bohr
d Bohrsches Atommodell *n*

691 BOHR MAGNETON, np
ELECTRONIC BOHR MAGNETON
A unit of magnetic moment used in specifying the magnetic moment of an atomic particle or system of particles, when $\mu_\rho = eh/4\pi m_e c = 9.27 \times 10^{-21}$ erg gauss^{-1} where m_e is the rest mass of the electron.
f magnéton *m* de Bohr
e magnetón *m* de Bohr
i magnetone *m* di Bohr
n bohrmagneton *n*
d Bohrsches Magneton *n*

692 BOHR ORBIT np
An allowed electron orbit in the Bohr model of the hydrogen atom.
f orbite *f* de Bohr
e órbita *f* de Bohr
i orbita *f* di Bohr
n elektronenbaan volgens Bohr
d Bohrsche Bahn *f*

693 BOHR RADIUS np
The radius of the lowest-energy electron orbit in the Bohr model of the hydrogen atom.
f rayon *m* de Bohr
e radio *m* de Bohr
i raggio *m* di Bohr
n bohrradius, straal van de s-baan
d Bohrscher Wasserstoffradius *m*

694 BOHR-SOMMERFELD ATOM np
A modification of the Bohr model of the atom, in which quantized elliptic orbits, as well as circular orbits, are allowed.
f atome *m* de Bohr-Sommerfeld
e átomo *m* de Bohr-Sommerfeld
i atomo *m* di Bohr-Sommerfeld
n atoommodel *n* volgens Bohr en Sommerfeld
d Bohr-Sommerfeldsches Atommodell *n*

695 BOHR-SOMMERFELD np, qm
QUANTUM THEORY,
OLD QUANTUM MECHANICAL
THEORY
A theory developed from quantum ideas of Planck, Einstein, Sommerfeld and others, culminating just before the present theory of quantum mechanics developed with the ideas of de Broglie, Schrödinger and Heisenberg.
f théorie *f* des quanta de Bohr-Sommerfeld
e teoría *f* cuántica de Bohr-Sommerfeld
i teoria *f* dei quanti di Bohr-Sommerfeld
n quantumtheorie volgens Bohr en Sommerfeld
d Bohr-Sommerfeldsche Quantentheorie *f*

696 BOHR-WHEELER THEORY np
An analytical theory of nuclear fission and the condition under which it might occur, based upon a liquid-drop analogy.
f théorie *f* de Bohr-Wheeler
e teoría *f* de Bohr-Wheeler
i teoria *f* di Bohr-Wheeler
n theorie van Bohr en Wheeler
d Bohr-Wheelersche Theorie *f*

697 BOILING HEAT TRANSFER cl
Heat transfer by the actual boiling of the coolant.
f transfert *m* de chaleur d'ébullition
e transferencia *f* de calor de ebullición
i trasferimento *m* di calore d'ebullizione
n kookhitteoverdracht
d Siedehitzeübertragung *f*

698 BOILING POINT gp
The temperature of a liquid at which its maximum or saturated vapo(u)r pressure is equal to the normal atmospheric pressure.
f point *m* d'ébullition
e punto *m* de ebullición
i punto *m* d'ebollizione
n kookpunt *n*
d Siedepunkt *m*

699 BOILING WATER nue
IRRADIATION LOOP
A closed pipe system in a nuclear reactor in which the boiling water through the pipe system is irradiated.
f circuit *m* fermé à irradiation d'eau bouillante

e circuito *m* cerrado de irradiación de agua
   hirviente
i circuito *m* chiuso ad irradiazione d'acqua
   bollente
n gesloten circuit *n* voor het bestralen van
   kokend water
d Kreislauf *m* zum Bestrahlen von siedendem
   Wasser

700 BOILING WATER REACTOR,   rt
      WATER BOILING REACTOR
      A nuclear reactor in which water is used
      as the coolant.
f réacteur *m* à eau bouillante
e reactor *m* de agua hirviente
i reattore *m* ad acqua bollente
n kokend-waterreactor
d Siedewasserreaktor *m*

701 BOLTZMANN CONSTANT   ma
      The constant K in the calculations arising
      in the development of the Boltzmann
      equation.
f constante *f* de Boltzmann
e constante *f* de Boltzmann
i costante *f* di Boltzmann
n constante van Boltzmann
d Boltzmann-Konstante *f*

702 BOLTZMANN DISTRIBUTION   np
      PRINCIPLE
      A somewhat general law relating to the
      statistical distribution of large numbers
      of minute particles subject to thermal
      agitation and acted upon by a magnetic
      field, an electric field or a gravitational
      field, etc.
f principe *m* de répartition de Boltzmann
e principio *m* de repartición de Boltzmann
i principio *m* di ripartizione di Boltzmann
n verdelingsprincipe *n* van Boltzmann
d Boltzmann-Verteilungsprinzip *n*

703 BOLTZMANN EQUATION   ma
      The algebraic equation stating that the
      entropy of a system of particles is
      proportional to the Napierian logarithm
      of the probability of its macroscopic
      state.
f équation *f* de Boltzmann
e ecuación *f* de Boltzmann
i equazione *f* di Boltzmann
n boltzmannvergelijking
d Boltzmannsche Gleichung *f*

704 BOLTZMANN TRANSPORT   np
      EQUATION
      The fundamental equation describing the
      conservation of particles which are
      diffusing in a scattering, absorbing and
      multiplying medium.
f équation *f* de transport de Boltzmann
e ecuación *f* de transporte de Boltzmann
i equazione *f* di trasporto di Boltzmann
n transportvergelijking van Boltzmann
d Boltzmannsche Stossgleichung *f*

705 BOLUS MATERIAL   xr
      Material usually having the density and
      effective atomic number of tissue, used in
      radiation therapy to fill up void spaces,
      thus reducing the irradiated volume to
      a simple geometric form.
f bolus *m*
e bolus *m*
i bolo *m*
n bolusmateriaal *n*
d Bolusmaterial *n*

706 BOMB CORE   nw
      The active central part of a nuclear bomb.
f noyau *m* de bombe
e núcleo *m* de bomba
i nocciolo *m* di bomba
n bomkern
d Bombenkern *m*

707 BOMB DEBRIS   nw
      The residue of a nuclear bomb after it
      has been exploded.
f débris *pl* de bombe
e derribo *m* de bomba
i macerie *pl* di bomba
n bomresten *pl*
d Bombenreste *pl*

708 BOMBARD (TO)   np
      To direct a stream of high energy particles
      against a target.
f bombarder v
e bombardeaɾ v
i bombardare v
n beschieten v, bombarderen v
d beschiessen v, bombardieren v

709 BOMBARDED PARTICLE,   np
      STRUCK PARTICLE,
      TARGET PARTICLE
      A particle that is hit by an incident
      particle in a nuclear process.
f particule *f* bombardée
e partícula *f* bombardeada
i particella *f* bombardata
n getroffen deeltje *n*
d beschossenes Teilchen *n*,
   getroffenes Teilchen *n*

710 BOMBARDING PARTICLE,   np
      COLLIDING PARTICLE,
      INCIDENT PARTICLE
      A particle that hits another particle.
f particule *f* incidente,
   particule *f* projectile
e partícula *f* incidente
i particella *f* incidente
n botsend deeltje *n*, invallend deeltje *n*
d Beschussteilchen *n*, einfallendes Teilchen *n*

711 · BOMBARDING VOLTAGE   pa
      The voltage which would be required to
      accelerate the particles to the velocity
      attainable by a particle accelerator.
f tension *f* d'accélération
e tensión *f* de aceleración

i   tensione *f* d'accelerazione
n   doorlopen spanning
d   Beschleunigungsspannung *f*

712 BOMBARDMENT   np
The process of directing high-speed
electrons at an object.
f   bombardement *m*
e   bombardeo *m*
i   bombardamento *m*
n   beschieting, bombardement *n*
d   Beschiessung *f*, Bombardement *n*

713 BOMBARDMENT DAMAGE,   ra
RADIATION DAMAGE
A general term for the effects of
radiation upon substances.
f   dégâts *pl* par rayonnement
e   daño *m* por radiación,
deterioro *m* por radiación
i   danneggiamento *m* da radiazione
n   stralingsschade
d   Strahlenschaden *m*

714 BOND   rt
In reactor technology, the intimate material
contact between fuel and can or cladding.
f   liaison *f*
e   ligazón *f*
i   contatto *m*, legamento *m*, saldatura *f*
n   binding
d   Verbund *m*

715 BOND   rt
In reactor technology, a material effecting
the intimate contact between fuel and
can or cladding.
f   matériel *m* de liaison
e   material *m* de ligazón
i   materiale *m* di contatto,
materiale *m* di legamento,
materiale *m* di saldatura
n   bindingsmateriaal *n*
d   Verbundmaterial *n*

716 BOND DIRECTION,   ch
ORIENTATION
Certain covalent bonds prefer to lie in
particular directions with respect to
the bonded atoms.
f   direction *f* de la valence
e   dirección *f* de la valencia
i   direzione *f* della valenza
n   valentierichting
d   Valenzrichtung *f*

717 BOND MOMENT   ch, gp
The electromagnetic dipole moment of
a chemical bond between them.
f   moment *m* de liaison
e   momento *m* de enlace
i   momento *m* di legame
n   bindingsmoment *n*
d   Bindungsmoment *m*

718 BOND ORBITAL   ch
An orbital which may be associated with
a definite chemical bond.

f   orbite *f* de liaison
e   órbita *f* de enlace
i   orbita *f* di legame
n   bindingsbaan
d   Bindungsbahn *f*

719 BOND STRENGTH   ch, gp
The energy required to rupture a given
valence bond in a given molecule.
f   force *f* de liaison
e   fuerza *f* de enlace
i   forza *f* di legame
n   bindingssterkte
d   Bindekraft *f*

720 BONDED FUEL ELEMENT   fu
A fuel element which has a true
metallurgical bond between the fuel and
the canning material.
f   élément *m* combustible à liaison
métallurgique
e   elemento *m* combustible de ligazón
metalúrgica
i   elemento *m* combustibile a saldatura
per diffusione
n   splijtstofelement *n* met metallurgische
binding
d   Brennelement *n* mit metallurgischer
Verbindung

721 BONDING ELECTRON   ec, np
An electron in a molecule which serves to
hold two adjacent nuclei together.
f   électron *m* de valence
e   electrón *m* de valencia
i   elettrone *m* di valenza
n   valentie-elektron *n*
d   Valenzelektron *n*

722 BONDING ORBITAL   ch
A molecular orbital coupling two atoms in
such a way that the energy has a minimum
value when the interatomic distance is
small, thus favo(u)ring a bond between them.
f   orbite *f* électronique commune
e   órbita *f* electrónica común
i   orbita *f* elettronica comune
n   gemeenschappelijke elektronenbaan
d   gemeinsame Elektronenbahn *f*

723 BONE AGE   np
The age determined by means of isotopes
of bones found during excavations, etc.
f   âge *m* des os
e   edad *f* ósea
i   età *f* ossea
n   gebeenteleeftijd
d   Knochenalter *n*

724 BONE MARROW   bi, md
Soft tissue which fills the cavity of most
bones.
f   moelle *f* osseuse
e   médula *f* ósea
i   midollo *m* osseo
n   beenmerg *n*
d   Knochenmark *n*

725 BONE MAXIMUM   ra
     PERMISSIBLE DOSE
     The maximum dose that can safely be
     delivered to the bone.
f dose *f* maximale admissible dans les os
e dosis *f* máxima permisible en los huesos
i dose *f* massima ammissibile nelle ossa
n maximaal toegestane dosis in beenderen
d höchstzugelassene Dosis *f* in Knochen

726 BONE SCANNING   mc
     The sequential examination of the bone by
     means e.g. of strontium-87m.
f exploration *f* d'os
e exploración *f* de huesos
i scansione *f* d'ossa
n beenderaftasting
d Knochenabtastung *f*

727 BONE SEEKER,   bi
     BONE SEEKING ELEMENT
     Any element which in vivo is incorporated
     in bone, in preference to other tissue.
f substance *f* ostéophile
e substancia *f* osteófila
i sostanza *f* osseofila
n botzoeker
d Knochensucher *m*

728 BOOSTER,   np, sp
     NEUTRON BOOSTER
     A fast-neutron multiplying assembly,
     composed of a subcritical mass of fertile
     material, which is used in conjunction
     with a pulsed neutron source to increase
     the neutron output for time-of-flight
     spectroscopy.
f amplificateur *m* du rendement neutronique,
   multiplicateur *m* de neutrons
e amplificador *m* del rendimiento neutrónico
i amplificatore *m* del rendimento neutronico
n neutronenrendementversterker
d Neutronenbooster *m*

729 BOOSTER ELEMENT   rt
     A fuel element temporarily inserted in a
     reactor to provide xenon override.
f élément *m* de surréactivité
e elemento *m* de sobrerreactividad
i elemento *m* d'avviamento
n opjaagelement *n*
d Anfahrtbrennelement *n*

730 BOOTING   cd
     A plastics sleeve between parts of a
     manipulator isolating the handling parts
     from contamination.
f manche *f* de protection
e funda *f* protectora,
   manguito *m* de protección
i manicotto *m* di protezione
n beschermingsmanchet
d Schutzmanschette *f*

731 BORAL   mt
     An alloy of alumin(i)um containing
     boron carbide and used i.a. as a material
     for control rods.

f boral *m*
e boral *m*
i boral *m*
n boral *n*
d Boral *n*

732 BORAL SHIELD   sa
     A special shielding against thermal neutrons
     consisting of rolled boral sheets.
f blindage *m* en boral
e blindaje *m* en boral
i schermo *m* in boral
n boralscherm *n*
d Boralschild *m*

733 BORATED GRAPHITE   mt
     Shielding material used in reactor design.
f graphite *m* boraté
e grafito *m* boratado
i grafite *f* borata
n geborateerd grafiet *n*
d borierter Graphit *m*

734 BORAZOLE   mt
     Inorganic compound, isoelectronic with
     benzene and useful as a solvent to give
     a neutron sensitive liquid scintillation
     detector.
f borazole *m*
e borazol *m*
i borazolo *m*
n borazol *n*
d Borazol *n*

735 BORE-HOLE RADIO-LOG,   ma
     RADIOMETRIC BORE-HOLE
     LOGGING ASSEMBLY
     An assembly designed to measure the
     radiation as a function of the depth.
f ensemble *m* d'exploration radiométrique
   de forage,
   ensemble *m* de radiocarottage
e conjunto *m* de radiosondeo,
   explorador *m* radiométrico para sondeo
i complesso *m* per radiocarotaggio
n opstelling voor activiteitsmeting in
   boorgaten
d Anordnung *f* zur radiometrischen
   Bohrlochvermessung

736 BORN APPROXIMATION   qm
     A quantum mechanical approximation
     method used particularly for the
     computation of cross sections in scattering
     problems.
f approximation *f* de Born
e aproximación *f* de Born
i approssimazione *f* di Born
n benadering van Born
d Bornsche Näherung *f*

737 BORON   ch
     Non-metallic element, symbol B, atomic
     number 5.
f bore *m*
e boro *m*
i boro *m*

n   borium *n*
d   Bor *n*

738 BORON CARBIDE   mt
A refractory material used in nuclear
reactor control rods.
f   carbure *m* de bore
e   carburo *m* de boro
i   carburo *m* di boro
n   boriumcarbide *n*
d   Borkarbid *n*

739 BORON COUNTER TUBE   ct
A counter tube filled with BF_3 and/or

having electrodes coated with boron or
boron compounds.
f   tube *m* compteur à bore
e   tubo *m* contador al boro
i   tubo *m* contatore al boro
n   boriumtelbuis
d   Boriumzählrohr *n*,
Bortrifluoridrhythmusmesser *m*

740 BORON FLUORIDE,   ch, ct
BORON TRIFLUORIDE,
FLUORBORATE
The anion $BO_2F_2^{3-}$ or BF_4^{-1}, or a

compound containing one of these anions.
f   fluorure *f* de bore
e   fluoruro *m* de boro
i   fluoruro *m* di boro
n   boriumfluoride *n*
d   Borfluorid *n*

741 BORON IONIZATION CHAMBER   ic
An ionization chamber containing boron
or boron compounds, which is used for
detecting neutrons, mainly slow neutrons.
f   chambre *f* d'ionisation à bore
e   cámara *f* de ionización al boro
i   camera *f* d'ionizzazione al boro
n   boriumionisatievat *n*
d   Boriumionisationskammer *f*,
Bortrifluoridionisationskammer *f*

742 BORON POISONING   sa
The changes of the boron poisoning in the
heavy water moderator compensate
long-term changes of reactivity in the
core.
f   empoisonnement *m* de bore
e   envenenamiento *m* de boro
i   avvelenamento *m* da boro
n   boriumeffect *n*
d   Borvergiftung *f*

743 BORON STEEL   mt
A steel used for reactor control and
shielding.
f   acier *m* au bore
e   acero *m* al boro
i   acciaio *m* al boro
n   boriumstaal *n*
d   Borstahl *m*

744 BORON THERMOPILE   me
A device for measuring neutron flux.
f   thermopile *f* au bore
e   termopila *f* al boro
i   termopila *f* al boro
n   boriumthermozuil
d   Borthermosäule *f*

745 BOSE-EINSTEIN STATISTICS,   ma
BOSE STATISTICS
The kind of statistics to be used with
systems of identical particles which have
the property that the wave function remains
unchanged if any two particles are interchanged.
f   statistique *f* de Bose-Einstein
e   estadística *f* de Bose-Einstein
i   statistica *f* di Bose-Einstein
n   statistiek van Bose en Einstein
d   Bose-Einstein-Statistik *f*

746 BOSON   np
Particle described by the Bose-Einstein
statistics.
f   boson *m*
e   bosón *m*
i   bosone *m*
n   boson *n*
d   Boson *n*

747 BOTTOM REFLECTOR   rt
Reflector placed at the lower part of a
reactor installation.
f   réflecteur *m* de fond
e   reflector *m* de fondo
i   riflettore *m* di fondo
n   bodemreflector
d   Bodenreflektor *m*

748 BOTTOMS,   ch
HEEL, TAILINGS
The material drawn off from the bottom
of a tower or still.
f   résidus *pl* de distillation
e   residuos *pl* de destilación
i   residui *pl* di distillazione
n   destillatieresten *pl*
d   Destillationsrückstände *pl*,
hochsiedender Rückstand *m*

749 BOUND ELECTRON   ec, np
An electron bound to the nucleus of an
atom by electrostatic attraction.
f   électron *m* lié
e   electrón *m* ligado
i   elettrone *m* legato
n   gebonden elektron *n*
d   gebundenes Elektron *n*

750 BOUND PARTICLE   np
A particle surrounded by a potential
barrier and having insufficient kinetic
energy to pass over the top of the barrier.
f   particule *f* liée
e   partícola *f* ligada
i   particella *f* legata
n   gebonden deeltje *n*
d   gebundenes Teilchen *n*

751 BOUNDARY,   cr
   BOUNDARY SURFACE
   In crystallography, the surface of contact
   between adjacent crystals in a metal.
f surface *f* limite
e superficie *f* límite
i superficie *f* limite
n grensvlak *n*
d Grenzfläche *f*

752 BOUNDARY LAYER   ch, gp
   An extremely thin layer between e.g. a
   fluid of low viscosity moving around a
   stationary body or through a stationary
   conduit.
f couche *f* limite
e capa *f* límite
i strato *m* limite
n grenslaag
d Grenzschicht *f*

753 BOUNDARY WAVELENGTH,   sp
   MINIMUM WAVELENGTH,
   QUANTUM LIMIT
   The shortest wavelength present in a
   spectrum, corresponding with the
   potential applied.
f limite *f* quantique,
   longueur *f* d'onde limite
e límite *m* cuántico,
   longitud *f* de onda límite
i limite *m* quantico,
   lunghezza *f* d'onda limite
n grensgolflengte, kortste golflengte
d Grenzwellenlänge *f*,
   kürzeste Wellenlänge *f*

754 BOWING UNDER LOAD   rt
   In nuclear power plant an effect in fuel
   elements due to the stacking of one on
   top of the other.
f flexion *f*
e flexión *f*
i flessione *f*
n doorbuiging
d Durchbiegung *f*

755 BRAGG ANGLE   cr
   The angle θ appearing in the Bragg law,
   used in X-ray orientation of quartz
   crystals being cut for radio use.
f angle *m* de Bragg
e ángulo *m* de Bragg
i angolo *m* di Bragg
n hoek van Bragg
d Braggscher Winkel *m*

756 BRAGG CURVE   ra
   A curve showing the average number of
   ions per unit distance along a beam of
   initially monoenergetic ionizing particles
   passing through a gas.
f courbe *f* de Bragg
e curva *f* de Bragg
i curva *f* di Bragg
n kromme van Bragg
d Braggsche Kurve *f*

757 BRAGG-GRAY CAVITY,   ic, np
   GRAY CAVITY
   An ideal cavity containing gas within a solid
   medium, the cavity being sufficiently small
   not to disturb the distribution of either
   the primary or the secondary radiation in
   the medium.
f cavité *f* de Bragg-Gray
e cavidad *f* de Bragg-Gray
i cavità *f* di Bragg-Gray
n holte van Bragg-Gray
d Bragg-Grayscher Hohlraum *m*

758 BRAGG-GRAY CAVITY   ic
   IONIZATION CHAMBER
   An ionization chamber whose characteristics
   are such that the conditions defining the
   Bragg-Gray cavity are met in practice.
f chambre *f* d'ionisation à cavité,
   chambre *f* d'ionisation de Bragg-Gray
e cámara *f* de ionización de cavidad de
   Bragg-Gray
i camera *f* d'ionizzazione a cavità di
   Bragg-Gray
n ionisatievat *n* volgens Bragg en Gray
d Ionisationskammer *f* nach Bragg-Gray

759 BRAGG-GRAY PRINCIPLE   ic
   A principle saying that the conditions of a
   free air ionization chamber are reproduced
   by a small cavity in an infinite medium of
   a material where both the cavity gas and
   medium have essentially the same effective
   atomic number as air.
f principe *m* de Bragg-Gray
e principio *m* de Bragg-Gray
i principio *m* di Bragg-Gray
n principe *n* van Bragg en Gray
d Bragg-Graysches Prinzip *n*

760 BRAGG LAW   cr
   The statement of the condition under which
   a crystal will reflect a beam of X-rays
   with maximum intensity.
f loi *f* de Bragg
e ley *f* de Bragg
i legge *f* di Bragg
n wet van Bragg
d Braggsche Formel *f*, Braggsches Gesetz *n*

761 BRAGG REFLECTION   cr
   The diffracted beam produced by reinforcement of the contributions from successive
   members of a set of crystal planes when
   their common normal is suitably oriented
   with respect to an incident beam of monochromatic radiation.
f réflexion *f* de Bragg
e reflexión *f* de Bragg
i riflessione *f* di Bragg
n reflectievoorwaarde van Bragg
d Reflexionsbedingung *f* von Bragg

762 BRAGG RULE   np, ra
   An empirical rule according to which the
   atomic stopping power is directly
   proportional to the square root of the
   atomic weight.

f   règle de Bragg
e   regla *f* de Bragg
i   regola *f* di Bragg
n   regel van Bragg
d   Braggsche Regel *f*

763   BRAGG SCATTERING,   cr
         ORDERED SCATTERING
Elastic scattering from a crystal, in which
individual waves reinforce each other to
give constructive interference.
f   diffusion *f* de Bragg
e   dispersión *f* de Bragg
i   deviazione *f* di Bragg
n   verstrooiing van Bragg
d   Braggsche Streuung *f*

764   BRAGG SPECTROMETER,   sp
         CRYSTAL SPECTROMETER,
         IONIZATION SPECTROMETER
A spectrometer in which diffraction by a
crystal is used to obtain the spectra of
X-rays and soft gamma rays as well as
of slow neutrons.
f   spectromètre *m* à cristal
e   espectrómetro *m* de cristal
i   spettrometro *m* a cristallo
n   kristalspectrometer
d   Kristallspektrometer *n*

765   BRANCH   np
A nuclear product or series of nuclear
products resulting from one mode of decay
of a radioactive nuclide exhibiting two or
more such modes.
f   produit *m* d'embranchement
e   producto *m* de bifurcación
i   prodotto *m* di ramificazione
n   vertakkingsprodukt *n*
d   Unterfamilie *f*, Zerfallsanteil *m*,
    Zweig *m*

766   BRANCHING   np
The occurrence of two or more modes by
which a radionuclide can undergo radioactive decay.
f   embranchement *m*
e   bifurcación *f*
i   ramificazione *f*
n   vertakking
d   Verzweigung *f*

767   BRANCHING DECAY,   np
         BRANCHING DISINTEGRATION,
         MULTIPLE DECAY,
         MULTIPLE DISINTEGRATION
Radioactive decay of a nuclide by two or
more modes.
f   désintégration *f* multiple,
    embranchement *m*
e   desintegración *f* de bifurcación
i   ramificazione *f* d'un decadimento
n   vertakt verval *n*
d   verzweigter Zerfall *m*

768   BRANCHING FRACTION   np
In branching decay, the fraction of nuclei
which disintegrate in a specified way.

f   fraction *f* d'embranchement
e   fracción *f* de bifurcación
i   frazione *f* di ramificazione
n   vertakkingsfractie
d   Verzweigungsanteil *m*

769   BRANCHING RATIO   np
The ratio of the branching fractions for
two specified modes of disintegration.
f   rapport *m* d'embranchement
e   relación *f* de bifurcación
i   rapporto *m* di ramificazione
n   vertakkingsverhouding
d   Verzweigungsverhältnis *n*

770   BRANNERITE   mi
A metatitanate of iron, calcium, thorium,
zirconium and uranium.
f   brannerite *f*
e   branerita *f*
i   brannerite *f*
n   branneriet *n*
d   Brannerit *m*

771   BRAVAIS LATTICE,   cr
         CRYSTAL LATTICE,
         SPACE LATTICE
In a crystal, the arrangement of the
constituent atoms or radicals which is
determined from the diffraction patterns
obtained by X-ray photography and other
methods.
f   réseau *m* cristallin
e   red *f* del cristal, red *f* cristalina
i   reticolo *m* cristallino
n   kristalrooster *n*
d   Kristallgitter *n*

772   BRAZING   mg
The process of joining metals by the fusion
and solidification of a metal or alloy
whose melting point is lower than those
of the metals being joined.
f   brasage *m*
e   soldadura *f* fuerte
i   brasatura *f*, saldatura *f* forte
n   hardsolderen *n*
d   Hartlöten *n*

773   BREAK POINT,   ab, rw
         BREAKTHROUGH POINT
That point in an absorption cycle at which
the effluent begins to show a marked
increase in concentration of the substance
being absorbed.
f   point *m* de rupture
e   punto *m* de ruptura
i   punto *m* di rottura
n   doorbraakpunt *n*
d   Durchbruchpunkt *m*

774   BREAKAWAY   nw
The onset of a condition in which the shock
front in the air moves away from the
exterior ball of fire produced by the
explosion of a nuclear weapon.
f   sécession *f*
e   secesión *f*

i secessione *f*
n afscheiding
d Abscheidung *f*

775 BREAKDOWN   rt
Operation incident leading to partial or
complete shut-down of a nuclear reactor.
f panne *f*
e avería *f*
i avaria *f*, collasso *m*, guasto *m*
n panne
d Panne *f*

776 BREAK-SEAL AMPOULE   is, md
An ampoule of which the top can be easily
broken off to enable the contents to be
discharged.
f ampoule *f* à rupture de pointe
e ampolla *f* de ruptura de punta
i ampolla *f* a rottura di punta
n ampul met afbreekbare punt
d Ampulle *f* mit Abbrechspitze

777 BREAKTHROUGH CURVE   rw
A graph used to study the ability of soil
or other solids to remove radioisotopes
from a flowing stream.
f courbe *f* de rupture
e curva *f* de ruptura
i curva *f* di rottura
n doorbraakgrafiek, doorbraakkromme
d Durchbruchkurve *f*

778 BREEDER END   rt
A tube of pure uranium at the top and
bottom of tubular enriched uranium
elements and separated therefrom by
molybdenum washers.
f anneau *m* en uranium
e anillo *m* en uranio
i anello *m* in uranio
n uraniumring
d Uranring *m*

779 BREEDER REACTOR,   rt
    NUCLEAR BREEDER
A nuclear reactor which produces more
fissile (fissionable) material than it
consumes, i.e., has a conversion ratio
greater than unity.
f réacteur *m* surrégénérateur
e reactor *m* regenerador
i reattore *m* rigeneratore
n kweekreactor
d Brutreaktor *m*

780 BREEDING   np, rt
Conversion, when the conversion factor
is greater than unity.
f surrégénération *f*
e cría *f*, regeneración *f*
i rigenerazione *f*
n kweken *n*
d Brüten *n*

781 BREEDING GAIN   np, rt
Breeding ratio minus one.

f gain *m* de surrégénération
e ganancia *f* de regeneración,
  rendimiento *m* de reproducción
i guadagno *m* di rigenerazione
n kweekwinst
d Brutgewinn *m*

782 BREEDING RATIO   np, rt
The conversion factor, when it is greater
than unity.
f rapport *m* de surrégénération
e relación *f* de regeneración,
  relación *f* de reproducción
i rapporto *m* integrale di rigenerazione
n kweekverhouding
d Brutverhältnis *n*

783 BREIT-WIGNER FORMULA   ma
An expression for the absorption cross
section of a nucleus as a function of the
energy E of the bombarding particle,
e.g. neutron, near a resonance.
f formule *f* de Breit-Wigner
e fórmula *f* de Breit-Wigner
i formula *f* di Breit-Wigner
n formule van Breit en Wigner
d Breit-Wigner-Formel *f*

784 BREMSSTRAHLEN   xr
The radiation produced by the bremsstrahlung process.
f rayons *pl* de freinage
e rayos *pl* de enfrenamiento,
  rayos *pl* de frenado
i raggi *pl* di rallentamento
n remstralen *pl*
d Bremsstrahlen *pl*

785 BREMSSTRAHLUNG   xr
The electromagnetic radiation arising
from the deceleration, or acceleration,
of charged particles.
f rayonnement *m* de freinage
e radiación *f* de enfrenamiento,
  radiación *f* de frenado
i radiazione *f* di rallentamento
n remstraling
d Bremsstrahlung *f*

786 BREVIUM   ch
Uranium X_2, one of the decay products
in the uranium series.
f brévium *m*
e brevio *m*
i brevio *m*
n brevium *n*
d Brevium *n*

787 BRITTLE CRACK   mt
An effect occurring in mild steel linings
of a reactor vessel.
f tapure *f* due à la fragilité
e grieta *f* de fragilidad
i cricca *f* da fragilità
n brosheidsscheur
d Sprödigkeitsriss *m*

788 BRITTLE FRACTURE   mt, te
The consequence eventually of a brittle
crack.
f cassure *f* due à la fragilité
e fractura *f* de fragilidad
i frattura *f* da fragilità
n brosheidsbreuk
d Sprödigkeitsbruch *m*

789 BROAD BEAM   ec, ra
Term used to determine a radiation beam
emitted without a collimator and in which
the influence of the diffused radiation is
large.
f faisceau *m* large
e haz *m* ancho
i fascio *m* largo
n brede bundel
e breiter Strahl *m*, breites Bündel *n*

790 BROAD BEAM ABSORPTION   ab, ra
In radiology, the absorption of a wide
beam of radiation in a given material,
measured under conditions such that
scattered radiation is detected.
f absorption *f* de faisceau large
e absorción *f* de haz ancho
i assorbimento *m* di fascio largo
n brede-bundelabsorptie
d Breitbündelabsorption *f*

791 BROAD BEAM ATTENUATION   ab
Due to the presence of scattered particles
in the beam, attenuation of a broad beam
is much less than that of a narrow beam
in the same material.
f atténuation *f* de faisceau large
e atenuación *f* de haz ancho
i attenuazione *f* di fascio largo
n brede-bundelverzwakking
d Breitbündelschwächung *f*

792 BROAD BEAM GEOMETRY   rt
In shielding calculation, the beam condition
in which many of the particles or photons
initially scattered out of the beam may be
scattered back in because of a second
event or where one scattering event does
not necessarily scatter the particle out
of the beam.
f géométrie *f* à champ large,
   géométrie *f* à faisceau large
e geometría *f* de haz ancho
i geometria *f* a fascio largo
n geometrie bij groot veld
d Grossfeldgeometrie *f*

793 BRÖGGERITE   mi
A uranium mineral containing a large
percentage of thorium.
f broeggerite *f*
e bröggerita *f*
i bröggerite *f*
n bröggeriet *n*
d Bröggerit *m*

794 BROMINE   ch
Non-metallic liquid element, symbol Br,
atomic number 35.
f brome *m*
e bromo *m*
i bromo *m*
n bromium *n*, broom *n*
d Brom *n*

795 BROMINE TRIFLUORIDE   mt
An interhalogen compound used as a
fluorinating agent in nuclear technology.
f trifluorure *m* de brome
e trifluoruro *m* de bromo
i trifluoruro *m* di bromo
n broomtrifluoride *n*
d Bromtrifluorid *n*

796 BRONCHOGRAPHY   xr
The radiological examination of the trachea,
bronchi or the bronchial tree following the
injection of air or other contrast medium.
f bronchographie *f*
e broncografía *f*
i broncografia *f*
n bronchografie
d Bronchographie *f*

797 BROWN URANIUM OXIDE   ch
The uranium oxide with formula UO_2.
f bioxyde *m* d'uranium.
e bióxido *m* de uranio
i biossido *m* d'uranio,
   ossido *m* bruno d'uranio
n uraniumdioxyde *n*
d Urandioxyd *n*

798 BROWNIAN MOVEMENT   ch, gp
A proof for the truth of the kinetic theory,
asserting that heat has to be explained as
uncontrolled movement of atoms and
molecules.
f mouvement *m* brownien
e movimiento *m* browniano
i movimento *m* browniano
n brownbeweging
d Brownsche Bewegung *f*

799 BUBBLE CAP   ch, is
A device resembling an inverted cup with
serrated or slotted edges employed in
connection with a bubble plate in a
distillation or absorption column.
f calotte *f* à barbotage
e platillo *m* de burbujeo
i campanella *f* di gorgogliamento
n borrelklok
d Bodenglocke *f*, Fraktionierbodenglocke *f*

800 BUBBLE CHAMBER   ic
An instrument for making the tracks of
ionizing particles visible as rows of
little bubbles.
f chambre *f* à bulles
e cámara *f* de burbujas
i camera *f* a bolle
n bellenvat *n*
d Blasenkammer *f*

801 BUBBLE PLATE, ch, is
BUBBLE TRAY
In a distillation or absorption column, a
tray or plate equipped with bubble caps
and an overflow pipe.
f plateau *m* de barbotage
e plato *m* de burbujeo
i piatto *m* a campanelle,
vassoio *m* di gorgogliamento
n borrelplaat, klokschotel
d Glockenboden *m*

802 BUBBLE PLATE COLUMN, ch, is
BUBBLE TRAY COLUMN
A column for distillation using a series
of bubble plates (trays).
f colonne *f* à plateaux de barbotage
e columna *f* de burbujeo
i colonna *f* di gorgogliamento
n borrelplaatkolom, klokschotelkolom
d Glockenbodenkolonne *f*

803 BUBBLE TEST te
A test whereby the source is immersed
in ethanediol or other suitable liquid
to check the appearance of fine bubbles.
f essai *m* de bulles
e prueba *f* de burbujas
i prova *f* di bolle
n blaasjesproef
d Bellenprüfung *f*

804 BUBBLING, cg, ch
SPARGING
Introduction of a gas or a vapo(u)r beneath
the surface of a liquid for the purpose of
agitation, heat transfer or stripping.
f barbotage *m*
e burbujeo *m*
i gorgogliamento *m*
n doorblazen *n*
d Durchblasen *n*, Hindurchperlen *n*

805 BUCKLING, rt
LAPLACIEN
The common value of the geometric and
material bucklings for a critical reactor.
f laplacien *m*
e laplaciano *m*
i parametro *m* di criticità
n bolling, welving
d Flussdichtewölbung *f*

806 BUCKY (US), xr
BUCKY GRID (GB), MOVING GRID,
POTTER-BUCKY GRID,
RECÍPROCATING GRID
An anti-diffusion grid which is kept in
motion during röntgen exposure so as to
eliminate the line pattern.
f grille *f* mobile, grille *f* oscillante,
Potter-Bucky *m*
e rejilla *f* de Potter-Bucky,
rejilla *f* móvil, rejilla *f* oscilante
i griglia *f* mobile, griglia *f* oscillante
n potter-bucky-raster *n*
d bewegliche Streustrahlenblende *f*,
Potter-Bucky-Blende *f*

807 BUILD-UP gr, xr
In a material irradiated by a beam of
X-rays or gamma rays, the increase in
absorbed dose-rate or ionization with depth
beneath the surface due to the production
of secondary electrons in the material.
f accroissement *m*, accumulation *f*
e crecimiento *m*
i accumulo *m*, aggiuntatura *f*
n opbouw, verhoging
d Aufbau *m*, Zuwachs *m*

808 BUILD-UP DOSE ra, xr
The specific absorbed dose in a point by
interaction of the irradiated material with
the secondary radiation produced in the
material itself by the primary beam.
f dose *f* d'accumulation
e dosis *f* de acumulación
i dose *f* aggiunta, dose d'accumulo
n opgebouwde dosis
d Aufbaudosis *f*

809 BUILD-UP FACTOR ra
In the passage of radiation through a medium,
the ratio of the total value of a specified
radiation quantity at any point, to the
contribution to that value from radiation
reaching the point without having undergone
a collision.
f facteur *m* d'accumulation
e factor *m* de acumulación
i fattore *m* d'accumulazione
n accumulatiefactor, opbouwfactor
d Aufbaufaktor *m*, Zuwachsfaktor *m*

810 BUILDING-UP PRINCIPLE np
The postulate that the totality of the
electronic terms of an atom or molecule
can be obtained by successive bringing
together of the parts.
f principe *m* de constitution
e principio *m* de constitución
i principio *m* di costituzione
n opbouwprincipe *n*
d Aufbauprinzip *n*

811 BUILT-IN REACTIVITY rt
The excess reactivity of the clean, cold
core.
f réserve *f* de réactivité
e reactividad *f* inherente
i reattività *f* intrinseca
n ingebouwde reactiviteit
d inhärente Reaktivität *f*,
Überschussreaktivität *f*

812 BULK SHIELDING FACILITY, rt, te
SHIELD TEST POOL FACILITY
A facility for testing shields at the Oak
Ridge National Laboratory.
f installation *f* d'essai de blindages en masse
e equipo *m* de prueba de blindajes en masa
i impianto *m* di prova di schermatura in
massa
n massabeproevingsinstallatie voor afschermingen
d Massenprüfanlage *f* für Abschirmungen

813 BULK SHIELDING REACTOR rt
A nuclear reactor capable of operating at
a wide variety of power levels.
f réacteur *m* à blindages différents
e reactor *m* de blindajes diferentes
i reattore *m* a schermature differenti
n reactor met verschillende afschermingen
d Reaktor *m* mit verschiedenen Abschirmungen

814 BULK TEST te
Test of a sample in a large quantity.
f essai *m* de volume
e prueba *f* de volumen
i prova *f* di volume
n bulktest, volumebeproeving
d Massenprüfung *f*

815 BUNCHING OF IONS ec
The joining together of ions in a number
of groups.
f groupement *m* d'ions
e agrupamiento *m* de iones
i accumulo *m* d'ioni
n ionenbundeling
d Ionenpaketbildung *f*

816 BUNDLE OF FUEL ELEMENTS, fu
CLUSTER OF FUEL ELEMENTS,
STRINGER
Rod shaped fuel elements, mounted
parallel-wise.
f faisceau *m* d'éléments combustibles,
grappe *m* d'éléments combustibles
e haz *m* de elementos combustibles
i fascio *m* d'elementi combustibili
n splijtstofelementenbundel
d Brennelementenbündel *n*

817 BUOYANCY nw
Upward drift experienced by a mass of
hot air when entering a colder atmosphere.
f carène *f*
e fuerza *f* ascensional
i spinta *f* ascendente
n opwaartse druk
d Auftrieb *m*

818 BURGERS VECTOR cr
A vector representing the displacement
of the material of the lattice required
to create a dislocation.
f vecteur *m* de Burgers
e vector *m* de Burgers
i vettore *m* di Burgers
n burgersvector
d Burgers-Vektor *m*

819 BURIAL GROUND, rt
GRAVEYARD
A place for burying unwanted radioactive
objects to prevent escape of radiations.
f cimetière *m*
e cementerio *m*
i cimitero *m*
n begraafplaats
d Friedhof *m*

820 BURN-OUT np
Describes the condition achieved in a
thermonuclear apparatus when the rate at
which neutral gas atoms are converted into
ions either by charge exchange or by
ionization, exceeds the rate at which
neutral atoms enter the working volume.
f dépassement *m* de débit,
épuisement *m* de particules neutres
e traspaso *m* de ritmo
i eccedenza *f* di velocità
n tempo-overschrijding
d Rateüberschreitung *f*

821 BURN-OUT rt
FUEL BURN-OUT
Severe local damage of a fuel element,
due to failure of the coolant to dissipate
all the heat produced in the element.
f brûlage *m*
e abrasamiento *m*
i bruciatura *f*
n doorsmelting
d Durchbrennen *n*

822 BURN-OUT FLUX, rt
BURN-OUT HEAT FLUX
The local heat flux density at which fuel
burn-out takes place.
f flux *m* de brûlage
e flujo *m* de abrasamiento
i flusso *m* termico critico
n warmteflux bij doorsmelting
d Wärmestromdichte *f* beim Durchbrennen

823 BURN-OUT POINT rt
For a liquid-cooled reactor, any combination of values of heat transfer parameters
which result in fuel burn-out.
f point *m* de caléfaction
e punto *m* de calefacción
i punto *m* di riscaldamento
n doorsmeltingspunt *n*
d Durchbrennpunkt *m*

824 BURN-OUT POISON, rt
BURNABLE POISON
Poison purposely included in a nuclear
reactor to help control long-term reactivity
changes by its progressive burn-up.
f poison *m* consommable
e veneno *m* de compensación
i veleno *m* di compensazione
n slijtend gif *n*, verdwijnend gif *n*
d abbrennbares Reaktorgift *n*

825 BURN-UP np, rt
Induced nuclear transformation of atoms
during reactor operation. The term may
be applied to fuel or other materials.
f combustion *f* nucléaire
e combustión *f* nuclear, consumo *m*
i consumo *m*
n verslijting, versplijting
d Abbrand *m*

826 BURN-UP FRACTION,							np, rt
    BURN-UP RATIO,
    FUEL UTILIZATION
    The fraction, usually expressed as a
    percentage, of an initial quantity of nuclei
    of a given type which has undergone
    burn-up.
f   taux *m* d'épuisement
e   fracción *f* de combustión nuclear,
    fracción *f* de consumo
i   frazione *f* di consumo,
    rapporto *m* di combustione
n   verslijtingsfractie, versplijtingsfractie
d   relativer Abbrand *m*

827 BURN-UP LEVEL,							rt
    IRRADIATION LEVEL
    The total amount of energy liberated per
    unit mass of fuel or fissile (fissionable)
    material.
f   consommation *f*
e   rendimiento *m* energético específico
    de un combustible nuclear
i   resa *f* energetica specifica d'un combus-
    tibile nucleare
n   energetisch versplijtingsrendement *n*
d   Abbrandtiefe *f*

828 BURNABLE POISON PRESSURIZED			rt
    WATER REACTOR
    A small-size reactor contemplated for
    small power outputs, e.g. for container
    ships.
f   réacteur *m* à eau sous pression et poison
    consommable
e   reactor *m* de agua presurizada y veneno
    de compensación
i   reattore *m* ad acqua in pressione a veleno
    di compensazione
n   drukwaterreactor met slijtend gif
d   Druckwasserreaktor *m* mit abbrennbarer
    Vergiftung

829 BURNER REACTOR							rt
    A reactor in which no significant conver-
    sion takes place.
f   réacteur *m* à conversion négligeable
e   reactor *m* de conversión insignificante
i   reattore *m* a conversione senza importanza
n   versplijtingsreactor
d   Zerspaltungsreaktor *m*

830 BURST									fu
    A defect in a fuel element which allows
    fission products to escape.
f   rupture *f*
e   hendidura *f*
i   fenditura *f*
n   barst, lek
d   Hüllenfehler *m*

831 BURST									ic, np
    A large number of ions produced in an
    ionization chamber if several cosmic-ray
    particles pass through it simultaneously.
f   gerbe *f* explosive
e   estallido *m*

i   fiotto *m*
n   ionenuitbarsting
d   Hoffmannscher Stoss *m*, Ionisationsstoss *m*

832 BURST CAN DETECTOR,					rt, sa
    BURST CARTRIDGE DETECTOR,
    DUNKOMETER (US)
    An instrument for the early detection of
    ruptures of the sheaths of fuel elements
    inside a reactor.
f   détecteur *m* d'éléments défectueux,
    détecteur *m* de rupture de gaine
e   detector *m* de elementos defectuosos
i   rivelatore *m* d'elementi difettosi
n   detector van lekke splijtstofelementen
d   Detektor *m* von schadhaften
    Brennelementen

833 BURST DETECTOR							rt
    Apparatus for detecting burst fuel elements
    by means of the escaping radiation of
    short-lived fission products.
f   détecteur *m* de ruptures
e   detector *m* de avarías
i   rivelatore *m* di rotture
n   breukdetector, scheurdetector
d   Rissdetektor *m*, Schadendetektor *m*

834 BURST SLUG,							rt
    FAILED ELEMENT
    A fuel element with a defect which allows
    fission products to escape.
f   élément *m* défectueux
e   elemento *m* defectuoso
i   elemento *m* difettoso
n   lek splijtstofelement *n*
d   Brennelementfleck *m*,
    schadhaftes Brennelement *n*

835 BUTEX,								mt
    DIBUTOXYDIETHYL ETHER,
    DIBUTYL CARBITOL,
    DIETHYLENE GLYCOL DIBUTYL
    ETHER
    An organic solvent used in the solvent
    extraction of plutonium from irradiated
    uranium.
f   éther *m* diéthylique du diéthylèneglycol
e   éter *m* dietílico del dietilenglicol
i   etere *m* dietil-glicol-dietilenico
n   diëthyleenglycol-diëthyleenether
d   Diäthylenglykoldiäthylenäther *m*

836 BY-PRODUCT							ch, ge
    A product in a chemical process which is
    not the main object but useful for other
    purposes.
f   sous-produit *m*
e   coproducto *m*
i   sottoprodotto *m*
n   bijprodukt *n*
d   Nebenprodukt *n*

837 BY-PRODUCT MATERIAL   mt

In atomic energy law, any radioactive material, except source of fissile (fissionable) material, obtained in the process of producing or using source or fissile (fissionable) material.

f   matériel *m* de sous-produit
e   material *m* de coproducto
i   materiale *m* di sottoprodotto
n   bijproduktmateriaal *n*
d   Nebenproduktmaterial *n*

C

838 CADMIUM   ch
Metallic element, symbol Cd, atomic
number 48.
f   cadmium *m*
e   cadmio *m*
i   cadmio *m*
n   cadmium *n*
d   Cadmium *n*, Kadmium *n*

839 CADMIUM CURTAIN,   co
CADMIUM STRIP
A control element in nuclear reactors
made of cadmium in strip form.
f   bande *f* de cadmium
e   hoja *f* de cadmio
i   striscia *f* di cadmio
n   cadmiumstrip
d   Kadmiumstreifen *m*

840 CADMIUM CUT-OFF,   rt
CADMIUM CUT-OFF ENERGY
In nuclear engineering, the neutron
energy level at which cadmium ceases to
show a high cross section of absorption.
f   limite *f* de capture pour le cadmium
e   límite *m* de captura para cadmio
i   limite *m* di cattura per cadmio
n   vangstgrens voor cadmium
d   Einfanggrenze *f* im Kadmium

841 CADMIUM RATIO   ct
The ratio of the neutron-induced saturated
activity in an unshielded foil to the
saturated activity of the same foil when
it is covered with cadmium.
f   rapport *m* cadmique
e   relación *f* cádmica
i   rapporto *m* cadmico
n   cadmiumverhouding
d   Kadmiumverhältnis *n*

842 CADMIUM REGULATOR   co
A regulating rod consisting of or
containing cadmium.
f   barre *f* de commande en cadmium
e   barra *f* de regulación en cadmio
i   barra *f* di regolazione in cadmio
n   cadmiumregelstaaf
d   Kadmiumsteuerstab *m*

843 CAESIUM (GB),   ch
CESIUM (US)
Metallic element, symbol Cs, atomic
number 55.
f   caesium *m*, césium *m*
e   cesio *m*
i   cesio *m*
n   cesium *n*
d   Caesium *n*, Zäsium *n*

844 CAESIUM-137 (GB),   ch, is
CESIUM-137 (US), RADIOCESIUM (US)
A radioisotope recovered from the waste
of nuclear reactors. Used in sterilizing
food and as a substitute for radium.
f   césium-137 *m*
e   cesio-137 *m*
i   cesio-137 *m*
n   cesium-137 *n*
d   Zäsium-137 *n*

845 CAESIUM-137 PELLET (GB), gr, md, ra
CESIUM-137 PELLET (US)
Pellet used in gamma sources for
teletherapy.
f   boulette *f* en césium-137
e   bolita *f* de cesio-137
i   pallottolina *f* in cesio-137
n   cesium-137-kogeltje *n*
d   Zäsium-137-Kügelchen *n*

846 CALANDRIA   rt
A closed reactor vessel with internal
tubes or channels arranged so as to keep
the liquid moderator separate from the
coolant, to provide irradiation facilities,
or to contain pressure tubes.
f   calandre *m*
e   calandria *f*,
tanque *m* de tubos internos
i   calandria *f*,
recipiente.*m* a pressione
n   calandria, pijpenvat *n*
d   Kalandriagefäss *n*, Röhrengefäss *n*

847 CALCIOSAMARSKITE   mi
Rare mineral containing about 9.4 to
11.3 % of U and 1.9 to 2.9 % of Th.
f   calciosamarskite *f*
e   calciosamarskita *f*
i   calciosamarskite *f*
n   calciumsamarskiet *n*
d   Kalziumsamarskit *m*

848 CALCIOTHERMY   mg
The process of manufacturing metallic
uranium or plutonium by reducing a
compound, usually a fluoride, by calcium.
f   calciothermie *f*
e   calciotermia *f*
i   calciotermia *f*
n   calciothermie
d   Kalziothermie *f*

849 CALCIOTHORITE   mi
Radioactive decomposition product of
thorite containing uranium and thorium.
f   calciothorite *f*
e   calciotorita *f*
i   calciotorite *f*
n   calciumthoriet *n*
d   Kalziumthorit *m*

850 CALCIUM   ch
Metallic element, symbol Ca, atomic
number 20.

f   calcium *m*
e   calcio *m*
i   calcio *m*
n   calcium *n*
d   Calcium *n*, Kalzium *n*

851 CALCIUM AGE   ma, np
The age calculated from the relative
number of atoms of a stable radiogenic
calcium product and radioactive parent
present.
f   âge *m* de calcium
e   edad *f* de calcio
i   età *f* di calcio
n   calciumleeftijd
d   Kalziumalter *n*

852 CALCIUM URANYL CARBONATE   ch
Compound present in the mineral randite.
f   carbonate *m* uranylique de calcium
e   carbonato *m* uranílico de calcio
i   carbonato *m* uranilico di calcio
n   calciumuranylcarbonaat *n*
d   Kalziumuranylkarbonat *n*

853 CALIBRATOR   me, co
Used for the calibration and periodic
testing of portable ionization chambers.
f   bloc *m* d'étalonnage
e   bloque *m* de muestreo
i   blocco *m* di campionatura
n   ijktoestel *n* voor ionisatievaten
d   Eichgerät *n* für Ionisationskammer

854 CALIFORNIUM   ch
The tenth element of the actinides,
symbol Cf, atomic number 98.
f   californium *m*
e   californio *m*
i   californio *m*
n   californium *n*
d   Californium *n*, Kalifornium *n*

855 CALORIE,   un
      CALORY
A unit for the measurement of the quantity
of heat and defined as 4.1840 absolute
joules.
f   calorie *f*
e   caloría *f*
i   caloria *f*
n   calorie, kalorie
d   Kalorie *f*

856 CALORIFIC POWER (US),   gp
      CALORIFIC VALUE (GB)
The number of units of heat obtained by
complete combustion of unit mass of a
substance.
f   puissance *f* calorifique
e   potencia *f* calorífica
i   potenza *f* calorifica
n   bovenste verbrandingswaarde,
    calorische waarde
d   Heizkraft *f*

857 CALUTRON   is
An isotope separator of the electromagnetic
type, based on the 180° single-focusing
Dempster mass spectrograph, for separating isotopes on the production scale.
f   calutron *m*
e   calutrón *m*
i   calutrone *m*
n   calutron *n*
d   Calutron *n*

858 CAN,   rt
      CLADDING, JACKET, SHEATH
An external layer of material applied,
usually to a nuclear fuel, to provide
protection from a chemically reactive
environment, to provide containment of
radioactive products produced during the
irradiation of the composite, or to provide
structural support.
f   gaine *f*
e   revestimiento *m* metálico, vaina *f*
i   guaina *f*, placcatura *f*,
    rivestimento *m* metallico
n   mantel, metaalbekleding, omhulling
d   Brennstoffhülle *f*, Brennstoffhülse *f*

859 CANAL   rt
A water-filled channel, leading to or
serving as a fuel cooling installation, into
which radioactive objects are discharged
from a reactor.
f   canal *m*
e   canal *m*
i   canale *m*
n   kanaal *n*
d   Kanal *m*

860 CANAL RAYS,   ra
      POSITIVE ION RAYS
Positive particles in a vacuum tube which
escape through holes or tunnels bored in
the cathode.
f   rayons *pl* canaux
e   rayos *pl* canales
i   raggi *pl* canale, raggi *pl* di Goldstein
n   kanaalstralen *pl*, positieve ionenstralen *pl*
d   Kanalstrahlen *pl*, positive Ionenstrahlen *pl*

861 CANCER,   md
      CARCINOMA
Any malignant neoplasm.
f   cancer *m*, carcinome *m*
e   cáncer *m*, carcinoma *m*
i   cancro *m*, carcinoma *m*
n   carcinoom *n*, kanker
d   Karzinom *n*, Krebs *m*

862 CANNED PUMP,   rt
      CANNED ROTOR PUMP,
      SHEATHED PUMP
A pump used in nuclear engineering of
which the rotor is surrounded by a
protective housing.
f   pompe *f* à roteur gainé
e   bomba *f* de rotor hermético
i   pompa *f* a rotore stagno

n pomp met ingekapselde rotor
d Pumpe *f* mit eingekapseltem Läufer

863 CANNING, rt
 CLADDING, JACKETING
 The process of providing a material
 with a can.
f gainage *m*
e camisadura *f*, envainadura *f*,
 revestimiento *m*
i ricopertura *f*, rivestimento *m*
n bekleden *n*, ommanteling
d Einhüllen *n*, Einhülsen *n*

864 CANYON fu
 A long, narrow space enclosed with
 heavy shields, constituting the major part
 of a building used for certain types of
 radiochemical plants, such as plants for
 reprocessing used fuel.
f canyon *m*, enceinte *f* blindé
e pasillo *m* blindado
i trincea *f* schermata
n afgeschermde tunnelvormige ruimte
d abgeschirmter tunnelförmiger Raum *m*

865 CAPACITOR BANK, ec, pp
 CAPACITOR BOX (GB),
 SUBDIVIDED CAPACITOR (US)
 The energy source of the Columbus
 experimental rig.
f boîte *f* de condensateurs
e caja *f* de condensadores
i cassa *f* di condensatori
n condensatorbank
d Kondensatorenkasten *m*

866 CAPACITOR DOSEMETER ma
 A dosemeter measuring the discharge of
 a capacitor under the effects of
 irradiation.
f dosimètre *m* à condensateur
e dosímetro *m* de condensador
i dosimetro *m* a condensatore
n capacitieve dosismeter,
 condensatordosismeter
d Kondensatordosismesser *m*

867 CAPACITOR EXPOSUREMETER me
 An exposure dosemeter measuring the
 discharge of a capacitor under the effect
 of irradiation.
f dosimètre *m* à condensateur,
 exposimètre *m* à condensateur
e exposímetro *m* de condensador
i esposimetro *m* a condensatore
n capacitieve exposiemeter,
 condensatorexposiemeter
d Exponierungsmesser *m*,
 Expositionsmesser *m*

868 CAPACITOR IONIZATION CHAMBER ic
 An ionization chamber in which the discharge caused by radiation induces a
 variation of the potential difference, which
 is the quantity measured, between the
 electrodes constituting a capacitor.

f chambre *f* d'ionisation à condensateur
e cámara *f* de ionización de condensador
i camera *f* d'ionizzazione a condensatore
n capacitief ionisatievat *n*,
 condensatorionisatievat *n*
d Kondensatorionisationskammer *f*

869 CAPACITOR-R-METER me
 An instrument for measuring radiations
 consisting of an electrometer and a
 detachable ionization chamber embodying
 a capacitor of suitable capacitance.
f radiomètre *m* à condensateur
e radiómetro *m* de condensador
i radiometro *m* a condensatore
n condensatorstralingsmeter
d Kondensatorstrahlungsmesser *m*

870 CAPILLARY THEORY OF ch
 SEPARATION
 A theory of the separation of gases by
 flow through a porous medium, based on
 the concept of momentum transfer.
f théorie *f* de séparation capillaire
e teoría *f* de separación capilar
i teoria *f* di separazione capillare
n capillariteitstheorie van de gasscheiding
d Kapillaritätstheorie *f* der Gastrennung

871 CAPSULE md, ra
 In safety regulations with respect to
 radioactive materials, a receptacle
 immediately surrounding the radioactive
 material, intended to contain it in a
 permanent manner and used for unsealed
 sources.
f capsule *f*
e cápsula *f*
i capsula *f*
n capsule
d Kapsel *f*

872 CAPSULE IRRADIATION ra
 The irradiation by means of a radioactive
 material which is enclosed in a capsule.
f irradiation *f* à capsule
e irradiación *f* por cápsula
i irradiazione *f* a capsula
n capsulebestraling
d Kapselbestrahlung *f*

873 CAPTURE np
 In general a process in which an atomic
 or nuclear system acquires an additional
 particle, e.g. the capture of electrons by
 positive ions.
f capture *f*
e captura *f*
i cattura *f*
n vangst
d Einfang *m*

874 CAPTURE CROSS SECTION cs, np
 The cross section for radiative capture.
f section *f* efficace de capture
e sección *f* eficaz de captura

i sezione *f* d'urto di cattura
n vangstdoorsnede
d Einfangquerschnitt *m*

875 CAPTURE EFFICIENCY,   pa
    SYNCHROTRON CAPTURE
    EFFICIENCY
    The ratio of the time available during a
    frequency modulation cycle for starting
    particles into stable orbits to the total
    time for repetition of the frequency
    modulation .cycle.
f rendement *m* de capture
e rendimiento *m* de captura
i rendimento *m* di cattura
n vangstopbrengst, vangstrendement *n*
d Einfangausbeute *f*

876 CAPTURE GAMMA,   np
    CAPTURE GAMMA RADIATION
    The gamma radiation emitted in
    radiative capture.
f rayonnement *m* gamma de capture
e radiación *f* gamma de captura
i radiazione *f* gamma di cattura
n vangstgammastraling
d Einfanggammastrahlung *f*

877 CAPTURE REACTION   np
    A nuclear reaction produced by the
    capture of a bombarding particle by an
    atomic nucleus.
f réaction *f* de capture
e reacción *f* de captura
i reazione *f* di cattura
n vangstreactie
d Einfangreaktion *f*

878 CAPTURE TO FISSION RATIO   np
    The relation between the number of
    neutrons captured to the number of these
    neutrons able to produce fission.
f relation *f* capture/fission
e relación *f* captura/fisión
i rapporto *m* cattura/fissione
n vangst/splijtingverhouding
d Einfang/Spaltverhältnis *n*

879 CARBON   ch
    Non-metallic element, symbol C, atomic
    number 6.
f carbone *m*
e carbono *m*
i carbonio *m*
n koolstof
d Kohlenstoff *m*

880 CARBON CYCLE   np
    A cycle of six consecutive reactions,
    resulting in the formation of a helium
    nucleus from 4 protons.
f cycle *m* du carbone
e ciclo *m* del carbono
i ciclo *m* del carbonio
n koolstofcyclus
d Kohlenstoffzyklus *m*

881 CARBON DIOXIDE   ch, co
    Gas used as a coolant in nuclear reactors.
f acide *m* carbonique
e ácido *m* carbónico, anhídrido *m* carbónico,
  gas *m* carbónico
i anidride *f* carbonica,
  biossido *m* di carbonio
n koolzuur *n*
d Kohlensäure *f*

882 CARBON DIOXIDE COOLED   cl, rt
    REACTOR
    A nuclear reactor cooled by carbon
    dioxide.
f réacteui *m* à refroidissement par acide
  carbonique
e reactor *m* enfriado al ácido carbónico
i reattore *m* con raffreddamento ad acido
  carbonico
n met koolzuur gekoelde reactor
d kohlensäuregekühlter Reaktor *m*

883 CARBON DIOXIDE LEAKAGE   rt
    An undesirable event sometimes occurring
    in nuclear power plants especially at
    connecting points.
f fuite *f* d'acide carbonique
e fuga *f* de ácido carbónico
i fuga *f* d'anidride carbonica
n koolzuurlek *n*
d Kohlensäureleck *m*

884 CARBON-14, see 73

885 CARBON-14 AGE,   ms, np
    RADIOCARBON AGE,
    RADIOCARBON DATING
    The age calculated from the specific
    activity, due to C-14, of the carbon in a
    once living object.
f datation *f* C-14
e determinación *f* de la fecha por radiocarbón
i datazione *f* C-14,
  datazione *f* per radiocarbonio
n C-14-ouderdom, koolstofouderdom
d C-14-Alter *n*, C-14-Datierung *f*

886 CARBON/HYDROGEN RATIO   ma
    MEASURING ASSEMBLY FOR
    HYDROCARBONS
    A measuring assembly including a beta
    radiation source and designed to determine
    carbon/hydrogen ratio in hydrocarbon
    samples of known densities, by measure-
    ment of the radiation transmitted through
    the sample.
f ensemble *m* de détermination du rapport
  carbone/hydrogène d'hydrocarbures
e medidor *m* carbono/hidrógeno para
  hidrocarburos
i complesso *m* di misura del rapporto
  carbonio/idrogeno d'un idrocarburo
n meter voor kool/waterstofverhouding
  in koolwaterstoffen
d Messanordnung *f* zur Bestimmung des
  C/H-Verhältnisse in Kohlenwasserstoffen

887 CARBON ISOTOPE RATIO												is
The relative amount of a particular isotope in a sample of carbon.
f   rapport *m* isotopique de carbone
e   relación *f* isotópica de carbono
i   rapporto *m* isotopico di carbonio
n   isotopenverhouding van koolstof
d   Isotopenverhältnis *n* von Kohlenstoff

887A CARBON-NITROGEN CYCLE,
see 631

888 CARBON STEEL VESSEL												rt
A closed vessel containing the reactor core and consisting of carbon steel.
f   récipient *m* en acier au carbone
e   recipiente *m* de acero al carbono
i   serbatoio *m* all acciaio al carbonio
n   vat *n* uit koolstofstaal
d   Kohlenstoffstahlgefäss *n*

889 CARBON-TRITIUM METER										me
A device for measuring the low energy of beta emitters with liquid scintillators, particularly of C-14 and tritium.
f   carbotrimètre *m*
e   medidor *m* de actividad de C-14 y tritio
i   misuratore *m* d'attività di C-14 e tritio
n   activiteitsmeter voor C-14 en tritium
d   C-14- und Tritiumaktivitätsmesser *m*

890 CARBURAN														mt
A mixture of uranium-lead compound.
f   carburane *m*
e   carburán *m*
i   carburano *m*
n   carburaan *n*
d   Karburan *m*

891 CARCINOGENESIS (US),											md
CARCINOSIS (GB)
The production and development of cancer.
f   carcinose *f*
e   carcinomatosis *f*, carcinosis *f*
i   carcinomatosi *f*, carcinosi *f*
n   carcinose
d   Karzinose *f*

892 CARCINOMA, see 861

893 CARLSON METHOD											is, ma
Method of digital integration of the transport equation.
f   méthode *f* de Carlson
e   método *m* de Carlson
i   metodo *m* di Carlson
n   carlsonmethode
d   Carlson-Verfahren *n*

894 CARNOTITE														mi
A common secondary mineral containing 52.8 to 55.0 % of U.
f   carnotite *f*
e   carnotita *f*
i   carnotite *f*
n   carnotiet *n*
d   Carnotit *m*

895 CARRIER,														is
CARRIER ELEMENT
An element associated with traces of isotopes of the same element, or of an analogous element, giving a quantity of material sufficiently large to ensure that it follows chemical or physical processes in the manner characteristic of bulk matter.
f   élément *m* porteur, entraîneur *m*
e   elemento *m* portador
i   elemento *m* portatore, trascinatore *m*
n   dragerelement *n*
d   Trägerelement *n*

896 CARRIER COMPOUND											np
Bulk material containing the radioactive atoms, especially the mother mass not converted by irradiation.
f   composé *m* porteur
e   compuesto *m* portador
i   composto *m* portatore
n   dragerverbinding
d   Trägerverbindung *f*

897 CARRIER-FREE													is
A preparation in which all the atoms of the activated element present are radioactive.
f   sans entraîneur, sans porteur
e   sin portador
i   senza portatore, senza trascinatore
n   dragervrij adj, zonder drager
d   trägerfrei adj

898 CARRIER MOBILITY,											np
HALL MOBILITY
The quantity μ^H in the relation $\mu^H = R\delta$, where R = Hall constant and δ = conductivity.
f   mobilité *f* du porteur
e   movilidad *f* del portador
i   mobilità *f* del portatore
n   dragerbeweeglijkheid
d   Trägerbeweglichkeit *f*

899 CARTRIDGE													ab, rt
An absorber, other than a control element, with any necessary container, for insertion in a reactor to irradiate material, or to limit reactivity, or to change the flux distribution.
f   cartouche *f* à absorption
e   cartucho *m* de absorción
i   cartuccio *m* ad assorbimento
n   absorptiepatroon
d   Absorptionspatrone *f*

900 CARYOTIN,														md
CHROMATIN, CHROMOPLASM
The material in the nucleus which stains deeply with basic dye.
f   chromatine *f*, chromoplasma *m*, substance *f* nucléaire
e   cariotina *f*, cromatina *f*, cromoplasma *m*
i   carioplasma *m*, cromatina *f*
n   chromatine *n*, karyoplasma *n*
d   Chromatin *n*, Kernsubstanz *f*

901 CASCADE,   is
       STAGE CASCADE
       An arrangement of separative elements
       or stages connected so as to multiply the
       separation produced by a single element
       or stage.
f cascade f, cascade f d'étages
e cascada f, cascada f de etapas
i cascata f, cascata f di stadi
n cascade, trappencascade
d Kaskade f, Stufenkaskade f

902 CASCADE GAMMA RAYS,   gr
       GAMMA CASCADE,
       SUCCESSIVE GAMMA RAYS
       The process by which two or more different
       gamma rays are emitted successively
       from one nucleus when it passes through
       one or more energy levels.
f cascade f gamma,
       émission f gamma en cascade
e cascada f gamma
i cascata f gamma
n gammacascade
d Gammakaskade f

903 CASCADE GENERATOR   pa
       A high-voltage apparatus for obtaining
       high-velocity particles.
f générateur m en cascade
e generador m en cascada
i generatore m in cascata
n cascadegenerator
d Kaskadengenerator m

904 CASCADE ISOTOPE SEPARATION ch, is
       PLANT
       An isotope separation plant in which
       cascade distillation is carried out.
f installation f de séparation d'isotopes
       en cascade
e equipo m de separación de isótopos en
       cascada
i impianto m di separazione d'isotopi
       in cascata
n cascade-isotopenscheidingsinstallatie
d Kaskadenisotopentrennungsanlage f

905 CASCADE OF CASCADES   is
       A system in which one or more cascades
       are fed by the product or rejected from
       others.
f unité f de cascade collectrice
e unidad f de cascada colectora
i unità f di cascata collettrice
n verzamelcascade-eenheid
d Sammelkaskadeneinheit f

906 CASCADE OF SEPARATING UNITS ch, is
       A series of isotope separating units.
f cascade f d'étages de séparation,
       groupe m d'étages
e cascada f de etapas de separación,
       grupo m de etapas
i cascata f di stadi di separazione,
       gruppo m di stadi
n scheidingstrappencascade
d Trennungsstufenkaskade f

907 CASCADE PARTICLE   np
       A hyperon with a mean life $\simeq 10^{-9}$ s,
       decaying into a Λ° and a negative pion.
f particule f cascade
e partícula f cascada
i particella f cascata
n cascadedeeltje n
d Kaskadenteilchen n

908 CASCADE SHOWER,   np, ra
       SOFT SHOWER
       A cosmic-ray shower that is initiated when
       a high-energy electron, in passage
       through matter, produces one or more
       photons having energies comparable
       with its own.
f gerbe f électrophotonique,
       gerbe f en cascade
e chaparrón m en cascada
i sciame m in cascata
n cascadebui
d Kaskadenschauer m

909 CASCADE THEORY OF   np
       COSMIC RADIATION
f Theory of multiplication of electrons,
       positrons and gamma rays in the passage
       of a cosmic-ray particle through matter.
f théorie f des cascades du rayonnement
       cosmique
e teoría f de las cascadas de la radiación
       cósmica
i teoria f delle cascate della radiazione
       cosmica
n cascadetheorie van de kosmische straling
d Kaskadentheorie f der kosmischen
       Strahlung

910 CASIMIR-DU PRÉ THEORY   ma
       OF SPIN-LATTICE RELAXATION
       A theory in which the spin and lattice
       systems are treated as separate thermo-
       dynamic systems, in thermal contact with
       each other.
f théorie f de relaxation spin-réseau de
       Casimir et Du Pré
e teoría f de relajación espín-celosía de
       Casimir y Du Pré
i teoria f di rilasciamento spin-reticolo
       di Casimir e Du Pré
n spin-rooster-relaxatietheorie van
       Casimir en Du Pré
d Spin-Gitter-Relaxationstheorie f von
       Casimir und Du Pré

911 CASK,   ge
       COFFIN, FLASK
       A shielded container used to store or
       transport radioactive material.
f château m de transport
e recipiente m blindado
i bara f, cofano m
n opslagvat n, transportvat n
d Transportgefäss n, Umladebehälter m

912 CASSETTE   xr
       A container with a cover transparent to
       röntgen rays and opaque to ordinary light,

in which the film used for radiography is enclosed.
f cassette *f*
e cajita *f*
i cassetta *f*
n cassette
d Kassette *f*

913 CASTLE ct
A housing for a counter and the radio-active material for assay, so designed as to reduce extraneous radiation.
f chambre *f* à parois absorbantes
e cámara *f* de paredes absorbentes
i camera *f* a pareti assorbenti, castello *m*
n kasteel *n*
d Absorptionskammer *f*

914 CATALYST, ch
CATALYZER
A substance which promotes a chemical reaction without taking part itself.
f catalyseur *m*
e catalizador *m*
i catalizzatore *m*
n katalysator
d Katalysator *m*

915 CATALYTIC EXCHANGE IN is
AQUEOUS MEDIA
A method of tritium labelling (tagging) suitable only for organic compounds which are stable in aqueous solution at temperatures up to $120^{\circ}C$ in the presence of a platinum catalyst.
f échange *m* catalytique en milieux aqueux
e cambio *m* catalítico en medios acuosos
i scambio *m* catalitico in mezzi acquosi
n katalytische uitwisseling in waterige media
d katalytischer Austausch *m* in wässerigen Medien

916 CATALYTIC EXCHANGE ch
REACTION
A chemical exchange reaction in which use is made of a catalyst.
f réaction *f* à échange catalytique
e reacción *f* de cambio catalítico
i reazione *f* a scambio catalitico
n katalytische uitwisselingsreactie
d katalytische Austauschreaktion *f*

917 CATALYTIC GAS EXPOSURE is
METHOD
A method of tritium labelling (tagging) involving the exposure of organic compounds to tritium gas for an agreed period.
f méthode *f* catalytique d'exposition au tritium
e método *m* catalítico de exposición al tritio
i metodo *m* catalitico d'esposizione al tritio
n katalytische methode met blootstelling aan tritium
d katalytisches Verfahren *n* mit Tritium-exposition

918 CATALYTIC RECOMBINER rt
A recombiner in which the radiolytic gases are led over a heated catalyst.
f recombinateur *m* catalytique
e recombinador *m* catalítico
i ricombinatore *m* catalitico
n katalytische recombinator
d katalytischer Rekombinator *m*

919 CATALYZED NUCLEAR np
REACTION,
COLD FUSION REACTION
A nuclear fusion reaction catalyzed by mu mesons at liquid hydrogen temperature.
f réaction *f* nucléaire à catalyseur mésonique
e reacción *f* nuclear con catalizador mesónico
i reazione *f* nucleare a catalizzatore mesonico
n kernreactie met mesonkatalysator
d Kernreaktion *f* mit Mesonenkatalysator

920 CATAPHORESIS ch
The migration of electronegative particles through a fluid under influence of an electric field.
f cataphorèse *f*
e cataforesis *f*
i cataforesi *f*
n kataforese
d Kataphorese *f*

921 CATCHER FOIL rt
A thin, accurately machined alumin(i)um foil used in measuring power levels at given points in a nuclear reactor.
f bande *f* de mesure
e banda *f* de medida
i banda *f* di misura
n meetstrook
d Messstreifen *m*

922 CATHODE xr
An electrode, usually of incandescent tungsten, which emits electrons in a röntgen tube.
f cathode *f*
e cátodo *m*
i catodo *m*
n katode
d Katode *f*

923 CATHODE BEAM, ec
CATHODE RAYS
High-speed electrons which have been artificially accelerated in an electric field.
f faisceau *m* cathodique,
rayons *pl* cathodiques
e haz *m* catódico, rayos *pl* catódicos
i fascio *m* catodico, raggi *pl* catodici
n elektronenbundel, elektronenstraal
d Katodenstrahlenbündel *n*,
Katodenstrahlung *f*

924 CATHODE EVAPORATION ec, xr
Detachment of particles from the cathode.

f   évaporation *f* cathodique
e   evaporación *f* catódica
i   evaporazione *f* catodica
n   katodeverstuiving
d   Katodenzerstäubung *f*

925 CATHODE-RAY TUBE   ec
A vacuum tube in which a beam of electrons
is produced by acceleration by an
electric field.
f   tube *m* à rayons cathodiques
e   tubo *m* de rayos catódicos
i   tubo *m* a raggi catodici
n   elektronenstraalbuis
d   Elektronenstrahlröhre *f*

926 CATION,   ec
POSITIVE ION
An ion having a positive charge.
f   cation *m*, ion *m* positif
e   catión *m*, ión *m* positivo
i   catione *m*, ione *m* positivo
n   kation *n*, positief ion *n*
d   Kation *n*, positives Ion *n*

927 CATIONIC CURRENT   ec
That part of an electric current in which
the charge is carried by cations.
f   courant *m* cationique
e   corriente *f* catiónica
i   corrente *f* cationica
n   positieve ionenstroom
d   positiver Ionenstrom *m*

928 CAULKING COMPOUND   ch, rt
A compound used to ensure a perfect seal
between parts of an installation.
f   composé *m* de calfatage
e   compuesto *m* para calafatear
i   composto *m* da calafatare
n   kalfaterverbinding
d   Kalfaterverbindung *f*

929 CAUSTIC SCRUBBER   cg
A vessel in which a contaminated gas is
washed with a solution of sodium hydroxide
to remove the contaminant.
f   récipient *m* de lavage caustique de gaz
e   recipiente *m* de lavado cáustico de gas
i   recipiente *m* di lavaggio caustico di gas
n   gaswasser met natriumhydroxyde
d   Gaswäscher *m* mit Natriumhydroxyd

930 CAUSTIC SCRUBBING   cg, rw
The washing of a contaminated gas.
f   lavage *m* caustique
e   lavado *m* cáustico
i   lavaggio *m* caustico
n   wassing met natriumhydroxyde
d   Natriumhydroxydwäsche *f*

931 CAVERN   rw
Gravel bed buried in the ground from
which a liquid waste stream can percolate
into the soil.
f   caverne *f* à sol de gravier
e   caverna *f* de suelo de cascajo

i   caverna *f* a suolo di ghiaia
n   met grind gevulde ondergrondse ruimte
d   unterirdische Sickerhöhle *f* mit
    Kiesboden

932 CAVITATION   fu
A term loosely applied to describe the
formation of intergranular holes in fuel
cans or material during deformation.
f   cavitation *f*
e   cavitación *f*
i   cavitazione *f*
n   cavitatie
d   Kavitation *f*

933 CELL,   rt
LATTICE CELL, REACTOR CELL
One of a set of elementary regions in a
heterogeneous reactor each of which
has the same geometrical form and
neutron characteristics.
f   cellule *f* de réacteur
e   célula *f* de reactor
i   cella *f* di reattore
n   reactorcel
d   Reaktorzelle *f*

934 CELL CORRECTION FACTOR   ma, rt
A factor introduced to correct for the
effect of idealizing the shape of actual
reactor cells in the calculation of
reactor parameters.
f   facteur *m* de correction de cellule
e   factor *m* de corrección de celda
i   fattore *m* di correzione di cella
n   correctiefactor voor de celvorm
d   Zellkorrekturfaktor *m*

935 CELL SIZE   me
In the photographic emulsion technique,
the regular intervals into which a track in
the emulsion is divided when measuring
the multiple scattering of a particle in
the emulsion.
f   intervalle *m* entre traces
e   intervalo *m* entre trazas
i   intervallo *m* tra traccie
n   baandeel *n*, spoordeel *n*
d   Bahnteil *m*, Spurteil *m*

936 CENT   un
A nuclear unit of reactivity equal to one
hundredth of a dollar.
f   cent *m*
e   cent *m*
i   cent *m*
n   cent
d   cent-Einheit *f*

937 CENTER OF AREA (US),   gp, ma
CENTRE OF AREA (GB), CENTROID
The centre(er) of mass of a two-dimensional
homogeneous surface.
f   centre *m* de surface
e   centro *m* de superficie
i   centro *m* di superficie
n   oppervlakmiddelpunt *n*
d   Flächenmittelpunkt *m*

938 CENTER OF GRAVITY (US), gp, ma
   CENTER OF INERTIA (US),
   CENTRE OF GRAVITY (GB),
   CENTRE OF INERTIA (GB)
The point in a body at which its weight
may be taken to act, and at which the
body may be supported in neutral
equilibrium. It is coincident with the
centre(er) of mass.
f centre m de gravité
e centro m de gravedad
i centro m di gravità
n zwaartepunt n
d Schwerpunkt m

939 CENTER OF MASS (US), gp, ma
   CENTER OF MASS (GB)
That point in a collection of mass particles
which moves as if the total mass of the
collection were concentrated there and
the resultant of all the external forces
were acting there. It is coincident with
the centre(er) of gravity.
f centre m de masse
e centro m de masa
i centro m di massa
n massamiddelpunt n
d Massenmittelpunkt m

940 CENTER OF MASS SYSTEM (US), gp, ma
   CENTRE OF MASS SYSTEM (GB)
A frame of reference of which the origin
is taken from the centre(er) of mass of the
particles considered or more generally,
for which the total pulse of the considered
particles is zero.
f système m du centre de masse
e sistema m del centro de masa
i sistema m del centro di massa
n massamiddelpuntsysteem n
d Massenmittelpunktsystem n

941 CENTER OF VOLUME (US), gp, ma
   CENTRE OF VOLUME (GB)
The centre(er) of mass of a homogeneous
medium.
f centre m de volume
e centro m de volumen
i centro m di volume
n volumemiddelpunt n
d Volumenmittelpunkt m

942 CENTRAL BARRIER np
A potential barrier which shows a high
central elevation.
f barrière f à centre élevé
e barrera f de centro elevado
i barriera f a centro elevato
n potentiaalstoep met centrale verhoging
d Potentialberg m mit hohem Zentrum

943 CENTRAL COLLISION, np
   HEAD-ON COLLISION
A collision of two particles moving toward
each other along a straight line.
f collision f frontale
e colisión f de frente

i collisione f frontale
n frontale botsing, rechtlijnige botsing
d geradlinige Kollision f, Zentralstoss m

944 CENTRAL FORCE, np
   CENTRAL NUCLEAR FORCE
A nuclear force which is a simple
attraction or repulsion directed along the
line joining a pair of nucleons.
f force f centrale
e fuerza f central
i forza f centrale
n centrale kracht
d Zentralkraft f

945 CENTRAL POTENTIAL np
A nuclear potential that is spherically
symmetric.
f potentiel m central
e potencial m central
i potenziale m centrale
n centrale potentiaal
d kugelsymmetrisches Potential n

946 CENTRAL RAY ra, xr
Radiation which is propagated along the
axis of the useful beam of radiation.
f rayon m central
e rayo m central
i raggio m centrale
n centrale straal
d Zentralstrahl m

947 CENTRIFUGAL BARRIER np
The nuclear barrier due to centrifugal
forces associated with angular momentum.
f barrière f centrifuge
e barrera f centrífuga
i barriera f centrifuga
n centrifugale stoep
d Zentrifugalberg m, Zentrifugalschwelle f,
   Zentrifugalwall m

948 CENTRIFUGAL EFFECT ch, gp
The effect produced by a centrifugal force.
f effet m centrifuge
e efecto m centrífugo
i effetto m centrifugo
n centrifugaal effect n
d Zentrifugaleffekt m

949 CENTRIFUGAL FORCE gp
A radially outward force experienced by
an observer in a reference frame which is
rotating at an angular velocity with
respect to an internal frame.
f force f centrifuge
e fuerza f centrífuga
i forza f centrifuga
n centrifugaalkracht, middelpuntvliedende
   kracht
d Zentrifugalkraft f

950 CENTRIFUGAL PUMP vt
A form of pump which displaces liquid by
whirling it around and outwardly by vanes
rotating rapidly in a closed case.

f   pompe *f* centrifuge
e   bomba *f* centrífuga
i   pompa *f* centrifuga
n   centrifugaalpomp
d   Zentrifugalpumpe *f*

951 CENTRIFUGAL SEPARATION   is
Separation of isotopes by spinning a
mixture in gas or vapo(u)r form at high
speed.
f   séparation *f* centrifuge
e   separación *f* centrífuga
i   separazione *f* centrifuga
n   centrifugaalscheiding
d   Zentrifugaltrennung *f*

952 CENTRIFUGE   ch
An apparatus for separating substances
by the application of centrifugal force.
f   centrifugeur *m*
e   centrífuga *f*·
i   centrifuga *f*
n   centrifuge
d   Zentrifuge *f*

953 CENTRIFUGE METHOD   is
The use of centrifuges in isotope separation.
f   méthode *f* à centrifuger
e   método *m* de centrífuga
i   metodo *m* a centrifuga
n   centrifugemethode
d   Zentrifugeverfahren *n*

954 CENTRIPETAL ACCELERATION   pa
That part of the radial component of the
acceleration of a particle moving in any
curved path which is equal to the magnitude of the radius vector from the
instantaneous centre(er) of rotation multiplied by the square of the instantaneous
angular velocity about that centre(er).
f   accélération *f* centripète
e   aceleración *f* centrípeta
i   accelerazione *f* centripeta
n   centripetale versnelling,
middelpuntzoekende versnelling
d   Zentripetalbeschleunigung *f*

955 CENTRIPETAL FORCE   gp
The force required to keep a moving mass
in a circular path.
f   force *f* centripète
e   fuerza *f* centrípeta
i   forza *f* centripeta
n   centripetale kracht,
middelpuntzoekende kracht
d   Zentripetalkraft *f*

956 CERAMIC BETA SOURCE   br, md, ra
A source consisting of a ceramic cylinder
of strontium titanate, protected by a layer
of high-melting point ceramic glaze, into
the surface of which cylinder strontium-90
is fired.
f   source *f* bêta céramique
e   fuente *f* beta cerámica
i   sorgente *f* beta ceramica

n   keramische bêtabron
d   keramische Betaquelle *f*

957 CERAMIC FUEL   rt
Nuclear fuel consisting of refractory
compounds, e.g., oxides and carbides.
f   combustible *m* céramique
e   combustible *m* cerámico
i   combustibile *m* ceramico
n   keramische splijtstof
d   keramischer Brennstoff *m*

958 CERAMIC FUELLED   cl, rt
GAS-COOLED REACTOR
A nuclear reactor with ceramic fuel and
cooled by a gas.
f   réacteur *m* à refroidissement à gaz avec
combustible céramique
e   reactor *m* de enfriamiento por gas con
combustible cerámico
i   reattore *m* a raffreddamento a gas con
combustibile ceramico
n   met gas gekoelde keramieksplijtstofreactor
d   gasgekühlter Keramikbrennstoffreaktor *m*

959 CERAMIC FUELLED REACTOR,   rt
CERAMIC REACTOR
A nuclear reactor constructed of fuel and
moderator assemblies of high-temperature
resistance ceramic materials.
f   réacteur *m* céramique
e   reactor *m* cerámico
i   reattore *m* a combustibile e moderatore
ceramico
n   keramische reactor
d   keramischer Reaktor *m*

960 CERAMIC URANIUM   fu
DIOXIDE FUEL
f   combustible *m* de dioxyde d'uranium
céramique
e   combustible *m* de dióxido de uranio
cerámico
i   combustibile *m* di diossido d'uranio
ceramico
n   splijtstof uit keramisch uraniumdioxyde
d   Brennstoff *m* aus keramischem Urandioxyd

961 CEREBRAL ANGIOGRAPHY   xr
The radiological examination of the
cerebral blood vessels following direct
injection of a contrast medium.
f   angiographie *f* cérébrale
e   angiografía *f* cerebral
i   angiografia *f* cerebrale
n   hersenangiografie
d   Gehirnangiographie *f*

962 CERENKOV COUNTER,   ct
CERENKOV DETECTOR
An instrument for detecting radiation from
high-energy particles traversing a medium
whose refractive index exceeds unity.
f   compteur *m* Cerenkov,
détecteur *m* Cerenkov
e   contador *m* Cerenkov, detector *m* Cerenkov
i   contatore *m* Cerenkov,
rivelatore *m* Cerenkov

n   cerenkovdetector, cerenkovteller
d   Cerenkov-Zähler *m*

963 CERENKOV EFFECT,   np, ra
      CERENKOV RADIATION
      Visible light produced when charged
      particles traverse a transparent medium
      with a velocity exceeding the velocity of
      light in the medium.
f   effet *m* Cerenkov, rayonnement *m* de
      Cerenkov
e   efecto *m* Cerenkov,
      radiación *f* de Cerenkov
i   effetto *m* Cerenkov,
      radiazione *f* di Cerenkov
n   cerenkoveffect *n*, cerenkovstraling
d   Cerenkov-Effekt *m*, Cerenkov-Strahlung *f*

964 CERENKOV EFFECT   ma
      FAILED ELEMENT MONITOR
      A failed element monitor using the
      Cerenkov effect caused, in water, by the
      beta radiation of the fission radionuclides.
f   moniteur *m* de rupture de gaine à
      effet Cerenkov
e   vigía *f* de averías por efecto Cerenkov
i   monitore *m* di rottura di guaine ad
      effetto Cerenkov
n   monitor voor beschadigde splijtstof-
      elementen berustend op het cerenkoveffect
d   Warngerät *n* für schadhafte Brennelemente
      nach dem Cerenkov-Effekt

965 CERENKOV RAYS   np, ra
      Röntgen rays or light rays produced by
      passage of electrons or other charged
      particles through a substance at speeds
      greater than the speed of light in that
      substance.
f   rayons *pl* de Cerenkov
e   rayos *pl* de Cerenkov
i   raggi *pl* di Cerenkov
n   cerenkovstralen *pl*
d   Cerenkov-Strahlen *pl*

966 CERITE   mi
      A cerium silicate with small percentages
      of uranium and thorium.
f   cérite *f*
e   cerita *f*
i   cerite *f*
n   ceriet *n*
d   Zerit *m*

967 CERIUM   ch
      Rare earth element, symbol Ce, atomic
      number 58.
f   cérium *m*
e   cerio *m*
i   cerio *m*
n   cerium *n*
d   Zer *n*

968 CERMET FUEL   rt
      Nuclear fuel consisting of an intimate
      mixture of metallic materials and
      refractory compounds. Either phase or

both may contain fissile (fissionable)
elements.
f   combustible *m* cermet
e   combustible *m* cermet
i   combustibile *m* cermet
n   cermetsplijtstof
d   Kermetbrennstoff *m*

969 CERTIFICATE OF   me
      MEASUREMENT
      A certificate issued when a radiation
      source has been individually calibrated.
f   certificat *m* de mesure
e   certificado *m* de medida
i   certificato *m* di misura   .
n   meetcertificaat *n*
d   Messzertifikat *n*

970 CERTIFICATE OF RADIUM   ra, te
      CONTENT
      A certificate provided for all
      radium-226 beryllium sources.
f   certificat *m* de teneur en radium
e   certificado *m* de contenido de radio
i   certificato *m* di contenuto in radio
n   radiumgehaltecertificaat *n*
d   Radiumgehaltzertifikat *n*

971 CERTIFIED EMISSION   te
      The emission guaranteed by the manufac-
      turer of the neutron emitter.
f   émission *f* garantie
e   emisión *f* garantizada
i   emissione *f* garantita
n   gegarandeerde emissie
d   garantierte Emission *f*

972 CESIUM (US), see 843

973 CESIUM-137 (US), see 844

974 CESIUM-137 PELLET (US), see 845

975 C-14 MONITOR   sa
      An air monitor for determining the
      pressure of C-14 in the atmosphere.
f   moniteur *m* atmosphérique C-14
e   monitor *m* atmosférico C-14
i   monitore *m* atmosferico C-14
n   luchtmonitor voor C-14
d   C-14-Luftmonitor *m*

976 CHADWICK-GOLDHABER EFFECT   np
      A term applied to nuclear reactions
      brought about by bombardment with
      gamma radiation.
f   effet *m* Chadwick-Goldhaber
e   efecto *m* Chadwick-Goldhaber
i   effetto *m* Chadwick-Goldhaber
n   effect *n* van Chadwick en Goldhaber
d   Chadwick-Goldhaber-Effekt *m*

977 CHAIN   ge
      A succession of phenomena of analog(ue)
      character each of which creates the
      necessary elements for the development of
      the next phenomenon.

f   chaîne *f*
e   cadena *f*
i   catena *f*
n   keten, ketting
d   Kette *f*

978   CHAIN DECAY,   np
        CHAIN DISINTEGRATION,
        SERIES DECAY,
        SERIES DISINTEGRATION
   The process of successive radioactive
   transformations in radioactive series.
f   désintégration *f* en chaîne
e   desintegración *f* ˙en cadena
i   disintegrazione *f* in catena
n   kettingdesintegratie, kettingverval *n*
d   Kettenzerfall *m*

979   CHAIN FISSION YIELD   np
   The fraction of fissions giving rise to
   nuclei of a particular mass number.
f   rendement *m* de fission pour nucléides
    isobares
e   rendimiento *m* de fisión para isobaros
    nucleares
i   rendimento *m* di fissione per isobari
    nucleari
n   splijtingsopbrengst voor gespecificeerde
    isobaren
d   Spaltausbeute *f* für spezifizierte Kern-
    isobaren

980   CHAIN FISSIONS   np, rt
   Fissions producing particles of the same
   nature as the initiating particles and which
   give rise to new fissions.
f   fissions *pl* en chaîne
e   fisiones *pl* en cadena
i   fissioni *pl* in catena
n   kettingsplijtingen *pl*
d   Kettenspaltungen *pl*

981   CHAIN REACTION,   np, rt
        NUCLEAR CHAIN REACTION
   A series of nuclear reactions in which one
   of the agents necessary to the series is
   itself produced by the reactions so as to
   cause like reactions.
f   réaction *f* nucléaire en chaîne
e   reacción *f* nuclear de cadena
i   reazione *f* nucleare a catena
n   nucleaire kettingreactie
d   Kernkettenreaktion *f*

982   CHAIN YIELD   np
   The sum of the direct yields of all the
   isobars of a given mass number leading
   to the final stable nuclide at the end of
   the chain.
f   rendement *m* total de chaîne
e   rendimiento *m* global de cadena
i   rendimento *m* globale di catena
n   totaal rendement *n* van de keten
d   Gesamtausbeute *f* der Kette

983   CHANCE COINCIDENCE, see 40

984   CHANCE VARIABLE,   ma
        RANDOM VARIABLE,
        STOCHASTIC VARIABLE
   Any variable that may assume each of its
   possible values X_1 with definite probability,
   not necessarily the same for all values.
f   variable *f* stochastique
e   variable *f* estocástica
i   variabile *f* aleatoria
n   stochastische variabele
d   aleatorische Variable *f*,
    Zufallsvariable *f*

985   CHANNEL,   rt
        HOLE
   A passage through the core of a reactor.
f   canal *m*
e   canal *m*
i   canale *m*
n   kanaal *n*
d   Kanal *m*

986   CHANNEL,   np, rt
        REACTION CHANNEL
   The way in which a nuclear reaction
   proceeds.
f   parcours *m* de réaction, voie *f* de réaction
e   trayectoria *f* de reacción
i   percorso *m* di reazione
n   reactiebaan
d   Reaktionsweg *m*

987   CHANNEL ACTIVITY   ma
        COMPARATOR
   A measuring assembly which automatically
   compares the concentration of fission
   products in each fuel channel or groups
   of channels with a preceding concentration
   measured in the same channel or group
   of channels, taken as reference.
f   évolumètre *m*
e   activímetro *m* comparador de canal
i   comparatore *m* d'attività di canale
n   activiteitsmeter voor koelkanalen
d   Kanalaktivitätsmesser *m*

988   CHANNEL CONFIGURATION   fu, rt
   The arrangement of the fuel containing
   channels in the reactor core.
f   configuration *f* des canaux
e   configuración *f* de los canales
i   configurazione *f* dei canali
n   kanaalopstelling
d   Kanalanordnung *f*

989   CHANNEL FLOW ADJUSTMENT   cl, rt
f   réglage *m* du courant dans les canaux
e   regulación *f* del flujo en los canales
i   regolazione *f* del flusso nei canali
n   kanaalstromingsregeling
d   Kanalstromregelung *f*

990   CHANNEL OUTLET   co, me
        COOLANT TEMPERATURE METER
   An assembly for measuring the temperature
   of the coolant at the outlet.

f thermomètre *m* de liquide de refroidissement à la sortie d'un canal
e termómetro *m* del canal de refrigeración
i termometro *m* per canali di raffreddamento
n thermometer voor de uitlaat van het koelmiddel
d Messgerät *n* für die Austrittstemperatur des Kühlmittels

991 CHANNEL PITCH PLATE rt
A holder for a number of fuel rods in groups of six or twelve to facilitate transport.
f plaque *f* de support
e placa *f* de soporte
i piastra *f* di sopporto
n stelplaat
d Halterplatte *f*

992 CHANNEL SPIN np
In a nuclear reaction, the vector sum of the spins of either the initial particles or the resulting particles.
f spin *m* de voie de réaction
e espín *m* de trayectoria de reacción
i spin *m* di percorso di reazione
n reactiebaanspin
d Reaktionswegsspin *m*

993 CHANNEL WIDTH, ec
WINDOW
In a channel width or window amplifier, the amplitude of a voltage pulses greater than a preset level.
f fenêtre *f*, largeur *f* de canal
e ancha *f* de canal, ventana *f*
i finestra *f*, larghezza *f* di canale
n kanaalbreedte, venster *n*
d Fenster *n*, Kanalbreite *f*

994 CHANNEL WIDTH AMPLIFIER, ec
WINDOW AMPLIFIER
An electronic circuit which accepts only those voltage pulses having amplitudes greater than a preset level and amplifies them linearly before passing them to a pulse height selector.
f amplificateur *m* à fenêtre
e amplificador *m* de ventana
i amplificatore *m* a finestra
n kanaalversterker, vensterversterker
d Fensterverstärker *m*, Kanalverstärker *m*

995 CHANNELLING EFFECT, ra
STREAMING
The increased transmission of electromagnetic or particulate radiation through a medium resulting from the presence of extended voids or other regions of low attenuation.
f effet *m* de canalisation
e efecto *m* de encanalamiento
i effetto *m* d'incanalamento
n stroming
d Kanaleffekt *m*, Kanalwirkung *f*

996 CHANNELLING EFFECT np
FACTOR
Attenuation per unit length of equivalent homogeneous material divided by attenuation per unit length of the material containing voids.
f facteur *m* d'effet de canalisation
e factor *m* de efecto de encanalamiento
i fattore *m* d'effetto d'incanalamento
n stromingsfactor
d Kanaleffektfaktor *m*, Kanalwirkungsfaktor *m*

997 CHAOUL TUBE xr
An X-ray tube used for röntgen ray therapy operating at low potential with a very short anode-object distance.
f tube *m* de Chaoul
e tubo *m* de Chaoul
i tubo *m* di Chaoul
n buis van Chaoul
d Chaoul-Röhre *f*

998 CHARACTERISTIC RADIATION, ra, sp
FLUORESCENT RADIATION,
SELF RADIATION
The line spectrum emitted by a substance, resulting from ionization or excitation of atoms.
f rayonnement *m* caractéristique, rayonnement *m* de fluorescence
e radiación *f* característica, radiación *f* de fluorescencia
i radiazione *f* caratteristica, radiazione.*f* di fluorescenza
n karakteristieke straling
d charakteristische Strahlung *f*, Fluoreszenzstrahlung *f*

999 CHARACTERISTIC X-RAYS np, ra, xr
Electromagnetic radiation emitted as a result of rearrangement of the electrons in the inner shell of atoms.
f rayons *pl* X caractéristiques
e rayos *pl* X característicos
i raggi *pl* X caratteristici
n karakteristieke röntgenstralen *pl*
d charakteristische Röntgenstrahlen *pl*

1000 CHARGE, rt
LOAD
The fuel placed in a reactor.
f charge *f*
e carga *f*
i carica *f*
n lading
d Beschickung *f*, Ladung *f*

1001 CHARGE (TO), rt
LOAD (TO)
To place the fuel in a reactor.
f charger v
e cargar v
i caricare v
n laden v
d beschicken v, laden v

1002 CHARGE CONJUGATION   np
A principle of nuclear physics that postulates that a positively charged particle
must have a negatively charged twin.
f   conjugaison *f* de charge
e   conjugación *f* de carga
i   coniugazione *f* di carica
n   ladingsconjugatie
d   Ladungskonjugation *f*, Ladungszuordnung *f*

1003 CHARGE DISCRIMINATOR UNIT   ec
A basic function unit which gives an
output signal when the electrical charge
exceeds a given-threshold value.
f   élément *m* discriminateur de charge
e   unidad *f* discriminadora de carga
i   elemento *m* discriminatore di carica
n   ladingsdiscriminator
d   Ladungsdiskriminator *m*

1004 CHARGE EXCHANGE   np
The transfer of electrical charge between
two colliding particles, e.g. between an
ion and a neutral atom in a plasma.
f   échange *m* de charge
e   cambio *m* de carga
i   scambio *m* di carica
n   ladingsverwisseling
d   Ladungsaustausch *m*

1005 CHARGE EXCHANGE   np
       PHENOMENON
The phenomenon in which a positive ion
possessing sufficient kinetic energy is
neutralized by colliding with a molecule and
capturing an electron from it. The
molecule is transformed into a positive ion.
f   phénomène *m* d'échange
e   fenómeno *m* de cambio
i   fenomeno *m* di scambio di carica
n   ladingsverwisselingsverschijnsel *n*
d   Ladungsaustauschereignis *n*

1006 CHARGE HOIST,   rt
       LOAD HOIST
The lift used to make the loading and
unloading machinery charge or discharge
the holes in a reactor.
f   montecharge *m*
e   cabina *f* de carga, montacargas *m*
i   montacarichi *m*
n   laadbok, laadlift
d   Ladeaufzug *m*, Ladewinde *f*

1007 CHARGE HOLE,   rt
       CHARGE TUBE, LOAD HOLE,
       LOAD TUBE
A hole in the charge face of a nuclear
reactor for inserting or withdrawing fuel
elements.
f   trou *m* de charge
e   hueco *m* de carga
i   foro *m* di carica
n   laadgat *n*
d   Ladeloch *n*

1008 CHARGE INDEPENDENCE   np
Hypothesis that the neutron-proton,
proton-proton and neutron-neutron forces
are all equal for the same spin states,
when the parts of the interactions that
are explicitly electromagnetic have been
subtracted out.
f   indépendance *f* de charge
e   independencia *f* de carga
i   indipendenza *f* di carica
n   ladingsonafhankelijkheid
d   Ladungsunabhängigkeit *f*

1009 CHARGE INVARIANCE   np
Hypothesis that nucleon-nucleon interactions
are invariant under rotations in isotopic
spin space.
f   invariance *f* de charge
e   invariancia *f* de carga
i   invarianza *f* di carica
n   ladingsonveranderlijkheid
d   Ladungsunveränderlichkeit *f*

1010 CHARGE-MASS RATIO,   np
       SPECIFIC CHARGE
The quotient of the electric charge by the
mass.
f   charge *f* spécifique d'un porteur
e   carga *f* específica de un portador
    electrizado
i   carica *f* specifica d'una particella
n   verhouding van lading tot massa
d   spezifische Ladung *f*

1011 CHARGE MEASURING   ma
       INTEGRATOR,
       RADIATION CHARGE METER
An assembly designed to measure, by
integrating the electrical output current
of a radiation detector, the electrical
charges collected, e.g. at the target of
an accelerator.
f   chargemètre *m* de rayonnement,
    intégrateur *m* pour mesure de charge
e   integrador *m* para la medida de carga
i   complesso *m* di misura di carica
n   opstelling met ladingsmeting
d   Ladungsmessanordnung *f* nach dem
    Integrationsverfahren

1012 CHARGE MULTIPLET   np
A series of several elementary particles
having the same baryon number and
practically the same mass but differing in
charge.
f   charge *f* multiplet
e   carga *f* multiplete
i   carica *f* multipletto
n   ladingsmultiplet
d   Ladungsmultiplett *n*

1013 CHARGE MULTIPLICATION   np
Caused mainly by collision of electrons
with gas molecules under the action of an
electric field.
f   multiplication *f* de charge
e   multiplicación *f* de carga

i   moltiplicazione *f* di carica
n   ladingsvermenigvuldiging
d   Ladungsvervielfachung *f*

1014   CHARGE NEUTRALIZATION   np
The coming together of ions of opposite
signs.
f   neutralisation *f* de charge
e   neutralización *f* de carga
i   neutralizzazione *f* di carica
n   ladingsneutralisatie
d   Ladungsausgleich *m*

1015   CHARGE OF THE ELECTRON,   np
ELECTRONIC CHARGE,
ELEMENTARY CHARGE
Elementary quantity of negative
electricity associated with the electron.
f   charge *f* de l'électron
e   carga *f* del electrón
i   carica *f* dell'elettrone
n   elektronenlading
d   Elektronenladung *f*

1016   CHARGE PULSE AMPLIFIER   ec
An amplifier designed to provide an output
pulse whose amplitude is proportional
to the input pulse charge.
f   amplificateur *m* de charge
e   amplificador *m* de carga
i   amplificatore *m* di carica
n   ladingsversterker
d   ladungsempfindlicher Impulsverstärker *m*

1017   CHARGE SINGLET   np
A fundamental particle which does
not belong to a charge multiplet.
f   singulet *m*
e   singulete *m*
i   singuletto *m*
n   singlet
d   Singulett *n*

1018   CHARGE SYMMETRY   np
A hypothesis which postulates that the
nuclear forces between two neutrons are
the same as those between two protons,
of the same angular momentum and spin.
f   symétrie *f* des charges
e   simetría *f* de las cargas
i   simmetria *f* delle cariche
n   ladingssymmetrie
d   Ladungssymmetrie *f*

1019   CHARGE TRANSFER   np
Transition of an electron between a
bonding and an antibonding orbital.
f   °transfert *m* de charge
e   transferencia *f* de carga
i   trasferimento *m* di carica
n   ladingsoverdracht
d   Ladungsübertragung *f*

1020   CHARGE-TRANSFER   np, sp
SPECTRUM
The spectrum caused by transition of an
electron from a bonding orbital to an
antibonding orbital.
f   spectre *m* du transfert de charge
e   espectro *m* de la transferencia de carga
i   spettro *m* del trasferimento di carica
n   ladingsoverdrachtspectrum *n*
d   Ladungsübertragungsspektrum *n*

1021   CHARGED NUCLEUS   np
Said of a nucleus when it is positively
charged due to the protons it contains.
f   noyau *m* chargé
e   núcleo *m* cargado
i   nucleo *m* caricato
n   geladen kern
d   geladener Kern *m*

1022   CHARGED PARTICLE   np
A particle which carries a positive or a
negative electric charge.
f   particule *f* chargée
e   partícula *f* cargada
i   particella *f* caricata
n   geladen deeltje *n*
d   geladenes Teilchen *n*

1023   CHARGED PARTICLE CARRIERS   np
The general name given to electrically
charged particles such as electrons,
positrons, ions, mesons.
f   porteurs *pl* de charge,
porteurs *pl* électrisés
e   portadores *pl* de carga,
portadores *pl* electrizados
i   particelle *pl* caricate,
portatori *pl* di carica
n   geladen deeltjes *pl*, ladingsdragers *pl*
d   geladene Teilchen *pl*, Ladungsträger *pl*

1024   CHARGED PARTICLE   np
EQUILIBRIUM,
CPE
The condition existing at a point within a
medium under irradiation when, for every
charged particle leaving a volume element
surrounding the point, another particle
of the same kind and energy enters.
f   équilibre *m* de particules chargées
e   equilibrio *m* de partículas cargadas
i   equilibrio *m* di particelle caricate
n   evenwicht *n* voor geladen deeltjes
d   Sekundärteilchengleichgewicht *n*

1025   CHARGED POINT DETECTOR   ct
A radiation counter which utilizes the
principle that the applied voltage necessary
to produce a breakdown discharge from a
charged conducting-point can be greatly
varied by irradiating the gas in the
neighbo(u)rhood of the point with e.g.
X-rays or -particles.
f   tube *m* compteur à pointe irradiée
e   tubo *m* contador de punta irradiada
i   tubo *m* contatore a punta irradiata
n   telbuis met bestraalde puntelektrode
d   bestrahltes Spitzenzählrohr *n*

1026 CHARGER-READER   me
   Equipment for charging a capacitive
   dosemeter and measuring its residual
   charge.
f chargeur-lecteur *m*
e cargador-indicador *m*, cargador-lector *m*,
   lector *m* de cargas
i caricatore-indicatore *m*,
   caricatore-lettòre *m*
n ladende meter
d Lade- und Ablesegerät *n*,
   ladendes Messgerät *n*

1027 CHARGING,   rt
       LOADING
   The act of placing the fuel in a reactor.
f chargement *m*
e cargo *m*
i caricamento *m*
n belading, lading
d Beschickung *f*, Ladung *f*

1028 CHARGING FACE,   rt
       LOADING FACE
   That side of a reactor where fuel
   charging is carried out.
f front *m* de chargement
e frente *f* de cargo
i fronte *f* di caricamento
n laadzijde
d Ladefläche *f*, Ladeseite *f*

1029 CHARGING MACHINE,   rt
       LOADING MACHINE
   A machine used for inserting into or
   withdrawing fuel elements from the core
   of a reactor.
f appareil *m* de chargement
e aparato *m* de cargo
i macchina *f* di caricamento
n laadmachine, laadmechanisme *n*
d Lademaschine *f*, Lademechanismus *m*

1030 CHARGING PROCEDURE,   rt
       LOADING PROCEDURE
f procédé *m* de chargement
e procedimiento *m* de cargo
i processo *m* di caricamento
n laadproces *n*
d Ladevorgang *m*

1031 CHARPY IMPACT TEST,   mg, te
       CHARPY TEST
   A type of impact test in which a notched
   specimen is supported freely at both ends
   and broken by the impact of a falling
   pendulum.
f essai *m* de choc de Charpy
e prueba *f* de impacto de Charpy
i prova *f* di resilienza di Charpy
n schokproef van Charpy
d Charpy-Prüfung *f*

1032 CHECK POINT   me
   The point in an electronic measuring
   assembly accessible to carry out the
   verification of the correct operation.

f point *m* de contrôle
e punto *m* de verificación
i punto *m* d'ispezione
n controlepunt *n*
d Kontrollpunkt *m*

1033 CHELATE COMPOUND   ch
   A compound containing a chelate ring,
   i.e. a ring structure formed by hydrogen
   bonding. Used in solvent extraction
   processes.
f chélate *m*
e quelato *m*
i chelato *m*
n chelaat *n*, klauwverbinding
d Chelat *n*

1034 CHELATING AGENT   ch
   The attached molecule of a chelate
   compound.
f atome *m* donneur de chélate
e átomo *m* donador de quelato
i atomo *m* donatore di chelato
n donoratoom *n* van een chelaat
d Donoratom *n* eines Chelats

1035 CHELATION   ch
   Formation of complex chemical structures,
   usually an organometallic nature.
f chélation *f*
e quelación *f*
i chelazione *f*
n chelaatvorming, chelatie
d Chelatbildung *f*, Chelierung

1036 CHELOID,   ra
       CHELOMA, KELOID, KELOMA,
       KELOS
   Excessive scar formation following
   irradiation or other forms of skin injury.
f chélóide *f*
e queloide *m*, queloma *m*, quelos *m*
i cheloide *m*
n kelóide *n*
d Keloid *n*, Wulstmarke *f*

1037 CHEMICAL AGE   ch, mi
   For a uranium containing mineral, the age
   calculated from chemical analysis only,
   assuming that all of the lead present is
   radiogenic.
f âge *m* chimique
e edad *f* química
i età *f* chimica
n chemische leeftijd
d chemisches Alter *n*

1038 CHEMICAL ASSAY   ch, te
   The analysis and calibration of substances
   by chemical means.
f essai *m* chimique
e ensayo *m* químico
i assaggio *m* chimico
n chemisch onderzoek *n*
d chemische Prüfung *f*

1039 CHEMICAL BINDING   ch
  EFFECT
  The dependence of the neutron cross
  sections of a material on the chemical
  binding of the atoms composing the
  material.
f effet *m* de liaison chimique
e efecto *m* de enlace químico
i effetto *m* di legame chimico
n invloed van de chemische binding
d Einfluss *m* der chemischen Bindung

1040 CHEMICAL BOND   ch
  The directed attractive force between two
  atoms in a molecule.
f liaison *f* chimique
e enlace *m* químico
i legame *m* chimico
n chemische binding
d chemische Bindung *f*

1041 CHEMICAL CLADDING REMOVAL, fu
  CHEMICAL DECANNING,
  CHEMICAL JACKET REMOVAL
  In the head treatment of irradiated
  combustibles, the operation of removing
  the cladding by chemical means.
f dégainage *m* chimique, pelage *m*
e descamisadura *f* química
i spogliatura *f* chimica
n chemische ontmanteling
d chemische Enthülsung *f*

1042 CHEMICAL COMBUSTION   ch
  Combustion by which a chemical reaction
  takes place.
f combustion *f* chimique
e combustión *f* química
i combustione *f* chimica
n chemische verbranding
d chemische Verbrennung *f*

1043 CHEMICAL CONSTITUTION,   ch
  ELEMENTAL COMPOSITION
  The composition of a mixture or compound
  expressed in terms of the constituent
  elements.
f constitution *f* chimique
e constitución *f* química
i costituzione *f* chimica
n chemische samenstelling
d chemische Zusammensetzung *f*

1044 CHEMICAL DOSEMETER   ch, me, ra
  A dosemeter in which the dose is measured
  by observing the extent, under specified
  conditions, of a chemical reaction
  caused by the ionizing radiation to be
  measured.
f dosimètre *m* chimique
e dosímetro *m* químico
i dosimetro *m* chimico
n chemische dosismeter
d chemischer Dosismesser *m*

1045 CHEMICAL ELEMENT   ch
  1. A substance all of whose atoms have
  the same atomic number.

2. A naturally occurring mixture of isotopes with the same atomic number.
  3. A class of atoms having a particular
  atomic number as the distinguishing
  characteristic.
f élément *m* chimique
e elemento *m* químico
i elemento *m* chimico
n chemisch element *n*
d chemisches Element *n*

1046 CHEMICAL ENERGY   ch
  The energy liberated in a chemical
  reaction.
f énergie *f* chimique
e energía *f* química
i energia *f* chimica
n chemische energie
d chemische Energie *f*

1047 CHEMICAL EQUIVALENT   ch
  1. A gram atomic weight of an element.
  2. The basic weight of an ion divided by
  its valence.
f équivalent *m* chimique
e equivalente *m* químico
i equivalente *m* chimico
n chemisch equivalent *n*
d chemisches Äquivalent *n*

1048 CHEMICAL EXCHANGE   ch, is
  A method of isotope separation in which
  two molecules of a compound containing
  different isotopes of an element interact.
f échange *m* chimique
e intercambio *m* químico
i scambio *m* chimico
n chemische uitwisseling
d chemischer Austausch *m*

1049 CHEMICAL EXCHANGE METHOD   is
  The separation of isotopes by means of
  chemical exchange.
f méthode *f* à échange chimique
e método *m* de cambio químico
i metodo *m* a scambio chimico
n chemische uitwisselingsmethode
d chemisches Austauschverfahren *n*

1050 CHEMICAL EXCHANGE   ch, is
  REACTION
  A reaction by which the isotopic composition of the element in the two chemical
  forms of a compound eventually becomes
  almost the same.
f réaction *f* à échange chimique
e reacción *f* de cambio químico
i reazione *f* di scambio chimico
n chemische uitwisselingsreactie
d chemische Austauschreaktion *f*

1051 CHEMICAL IMPURITY   ch, ra
  The amount of non-radioactive element
  present in a specified chemical form
  expressed as per cent of the total amount
  of the substances in the specified chemical
  form.
f quantité *f* d'élément non-actif

e   cuantidad *f* de elemento no activo
i   quantità *f* d'elemento non attivo
n   hoeveelheid niet-actief element
d   Begleitelementmenge *f*

1052 CHEMICAL IMPURITY,   ec, np
    CRYSTAL IMPURITY,
    FOREIGN ATOM
    An atom within a crystal which is foreign
    to the crystal.
f   atome *m* d'impureté
e   átomo *m* de. impureza
i   atomo *m* estraneo, impurezza *f* chimica
n   verontreiniging, vreemdatoom *n*
d   Fremdatom *n*

1053 CHEMICAL INDICATOR,   ch
    CHEMICAL TRACER
    A tracer having chemical properties
    which are similar to those of the substance
    being traced' and with which it is mixed
    homogeneously.
f   indicateur *m* chimique, traceur *m* chimique
e   indicador *m* químico, trazador *m* químico
i   tracciante *m* chimico
n   chemische indicator, chemische tracer
d   chemischer Indikator *m*,
    chemischer Tracer *m*

1054 CHEMICAL PLANT   ch
    A plant for carrying out chemical
    processes.
f   usine *f* chimique
e   fábrica *f* química
i   stabilimento *m* chimico
n   chemische fabriek
d   chemische Fabrik *f*

1055 CHEMICAL PLANT PROCESS -   ma
    CONTROL EQUIPMENT
    An equipment designed to determine the
    nature and the content of radionuclides
    in a mixture or in a solution and based
    on the selective measurement of the
    radiation emitted by the radionuclides
    present.
f   équipement *m* de contrôle des usines
    chimiques
e   equipo *m* para regulación en fábricas
    químicas
i   apparecchiatura *f* per controllo
    d'impianti chimici
n   uitrusting voor procesregeling in
    chemische fabrieken
d   Ausrüstung *f* für die Überwachung der
    chemischen Aufbereitung

1056 CHEMICAL PROCESS   ch
    A process whereby use is made of chemi-
    cal reaction.
f   procédé *m* chimique
e   procedimiento *m* químico
i   processo *m* chimico
n   chemisch proces *n*, chemische werkwijze
d   chemischer Prozess *m*,
    chemisches Verfahren *n*

1057 CHEMICAL PROCESSING,   ch
    CHEMICAL TREATMENT
    The treatment of substances by means of
    chemical processes.
f   traitement *m* chimique
e   elaboración *f* química,
    tratamiento *m* químico
i   lavorazione *f* chimica,
    trattamento *m* chimico
n   chemische bewerking
d   chemische Bearbeitung *f*

1058 CHEMICAL PROCESSING   rt
    REACTOR,
    CHEMONUCLEAR REACTOR
    A nuclear reactor designed as a source
    of radiation for effecting chemical
    transformation on an industrial scale.
f   réacteur *m* de radiochimie
e   reactor *m* quimionuclear
i   reattore *m* chimiconucleare,
    reattore *m* di radiochimica
n   chemonucleaire reactor
d   chemotechnischer Reaktor *m*,
    kernchemischer Reaktor *m*

1059 CHEMICAL PROPERTY   ch
    Any property involving the notion of
    chemical change.
f   propriété *f* chimique
e   propiedad *f* química
i   proprietà *f* chimica
n   chemische eigenschap
d   chemische Eigenschaft *f*

1060 CHEMICAL PROTECTION   ch, sa
    For a given dose of radiation to a chemical
    or biological system, the reduction of a
    particular effect of the radiation by the
    addition of a specified chemical material
    to that system.
f   addition *f* de matériaux protecteurs
e   adición *f* de materiales protectores
i   addizione *f* di materiali protettivi
n   toevoeging van beschermende materialen
d   Schutzstoffzugabe *f*

1061 CHEMICAL PROTECTOR   ch, sa
    A chemical product which added to a
    chemical or biological system reduces
    a particular effect of the irradiation to
    which the system is subjected.
f   protecteur *m* chimique,
    radioprotecteur *m*
e   substancia *f* protectora química
i   sostanza *f* protettrice chimica
n   chemisch beschermmiddel *n*
d   chemischer Schutzstoff *m*

1062 CHEMICAL PURITY   ch
    Absence of foreign elements in traceable
    quantities.
f   pureté *f* chimique
e   pureza *f* química
i   purezza *f* chimica
n   chemische zuiverheid
d   chemische Reinheit *f*

1063 CHEMICAL REACTION ch
A reaction between two or more chemical
substances.
f réaction f chimique
e reacción f química
i reazione f chimica
n chemische reactie
d chemische Reaktion f

1064 CHEMICAL REPROCESSING ch, fu
The treatment of used chemical substances,
e.g. reactor fuel elements, to make them
ready for use again.
f traitement m de régénération chimique,
traitement m nouveau
e reprocesamiento m químico
i rigenerazione f chimica,
trattamento m chimico
n chemische opwerking
d chemische Aufarbeitung f

1065 CHEMICAL SEPARATION ch
The separation of chemical substances
by means of a chemical process.
f séparation f chimique
e separación f química
i separazione f chimica
n chemische scheiding
d chemische Trennung f

1066 CHEMICAL SHIM rt
A substance, such as boric acid, placed
in a reactor coolant to control the reactor
by absorbing neutrons.
f compensateur m chimique
e ajustador m químico
i regolatore m chimico a lungo termine
n chemische regelaar
d chemischer Trimmer m

1067 CHEMICAL SHIMMING rt
The use of neutron absorbing chemicals
in the primary coolant, a fluid moderator,
or some special fluid component, for the
purpose of poison control.
f compensation f chimique
e ajuste m químico
i regolazione f chimica a lungo termine
n chemische regeling
d chemisches Trimmen n

1068 CHEMICAL WASTE ch
The waste resulting from the chemical
processes carried out in a chemical plant.
f déchets pl chimiques
e desechos pl químicos
i rifiuti pl chimici
n chemische afvalprodukten pl
d chemischer Abfall m

1069 CHEMONUCLEAR ch
Refers to chemical processes induced
by nuclear radiation.
f radiochimique adj
e quimionuclear adj
i chimiconucleare adj
n chemonucleair adj
d kernchemisch adj

1070 CHERALITE mi
A rare, uranium and thorium containing
ore.
f chéralite f
e cheralita f
i cheralite f
n cheraliet n
d Cheralit m

1071 CHEVKINITE, mi
TSCHEFFKINITE
A complex silicate containing uranium and
thorium.
f chevkinite f
e chevkinita f
i chevkinite f
n tscheffkiniet n
d Tscheffkinit m

1072 CHILD-LANGMUIR EQUATION, ec
CHILD-LANGMUIR LAW
An equation representing the cathode
current of a thermionic diode in a
space-charge-limited-current state.
f loi f de Child-Langmuir
e ley f de Child-Langmuir
i legge f di Child-Langmuir
n vergelijking van Child en Langmuir
d Child-Langmuirsche Gleichung f

1073 CHINGLUSUITE mi
A complex silicate of Na, Mn, Ca and Ti,
containing small amounts of Th.
f chinglusuite f
e chinglusuita f
i chinglusuite f
n chinglusuïet n
d Chinglusuit m

1074 CHINKOLOBWITE, mi
SHINKOLOBWITE, SKLODOWSKITE
A rare secondary mineral containing
about 55.6 % of U.
f sklodowskite f
e sklodowskita f
i sklodowskite f
n sklodowskiet n
d Sklodowskit m

1075 CHLOPINITE, mi
HLOPINITE, KHLOPINITE
A material probably related to euxenite
and containing thorium and uranium.
f chlopinite f
e clopinita f
i clopinite f
n chlopiniet n
d Chlopinit m

1076 CHLORINE ch
Non-metallic element, gaseous at ordinary
temperatures, symbol Cl, atomic number
17.
f chlore m
e cloro m
i cloro m
n chloor n
d Chlor n

1077 CHLORINE TRIFLUORIDE   mt
A chlorine compound which to some extent
has superseded fluorine as a fluorinating
agent, e.g. in the chemical processing
of nuclear fuel.
f   trifluorure *m* de chlore
e   trifluoruro *m* de cloro
i   trifluoruro *m* di cloro
n   chloortrifluoride *n*
d   Chlortrifluorid *n*

1078 CHOLANGIOGRAPHY   xr
Radiological examination of the gall ways
after oral administration of an iodine
compound or injection of a contrast medium.
f   cholangiographie *f*
e   colangiografía *f*
i   colangiografia *f*
n   cholangiografie
d   Cholangiographie *f*

1079 CHOLECYSTOGRAPHY   xr
Radiological examination of the gall
bladder.
f   cholécystographie *f*
e   colecistografía *f*
i   colecistografia *f*
n   cholecystografie
d   Cholezystographie *f*

1080 CHOP AND LEACH PROCESS,   fu
      CHOPPING-LEACHING PROCESS
Treatment process of irradiated fuel
pins consisting in chopping the pins into
small pieces and subjecting these pieces
to the action of nitric acid.
f   procédé *m* tronconnage-dissolution
e   procedimiento *m* de recortado y
    disolución
i   processo *m* di taglio e dissoluzione
n   hak- en uitloogproces *n*
d   Hack- und Auslaugeverfahren *n*

1081 CHOPPER,   ge
      VIBRATING CONTACTOR,
      VIBRATOR
A device for modulating an electric
current or a beam.
f   hacheur *m*
e   modulador *m*
i   modulatore *m*
n   stralenhakker, stroomhakker
d   Zerhacker *m*

1082 CHOPPER AMPLIFIER   ec
A device in which a direct voltage is
periodically interrupted by a mechanical
or electronic switch, thus providing a
pulsating voltage which is applied to an
alternating-current amplifier.
f   amplificateur *m* à hacheur
e   amplificador *m* de modulador
i   amplificatore *m* a modulatore
n   stroomhakkerversterker
d   Zerhackerverstärker *m*

1083 CHOPPER SPECTROMETER   sp
Radiation spectrometer designed to
measure the time of flight of neutrons
grouped in packets beforehand by means
of a neutron chopper.
f   spectromètre *m* mécanique de neutrons
e   espectrómetro *m* mecánico de neutrones
i   spettrometro *m* meccanico di neutroni
n   mechanische neutronenspectrometer
d   mechanisches Neutronenspektrometer *n*

1084 CHROMATID   md
One of the sister threads of the
longitudinally divided chromosome prior
to nuclear division.
f   chromatide *f*
e   cromatida *f*
i   cromatide *m*
n   chromatide *n*
d   Chromatid *n*

1085 CHROMATID BREAK   md
The transverse break of one or both of
the chromatid threads after the longitudinal
division of the chromosome.
f   division *f* de la chromatide
e   división *f* de la cromatida
i   divisione *f* del cromatide
n   chromatidedeling
d   Chromatidenteilung *f*

1086 CHROMATIN,
      CHROMOPLASM, see 900

1087 CHROMATOGRAPHIC ANALYSIS,   an
      CHROMATOGRAPHY
Analysis of mixtures by passage of their
solutions through an adsorbent medium.
f   chromatographie *f*
e   cromatografía *f*
i   cromatografia *f*
n   chromatografie
d   Chromatographie *f*

1088 CHROMIUM   ch
Metallic element, symbol Cr, atomic
number 24.
f   chrome *m*
e   cromo *m*
i   cromo *m*
n   chroom *n*
d   Chrom *n*

1089 CHROMOPHORIC ELECTRONS   ec
The electrons which form the double
bonds of a group of atoms with special
ability to absorb radiations in the
visible spectrum.
f   électrons *pl* chromophores
e   electrones *pl* cromóforos
i   elettroni *pl* cromofori
n   chromofore elektronen *pl*
d   chromofore Elektronen *pl*

1090 CHROMOSOME   md
One of a definite number of small
dark-staining and more or less rod-shaped

bodies situated in the nucleus of a cell.
f anse *f* chromatique, chromosome *m*
e cromosoma *m*
i cromosoma *m*
n chromosoom *n*, kernlis
d Chromosom *n*, Kernschleife *f*,
  Kernsegment *n*

1091 CHROMOSOME ABERRATION   md
Any rearrangement of chromosome parts
as a result of breakage and reunion of
broken ends.
f mutation *f* chromosomatique
e mutación *f* cromosomática
i mutazione *f* cromosomatica
n chromosoommutatie
d Chromosomenunregelmässigkeit *f*

1092 CHROMOSOME BREAK   md
The transverse break of a chromosome
before longitudinal division into chromatid
threads.
f contraction *f* chromosomatique
e contracción *f* cromosomática
i contrazione *f* cromosomatica
n chromosoombreuk, chromosoominsnoering
d Chromosomeneinschnürung *f*

1093 CHROMOSOME DELETION   md
Loss of a section of a chromosome.
f déficience *f* chromosomatique
e deficiencia *f* cromosomática
i deficienza *f* cromosomatica
n chromosoomdeletie
d Chromosomenausfall *m*,
  Chromosomendefizienz *f*

1094 CHROMOSOME DIVISION   md
The longitudinal division of a chromosome
into chromatid threads.
f division *f* chromosomatique
e división *f* cromosomática
i divisione *f* cromosomatica
n chromosoomdeling
d Chromosomenteilung *f*

1095 CHROMOSOME EXCHANGE,   md
     CHROMOSOME INTERCHANGE
Joining in other than the virginal
arrangement of the four ends resulting
from two breaks in the chromosome.
f échange *m* de chromosomes
e intercambio *m* de cromosomas
i scambio *m* di cromosomi
n chromosoomuitwisseling
d Chromosomenaustausch *m*

1096 CHROMOSOME INVERSION   md
The appearance in a chromosome of a
segment in which the normal linear order
of the genes is reversed.
f inversion *f* de chromosomes
e inversión *f* de cromosomas
i inversione *f* di cromosomi
n chromosoominversie
d Chromosomeninversion *f*

1097 CHROMOSOME TRANSLOCATION   md
Change in position of a portion of a
chromosome either to different regions
of the same chromosome or to another
chromosome.
f translocation *f* de chromosomes
e translocación *f* de cromosomas
i traslocazione *f* di cromosomi
n chromosoomtranslocatie
d Chromosomentranslokation *f*

1098 CHRONIC EXPOSURE   xr
An irradiation over a long period of time,
either fractionated or continuous.
f exposition *f* chronique
e exposición *f* crónica
i esposizione *f* cronica
n langdurige bestraling
d ständige Bestrahlung *f*

1099 CHRONOTRON   ec
A device which utilizes a measurement of
the position of the superposed loci of a
pair of pulses on a transmission line to
determine the time between the events
which initiate the pulses.
f chronotron *m*
e cronotrón *m*
i cronotrone *m*
n chronotron *n*
d Chronotron *n*

1100 CHRYSOBERYL   mg, mi
A mineral consisting mainly of
beryllium-alumin(i)um oxide.
f chrysobéryl *m*
e crisoberilo *m*
i crisoberillio *m*
n chrysoberil *n*
d Chrysoberyll *n*, Goldberyll *n*

1101 CHUGGING   rt
An oscillatory instability in water-moderated reactors caused by the formation
of steam bubbles in the core.
f cahotement *m*
e traqueteo *m*
i traballamento *m*
n schokken *n*
d ruckweise Bewegung *f*

1102 CINEFLUOROGRAPHY,   xr
     CINERADIOGRAPHY,
     RÖNTGENCINEMATOGRAPHY
The making of motion pictures by photofluorography or radiography.
f cinématographie *f* aux rayons X,
  cinéradiographie *f*,
  radiocinématographie *f*
e cinematografía *f* de rayos X;
  cinerradiografía *f*,
  radiocinematografía *f*
i cinematografia *f* con raggi X,
  cineradiografia *f*, radiocinematografia *f*
n röntgencinematografie
d Röntgenkinematographie *f*

1103 CIRCULATING FUEL REACTOR,   rt
     CIRCULATING REACTOR
   A nuclear reactor in which the fissile
   (fissionable) material circulates through
   the core.
f réacteur *m* à circulation du combustible
e reactor *m* de circulación del combustible
i reattore *m* a circolazione del combustibile
n reactor met circulerende splijtstof
d Kreislaufreaktor *m*,
   Reaktor *m* mit umlaufendem Brennstoff

1104 CIRCULATION LOOP   rt
   A loop in a reactor system through
   which a liquid circulates in a closed cycle.
f circuit *m* de circulation
e circuito *m* de circulación
i circuito *m* di circolazione
n circulatiecircuit *n*
d Kreislaufleitung *f*

1105 CIRCUMFERENTIAL FINNING,   cl
     TRANSVERSE FINNING
   Method of arranging cooling fins to
   improve the heat transfer characteristics
   of a nuclear reactor.
f à ailettes transversales
e con aletas transversales
i ad alette trasversali
n van dwarsribben voorzien
d Querberippung *f*, mit Querrippen versehen

1106 CIRCUMFERENTIALLY-FINNED cl, rt
     ELEMENT,
       TRANSVERSELY-FINNED FUEL
       ELEMENT
f élément *m* combustible à ailettes
   transversales
e elemento *m* combustible con aletas
   transversales
i elemento *m* combustibile ad alette
   trasversali
n van dwarsribben voorzien splijtelement *n*
d Brennelement *n* mit Querberippung

1107 CLADDING (CAN), see 858

1108 CLADDING (CANNING), see 863

1109 CLADDING TEMPERATURE   ma
     COMPUTER
   For a nuclear reactor, a computer that
   calculates the temperature reached by
   the hottest cladding inside a nuclear
   reactor.
f calculateur *m* de la température de gaine
e calculadora *f* de la temperatura en la
   vaina
i calcolatore *m* della temperatura di
   guaina
n machine ter berekening van de temperatuur
   van de omhulling
d Hüllentemperaturrechner *m*

1110 CLAMPING CIRCUIT,   ec
     CROWBAR
   In thermonuclear research, an auxiliary
   circuit used to shortcircuit an inductive

load at peak current.
f circuit *m* limiteur
e circuito *m* enclavador
   circuito *m* limitador
i circuito *m* di vincolamento,
   circuito *m* limitatore
n begrenzerschakeling, klemschakeling
d Klemmschaltung *f*

1111 CLARKEITE   mi
   A mineral containing potassium, sodium,
   lead and uranium.
f clarkéite *f*
e clarkeita *f*
i clarkeite *f*
n clarkeïet *n*
d Clarkeit *m*

1112 CLASSICAL   gp
   Pertaining to the system of physics which
   preceded the quantum theory.
f classique adj
e clásico adj
i classico adj
n klassiek adj
d klassisch adj

1113 CLASSICAL ELECTRON RADIUS   np
   Usually the radius given by the expression:
   $r = e^2/m_e c^2 = 2.82 \times 10^{-13}$ cm, where
   m_e is the rest mass of the electron.
f rayon *m* classique d'électron
e radio *m* clásico de electrón
i raggio *m* classico d'elettrone
n klassieke straal van het elektron
d klassischer Elektronenradius *m*

1114 CLASSICAL MECHANICS,   gp
     NEWTONIAN MECHANICS
   System of mechanics developed from
   Newton's laws of motion.
f mécanique *f* classique,
   mécanique *f* de Newton
e mecánica *f* clásica,
   mecánica *f* newtoniana
i meccanica *f* classica,
   meccanica *f* di Newton
n klassieke mechanica, newtonmechanica
d Newtonsche Mechanik *f*

1115 CLASSICAL SCATTERING   cs
     CROSS SECTION,
       SCATTERING CROSS SECTION,
       THOMSON CROSS SECTION
   The cross section for the scattering process.
f section *f* efficace de diffusion
e sección *f* eficaz de dispersión
i sezione *f* d'urto per deviazione
n verstrooiingsdoorsnede
d Streuquerschnitt *m*

1116 CLASSICAL SYSTEM,   np
     NON-QUANTIZED SYSTEM
   System of particles whose energies are
   assumed to be capable of varying in a
   continuous manner.

f   système *m* classique,
    système *m* non-quantifié
e   sistema *m* clásico,
    sistema *m* no cuantificado
i   sistema *m* classico,
    sistema *m* non quantizzato
n   klassiek systeem *n*,
    niet-gequantiseerd systeem *n*
d   klassisches System *n*,
    nichtquantisiertes System *n*

1117 CLEAN rt
Of a reactor, having no induced radioactivity and no fission products.
f   propre adj
e   limpio adj
i   vergine adj
n   schoon adj
d   sauber adj

1118 CLEAN BOMB nw
A nuclear bomb designed to produce
significantly less radioactivity than a
bomb of the same explosive designed
without regard to this requirement.
f   bombe *f* propre
e   bomba *f* limpia
i   bomba *f* vergine
n   schone bom
d   saubere Bombe *f*

1119 CLEAN BUBBLE CHAMBER ic
A bubble chamber with smooth walls.
f   chambre *f* à bulles à parois lisses
e   cámara *f* de burbujas con paredes lisas
i   camera *f* a bolle con pareti liscie
n   bellenvat *n* met gladde wanden
d   Blasenkammer *f* mit glatten Wänden

1120 CLEAN DISTANCE, nw
      SKIP DISTANCE
The distance between the source of
particulates and the spot where the
particulates reach ground level.
f   zone *f* de non-activité
e   zona *f* de no actividad
i   zona *f* di nonattività
n   niet-actieve zone
d   nichtaktive Zone *f*

1121 CLEAN REACTIVITY rt
Intrinsic reactivity in a multiplying
system in which no poison products of
the system itself are present.
f   réactivité *f* propre
e   reactividad *f* limpia
i   reattività *f* a sistema vergine
n   schone reactiviteit
d   saubere Reaktivität *f*

1122 CLEAN REACTOR rt
A nuclear reactor having no poisons other
than those present when it was constructed.
f   reacteur *m* propre
e   reactor *m* limpio
i   reattore *m* vergine
n   schone reactor
d   jungfräulicher Reaktor *m*

1123 CLEAN-UP CIRCUIT cl
Circuit used to remove residual radioactivity from the coolant.
f   circuit *m* de purification
e   circuito *m* de depuración,
    circuito *m* de purificación
i   circuito *m* di depurazione
n   reinigingskring
d   Reinigungskreis *m*

1124 CLEARANCE RATE md
Of a specified organ or tissue, the
efficiency with which it removes a particular substance, which may be isotopically
labelled, from a given fluid in the body.
f   vitesse *f* d'élimination
e   velocidad *f* de eliminación
i   velocità *f* d'eliminazione
n   afscheidingssnelheid
d   Ausscheidungsgeschwindigkeit *f*

1125 CLEARING FIELD ic
An electrostatic field applied across the
gas space in a cloud chamber for
clearing the chamber of ions formed at
times other than the desired sensitive
time.
f   champ *m* clarificateur,
    champ *m* d'effacement
e   campo *m* clarificador
i   campo *m* chiarificatore
n   schoonmaakveld *n*
d   Reinigungsfeld *n*, Ziehfeld *n*

1126 CLEVEITE mi
A uranium compound with rare earth and
helium contents.
f   clévéite *f*
e   cleveita *f*
i   cleveite *f*
n   cleveïet *n*
d   Cleveit *m*

1127 CLINICAL RADIATION SOURCE md
A beta and/or gamma emitting source
designed specially for clinical use.
f   source *f* de rayonnement pour cliniques
e   fuente *f* de radiación para clínicas
i   sorgente *f* di radiazione per cliniche
n   stralingsbron voor klinieken
d   Strahlungsquelle *f* für Kliniken

1128 CLIPPER, ec
      CLIPPING CIRCUIT
A circuit for preventing the peak amplitude
of an electrical signal to exceed a certain
level.
f   circuit *m* écrêteur
e   circuito *m* recortador
i   circuito *m* limitatore
n   afsnijschakeling
d   Abkapperkreis *m*

1129 CLIPPING TIME ec
The time constant of the clipping circuit.
f   constante *f* de temps de l'écrêteur
e   constante *f* de tiempo del recortador
i   costante *f* di tempo del limitatore

n tijdconstante van de afsnijschakeling
d Zeitkonstante f des Abkapperkreises

1130 CLOCK GENERATOR,
CLOCK PULSE GENERATOR, see 128

1131 CLOSE COLLISION np
In studying the binary collisions between
charged particles, the collision for which
the relative velocity of the two particles
undergoes an angular deviation of 90°
or more.
f collision f proche
e colisión f próxima
i collisione f vicina
n nabije botsing
d naher Stoss m

1132 CLOSE-IN FALL-OUT nw
The larger particles of the fall-out of a
nuclear explosion falling under gravity at
an appreciable speed.
f dépôt m radioactif primaire
e depósito m radiactivo primario
i deposito m radioattivo primario
n primaire radioactieve neerslag
d primärer radioaktiver Niederschlag m

1133 CLOSED CIRCUIT, gp
CLOSED CYCLE
Cycle of operation of a heat engine in
which the same power fluid is used repeatedly.
f cycle m fermé
e ciclo m cerrado
i ciclo m chiuso
n kringloop
d Kreislauf m

1134 CLOSED COOLING SYSTEM cl
f système m de refroidissement fermé
e sistema m de enfriamiento cerrado
i sistema m di raffreddamento chiuso
n gesloten koelsysteem n
d geschlossenes Kühlsystem n

1135 CLOSED-CYCLE cl
REACTOR SYSTEM
A system in which the primary coolant
flows to a heat exchanger and then re-
circulates through the core in a completely
closed circuit.
f système m de réacteur à cycle fermé
e sistema m de reactor de ciclo cerrado
i sistema m di reattore a ciclo chiuso
n reactorsysteem n met koelmiddel-
kringloop
d Reaktorsystem n mit Kühlmittelkreislauf

1136 CLOSED SHELL, np
COMPLETED SHELL
A shell containing the full number of
electrons according to the Bohr theory.
f couche f électronique saturée
e capa f electrónica saturada
i strato m elettronico saturato
n afgesloten schil, complete schil
d abgeschlossene Elektronenschale f

1137 CLOTHING MONITOR sa
An instrument for monitoring radioactive
contamination in clothing.
f moniteur m d'habits
e monitor m para ropas
i monitore m d'abiti
n kledingmonitor
d Kleidungsmonitor m

1138 CLOUD CHAMBER, ic
EXPANSION CLOUD CHAMBER,
WILSON CHAMBER
A cloud chamber in which the supersatura-
tion of the vapo(u)r is produced for a short
time by a rapid expansion.
f chambre f à brouillard,
chambre f à nuage, chambre f de Wilson
e cámara f de niebla, cámara f de Wilson
i camera f a nebbia, camera f di Wilson
n expansienevelvat n, nevelvat n,
wilsonvat n
d Expansionsnebelkammer f,
Nebelkammer f, Wilson-Kammer f

1139 CLOUD CHAMBER ic
EXPANSION RATIO,
EXPANSION RATIO
The ratio of the volume of the cloud
chamber gas after expansion to its
volume before expansion.
f rapport m d'expansion
e relación f de expansión
i rapporto m d'espansione
n expansieverhouding
d Expansionsverhältnis n

1140 CLOUD COLUMN nw
The visible column of smoke extending
upward from the point of burst of a
nuclear weapon.
f colonne f de fumée
e columna f de humo
i colonna f di fumo
n rookzuil
d Rauchsäule f

1141 CLOUD TRACK, ic
CLOUD TRAIL, FOG TRACK
A streak of droplets formed along the
path of an ionizing particle in a cloud
chamber.
f ligne f de brouillard, trace f nébuleuse
e traza f de niebla, traza f nebulosa
i traccia f di nebbia, traccia f nebulosa
n nevelspoor n
d Nebelspur f

1142 CLOUD TRACK ic, ph
INTERPRETATION
The study of cloud tracks in a cloud
chamber so as to obtain information on the
sign and on the amount of charge, mass,
energy, etc. of the particle.
f étude f des traces
e interpretación f de las trazas
i interpretazione f delle traccie
n spooronderzoek n
d Spurauswertung f

1143 CLOUDY CRYSTAL np
 BALL MODEL,
 OPTICAL MODEL
A nuclear model, where the nucleus is
treated as a sphere with a certain refractive index and absorption coefficient.
f modèle *m* optique du noyau
e modelo *m* óptico del núcleo
i modello *m* ottico del nucleo
n optisch kernmodel *n*
d optisches Kernmodell *n*

1144 CLOVERLEAF CYCLOTRON pa
A cyclotron with three polar pieces with
three sectors and using the same
focusing methods as the fixed field alternating gradient accelerator.
f cyclotron *m* en feuille de trèfle
e ciclotrón *m* trébol
i ciclotrone *m* trifoglio
n klaverbladcyclotron *n*
d Kleeblattzyklotron *n*

1145 CLUSIUS AND STARKE ch, rt
 HEAVY WATER PRODUCTION
 PROCESS
An example of the methods of heavy
water production which involve the
distillation of hydrogen.
f procédé *m* de production d'eau lourde
de Clusius et Starke
e procedimiento *m* de producción de agua
pesada de Clusius y Starke
i processo *m* di produzione d'acqua
pesante di Clusius e Starke
n zwaar-waterproduktieproces *n* volgens
Clusius en Starke
d Schwerwasserherstellungsverfahren *n*
nach Clusius und Starke

1146 CLUSIUS COLUMN is
A column used in the separation of
isotopes by thermal diffusion, in which a
temperature gradient is established
between two concentric vertical cylinders.
f colonne *f* de Clusius
e columna *f* de Clusius
i colonna *f* di Clusius
n clusiuskolom
d Clusiussches Trennrohr *n*

1147 CLUSTER cr
Complex lattice fault.
f aggrégat *m* de défauts
e agregado *m* de defectos
i aggregato *m* di difetti
n defectenopeenhoping
d Defektensammlung *f*, Fehlernest *n*

1148 CLUSTER COUNTING, see 671

1149 CLUSTER IONS, np
 COMPLEX IONS
A complex electrically charged radical
or group of atoms.
f ions *pl* complexes
e iones *pl* complejos
i ioni *pl* complessi
n complexe ionen *pl*
d Cluster-Ionen *pl*, Komplexionen *pl*

1150 CLUSTER LATTICE fu
A fuel lattice in which the elements are
united in a cluster-shaped configuration.
f réseau *m* à faisceau, réseau *m* à grappe
e celosía *f* de haz
i reticolo *m* a fascio
n bundelrooster *n*
d Bündelgitter *n*

1151 CLUSTER OF FUEL ELEMENTS,
 see 816

1151A C-N CYCLE, see 631

1152 COAL ASH CONTENT METER ma
A content meter designed to determine
ash content in coal by measurement of the
radiation back-scattered or transmitted
by the coal.
f teneurmètre *m* en cendres de charbon
e valorímetro *m* de cenizas de carbón
i tenorimetro *m* in cenere di carbone
n gehaltemeter voor as in steenkool
d Strahlungsmessgerät *n* zur Bestimmung
des Aschengehalts der Kohle

1153 COAL ASH MONITOR me
An apparatus using X-rays produced by
radioisotopes to monitor continuously and
automatically the mineral content of coal.
f moniteur *m* de cendres de charbon
e monitor *m* de cenizas de carbón
i monitore *m* di cenere di carbone
n steenkoolasmonitor
d Steinkohlenaschenmonitor *m*

1154 COARSE CONTROL MEMBER, co, rt
 DEPLETION MEMBER
A control member used for gross
adjustment of the reactivity of a nuclear
reactor or for altering flux density
distribution.
f élément *m* de réglage grossier
e elemento *m* de regulación gruesa
i elemento *m* di regolazione approssimata
n grofregellichaam *n*
d Grobsteuerelement *n*

1155 COASTAL SEA-BED MOVEMENT is
One of the problems to be studied in using
isotopic tracers.
f mouvement *m* du fond de la mer côtier
e movimiento *m* del fondo del mar costero
i movimento *m* del fondo marino costiero
n beweging van de zeebodem aan de kust
d Bewegung *f* des Meeresbodens an die
Küste

1156 COASTAL WATER POLLUTION is
One of the problems to be studied by using
isotopic tracers.
f pollution *f* de l'eau côtière
e contaminación *f* del agua costera

i   inquinamento *m* dell'acqua costiera
n   kustwaterverontreiniging
d   Küstenwasserverunreinigung *f*

1157 COATED FUEL PARTICLE   fu
A fuel pallet of relatively large diameter
with a coating thickness of 150 microns.
f   particule *f* combustible enrobée
e   partícula *f* combustible recubierta
i   particella *f* combustibile ricoperta
n   bekleed splijtstofdeeltje *n*
d   beschichtetes Brennstoffteilchen *n*

1158 COATING   fu
Integral protective envelope deposited on
a fuel particle by plating, dipping, etc.
f   enrobage *m*
e   revestimiento *m*
i   rivestimento *m*
n   bekleding
d   Bekleidung *f*, Beschichtung *f*

1159 COBALT   ch
Metallic element, symbol Co, atomic
number 27.
f   cobalt *m*
e   cobalto *m*
i   cobalto *m*
n   kobalt *n*
d   Kobalt *n*

1160 COBALT BEAM THERAPY   gr, md
Therapy involving the use of gamma
radiation from a cobalt-60 source mounted
in a cobalt bomb.
f   thérapie *f* par isotopes de cobalt
e   terapia *f* por isótopos de cobalto
i   terapia *f* per isotopi di cobalto
n   kobalttherapie
d   Kobalttherapie *f*

1161 COBALT BOMB   nw
A hypothetic nuclear weapon with a shell
of cobalt around it.
f   bombe *f* à gaine de cobalt
e   bomba *f* de vaina de cobalto
i   bomba *f* a guaina di cobalto
n   waterstofbom met kobaltmantel
d   Wasserstoffbombe *f* mit Kobaltmantel

1162 COBALT BOMB, COBALT GUN gr, md
A housing containing a quantity of cobalt.
f   bombe *f* au cobalt,
    caisse *f* d'irradiation à cobalt
e   cañon *m* de cobalto,
    disparador *m* de cobalto
i   bomba *f* al cobalto
n   kobaltbom
d   Kobaltkanone *f*

1163 COBALT DISK   gr, md, ra
Disk used in gamma sources for
teletherapy.
f   disque *m* en cobalt
e   disco *m* en cobalto
i   disco *m* in cobalto
n   kobaltschijf
d   Kobaltscheibe *f*

1164 COBALT PELLET   gr, md, ra
Pellet used in gamma sources for
teletherapy.
f   boulette *f* en cobalt
e   bolita *f* en cobalto
i   pallottolina *f* in cobalto
n   kobaltkogeltje *n*
d   Kobaltkügelchen *n*

1165 COBALT-60   is
The most important gamma source for
medical and industrial radiography.
f   cobalt-60 *m*
e   cobalto-60 *m*
i   cobalto-60 *m*
n   kobalt-60 *n*
d   Kobalt-60 *n*

1166 COBALT-60 PEARLS   md, ra
Small pellets of cobalt-60 enclosed in a
cobalt-60 gun.
f   perles *pl* à cobalt-60
e   perlas *pl* de cobalto-60
i   perle *pl* di cobalto-60
n   kobalt-60 parels *pl*
d   Kobalt-60 Perlen *pl*

1167 COCKCROFT-WALTON   np
        EXPERIMENT
The first successful nuclear disintegration
produced by artificially accelerated protons
(1930).
f   expérience *f* Cockcroft-Walton
e   experimento *m* Cockcroft-Walton
i   esperimento *m* Cockcroft-Walton
n   experiment *n* van Cockcroft en Walton
d   Versuch *m* von Cockcroft und Walton

1168 COCKCROFT-WALTON LINEAR   pa
        DIRECT CURRENT ACCELERATOR
A device for accelerating charged particles
to very high velocities by application of
direct current voltage along a straight
insulated tube.
f   accélérateur *m* Cockcroft-Walton,
    accélérateur *m* Greinacher
e   acelerador *m* Cockcroft-Walton
i   acceleratore *m* Cockcroft-Walton
n   Cockcroft-Waltonversneller
d   Cockcroft-Walton-Beschleuniger *m*

1169 COCKCROFT-WALTON   ec, pa
        RECTIFIER,
        VOLTAGE MULTIPLIER RECTIFIER
A rectifier designed for developing a high
potential direct current voltage for
accelerating nuclear particles.
f   redresseur *m* multiplicateur de tension
e   rectificador *m* multiplicador de tensión
i   raddrizzatore *m* moltiplicatore di tensione
n   spanningsvermenigvuldigingsgelijkrichter
d   Spannungsvervielfachungsgleichrichter *m*

1170 COCURRENT FLOW,   ch
        PARALLEL FLOW
The flow in the same direction of two
streams within a system.
f   écoulement *m* parallèle

e flujo m paralelo
i flusso m parallelo
n gelijkstroming
d Gleichstrom m

1171 COEFFICIENT ge
The numerical expression for the extent
of a physical change with changing
conditions.
f coefficient m
e coeficiente m
i coefficiente m
n coëfficiënt
d Koeffizient m

1172 COEFFICIENT OF ELASTICITY gp
IN SHEAR,
MODULUS OF RIGIDITY
The modulus of elasticity in shear.
f module m de cisaillement,
module m de glissement
e módulo m de elasticidad al corte
i modulo m d'elasticità tangenziale
n glijdingsmodulus
d Gleitmodul m, Schubmodul m

1173 COEFFICIENT OF HEAT gp
TRANSFER (GB),
FILM COEFFICIENT OF HEAT
TRANSFER (US)
The rate of transfer of heat through unit
area of gas or liquid film under unit
driving force.
f coefficient m de transmission superficielle
de chaleur
e coeficiente m de transmisión superficial
de calor
i coefficiente m di trasmissione superficiale di calore
n warmteoverdrachtscoëfficiënt
d Wärmeübertragungskoeffizient m

1174 COEFFICIENT OF np
RECOMBINATION
The coefficient that appears in the law
expressing the recombination of ions in
a gas.
f coefficient m de recombinaison
e coeficiente m de recombinación
i coefficiente m di ricombinazione
n recombinatiecoëfficiënt
d Rekombinationskoeffizient m

1175 COEFFICIENT OF RESTITUTION, np
COLLISION COEFFICIENT
In a two-body collision involving particles
1 and 2, moving in the same straight line,
this coefficient is defined by

$$e = \frac{v_2 - v_1}{u_1 - u_2}$$

f coefficient m de restitution
e coeficiente m de restitución
i coefficiente m di restituzione
n botsingsgetal n
d Stosszahl f

1176 COFFIN, see 911

1177 COFFINITE mi
A rare uranium-containing ore, mostly
associated with carbonaceous material.
f coffinite f
e cofinita f
i coffinite f
n coffiniet n
d Koffinit m

1178 COHERENT np
Said of a scattered wave, possessing a
non-random phase relation with the
incoming wave.
f cohérent adj
e coherente adj
i coerente adj
n coherent adj
d kohärent adj

1179 COHERENT RADIATION ra
Radiation in which there are definite
phase relationships between different
points in a cross section of the beam.
f rayonnement m cohérent
e radiación f coherente
i radiazione f coerente
n coherente straling
d kohärente Strahlung f

1180 COHERENT SCATTERING np
A process in which radiation is scattered
in such a manner that a definite phase
relation exists between the scattered and
incident waves.
f diffusion f cohérente
e dispersión f coherente
i deviazione f coerente
n coherente verstrooiing
d kohärente Streuung f

1181 COHERENT SCATTERING cs
CROSS SECTION
The cross section for the coherent
scattering process.
f section f efficace de diffusion cohérente
e sección f eficaz de dispersión coherente
i sezione f d'urto di deviazione coerente
n doorsnede voor coherente verstrooiing
d Wirkungsquerschnitt m für kohärente
Streuung

1182 COHESION, gp
COHESIVE FORCE
A force between the particles of any given
mass by virtue of which it resists physical
disintegration.
f cohésion f
e cohesión f
i coesione f
n cohesie
d Kohäsion f

1183 COINCIDENCE ct
The occurrence of counts in two or more

detectors simultaneously or within an assignable time interval.
f coïncidence *f*
e coincidencia *f*
i coincidenza *f*
n coïncidentie
d Koinzidenz *f*

1184 COINCIDENCE AMPLIFIER   ct, ec
An amplifier whose output is different from zero only when two or more input pulses are applied simultaneously. Used in coincidence counters.
f amplificateur *m* de coïncidence
e amplificador *m* de coincidencia
i amplificatore *m* di coincidenza
n coïncidentieversterker
d Koinzidenzverstärker *m*

**1185 COINCIDENCE AND ANTICOINCI-   ma
     DENCE PULSE COUNTING ASSEMBLY**
A pulse counting assembly possessing several coincidence and anticoincidence channels, the circuits of which may be connected to obtain various combinations of coincidence, with or without anti-coincidence.
f ensemble *m* de mesure à coïncidences et anticoïncidences
e conjunto *m* contador por coincidencias y anticoincidencias
i complesso *m* di misura a coincidenze ed anticoincidenze
n telopstelling voor coïncidenties en anti-coïncidenties
d Koinzidenz- und Antikoinzidenzzähl-anordnung *f*

**1186 COINCIDENCE CIRCUIT,   ct
     COINCIDENCE COUNTER,
     COINCIDENCE GATE**
An electronic circuit that produces a suitable output pulse only when each of two or more input circuits receives pulses simultaneously or within an assignable time interval.
f circuit *m* de coïncidence
e circuito *m* de coincidencia
i circuito *m* di coincidenza
n coïncidentieschakeling
d Koinzidenzschaltung *f*

1187 COINCIDENCE CORRECTION   ct
A correction to be applied in the counting rate in order to take account of coincidence loss.
f correction *f* de coïncidence
e corrección *f* de coincidencia
i correzione *f* di coincidenza
n coïncidentiecorrectie
d Koinzidenzkorrektion *f*

1188 COINCIDENCE COUNTER   ct
An arrangement of counters and circuits which records the occurrence of counts in two or more detectors simultaneously or within an assignable time interval.

f compteur *m* de coïncidences
e contador *m* de coincidencias
i contatore *m* di coincidenze
n coïncidentieteller
d Koinzidenzzähler *m*

1189 COINCIDENCE COUNTING   ct
An experimental technique in which particular types of events are distinguished from background events by means of coincidence circuits so designed or employed as to register coincidences caused by the type of events under consideration.
f comptage *m* par coïncidences
e recuento *m* por coincidencias
i conteggio *m* per coincidenze
n coïncidentietelling
d Koinzidenzzählung *f*

1190 COINCIDENCE LOSS   ct
Loss of counts due to the occurrence of ionizing events at intervals less than the resolution time of the counting system.
f pertes *pl* par coïncidence
e pérdidas *pl* per coincidencia
i perdite *pl* per coincidenza
n coïncidentieverliezen *pl*
d Koinzidenzausfall *m*, Koinzidenzverluste *pl*

1191 COINCIDENCE OF PULSES   ct
The occurrence of two or more channels of pulses separated by a time interval which is less than a specified value.
f coïncidence *f* d'impulsions
e coincidencia *f* de impulsos
i coincidenza *f* d'impulsi
n coïncidentie van pulsen
d Impulskoinzidenz *f*

**1192 COINCIDENCE RESOLUTION   ct, ec
     TIME,
     COINCIDENCE RESOLVING TIME**
The greatest time interval which can elapse between the occurrence of pulses in the respective input shaping circuits associated with a coincidence circuit in order that the latter produces a usable output pulse.
f temps *m* de résolution de coïncidence
e tiempo *m* de resolución de coincidencia
i tempo *m* di risolvenza di coincidenza
n scheidingstijd van een coïncidentie-schakeling
d Koinzidenzauflösungszeit *f*

1193 COINCIDENCE SELECTOR   ct
A selector having a coincidence circuit and pulse shaping circuits which are placed between each radiation detector and the corresponding input of the coincidence circuit.
f sélecteur *m* de coïncidences
e selector *m* de coincidencias
i selettore *m* di coincidenze
n coïncidentiekiezer
d Koinzidenzwähler *m*

1194 COINCIDENCE SELECTOR UNIT   ct
A basic function unit having a coincidence
circuit and pulse shaping circuits which
are placed between each radiation detector
and the corresponding input of the
coincidence circuit.
f élément *m* sélecteur de coïncidences
e unidad *f* selectora de coincidencias
i elemento *m* selettore di coincidenze
n coïncidentieschakel
d Koinzidenzeinheit *f*

1195 COLD   np
Said of a material or atmosphere in which
there is little or no radioactivity.
f froid adj, non-radioactif adj,
   sans radioactivité
e frío adj, no radiativo adj
i freddo adj, non radioattivo adj
n koud adj, niet-radioactief adj
d aktivitätsfrei adj, kalt adj,
   nichtradioaktiv adj

1196 COLD AREA   rt
An area in a nuclear plant or laboratory
in which there is little or no danger of
radioactivity.
f région *f* de peu de radioactivité
e región *f* de baja radiactividad
i regione *f* di bassa radioattività
n niet-actieve ruimte
d aktivitätsfreier Raum *m*

1197 COLD CATHODE   ec
An electrode used to furnish electrons
by secondary emission, sometimes used
in the ion source of a mass spectrometer.
f cathode *f* froide
e cátodo *m* frío
i catodo *m* freddo
n koude katode
d kalte Katode *f*

1198 COLD-CATHODE   ct
   COUNTER TUBE,
      DECADE COUNTER TUBE,
      DECADE GLOW COUNTER TUBE
A counter tube having one anode and three
sets of cold cathodes.
f tube *m* compteur à décades,
   tube *m* compteur décimal
e tubo *m* contador de decadas,
   tubo *m* contador decimal
i tubo *m* contatore a base decimale
n decimale telbuis
d Dekadenzählrohr n,
   dekadisches Zählrohr *n*

1199 COLD-CATHODE TUBE (US),   ec
   COLD-CATHODE VALVE (GB)
An electron tube or valve containing a
cold cathode.
f tube *m* à cathode froide
e válvula *f* de cátodo frío
i valvola *f* a catodo freddo
n koude-katodebuis
d Kaltkatodenröhre *f*

1200 COLD CLEAN REACTOR   rt
A nuclear reactor having no induced
radioactivity and no poisons other than
those present when it was constructed.
f réacteur *m* froid et propre
e reactor *m* frío y limpio
i reattore *m* freddo e virgine
n koude schone reactor
d kalter jungfräulicher Reaktor *m*

1201 COLD EMISSION, see 446

1202 COLD FUSION REACTION, see 919

1203 COLD LABORATORY   sa
A laboratory where a radiation survey
meter will not indicate more than the
background radiation.
f laboratoire *m* de peu de radioactivité
e laboratorio *m* de baja radiactividad
i laboratorio *m* di bassa radioattività
n niet-actief laboratorium *n*
d aktivitätsfreies Laboratorium *n*

1204 COLD LAPS,   mg
   COLD SHUT
A discontinuity on or near the surface of
a casting which results from the failure
of two streams of metal to unite.
f reprise *f*
e junta *f* fría
i ripresa *f*
n koudloop
d Kaltschweisse *f*

1205 COLD NEUTRONS   np
Neutrons whose most probable energy
corresponds to a temperature considerably
lower than room temperature.
f neutrons *pl* froids
e neutrones *pl* fríos
i neutroni *pl* freddi
n koude neutronen *pl*
d kalte Neutrone *pl*,
   unterthermische Neutrone *pl*

1206 COLD PLASMA   pp
A plasma in which in order to facilitate
calculations, the thermal agitation of the
particles is neglected.
f plasma *m* froid
e plasma *m* frío
i plasma *m* freddo
n koud plasma *n*
d kaltes Plasma *n*

1207 COLD REACTIVITY   rt
Intrinsic reactivity of a multiplying
system at normal temperature.
f réactivité *f* froide
e reactividad *f* fría
i reattività *f* a sistema freddo
n koude reactiviteit
d kalte Reaktivität *f*

1208 COLD REACTOR   rt
A nuclear reactor having no induced
radioactivity.

f   réacteur *m* froid
e   reactor *m* frío
i   reattore *m* freddo
n   koude reactor
d   kalter Reaktor *m*

1209   COLD TESTING   te
Preliminary test, carried out with non-
radioactive products or with radioactive
tracers, of a chemical process which
must be applied to highly radioactive
bodies.
f   essai *m* en inactif
e   prueba *f* en no activo
i   prova *f* in non attivo
n   nonactiviteitsonderzoek *n*
d   Nonaktivitätsprüfung *f*

1210   COLD TRAPPING   cl, rt
The trapping of sodium oxide in a refrig-
erated part of the cooling circuit.
f   piégeage *m* à froid
e   captura *f* al frío, extracción *f* al frío
i   cattura *f* al freddo, estrazione *f* al freddo
n   koude extractie, koude vangst
d   kalte Extraktion *f*, kalter Einfang *m*

1211   COLLAPSIBLE CLADDING   rt
A fuel element cladding which is designed
to achieve direct contact with the fuel
under pressure of the coolant.
f   gaine *f* non résistante
e   revestimiento *m* abatible
i   rivestimento *m* non resistente
n   vervormbare bekleding
d   Andrückhülle *f*

1212   COLLATERAL   np
Describes nuclides which decay into one
of the main radioactive series but not in
the direct line of the series.
f   collatéral adj
e   colateral adj
i   collaterale adj
n   collateraal adj
d   zugehörig adj

1213   COLLATERAL CHAIN,   np
COLLATERAL FAMILIES,
COLLATERAL SERIES
A radioactive decay series, initiated by
transmutation, which eventually joins into
one of the four radioactive decay series.
f   familles *pl* collatérales
e   familias *pl* colaterales
i   famiglie *pl* collaterali
n   collaterale reeksen *pl*
d   zugehörige Zerfallsreihen *pl*

1214   COLLECTING ELECTRODE,   ct, ic
COLLECTOR
In an ionization chamber or in a counter
tube, the electrode which is intended to be
connected to the measuring electrode.
f   électrode *f* collectrice
e   electrodo *m* colector
i   elettrodo *m* collettore

n   collectorelektrode, vergaarelektrode,
verzamelelektrode
d   Auffangelektrode *f*, Sammelektrode *f*

1215   COLLECTING POTENTIAL,   ic
POLARIZING VOLTAGE
The potential difference applied between the
electrodes of an ionization chamber to
provide the electric field required for ion
collection.
f   tension *f* de l'électrode collectrice
e   tensión *f* del electrodo colector
i   tensione *f* dell'elettrodo collettore
n   collectorspanning
d   Auffangelektrodenspannung *f*

1216   COLLECTION TIME   ct, ic
For ions or electrons, the time interval
between the almost instantaneous creation
of ions by ionizing radiation and the total
collection on the collector electrode of
these ions (positive ions or electrons,
according to the polarity of that electrode).
f   temps *m* de collection
e   tiempo *m* de colección
i   tempo *m* di collezione
n   vergaartijd, verzameltijd
d   Auffangzeit *f*, Sammelzeit *f*

1217   COLLECTIVE INTERACTIONS   pp
In plasma physics those interactions which
involve the combined effect of two or more
interacting particles.
f   interactions *pl* collectives
e   interacciones *pl* colectivas
i   interazioni *pl* collettive
n   collectieve wisselwerkingen *pl*
d   kollektive Wechselwirkungen *pl*

1218   COLLECTIVE MODEL   np
The modern version of the liquid drop
model.
f   modèle *m* collectif
e   modelo *m* colectivo
i   modello *m* collettivo
n   collectief model *n*
d   kollektives Modell *n*

1219   COLLIDE (TO)   gp, np
To come into collision.
f   entrechoquer v
e   chocar v
i   collidere v, urtare v
n   botsen v
d   aufprallen v, stossen v, zusammenstossen v

1220   COLLIDING BEAM ACCELERATOR pa
An accelerator consisting of two synchro-
trons or of one synchrotron and its storage
ring in which the two corresponding beams
are forced to meet, directed in such a way
as to produce very efficient collisions.
f   accélérateur *m* à recoupement de faisceaux,
accélérateur *m* de collisions
e   acelerador *m* de colisiones
i   acceleratore *m* di collisioni
n   versneller met bundelcollisie
d   Beschleuniger *m* mit Strahlenkollision

1221 COLLIDING PARTICLE, see 710

1222 COLLIMATION   xr
In radiology, the limiting of a beam of
radiation to the required dimensions and
angular spread.
f collimation *f*
e colimación *f*
i collimazione *f*
n collimatie
d Bündelung *f*, Kollimation *f*

1223 COLLIMATOR   xr
An arrangement of absorbers for the
collimation of a beam of radiation.
f collimateur *m*
e colimador *m*
i collimatore *m*
n collimator
d Blende *f*, Kollimator *m*

1224 COLLISION   np
A close approach of two or more objects
(particles, photons, etc.) during which
there occurs an interchange of quantities
such as energy, momentum and charge.
f choc *m*, collision *f*
e colisión *f*, choque *m*
i collisione *f*, urto *m*
n botsing, collisie
d Kollision *f*, Stoss *m*

1225 COLLISION COEFFICIENT, see 1175

1226 COLLISION DAMPING   sp
A phenomenon of line broadening in
spectral lines of a spectral series,
produced by collisions between atoms or
other radiation and the spectral radiation.
f enlargissement *m* de choc,
   enlargissement *m* de collision
e ensanchamiento *m* de colisión,
   ensanchamiento *m* de choque
i allargamento *m* di collisione,
   allargamento *m* d'urto
n botsingsverbreding
d Druckverbreiterung *f*,
   Stossverbreiterung *f*

1227 COLLISION DENSITY   np
The number of neutron collisions which
matter per unit volume per unit time.
f densité *f* de chocs, densité *f* de collisions
e densidad *f* de colisiones,
   densidad *f* de choques
i densità *f* di collisioni, densità *f* d'urti
n botsingsdichtheid, collisiedichtheid
d Kollisionsdichte *f*, Stossdichte *f*

1228 COLLISION EXCITATION   np
The change of structure of certain atoms
or molecules of a gas characterized by
the passage of an electron from one energy
level to a higher level.
f excitation *f* par choc,
   excitation *f* par collision

e excitación *f* por colisión,
   excitación *f* por choque
i eccitazione *f* per collisione,
   eccitazione *f* per urto
n aanslag door botsing, aanslag door collisie
d Kollisionsanregung *f*, Stossanregung *f*

1229 COLLISION FREQUENCY   np
The number of collisions between an
electron and a molecule of a gas per unit
time.
f fréquence *f* de chocs,
   fréquence *f* de collisions
e frecuencia *f* de colisiones,
   frecuencia *f* de choques
i frequenza *f* di collisioni,
   frequenza *f* d'urti
n botsingsfrequentie, collisiefrequentie
d Kollisionsfrequenz *f*, Stossfrequenz *f*

1230 COLLISION INTEGRAL   np
Integral used in studying collision
phenomena.
f intégrale *f* de choc,
   intégrale *f* de collision
e integral *f* de colisión, integral *f* de choque
i integrale *m* di collisione,
   integrale *m* d'urto
n botsingsintegraal, collisie-integraal
d Kollisionsintegral *n*, Stossintegral *n*

1231 COLLISION IONIZATION,   np
     IMPACT IONIZATION,
     IONIZATION BY COLLISION
The ionization of atoms or molecules of
a gas or vapo(u)r by collision with other
particles.
f ionisation *f* par choc,
   ionisation *f* par collision
e ionización *f* por colisión,
   ionización *f* por choque
i ionizzazione *f* per collisione,
   ionizzazione *f* per urto
n ionisatie door collisie, stootionisatie
d Kollisionsionisation *f*, Stossionisation *f*

1232 COLLISION NUMBER   np
The average number N_c of collisions
suffered by an ion in diffusing out of the
cavity.
f nombre *m* de chocs, nombre *m* de collisions
e número *m* de colisiones,
   número *m* de choques
i numero *m* di collisioni, numero *m* d'urti
n botsingsgetal *n*, collisiegetal *n*
d Kollisionszahl *f*, Stosszahl *f*

1233 COLLISION OF THE FIRST KIND   np
The collision of an accelerated particle
with an atom resulting in a transfer of
energy, whereby the atom becomes excited
and the electron is slowed.
f choc *m* de première espèce,
   collision *f* de première espèce
e colisión *f* de primera clase,
   choque *m* de primera clase

i   collisione *f* di prima specie,
    urto *m* di prima specie
n   botsing van de eerste soort,
    collisie van de eerste soort
d   Kollision *f* erster Art,
    Stoss *m* erster Art

**1234 COLLISION OF THE SECOND   np
        KIND**
The collision of an excited atom with a
slow particle whereby the atom undergoes
transition to a lower energy-state and
the other particle is accelerated.
f   choc *m* de deuxième espèce,
    collision *f* de deuxième espèce
e   colisión *f* de segunda clase,
    choque *m* de segunda clase
i   collisione *f* di seconda specie,
    urto *m* di seconda specie
n   botsing van de tweede soort,
    collisie van de tweede soort
d   Kollision *f* zweiter Art,
    Stoss *m* zweiter Art

1235 COLLISION REACTION   np
Any reaction produced when a neutron
strikes a nucleus, including elastic
scattering, inelastic scattering, radiative
capture and fission.
f   réaction *f* de choc, réaction *f* de collision
e   reacción *f* de colisión,
    reacción *f* de choque
i   reazione *f* di collisione,
    reazione *f* d'urto
n   botsingsreactie, collisiereactie
d   Kollisionsreaktion *f*, Stossreaktion *f*

1236 COLLISION REPLACEMENT   np
The phenomenon that an atom displaced
from its lattice site collides with
another atom and takes its place.
f   substitution *f* par collision
e   substitución *f* por colisión
i   sostituzione *f* per urto
n   botsingssubstitutie
d   Stossubstitution *f*

1237 COLLISION TIME   np
The time interval between two close
collisions.
f   temps *m* de collision
e   tiempo *m* de colisión
i   tempo *m* d'urto
n   botsingstijd
d   Stosszeit *f*

1238 COLLISIONLESS PLASMA   pp
A plasma of which the density is so low
that the binary close collisions between
particles practically play no role.
f   plasma *m* sans collision
e   plasma *m* sin colisión
i   plasma *m* senza collisione
n   collisievrij plasma *n*
d   kollisionsfreies Plasma *n*

**1239 COLOR CENTERS (US),   cr
        COLOUR CENTRES (GB)**
Crystal defects which interact with
electrons in such a way as to cause intense
light absorption, thus enabling small
numbers of such defects to change the
colo(u)r of the crystal.
f   centres *pl* de couleur
e   centros *pl* de color
i   centri *pl* di colore
n   kleurcentra *pl*
d   Farbzentren *pl*

**1240 COLUMBITE,   mi
        NIOBITE**
An iron-manganese-niobium-tantalum
ore containing about 8 % of U.
f   columbite *f*, niobite *f*
e   columbita *f*, niobita *f*
i   colombite *f*, niobite *f*
n   columbiet *n*, niobiet *n*
d   Columbit *m*, Niobit *m*

**1241 COLUMN,   nw
        PLUME, WATER COLUMN**
A hollow cylinder of water and spray
thrown up from an underwater burst of a
nuclear weapon, through which the hot
high-pressure gases formed in the
explosion are vented to the atmosphere.
f   colonne *f* d'eau
e   columna *f* de agua
i   colonna *f* d'acqua
n   waterzuil
d   Wassersäule *f*

**1242 COLUMN,   ch
        TOWER**
A facility in which various chemical
processes such as distillation, rectification,
extraction, etc. are carried out.
f   colonne *f*, tour *f*
e   columna *f*, torre *f*
i   colonna *f*, torre *f*
n   kolom, toren, zuil
d   Kolonne *f*, Säule *f*, Turm *m*

1243 COLUMNAR IONIZATION   ic, np
The cylinder of dense ionization existing
along the track of a heavy charged particle
immediately after its passage through
a material.
f   ionisation *f* colonnaire
e   ionización *f* columnaria
i   ionizzazione *f* colonnare
n   zuilionisatie
d   Kolonnenionisation *f*, Säulenionisation *f*

1244 COLUMNAR RECOMBINATION   ic, np
Recombination which takes place before the
ions have left the track, in the case where
the ionization takes place in along a column.
f   recombinaison *f* colonnaire
e   recombinación *f* columnaria
i   ricombinazione *f* colonnare
n   zuilrecombinatie
d   Säulenrekombination *f*

1245 COMBUSTION   gp
 Any chemical process accompanied by
 the evolution of light and heat.
f combustion f
e combustión f
i combustione f
n verbranding
d Verbrennung f

1246 COMBUSTION HEAT,   gp
  HEAT OF COMBUSTION
 The increase in the heat content when
 one mole of a substance undergoes
 oxidation, whereby the products obtained
 in complete combustion are produced.
f chaleur f de combustion
e calor m de combustión
i calore m di combustione
n verbrandingswarmte
d Verbrennungswärme f

1247 COMPACT   ch, mg
 A body produced by the compression of
 metallic or non-metallic powders in a
 die.
f comprimé m, pastille f
e comprimido m, pieza f aglomerada,
 pieza f prensada
i compressa f, pastiglia f
n pastille, persstuk n
d Pastille f, Presskörper m, Pressling m

1248 COMPARATIVE LIFETIME   np
 The product of the half life t of a beta
 disintegration and a function f that
 expresses the probability per unit of
 time that a beta transition will take place
 in a given nucleus.
f période f comparative, valeur f ft
e valor m ft, vida f comparativa
i valore m ft, vita f comparativa
n ft-waarde
d ft-Wert m, vergleichbares Alter n

1249 COMPARATOR   ma
 An electronic instrument that measures
 a quantity and compares it with a
 precision standard.
f comparateur m
e comparador m,
i comparatore m
n comparator
d Komparator m

1250 COMPATIBILITY   ch, gp
 The lack of interaction at significant rates
 between one solid and other solids,
 liquids and gases; it may be affected by
 radiation.
f compatibilité f
e compatibilidad f
i compatibilità f
n compatibiliteit
d Verträglichkeit f

1251 COMPATIBILITY PROBLEM   rt
 The problem of selecting the right
 chemical substances when constructing

 nuclear reactors in view of their mutual
 reactions.
f problème m du choix des matériaux
e problema m de la selección de los
 materiales
i problema m della selezione dei materiali
n materiaalkeuzeprobleem n
d Problem n der Werkstoffswahl,
 Verträglichkeitsproblem n

1252 COMPENSATED   ic
  IONIZATION CHAMBER
 Differential ionization chamber designed
 in such a manner as to eliminate by
 compensation the effect of another
 radiation superimposed on that of the
 radiation which it is desired to measure.
f chambre f d'ionisation compensée
e cámara f de ionización compensada
i camera f d'ionizzazione compensata
n compensatie-ionisatievat n,
 ionisatievat n met compensatie
d Kompensationsionisationskammer f

1253 COMPENSATED SEMICONDUCTOR, ec
  P.I.N. SEMICONDUCTOR
 A semiconductor in which one type of
 impurity or imperfection partially
 cancels the effect of the other type of
 impurity or imperfection.
f semiconducteur m compensé,
 semiconducteur m P.I.N. compensé
e semiconductor m compensado,
 semiconductor m tipo P.I.N. compensado
i semiconduttore m compensato,
 semiconduttore m tipo P.I.N. compensato
n gecompenseerde halfgeleider,
 gecompenseerde P.I.N.-halfgeleider
d kompensierter Halbleiter m,
 kompensierter P.I.N.Typ-Halbleiter m

1254 COMPENSATED SEMI-   ra
  CONDUCTOR DETECTOR,
  P.I.N. SEMICONDUCTOR DETECTOR
 A semiconductor detector consisting of
 a compensated region between a P and
 an N region.
f détecteur m semiconducteur compensé,
 détecteur m semiconducteur P.I.N.
e detector m semiconductor tipo P.I.N.
i rivelatore m semiconduttore tipo P.I.N.
n P.I.N.-halfgeleiderdetector
d P.I.N.-Typ Halbleiterdetektor m

1255 COMPENSATING COILS   pa
 A set of coils so distributed as to alter
 a magnetic flux distribution in a desired
 manner.
f bobines pl de compensation,
 enroulements pl correcteurs
e bobinas pl de compensación
i bobine pl di compensazione
n beïnvloedingsspoelen pl
d Kompensationsspulen pl

1256 COMPENSATION FACTOR   ic
 Of a compensated ionization chamber, the
 ratio of the undesired sensitivity to gamma

radiation of the compensated chamber,
to the sensitivity to gamma radiation of the
same chamber, if it were not to be
compensated.
f facteur *m* de compensation
e factor *m* de compensación
i fattore *m* di compensazione
n compensatiefactor
d Kompensationsfaktor *m*

1257 COMPENSATION FILTER   xr
f filtre *m* de compensation
e filtro *m* de compensación
i filtro *m* di compensazione
n compensatiefilter *n*
d Ausgleichsfilter *m*

1258 COMPENSATION RATIO   ic
Of a compensated ionization chamber,
the inverse of the compensation factor,
and used to indicate the quality of a
compensated ionization chamber.
f rapport *m* de compensation
e relación *f* de compensación
i rapporto *m* di compensazione
n compensatieverhouding
d Kompensationsverhältnis *n*

1259 COMPLEMENTARITY   qm
In quantum physics, the relation between
two aspects of a phenomenon where these
aspects can be studied separately, but not
at the same time.
f complémentarité *f*
e complementaridad *f*
i complementarità *f*
n complementariteit
d Komplementarität *f*

1260 COMPLEMENTARITY   gp
     PRINCIPLE
Particle motion in terms of momentum
and energy may be expressed as a wave
motion in terms of wavelength and
frequency and vice-versa. The transforming equations are: $p = h \lambda$ (1)
                        $E = h\nu$ (2), where
p is momentum descriptive of particle
motion, E is particle energy, λ is
wavelength and h is Planck's constant.
f principe *m* de complémentarité
e principio *m* de complementaridad
i principio *m* di complementarità
n complementariteitsprincipe *n*
d Komplementaritätsprinzip *n*

1261 COMPLETED SHELL, see 1136

1262 COMPLEX COMPOUND   ch
A compound which is made up structurally
of two or more compounds or ions.
f composé *m* complexe
e compuesto *m* complejo
i composto *m* complesso
n complexe verbinding
d Komplexverbindung *f*

1263 COMPLEX IONS, see 1149

1264 COMPLEXING AGENT,   ch
     SEQUESTERING AGENT
Compounds capable of binding metal ions
so that they no longer exhibit their normal
reactions in the presence of precipitating
agents. Used for decontamination of
radioactive surfaces.
f agent *m* complexant, agent *m* séquestrant
e agente *m* secuestrante
i agente *m* sequestrante
n afscheidingsmiddel *n*
d Abscheidemittel *n*, Trennmittel *n*

1265 COMPOUND   ch
A pure substance which can be decomposed
into other different pure substances.
f composé *m*
e compuesto *m*
i composto *m*
n verbinding
d Verbindung *f*

1266 COMPOUND CYCLE   cl, gp
A thermal cycle employing more than one
circuit.
f cycle *m* multiple
e ciclo *m* múltiple
i ciclo *m* multiplo
n meervoudige kringloop
d Mehrfachkreislauf *m*

1267 COMPOUND FILTER   ra, sp
A filter composed of more than one
material designed to transmit preferentially
a part of the spectrum.
f filtre *m* composé
e filtro *m* compuesto
i filtro *m* composto
n samengesteld filter *n*
d zusammengesetzter Filter *m*

1268 COMPOUND NUCLEUS,   np
     INTERMEDIATE NUCLEUS
The highly excited nucleus formed as the
immediate result of a nuclear collision.
f noyau *m* composé, noyau *m* intermédiaire
e núcleo *m* compuesto
i nucleo *m* composto
n tussenkern
d Verbundkern *m*, Zwischenkern *m*

1269 COMPRESSIBILITY FACTOR   gp
The ratio of the actual volume of a given
mass of gas to the volume it would have
at the same temperature and pressure
if it were a perfect gas.
f facteur *m* de compressibilité
e factor *m* de compresibilidad
i fattore *m* di compressibilità
n compressiefactor
d Kompressionsfaktor *m*

1270 COMPRESSIBLE FLOW   gp
The flow of fluid under conditions such that
its density changes significantly.

f débit *m* compressible
e flujo *m* compresible
i flusso *m* compressibile
n stroming met sterk veranderlijke dichtheid
d kompressibele Strömung *f*

1271 COMPRESSION CONE xr
A device used to exert pressure upon the body irradiated to immobilize the object or to depress the surface of the body near the underlying tumor in radiotherapy.
f cône *m* de compression
e cono *m* de compresión, localizador *m* de compresión
i localizzatore-compressore *m*
n compressieconus, compressor
d Kompressionstubus *m*

1272 COMPRESSIONAL MIRROR pp
Instrument used in plasma containment experiments.
f miroir *m* de compression
e espejo *m* de compresión
i specchio *m* di compressione
n comprimerende spiegel
d komprimierender Spiegel *m*

1273 COMPTON ABSORPTION. ab, gr, xr
The absorption of an X-ray or gamma-ray photon in the Compton effect.
f absorption *f* Compton
e absorción *f* Compton
i assorbimento *m* Compton
n comptonabsorptie
d Compton-Absorption *f*

1274 COMPTON EFFECT, np
COMPTON SCATTERING
The elastic scattering of photons by electrons.
f diffusion *f* de Compton, effet *m* Compton
e dispersión *f* de Compton, efecto *m* Compton
i deviazione *f* di Compton, effetto *m* Compton
n comptoneffect *n*, comptonverstrooiing
d Compton-Effekt *m*, Compton-Streuung *f*

1275 COMPTON ELECTRON, np
COMPTON RECOIL ELECTRON
An electron that has been set in motion through interaction with a photon in the Compton effect.
f électron *m* Compton, électron *m* de recul
e electrón *m* Compton, electrón *m* de rebote
i elettrone *m* Compton, elettrone *m* di rimbalzo, elettrone *m* di ripulsione
n comptonelektron *n*, terugslagelektron *n*
d Compton-Elektron *n*, Rückstosselektron *n*

1276 COMPTON METER cr, ic
A compensating ionization chamber with a uranium source which is adjusted until it balances out the normal cosmic radiation.
f chambre *f* d'ionisation de Compton

e cámara *f* de ionización de Compton
i camera *f* d'ionizzazione di Compton
n comptonionisatievat *n*
d Compton-Ionisationskammer *f*

1277 COMPTON RECOIL PARTICLE np
Any particle that has acquired its momentum in a scattering process similar in type to the Compton effect.
f particule *f* Compton
e partícula *f* Compton
i particella *f* Compton
n comptondeeltje *n*
d Compton-Teilchen *n*

1278 COMPTON SCATTERER np
A fictitious medium consisting of a free-electron gas.
f milieu *m* de diffusion de Compton
e medio *m* de dispersión de Compton
i mezzo *m* di deviazione di Compton
n verstrooiingsmedium *n* volgens Compton
d Comptonsches Streuungsmedium *n*

1279 COMPTON SHIFT np
The difference between the wavelengths of the scattered and the incident photons on the Compton effect.
f déplacement *m* Compton
e desplazamiento *m* Compton
i spostamento *m* Compton
n comptonverschuiving
d Compton-Verschiebung *f*

1280 COMPTON-SIMON EXPERIMENT xr
Fundamental experiment demonstrating the quantum nature of X-rays by scattering them by electrons in a Wilson cloud chamber.
f expérience *f* Compton-Wilson
e experimento *m* Compton-Simon
i esperimento *m* Compton-Simon
n proef van Compton en Simon
d Versuch *m* von Compton und Simon

1281 COMPTON WAVELENGTH np
A wavelength λ_0 characteristic of a particle of mass m and equal to $h/m_0 c$.
f longueur *f* d'onde de Compton
e longitud *f* de onda de Compton
i lunghezza *f* d'onda di Compton
n comptongolflengte
d Compton-Wellenlänge *f*

1282 CONCENTRATED IRRADIATION xr
f irradiation *f* concentrée
e irradiación *f* concentrada
i irradiazione *f* concentrata
n geconcentreerde bestraling
d konzentrierte Bestrahlung *f*

1283 CONCENTRATING CUP, xr
FOCUSING CUP
A metal cup in which the incandescent cathode of a röntgen tube is mounted, which electrostatically focuses the electron beam

upon the focal spot on the surface of the
anode.
f cupule *f* de concentration,
   cupule *f* de focalisation
e cúpula *f* de concentración,
   cúpula *f* de enfoque
i cupola *f* di concentrazione,
   cupola *f* di focalizzazione
n focusseerkap, katodekap
d Katodenbecher *m*

1284 CONCENTRATION, see 25

1285 CONCENTRATION CONTROL   ch
   A method of controlling a unit by
   measuring the concentration in a feed or
   product stream.
f contrôle *m* de la concentration
e control *m* de la concentración
i controllo *m* della concentrazione
n concentratiecontrole
d Konzentrationskontrolle *f*

1286 CONCENTRATION FACTOR,   rw
   RATIO OF ACTIVITY DENSITIES
   A number which indicates the presence
   of radioactivity of a substance relative
   to known or standard conditions.
f rapport *m* d'activité
e relación *f* de actividad
i rapporto *m* d'attività
n activiteitsverhouding
d Aktivitätsverhältnis *n*

1287 CONCRETE   mt
   A mixture of cement, sand and gravel,
   with water in varying proportions according
   to the use which is to be made of it.
f béton *m*
e hormigón *m*
i calcestruzzo *m*
n beton *n*
d Beton *m*

1288 CONCRETE BIOLOGICAL   sa
   SHIELD
   A biological shield made of concrete.
f bouclier *m* biologique en béton
e pantalla *f* biológica de hormigón
i schermo *m* biologico in calcestruzzo
n biologisch betonscherm *n*
d biologische Betonabschirmung *f*

1289 CONCRETE PIT   rw
   Underground burying space for containers
   of radioactive waste.
f fosse *f* cimentée
e fosa *f* de hormigón
i fossa *f* in calcestruzzo,
   vasca *f* in calcestruzzo
n betonput
d Betongrube *f*

1290 CONCRETE SHIELD   nue, sap
   A shield utilized for protection consisting
   of concrete.
f carapace *f* en béton

e blindaje *m* de hormigón
i schermo *m* in calcestruzzo
n betonpantser *n*
d Betonpanzer *m*

1291 CONCURRENT CENTRIFUGE,   is
   FLOW-THROUGH CENTRIFUGE
   A centrifuge in which one or more
   streams of gas enter in one end and the
   partially separated isotopes are removed
   in two or more streams at the other end.
f centrifugeur *m* à courant parallèle
e centrífuga *f* de corriente paralela
i centrifuga *f* a corrente parallela
n gelijkstroomcentrifuge
d Gleichstromzentrifuge *f*

1292 CONDENSATION   ch
   The formation of a liquid or a solid from
   a vapo(u)r.
f condensation *f*
e condensación *f*
i condensazione *f*
n condensatie, verdichting
d Kondensation *f*, Verdichtung *f*

1293 CONDENSATION CLOUD   nw
   A mist or fog of minute water droplets
   which temporarily surrounds the ball of
   fire following a nuclear explosion in a
   comparatively humid atmosphere.
f nuage *m* de condensation
e nube *m* de condensación
i nuvolone *m* di condensazione
n condenswolk
d Kondenswolke *f*

1294 CONDENSATION COLUMN,   ch
   CONDENSATION TOWER
   A component part of a chemical plant,
   a.o. of a distillation plant.
f colonne *f* de condensation
e columna *f* de condensación
i colonna *f* di condensazione
n condensatiekolom, condensatiezuil
d Kondensationssäule *f*, Kondensationsturm *m*

1295 CONDENSER   ch
   An apparatus for condensing vapo(u)rs to
   a liquid or solid state.
f condenseur *m*
e condensador *m*
i condensatore *m*
n condensor
d Kondensator *m*

1296 CONDITIONAL RELEASE   rw
   The provisional release of radioactively
   contaminated materials for future use or
   disposal under stated conditions.
f libération *f* conditionnelle
e liberación *f* condicional
i liberazione *f* condizionale
n voorwaardelijke vrijgeving
d bedingte Freigabe *f*

1297 CONDUCTION,   gp
    HEAT CONDUCTION
The flow of heat through a body by the
transfer of kinetic energy from molecule
to molecule without gross mixing.
f conduction f de chaleur
e conducción f de calor
i conduzione f di calore
n warmtegeleiding
d Wärmeleitung f

1298 CONDUCTION BAND   ec, np
A range of states in the energy spectrum
of a solid in which electrons can move
freely.
f bande f de conduction
e banda f de conducción
i banda f di conduzione
n geleidingsband
d Leitungsband n

1299 CONDUCTION CURRENT   ec, np
Continuous movement of charges in a
body due to a flow of conduction electrons,
created by the contribution of external
energy.
f courant m de conduction
e corriente f de conducción
i corrente f di conduzione
n geleidingsstroom
d Leitungsstrom m

1300 CONDUCTION ELECTRON,   ec, np
    OUTER-SHELL ELECTRON,
    PERIPHERAL ELECTRON,
    VALENCE ELECTRON
An electron belonging normally to the outer
shell and which is concerned in light
phenomena, conduction phenomena and
also in the chemical properties of the
atom.
f électron m de conduction,
    électron m de valence,
    électron m optique,
    électron m périphérique
e electrón m de conducción,
    electrón m de valencia, electrón m óptico,
    electrón m periférico
i elettrone m di conduzione,
    elettrone m di valenza, elettrone m ottico,
    elettrone m periferico
n elektron n van de buitenschil,
    geleidingselektron n, valentie-elektron n
d kernfernes Elektron n, Valenzelektron n

1301 CONDUCTION PUMP   cl
Electromagnetic pump for liquid metal
transport in which the current is conducted
from some outside source into the liquid
metal.
f pompe f à courant conductif
e bomba f de corriente conductiva
i pompa f a corrente conduttiva
n geleidingsstroompomp
d Leitungsstrompumpe f

1302 CONDUCTIVITY METER   ma
An assembly for measuring the conductivity
of a liquid, generally water, associated
with the operation of a nuclear reactor.
f conductivimètre m
e conductímetro m
i conduttivimetro m
n geleidingsmeter
d Leitfähigkeitsmessgerät n

1303 CONE   xr
A conical attachment to the X-ray tube
for concentrating the emitted rays.
f cône m, localisateur m
e cono m, localizador m
i cono m
n conus
d Tubus m

1304 CONE HOLDER   xr
f support m de cône
e soporte m de cono
i sopporto m di cono
n conushouder
d Tubusfassung f

1305 CONE OF RADIATION,   xr
    TREATMENT CONE
In the medical use of X-rays the X-ray
beam is frequently restricted to a narrow
cone by the use of diaphragms.
f cône m de rayons
e cono m de rayos
i cono m di raggi
n stralenkegel
d Strahlenkegel m

1306 CONFIGURATION   ge
Arrangement of parts in a form or figure
and the form resulting from such arrangement.
f configuration f
e configuración f
i configurazione f
n configuratie, ruimtelijke rangschikking
d Konfiguration f, räumliche Anordnung f

1307 CONFIGURATION CONTROL,   co
    LEAKAGE CONTROL
A method of control based on the principles
of varying either the rate of production
or the rate of leakage of neutrons, e.g. by
varying the amount and/or the configuration
of the fuel, reflector, coolant or moderator
if any.
f commande f par configuration
e regulación f por configuración
i regolazione f per configurazione
n configuratieregeling
d Konfigurationssteuerung f

1308 CONFINING END MIRROR   pp
A mirror at both ends of a containment
space.
f miroir m de confinement terminal
e espejo m de confinamiento terminal
i specchio m di confinamento terminale

n opsluitende eindspiegel
d einschliessender Endspiegel *m*

1309 CONJUGATE LAYERS   ch
In solvent extraction, the two inmiscible
solutions.
f couches *pl* conjuguées
e capas *pl* conjugadas
i strati *pl* coniugati
n geconjugeerde lagen *pl*, lagen *pl* in evenwicht
d konjugierte Schichten *pl*

1310. CONJUGATE VARIABLES   ma
Variables occurring in pairs.
f grandeurs *pl* conjuguées,
variables *pl* conjuguées
e magnitudes *pl* conjugadas,
variables *pl* conjugadas
i grandezze *pl* coniugate,
variabili *pl* coniugate
n geconjugeerde grootheden *pl*
d konjugierte Grössen *pl*

1311 CONSERVATION OF ENERGY   gp
The total quantity of energy in a closed
system is constant.
f conservation *f* de l'énergie
e conservación *f* de la energía
i conservazione *f* dell'energia
n behoud *n* van het arbeidsvermogen
d Erhaltung *f* des Arbeitsvermögens

1312 CONSERVATION OF MASS   gp
According to the law of conservation of
mass, matter can neither be created nor
destroyed.
f conservation *f* de la masse
e conservación *f* de la masa
i conservazione *f* della massa
n behoud *n* van de massa
d Erhaltung *f* der Masse

1313 CONSERVATION OF MOMENTUM   gp
For any collision, the vector sum of the
moments of the colliding bodies after
collision equals the vector sum of their
moments before collision.
f conservation *f* de l'impulsion
e conservación *f* del impulso
i conservazione *f* dell'impulso
n behoud *n* van de impuls
d Erhaltung *f* des Impulses

1314 CONSTANT POTENTIAL   pa
ACCELERATOR
A device in which a direct current potential
is applied to an accelerating tube to
produce high-energy ions or electrons.
f accélérateur *m* à tension continue
e acelerador *m* de tensión continua
i acceleratore *m* a tensione continua
n gelijkspanningsversneller
d Gleichspannungsbeschleuniger *m*

1315 CONSTANT RATE   ch, gp
DRYING PERIOD
In the drying of solids, a period in the

initial stages of drying in which the rate
of moisture loss is essentially constant.
f temps *m* de séchage à vitesse constante
e tiempo *m* de secado con velocidad
constante
i tempo *m* di disseccamento a velocità
costante
n constante-verdampingstijd
d Zeit *f* der konstanten Trocknungs-
geschwindigkeit

1316 CONSTITUENT PARTICLE,   gp
MATERIAL PARTICLE
A particle characterized by the properties
of mass and an observable position in
public space and time.
f particule *f* constituante
e partícula *f* constitutiva
i particella *f* costitutiva
n stoffelijk deeltje *n*
d materielles Teilchen *n*

1317 CONSTITUTION DIAGRAM,   ch, gp
EQUILIBRIUM DIAGRAM,
PHASE DIAGRAM
A representation of the equilibrium
temperature, pressure and composition
limits of the phase regions of a system.
f diagramme *m* de phases
e diagrama *m* de fases
i diagramma *m* di fasi
n fazediagram *n*
d Zustandsdiagramm *n*

1318 CONSTRAINT   np
The condition wherein a particle or group
of particles has less than 3 N degrees of
freedom, where N is the number of
particles in the group.
f contrainte *f*
e constreñimiento *m*
i costrizione *f*
n beperking
d Beschränkung *f*

1319 CONSUMED PARTICLE   np
In a nuclear reaction a particle that has
done the work for which it was destined.
f particule *f* consumée
e partícula *f* consumida
i particella *f* consumata
n verbruikt deeltje *n*
d verbrauchtes Teilchen *n*

1320 CONTACT BETA-RADIOGRAPHY
DATING, see 609

1321 CONTACT RADIOGRAM   xr
A radiogram obtained by putting the X-ray
tube in direct contact with the object.
f radiogramme *m* de contact
e radiograma *m* de contacto
i radiogramma *m* di contatto
n contactröntgenfoto
d Kontaktradiogramm *n*,
Röntgenkontaktaufnahme *f*

1322 CONTACT RÖNTGENTHERAPY, xr
SHORT FOCAL DISTANCE THERAPY
X-Ray therapy with specially constructed
tubes in which the target-skin distance is
very short, i.e. less than 2 cm.
f contact-thérapie *f*, plésiothérapie *f*,
röntgenthérapie *f* de contact
e plesioterapia *f*,
radioterapia *f* de contacto,
röntgenoterapia *f* de contacto
i contatto-terapia *f*,
röntgenterapia *f* di contatto
n contactröntgentherapie
d Kontaktröntgentherapie *f*

1323 CONTAINER md, ra
For a radioactive source, an enclosure
used as a shielding against radiation.
f récipient *m*
e recipiente *m*
i recipiente *m*
n houder
d Behälter *m*

1324 CONTAINER LOAD ma
ACTIVITY METER
An activity meter which includes detectors
associated with an electronic sub-assembly
and designed to measure and possibly to
record the activity of a container load.
f activimètre *m* par unité d'extraction,
ensemble *m* de mesure de l'activité par
unité d'extraction
e radiámetro *m* valorador de carga
i attivimetro *m* per unità d'estrazione
n activiteitsmeter voor vrachtladingen
d Gerät *n* zur Bestimmung der Behälter-
ladung

1325 CONTAINER SORTING ma
MINE-HEAD GRADING EQUIPMENT
An equipment which includes a container
load activity meter associated with an
automatic sorting device.
f équipement *m* d'estimation et de triage
par unité d'extraction
e equipo *m* selector clasificador de carga
a bocamina
i apparecchiatura *f* per stima e cernita
di minerali per unità d'estrazione
n uitrusting voor sortering van vrachtladingen
d Klassiereinrichtung *f* bei der automa-
tischen Behältersortierung

1326 CONTAINING FIELD pp
The field generated by powerful magnets
for preventing a plasma from reaching
the wall of the vessel in which it is
produced.
f champ *m* de confinement
e campo *m* de confinamiento
i campo *m* di confinamento
n opsluitingsveld *n*
d Einschliessungsfeld *n*

1327 CONTAINMENT, pp
PLASMA CONTAINMENT
In the controlled thermonuclear
program(me), a term which refers to the
condition in which an extremely hot
ionized gas or plasma is effectively
separated from material walls by means
of a magnetic field.
f confinement *m* du plasma
e confinamiento *m* del plasma
i confinamento *m* del plasma
n plasmaopsluiting
d Plasma-Einschliessung *f*

1328 CONTAINMENT SPHERE, rt
SPHERICAL CONTAINMENT VESSEL
f enveloppe *f* de sécurité sphérique
e envoltura *f* de seguridad esférica
i involucro *m* di contenimento sferico
n bolvormig omhulsel *n*
d kugelförmiger Sicherheitsbehälter *m*

1329 CONTAINMENT TIME pp
The approximate time for which ions
remain trapped by the containing field.
f durée *f* de confinement
e duración *f* de confinamiento
i durata *f* di confinamento
n opsluitingstijd
d Einschliessungszeit *f*

1330 CONTAINMENT VESSEL, rt
REACTOR CONTAINMENT
The system or device preventing the release,
even under the conditions of a reactor accident, of unacceptable quantities of radioactive materials beyond a controlled zone.
f enveloppe *f* de sécurité
e envoltura *f* de seguridad
i involucro *m* di contenimento
n insluitsysteem *n*, veiligheidsomhulsel *n*
d Reaktorsicherheitshülle *f*

1331 CONTAMINANT, np
RADIOACTIVE CONTAMINANT
A radioactive substance dispersed in
materials or places where it is undesirable.
f matière *f* contaminante
e contaminante *m* radiactivo
i contaminante *m* radioattivo
n radioactieve smetstof
d radioaktiver Kontaminationsstoff *m*,
Verseuchungsstoff *m*

1332 CONTAMINATED ge, rt
Made radioactive by addition or adsorption
of more or less minute quantities of
radioactive material.
f contaminé adj
e contaminado adj
i contaminato adj
n besmet adj
d kontaminiert adj, verseucht adj

1333 CONTAMINATED STEAM ra, sa
Steam developed by underground explosion
in rock formations and used for industrial
heating purposes.

f   vapeur *f* radioactive
e   vapor *m* radiactivo
i   vapore *m* radioattivo
n   besmette stoom
d   kontaminierter Dampf *m*

1334 CONTAMINATION,   np
      RADIOACTIVE CONTAMINATION
      The presence of a radioactive substance
      dispersed in materials or places where it
      is undesirable.
f   contamination *f* radioactive
e   contaminación *f* radiactiva
i   contaminazione *f* radioattiva
n   radioactieve besmetting
d   radioaktive Kontamination *f*

1335 CONTAMINATION DOSE,   nw, sa
      DEPOSIT DOSE
      Gamma dose from the residual radio-
      activity deposited on a surface after a
      nuclear explosion, as by fallout particles,
      or water falling as rain from the base
      surge of an underwater explosion.
f   dose *f* de dépôt
e   dosis *f* depositada
i   dose *f* di deposito
n   neerslagdosis
d   Niederschlagsdosis *f*

1336 CONTAMINATION HAZARDS   sa
f   dangers *pl* de contamination
e   riesgos *pl* de contaminación
i   rischi *pl* di contaminazione
n   besmettingsgevaren *pl*
d   Kontaminationsgefährdung *f*

1337 CONTAMINATION METER   me
f   appareil *m* de mesure de la contamination,
    contaminamètre *m*
e   aparato *m* de medida de la contaminación
i   apparecchio *m* di misura della
    contaminazione
n   besmettingsmeter
d   Kontaminationsmeter *n*,
    Verseuchungsmessgerät *n*

1338 CONTAMINATION MONITORING   co
      Periodical and continuous control of
      contamination of an area, fluid, object
      or person.
f   contrôle *m* de contamination
e   control *m* de contaminación
i   controllo *m* di contaminazione
n   besmettingscontrole
d   Kontaminationsüberwachung *f*

1339 CONTENT METER   ma
      A measuring assembly that includes an
      ionizing radiation source and is designed
      to determine the content in one or several
      components of a gaseous, liquid or solid
      substance, using the measurement of
      characteristics, within a defined geometry,
      of radiation resulting of the utilized
      process.
f   teneurmètre *m*
e   valorímetro *m*

i   tenorimetro *m*
n   gehaltemeter
d   Gehaltmessgerät *n*

1340 CONTENT METER BY   ma
      U.V. FLUORESCENCE,
         U.V. EXCITATION FLUORIMETER
      A content meter usually used to determine
      the uranium content of a solution by
      measuring the fluorescent light intensity
      of uranyl salts excited by ultraviolet
      radiation.
f   fluorimètre *m* à excitation U.V.,
    teneurmètre *m* à fluorescence U.V.
e   fluorímetro *m* por excitación ultravioleta,
    uraniómetro *m* por fluorescencia
i   tenorimetro *m* a fluorescenza
n   U.V. fluorimeter
d   UV-Fluoreszenzanalysegerät *n*

1341 CONTINUITY EQUATION,   ma
      EQUATION OF CONTINUITY,
      PRINCIPLE OF CONTINUITY
      An equation for continuous forces
      occurring in motions of ions and electrons.
f   équation *f* de continuité
e   ecuación *f* de continuidad
i   equazione *f* di continuità
n   continuiteitsvergelijking
d   Kontinuitätsgleichung *f*

1342 CONTINUOUS AIR MONITOR   sa
      An air monitor in which the filter moves
      in front of the built-in detectors which
      give a continuous measure of the
      contamination.
f   moniteur *m* atmosphérique en continu
e   vigía *f* continua del aire
i   monitore *m* atmosferico continuo
n   continue luchtmonitor
d   kontinuierlich arbeitender Luftmonitor *m*

1343 CONTINUOUS CREATION   cr, np
      HYPOTHESIS,
         STEADY STATE COSMOLOGY
      Hypothesis that the universe is being
      created continuously, by the formation,
      without apparent mechanism, of one

      nucleon per 10^9 year-litre(er).
f   hypothèse *f* de la création continue
e   hipótesis *f* de la creación continua
i   ipotesi *f* della creazione continua
n   continue scheppingshypothese
d   Hypothese *f* der kontinuierlichen
    Schöpfung

1344 CONTINUOUS GRAIN   ms
      IRRADIATION PLANT
      A large-scale plant using cobalt-60 as
      the radiation source.
f   installation *f* d'irradiation continue du blé
e   equipo *m* de irradiación continua de
    cereales
i   impianto *m* d'irradiazione continua di
    cereali
n   continue graanbestralingsinstallatie
d   kontinuierliche Getreidebestrahlungsan-
    lage *f*

1345 CONTINUOUS PROCESS   ch
A process in which for extended periods
uninterrupted flows of proportional components enter and products leave a system.
f procédé *m* continu
e procedimiento *m* continuo
i processo *m* continuo
n continu proces *n*
d kontinuierliches Verfahren *n*

1346 CONTINUOUS SLOWING-DOWN   np
    MODEL
A treatment of the slowing-down process
which replaces the step-wise decrease of
energy due to collisions by a continuous
curve.
f modèle *m* de ralentissement continu
e modelo *m* de retardamiento continuo
i modello *m* di rallentamento continuo
n model *n* van continue verlangzaming
d Modell *n* der kontinuierlichen Verlangsamung

1347 CONTINUOUS SPECTRUM   sp
A spectrum in which the radiation varies
slowly along the spectrum.
f spectre *m* continu
e espectro *m* continuo
i spettro *m* continuo
n continu spectrum *n*
d kontinuierliches Spektrum *n*, Kontinuum *n*

1348 CONTINUOUS X-RAY   sp, xr
    SPECTRUM
Röntgen radiation from an X-ray tube,
excluding characteristic radiation.
f spectre *m* continu de rayons X
e espectro *m* continuo de rayos X
i spettro *m* continuo di raggi X
n continu röntgenspectrum *n*
d kontinuierliches Röntgenspektrum *n*

1349 CONTINUOUS X-RAYS   xr
The electromagnetic radiation of
continuous spectral distribution produced by
bremsstrahlung, e.g. when electrons
strike a target.
f rayons *pl* X continus
e rayos *pl* X continuos
i raggi *pl* X continui
n continue röntgenstralen *pl*
d kontinuierliche Röntgenstrahlen *pl*

1350 CONTRACTION CAVITY,   mg
    PIPE, SHRINKAGE CAVITY
A cavity, usually along the centre(er) line
of an ingot, created by the contraction which
accompanies solidification.
f retassure *f*
e rechupe *m*
i risucchio *m*
n slinkholte
d Lunker *m*

1351 CONTRAST MEDIUM   xr
Metal suspension or metallic salt, e.g.
bismuth, barium, which makes visible the

parts of the body filled with it in the radiogram.
f moyen *m* de contraste
e medio *m* de contraste
i mezzo *m* di contraste
n contrastmiddel *n*
d Kontrastmittel *n*

1352 CONTRAST RADIOGRAPHY   xr
Röntgenography of organs with the aid of
a contrast medium.
f radiographie *f* à contraste
e radiografía *f* de contraste
i radiografia *f* di contrasto
n contraströntgenfotografie
d Röntgenkontrastdarstellung *f*

1353 CONTROL   co
Of a nuclear reactor, the intentional
variation of the reaction rate or the
adjustment made to prevent variation.
f commande *f*
e regulación *f*
i regolazione *f*
n regeling
d Steuerung *f*

1354 CONTROL ARM   co
A control member, usually of cylindrical
shape, capable of being translated along
its axis.
f barre *f* de commande rotative,
e barra *f* de regulación giratoria
i barra *f* di regolazione rotativa
n draaibare regelarm
d drehbarer Steuerstab *m*

1355 CONTROL CIRCUIT   co
The electrical circuits and the pneumatic
and electromechanical organs which effect
the regulation and control of an installation.
f circuit *m* de commande,
   circuit *m* de contrôle, circuit *m* de réglage
e circuito *m* de control, circuito *m* de mando,
   circuito *m* de regulación
i circuito *m* di comando,
   circuito *m* di controllo,
   circuito *m* di regolazione
n regelcircuit *n*
d Regelkreis *m*, Steuerkreis *m*

1356 CONTROL DRIVE   co, rt
A device used for moving a control member
in the course of reactor control.
f mécanisme *m* de commande
e mecanismo *m* de regulación
i meccanismo *m* di regolazione
n aandrijving van een regellichaam
d Steuerantrieb *m*

1357 CONTROL ELEMENT,   co, rt
    CONTROL MEMBER
A movable part of a nuclear reactor which
itself affects reactivity and is used for
reactor control.
f élément *m* de commande
e elemento *m* de regulación

i elemento *m* di regolazione
n regellichaam *n*
d Steuerelement *n*

1358 CONTROL INSTALLATION   co
A set of apparatus for controlling the rate
of reaction in a nuclear reactor.
f appareillage *m* de commande,
appareillage *m* de contrôle,
appareillage *m* de réglage
e aparatos *pl* de control,
aparatos *pl* de mando,
aparatos *pl* de regulación
i apparecchiatura *f* di comando,
apparecchiatura *f* di controllo,
apparecchiatura *f* di regolazione
n regelapparatuur
d Regelapparatur *f*, Steuerapparatur *f*

1359 CONTROL PANEL   co
A panel which contains the various control
knobs, switches, etc.
f tableau *m* de commande
e tablero *m* de control
i quadro *m* di comando
n schakelbord *n*
d Schaltbrett *n*

1360 CONTROL ROD, see 10

1361 CONTROL ROD CALIBRATION   co
The carrying out of checking operations
in order to calibrate the control rod.
f calibrage *m* d'une barre de commande
e calibración *f* de una barra de regulación
i taratura *f* d'una barra di regolazione
n ijking van een regelstaaf
d Eichung *f* eines Steuerstabs

1362 CONTROL ROD DRIVE   co
The means for operating any rod to control
the reactivity of a nuclear reactor.
f mécanisme *m* de la barre de commande
e mecanismo *m* de la barre de regulación
i comando *m* della barra di regolazione
n regelstaafmechanisme *n*
d Steuerstabantrieb *m*,
Steuerstabmechanismus *m*

1363 CONTROL ROD FOLLOWER,   co
FOLLOWER
Appendix to the end of a control rod of
suitable material which will occupy the
vacancy created by withdrawing the fuel
element.
f allongement *m* d'une barre de commande
e prolongación *f* de barra de regulación
i prolungamento *m* d'una barra di
regolazione
n regelstaafverlengstuk *n*
d Steuerstabansatz *m*

1364 CONTROL ROD FUEL   co, fu
FOLLOWER
Appendix to a control rod containing nuclear
fuel, applied to some reactors for reducing
local spikes.
f allongement *m* en combustible d'une barre
de commande

e prolongación *f* de combustible de una
barra de regulación
i prolungamento *m* di combustibile d'una
barra di regolazione
n splijtstofhoudend regelstaafverlengstuk *n*
d brennstoffhaltiger Steuerstabansatz *m*

1365 CONTROL ROD GEAR BOX   co
The gear box containing the handling
connection for the control rods.
f carter *m* des barres de commande
e cárter *m* de las barras de regulación
i scatola *f* delle barre di regolazione
n regelstaafcarter *n*
d Steuerstabgetriebekasten *m*

1366 CONTROL ROD SHADOWING,   co
SHADOWING
Shadow effect between control rods whereby
the efficiency of one rod is diminished when
inserting another rod.
f interaction *f* négative entre barres de
commande
e interacción *f* negativa entre barras de
regulación
i interazione *f* negativa fra barre di
regolazione
n nadelig schaduweffect *n*
d nachteilige Schattenwirkung *f*,
nachteiliger Schatteneffekt *m*

1367 CONTROL ROD WORTH   rt
The reactivity change resulting from the
complete insertion of a fully withdrawn
control rod into a critical reactor under
specified conditions.
f efficacité *f* d'une barre de commande
e eficacia *f* de una barra de regulación
i efficacia *f* d'una barra di regolazione
n regelstaafwaarde
d Leistungsfähigkeit *f* eines Steuerstabs

1368 CONTROL ROOM   co
A room from which engineers and production men control direct the reactor.
f salle *f* de commande
e sala *f* de regulación
i sala *f* di regolazione
n regelkamer
d Regelraum *m*, Steuerwarte *f*

1369 CONTROLLABLE REACTION   rt
A nuclear reaction which can be initiated,
controlled or stopped by personnel of the
reactor.
f réaction *f* contrôlable
e reacción *f* controlable
i reazione *f* controllabile
n beheerste reactie
d beherrschte Reaktion *f*

1370 CONTROLLED AREA, see 71

1371 CONTROLLED CONDITION,   co
CONTROLLED VARIABLE
That quantity or condition which is
measured and controlled.
f grandeur *f* réglée

e magnitud f regulada
i grandezza f regolata
n geregelde grootheid
d Regelgrösse f

1372 CONTROLLED FUSION REACTOR,   rt
      CONTROLLED THERMONUCLEAR
      REACTOR
A hypothetical device which harnesses
the nuclear fusion process to produce
electrical power.
f réacteur m à fusion contrôlée
e reactor m de fusión controlada
i reattore m a fusione controllata
n beheerste fusiereactor
d Reaktor m für kontrollierte Fusion

1373 CONTROLLED INJURY ZONE   rw
A radiation zone wherein there are radioactive substances available to such an
extent that an accident involving the
ingestion, inhalation or injection of
foreign matter into the human body could
result in a significant deposition of the
radioactive substance(s).
f zone f dangereuse surveillée
e zona f de peligro vigilada
i zona f di pericolo sorvegliata
n bewaakt gevarengebied n
d überwachte Gefahrenzone f

1374 CONVECTION   gp
The transfer of heat by circulation or
mixing.
f convection f
e convección f
i convezione f
n convectie
d Konvektion f

1375 CONVECTION CURRENT   ch
The current resulting from the flow of
charged particles in electrolytes, insulating liquids, gases or in a vacuum.
f courant m de convection
e corriente f de convección
i corrente f di convezione
n convectiestroom
d Konvektionsstrom m

1376 CONVENTIONAL FLUX DENSITY,   np
      2200 METRE(ER) PER SECOND
      FLUX DENSITY
A fictitious flux density equal to the product
of the total number of neutrons per cubic
centimetre(er) and a neutron speed of
2.2×10^5 centimetre(er)s per second.
f flux m conventionnel,
   flux m de 2200 mètres par seconde
e flujo m convencional,
   flujo m de 2200 metros por segundo
i flusso m convenzionale,
   flusso m di 2200 metri per secondo
n conventionele fluxdichtheid,
   2200 m/s fluxdichtheid
d konventionelle Flussdichte f,
   2200 m/s Flussdichte f

1377 CONVERGENCY IRRADIATION,   xr
      CONVERGENT BEAM THERAPY
A technique whereby convergent beams of
röntgen or gamma rays are brought to
bear upon a certain region.
f radiothérapie f convergente
e radioterapia f convergente
i radioterapia f convergente
n convergentiebestraling
d Konvergenzbestrahlung f

1378 CONVERGENT BEAM UNIT   xr
An X-ray apparatus for carrying out
convergent beam therapy.
f dispositif m d'irradiation convergente
e dispositivo m de irradiación convergente
i dispositivo m d'irradiazione convergente
n apparaat n voor convergentiebestraling
d Konvergenzbestrahlungsgerät n

1379 CONVERGENT REACTION   np
A nuclear chain reaction in which the
number of reactions caused directly by one
reaction is on the average less than unity.
f réaction f convergente
e reacción f convergente
i reazione f convergente
n convergente reactie
d abklingende Reaktion f,
   konvergente Reaktion f

1380 CONVERSION   rt
In reactor technology, nuclear transformation of a fertile substance into a fissile
substance.
f conversion f
e conversión f
i conversione f
n conversie
d Konversion f, Umwandlung f

1381 CONVERSION COEFFICIENT,   np
      CONVERSION FRACTION
The ratio of the number of internal conversion electrons to the total number of quanta
plus the number of conversion electrons
emitted in a given mode of de-excitation of
a nucleus.
f fraction f de conversion
e fracción f de conversión
i frazione f di conversione
n conversiefactor, conversiefractie
d Konversionsanteil m, Konversionsfaktor m

1382 CONVERSION EFFICIENCY   ra
In bremsstrahlung sources, the number of
photons per beta particle.
f rendement m de conversion
e rendimiento m de conversión
i rendimento m di conversione
n conversierendement n
d Konversionsausbeute f

1383 CONVERSION ELECTRON   np
An electron emitted by internal conversion
during de-excitation of a nucleus.
f électron m de conversion

e electrón *m* de conversión
i elettrone *m* di conversione
n conversie-elektron *n*
d Konversionselektron *n*

1384 CONVERSION FACTOR   np, rt
In a converter reactor, the number of
fissile (fissionable) atoms produced from
the fertile material per fissile (fissionable) atom destroyed in the fuel.
f facteur *m* de conversion
e factor *m* de conversión
i fattore *m* di conversione
n conversiefactor
d Konversionsfaktor *m*

1385 CONVERSION GAIN   np, rt
The conversion factor minus one.
f gain *m* de conversion
e ganancia *f* de conversión
i guadagno *m* di conversione
n conversiewinst
d Konversionsgewinn *m*

1386 CONVERSION QUANTUM   ec
EFFICIENCY
Of a photocathode, the ratio of the number
of electrons emitted to the number of
incident photons at the photocathode.
f rendement *m* quantique de conversion
e rendimiento *m* cuántico de conversión
i rendimento *m* quantico di conversione
n elektronenopbrengst van de quantumom-
   zetting
d Quantenausbeute *f* an einer Photokatode

1387 CONVERSION RATIO   np, rt
The ratio of the number of fissile
(fissionable) nuclei produced by conversion
to the number of fissile (fissionable)
nuclei destroyed.
f rapport *m* de conversion
e relación *f* de conversión
i rapporto *m* integrale di conversione
n conversieverhouding
d Konversionsverhältnis *n*

1388 CONVERSION TRANSITION   np
A transition in the nucleus.
f transition *f* par conversion
e transición *f* por conversión
i transizione *f* per conversione
n conversieovergang
d Konversionsübergang *m*

1389 CONVERTER,   np
NEUTRON CONVERTER
A device placed in a flux density of
slow neutrons to produce fast neutrons.
f convertisseur *m* de neutrons
e convertidor *m* de neutrones
i convertitore *m* di neutroni
n neutronenomzetter
d Neutronenkonverter *m*

1390 CONVERTER REACTOR,   rt
REGENERATIVE REACTOR
A nuclear reactor which produces a

fissile (fissionable) material identical
with the consumed material.
f réacteur *m* régénérateur
e reactor *m* de conversión
i reattore *m* convertitore
n conversiereactor
d Konverterreaktor *m*

1391 CONVEYOR GRADING   ma
EQUIPMENT
An equipment designed to grade ore or a
mixture of ores transported on a conveyor,
and also estimating the uranium content.
f équipement *m* de contrôle sur bande
   transporteuse
e equipo *m* transportador clasificador
i apparecchiatura *f* per controllo su nastro
   trasportatore
n uitrusting voor gehaltecontrole aan de band
d Klassiereinrichtung *f* am Förderband

1392 COOLANT,   cl
COOLING MEDIUM
f fluide *m* caloporteur,
   fluide *m* de refroidissement, réfrigérant *m*
e enfriador *m*, refrigerante *m*
i fluido *m* termovettore, refrigerante *m*
n koelmiddel *n*
d Kühlmittel *n*

1393 COOLANT ADDITIVE   cl, mt
A substance added to the coolant in a nuclear
reactor to counteract the radiolytic
attack.
f substance *f* d'addition au réfrigérant
e substancia *f* de adición al refrigerante
i sostanza *f* d'addizione al refrigerante
n toevoegsel *n* aan het koelmiddel
d Kühlmittelzusatz *m*

1394 COOLANT CIRCUIT   cl
f circuit *m* de refroidissement
e circuito *m* de enfriamiento
i circuito *m* di raffreddamento
n koelcircuit *n*, koelkring
d Kühlkreis *m*

1395 COOLANT CIRCUIT   co, sa
LEAKAGE INDICATOR
An indicator designed to detect leaks in
the cooling circuit.
f signaleur *m* de pertes du circuit de
   refroidissement
e indicador *m* de pérdidas del circuito de
   refrigeración
i segnalatore *m* di perdite del circuito di
   raffreddamento
n lekindicator voor het koelcircuit
d Kühlkreisleckanzeiger *m*

1396 COOLANT GROSS   co, sa
ACTIVITY MONITOR
A monitor designed to measure the activity
of the coolant of a nuclear reactor and to
give a warning when it exceeds a predetermined value.
f moniteur *m* d'activité globale du fluide
   de refroidissement

e monitor *m* de actividad global del
  refrigerante
i monitore *m* d'attività globale del fluido
  di raffreddamento
n monitor voor de totale activiteit van
  het koelmiddel
d Warngerät *n* für die Gesamtaktivität des
  Kühlmittels

1397 COOLANT LEAKAGE INDICATOR co, sa
  An indicator designed to detect the loss
  of coolant of a nuclear reactor.
f signaleur *m* de pertes du fluide de
  refroidissement
e indicador *m* de pérdidas del refrigerante
i segnalatore *m* di perdite del fluido di
  raffreddamento
n lekindicator voor het koelmiddel
d Kühlmittelleckanzeiger *m*

1398 COOLANT LEAKAGE   co, me
  MEASURING ASSEMBLY
  A measuring assembly designed to
  determine the loss of coolant from a
  nuclear reactor.
f ensemble *m* de mesure de pertes du fluide
  de refroidissement
e conjunto *m* de medida de pérdidas del
  refrigerante
i complesso *m* di misura di perdite del
  fluido di raffreddamento
n meetopstelling voor koelmiddellekkage
d Kühlmittelverlust-Messanordnung *f*

1399 COOLANT PUMP   co
  Pump used for circulating a cooling
  substance.
f pompe *f* à circulation du réfrigérant
e bomba *f* de circulación del refrigerante
i pompa *f* di circolazione del refrigerante
n koelmiddelpomp
d Kühlmittelpumpe *f*

1400 COOLIDGE TUBE   xr
  An X-ray tube in which the needed electrons
  are produced by an incandescent cathode.
f tube *m* Coolidge
e tubo *m* Coolidge
i tubo *m* Coolidge
n coolidgebuis
d Coolidge-Röhre *f*

1401 COOLING   cl, ge
f réfrigération *f*, refroidissement *m*
e enfriamiento *m*, refrigeración *f*
i raffreddamento *m*, refrigerazione *f*
n afkoeling, koeling
d Abkühlung *f*, Kühlung *f*

1402 COOLING   rt
  Of a strongly radioactive material, the
  decrease of its radioactivity through
  radioactive decay.
f désactivation *f*, refroidissement *m*
e desactivación *f*, enfriamiento *m*
i decadimento *m*

n activiteitsvermindering, afkoeling,
  versterven *n*
d Abklingen *n*, Aktivitätsverminderung *f*,
  Auskühlung *f*

1403 COOLING AIR, see 151

1404 COOLING COLUMN,   cl
  COOLING TOWER
f colonne *f* de refroidissement,
  tour *f* de refroidissement
e columna *f* de enfriamiento,
  torre *f* de enfriamiento
i colonna *f* di raffreddamento,
  torre *f* di raffreddamento
n koeltoren
d Kühlturm *m*

1405 COOLING PERIOD,   sa
  COOLING TIME
  The interval between the removal of a
  radioactive product from the reactor and
  its chemical treatment.
f durée *f* de désactivation,
  durée *f* de refroidissement
e duración *f* de desactivación,
  duración *f* de enfriamiento
i durata *f* di dedicamento
n versterftijd
d Abklingzeit *f*

1406 COOLING PIT,   rt
  COOLING POND,
  FUEL COOLING INSTALLATION
  A large container or cell, usually filled
  with water, in which spent nuclear fuel is
  set aside until its radioactivity has
  decreased to a desired level.
f installation *f* de refroidissement du
  combustible, piscine *f* de désactivation
e equipo *m* de enfriamiento del combustible
i bacino *m* di decadimento,
  vasca *f* di decadimento
n afkoelbassin *n* voor splijtstoffen
d Abklingbecken *n*

1407 CO-ORDINATE (GB),   ma
  COORDINATE (US)
  Any one of two or more magnitudes that
  determine position relative to the
  reference axes of a co-ordinate system.
f coordinée *f*
e coordenada *f*
i coordinata *f*
n coördinaat
d Koordinate *f*

1408 CO-ORDINATE SYSTEM (GB),   ma
  COORDINATE SYSTEM (US)
  A set of numbers or surfaces which may
  be used to locate a point or geometric
  element in space.
f système *m* de coordinées
e sistema *m* de coordenadas
i sistema *m* di coordinate
n coördinatensysteem *n*
d Koordinatensystem *n*

1409 COPPER   ch
    Metallic element,symbol Cu, atomic
    number 29.
f cuivre *m*
e cobre *m*
i rame *m*
n koper *n*
d Kupfer *n*

1410 COPPER URANITE,   mi
    TORBERNITE
    A common secondary mineral, containing
    about 47 % of U.
f chalcolite *f*, torbernite *f*
e torbernita *f*
i torbernite *f*
n torberniet *n*
d Kupferphosphoruranit *m*, Torbernit *m*

1411 COPRECIPITATION,   ch
    COSEPARATION
    In radiochemistry, the precipitation of a
    material from its unsaturated solution,
    or from a highly dispersed suspension,
    promoted by the precipitation of another
    material.
f coprécipitation *f*
e coprecipitación *f*
i coprecipitazione *f*
n coprecipitatie
d Mitfällung *f*

1412 CORACITE   mi
    An impure gummite.
f coracite *f*
e coracita *f*
i coracite *f*
n coraciet *n*
d Corazit *m*

1413 CORDYLITE   mi
    A cerium lanthanum barium mineral
    containing a small percentage of Th.
f cordylite *f*
e cordilita *f*
i cordilite *f*
n cordyliet *n*
d Cordylit *m*

1414 CORE,   rt
    REACTOR CORE
    That region of a reactor in which a chain
    reaction can take place.
f coeur *m* du réacteur
e alma *f* del reactor
i nocciolo *m* del reattore
n reactorkern
d Spaltzone *f*

1415 CORE CONVERSION RATIO,   rt
    INTERNAL CONVERSION RATIO
    The conversion ratio relating only to the
    fissile (fissionable) nuclei produced in the
    interior of the core of a converter reactor.
f rapport *m* de conversion interne
e relación *f* de conversión interna
i rapporto *m* di conversione interna

n interne conversieverhouding
d inneres Konversionsverhältnis *n*

1416 CORE PRESSURE VESSEL,   rt
    PRESSURE VESSEL,
    REACTOR VESSEL
    A vessel for containing the core and
    reflector of a reactor, in which the cooling
    medium is maintained at a high pressure.
f caisson *m* de réacteur, cuve *f* de réacteur,
    récipient *m* à pression,
    récipient *m* du réacteur
e recipiente *m* de presión,
    recipiente *m* del reactor
i recipiente *m* a pressione,
    recipiente *m* del reattore
n drukvat *n*, reactorvat *n*
d Reaktorbehälter *m*

1417 CORE TANK,   rt
    SOLUTION TYPE REACTOR TANK
    The tank holding the uranium compound
    solution in a solution type tank.
f récipient *m* du coeur de réacteur
e recipiente *m* de la alma del reactor
i recipiente *m* del nocciolo
n kernvat *n*
d Kernbehälter *m*

1418 CORING,   mg
    INTERDENDRITIC SEGREGATION,
    MICROSCOPIC SEGREGATION
    A variation in composition within the
    dendrites or grains of a cast solution
    phase caused by failure to achieve
    equilibrium during solidification.
f ségrégation *f* mineure
e microsegregación *f*,
    segregación *f* interdendrítica
i microsegregazione *f*
n microsegregatie
d interdendritische Seigerung *f*,
    Kornseigerung *f*, Mikroseigerung *f*

1419 CORONA LOSS   pa
    Loss of charge occurring sometimes in the
    Van de Graaff and Cockcroft-Wilson
    particle accelerator.
f perte *f* par effet de couronne
e pérdida *f* por efecto corona
i perdita *f* per effetto corona
n coronaverlies *n*
d Koronaverlust *m*

1420 CORPUSCLE   md
    A protoplasmic cell, floating freely in a
    fluid, or embedded in a matrix.
f corpuscule *m*
e corpúsculo *m*
i corpuscolo *m*
n lichaampje *n*
d Körperchen *n*

1421 CORPUSCULAR EMISSION,   gr, xr
    PARTICLE EMISSION
    The full complement of secondary charged
    particles, usually electrons, associated

with an X-ray or gamma-ray beam in its
passage through air.
f émission *f* corpusculaire,
émission *f* de particules
e emisión *f* corpuscular,
emisión *f* de partículas
i emissione *f* corpuscolare,
emissione *f* di particelle
n corpusculaire emissie,
emissie van deeltjes
d Korpuskularemission *f*

1422 CORPUSCULAR RADIATION, ra
PARTICLE RADIATION
A stream of atomic or sub-atomic
particles which may be charged positively
or negatively or not at all.
f rayonnement *m* corpusculaire
rayonnement *m* de particules
e radiación *f* corpuscular,
radiación *f* de partículas
i radiazione *f* corpuscolare
n corpusculaire straling
d Korpuskularstrahlung *f*

1423 CORRECTION TIME, co
RECOVERY TIME,
SETTLING TIME
The time required for the controlled
variable to reach and stay within a
predetermined band about the control point
following any change of the independent
variable or operating condition.
f durée *f* de réglage,
temps *m* de réglage
e tiempo *m* de recuperación,
tiempo *m* de regulación
i tempo *m* di correzione,
tempo *m* di ripristino
n regelduur
d Regeldauer *f*

1424 CORRELATION FUNCTION, ma
LANGEVIN FUNCTION
An equation used in studying the Langevin
effect.
f fonction *f* de Langevin
e función *f* de correlación
i funzione *f* di correlazione
n correlatiefunctie
d Langevin-Funktion *f*

1425 CORRESPONDENCE PRINCIPLE qm
The principle that, in the limit of high
quantum numbers, the predictions of
quantum theory correspond with those of
classical mechanics.
f principe *m* de correspondance
e principio *m* de correspondencia
i principio *m* di correspondenza
n correspondentieprincipe *n*
d Korrespondenzprinzip *n*

1426 CORROSION mg
The conversion of iron and other metals
and alloys into oxides and carbonates by
the action of air and/or water.

f corrosion *f*
e corrosión *f*
i corrosione *f*
n corrosie
d Korrosion *f*

1427 CORROSION FATIGUE mg
Failure in service of fabricated parts that
are subjected simultaneously to alternating
stresses and to corrosion.
f fatigue *f* par corrosion
e fatiga *f* por corrosión
i fatica *f* per corrosione
n corrosievermoeiing
d Korrosionsermüdung *f*

1428 CORROSION RESISTANCE mg
f résistance *f* à la corrosion
e resistencia *f* a la corrosión
i resistenza *f* alla corrosione
n corrosievastheid
d Korrosionsfestigkeit *f*

1429 CORROSIVE FLUID mg
f fluide *m* corrosif
e flúido *m* corrosivo
i fluido *m* corrosivo
n corroderend gas *n*, corroderende vloeistof
d Korrosionsflüssigkeit *f*, Korrosionsgas *n*

1430 CORVUSITE mi
A vanadium ore containing about 1.45 %
of U.
f corvusite *f*,
e corvusita *f*
i corvusite *f*
n corvusiet *n*
d Corvusit *m*

1431 COSMIC ABUNDANCE cm
The relative abundance of nuclides or
elements in the universe, expressed as a
fraction of the total.
f abondance *f* cosmique
e abundancia *f* cósmica
i abbondanza *f* cosmica
n kosmische abondantie
d kosmische Häufigkeit *f*

1432 COSMIC RADIATION cm
Ionizing radiation from extraterrestrial
sources.
f rayonnement *m* cosmique
e radiación *f* cósmica
i radiazione *f* cosmica
n kosmische straling
d kosmische Strahlung *f*

1433 COSMIC-RAY DECAY ra
ELECTRONS
Electrons in the soft component of cosmic
rays which originate from the decay of
mesons.
f électrons *pl* de désintégration du
rayonnement cosmique
e electrones *pl* de desintegración de la
radiación cósmica

i   elettroni *pl* di disintegrazione della
    radiazione cosmica
n   vervalselektronen *pl* van de kosmische
    straling
d   Zerfallselektronen *pl* der kosmischen
    Strahlung

1434 COSMIC-RAY KNOCK-ON   ra
      ELECTRONS
      Electrons in the soft components of
      cosmic rays which originate in the direct
      impact of fast mesons with the orbital
      electrons of the oxygen and nitrogen atoms
      of the atmosphere.
f   électrons *pl* cosmiques de choc
e   electrones *pl* cósmicos de choque
i   elettroni *pl* cosmici d'urto
n   kosmische botsingselektronen *pl*
d   kosmische Stosselektronen *pl*

1435 COSMIC-RAY SHOWER   ra
      The simultaneous appearance of a number
      of downward-directed ionizing particles
      caused by a single cosmic ray.
f   gerbe *f* cosmique
e   chaparrón *m* cósmico
i   sciame *m* cosmico
n   kosmische. bui
d   kosmischer Schauer *m*

1436 COSMIC-RAY TELESCOPE   cr, me
      An array of cosmic counters connected
      in coincidence, anticoincidence, or some
      combination of the two and so arranged
      that a count will be recorded only for a
      cosmic ray particle that is incident from
      a given direction.
f   ensemble *m* compteur de rayonnement
    cosmique
e   conjunto *m* contador de radiación
    cósmica
i   complesso *m* contatore di radiazione
    cosmico
n   telleropstelling voor kosmische straling
d   Zähleranordnung *f* für kosmische
    Strahlung

1437 COSMIC RAYS   ra
      Radiation that has its ultimate origin
      outside of the earth's atmosphere,
      capable of producing ionizing events and
      of penetrating many feet of material such
      as rock.
f   rayons *pl* cosmiques
e   rayos *pl* cósmicos
i   raggi *pl* cosmici
n   kosmische stralen *pl*
d   Höhenstrahlung *f*, kosmische Strahlen *pl*

1438 COSMOTRON,   pa
      PROTON-SYNCHROTRON
      A synchrotron modified to permit the
      acceleration of protons by frequency
      modulation of the r.f. accelerating voltage.
f   cosmotron *m*, proton-synchrotron *m*
e   cosmotrón *m*, protón-sincrotrón *m*
i   cosmotrone *m*, protone-sincrotrone *m*

n   cosmotron *n*, proton-synchrotron *n*
d   Cosmotron *n*, Protonen-Synchrotron *n*

1439 COTTON ASBESTOS, see 333

1440 COTTRELL EFFECT,   mg, rt
      URANIUM CREEP
      The property of uranium under irradiation,
      e.g. in a nuclear reactor core, to lose its
      elasticity and to flow like an extremely
      viscous liquid when stress is applied.
f   effet *m* Cottrell, fluage *m* d'uranium
e   efecto *m* Cottrell, fluencia *f* de uranio
i   effetto *m* Cottrell, scorrimento *m* d'uranio
n   cottrelleffect *n*, uraniumkruip
d   Cottrell-Effekt *m*, Urankriechen *n*

1441 COULOMB BARRIER   np
      The nuclear barrier due to electrostatic
      forces.
f   barrière *f* coulombienne,
    barrière *f* de Coulomb
e   barrera *f* de Coulomb
i   barriera *f* di Coulomb
n   coulombstoep
d   Coulomb-Berg *m*, Coulomb-Schwelle *f*,
    Coulomb-Wall *m*

1442 COULOMB BARRIER RADIUS   np
      The radius deduced from the rate of alpha
      disintegration or from cross sections of
      nuclear reactions involving charged
      particles.
f   rayon *m* de la barrière de Coulomb
e   radio *m* de la barrera de Coulomb
i   raggio *m* della barriera di Coulomb
n   straal van de potentiaalstoep van Coulomb
d   Radius *m* des Coulombschen Potentialwalls

1443 COULOMB DEGENERACY   np
      Identity of the energy levels of a charged
      particle bound in a Coulomb field for
      different values of the orbital angular
      momentum, provided that the principal
      quantum number and spin state are the same.
f   dégénérescence *f* coulombienne
e   degeneración *f* de Coulomb
i   degenerazione *f* di Coulomb
n   coulombdegeneratie
d   Coulombsche Degenerierung *f*

1444 COULOMB ENERGY,   np
      COULOMB FORCE
      The contribution to the atomic binding
      energy of a solid which is due to the
      electrostatic attraction between electrons
      and ions.
f   force *f* coulombienne
e   fuerza *f* de Coulomb
i   forza *f* di Coulomb
n   coulombkracht
d   Coulomb-Kraft *f*

1445 COULOMB EXCITATION   np
      The excitation of a nucleus by the electric
      field of a passing charged particle.
f   excitation *f* coulombienne

e excitación f de Coulomb
i eccitazione f di Coulomb
n coulombaanslag
d Coulomb-Anregung f

1446 COULOMB FIELD np
An electric field due to a charge acting
as if concentrated at a point, so that the
field intensity is inversely proportional
to the square of the radial distance from
that point.
f champ m coulombien
e campo m de Coulomb
i campo m di Coulomb
n coulombveld n
d Coulomb-Feld n

1447 COULOMB INTERACTION, np
 ELECTROSTATIC INTERACTION
That part of the total interaction between
two charged particles that is due to the
Coulomb force between them.
f interaction f coulombienne,
 interaction f électrostatique
e interacción f de Coulomb,
 interacción f electrostática
i interazione f di Coulomb,
 interazione f elettrostatica
n elektrostatische wisselwerking,
 wisselwerking van Coulomb
d Coulombsche Wechselwirkung f,
 elektrostatische Wechselwirkung f

1448 COULOMB POTENTIAL np
A scalar point function equal to the work
per unit charge done against or by the
Coulomb force in transferring a particle
bearing an infinitesimal positive charge
from infinity to the field of a charged
particle in a vacuum.
f potentiel m coulombien
e potencial m de Coulomb
i potenziale m di Coulomb
n coulombpotentiaal
d Coulomb-Potential n

1449 COULOMB SCATTERING np
The scattering of charged particles by an
inverse square attractive or repulsive
field.
f diffusion f coulombienne
e dispersión f de Coulomb
i deviazione f di Coulomb
n coulombverstrooiing
d Coulomb-Streuung f

1450 COUNT ct
Number of pulses recorded during a
measurement.
f compte m
e cuenta f, recuento m
i conteggio m
n telling
d Zählung f

1451 COUNT ct
Information corresponding to a pulse
processed for counting.

f choc m, coup m
e impulso m de cuenta, impulso m de recuento
i impulso m di conteggio
n telpuls, telsignaal n
d Zählimpuls m

1452 COUNTER ct
1. A device for counting ionizing events.
2. An electronic device for counting
electric pulses.
f compteur m
e contador m
i contatore m
n teller
d Zähler m

1453 COUNTER AMPLIFIER ct, ec
f amplificateur m de compteur
e amplificador m de contador
i amplificatore m di contatore
n tellerversterker
d Zählerverstärker m

1454 COUNTER CHARACTERISTIC ct
 CURVE
The curve of the counting rate of a counter
of the Geiger type, against applied voltage,
all pulses being counted being greater than
a minimum size, determined by the sensitivity of the circuit.
f courbe f caractéristique du compteur
e curva f característica del contador
i curva f caratteristica del contatore
n karakteristieke kromme van de teller
d Zählerkennlinie f

1455 COUNTER-CONTROLLED ct, ic
 CLOUD CHAMBER
A cloud chamber whose expansion is
triggered by a counter or counters.
f chambre f d'ionisation à compteur pilote
e cámara f de niebla con contador regulador
i camera f di nebbia con contatore
 regolatore
n door teller gestuurd nevelvat n
d Nebelkammer f mit Zählersteuerung

1456 COUNTER DEAD TIME ct
The time interval between the start of a
counted event and the earliest instant at
which a new event can be counted by a
radiation counter.
f temps m mort d'un compteur
e tiempo m muerto de un contador
i tempo m morto d'un contatore
n dode tijd van een teller
d Zählertotzeit f

1457 COUNTER EFFICIENCY, ct
 COUNTING EFFICIENCY,
 COUNTING YIELD
The ratio of the average number of photons
or ionizing particles that produce counts
to the average number incident on the
sensitive area of a radiation counter.
f sensibilité f du compteur
e sensibilidad f del contador
i sensibilità f del contatore

n tellergevoeligheid, tellerrendement *n*
d Ansprechvermögen *n*, Zählerausbeute *f*

1458 COUNTER FIELD EMISSION   ct
A radiation counter phenomenon in which
an approaching ionic charge releases an
electron from the counter cathode surface.
f émission *f* froide d'un compteur
e emisión *f* fría de un contador
i emissione *f* fredda d'un contatore
n koude emissie van een teller
d Feldemission *f* eines Zählers

1459 COUNTER FILLING SYSTEM   ct
A system consisting of a gas and/or
vapo(u)r reservoir, a manifold for
attaching counters, and a vacuum system,
used for the evacuation and filling of
the counter tubes.
f système *m* de remplissage d'un tube
   compteur
e sistema *m* de llenado de un tubo contador
i sistema *m* di ripieno d'un tubo contatore
n gasvullingssysteem *n* van een telbuis
d .Gasfüllungssystem *n* eines Zählrohrs.

1460 COUNTER GAS AMPLIFICATION   ct
The ratio of the charge collected to the
charge liberated by the initial ionizing
event.
f amplification *f* due au gaz du tube compteur
e amplificación *f* debida al gas del tubo
   contador
i amplificazione *f* dovuta al gas del tubo
   contatore
n door het gas van de telbuis veroorzaakte
   versterking
d Gasverstärkung *f* des Zählrohrs

1461 COUNTER LAG-TIME,   ct
        COUNTER TIME-LAG,
        STATISTICAL COUNTER TIME-LAG
The time between the occurrence of the
primary ionizing event and the occurrence
of the count in the counter.
f retard *m* du compteur
e retardo *m* del contador
i ritardo *m* del contatore
n tellerachterstand, tellervertraging
d Zählernacheilung *f*, Zählerverzögerung *f*

1462 COUNTER LIFETIME   ct
The number of counts which a counter
is capable of detecting before becoming
useless.
f vie *f* du compteur
e vida *f* del contador
i vita *f* del contatore
n tellerlevensduur
d Zählerlebensdauer *f*

1463 COUNTER OPERATING VOLTAGE   ct
The voltage across a radiation counter,
when operating, measured between the
anode and the cathode.
f tension *f* de régime du compteur
e tensión *f* de funcionamiento del contador
i tensione *f* di funzionamiento del contatore

n tellerbedrijfsspanning
d Zählerbetriebsspannung *f*

1464 COUNTER OVERSHOOTING   ct
Said of a counter when an increase of the
anode voltage is greater than the counter
overvoltage.
f dépassement *m* du compteur
e sobreexcitación *f* del contador
i sovraeccitazione *f* del contatore
n doorschieten *n* van een teller
d Überanregung *f* eines Zählers

1465 COUNTER OVERVOLTAGE   ct
In a radiation counter, the amount by which
the applied voltage exceeds the Geiger-
Müller threshold.
f surtension *f* de compteur
e sobretensión *f* de contador
i sovratensione *f* di contatore
n telleroverspanning
d Zählerüberspannung *f*

1466 COUNTER PLATEAU   ct
The region of the counter characteristic
curve in which the counting rate is
substantially independent of voltage.
f palier *m* de compteur,
   plateau *m* de compteur
e meseta *f* de contador, plato *m* de contador
i ripiano *m* di contatore
n tellerplateau *n*
d Zählerplateau *n*

1467 COUNTER RANGE   ct, rt
The range of reactor power level within
which a particle counter is required for
adequate measurement of the neutron flux
density.
f domaine *m* de comptage
e régimen *m* del contador
i campo *m* del contatore,
   intervallo *m* del contatore
n tellergebied *n*
d Zählerbereich *m*

1468 COUNTER RECOVERY TIME,   ct
        RECOVERY TIME
The time interval, after the initiation of
a count, which must elapse before a counter
is capable of delivering, at the next ionizing
event, a pulse of substantially full size.
f temps *m* de récupération,
   temps *m* de restitution
e tiempo *m* de restitución
i tempo *m* di riassetto
n hersteltijd
d Erholungszeit *f*, innere Totzeit *f*

1469 COUNTER REIGNITION,   ct
        REIGNITION
In radiation counter tubes, a process by
which multiple counts are generated within
a counter tube by atoms or molecules
excited or ionized in the discharge
accompanying a tube count.
f amorçage *m* parasitaire

e encendido *m* parásito
i accensione *f* parassitaria
n parasitaire ontsteking
d Störzündung *f*

1470 COUNTER RESOLUTION TIME, ct
COUNTER RESOLVING TIME,
RESOLUTION TIME,
RESOLVING TIME
The time from the start of a counted
pulse to the instant a succeeding pulse
can assume the minimum strength to be
detected by the counting circuit.
f temps *m* de résolution
e tiempo *m* de resolución
i tempo *m* di risolvenza
n scheidingstijd
d Auflösungszeit *f*

1471 COUNTER STARTING POTENTIAL ct
The voltage which must be applied to a
radiation counter of the Geiger type
to cause it to count.
f tension *f* d'amorçage du compteur
e tensión *f* de cebado del contador
i potenziale *m* d'accensione del contatore
n tellerontsteekspanning
d Zählerzündspannung *f*

1472 COUNTER TUBE ct
Radiation detector consisting of a
gas-filled tube or valve whose gas
amplification factor is much greater than
one, and in which the individual ionizing
events give rise to discrete electrical
pulses.
f tube *m* compteur
e tubo *m* contador
i tubo *m* contatore
n telbuis
d Zählrohr *n*

1473 COUNTER TUBE FAST ma
NEUTRON FLUX-METER
An assembly designed to measure fast
neutron flux density, in which the detector
is a boron counter tube surrounded by a
moderating material.
f fluxmètre *m* de neutrons rapides à tube
compteur
e flujómetro *m* de neutrones rápidos con
tubo contador
i flussometro *m* di neutroni veloci a tubo
contatore
n fluxdichtheidsmeter voor snelle neutronen
met telbuis
d Gerät *n* zur Messung der Fluenz
schneller Elektronen mit Zählrohr

1474 COUNTER TUBE HYSTERESIS ct
Of a radiation counter tube, a temporary
change in the counting rate versus voltage
characteristic caused by previous operation.
f hystérésis *f* de tube compteur
e histéresis *f* de tubo contador
i isteresi *f* di tubo contatore
n telbuishysteresis
d Zählrohrhysteresis *f*

1475 COUNTER TUBE WITH INTERNAL ct
GAS SOURCE
A counter tube in which the filling gas
consists in all or in part of the radioactive
gas, whose activity is to be measured.
f tube *m* compteur à source interne gazeuse
e tubo *m* contador de fuente interna de gas
i tubo *m* contatore a sorgente interna di gas
n telbuis met interne gasvormige bron
d Gasfüllzählrohr *n*,
Zählrohr *n* mit innerer Gasquelle

1476 COUNTERCURRENT is
CENTRIFUGE
A centrifuge in which countercurrent
circulation of a gas is established either·
thermally or mechanically.
f centrifugeur *m* à contre-courant
e centrifuga *f* de contracorriente
i centrifuga *f* a controcorrente
n tegenstroomcentrifuge
d Gegenstromzentrifuge *f*

1477 COUNTERCURRENT ch
DISTILLATION
A distillation method in which use is made
of countercurrent flow.
f distillation *f* à contre-courant
e destilación *f* por contracorriente
i distillazione *f* a controcorrente
n tegenstroomdestillatie
d Gegenstromdestillation *f*

1478 COUNTERCURRENT is
ELECTROMIGRATION
The process used for the separation of the
chlorine isotopes.
f électromigration *f* à contre-courant,
migration *f* ionique à contre-courant
e electromigración *f* de contracorriente,
migración *f* iónica de contracorriente
i elettromigrazione *f* a controcorrente,
migrazione *f* ionica a controcorrente
n ionenmigratie met tegenstroom
d Gegenstromionenmigration *f*

1479 COUNTERCURRENT FLOW, ch
COUNTERFLOW
The flow in opposite directions of two
streams within a system.
f contre-courant *m*
e contracorriente *f*
i controcorrente *f*
n tegenstroom
d Gegenstrom *m*

1480 COUNTERCURRENT ch
PACKED COLUMN
A column used in solvent extraction
processes in which the feed solution enters
at the top and the aqueous raffinate leaves
at the bottom, while the solvent phase runs
in at the bottom and out through the top.
f colonne *f* garnie à contre-courant
e columna *f* atestado de contracorriente
i colonna *f* a riempimento ed a controcorrente
n gevulde kolom met tegenstroom

d Gegenstromfüllkörperkolonne f

**1481 COUNTERCURRENT   ch
PULSED COLUMN**
Column used in solvent extraction
processes, in which increased mixing is
obtained by means of pulses transmitted
by a plunger.
f colonne f à pulsations et à contre-courant
e columna f de pulsaciones y de contracorriente
i colonna f a pulsazioni ed a controcorrente
n gepulsde kolom met tegenstroom
d pulsierte Gegenstromkolonne f

**1482 COUNTERCURRENT   ch
ROTARY COLUMN**
Countercurrent column without packing
used in solvent extraction processes,
carrying down its centre(er) a rapidly
rotating rod which has the effect of dividing
up the column into areas of mixed and
settled solvent-aqueous phases.
f colonne f à rotation et à contre-courant
e columna f de rotación y de contracorriente
i colonna f a rotazione ed a controcorrente
n draaiingskolom met tegenstroom
d Gegenstromdrehungskolonne f

1483 COUNTING   ct
Determination of the number of electrical
impulses delivered by a radiation detector.
f comptage m
e cuenta f
i conteggio m
n tellen n
d Zählen n

**1484 COUNTING CELL,   md
HEMACYTOMETER,
HEMATIMETER**
An apparatus for counting blood corpuscles.
f hématimètre m
e hematímetro m
i ematimetro m
n bloedlichaampjesteller
d Blutkörperchenzählapparat m

1485 COUNTING CHAMBER   md
Component part of a hemacytometer.
f cellule f quadrillée
e cámara f cuentaglóbulos
i spazio m quadrato d'un ematimetro
n telkamer
d Zählkammer f

**1486 COUNTING ERROR,   ct
STATISTICAL COUNTING ERROR**
Error in counting ionizing events, due to
the random time distributions of
disintegrations.
f erreur f de comptage
e error m de recuento
i errore m di conteggio
n telfout
d Zählfehler m

1487 COUNTING GEOMETRY   ct
The dimensions of the samples and of the
distances source-counter between two
sources symmetrically arranged.
f géométrie f de comptage
e geometría f de recuento
i geometria f di conteggio
n tellingsgeometrie
d Zählungsgeometrie f

**1488 COUNTING IONIZATION   ic
CHAMBER,
PULSE IONIZATION CHAMBER**
A type of ionization chamber designed to
detect individually the pulses due to
ionizing particles.
f chambre f d'ionisation à impulsions
e cámara f de ionización de impulsos
i camera f d'ionizzazione ad impulsi
n ionisatievat n met pulstelling
d Impulsionisationskammer f

1489 COUNTING LOSS   ct
Of a pulse counting assembly the reduction
of the counting rate resulting from
phenomena such as the resolving time,
the paralysis time, or the G.M. counter
tube dead time.
f perte f de comptage
e pérdida f de cuenta, pérdida f de recuento
i perdita f di conteggio
n telverlies n
d Zählverlust m

**1490 COUNTING PROSPECTING   ma
RADIATION METER WITH
G.M. COUNTER TUBE**
A portable prospecting assembly containing
its own power source, and designed for
counting the photons detected by means
of one or more Geiger-Müller counter tubes.
f radiamètre m de comptage pour prospection
à tube compteur Geiger-Müller
e radiámetro m contador para exploración
con tubo contador Geiger-Müller
i radiametro m di prospezione a conteggio
a tubo contatore Geiger-Müller
n stralingsmeter met pulstelling voor
prospectie met geigertelbuis
d Lagerstättensuchgerät n mit G.M.-Zählrohr

1491 COUNTING RATE   ct
Number of counts occurring in unit time.
f taux m de comptage
e ritmo m de recuento, velocidad f de cuenta
i velocità f di conteggio
n telsnelheid, teltempo n
d Zählgeschwindigkeit f, Zählrate f

**1492 COUNTING RATE   ct
CHARACTERISTIC,
COUNTING RATE CURVE**
The relation between counting rate and
voltage applied to a counter tube for a
given constant source of radiation.
f caractéristique f de la vitesse de comptage
e característica f de la velocidad de recuento

i caratteristica f della velocità di
conteggio
n telsnelheidskarakteristiek,
teltempokarakteristiek
d Zählrate-Kennlinie f

1493 COUNTING RATEMETER, ct
RATEMETER
An electronic sub-assembly which gives an
indication of the average counting rate in
a given time.
f ictomètre m
e impulsímetro m
i frequenzimetro m statistico,
rateometro m di conteggio
n telsnelheidsmeter, teltempometer
d Zählratenmessgerät n

1494 COUNTRY ROCK, mi
GANGUE, NATIVE ROCK
The valueless rock forming the walls
of a reef or lode.
f gangue f
e ganga f
i ganga f
n ganggesteente n
d Ganggestein n, Gangmasse f

1495 COUPLING np
An interaction between different
properties of a system, or an interaction
between two or more systems, e.g. the
interaction between the lepton field and a
nucleon involved in a beta disintegration
process.
f couplage m
e acoplamiento m
i accoppiamento m
n koppeling
d Kopplung f

1496 COUPLING CONSTANT gp
A constant expressing the strength of a
particular coupling.
f constante f de couplage
e constante f de acoplamiento
i costante f d'accoppiamento
n koppelingsconstante
d Kopplungskonstante f

1497 COUPLING MEDIUM ec
A substance sometimes used between the
photomultiplier surface and the scintillator
or light guide surface to reduce light losses
due to total reflections.
f joint m optique
e junta f óptica
i giunzione f ottica
n optische koppeling
d optische Kupplung f

1498 COVALENCE, ch
COVALENCY
A chemical linkage in which the sharing
of electrons occurs in pairs, each pair
being equivalent to one conventional
chemical bond.

f covalence f
e covalencia f
i covalenza f
n covalentie
d Kovalenz f

1499 COVALENT BOND, ch
HOMOPOLAR BOND
A type of linkage between two atoms,
wherein each atom contributes one electron
to a shared pair that constitutes an
ordinary chemical bond.
f liaison f covalente
e enlace m covalente
i legame m covalente
n covalente binding
d kovalente Bindung f

1500 COVALENT COMPOUND ch
A compound formed by the sharing of
electrons between atoms.
f composé m covalent
e compuesto m covalente
i composto m covalente
n covalente verbinding
d kovalente Verbindung f

1501 CPE, see 1024

1502 CRANE MANIPULATOR, co, rt
RECTILINEAR MANIPULATOR
A manipulator for remote handling of
radioactive materials.
f manipulateur m rectiligne
e manipulador m rectilineo
i manipolatore m rettilineo
n coördinatenmanipulator
d Koordinatenmanipulator m

1503 CRATERING EXPLOSION mi
A proposed technique i.a. for the exploration of the sea-bed for locating minerals.
f explosion f à formation de cratères
e explosión f de craterización
i esplosione f a formazione di crateri
n kratervormende explosie
d kraterbildende Explosion f

1504 CREATION RATE, ec, np
FORMATION RATE,
PRODUCTION RATE
The time rate of creation of electron-hole
pairs.
f taux m de formation d'une paire
e velocidad f de formación de un par
i velocità f di formazione d'una coppia
n paarvormingstempo n
d Paarbildungsgrad m, Paarbildungsrate f

1505 CREEP mg
The plastic deformation of metals that
occurs with time upon application of load.
f fluage m
e fluencia f
i scorrimento m
n kruip
d Kriechen n

1506 CRIB   rw
Open work box buried in the ground from
which under a given set of conditions is
soil.
f caisse f à clairevoie
e jaula f
i cassa f a trafori
n krat
d Senkkiste f, Sickerkiste f

1507 CRIT   rt
The mass of fissile (fissionable) material
which under a given set of conditions is
critical.
f masse f critique spécifique
e masa f crítico específica
i massa f critica specifica
n specifieke kritieke massa,
   specifieke kritische massa
d spezifische kritische Masse f

1508 CRITERION OF DEGENERACY   np
The criterion defined by the value of
$Ah^3/2m^3$ which determines the degeneracy
of a system of particles.
f critère m de dégénérescence
e criterio m de degenerescencia
i criterio m di degenerescenza
n ontaardingscriterium n
d Entartungskriterium n

1509 CRITICAL   rt
Fulfilling the condition that a nuclear
chain reacting medium has an effective
multiplication constant equal to unity.
f critique adj
e crítico adj
i critico adj
n kritiek adj, kritisch adj
d kritisch adj

1510 CRITICAL ABSORPTION   ra
      WAVELENGTH
The wavelength, characteristic of a given
electron energy level in an atom of a
specified element, at which an absorption
discontinuity occurs.
f longueur f d'onde critique d'absorption
e longitud f de onda crítica de absorción
i lunghezza f d'onda critica d'assorbimento
n kritieke absorptiegolflengte,
   kritische absorptiegolflengte
d kritische Absorptionswellenlänge f

1511 CRITICAL ASSEMBLY   rt
An assembly of fissile (fissionable)
material plus moderator which is capable
of maintaining a fission chain reaction at
very low power level.
f ensemble m critique
e conjunto m crítico
i complesso m critico
n kritieke opstelling, kritische opstelling
d kritische Anordnung f

1512 CRITICAL CONCENTRATION   rt
Concentration for which a homogeneous
solution of fissile (fissionable) material

becomes critical in an infinite medium.
f concentration f critique
e concentración f crítica
i concentrazione f critica
n kritieke concentratie, kritische concentratie
d kritische Konzentration f

1513 CRITICAL CONDITION   np, rt
The condition of a nuclear reactor in
which the reaction can just support itself.
f condition f critique
e condición f crítica
i condizione f critica
n kritieke toestand, kritische toestand
d kritischer Zustand m

1514 CRITICAL DIMENSIONS,   rt
      CRITICAL SIZE
The minimum physical dimensions of a
reactor core or an assembly which can
be made critical for a specified geometrical
arrangement and material composition.
f taille f critique
e dimensiones pl críticas
i dimensioni pl critiche
n kritieke afmetingen pl,
   kritische afmetingen pl
d kritische Abmessungen pl,
   kritische Grösse f

1515 CRITICAL EQUATION,   ma
      PILE EQUATION
Any equation relating parameters of an
assembly which must be satisfied for the
assembly to be critical.
f équation f critique
e ecuación f crítica
i equazione f di criticità
n kriticiteitsvergelijking
d kritische Gleichung f

1516 CRITICAL EXPERIMENT   rt
A test or series of tests performed with
an assembly of reactor materials which
can be gradually brought to the critical
state for the purpose of determining the
nuclear characteristics of a reactor.
f expérience f critique
e experimento m crítico
i esperimento m di criticità
n kritiek experiment n, kritisch experiment n
d kritisches Experiment n

1517 CRITICAL FACILITY   rt
A facility where critical experiments are
conducted.
f salle f d'expériences critiques
e sala f de experimentos críticos
i sala f d'esperimenti critici
n kritieke-proefruimte,
   kritische-proefruimte
d kritischer Versuchsraum m

1518 CRITICAL FIELD OF   ct
      A COUNTER TUBE
The minimum electric field strength
necessary for gas multiplication to be
initiated.

f   champ *m* critique d'un tube compteur
e   campo *m* crítico de un tubo contador
i   campo *m* critico d'un tubo contatore
n   kritieke veldsterkte van een telbuis,
    kritische veldsterkte van een telbuis
d   kritische Feldstärke *f* eines Zählrohrs

1519 CRITICAL IONIZATION POTENTIAL np
A measure of the quantity of energy per
unit charge required to move an electron
from the lowest energy level of a normal
atom, to a sufficient distance that the
atom remains positively ionized.
f   potentiel *m* critique d'ionisation
e   potencial *m* crítico de ionización
i   potenziale *m* critico d'ionizzazione
n   kritieke ionisatiespanning,
    kritische ionisatiespanning
d   kritische Ionisierungsspannung *f*

1520 CRITICAL MASS rt
The minimum mass of fissile(fissionable)
material which will sustain a nuclear chain
reaction for a specified geometrical
arrangement and material composition.
f   masse *f* critique
e   masa *f* crítica
i   massa *f* critica
n   kritieke massa, kritische massa
d   kritische Masse *f*

1521 CRITICAL MOISTURE CONTENT cp
In the drying of solids, the moisture
content at the end of the constant rate
drying period and the beginning of the
falling rate drying period.
f   humidité *f* critique
e   humedad *f* crítica
i   umidità *f* critica
n   kritieke vochtigheid, kritische vochtigheid
d   kritische Feuchtigkeit *f*

1522 CRITICAL ORGAN xr
That part of the body or of an animal
deemed to be particularly liable to injury
by irradiation of a given form.
f   organe *m* critique
e   órgano *m* crítico
i   organo *m* critico
n   gevoelig orgaan *n*, kritisch orgaan *n*
d   kritisches Organ *n*,
    strahlungsempfindliches Organ *n*

1523 CRITICAL PLASMA pp
    TEMPERATURE,
    IGNITION TEMPERATURE
In a plasma of given composition, the
temperature at which the power lost by
radiative collisions is balanced by the
power deposited in the plasma by thermonuclear reactions.
f   température *f* d'ignition
e   temperatura *f* de encendido,
    temperatura *f* de ignición
i   temperatura *f* d'accensione
n   ontstekingstemperatuur
d   Zündtemperatur *f*

1524 CRITICAL POINT ch, gp
A point where two phases, which are
continually approximating each other,
become identical and form but one phase.
f   point *m* critique
e   punto *m* crítico
i   punto *m* critico
n   kritiek punt *n*, kritisch punt *n*
d   kritischer Punkt *m*

1525 CRITICAL POTENTIAL np
A measure of the energy required to raise
the energy level of an orbital electron to
a higher energy band in the atom.
f   potentiel *m* critique
e   potencial *m* crítico
i   potenziale *m* critico
n   kritieke potentiaal, kritische potentiaal
d   kritisches Potential *n*

1526 CRITICAL REACTOR rt
A nuclear reactor in the critical state.
f   réacteur *m* critique
e   reactor *m* crítico
i   reattore *m* critico
n   kritieke reactor, kritische reactor
d   kritischer Reaktor *m*

1527 CRITICAL SHEAR STRESS mg
The resolved shear stress which is required
to initiate slip in a given crystallographic
direction along a given crystallographic
plane of a single crystal of a metal.
f   effort *m* critique de cisaillement,
    résistance *f* critique au cisaillement
e   resistencia *f* crítica al empuje
i   sollecitazione *f* critica di taglio
n   kritieke schuifspanning,
    kritische schuifspanning
d   kritische Schubfestigkeit *f*

1528 CRITICAL STATE rt
The state of a reactor system in which
just as many neutrons are being produced
as are lost by absorption and leakage.
f   état *m* critique
e   estado *m* crítico
i   stato *m* critico
n   kritieke toestand, kritische toestand
d   kritischer Zustand *m*

1529 CRITICAL THICKNESS OF rt
    AN INFINITE SLAB
In a reactor, the thickness which makes
the geometric buckling equal to the
material buckling.
f   épaisseur *f* critique d'une plaque infinie
e   espesor *m* crítico de una placa infinita
i   spessore *m* critico d'una piastra infinita
n   kritieke dikte van een oneindige plaat,
    kritische dikte van een oneindige plaat
d   kritische Dicke *f* einer unendlichen Platte

1530 CRITICAL VELOCITY ch, gp
That average linear velocity below which
a given fluid, at a given temperature and
pressure, flowing in a given apparatus,

will move in stream line flow and above
which the motion is usually turbulent.
f vitesse *f* critique
e velocidad *f* crítica
i velocità *f* critica
n kritieke snelheid, kritische snelheid
d kritische Geschwindigkeit *f*

1531 CRITICALITY rt
The condition of being critical.
f criticité *f*
e criticidad *f*
i criticità *f*
n kriticiteit
d Kritikalität *f*, Kritizität *f*

1532 CRITICALITY MONITOR me, sa
A monitor designed to measure a quantity
connected with a possible criticality
accident and to give a warning when it
exceeds a predetermined value.
f moniteur *m* de criticité
e monitor *m* de criticidad
i monitore *m* di criticità
n kriticiteitsmonitor
d Kritikalitätswarngerät *n*,
Kritizitätsmonitor *m*

1533 CRITICALITY MONITOR BASED me, sa
ON GAMMA RADIATION
f moniteur *m* gamma de criticité
e monitor *m* gamma de criticidad
i monitore *m* gamma di criticità
n gammakriticiteitsmonitor
d Kritikalitätswarngerät *n* mit Nachweis
der Gammastrahlung,
Kritizitätswarngerät *n* mit Nachweis
der Gammastrahlung

1534 CRITICALITY MONITOR me, sa
BASED ON NEUTRON RADIATION
f moniteur *m* neutronique de criticité
e monitor *m* neutrónico de criticidad
i monitore *m* neutronico di criticità
n neutronkriticiteitsmonitor
d Kritikalitätswarngerät *n* mit Nachweis
von Neutronen,
Kritizitätswarngerät *n* mit Nachweis
von Neutronen

1535 CROSS-BOMBARDMENT np
A method for assigning the mass to a
radioactive nuclide by producing it in
different nuclear reactions.
f bombardements *pl* croisés,
recoupement *m* par bombardement
e bombardeos *pl* cruzados
i bombardamenti *pl* incrociati
n kruisvuur *n*
d gekreuzte Kernreaktionen *pl*,
Kreuzbeschuss *m*

1536 CROSS-CONTAMINATION rw
The spread of certain isotopes to systems
where they interfere with the measurement
or handling of other isotopes, e.g.
plutonium in an intended uranium solution.

f contamination *f* croisée
e contaminación *f* cruzada
i contaminazione *f* incrociata
n kruisbesmetting
d Kreuzkontamination *f*, Kreuzverseuchung *f*

1537 CROSS-DRIFT np
A drift in electron and ion orbits which
produces a mass motion with no electric
current.
f mouvement *m* transversal
e movimiento *m* transversal
i movimento *m* trasversale
n dwarsdrift
d Querdrift *f*

1538 CROSS-FIRING xr
A radiation-therapeutic technique by which
a lesion is subjected to radiation entering
the body through several portals.
f feux *pl* croisés
e destellos *pl* cruzados
i fuochi *pl* incrociati
n kruisvuur *n*
d Kreuzfeuerbestrahlung *f*

1539 CROSS-LINKING ch
The formation of additional links between
the chains of atoms in polymerized
material.
f formation *f* d'une structure moléculaire
en réseau,
réticulation *f*
e enlace *m* cruzado, entrelazamiento *m*
i legame *m* in forma di rete, reticolazione *f*
n netvorming
d Verknüpfung *f*, Vernetzung *f*

1540 CROSS-OVER TRANSITION, ra
CROSSING OVER
In atomic or nuclear spectroscopy, a
radiative transition between two states
may be referred to as a cross-over
transition in comparison with the two-step
transition through an intermediate energy
level.
f transition *f* radiative
e transición *f* radiactiva
i transizione *f* radiativa
n stralingsovergang
d Strahlungsübergang *m*

1541 CROSS SECTION cs
A measure of the probability of a specified
interaction between an incident radiation
and a target particle or system of
particles.
f section *f* efficace
e sección *f* eficaz
i sezione *f* d'urto, sezione *f* efficace
n doorsnede, werkzame doorsnede,
d Wirkungsquerschnitt *m*

1542 CROSS SECTION DENSITY, cs
MACROSCOPIC CROSS SECTION
The cross section per unit volume of a
given material for a specified process.

f   section f efficace macroscopique,
    section f efficace volumique
e   sección f eficaz macroscópica
i   sezione f d'urto macroscopica
n   macroscopische doorsnede
d   makroskopischer Querschnitt m

1543 CROSSING OVER   md
   The mutual exchange, during prophase
   of meiosis, of corresponding segments
   of homologous chromatids.
f   entrecroisement m chromosomique
e   cruzamiento m cromosómico
i   incrociamento m cromosomico
n   crossing-over, overkruisen n
d   Crossing-over n, Überkreuzung f

1544 CROSSING OVER   np
   The order of single particle levels in the
   mean nuclear potential may change from
   nucleus to nucleus, due to changes in the
   relative importance of the surface.
f   échange m de niveau
e   cambio m de nivel
i   scambio m di livello
n   niveauverwisseling
d   Niveauaustausch m

1545 CROWBAR, see 1110

1546 CROWDION   cr
   A crystal defect due to the crowding of
   neighbo(u)ring ions along the row in which
   an interstitial ion is placed.
f   presse f d'ions
e   hacinamiento m de iones
i   calca f d'ioni
n   ionengedrang n
d   Ionengedränge n

1547 CRUD   ch
   Undesirable solid material of uncertain
   composition arising in chemical processes.
f   infection f
e   materia f extrañea
i   feccia f
n   smurrie
d   Fremdstoff m

1548 CRUDE ORE   mi
   Ore as it is found in natural deposits.
f   minerai m cru, minerai m tout-venant
e   mineral m complejo, mineral m tosco
i   minerale m grezzo
n   ruw erts n
d   Roherz n

1549 CRUSHING   mi
   Ore treatment in order to facilitate the
   recovery of the metal contained.
f   broyage m, concassage m
e   machaqueo m, trituración f
i   frantumazione f, triturazione f
n   breken n
d   Zerkleinerung f

1550 CRYOGENIC COIL   pp
   Coil working at very low temperatures
   used in plasma research.
f   bobine f en circuit cryogène
e   bobina f en circuito criógeno
i   bobina f in circuito criogeno
n   spoel in cryogeen circuit
d   Spule f in kryogenem Kreis

1551 CRYSTAL ANALYSIS,   cr
      RADIOCRYSTALLOGRAPHY
   A study of the structure of crytals by the
   processes of diffraction of X-rays,
   electrons, neutrons,etc.
f   radiocristallographie f
e   radiocristalografía f
i   analisi f cristallografica,
    radiocristallografia f
n   röntgenkristallografie
d   Kristallanalyse f, Röntgenstrukturanalyse f

1552 CRYSTAL COUNTER,   ct
      DIAMOND COUNTER
   In nucleonics, a counter utilizing one of
   several known crystals that are rendered
   momentarily conducting by ionizing events.
f   compteur m à cristaux
e   contador m de cristales
i   contatore m a cristalli
n   kristalteller
d   Kristallzähler m

1553 CRYSTAL EFFECTS   cr, np
   In nucleonics, dependence of the microscopic neutron cross section of the
   material upon the crystalline structure of
   the material.
f   influence f exercée sur le réseau
    cristallin
e   influencia f sobre la red cristalina
i   influenzamento m del reticolo cristallino
n   kristalroosterbeïnvloeding
d   Kristalgitterbeeinflüssung f

1554 CRYSTAL GROWTH   cr
   Growth of a crystal proceeds only if
   there is a screw dislocation present; then
   it proceeds in a spiral fashion by the
   accretion of atoms at the edge of growth
   steps.
f   croissance f des cristaux
e   crecimiento m de cristales
i   crescita f di cristalli
n   kristalgroei
d   Kristallwachstum n

1555 CRYSTAL IMPERFECTION,   cr
      IMPERFECTION
   Any deviation in structure from that of an
   ideal crystal.
f   défaut m de cristal
e   defecto m de cristal
i   difetto m di cristallo
n   kristalfout, structuurafwijking
d   Baufehler m, Strukturfehler m

1556 CRYSTAL IMPURITY, see 1052

1557 CRYSTAL LATTICE, see 771

1558 CRYSTAL SCINTILLATOR   ct
A scintillator consisting of an inorganic
or organic crystal.
f scintillateur *m* cristallin
e centelleador *m* cristalino
i scintillatore *m* a cristallo
n kristalscintillator
d Kristalszintillator *m*

1559 CRYSTAL SPECTROGRAPH   sp
In crystallography, an instrument used to
photograph spectra produced by radiation
transmitted through a crystal.
f spectrographe *m* à cristal
e espectrógrafo *m* de cristal
i spettrografo *m* a cristallo
n kristalspectrograaf
d Kristallspektrograph *m*

1560 CRYSTAL SPECTROMETER, see 764

1561 CRYSTAL STRUCTURE   cr
The internal structure of a particular
crystal.
f structure *f* cristalline
e estructura *f* cristalina
i struttura *f* cristallina
n kristalstructuur
d Kristallstruktur *f*

1562 CRYSTALLINE MATERIAL   cr
f matière *f* cristalline
e materia *f* cristalina
i materia *f* cristallina
n kristallijn materiaal *n*
d kristallines Material *n*

1563 CRYSTALLITE,   cr
      GRAIN
The individual crystal in a polycrystalline
substance.
f cristallite *m*, grain *m*
e cristalito *m*, grano *m*
i cristallito *m*, grano *m*
n kristalliet *n*
d Korn *n*, Kristallit *n*

1564 CUBE ROOT LAW   ma, nw
A scaling law applicable to many blast
phenomena.
f loi *f* de racine cubique
e ley *f* de raíz cúbica
i legge *f* di radica cubica
n wet van de derdemachtswortel
d Kubikwurzelgesetz *n*

1565 CUBIC LATTICE   cr
The lattice of a crystal belonging to the
cubic or regular system.
f réseau *m* cubique
e red *f* cúbica
i reticolo *m* cubico
n kubisch rooster *n*
d kubisches Gitter *n*

1566 CUBICAL REACTOR   rt
A nuclear reactor in which the core
approximately has the shape of a cube.
f réacteur *m* cubique
e reactor *m* cúbico
i reattore *m* cubico
n kubische reactor
d kubischer Reaktor *m*

1567 CUMULATIVE ABSORBED   ra, xr
      DOSE,
      CUMULATIVE DOSE
The total dose resulting from repeated
exposures to radiations of the same region
or of the whole body.
f dose *f* absorbée cumulée
e dosis *f* absorbida acumulada
i dose *f* assorbita unica equivalente,
   dose *f* totale assorbita
n cumulatieve geabsorbeerde dosis
d akkumulierte Energiedosis *f*

1568 CUMULATIVE EXCITATION   np
The process by which an atom is raised
by collision from one excited state to
higher states.
f excitation *f* cumulative
e excitación *f* acumulativa
i eccitazione *f* cumulativa
n cumulatieve aanslag
d kumulative Anregung *f*

1569 CUMULATIVE FISSION YIELD   np
The fraction of fissions which have
resulted in the production of a given
nuclide either directly or indirectly up
to a specified time.
f rendement *m* de fission cumulé
e rendimiento *m* de fisión acumulativo
i resa *f* di fissione cumulativa
n cumulatieve splijtingsopbrengst
d kumulative Spaltausbeute *f*

1570 CUMULATIVE IONIZATION   np
Of a gas or vapo(u)r: a rapidly growing
ionization due to a first electron ionizing
by successive collisions a number of
atoms (or molecules), the electrons thus
freed ionizing further atoms (or molecules)
in their turn.
f ionisation *f* cumulative
e ionización *f* acumulativa
i ionizzazione *f* cumulativa
n cumulatieve ionisatie
d lawinenartige Ionisation *f*

1571 CUPROAUTUNITE   mi
A calcium-uranium compound.
f cuproautunite *f*
e cuproautunita *f*
i cuproautunite *f*
n cuproautuniet *n*
d Cuproautunit *m*

1572 CUPROSKLODOWSKITE   mi
A copper-uranium material, containing
about 54.1 % of U.

f cuprosklodowskite *f*
e cuprosklodowskita *f*
i cuprosklodowskite *f*
n cuprosklodowskiet *n*
d Cuprosklodowskit *m*

1573 CUPROZIPPEITE mi
A uranium-containing ore.
f cupro-zippéite *f*
e cuprozippeita *f*
i cuprozippeite *f*
n cuprozippeïet *n*
d Cuprozippeit *m*

1574 CURIE un
The unit of activity, defined as 3.7×10^{10} disintegrations per second exactly.
f curie *m*
e curie *m*
i curie *m*
n curie
d Curie *n*

1575 CURIETHERAPY, md, ra
RADIO-ISOTOPE THERAPY
Treatment of diseases by use of radioactive nuclides.
f curiethérapie *f*,
thérapie *f* par radionucléides
e curieterapia *f*,
terapia *f* por isótopos radiactivos
i curieterapia *f*,
terapia *f* con radioisotopi
n curietherapie, isotopentherapie
d Curietherapie *f*,
Therapie *f* mit Radioisotopen

1576 CURITE mi
A mineral containing lead and uranium.
f curite *f*
e curita *f*
i curite *f*
n curiet *n*
d Curit *m*, Lebererz *n*

1577 CURIUM ch
Transuranic element, symbol Cm, atomic number 96.
f curium *m*
e curio *m*
i curio *m*
n curium *n*
d Curium *n*

1578 CURRENT DENSITY OF ma
PARTICLES,
PARTICLE CURRENT DENSITY
A vector quantity the integral of whose normal component over any surface is equal to the current of particles through that surface.
f densité *f* de courant de particules
e densidad *f* de corriente de partículas
i densità *f* di corrente di particelle
n stroomdichtheid van deeltjes
d Teilchenstromdichte *f*

1579 CURRENT DENSITY OF ma
QUANTA,
QUANTA CURRENT DENSITY
A vector quantity the integral of whose normal component over any surface is equal to the current of quanta through that surface.
f densité *f* de courant de quanta
e densidad *f* de corriente de cuantos
i densità *f* di corrente di quanti
n stroomdichtheid van quanten
d Quantenstromdichte *f*

1580 CURRENT IONIZATION CHAMBER ic
An ionization chamber used in such a manner that the average value of the ionization current in the chamber is measured.
f chambre *f* d'ionisation à courant
e cámara *f* de ionización de corriente
i camera *f* d'ionizzazione a corrente
n ionisatievat *n* met stroommeting
d Ionisationskammer *f* mit Strommessung, Stromionisationskammer *f*

1581 CURRENT PULSE AMPLIFIER ec
Electrical pulse amplifier, designed to supply an output signal as a function of the current through an associated detector.
f amplificateur *m* à impulsion de courant
e amplificador *m* de impulso de corriente
i amplificatore *m* ad impulso di corrente
n stroomversterker
d stromempfindlicher Pulsverstärker *m*

1582 CURTAIN, rt
NEUTRON CURTAIN, STRIP
A thin shield, interposed in a nuclear reactor to shut off a flow of slow neutrons.
f bande *f*
e hoja *f*
i striscia *f*
n strip
d Streifen *m*

1583 CUSP INJECTION EXPERIMENT pp
An experiment in which two conical thetatrons are used injecting plasma blobs simultaneously towards each other and into steady magnetic fields which guide the plasma into a central cusp-shaped magnetic cage.
f expérience *f* d'injection à cuspide
e experimento *m* de inyección de cúspide
i esperimento *m* d'iniezione a cuspide
n cuspidaal injectie-experiment *n*
d Cusp-Injektionsexperiment *n*

1584 CUSP MIRROR pp
Instrument used in plasma containment experiments.
f miroir *m* en forme de cuspide
e espejo *m* en forma de cúspide
i specchio *m* in forma di cuspide
n cuspidale spiegel
d Cusp-Spiegel *m*

1585 CUSP THERMONUCLEAR   rt
        REACTOR,
           PICKET FENCE THERMONUCLEAR
           REACTOR
    A controlled thermonuclear device with
    a plasma configuration in such a way that
    the device is theoretically stable against
    hydromagnetic instabilities.
f   réacteur *m* thermonucléaire à plasma en
    forme de cuspide
e   reactor *m* termonuclear con plasma en
    forma de cúspide
i   reattore *m* termonucleare con plasma in
    forma di cuspide
n   thermonucleaire reactor met cuspidaal
    plasma
d   thermonuklearer Cusp-Reaktor *m*

1586 CUSPED GEOMETRY   pp
    Elementary shape of the configuration of
    the containing magnetic field in thermo-
    nuclear research.
f   configuration *f* cuspidée
e   geometría *f* de forma de cúspide
i   geometria *f* in forma di cuspide
n   cuspidale geometrie
d   Cusp-Geometrie *f*

1587 CUT   is
    In isotope separation, the fraction of the
    feed into the separative element which is
    either removed as product or advanced to
    the next separative element in the direction
    of the product.
f   coupe *f*
e   corte *m*
i   taglio *m*
n   aftap
d   Schnitt *m*

1588 CUT-OFF ENERGY   np
    For a specific absorbing cover surrounding
    a given detector in a given experimental
    configuration, that energy value which
    satisfies the condition that if the cover
    were replaced by a hypothetical cover
    black to neutrons with energy below this
    value and transparent to neutrons with
    energy above this value, the observed
    detector response would be unchanged.
f   énergie *f* de coupure, seuil *m* d'énergie
e   energía *f* de corte,
    umbral *m* de energía
i   energia *f* di taglio, soglia *f* d'energia
n   grensenergie
d   Grenzenergie *f*

1589 CUTANEOUS DROPSY,   md
        EDEMA, OEDEMA
    Presence of abnormally large amounts
    of fluid in the intercellular tissue spaces
    of the body or part of the body.
f   oedème *m*
e   edema *m*, hidropesia *f* cutánea
i   edema *m*
n   oedeem *n*
d   Gewebswassersucht *f*, Ödem *n*

1590 CUTIE PIE, see 491

1591 CYCLE   ge
    A series of changes executed in orderly
    sequence, by means of which a mechanism,
    a working substance, or a system is caused
    periodically to return to the same initial
    conditions, constitutes a cycle.
f   cycle *m*
e   ciclo *m*
i   ciclo *m*
n   cyclus, kringloop
d   Kreislauf *m*, Zyklus *m*

1592 CYCLING   rt
    Periodic change of power of a nuclear
    reactor.
f   cyclage *m*
e   ciclaje *m*
i   variazione *f* periodica di potenza
n   periodieke vermogensvariatie
d   periodische Leistungsvariation *f*

1593 CYCLING (US),   co
        HUNTING (GB)
    A periodic change of the controlled variable.
f   pompage *m*
e   bombeo *m*, pendulación *f*
i   pendolazione *f*
n   slingeren *n*
d   Pendelung *f*

1594 CYCLOTRON   pa
    A device for accelerating charged
    particles to high energies by means of an
    alternating current placed in a constant
    magnetic field.
f   accélérateur *m* magnétique à résonance,
    cyclotron *m*
e   ciclotrón *m*
i   ciclotrone *m*
n   cyclotron *n*
d   Cyclotron *n*, Zyklotron *n*

1595 CYCLOTRON ANGULAR   ma
        FREQUENCY
    $\omega_c = \frac{q}{m}$. B, where $\frac{q}{m}$ is the charge to mass
    ratio of the particle and B is the magnetic
    flux density.
f   fréquence *f* angulaire de cyclotron
e   frecuencia *f* angular de ciclotrón
i   frequenza *f* angolare di ciclotrone
n   cyclotronfrequentie
d   Zyklotronkreisfrequenz *f*,
    Zyklotronwinkelfrequenz *f*

1596 CYCLOTRON ANGULAR   ma
        PRECESSION FREQUENCY,
        NUCLEAR ANGULAR PRECESSION
        FREQUENCY
    $\omega_n = g \frac{e}{2m_p}$ B, where g is the g-factor and
    B is the magnetic flux density.
f   fréquence *f* angulaire de précession
    nucléaire
e   frecuencia *f* angular de precesión nuclear

i frequenza *f* angolare di precessione nucleare
n hoekfrequentie voor cyclotronprecessie, hoekfrequentie voor kernprecessie
d Kreisfrequenz *f* für Zyklotronpräzession, Winkelfrequenz *f* für Zyklotronpräzession

1597 CYCLOTRON FREQUENCY pa
The frequency at which an electron traverses an orbit in a steady, uniform, magnetic field and zero field.
f fréquence *f* de cyclotron
e frecuencia *f* de ciclotrón
i frequenza *f* di ciclotrone
n cyclotronfrequentie
d Zyklotronfrequenz *f*

1598 CYCLOTRON RADIATION pa, ra
Radiation emitted by a charged particle in a magnetic field.
f rayonnement *m* cyclotron
e radiación *f* ciclotrón
i radiazione *f* ciclotrone
n cyclotronstraling
d Zyklotronstrahlung *f*

1599 CYCLOTRON RESONANCE, ec, pp
GYROMAGNETIC RESONANCE
In a plasma, submitted to a uniform magnetic field and to an alternating and uniform electric field, a phenomenon which occurs when the electric field frequency is equal to the gyromagnetic frequency of the electrons or ions.
f résonance *f* cyclotron, résonance *f* gyromagnétique
e resonancia *f* giromagnética
i risonanza *f* giromagnetica
n gyromagnetische resonantie
d gyromagnetische Resonanz *f*

1600 CYCLOTRON RESONANCE pp
HEATING
A method of heating the plasma in the stellarator thermonuclear device based upon the unique rotational frequency of a deuteron in a magnetic field.
f chauffage *m* par la résonance du cyclotron
e calentamiento *m* por la resonancia del ciclotrón
i riscaldamento *m* per la risonanza del ciclotrone
n verhitting door cyclotronresonantie
d Heizung *f* durch Zyklotronresonanz

1601 CYLINDRICAL COUNTER ct
CHAMBER,
CYLINDRICAL RADIATION COUNTER
A counter chamber consisting of a cylinder acting as one electrode and fine wire coaxial with the cylinder acting as the other electrode.
f compteur *m* cylindrique
e contador *m* cilíndrico
i contatore *m* cilindrico
n cilindrische teller
d Zylinderzähler *m*

1602 CYLINDRICAL REACTOR rt
A nuclear reactor in which the core is of cylindrical shape and has a nearly circular cross section.
f réacteur *m* cylindrique
e reactor *m* cilíndrico
i reattore *m* cilindrico
n cilindrische reactor
d zylindrischer Reaktor *m*

1603 CYRTOLITE mi
A zirconium orthosilicate.
f cyrtolite *f*
e cirtolita *f*
i cirtolite *f*
n cyrtoliet *n*
d Cyrtolit *m*

1604 CYSTAMINE ch
An example of a chemical protector.
f cystéamine *f*
e quistamina *f*
i cistamina *f*
n cystamine *n*
d Zystamin *n*

1605 CYSTOGRAPHY xr
Radiological examination of the bladder by means of a contrast medium.
f cystographie *f*
e cistografía *f*
i cistografia *f*
n cystografie
d Harnblasendarstellung *f*

1606 CYTOPLASM, md
KARYOPLASM, NUCLEOPLASM
The protoplasma of a cell, exclusive of the nucleus.
f cytoplasma *m*, protoplasma *m*
e citoplasma *m*, protoplasma *m*
i citoplasma *m*, protoplasma *m*
n cytoplasma *n*, protoplasma *n*
d Karyoplasma *n*, Protoplasma *n*, Zellplasma *n*

1607 CYTOPLASMIC INHERITANCE md
Inheritance not controlled by the genes in the nucleus, but by the cytoplasm.
f hérédité *f* protoplasmique, héritage *m* protoplasmique
e heredabilidad *f* protoplásmica, herencia *f* protoplásmica
i eredità *f* protoplasmica, ereditarietà *f* protoplasmica
n protoplasmatische overerving
d plasmatische Vererbung *f*.

D

1608 d ELECTRON   np
An electron having an orbital momentum
quantum number of two.
f électron *m* d
e electrón *m* d
i elettrone *m* d
n d-elektron *n*
d d-Elektron *n*

1609 DACRYOCYSTOGRAPHY   xr
The radiological examination of the
canilucili, lachrymal sac and nasal duct
following direct injection of a contrast
medium.
f dacryocystographie *f*
e dacriocistografía *f*
i dacriocistografia *f*
n dacryocystografie
d Dacryocystographie *f*

1610 DAMAGE CRITERIA   sa
Standards or measures used in estimating
specific levels of damage.
f critériums *pl* de lésion
e criterios *pl* de lesión
i criteri *pl* di lesione
n schademaatstaven *pl*
d Schadenmassstäbe *pl*

1611 DANGER COEFFICIENT   rt
Of a substance, for a particular reactor,
the change in reactivity caused by inserting
that substance in the reactor.
f coefficient *m* de danger,
   coefficient *m* d'empoisonnement
e coeficiente *m* de envenenamiento,
   coeficiente *m* de peligro
i coefficiente *m* d'avvelenamento,
   coefficiente *m* di pericolo
n gevaarscoëfficiënt, vergiftigingscoëfficiënt
d Gefährdungskoeffizient *m*,
   Vergiftungskoeffizient *m*

1612 DANGER RANGE   sa
The distance from a radiation source
beyond which the exposure dose is inferior
to the prescribed value of 100 mr.
f distance *f* de danger
e distancia *f* de peligro
i distanza *f* di pericolo
n gevaarsafstand
d Gefährdungsabstand *m*

1613 DAPEX PROCESS,   ch
       DIALKYL PHOSPHORIC ACID PROCESS
A solvent extraction process using dialkyl
phosphoric acid as a solvent.
f procédé *m* à l'acide dialcoylphosphorique
e procedimiento *m* con ácido dialquilfosfórico
i processo *m* con acido dialchilfosforico
n dialkylfosforzuurproces *n*
d Dialkylphosphorsäureverfahren *n*

1614 DARK CONDUCTION   ec
Residual conduction in a photosensitive
substance that is not illuminated.
f conduction *f* d'obscurité
e conducción *f* de obscuridad
i conduzione *f* d'oscurità
n donkergeleiding
d Dunkelleitung *f*

1615 DARK CURRENT   ec
The current flowing in the external circuit
of a photoelectronic cell in the absence of
irradiation.
f courant *m* d'obscurité
e corriente *f* de obscuridad
i corrente *f* d'oscurità
n donkerstroom
d Dunkelstrom *m*

1616 DARK-CURRENT PULSE   ec
A phototube dark-current excursion that
can be resolved by the system employing
the photocathode.
f impulsion *f* de courant d'obscurité
e impulsión *f* de corriente de obscuridad
i impulso *m* di corrente d'oscurità
n donkerstroompuls
d Dunkelstromimpuls *m*

1617 DARK RESISTANCE   ec
The resistance of a photoelectric device in
total darkness.
f résistance *f* d'obscurité
e resistencia *f* de obscuridad
i resistenza *f* d'oscurità
n donkerweerstand
d Dunkelwiderstand *m*

1618 DATING   np
The determination of the radioactive
age of an object from its content of
radioactive substances.
f datation *f*, détermination *f* de l'âge
e datación *f*
i datazione *f*
n leeftijdsbepaling
d Altersbestimmung *f*, Datierung *f*

1619 DAUGHTER PRODUCT   np
Any nuclide which follows a specified
radionuclide in a decay chain.
f descendant *m* radioactif,
   produit *m* de filiation
e hija *f* radiactiva, hijo *m* radiactivo
i figlio *m* radioattivo
n dochternuclide *n*
d Folgenuklid *n*, Tochternuklid *n*,
   Tochterprodukt *n*

1620 DAVIDITE   mi
A mixture of ilmenite, carnotite and
chevkenite.

f davidite *f*
e davidita *f*
i davidite *f*
n davidiet *n*
d Davidit *m*

1621 DE BROGLIE ATOM   np
Similar in all respects to the Bohr model
except that the quantum condition on the
angular momentum electron, mvr = nh,
results not as an ad hoc assumption but
as a result of imposing a condition on the
wavelength of the de Broglie wave
associated with the electron.
f atome *m* de de Broglie
e átomo *m* de de Broglie
i atomo *m* di de Broglie
n atoommodel *n* volgens de Broglie
d de Broglie-Atommodell *n*

1622 DE BROGLIE HYPOTHESIS,   ma
DE BROGLIE RELATION
The expression $\lambda = h/p$ for the de Broglie
wavelength λ ascribed by wave or
quantum mechanics to any particle
having momentum p.
f relation *f* de de Broglie
e ecuación *f* de de Broglie
i equazione *f* di de Broglie
n hypothese van de Broglie
d de Broglie-Beziehung *f*

1623 DE BROGLIE WAVELENGTH   np
The length of the waves which, according
to de Broglie, are associated with a
moving particle.
f longueur *f* d'onde de de Broglie,
longueur *f* d'onde pilote
e longitud *f* de onda de de Broglie
i lunghezza *f* d'onda di de Broglie
n golflengte van de Broglie
d de Broglie-Wellenlänge *f*

1624 DEAD BAND (US),   co
DEAD ZONE (GB)
The portion of the operating range of a
control device, over which there is no
change in output.
f zone *f* morte
e zona *f* muerta
i campo *m* d'insensibilità,
intervallo *m* d'insensibilità, zona *f* morta
n dood gebied *n*
d Totzone *f*

1625 DEAD TIME,   ge
INSENSITIVE TIME
In an electrical circuit, tube, valve or
instrument, the time immediately after
receiving a stimulus during which it is
insensitive to another impulse or
stimulus.
f temps *m* mort
e tiempo *m* muerto
i tempo *m* d'inazione, tempo *m* morto
n dode tijd
d Totzeit *f*

1626 DEAD TIME CORRECTION   ct
A correction to the observed counting rate
to allow for the probability of the occurrence
of events within the dead time of the system.
f correction *f* pour le temps mort
e corrección *f* para el tiempo muerto
i correzione *f* per il tempo morto
n correctie voor dode tijd
d Totzeitkorrektion *f*

1627 DEAD TIME OF A   ct
GEIGER-MÜLLER COUNTER TUBE
Time following the initiation of a pulse
caused by an ionizing event, during which
a Geiger-Müller counter tube is incapable
of responding to a further ionizing event.
f temps *m* mort d'un tube compteur
Geiger-Müller
e tiempo *m* muerto de un tubo contador
Geiger-Müller
i tempo *m* morto d'un tubo contatore
Geiger-Müller
n dode tijd van een geigermüllertelbuis
d Totzeit *f* eines Geiger-Müller-Zählrohrs

1628 DEBYE LENGTH   np, pp
A characteristic distance beyond which the
surrounding electrons in a plasma, by
their collective motion, effectively screen
the coulomb field of a particular charged
particle from that of another moving past
the first.
f longueur *f* de Debye
e longitud *f*, de Debye
i lunghezza *f* di Debye
n debijelengte
d Debye-Länge *f*

1629 DEBYE-SCHERRER METHOD,   cr
POWDER METHOD
The examination of the diffraction holes
produced when a crystalline powder
specimen is irradiated in a monochromatic
beam radiation.
f méthode *f* Debye-Scherrer
e método *m* Debye-Scherrer
i metodo *m* Debye-Scherrer
n methode van Debije en Scherrer,
poedermethode
d Debye-Scherrer-Verfahren *n*,
Pulvermethode *f*

1630 DEBYE SPHERE   np
A sphere whose radius is a Debye length.
f sphère *f* de Debye
e esfera *f* de Debye
i sfera *f* di Debye
n debijebol
d Debye-Sphäre *f*

1631 DECADE COUNTER TUBE,
DECADE GLOW COUNTER TUBE,
see 1198

1632 DECADE SCALER   ct
A scaler whose scaling factor is 10.
f démultiplicateur *m* décimal,
échelle *f* décimale

e desmultiplicador *m* decimal,
   escalímetro *m* decimal
i demoltiplicatore *m* decimale
n decimale deelschakeling
d Dezimalteiler *m*

1633 DECALESCENCE   mg
Absorption of heat, usually by an alloy
without rise of temperature, due to an
allotropic transformation.
f décalescence *f*
e decalescencia *f*
i decalescenza *f*
n decalescentie
d Dekaleszenz *f*

1634 DECALESCENT POINT   mg
The point at which there is a sudden
absorption of heat as the metal is
raised in temperature.
f point *m* de décalescence
e punto *m* de decalescencia
i punto *m* di decalescenza
n decalescentiepunt *n*
d Dekaleszenzpunkt *m*

1635 DECANNING,   fu, rt
   DECLADDING, DEJACKETING
The removal of the envelope of a fuel
element.
f dégainage *m*
e descamisadura *f*, desvainadura *f*
i spogliatura *f*
n ontmanteling
d Enthülsung *f*

1636 DECANNING PLANT,   fu
   DECLADDING PLANT,
   DEJACKETING PLANT
An installation for the removal of the
cladding of nuclear fuel rods.
f installation *f* à dégainer
e equipo *m* descamisador,
   equipo *m* desvainador
i impianto *m* di spogliatura
n ontmantelingsinstallatie
d Enthülsungsanlage *f*

1637 DECAY, see 134

1638 DECAY,   np
   DISINTEGRATION
Of a radioactive substance, the gradual
decrease of its activity, or its transformation into its daughter products.
f décroissance *f*, désintégration *f*
e decaimiento *m*, desintegración *f*
i decadimento *m*, disintegrazione *f*
n desintegratie, verval *n*
d Umwandlung *f*, Zerfall *m*

1639 DECAY CHAIN,   np
   DISINTEGRATION CHAIN,
   RADIOACTIVE CHAIN,
   RADIOACTIVE SERIES
A series of successive radioactive trans-
formations which occurs when the nucleus

deviates so much from a stable configuration of protons and neutrons that more than
one transformation must occur.
f chaîne *f* de désintégrations
e cadena *f* de desintegraciones,
   serie *f* de desintegraciones
i catena *f* di decadimenti
n vervalserie
d Zerfallskette *f*, Zerfallsreihe *f*

1640 DECAY CONSTANT,   ma
   DISINTEGRATION CONSTANT
$\frac{dN}{dt} = -\lambda N$, where N is the number of radioactive atoms at time t.
f constante *f* de désintégration
e constante *f* de desintegración
i costante *f* di decadimento
n desintegratieconstante,
   vervalconstante
d Zerfallskonstante *f*

1641 DECAY CURVE, see 104

1642 DECAY ENERGY,   np
   DISINTEGRATION ENERGY
For a given nuclear disintegration, the
amount of energy released.
f énergie *f* de désintégration
e energía *f* de desintegración
i energia *f* di disintegrazione
n desintegratie-energie
d Zerfallsenergie *f*

1643 DECAY HEAT,   np
   DISINTEGRATION HEAT
That part of the energy developed in a
radioactive disintegration which manifests
itself in a rise of temperature in the
medium in which the disintegration takes
place.
f chaleur *f* de désintégration
e calor *m* de desintegración
i calore *m* da disintegrazione
n vervalwarmte
d Zerfallswärme *f*

1644 DECAY LAW   np
Of a radioactive nuclide, the law which
states that the number of atoms decaying
in unit time is proportional to the number
present.
f loi *f* de décroissance
e ley *f* de desintegración
i legge *f* di decadimento
n vervalwet
d Zerfallsgesetz *n*

1645 DECAY PARTICLE,   np
   DISINTEGRATION PARTICLE
A particle that undergoes one or more
modes of disintegration.
f particule *f* de désintégration
e partícula *f* de desintegración
i particella *f* di disintegrazione
n desintegratiedeeltje *n*, vervaldeeltje *n*
d Zerfallsteilchen *n*

1646 DECAY PRODUCT,   np
      DISINTEGRATION PRODUCT
      Any nuclide, radioactive or stable,
      resulting from the radioactive disintegra-
      tion of a radionuclide.
f produit *m* de décroissance
e producto *m* de desintegración
i prodotto *m* di decadimento
n volgprodukt *n*
d Folgeprodukt *n*

1647 DECAY RATE,   np
      DISINTEGRATION RATE
      The absolute rate of decay of a radio-
      active substance, usually expressed in
      terms of disintegrations per unit of time.
f taux *m* de désintégration
e velocidad *f* de desintegración
i velocità *f* di disintegrazione
n desintegratietempo *n*
d Zerfallsrate *f*

1648 DECAY SCHEME,   np
      DISINTEGRATION SCHEME
      The modes of decay of a radioactive
      nuclide generally expressed in the form
      of a diagram.
f diagramme *m* de désintégration
e diagrama *m* de desintegración
i diagramma *m* di disintegrazione
n desintegratieschema *n*, vervalschema *n*
d Zerfallsschema *n*

1649 DECAY SEQUENCE,   np
      DISINTEGRATION SEQUENCE
      The series of disintegrations occurring
      before uranium finally becomes lead.
f succession *f* de désintégration
e sucesión *f* de desintegración
i successione *f* di disintegrazione
n desintegratievolgorde, vervalvolgorde
d Zerfallsfolge *f*

1650 DECAY STORAGE   rw
      A convenience where radioactive material
      is stored to decay.
f dépôt *m* de décroissance
e depósito *m* de desintegración
i deposito *m* di decadimento
n versterfruimte
d Abkühlungsraum *m*, Auskühlungsraum *m*

1651 DECAY TABLES   is
      Tables giving the activity level as a
      function of time of the greater part of
      radioisotopes now used in medicine.
f tables *pl* de décroissance radioactive
e cuadros *pl* de desintegración
i tavole *pl* di decadimento
n vervaltabellen *pl*
d Zerfallstabellen *pl*

1652 DECELERATION   pa
      Negative acceleration.
f décélération *f*
e deceleración *f*
i decelerazione *f*

n verlangzaming
d negative Beschleunigung *f*,
   Verlangsamung *f*

1653 DECONTAMINATION,   sa
      RADIOACTIVE DECONTAMINATION
      Removal or reduction of radioactive
      contamination.
f décontamination *f*
e descontaminación *f*
i decontaminazione *f*
n decontaminatie, ontsmetting
d Dekontamination *f*, Entseuchung

1654 DECONTAMINATION FACTOR   np
      The ratio of the initial concentration of
      contaminating radioactive material to the
      final concentration resulting from a
      process.
f facteur *m* de décontamination
e factor *m* de descontaminación
i fattore *m* di decontaminazione
n decontaminatiefactor, ontsmettingsfactor
d Dekontaminationsfaktor *m*

1655 DECONTAMINATION INDEX   cg
      The logarithm of the ratio of initial
      specific radioactivity resulting from a
      separation process.
f indice *m* de décontamination
e índice *m* de descontaminación
i indice *m* di decontaminazione
n ontsmettingsindex
d Dekontaminationsindex *m*,
   Entseuchungsindex *m*

1656 DECONTAMINATION PLANT   cg
f installation *f* de décontamination
e instalación *f* de descontaminación
i impianto *m* di decontaminazione
n ontsmettingsinstallatie
d Dekontaminationsanlage *f*,
   Entseuchungsanlage *f*

1657 DECONTAMINATION ROOM,   cl
      DETOXIFICATION ROOM
f chambre *f* de décontamination,
   chambre *f* de détoxication
e cámara *f* de descontaminación,
   cámara *f* de detoxificación
i camera *f* di decontaminazione,
   camera *f* di disintossicazione
n ontsmettingsruimte
d Dekontaminationsraum *m*,
   Entgiftungsraum *m*

1658 DECONTAMINATION SQUAD   cg
      A special group of men for removing
      the radioactive waste in laboratories,
      etc.
f équipe *m* de décontamination
e equipo *m* de descontaminación
i squadra *f* di decontaminazione
n ontsmettingsploeg
d Dekontaminationstrupp *m*

1659 DECOUPLING   nw
A method for decreasing the seismic
effect of an underground explosion by
creating a certain distance between the
bomb and the wall of the cavern.
f méthode *f* sans contact
e método *m* sin contacto
i metodo *m* senza contatto
n contactloze methode
d kontaktfreies Verfahren *n*

1660 DEE   pa
A hollow accelerating D-shaped electrode
in a cyclotron.
f dé *m*
e electrodo *m* en D,
   electrodo *m* hueco semicilíndrico
i elettrodo *m* in D
n d-elektrode
d D-Elektrode *f*, Duante *f*

1661 DEE LINES   pa
Structural members which support the
dees of a cyclotron.
f supports *pl* des dés
e soportes *pl* de los electrodos en D
i sostegni *pl* degli elettrodi in D
n d-stelen *pl*
d Duantenstützen *pl*

1662 DEEP RÖNTGEN THERAPY   xr
Röntgen therapy directed towards lesions
situated within the depths of the body
using quantum energies about 200kV
or more.
f röntgenthérapie *f* profonde
e radioterapia *f* profunda,
   röntgenoterapia *f* profunda
i röntgenterapia *f* profonda
n röntgendieptetherapie
d Röntgentiefentherapie *f*

1663 DEEP-THERAPY TUBE   xr
Röntgen tube for deep therapy.
f tube *m* pour röntgenthérapie profonde
e tubo *m* para röntgenoterapia profunda
i tubo *m* per röntgenterapia profonda
n dieptetherapiebuis
d Tiefentherapieröhre *f*

1664 DE-EXCITATION   np
The return of e.g. an atom from the
excited state to the ground state.
f désexcitation *f*
e desexcitación *f*
i diseccitazione *f*
n de-excitatie
d Entregung *f*

1665 DEFECT   cr, ec
A term used to include various types of
point imperfections in solids, such as
vacancies, interstitial atoms, etc., as
distinct from extended imperfections such
as dislocations.
f défaut *m*
e defecto *m*

i difetto *m*
n defect *n*
d Defekt *m*

1666 DEFECT CONDUCTION,   ec
        HOLE CONDUCTION
Conduction by holes in the valence band
of a semiconductor.
f conduction *f* par trous
e conducción *f* por huecos
i conduzione *f* per buche,
   conduzione *f* per difetti
n gatengeleiding
d Defektelektronenleitung *f*, Löcherleitung *f*

1667 DEFINITION   gp, ph
A term which refers to the sharpness of
delineation of image details in a radiograph,
photographic reproduction or fluorescent-
screen image.
f définition *f*
e definición *f*
i definizione *f*
n definitie
d Definition *f*

1668 DEFLECTION   ra
A bending of rays from a straight line.
f déviation *f*
e desviación *f*
i deviazione *f*
n afbuiging, deflectie
d Ablenkung *f*

1669 DEFLECTION ANGLE   np, ra
The total angle through which a beam is
bent or deflected horizontally, vertically
or diagonally.
f angle *m* de déviation
e ángulo *m* de desviación
i angolo *m* di deviazione
n afbuighoek
d Ablenkwinkel *m*

1670 DEFLECTOR   pa
A device which applies electrostatic
and/or magnetic forces to cause accelerated particles to depart from their
normal path.
f déflecteur *m*
e deflector *m*
i deflettore *m*
n deflector
d Deflektor *m*

1671 DEFLOCCULATION   ch
Breakup of soil clusters into smaller units
or particles.
f défloculation *f*
e desfloculación *f*
i defloccolazione *f*
n ontvlokking
d Entflockung *f*

1672 DEFORMATION BANDS   cr
Regions within a metal crystal which have
assumed different orientations as a result
of slip.

f bandes *pl* de déformation
e bandas *pl* de deformación
i bande *pl* di deformazione
n glijbanden *pl*
d Deformationsbänder *pl*, Gleitbänder *pl*

1673 DEFORMATION ENERGY np
In the liquid-drop model of fission the
deformation energy is the energy required
to deform the nucleus from a spherical
shape to an ellipsoid.
f énergie *f* de déformation
e energía *f* de deformación
i energia *f* di deformazione
n vervormingsenergie
d Verformungsenergie *f*

1674 DEFORMATION POTENTIAL ec
The effective electric potential acting on
a free electron in a metal or semiconductor as a result of a local deformation in
the crystal lattice.
f potentiel *m* de déformation
e potencial *m* de deformación
i potenziale *m* di deformazione
n deformatiepotentiaal
d Deformationspotential *n*

1675 DEGENERACY, qm
DEGENERATE SYSTEM
The term used to describe the situation
in which a quantum mechanical system
possesses more than one eigenstate
characteristic of the same eigenvalue(s)
of the operator(s) corresponding to a
given (set of) dynamical variable(s).
f dégénérescence *f*
e degenerescencia *f*
i degenerescenza *f*
n degeneratie, ontaarding
d Entartung *f*

1676 DEGENERATE GAS gp
A gas formed by a system of particles
whose concentration is very great, with
the result that the Maxwell-Boltzmann
law does not apply.
f gaz *m* dégénéré
e gas *m* degenerado
i gas *m* degenere
n ontaard gas *n*
d entartetes Gas *n*

1677 DEGENERATE STATE qm
In quantum mechanics, when different
states of motion correspond to the same
energy level, the states are said to be
degenerate.
f état *m* dégénéré
e estado *m* degenerado
i stato *m* degenere
n ontaarde toestand
d entarteter Zustand *m*

1678 DEGRADATION, np
THIN-DOWN
Loss of energy by particles or photons
as the result of collision.

f dégradation *f*, perte *f* d'énergie
e degradación *f*, pérdida *f* de energía
i degradazione *f*, perdita *f* d'energia
n energieverlies *n*, verlangzaming
d Energieverlust *m*

1679 DEGRADATION OF ch, ms, ra
POLYMERS
A phenomenon induced by irradiation in
all categories of polymers.
f dégradation *f* de polymères
e degradación *f* de polímeros
i degradazione *f* di polimeri
n degradatie van polymeren
d Abbau *m* von Polymeren

1680 DEGREE OF ENRICHMENT rt
Enrichment factor minus one.
f degré *m* d'enrichissement
e grado *m* de enriquecimiento
i grado *m* d'arricchimento
n verrijkingsgraad
d Anreicherungsgrad *m*

1681 DEGREE OF FREEDOM ma
The least number of independent variables
which determine the state of a system and
which must be defined in order to give a
complete description of that particular
state.
f degré *m* de liberté
e grado *m* de libertad
i grado *m* di libertà
n vrijheidsgraad
d Freiheitsgrad *m*

1682 DEGREE OF MODERATION, see 395

1683 DE-IONIZATION np
The disappearance of ions in an ionized gas.
f désionisation *f*
e desionización *f*
i deionizzazione *f*
n desionisatie, ontionisatie
d Entionisierung *f*

1684 DE-IONIZATION POTENTIAL ec
The potential at which the ionization of the
gas within a gasfilled tube or valve ceases
and conduction is interrupted.
f potentiel *m* de désionisation
e potencial *m* de desionización
i potenziale *m* di deionizzazione
n desionisatiepotentiaal,
ontionisatiepotentiaal
d Entionisierungspotential *n*

1685 DE-IONIZATION RATE np
The number of pairs of ions with opposite
charges disappearing per unit volume of a
gas in unit time.
f vitesse *f* spécifique de désionisation
e velocidad *f* específica de desionización
i velocità *f* specifica di deionizzazione
n desionisatiesnelheid, ontionisatiesnelheid
d Entionisierungsgeschwindigkeit *f*

1686 DEKATRON   ct
A gas-filled multi-electrode tube in which
incoming pulses are counted by causing a
glow discharge to be transferred sequenti-
ally to cathodes arranged in a circle
around a central anode.
f décatron *m*
e decatrón *m*
i decatrone *m*
n decatron *n*
d dekadisches Kaltkatodenzählrohr *n*,
   Dekatron *n*

1687 DELAY,   co, rt
      DELAY TIME
The time between the establishment of an
undesired condition and the start of
corrective motion of a control rod.
f retard *m*
e retraso *m*
i ritardo *m*
n vertraging
d Verzugszeit *f*

1688 DELAY COINCIDENCE CIRCUIT   ct
A coincidence circuit that is actuated by
two pulses, one of which is delayed by a
specified time interval with respect to the
other.
f circuit *m* de coïncidence à retard partiel
e circuito *m* de coincidencia de retraso
   parcial
i circuito *m* di coincidenza a ritardo
   parziale
n coïncidentieschakeling met gedeeltelijke
   vertraging
d Koinzidenzschaltung *f* mit teilweiser
   Verzögerung

1689 DELAY TANK   rt, sa
A tank or reservoir for the temporary
hold-up of radioactive fluids to permit
their activity to decay.
f réservoir *m* à diminution de radioactivité,
   réservoir *m* de désactivation
e tanque *m* de desactivación,
   tanque *m* de diminución de radiactividad
i cisterna *f* di decadimento,
   serbatoio *m* a diminuzione di radioattività
n versterftank, vertraagtank
d Abklinggefäss *n*, Verweiltank *m*

1690 DELAY UNIT   ec
A basic function unit comprising an
electronic circuit which delivers an output
pulse after a given time interval in
response to an input pulse.
f élément *m* retard
e unidad *f* de impulso definido
i elemento *m* di ritardo
n vertraagschakel
d Verzögerungsgerät *n*,
   Verzögerungsstufe *f*

1691 DELAYED ALPHA PARTICLES   ar, np
Alpha particles emitted promptly by excited nuclei formed in a beta disintegration
process.

f particules *pl* alpha retardées
e partículas *pl* alfa retrasadas
i particelle *pl* alfa ritardate
n nakomende alfadeeltjes *pl*
d verzögerte Alphateilchen *pl*

1692 DELAYED COINCIDENCE   ct
The occurrence of a count in one detector
at a short, but measurable, time later than
a count in another detector.
f coïncidence *f* retardée
e coincidencia *f* diferida
i coincidenza *f* ritardata
n vertraagde coïncidentie
d verzögerte Koinzidenz *f*

1693 DELAYED COINCIDENCE   ct
      COUNTING
The recording of such pairs of ionizing
events as have a specified time interval
between them.
f comptage *m* de coïncidence retardée
e recuento *m* de coincidencia diferida
i conteggio *m* di coincidenza ritardata
n telling bij vertraagde coïncidentie
d Zählung *f* bei verzögerter Koinzidenz

1694 DELAYED COINCIDENCE UNIT   ct
A coincidence selector unit designed to
produce an output signal when one or more
pulses are delayed by a specified time
interval with respect to one another.
f élément *m* sélecteur de coïncidences
   retardées
e unidad *f* selectora de coincidencias
   diferidas
i elemento *m* selettore di coincidenze
   ritardate
n uitgestelde-coïncidentie-eenheid
d Koinzidenzstufe *f* mit Verzögerung

1695 DELAYED CRITICAL   rt
Term used to emphasize that the delayed
neutrons are necessary to achieve the
critical state.
f critique différé
e crítico con neutrones retrasados
i critico ritardato
n kritiek met nakomende neutronen,
   kritisch met nakomende neutronen
d verzögert-kritisch

1696 DELAYED CRITICALITY   rt
Condition of criticality which includes
the effect of the prompt neutrons or of
the delayed neutrons.
f criticité *f* différée
e criticidad *f* retrasada
i criticità *f* ritardata
n vertraagde kriticiteit
d verzögerte Kritizität *f*

1697 DELAYED NEUTRON EMITTER   np
A neutron emitter in which the excitation
energy of the product nucleus exceeds the
neutron binding energy for that nucleus.
f émetteur *m* de neutrons retardés
e emisor *m* de neutrones retrasados

i emettitore *m* di neutroni ritardati
n vertraagde-neutronenstraler
d verzögerter Neutronenstrahler *m*

1698 DELAYED NEUTRON FAILED   ma
      ELEMENT MONITOR
A failed element monitor based on detection of delayed neutrons emitted by certain
fission products in the coolant.
f moniteur *m* de rupture de gaine par
   détection de neutrons retardés
e vigía *f* de averías por neutrones
   retrasados
i monitore *m* di rottura di guaine a
   neutroni ritardati
n monitor voor beschadigde splijtstofelementen berustend op nakomende neutronen.
d Warngerät *n* für schadhafte Brennelemente,
   das auf verzögerte Neutronen anspricht.

1699 DELAYED NEUTRON   np
      FRACTION
The ratio of the mean number of delayed
neutrons per fission to the mean total
number of neutrons (prompt plus
delayed) per fission.
f fraction *f* de neutrons retardés
e fracción *f* de neutrones retrasados
i frazione *f* di neutroni ritardati
n fractie van nakomende neutronen
d Anteil *m* der verzögerten Neutronen

1700 DELAYED NEUTRON   np
      PRECURSOR
A nuclide whose nuclei undergo beta decay
followed by neutron emission.
f précurseur *m* de neutrons retardés
e precursor *m* de neutrones retrasados
i precursore *m* di neutroni ritardati
n moedernuclide *n* van nakomende
   neutronen
d Mutterkern *m* verzögerter Neutronen

1701 DELAYED NEUTRONS   np
Neutrons emitted by nuclei in excited
states which have been formed in the
process of a beta decay.
f neutrons *pl* retardés
e neutrones *pl* retrasados
i neutroni *pl* ritardati
n nakomende neutronen *pl*
d verzögerte Neutronen *pl*

1702 DELAYED REACTIVITY   rt
Reactivity due to the effect of delayed
neutrons.
f réactivité *f* retardée
e reactividad *f* retrasada
i reattività *f* ritardata
n vertraagde reactiviteit
d verzögerte Reaktivität *f*

1703 DELINEATION OF FALL-OUT   nw
      CONTOURS
The drawing of the contour lines of the
fall-out concentrations after a nuclear
explosion.

f délinéation *f* des contours des retombées
   radioactives
e delineación *f* de los contornos de los
   depósitos radiactivos
i delineazione *f* dei contorni della ricaduta
   radioattiva
n tekenen *n* van de omtreklijnen van de radioactieve neerslag
d Zeichnen *n* der Umrisslinien des radioaktiven Ausfalls

1704 DELORENZITE   mi
Mineral containing ytterbium, uranium,
iron and titanium.
f delorenzite *f*
e delorenzita *f*
i delorenzite *f*
n delorenziet *n*
d Delorenzit *m*

1705 DELTA ELECTRON,   np
      DELTA RAY
An electron that is ejected by recoil
when a rapidly moving charged particle
passes through matter.
f rayon *m* delta
e rayo *m* delta
i raggio *m* delta
n deltastraal
d Deltastrahl *m*

1706 DELTA RAY COUNTING   ct, ph
In the photographic emulsion technique,
a method of determining the charge of a
particle by counting the number of secondary electron tracks branching off the main
track of the particle.
f comptage *m* des rayons delta
e recuento *m* de los rayos delta
i conteggio *m* dei raggi delta
n telling van de deltastralen
d Zählung *f* der Deltastrahlen

1707 DEMATERIALIZATION   np
In nuclear physics, the process in which a
positively and a negatively charged
material particle unite and cease to exist
as such, whereas the energy corresponding
to their material masses is reproduced
as electromagnetic radiation.
f dématérialisation *f*
e desmaterialización *f*
i dematerializzazione *f*
n dematerialisatie
d Dematerialisation *f*

1708 DEMONSTRATION REACTOR   rt
A nuclear reactor designed to demonstrate
the technical feasibility and to explore the
economic potential of a given reactor type.
f réacteur *m* de démonstration
e reactor *m* de demostración
i reattore *m* di dimostrazione
n demonstratiereactor
d Demonstrationsreaktor *m*,
   Veranschaulichungsreaktor *m*

1709 DEMPSTER MASS   sp
      SPECTROGRAPH
      A mass spectrograph in which ions are
      accelerated through a slit by an electric
      field, then deflected by a magnetic field
      so that all ions of the same charge-to-mass
      ratio pass through a second slit.
f spectrographe *m* de masse(s) de Dempster
e espectrógrafo *m* de masa de Dempster
i spettrografo *m* di massa di Dempster
n massaspectrograaf volgens Dempster
d Massenspektrograph *m* nach Dempster

1710 DEMPSTER POSITIVE   an
      RAY ANALYSIS
      A method of determining the charge-to-
      mass ratio by separating particles by
      means of a Dempster mass spectrograph.
f méthode *f* de détermination du rapport
   charge/masse de Dempster
e método *m* de determinación de la relación
   carga-masa de Dempster
i metodo *m* di determinazione del rapporto
   carica-massa di Dempster
n methode voor het bepalen van de verhouding lading-massa volgens Dempster
d Verfahren *n* zur Bestimmung des
   Verhältnisses Ladung-Masse nach
   Dempster

1711 DENATURANT   nw
      An isotope added to fissile (fissionable)
      material to make it unsuitable for use in
      atomic weapons except after extensive
      processing.
f dénaturant *m*
e desnaturalizante *m*
i denaturante *m*
n denatureermiddel *n*
d Denaturierungsmittel *n*

1712 DENATURATION   nw
      In nuclear technology, the adding of an
      isotope to fissile (fissionable) material to
      make it unsuitable for use in nuclear
      weapons except after extensive processing.
f dénaturation *f*
e desnaturalización *f*
i denaturazione *f*
n denaturatie
d Denaturierung *f*

1713 DENDRITE   cr
      A crystal, usually produced by solidification
      of liquid, and characterized by a tree-like
      structure with many branches.
f dendrite *f*
e dendrita *f*
i dendrite *f*
n dendriet *n*
d Dendrit *m*, Tannenbaumkristall *m*

1714 DENSITOMETER   me, ph
      Instrument for measuring photographic
      density.
f densitomètre *m*
e densitómetro *m*

i densitometro *m*
n densitometer, zwartingsmeter
d Densitometer *n*, Schwärzungsmesser *m*

1715 DENSITY,   ph
      PHOTOGRAPHIC DENSITY
      The degree of darkening of photographic
      film.
f densité *f* photographique
e densidad *f* fotográfica
i densità *f* fotografica
n zwarting
d photographische Dichte *f*

1716 DENSITY EFFECT   np
      The reduction in the stopping power of a
      dense material relative to the stopping
      power of the same mass per unit area of
      the same material in a rarefied state.
f effet *m* de densité
e efecto *m* de densidad
i effetto *m* di densità
n dichtheidseffect *n*
d Dichteeffekt *m*

1717 DENSITY OF AN   ec
      ELECTRON BEAM,
      ELECTRON BEAM DENSITY
      The density of the electron current of the
      beam at any given point.
f densité *f* d'un faisceau électronique
e densidad *f* de un haz electrónico
i densità *f* d'un fascio elettronico
n elektronenbundeldichtheid
d Elektronenstrahldichte *f*

1718 DENSITY OF ELECTRONS,   np
      ELECTRON DENSITY
      The number of electrons per unit volume.
f densité *f* d'électrons
e densidad *f* de electrones
i densità *f* d'elettroni
n elektronendichtheid
d Elektronendichte *f*

1719 DENSITY OF ENERGY,   gp
      ENERGY DENSITY
      The ratio of the total amount of energy
      carried by or contained in a volume to that
      volume.
f densité *f* d'énergie
e densidad *f* de energía
i densità *f* d'energia
n energiedichtheid
d Energiedichte *f*

1720 DEPENDENT VARIABLE   co
      Of a mathematical or physical system,
      that dependent quantity which, through the
      action of the independent variable(s),
      changes according to some established
      relationship.
f variable *f* dépendante
e variable *f* dependiente
i variabile *f* dipendente
n afhankelijke variabele
d abhängige Variable *f*

1721 DEPHLEGMATION,   ch
     PARTIAL CONDENSATION
The partial condensation of a mixed
vapo(u)r thereby increasing the concentration of the more volatile constituents in
the remaining vapo(u)r.
f déflegmation *f*
e deflegmación *f*
i deflemmazione *f*
n deflegmatie
d Dephlegmieren *n*

1722 DEPILATION,   md, xr
     EPILATION
The temporary or permanent removal of
háir, e.g. by electrolysis or X-ray
irradiation.
f dépilation *f*, épilation *f*
e depilación *f*
i depilazione *f*
n epileren *n*, ontharen *n*
d Depilation *f*, Enthaarung *f*

1723 DEPILATION DOSE,   xr
     EPILATION DOSE
Dose of radiation that produces temporarily
epilation after a latent period.
f dose *f* de dépilation, dose *f* d'épilation
e dosis *f* de depilación
i dose *f* di depilazione
n epilatiedosis
d Depilationsdosis *f*

1724 DEPLETE (TO),   fu, is
     STRIP (TO)
1. To decrease the abundance of a
particular isotope in a mixture of the
isotopes of an element.
2. Applied to nuclear fuel, to decrease
the abundance of fissile (fissionable)
isotopes.
f épuiser v, extraire v fractions légères
e agotar v
i estrarre v frazioni leggeri, impoverire v
n uitdrijven v
d abreichern v, abtreiben v

1725 DEPLETED FRACTION   ch, is
A fraction which has a smaller abundance
of the desired isotope.
f fraction *f* épuisée
e fracción *f* agotada
i frazione *f* impoverita
n uitgedreven fractie
d abgereicherter Anteil *m*

1726 DEPLETED FUEL,   fu, rt
     IMPOVERISHED FUEL
Nuclear fuel which has lost a percentage
of its quantity of fissile (fissionable)
atoms due to the reactor operation or to
an isotope separation process.
f combustible *m* appauvri
e combustible *m* agotado
i combustibile *m* impoverito
n verarmd splijtmateriaal *n*
d abgereicherter Brennstoff *m*

1727 DEPLETED MATERIAL   fu, rt
Material which has undergone depletion.
f matière *f* appauvrie
e material *m* agotado
i materiale *m* impoverito
n verarmd materiaal *n*
d abgereichertes Material *n*,
   verarmtes Material *n*

1728 DEPLETED URANIUM   fu
f uranium *m* appauvri
e uranio *m* agotado
i uranio *m* impoverito
n verarmd uranium *n*
d abgereichertes Uran *n*

1729 DEPLETED URANIUM   ra, sa
     SHIELDING
Shielding by using depleted uranium
elements in medical Co-60 teletherapy
units, industrial radiography units, etc.
f blindage *m* par uranium appauvri
e blindaje *m* por uranio agotado
i schermatura *f* per uranio impoverito
n afscherming door middel van verarmd
   uranium
d Abschirmung *f* durch abgereichertes Uran

1730 DEPLETED WATER   np
Water which after an isotopic exchange
treatment of natural water contains a
smaller percentage of D_2O than natural
water.
f eau *f* appauvrie en eau lourde
e agua *f* agotada
i acqua *f* impoverita
n verarmd water *n*
d abgereichertes Wasser *n*,
   verarmtes Wasser *n*

1731 DEPLETION   fu, rt
Reduction of the concentration of one or
more specified isotopes in a material or
one of its constituents.
f appauvrissement *m*
e agotamiento *m*
i impoverimento *m*
n verarming
d Abreicherung *f*, Verarmung *f*

1732 DEPLETION MEMBER, see 1154

1733 DEPOLARIZATION   gp
The removal of any alignment in space of
the elementary electric or magnetic dipoles
in a substance, or in a beam of particles.
f dépolarisation *f*
e despolarización *f*
i depolarizzazione *f*
n depolarisatie
d Depolarisation *f*

1734 DEPOSIT   mi
A natural occurrence or accumulation of
ore.
f gisement *m*
e yacimiento *m*

i giacimento *m*
n afzetting
d Ablagerung *f*, Lagerstätte *f*

1735 DEPOSIT DOSE, see 1335

1736 DEPOSIT OF SECOND ORIGIN   mi
Ore deposit supposed to be formed in the
Mesozoin era.
f gisement *m* d'origine secondaire
e yacimiento *m* de origen secundario
i giacimento *m* d'origine secondario
n afzetting uit het mesozoïsche tijdperk
d Lagerstätte *f* aus der mesozoischen
    Periode

1737 DEPRESSURIZATION   rt, sa
In off-load refuelling, the letting escape
of all gases and fluids under pressure.
f élimination *f* de la pression
e despresurización *f*
i rimozione *f* della pressione
n wegnemen *n* van de druk
d Druckbeseitigung *f*, Druckminderung *f*

1738 DEPTH ABSORBED DOSE,   ra
          DEPTH DOSE
The dose of radiation delivered at a
particular depth beneath the surface of a
body or other irradiated material.
f dose *f* absorbée en profondeur,
    dose *f* en profondeur
e ʳdosis *f* en profundidad
i dose *f* specifica assorbita in profondità
n dieptedosis
d Tiefendosis *f*

1739 DERIVATIVE ACTION (GB),   co
          RATE ACTION (US)
That action in which there is a continuous
relation between rate of change of the
controlled variable and position of a final
element.
f action *f* D, action *f* par dérivation
e acción *f* D, acción *f* derivada
i azione *f* D, azione *f* derivatrice
n D-werking, differentiële werking
d D-Verhalten *n*, Differentialverhalten *n*

1740 DE-SHRINKAGE   ph
Of a photographic emulsion, the
restoration of the original thickness of the
emulsion, after the loss of silver halides
during fixation, by introducing some filler
such as resin.
f regonflement *m*,
    restauration *f* de l'épaisseur de la couche
    d'émulsion
e restablecimiento *m* del espesor de la
    capa de emulsión
i ricostituzione *f* dello spessore dello
    strato d'emulsione
n dikteherstel *n* van de emulsielaag
d Wiederherstellung *f* der Emulsions-
    schichtdicke

1741 DESIGN POWER   rt
The power for which an energy producing
reactor is planned.
f puissance *f* calculée
e potencia *f* de diseño, potencia *f* presumida,
    potencia *f* teórica
i potenza *f* designata
n gepland vermogen *n*
d projektierte Leistung *f*, Solleistung *f*

1742 DESIGN TEMPERATURE RISE   cl, rt
Results from a compromise between
1. proportion of coolant in the reactor core,
2. pressure drop and pumping power,
3. allowable temperature difference.
f augmentation *f* admissible calculée de
    la température
e aumento *m* permisible presumida de la
    temperatura
i aumento *m* ammissibile designato della
    temperatura
n geplande toelaatbare temperatuurstijging
d projektierter zugelassener Temperatur-
    anstieg *m*

1743 DESIRED VALUE (GB),   co
          IDEAL VALUE (US)
The specified value of the controlled
condition.
f valeur *f* prescrite
e valor *m* prescrito
i valore *m* desiderato
n gewenste waarde
d Aufgabenwert *m*

1744 DESORPTION   ch
The reverse process of adsorption whereby
adsorbed matter is removed from the
adsorbant. The term is also used as the
reverse process of absorption.
f désorption *f*
e desorción *f*
i dissorbimento *m*
n desorptie
d Desorption *f*

1745 DESQUAMATION,   md
          ECDYSIS, MOULTING, SHEDDING
An undesired effect of radiation on the skin.
f désquamation *f*
e descamación *f*
i desquamazione *f*
n afschilfering, desquamatie
d Abschuppung *f*, Häutung *f*, Schälung *f*,
    Schuppung *f*

1746 DETECTABLE ACTIVITY   te
In testing neutron sources by the immersion
method, the detectable activity should be
smaller than 10^{-8} c.
f activité *f* discernable
e actividad *f* discernible
i attività *f* discernibile
n aantoonbare activiteit
d nachweisbare Aktivität *f*

1747 DETECTION,   ra
      RADIATION DETECTION
f détection f d'un rayonnement
e detección f de una radiación
i rivelazione f d'una radiazione
n stralingsdetectie
d Strahlungsdetektion f

1748 DETECTION EFFICIENCY   ra
      The ratio between the number of particles
 or photons detected to the number of
 similar particles or photons which have
 entered the sensitive volume of the
 detector.
f rendement m de détection
e rendimiento m de detección
i rendimento m di rivelazione
n detectierendement n
d Nachweisausbeute f

1749 DETECTOR   me, sa
      An instrument for determining presence
 and sometimes measuring the amount of
 radiation or neutron flux.
f détecteur m
e detector m
i rivelatore m
n detector
d Detektor m

1750 DETECTOR EFFICIENCY   ra
      The ratio of the number of particles or
 photons detected to the number of similar
 particles or photons which have entered
 the sensitive volume of the detector.
f rendement m d'un détecteur
e rendimiento m de un detector
i rendimento m d'un rivelatore
n detectorrendement n
d Ausbeute f eines Nachweisgeräts

1751 DETECTOR NOISE   ct
      The variation in integrated output of a
 radiation detectur due to the random
 arrival in time of the ionizing events
 within the detector.
f bruit m de détecteur de rayonnement
e ruido m de detector de radiación
i rumore m di rivelatore di radiazione
n stralingsdetectorruis
d Rauschen n des Strahlungsnachweisgeräts

1752 DETECTOR 1/V   me
      A neutron detector for which the cross
 section of the detection reaction varies
 inversely with neutron speed.
f détecteur m en $1/v$
e detector m en $1/v$
i rivelatore m in $1/v$
n $1/v$-detector
d $1/v$-Detektor m

1753 DETECTOR WITH INTERNAL   ra
      GAS SOURCE
      A radiation detector in which the filling
 gas consists in all or in part of the radioactive gas, whose activity is to be
 measured.

f détecteur m à source interne gazeuse
e detector m de fuente interna de gas
i rivelatore m a sorgente interna di gas
n detector met inwendige gasvormige bron
d Nachweisgerät n mit innerer Gasquelle

1754 DEUTERIDE,   ch
      HEAVY HYDRIDE
f hydrure m lourd
e deuteruro m, hidruro m pesado
i deuteruro m, idruro m pesante
n deuteride n, zwaar hydride n
d Deuterid n, schweres Hydrid n

1755 DEUTERIUM,   ch, is
      HEAVY HYDROGEN
      An isotope of hydrogen having the mass
 number 2.
f deutérium m, hydrogène m lourd
e deuterio m, hidrógeno m pesado
i deuterio m, idrogeno m pesante
n deuterium n, zware waterstof
d Deuterium n, schwerer Wasserstoff m

1756 DEUTERIUM COOLED REACTOR,   rt
      DEUTERIUM OXIDE COOLED
      REACTOR,
      HEAVY WATER COOLED REACTOR
      A nuclear reactor in which heavy water
 is the cooling material.
f réacteur m refroidi à eau lourde
e reactor m enfriado de agua pesada
i reattore m raffreddato ad acqua pesante
n met zwaar water gekoelde reactor
d schwerwassergekühlter Reaktor m

1757 DEUTERIUM LEAK DETECTOR   sa
      The detector used in deuterium-moderated
 reactor to detect leaks.
f détecteur m de fuites de deutérium
e detector m de fugas de deuterio
i rivelatore m di fughe di deuterio
n deuteriumlekdetector
d Deuteriumleckdetektor m

1758 DEUTERIUM MODERATED   rt
      REACTOR,
      DEUTERIUM OXIDE MODERATED
      REACTOR,
      HEAVY WATER MODERATED
      REACTOR
      A nuclear reactor in which heavy water
 is the moderating material.
f réacteur m modéré à eau lourde
e reactor m moderado de agua pesada
i reattore m moderato ad acqua pesante
n met zwaar water gemodereerde reactor
d D_2O-Reaktor m,

      schwerwassermoderierter Reaktor m

1759 DEUTERIUM TARGET   ra
      A convenient source of monoenergetic
 neutrons consisting of absorbed deuterium
 in a thin layer of titanium or zirconium
 deposited on suitable backing metals.
f cible f en deutérium
e blanco m en deuterio

i   bersaglio *m* in deuterio
n   deuteriumtrefplaat
d   Deuterium-Treffplatte *f*

1760 DEUTERON   np
A nucleus of deuterium.
f   deutéron *m*
e   deuterón *m*
i   deuterone *m*
n   deuteron *n*
d   Deuteron *n*

1761 DEUTERON-ALPHA REACTION   np
The capture of a bombarding deuteron
by the target nucleus; the compound
nucleus thus formed ejects an alpha
particle and thereby changes into a new
nucleus.
f   réaction *f* deutéron-alpha
e   reacción *f* deuterón-alfa
i   reazione *f* dèuterone-alfa
n   deuteron-alfa-reactie
d   Deuteron-Alpha-Reaktion *f*

1762 DEUTERON-NEUTRON   np
REACTION
The capture of a bombarding deuteron by
the target nucleus; the compound nucleus
thus formed ejects a neutron and thereby
changes into a new nucleus.
f   réaction *f* deutéron-neutron
e   reacción *f* deuterón-neutrón
i   reazione *f* deuterone-neutrone
n   deuteron-neutron-reactie
d   Deuteron-Neutron-Reaktion *f*

1763 DEUTERON-PROTON REACTION   np
The capture of a bombarding deuteron
by the target nucleus; the compound
nucleus thus formed ejects a proton and
thereby changes into a new nucleus.
f   réaction *f* deutéron-proton
e   reacción *f* deuterón-protón
i   reazione *f* deuterone-protone
i   deuteron-proton-reactie
d   Deuteron-Proton-Reaktion *f*

1764 DEVIATION FROM THE   co
DESIRED VALUE
The difference between the value of the
controlled condition and the desired value
at the instant under consideration.
f   écart *m*
e   desviación *f*
i   deviazione *f*
n   afwijking
d   Abweichung *f*

1765 DEVIATION FROM THE   co
INDEX VALUE
The difference between the true value of
the controlled variable and the value of
the controlled variable corresponding with
the set value.
f   écart *m* de consigne, écart *m* de réglage
e   desviación *f* de consigna,
desviación *f* de referencia

i   scarto *m* di regolazione
n   regelafwijking
d   Regelabweichung *f*

1766 DEWINDTITE   mi
A mineral containing lead, uranium and
phosphorus.
f   dewindtite *f*
e   dewindtita *f*
i   dewindtite *f*
n   dewindtiet *n*
d   Dewindtit *m*

1767 DIAGNOSTIC RADIOLOGY   xr
Radiology applied to medical diagnosis.
f   radiodiagnostic *m*
e   diagnóstico *m* radiológico
i   radiodiagnostica *f*
n   radiologiediagnostiek
d   Strahlendiagnostik *f*

1768 DIAGNOSTIC RÖNTGENOLOGY (GB),xr
RADIODIAGNOSIS (US)
Röntgenology applied to medical diagnosis.
f   röntgendiagnostic *m*
e   röntgendiagnóstico *m*
i   röntgendiagnostica *f*
n   röntgendiagnostiek
d   Röntgendiagnostik *f*

1769 DIAGNOSTIC TUBE   xr
An X-ray tube designed for use in
diagnostic radiology.
f   tube *m* à diagnostic
e   tubo *m* de diagnóstico
i   tubo *m* a diagnostica
n   diagnostiekbuis
d   Diagnostikröhre *f*

1770 DIAGNOSTIC TYPE PROTECTIVE   xr
TUBE HOUSING
Housing for which the leakage radiation
1 metre(er) from X-ray source when the
tube is continuously operated, with closed
window, at its maximum rated current for
the maximum rated voltage, is reduced to
a value considered safe for diagnostic
applications.
f   gaine *f* protectrice de tube à diagnostic
e   envuelta *f* protectora de tubo de diagnóstico
i   guaina *f* protettiva per tubo a diagnostica
n   omhulling voor diagnostiekbuis
d   diagnostisches Röhrenschutzgehäuse *n*

1771 DIALKYL PHOSPHORIC ACID
PROCESS, see 1613

1772 DIAMAGNETISM OF THE   pp
PLASMA PARTICLES,
PLASMA DIAMAGNETISM
The diamagnetic effect exhibited by a
plasma in a magnetic field as a result of
the Larmor motion of the charged
particles in the magnetic field.
f   diamagnétisme *m* des particelles du
plasma

e diamagnetismo *m* de las partículas del
   plasma
i diamagnetismo *m* delle particelle del
   plasma
n diamagnetisme *n* van de plasmadeeltjes
d Diamagnetismus *m* der Plasmateilchen

1773 DIAMOND COUNTER, see 1552

1774 DIAPHRAGM   ge, ph, ra
An opening which can be regulated so as
to limit the beam of rays used.
f diaphragme *m*
e diafragma *m*
i diaframma *m*
n diafragma *n*
d Blende *f*

1775 DIBUTOXYDIETHYL ETHER,
      DIBUTYL CARBITOL,
      DIETHYLENE GLYCOL DIBUTYL
      ETHER,
      see 835

1776 DIBUTOXYDIMETHYL ETHER   mt
An organic solvent used in separating
uranium and plutonium from fission
products during processing chemically
dissolved irradiated fuel elements.
f dibutoxydiméthyléther *m*
e dibutoxidimetiléter *m*
i dibutossidimetiletere *m*
n dibutoxydimethylether
d Dibutoxydimethyläther *m*

1777 DIDERICHITE,   mi
      RUTHERFORDINE
A rare alteration product of uraninite.
f didérichite *f*, rutherfordine *f*
e dideriquita *f*, rutherfordina *f*
i diderichite *f*, rutherfordine *f*
n diderichiet *n*, rutherfordine *n*
d Diderichit *m*, Rutherfordin *n*

1778 DIFFERENCE APPARATUS   ge
Apparatus in which the output signal is
a function of the difference between two
input signals.
f appareil *m* à différence
e aparato *m* por diferencia
i apparecchio *m* a differenza
n verschilapparaat *n*
d differenzbildendes Gerät *n*

1779 DIFFERENCE IONIZATION   ic
      CHAMBER,
      DIFFERENTIAL IONIZATION
      CHAMBER
An ionization chamber composed of two
portions designed in such a manner that
the output signal corresponds to the
difference between the ionization currents
of the two portions.
f chambre *f* d'ionisation à différence,
   chambre *f* d'ionisation différentielle
e cámara *f* de ionización de diferencia
   cámara *f* de ionización diferencial

i camera *f* d'ionizzazione a differenza,
   camera *f* d'ionizzazione differenziale
n differentieel ionisatievat *n*,
   verschilionisatievat *n*
d differenzbildende Ionisationskammer *f*

1780 DIFFERENCE LINEAR   ec, ma
      RATEMETER
A linear ratemeter in which the output
signal is a function of the difference
between two input signals.
f ictomètre *m* linéaire à différence
e impulsímetro *m* lineal por diferencia
i rateometro *m* di conteggio lineare a
   differenza
n lineaire verschilsnelheidsmeter,
   lineaire verschiltempometer
d linearer Differenz-Zählratenmesser *m*

1781 DIFFERENCE MEASURING   me
      ASSEMBLY
A measuring assembly in which the output
signal is a function of the difference
between two input signals.
f ensemble *m* de mesure à différence
e conjunto *m* de medida por diferencia
i complesso *m* di misura a differenza
n verschilmeetopstelling
d differenzbildende Messanordnung *f*

1782 DIFFERENCE NUMBER,   np
      ISOTOPIC NUMBER,
      NEUTRON EXCESS
Of a nuclide, the difference I between the
neutron number and the number of protons
Z.
f excès *m* de neutrons
e exceso *m* de neutrones
i eccesso *m* di neutroni
n neutronenoverschot *n*
d Neutronenüberschuss *m*

1783 DIFFERENCE SCALER   ct
A scaler having two inputs, which adds one
to its contents for each pulse arriving at
one input and subtracts one from its
contents for each pulse arriving at the
other input.
f échelle *f* de comptage à différence
e escala *f* de recuento por diferencia
i unità *f* per conteggio a differenza
n verschilpulsteller
d Differenzzähler *m*

1784 DIFFERENTIAL ABSORPTION   ab, ra
The variation in absorbed dose in
different tissues when irradiated in the
same way, due to difference in atomic
composition.
f absorption *f* différentielle
e absorción *f* diferencial
i assorbimento *m* differenziale
n differentiële absorptie
d differentielle Konzentration *f*

1785 DIFFERENTIAL ABSORPTION   ab, ra
      RATIO
      The ratio of the activity per unit mass in
      a particular tissue, e.g. a malignant tissue,
      to the activity per unit mass in some other
      particular tissue, e.g.adjacent normal
      tissue, measured at a given time after
      administration of an element which is
      isotopically labelled.
f facteur m d'absorption différentielle
e factor m de absorción diferencial,
   tasa f de absorción diferencial
i fattore m d'assorbimento differenziale
n differentiële-absorptie-verhouding
d differentielles Konzentrationsverhältnis n

1786 DIFFERENTIAL BACKWASHING,   ch
      SPLITTING
      A method of separating uranium and
      plutonium by using differential back-
      washing following either one or two cycles
      of primary decontamination.
f réextraction f différentielle
e reextracción f diferencial
i riestrazione f differenziale
n differentiële herextractie
d differentielle Rückwaschung f

1787 DIFFERENTIAL CONTROL   co, rt
      ROD WORTH
      Effect of the displacement of a control rod
      on the reactivity of a critical nuclear
      reactor.
f efficacité f différentielle
e efecto m de desplazamiento
i efficacità f differenziale
n verplaatsingseffect n
d Verschiebungseffekt m

1788 DIFFERENTIAL CROSS SECTION   cs
      The cross section for an interaction
      process involving one or more outgoing
      particles with specified direction or energy
      per unit interval of solid angle or energy.
f section f efficace différentielle
e sección f eficaz diferencial
i sezione f d'urto differenziale
n differentiële doorsnede
d differentieller Wirkungsquerschnitt m

1789 DIFFERENTIAL   ct, ec
      DISCRIMINATOR,
      SINGLE CHANNEL PULSE
      HEIGHT ANALYZER
      An instrument incorporating a pulse
      height selector in which the channel width
      is preset and the threshold varied, either
      manually or automatically, to scan the
      amplitude spectrum of the incoming
      pulses .
f analysateur m d'amplitude à canal mobile
e analizador m de amplitud con canal móvil
i analizzatore m d'ampiezza a canale
   mobile
n eenkanaal- amplitudeanalysator
d Einkanalamplitudenanalysator m

1790 DIFFERENTIAL ENERGY   np
      FLUX DENSITY
      That part of the energy flux density
      resulting from particles or photons
      having a specified direction, energy or
      both per unit interval of solid angle,
      energy or both.
f densité f de flux énergétique différentielle
e densidad f de flujo de energía
   diferencial
i densità f di flusso d'energia differenziale
n differentiële energiefluxdichtheid
d Differentialflussdichte f der Energie

1791 DIFFERENTIAL PARTICLE   np
      FLUX DENSITY
      That part of the particle flux density
      resulting from particles or photons
      having a specified direction, energy or
      both per unit interval of solid angle,
      energy or both.
f densité f de flux particulaire
   différentielle
e densidad f de flujo de partículas
   diferencial
i densità f di flusso di particelle
   differenziale
n differentiële deeltjesfluxdichtheid
d differentielle Teilchenflussdichte f

1792 DIFFERENTIAL RECOVERY   md
      RATE
      Among tissues recovering at different
      rates, those having slower rates will
      ultimately suffer greater damage from
      a series of irradiations.
f vitesse f différentielle de guérison
e velocidad f diferencial de curación
i velocità f differenziale di guarigione
n differentieel hersteltempo n
d differentielle Erholungsrate f

1793 DIFFERENTIAL SCATTERING   cs
      CROSS SECTION
      Cross section for scattering of a particle
      from its initial velocity to a new velocity
      per unit solid angle per unit speed at the
      new velocity.
f section f différentielle pour diffusion
e sección f diferencial para dispersión
i sezione f differenziale per. deviazione
n differentiële doorsnede voor verstrooiing
d Differentialquerschnitt m für Streuung

1794 DIFFERENTIAL SHRINKAGE   mt
      Due to a gradient of flux and temperature
      from the inside of a reactor brick to the
      outside.
f différence f de contraction
e diferencia f de contracción
i differenza f di contrazione
n krimpverschil n
d Schwundunterschied m

1795 DIFFERENTIAL WHITE COUNT,   md
      TOTAL WHITE COUNT
      The number of each variety of white
      corpuscles in a count of 100.

f   formule *f* leucocytaire
e   cuadro *m* hemático blanco,
    fórmula *f* leucocitaria
i   quadro *m* leucocitario
n   aantal *n* van de verschillende witte
    bloedlichaampjes
d   Anzahl *f* der verschiedenen weissen
    Blutkörperchen

1796 DIFFRACTION   ge, ra
    The phenomenon of the preferential
    scattering of a beam of radiation in
    certain directions.
f   diffraction *f*
e   difracción *f*
i   diffrazione *n*
n   buiging, diffractie
d   Beugung *f*

1797 DIFFRACTION ANALYSIS   xr
    The study of atomic arrangement by
    means of X-rays or material particles.
f   analyse *f* de cristaux par diffraction
e   análisis *f* de cristales por difracción
i   analisi *f* dei cristalli per diffrazione
n   buigingsanalyse
d   Beugungsanalyse *f*

1798 DIFFRACTION GRATING   cr
    A polished metal or glass surface having
    closely spaced parallel reflecting
    grooves that produce a spectrum by
    interference between different colo(u)rs of
    light when white light arrives at a certain
    angle.
f   réseau *m* de diffraction
e   red *f* de difración
i   reticolo *m* di diffrazione
n   buigingsrooster *n*, diffractierooster *n*
d   Beugungsgitter *n*

1799 DIFFRACTION INSTRUMENT,   me, ra
        DIFFRACTOMETER
    Any device employed for studying the
    structure of matter or the properties of
    radiation by means of the diffraction of
    waves.
f   diffractomètre *m*, instrument *m* à diffraction
e   difractómetro *m*,
    instrumento *m* de difracción
i   diffrattometro *m*,
    strumento *m* a diffrazione
n   diffractie-instrument *n*, diffractometer
d   Beugungsmessgerät *n*

1800 DIFFRACTION OF X-RAYS,   xr
        X-RAY DIFFRACTION
f   diffraction *f* de rayons X
e   difracción *f* de rayos X
i   diffrazione *f* di raggi X
n   buiging van röntgenstralen
d   Beugung *f* von Röntgenstrahlen

1801 DIFFRACTION SCATTERING,   np
        SHADOW SCATTERING
    The result of interference between the
    incident wave and scattered waves.

f   diffusion *f* diffractive
e   dispersión *f* difrangente
i   deviazione *f* diffrangente
n   diffractieverstrooiing
d   Schattenstreuung *f*

1802 DIFFUSATE   ch
    Material which, in the process of dialysis,
    has diffused or passed through the
    separating membrane.
f   diffusat *m*
e   difusato *m*
i   diffusato *m*
n   diffusaat *n*
d   Diffusat *n*

1803 DIFFUSE SCATTERING   cr, np
    A special case of incoherent scattering.
f   diffusion *f* diffuse
e   dispersión *f* difusa
i   deviazione *f* diffusa
n   diffuse verstrooiing
d   diffuse Streuung *f*

1804 DIFFUSED JUNCTION   ra
        SEMICONDUCTOR DETECTOR
    A semiconductor detector in which the
    P-N or N-P junction is produced by
    diffusion of donor or acceptor purities.
f   détecteur *m* semiconducteur à jonction
    diffusée
e   detector *m* semiconductor por junta de
    difusión
i   rivelatore *m* a semiconduttore a
    giunzione diffusa
n   halfgeleiderdetector met gediffundeerde
    overgangslaag
d   Diffusions-Halbleiterdetektor *m*

1805 DIFFUSER   rt
    A conduit in which the velocity of liquids
    or gases in converted into pressure.
f   diffuseur *m*
e   difusor *m*
i   diffusore *m*
n   diffusor
d   Ausströmraum *m*, Diffusor *m*

1806 DIFFUSION   np
    In nuclear physics, the passage of particles
    through matter in such a way that the
    probability of scattering is large compared
    with that of capture.
f   diffusion *f*
e   difusión *f*
i   diffusione *f*
n   diffusie
d   Diffusion *f*

1807 DIFFUSION APPROXIMATION   ma, np
    Consists in describing the behavio(u)r of
    the neutron population in a medium by
    means of the diffusion theory.
f   approximation *f* de diffusion
e   aproximación *f* de difusión
i   approssimazione *f* di diffusione
n   diffusiebenadering
d   Diffusionsnäherung *f*

1808 DIFFUSION AREA   np
In an infinite homogeneous medium,
one-sixth of the mean square distance
travel(l)ed by a particle of a given type
and class from appearance to disappearance
in an infinite homogeneous medium.
f aire f de diffusion
e área f de difusión
i area f di diffusione
n diffusieoppervlak n
d Diffusionsfläche f

1809 DIFFUSION BARRIER (US),   is
     DIFFUSION MEMBRANE (GB)
A device used in the process of separating
isotopes by means of diffusion.
f barrière f de diffusion
e barrera f de difusión
i barriera f antidiffusiva
n diffusiewand
d Diffusionswand f

1810 DIFFUSION CASCADE   ch, is
An isotope separation plant consisting of
an arrangement of diffusion membranes
and compressors such that the small
change in isotope ratio effected by the
diffusion of the process gas through a
single membrane is multiplied to give the
desired separation.
f cascade f de diffusion
e cascada f de difusión
i cascata f di diffusione
n diffusiecascade
d Diffusionskaskade f

1811 DIFFUSION CLOUD CHAMBER   ic, np
A cloud chamber that produces a supersaturated condition by means of diffusion
of vapo(u)r acted upon by a large
temperature gradient.
f chambre f à détente à diffusion
e cámara f de expansión de difusión
i camera f d'ionizzazione a diffusione
n diffusienevelvat n
d Diffusionsnebelkammer f

1812 DIFFUSION COEFFICIENT,
     DIFFUSIVITY, see 540

1813 DIFFUSION COEFFICIENT   np
     FOR NEUTRON FLUX DENSITY
The ratio of the neutron current density at
a particular energy to the negative gradient
of the neutron flux density at the same
energy in the direction of that current.
f coefficient m de diffusion pour la densité
  de flux de neutrons
e coeficiente m de difusión para la densidad
  de flujo de neutrones
i coefficiente m di diffusione per la densità
  di flusso di neutroni
n diffusiecoëfficiënt voor neutronenfluxdichtheid
d Diffusionskoeffizient m für Neutronenflussdichte

1814 DIFFUSION COEFFICIENT   np
     FOR NEUTRON NUMBER DENSITY
$J_x = - D_m \frac{\delta n}{\delta x}$, where J is the x-component
of the neutron current density and n is the
number density of neutrons.
f coefficient m de diffusion pour le nombre
  volumique
e coeficiente m de difusión para la densidad
  de neutrones
i coefficiente m di diffusione per la densità
  neutronica
n diffusiecoëfficiënt voor neutronendichtheid
d Diffusionskoeffizient m für Neutronenzahldichte

1815 DIFFUSION COLUMN   ch, is
A vertical tube, within which a radial
temperature gradient is maintained; used in
the separation of isotopes.
f colonne f de diffusion
e columna f de difusión
i colonna f di diffusione
n diffusiekolom
d Diffusionssäule f

1816 DIFFUSION CROSS SECTION   np
The effective target area presented by an
atom or a subatomic particle to an incident
particle or photon for the process of
diffusion.
f section f efficace pour diffusion
e sección f eficaz para difusión
i sezione f d'urto per diffusione
n werkzame doorsnede voor diffusie
d Wirkungsquerschnitt m für Diffusion

1817 DIFFUSION CURRENT DENSITY   ch
The density of the limiting current that is
reached by electrolytic migration of the
ions in a solution under the application of
a potential difference of the electrodes.
f densité f de courant pour diffusion
e densidad f de corriente para difusión
i densità f di corrente per diffusione
n stroomdichtheid voor diffusie
d Stromdichte f für Diffusion

1818 DIFFUSION ENERGY   np
Activation energy for diffusion.
f énergie f d'activation pour diffusion.
e energía f de activación para difusión
i energia f d'attivazione per diffusione
n activeringsenergie voor diffusie
d Aktivierungsenergie f für Diffusion

1819 DIFFUSION EQUATION,   ma
     ELEMENTARY DIFFUSION
     EQUATION
An equation used in studying the energy
distribution in motion of ions and electrons.
f équation f de diffusion
e ecuación f de difusión
i equazione f di diffusione
n diffusievergelijking
d Diffusionsgleichung f

1820 DIFFUSION FREQUENCY ma
Expression used in theoretical discussion
on the gain equation.
f   fréquence f de diffusion
e   frecuencia f de difusión
i   frequenza f di diffusione
n   diffusiefrequentie
d   Diffusionsfrequenz f

1821 DIFFUSION KERNEL, ma
YUKAWA KERNEL
A Green's function of the elementary
diffusion equation for a nuclear reactor.
f   noyau m de diffusion de Yukawa,
noyau m de l'intégrale de diffusion
e   núcleo m de difusión de Yukawa
i   nucleo m di diffusione di Yukawa
n   integraalkern voor diffusie, yukawakern
d   Diffusionskern m,
Integralkern m nach Yukawa

1822 DIFFUSION LENGTH ma
The mean distance travel(l)ed by a diffusing
particle from the point of its formation
to the point at which it is desired.
f   longueur f de diffusion
e   longitud f de difusión
i   lunghezza f di diffusione
n   diffusielengte
d   Diffusionslänge f

1823 DIFFUSION OF A PLASMA pp
ACROSS A MAGNETIC FIELD
The diffusion by a series of collisions of
individual charged particles of a plasma
across the magnetic field.
f   diffusion f du plasma à travers un champ
magnétique
e   difusión f del plasma a través de un
campo magnético
i   diffusione f del plasma attraverso un
campo magnetico
n   plasmadiffusie loodrecht op het
magneetveld
d   Plasmadiffusion f quer zum Magnetfeld

1824 DIFFUSION PARAMETERS np
Parameters which belong to the atmospheric
eddy diffusion equation for characterizing
the particular micrometeorological
situation.
f   paramètres pl de diffusion
e   parámetros pl de difusión
i   parametri pl di diffusione
n   diffusieparameters pl
d   Diffusionsparameter pl

1825 DIFFUSION PLANT ch, is
An isotopic separation plant based on the
diffusion of a gas through a membrane.
f   installation f de diffusion
e   instalación f de difusión
i   impianto m di diffusione
n   diffusie-installatie
d   Diffusionsanlage f

1826 DIFFUSION PUMP vt
A vacuum pump in which a stream of heavy
molecules, such as mercury vapo(u)r,
carries gas molecules out of the volume
being evacuated.
f   pompe f à diffusion
e   bomba f de difusión
i   pompa f a diffusione·
n   diffusiepomp
d   Diffusionspumpe f

1827 DIFFUSION STACK, rt
SIGMA PILE
An assembly of moderating material
containing a neutron, used to study the
neutron properties of the moderator.
f   colonne f sigma
e   pila f sigma
i   pila f sigma
n   sigmazuil
d   Sigmaanordnung f

1828 DIFFUSION THEORY np
An approximate theory for the diffusion of
particles, especially neutrons, based on
the assumption that in a homogeneous
medium the current density is proportional
to the gradient of the flux density.
f   théorie f de la diffusion
e   teoría f de la difusión
i   teoria f della diffusione
n   diffusietheorie
d   Diffusionstheorie f

1829 DIFFUSION TIME pp
Of a particle in a plasma, the mean time
taken by a particle, diffusing across the
magnetic field, to pass out of the confined
volume.
f   temps m de diffusion
e   tiempo m de difusión
i   tempo m di diffusione
n   diffusietijd, ontsnappingstijd
d   Diffusionszeit f

1830 DIFFUSIONAL BOND, rt
METALLURGICAL BOND
A bond in which the materials are so close
that interatomic forces are operative.
f   liaison f métallurgique
e   ligazón f metalúrgica
i   legamento f metallurgico,
saldatura f per diffusione
n   metallurgische binding
d   metallurgische Verbindung f,
metallurgischer Verband m

1831 DIGESTIVE CANAL, see 180

1832 DIGITAL SIGNAL ec
A signal which, in a quantized (numeral)
form represents the value of a physical
quantity.
f   signal m numérique
e   señal f digital, señal f numérica
i   segnale m digitale, segnale m numerico

n digitaal signaal *n*, numeriek signaal *n*
d digitales Signal *n*, numerisches Signal *n*

1833 DIGITAL TIME-CONVERTER   ma
A sub-assembly designed to provide an
output signal which is the digital representation of the time interval between two
input signals.
f convertisseur *n* numérique de temps
e convertidor *m* numérico de tiempo
i convertitore *m* numerico di tempo
n tijd-digitaalomzetter
d digitaler Zeitintervallwandler *m*

1834 DILUENT   ch
Chemically inert solvent added to
complexing solvent to facilitate solvent
extraction.
f diluant *m*
e diluente *m*
i diluente *m*
n verdunningsmiddel *n*
d Verdünnungsmittel *n*

1835 DILUTION   ch
The act of increasing the proportion of
solvent to solute in any solution.
f dilution *f*
e dilución *f*
i diluizione *f*
n verdunning
d Verdünnung *f*

1836 DILUTION ANALYSIS   an, is
A process used in determining the radiochemical purity of a labelled (tagged)
compound.
f analyse *f* par dilution
e análisis *f* por dilución
i analisi *f* per diluizione
n oplossingsanalyse
d Lösungsanalyse *f*

1837 DILUTION EFFECT   rw
Effect of the mixing of radioactive effluent
with e.g. seawater.
f effet *m* de dilution
e efecto *m* de dilución
i effetto *m* di diluizione
n verdunningseffect *n*
d Verdünnungseffekt *m*

1838 DILUTION RATIO   ch
The dilution equation $v = v_o(s_o(s - 1))$

which applies when a small volume v_o with

a specific activity s_o is mixed with a volume

v of inert material resulting in a specific
activity s provides a valuable method of
analysing complex mixtures.
f rapport *m* de dilution
e relación *f* de dilución
i rapporto *m* di diluizione
n verdunningsverhouding
d Verdünnungsverhältnis *n*

1839 DIMENSIONAL CHANGE   rt
A change occurring with the graphite in
an advanced gas-cooled reactor.
f altération *f* de dimensions
e alteración *f* de dimensiones
i variazione *f* di dimensioni
n wijziging van de afmetingen
d Abänderung *f* der Abmessungen

1840 DIMINUTION OF CONTRAST   xr
f réduction *f* de contraste
e disminución *f* de contraste
i diminuzione *f* di contrasto
n contrastvermindering
d Kontrastverminderung *f*

1841 DI-NEUTRON   np
An unstable system composed of two
neutrons.
f di-neutron *m*
e di-neutrón *m*
i di-neutrone *m*
n di-neutron *n*
d Di-Neutron *n*

1842 DIP COUNTER TUBE   ct
A counter tube especially developed to be
dipped into the liquid whose activity is to
be measured.
f tube *m* compteur à immersion
e tubo *m* contador de inmersión
i tubo *m* contatore ad immersione
n dompeltelbuis
d Tauchzählrohr *n*

1843 DIPHENYL, see 651

1844 DIPLET,   sp
     DOUBLET
A spectrum line which, when examined
under high resolution, is composed of two
closely-packed fine lines.
f doublet *m*
e doblete *m*
i doppietto *m*
n doublet *n*
d Dublett *n*

1845 DIPLOID CELL   md
Having two complete sets of chromosomes,
the normal somatic number of chromosomes in higher organisms, as opposed to
the half number formed in mature germ
cells.
f cellule *f* diploïde
e célula *f* diploide
i cellula *f* diploide
n diploïde cel
d diploide Zelle *f*

1846 DIPOLE,   np
     ELECTRIC DIPOLE
A combination of two electrically or
magnetically charged particles of opposite
sign which are separated by a very small
distance.
f dipôle *m*

e dipolo *m*
i dipolo *m*
n dipool
d Dipol *m*

1847 DIPOLE LAYER,   np
      ELECTRIC DOUBLE LAYER
      A hypothetical distribution of charge
      comprising a layer of positive charge
      and a layer of negative charge.
f couche *f* dipôle
e capa *f* dipolo
i strato *m* dipolare
n dipoollaag
d Dipolschicht *f*, Dipolzone *f*

1848 DIPOLE MOMENT   np
      The product of one of the charges of a
      dipole unit by the distance separating the
      two dipolar charges.
f moment *m* dipôle
e momento *m* dipolo
i momento *m* dipolare
n dipoolmoment *n*
d Dipolmoment *m*

1849 DI-PROTON   np
      An unstable system composed of two
      protons.
f di-proton *m*
e di-protón *m*
i di-protone *m*
n di-proton *n*
d Di-Proton *n*

1850 DIRAC CLASSICAL ELECTRON   np
      THEORY
      Method of defining the force on an electron
      by subtracting the proper field from the
      external field.
f théorie *f* classique des électrons de Dirac
e teoría *f* clásica de los electrones de
   Dirac
i teoria *f* classica degli elettroni di Dirac
n klassieke elektronentheorie van Dirac
d Diracsche Theorie *f* des Elektrons

1851 DIRAC $\hbar$ ma
      h BAR
      The symbol $\hbar$, usually called h bar, for
      the universal $h/2\pi$, where h is the
      Planck constant.
f constante *f* réduite de Planck,
   fonction *f* de Dirac
e función *f* de Dirac
i funzione *f* di Dirac
n diracfunctie
d Dirac-Funktion *f*

1852 DIRECT CURRENT   ec
      AMPLITUDE DISCRIMINATOR UNIT
      A basic function unit which gives an output
      signal when the d.c. input voltage or
      current exceeds a given threshold value.
f élément *m* discriminateur d'amplitude
   à courant continu

e unidad *f* discriminadora de amplitud por
   corriente continua
i elemento *m* discriminatore d'ampiezza
   a corrente continua
n gelijkstroomamplitudediscriminator
d Spannungsdiskriminator *m*

1853 DIRECT-CYCLE INTEGRAL   rt
      BOILING REACTOR
      A boiling water reactor with direct
      cooling and a built-in heat exchanger.
f réacteur *m* à eau bouillante à refroidissement direct et échangeur de chaleur
   incorporé
e reactor *m* de agua hirviente de enfriamiento
   directo y cambiador de calor encerrado
i reattore *m* ad acqua bollente a raffreddamento diretto e scambiatore di calore
   incorporato.
n direct gekoelde kokend-water-reactor met
   ingebouwde warmte-uitwisselaar
d direktgekühlter Siedewasserreaktor *m*
   mit eingebautem Wärmeaustaucher

1854 DIRECT-CYCLE REACTOR   rt
      A nuclear reactor in which the primary
      coolant is used directly to produce useful
      power.
f réacteur *m* à cycle direct
e reactor *m* de ciclo directo
i reattore *m* a ciclo diretto
n reactor met directe kringloop
d Reaktor *m* mit direktem Kreislauf

1855 DIRECT FISSION YIELD,   np
      INDEPENDENT FISSION YIELD,
      PRIMARY FISSION YIELD
      The fraction of fissions giving rise to a
      particular nuclide before any beta or
      gamma decay has occurred.
f rendement *m* de fission primaire
e rendimiento *m* de fisión primario
i resa *f* diretta di fissione
n primaire splijtingsopbrengst
d Fragmentausbeute *f*,
   primäre Spaltausbeute *f*,
   unabhängige Spaltausbeute *f*

1856 DIRECT INTERACTION   np
      A nuclear reaction in which a compound
      nucleus is not formed.
f interaction *f* directe
e interacción *f* directa
i interazione *f* diretta
n directe wisselwerking
d direkte Wechselwirkung *f*

1857 DIRECT RADIATION   me
      PROXIMITY INDICATOR
      A proximity indicator using direct
      ionizing radiation.
f signaleur *m* de proximité par rayonnement
   ionisant direct
e indicador *m* de proximidad por radiación
   ionizante directa
i segnalatore *m* di prossimità a radiazione
   ionizzante diretta

n   nabijheidsindicator berustend op directe
    ioniserende straling
d   Wegindikator *m* durch direkte ionisierende
    Strahlung

1858 DIRECT X-RAY ANALYSIS,   cr
      HEAVY ATOM METHOD
      In some special cases, the crystal
      structure can be found, when there is a
      heavy atom at the centre(er) of symmetry of
      the unit cell, outweighing the contri-
      butions of the other atoms to be
      scattering.
f   analyse *f* par radiocristallographie
    directe
e   análisis *f* por radiocristalografía directa
i   analisi *f* per radiocristallografia diretta
n   directe röntgenstructuuranalyse
d   direkte Röntgenstrukturanalyse *f*

1859 DIRECTIONAL CORRELATION   gr
      OF SUCCESSIVE GAMMA RAYS
      When two gamma rays are successively
      emitted from the same nucleus, their
      directions and planes of polarization are
      not entirely independent. The directional
      correlation is experimentally observed as
      a chance in coincidence counting rate as
      the angle between the lines joining the
      source with the two counters is varied.
f   corrélation *f* directionnelle de rayons
    gamma successifs
e   correlación *f* direccional de rayos gamma
    sucesivos
i   correlazione *f* direzionale di raggi
    gamma successivi
n   richtingscorrelatie van elkaar volgende
    gammastralen
d   Richtungskorrelation *f* von einander
    folgenden Gammastrahlen

1860 DIRECTIONAL COUNTER   ct
      A counter so arranged as to be more
      sensitive to radiation from some directions
      than from others.
f   compteur *m* directionnel, compteur *m* dirigé
e   contador *m* direccional, contador *m* dirigido
i   contatore *m* direzionale
n   gerichte teller
d   Richtzähler *m*

1861 DIRECTLY IONIZING   np
      PARTICLES
      Charged particles (electrons, protons,
      alpha particles, etc.) having sufficient
      kinetic energy to produce ionization by
      collision.
f   particules *pl* directement ionisantes
e   partículas *pl* directamente ionizantes
i   particelle *pl* direttamente ionizzanti
n   direct ioniserende deeltjes *pl*
d   direkt ionisierende Teilchen *pl*

1862 DIRECTLY IONIZING RADIATION   ra
      Radiation consisting of directly
      ionizing particles.
f   rayonnement *m* directement ionisant

e   radiación *f* directamente ionizante
i   radiazione *f* direttamente ionizzante
n   direct ioniserende straling
d   direkt ionisierende Strahlung *f*

1863 DIRTY BOMB   nw
      A nuclear bomb with a large amount of
      radioactive fall-out.
f   bombe *f* sale
e   bomba *f* sucia
i   bomba *f* sudicia
n   vuile bom
d   schmutzige Bombe *f*

1864 DIRTY BUBBLE CHAMBER   ic
      A bubble chamber containing gaskets,
      joints, etc. inside.
f   chambre *f* à bulles à parois rugueuses
e   cámara *f* de burbujas con paredes rugosas
i   camera *f* a bolle con pareti ruvide
n   bellenvat *n* met ruwe wanden
d   Blasenkammer *f* mit rohen Wänden

1865 DISADVANTAGE FACTOR   rt
      In a reactor cell, the ratio of the average
      neutron flux density in a material to that
      in the fuel.
f   facteur *m* de désavantage
e   factor *m* de desventaja
i   fattore *m* di svantaggio
n   nadeelfactor, ongunstfactor
d   Absenkungsfaktor *m*,
    Absenkungsverhältnis *n*

1866 DISCHARGE CHAMBER   pp
      Component part of a thermonuclear
      device.
f   chambre *f* de décharge,
    espace *f* de décharge
e   cámara *f* de descarga,
    espacio *m* de descarga
i   camera *f* di scarica, spazio *m* di scarica
n   ontladingskamer, ontladingsruimte
d   Entladungskammer *f*, Entladungsraum *m*

1867 DISCOMPOSITION,   cr
      KNOCKING OUT
      The process in which an atom is knocked
      out of its position in a crystal lattice by
      direct nuclear impact.
f   déplacement *m* d'un atome
e   dislocación *f* de un átomo,
    expulsión *f* de un átomo
i   dislocazione *f* d'un atomo
n   atoomdislocatie, wegslaan *n*
d   Atomumlagerung *f*,
    Besetzung *f* von Zwischengitterplätzen

1868 DISCOMPOSITION EFFECT,   rt
      WIGNER EFFECT
      In reactor operation, the change in physical
      properties of graphite resulting from
      displacement of lattice atoms by high
      energy neutrons and other energetic
      particles.
f   effet *m* Wigner
e   efecto *m* Wigner

i   effetto *m* Wigner
n   wignereffect *n*
d   Wigner-Effekt *m*

1869 DISCRIMINATING UNIT   ec
A basic function unit comprising an
electronic circuit which gives an output
pulse for each pulse whose amplitude
lies above a given threshold value.
f   élément *m* discriminateur
e   discriminador *m*
i   discriminatore *m*
n   discriminator
d   Diskriminator *m*

1870 DISCRIMINATION CURVE   ct
A curve showing the counting rate as a
function of the discrimination level.
f   courbe *f* de discrimination
e   curva *f* de discriminación
i   curva *f* di discriminazione
n   discriminatiekromme
d   Diskriminationskurve *f*

1871 DISCRIMINATION FACTOR   bi, is
In a given biological process the relative
transmission or utilization rate of an
element A in relation to an element B.
f   facteur *m* de discrimination
e   factor *m* de discriminación
i   fattore *m* di discriminazione
n   discriminatiefactor
d   Diskriminationsfaktor *m*

1872 DISINTEGRATION, see 1638

1873 DISINTEGRATION CHAIN, see 1639

1874 DISINTEGRATION CONSTANT, see 1640

1875 DISINTEGRATION CURVE, see 104

1876 DISINTEGRATION ENERGY, see 1642

1877 DISINTEGRATION HEAT, see 1643

1878 DISINTEGRATION OF   np
ELEMENTARY PARTICLES
Spontaneous transformation of unstable
elementary particles into other elementary
particles.
f   désintégration *f* des particules
élémentaires
e   desintegración *f* de las partículas
elementales
i   disintegrazione *f* delle particelle
elementari
n   verval *n* van elementaire deeltjes
d   Zerfall *m* von Elementarteilchen

1879 DISINTEGRATION PARTICLE, see 1645

1880 DISINTEGRATION PRODUCT, see 1646

1881 DISINTEGRATION RATE, see 1647

1882 DISINTEGRATION SCHEME, see 1648

1883 DISINTEGRATION SEQUENCE, see 1649

1884 DISK SOURCE   ar, md, rd
A source in which the radioisotope is
deposited on a disk of platinum or gold.
f   source *f* discoïde
e   fuente *f* discoidal
i   sorgente *f* in forma di disco
n   schijfvormige bron
d   scheibenförmige Quelle *f*

1885 DISLOCATION   cr
The result of a slip along a surface which
ends somewhere inside the crystal.
f   dislocation *f*
e   dislocación *f*
i   dislocazione *f*
n   dislocatie
d   Versetzung *f*

1886 DISLOCATION EDGE   cr
A point on a dislocation line at which the
line passes from one plane to an adjacent
higher one, parallel to the first plane.
f   coin *m* de dislocation
e   esquina *f* de dislocación
i   scalino *m* in una dislocazione
n   dislocatiekant
d   Versetzungsstufe *f*

1887 DISLOCATION LINE   cr
The edge where the slip of a dislocation
ends.
f   ligne *f* de dislocation
e   línea *f* de dislocación
i   linea *f* di dislocazione
n   dislocatielijn
d   Versetzungslinie *f*

1888 DISMANTLING   rt
Taking apart a fuel assembly with the
object of reprocessing it.
f   désassemblage *m*
e   desmantelamiento *m*
i   smontaggio *m*
n   slopen *n*
d   Abmontieren *n*

1889 DISORDERED SCATTERING   np
Diffuse scattering due to crystalline
imperfection.
f   diffusion *f* désordonnée
e   dispersión *f* en desorden
i   deviazione *f* in disordine
n   ongeordende verstrooiing
d   ungeordnete Streuung *f*

1890 DISORDERING   cr
Any process by which atoms are displaced
from or rearranged among their positions
in a crystal lattice, e.g., by ionizing
radiation.
f   création *f* de défaut
e   desorden *m*
i   creazione *f* di difetto
n   roosterverstoring, verstoring
d   Gitterstörung *f*

1891 DISPERSAL EFFECT   rw
Effect of discharging radioactive effluent
e.g. into the sea on the radioactivity of
the water.
f effet *m* de décharge, effet *m* d'écoulement
e efecto *m* de descarga, efecto *m* de salida
i effetto *m* di scarico,
   effetto *m* di spargimento
n lozingseffect *n*
d Abflusseffekt *m*, Entleerungseffekt *m*

1892 DISPERSION   gp
Colloidal system of one phase dispersed in
another, e.g. smoke from solid particles
in gas phase or emulsion formed from two
immiscible liquid phases.
f dispersion *f*
e dispersión *f*
i dispersione *f*
n dispersie
d Dispersion *f*

1893 DISPERSION FUEL   rt
A nuclear fuel in the form of fine particles
dispersed in another material.
f combustible *m* en dispersion
e combustible *m* de dispersión
i combustibile *m* in dispersione
n gedispergeerde splijtstof
d dispergierter Brennstoff *m*

1894 DISPERSION FUEL ELEMENT   fu
A fuel element in which the combustible
consists of a dispersion solution.
f élément *m* combustible en dispersion
e elemento *m* combustible de dispersión
i elemento *m* combustibile in dispersione
n dispersiesplijtstofelement *n*
d Dispersionsbrennelement *n*

1895 DISPERSION RELATIONS   pp
As used in plasma physics, the term refers
to the relation between the radian
frequency ω and the wave factor k for the
various kinds of wave motions and
instabilities which can occur in a plasma.
f relations *pl* de dispersion
e relaciones *pl* de dispersión
i relazioni *pl* di dispersione
n dispersieverhoudingen *pl*
d Dispersionsverhältnisse *pl*

1896 DISPLACEMENT   ma
The vector representing the change in
position of a particle.
f déplacement *m*
e desplazamiento *m*
i spostamento *m*
n verschuiving
d Verschiebung *f*

1897 DISPLACEMENT KERNEL   ma
A kernel in which the spatial dependence
is upon distance only between two points,
not upon direction.
f noyau *m* de déplacement

e núcleo *m* de desplazamiento
i nucleo *m* di spostamento
n verschuivingsintegraalkern
d Verschiebungsintegralkern *m*

1898 DISPLACEMENT LAWS,   ma, np
       DISPLACEMENT RULES
The rules which connect the type of radioactive decay with the displacement of the
daughter relative to the parent in the
periodic table of elements.
f lois *pl* de déplacement
e leyes *pl* de desplazamiento
i leggi *pl* di spostamento
n verschuivingswetten *pl*
d Verschiebungssätze *pl*

1899 DISPLACEMENT SPIKE   np
The zone of temporarily or permanently
displaced atoms produced in a solid or
liquid caused by the passage of a heavy
ion.
f zone *f* de déplacements
e zona *f* de átomos desplazados
i zona *f* d'atomi dislocati
n gebied *n* van verschoven atomen
d Umlagerungsbereich *m*,
   Umordnungsbereich *m*

1900 DISRUPTIVE VOLTAGE   pa
The voltage necessary to produce a
disruptive discharge between two conductors.
f tension *f* disruptive
e tensión *f* disruptiva
i tensione *f* disruttiva
n doorslagspanning
d Durchschlagsspannung *f*

1901 DISSIPATION OF ENERGY,   gp
       ENERGY DISSIPATION
A waste of energy.
f dissipation *f* d'énergie
e disipación *f* de energía
i dissipazione *f* d'energia
n energieverlies *n*
d Energieverlust *m*

1902 DISSOCIATIVE CAPTURE   np
A specific case of ion recombination.
f capture *f* dissociative
e captura *f* disociativa
i cattura *f* dissociativa
n dissociatieve vangst
d dissoziativer Einfang *m*

1903 DISSOCIATIVE IONIZATION   np
Ionization in which dissociation yields an
ion and an atom which may be in an
excited state.
f ionisation *f* dissociative
e ionización *f* disociativa
i ionizzazione *f* dissociativa
n dissociatieve ionisatie
d dissoziative Ionisation *f*

1904 DISSOLUTION,   ch
    SOLUTION
An extremely intimate mixture of variable
composition, of two or more.substances.
f dissolution f, solution f
e disolución f, solución f
i dissoluzione f, soluzione f
n oplossing
d Lösung f

1905 DISSOLVER GAS   rw
The term used with reference to the gases
and vapo(u)rs evolved when irradiated
reactor fuel elements are dissolved in
nitric acid prior to solvent extraction
processing.
f gaz m de dégagement
e gas m de reacción
i gas m di sviluppo
n ontwijkend gas n
d entweichendes Gas n

1906 DISSOLVER SOLUTION   rw
The solution obtained by dissolving
irradiated reactor fuel elements in nitric
or other acids in the first stage or head
end of the chemical processing of the fuels.
f solution f de combustible irradié
e solución f de combustible irradiado
i soluzione f di combustibile irradiato
n oplossing van bestraalde splijtstof
d Lösung f bestrahlter Brennstoff

1907 DISTANT COLLISION   np
When studying binary collisions between
charged particles, the collision during
which the relative velocity undergoes
only a small deviation, the value of the
impact parameter being much larger
than the critical impact parameters.
f. collision f lointaine
e colisión f a distancia
i urto m distante
n afstandscollisie
d entfernte Kollision f

1908 DISTILLATE   ch, is
The volatile product of a distillation
system.
f distillat m
e destilado m
i distillato m
n destillaat n
d Destillat n

1909 DISTILLATION   ch, is
The separation of the components of a
liquid mixture by partial vaporization of
the mixture and separate recovery of
vapo(u)r and residue.
f distillation f
e destilación f
i distillazione f
n destillatie
d Destillation f

1910 DISTILLATION COLUMN,   ch, is
    DISTILLATION TOWER
f colonne f de distillation
e columna f de destilación
i colonna f di distillazione
n destilleerkolom
d Destillationssäule f

1911 DISTILLATION METHOD   is
The separation of isotopes by means of
distillation.
f méthode f à distillation
e método m de destilación
i metodo m a distillazione
n destillatiemethode
d Destillationsverfahren n

1912 DISTRIBUTION COEFFICIENT,   ch
    DISTRIBUTION RATIO
A constant defining the distribution of an
ion between a solution and an ion exchange
resin in equilibrium under a fixed set of
conditions.
f coefficient m de distribution
e coeficiente m de distribución
i coefficiente m di distribuzione
n verdelingscoëfficiënt
d Verteilungskoeffizient m

1913 DISTRIBUTION FACTOR   ma, sa
The factor used in computing dose
equivalent to account for the non-uniform
distribution of internally deposited
radionuclides.
f facteur m de distribution
e factor m de distribución
i fattore m di distribuzione
n distributiefactor, verdelingsfactor
d Verteilungsfaktor m

1914 DISTRIBUTION FUNCTION   ma
A function used in theoretical treatise on
the Boltzmann equation.
f fonction f de distribution
e función f de distribución
i funzione f di distribuzione
n functie van de verdeling, verdelingsfunctie
d Verteilungsfunktion f

1915 DISTRIBUTION OF DOSE   xr
f distribution f de la dose
e distribución f de la dosis
i distribuzione f della dose
n dosisverdeling
d Dosisverteilung f

1916 DISTRIBUTION RATIO,   ch
    PARTITION COEFFICIENT
The ratio of the equilibrium concentrations
of a given component in two immiscible
solutions.
f coefficient m de partition,
   rapport m de distribution
e relación f de distribución
i rapporto m di distribuzione
n concentratieverhouding,
   verdelingscoëfficiënt
d Konzentrationsverhältnis n

1917 DIVERGENCE   rt
Growth of a reaction with time.
f   divergence *f*
e   divergencia *f*
i   divergenza *f*
n   divergentie
d   Anstieg *m*, Divergenz *f*

1918. DIVERGENT ·   np
Said of a multiplying system when the
effective multiplication constant is
equal to 1 or greater than 1.
f   divergent adj
e   divergente adj
i   divergente adj
n   divergent adj
d   divergent adj

1919 DIVERGENT REACTION   rt
A nuclear chain reaction in which the
number of reactions caused directly
by one reaction is greater than unity.
f   réaction *f* divergente
e   reacción *f* divergente
i   reazione *f* divergente
n   divergente reactie
d   ansteigende Reaktion *f*,
Divergenzreaktion *f*

1920 DIVERSION BOX   rw
An underground concrete box used to
route liquid wastes between separation
processing plants and storage tanks,
with desired connections made by pipes
between wall terminals.
f   caisse *f* de diversion
e   caja *f* de diversión
i   cassa *f* di diversione
n   afvoerkist
d   Ableitungskasten *m*

1921 DIVERTOR   pp
An auxiliary apparatus in thermonuclear
apparatus, intended to divert impurity
atoms leaving the wall of the discharge
chamber into a cooled trap and so to
prevent them from reaching the hot
plasma.
f   déviateur *m*, écorceur *m*
e   desviador *m*
i   deviatore *m*
n   afleider
d   Ableiter *m*

1922 DJALMAITE   mi
A tantalate of calcium and uranium.
f   djalmaite *f*
e   dialmaita *f*
i   dialmaite *f*
n   djalmaiet *n*
d   Djalmait *m*

1923 DOERNER-HOSKINS   ch, is
DISTRIBUTION LAW
The logarithmic law governing the
distribution of activity between precipitate
and solution in many co-precipitations of

radioactive tracers with carrier
precipitates.
f   loi *f* de distribution de Doerner-Hoskins
e   ley *f* de distribución de Doerner-Hoskins
i   legge *f* di distribuzione di Doerner-
Hoskins
n   verdelingswet van Doerner en Hoskins
d   Verteilungsgesetz *n* nach Doerner und
Hoskins

1924 DOLLAR OF REACTIVITY   un
One dollar is defined as the amount of
reactivity equal to the delayed neutron
fraction.
f   dollar *m*
e   dólar *m*
i   dollaro *m*
n   dollar
d   dollar-Einheit *f*

1925 DOME   nw
The mound of water spray thrown up into
the air when the shock wave from an
underwater detonation of a nuclear weapon
reaches the surface.
f   coupole *f* d'eau
e   cúpula *f* de agua
i   cupola *f* d'acqua
n   waterberg
d   Wasserberg *m*

1926 DOMINANT CHARACTER   md
Of a pair of contrasted characteristics
the one which will appear in a hybrid
resulting from cross-breeding of
homozygous parents unlike with respect
to that characteristic.
f   caractère *m* dominant
e   carácter *m* predominante
i   carattere *m* dominante
n   overheersende eigenschap
d   dominierende Eigenschaft *f*

1927 DONOR,   ec
DONOR IMPURITY
In a semiconductor, an impurity which may
induce electronic conduction.
f   donneur *m*
e   donador *m*
i   donatore *m*
n   donor
d   Donator *m*, Donor *m*

1928 DONOR LEVEL   ec
In the energy diagram of an extrinsic
semiconductor, the intermediate level close
to the conduction band.
f   niveau *m* de donneur
e   nivel *m* de donador
i   livello *m* donatore di carica
n   donorniveau *n*
d   Donatorniveau *n*

1929 DONUT (US),   rt
DOUGHNUT (GB)
Assembly of fissile (fissionable) material
of higher enrichment than that of the basic

reactor, used in a thermal reactor to
provide for experimental purposes a local
increase in fast neutron flux.
f amplificateur *m* de flux
e amplificador *m* de flujo
i amplificatore *m* di flusso
n fluxversterker
d Fluenzverstärker *m*

1930 DONUT (US),					pa
	DOUGHNUT (GB), TOROID
The toroidal vacuum chamber of a betatron
or synchrotron, in which electrons are
accelerated.
f chambre *f* à vide
e cámara *f* toroidal de vacío
i ciambella *f*
n ringvormige buis
d Ringröhre *f*

1931 DOPING CONTROL OF				ms, sc
	SEMICONDUCTORS
Carried out by means of ion implantation.
f contrôle *m* de dopage de semiconducteurs
e control *m* de adulteración de semi-
	conductores
i controllo *m* di drogaggio di semiconduttori
n dopingcontrole voor halfgeleiders
d Halbleiterdotierungskontrolle *f*

1932 DOPPLER AVERAGED CROSS			cs
	SECTION
A cross section averaged over energy,
employing appropriate weighting factors,
to take into account the effect of thermal
motion of the target particles such that
the product of the average cross section
so obtained and the flux density in the
laboratory system gives the correct
reaction rate.
f section *f* efficace moyenne Doppler
e sección *f* eficaz media Doppler
i sezione *f* d'urto media Doppler,
	sezione *f* efficace media Doppler
n doorsnede met dopplercorrectie
d Doppler-Wirkungsquerschnitt *m*

1933 DOPPLER BROADENING			np
Phenomenon occurring in electron-ion
recombination.
f élargissement *m* Doppler
e ensanchamiento *m* Doppler
i allargamento *m* Doppler
n dopplerverbreding
d Doppler-Verbreiterung *f*

1934 DOPPLER EFFECT,				ra
	DOPPLER SHIFT
The change in the observed length of a
radiation which results from the motion
of its source relative to the observer.
f effet *m* Doppler-Fizeau
e efecto *m* Doppler
i effetto *m* Doppler
n dopplereffect *n*
d Doppler-Effekt *m*

1935 DORSAL PROJECTION, see 283

1936 DOSAGE					xr
f dosage *m*
e dosificación *f*
i dosaggio *m*
n dosering
d Dosierung *f*

1937 DOSE					ra
A general term denoting the quantity of
radiation or energy absorbed.
f dose *f*
e dosis *f*
i dose *f*
n dosis
d Dosis *f*

1938 DOSE BUILD-UP FACTOR			ra
The build-up factor used for calculating
true biological dose rates through thick
shields.
f facteur *m* d'accroissement de dose
e factor *m* de incremento de dosis
i fattore *m* di dose aggiunta,
	fattore *m* di dose d'accumulo
n dosisaanwasfactor
d Dosiszuwachsfaktor *m*,
	Fano-Faktor *m*

1939 DOSE-EFFECT RELATION			ra
Relation between radiation dose and a
particular effect, possibly expressed
graphically.
f relation *f* dose-effet
e relación *f* dosis-efecto
i rapporto *m* dose-effetto
n dosis-effect-verhouding
d Dosis-Effekt-Verhältnis *n*

1940 DOSE EQUIVALENT				ra
When biological tissue is irradiated the
dose equivalent is an estimate of the
absorbed dose of medium energy
X-radiation which would give the same
biological or therapeutic effect.
f équivalent *m* de dose
e equivalente *m* de dosis
i equivalente *m* di dose
n dosisequivalent *n*
d Äquivalentdosis *f*

1941 DOSE EQUIVALENT RATE			xr
The quotient of the increase of the dose
equivalent during a certain time by that
time.
f débit *m* d'équivalent de dose
e razón *m* de equivalente de dosis
i ritmo *m* d'equivalente di dose
n dosisequivalenttempo
d Äquivalentdosisrate *f*

1942 DOSE FRACTIONATION,			ra
	FRACTIONATION
The prolongation of a given radiation dose
by the delivery of a succession of small
doses separated in time.

f fractionnement *m* de la dose
e fraccionamiento *m* de la dosis
i frazionamento *m* della dose
n fractionering
d Dosisfraktionierung *f*

1943 DOSE OF AN ISOTOPE   is
The administered quantity expressed in
any convenient unit, which in the case of
radioisotopes is usually the millicurie
or microcurie.
f dose *f* d'isotope
e dosis *f* de isótopo
i dose *f* d'isotopo
n isotoopdosis
d Isotopendosis *f*

1944 DOSE PROTRACTION,   ra
      PROTRACTION
The prolongation of the total time during
which a given radiation dose is delivered.
f étalement *m* de la dose,
   protraction *f* de la dose
e dosis *f* prolongada
i protrazione *f* della dose
n dosisprotrahering
d Dosisprotrahierung *f*

1945 DOSE RATE   ra
Dose per unit time.
f taux *m* de dose
e intensidad *f* de dosis
i intensità *f* di dose
n doseringssnelheid, doseringstempo *n*
d Dosisleistung *f*

1946 DOSE RATEMETER   me, ra
An assembly designed to measure radiation
so as to permit evaluation of absorbed
dose rate.
f débitmètre *m* de dose
e debitómetro *m* de dosis,
   medidor *m* de dosis
i rateometro *m* di dose
n doseringssnelheidsmeter, dosistempometer
d Dosisleistungsmesser *m*

1947 DOSEMETER,   me, ra
      RADIATION DOSEMETER
An instrument used for measuring or
evaluating the absorbed dose, exposure,
or similar radiation quantity.
f dosimètre *m*
e dosímetro *m*
i dosimetro *m*
n dosismeter
d Dosimeter *n*, Dosismesser *m*

1948 DOSEMETER CHARGER,   ra
      MINOMETER
A device used to charge dosemeters of
the quartz fibre(er) or other capacitor
ionization chamber type.
f chargeur *m* de dosimètre
e cargador *m* de dosímetro
i caricatore *m* di dosimetro
n dosismeterlader
d Dosismesserlader *m*

1949 DOSIMETRY   me, ra
The methods for measuring directly, or
measuring indirectly and computing,
absorbed dose or exposure dose.
f dosimétrie *f*
e dosimetría *f*
i dosimetria *f*
n dosimetrie
d Dosimetrie *f*

1950 DOUBLE BETA DECAY   br
A rare process in which two beta rays are
emitted simultaneously.
f désintégration *f* bêta double
e desintegración *f* beta doble
i disintegrazione *f* beta doppia
n dubbel bêtaverval *n*
d doppelter Betazerfall *m*

1951 DOUBLE CLAD VESSEL   rt
f récipient *m* à deux chemises
e recipiente *m* de doble chapado
i serbatoio *m* a due camicie
n dubbelbekleed vat *n*
d doppeltbekleidetes Gefäss *n*

1952 DOUBLE COATED FILM,   ph, xr
      DOUBLE EMULSION FILM
A radiographic film with photographic
emulsion on both sides.
f film *m* à double couche,
   film *m* à double émulsion
e película *f* de doble capa,
   película *f* de doble emulsión
i pellicola *f* con doppia emulsione,
   pellicola *f* con doppio strato
n dubbel begoten film, tweelaagsfilm
d beidseitig begossener Film *m*

1953 DOUBLE COMPTON   np
      SCATTERING
The process in which a photon is incident
on a charged particle and two protons are
given.
f diffusion *f* Compton double
e dispersión *f* Compton doble
i deviazione *f* Compton doppia
n dubbele comptonverstrooiing
d doppelte Compton-Streuung *f*

1954 DOUBLE-CONTRAST   xr
      ENEMA TECHNIQUE
The outlining, during radiological
examination, of the large intestine by the
injection of a quantity of air following the
evacuation of an opaque enema usually
containing barium.
f technique *f* du lavement à double contraste
e técnica *f* de la enema de doble contraste
i tecnica *f* del serviziale di doppio
   contrasto
n dubbelcontrastmethode
d Doppelkontrasteinlauftechnik *f*

1955 DOUBLE FOCUS TUBE   xr
A röntgen tube with two focal spots,
usually of different size.
f tube *m* à double foyer

e tubo *m* de doble foco
i tubo *m* con doppio fuoco
n dubbelfocusbuis
d Doppelfokusröhre *f*

1956 DOUBLE FOCUSING   me, sp
      MASS SPECTROGRAPH
      A mass spectrograph in which both
      velocity focusing and direction (or
      deflection) focusing are used.
f spectrographe *m* de masse(s) à double
   focalisation
e espectrógrafo *m* de masa de doble
   enfoque
i spettrografo *m* di massa di doppia
   focalizzazione
n massaspectrograaf met dubbelfocus
d Doppelfokusmassenspektrograph *m*

1957 DOUBLE SLIT   cr, me
      Two long narrow parallel openings used
      in certain diffraction and interference
      instruments.
f fente *f* double
e lumbrera *f* doble
i fenditura *f* doppia, fessura *f* doppia
n dubbele spleet
d doppelter Spalt *m*

1958 DOUBLET, see 1844

1959 DOUBLET,   np
      DUPLET, ELECTRON DUPLET
      Two electrons which are shared by two
      atoms so as to form a non-polar
      valence bond.
f doublet *m*
e doblete *m*
i doppietto *m*
n doublet *n*
d Dublett *n*

1960 DOUBLING TIME   rt
      1. The period of a reactor multiplied by
      ln 2.
      2. If the neutron flux is changing or is
      assumed to change exponentially, the time
      required for the neutron flux to be doubled.
      3. Of a breeder system, the time required
      to double the total amount of fissile
      (fissionable) material in the reactor.
f temps *m* de doublement
e tiempo *m* de doblado
i tempo *m* di raddoppio
n verdubbelingstijd
d Verdopplungszeit *f*

1961 DOUBLING TIME METER   me
      An instrument which measures the
      doubling time of a nuclear reactor at
      suitable intervals during the start-stop.
f appareil *m* de mesure du temps de
   doublement
e medidor *m* del tiempo de doblado
i misuratore *m* del tempo di raddoppio
n verdubbelingstijdmeter
d Verdopplungszeitmesser *m*

1962 DOUGHNUT (GB), see 1929

1963 DOUGHNUT (GB), see 1930

1964 DOWN TIME   ge
      The time during which a piece of equipment
      is not in operation because of a breakdown.
f temps *m* de non-opération
e tiempo *m* de parada
i tempo *m* di guasto
n stilstand
d Stillstandzeit *f*

1965 DOWNWARDS COOLANT   cl, rt
      FLOW
      A coolant flow which provides a
      relatively cool environment at the top
      of a reactor core.
f courant *m* de refroidissement de haut en
   bas
e flujo *m* de enfriamiento hacia abajo
i flusso *m* di raffreddamento verso il basso
n benedenwaartse koelmiddelstroom
d abwärtsfliessendes Kühlmittel *n*

1966 DOWNWIND FALL-OUT,   nw
      KATABATIC WIND FALL-OUT
f retombées *pl* radioactives de vent
   katabatique
e depósito *m* radiactivo de viento
   catabático
i ricaduta *f* radioattiva di vento catabatico,
   ricaduta *f* radioattiva di vento discendente
n radioactieve valwindneerslag
d radioaktiver Fallwindausfall *m*

1967 DRAG LOADING   nw
      The force on an object or structure due
      to the transient winds accompanying the
      passage of a blast wave.
f force *f* d'entraînement
e fuerza *f* de arrastre
i forza *f* trascinante
n meesleepkracht
d Mitschleppkraft *f*

1968 DRAG PRESSURE   nw
      The product of the dynamic pressure and
      a coefficient which is dependent upon the
      shape or geometry of the structure or
      object.
f pression *f* d'entraînement
e presión *f* de arrastre
i pressione *f* trascinante
n meesleepdruk
d Mitschleppdruck *m*

1969 DRESSING   mi
      The treatment of ore.
f triage *m*
e preparación *f*
i preparazione *f*
n bewerking
d Aufbereitung *f*

1970 DRIFT ENERGY   np
The energy of mobility of ions and
electrons.
f énergie *f* de mobilité
e energía *f* de movilidad
i energia *f* di mobilità
n driftenergie
d Driftenergie *f*

1971 DRIFT MOBILITY   np
The average drift velocity of carriers per
unit electric field.
f mobilité *f* de dérive, mobilité *f* moyenne
e movilidad *f* media
i mobilità *f* di spostamento
n driftsnelheid
d Driftgeschwindigkeit *f*

1972 DRIFT SPEED,   np
       DRIFT VELOCITY
The mean speed with which electrons or
ions progress through a medium where
they are continually experiencing
collisions.
f vitesse *f* moyenne de migration,
   vitesse *f* moyenne de pénétration
e velocidad *f* media de migración,
   velocidad *f* media de penetración
i velocità *f* media di migrazione,
   velocità *f* media di penetrazione
n gemiddelde doordringsnelheid,
   gemiddelde zwerfsnelheid
d mittlere Durchdringungsgeschwindigkeit *f*,
   mittlere Wandergeschwindigkeit *f*

1973 DRIFT TUBE   pa
In linear accelerators and certain
klystrons, a conducting tube through
which the charged particles drift while
the accelerating voltage changes.
f tube *m* de propagation
e tubo *m* de corrimiento
i tubo *m* di propagazione
n voortplantingsbuis
d Driftrohr *n*

1974 DROOGMANSITE   mi
A uranium ore probably related to
sklodowskite.
f droogmansite *f*
e droogmansita *f*
i droogmansite *f*
n droogmansiet *n*
d Droogmansit *m*

1975 DRUDE THEORY OF   np
       ELECTRONS IN METALS
The original from the free electron
theory of metals in which the electrons
were treated as a gas of classical
particles.
f théorie *f* électronique de Drude en
   métaux
e teoría *f* electrónica de Drude en metales
i teoria *f* elettronica di Drude in metalli
n elektronentheorie van Drude in metalen
d Drudesche Elektronentheorie *f* in Metalle

1976 DRY BOX   rt
A box for manipulating materials under
conditions in which moisture is excluded.
f boîte *f* sèche
e cámara *f* seca
i scatola *f* secca
n vochtvrije kast
d Trockenschrank *m*

1977 DRY CRITICALITY   rt
Reactor criticality achieved without a
coolant.
f criticité *f* en absence de réfrigérant
e criticidad *f* sin refrigerante
i criticità *f* senza refrigerante
n koelingsloze criticiteit
d kühlungsfreie Kritizität *f*

1978 DRY DEPOSIT   nw
The deposit of small particles resulting
from a nuclear explosion, not settling at
a significant rate.
f dépôt *m* radioactif secondaire
e depósito *m* radiactivo segundario
i deposito *m* radioattivo secondario
n secondaire radioactieve neerslag
d sekundärer radioaktiver Niederschlag *m*

1979 DRY-OUT MARGIN   fu, sa
The allowable percentages of humidity
of fuel elements.
f limites *pl* de séchage
e límites *pl* de secado
i limiti *pl* d'essicazione
n uitdrogingsgrenzen *pl*
d Austrocknungsgrenzen *pl*

1980 DRY REPROCESSING,   fu, mg
       NON-AQUEOUS REPROCESSING,
       PYROMETALLURGICAL REPRO-
       CESSING
The reprocessing of irradiated fuel
elements by metallurgical means.
f traitement *m* pyrométallurgique du
   combustible irradié
e reprocesamiento *m* pirometalúrgico
i trattamento *m* pirometallurgico del
   combustibile irradiato
n pyrometallurgische opwerking van
   splijtstof
d pyrometallurgische Brennstoffaufarbeitung *f*

1981 DRY-WAY PROCESS   ch
A non-aqueous chemical process carried
out with solids or powder.
f procédé *m* sec
e procedimiento *m* seco
i processo *m* secco
n droog proces *n*
d Trockenverfahren *n*

1982 DRYING   ch
f séchage *m*
e secado *m*
i disseccamento *m*
n drogen *n*
d Trocknen *n*

1983 DUAL BETA DECAY np
A type of branching in which a radioactive
nuclide may decay by the emission of
either a negative or positive electron or
by electron capture.
f désintégration *f* bêta composée
e desintegración *f* beta dual
i decadimento *m* beta duale
n samengesteld bêtaverval *n*
d dualer Betazerfall *m*

1984 DUAL CYCLE REACTOR rt
A nuclear reactor from which useful
power is produced by utilization of heat
from both the primary and secondary
coolant circuits.
f réacteur *m* à double cycle
e reactor *m* de doble ciclo
i reattore *m* a doppio ciclo
n reactor met tweevoudige kringloop
d Reaktor *m* mit Doppelkreislauf

1985 DUAL PURPOSE REACTOR, rt
PRIMARY REACTOR
Any nuclear reactor designed to produce
fissile (fissionable) material and to act
as a heat-energy source for a power plant.
f réacteur *m* à double dessin
e reactor *m* de doble aprovechamiento
i reattore *m* a duplice scopo
n reactor met tweeledig doel
d Zweizweckreaktor *m*

1986 DUAL TEMPERATURE ch, cs
PROCESS
A modification of the chemical exchange
process making use of the difference in
the equilibrium constant of an exchange
reaction at two temperatures.
f procédé *m* d'échange à deux températures
e procedimiento *m* de cambio con dos
temperaturas
i processo *m* di scambio a due temperature
n uitwisselingsproces *n* met twee
temperaturen
d Zweitemperaturenaustauschverfahren *n*

1987 DUANE AND HUNT LAW xr
X-Rays generated by electrons striking
a target cannot have a frequency greater
than eV/h where e is the electronic
charge, V the exciting voltage and h is the
Planck constant.
f loi *f* de Duane et Hunt
e ley *f* de Duane y Hunt
i legge *f* di Duane e Hunt
n wet van Duane en Hunt
d Duane-Hunt-Gesetz *n*

1988 DUCTILE-BRITTLE mg
TRANSITION TEMPERATURE,
EMBRITTLEMENT TEMPERATURE
The temperature above which a metal or
alloy becomes brittle.
f température *f* de fragilisation,
température *f* de transition de résilience
e temperatura *f* de aquebradización

i temperatura *f* d'infragilimento
n verbrozingstemperatuur
d Versprödungstemperatur *f*

1989 DUMONTITE mi
A mineral containing lead, uranium and
phosphorus.
f dumontite *f*
e dumontita *f*
i dumontite *f*
n dumontiet *n*
d Dumontit *m*

1990 DUMP rw
In reactor technology, a place where
burnt-out fuel can be temporarily or
definitely stored.
f chantier *m* de dépôt
e escombrera *f*, vertedero *m*
i magazzino *m* di scarico
n dump, tijdelijk depot *n*
d Abladeplatz *m*

1991 DUMP CONDENSER rt
A water-cooled steam condenser for
taking the heat output power of a reactor
in case the normally used turbine system
should become inoperative.
f condensateur *m* de réserve,
condensateur *m* de sécurité
e condensador *m* de seguridad
i condensatore *m* di supero
n reservecondensor
d Überschusskondensator *m*

1992 DUMP TANK rt, sa
A tank for accepting the moderator in case
of emergency shut-down.
f réservoir *m* de vidange
e tanque *m* vaciador
i vasca *f* di raccolta
n opslagtank
d Ablassbehälter *m*

1993 DUMP VALVE cd
An automatic valve which deviates in case
of emergency the vapo(u)r generated in a
nuclear plant to a dump condenser.
f soupape *f* de réserve
e válvula *f* de seguridad
i valvola *f* di supero
n reserveklep
d Überschussventil *n*

1994 DUNKOMETER (US), see 832

1995 DUPLET, see 1959

1996 DURATION OF A ec
SCINTILLATION
The interval of time between the beginning
of a scintillation and the instant at which
90 % of the photons of the scintillation
have been emitted.
f durée *f* d'une scintillation
e duración *f* de escintilación
i durata *f* di scintillazione

n   scintillatieduur
d   Szintillationsdauer *f*

1997 DUST CLOUD   nw
The cloud formed by the hot ball of gas
sucking up vast quantities of air and sand,
spreading out to the mushroom head.
f   nuage *m* de poussière,
tourbillon *m* de poussière
e   nube *m* de polvo, polvareda *f*
i   nuvolone *m* di polvere, polverone *m*
n   stofwolk
d   Staubwolke *f*

1998 DUST-COOLED REACTOR   cl, rt
A reactor cooled by a suspension of solid
particles in a gas, e.g. graphite in carbon
dioxide.
f   réacteur *m* refroidi par poussière
e   reactor *m* de enfriamiento por polvo
i   reattore *m* a·raffreddamento per polvere
n   door poedersuspensie gekoelde reactor
d   mit Pulversuspension gekühlter
Reaktor *m*

1999 DUST MONITOR   me, sa
A radiation monitor for measuring the
activity of airborne dust.
f   moniteur *m* de poussière radioactive
transportée par l'air
e   monitor *m* de radiactividad aeroportada
i   avvisatore *m* di radioattività aeroportata
n   luchtstofmonitor
d   Monitor *m* für durch die Luft getragenen
Staub

2000 DUST TRAP   cg, sa
f   nid *m* à poussière
e   nido *m* de polvo
i   nido *m* a polvere
n   stofnest *n*
d   Staubfänger *m*

2001 DUTY CYCLE,   gen
LENGTH OF WORKING CYCLE
The time interval occupied by a device on
intermittent duty in starting, running,
stopping or idling.
f   cycle *m* de fonctionnement,
facteur *m* d'utilisation,
fréquence *f* de travail
e   ciclo *m* de servicio, ciclo *m* de trabajo,
duración *f* de funcionamiento
i   ciclo *m* di lavoro,
durata *f* di funzionamento,
durata *f* d'inserimento
n   arbeidscyclus, schakelduur, werkfractie
d   Arbeitsphase *f*, Arbeitszyklus *m*,
Belastungsverhältnis *n*, Einschaltdauer *f*,
Tastverhältnis *n*

2002 DYNAMIC CAPACITOR   me
ELECTROMETER,
VIBRATING REED ELECTROMETER
An instrument using a vibrating reed
capacitor for the measurement of a small
charge or a small direct current at a high
impedance, by the conversion of a direct
voltage into an alternating voltage which
can be more easily amplified.
f   électromètre *m* à lame vibrante
e   electrómetro *m* de lengüeta vibrante
i   elettrometro *m* a lamina vibrante
n   trilcondensatorelektrometer
d   Schwingkondensatorelektrometer *m*

2003 DYNAMIC EQUILIBRIUM RATIO   np
A ratio between electrons of concentration
n_1 and negative ions of concentration n_2.
f   rapport *m* dynamique d'équilibre
e   relación *f* dinámica de equilibrio
i   rapporto *m* dinamico d'equilibrio
n   dynamische evenwichtsverhouding
d   dynamisches Gleichgewichtsverhältnis *n*

2004 DYNAMIC PINCH   pp
An experimental device using the pinch
effect to heat and confine an ionized gas
along the axis of a cylinder or along the
minor axis of a torus.
f   striction *f* dynamique
e   estricción *f* dinámica
i   contrazione *f* dinamica
n   dynamische insnoering
d   dynamische Einschnürung *f*

2005 DYNAMIC PRESSURE   nw
The air pressure which results from the
mass air flow behind the shock front of a
blast wave.
f   pression *f* dynamique
e   presión *f* dinámica
i   pressione *f* dinamica
n   dynamische druk
d   dynamischer Druck *m*

2006 DYNAMICAL FRICTION   np
Friction acting on ions when the drift
velocity is small and there is little random
velocity about this drift velocity.
f   friction *f* dynamique
e   fricción *f* dinámica
i   attrito *m* dinamico
n   dynamische wrijving
d   dynamische Reibung *f*

2007 DYSPROSIUM   ch
Rare earth metallic element, symbol Dy,
atomic number 66.
f   dysprosium *m*
e   disprosio *m*
i   disprosio *m*
n   dysprosium *n*
d   Dysprosium *n*

E

2008 E LAYER, pp
     HIGH-ENERGY ELECTRON SHEET
A cylindrical sheet of high-energy electrons
providing both containment and heating
of the plasma.
f conche *f* électronique d'énergie élevée
e capa *f* electrónica de alta energía
i strato *m* elettronico d'alta energia
n hoogenergetische elektronenlaag
d hochenergetische Elektronenschicht *f*

2009 EARTH'S ATMOSPHERE, see 350

2010 EARTHQUAKE-PROOF SITE, rt
     SEISMIC SITE
An area on which a reactor is tested to
measure its resistance to vibratory
motion and differential movements of
the ground.
f site *f* résistante aux tremblements de
   terre
e ubicación *f* antisísmica
i sito *m* antiterremoto
n tegen aardbevingen veilige bouwplaats
d erdbebenfeste Lage *f*,
   erdbebensichere Lage *f*

2011 EASEMENT CURVE, mi
     TRANSITION CURVE
In area survey, a curve of special form
connecting a straight and a circular arc.
f courbe *f* de transition
e curva *f* de transición
i curva *f* di transizione
n overgangskromme
d Übergangskurve *f*

2012 EAST-WEST EFFECT cr
The east-west asymmetry in the number
of charged cosmic-ray particles observed
on the earth caused by the deflection of
the cosmic rays by the earth's magnetic
field in combination with the rotation of
the earth from west to east.
f effet *m* est-ouest
e efecto *m* este-oeste
i effetto *m* est-ovest
n oost-west-effect *n*
d Ost-West-Effekt *m*

2013 ECCENTRIC PENDULUM xr
     THEORY
f traitement *m* pendulaire excentrique
e terapia *f* pendular excéntrica
i terapia *f* pendolare eccentrica
n excentrische slingertherapie
d exzentrische Pendeltherapie *f*

2014 ECDYSIS, see 1745

2015 ECOLOGY ge
The science of the relations between
animals and plants and the space wherein
they live.
f écologie *f*
e ecología *f*
i ecologia *f*
n ecologie
d Ökologie *f*

2016 EDDINGTON THEORY np
Theory proposed by A.S. Eddington to
describe the observed values of the
proton-electron mass ratio, the fine
structure constant and other dimensionless
ratios.
f théorie *f* d'Eddington
e teoría *f* de Eddington
i teoria *f* d'Eddington
n theorie van Eddington
d Eddingtonsche Theorie *f*

2017 EDDY DIFFUSION gp
Transport process in which the transport
itself is due to a convective phenomenon
and not to a diffusion phenomenon.
f diffusion *f* à tourbillonnement
e difusión *f* turbulenta
i diffusione *f* turbolenta
n turbulente diffusie
d turbulenzüberlagerte Diffusion *f*

2018 EDEMA, see 1589

2019 EDGE BREAKS, mg
     EDGE CRACKS
The cracks that occur at the edges of
metal sheets during rolling.
f criques *pl* du bord, fissures *pl* du bord
e roturas *pl* de los cantos
i cricche *pl* sullo spigolo,
   fessure *pl* sullo spigolo
n randscheuren *pl*
d Kantenrisse *pl*

2020 EDGE DISLOCATION cr
A dislocation whose Burgers vector is
normal to the line of dislocation.
f dislocation *f* coin
e dislocación *f* en esquina
i dislocazione *f* a scalino
n kantdislocatie
d Stufenversetzung *f*

2021 EFFECTIVE ABSORPTION ra
     COEFFICIENT,
     REDUCTION COEFFICIENT
Of a substance for a beam of homogeneous
radiation, the ratio of the rate of decrease
in the direction of propagation of the
intensity at any point to the intensity at
that point.

f   coefficient *m* d'absorption efficace
e   coeficiente *m* de absorción eficaz
i   coefficiente *m* d'assorbimento efficace
n   effectieve absorptiecoëfficiënt
d   effektiver Absorptionskoeffizient *m*

2022   EFFECTIVE ATOMIC CHARGE   np
The spectra of monovalent atoms may be
approximated by a Balmer-like formula
in which the charge on the nucleus is
reduced by a shielding constant due to
the other electrons in the atom.
f   charge *f* atomique efficace
e   carga *f* atómica eficaz
i   carica *f* atomica efficace
n   effectieve atoomlading
d   effektive Atomladung *f*

2023   EFFECTIVE ATOMIC   ch, ra
NUMBER
For a material containing two or more
elements, the number which replaces
the atomic number in the calculation of
the interaction of that material with a
given radiation.
f   numéro *m* atomique effectif
e   número *m* atómico efectivo
i   numero *m* atomico effettivo
n   effectief atoomgetal *n*
d   effektive Ordnungszahl *f*

2024   EFFECTIVE CADMIUM   np
CUT-OFF
That energy value which, for a given
.experimental configuration, is determined
by the condition that, if a cadmium cover
surrounding a detector were replaced by
a fictitious cover opaque to neutrons with
energy below this value and transparent
to neutrons above this value, the observed
detector response would be unchanged.
f   seuil *m* cadmium effectif
e   umbral *m* cadmio efectivo
i   soglia *f* cadmio effettiva
n   effectieve cadmiumgrens
d   effektive Kadmiumschwellenenergie *f*

2025   EFFECTIVE CAPTURE   cs
CROSS SECTION
The effective cross section for radioactive capture.
f   section *f* efficace de capture
e   sección *f* eficaz de captura
i   sezione *f* d'urto di cattura
n   effectieve vangstdoorsnede
d   Einfangwirkungsquerschnitt *m*

2026   EFFECTIVE COLLISION   cs, np
CROSS SECTION
The quotient of the probability of collision
by the concentration of the gas.
f   section *f* efficace de choc
e   sección *f* eficaz de choque
i   sezione *f* d'urto
n   werkzame botsingsdoorsnede
d   Wirkungsquerschnitt *m* für Stoss

2027   EFFECTIVE CROSS   cs
SECTION FOR RESONANCE,
RESONANCE CROSS SECTION
A cross section of which the value at any
one of the maxima is attributed to
resonance.
f   section *f* efficace de résonance
e   sección *f* eficaz de resonancia
i   sezione *f* d'urto di risonanza
n   werkzame doorsnede voor resonantie
d   wirksamer Resonanzquerschnitt *m*

2028   EFFECTIVE DELAYED   np
NEUTRON FRACTION
The ratio of the mean number of fissions
caused by delayed neutrons to the mean
total number of fissions caused by
delayed plus prompt neutrons.
f   fraction *f* efficace de neutrons retardés
e   fracción *f* eficaz de neutrones retrasados
i   frazione *f* efficace di neutroni ritardati
n   effectieve fractie nakomende neutronen
d   effektiver Anteil *m* der verzögerten
Neutronen

2029   EFFECTIVE DOSE   xr
f   dose *f* efficace
e   dosis *f* efectiva
i   dose *f* effettiva
n   werkzame dosis
d   Wirkungsdosis *f*

2030   EFFECTIVE ENERGY   ra
Of heterogeneous radiation, the quantum
energy of that beam of homogeneous
radiation which under the same specified
conditions is absorbed or scattered to the
same extent as the given beam of heterogeneous radiation.
f   énergie *f* effective
e   energía *f* efectiva
i   energia *f* effettiva
n   effectieve energie
d   effektive Energie *f*

2031   EFFECTIVE FOCAL SPOT,   xr
PROJECTED FOCAL SPOT
The geometric projection of the focal
spot on to a plane perpendicular to the
central ray.
f   foyer *m* optique, spot *m* lumineux effectif
e   foco *m* óptico, mancha *f* focal efectiva
i   fuoco *m* ottico,
macchia *f* focale effettiva
n   effectieve brandvlek
d   effektiver Brennfleck *m*

2032   EFFECTIVE HALF-LIFE   bi, np
The time required for the amount of a
particular specimen of a radioactive
nuclide in a system to be reduced to half
value as a consequence of both radioactive
decay and other processes such as
biological elimination and burn-up.
f   demi-vie *f* résultante, période *f* résultante
e   período *m* efectivo

i   tempo *m* effettivo di dimezzamento
n   effectieve halveringstijd
d   effektive Halbwertzeit *f*

2033 EFFECTIVE MASS   ec, np
A parameter of the dimensions of a mass
which is often used in the band theory of
solids.
f   masse *f* effective
e   masa *f* efectiva
i   massa *f* effettiva
n   werkzame massa
d   wirksame Masse *f*

2034 EFFECTIVE MULTIPLICATION   np
FACTOR,
MULTIPLICATION CONSTANT
The multiplication factor evaluated for
a finite medium.
f   facteur *m* de multiplication effectif
e   factor *m* de multiplicación efectivo
i   coefficiente *m* di moltiplicazione
effettivo
n   effectieve vermenigvuldigingsfactor
d   effektiver Vermehrungsfaktor *m*

2035 EFFECTIVE PARTICLE   pa
VELOCITY
The root-mean square value of the
instantaneous particle velocities at a point.
f   vitesse *f* effective de particule
e   velocidad *f* eficaz de partícula
i   velocità *f* effettiva di particella
n   effectieve deeltjessnelheid
d   effektive Teilchengeschwindigkeit *f*

2036 EFFECTIVE RADIUM CONTENT   ra
Of a radium container, that quantity of
radium element which, in the absence of
self absorption and wall absorption,
would produce the same effect in a
specified medium or apparatus as the
given container.
f   contenu *m* efficace de radium
e   contenido *m* eficaz de radio
i   contenuto *m* efficace di radio
n   effectief radiumgehalte *n*
d   effektiver Radiumgehalt *m*

2037 EFFECTIVE RADIUS OF   co
A CONTROL ROD
The radius of the ideal cylinder in which
the thermal neutron flux is annulled.
f   diamètre *m* efficace d'une barre de
commande
e   diámetro *m* eficaz de una barra de
regulación
i   diametro *m* efficace d'una barra di
regolazione
n   effectieve doorsnede van een regelstaaf
d   effektiver Durchmesser *m* eines
Steuerstabs

2038 EFFECTIVE RANGE   np
The radius of a spherical well which
possesses the same effect, in first
approximation, as the force under study.

f   champ *m* de force effectif
e   campo *m* de fuerza efectivo
i   campo *m* di forza effettivo
n   effectief krachtbereik *n*
d   effektive Kraftwirkung *f*

2039 EFFECTIVE RELAXATION   ra
LENGTH
The relaxation length determined taking
into account the geometric attenuation.
f   longueur *f* de relaxation effective
e   longitud *f* de relajación eficaz
i   lunghezza *f* d'attenuazione efficace
n   effectieve relaxatielengte
d   effektive Relaxationslänge *f*

2040 EFFECTIVE REMOVAL   cs
CROSS SECTION,
REMOVAL CROSS SECTION
A modified total cross section used in
considerations of neutron flux in thick
shields, in order to calculate the
penetration of the higher energy neutrons
in the incident beam.
f   section *f* efficace de déplacement
e   sección *f* eficaz de remoción
i   sezione *f* d'urto di rimozione
n   verplaatsingsdoorsnede
d   Ausscheidquerschnitt *m*,
Removalquerschnitt *m*

2041 EFFECTIVE RESONANCE   ma, np
INTEGRAL
The resonance integral where the cross
section has been replaced by an effective
cross section which will give the true
reaction rate when the flux density does
not vary inversely as the neutron energy.
f   intégrale *f* effective de résonance
e   integral *f* efectiva de resonancia
i   integrale *m* effettivo di risonanza
n   effectieve resonantie-integraal
d   effektives Resonanzintegral *n*

2042 EFFECTIVE SIMPLE   is
PROCESS FACTOR,
STAGE SEPARATION FACTOR
The separation factor obtained from a
single stage of a cascade.
f   facteur *m* efficace de séparation d'un étage
e   factor *m* eficaz de separación de una etapa
i   fattore *m* efficace di separazione d'uno
stadio
n   effectieve scheidingsfactor van een trap
d   effektiver Trennfaktor *m* einer Stufe

2043 EFFECTIVE SOURCE AREA   ra
Defined as the part of a plane source
contained within a source centred at the
base of the perpendicular from detector
to source.
f   aire *f* effective de source
e   superficie *f* eficaz de fuente
i   superficie *f* effettiva di sorgente
n   werkzaam bronoppervlak *n*
d   wirksame Quellenfläche *f*

2044 EFFECTIVE STACK HEIGHT   rw
The physical height of a stack or chimney
carrying effluent gases, plus the height
to which the gases are carried by
momentum and buoyancy effects.
f altitude f effective de cheminée
e altitud f eficaz de chimenea
i altezza f effettiva di fumaiuolo
n werkzame schoorsteenhoogte
d wirksame Schlothöhe f

2045 EFFECTIVE STANDARD   me
      DEVIATION
The standard deviation of a neutron
emitter less than $\pm$ 3 %.
f déviation f normale effective
e desviación f normal efectiva
i deviazione f normale effettiva
n effectieve normale afwijking
d effektive normale Abweichung f

2046 EFFECTIVE TARGET AREA   pa
That surface of the target that is
accessible to the incident particles.
f surface f efficace de cible
e superficie f eficaz de blanco
i superficie f efficace di bersaglio
n werkzaam trefplaatoppervlak n
d wirksame Auftrefferfläche f,
   wirksame Treffplattenfläche f

2047 EFFECTIVE THERMAL   cs
      CROSS SECTION,
      WESTCOTT CROSS SECTION
A fictitious cross section for a specified
interaction which, when multiplied by
the conventional flux density, gives the
correct reaction rate.
f section f efficace thermique effective
e sección f eficaz térmica efectiva
i sezione f d'urto termica
n effectieve thermische doorsnede,
   westcottdoorsnede
d effektiver thermischer Wirkungsquerschnitt m

2048 EFFECTIVE VALUE   ma
The effective amount or magnitude of a
quantity or property.
f valeur f effective
e valor m efectivo
i valore m effettivo
n effectieve waarde
d Effektivwert m

2049 EFFECTIVE WAVELENGTH   ra
Of heterogeneous radiation, the wavelength
of that beam of homogeneous radiation
which, under the same specified
conditions, is absorbed to the same extent
as the given beam of heterogeneous
radiation.
f longueur f d'onde effective
e longitud f de onda efectiva
i lunghezza f d'onda effettiva
n effectieve golflengte
d effektive Wellenlänge f

2050 EFFICIENCY   ge
The ratio of useful output of a device to
total input, generally expressed as a
percentage.
f rendement m
e rendimiento m
i rendimento m
n rendement n
d Ausbeute f, Wirkungsgrad m

2051 EFFLUENT   ch
In chemical engineering, a stream leaving
a process or a unit of process equipment.
f effluent m
e efluente m, escurente m
i affluente m, scarico m attivo
n afvoer, afvoerstroom
d Ausfluss m

2052 EFFLUENT ACTIVITY   rt, rw
      METER,
      EFFLUENT MONITOR
An instrument for measuring the radioactivity of effluent.
f activimètre m d'effluent,
   moniteur m d'effluent
e monitor m de efluente
i avvisatore m di scarico attivo
n afvoermonitor, afvoerstroommonitor
d Ausflussmonitor m

2053 EHRENFEST ADIABATIC   gp, np, qm
      LAW
For a virtual and infinitely slow
alteration of the coupling conditions, the
quantum numbers of the atomic system
do not change, and the number of terms
also does not change.
f loi f adiabatique d'Ehrenfest
e ley f adiabática de Ehrenfest
i legge f adiabatica d'Ehrenfest
n adiabatenwet van Ehrenfest
d Adiabatensatz m von Ehrenfest

2054 EINSTEIN EQUATION,   gp
      MASS-ENERGY RELATION
The formula $E = mc^2$, relating the change
E in the energy of a system with the
change m in its mass which follows from
Einstein's theory of relativity.
f équation f d'Einstein,
   relation f masse-énergie
e ecuación f de Einstein,
   relación f masa-energía
i equazione f d'Einstein,
   rapporto m massa-energia
n betrekking tussen massa en energie,
   vergelijking van Einstein
d Einstein-Gleichung f,
   Masse-Energieverhältnis n

2055 EINSTEIN PHOTO-   ma
      ELECTRIC EQUATION
An equation giving the kinetic energy of an
electron ejected from a system in the photoelectric effect, where the electron is

ejected by an incident photon, absorbing
all the energy of the latter.
f équation f photoélectrique d'Einstein
e ecuación f fotoeléctrica de Einstein
i equazione f fotoelettrica d'Einstein
n vergelijking van Einstein voor het
   foto-effect
d Einsteinsche Gleichung f für den
   Photoeffekt

2056 EINSTEINIUM   ch
   Radioactive element, symbol E, atomic
   number 99.
f einsteinium m
e einsteinio m
i einsteinio m
n einsteinium n
d Einsteinium n

2057 EJECTED BEAM,   pa
      EXTRACTED BEAM
   Term used for the beam of accelerated
   particles which leaves the accelerator
   or a reactor to be directed towards a
   target.
f faisceau m sorti
e haz m de salida
i fascio m d'uscita
n uitgezonden bundel
d emittierter Strahl m

2058 EJECTED PARTICLE,   np
      EMITTED PARTICLE
   A particle that has been emitted from a
   solid or liquid.
f particule f émise
e partícula f emitida, partícula f expulsada
i particella f emessa
n geëmitteerd deeltje n,
   uitgestoten deeltje n
d emittiertes Teilchen n

2059 ELASTIC COLLISION   np
   A collision in which there is no change
   either in the internal energy of each
   participating system or in the sum of
   their kinetic energies of translation.
f collision f élastique
e colisión f elástica
i collisione f elastica, urto m elastico
n elastische botsing, elastische collisie
d elastische Kollision f, elastischer Stoss m

2060 ELASTIC RANGE   gp
   The stress range in which a material will
   recover its original form when the force
   or loading is removed.
f domaine m élastique
e zona f de deformaciones elásticas
i campo m elastico
n elastisch gebied n
d Elastizitätsgebiet n

2061 ELASTIC SCATTERING   np
   A scattering process in which the total
   kinetic energy is unchanged.
f diffusion f élastique

e dispersión f elástica
i deviazione f elastica
n elastische verstrooiing
d elastische Streuung f

2062 ELASTIC SCATTERING   cs
      CROSS SECTION
   The cross section for the elastic scattering
   process.
f section f efficace de diffusion élastique
e sección f eficaz de dispersión elástica
i sezione f d'urto di deviazione elastica
n doorsnede voor elastische verstrooiing
d Wirkungsquerschnitt m für elastische
   Streuung

2063 ELASTIC STRAIN   mg
   The process of deformation of a solid
   characterized by a ratio of proportionality
   between the deformation and the forces and
   by the fact that the deformations are
   reversible.
f déformation f élastique
e deformación f elástica
i deformazione f in regime elastico
n elastische vervorming
d Beanspruchung f unterhalb der
   Elastizitätsgrenze

2064 ELASTIC THERMAL STRESS   rt
   Stresses set up in core components as
   a result of the temperature gradient
   produced by fission heating or gamma
   heating which do not produce permanent
   changes in dimensions.
f effort m thermique élastique
e esfuerzo m térmico elástico
i sollecitazione f termica elastica
n elastische thermische spanning
d elastische thermische Beanspruchung f

2065 ELASTICITY   gp
f élasticité f
e elasticidad f
i elasticità f
n elasticiteit, veerkracht
d Elastizität f

2066 ELECTRIC CONDUCTION   ge
   The conduction of electricity by means of
   electrons, ionized atoms, ionized molecules or semiconductor holes.
f conductibilité f électrique
e conductibilidad f eléctrica
i conducibilità f elettrica
n elektriciteitsgeleiding
d Elektrizitätsleitung f

2067 ELECTRIC DIPOLE, see 1846

2068 ELECTRIC DOUBLE LAYER, see 1847

2069 ELECTRIC FIELD,   ec
      FIELD OF FORCE
   The space in the neighbo(u)rhood of a
   charged body, or of a varying magnetic
   field, throughout which an electric charge

would experience a mechanical force.
f champ *m* électrique
e campo *m* eléctrico
i campo *m* elettrico
n elektrisch veld *n*
d elektrisches Feld *n*

2070 ELECTRIC FIELD STRENGTH   ec
Measured in magnitude and direction by
the mechanical force per unit charge
experienced by a very small charged
body placed at the point.
f intensité *f* du champ électrique
e intensidad *f* del campo eléctrico
i intensità *f* del campo elettrico
n veldsterkte
d Feldstärke *f*

2071 ELECTRIC STRENGTH   ec, gp
The property of a dielectric which opposes
a disruptive discharge.
f rigidité *f* diélectrique
e rigidez *f* dieléctrica
i rigidità *f* dielettrica
n doorslagvastheid
d dielektrische Festigkeit *f*,
   Durchschlagsfestigkeit *f*

2072 ELECTRICAL CONDUCTIVITY   pp
      IN A PLASMA
The ratio of the current density to the
rate at which electrons in a unit volume
gain momentum from collisions with
positive ions.
f conductivité *f* de plasma
e conductividad *f* de plasma
i conduttività *f* di plasma
n plasmageleidendheid
d Plasmaleitfähigkeit *f*

2073 ELECTRICITY   rt
      PRODUCTION REACTOR
A nuclear reactor designed primarily
to supply electrical energy for industrial
purposes.
f réacteur *m* de production d'électricité
e reactor *m* de producción de electricidad
i reattore *m* di produzione d'elettricità
n reactor voor elektriciteitsproduktie
d Reaktor *m* zur Elektrizitätsproduktion

2074 ELECTRIFICATION OF A GAS   np
The state of a gas made conducting by the
introduction of charged particles,
generally electrons, without ionizing the
gas.
f électrisation *f* d'un gaz
e electrización *f* de un gas
i elettrizzazione *f* d'un gas
n elektrisch geleidend maken *n* van een gas
   zonder ionisatie
d Elektrisierung *f* eines Gases

2075 ELECTRODE   ec, ge
A conductor fulfilling one or several
functions: emission, capture or control of
electrons or ions by means of an electric
field.
f électrode *f*
e electrodo *m*
i elettrodo *m*
n elektrode
d Elektrode *f*

2076 ELECTRODISINTEGRATION   np
The disintegration of a nucleus under
electron bombardment.
f électrodésintégration *f*
e electrodesintegración *f*
i elettrodisintegrazione *f*
n elektrodesintegratie
d Elektrodesintegration *f*

2077 ELECTROKINETIC EFFECTS   pa
Movement of particles under the influence
of an applied electric field.
f effets *pl* électrocinétiques
e efectos *pl* electrocinéticos
i effetti *m* elettrocinetici
n elektrokinetische effecten *pl*
d elektrokinetische Effekte *pl*

2078 ELECTROKYMOGRAPH   xr
An instrument for recording the time rate
of motion of the shadow on a fluoroscopic
screen.
f électrokymographe *m*
e electroquimógrafo *m*
i elettrochimografo *m*
n elektrokymograaf
d Elektrokymograph *m*

2079 ELECTROLYTIC METHOD   is
The separation of isotopes by means of
motion of ions in liquids.
f méthode *f* électrolytique
e método *m* electrolítico
i metodo *m* elettrolitico
n elektrolytische methode
d Elektrolyseverfahren *n*

2080 ELECTROLYTIC SEPARATION,   is
      SEPARATION BY MEANS OF
      IONS IN LIQUIDS
Isotope separation by electrolytic
decomposition of a solution or melt.
f séparation *f* électrolytique
e separación *f* electrolítica
i separazione *f* elettrolitica
n elektrolytische scheiding
d elektrolytische Trennung *f*

2081 ELECTROMAGNETIC FIELD   gp
According to the theory of Maxwell, a
variable electric field is always accompanied by a magnetic field and conversely,
a variable magnetic field is accompanied
by an electric field. The joint interplay
of electric and magnetic forces is what
is called an electromagnetic field.
f champ *m* électromagnétique
e campo *m* electromagnético
i campo *m* elettromagnetico
n elektromagnetisch veld *n*
d elektromagnetisches Feld *n*

2082 ELECTROMAGNETIC   me
      FLOWMETER
A flowmeter operating on the same
principle as an electric generator, as
symbolized by Fleming's right hand rule.
f débitmètre *m* électromagnétique
e flujómetro *m* electromagnético
i flussometro *m* elettromagnetico
n elektromagnetische debietmeter,
   elektromagnetische stromingsmeter
d elektromagnetischer Strömungsmesser *m*

2083 ELECTROMAGNETIC ISOTOPE   is
      SEPARATION UNIT
A set of apparatus for separating isotopes
by electromagnetic methods.
f ensemble *m* d'appareils pour la séparation
   électromagnétique d'isotopes
e conjunto *m* de aparatos para la separación
   electromagnética de isótopos
i complesso *m* d'apparecchi per la
   separazione elettromagnetica d'isotopi
n samenstel *n* van apparaten voor de
   elektromagnetische scheiding van
   isotopen
d Apparatengruppe *f* für die elektromagnetische Trennung von Isotopen

2084 ELECTROMAGNETIC   is
      ISOTOPE SEPARATOR
An apparatus used in the electromagnetic
isotope separation method.
f séparateur *m* électromagnétique
   d'isotopes
e separador *m* electromagnético de
   isótopos
i separatore *m* elettromagnetico d'isotopi
n elektromagnetische isotopenscheider
d Gerät *n* zur elektromagnetischen
   Isotopentrennung

2085 ELECTROMAGNETIC MASS   gp
A quantity postulated by the Maxwell
theory on the electromagnetic field.
f masse *f* électromagnétique
e masa *f* electromagnética
i massa *f* elettromagnetica
n elektromagnetische massa
d elektromagnetische Masse *f*

2086 ELECTROMAGNETIC METHOD   is
      OF ISOTOPE SEPARATION,
      MASS-SPECTROGRAPHIC METHOD
      OF ISOTOPE SEPARATION
A separation method in which the ions of
varying mass are separated by a combination of electric and magnetic fields.
The method is based on the principle of
the mass spectrograph.
f méthode *f* électromagnétique,
   méthode *f* masse-spectrographique de
   séparation d'isotopes
e método *m* de separación de isótopos por
   espectrografía de masa,
   método *m* electromagnético
i metodo *m* di separazione d'isotopi mediante spettrografia di massa,
   metodo *m* elettromagnetico

n elektromagnetische methode,
   isotopenscheiding volgens de spectro-
   grafische methode
d elektromagnetisches Verfahren *n*,
   Isotopentrennung *f* nach dem
   Spektrographenverfahren

2087 ELECTROMAGNETIC POSITION   co
      MEASURING ASSEMBLY
A position measuring assembly, using the
reluctance variation of a magnetic circuit,
part of which is connected to the control
element.
f ensemble *m* électromagnétique de mesure
   de position
e conjunto *m* medidor de posición electromagnético
i posiziometro *m* elettromagnetico
n elektromagnetische opstelling voor
   plaatsbepaling
d Stellungsmessanordnung *f* mit elektro-
   magnetischem Geber

2088 ELECTROMAGNETIC PULSE,   nw, sa
      EMP
A pulse created when gamma rays strike
electrons in the air surrounding a nuclear
weapon explosion.
f impulsion *f* électromagnétique
e impulsión *f* electromagnética
i impulso *m* elettromagnetico
n elektromagnetische puls
d elektromagnetischer Impuls *m*

2089 ELECTROMAGNETIC PUMP   pa
A pump for liquid metals operating on the
principle of the electric motor.
f pompe *f* électromagnétique
e bomba *f* electromagnética
i pompa *f* elettromagnetica
n elektromagnetische pomp
d elektromagnetische Pumpe *f*,
   Konduktionspumpe *f*

2090 ELECTROMAGNETIC   ra
      RADIATION
The propagation of energy through space
or through material media in the form of
electromagnetic waves, but subdivided in
some manner into discrete portions or
quanta.
f rayonnement *m* électromagnétique
e radiación *f* electromagnética
i radiazione *f* elettromagnetica
n elektromagnetische straling
d elektromagnetische Strahlung *f*

2091 ELECTROMAGNETIC SAFETY   sm
      MECHANISM
A safety mechanism in which the safety
member is actuated by an electromagnetic
device.
f mécanisme *m* électromagnétique de
   sécurité
e mecanismo *m* electromagnético de
   seguridad
i meccanismo *m* elettromagnetico di
   sicurezza

n elektromagnetisch veiligheidsmechanisme *n*
d elektromagnetische Schnellschluss-
   vorrichtung *f*

2092 ELECTROMECHANICAL   ct
      REGISTER
      An electromechanical instrument for
      counting electrical pulses.
f numéroteur *m* électromécanique
e registrador *m* electromecánico
i numeratore *m* elettromeccanico
n elektromechanische teller
d elektromechanischer Zähler *m*

2093 ELECTROMECHANICAL   ct
      REGISTER UNIT
      A basic function unit containing an
      electromechanical register.
f élément *m* numéroteur électromécanique
e unidad *f* registradora electromecánica
i elemento *m* numeratore elettromeccanico
n elektromechanisch telwerk *n*
d elektromechanische Zählstufe *f*

2094 ELECTROMETER   me
      An instrument for the measurement of
      small electrical charges or currents.
f électromètre *m*
e electrómetro *m*
i elettrometro *m*
n elektrometer
d Elektrometer *n*

2095 ELECTROMETER AMPLIFIER   ec
f amplificateur *m* d'électromètre
e amplificador *m* de electrómetro
i amplificatore *m* elettrometrico
n elektrometerversterker
d Elektrometerverstärker *m*

2096 ELECTROMETER DOSEMETER   ra
      Dosemeter utilizing the discharge of the
      self-capacity of an ionization chamber
      under the action of radiation, this dis-
      charge being controlled by the position
      of an electrometer.
f dosimètre *m* à électromètre
e dosímetro *m* de electrómetro
i dosimetro *m* ad elettrometro
n elektrometerdosismeter
d Elektrometerdosismesser *m*

2097 ELECTROMETER TUBE (US),   ec, me
      ELECTROMETER VALVE (GB)
      Electron tube or valve with a high input
      resistance generally used for the indirect
      measurement of very small currents
      from sources of high internal resistance
      by means of the measurement of voltages.
f tube *m* d'électromètre
e válvula *f* de electrómetro
i valvola *f* d'elettrometro
n elektrometerbuis
d Elektrometerröhre *f*

2098 ELECTROMIGRATION PROCESS,   is
      IONIC MIGRATION METHOD
      An isotope separation process whereby
      separation is achieved due to the differ-
      ence in mobility of isotopic ions moving
      in an aqueous solution or in a molten salt,
      under the influence of an electric field.
f procédé *m* d'électromigration,
   procédé *m* de migration ionique
e procedimiento *m* de electromigración,
   procedimiento *m* de migración iónica
i processo *m* d'elettromigrazione,
   processo *m* di migrazione ionica
n ionenmigratieproces *n*
d Ionenmigrationsverfahren *n*

2099 ELECTRON,   np
      NEGATIVE ELECTRON, NEGATRON
      An elementary particle of rest mass
      m_e equal to 9.107 x 10^{-28} g and charge
      equal to 4.302 x 10^{-10} statcoul.
f électron *m*, négaton *m*
e electrón *m*, negatón *m*
i elettrone *m*, negatone *m*
n elektron *n*, negaton *n*
d Elektron *n*, Negatron *n*

2100 ELECTRON AFFINITY,   np
      WORK FUNCTION
      The minimum energy that it is necessary
      to give to an electron for it to be able
      to pass through the potential barrier.
      In a metal it is the energy gap between
      the crest of the potential barrier and the
      Fermi characteristic energy level.
f travail *m* d'extraction, travail *m* de sortie
e trabajo de extracción, trabajo *m* de salida
i lavoro *m* d'estrazione
n uittreearbeid
d Austrittsarbeit *f*

2101 ELECTRON ATOMIC MASS   np
      The atomic mass of the electron is the
      product of Avogadro's constant N and the
      electron rest mass *m*.
f masse *f* de l'électron
e masa *f* del electrón
i massa *f* dell'elettrone
n massa van het elektron
d Elektronenmasse *f*

2102 ELECTRON ATTACHMENT   np
      The formation of a negative ion when a
      free electron becomes attached to an atom
      or molecule.
f attachement *m* de l'électron
e atracción *f* del electrón
i attaccamento *m* dell'elettrone
n elektronaanhechting
d Elektronenanlagerung *f*

2103 ELECTRON AVALANCHE   np
      A group of electrons freed by cumulative
      ionization.
f avalanche *f* électronique

e   avalancha f electrónica
i   valanga f elettronica
n   elektronenlawine
d   Elektronenlawine f

2104 ELECTRON BEAM   ec
A narrow stream of electrons moving in
the same direction, all having about the
same velocity.
f   faisceau m d'électrons
e   haz m de electrones
i   fascio m d'elettroni
n   elektronenbundel, elektronenstraal
d   Elektronenbündel n, Elektronenstrahl m

2105 ELECTRON BEAM DENSITY, see 1717

2106 ELECTRON CAPTURE   np
A radioactive transformation whereby a
nucleus captures one of its orbital
electrons.
f   capture f d'électron
e   captura f de electrón
i   cattura f d'elettrone
n   elektronvangst
d   Elektroneneinfang m

2107 ELECTRON CATCHER   pa
Consists of positively charged points on
the surface of the tube of a storage ring,
which extracts undesired electrons
from the particle beam.
f   capteur m d'électrons
e   captador m de electrones
i   prenditore m d'elettroni
n   elektronenborstel
d   Elektronenfänger m

2108 ELECTRON CHARGE-TO-MASS   np
RATIO,
SPECIFIC ELECTRONIC CHARGE
The ratio e/m_c of the electronic charge
to the rest mass of the electron.
f   charge f spécifique de l'électron
e   carga f específica del electrón
i   carica f specifica dell'elettrone
n   specifieke elektronenlading,
verhouding van lading tot massa van
het elektron
d   spezifische Ladung f des Elektrons

2109 ELECTRON CLOUD   np
The total of satellite electrons moving
around the nucleus.
f   cortège m électronique,
nuage m électronique
e   nube f de electrones
i   nuvolo m d'elettroni
n   elektronenwolk
d   Elektronenwolke f

2110 ELECTRON COLLECTION   ic
A technique for obtaining a signal from an
ionization chamber, which takes advantage
of the high mobility of electrons as
compared with that of ions.

f   collection f d'électrons
e   recolección f de electrones
i   collezione f d'elettroni
n   elektronenextractie, elektronenvergaring
d   Elektronensammlung f

2111 ELECTRON COLLECTION TIME   ic
The time between the quasi-instantaneous
creation of ions by ionizing particles and
the total collection by the collecting
electrode of the corresponding electrons.
f   temps m de collection d'électrons
e   tiempo m de recolección de electrones
i   tempo m di collezione d'elettroni
n   elektronenvergaringstijd
d   Elektronensammlungszeit f

2112 ELECTRON CONCENTRATION   ec
The ratio of the number of valence
electrons to the number of atoms in a
molecule.
f   concentration f électronique
e   concentración f electrónica
i   concentrazione f elettronica
n   elektronenconcentratie
d   Elektronenkonzentration f

2113 ELECTRON COUPLING   np
The combination of the orbital and spin
angular momentum vectors for a group
of electrons.
f   couplage m électronique
e   acoplamiento m electrónico
i   accoppiamento m elettronico
n   elektronenkoppeling
d   Elektronenkopplung f

2114 ELECTRON CRACK,   np
ELECTRON TRANSITION,
JUMP OF ELECTRONS
The transition of extranuclear electrons
from one energy level to another.
f   transition f électronique
e   transición f electrónica
i   transizione f elettronica
n   elektronenovergang
d   Elektronenübergang m

2115 ELECTRON CURRENT   ec, np
Current resulting from a flow of electrons.
f   courant m électronique
e   corriente f electrónica
i   corrente f elettronica
n   elektronenstroom
d   Elektronenstrom m

2116 ELECTRON CYCLOTRON   pa
FREQUENCY
The frequency at which an electron of
charge q (in esu) and mass m rotates in
a steady uniform magnetic field of flux
density B (in emu).
f   fréquence f de cyclotron d'un électron
e   frecuencia f de ciclotrón de un electrón
i   frequenza f di ciclotrone d'un elettrone
n   cyclotronfrequentie van een elektron
d   Elektron n mit Zyklotronfrequenz

2117 ELECTRON DENSITY, see 1718

2118 ELECTRON DETACHMENT,   np
ELECTRON REMOVAL
Occurring in collision processes of
negative ions with gas molecules in high
electrical fields or by photodissociation.
f arrachement *m* d'un électron
e destacamiento *m* de un electrón
i distacco *m* d'un elettrone
n losmaking van een elektron
d Ablösung *f* eines Elektrons

2119 ELECTRON DIFFRACTION   ms
Diffraction method used to study crystal
structure in a manner very similar to
X-ray diffraction.
f diffraction *f* à l'aide d'électrons,
diffraction *f* électronique
e difracción *f* con electrones,
difracción *f* electrónica
i diffrazione *f* con elettroni,
diffrazione *f* elettronica
n diffractie door middel van elektronen,
elektronendiffractie
d Beugung *f* mittels Elektronen,
Elektronenbeugung *f*

2120 ELECTRON DISTRIBUTION   np
An arrangement of electrons, especially
the arrangement of electrons in orbits
or shells around the nucleus of an atom
or an ion.
f distribution *f* électronique
e distribución *f* electrónica
i distribuzione *f* elettronica
n elektronenverdeling
d Elektronenverteilung *f*

2121 ELECTRON DONOR   ec, np
When the type of bond between two atoms
is of the dative type, the atom supplying
the duplet is the electron donor.
f donneur *m* d'électrons
e donador *m* de electrones
i donatore *m* d'elettroni
n elektronendonor
d Elektronendonator *m*

2122 ELECTRON DUPLET, see 1959

2123 ELECTRON-ELECTRON   np
SCATTERING
One of the basic models of scattering.
f diffusion *f* électron-électron
e dispersión *f* electrón-electrón
i deviazione *f* elettrone-elettrone
n elektron-elektron-verstrooiing
d Elektron-Elektron-Streuung *f*

2124 ELECTRON EMISSION   ec
The liberation of electrons from an electrode into the surrounding space.
f émission *f* électronique
e emisión *f* electrónica
i emissione *f* elettronica
n elektronenemissie
d Elektronenemission *f*

2125 ELECTRON EQUILIBRIUM,   np
ELECTRONIC EQUILIBRIUM
The condition set up at a point in an
irradiated material when the energy
transferred to a small volume about that
point by secondary electrons crossing it
is the same as the total energy dissipated
outside the volume by secondary electrons
produced in it.
f équilibre *m* électronique
e equilibrio *m* electrónico
i equilibrio *m* elettronico
n elektronenevenwicht *n*,
elektronisch evenwicht *n*
d Elektronengleichgewicht *n*,
elektronisches Gleichgewicht *n*

2126 ELECTRON FLOW   ec
A current produced by the movement of
free electrons toward a positive terminal.
f courant *m* électronique
e corriente *f* de electrones
i corrente *f* elettronica
n elektronenstroom
d Elektronenstrom *m*

2127 ELECTRON FLUX   np
For electrons of a given energy the
product of electron density with speed.
f flux *m* électronique
e flujo *m* electrónico
i flusso *m* elettronico
n elektronenflux
d Elektronenfluss *m*

2128 ELECTRON FLUX DENSITY   np
At a given point in space, the number of
electrons incident on a small sphere in
a time interval divided by the cross
sectional area of that sphere and by the
time interval.
f densité *f* de flux d'électrons
e densidad *f* de flujo de electrones
i densità *f* di flusso d'elettroni
n elektronenfluxdichtheid,
fluxdichtheid van elektronen
d Elektronenfluenz *f*

2129 ELECTRON FLUX   me, np
DENSITY INDICATOR
An indicator designed to give an estimate
of electron flux density.
f signaleur *m* de la densité de flux
d'électrons
e indicador *m* de la densidad de flujo de
electrones
i segnalatore *m* della densità di flusso
d'elettroni
n indicator voor elektronenfluxdichtheid
d Elektronenfluenzanzeiger *m*

2130 ELECTRON FLUX   me
DENSITY METER
A measuring assembly for electron flux
density.
f fluxmètre *m* d'électrons
e fluxímetro *m* de electrones
i flussometro *m* d'elettroni

n meter voor elektronenfluxdichtheid
d Gerät n zur Bestimmung der
  Elektronenfluenz

2131 ELECTRON FLUX DENSITY	np, me
  MONITOR
  A monitor designed to measure and respond
  to electron flux density.
f moniteur m de la densité de flux d'électrons
e monitor m de la densidad de flujo de
  electrones
i monitore m della densità di flusso
  d'elettroni
n monitor voor elektronenfluxdichtheid
d Warngerät n für Elektronenfluenz

2132 ELECTRON GAS	ec, np
  The aggregate of free electrons moving
  in a vacuous or gaseous space or within
  a conductor or semiconductor.
f gaz m électronique
e gas m electrónico
i gas m elettronico
n elektronengas n
d Elektronengas n

2133 ELECTRON GUN	ec
  A device for producing a collimated beam
  of electrons.
f canon m électronique
e cañon m electrónico
i cannone m elettronico
n elektronenkanon n
d Elektronenkanone f

2134 ELECTRON IMPACT	np
  The action of two electrons in collision.
f impact m d'électrons
e choque m de electrones
i impatto m d'elettroni
n elektronentreffen n
d Elektronenaufprall m

2135 ELECTRON INJECTOR	pa
  The electron gun of a betatron.
f injecteur m d'électrons
e inyector m de electrones
i iniettore m d'elettroni
n elektroneninjector
d Elektroneninjektor m

2136 ELECTRON-ION RECOMBINATION	np
  One of the basic models of recombination.
f recombinaison f électron-ion
e recombinación f electrón-ión
i ricombinazione f elettrone-ione
n elektron-ion-recombinatie
d Elektron-Ion-Rekombination f

2137 ELECTRON-ION WALL	np
  RECOMBINATION
  A recombination of electrons and ions at
  a wall.
f recombinaison f électron-ion à la paroi
e recombinación f electrón-ión de pared
i ricombinazione f elettrone-ione alla
  parete

n elektron-ion-wandrecombinatie
d Elektron-Ion-Wandrekombination f

2138 ELECTRON MICROSCOPE	me
  An instrument employing electron beams
  for making enlarged images of objects.
f microscope m électronique
e microscopio m electrónico
i microscopio m elettronico
n elektronenmicroscoop
d Elektronenmikroskop n

2139 ELECTRON MIRROR	ec
  Electronic device realizing the total
  reflection of an electron beam.
f miroir m électronique
e espejo m electrónico
i specchio m elettronico
n elektronenspiegel
d Elektronenspiegel m

2140 ELECTRON MULTIPLICATION	ec
  The ratio of the number of electrons
  reaching the anode to the number emitted
  at the cathode.
f multiplication f d'électrons
e multiplicación f de electrones
i moltiplicazione f d'elettroni
n elektronenvermenigvuldiging
d Elektronenvervielfachung f

2141 ELECTRON MULTIPLIER	ec
  A tube or valve or a section of a tube or
  valve in which an electron current is
  amplified in a cascade process by means
  of secondary emission at electrodes
  called dynodes.
f multiplicateur m d'électrons
e multiplicador m de electrones
i moltiplicatore m d'elettroni
n elektronenvermenigvuldiger
d Elektronenvervielfacher m

2142 ELECTRON-NEUTRINO FIELD,	np
  LEPTON FIELD
  Mathematically described as being the sum
  of products of electron and antineutrino
  wave functions plus a sum of products
  of posit(r)on and neutrino wave functions.
f champ m électron-neutrino,
  champ m lepton
e campo m electrón-neutrino,
  campo m leptón
i campo m elettrone-neutrino,
  campo m leptone
n elektron-neutrino-veld n, leptonveld n
d Elektron-Neutrino-Feld n, Leptonfeld n

2143 ELECTRON OCTET	np
  A group of eight valence electrons which
  constitutes the most stable configuration
  of the outermost electron shell of the atom.
f octet m électronique
e octete m electrónico
i ottetto m elettronico
n elektronenoctet n
d Elektronenoktett n

2144 ELECTRON ORBIT,   np
        ELECTRON SHELL
     The groups of electrons which have
     adjacent energy levels.
f   couche f électronique,
     orbite f électronique
e   capa f electrónica
i   strato m d'elettroni
n   elektronenschil
d   Elektronenhülle f

2145 ELECTRON PAIR   np
     A general feature of the architecture of
     many molecular structures, in which
     neighbo(u)ring atoms or nuclei are
     bonded by sharing a pair of valence
     electrons, forming a non-polar bond.
f   paire f d'électrons
e   par m de electrones
i   coppia f d'elettroni
n   elektronenpaar n
d   Elektronenpaar n

2146 ELECTRON PARAMAGNETIC   np
        RESONANCE
     Paramagnetic resonance in which the
     resonance effect is due to conduction
     electrons in metals or semiconductors.
f   résonance f paramagnétique électronique
e   resonancia f paramagnética electrónica
i   risonanza f paramagnetica elettronica
n   paramagnetische elektronenresonantie
d   paramagnetische Elektronenresonanz f

2147 ELECTRON PLASMA FREQUENCY, pp
        LANGMUIR FREQUENCY,
        PLASMA FREQUENCY
     The oscillation frequency of plasma
     electrons about an equilibrium charge
     distribution.
f   fréquence f de Langmuir,
     fréquence f de plasma des électrons
e   frecuencia f de Langmuir,
     frecuencia f de plasma de los electrones
i   frequenza f di Langmuir,
     frequenza f di plasma degli elettroni
n   langmuirfrequentie,
     plasmafrequentie van de elektronen
d   Langmuir-Frequenz f,
     Plasmafrequenz f der Elektronen

2148 ELECTRON POLARIZATION   np
     The part of the total induced polarization
     of a molecule that is due to the distortion
     or deformation of the electron shells
     under the influence of external electric
     fields.
f   polarisation f électronique
e   polarización f electrónica
i   polarizzazione f elettronica
n   elektronenpolarisatie
d   Elektronenpolarisation f

2149 ELECTRON-POSITON PAIR,   np
        ELECTRON-POSITRON PAIR
     The electron and posit(r)on simultaneously
     created by the process of pair production.

f   paire f électron-positon
e   par m electrón-positón
i   coppia f elettrone-positone
n   elektron-positon-paar n
d   Elektron-Positron-Paar n

2150 ELECTRON PROBE   te
        MICROANALYZER
     Apparatus for focusing the electrons by
     means of magnetic or electronic lenses.
f   microsonde f de Castaing
e   microsonda f de Castaing
i   microsonda f di Castaing
n   microsonde van Castaing
d   Castaingsche Mikrosonde f

2151 ELECTRON RADIOGRAPHY   me
     Radiography by means of electrons.
f   radiographie f électronique
e   radiografía f electrónica
i   radiografia f elettronica
n   elektronenradiografie
d   Elektronenradiographie f

2152 ELECTRON RADIUS   np
     The length $e^2/m_e c^2$, equal to
     2.82×10^{-13} cm approximately.
f   rayon m de l'électron
e   radio m del electrón
i   raggio m dell'elettrone
n   straal van het elektron
d   Elektronenradius m

2153 ELECTRON REDUCED MASS   np
     The product of the rest mass of the
     electron m_o, and the atomic mass of the
     proton M, divided by the atomic mass of
     hydrogen, m_o+M.
f   masse f réduite de l'électron
e   masa f reducida del electrón
i   massa f ridotta dell'elettrone
n   gereduceerde massa van het elektron
d   reduzierte Masse f des Elektrons

2154 ELECTRON REMOVAL, see 2118

2155 ELECTRON REST MASS   np
     The rest mass of the electron at zero
     velocity relative to an inertial frame of
     reference.
f   masse f de repos de l'électron
e   masa f del electrón en reposo
i   massa f di riposo dell'elettrone
n   rustmassa van het elektron
d   Ruhemasse f des Elektrons

2156 ELECTRON SCATTERING   np
     The scattering of electrons in solids, e.g.
     by thermal vibration.
f   diffusion f d'électrons
e   dispersión f de electrones
i   deviazione f d'elettroni
n   elektronenverstrooiing
d   Elektronenstreuung f

2157 ELECTRON SCREENING,   np
     SCREENING OF NUCLEUS
The reduction of the electric field about a
nucleus by the space charge of the
surrounding electron.
f effet *m* d'écran du noyau
e efecto *m* de pantalla del núcleo
i effetto *m* di schermo del nucleo
n afscherming van de kernlading,
   kernafscherming
d Abschirmung *f* der Kernladung,
   Kernabschirmung *f*

2158 ELECTRON SPIN   np
The rotation of the electron about its
own axis.
f spin *m* d'un électron
e espín *m* de un electrón
i spin *m* d'un elettrone
n elektronenspin
d Elektronendrall *m*,
   Elektroneneigendrehimpuls *m*,
   Elektronenspin *m*

2159 ELECTRON-SPIN RESONANCE   np
Interaction of electric and magnetic fields
with the spin of an electron about its own
axis.
f résonance *f* du spin d'un électron
e resonancia *f* del espín de un electrón
i risonanza *f* dello spin d'un elettrone
n elektronenspinresonantie
d Elektronenspinresonanz *f*

2160 ELECTRON SYNCHROTRON   pa
A synchrotron designed to accelerate
electrons.
f synchrotron *m* d'électrons
e sincrotrón *m* de electrones
i sincrotrone *m* d'elettroni
n elektronensynchrotron *n*
d Elektronensynchrotron *n*

2161 ELECTRON TRAJECTORY   ec
The path of one electron in a beam.
f parcours *m* d'un électron
e trayectoria *f* de un electrón
i percorso *m* d'un elettrone
n elektronenbaan
d Elektronenbahn *f*

2162 ELECTRON TRANSFER   np
The transfer of n electrons from a lower
oxidation state of an element to a state
n units higher in oxidation level.
f transfert *m* d'électrons
e transferencia *f* de electrones
i trasferimento *m* d'elettroni
n elektronenoverdracht
d Elektronenübertragung *f*

2163 ELECTRON TRANSIT TIME   ec
Time taken by an electron in moving
between two specified points.
f temps *m* de transit d'un électron
e tiempo *m* de tránsito de un electrón
i tempo *m* di transito d'un elettrone

n looptijd van een elektron
d Elektronenlaufzeit *f*

2164 ELECTRON TRANSITION, see 2114

2165 ELECTRON TRAP (GB),   ec
     TRAPPING CENTER (US)
A localized energy state in the energy gap
of a semiconductor caused by the presence
of an imperfection.
f centre *m* de capture, piège *m*
e centro *m* de captura, trampa *f*
i centro *m* di cattura, trappola *f*
n pleisterplaats, vangstcentrum *n*
d Fangstelle *f*, Haftstelle *f*

2166 ELECTRON VACANCY   ec, np
An unoccupied electronic site in an atomic
structure.
f électron-trou *m*, lacune *f* électronique,
   trou *m* électronique
e hueco *m* electrónico
i buca *f* elettronica, lacuna *f* elettronica
n elektronengat *n*
d Defektelektron *n*

2167 ELECTRON-VELOCITY   np
     SORTING
Any process of selecting electrons
according to their velocities.
f distribution *f* de la vitesse des électrons
e distribución *f* de la velocidad de los
   electrones
i distribuzione *f* della velocità degli
   elettroni
n snelheidsverdeling van elektronen
d Elektronengeschwindigkeitsverteilung *f*

2168 ELECTRON WAVELENGTH   np
The wavelength, λ, of the wave train
which characterizes electrons moving
with momentum p.
f longueur *f* d'onde électronique
e longitud *f* de onda electrónica
i lunghezza *f* d'onda elettronica
n elektronengolflengte
d Elektronenwellenlänge *f*

2169 ELECTRONEGATIVE ELEMENT   ch
An element which has a relatively great
tendency to attract electrons.
f élément *m* électronégatif
e elemento *m* electronegativo
i elemento *m* elettronegativo
n elektronegatief element *n*
d elektronegatives Element *n*

2170 ELECTRONEGATIVITY   ch
The extent, relative to other atoms, to
which a given atom or group of atoms
tends to attract and hold valence
electrons in its immediate
neighbo(u)rhood.
f électronégativité *f*
e electronegatividad *f*
i elettronegatività *f*
n elektronegativiteit
d Elektronegativität *f*

2171 ELECTRONIC ABSORPTION   np, ra
      COEFFICIENT
      The fractional decrease in the intensity
      of a beam of radiation per number of
      electrons per unit area.
f   coefficient *m* d'absorption électronique
e   coeficiente *m* de absorción electrónica
i   coefficiente *m* d'assorbimento elettronico
n   elektronische absorptiecoëfficiënt
d   elektronischer Absorptionskoeffizient *m*

2172 ELECTRONIC ACCELEROMETER   me
      An instrument for measuring one or more
      components of acceleration and including
      some form of transducer which usually
      produces a voltage proportional to the
      acceleration.
f   accéléromètre *m* électronique
e   acelerómetro *m* electrónico
i   accelerometro *m* elettronico
n   elektronische versnellingsmeter
d   elektronischer Beschleunigungsmesser *m*

2173 ELECTRONIC BAND SPECTRA   np, sp
      Spectra arising from electronic
      transitions in molecules.
f   spectres *pl* électroniques de bandes
e   espectros *pl* electrónicos de bandas
i   spettri *pl* elettronici di bande
n   elektronenbandenspectra *pl*
d   Elektronenbandenspektren *pl*

2174 ELECTRONIC BOHR MAGNETON,
      see 691

2175 ELECTRONIC CHARGE, see 1015

2176 ELECTRONIC CONFIGURATION   np
      An assignment of electrons to states or
      orbitals in an atom, molecule or
      crystal.
f   configuration *f* d'électrons
e   configuración *f* de electrones
i   configurazione *f* d'elettroni
n   elektronenconfiguratie
d   Elektronenkonfiguration *f*

2177 ELECTRONIC COUNTER   ct
      A circuit using electron tubes or
      equivalent devices for counting electric
      pulses.
f   compteur *m* électronique
e   contador *m* electrónico
i   contatore *m* elettronico
n   elektronische teller
d   elektronischer Zähler *m*

2178 ELECTRONIC CROSS SECTION   cs, np
      The cross section of an electron for a
      particular process.
f   section *f* électronique
e   sección *f* electrónica
i   sezione *f* elettronica
n   elektronendoorsnede
d   Elektronenquerschnitt *m*

2179 ELECTRONIC EQUILIBRIUM, see 2125

2180 ELECTRONIC FORMULA   ec, np
      A formula that shows the electronic state
      of each atom in a compound.
f   formule *f* électronique
e   fórmula *f* electrónica
i   formula *f* elettronica
n   elektronenformule
d   Elektronenformel *f*

2181 ELECTRONIC GREY WEDGE   ec
      Electronic part of instrument used in
      pulse amplitude analysis.
f   échelle *f* de gris électronique
e   escala *f* de gris electrónica
i   scala *f* di grigi elettronica
n   elektronische grijstrap
d   elektronischer Graukeil *m*

2182 ELECTRONIC STOPPING   np
      POWER
      The energy loss per electron and per unit
      area normal to the motion of the particle.
f   pouvoir *m* d'arrêt électronique
e   poder *m* de frenado electrónico
i   potere *m* di rallentamento elettronico
n   elektronisch stoppend vermogen *n*
d   elektronisches Bremsvermögen *n*

2183 ELECTRONIC STRUCTURE   np
      A term sometimes used for electronic
      configuration.
f   structure *f* électronique
e   estructura *f* electrónica
i   struttura *f* elettronica
n   elektronstructuur
d   Elektronenaufbau *m*

2184 ELECTRONIC THEORY   ch
      OF VALENCE
      The explanation of the nature of chemical
      bonds.
f   théorie *f* électronique de la valence
e   teoría *f* electrónica de la valencia
i   teoria *f* elettronica della valenza
n   elektronentheorie van de valentie,
    elektronentheorie van de waardigheid
d   Elektronentheorie *f* der Valenz,
    Elektronentheorie *f* der Wertigkeit

2185 ELECTRONIC TIMER   xr
      A timer using an electronic circuit to
      operate a relay at a predetermined
      interval of time after the circuit is
      energized.
f   interrupteur *m* à temps électronique,
    minuterie *f* électronique
e   contador *m* de tiempo electrónico
i   interruttore *m* a tempo elettronico
n   elektronische tijdschakelaar
d   elektronischer Zeitschalter *m*

2186 ELECTRONICS   ec
      That field of science and engineering which
      deals with electronic devices and their
      utilization.
f   électronique *f*
e   electrónica *f*

i   elettronica *f*
n   elektronenwetenschap, elektronica
d   Elektronik *f*

2187 ELECTRONOGEN,   ra
      LUMINOPHORE
      A molecule which emits electrons when
      illuminated.
f   électronogène adj
e   electronógeno adj
i   elettronogeno adj
n   elektronen genererend adj
d   elektronenerzeugend adj

2188 ELECTRONVOLT,   un
      eV
      A unit of energy equal to the change of
      energy of an electron which passes through
      a potential difference of one volt.
f   électron-volt *m*, eV
e   electrón-voltio *m*, eV
i   elettronevolt *m*, eV
n   elektronvolt, eV
d   Elektronenvolt *n*, eV

2189 ELECTROPHORESIS   ch
      The migration of charged particles
      through a fluid under the influence of an
      electric field.
f   électrophorèse *f*
e   electroforesis *f*
i   elettroforesi *f*
n   elektroforese
d   Elektrophorese *f*

2190 ELECTROSCOPE   me
      An instrument for indicating an electric
      charge by means of mechanical forces
      exerted between electrically charged
      bodies.
f   électroscope *m*
e   electroscopio *m*
i   elettroscopio *m*
n   elektroscoop
d   Elektroskop *n*

2191 ELECTROSTATIC ACCELERATOR   pa
      An accelerator which depends on an
      electrostatic field due to a high direct
      voltage.
f   accélérateur *m* électrostatique
e   acelerador *m* electrostático
i   acceleratore *m* elettrostatico
n   elektrostatische versneller
d   elektrostatischer Beschleuniger *m*

2192 ELECTROSTATIC BOND,   ch
      IONIC BOND
      A valence linkage in which two atoms
      are held together by electrostatic forces
      resulting from the transfer of one or
      more electrons from the outer shell to
      that of the other.
f   liaison *f* ionique
e   enlace *m* iónico
i   legame *m* ionico
n   ionbinding
d   Ionenbindung *f*

2193 ELECTROSTATIC COLLECTOR   ma
      FAILED ELEMENT MONITOR
      A failed element monitor using the
      measurement of the activity of the fission
      gas daughters such as rubidium and
      c(a)esium after collecting them on a
      negative electrode.
f   moniteur *m* de rupture de gaine à
      collection électrostatique
e   vigía *f* de colector electrostático
i   monitore *m* di rottura di guaine a
      raccolta elettrostatica
n   monitor voor beschadigde splijtstof-
      elementen berustend op elektrostatische
      opzameling
d   elektrostatisches Warngerät *n* für
      schadhafte Brennelemente

2194 ELECTROSTATIC FIELD   ec
      An electric field that is constant in time,
      i.e. an electric field produced by
      stationary charges.
f   champ *m* électrostatique
e   campo *m* electrostático
i   campo *m* elettrostatico
n   elektrostatisch veld *n*
d   elektrostatisches Feld *n*

2195 ELECTROSTATIC FILTER   cg
      A filter used, e.g., in nuclear energy
      plants to collect active dust particles.
f   filtre *m* électrostatique
e   filtro *m* electrostático
i   filtro *m* elettrostatico
n   elektrostatisch filter *n*
d   elektrostatischer Filter *m*

2196 ELECTROSTATIC FORCE   ec
      Force due to the interaction of electric
      charges.
f   force *f* électrostatique
e   fuerza *f* electrostática
i   forza *f* elettrostatica
n   elektrostatische kracht
d   elektrostatische Kraft *f*

2197 ELECTROSTATIC GENERATOR,   pa
      VAN DE GRAAFF ACCELERATOR
      An accelerator in which a conveyor belt
      is used to charge an insulated electrode
      to the desired potential.
f   accélérateur *m* électrostatique,
      générateur *m* Van de Graaff
e   acelerador *m* electrostático,
      generador *m* Van de Graaff
i   acceleratore *m* elettrostatico,
      generatore *m* Van de Graaff
n   elektrostatische versneller,
      vandegraaffgenerator
d   Statitron *n*, Van-de-Graaff-Generator *m*

2198 ELECTROSTATIC INTERACTION,
      see 1447

2199 ELECTROSTATIC METHOD,   is
      ISOTRON METHOD
      Isotope separation by using radio-
      frequency voltages.

f   méthode *f* électrostatique
e   método *m* electrostático
i   metodo *m* elettrostatico
n   elektrostatische methode
d   elektrostatisches Verfahren *n*

2200 ELECTROSTATIC PRECIPITATOR   cg
The apparatus used for purifying air or
a gas by removing unwanted particles
after charging them electrostatically and
attracting them to an electrode charged
with the opposite sign.
f   séparateur *m* électrostatique
e   precipitador *m* electrostático
i   precipitatore *m* elettrostatico
n   elektrostatische afscheider
d   elektrostatischer Abscheider *m*

2201 ELECTROSTATIC RADIUS   np
The radius deduced from an analysis of
nuclear binding energies, especially
of mirror nuclides.
f   rayon *m* électrostatique
e   radio *m* electrostático
i   raggio *m* elettrostatico
n   elektrostatische radius
d   elektrostatischer Radius *m*

2202 ELECTROSTATIC SEPARATION   mi
A method by which materials having
different permittivities are deflected
by different amounts when falling between
charged electrodes.
f   séparation *f* électrostatique
e   separación *f* electrostática
i   separazione *f* elettrostatica
n   elektrostatische scheiding
d   elektrostatische Trennung *f*

2203 ELECTROSTATIC UNIT,   un
.esu
Unit of electric charge.
f   u.e.s., unité *f* électrostatique
e   u.e.s., unidad *f* electrostática
i   u.e.s., unità *f* elettrostatica
n   elektrostatische eenheid
d   elektrostatische Einheit *f*, esE

2204 ELECTROSTATIC VALENCE,   ch
ELECTROVALENCE,
IONIC VALENCE
The type of valence which involves
electron transfer.
f   électrovalence *f*
e   valencia *f* electrónica,
valencia *f* electrostática,
valencia *f* iónica
i   valenza *f* elettronica,
valenza *f* elettrostatica, valenza *f* ionica
n   ionwaardigheid
d   Elektrovalenz *f*, heteropolare Valenz *f*,
Ionenvalenz *f*

2205 ELECTROSTATIC WAVES,   pp
LANGMUIR WAVES
Longitudinal waves appearing in a plasma
due to a perturbation of the electric
neutrality.

f   ondes *pl* de Langmuir,
ondes *pl* électrostatiques
e   ondas *pl* de Langmuir,
ondas *pl* electrostáticas
i   onde *pl* di Langmuir,
onde *pl* elettrostatiche
n   elektrostatische golven *pl*,
langmuirgolven *pl*
d   elektrostatische Wellen *pl*,
Langmuir-Wellen *pl*

2206 ELECTROVALENT COMPOUND   ch
A compound in which there occurs a
transfer of electrons.
f   composé *m* électrovalent
e   compuesto *m* electrovalente
i   composto *m* elettrovalente
n   ionwaardigheidsverbinding
d   elektronenvalente Verbindung *f*

2207 ELEMENT   ch, is
1. A substance all of whose atoms have
the same atomic number.
2. A naturally occurring mixture of
isotopes.
3. A class of atom having a particular
atomic number as the distinguishing
characteristic.
f   élément *m*
e   elemento *m*
i   elemento *m*
n   element *n*
d   Element *n*

2208 ELEMENTAL COMPOSITION, see 1043

2209 ELEMENTARY CELL,   cr
LATTICE CELL, UNIT CELL
The simplest geometric figure which
includes all the characteristics of the
lattice structure of a crystal.
f   cellule *f* élémentaire
e   célula *f* elemental
i   cella *f* elementare
n   elementaire cel
d   Elementarparallelepiped *n*,
Elementarzelle *f*

2210 ELEMENTARY CHARGE, see 1015

2211 ELEMENTARY DIFFUSION
EQUATION, see 1819

2212 ELEMENTARY PARTICLE,   np
FUNDAMENTAL PARTICLE
A term sometimes used in connection
with an electron, proton, etc.
f   particule *f* élémentaire,
particule *f* fondamentale
e   partícula *f* elemental,
partícula *f* fundamental
i   particella *f* elementare,
particella *f* fondamentale
n   elementair deeltje *n*
d   Elementarteilchen *n*

2213 ELIASITE   mi
An impure gummite.

f éliasite *f*
e eliasita *f*
i eliasite *f*
n eliasiet *n*
d Eliasit *m*

2214 ELLSWORTHITE   mi
A mineral containing uranium, calcium,
niobium and titanium.
f ellsworthite *f*
e ellsworthita *f*
i ellsworthite *f*
n ellsworthiet *n*
d Ellsworthit *m*

2215 ELUTRIATION   ch
The separation of the lighter particles
of a powder from the heavier ones by
means of an upward stream of fluid.
f élutriation *f*, lévigation *f*
e elutriación *f*, levigación *f*
i elutriazione *f*, levigazione *f*
n elutreren *n*, opslibbing
d Schlämmung *f*

2216 EMANATING POWER   ch, ra
Of a material, the rate of emission of
inert gas atoms from the surface of a
given material expressed as a fraction of
the rate of production in the material.
f pouvoir *m* d'émanation
e poder *m* de emanación
i potere *m* d'emanazione
n emanerend vermogen *n*
d Emaniervermögen *n*

2217 EMANATION   ch, np
The inert gases escaping from irradiated
or radioactive material.
f émanation *f*
e emanación *f*
i emanazione *f*
n emanatie
d Emanation *f*

2218 EMANATION METHOD   te
A method used in leak testing of
neutron sources.
f méthode *f* à émanation
e método *m* de emanación
i metodo *m* ad emanazione
n emanatiemethode
d Emanationsverfahren *n*

2219 EMANATION PROSPECTING   mi
Prospection technique based on the
measure of radon content variations in the
air contained in the soil.
f prospection *f* émanométrique
e exploración *f* emanométrica
i prospezione *f* emanometrica
n emanatieprospectie
d Emanationsschürfung *f*

2220 EMBRITTLEMENT   mg
An increase in the susceptibility of a metal
to fracture under stress caused by the

introduction of gas or foreign atoms, etc.
f accroissement *m* de la fragilité,
   fragilisation *f*
e aquebradización *f*
i infragilimento *m*
n verbrozing
d Versprödung *f*

2221 EMBRITTLEMENT TEMPERATURE,
see 1988

2222 EMERGENCY COOLING   cl
f refroidissement *m* de sécurité
e enfriamiento *m* de seguridad
i raffreddamento *m* di sicurezza
n noodkoeling
d Notkühlung *f*

2223 EMERGENCY DOSE   ra
The absorbed dose incurred when the
maximum permissible dose equivalent is
knowingly exceeded in the performance
of an unusual task to protect individuals
or valuable property.
f dose *f* d'urgence
e dosis *f* de emergencia
i dose *f* d'emergenza
n nooddosis
d Notstandäquivalentdosis *f*

2224 EMERGENCY EXPOSURE   ra
TO EXTERNAL RADIATIONS,
PLANNED EMERGENCY EXPOSURE
An exposure calculated and laid down
beforehand to be applied in emergency
cases.
f exposition *f* externe exceptionnelle
   concertée
e exposición *f* externa excepcional
   planeada
i esposizione *f* concordata
n geplande exposie in noodgeval
d für Notfälle geplante Exposition *f*

2225 EMERGENCY HIGH EXPOSURE   sa
TO RADIOACTIVE MATERIALS
According to the terms of regulations
in health physics, an event studied and
accepted beforehand as a risk, which
results in an internal contamination to
such a degree that it exceeds the one that
would result from a continuous exposure
during 3 months to the maximum
admissible concentration for personnel
directly concerned with work in radiation
environment.
f contamination *f* interne exceptionnelle
   concertée
e contaminación *f* interna excepcional
   planeada
i contaminazione *f* eccezionale concordata
n voorziene interne besmettingen in nood-
   gevallen
d für Notfälle vorgesehene innere
   Kontamination *f*

2226 EMERGENCY SHUT-DOWN,   rt
        SCRAM
      The act of shutting down a reactor suddenly
      to prevent or minimize a dangerous
      condition.
f   arrêt *m* d'urgence
e   paro *m* de emergencia, retiro *m* rápido
i   arresto *m* d'emergenza, arresto *m* rapido
n   noodstop
d   Notabschaltung *f*

2227 EMERGENCY SHUT-DOWN   sa
        MEMBER,
          SAFETY ELEMENT,
          SAFETY MEMBER
      A control member which, singly or in
      concert with others, provides a reserve
      of negative reactivity for the purpose of
      emergency shut-down of a reactor.
f   élément *m* d'arrêt d'urgence,
      élément *m* de sécurité
e   dispositivo *m* de seguridad,
      elemento *m* de paro para emergencia
i   dispositivo *m* di sicurezza,
      elemento *m* d'arresto d'emergenza
n   noodstopelement *n*, veiligheidslichaam *n*
d   Sicherheitselement *n*

2228 EMERGENCY SHUT-DOWN   sa
        SAFETY ASSEMBLY
      A safety assembly which performs the
      rapid shut-down of a nuclear reactor by
      an action which is not necessarily
      reversible.
f   ensemble *m* de sécurité d'arrêt de secours
e   conjunto *m* de seguridad de paro para
      emergencia
i   complesso *m* di sicurezza d'arresto
      d'emergenza
n   veiligheidsopstelling voor een noodstop
d   Not-Schnellschlusseinrichtung *f*

2229 EMERGENCY TRIP   rt, sa
      The fastest possible reduction of reactor
      power to a negligible level achieved by
      fully inserting all control members at
      their emergency speed, and the closing of
      the reactor building, when fitted.
f   arrêt *m* brusque, interruption *f* d'urgence
e   interrupción *f* de emergencia
i   arresto *m* rapido
n   noodstop
d   Schnellabschaltung *f*, Schnellschluss *m*

2230 EMERGENT RAY POINT   xr
      The centre(er) of the exit.
f   point *m* de sortie
e   punto *m* de rayo emergente
i   punto *m* d'uscita
n   uittreepunt *n*
d   Ausgangspunkt *m* des Zielstrahls,
      Zentrum *n* der Ausgangsblende

2231 EMISSION BAND   ec
      The energy or wavelength band of the
      photons whose emission by a scintillating
      material is most intense.

f   bande *f* d'émission
e   banda *f* de emisión
i   banda *f* d'emissione
n   emissieband
d   Emissionsbande *f*

2232 EMISSION SPECTRUM   sp
      Of a scintillating material, the curve
      representing the distribution of emitted
      photons as a function of wavelength or
      energy.
f   spectre *m* d'émission
e   espectro *m* de emisión
i   spettro *m* d'emissione
n   emissiespectrum *n*
d   Emissionsspektrum *n*

2233 EMITTED PARTICLE, see 2058

2234 EMITTER   ra
      A source of radiation, e.g. beta emitter,
      meaning a source of beta particles,
      gamma emitter, meaning a source of
      gamma radiation.
f   émetteur *m*
e   emisor *m*
i   emettitore *m*
n   straler
d   Strahler *m*

2235 EMP, see 2088

2236 EMPIRICAL MASS FORMULA   ma
      A mass formula that has less theoretical
      foundation than the semi-empirical mass
      formula, but greater freedom for
      adjustment to fit empirical data.
f   formule *f* empirique de masse
e   fórmula *f* empírica de masa
i   formula *f* empirica di massa
n   proefondervindelijke massaformule
d   empirische Massenformel *f*

2237 EMPTY BAND   ec, np
      A band of possible energy levels none of
      which corresponds to the energy of any
      electron in the given substance or in the
      given state.
f   bande *f* vide
e   banda *f* vacía
i   banda *f* vuota
n   lege band
d   leeres Energieband *n*

2238 ENABLING PULSE   ec
      A pulse which opens an electrical gate
      normally closed.
f   impulsion *f* d'ouverture
e   impulsión *f* de abertura
i   impulso *m* d'apertura
n   openingspuls
d   Öffnungsimpuls *m*

2239 ENALITE   mi
      A mineral containing zirconium and
      uranium.
f   énalite *f*

e enalita *f*
i enalite *f*
n enaliet *n*
d Enalit *m*

2240 ENCAPSULATED FUEL   fu
    UNIT (US),
      SEALED-IN FUEL ELEMENT (GB)
A fuel element hermetically sealed in
an envelope.
f élément *m* combustible scellé°
e elemento *m* combustible encamisado
i elemento *m* combustibile incamiciato
n ingekapseld splijtstofelement *n*
d eingehülstes Brennelement *n*,
   eingekapseltes Brennelement *n*

2241 ENCAPSULATION   sa
The covering and enclosing of neutron
sources in such a way that no leakage can
occur under normal circumstances.
f scellement *m*
e encapsulación *f*
i incamiciatura *f*
n inkapselen *n*
d Einkapselung *f*

2242 ENCEPHALOGRAM   xr
A radiograph made by passing X-rays
through the brain to a sensitive film.
f encéphalogramme *m*
e encefalograma *m*
i encefalogramma *m*
n encefalogram *n*
d Enzephalogramm *n*

2243 ENCEPHALOGRAPHY   xr
The radiological examination of the
ventricles and subarachnoid space
following the injection of air by cisternal
or lumbar puncture.
f encéphalographie *f*
e encefalografía *f*
i encefalografia *f*
n encefalografie
d Enzephalographie *f*

2244 END EFFECT   ct
Disturbance of the electrical field near the
ends of the collector electrode of a
counter tube resulting in count losses.
f effet *m* de bout
e efecto *m* de extremidad
i effetto *m* d'estremità
n eindeffect *n*
d Endeffekt *m*

2245 END PRODUCT,   ch, np
    FINAL PRODUCT
Of a radioactive series, the stable nuclide
that is its final member.
f produit *m* final, produit *m* terminal
e producto *m* final, producto *m* terminal
i prodotto *m* finale, prodotto *m* terminale
n eindprodukt *n*
d Schlussglied *n* einer Zerfallsreihe,
   stabiler Kern *m*

2246 END-WINDOW   ct
    COUNTER TUBE
A thin window counter tube with the
window situated at one end.
f tube *m* compteur à fenêtre en bout
e tubo *m* contador con ventana apical
i tubo *m* contatore a finestra frontale
n eindvenstertelbuis
d Endfensterzählrohr *n*

2247 ENDOERGIC,   gp
    ENDOTHERMIC
Involving an absorption of heat or energy.
f endothermique adj
e endotérmico adj
i endotermico adj
n endotherm adj
d endotherm adj

2248 ENDOERGIC PROCESS,   rt
    ENDOTHERMIC PROCESS
A nuclear process in which energy is
consumed.
f procédé *m* endothermique
e procedimiento *m* endotérmico
i processo *m* endotermico
n endotherm proces n
d endothermes Verfahren *n*

2249 ENDOSTEUM,   md
    MEDULLARY MEMBRANE
The tissue lining the internal cavity of
the bone.
f périoste *m* interne
e endostio *m*, periostio *m* interno
i endostio *m*
n endostium *n*
d Endost *m*, Endosteum *n*

2250 ENDOTHELIUM   md
The layer of simple squamous cells lining
the inner surface of the circulatory
organs and certain other closed bone
cavities.
f endothélium *m*
e endotelio *m*
i endotelio *m*
n endothelium *n*
d Endothel *n*, Endothelium *n*

2251 ENDURANCE LIMIT,   mg
    FATIGUE LIMIT
The stress below which failure will
presumably not occur in an infinite number
of cycles.
f limite *f* de résistance à la fatigue
e límite *m* de resistencia a la fatiga
i limite *m* di resistenza alla fatica
n vermoeiingsgrens
d Ermüdungsgrenze *f*

2252 ENDURANCE RATIO,   mg
    FATIGUE RATIO
The endurance limit divided by the tensile
strength.
f rapport *m* de fatigue,
   rapport *m* de résistance

e   relación f de fatiga,
    relación f de resistencia
i   rapporto m di fatica,
    rapporto m di resistenza
n   vermoeiingsverhouding
d   Ermüdungsverhältnis n

2253   ENDURANCE TEST,   mg
       FATIGUE TEST
       A test to determine for a definite stress
       cycle, applied at a definite frequency,
       the number of cycles that a material can
       withstand without rupture.
f   essai m de résistance à la fatigue
e   prueba f de fatiga
i   prova f di fatica
n   vermoeiingsproef
d   Dauerfestigkeitsversuch m,
    Dauerschwingversuch m,
    Ermüdungsversuch m

2254   ENERGY   ge
       The ability to do work.
f   énergie f
e   energía f
i   energia f
n   energie
d   Energie f

2255   ENERGY ABSORPTION   ra
       A phenomenon in which the incident
       radiation imparts some or all of its
       energy to the material it traverses.
f   absorption f d'énergie
e   absorción f de energía
i   assorbimento m d'energia
n   energieabsorptie
d   Energieabsorption f

2256   ENERGY ABSORPTION COEFFICIENT,
       see 14

2257   ENERGY BALANCE   ge
       The amount of energy released in each
       individual reaction, designated by Q.
f   bilan m énergétique
e   balance m energético
i   bilancia f energetica
n   energiebalans
d   Energiebilanz f

2258   ENERGY BAND, see 672

2259   ENERGY BAND STRUCTURE   ec, sc
       Structure representing the physical values
       of the different bands of a specific material.
f   structure f de la bande d'énergie
e   estructura f de la banda de energía
i   struttura f della banda d'energia
n   energiebandstructuur
d   Energiebandstruktur f

2260   ENERGY BUILD-UP FACTOR   ra
       The build-up factor used when calculating
       the heating effect of the gamma rays on the
       shield.
f   facteur m d'accroissement d'énergie

e   factor m de incremento de energía
i   fattore m d'accumulazione d'energia
n   energieaanwasfactor
d   Energiezuwachsfaktor m

2261   ENERGY CONVERSION   ec
       EFFICIENCY OF A SCINTILLATOR,
       LUMINOUS EFFICIENCY OF A
       SCINTILLATOR
       Ratio of the total energy of the photons
       emitted by a scintillator to the fraction
       absorbed of the incident energy.
f   rendement m énergétique de conversion
    d'un scintillateur
e   rendimiento m energético de conversión
    de un centelleador
i   rendimento m energetico di conversione
    d'uno scintillatore
n   rendement n van de energieomzetting
    van een scintillator
d   Energieumwandlungsausbeute f eines
    Szintillators

2262   ENERGY CONVERSION FACTOR   ma
       The ratio of two measures of the same
       quantity of energy, in different units.
f   facteur m de conversion d'énergie
e   factor m de conversión de energía
i   fattore m di conversione d'energia
n   energieomrekeningsfactor
d   Energieumrechnungsfaktor m

2263   ENERGY DECREMENT   gp
       The diminution of the energy in a system.
f   décrément m d'énergie
e   decremento m de energía
i   decremento m d'energia
n   energiedecrement n, energievermindering,
    geleidelijke energieafneming
d   Energieabnahme f, Energiedekrement n

2264   ENERGY DENSITY, see 1719

2265   ENERGY DEPENDENCE   ct, me
       Of a radiation detector, the dependence
       of its response on the energy of the
       incident radiation.
f   dépendance f de l'énergie
e   dependencia f de la energía
i   dipendenza f dell'energia
n   energieafhankelijkheid
d   Energieabhängigkeit f

2266   ENERGY DEPOSITION   ra
       When a solid, liquid or gas is irradiated,
       it may absorb energy by interaction with
       the incident radiation; this process is
       described as energy deposition.
f   acquisition f d'énergie
e   adquisición f de energía
i   acquisto m d'energia
n   energieopneming
d   Energieaufnahme f

2267   ENERGY DEPOT   rt
       A nuclear power system concept in which
       the electrical power would be used to

generate a chemical fuel such as
hydrogen by the dissociation of water into
oxygen and hydrogen or to regenerate a
spent fuel.
f   dépôt *m* d'énergie
e   depósito *m* de energía
i   deposito *m* d'energia
n   energiedepot *n*
d   Energieablagerung *f*

2268   ENERGY DISSIPATION, see 1901

2269   ENERGY DISTRIBUTION   ec, np
The distribution of the electrons emitted
from a surface.
f   distribution *f* de l'énergie
e   distribución *f* de la energía
i   distribuzione *f* dell'energia
n   energieverdeling
d   Energieverteilung *f*

2270   ENERGY EQUIVALENCE   un
Approximately: 1 mass unit = 930 million
electronvolts.
f   équivalence *f* d'énergie
e   equivalencia *f* de energía
i   equivalenza *f* d'energia
n   energie-equivalentie
d   Energieäquivalenz *f*

2271   ENERGY FLUENCE   np
At a given point in space, the sum of the
energies, exclusive of rest energies, of
all the particles or quanta incident on
a small sphere in a time interval, divided
by the cross sectional area of that
sphere.
f   fluence *f* d'énergie, fluence *f* énergétique
e   fluencia *f* de energía
i   fluenza *f* d'energia
n   energiefluentie
d   Energiefluenz *f*

2272   ENERGY FLUENCE RATE,   np
ENERGY FLUX DENSITY
At a given point in space, the sum of the
energies, exclusive of rest energies, of
all the particles incident on a small
sphere in a time interval, divided by the
cross sectional area of that sphere and by
the time interval.
f   débit *m* de fluence énergétique
e   densidad *f* de flujo de energía
i   densità *f* di flusso d'energia
n   energiefluxdichtheid,
fluxdichtheid van de energie
d   Energieflussdichte *f*

2273   ENERGY GAP   ec, np
The energy range between the bottom of
the conduction band and the top of the
valence band.
f   écart *m* énergétique entre deux bandes
e   separación *f* energética entre dos bandas
i   intervallo *m* energetico interbanda
n   bandafstand, energieverschil *n*,
niveauverschil *n*
d   Bandabstand *m*

2274   ENERGY IMPARTED TO   np, ra
MATTER,
IMPARTED ENERGY
The difference between the sum of the
energies of all ionizing particles or photons
which have entered a volume and the sum
of the energies of all those which have left
it, minus the energy equivalent of any
increase in rest mass that took place in
nuclear or elementary particle reactions
within the volume.
f   énergie *f* communiquée à la matière
e   energía *f* comunicada a la materia
i   energia *f* comunicata alla materia
n   aan materie overgedragen energie
d   übertragene Energie *f*

2275   ENERGY LEVEL   ec
A stationary state of energy of any
physical system.
f   niveau *m* d'énergie
e   nivel *m* de energía
i   livello *m* d'energia
n   energieniveau *n*
d   Energieniveau *n*

2276   ENERGY-LEVEL DIAGRAM   ec
A diagram representing the energy levels
of the particles of a quantized system by
horizontal lines, having for ordinates the
energy of these particles.
f   diagramme *m* énergétique
e   diagrama *m* energético
i   diagramma *m* energetico
n   energieniveauschema *n*
d   Energieschema *n*, Termschema *n*

2277   ENERGY-LEVEL WIDTH,   np
LEVEL WIDTH, TOTAL WIDTH
A level of the spread in excitation energy
of an unstable state of a quantized system.
f   largeur *f* du niveau d'énergie
e   ancho *m* del nivel de energía
i   larghezza *f* del livello d'energia
n   breedte van het energieniveau
d   Breite *f* des Energieniveaus

2278   ENERGY LOSS PER ATOM   np
The amount of energy lost by an atom per
unit area normal to the particle's motion.
f   perte *f* d'énergie par atome
e   pérdida *f* de energía por átomo
i   perdita *f* d'energia per atomo
n   energieverlies *n* per atoom
d   Energieverlust *m* je Atom

2279   ENERGY LOSS PER ION PAIR   np
The mean energy lost by all processes
per ion pair produced in a given material
by a particular type of radiation.
f   perte *f* d'énergie par paire d'ions formée
e   pérdida *f* de energía por par de iones
generado
i   perdita *f* d'energia per coppia d'ioni
formata
n   energieverlies *n* per gevormd ionenpaar
d   Energieverlust *m* je gebildetes Ionenpaar

2280 ENERGY LOSS TIME,   pp
      ENERGY REPLACEMENT TIME
      The time needed to make a plasma lose
      its average kinetic energy.
f temps m de vie de l'énergie du plasma
e tiempo m de vida de la energía del plasma
i tempo m di vita dell'energia del plasma
n levensduur van de plasmaenergie
d Lebensdauer f der Plasmaenergie

2281 ENERGY OF DISLOCATION   cr
      In crystallography, the energy measured
      per unit length of dislocation.
f énergie f de dislocation
e energía f de dislocación
i energia f di dislocazione
n dislocatie-energie
d Versetzungsenergie f

2282 ENERGY OF RESONANCE   np
      ABSORPTION,
      RESONANCE ABSORPTION ENERGY
      The amount of energy connected with the
      resonance absorption in a nuclear
      reaction.
f énergie f d'absorption par résonance
e energía f de absorción por resonancia
i energia f d'assorbimento per risonanza
n energie van de resonantieabsorptie
d Energie f der Resonanzabsorption,
   Resonanzabsorptionsenergie f

2283 ENERGY PER UNIT MASS   gp, np
f énergie f par unité de masse
e energía f por unidad de masa
i energia f per unità di massa
n energie per massaeenheid
d Energie f je Masseneinheit

2284 ENERGY QUANTUM,   qm
      QUANTUM
      A definite amount of radiant energy
      which is related quantitatively to the
      frequence of the radiation.
f quantum m, quantum m d'énergie
e cuanto m, cuanto m de energía
i quanto m, quanto m d'energia
n quantum n
d Energiequantum n, Quantum n

2285 ENERGY RANGE,   np
      ENERGY REGION, RANGE, REGION
      The distance that a particle will penetrate
      a given substance before its kinetic energy
      is reduced to a value below which it can
      no longer produce ions.
f portée f
e alcance m
i portata f
n dracht
d Reichweite f

2286 ENERGY RELEASE,   ch, gp
      RELEASE OF ENERGY
      The liberation of energy during or after
      a process.
f libération f d'énergie

e liberación f de energía
i liberazione f d'energia,
   rilascio m d'energia
n energieontwikkeling,
   vrijmaking van energie
d Energieabgabe f, Energiefreigabe f

2287 ENERGY RESOLUTION   sp
      A measure, at a given energy, of the
      smallest relative difference between the
      energies of two particles or two photons
      capable of being distinguished by a
      radiation spectrometer.
f résolution f en énergie
e resolución f en energía
i risoluzione f in energia
n energieresolutie, energiescheiding
d Energieauflösung f

2288 ENERGY SPECTROMETER,   sp
      SPECTROMETER
      An instrument designed for measuring the
      spectrum of a given kind of radiation.
f spectromètre m
e espectrómetro m
i spettrometro m
n spectrometer
d Spektrometer n

2289 ENERGY SPECTRUM,   sp
      RADIATION SPECTRUM
      A spectral representation of the energy
      distribution.
f spectre m énergétique
e espectro m energético
i spettro m energetico
n energiespectrum n
d Energiespektrum n

2290 ENERGY TRANSFER   gp
      The transfer of energy from one system
      to another.
f transfert m d'énergie
e transferencia f de energía
i trasferimento m d'energia
n energieoverdracht
d Energieübertragung f

2291 ENERGY TRANSFER   np
      COEFFICIENT
      The quotient of the kerma by the energy
      fluence.
f coefficient m de transfert d'énergie
e coeficiente m de transferencia de energía
i coefficiente m di trasferimento d'energia
n energieoverdrachtscoëfficiënt
d Energieübertragungskoeffizient m

2292 ENERGY UNIT   un
      A proposed unit for the dose received
      from ionizing radiation at a point in a
      tissue when the energy absorbed is
      93 ergs per gram(me) at that point.
f unité f d'énergie
e unidad f de energía
i unità f d'energia
n energie-eenheid
d Energie-Einheit f

2293 ENERGY YIELD,   ch
      g VALUE
The number of molecules of a specified
type altered per 100eV of energy absorbed
in a system.
f coefficient m G,
   rendement m énergétique
é coeficiente m G,
   rendimiento m energético
i coefficiente m G,
   rendimento m energetico
n energierendement n, G-factor
d Energieausbeute f, G-Faktor m

2294 ENERGY YIELD,   gp, nw
      YIELD
The total effective energy released in a
nuclear explosion.
f rendement m énergétique
e rendimiento m energético
i rendimento m energetico
n energieopbrengst, energierendement n
d Energieausbeute f

2295 ENLARGEMENT LOSS   gp
The frictional energy loss due to sudden
and sharp enlargement of the flow cross
section.
f perte f par enlargement
e pérdida f por ensanchamiento
i perdita f per allargamento
n verwijdingsverlies n
d Aufweitungsverlust m

2296 ENRICH (TO)   np, is
1. To increase the abundance of a
particular isotope in a mixture of the
isotopes of an element.
2. Applied to other fuel, to increase the
abundance of fissile (fissionable)
isotopes.
f enrichir v
e enriquecer v
i arricchire v
n verrijken v
d anreichern v

2297 ENRICHED FRACTION   ch, is
A fraction which has a greater abundance
of a desired isotope.
f fraction f enrichie
e fracción f enriquecida
i frazione f arricchita
n verrijkte fractie
d angereicherte Fraktion f

2298 ENRICHED FUEL   fu
Nuclear fuel containing uranium which has
been enriched in one or more of its
fissile (fissionable) isotopes or to which
chemically different fissile (fissionable)
nuclides have been added.
f combustible m enrichi
e combustible m enriquecido
i combustibile m arricchito
n verrijkte splijtstof
d angereicherter Brennstoff m

2299 ENRICHED MATERIAL   fu, rt
Material in which the concentration of
one or more specified isotopes of a
constituent is greater than its natural
value.
f matière f enrichie
e material m enriquecido
i materiale m arricchito
n verrijkt materiaal n
d angereichertes Material n

2300 ENRICHED REACTOR,   rt
      ENRICHED URANIUM REACTOR
A nuclear reactor in which the fuel
consists of enriched uranium.
f réacteur m à uranium enrichi,
   réacteur m enrichi
e reactor m de uranio enriquecido,
   reactor m enriquecido
i reattore m ad uranio arricchito,
   reattore m arricchito
n reactor met verrijkt uranium,
   verrijkte reactor
d angereicherter Reaktor m,
   Reaktor m mit angereichertem Uran

2301 ENRICHED URANIUM   fu
Uranium containing a higher percentage
of U-235 than in the normal state.
f uranium m enrichi
e uranio m enriquecido
i uranio m arricchito
n verrijkt uranium n
d angereichertes Uran n

2302 ENRICHED WATER   np, rt
Water which after an isotopic exchange
treatment contains a higher percentage
of D_2O than natural water.
f eau f enrichie
e agua f enriquecida
i acqua f arricchita
n verrijkt water n
d angereichertes Wasser n

2303 ENRICHMENT   fu, rt
A process that changes the isotopic ratio
in a material.
f enrichissement m
e enriquecimiento m
i arricchimento m
n verrijking
d Anreicherung f

2304 ENRICHMENT FACTOR,   fu, rt
      SEPARATION FACTOR
The ratio of the fraction of atoms of a
particular isotope in a mixture enriched
in that isotope, to the fraction of atoms
in that isotope in a mixture of natural
composition.
f facteur m d'enrichissement
e factor m de enriquecimiento
i fattore m d'arricchimento
n verrijkingsfactor
d Anreicherungsfaktor m

2305 ENRICHMENT PLANT   fu
    A chemical plant in which fissile (fission-
    able) material is enriched.
f   installation f d'enrichissement
e   instalación f de enriquecimiento
i   impianto m d'arricchimento
n   verrijkingsinstallatie
d   Anreicherungsanlage f

2306 ENRICHMENT PROCESS   fu, is
    The process in which a specific isotope
    content of an element is increased.
f   procédé m d'enrichissement
e   procedimiento m de enriquecimiento
i   processo m d'arricchimento
n   verrijkingsproces n
d   Anreicherungsverfahren n

2307 ENTRAINMENT   ch
    The suspension of liquid droplets, gas
    bubbles or fine solid particles being
    carried by a stream of fluids.
f   entraînement m
e   arrastre m
i   trascinamento m
n   meeslepen n
d   Mitreissen n, Mitschleppen n

2308 ENTRAINMENT FILTER   ch
    A special type of filter used in distillation.
f   filtre m d'entraînement
e   filtro m de arrastre
i   filtro m di trascinamento
n   meesleepfilter n
d   Mitreissfilter m

2309 ENTRANCE CHANNEL SPIN   np
    In a nuclear reaction, the vector sum of
    the spins of the initial particles.
f   spin m de voie d'entrée, voie f d'entrée
e   espín m de trayectoria de entrada
i   spin m di percorso d'ingresso
n   ingansbaanspin
d   Eingangswegspin m

2310 ENTRANCE PRESSURE DROP   cl, rt
    The pressure loss when the coolant for a
    reactor plant enters a channel.
r   chute f de pression d'entrée
e   caída f de presión de entrada
i   caduta f di pressione d'ingresso
n   ingangsdrukverval n
d   Eingangsdruckabfall m

2311 ENTROPY TRAPPING   pp
    Capture of an ordered particle beam or
    of a plasma jet in a magnetic configuration
    by means of a disordering process of the
    ordered movement of the particles,
    thereby enhancing the entropy of the system.
f   piégeage m entropique
e   captura f entrópica
i   cattura f entropica
n   entropievangst
d   Entropieeinfang m

2312 ENTRY PORTAL,   xr
        FIELD OF INCIDENCE
    The area through which a beam of
    radiation enters a patient's body.
f   champ m d'incidence, porte f d'entrée
e   campo m de entrada
i   porta f d'ingresso
n   invalsveld n
d   Einfallsfeld n

2313 ENVELOPE   ge
    A curve that is tangent to each of a
    family of curves.
f   enveloppante f
e   envolvente m
i   inviluppo m
n   omhullende
d   Einhüllende f, Hüllkurve f

2314 ENVIRONMENT   ge
    The aggregate of all the conditions and
    influences that affect the operation of
    equipment and components.
f   ambiance f
e   ambiente m
i   ambiente m
n   omgeving
d   Umgebung f

2315 ENVIRONMENTAL   ra, rt
        CONTAMINATION
    Contamination of the direct surroundings
    of an area where, e.g., a nuclear reaction
    takes place.
f   contamination f de l'ambience
e   contaminación f del ambiente
i   contaminazione f dell'ambiente
n   omgevingsbesmetting
d   Umgebungskontamination f

2316 EOSINOPHIL   md
    Showing affinity to eosin dyes.
f   éosinophile adj
e   eosinófilo adj
i   eosinofilo adj
n   eosinofiel adj
d   eosinofiel adj

2317 EPHANTINITE   mi
    A rare uranium ore containing an alteration
    product of ianthinite.
f   éphantinite f
e   efantinita f
i   efantinite f
n   efantiniet n
d   Ephantinit m

2318 EPICADMIUM   np
    Energies above the cadmium cut-off.
f   épicadmium m
e   epicadmio m
i   epicadmio m
n   epicadmium n
d   Epikadmium n

2319 EPICADMIUM NEUTRONS   np
     Neutrons of kinetic energy greater than
     the cadmium cut-off energy.
f neutrons *pl* épicadmiques
e neutrones *pl* epicádmicos
i neutroni *pl* epicadmici
n epicadmiumneutronen *pl*
d Epikadmiumneutronen *pl*

2320 EPICADMIUM RESONANCE   ma, np
     INTEGRAL
     The resonance integral which has the
     effective cadmium cut-off energy as the
     lower energy unit.
f intégrale *f* de résonance épicadmique
e integral *f* de resonancia epicádmica
i integrale *m* di risonanza epicadmica
n epicadmiumresonantie-integraal
d Epikadmiumresonanzintegral *n*

2321 EPICENTER (US),   nw
     EPICENTRE (GB)
     The point vertically above or below the
     centre(er) of a burst of a nuclear weapon.
f épicentre *m*
e epicentro *m*
i epicentro *m*
n epicentrum *n*
d Epizentrum *n*

2322 EPIDUROGRAPHY   xr
     Radiological examination of a blood
     eruption between the dura and the brains.
f épidurographie *f*
e epidurografía *f*
i epidurografia *f*
n epidurografie
d Epidurographie *f*

2323 EPILANTHINITE   mi
     A rare alteration product of lanthinite,
     containing about 87.8 % of U.
f épilanthinite *f*
e epilantinita *f*
i epilantinite *f*
n epilanthiniet *n*
d Epilanthinit *m*

2324 EPILATION, see 1722

2325 EPILATION DOSE, see 1723

2326 EPIPHARYNGOGRAPHY   xr
     Radiological examination of the nasal
     part of the pharynx.
f épipharyngographie *f*
e epifaringografía *f*
i epifaringografia *f* .
n epifaryngografie
d Epipharyngographie *f*

2327 EPIPHYSIS   md
     A part or process of a bone which ossifies
     separately and subsequently becomes
     joined to the main part of the bone.
f épiphyse *f*
e epifisis *f*

i epifisi *f*
n epifyse
d Epiphyse *f*

2328 EPITHELIAL CELLS   md
     Cells in the cellular tissue covering a
     free surface or lining a tube or cavity.
f cellules *pl* épithéliales
e células *pl* epiteliales
i cellule *pl* epiteliali
n epitheelcellen *pl*
d Epithelzellen *pl*

2329 EPITHELIOMA   md
     Malignant neoplasm derived from
     epithelial cells lining the internal and
     external body surfaces.
f épithélioma *m*
e epitelioma *m*
i epitelioma *m*
n epithelioom *n*
d Epitheliom *n*, Epithelzellengeschwulst *f*

2330 EPITHERMAL   np
     Having energy just above the energy of
     thermal agitation and comparable with
     chemical bond energies.
f épithermique adj
e epitérmico adj
i epitermico adj
n epithermisch adj
d epithermisch adj

2331 EPITHERMAL LEAKAGE   np
f fuite *f* de neutrons épithermiques
e fuga *f* de neutrones epitérmicos
i fuga *f* di neutroni epitermici
n lekken *n* van epithermische neutronen
d Abfluss *m* von epithermischen Neutronen

2332 EPITHERMAL NEUTRONS   np
     Neutrons having a kinetic energy above
     that of thermal agitation.
f neutrons *pl* épithermiques
e neutrones *pl* epitérmicos
i neutroni *pl* epitermici
n epithermische neutronen *pl*
d epithermische Neutronen *pl*

2333 EPITHERMAL REACTOR   rt
     A nuclear reactor in which fission is
     induced predominantly by epithermal
     neutrons.
f réacteur *m* épithermique
e reactor *m* epitérmico
i reattore *m* epitermico
n epithermische reactor
d epithermischer Reaktor *m*

2334 EQUATION OF CONTINUITY, see 1341

2335 EQUATION OF STATE   gp, ma
     An equation relating the volume, pressure
     and temperature of a given system.
f équation *f* d'état
e ecuación *f* de estado
i equazione *f* di stato

n   toestandsvergelijking
d   Zustandsgleichung f

2336 EQUILIBRATION   is
The process of establishing isotopic
homogeneity throughout a system in which
allobars have been mixed.
f   homogénéisation f isotopique
e   homogenización f isotópica
i   omogenizzazione f isotopica
n   isotopische homogenisatie
d   Abgleichung f,
isotopische Homogenisierung f

2337 EQUILIBRIUM CONCENTRATION   ch
That concentration of a chemical solution
which will remain unchanged by further
exposure to air of a given humidity,
temperature and pressure.
f   concentration f équilibrée
e   concentración f equilibrada
i   concentrazione f equilibrata
n   evenwichtsconcentratie
d   Gleichgewichtskonzentration f

2338 EQUILIBRIUM DIAGRAM, see 1317

2339 EQUILIBRIUM DISTILLATION   ch, is
A distillation method based on the slight
difference which exists between the
relative concentrations of two isotopes in
a liquid and in its vapo(u)r, when liquid
and vapo(u)r are in equilibrium.
f   distillation f globalement équilibrée
e   destilación f cerrada,
destilación f de equilibrio
i   distillazione f in equilibrio
n   evenwichtsdestillatie
d   geschlossene Destillation f,
Gleichgewichtsdestillation f

2340 EQUILIBRIUM ENERGY   ge, np
The energy of a condition in which all the
forces or tendencies present are exactly
counterbalanced by equal and opposite
forces and tendencies.
f   énergie f d'équilibre
e   energía f de equilibrio
i   energia f d'equilibrio
n   evenwichtsenergie
d   Gleichgewichtsenergie f

2341 EQUILIBRIUM ENRICHMENT   is
FACTOR
The ratio of isotopic ratios after
enrichment to that before enrichment
that will remain unchanged.
f   facteur m d'enrichissement équilibré
e   factor m de enriquecimiento equilibrado
i   fattore m d'arricchimento equilibrato
n   evenwichtsverrijkingsfactor
d   Gleichgewichtsanreicherungsfaktor m

2342 EQUILIBRIUM METHOD   ma
One of the three methods of calculating
the coefficient of recombination.
f   méthode f d'équilibre

e   método m de equilibrio
i   metodo m d'equilibrio
n   evenwichtsmethode
d   Gleichgewichtsverfahren n

2343 EQUILIBRIUM OF A PARTICLE   np
A particle is said to be in equilibrium
if the vector sum of all forces through
the particle is equal to zero.
f   particule f en équilibre
e   partícula f en equilibrio
i   particella f in equilibrio
n   deeltje n in evenwicht
d   Teilchen n in Gleichgewicht

2344 EQUILIBRIUM ORBIT,   pa
STABLE ORBIT
The constant-radius circular path of
accelerated particles in a betatron or
synchrotron.
f   orbite f d'équilibre,
orbite f stable
e   órbita f de equilibrio,
órbita f estable
i   orbita f d'equilibrio,
orbita f stabile
n   stabiele baan
d   stabile Bahn f

2345 EQUILIBRIUM STRESS   mt
The stress corresponding to the best
value for the minimum creep rate.
f   effort m d'équilibre
e   esfuerzo m de equilibrio
i   sollecitazione f d'equilibrio
n   evenwichtsspanning
d   Gleichgewichtsbeanspruchung f

2346 EQUILIBRIUM TIME,   is
STARTING TIME
In an isotope separation plant, the time
interval between start-up and full-rate
production.
f   période f de mise en train
e   tiempo m de puesta en marcha
i   tempo m d'avviamento
n   aanlooptijd, insteltijd
d   Anlaufzeit f

2347 EQUILIBRIUM WATER   ch
That water content of a solid which will
remain unchanged by further exposure to
air of a given humidity, temperature and
pressure.
f   humidité f d'équilibre
e   humedad f de equilibrio
i   umidità f d'equilibrio
n   evenwichtsvochtigheid,
vochtigheidsevenwicht n
d   Feuchtigkeitsgleichgewicht n,
Gleichgewichtsfeuchtigkeit f

2348 EQUIPMENT   ge
An association of assemblies associated
to attain a determined final objective.
f   équipement m
e   equipo m

i apparecchiatura f
n uitrusting
d Einrichtung f

2349 EQUIVALENT ACTIVITY md, ra
OF A SOURCE
The activity equal to the activity of a
point source of the same radionuclide
which would give the same exposure rate
at the same distance from the centre(er)
of the source.
f activité f équivalente d'une source
e actividad f equivalente de una fuente
i attività f equivalente d'una sorgente
n equivalente activiteit van een bron
d Äquivalentaktivität f einer Quelle

2350 EQUIVALENT CONSTANT xr
POTENTIAL
The constant potential which must be
applied to an X-ray tube to produce
radiation having an absorption curve in a
given material closely similar to that
of the beam under consideration.
f potentiel m constant équivalent
e potencial m constante equivalente
e potenziale m costante equivalente
n equivalente constante spanning
d konstante Äquivalentspannung f

2351 EQUIVALENT CURIE un
The activity of a rod which would give
the same dose rate at 1 metre(er) in a
direction perpendicular to the axis of the
rod as would be obtained from a point
source of cobalt-60 at the same distance.
f curie m équivalent
e curie m equivalente
i curie m equivalente
n equivalentcurie
d Äquivalentcurie n

2352 EQUIVALENT DARK ec
CURRENT INPUT
The incident luminous flux required to
give a signal output current equal to the
dark current in a phototube.
f entrée f de courant équivalent
d'obscurité
e entrada f de corriente equivalente de
obscuridad
i ingresso m di corrente equivalente
d'oscurità
n equivalenter donkerstroomingang
d äquivalenter Dunkelstromeingang m

2353 EQUIVALENT EFFECTIVE co
RADIUS OF A CONTROL ROD
For a non-cylindrical rod the radius of
a cylindrical rod with the same efficiency.
f diamètre m équivalent d'une barre de
commande
e diámetro m equivalente de una barra de
regulación
i raggio m equivalente d'una barra di
regolazione
n equivalente doorsnede van een regelstaaf

d Äquivalentdurchmesser m eines
Steuerstabs

2354 EQUIVALENT ELECTRON ch, np
DENSITY
In an ionized gas, the product of ion
density and the ratio of the mass of an
electron to an ion of the gas.
f densité f électronique équivalente
e densidad f electrónica equivalente
i densità f elettronica equivalente
n equivalente elektronendichtheid
d äquivalente Elektronendichte f

2355 EQUIVALENT ELECTRONS np
Electrons which occupy the same orbit
in an atom.
f électrons pl équivalents
e electrones pl equivalentes
i elettroni pl equivalenti
n equivalente elektronen pl
d Äquivalentelektronen pl

2356 EQUIVALENT NUCLEI np
Sets of those nuclei in a molecule which
can be transformed into one another by
the symmetry operations permitted by
the molecule.
f noyaux pl équivalents
e núcleos pl equivalentes
i nuclei pl equivalenti
n equivalente kernen pl
d äquivalente Kerne pl

2357 EQUIVALENT RADIUM CONTENT ra
Of a radioactive specimen that quantity
of radium which, under similar
conditions of screenage, produces the
same effect in a specified medium or
apparatus.
f contenu m équivalent de radium
e contenido m equivalente de radio
i contenuto m equivalente di radio
n equivalent radiumgehalte n
d äquivalenter Radiumgehalt m

2358 EQUIVALENT STOPPING POWER np
Sometimes used synonymously with
relative stopping power and sometimes
with stopping equivalent.
f pouvoir m d'arrêt équivalent
e poder m de frenado equivalente
i potere m di rallentamento equivalente
n equivalent stoppend vermogen n
d äquivalentes Bremsvermögen n

2359 ERBIUM ch
Rare earth metallic element, symbol Er,
atomic number 68.
f erbium m
e erbio m
i erbio m
n erbium n
d Erbium n

2360 ERG   un
A unit of work or energy in the cgs system
of units.
f erg *m*
e ergio *m*
i erg *m*
n erg
d erg-Einheit *f*

2361 EROSION   ch, gp
The wearing away of surfaces by
weathering, corrosion and transportation.
f érosion *f*
e erosión *f*
i erosione *f*
n erosie
d Erosion *f*

2362 ERROR   co
The difference at any time between the
reference input and the value of the
controlled variable.
f erreur *f*
e error *m*
i errore *m*
n fout
d Fehler *m*

2363 ERYTHEMA,   md
ERYTHEMATOUS ERUPTION
An abnormal redness of the skin due to
distention of the capillaries with blood.
f éruption *f* érythémateuse, érythème *m*
e eritema *m*, erupción *f* eritematosa
i eritema *m*, eruzione *f* eritematosa
n erytheem *n*, erythemateuze uitslag
d Erythem *n*, erythematöser Ausschlag *m*

2364 ERYTHEMA DOSE   xr
Dose of radiation that produces a reddening
of the skin followed by pigmentation.
f dose *f* d'érythème
e dosis *f* de eritema
i dose *f* d'eritema
n erytheemdosis
d Erythemdosis *f*

2365 ERYTHROCYTE,   md
RED CORPUSCLE
A colo(u)r red blood corpuscle of
vertebrates, containing haemoglobin.
f érythrocyte *m*, globule *m* rouge
e eritrocito *m*, glóbulo *m* rojo
i eritrocito *m*, globulo *m* rosso
n erythrocyt *n*, rood bloedlichaampje *n*
d Erythrozyt *m*, rotes Blutkörperchen *n*

2366 ESCAPE FACTOR   ph
In photographic emulsion technique, the
fraction of particles whose tracks in the
emulsion reach the top or the bottom
before completing their range.
f facteur *m* de fuite
e factor *m* de fuga
i fattore *f* di fuga
n ontsnappingsfactor
d Entkommfaktor *m*, Fluchtfaktor *m*

2367 ESCHYNITE   mi
A thorium containing ore from granite
pegmatites and nepheline syenites.
f éschynite *f*
e eschinita *f*
i eschinite *f*
n eschyniet *n*
d Eschynit *m*

2368 ESTIMATED ACTUAL   ra
RADIUM CONTENT
An estimate of the radium content of a
given container derived from its effective
radium content, after correction for
self-absorption and wall-absorption.
f contenu *m* réel de radium estimé
e contenido *m* real de radio estimado
i contenuto *m* reale di radio stimato
n geschat absoluut radiumgehalte *n*
d abgeschätzter absoluter Radiumgehalt *m*

2369 ESTIMATED MAXIMUM   ge
SYSTEM ERROR,
ESTIMATED MAXIMUM SYSTEMATIC
ERROR
f erreur *f* systématique estimée maximale
e error *m* sistemático máximo estimado
i errore *m* sistematico massimo stimato
n maximale geschatte systematische fout
d höchstzugelassener abgeschätzter
systematischer Fehler *m*

2370 esu, see 2203

2371 ETA FACTOR,   np
NEUTRON YIELD PER ABSORPTION
The average number of primary fission
neutrons, including delayed neutrons,
emitted per neutron absorbed by a fissile
(fissionable) nuclide or by a nuclear fuel
as specified.
f facteur *m* êta
e factor *m* eta
i fattore *m* eta,
resa *f* neutronica per assorbimento
n êtafactor,
neutronenopbrengst per absorptie
d Eta-Faktor *m*,
Neutronenausbeute *f* je Absorption

2372 ETCHING   ch, gp
The process of revealing structural
details by preferential attack of a metal
surface.
f attaque *f* chimique
e ataque *m* qufmico
i attacco *m* chimico
n etsen *n*
d Ätzung *f*

2373 ETHANE IONIZATION   ic
CHAMBER
An ionization chamber in which the
organic gas filling consists of ethane
(C_2H_6).
f chambre *f* d'ionisation à éthane

e   cámara ƒ de ionización con etano
i   camera ƒ d'ionizzazione ad etano
n   met ethaan gevuld ionisatievat n
d   äthangefüllte Ionisationskammer ƒ

2374   ETHYLENE-DIAMINE-   ch
       TETRA-ACETIC ACID
       An efficient sequestering agent.
f   acide m éthylène-diamine-tétracétique,
    acide m versinique
e   ácido m etilenodiaminatetraacético
i   acido m etilendiaminatetraacetico
n   ethyleendiamintetra-azijnzuur n
d   Äthylendiamintetraessigsäure ƒ

2375   ETIOLOGY   md
       Sum of knowledge regarding a disease.
f   étiologie ƒ
e   etiología ƒ
i   etiologia ƒ
n   etiologie
d   Ätiologie ƒ

2376   EUCKEN EQUATION   gp
       The equation for calculating the thermal
       conductivity of pure gases and gaseous
       mixtures at elevated temperatures.
f   équation ƒ d'Eucken
e   ecuación ƒ de Eucken
i   equazione ƒ d'Eucken
n   vergelijking van Eucken
d   Euckensche Gleichung ƒ

2377   EUROPIUM   ch
       Rare earth metal element, symbol Eu,
       atomic number 63.
f   europium m
e   europio m
i   europio m
n   europium n
d   Europium n

2378   EUTECTIC   mg
       The composition of a binary or ternary
       alloy system that has a lower melting
       point than neighbo(u)ring compositions.
f   eutectique adj
e   eutéctico adj
i   eutettico adj
n   eutectisch adj
d   eutektisch adj

2379   EUXENITE,   mi
       POLYCRASE, POLYCRASITE
       A mineral containing a large number of
       metallic elements, one of which is
       uranium.
f   euxénite ƒ, polycrasite ƒ
e   euxenita ƒ, policrasita ƒ
i   eusenite ƒ, policrasite ƒ
n   euxeniet n, polycrasiet n
d   Euxenit m, Polycrasit m

2380   eV, see 2188

2381   EVACUATED GLASS AMPOULE   is, md
f   ampoule ƒ en verre évacuée

e   ampolla ƒ de vidrio evacuada
i   ampolla ƒ in vetro vuotata
n   luchtledige glazen ampul
d   luftleere Glasampulle ƒ

2382   EVACUATED SEALED AMPOULE   tr
       A container for, e.g., labelled (tagged)
       compounds.
f   ampoule ƒ évacuée scellée
e   ampolla ƒ evacuada sellada
i   ampolla ƒ vuotata sigillata
n   luchtledige dichtgesmolten ampul
d   luftleere verschmolzene Ampulle ƒ

2383   EVANSITE   mi
       An alumin(i)um phosphate, reported to
       contain small amounts of U.
f   évansite ƒ
e   evansita ƒ
i   evansite ƒ
n   evansiet n
d   Evansit m

2384   EVAPORATED SIMPLE   me
       LIQUID ACTIVITY METER
       An assembly designed to measure the
       specific activity of a liquid and utilizing
       for this purpose residues of evaporated
       samples.
f   activimètre m de liquide par évaporation
e   activímetro m de líquido por evaporación
i   attivimetro m per campioni liquidi ad
    evaporazione
n   activiteitsmeter voor ingedampte
    vloeistoffen
d   Gerät n zur Bestimmung der Aktivität
    von Flüssigkeiten mittels eingedampfter
    Proben

2385   EVAPORATION,   ch
       VAPORIZATION
       The removal by vaporization of relatively
       large amounts of liquids from solutions
       or suspensions.
f   évaporation ƒ, vaporisation ƒ
e   evaporación ƒ, vaporización ƒ
i   evaporazione ƒ, vaporizzazione ƒ
n   verdamping
d   Verdampfung ƒ, Verdunstung ƒ

2386   EVAPORATION HEAT,   ch, gp
       HEAT OF EVAPORATION,
       HEAT OF VAPORIZATION,
       VAPORIZATION HEAT
       The increase of heat content when unit
       mass, or one mole, of a liquid is converted
       into a vapo(u)r at the boiling point,
       without change of temperature.
f   chaleur ƒ d'évaporation
e   calor m de evaporación
i   calore m d'evaporazione
n   verdampingswarmte
d   Verdampfungswärme ƒ

2387   EVAPORATION NUCLEAR MODEL   np
       A nuclear model in which the analogy
       with the evaporation of a liquid drop is
       stressed.

f   modèle *m* nucléaire d'évaporation
e   modelo *m* nuclear de evaporación
i   modello *m* nucleare d'evaporazione
n   verdampingsmodel *n*
d   Verdampfungsmodell *n*

2388 EVAPORATION NUCLEON   np
Name given to a proton or a neutron
emitted by a nucleus during the cooling
down of the latter after certain collisions
which have the effect of heating it.
f   nucléon *m* d'évaporation
e   nucleón *m* de evaporación
i   nucleone *m* d'evaporazione
n   verdampingsnucleon *n*
d   Verdampfungsnukleon *n*

2389 EVAPORATIVE CENTRIFUGE   is
A centrifuge in which the mixture to be
separated is in the liquid phase at the
periphery of the centrifuge.
f   centrifugeur *m* à évaporation
e   centrifuga *f* de evaporación
i   centrifuga *f* ad evaporazione
n   verdampingscentrifuge
d   Verdampfungszentrifuge *f*

2390 EVEN-EVEN NUCLEUS   np
Nucleus in which there are an even number
of protons and an even number of
neutrons.
f   noyau *m* pair-pair
e   núcleo *m* par-par
i   nucleo *m* pari-pari
n   even-even-kern
d   gerade-gerade-Kern *m*, gg-Kern *m*

2391 EVEN NUMBER   ma
Capable of being divided by two without
a remainder.
f   nombre *m* pair
e   número *m* par
i   numero *m* pari
n   even getal *n*
d   gerade Zahl *f*

2392 EVEN-ODD NUCLEUS   np
A nucleus in which there are an even
number of protons and an odd number of
neutrons.
f   noyau *m* pair-impair
e   núcleo *m* par-impar
i   nucleo *m* pari-dispari
n   even-oneven-kern
d   gerade-ungerade-Kern *m*, gu-Kern *m*

2393 EVEN TERM OF ATOM   ma, np
A term for which $\sum_i 1_i$, summed over all
the electrons of the atom, is even.
f   terme *m* pair de l'atome
e   término *m* par del átomo
i   termine *m* pari dell'atomo
n   atoomterm met even spin
d   Atomterm *m* mit gradzahligem Spin

2394 EVERSAFE   rt
Generally used as an adjective to describe
vessels, containing solutions of fissile
(fissionable) materials, whose dimensions
are such that under no conditions of
concentration or total mass can nuclear
criticality arise.
f   absolument sûr,
    nucléairement sûr
e   absolutamente seguro
i   assolutamente sicuro
n   absoluut veilig
d   narrensicher

2395 EXCESS ABSORPTION   ab
Absorption possessing a value above a
certain stated value.
f   excédent *m* d'absorption
e   absorción *f* excedente
i   eccesso *m* d'assorbimento
n   overabsorptie
d   Überabsorption *f*

2396 EXCESS MULTIPLICATION   ma, np
        CONSTANT
The multiplication constant minus one.
f   facteur *m* de multiplication excédentaire
e   factor *m* de multiplicación excedente
i   fattore *m* di moltiplicazione eccessivo
n   vermenigvuldigingsfactor min 1
d   Vermehrungsfaktor *m* minus 1

2397 EXCESS NEUTRON   rt
        FLUX SHUT-DOWN
Shut-down of a reactor automatically
caused by suitable pre-arranged measures
when the neutron flux exceeds a prescribed
level.
f   arrêt *m* par flux neutronique excessif
e   paro *m* por flujo neutrónico excesivo
i   arresto *m* per eccesso di flusso
    neutronico
n   uitschakeling door overmatige
    neutronenflux
d   Abschaltung *f* durch übermässigen
    Neutronenfluss

2398 EXCESS REACTIVITY,   rt
        REACTIVITY EXCESS
The maximum reactivity attainable at
any time by adjustment of the control
members.
f   excédent *m* de réactivité
e   reactividad *f* excedente
i   eccesso *m* di reattività
n   overreactiviteit
d   Überreaktivität *f*

2399 EXCESS RESONANCE   ma, np
        INTEGRAL
The resonance integral when the cross
section excludes that part which varies
inversely with neutron speed.
f   excès *m* de l'intégrale de résonance
e   exceso *m* de la integral de resonancia

i   eccesso *m* dell'integrale di risonanza
n   resonantiepiekintegraal
d   Überschussresonanzintegral *n*

2400 EXCESS TEMPERATURE   rt
      SHUT-DOWN
      Shut-down of a reactor caused automatic-
      ally by suitable pre-arranged measures
      when the temperature at a specified
      spot of the reactor exceeds a prescribed
      value.
f   arrêt *m* par température excessive
e   paro *m* por temperatura excesiva
i   arresto *m* per eccesso di temperatura
n   uitschakeling door te hoge temperatuur
d   Abschaltung *f* durch Temperaturüber-
    höhung

2401 EXCESSIVE NEUTRON FLUX   rt
      The result of a nuclear reaction in which
      an excess of neutrons is produced.
f   flux *m* neutronique excessif
e   flujo *m* neutrónico excesivo
i   flusso *m* neutronico eccessivo
n   overmatige neutronenflux
d   übermässiger Neutronenfluss *m*

2402 EXCHANGE CAPACITY   ch
      A measure of the capacity of a solid to
      capture and hold ions from solution.
f   capacité *f* d'échange
e   capacidad *f* de intercambio
i   capacità *f* di scambio
n   uitwisselingscapaciteit
d   Austauschkapazität *f*

2403 EXCHANGE COLLISION   gp, np
      A collision in which exchange of energy
      takes place.
f   collision *f* à échange d'énergie
e   colisión *f* con intercambio de energía
i   collisione *f* a scambio d'energia
n   botsing met energieuitwisseling
d   Kollision *f* mit Energieaustausch

2404 EXCHANGE ENERGY   qm
      A specifically quantum-mechanical effect
      which has no classical analog(ue).
f   énergie *f* d'échange
e   energía *f* de intercambio
i   energia *f* di scambio
n   uitwisselingsenergie
d   Austauschenergie *f*

2405 EXCHANGE FORCE   ma, np
      A type of force, acting between two
      particles, the mathematical expression
      of which involves an interchange of their
      co-ordinates.
f   force *f* d'échange
e   fuerza *f* de intercambio
i   forza *f* di scambio
n   uitwisselingskracht
d   Austauschkraft *f*

2406 EXCHANGE OF   ge
      KINETIC ENERGY
      Exchange of energy which a body possesses
      by virtue of its motion..
f   échange *m* d'énergie cinétique
e   cambio *m* de energía cinética
i   scambio *m* d'energia cinetica
n   uitwisseling van kinetische energie
d   Austausch *m* der kinetischen Energie

2407 EXCHANGE REACTION   ch
      A reaction in which exchange of energy
      takes place.
f   réaction *f* à échange d'énergie
e   reacción *f* con intercambio de energía
i   reazione *f* a scambio d'energia
n   reactie met energieuitwisseling
d   Reaktion *f* mit Energieaustausch

2408 EXCITATION   np
      The addition of energy to a system,
      transferring it from its ground state
      to an excited state.
f   excitation *f*
e   excitación *f*
i   eccitazione *f*
n   aanslag
d   Anregung *f*

2409 EXCITATION BAND   np
      A range of possible neighbo(u)ring values
      to which the energies of the electrons
      of an atom can be raised by excitation.
f   bande *f* d'excitation
e   banda *f* de excitación
i   banda *f* d'eccitazione
n   aanslagband,
    band van de aangeslagen toestanden
d   Anregungsband *n*

2410 EXCITATION CURVE,   np
      EXCITATION FUNCTION
      The yield of a specified nuclear reaction
      related to the energy of the particle or
      photon which causes it.
f   fonction *f* d'excitation
e   función *f* de excitación
i   funzione *f* d'eccitazione
n   aanslagfunctie
d   Anregungsfunktion *f*

2411 EXCITATION ENERGY   np
      The minimum energy necessary to carry
      a non-excited atom to a certain degree of
      excitation.
f   énergie *f* d'excitation
e   energía *f* de excitación
i   energie *f* d'eccitazione
n   aanslagenergie
d   Anregungsenergie *f*

2412 EXCITATION FREQUENCY   np
      A physical value used in theoretical
      electronics and represented by V_x.
f   fréquence *f* d'excitation
e   frecuencia *f* de excitación

i frequenza f d'eccitazione
n aanslagfrequentie
d Anregungsfrequenz f

2413 EXCITATION NUMBER   np
A physical value used in theoretical
electronics and represented by N_x.
f nombre m d'excitation
e número m de excitación
i numero m d'eccitazione
n aanslaggetal n
d Anregungszahl f

2414 EXCITATION OF A GAS   np
The change of structure of certain atoms
or molecules of a gas characterized by
the passage of an electron from one
energy level to a higher level.
f excitation f d'un gaz
e excitación f de un gas
i eccitazione f d'un gas
n aanslag van een gas
d Anregung f eines Gases

2415 EXCITATION POTENTIAL   np
The difference of potential required to
give an electron starting from rest the
necessary minimum energy to permit it
to excite by collision an atom or a
molecule initially in a normal state.
f potentiel m d'excitation
e potencial m de excitación
i potenziale m d'eccitazione
n aanslagpotentiaal
d Anregungsspannung f

2416 EXCITATION PROBABILITY   np
Dependent on the difference between the
energy of the electron and the critical
energy.
f probabilité f d'excitation
e probabilidad f de excitación
i probabilità f d'eccitazione
n aanslagkans, aanslagwaarschijnlijkheid
d Anregungswahrscheinlichkeit f

2417 EXCITED ATOM,   np
        HOT ATOM
An atom which possesses more energy
than a normal atom.
f atome m excité
e átomo m excitado
i atomo m eccitato
n aangeslagen atoom n
d angeregtes Atom n

2418 EXCITED-ATOM DENSITY   np
The number of excited atoms in a gas
per unit volume.
f concentration f des atomes excités,
    densité f des atomes excités
e concentración f de los átomos excitados,
    densidad f de los átomos excitados
i concentrazione f degli atomi eccitati,
    densità f degli atomi eccitati
n concentratie van de aangeslagen atomen,
    dichtheid van de aangeslagen atomen
d Dichte f der angeregten Atome

2419 EXCITED ION   np
An ion resulting from the loss by an atom
of one valence electron and the transition
of another valence electron to a higher
energy level.
f ion m excité
e ión m excitado
i ione m eccitato
n aangeslagen ion n
d angeregtes Ion n

2420 EXCITED LEVEL   np
The level of the energy of a particle
higher than that of the ground state.
f niveau m d'excitation
e nivel m de excitación
i livello m d'eccitazione
n aanslagniveau n
d Anregungsniveau n

2421 EXCITED NUCLEUS   np
A nucleus having a higher energy than in
the ground state.
f noyau m excité
e núcleo m excitado
i nucleo m eccitato
n aangeslagen kern
d angeregter Kern m

2422 EXCITED STATE   np
A state of a nucleus atom, or molecule,
having a higher energy than the ground
state energy.
f état m d'excitation
e estado m de excitación
i stato m d'eccitazione
n aangeslagen toestand, aanslagtoestand
d angeregter Zustand m

2423 EXCITON   cr
A free electron attached to a hole.
f exciton m
e excitón m
i eccitone m
n exciton n
d Exciton n

2424 EXCLUSION AREA   sa
The area surrounding a nuclear installation
in which no building not belonging to
that installation may be erected, the
boundary being determined as a function of
danger arising from accidents.
f zone f d'exclusion
e zona f de exclusión
i zona f d'esclusione
n bouwverbodgebied n
d Bauverbotsgebiet n

2425 EXCRETION MONITORING   rw, sa
A method of detecting the presence of
radioactive material and of estimating
the quantity in the human body.
f contrôle m de l'activité de l'excrétion
e reconocimiento m de la radiactividad de
    la excreción
i controllo m di radioattività dell'escrezione
n activiteitsonderzoek n van de excretie

d Aktivitätskontrolle f der Auswurf-
stoffen

2426 EXFOLIATION,   mg
FLAKING, SPALLING
Flaking of the surface of metals or
refractories leaving new surfaces exposed.
f affouillement m, écaillement m
e desconchado m
i scagliatura f, screpolatura f
n afbladderen n
d Abblättern n, Häutung f

2427 EXIT CHANNEL SPIN   np
In a nuclear reaction, the vector sum of
the spins of the resulting particles.
f spin m de voie de sortie, voie f de sortie
e espín m de trayectoria de salida
i spin m di percorso d'uscita
n uitgangsbaanspin
d Ausgangswegspin m

2428 EXIT DOSE   xr
Dose of radiation of surface of body
opposite to that on which the beam is
incident.
f dose f à la sortie
e dosis f de salida
i dose f all'uscita
n uittreedosis
d Austrittsdosis f

2429 EXIT PORTAL,   xr
FIELD OF EXIT
The area through which a beam of radiation
leaves a patient's body.
f champ m de sortie, porte f de sortie
e campo m de salida
i porta f d'uscita
n uittreeveld n
d Austrittsfeld n

2430 EXIT PRESSURE DROP   cl, rt
The pressure loss when the coolant for
a reactor plant suffers an abrupt
contraction in fluid flow when leaving a
fuel-element channel.
f chute f de pression de sortie
e caída f de presión de salida
i caduta f di pressione d'uscita
n uitgangsdrukafval
d Ausgangsdruckabfall m

2431 EXOERGIC,   gp
EXOTHERMIC
Involving an evolution of heat or other
energy.
f exothermique adj
e exotérmico adj
i esotermico adj
n exotherm adj
d exotherm adj

2432 EXOERGIC PROCESS,   np
EXOTHERMIC PROCESS
A nuclear process in which energy is
liberated.

f procédé m exothermique
e procedimiento m exotérmico
i processo m esotermico
n exotherm proces n
d exothermes Verfahren n

2433 EXOTHERMIC NUCLEAR   np
DISINTEGRATION
A nuclear disintegration proceeding with
the development of heat.
f désintégration f nucléaire exothermique
e desintegración f nuclear exotérmico
i disintegrazione f nucleare esotermica
n exotherme kerndesintegratie
d exothermer Kernzerfall m

2434 EXOTHERMIC NUCLEAR   rt
REACTION
Nuclear reaction which proceeds with the
development of heat.
f réaction f nucléaire exothermique
e reacción f nuclear exotérmica
i reazione f nucleare esotermica
n exotherme kernreactie
d exotherme Kernreaktion f

2435 EXPANSION CLOUD CHAMBER,
see 1138

2436 EXPANSION ORBIT   np
The ultimate part of the electron path
which terminates at the target.
f orbite f d'expansion
e órbita f de expansión
i orbita f d'espansione
n expansiebaan, uitloopbaan
d Auslaufbahn f

2437 EXPANSION RATIO, see 1139

2438 EXPERIMENT   ge
An operation conducted for the purpose
of determining some unknown fact or of
obtaining knowledge concerning a general
principle of truth.
f essai m, expérience f
e experimento m, prueba f
i esperimento m, prova f
n experiment n, proef
d Experiment n, Versuch m

2439 EXPERIMENT RIG,   rt, te
TEST RIG
The complete set of apparatus for carrying
out experiments in a nuclear reactor.
f appareillage m expérimental,
équipement m expérimental
e equipo m experimental
i apparecchiatura f pilota
n experimenteeropstelling
d Experimentieraufstellung f

2440 EXPERIMENT THIMBLE   rt, te
A long narrow tube for carrying out
various tests in a reactor.
f tube m d'expérimentation
e tubo m de experimentación

i tubo *m* d'esperimento
n experimenteerbuis
d Experimentierrohr *n*

2441 EXPERIMENTAL BREEDER   rt
      REACTOR
      A fast heterogeneous nuclear reactor used
      for research and breeding.
f réacteur *m* surrégénérateur expérimental
e reactor *m* regenerador experimental
i reattore *m* rigeneratore sperimentale
n experimentele kweekreactor
d experimentéller Brutreaktor *m*

2442 EXPERIMENTAL CELL   rc, te
      In a nuclear reactor, an element of an
      experimental circuit comprising a channel
      and a heat producing element.
f cellule *f* d'essai
e célula *f* experimental
i cella *f* sperimentale
n experimenteercel
d Experimentierzelle *f*

2443 EXPERIMENTAL HOLE,   rt
      IRRADIATION CHANNEL,
      TEST HOLE
      A hole through a reactor shield into the
      interior of the reactor in which irra-
      diations are carried out.
f canal *m* expérimental
e canal *m* de irradiación
i canale *m* d'irradiazione
n bestralingskanaal *n*
d Bestrahlungskanal *m*, Versuchskanal *m*

2444 EXPERIMENTAL LOOP,   rt
      LOOP
      In reactor technology, a tube forming a
      closed circuit, through which various
      coolants may be pumped, in order to
      measure mass transfer rates, etc.
f boucle *f*, circuit *m* expérimental
e circuito *m* experimental, lazo *m*
i cappio *m*, circuito *m* sperimentale
n experimenteerkringloop, kringloop, lus
d Experimentierkreislauf *m*,
   Versuchskreislauf *m*

2445 EXPERIMENTAL REACTOR   rt
      A nuclear reactor designed primarily to
      obtain reactor physics or engineering data
      for the design or development of a
      reactor type.
f réacteur *m* expérimental
e reactor *m* experimental
i reattore *m* pilota
n experimentele reactor
d Experimentierreaktor *m*

2446 EXPLOSION   nw
f explosion *f*
e explosión *f*
i esplosione *f*
n explosie, uitbarsting
d Explosion *f*

2447 EXPLOSION TRAP   sa
      A vessel intended to receive the contents
      of another vessel should they be displaced
      by an explosion.
f récipient *m* de receuil
e recipiente *m* evitador de explosión
i recipiente *m* da attutire
n opvangvat *n*
d Auffanggefäss *n*

2448 EXPLOSION TRAP,
      ANTI-EXPLOSION VALVE,
      see 297

2449 EXPLOSIVE FISSION   nw
      The type of fission used in nuclear
      weapons.
f fission *f* explosive
e fisión *f* explosiva
i fissione *f* esplosiva
n explosieve splijting
d explosive Spaltung *f*

2450 EXPLOSIVE MATERIAL   nw
      Material that can either be exploded or
      detonated.
f matière *f* explosive
e materia *f* explosiva
i materia *f* esplosiva
n explosief materiaal *n*
d explosives Material *n*

2451 EXPLOSIVE REACTION   nw
      The type of reaction used in nuclear
      weapons.
f réaction *f* explosive
e reacción *f* explosiva
i reazione *f* esplosiva
n explosieve reactie
d explosive Reaktion *f*

2452 EXPONENTIAL ABSORPTION   np
      The removal of particles or photons from
      a beam, as it passes through matter.
f absorption *f* exponentielle
e absorción *f* exponencial
i assorbimento *m* esponenziale
n exponentiële absorptie
d exponentielle Absorption *f*

2453 EXPONENTIAL ASSEMBLY   rt
      A subcritical assembly used for an
      exponential experiment.
f ensemble *m* exponentiel
e conjunto *m* exponencial
i complesso *m* esponenziale
n exponentiële opstelling
d exponentielle Anordnung *f*

2454 EXPONENTIAL DECAY,   np
      EXPONENTIAL DISINTEGRATION
      Variation of a quantity a according to the
      law $A = A_o e^{-\lambda t}$, where A and A_o are the
      values of the quantity being considered at
      time t and zero, respectively, and λ is an
      appropriate constant.

f   décroissance f exponentielle
e   desintegración f exponencial
i   decadimento m esponenziale,
    disintegrazione f esponenziale
n   exponentieel verval n
d   exponentieller Zerfall m

2455 EXPONENTIAL EXPERIMENT   rt
    An assembly of a neutron source and a
    part of a possible reactor core, used to
    determine the buckling of the core.
f   expérience f exponentielle
e   experimento m exponencial
i   esperimento m esponenziale
n   exponentieel experiment n
d   Exponentialexperiment n,
    Exponentialversuch m

2456 EXPONENTIAL LAW OF   ra
      ATTENUATION
    The law of attenuation which applies to all
    processes where a constant fraction of
    the quantity being measured is lost per
    unit of distance travelled.
f   loi f exponentielle d'atténuation
e   ley f exponencial de atenuación
i   legge f esponenziale d'attenuazione
n   exponentiële verzwakkingswet
d   exponentielles Schwächungsgesetz n

2457 EXPONENTIAL METHOD   rt
    The method of building an assembly of
    fissile (fissionable) material and moderator
    which contains too little of the first
    mentioned material to sustain a chain
    reaction.
f   méthode f exponentielle
e   método m exponencial
i   metodo m esponenziale
n   exponentiële methode
d   Exponentialverfahren n

2458 EXPONENTIAL REACTOR   rt
    A nuclear reactor in which the fissile
    (fissionable) material is assembled
    according to the exponential method.
f   réacteur m exponentiel
e   reactor m exponencial
i   reattore m esponenziale
n   exponentiële reactor
d   Exponentialreaktor m

2459 EXPONENTIAL WELL   ma, np
    A potential well in which the potentials
    are proportional to $(1/r)$ exp-kr.
f   puits m exponentiel
e   pozo m exponencial
i   pozzo m esponenziale
n   exponentiële kuil
d   Exponentialtopf m

2460 EXPOSURE   np
    Of X- or gamma radiation at a certain
    place, a measure of the radiation that is
    based upon its ability to produce ionization.
f   exposition f
e   exposición f

i   esposizione f
n   exposie
d   Exponierung f, Exposition f

2461 EXPOSURE CHART   xr
    A chart for indicating the radiographic
    exposures appropriate for different
    thicknesses of a specified material or for
    different parts of the human body.
f   tableau m d'exposition
e   tabla f de tiempos de exposición
i   tabella f dei tempi d'esposizione
n   exposietabel
d   Expositionstabelle f

2462 EXPOSURE CONTAINER   ra
    A container for a sealed source for
    immediate use.
f   récipient m d'exposition
e   recipiente m de exposición
i   recipiente m d'esposizione
n   exposievat n
d   Expositionsbehälter m

2463 EXPOSURE DIMMER   xd
    A device for regulating the time and/or
    intensity of the exposure.
f   atténuateur m d'exposition
e   atenuador m de exposición
i   attenuatore m d'esposizione
n   exposieregelaar
d   Expositionsregler m

2464 EXPOSURE INDICATOR,   gr, ra
      RADIATION EXPOSURE INDICATOR
    An indicator designed to detect the
    presence of gamma radiation and to give
    an estimate of the exposure.
f   signaleur m d'exposition
e   indicador m de exposición
i   segnalatore m d'esposizione
n   exposie-indicator
d   Expositionsanzeiger m

2465 EXPOSURE RATE   me
    Exposure in a certain time interval,
    divided by that interval.
f   débit m d'exposition
e   intensidad f de exposición
i   intensità f d'esposizione
n   exposiesnelheid, exposietempo n
d   Exponierungsstärke f, Expositionsstärke f

2466 EXPOSURE RATEMETER   me
    A measuring assembly for the exposure
    rate.
f   débitmètre m d'exposition
e   dosímetro m de exposición
i   rateometro m d'esposizione
n   exposietempometer
d   Dosisleistungsmesser m

2467 EXPOSURE TIME RANGE   xr
f   plage f de temps d'exposition
e   campo m de tiempo de exposición
i   campo m di tempo d'esposizione
n   exposietijdgebied n
d   Expositionszeitgebiet n

2468 EXPOSURE TO RADIATION   ra
Voluntary or accidental exposure to
radiation of a living organism or a
material substance.
f irradiation *f*
e irradiación *f*
i irradiazione *f*
n bestraling
d Bestrahlung *f* ·

2469 EXPOSUREMETER   xr
An instrument for predicting the exposure
time required for specified radiographic
conditions.
f exposimètre *m*
e exposímetro *m*
i esposimetro *m*
n exposiemeter
d Expositionsmesser *m*

2470 EXTENSIVE SHOWER, see 440

2471 EXTERNAL BREEDING RATIO   rt
Breeding ratio concerning only the
fissile (fissionable) nuclei produced in
the blanket surrounding the core of a
breeder reactor.
f rapport *m* de régénération externe
e relación *f* de regeneración externa
i rapporto *m* di rigenerazione esterna
n uitwendige kweekverhouding
d äusseres Brutverhältnis *n*

2472 EXTERNAL CATHODE COUNTER   ct
TUBE,
MAZE COUNTER TUBE
A counter tube possessing an envelope
generally of glass, and a cathode which
is a carbon or metal coating on the
external surface of the envelope.
f tube *m* compteur à cathode externe,
tube *m* compteur de Maze
e tubo *m* contador de cátodo externo, ·
tubo *m* contador de Maze
i tubo *m* contatore a catodo esterno,
tubo *m* contatore di Maze
n telbuis met uitwendige katode
d Zählrohr *n* mit Aussenkatode,
Zählrohr *n* nach Maze

2473 EXTERNAL CIRCULATING   cl, su
SYSTEM
A circulating system of special design
because of the use of high pressures and
temperatures, the desirability of small
pressure drop and the potential hazard
of radioactivity from the coolant fluid.
f système *m* de circulation externe
e sistema *m* de circulación externa
i sistema *m* di circolazione esterna
n uitwendig circulatiesysteem *n*
d äusseres Umlaufsystem *n*

2474 EXTERNAL CONVERSION RATIO,
see 661

2475 EXTERNAL EXPOSURE,   ra, xr
EXTERNAL IRRADIATION
f exposition *f* extérieure,
irradiation *f* externe
e exposición *f* exterior,
irradiación *f* externa
i esposizione *f* ad irradiazione esterna,
irradiazione *f* esterna
n uitwendige exposie
d äussere Bestrahlung *f*,
Fremdbestrahlung *f*

2476 EXTERNAL FIELD   ec
An electric field applied externally.
f champ *m* extérieur
e campo *m* exterior
i campo *m* esteriore
n uitwendig veld *n*
d äusseres Feld *n*

2477 EXTERNAL QUENCHING   ct
The quenching of, e.g., a Geiger–Müller-
counter tube by momentarily reducing the
applied potential difference.
f coupure *f* externe
e extinción *f* externa
i spegnimento *m* esterno
n uitwendige doving
d Fremdlöschung *f*

2478 EXTERNAL RADIATION   ra
That radiation reaching a given point in
the body or irradiated material which
is due directly or indirectly to a source
of radiation outside that body or material.
f rayonnement *m* externe
e radiación *f* externa
i radiazione *f* esterna
n uitwendige bestraling
d äussere Bestrahlung *f*

2479 EXTERNALLY QUENCHED   ct
COUNTER TUBE
A radiation counter tube quenched by
momentary reduction of the applied
potential difference.
f tube *m* compteur à coupure externe
e tubo *m* contador de extinción externa
i tubo *m* contatore a spegnimento esterno
n telbuis met uitwendige doving
d Zählrohr *n* mit externer Löschung

2480 EXTINCT NATURAL   mi, np
RADIONUCLIDES
Natural radionuclides which have lifetimes
that are too short for survival from the
time of nucleogenesis to the present, but
long enough for persistence into early
geologic times with measurable effect.
f radionucléides *pl* naturels éteints
e radionúclidos *pl* naturales extintos
i radionuclidi *pl* naturali estinti
n uitgestorven natuurlijke radionucliden *pl*
d zerfallene natürliche Radionuklide *pl*

2481 EXTINCTION   ra
The quenching of a beam of radiation
by diffraction as distinct from absorption.
f extinction f
e extinción f
i estinzione f, spegnimento m
n doving
d Löschung f

2482 EXTRACORPOREAL BLOOD
IRRADIATOR, see 612

2483 EXTRACT LAYER   ch
The liquid layer in a solvent extraction
system into which the desired solvent is
extracted.
f couche f d'extraction
e capa f de extracción
i strato m d'estrazione
n extractielaag, extraherende laag
d extrahierende Schicht f,
Extraktionsschicht f

2484 EXTRACTANT,   ch
EXTRACTING AGENT
A substance other than water used in
extraction processes.
f agent m extracteur
e extractante m, extraente m
i agente m estrattore
n extractiemiddel n, extraheermiddel n
d Extraktionsmittel n

2485 EXTRACTED BEAM, see 2057

2486 EXTRACTION,   ch
RECOVERY
The operation wherein a liquid or solid
mixture is brought into contact with an
immiscible or partially immiscible
liquid to achieve a redistribution of
solute between the phases.
f extraction f
e extracción f
i estrazione f
n extractie
d Extraktion f

2487 EXTRACTION CYCLE   ch
The total of the operations in any
extraction process.
f cycle m d'extraction
e ciclo m de extracción
i ciclo m d'estrazione
n extractiecyclus
d Extraktionszyklus m

2488 EXTRACTION PLANT   ch
f installation f d'extraction
e instalación f de extracción
i impianto m d'estrazione
n extractie-installatie
d Extraktionsanlage f

2489 EXTRACTION POTENTIAL   np
The minimum electric potential needed
to extract an electron in a given manner at

a given temperature from a substance.
f potentiel m d'extraction
e potencial m de extracción
i potenziale m d'estrazione
n uittreepotentiaal
d Austrittspotential n

2490 EXTRACTION TOWER   ch
f tour f d'extraction
e torre f de extracción
i torre f d'estrazione
n extractietoren
d Extraktionsturm m

2491 EXTRACTIVE DISTILLATION   ch
Separation by distillation of a relatively
non-volatile azeotrope formed from an
added compound and one of the constituents
of the original mixture.
f distillation f extractive
e destilación f extractiva
i distillazione f estrattiva
n extractieve destillatie
d extraktive Destillation f

2492 EXTRACTIVE METALLURGY   mg
A method used for obtaining pure uranium.
f métallurgie f extractive
e metalurgia f extractiva
i metallurgia f estrattiva
n extractiemetallurgie,
extractieve metallurgie
d Extraktionsmetallurgie f,
extraktive Metallurgie f

2493 EXTRANUCLEAR ELECTRON   np
An electron of the outer shells.
f électron m extranucléaire
e electrón m extranuclear
i elettrone m estranucleare
n schaalelektron n
d Hüllenelektron n

2494 EXTRANUCLEAR PROCESS   np
A nuclear process restricted to an electron
of the outer shell.
f procédé m extranucléaire
e procedimiento m extranuclear
i processo m estranucleare
n extranucleair proces n
d extranuklearer Prozess m

2495 EXTRANUCLEAR STRUCTURE   np
The structure of an atom outside the
nucleus.
f structure f extranucléaire
e structura f extranuclear
i struttura f estranucleare
n extranucleaire structuur
d extranukleare Struktur f

2496 EXTRAPOLATED BOUNDARY   np
A hypothetical surface outside an assembly
on which the neutron flux density would be
zero if extrapolated from the flux
distribution neglecting the distribution
within a few mean-free paths of the
physical surface.

f   limite *f* extrapolée
e   lîmite *m* extrapolado
i   contorno *m* d'estrapolazione
n   geëxtrapoleerde grens
d   extrapolierte Grenze *f*,
     extrapolierte Reaktorbegrenzung *f*

2497 EXTRAPOLATED RANGE,   np
      VISUAL RANGE
      The value of the range in matter, for a
      given type of charged particle, obtained
      by an extrapolation of the absorption
      curve according to some prescription.
f   portée *f* extrapolée
e   alcance *m* extrapolado
i   portata *m* estrapolata
n   geëxtrapoleerde dracht
d   extrapolierte Reichweite *f*

2498 EXTRAPOLATION DISTANCE, see 443

2499 EXTRAPOLATION IONIZATION   ic
      CHAMBER
      An ionization chamber in which one of the
      characteristics can be varied - most often
      the spacing between electrodes - in order
      to extrapolate its readings to zero chamber
      value.
f   chambre *f* d'ionisation à extrapolation
e   cámara *f* de ionización de extrapolación
i   camera *f* d'ionizzazione ad estrapolazione
n   extrapolatie-ionisatievat *n*
d   Extrapolationsionisationskammer *f*

2500 EXTREME VACUUM,   vt
      ULTRA-HIGH VACUUM
      A vacuum in which the pressure range
      covers under 10^{-13} Torr.
f   vide *m* ultra-élevé
e   vacío *m* ultraelevado
i   vuoto *m* superelevato
n   zeer hoog vacuüm *n*
d   Höchstvakuum *n*

2501 EXTRUSION   mg
      The forming of metal parts of continuous
      cross section by plastically forcing the
      metal through a die in an extrusion press.
f   extrusion *f*
e   extrusión *f*
i   estrusione *f*
n   extrusie, spuitgieten *n*
d   Extrusion *f*, Fliesspressen *n*,
     Kaltspritzen *n*

2502 EYE SHIELD   sa, xr
      A shield of protective material arranged
      to fit into the conjunctival sac so as to
      protect the eye, and in particular the lens,
      from harmful quantities of an ionizing
      radiation.
f   oeuilleton *m* protecteur
e   protector *m* ocular de contacto
i   protettore *m* oculare
n   oogscherm *n*
d   Augenmuschel *f*

F

2503 F CENTERS (US), ec, np
    F CENTRES (GB)
Colo(u)r centre(er)s formed by electrons
trapped in anion vacancies.
f centres *pl* F
e centros *pl* F
i centri *pl* F
n F-centra *pl*
d F-Zentren *pl*

2504 F CURVE np
The curve showing the variation of the
atomic scattering factor f, with $(\sin \theta)\lambda$,
where θ is the Bragg angle and λ the
incident wavelength.
f courbe *f* f
e curva *f* f
i curva *f* f
n f-kromme
d f-Kurve *f*

2505 FACE-CENTRED CUBIC cr
    LATTICE (GB),
    FACE-CENTERED
    CUBIC LATTICE (US),
    FACE-CENTRED STRUCTURE (GB),
    FACE-CENTERED STRUCTURE (US)
A crystal structure characterized by a
cubic unit cell with an atom at each
corner and an atom in the centre(er) of each
of the six faces.
f réseau *m* cubique à faces centrées
e structura *f* cúbica de caras centradas
i struttura *f* cubica a faccie centrate
n in de vlakken gecentreerd kubisch
   rooster *n*,
   midvlaks kubisch gecentreerd rooster *n*
d flächenzentriertes kubisches Gitter *n*

2506 FACTOR ge
Generally a proportionality constant in a
physical equation.
f facteur *m*
e factor *m*
i fattore *m*
n factor
d Faktor *m*

2507 FACTOR OF SAFETY sa
A number expressing the relation between
the utmost endurance of a structural part,
or of a complete structure, to the maximum
actual demand that may be expected ever
to be made upon it.
f facteur *m* de sûreté
e factor *m* de seguridad
i fattore *m* di sicurezza
n veiligheidsfactor
d Sicherheitsfaktor *m*

2508 FAILED ELEMENT, see 834

2509 FAILED ELEMENT DETECTION ma
    AND LOCALIZATION EQUIPMENT
An equipment for detection and localization
of failed fuel elements and for following
the progress of the failure.
f équipement *m* d'avertissement et de
   localisation des ruptures de gaines.
e equipo *m* detector y localizador de un
   elemento combustible averiado
i apparecchiatura *f* di rivelazione e
   localizzazione di rottura di guaine
n uitrusting voor vaststelling en lokalisatie
   van beschadigde splijtstofelementen
d Einrichtung *f* zum Nachweis und zur
   Lokalisierung schadhafter Brennelemente

2510 FAILED ELEMENT INDICATOR ma
An indicator whose detector is situated in
the main coolant circuit and which
measures the concentration of fission
products for fast indication of element
failure.
f signaleur *m* de rupture de gaine
e indicador *m* de averías
i segnalatore *m* di rottura di guaine
n indicator voor beschadigde splijtstofelementen
d Anzeiger *m* für schadhafte Brennelemente

2511 FAILED ELEMENT MONITOR ma
An assembly designed for the detection of
failures likely to occur in the clads which
seal off the fuel from the coolant of a
nuclear reactor.
f moniteur *m* de rupture de gaine
e vigía *f* de averías
i monitore *m* di rottura di guaine
n monitor voor beschadigde splijtstofelementen
d Warngerät *n* für schadhafte Brennelemente

2512 FAILSAFE sa
Safe under any circumstances.
f à autoprotection, à sûreté absolue
e a prueba de fallas, de autoprotección
i ad autoprotezione, a sicurezza assoluta
n veilig falend
d fehlsicher, gesichert gegen Versagen

2513 FALL-BACK nw
The material that drops back to the earth
or water at the site of a surface or
subsurface nuclear explosion.
f retombée *f*
e caída *f*, materia *f* descente
i materia *f* cadente, ricaduta *f*
n neervallend materiaal *n*
d Rückfallstoffe *pl*

2514 FALL-OUT, see 82

2515 FALLING RATE DRYING PERIOD ch
In the drying of solids, a period in the later
stages of drying during which the rate of
moisture loss is decreasing.

f intervalle *m* de temps de séchage à
   vitesse décroissante
e intervalo *m* de tiempo de secado de
   velocidad decreciente
i intervallo *m* di tempo d'essicazione a
   velocità decrescente
n tijdsspan *n* van afnemende droogsnelheid
d Zeitspanne *f* abnehmender Trocken-
   geschwindigkeit

2516 FALSE CURVATURE   ic
   The curvature of electron tracks in an
   ionization chamber in the absence of an
   applied magnetic field, caused by
   scattering collisions between electrons
   and gas atoms.
f trace *f* fictive
e traza *f* fictiva
i traccia *f* finta
n loos spoor *n*
d Blindspur *f*

2517 FAMILY,   ch
       GROUP
   A group of elements characterized by
   similar chemical properties such as
   valence, solubility of salts,behavio(u)r
   towards reagents, etc.
f groupe *m*
e grupo *m*
i gruppo *m*
n groep
d Gruppe *f*

2518 FAMILY OF CHARACTERISTICS   ge
   A collection of characteristics showing
   the interrelation of dependent and
   independent variables of a device.
f faisceau *m* de caractéristiques,
   faisceau *m* de courbes
e familia *f* de características,
   familia *f* de curvas
i famiglia *f* di caratteristiche,
   famiglia *f* di curve
n karakteristiekenschaar, krommenschaar
d Kennlinienfeld *n*, Kurvenschar *f*

2519 FAMILY OF REACTORS,   rt
       REACTOR FAMILY,
       REACTOR SYSTEM
   A set of characteristics defining a
   category of nuclear reactors which can be
   built, e.g., the nature of the fuel, the
   nature of the moderator, the nature of the
   cooling system, etc. By extension, a
   group of reactors having this set of
   characteristics.
f filière *f* de réacteurs
e familia *f* de reactores
i famiglia *f* di reattori, serie *f* di reattori
n reactorfamilie
d Reaktorbaulinie *f*

2520 FAN-IN UNIT,   cd
       SEQUENCING-MIXING UNIT
   Component part of the 400 channel pulse
   height analyzer.

f aiguilleur-mélangeur *m*
e mezclador *m* formulador en serie
i mescolatore *m* ordinatore di sequenza
n volgorde bepalende menger
d reihenfolgebestimmender Mischer *m*

2521 FAST BREEDER,   rt
       FAST BREEDING REACTOR
   A reactor of the breeder type which
   produces fast neutrons.
f réacteur *m* surrégénérateur à neutrons
   rapides
e reactor *m* regenerador de neutrones
   rápidos
i reattore *m* rigeneratore a neutroni
   veloci
n kweekreactor met snelle neutronen
d Brutreaktor *m* mit schnellen Neutronen

2522 FAST CHOPPER   ec
   A chopper operating at a high speed.
f hacheur *m* rapide
e modulador *m* rápido
i modulatore *m* veloce
n snelle hakker
d schneller Zerhacker *m*

2523 FAST COINCIDENCE UNIT   ct
   Coincidence counting apparatus with three
   channels with one anticoincidence channel.
f sélecteur *m* de coïncidences rapides
e selector *m* de coincidencias rápidas
i selettore *m* di coincidenze veloci
n snelle coïncidentiekiezer
d schneller Koinzidenzwähler *m*

2524 FAST EFFECT,   rt
       FAST FISSION EFFECT,
       FAST MULTIPLICATION EFFECT
   In a thermal reactor, the increase in
   reactivity, due to fast fission.
f effet *m* de fission rapide
e efecto *m* de fisión rápida
i effetto *m* di fissione veloce
n snelsplijtingseffect *n*
d Schnellspalteffekt *m*

2525 FAST EXPONENTIAL   rt
       EXPERIMENT,
       FEE
   A critical facility built and operated by
   Argonne National Laboratory at Lemont,
   Ill.
f installation *f* d'expériences exponentielles
   rapides
e instalación *f* de experimentos exponenci-
   ales rápidos
i impianto *m* d'esperimenti esponenziali
   veloci
n installatie voor snelle exponentiële
   experimenten
d Anlage *f* für schnelle exponentielle
   Experimente

2526 FAST FISSION,   np
    FAST NEUTRON FISSION,
    HIGH-ENERGY FISSION
  Fission caused by fast neutrons.
f fission *f* rapide
e fisión *f* rápida
i fissione *f* veloce
n snelsplijting, splijting met snelle neutronen
d Schnellspaltung *f*,
   Spaltung *f* durch schnelle Neutronen

2527 FAST FISSION EFFECT   np, rt
    FACTOR,
    FAST FISSION FACTOR,
    FAST MULTIPLICATION FACTOR
  The ratio of the number of fissions due to
  neutrons of all energies, to that due to
  thermal neutrons only.
f facteur *m* de fission rapide
e factor *m* de fisión rápida
i fattore *m* di fissione veloce
n factor voor snelsplijtingen,
   snelsplijtingsfactor
d Schnellspaltfaktor *m*

2528 FAST FRAGMENT   np, rt
  A fragment due to fast neutrons in a
  thermal reactor.
f fragment *m* rapide
e fragmento *m* rápido
i frammento *m* veloce
n snel brokstuk *n*
d schnelles Bruchstück *n*

2529 FAST MEDIUM   np
  A medium containing a fissile (fissionable)
  substance in which the proportion of
  slowing-down neutrons is low enough to
  make it possible that, in absence of a
  reflector, there is predominance of fissions
  induced by fast neutrons.
f milieu *m* rapide
e medio *m* de fisión rápida
i mezzo *m* di fissione veloce
n snelsplijtmilieu *n*
d Schnellspaltmedium *n*

2530 FAST NEUTRON COUNTER TUBE   ct
  A tube consisting of a series of very pure
  polyethylene cavities, their metal-cooled
  surface forming the cathode.
f tube *m* compteur de neutrons rapides
e tubo *m* contador de neutrones rápidos
i tubo *m* contatore di neutroni veloci
n telbuis voor snelle neutronen
d Zählrohr *n* für schnelle Neutronen

2531 FAST NEUTRON RANGE,   np
    FAST NEUTRON REGION
  The distance that a fast neutron will
  penetrate a given substance before its
  kinetic energy is reduced below which
  it can no longer produce ionization.
f portée *f* de neutrons rapides
e alcance *m* de neutrones rápidos
i portata *f* di neutroni veloci
n dracht van snelle neutronen
d Reichweite *f* von schnellen Neutronen

2532 FAST NEUTRON REACTOR,   rt
    FAST REACTOR
  A nuclear reactor in which fission is
  induced predominantly by fast neutrons.
f réacteur *m* rapide
e reactor *m* rápido
i reattore *m* veloce
n reactor met snelle neutronen,
   snelle reactor
d schneller Reaktor *m*

2533 FAST NEUTRONS,   np
    HIGH-SPEED NEUTRONS
  Neutrons of kinetic energy greater than
  some specified value.
f neutrons *pl* rapides
e neutrones *pl* rápidos
i neutroni *pl* veloci
n snelle neutronen *pl*
d schnelle Neutronen *pl*

2534 FAST REACTION   np
  In nuclear physics, a term applied to a
  nuclear reaction which takes place
  within 10^{-22} sec.
f réaction *f* rapide
e reacción *f* rápida
i reazione *f* veloce
n snelle reactie
d schnelle Reaktion *f*

2535 FAST THERMAL COUPLED   rt
    REACTOR
  A nuclear reactor with two separate core
  regions, one operating as a thermal reactor
  and the other as a fast reactor.
f réacteur *m* thermique-rapide
e reactor *m* térmico-rápido
i reattore *m* termico-veloce
n gecombineerde thermische en snelle
   reactor
d kombinierter thermischer und schneller
   Reaktor *m*

2536 FATIGUE   mg
  The tendency for a metal to fracture under
  repeated stressing considerably below the
  ultimate tensile strength.
f fatigue *f*
e fatiga *f*
i fatica *f*
n vermoeiing
d Ermüdung *f*

2537 FATIGUE FRACTURE   mg
  A brittle fracture due to fatigue caused
  by cyclic stressing of a metal.
f fracture *f* de fatigue
e fractura *f* de fatiga
i frattura *f* da fatica
n vermoeiingsbreuk
d Ermüdungsbruch *m*

2538 FATIGUE LIMIT, see 2251

2539 FATIGUE RATIO, see 2252

2540 FATIGUE TEST, see 2253

2541 FAVOURABLE GEOMETRY   rt
A geometry in which all risk of criticality
is excluded.
f géométrie *f* favorable
e geometría *f* favorable
i geometria *f* favorevole
n gunstige geometrie
d günstige Geometrie *f*

2542 FEATHER ANALYSIS   ab
A method, due to Feather, of determining
the approximate range, in an absorber,
of beta rays from an active material.
f analyse *f* de Feather
e análisis *f* de Feather
i analisi *f* di Feather
n featheranalyse
d Feather-Analyse *f*

2543 FEATHER RULE   br
An empirical relation between the range
and energy of continuous beta rays.
f formule *f* de Feather
e fórmula *f* de Feather
i formula *f* di Feather
n formule van Feather
d Feather-Formel *f*

2544 FECUNDATION,   md
FERTILIZATION
Fusion of two gametes, male and female,
to form a zygote.
f fécondation *f*, fertilisation *f*
e fecondación *f*
i fecondazione *f*, fertilizzazione *f*
n bevruchting, fecundatie
d Befruchtung *f*, Fekundation *f*

2545 FEE, see 2525

2546 FEED,   ch, is
INPUT
The product introduced in a cascade in
order to realize the separation into two
fractions.
f alimentation *f*
e alimentación *f*
i alimentazione *f*
n inbrengst
d Einlauf *m*

2547 FEEDBACK SIGNAL,   co
RETURN SIGNAL
A signal responsible for the value of the
controlled variable.
f signal *m* de réaction
e señal *f* de retorno
i segnale *m* di reazione
n terugvoersignaal *n*
d Rückführsignal *n*

2548 FERGHANITE   mi
A secondary mineral containing about
68 % of U and related to tynyamunite.
f ferghanite *f*

e ferganita *f*
i ferganite *f*
n ferghaniet *n*
d Ferghanit *m*

2549 FERGUSONITE   mi
Mineral containing ytterbium, erbium,
cerium and iron and 0.8 - 7.2 % of U,
0.7 - 2.5 % of Th.
f fergusonite *f*
e fergusonita *f*
i fergusonite *f*
n fergusoniet *n*
d Fergusonit *m*

2550 FERMI AGE (GB), see 139

2551 FERMI AGE EQUATION, see 140

2552 FERMI AGE MODEL   np
A model for the study of the slowing down
of neutrons by elastic collisions.
f modèle *m* de l'âge de Fermi
e modelo *m* de la edad de Fermi
i modello *m* dell'età di Fermi
n leeftijdmodel *n* van Fermi
d Fermi-Altermodell *n*

2553 FERMI AGE THEORY   np
An approximate method, due to Fermi, for
calculating the slowing down density in
a medium by means of the age equation
and also for calculating the Fermi age
or slowing-down length.
f théorie *f* de l'âge de Fermi
e teoría *f* de la edad de Fermi
i teoria *f* dell'età di Fermi
n leeftijdstheorie van Fermi
d Fermische Alterstheorie *f*

2554 FERMI BETA DECAY THEORY   br
Theory proposed in 1934 on beta decay.
f théorie *f* de désintégration bêta de Fermi
e teoría *f* de desintegración beta de Fermi
i teoria *f* di disintegrazione beta di Fermi
n bêtavervaltheorie van Fermi
d Betazerfallstheorie *f* nach Fermi

2555 FERMI BRIM,   np
FERMI LEVEL
The value of the electron energy at which
the Fermi-Dirac distribution function
has the value one half.
f niveau *m* de Fermi
e nivel *m* de Fermi
i livello *m* di Fermi
n ferminiveau *n*
d Fermi-Niveau *n*

2556 FERMI CHARACTERISTIC   np
ENERGY LEVEL
The inner work function in the case of a
metal.
f niveau *m* caractéristique de Fermi
e nivel *m* característico de Fermi
i livello *m* caratteristico di Fermi
n karakteristiek ferminiveau *n*
d Fermi-Kante *f*

2557 FERMI CONSTANT np
A universal constant, introduced in beta disintegration theory and often denoted by the symbol g or G, that expresses the strength of the interaction between the transforming nucleon and the electron-neutrino field.
f constante f de Fermi
e constante f de Fermi
i costante f di Fermi
n fermiconstante
d Fermi-Konstante f

2558 FERMI-DIRAC np, sc
DISTRIBUTION FUNCTION
A function having a value in the range from zero to unity, specifying the probability that an electron in a semiconductor will occupy a certain quantum state of energy when thermal equilibrium exists.
f fonction f de distribution de Fermi-Dirac
e función f de distribución de Fermi-Dirac
i funzione f di distribuzione di Fermi-Dirac
n verdelingsfunctie van Fermi en Dirac
d Fermi-Diracsche Verteilungsfunktion f

2559 FERMI-DIRAC GAS np
An assembly of independent particles obeying Fermi-Dirac statistics.
f gaz m de Fermi-Dirac
e gas m de Fermi-Dirac
i gas m di Fermi
n fermigas n
d Fermi-Gas n

2560 FERMI-DIRAC STATISTICS np
Quantum statistics in which no more than one of a set of identical particles may occupy a particular quantum state.
f statistique f de Fermi-Dirac
e estadística f de Fermi-Dirac
i statistica f di Fermi-Dirac
n statistiek van Fermi en Dirac
d Fermi-Diracsche Statistik f

2561 FERMI DISTRIBUTION np
The energy distribution of the electrons in a metal.
f distribution f de Fermi
e distribución f de Fermi
i distribuzione f di Fermi
n fermiverdeling
d Verteilung f nach Fermi

2562 FERMI ENERGY np
The energy E_F occurring as a parameter
in the Fermi distribution function and measuring the highest occupied level at very low temperatures or to be the average energy of the electrons, which is $\frac{3}{5} E_F$.
f énergie f de Fermi
e energía f de Fermi
i energia f di Fermi
n fermi-energie
d Fermi-Energie f

2563 FERMI HOLE ec, np
The depletion region surrounding an electron in the energy band theory of solids.
f trou m de Fermi
e hueco m de Fermi
i buca f di Fermi
n fermigat n
d Fermi-Loch n

2564 FERMI INTERCEPT, ma, np
SCATTERING LENGTH
A parameter appearing in the analysis of neutron-proton scattering at low energies.
f parcours m de diffusion
e trayectoria f de dispersión
i percorso m di deviazione
n verstrooiingsweglengte
d Streuweglänge f

2565 FERMI PERTURBATIONS, ch, np
FERMI RESONANCE
In polyatomic molecules, two vibrational levels belonging to different vibrations may happen to have nearly the same energy and therefore be accidentally degenerate.
f résonance f de Fermi
e resonancia f de Fermi
i risonanza f di Fermi
n fermiresonantie
d Fermi-Resonanz f

2566 FERMI PLOT, br, sp
KURIE PLOT
Of a beta particle spectrum, a graph in which a suitable function of the observed intensity is plotted against the particle energy, the function being chosen so that the graph is a straight line for allowed beta transitions.
f diagramme m de Fermi,
diagramme m de Kurie, droite f de Fermi, droite f de Kurie
e diagrama m de Fermi,
diagrama m de Kurie
i diagramma m di Fermi,
diagramma m di Kurie
n fermidiagram n, kuriediagram n
d Fermi-Diagramm n, Kurie-Diagramm n

2567 FERMI POTENTIAL np
The energy of the Fermi level interpreted as an electric potential, i.e. the ratio of the Fermi level to the electronic charge.
f potentiel m de Fermi
e potencial m de Fermi
i potenziale m di Fermi
n fermipotentiaal
d Fermi-Potential n

2568 FERMI SELECTION RULES br
In beta theory, a particular set of selection rules which follow from a definite assumption about the coupling term.
f lois pl de sélection de Fermi
e leyes pl de selección de Fermi
i leggi pl di selezione di Fermi

n selectieregels *pl* van Fermi
d Fermische Auswahlregeln *pl*

2569 FERMI STATISTICS   np
The study of the probability of the
macroscopic states of a quantized system
of particles, taking account of the
Pauli-Fermi principle.
f statistique *f* de Fermi
e estadística *f* de Fermi
i statistica *f* di Fermi
n fermistatistiek
d Fermi-Statistik *f*

2570 FERMI SURFACE   np
The constant energy surface corresponding
to the Fermi level of which it is, in a
sense, the three-dimensional analog(ue).
f surface *f* de Fermi
e superficie *f* de Fermi
i superficie *f* di Fermi
n fermivlak *n*
d Fermi-Fläche *f*

2571 FERMI TEMPERATURE   np
The degeneracy temperature of a
Fermi-Dirac gas defined by $E_F = k$, where

E_F is the energy of the Fermi level,

occurring as a parameter in the
Fermi-Dirac distribution function.
f température *f* de Fermi
e temperatura *f* de Fermi
i temperatura *f* di Fermi
n fermitemperatuur
d Fermi-Temperatur *f*

2572 FERMI THEORY OF COSMIC   cr, np
RAY ACCELERATION
A theory in which collisions are assumed
to take place between charged cosmic
ray particles and the magnetic fields
associated with turbulent interstellar
clouds.
f théorie *f* de Fermi de l'accélération de
rayons cosmiques
e teoría *f* de Fermi de aceleración de
rayos cósmicos
i teoria *f* di Fermi d'accelerazione di
raggi cosmici
n theorie van Fermi voor de versnelling
van kosmische stralen
d Fermische Theorie *f* für die
Beschleunigung der kosmischen Strahlen

2573 FERMION   np
A particle which conforms to Fermi-Dirac
statistics.
f fermion *m*
e fermión *m*
i fermione *m*
n fermion *n*
d Fermion *n*

2574 FERMIUM   ch
Radioactive element, symbol Fm, atomic
number 100.

f fermium *m*
e fermio *m*
i fermio *m*
n fermium *n*
d Fermium *n*

2575 FERROUS SULFATE   me, ra
DOSEMETER (US),
FERROUS SULPHATE DOSEMETER
(GB)
f dosimètre *m* à sulfate ferreux
e dosímetro *m* de sulfato ferroso
i dosimetro *m* a solfato ferroso
n ferrosulfaatdosismeter
d Ferrosulphatdosismesser *m*

2576 FERTILE   rt
Of a nuclide, capable of being transformed,
directly or indirectly, into a fissile (fissionable) nuclide by neutron capture of a
material, containing one or more fertile
nuclides.
f fertile adj
e fértil adj
i fertile adj
n vruchtbaar adj
d brütbar adj

2577 FERTILE ELEMENT   rt
Element capable of being transformed
into fissile (fissionable) material in a
reactor.
f élément *m* fertile
e elemento *m* fértil
i elemento *m* fertile
n vruchtbaar element *n*
d Brutstoff *m*

2578 FERTILE MATERIAL   np
Material which in itself is not fissile
(fissionable) but which, in a reactor, may
be transformed into fissile (fissionable)
material through a nuclear transformation.
f matière *f* fertile
e materia *f* fértil
i materia *f* fertile
n splijtgrondstof
d Spaltrohstoff *m*

2579 FERTILIZATION, see 2544

2580 FEYNMAN DIAGRAM   np, qm
Sketch in Minkowski space of the
development of a process involving the
interaction of particles and/or electromagnetic radiation.
f diagramme *m* de Feynman
e diagrama *m* de Feynman
i diagramma *m* di Feynman
n diagram *n* van Feynman
d Feynman-Diagramm *n*

2581 FEYNMAN POSITON THEORY,   np
FEYNMAN POSITRON THEORY
A theory the purpose of which is to
simplify all calculations involving
posit(r)ons and electrons interacting with
external fields.

f   théorie *f* de Feynman des positons
e   teoría *f* de Feynman de los positones
i   teoria *f* di Feynman dei positoni
n   positonentheorie van Feynman
d   Positronentheorie *f* von Feynman

2582   FICK'S LAW   np
The law, stating that the rate of flow of
molecules in gas is proportional to the
concentration gradient, extended to
nuclear fission.
f   loi *f* de Fick
e   ley *f* de Fick
i   legge *f* di Fick
n   wet van Fick
d   Ficksches Gesetz *n*

2583   FIELD DESORPTION   np
One of the forms of field emission of
positive ions.
f   désorption *f* de champ
e   desorción *f* de campo
i   dissorbimento *m* di campo
n   velddesorptie
d   Felddesorption *f*

2584   FIELD DOSE,   xd
       GIVEN DOSE, SURFACE DOSE
The dose of radiation given to an irradiated
surface area, measured with backscatter
unless otherwise specified.
f   dose *f* en surface
e   dosis *f* de superficie
i   dose *f* di superficie
n   oppervlakdosis
d   Oberflächendosis *f*

2585   FIELD EMISSION, see 446

2586   FIELD-EMISSION COUNTER   ct
A radiation counter in which an approaching
ionic charge releases an electron from the
counter cathode surface.
f   compteur *m* à émission froide
e   contador *m* de emisión fría
i   contatore *m* ad emissione fredda
n   teller met koude emissie
d   Zähler *m* mit Feldemission

2587   FIELD-FREE EMISSION CURRENT, ec
       ZERO-FIELD EMISSION CURRENT
The emission current from an emitter
when the electric gradient at the surface
is zero.
f   courant *m* d'émission à champ nul
e   corriente *f* de emisión en campo nulo
i   corrente *f* d'emissione in campo nullo,
    corrente *f* di lancio
n   veldvrije emissiestroom
d   feldfreier Emissionsstrom *m*

2588   FIELD INDEX,   pa
       MAGNETIC FIELD INDEX
Of a magnetic field H, designed to make
charged particles move approximately in
a circle the value $n = - \dfrac{d(\ln H)}{d(\ln r)}$, where r is

the distance from the centre(er) of the
design orbit.
f   indice *m* de champ
e   índice *m* de campo
i   indice *m* di campo
n   veldindex
d   Feldindex *m*

2589   FIELD-INTENSIFIED DISCHARGE,   np
       NON-SELF-MAINTAINED DISCHARGE,
       TOWNSEND DISCHARGE
A conduction current in a gas which is due
to ionization of the gas from an external
source other than the applied voltage.
f   décharge *f* de Townsend,
    décharge *f* non autonome
e   descarga *f* de Townsend,
    descarga *f* no autónoma
i   scarica *f* di Townsend,
    scarica *f* non autonoma
n   onzelfstandige ontlading,
    townsendontlading
d   Townsend-Entladung *f*,
    unselbständige Entladung *f*

2590   FIELD-ION EMISSION   np
Field emission realized by ionization of
gas molecules impinging on the emitter.
f   émission *f* ionique par champ électrique
e   emisión *f* iónica de campo
i   emissione *f* ionica di campo
n   ionenemissie in een veld
d   Feldionenemission *f*

2591   FIELD OF EXIT, see 2429

2592   FIELD OF FORCE, see 2069

2593   FIELD OF INCIDENCE, see 2312

2594   FIELD OPERATOR   ma
In quantized field theory, an operator
which represents the creation or
annihilation of a particle.
f   opérateur *m* de champ
e   operador *m* de campo
i   operatore *m* di campo
n   veldoperator
d   Feldoperator *m*

2595   FIELD QUANTUM,   qm
       FUNDAMENTAL FIELD PARTICLE
In field theories, a field is quantized by
application of a proper quantum mechanical
procedure and thus results in the existence
of a fundamental field particle, which may
be called field quantum.
f   quantum *m* de champ
e   cuanto *m* de campo
i   quanto *m* di campo
n   veldquantum *n*
d   Feldquant *m*

2596   FIELD TUBE   ct
A device used in a proportional counter to
restrict the counting volume to that part
of the counter where the lines of force

are effectively normal to the central wire.
f tube *m* limiteur du champ
e tubo *m* limitador del campo
i tubo *m* limitatore del campo
n veldbegrenzingsbuis
d Feldbegrenzungsrohr *n*

2597 FIGURE EIGHT   pp
A term used in 'the controlled fusion
project and which refers to an early
model of the stellarator.
f figure *f* en huit
e figura *f* en ocho
i figura *f* in otto
n achtvorm
d Achtform *f*

2598 FILLED BAND   ec, np
An energy band in which each energy level
is occupied by an electron.
f bande *f* remplic
e banda *f* llena, banda *f* ocupada
i banda *f* interamente occupata
n bezette band, volle band
d vollbesetztes Energieband *n*

2599 FILM,   gp
LAMINAR BOUNDARY LAYER
That portion of the boundary layer in which
the motion approaches streamline flow.
f couche *f* limite laminaire
e capa *f* límite laminar
i strato *m* limite laminare
n laminaire grenslaag
d laminare Grenzschicht *f*

2600 FILM BADGE   ph, sa
A photographic film in a suitable container
worn by personnel and used as a radiation
monitor.
f film-témoin *m*, plaquette *f* de film
e film *m* dosimétrico, fotactímetro *m*
i pellicola *f* dosimetrica
n filmplak
d Filmdosimeter *n*, Filmplakete *f*

2601 FILM COEFFICIENT OF HEAT
TRANSFER (US), see 1173

2602 FILM ILLUMINATOR,   xr
NEGATOSCOPE, VIEWING SCREEN
Equipment incorporating a suitable source
of illumination for viewing radiographs
or other transparencies.
f négatoscope *m*
e negatoscopio *m*
i negatoscopio *m*
n negatoscoop
d Negativschaukasten *m*

2603 FILM MARKER,   xr
IDENTIFICATION MARKER
A marker, usually of heavy metal, used to
provide a reference point or identification
mark in a radiograph.
f marqueur *m* de cliché,
numéroteur *m* de cliché

e numerador *m* de clisé
i marcatore *m* di pellicola,
numeratore *m* di pellicola
n filmmarkeerapparaat *n*
d Filmmarkiergerät *n*

2604 FILM RING   sa
A film badge in the form of a finger ring.
f dosimètre *m* photographique annulaire
e fotactímetro *m* anular
i pellicola *f* dosimetrica anulare
n filmring, ringfilm
d Filmring *m*

2605 FILM TRACK DARKENING   ph
The blackening of the emulsion by the
conversion of the silver bromide to silver.
f noircissement *m* de l'émulsion
e ennegrecimiento *m* de la emulsión
i annerimento *m* dell'emulsione
n emulsiezwarting
d Emulsionsschwärzung *f*

2606 FILTER   ra
Material interposed in the path of radiation
in order to modify the spectral or the
spatial distribution of the radiation, or both.
f filtre *m*
e filtro *m*
i filtro *m*
n filter *n*
d Filter *m*

2607 FILTER AID   ch
A finely divided solid material used either
as a precoat on the filter medium or
incorporated in the suspension to be
filtered.
f adjuvant *m* de filtration
e ayudo *m* filtrante,
coadyuvante *m* de filtración
i materiale *m* ausiliario di filtrazione
n filtreerhulpmiddel *n*
d Filterhilfsmittel *n*, Filterzusatz *m*

2608 FILTER CAKE   cg
The compacted aggregate of solids
collected on the filter cloth or other filter
medium.
f tourteau *m*
e torta *f* de filtro
i torta *f* di filtro
n filterkoek
d Filterkuchen *m*

2609 FILTER CLOTH   cg
The material upon which the solids to be
removed are collected or in which they
are entrapped.
f blanchet *m*, toile *f* de filtre
e tejido *m* de filtro
i tessuto *m* per filtro
n filtreerdoek
d Filtertuch *n*

2610 FILTER HOUSE   cg
The container in which the filter is housed.
f boîte *f* de filtre
e caja *f* de filtro
i cassa *f* di filtro
n filterhuis *n*
d Filtergehäuse *n*

2611 FILTRATE   cg
The liquid that has passed through the
filter.
f filtrat *m*
e filtrado *m*
i filtrato *m*
n filtraat *n*
d Filtrat *n*

2612 FILTRATION   cg
The separation of finely divided solids
from a fluid by passing the fluid through
a porous medium which retains the
particles.
f filtration *f*
e filtración *f*
i filtrazione *f*
n filtratie, filtreren *n*
d Filtration *f*, Filtrieren *n*, Filtrierung *f*

2613 FILTRATION   xr
The preferential absorption of the less
penetrating components of a beam of
heterogeneous röntgen or gamma rays
caused by passage of the beam through
a sheet of material called a filter.
f filtration *f*
e filtración *f*
i filtrazione *f*
n filteren *n*, filtering
d Filterung *f*

2614 FINAL CONTROLLING DRIVE,   co
FINAL CONTROLLING ELEMENT
That portion of the controlling means
which directly changes the value of the
manipulated variable.
f organe *m* de réglage final
e órgano *m* de mando final
i elemento *m* di controllo finale
n eindregelorgaan *n*
d Endstellglied *n*

2615 FINAL PRODUCT, see 2245

2616 FINE CONTROL,   co
REGULATING
A control designed to correct low
amplitude reactivity variations.
f pilotage *m*
e regulación *f* fina
i regolazione *f* di precisione
n fijnregeling
d Feinsteuerung *f*

2617 FINE-CONTROL CHANNEL,   cp
FINE-CONTROL HARDWARE
The whole set of apparatus, connecting
cables, and electronic tubes, allowing to

transfer to the control desk of a nuclear
reactor the signals intended for handling
the fine control elements.
f chaîne *f* de pilotage
e equipo *m* de regulación fina
i apparecchiatura *f* di regolazione di
precisione
n fijnregelapparatuur
d Feinsteuergeräte *pl*

2618 FINE-CONTROL MEMBER,   co, rt
REGULATING MEMBER
A control member used for small and
precise adjustment of the reactivity of
a nuclear reactor.
f élément *m* de pilotage,
élément *m* de réglage fin
e elemento *m* de regulación fina
i elemento *m* di regolazione di precisione
n fijnregellichaam *n*
d Feinsteuerelement *n*

2619 FINE-CONTROL ROD,   co
REGULATING ROD
A control rod for effecting fine regulation.
f barre *f* de pilotage,
barre *f* de réglage fin
e barra *f* de regulación fina
i barra *f* di regolazione di precisione
n fijnregelstaaf
d Feinsteuerstab *m*

2620 FINE DISTRIBUTION,   rt
FINE FLUX
Distribution of the neutron flux in the lattice
cell constituting the system.
f distribution *f* fine
e distribución *f* fina
i distribuzione *f* fina, flusso *m* fino
n fijnverdeling
d Feinverteilung *f*

2621 FINE FOCUS   xr
f petit foyer *m*
e foco *m* fino
i fuoco *m* fino
n fijnfocus
d Feinfokus *n*

2622 FINE GRID   xr
f grille *f* fine
e diafragma *m* de rejilla fina
i diaframma *m* a griglia fina
n fijnraster *n*
d Feinrasterblende *f*

2623 FINE STRUCTURE   sp
Said of the distribution of an alpha
particle spectrum when more than one
group is present.
f structure *f* fine
e estructura *f* fina
i struttura *f* fina
n fijnstructuur
d Feinstruktur *f*

2624 FINE-STRUCTURE CONSTANT   np
A constant given by the square of the
electronic charge e, divided by the product
of the modified Planck's constant h, and the
velocity of light.
f   constante f de structure fine
e   constante f de estructura fina
i   costante f di struttura fina
n   fijnstructuurconstante
d   Feinstrukturkonstante f

2625 FINELY GROUND   mi
          CONCENTRATES
Ore concentrates pulverized by mechanical
means.
f   concentrés pl pulvérisés
e   concentrados pl pulverizados
i   concentrati pl polverizzati
n   fijngemalen concentraten pl
d   feingemahlene angereicherte Erze pl

2626 FINITE CYLINDRICAL REACTOR   rt
A theoretical reactor of cylindrical form
and finite dimensions.
f   réacteur m cylindrique fini
e   reactor m cilíndrico finito
i   reattore m cilindrico finito
n   eindige cilindrische reactor
d   zylindrischer Reaktor m endlicher
    Abmessungen

2627 FINNING   cl, fu, rt
The application of cooling fins to the can
of the fuel element.
f   application f d'ailettes
e   aplicación f de aletas
i   applicazione f d'alette
n   aanbrengen n van koelvinnen
d   Berippung f

2628 FIR,   rt
          FOOD IRRADIATION REACTOR
f   réacteur m d'irradiation de denrées
e   reactor m de irradiación de alimentos
i   reattore m d'irradiazione di prodotti
    alimentari
n   reactor voor het bestralen van levens-
    middelen
d   Reaktor m zur Lebensmittelbestrahlung

2629 FIRE STORM   nw
Stationary mass fire, in built-up urban
areas, generating strong in-rushing winds
from all sides, which keep the fires
from spreading while adding fresh
oxygen to increase their intensity.
f   tempête f de feu
e   tempestad f de fuego
i   tempesta f di fuoco
n   vuurstorm
d   Feuersturm m

2630 FIREBALL, see 520

2631 FIRST COLLISION KERMA   np
A value of kerma obtained when limited
to the charged particles created in the
course of the first interaction possible
for each individual ionizing particle.
f   kerma m de première collision
e   kerma m de primera colisión
i   kerma m di primo urto
n   kerma bij eerste botsing
d   Erststosskerma m

2632 FIRST IONIZATION POTENTIAL   np
Needed for the removal of the most
loosely bound electron from an initially
neutral atom.
f   premier potentiel m d'ionisation
e   primer potencial m de ionización
i   primo potenziale m d'ionizzazione
n   eerste ionisatiepotentiaal
d   erstes Ionisationspotential n

2633 FIRST QUANTUM NUMBER,   np, qm
          MAIN QUANTUM NUMBER
The number which gives the size of the
electron orbit, the discontinuous changes
of which lead to important changes of
energy.
f   nombre m quantique principal
e   número m cuántico principal
i   numero m quantico principale,
    primo numero m quantico
n   hoofdquantumgetal n
d   Hauptquantenzahl f

2634 FIRST RADIATION CONSTANT   np
A constant c_1, equal to 8π times the
product of Planck's constant, h, and the
velocity of light c.
f   première constante f de rayonnement
e   primera constante f de radiación
i   prima costante f d'irradiazione
n   eerste stralingsconstante
d   erste Strahlungskonstante f

2635 FIRST TOWNSEND COEFFICIENT   ic
The number of ion pairs formed per
electron and per centimetre(er) of drift
towards the central wire of a counter.
f   premier coefficient m de Townsend
e   primer coeficiente m de Townsend
i   primo coefficiente m di Townsend
n   eerste townsendcoëfficiënt
d   erster Townsend-Koeffizient m

2636 FIRST TOWNSEND DISCHARGE   np
A semi-self-maintained discharge in which
the additional which appear are solely
due to the ionization of the gas by electron
collision.
f   première décharge f de Townsend
e   primera descarga f de Townsend
i   prima scarica f di Townsend
n   eerste townsendontlading
d   primäre Townsend-Entladung f

2637 FISH-POLE PROBE, see 598

2638 FISH TRACKS   ic
In a cloud chamber, short tracks caused
by scattering of X-ray photons.

f   traces *pl* de poisson
e   huellas *pl* de pescado
i   traccie *pl* di pesce
n   vissporen *pl*
d   Fischspuren *pl*

2639 FISSILE   np
Of a nuclide, capable of undergoing fission
by interaction with slow neutrons.
f   fissile par neutrons lents
e   físil por neutrones lentos
i   fissile per neutroni lenti
n   fissiel,
    thermisch splijtbaar door langzame
    neutronen
d   thermisch spaltbar durch langsame
    Neutronen

2640 FISSILE (GB),   np
      FISSIONABLE (US)
Of a nuclide, capable of undergoing fission
by any process.
f   fissile adj
e   físil adj
i   fissile adj
n   splijtbaar adj
d   spaltbar adj

2641 FISSILE ATOM (GB),   np
      FISSIONABLE ATOM (US)
An atom capable of undergoing fission.
f   atome *m* fissile
e   átomo *m* físil
i   atomo *m* fissile
n   splijtbaar atoom *n*
d   spaltbares Atom *n*

2642 FISSILE DERIVATIVE (GB),   np
      FISSIONABLE DERIVATIVE (US)
A fission product capable of undergoing
fission again.
f   dérivé *m* fissile
e   derivado *m* físil
i   derivato *m* fissile
n   splijtbaar derivaat *n*
d   spaltbares Derivat *n*

2643 FISSILE MATERIAL (GB),   rt
      FISSIONABLE MATERIAL (US)
A material having the property of capturing
neutrons and thereupon splitting into two
particles with great kinetic energy.
f   matière *f* fissile
e   material *m* físil
i   materiale *m* fissile
n   splijtbaar materiaal *n*, splijtstof
d   spaltbares Material *n*, Spaltstoff *m*

2644 FISSILE MATERIAL PRODUCTION   rt
      REACTOR (GB),
      FISSIONABLE MATERIAL
      PRODUCTION REACTOR (US)
A reactor whose primary purpose is to
produce fissile (fissionable) material.
f   réacteur *m* de production des matériaux
    fissiles
e   reactor *m* de producción del material físil

i   reattore *m* di produzione del materiale
    fissile
n   reactor voor het produceren van splijt-
    baar materiaal
d   Reaktor *m* zur Produktion von spaltbarem
    Material

2645 FISSILE NUCLEUS (GB),   np
      FISSIONABLE NUCLEUS (US)
A nucleus which may be used in a fission
process.
f   noyau *m* fissile
e   núcleo *m* físil
i   nucleo *m* fissile
n   splijtbare kern
d   spaltbarer Kern *m*

2646 FISSION   np
The splitting of a nucleus into two more
or less equal fragments.
f   fission *f*
e   fisión *f*
i   fissione *f*, scissione *f*
n   splijting
d   Spaltung *f*

2647 FISSION BOMB, see 1

2648 FISSION CAPTURE,   np
      USEFUL CAPTURE
The capture of an incident particle by the
target nucleus, resulting in nuclear fission.
f   capture *f* utile
e   captura *f* de fisión
i   cattura *f* a fissione
n   splijtingsvangst
d   Spaltungseinfang *m*

2649 FISSION CHAIN,   np
      FISSION DECAY CHAIN
The successive beta decays starting with
a fission product and ending in a stable
nucleus of the same mass number.
f   chaîne *f* de fission
e   cadena *f* de fisión
i   catena *f* di fissione
n   splijtingsketting
d   Spaltungskette *f*

2650 FISSION CHAIN REACTION   np
A reaction whereby the fission of an
atomic nucleus causes the fission of other
atomic nuciei and so on.
f   réaction *f* en chaîne de fission
e   reacción *f* en cadena de fisión
i   reazione *f* in catena di fissione
n   splijtingskettingreactie
d   Spaltungskettenreaktion *f*

2651 FISSION CHANNEL   np
The way in which fission occurs.
f   progrès *m* de la fission,
    voie *f* de fission
e   progreso *m* de la fisión
i   progresso *m* della fissione
n   splijtingsverloop *n*
d   Spaltverlauf *m*

2652 FISSION COUNTER TUBE   ct
    A counter tube for detecting neutrons,
    containing fissile(fissionable) materials
    and in which the initial ionization is
    caused mainly by fission fragments
    produced by these neutrons.
f tube *m* compteur à fission
e tubo *m* contador de fisión
i tubo *m* contatore a fissione
n splijtingstelbuis
d Spaltzählrohr *n*

2653 FISSION CROSS SECTION   cs
    The cross section for the fission process.
f section *f* efficace de fission
e sección *f* eficaz de fisión
i sezione *f* d'urto di fissione
n doorsnede voor splijting
d Wirkungsquerschnitt *m* für Kernspaltung

2654 FISSION ENERGY   np
    The energy released in a fission process.
f énergie *f* de fission
e energía *f* de fisión
i energia *f* di fissione
n splijtingsenergie
d Spaltenergie *f*, Spaltungsenergie *f*

2655 FISSION ENERGY SPECTRUM,   np, sp
    FISSION SPECTRUM
    Of a fissile (fissionable) material, the
    energy distribution of the neutrons
    produced by the fission.
f spectre *m* énergétique de fission
e espectro *m* energético de fisión
i spettro *m* energetico di fissione
n splijtingsenergiespectrum *n*
d Spaltungsenergiespektrum *n*

2656 FISSION FRAGMENTS   np
    The nuclei resulting from fission and
    possessing kinetic energy acquired
    from that fission.
f fragments *pl* de fission
e fragmentos *pl* de fisión
i frammenti *pl* di fissione
n splijtingsfragmenten *pl*
d Kernspaltungsfragmente *pl*,
    Spaltbruchstücke *pl*

2657 FISSION-FUSION-FISSION-BOMB,   nw
    THREE-F BOMB
    A thermonuclear bomb, in which the
    energy is partly derived from fission and
    partly from fusion.
f bombe *f* à fission-fusion-fission
e superbomba *f* de fisión-fusión-fisión
i bomba *f* a fissione-fusione-fissione
n splijtings-versmeltings-splijtings-bom
d Spaltungs-Verschmelzungs-Spaltungs-
    Bombe *f*

2658 FISSION GAS,   np
    VOLATILE FISSION PRODUCT
    A fission product in gaseous form.
f gaz *m* de fission
e gas *m* de fisión

i gas *m* di fissione
n splijtingsgas *n*
d Spaltgas *n*

2659 FISSION GAS RELEASE   rt
    An undesired effect in nuclear reactors.
f libération *f* de gaz de fission
e liberación *f* de gas de fisión
i liberazione *f* di gas di fissione
n splijtgasontwikkeling
d Spaltgasentwicklung *f*

2660 FISSION HEAT   rt
f chaleur *f* de fission
e calor *m* de fisión
i calore *m* di fissione
n splijtingswarmte
d Spaltungswärme *f*

2661 FISSION IONIZATION CHAMBER   ic
    An ionization chamber for detecting
    neutrons, containing fissile(fissionable)
    materials, in which the ionization is
    caused mainly by fission fragments
    produced by these neutrons.
f chambre *f* d'ionisation à fission
e cámara *f* de ionización de fisión
i camera *f* d'ionizzazione a fissione
n splijtingsionisatievat *n*
d Spaltkammer *f*,
    Spaltungsionisationskammer *f*

2662 FISSION NEUTRONS   np
    Neutrons originating in the fission process
    which have retained their original energy.
f neutrons *pl* de fission
e neutrones *pl* de fisión
i neutroni *pl* di fissione
n splijtingsneutronen *pl*
d Spaltneutronen *pl*

2663 FISSION POISONS   np
    Fission products which have appreciable
    cross sections for neutrons.
f poisons *pl* de fission
e venenos *pl* de fisión
i veleni *pl* di fissione
n splijtingsgiffen *pl*
d Spaltgifte *pl*

2664 FISSION PROBABILITY   np
    In a nuclear process, the probability
    that fission will occur.
f probabilité *f* de fission
e probabilidad *f* de fisión
i probabilità *f* di fissione
n splijtingskans, splijtingswaarschijnlijkheid
d Spaltungswahrscheinlichkeit *f*

2665 FISSION-PRODUCT-RETAINING   fu, rt
    COATED FUEL PARTICLE
    The fuel used in the fuel tubes inside the
    removable graphite blocks in the DRAGON
    project.
f particule *f* combustible enrobée à
    rétention de produits de fission

e partícula ƒ combustible recubierta de
retención de productos de fisión
i particella ƒ combustibile ricoperta da
ritenere prodotti di fissione
n splijtingsprodukten vasthoudend bekleed
splijtstofdeeltje *n*
d Spaltprodukte zurückhaltendes umhülltes
Brennstoffteilchen *n*

**2666 FISSION PRODUCT SEPARATOR   ma
FAILED ELEMENT MONITOR**
A failed element monitor using the
separation of one or several fission
products from the reactor coolant for
their determination by measuring their
activities.
f moniteur *m* de rupture de gaine à séparation
des produits de fission
e vigía ƒ de averías por separación de los
productos de fisión
i monitore *m* di rottura di guaine a
separazione dei prodotti di fissione
n monitor voor beschadigde splijtstofelementen berustend op de afscheiding van
splijtingsprodukten
d Warngerät *n* für schadhafte Brennelemente,
das auf abgeschiedene Spaltprodukte anspricht

2667 FISSION PRODUCTS   np
The nuclides produced either by fission
or by the subsequent radioactive disintegration of the nuclides thus formed.
f produits *pl* de fission
e productos *pl* de fisión
i prodotti *pl* di fissione
n splijtingsprodukten *pl*
d Spaltprodukte *pl*

**2668 FISSION PRODUCTS   rt, sa
POISONING PREDICTOR**
A computer designed to determine the
evolution, due to the fission products
poisoning, of the reactivity of a nuclear
reactor.
f prédicteur *m* d'empoisonnement par
produits de fission
e predictor *m* de envenenamiento por
productos de fisión
i calcolatore *m* per calcolo preliminare
d'avvelenamento per prodotti di fissione,
previsore *m* dell'avvelenamento per
prodotti di fissione
n machine voor voorafgaande berekening van
splijtingsproduktenvergiftiging
d Spaltproduktvergiftungsrechner *m*

**2669 FISSION PRODUCTS   np, rt
POISONING PROCESS**
When a nuclear reactor operates, certain
nuclei with larger cross sections for the
capture of thermal neutrons are produced
either as direct fission products or from
decay of the latter. These act as poisons
in the reactor.
f procédé *m* d'empoisonnement par produits
de fission

e procedimiento *m* de envenenamiento por
productos de fisión
i processo *m* d'avvelenamento per prodotti
di fissione
n proces *n* van splijtingsproduktenvergiftiging
d Spaltproduktvergiftungsprozess *m*

2670 FISSION PRODUCTS RELEASE   rt
Escape of fission products from the fuel
elements towards the coolant.
f fuite ƒ de produits de fission
e fuga ƒ de productos de fisión
i fuga ƒ di prodotti di fissione
n ontsnapping van splijtingsprodukten
d Entweichung ƒ von Spaltprodukten

2671 FISSION RATE   np
The number of fissions per unit time.
f débit *m* de fission
e velocidad ƒ de fisión
i velocità ƒ di fissione
n splijtingstempo *n*
d Spalthäufigkeit ƒ, Spaltrate ƒ

2672 FISSION RECOILS   np
The fission fragments at the moment of
separation.
f particules *pl* de recul à la fission
e partículas *pl* de retroceso a la fisión
i particelle *pl* di rinculo alla fissione
n splijtingsterugslagdeeltjes *pl*,
verse brokstukken *pl*
d Rückstossteilchen *pl* bei der Spaltung,
Spaltbruchstücke *pl* im Augenblick der
Spaltung

2673 FISSION RESONANCES   np
Resonances in the fission cross section
of neutrons with nuclei.
f résonances *pl* de fission
e resonancias *pl* de fisión
i risonanze *pl* di fissione
n splijtingsresonanties *pl*
d Spaltungsresonanzen *pl*

2674 FISSION SPECIES   np
Collective expression for the various kinds
of fission.
f espèces *pl* de fission
e especies *pl* de fisión
i specie *pl* di fissione
n splijtingsprincipes *pl*
d Spaltungsarten *pl*

2675 FISSION SPECTRUM   np, sp
For a specified fissionable nuclide, the
energy distribution of its prompt neutrons.
f spectre *m* de fission
e espectro *m* de fisión
i spettro *m* di fissione
n splijtingsspectrum *n*
d Spaltspektrum *n*

2676 FISSION SPIKE   np
A displacement spike in the special case of
the damage track being caused by fission
fragments.

f pic *m* de déplacement de fragments de
fission
e zona *f* de fragmentos de fisión desplazados
i zona *f* di frammenti di fissione dislocati
n gebied *n* van verplaatste splijtingsfragmenten
d Umlagerungsbereich *m* von versetzten
Spaltbruchstücken

2677 FISSION THRESHOLD, np
THRESHOLD ENERGY
The energy limit for an incident particle
or photon below which a particular
endothermic reaction will not occur.
f énergie *f* de seuil, seuil *m* d'énergie
e energía *f* de umbral
i energia *f* di soglia
n drempelenergie
d Energieschwelle *f*, Schwellenenergie *f*

2678 FISSION WEAPON nw
The type of bomb in which the active
element is at the heavy end of the
periodic table.
f arme *f* à fission nucléaire
e arma *f* de fisión nuclear
i arma *f* a fissione nucleare
n kernsplijtingswapen *n*
d Kernspaltungswaffe *f*

2679 FISSION YIELD np
The fraction of fissions leading to fission
products of a given type.
f rendement *m* de fission
e rendimiento *m* de fisión
i rendimento *m* di fissione,
resa *f* di fissione
n splijtingsopbrengst, splijtingsrendement *n*
d Spaltausbeute *f*

2680 FISSION YIELD CHARACTERISTIC, np
FISSION YIELD CURVE
A curve of the double hump type which
shows the variation of percentage fission
yield versus the mass number.
f courbe *f* de rendement de fission
e curva *f* de rendimiento de fisión
i curva *f* di rendimento di fissione
n splijtingsopbrengstkromme
d Spaltausbeutekurve *f*

2681 FISSIONING DISTRIBUTION, np, sp
FISSIONING SPECTRUM,
REACTOR SPECTRUM
For a given neutron spectrum the energy
distribution of those neutrons which
actually cause fission.
f distribution *f* de l'énergie des neutrons
dans le réacteur
e distribución *f* de la energía de los
neutrones en el reactor
i distribuzione *f* dell'energia dei
neutroni nel reattore
n energieverdeling van de neutronen in de
reactor
d Energieverteilung *f* der Neutronen im
Reaktor

2682 FISSIUM np
Fissile (fissionable) material artificially
mixed with fission product elements to
simulate the material resulting from
fission.
f fissium *m*
e fisio *m*
i fissio *m*
n fissium *n*
d Fissium *n*

2683 FIXED-FIELD pa
ALTERNATING-GRADIENT
ACCELERATOR
A particle accelerator employing a
fixed-field alternating-gradient radiofrequency system.
f accélérateur *m* à gradient alterné et
champ fixe
e acelerador *m* de gradiente alterno y
campo fijo
i acceleratore *m* a gradiente alternato e
campo fisso
n versneller met wisselende gradiënt en
vast veld
d Beschleuniger *m* mit alternierendem
Gradienten und ruhendem Feld

2684 FIXED REFLECTOR, rt
STATIC REFLECTOR
A neutron reflector which is designed to
be non-removable.
f réflecteur *m* fixe
e reflector *m* fijo
i riflettore *m* fisso
n vaste reflector
d feststehender Reflektor *m*,
ortsfester Reflektor *m*

2685 FLAKING, see 2426

2686 FLASH BOILER, see 572

2687 FLASH BURN, md, nw
THERMAL RADIATION BURN
A burn caused by excessive exposure of
bare skin to thermal radiation, such as
that produced by a nuclear weapon.
f lucite *f* par rayonnement thermique
e quemadura *f* por radiación térmica
i bruciatura *f* per radiazione termica
n warmtestralingsverbranding
d Strahlungsbrandwunde *f*,
Verbrennung *f* durch Wärmestrahlung

2688 FLASH GAMMA RADIATION, np
PROMPT GAMMA RADIATION
Gamma radiation accompanying the
fission process without measurable delay.
f rayonnement *m* gamma instantané
e radiación *f* gamma inmediata,
radiación *f* gamma instantánea
i radiazione *f* gamma istantanea
n prompte gammastraling
d prompte Gammastrahlung *f*

2689 FLASH RADIOGRAPHY xr
Radiography in which the exposure time is
extremely short, e.g. 1 microsecond.
f milligraphie *f*
e miligraffa *f*
i milligrafia *f*
n bliksemlichtröntgenfotografie
d Blitzlichtröntgenographie *f*

2690 FLASK, see 911

2691 FLAT COUNTER TUBE ct
Proportional counter tube formed by two
metallized plane sheets between which
several parallel wires are suspended,
parallel to these sheets.
f tube *m* compteur plan
e tubo *m* contador plano
i tubo *m* contatore piano
n platte telbuis
d Grossflächenzählrohr *n*

2692 FLAT LINEAR INDUCTION PUMP, vt
F.L.I.P.
An induction pump with polyphase
windings arranged on either side of a
flat tube.
f pompe *f* à induction linéaire à disposition
plate des enroulements
e bomba *f* de inducción lineal con
disposición plana de los devanados
i pompa *f* ad induzione lineare con
disposizione piana degli avvolgimenti
n lineaire inductiepomp met platte opstelling
van de windingen
d lineare Induktionspumpe *f* mit flacher
Anordnung der Windungen

2693 FLATTENED REGION, rt
FLATTENED ZONE,
FLUX FLATTENING REGION,
FLUX FLATTENING ZONE
The region in a nuclear reactor over
which the neutron flux is maintained
approximately uniform.
f zone *f* d'aplatissement
e zona *f* de aplanamiento
i zona *f* d'appiattimento
n afvlakkingsgebied *n*
d Abflachungsgebiet *n*

2694 FLATTENING, md, xr
ISODOSE EQUALIZING
The use of a specially mo(u)lded filtering
medium in order to obtain in a body submitted to an irradiating beam flat isodose
curves.
f égalisation *f* des isodoses
e aplanamiento *m* de isodosis
i appiattamento *m* d'isodosi
n isodosisafvlakking
d Isodosisabflächung *f*

2695 FLATTENING FILTER ra
The specially shaped filter used in
flattening a beam of radiation.
f filtre *m* de nivellement

e filtro *m* de nivelación
i filtro *m* di livellamento
n nivelleringsfilter *n*
d Nivellierungsfilter *m*

2696 FLATTENING MATERIAL, rt
FLUX FLATTENING MATERIAL
Neutron absorbers placed in a reactor to
flatten the flux.
f substance *f* d'aplatissement
e substancia *f* de aplanamiento
i sostanza *f* d'appiattimento
n afvlakkingsmiddel *n*
d Abflachungsmittel *n*

2697 FLATTENING RADIUS, rt
FLUX FLATTENING RADIUS
In a cylindrical reactor core, the radius
of the flattened region.
f rayon *m* d'aplatissement
e radio *m* de aplanamiento
i raggio *m* d'appiattimento
n afvlakkingsstraal
d Abflachungsstrahl *m*

2698 FLAW SENSITIVITY xr
In a radiograph the minimum detectable
flaw or defect, measured in the direction
of the primary beam of radiation,
expressed as a percentage of the total
thickness of the object irradiated.
f sensibilité *f* au défaut d'épaisseur,
sensibilité *f* aux contrastes
e sensibilidad *f* para defecto de espesor,
sensibilidad *f* para los contrastes
i sensibilità *f* ai contrasti,
sensibilità *f* al difetto di spessore
n contrastgevoeligheid,
gevoeligheid voor dikteonregelmatigheden
d Empfindlichkeit *f* für Dickenunregel-
mässigkeiten,
Kontrastempfindlichkeit *f*

2699 FLEXIBLE BETA-RAY br, md, ra
SHEETING
A sheet of polyethylene in which 20 % by
weight red phosphorus or yttrium oxide
are suspended. The surfaces are coated
with 0.16 mm inactive polyvinyl chloride.
f feuille *f* flexible à émission bêta
e hoja *f* flexible de emisión beta
i foglia *f* flessibile ad emissione beta
n buigzame foelie met bêtaemissie
d biegsame Folie *f* mit Beta-Emission

2700 FLINT GLASS ct, mt
Glass used for windows of counter tubes
to stop gamma rays.
f verre *m* de flint
e cristal *m* de roca, vidrio *m* flint
i vetro *m* flint
n flintglas *n*
d Flintglas *n*

2701 FLIP-OVER PROCESS, np
UMKLAPP PROCESS
A type of collision between phonons, or

between phonons and electrons, where
crystal momentum is not conserved.
f   processus *m* de fustigation
e   proceso *m* de fustigación
i   processo *m* di fustigazione
n   omslagproces *n*
d   Umklapprozess *m*

2702 FLOATING ACTION   co
That action wherein there is a predetermined relation between the value of the
controlled variable and the rate of motion
of a final control element.
f   action *f* flottante
e   acción *f* astática, acción *f* flotante
i   azione *f* astatica
n   astatische actie
d   gleitendes Verhalten *n*

2703 FLOATING POWER SUPPLY   ge
A power supply electrically isolated from
any common circuits.
f   alimentation *f* flottante
e   alimentación *f* flotante
i   alimentazione *f* isolata
n   zwevende voeding
d   potentialfreie Stromversorgung *f*

2704 FLOCCULATION   ch, rw
A method of separating radioisotopes
from large volumes of water in the
treatment of radioactive waste products.
f   floculation *f*
e   floculación *f*
i   flocculazione *f*
n   uitvlokking
d   Ausflockung *f*

2705 FLOODING POINT   ch
The point in two-phase countercurrent
flow beyond which the column becomes
inoperable due to irregular flow.
f   point *m* d'engorgement
e   punto *m* de inundación
i   portata *f* d'ingorgo,
    punto *m* d'ingolfamento
n   overstromingspunt *n*
d   Flutpunkt *m*, Überflutungspunkt *m*

2706 FLOODING POINT-RATE   ch
The limiting flow rate in two-phase
countercurrent flow through a column
above which the column is inoperable due
to irregular flow.
f   vitesse *f* au point d'engorgement
e   velocidad *f* al punto de inundación
i   velocità *f* al punto d'ingolfamento,
    velocità *f* della portata d'ingorgo
n   snelheid aan het overstromingspunt
d   Geschwindigkeit *f* am Flutpunkt

2707 FLOOR CONTAMINATION   me
      INDICATOR
A surface contamination indicator for a
specified area of a floor.
f   signaleur *m* de contamination surfacique
    des sols

e   indicador *m* de contaminación superficial
    de suelos
i   segnalatore *m* di contaminazione di
    pavimenti
n   oppervlakbesmettingsindicator voor
    vloeren
d   Fussbodenkontaminationsanzeiger *m*

2708 FLOOR CONTAMINATION   me
      METER
A surface contamination meter for
measuring the contamination on a specified
size of floor.
f   contaminamètre *m* surfacique des sols
e   medidor *m* de contaminación de suelos
i   contaminametro *m* per pavimenti
n   oppervlakbesmettingsmeter voor
    vloeren
d   Gerät *n* zur Bestimmung der Fussboden-
    oberflächenkontamination

2709 FLOOR CONTAMINATION   sa
      MONITOR
A surface contamination monitor for a
specified area of a floor.
f   moniteur *m* de contamination surfacique
    des sols
e   monitor *m* de contaminación superficial
    de suelos
i   monitore *m* di contaminazione per
    pavimenti
n   oppervlakbesmettingsmonitor voor
    vloeren
d   Kontaminationswarngerät *n* für
    Fussbodenflächen

2710 FLOOR-TYPE RADIATION   ra, sa
      BEACON
f   balise *f* de rayonnement de parquet
e   baliza *f* de radiación de suelo
i   faro *m* di radiazione di pavimento
n   vloerstralingsbaken *n*
d   Fussbodenstrahlungsbake *f*

2711 FLOW   ch
In nucleonics, the movement or passage
of a quantity of a material through, to or
from a tube or container.
f   courant *m*, flux *m*
e   corriente *f*, flujo *m*
i   corrente *f*, flusso *m*
n   stroming, stroom
d   Strom *m*, Strömung *f*

2712 FLOW CHART,   ge
      FLOW DIAGRAM, FLOW SHEET
A graphical representation of a sequence
of operations.
f   graphe *m* de fluence, organigramme *m*
e   ordinograma *m*
i   schema *m* di flusso
n   blokschema *n*, organigram *n*
d   Betriebsfolgediagramm *n*, Flussbild *n*

2713 FLOW OF MATERIALS   ge
The handling of the various materials
used, e.g., in building a nuclear reactor.

f   circulation ƒ du matériel
e   circulación ƒ de los materiales
i   circolazione ƒ dei materiali
n   materiaalomloop
d   Materialumlauf *m*

2714 FLOW-THROUGH CENTRIFUGE,
      see 1291

2715 FLOWMETER   me
      Any instrument for measuring the rate of
      movement of a fluid through a pipe.
f   débitmètre *m*
e   flujómetro *m*
i   flussometro *m*
n   debietmeter, stromingsmeter
d   Strömungsmesser *m*

2716 FLUCTUATING MAINS VOLTAGE   ge
f   oscillation ƒ de la tension du réseau
e   fluctuación ƒ del voltaje de la red
i   fluttuazione ƒ della tensione di rete
n   netspanningsschommeling,
    netspanningsvariatie
d   Netzspannungsschwankung ƒ

2717 FLUCTUATION TIME   ma
      Physical value used in studying the
      Langevin equation.
f   temps *m* de fluctuation
e   tiempo *m* de fluctuación
i   tempo *m* di fluttuazione
n   fluctuatietijd
d   Wanderzeit ƒ

2718 FLUENCE   np
      In a given point in space, the quotient of
      the number of particles penetrating in a
      small sphere centred in this point by the
      cross-sectional area of that sphere.
f   fluence ƒ
e   fluencia ƒ
i   fluenza ƒ
n   fluentie
d   Fluenz ƒ

2719 FLUENCE RATE,   np
      PARTICLE FLUX DENSITY
      At a given point in space, the number of
      particles incident on a small sphere in a
      time interval, divided by the cross
      sectional area of that sphere and by the
      time interval.
f   débit *m* de fluence,
    densité ƒ de flux de particules,
    densité ƒ de flux particulaire
e   densidad ƒ de flujo de partículas
i   densità ƒ di flusso di particelle
n   deeltjesfluxdichtheid,
    fluxdichtheid van deeltjes
d   Teilchenflussdichte ƒ

2720 FLUID FLOWMETER   ma
      An assembly for measuring the flow of a
      fluid in the cooling or moderating circuits
      of a nuclear reactor.
f   débitmètre *m* de fluide

e   contador *m* de flúido
i   complesso *m* di misura della portata di
    fluido
n   debietmeter voor vloeistof
d   Gerät *n* zur Messung des Durchflusse
    von Flüssigkeiten

2721 FLUID MIXING TANK   ch
f   cuve ƒ mélangeuse
e   tanque *m* mezclador
i   serbatoio *m* mescolatore
n   mengtank
d   Mischbehälter *m*

2722 FLUID POISON CONTROL   co, rt
      Control of a nuclear reactor by adjustment
      of the position or quantity of a fluid
      nuclear poison in such a way as to change
      the reactivity.
f   commande ƒ par poison fluide
e   regulacion por veneno flúido
i   regolazione ƒ con veleno fluido
n   regeling met gedispergeerd gif
d   Steuerung ƒ durch flüssige Neutronen-
    gifte

2723 FLUID POISON INJECTION   rt, sa
      SAFETY MECHANISM
      A safety mechanism for a nuclear reactor
      in which fluid poison is injected to
      rapidly decrease the reactivity.
f   mécanisme *m* de sécurité par injection
    de liquide comme poison
e   mecanismo *m* de seguridad por inyección
    de veneno líquido
i   meccanismo *m* di sicurezza mediante
    iniezione di veleno liquido
n   veiligheidsmechanisme *n* berustend op
    injectie van vloeibaar gif
d   Sicherheitsvorrichtung ƒ mittels
    Gifteinspritzung

2724 FLUID-TYPE INSTABILITY   pp
      Instability which may be deduced from
      macroscopic equations of a plasma.
f   instabilité ƒ fluide
e   inestabilidad ƒ flúido
i   instabilità ƒ fluido
n   fluïde-instabiliteit
d   Fluiduminstabilität ƒ

2725 FLUIDIZED-BED REACTOR   rt
      A nuclear reactor in which the fuel
      consists of small balls kept in motion by
      the current of the moderator and the
      coolant.
f   réacteur *m* à lit fluidisé
e   reactor *m* de lecho fluidificado
i   reattore *m* a letto fluidizzato
n   reactor met gefluïdiseerd bed
d   Wirbelbettreaktor *m*

2726 FLUIDIZED PASTE REACTOR,   rt
      PASTE REACTOR
      A nuclear reactor which makes use of a
      fuel system, consisting of a bed of
      compounds of fissile (fissionable) and

fertile materials in a fluidized state under laminar non-turbulent flow conditions.

f   réacteur *m* à pâte combustible
e   reactor *m* de pasta combustible
i   reattore *m* a pasta combustibile
n   deegreactor, pastareactor
d   Pastareaktor *m*, Teigreaktor *m*

**2727 FLUIDIZED REACTOR,   rt
       SLURRY REACTOR**
A nuclear reactor in which the fuel has the properties of a quasi- or semi-fluid, i.e. a suspension of fine particles in a carrying gas or liquid.

f   réacteur *m* à combustible en suspension, réacteur *m* à combustible fluidisé
e   reactor *m* de combustible fluidificado
i   reattore *m* a combustibile fluidizzato
n   reactor met gefluïdiseerde splijtstof, suspensiereactor
d   Flüssigsuspensionsreaktor *m*

2728 FLUORBORATE, see 740

2729 FLUORESCENCE   ra
Emission of light or other electromagnetic radiation by a material exposed to another type of radiation or to a beam of particles, with the luminescence ceasing within about 10^{-8} second after irradiation is stopped.

f   fluorescence *f*
e   fluorescencia *f*
i   fluorescenza *f*
n   fluorescentie
d   Fluoreszenz *f*

2730 FLUORESCENCE YIELD   ra, xr
The probability that an atom whose electronic structure has been excited will emit an X-ray photon, in the first transition, rather than an Auger electron.

f   rendement *m* de fluorescence
e   rendimiento *m* de fluorescencia
i   rendimento *m* di fluorescenza
n   fluorescentierendement *n*
d   Fluoreszenzausbeute *f*

2731 FLUORESCENT DYE   ch, rw
A dye used in experiments for determining dispersal and dilution effects when discharging radioactive waste into open water.

f   colorant *m* fluorescent
e   colorante *m* fluorescente
i   colorante *m* fluorescente
n   fluorescerende kleurstof
d   fluoreszierender Farbstoff *m*

**2732 FLUORESCENT MATERIAL,   ch, ec
       FLUORESCENT SUBSTANCE**
A substance possessing the property to absorb radiant energy of definite wavelength and of emitting it as waves of different wavelength, characteristic of the substance.

f   substance *f* fluorescente
e   substancia *f* fluorescente

i   sostanza *f* fluorescente
n   fluorescerende stof
d   fluoreszierender Stoff *m*

2733 FLUORESCENT RADIATION, see 998

**2734 FLUORESCENT RADIATION   me
       DETECTOR**
A detector which detects radiation by using the property of certain materials to fluoresce when subjected to bombardment by radiation.

f   détecteur *m* de rayonnement fluorescent
e   detector *m* de radiación fluorescente
i   rivelatore *m* di radiazione fluorescente
n   fluorescerende stralingsdetector
d   fluoreszierender Strahlungszeiger *m*

2735 FLUORESCENT SCREEN   ec, ra
A screen covered with a fluorescent substance which emits visible light when excited by ionizing radiation.

f   écran *m* fluorescent
e   pantalla *f* fluorescente
i   schermo *m* fluorescente
n   fluorescentiescherm *n*
d   Fluoreszenzschirm *m*, Leuchtschirm *m*

**2736 FLUORIMETER,   me
       FLUOROMETER,
       FLUOROPHOTOMETER**
An apparatus for the measurement of fluorescence.

f   fluorimètre *m*
e   fluorímetro *m*
i   fluorimetro *m*
n   fluorescentiemeter
d   Fluoreszenzmesser *m*

2737 FLUORIMETRIC ANALYSIS   an
Analysis by means of a fluorimeter.

f   analyse *f* fluorimétrique
e   análisis *f* fluorimétrica
i   analisi *f* fluorimetrica
n   fluorimetrische analyse
d   fluorimetrische Analyse *f*

**2738 FLUORIMETRY,   me
       FLUOROMETRY**
The study of the frequency, intensity and power of fluorescent radiation.

f   fluorimétrie *f*
e   fluorimetría *f*
i   fluorimetria *f*
n   fluorimetrie
d   Fluorimetrie *f*

2739 FLUORINATION METHOD   ch
A method of purifying uranium.

f   méthode *f* de fluorination
e   método *m* de fluorinación
i   metodo *m* di fluorinazione
n   fluorinatiemethode
d   Fluorinationsverfahren *n*

2740 FLUORINE   ch
Gaseous element, symbol F, atomic number 3.

f fluor *m*
e flúor *m*
i fluor *m*
n fluor *n*
d Fluor *n*

2741 FLUOROGRAPHY,   xr
       PHOTOFLUOROGRAPHY
Photography of image on a fluorescent
screen.
f radiophotographie *f*
e radiofotografía *f*
i radiofotografia *f*
n schermbeeldfotografie
d Schirmbildverfahren *n*

2742 FLUOROSCOPE,   xr
       RÖNTGENOSCOPE
A device, consisting of a fluorescent
screen suitably mounted, by means of
which X-ray shadows of objects interposed
between the tube and the screen are made
visible.
f appareil *m* de radioscopie,
   röntgenoscope *m*
e aparato *m* de radioscopia,
   röntgenoscopio *m*
i apparecchio *m* di radioscopia,
   röntgenoscopio *m*
n doorlichtapparaat *n*, fluoroscoop
d Durchleuchtgerät *n*, Fluoroskop *n*,
   Röntgenoskop *n*

2743 FLUOROSCOPIC SCREEN,   xr
       SALT SCREEN
A sheet of material coated with a fluorescent substance emitting visible light when
irradiated with ionizing radiation.
f écran *m* radioscopique
e pantalla *f* radioscópica
i schermo *m* radioscopico
n doorlichtscherm *n*
d Leuchtschirm *m*

2744 FLUOROSCOPY,   xr
       RADIOSCOPY, RÖNTGENOSCOPY
Röntgen rays examination by means of a
fluoroscope.
f radioscopie *f*, röntgenoscopie *f*
e radioscopia *f*, röntgenoscopia *f*
i radioscopia *f*, röntgenoscopia *f*
n doorlichting, fluoroscopie
d Durchleuchtung *f*, Radioskopie *f*,
   Röntgenoskopie *f*

2745 FLUTE-TYPE INSTABILITY,   pp
       INTERCHANGE INSTABILITY
Instability of the magnetodynamic type
in which the plasma tends to escape when
the containing magnetic field undergoes no
modification.
f instabilité *f* à cannelures,
   instabilité *f* d'échange,
   instabilité *f* en flûtes
e inestabilidad *f* de intercambio
i instabilità *f* d'intercambio
n uitwisselingsinstabiliteit
d Auswechslungsinstabilität *f*

2746 FLUX   np
The product of the number of particles
per unit volume and their average speed.
f flux *m*
e flujo *m*
i flusso *m*
n flux
d Fluss *m*

2747 FLUX DENSITY,   ec
       MAGNETIC FLUX DENSITY,
       MAGNETIC INDUCTION
The number of electric lines of force per
unit area at right angles to the lines.
f densité *f* de flux
e densidad *f* de flujo
i densità *f* di flusso
n fluxdichtheid
d Flussdichte *f*

2748 FLUX DENSITY,   np
       NEUTRON FLUX DENSITY
At a given point in space, the number of
neutrons incident on a small sphere in a
time interval divided by cross sectional
area of that sphere and by the time
interval.
f débit *m* de fluence de neutrons,
   densité *f* de flux de neutrons
e densidad *f* de flujo de neutrones
i densità *f* di flusso neutronico
n fluxdichtheid van neutronen,
   neutronenfluxdichtheid
d Neutronenflussdichte *f*, Neutronenfluenz *f*

2749 FLUX DISTRIBUTION,   np
       NEUTRON FLUX DISTRIBUTION
The distribution of the power, or energy
per unit time, of the neutron flux.
f distribution *f* de flux de neutrons
e distribución *f* de flujo de neutrones
i distribuzione *f* di flusso neutronico
n neutronenfluxverdeling
d Neutronenflussverteilung *f*

2750 FLUX FLATTENING   np
In a nuclear reactor, the achievement of
an approximately uniform neutron flux
over the central region by depressing the
thermal neutron flux over that region.
f aplatissement *m* du flux
e aplanamiento *m* del flujo
i appiattimento *m* del flusso
n fluxafvlakking
d Flussabflachung *f*

2751 FLUX FLATTENING MATERIAL,
       see 2696

2752 FLUX FLATTENING RADIUS, see 2697

2753 FLUX FLATTENING REGION,
       FLUX FLATTENING ZONE,
       see 2693

2754 FLUX LEVEL   np
The flux value at which a multiplying

system operates under constant operating
conditions.
f niveau *m* de flux
e nivel *m* de flujo
i livello *m* di flusso
n fluxniveau *n*
d Flussniveau *n*

2755 FLUX TRAP   rt
A device consisting of a moderator
nucleus surrounded by a nuclear fuel
and a reflector.
f piège *m* à flux
e trampa *f* de flujo
i trappola *f* a flusso
n fluxval
d Fluxfangstelle *f*

2756 FLUX TRAP REACTOR   rt
A nuclear reactor in which the configuration is arranged in such a manner
as to provoke the concentration of the
neutron flux in a portion of the core.
f réacteur *m* à flux concentré
e reactor *m* de flujo concentrado
i reattore *m* a flusso concentrato
n reactor met geconcentreerde neutronenflux
d Reaktor *m* mit konzentriertem
   Neutronenfluss

2757 FLY-OFF   rt
Upward movement of the fuel rods in a
nuclear reactor in which the cooling
liquid circulates from bottom to top.
f envol *m*
e movimiento *m* ascencional
i movimento *m* ascensionale
n opwaartse beweging
d Aufwärtsbewegung *f*

2758 F-M CYCLOTRON,   pa
        PHASOTRON,
        SYNCHROCYCLOTRON
A cyclotron in which the radiofrequency of
the electric field applied between the dees
is frequency-modulated to permit the
acceleration of particles to relativistic
energies.
f cyclotron *m* modulé en fréquence,
   synchrocyclotron *m*
e ciclotrón *m* modulado en frecuencia,
   sincrociclotrón *m*
i ciclotrone *m* modulato in frequenza,
   sincrociclotrone *m*
n cyclotron *n* met frequentiemodulatie,
   synchrocyclotron *n*
d frequenzmoduliertes Zyklotron *n*,
   Synchrozyklotron *n*

2759 FOCAL DISTANCE   xr
f distance *f* focale
e distancia *f* focal
i distanza *f* focale
n brandpuntsafstand
d Brennweite *f*

2760 FOCAL SPOT   xr
The part of the target of the X-ray tube
which is hit by the main electron stream.
f spot *m* lumineux
e mancha *f* focal
i macchia *f* focale
n brandvlek
d Brennfleck *m*

2761 FOCUS   ge
The point at which rays of light or
electrons of a beam converge to form a
minimum-diameter spot.
f foyer *m*
e foco *m*
i fuoco *m*
n brandpunt *n*, focus
d Brennpunkt *m*, Fokus *m*

2762 FOCUS TO FILM DISTANCE,   xr
        SOURCE TO FILM DISTANCE
The distance from the focus of a source
of radiation to a film set up for a radiographic camera.
f distance *f* foyer-film
e distancia *f* foco-película
i distanza *f* fuoco-pellicola
n focus-film-afstand
d Fokus-Film-Abstand *m*,
   Quelle-Film-Abstand *m*

2763 FOCUS TO SKIN DISTANCE   xr
The distance from the focus of an X-ray
tube to the surface of incidence on a
patient, usually measured along the
beam axis.
f distance *f* foyer-peau
e distancia *f* foco-piel
i distanza *f* fuoco-pelle
n focus-huid-afstand
d Fokus-Haut-Abstand *m*

2764 FOCUSING   pa, sp
1. In particle spectrometry, the lens-like
action of a suitably designed field.
2. In cyclic accelerators, the action of the
deflecting field in causing the particles
to oscillate about the design orbit so that
they are not lost.
f focalisation *f*
e enfoque *m*, focalización *f*
i focalizzazione *f*
n focussering
d Fokussierung *f*

2765 FOCUSING CUP, see 1283

2766 FOG   ph, xr
Deleterious photographic density in the
completed röntgenogram.
f flou *m*, voile *m*
e velo *m*
i velatura *f*
n sluier
d photographischer Schleier *m*

2767 FOG-COOLED REACTOR   cl, rt
A reactor cooled by a suspension of water
droplets in air.
f réacteur *m* refroidi par brouillard
e reactor *m* enfriado por niebla
i reattore *m* raffreddato per nebbia
n reactor met nevelkoeling
d Reaktor *m* mit Nebelkühlung

2768 FOG TRACK, see 1141

2769 FOIL DETECTOR,   ma, rt
METAL FOIL DETECTOR
Thin sheet of metal, e.g. gold or indium,
used for measuring thermal neutron flux
density.
f détecteur *m* à feuille
e detector *m* de hoja
i rivelatore *m* a foglia
n foeliedetector
d Foliendetektor *m*

2770 FOKKER-PLANCK EQUATION   ma
An equation similar to the Boltzmann
transport equation except that the
collision integral terms in the Boltzmann
equation are replaced by derivative terms
which describe the additive effect of
two-body encounters.
f équation *f* de Fokker-Planck
e ecuación *f* de Fokker-Planck
i equazione *f* di Fokker-Planck
n vergelijking van Fokker en Planck
d Fokker-Plancksche Gleichung *f*

2771 FOLLOWER, see 1363

2772 FOOD CONTAMINATION-   co, sa
MONITOR
f moniteur *m* de denrées,
moniteur *m* de vivres
e monitor *m* de alimentos,
monitor *m* de víveres
i monitore *m* di prodotti alimentari
n levensmiddelenmonitor, voedselmonitor
d Lebensmittelmonitor *m*

2773 FOOD IRRADIATION REACTOR,
see 2628

2774 FORBIDDEN BAND   ec, np
The range of energy separating two bands
of possible values of energy level.
f bande *f* interdite
e banda *f* reservada
i banda *f* interdetta, banda *f* proibita
n verboden gebied *n*, verboden zone
d verbotenes Energieband *n*

2775 FORBIDDEN TRANSITION   np, qm
In nucleonics, a transition between two
states of a quantum-mechanical system
for which the change in the quantum
number is such, under the appropriate
selection rules, as to make the transition
less probable than a competing allowed
transition, other things being equal.

f transition *f* interdite
e transición *f* prohibida
i transizione *f* proibita
n verboden overgang
d verbotener Übergang *m*

2776 FORBUSH DECREASE   cr
A decrease in cosmic radiation measured
at the ground due to a solar wind.
f décroissance *f* de Forbush
e disminución *f* de Forbush
i diminuzione *f* di Forbush
n stralingsachteruitgang van Forbush
d Forbush-Herabsetzung *f*

2777 FORCE CONSTANTS   np
OF LINKAGE
Expressions of the forces acting between
nuclei to restrain relative displacement.
f constantes *pl* de forces de liaison
nucléaire
e constantes *pl* de fuerzas de enlace nuclear
i costanti *pl* di forze di legame nucleare
n kernbindingskrachten *pl*
d Kernbindungskräfte *pl*

2778 FORCE-FREE FIELD   ec
A field in which the electric currents are
in all points parallel to the force lines
of the magnetic field.
f champ *m* sans force
e campo *m* sin fuerza
i campo *m* senza forza
n krachtvrij veld *n*
d kraftfreies Feld *n*

2779 FORCE OF ATTRACTION, see 436

2780 FORCE OF REPULSION,   np
REPULSIVE FORCE
Force between particles which tends to
keep them apart.
f force *f* de répulsion
e fuerza *f* de repulsión
i forza *f* di ripulsione
n afstotingskracht
d Abstossungskraft *f*

2781 FORE VACUUM PUMP   vt
A backing vacuum pump which maintains
a low enough pressure so that a diffusion
pump can operate into it.
f pompe *f* à vide préliminaire élevé
e bomba *f* de vacío preliminar elevado
i pompa *f* a vuoto preliminare elevato
n voorvacuümpomp
d Vorvakuumpumpe *f*

2782 FOREIGN ATOM, see 1052

2783 FORM FACTOR   ma
The ratio of the effective value of a
quantity to its half-period average value.
f facteur *m* de forme
e factor *m* de forma
i fattore *m* di forma
n vormfactor
d Formfaktor *m*

2784 FORMANITE   mi
A rare ore of the fergusonite-formanite
series containing small percentages of
uranium and thorium.
f formanite f
e formanita f
i formanite f
n formaniet n
d Formanit m

2785 FORMATION OF HEAT,   gp
      HEAT OF FORMATION
The increase of heat content in a system
when one mole of a substance is formed
from its elements.
f chaleur f de formation
e calor m de formación
i calore m di formazione
n vormingswarmte
d Bildungswärme f, Entstehungswärme f

2786 FORMATION RATE, see 1504

2787 FORMATIVE TIME   np
The time interval elapsing between the
appearance of the first Townsend discharge in a given gap and the formation
in it of a self-maintaining glow discharge.
f temps m d'établissement
e tiempo m de establecimiento
i tempo m di stabilimento
n opbouwtijd
d Aufbauzeit f

2788 FORWARD SCATTER   np
That part of the scattered radiation which
has a scattering angle of less than 90°.
f diffusion f en avant
e dispersión f hacia adelante
i deviazione f in avanti
n voorwaartse verstrooiing
d Vorwärtsstreuung f

2789 FOULING   rt
Deposit of products on the contact
surfaces of a fluid, which may alter the
resistance in heat transfer.
f engorgement m
e incrustación f
i incrostazione f
n korst
d Verkrustung f

2790 FOULING FACTOR   cl, mg
Correcting factor to be applied to load
losses and to the transmission coefficient
for the fouled surfaces compared with
clean surfaces.
f facteur m d'engorgement
e factor m de incrustación
i fattore m d'incrostazione
n korstfactor
d Verkrustungsfaktor m

2791 FOUNTAIN-PEN TYPE   me
      DOSEMETER,
          POCKET DOSEMETER
A term commonly used to denote a

particular type of personal dosemeter.
f stylo-dosimètre m
e dosímetro m de bolsillo
i penna f dosimetrica
n pendosismeter
d Stabdosismesser m,
   Taschendosismesser m

2792 FOUNTAIN-PEN TYPE   me
      EXPOSUREMETER,
          POCKET EXPOSUREMETER
A term commonly used to denote a
particular type of exposuremeter.
f stylo-exposimètre m
e exposímetro m de bolsillo
i esposimetro m tascabile
n penexposiemeter
d Taschenexponierungsmesser m,
   Taschenexpositionsmesser m

2793 FOUR-CHANNEL FAST   ct
      COINCIDENCE UNIT
An apparatus for coincidence counting,
whereas one of the channels may be used
for anticoincidence counting.
f sélecteur m de coïncidences rapides
   à quatre voies
e selector m de coincidencias rápidas de
   cuatro canales
i selettore m di coincidenze veloci a
   quattro canali
n snelle coïncidentiekiezer met vier kanalen
d schneller Koinzidenzwähler m mit vier
   Kanälen

2794 FOUR-FACTOR FORMULA   ma
A formula named for the four factors of
which it is composed applicable to a
thermal neutron reactor with no neutron
leakage.
f formule f des quatre facteurs,
   produit m des quatre facteurs
e fórmula f de los cuatro factores
i formula f dei quattro fattori
n vierfactorenformule
d Vierfaktorenformel f

2795 FOUR-MOMENTUM OF A   np
      CLASSICAL PARTICLE
The four-velocity times the rest-mass of
a particle.
f moment m quadratique
e momento m cuadrático
i momento m quadratico
n quadratisch moment n
d Vierermoment n

2796 4n + 3 SERIES, see 52

2797 4 π COUNTER   ct
A counter for measuring radiation emitted
in all directions by a specimen of radioactive material placed within.
f compteur m 4π
e contador m 4π
i contatore m 4π
n 4π-teller
d 4π-Zähler m

2798 4π IONIZATION CHAMBER, ic
RE-ENTRANT IONIZATION CHAMBER
An ionization chamber with which the
radiation emitted by a radioactive source
may be detected within a solid angle of
4π steradians.
f chambre f d'ionisation 4π
e cámara f de ionización 4π
i camera f d'ionizzazione 4π
n 4π -ionisatievat n
d 4π -Ionisationskammer f

2799 4π PULSE COUNTING ASSEMBLY ma
A pulse counting assembly which includes
a 4π radiation detector.
f ensemble m de mesure à impulsions 4π
e conjunto m contador por impulsos 4π
i complesso m di misura ad impulsi 4π
n 4π -telopstelling
d 4π -Impulszählanordnung f

2800 FOUR-STAGE MERCURY vt
VAPOR PUMP (US),
FOUR-STAGE MERCURY VAPOUR
PUMP (GB)
A pump used in recovering expensive gas
which is not usefully converted to ions.
f pompe f à vapeur de mercure à quatre
étages
e bomba f de vapor de mercurio de cuatro
etapas
i pompa f a vapore di mercurio a quattro
stadi
n viertrapskwikdamppomp
d Vierstufen-Quecksilberdampfpumpe f

2801 4096 CHANNEL MEMORY MODULE ct
Apparatus for analyzing detector output
data in experimental nuclear physics.
f bloc m d'exploitation numérique à 4096
canaux,
module m numérique à mémoire à 4096
canaux
e módulo m digital de memoria de 4096
canales
i modulo m digitale di memoria a 4096
canali
n digitaal geheugenmoduul n met 4096
kanalen
d digitaler Speichermodul m mit 4096
Kanälen

2802 FOUR-VELOCITY np
The four quantities $v_\mu = \dfrac{.dx_\mu}{ds}$
describing the velocity of a classical
particle in relativity theory, where
$\dfrac{d}{ds}$ denotes differentiation with respect to
the proper time of the particle.
f vitesse f quadratique
e velocidad f cuadrática
i velocità f quadratica
n kwadratische snelheid
d Vierergeschwindigkeit f

2803 FOURIER SYNTHESIS ma
A mathematical technique applied in
diffraction analysis.
f synthèse f de Fourier
e síntesis f de Fourier
i sintesi f di Fourier
n fouriersynthese
d Fourier-Synthese f

2804 FOURMARIERITE mi
A mineral containing uranium and lead.
f fourmariérite f
e fourmarierita f
i fourmarierite f
n fourmarieriet n
d Fourmarierit m

2805 FRACTION ch, is
In a separation plant, each of the partial
flows, having different abundances,
produced by a stage.
f fraction f
e fracción f
i frazione f
n fractie
d Fraktion f

2806 FRACTION EXCHANGE is
A number, varying from 0 to 1.0, denoting
the progress of an isotope exchange
reaction.
f facteur m d'interchange
e factor m de intercambio
i fattore m di scambio
n uitwisselingsfactor
d Austauschfaktor m

2807 FRACTIONAL DISTILLATION ch, is
A method for the separation of several
volatile components or fractions of
different boiling points.
f distillation f fractionnée
e destilación f fraccionada
i distillazione f frazionata
n gefractioneerde destillatie
d fraktionierte Destillation f

2808 FRACTIONAL IONIZATION, ra
IONIZATION DEGREE
The number of ion pairs formed in relation
to the total number of atoms in a gas.
f degré m d'ionisation
e grado m de ionización
i grado m d'ionizzazione
n ionisatiegraad
d Ionisationsgrad m

2809 FRACTIONAL ISOTOPIC is
ABUNDANCE,
RELATIVE ISOTOPIC ABUNDANCE
The ratio of the number of atoms of a
particular isotope to the total number of
atoms of the element, both in a given
sample.
f abondance f isotopique relative
e abundancia f isotópica relativa

i   abbondanza *f* isotopica relativa
n   relatieve isotopenverhouding
d   relative Isotopenhäufigkeit *f*

2810   FRACTIONAL YIELD   ch, cr
       Intermediate yield in a step by step
       process, e.g. in fractional distillation or
       fractional crystallization.
f   rendement *m* par étage
e   rendimiento *m* por etapa
i   rendimento *m* per stadio
n   opbrengst per trap
d   Ausbeute *f* je Stufe

2811   FRACTIONATING COLUMN,   ch
       FRACTIONATING TOWER
       In fractional distillation, an apparatus
       designed to bring into intimate contact
       the rising vapo(u)r and falling liquid.
f   colonne *f* de fractionnement
e   columna *f* fraccionadora
i   colonna *f* di frazionamento
n   fractioneerzuil, rectificeerkolom
d   Fraktionierkolonne *f*, Fraktionierturm *m*,
    Rektifikationssäule *f*

2812   FRACTIONATION, see 1942

2813   FRAGMENTATION OF NUCLEUS   np
       The result of an explosive nuclear
       reaction.
f   fragmentation *f* du noyau
e   fragmentación *f* del núcleo
i   frammentazione *f* del nucleo
n   kernexplosie
d   Kernzertrümmerung *f*

2814   FRAME OF REFERENCE,   ma
       REFERENCE FRAME,
       REFERENCE SYSTEM
       Something that may be referred to as to
       give guidance or purpose in working out
       analysis or in solving a problem.
f   système *m* de référence
e   sistema *m* de referencia
i   sistema *m* di riferenza
n   verwijzingssysteem *n*
d   Bezugssystem *n*

2815   FRANCEVILLITE   mi
       A yellow uranium mineral consisting of a
       hydrated vanadium compound of uranium
       and barium.
f   francevillite *f*
e   francevilita *f*
i   francevillite *f*
n   francevilliet *n*
d   Francevillit *m*

2816   FRANCIUM   ch
       Metallic element, symbol Fr, atomic
       number 87.
f   francium *m*
e   francio *m*
i   francio *m*
n   francium *n*
d   Francium *n*

2817   FRANCK-CONDON PRINCIPLE   np, sp
       A theoretical interpretation of the relative
       intensity of spectral bands of a given
       system on the basis of electronic transi-
       tions within the molecule, and the
       vibrations which result from them.
f   principe *m* de Franck-Condon
e   principio *m* de Franck-Condon
i   principio *m* di Franck-Condon
n   principe *n* van Franck en Condon
d   Franck-Condon-Prinzip *n*

2818   FRANCK-HERTZ EXPERIMENT   np
       Determination of the excitation and
       ionization potentials of gases by the
       measurement of the losses of kinetic
       energy of electrons passing through a
       gas or vapo(u)r.
f   expérience *f* de Franck-Hertz
e   experimento *m* de Franck-Hertz
i   esperimento *m* di Franck-Hertz
n   proef van Franck en Hertz
d   Franck-Hertz-Versuch *m*

2819   FREE AIR   nw
       In contexts regarding nuclear explosions,
       air sufficiently remote from surfaces or
       objects that an explosive effect, a blast,
       is not modified by reflected shock or
       scattering objects.
f   air *m* libre
e   aire *f* libre
i   aria *f* libera
n   open lucht, vrije lucht
d   freie Luft *f*, offene Luft *f*

2820   FREE-AIR DOSE, see 156

2821   FREE AIR IONIZATION CHAMBER,   ic
       OPEN AIR IONIZATION CHAMBER
       An ionization chamber open to the air, in
       which a delimited beam of radiation passes
       between the electrodes without striking
       them or other internal parts of the
       instrument.
f   chambre *f* d'ionisation à air libre
e   cámara *f* de ionización de aire libre
i   camera *f* d'ionizzazione ad aria libera
n   openluchtionisatievat *n*
d   Freiluftionisationskammer *f*,
    offene Luftionisationskammer *f*

2822   FREE AIR OVERPRESSURE,   nw
       FREE AIR PRESSURE
       The unreflected pressure, in excess of the
       ambient atmospheric pressure, created in
       the air by the blast wave from an explosion
       of a nuclear weapon.
f   surpression *f* en air libre
e   sobrepresión *f* en aire libre
i   sovrappressione *f* in aria libera
n   overdruk in vrije lucht
d   Überdruck *m* in freier Luft

2823   FREE CONTENT   is
       Of a particular isotope in an enriched
       material, the difference between the amount

of the isotope in a specified molar quantity
of the enriched material and the amount in
an equal molar quantity of the feed or the
natural isotopic mixture.
f contenu *m* libre
e contenido *m* libre
i contenuto *m* libero
n vrij gehalte *n*
d freier Gehalt *m*

2824 FREE ELECTRON   np
An electron which is not restrained to
remain in the immediate neighbo(u)rhood
of a nucleus or an atom.
f électron *m* libre
e electrón *m* libre
i elettrone *m* libero
n vrij elektron *n*
d freies Elektron *n*

2825 FREE-FREE TRANSITION   np
The transition of free electrons from one
hyperbolic orbit to another one in the
field of atomic nuclei or atomic ions.
f transition *f* libre-libre
e transición *f* libre-libre
i transizione *f* libero-libero
n vrij-vrij-overgang
d frei-frei-Übergang *m*

2826 FREE ISOTOPE HOLD-UP   ch, is
The free content of the rectifier of a plant.
f charge *f* ouvrable libre
e carga *f* práctica libre
i carica *f* di materiale libera
n vrije werkinhoud
d freier Materialeinsatz *m*

2827 FREE MOLECULE DIFFUSION,   gp
KNUDSEN FLOW
Flow of a gas through a long tube at
pressures such that the mean free path is
much greater than the tube radius.
f courant *m* thermique moléculaire,
   écoulement *m* de Knudsen
e flujo *m* térmico molecular
i flusso *m* termico molecolare
n knudseneffect *n*.
   thermische moleculaire stroming
d Knudsen-Effekt *m*,
   thermische Molekularströmung *f*

2828 FREE NEUTRON   np
A neutron liberated from the nucleus of
an atom.
f neutron *m* libre
e neutrón *m* libre
i neutrone *m* libero
n vrij neutron *n*
d freies Neutron *n*

2829 FREE RADICAL   ch
An unsaturated molecular fragment in
which some of the valence electrons
remain free.
f radical *m* libre
e radical *m* libre

i radicale *m* libero
n vrij radicaal *n*
d freies Radikal *n*

2830 FREE VALENCE   ch
A valence that does not appear to be
satisfied, as the valence of a free radical.
f valence *f* libre
e valencia *f* libre
i valenza *f* libera
n vrije waardigheid
d freie Wertigkeit *f*

2831 FREE WATER   ch
The amount of water removed in drying a
solid to its equilibrium water content.
f eau *f* libre
e agua *f* sobrante
i acqua *f* libera
n niet-gebonden water *n*, vrij water *n*
d ungebundenes Wasser *n*

2832 FREEZER UNIT,   rt
SOLIDIFYING UNIT
The apparatus for solidifying or freezing
material used in a reactor.
f dispositif *m* de congélation,
   dispositif *m* de solidification
e dispositivo *m* de congelación,
   dispositivo *m* de solidificación
i dispositivo *m* di congelazione,
   dispositivo *m* di solidificazione
n stolruimte, vriesruimte
d Erstarrungsgerät *n*, Gefriergerät *n*

2833 FREEZING,   ch, gp
SOLIDIFICATION
The solidification of a non-solid substance,
especially at low temperatures.
f solidification *f*
e solidificación *f*
i solidificazione *f*
n stolling
d Erstarrung *f*

2834 FRENKEL DEFECT   ec, np
The combination of a vacancy and an
interstitial atom in a crystal lattice.
f défaut *m* de Frenkel
e defecto *m* de Frenkel
i difetto *m* di Frenkel
n frenkeldefect *n*
d Frenkel-Defekt *m*, Frenkel-Fehlstelle *f*

2835 FREQUENCY CONDITION   np, ra
In order to be able to emit radiation of a
given frequency an atom must undergo a
change of energy which is equal to hv,
where h is Planck's constant and v is the
frequency.
f condition *f* de la fréquence
e condición *f* de la frecuencia
i condizione *f* della frequenza
n frequentievoorwaarde
d Frequenzbedingung *f*

2836 FREQUENCY RANGE   ge
The range of frequencies over which a
device may be considered useful with
various circuit and operating conditions.
f gamme f de fréquences
e gama f de frecuencias,
   rango m de frecuencias
i campo m di frequenza,
   gamma f di frequenze
n frequentiegebied n
d Frequenzbereich m

2837 FRETTING CORROSION,   mg
     GALLING
The localized mutual seizure of two
metal surfaces during sliding friction
accompanied by the removal of metal
particles from one or both surfaces.
f corrosion f par frottement,
   écorchure f, grippage m
e corrosión f de rozamiento
i corrosione f di frizione, scorticatura f
n wrijvingscorrosie
d Reibungskorrosion f

2838 FREYALITE   mi
A mineral chiefly consisting of a thorium
silicate containing alumin(i)um, iron,
manganese and sodium.
f fréyalite f
e freialita f
i freialite f
n freyaliet n
d Freyalit m

2839 FRICTION FACTOR   gp
If the friction head loss for a fluid in any
pipe is written in the form $h_r = f \frac{1}{d} hv$,
then f is a function of the velocity v, the
diameter d and the kinematic viscosity of
the fluid provided that the pipes are all in
the same internal condition of roughness.
f facteur m de frottement
e factor m de rozamiento
i fattore m d'attrito
n wrijvingsfactor
d Reibungsfaktor m

2840 FRISCH IONIZATION CHAMBER,   ic
     GRID IONIZATION CHAMBER
A pulse ionization chamber with flat
electrodes and an additional electrode
(Frisch grid) used to measure the energy
of alpha particles or fission fragments.
f chambre f d'ionisation à grille,
   chambre f d'ionisation de Frisch
e cámara f de ionización con rejilla de
   Frisch
i camera f d'ionizzazione a griglia di
   Frisch
n roosterionisatievat n van Frisch
d Gitterionisationskammer f nach Frisch,
   Ionisationskammer f mit Frisch-Gitter

2841 FRITZSCHEITE   mi
A manganese-containing variety of autunite.

f fritzschéite f
e fritzscheita f
i fritzscheite f
n fritzscheïet n
d Fritzscheit m

2842 FROGMAN SUIT,   sa
     FROGSUIT,
     PROTECTIVE CLOTHES
f vêtements pl de sûreté
e traje m protector,
   vestimentas pl de seguridad
i scafandro m, vestiti pl protettivi
n beschermende kleding, veiligheidspak n
d Hautanzug m, Schutzkleidung f

2843 FRONTAL PROJECTION, see 284

2844 FROZEN MAGNETIC FIELD   pp
Term used to describe the phenomenon
that, as can be demonstrated in ideal
magnetodynamics, the movement of an
infinitely conducting liquid in the presence
of a magnetic field is accompanied by a
field deformation as if the lines of forces
were frozen in the matter and entrained
by it.
f champ m magnétique gelé
e campo m magnético gelado
i campo m magnetico gelato
n bevroren magnetisch veld n
d gefrorenes magnetisches Feld n

2845 FROZEN SHAFT SEAL   cd
A gas-tight seal made by allowing some
of the liquid to leak along the shaft to an
annular space around the shaft which can
be cooled and frozen.
f scellement m axial congélé
e cerrado m axial congelada
i tenuta f ermetica assiale congelata
n bevroren asafsluiting
d erfrorener Achsenverschluss m

2846 FROZEN STATIC SEAL   cd
Liquid flow prevention in a pipe by placing
a cooling jacket around the pipe to freeze
the liquid and so form a static seal.
f scellement m hermétique statique congélé
e cerrado m estático congelado
i tenuta f ermetica statica congelata
n bevroren statische afsluiting
d erfrorener statischer Verschluss m

2847 ft-VALUE   br
Quantity characterizing the degree of
forbiddeness of a beta process, defined
as the product of the half life t and a
certain function f of the beta energy limit
and the charge of the beta active nucleus.
f valeur f ft
e valor m ft
i valore m ft
n ft-waarde
d ft-Wert m

2848 FUEL, fu, rt
NUCLEAR FUEL
Material containing fissile (fissionable)
nuclides which, when placed in a reactor,
enables a chain reaction to be achieved.
f combustible *m* nucléaire
e combustible *m* nuclear
i combustibile *m* nucleare
n splijtstof
d Kernbrennstoff *m*

2849 FUEL ASSEMBLY, fu, rt
FUEL ELEMENT ASSEMBLY
A grouping of fuel elements which is not
taken apart during the charging and discharging of a reactor core.
f assemblage *m* combustible
e ensemblaje *m* combustible
i gruppo *m* d'elementi di combustibile
n splijtstofpakket *n*
d Brennstoffkassette *f*

2850 FUEL ASSEMBLY GRID, rt
GRID
A system of lattices and plates to support
the fuel assembly in a sodium cooled fast
neutron reactor.
f sommier *m*
e plataforma *f* de elementos combustibles
i piattaforma *f* d'elementi combustibili
n splijtelementenplatform *n*
d Brennstoffbühne *f*

2851 FUEL BOX fu
A container for one or more elements.
f caisse *f* de combustible
e caja *f* de combustible
i cassa *f* di combustibile
n splijtstofhouder
d Brennstoffbehälter *m*

2852 FUEL BURN-OUT, see 821

2853 FUEL CHANNEL rt
A duct through the moderator which is
designed to contain one or more fuel
assemblies and through which the coolant
circulates.
f canal *m* de combustible
e canal *m* de combustible
i canale *m* di combustibile
n splijtstofkanaal *n*
d Brennelementkanal *m*

2854 FUEL CLADDING fu, me
TEMPERATURE METER
An assembly for measuring the
temperature of the cladding of a fuel
element.
f thermomètre *m* de gaine de combustible
e termómetro *m* de la vaina del combustible
i termometro *m* per guaine
n thermometer voor splijtstofbekleding
d Hüllentemperaturmessgerät *n*

2855 FUEL COMPACT fu
A conglomerated bar, cylinder, pellet,
etc.

f comprimé *m* combustible
e comprimido *m* combustible
i compatto *m* combustibile
n splijtbaar persstuk *n*
d Brennstoffpressling *m*

2856 FUEL CONTROL co, rt
Control of a nuclear reactor by adjustment of the properties, position or
quantity of fuel in such a way as to change
the reactivity.
f commande *f* par combustible
e regulación *f* por combustible
i regolazione *f* per combustibile
n regeling met splijtstof
d Steuerung *f* durch Brennstoff

2857 FUEL COOLING INSTALLATION,
see 1406

2858 FUEL CYCLE rt
The sequence of steps, such as utilization,
reprocessing and refabrication, through
which nuclear fuel passes.
f cycle *m* du combustible
e ciclo *m* del combustible
i ciclo *m* del combustibile
n splijtstofcyclus, splijtstofkringloop
d Brennstoffkreislauf *m*

2859 FUEL DAMAGE fu
Alteration and deformation occurring to
the fuel in the reactor by various causes.
f détérioration *f* du combustible
e daño *m* del combustible
i danneggiamento *m* del combustibile
n splijtstofbeschadiging
d Brennstoffschaden *m*

2860 FUEL ECONOMY fu
In a nuclear reactor, the efficiency of the
fuel spent, be it by fission or by capture
without fission.
f économie *f* du combustible
e economía *f* del combustible
i grado *m* d'utilizzazione del combustibile
n splijtstofeconomie
d Brennstoffökonomie *f*

2861 FUEL ELEMENT rt
Elementary unit of the core of a reactor,
containing the nuclear fuel.
f élément *m* combustible
e elemento *m* combustible
i elemento *m* combustibile
n splijtstofelement *n*
d Brennelement *n*

2862 FUEL ELEMENT CORE fu
f noyau *m* d'un élément combustible
e núcleo *m* de un elemento combustible
i anima *f* d'un elemento combustibile
n splijtstofelementkern
d Brennelementkern *m*

2863 FUEL ELEMENT NOZZLE fu
The extremity of a fuel element shaped in
such a way as to facilitate the

circulation of the coolant.
f distributeur *m* du réfrigérant
e distribuidor *m* de refrigerante
i distributore *m* di refrigerante
n koelwaterverdeler
d Kühlwasserverteiler *m*

2864 FUEL HANDLING   fu
The process of loading a fuel in the reactor
and unloading it after irradiation.
f manipulation *f* du combustible
e manejo *m* del combustible
i manipolazione *f* del combustibile
n splijtstofmanipulatie
d Brennstoffbehandlung *f*

2865 FUEL INVENTORY   fu
The total amount of fuel present in a given
nuclear plant at a given date.
f charge *f* ouvrable de combustible
e carga *f* práctica de combustible
i dotazione *f* di combustibile,
impegno *m* di combustibile
n splijtstofinbrengst
d Brennstoffeinsatz *m*

2866 FUEL IRRADIATION LEVEL,   np, rt
SPECIFIC BURN-UP
The total energy released per unit mass
in a nuclear fuel.
f combustion *f* massique,
niveau *m* d'irradiation du combustible,
taux *m* de combustion
e combustión *f* de masa,
consumo *m* específico
i consumo *m* specifico,
consumo *m* unitario
n massieke versplijting,
specifieke versplijting
d spezifischer Abbrand *m*

2867 FUEL LIFETIME   fu
f vie *f* de combustible
e duración *f* de vida de combustible
i durata *f* di combustibile
n splijtstoflevensduur
d Brennstofflebensdauer *f*

2868 FUEL MAKE-UP,   fu
FUEL REFRESHMENT
The replenishing of the inventory of fuel
material.
f recharge *f* de combustible
e recarga *f* de combustible
i rifornimento *m* di combustibile
n splijtstofaanvulling, splijtstofoplading
d Brennstoffauffrischung *f*,
Brennstoffnachfüllung *f*

2869 FUEL PELLET   fu
Pellet-shaped fuel, ready for sheathing or
canning.
f comprimé *m* de combustible
e comprimido *m* de combustible
i pastiglia *f* di combustibile
n splijtstofkogel
d Brennstoffkugel *f*

2870 FUEL PENCIL,   fu
FUEL PIN
A thin fuel rod formed usually of a pile
of small pellets.
f aiguille *f* de combustible,
crayon *m* combustible
e perno *m* de combustible
i ago *m* di combustibile
n splijtstofpen
d Brennstoffstift *m*

2871 FUEL PLATE,   fu
FUEL SLAB
A sandwich structure, consisting of a thin
layer of nuclear fuel contained between
two layers of sheathing material.
f plaque *f* combustible
e placa *f* combustible
i piastra *f* combustibile
n splijtstofplaat
d Brennstoffplatte *f*

2872 FUEL PROCESS CELL,   rt
HOT CAVE, HOT CELL
A heavily shielded compartment used for
storing or processing highly radioactive
materials.
f casematte *f*, cellule *f* chaude
e casamata *f*, celda *f* caliente
i casamatta *f*, cella *f* calda
n afgeschermd gewelf *n*, hete cel, kazemat
d abgeschirmte Höhle *f*, heisse Zelle *f*,
Kasematte *f*

2873 FUEL PROCESSING   fu
The total of operations necessary to obtain
a fuel product.
f traitement *m* de combustible
e tratamiento *m* de combustible
i trattamento *m* di combustibile
n splijtstofbewerking
d Brennstoffbearbeitung *f*

2874 FUEL RATING   fu, rt
In a nuclear reactor, the rate of heat
production from fission, divided by the
weight of heavy atoms, initially present
in the fuel.
f puissance *f* spécifique
e potencia *f* específica
i potenza *f* specifica
n specifiek vermogen *n*
d spezifisches Vermögen *n*

2875 FUEL RECONDITIONING,   fu
FUEL REGENERATION,
FUEL REPROCESSING
The processing of nuclear fuel, after its
use in a reactor, to remove fission
products and recover fissile (fissionable)
and fertile material.
f traitement *m* du combustible irradié
e reprocesamiento *m* del combustible
agotado
i trattamento *m* del combustibile irradiato
n opwerken *n* van splijtstof
d Brennstoffaufarbeitung *f*

2876 FUEL ROD   fu
Rod-shaped nuclear fuel, ready for
sheathing or canning.
f barre *f* de combustible
e barra *f* de combustible,
   varilla *f* de combustible
i barra *f* di combustibile
n splijtstofstaaf
d Brennstoffstab *m*

2877 FUEL ROD COATING   fu
A coating, bonded to or in contact with the
fuel rod protecting it from reaction with
the canning material.
f barrière *f* de diffusion,
   couche *f* protectrice de barre de combustible
e capa *f* protectora de barra de combustible
i strato *m* protettivo di barra di combustibile
n beschermende laag op splijtstofstaaf
d Brennstoffstabbeschichtung *f*

2878 FUEL SLUG,   fu
      SLUG
A lump of nuclear fuel, to be inserted
into holes or channels in the active
lattice of a reactor.
f barreau *m* de combustible,
   bloc *m* de combustible
e bloqueo *m* de combustible
i blocco *m* di combustibile
n splijtstofblok *n*
d Brennstoffblock *m*

2879 FUEL SOLUTION   fu
The solution obtained by dissolving
irradiated solid fuels prior to
reprocessing.
f solution *f* de combustible
e solución *f* de combustible
i soluzione *f* di combustibile
n splijtstofoplossing
d Brennstofflösung *f*

2880 FUEL TRANSFER POND   fu, rt
An intermediate station in the loading and
unloading procedure.
f chambre *f* de transfert de combustible
e cámara *f* de transferencia de combustible
i camera *f* di trasferimento di combustibile
n splijtstofuitwisselingsruimte
d Brennstoffauswechslungsraum *m*

2881 FUEL TUBE   fu
Tube-shaped fuel ready for sheathing or
canning.
f tube *m* de combustible
e tubo *m* de combustible
i tubo *m* di combustibile
n splijtstofbuis
d Brennstoffrohr *n*

2882 FUEL UTILIZATION, see 826

2883 FUELLED LOOP   rt
A device in a reactor containing nuclear

fuel shaped in such a way as to permit
circulation of a coolant.
f circuit *m* actif
e circuito *m* activo
i circuito *m* attivo
n splijtstofelement *n* met koelmiddel-
   circulatie
d Brennelement *n* mit Kühlmittelumlauf

2884 FUELS FABRICATION FACILITY   fu
A plant at the Argonne National
Laboratory which is equipped to machine
plutonium and plutonium alloys in a
controlled atmosphere.
f installation *f* de traitement mécanique
   de combustibles
e instalación *f* de tratamiento mecánico
   de combustibles
i impianto *m* da trattamento meccanico
   di combustibili
n installatie voor de mechanische bewerking
   van splijtstoffen
d Anlage *f* zur mechanischen Bearbeitung
   von Brennstoffen

2885 FUELS TECHNOLOGY CENTRE(ER)
An installation at the Argonne National
Laboratory where metallurgical research
on the properties, structure and
behavio(u)r of plutonium and other metals
of interest in the nuclear field will be
carried out.
f centre *m* d'études métallurgiques
   des combustibles
e centro *m*'de investigaciones metalúrgicas
   de combustibles
i centro *m* di ricerche metallurgiche di
   combustibili
n centrale voor metallurgisch splijtstof-
   onderzoek
d Zentrum *n* für metallurgische Brennstoff-
   prüfungen

2886 FULL LOAD,   ge
      NOMINAL LOAD
The greatest load that a circuit or piece
of equipment is designed to carry under
specified conditions.
f charge *f* nominale, pleine charge *f*
e carga *f* nominal, plena carga *f*
i carico *m* nominale, pieno carico *m*
n nominale belasting, volle belasting
d Nennlast *f*, Vollast *f*

2887 FULL-LOAD REFUELLING,   fu, rt
      ON-LOAD REFUELLING
f rechargement *m* en opération
e reaprovisionamiento *m* en operación
i ricarica *f* in esercizio
n oplading tijdens bedrijf
d Aufladung *f* während des Betriebs

2888 FULL WIDTH AT HALF MAXIMUM   ma
In a distribution curve comprising a
single peak, the distance between the

abscissae of two points on the curve
determined by the ordinate which is 50 %
of the ordinate of the peak.
f largeur f de bande à mi-hauteur
e anchura f de banda de media de altura
i larghezza f di banda a mezza altezza
n piekbreedte op halve hoogte
d Halbhöhenspitzenbreite f

2889 FULLY CONTAINED EXPLOSION   mi
A technique proposed for the exploration
of the sea-bed for locating minerals.
f explosion f complètement enfermée
e explosión f completamente encerrado
i esplosione f completamente rinchiusa
n volledig ingesloten explosie
d vollständig eingeschlossene Explosion f

2890 FULLY PROTECTIVE TUBE   sa, xr
       HOUSING
Housing for which the leakage is reduced
at the surface of the tube housing when the
tube is continuously operated, with closed
window, at its maximum rated current
for the maximum rated voltage to a value
considered safe.
f gaine f de tube à protection totale
e envuelta f del tubo de protección total
i guaina f per tubo a protezione totale
n omhulling voor volledige bescherming
d Vollschutzgehäuse n

2891 FULLY REFLECTED REACTOR   rt
A reflected reactor in which the reflector
completely surrounds the core of the
reactor.
f réacteur m à réflecteur complet
e reactor m de reflector completo
i reattore m a riflettore completo
n reactor met volledige reflector
d Reaktor m mit vollständigem Reflektor

2892 FUMELESS DISSOLVING   fu
Accomplished by leading oxygen through
the dissolving nitric acid.
f solution f sans fumée
e solución f fumífuga, solución f sin humo
i soluzione f senza fumo
n rookvrij oplossen n
d rauchfreies Lösen n

2893 FUNCTION   ge
A quantity whose value depends on the
value of one or more other quantities.
f fonction f
e función f
i funzione f
n functie
d Funktion f

2894 FUNDAMENTAL FIELD PARTICLE,
       see 2595

2895 FUNDAMENTAL PARTICLE, see 2212

2896 FUNDAMENTAL-PARTICLE   np
       PHYSICS
The study of the fundamental particles
themselves.
f physique f des particules fondamentales
e física f de las partículas fundamentales
i fisica f delle particelle fondamentali
n fysica van de elementaire deeltjes
d Elementarteilchenphysik f

2897 FUSED-SALT REACTOR   rt
A type of nuclear reactor that uses molten
salts of uranium for both fuel and coolant.
f réacteur m à sels fondus
e reactor m de sales fundidas
i reattore m a sali fusi
n reactor met gesmolten zouten
d Reaktor m mit geschmolzenen Salzen

2898 FUSION   gp
Operation of melting or rendering fluid
by heat.
f fusion f
e fusión f
i fusione f
n smelting
d Schmelzung f

2899 FUSION BOMB,   nw
       H BOMB, HYDROGEN BOMB
A thermonuclear bomb in which the energy
is mainly derived from fusion.
f bombe f à hydrogène, bombe f H
e bomba f de fusión, bomba f de hidrógeno
i bomba f ad idrogeno
n H-bom, waterstofbom
d H-Bombe f, Wasserstoffbombe f

2900 FUSION ENERGY,   rt
       FUSION POWER
Energy derived from nuclear fusion in a
hot plasma.
f énergie f de fusion
e energía f de fusión
i energia f di fusione
n kernversmeltingsenergie
d Kernverschmelzungsenergie f

2901 FUSION FUEL   fu, np
The material usable for a fusion reaction.
f combustible m à fusion
e combustible m de fusión
i combustibile m nucleare per processo di
   fusione
n fusiebrandstof
d Fusionsbrennstoff m

2902 FUSION HEAT,   gp
       HEAT OF FUSION
The increase of heat content when unit
mass, or one mole of a solid, is converted
into liquid at its melting point.
f chaleur f de fusion
e calor m de fusión
i calore m di fusione
n smeltwarmte
d Schmelzwärme f

2903 FUSION REACTION,   np, rt
       NUCLEAR FUSION REACTION,
       THERMONUCLEAR REACTION
       A reaction between two light nuclei
       resulting in the production of at least one
       nuclear species heavier than either
       initial nucleus, together with excess
       energy.
f réaction f de fusion nucléaire,
   réaction f thermonucléaire
e reacción f de fusión nuclear,
   reacción f termonuclear
i reazione f di fusione nucleare,
   reazione f termonucleare
n kernfusiereactie, kernversmeltingsreactie,
   thermonucleaire reactie
d Kernfusionsreaktion f,
   Kernverschmelzungsreaktion f,
   thermonukleare Reaktion f

2904 FUSION REACTOR   rt
       A nuclear reactor in which a thermonuclear
       reaction may be produced and controlled.
f réacteur m à fusion
e reactor m de fusión
i reattore m a fusione
n kernversmeltingsreactor
d Fusionsreaktor m,
   Kernverschmelzungsreaktor m

2905 FUSION WEAPON   nw
       The type of bomb in which isotopes of
       hydrogen, deuterium or tritium combine
       to form helium, with the release of
       enormous energy, and neutrons.
f arme f à fusion nucléaire
e arma f de fusión nuclear
i arma f a fusione nucleare
n kernfusiewapen n
d Kernfusionswaffe f

G

2906 g FACTOR OF ATOM OR   ma
      ELECTRON,
      g VALUE OF ATOM OR ELECTRON

$$\gamma = g\,\frac{\mu_B}{\hbar} = g\frac{\gamma e}{2m_e}\ ,\ \text{where }\gamma\text{ is the}$$

gyromagnetic ratio and μ_B is the
Bohr magneton.
f facteur m g d'un atome ou d'un électron
e factor m g de un átomo o de un electrón
i fattore m g d'un atomo o d'un elettrone
n g-faktor van een atoom of elektron
d Atom-g-Faktor m, Elektron-g-Faktor m

2907 g FACTOR OF NUCLEUS,   ma
      g VALUE OF NUCLEUS

$$\gamma = \frac{\gamma\mu_N}{\hbar} = g\frac{2}{2mp}\ \ \text{where }\gamma\text{ is the}$$

gyromagnetic ratio and μ_N is the nuclear
magneton.
f facteur m d'un noyau
e factor m g de un núcleo
i fattore m g d'un nucleo
n g-factor van een kern
d Kern-g-Faktor m

2908 g VALUE, see 2293

2909 GADOLINIUM   ch
      Rare earth metallic element, symbol Gd,
      atomic number 64.
f gadolinium m
e gadolinio m
i gadolinio m
n gadolinium n
d Gadolinium n

2910 GAEDE ROTARY OIL PUMP   vt
      A pump consisting of a steel cylinder,
      traversed by a diametral slot, rotating
      about its axis, which is set eccentrically
      in a cylindrical steel casing.
f pompe f rotative à huile de Gaede
e bomba f giratoria de aceite de Gaede
i pompa f rotativa ad olio di Gaede
n roterende oliepomp van Gaede
d Gasballastpumpe f,
   rotierende Ölpumpe f nach Gaede

2911 GAGE (US),   me
      GAUGE (GB)
      An instrument or device for comparing
      some physical characteristic, such as
      force, radiation, intensity, etc.
f calibre m, jauge f
e calibrador m, galga f
i calibro m, indicatore m
n meetapparaat n, meter
d Lehre f, Messer m, Messgerät n

2912 GAIN   ec
      In an amplifier or multiplier the relation
      of the output current to the input current.
f gain m .
e ganancia f
i guadagno m
n versterking
d Verstärkung f

2913 GALLING, see 2837

2914 GALLIUM   ch
      Metallic element, symbol Ga, atomic
      number 31.
f gallium m
e galio m
i gallio m
n gallium n
d Gallium n

2915 GAMETE,   md
      GERM CELL, INITIAL CELL,
      SEXUAL CELL
      A mature germ cell such as an
      unfertilized ovum or spermatozoon.
f cellule f germinale, cellule f sexuelle,
   gamète m, gamonte m
e célula f germen, célula f inicial,
   célula f sexual, gameto m
i cellula f germinale, cellula f sessuale,
   gamete m
n gameet, geslachtscel, kiemcel
d Gamete f, Geschlechtszelle f,
   Keimzelle f

2916 GAMMA   gr
      Pertaining to gamma radiation.
f gamma
e gamma
i gamma
n gamma
d gamma

2917 GAMMA ABSORPTION   gr
      The absorption of gamma rays.
f absorption f gamma
e absorción f gamma
i assorbimento m gamma
n gamma-absorptie
d Gammaabsorption f

2918 GAMMA BACKSCATTER   ma
      THICKNESS METER
      A thickness meter including a gamma
      source and designed to determine material
      thickness by measurement of the radiation
      backscattered by this material.
f épaisseurmètre m à rétrodiffusion gamma
e calibrador m por retrodispersión gamma
i spessimetro m a sparpagliamento gamma
   all'indietro

n diktemeter berustend op gammaterug-
   verstrooiing
d Dickenmessgerät *n* nach der
   Gammarückstreumethode

2919 GAMMA CASCADE, see 902

2920 GAMMA COMPENSATED   ic
      IONIZATION CHAMBER
      A differential ionization chamber having
      one compartment sensitive to gamma
      radiation and the other compartment
      sensitive to both neutron flux and gamma
      radiation.
f chambre *f* d'ionisation à compensation
   gamma
e cámara *f* de ionización de compensación
   gamma
i camera *f* d'ionizzazione a compensazione gamma
n ionisatievat *n* met gammacompensatie
d Ionisationskammer *f* mit Gammakompensation

2921 GAMMA COMPENSATION   np
      Used in a gamma-compensated ionization
      chamber.
f compensation *f* du rayonnement gamma,
   compensation *f* gamma
e compensación *f* de la radiación gamma,
   compensación *f* gamma
i compensazione *f* della radiazione gamma,
   compensazione *f* gamma
n compensatie van de gammastraling,
   gammacompensatie
d Gammakompensation *f*,
   Gammastrahlungskompensation *f*

2922 GAMMA CONTAMINATION   sa
      INDICATOR
      An indicator designed to detect gamma
      surface contamination in which the output
      pulses control a warning signal.
f signaleur *m* de contamination gamma
e indicador *m* de contaminación gamma
i segnalatore *m* di contaminazione gamma
n indicator voor gammabesmetting
d Gammakontaminationsanzeiger *m*

2923 GAMMA CROSS SECTION   cs
      The effective cross section for gamma
      rays.
f section *f* efficace pour rayons gamma
e sección *f* eficaz para rayos gamma
i sezione *f* d'urto per raggi gamma
n werkzame doorsnede voor gammastralen
d Wirkungsquerschnitt *m* für Gammastrahlen

2924 GAMMA EMITTER   gr
      An atom whose radioactive decay process
      is associated with the emission of gamma
      rays.
f émetteur *m* gamma
e emisor *m* gamma
i emettitore *m* gamma
n gammastraler
d Gammastrahler *m*

2925 GAMMA FLUX   gr, np
      For gamma particles of a given energy,
      the product gamma particle density with
      speed.
f flux *m* gamma
e flujo *m* gamma
i flusso *m* gamma
n gammaflux
d Gammafluss *m*

2926 GAMMA FLUX DENSITY   gr, np
      At a given point in space, the number of
      gamma particles incident on a small
      sphere in a time interval divided by cross
      sectional area of that sphere and by the
      time interval.
f densité *f* de flux gamma
e densidad *f* de flujo gamma
i densità *f* di flusso gamma
n gammafluxdichtheid
d Gammafluenz *f*

2927 GAMMA FLUX DENSITY   me, np
      INDICATOR
      An indicator designed to give an estimate
      of gamma flux density.
f signaleur *m* de la densité de flux gamma
e indicador *m* de la densidad de flujo gamma
i segnalatore *m* della densità di flusso gamma
n indicator voor gammafluxdichtheid
d Gammafluenzanzeiger *m*

2928 GAMMA FLUX DENSITY   gr, me
      METER
      A measuring assembly for gamma flux
      density.
f fluxmètre *m* gamma
e fluxímetro *m* gamma
i flussometro *m* gamma
n meter voor gammafluxdichtheid
d Gerät *n* zur Bestimmung der Gammafluenz

2929 GAMMA FLUX DENSITY   gr, np, me
      MONITOR
      A monitor designed to measure and
      respond to gamma flux density.
f moniteur *m* de la densité de flux gamma
e monitor *m* de la densidad de flujo gamma
i monitore *m* della densità di flusso gamma
n monitor voor gammafluxdichtheid
d Warngerät *n* für Gammafluenz

2930 GAMMA-GAMMA COINCIDENCE   ma
      PULSE COUNTING ASSEMBLY
      A coincidence counting assembly used, in
      particular, for the direct measurement of
      the activity of certain radionuclides.
f ensemble *m* de mesure à coïncidences
   gamma-gamma
e conjunto *m* contador por coincidencias
   gamma-gamma
i complesso *m* di misura a coincidenze
   gamma-gamma
n telopstelling voor gamma-gamma-coïn-
   cidenties
d Gamma-Gamma-Koinzidenzzählanordnung
   f

2931 GAMMA HEATING   gr
The thermal effect of gamma rays in a
nuclear reactor.
f   chauffage m par rayonnement gamma,
    échauffement m gamma
e   calentamiento m por radiación gamma
i   riscaldamento m per radiazione gamma
n   gammaverhitting
d   Gammaaufheizung f

2932 GAMMA LEAKAGE PEAK,   sp
      LEAKAGE PEAK
For a gamma radiation, part of the
curve characteristic for the gamma
leakage spectrum corresponding to the
absorption in the detector medium of an
energy equal to that of the primary
photo-electrons.
f   pic m d'échappement, pic m de fuite
e   pico m de fuga
i   picco m di fuga
n   gammalekpiek
d   Gammaleckspitze f

2933 GAMMA LEAKAGE SPECTRUM   rt, sp
The spectrum of the gamma rays leaking
through the shielding of a nuclear reactor.
f   spectre m de fuites de rayons gamma
e   espectro m de fugas de rayos gamma
i   spettro m di fughe di raggi gamma
n   gammalekspectrum n
d   Gammaleckspektrum n

2934 GAMMA MILKER,   is
      ISOTOPE MILKER, MILKER
A device which separates a short half-life
isotope from a solution containing also
its longer-lived parent by repeated controlled injection into a cartridge
containing an ion-exchange resin, and
subsequent instantaneous flushing and
rinsing
f   séparateur m répétiteur d'isotope
e   separador m repetidor de isótopo
i   separatore m ripetitore d'isotopo
n   herhalingsisotopenscheider
d   Wiederholungsisotopenabscheider m

2935 GAMMA PULSE   me
      COUNTING ASSEMBLY
A pulse counting assembly which includes
a gamma radiation detector, whose output
pulses are applied to a counting sub-assembly and/or to a counting rate measuring
sub-assembly.
f   ensemble m de mesure à impulsions pour
    particules gamma
e   conjunto m contador por impulsos para
    partículas gamma
i   complesso m di misura ad impulsi per
    particelle gamma
n   telopstelling voor gammaquanta
d   Gammazählanordnung f

2936 GAMMA QUANTUM   gr
A photon of gamma radiation.
f   quantum-gamma m

e   cuanto-gamma m
i   quanto-gamma m
n   gammaquantum n
d   Gammaquant m

2937 GAMMA QUENCH   mg
The rapid cooling of uranium from a
temperature in the gamma phase either
to randomize the structure or to refine the
grain size.
f   refroidissement m rapide dans la phase
    gamma
e   enfriamiento m rápido en la fase gamma
i   raffreddamento m rapido nella fase gamma
n   afschrikken n in de gammafaze
d   Abschrecken n in der Gammaphase

2938 GAMMA RADIATION   gr
Electromagnetic radiation emitted in the
process of nuclear transition or particle
annihilation.
f   rayonnement m gamma
e   radiación f gamma
i   radiazione f gamma
n   gammastraling
d   Gammastrahlung f

2939 GAMMA-RAY ABSORPTION   ab, gr
A phenomenon in which gamma radiation
transfers to the matter which it traverses some or all of its energy.
f   absorption f gamma
e   absorción f gamma
i   assorbimento m gamma
n   gamma-absorptie
d   Gammastrahlenabsorption f

2940 GAMMA-RAY CAPSULE   gr
A capsule containing a gamma source.
f   capsule f à rayonnement gamma
e   cápsula f de radiación gamma
i   capsula f a radiazione gamma
n   gammacapsule
d   Gammakapsel f

2941 GAMMA-RAY DETECTORS   gr
A range of apparatus as counters, monitors,
dosemeters, etc.
f   détecteurs pl de rayonnement gamma
e   detectores pl de radiación gamma
i   rivelatori pl di radiazione gamma
n   gammadetectors pl
d   Gammadetektoren pl

2942 GAMMA-RAY SOURCE   gr
A quantity of matter emitting gamma
radiation in a form convenient for
radiology.
f   source f de rayonnement gamma
e   fuente f de radiación gamma
i   sorgente f di radiazione gamma
n   gammabron
d   Gammaquelle f

2943 GAMMA-RAY SOURCE CONTAINER gr
A container of dense material having a
wall thickness sufficient to allow of safe
handling for a specified limited time.

f bombe f gamma,
   support m de source gamma
e bomba f gamma,
   soporte m de fuente gamma
i bomba f gamma,
   sopporto m di sorgente gamma
n gammabom, gammabronhouder
d Gammabombe f,
   Gammaquellenhaltervorrichtung f

2944 GAMMA-RAY SOURCE   gr
      STRENGTH
      The output of gamma radiation from a
      gamma-ray source under specified con-
      ditions of filtration.
f intensité f de source gamma
e intensidad f de fuente gamma
i intensità f di sorgente gamma
n gammabronsterkte
d Gammaquellenstärke f

2945 GAMMA-RAY SPECTROMETER   sp
      An instrument for determining the energy
      spectrum of gamma rays.
f spectromètre m du rayonnement gamma
e espectrómetro m de la radiación gamma
i spettrometro m della radiazione gamma
n gammaspectrometer
d Gammaspektrometer n

2946 GAMMA-RAY SPECTRUM   gr, sp
      One or more sharp lines each correspond-
      ing to an intensity and energy characteris-
      tic of the source.
f spectre m du rayonnement gamma
e espectro m de radiación gamma
i spettro m di radiazione gamma
n gammaspectrum n
d Gammaspektrum n

2947 GAMMA RAYS   gr
      Electromagnetic rays emitted by radio-
      active substances and of frequencies
      generally higher than those of X-rays.
f rayons pl gamma
e rayos pl gamma
i raggi pl gamma
n gammastralen pl
d Gammastrahlen pl

2948 GAMMA SPACE,   ma
      PHASE SPACE
      A Euclidian hyperspace of 2f dimensions
      having 2f rectangular axes.
f espace m de phase, extension f en phase
e espacio m de fase, extensión f de fase
i spazio m delle fasi
n fazeruimte
d Phasenraum m

2949 GAMMA SPECTROSCOPY   sp
      Spectroscopy using gamma rays as a source
      of radiation.
f spectroscopie f au rayonnement gamma
e espectroscopia f con radiación gamma
i spettroscopia f a radiazione gamma
n gammaspectroscopie
d Gammaspektroskopie f

2950 GAMMA TELETHERAPY   gr, md, ra
f télégammathérapie f
e telegammaterapia f
i telegammaterapia f
n telegammatherapie
d Gammateletherapie f

2951 GAMMA URANIUM   mg
      That allotropic modification of uranium
      metal which is stable above approximately
      770ºC.
f uranium m gamma
e uranio m gamma
i uranio m gamma
n gamma-uranium n
d Gamma-Uran n

2952 GAMMAGRAPHY,   gr
      GAMMARADIOGRAPHY
      The art of making radiographs by making
      use of gamma rays.
f gammagraphie f, gammaradiographie f
e gammagrafía f, gammaradiografía f
i gammagrafia f, gammaradiografia f
n gammagrafie, gammaradiografie
d Gammagraphie f, Gammaradiographie f

2953 GAMOW BARRIER, see 541

2954 GAMOW FACTOR   np
      A term used for the penetration probability
      for low energy charged particles in a
      Coulomb field.
f facteur m de Gamow
e factor m de Gamow
i fattore m di Gamow
n gamowfactor
d Gamow-Faktor m

2955 GAMOW-TELLER   np
      SELECTION RULES
      In beta theory, a particular set of selection
      rules which follow from a definite assump-
      tion about the coupling term.
f lois pl de sélection de Gamow-Teller
e leyes pl de selección de Gamow-Teller
i leggi pl di selezione di Gamow-Teller
n selectieregels pl van Gamow en Teller
d Gamow-Tellersche Auswahlregeln pl

2956 GANGUE, see 1494

2957 GAS ACTIVITY METER,   ma
      GAS RADIOACTIVITY METER
      A (radio)activity meter designed to
      measure the activity in a gas, and equipped
      with an indicating and/or recording
      measuring instrument.
f activimètre m de gaz,
   radioactivimètre m de gaz
e activímetro m para gases,
   radiactivímetro m para gases
i attivimetro m per gas,
   radioattivimetro m per gas
n gasactiviteitsmeter
d Aktivitätsmessgerät n für Gase

2958 GAS AMPLIFICATION,   ct
       GAS MAGNIFICATION
       The ratio of the charge collected in a
       gas-filled radiation counter tube to the
       charge produced in the active volume by
       the preliminary ionizing event.
f amplification *f* due au gaz
e amplificación *f* por gas
i amplificazione'*f* mediante ionizzazione
   d'un gas
n gasversterking
d Gasionisationsverstärkung *f*,
   Gasverstärkung *f*

2959 GAS AMPLIFICATION FACTOR   ct, ic
       The ratio of the charge collected from the
       sensitive volume of an enclosure
       containing a gas to the charge produced
       in this volume by the initial ionization.
f coefficient *m* d'amplification due au gaz
e factor *m* de amplificación por gas
i fattore *m* d'amplificazione mediante
   ionizzazione d'un gas
n gasversterkingsfactor
d Gasionisationsverstärkungsfaktor *m*

2960 GAS BEARINGS   cd
       Special bearings in which a moving and a
       stationary surface are separated by a film
       of gas which serves as a lubricant.
f coussinets *pl* gazeux
e cojinetes *pl* gaseosos
i cuscinetti *pl* gassosı
n gaslagers *pl*
d Gaslager *pl*

2961 GAS CHROMATOGRAPHY,   an
       VAPOR CHROMATOGRAPHY (US),
       VAPOUR CHROMATOGRAPHY (GB)
       A method of separating mixtures of gases
       or vapo(u)rs by differential adsorption
       or solution.
f chromatographie *f* à gaz
e cromatografía *f* de gas
i cromatografia *f* a gas
n gaschromatografie
d Gaschromatographie *J*

2962 GAS CIRCULATION   cl
       The movement of, e.g., the coolant gas
       used in a nuclear reactor.
f circulation *f* du gaz
e circulación *f* del gas
i circolazione *f* del gas
n gascirculatie, gasomloop
d Gasumlauf *m*, Gaszirkulation *f*

2963 GAS CIRCULATOR   rt
f appareil *m* circulateur de gaz
e aparato *m* circulador de gas
i apparecchio *m* circolatore di gas
n gascirculatie-apparaat *n*,
   gascirculatiepomp
d Gaszirkulationsvorrichtung *f*

2964 GAS-COOLED REACTOR   rt
       A nuclear reactor in which the coolant is
       a gas, commonly air or carbon dioxide.

f réacteur *m* à refroidissement au gaz
e reactor *m* enfriado de gas
i reattore *m* refrigerato a gas
n met gas gekoelde reactor,
   reactor met gaskoeling
d gasgekühlter Reaktor *m*

2965 GAS COUNTER   ct
       A counter in which the sample is prepared
       in form of a gas and introduced into the
       counter tube itself.
f compteur *m* à gaz
e contador *m* a gas
i contatore *m* a gas
n gasteller
d Gaszähler *m*

2966 GAS CURRENT,   ch
       IONIZATION CURRENT
       The electric current resulting from the
       movement of electric charges in an
       ionized medium, under the influence of
       an applied electric field.
f courant *m* d'ionisation
e corriente *f* de ionización
i corrente *f* d'ionizzazione
n ionisatiestroom
d Ionisationsstrom *m*,
   Ionisierungsstrom *m*

2967 GAS ENTRAINMENT   cl, rt
       An inconvenience sometimes occurring in
       liquid-metal cooled reactors and
       effecting large reactivity changes.
f entraînement *m* de gaz
e arrastre *m* de gas
i trascinamento *m* di gas
n meeslepen *n* van gas
d Gasmitreissung *f*

2968 GAS-FILLED COUNTER   ct, ic
       A form of ionization chamber which, when
       operated under suitable conditions, may be
       used as a counter.
f compteur *m* à remplissage gazeux
e contador *m* gaseoso
i contatore *m* con atmosfera gassosa
n met gas gevulde teller
d gasgefüllter Zähler *m*

2969 GAS-FILLED COUNTER TUBE   ct
       A counter tube for detection by means of
       gas ionization.
f tube *m* compteur à remplissage gazeux
e tubo *m* contador gaseoso
i tubo *m* contatore con atmosfera gassosa
n met gas gevulde telbuis
d gasgefülltes Zählrohr *n*

2970 GAS-FLOW COUNTER TUBE   ct
       A counter tube in which an appropriate
       atmosphere is maintained by means of a
       slow flow of a suitable gas.
f tube *m* compteur à balayage gazeux,
   tube *m* compteur à courant gazeux
e tubo *m* contador de corriente de gas
i tubo *m* contatore a corrente di gas
n gasdoorstroomtelbuis

d durchströmtes Zählrohr *n*,
Gasdurchflusszählrohr *n*

2971 GAS-FLOW IONIZATION CHAMBER ic
An ionization chamber in which an
appropriate atmosphere is maintained by
means of a slow flow of a suitable gas.
f chambre *f* d'ionisation à courant gazeux
e cámara *f* de ionización de corriente de gas
i camera *f* d'ionizzazione a corrente di
gas
n gasdoorstroomionisatievat *n*
d durchströmte Ionisationskammer *f*,
Gasdurchflussionisationskammer *f*

2972 GAS-FLOW NEUTRON FLUX ma
DENSITY MEASURING ASSEMBLY
An assembly designed to measure the
neutron flux density in a nuclear reactor
and which consists of a target of fissile
(fissionable) material and a flow of inert
gas from target to detector.
f ensemble *m* de mesure de densité de flux
de neutrons à courant gazeux
e conjunto *m* medidor de la densidad del
flujo neutrónico por corriente gaseosa
i complesso *m* di misura della densità di
flusso neutronico a corrente gassosa
n meetopstelling voor de neutronenfluxdichtheid met gasstroom
d Gasdurchflusszählgerät *n* zur
Neutronenfluenzmessung

2973 GAS-GRAPHITE REACTOR rt
A nuclear reactor in which the coolant is
a gas and the moderator consists of
graphite.
f réacteur *m* à gaz-graphite
e reactor *m* de gas-grafito
i reattore *m* a gas-grafite
n gas-grafietreactor
d Gas-Graphitreaktor *m*

2974 GAS MULTIPLICATION ct
Multiplication, due to the action of an
electric field, of the initial number of ions
produced in the gas of the sensitive
volume of a counter tube.
f multiplication *f* due au gaz
e multiplicación *f* por gas
i moltiplicazione *f* dovuta al gas
n gasversterking
d Gasverstärkung *f*

2975 GAS PASSAGES cd
Usually ducts along which a gas flows and
which are of a geometrically complex
shape dictated by factors in the reactor or
process other than the optimum gas flow.
f tuyauterie *f* de gaz
e tubería *f* de gas
i gasdotto *m*, tubatura *f* di gas
n gasleidingen *pl*
d Gasleitungen *pl*

2976 GAS SCATTERING ct
In a counter tube, the scattering in the
filling gas.

f diffusion *f* dans le gaz
e dispersión *f* en el gas
i deviazione *f* nel gas
n verstrooiing in het gas
d Streuung *f* im Gas

2977 GAS SCINTILLATOR, ct
GASEOUS SCINTILLATOR
A scintillator containing a gas.
f scintillateur *m* gazeux
e centelleador *m* gaseoso
i scintillatore *m* gassoso
n gasvormige scintillator
d gasförmiger Szintillator *m*

2978 GAS SCINTILLATOR DETECTOR ra
A scintillation detector in which the
scintillator is a gas.
f détecteur *m* à scintillateur gazeux
e detector *m* de centelleador gaseoso
i rivelatore *m* a scintillatore gassoso
n scintillatiedetector met gas
d Detektor *m* mit gasförmigem Szintillator

2979 GAS SCRUBBING cg, ch
The contacting of a gaseous material with
a liquid for the purpose of removing gases
or entrained liquids or solids.
f lavage *m* de gaz
e lavado *m* de gas
i lavaggio *m* di gas
n gaswassing
d Gaswäsche *f*

2980 GAS SEPARATOR ch
A unit for separating gas or vapo(u)r from
entrained liquids.
f séparateur *m* de gaz
e separador *m* de gas
i separatore *m* di gas
n gasafscheider
d Gasabscheider *m*

2981 GAS SPARGING cd
1. Blowing a rapid stream of gas into a
reaction vessel to expel air before initiating a reaction.
2. Using a rapid stream of gas as a mixer
for liquids.
f purgation *f* à gaz
e rociadura *f* con gas
i soffiamento *m* a traverso con gas
n gas doorblazen *n*
d Durchblasen *n* von Gas

2982 GAS TARGET ge
A target consisting of a layer or jet of gas.
f cible *f* en gaz
e blanco *m* de gas
i bersaglio *m* di gas
n gastreflaag
d Gastreffschicht *f*

2983 GAS-TIGHT CASING ge
f enveloppe *f* étanche au gaz
e envoltura *f* hermética
i involucro *m* a tenuta di gas
n gasdicht omhulsel *n*
d gasdichte Hülle *f*

2984 GAS X-RAY TUBE   xr
An X-ray tube in which the emission of
electrons from the cold cathode is
produced by positive-ion bombardment
when the applied cathode-anode voltage is
made sufficiently high.
f tube *m* à rayons X à gaz
e tubo *m* de rayos X de gas
i tubo *m* a raggi X a gas
n met gas gevulde röntgenbuis
d gasgefüllte Röntgenröhre *f*

2985 GASEOUS DIFFUSION,   ch, is
        POROUS DIFFUSION
The kinetic process whereby gaseous
molecules intermingle and become
uniformly scattered throughout a volume.
f diffusion *f* gazeuse
e difusión *f* gaseosa
i diffusione *f* gassosa,
   diffusione *f* in mezzo poroso
n gasdiffusie
d Gasdiffusion *f*

2986 GASEOUS DIFFUSION METHOD   is
The separation of isotopes by means of
gaseous diffusion.
f méthode *f* à diffusion gazeuse
e método *m* de difusión gaseosa
i metodo *m* a diffusione in mezzo poroso
n gasdiffusiemethode
d Gasdiffusionsverfahren *n*

2987 GASEOUS PHASE   ch, gp
f phase *f* gazeuse
e fase *f* gaseosa
i fase *f* gassosa
n gasfaze
d Gasphase *f*

2988 GASEOUS PHASE   me
        RADIOCHROMATOGRAPH
A measuring device to draw a representative curve of the activity of different
labelled components of gaseous phase
mixture flowing past a radiation detector
after having been separated along a
chromatographic column.
f radiochromatographe *m* en phase gazeuse
e radiocromatógrafo *m* en fase gaseosa
i radiogascromatografo *m*
n radiochromatograaf voor de gasfaze
d Radiogaschromatograph *m*

2989 GASTROINTESTINAL TRACT, see 180

2990 GASTUNITE   mi
A rare, uranium, calcium and silicon
containing ore.
f gastunite *f*
e gastunita *f*
i gastunite *f*
n gastuniet *n*
d Gastunit *m*

2991 GATE   ec
An electronic circuit with more than one
input and only one output.

f circuit *m* de gâchette,
   circuit *m* de porte, gâchette *f*, porte *f*
e circuito *m* de puertas, entrada *f*,
   puerta *f*
i circuito *f* de puerta, entrada *f*,
   sblocco,
   porta *f* elettronica
n poort, poortschakeling
d Gatter *n*, Torschaltung *f*

2992 GATE,   rt, sa
        SHIELDING BARRIER
A movable barrier of shielding material
used for closing a hole.
f porte *f* de blindage
e puerta *f* de blindaje
i porta *f* di schermatura
n afschermdeur
d Abschirmtür *f*

2993 GATING   ec
The process of selecting those portions
of a wave which exist during one or more
selected time intervals or which have
magnitudes between selected limits.
f sélection *f* de signal,
   séparation *f* d'impulsions
e selección *f* de señal,
   separación *f* de impulsiones
i selezione *f* di segnali,
   separazione *f* d'impulsi
n poorten *n*, pulsscheiding, signaalselectie
d Ausblenden *n*, Signalauswertung *f*

2994 GATING UNIT   ec
A basic function unit comprising an
electronic circuit designed to either
reject or accept a pulse as a function of
a controlling signal.
f élément *m* porte
e elemento *m* puerta, unidad *f* selectora
i elemento *m* porta
n poort, poortschakel
d Torschalter *m*, Torstufe *f*

2995 GAUGE (GB), see 2911

2996 GAUNTLET GLOVE   pe
The glove used in a glove box.
f gant *m* à crispin
e guante *m* protector
i guanto *m* di protezione
n kaphandschoen
d Handschuh *m* mit Schutzkappe,
   Stulpenhandschuh *m*

2997 GAUSSIAN WELL   np
A well in a nuclear potential in which
$$V = - V_0 e^{-r/b^2}$$

f puits *m* de Gauss
e pozo *m* de Gauss
i pozzo *m* di Gauss
n gausskuil
d Gaussche Potentialmulde *f*,
   Gausscher Potentialtopf *m*

2998 GEIGER FORMULA   ar, ma
A relationship between the initial velocity

of alpha particles and their range.
f formule *f* de Geiger
e fórmula *f* de Geiger
i formula *f* di Geiger
n geigerformule
d Geigersche Formel *f*

2999 GEIGER-MÜLLER   ct
      COUNTER TUBE,
      G.M. COUNTER TUBE
      A counter tube operating in the
      Geiger-Müller region.
f tube *m* compteur Geiger-Müller
e tubo *m* contador Geiger-Müller,
   tubo *m* contador G.M.
i tubo *m* contatore Geiger-Müller
n geiger-müllertelbuis, geigertelbuis
d Geiger-Müller-Zählrohr *n*,
   G.M.-Zählrohr *n*

3000 GEIGER-MÜLLER COUNTER   ma
      TUBE EXPOSURE RATEMETER
      A measuring assembly for the exposure
      rate, in which the detector is a built-in
      or separate Geiger-Müller counter tube.
f débitmètre *m* d'exposition à tube compteur
   de Geiger-Müller
e dosímetro *m* con contador G.M.
i rateometro *m* d'esposizione a tubo contatore Geiger-Müller
n exposietempometer met geigertelbuis
d Dosisleistungsmesser *m* mit G.M.-Zähl-
   rohr

3001 GEIGER-MÜLLER REGION,   ct
      GEIGER REGION
      The range of operating voltage of a
      counter tube in which each ionizing event
      gives rise to an output pulse having an
      amplitude independent of the number of
      ions initially produced in the sensitive
      volume by that ionizing event.
f région *f* de Geiger-Müller
e región *f* de Geiger-Müller
i regione *f* di Geiger-Müller
n geiger-müllergebied *n*, geigergebied *n*
d Auslösebereich *m*,
   Geiger-Müller-Bereich *m*

3002 GEIGER-MÜLLER THRESHOLD,   ct
      GEIGER THRESHOLD
      The lowest voltage which must be applied
      to a counter tube for it to operate in the
      Geiger-Müller region.
f seuil *m* de Geiger-Müller
e umbral *m* de Geiger-Müller
i soglia *f* di Geiger-Müller
n geiger-müllerdrempel, geigerdrempel
d Geiger-Müller-Schwelle *f*

3003 GEIGER-NUTTALL RELATION   ar
      The empirical formula $\log R = A + B \log \lambda$,
      where λ is the disintegration constant
      of an alpha emitter, R the range of the
      alpha particles emitted and B is a constant.
f relation *f* de Geiger-Nuttall
e relación *f* de Geiger-Nuttall

i rapporto *m* di Geiger-Nuttall
n verhouding van Geiger en Nuttall
d Geiger-Nuttall-Verhältnis *n*

3004 GENE   md
      Fundamental unit of inheritance, which
      determines and controls hereditarily
      transmissible characteristics.
f gène *m*
e gen *m*
i gene *m*
n gen *n*
d Erbfaktor *m*, Gen *n*, Vererbungskeim *m*

3005 GENE MUTATION   md
      A sudden and permanent change in a gene.
f mutation *f* de gènes
e mutación *f* de genes
i mutazione *f* di geni
n genmutatie
d Genmutation *f*

3006 GENERAL DIFFUSION EQUATION   ma
      A generalization of the thermal neutron
      diffusion equation, in which the density
      of thermal neutrons may change with
      time, and in which scattering, leakage
      and absorption are taken into account.
f équation *f* générale de diffusion
e ecuación *f* general de difusión
i equazione *f* generale di diffusione
n algemene diffusievergelijking,
   tijdsafhankelijke diffusievergelijking
d allgemeine Diffusionsgleichung *f*,
   zeitabhängige Diffusionsgleichung *f*

3007 GENERAL PURPOSE   rt
      MANIPULATOR
      A device in a nuclear reactor for
      carrying out all kinds of mechanical
      operations.
f télémanipulateur *m* universel
e manipulador *m* para usos generales
i manipolatore *m* universale
n universele manipulator
d Universalmanipulator *m*

3008 GENERALIZED VELOCITIES   ma
      In particle mechanics, the time rates of
      change of the generalized co-ordinates.
f vitesses *pl* généralisées
e velocidades *pl* generalizadas
i velocità *pl* generalizzate
n gegeneraliseerde snelheden *pl*
d verallgemeinerte Geschwindigkeiten *pl*

3009 GENERATION TIME   np
      The mean time required for neutrons
      arising from fissions to produce other
      fissions.
f temps *m* de génération
e duración *f* de generación
i tempo *m* di generazione,
   tempo *m* medio d'una generazione
   neutronica
n generatietijd
d Generationsdauer *f*

3010 GENERATIONS OF NUCLEI   np
    The successive numbers of nuclei produced
    in a nuclear chain reaction.
f générations *pl* de noyaux
e generaciones *pl* de núcleos
i generazioni *pl* di nuclei
n kerngeneraties *pl*
d Kerngenerationen *pl*

3011 GENETIC EFFECT OF   md
    RADIATION
    Inheritable changes, chiefly mutations,
    produced by the absorption of ionizing
    radiation.
f effet *m* génétique du rayonnement
e efecto *m* genético de la radiación
i effetto *m* genetico della radiazione
n genetisch stralingseffect *n*
d genetischer Strahlungseffekt *m*

3012 GEOCHRONOLOGY   np
    The science of computing time or periods
    of time, and of assigning events to their
    true dates in respect of the earth.
f géochronologie *f*
e geocronología *f*
i geocronologia *f*
n geochronologie,
    tijdrekenkunde van de aarde
d Geochronologie *f*

3013 GEOLOGICAL AGE   mi, np
    The interval of time between some
    geological event, such as the formation
    of a mineral and the present time.
f âge *m* géologique
e edad *f* geológica
i età *f* geologica
n ouderdom der aarde
d Alter *n* der Erde

3014 GEOLOGICAL GROUND SURVEY   mi
    The survey of an area by geological
    means, e.g. for one location.
f levé *m* géologique terrestre
e levantamiento *m* geológico de terrenos
i rilevamento *m* geologico di terreni
n geologisch terreinonderzoek *n*
d geologische Geländeaufnahme *f*

3015 GEOMAGNETIC EFFECT,   cr
    LATITUDE EFFECT,
    MAGNETIC LATITUDE EFFECT
    An effect due to the earth's magnetic
    field by which the cosmic radiation flux
    received by the earth decreases when
    one moves from the poles to the
    magnetic equator.
f effet *m* de latitude,
    effet *m* géomagnétique
e efecto *m* de latitud,
    efecto *m* geomagnético
i effetto *m* di latitudine,
    effetto *m* geomagnetico
n breedte-effect *n*, geomagnetisch effect *n*
d Breiteneffekt *m*, geomagnetischer Effekt *m*

3016 GEOMETRIC ATTENUATION   ra
    The reduction of a radiation quantity due
    to the effect only of the distance between
    the point of interest and the source,
    excluding the effect of any matter present.
f atténuation *f* géométrique
e atenuación *f* geométrica
i attenuazione *f* geometrica
n geometrische verzwakking
d geometrische Schwächung *f*

3017 GEOMETRIC BUCKLING   np
    In reactor theory, the lowest eigenvalue
    that results from solving the equation
    $\nabla^2\phi(\mathrm{F}) + B g^2\phi(\mathrm{r}) = 0$ with the boundary
    condition that the thermal neutron flux
    at the extrapolated boundary of the
    system = 0.
f laplacien *m* géométrique
e laplaciano *m* geométrico
i parametro *m* geometrico di criticità
n geometrische bolling,
    geometrische welving
d geometrische Flussdichtewölbung *f*

3018 GEOMETRIC CROSS SECTION   cs
    Of a nucleus, the product πR^2, where
    R is the nuclear radius.
f section *f* géométrique
e sección *f* geométrica
i sezione *f* geometrica
n geometrische doorsnede
d geometrischer Querschnitt *m*

3019 GEOMETRICAL CONFIGURATION   rt
    Configuration of a mathematical model
    used in carrying out calculations for a
    nuclear reactor representing more or
    less the reactor envisaged.
f configuration *f* géométrique
e configuración *f* geométrica
i configurazione *f* geometrica
n geometrische configuratie
d geometrische Konfiguration *f*

3020 GEOMETRICALLY SAFE   rt
    Of a system containing fissile (fissionable)
    material, incapable of supporting a
    self-sustaining nuclear chain reaction by
    virtue of the geometric arrangement or
    shape of the components.
f géométriquement sûr
e geometricamente seguro
i geometricamente sicuro
n geometrisch veilig
d geometrisch sicher

3021 GEOMETRY   ge, ma
    The physical relationship and symmetry
    of the parts of an assembly.
f géométrie *f*
e geometría *f*
i geometria *f*
n geometrie *f*
d Geometrie *f*

3022 GEOMETRY CORRECTION   rt
The correction in geometry required to
allow for finite resolution in energy or
angle.
f   correction *f* de la géométrie
e   corrección *f* de la geometría
i   correzione *f* della geometria
n   geometriecorrectie
d   Geometriekorrektion *f*

3023 GEOMETRY FACTOR   ra
The average solid angle at a radiation
source that is subtended by the aperture
or sensitive volume of a radiation detector,
divided by the solid angle.
f   facteur *m* de géométrie
e   factor *m* de geometría
i   fattore *m* di geometria
n   geometriefactor
d   Geometriefaktor *m*

3024 GEOPHYSICAL PROSPECTING   mi
Prospecting by measuring differences in
the density, electrical resistance or
magnetic properties of the earth's crust.
f   prospection *f* géophysique
e   exploración *f* geofísica
i   prospezione *f* geofisica
n   geofysische prospectie
d   geophysikalische Schürfung *f*

3025 GEOTHERMIC ENERGY   ge
Energy released by rock formations after
being subjected to nuclear explosions.
f   énergie *f* géothermique
e   energía *f* geotérmica
i   energia *f* geotermica
n   geothermische energie
d   geothermische Energie *f*

3026 GERM CELL, see 2915

3027 GERMANIUM   ch
Metallic element, symbol Ge, atomic
number 32.
f   germanium *m*
e   germanio *m*
i   germanio *m*
n   germanium *n*
d   Germanium *n*

3028 GeV,
     GIGA-ELECTRON VOLT, see 632

3029 GIANT AIR SHOWER, see 440

3030 GIANT RESONANCE   cs
A resonance peak in the curve of photo-
nuclear cross section vs. photon energy,
which occurs at about 20 meV for heavy
nuclei.
f   crête *f* de résonance
e   cresta *f* de resonancia,
    pico *m* de resonancia
i   cresta *f* di risonanza
n   resonantiepiek
d   Resonanzspitze *f*

3031 GILPINITE   mi
A mineral containing copper and uranium.
f   gilpinite *f*
e   gilpinita *f*
i   gilpinite *f*
n   gilpiniet *n*
d   Gilpinit *m*

3032 G.I. TRACT, see 180

3033 GIVEN DOSE, see 2584

3034 GLANDULOGRAPHY   xr
Radiological examination of the glandulae.
f   glandulographie *f*
e   glandulografía *f*
i   glandulografia *f*
n   glandulografie
d   Glandulographie *f*

3035 GLASS DOSEMETER   me, ra
A dosemeter comprising a piece of glass
of a special kind which fluoresces under
ultraviolet light following gamma
irradiation.
f   dosimètre *m* à plaque en verre fluorescent,
    verre *m* dosimètre
e   dosímetro *m* con placa de vidrio
    fluorescente
i   dosimetro *m* a piastra di vetro
    fluorescente
n   dosismeter met fluorescerend glasplaatje
d   Dosismesser *m* mit fluoreszierender
    Glasplatte

3036 GLORY HOLE (US), see 561

3037 GLOVE BOX   pe
An enclosure in which material may be
manipulated in isolation from the operator's
environment.
f   boîte *f* à gants
e   caja *f* con guantes, caja *f* manipuladora
i   camera *f* a guanti, scatola *f* a guanti
n   handschoenkast
d   Handschuhbox *m*, Handschuhkasten *m*

3038 GLOVE PORT   pe
The hole in the wall of a glove box to
which the gauntlet of a glove is fastened.
f   entrée *f* de boîte, rond *m* de gant
e   orificio *m* para guantes
i   orificio *m* per guanti
n   kastinlaat
d   Kasteneinlass *m*

3039 GOLD   ch
Metallic element, symbol Au, atomic
number 79.
f   or *m*
e   oro *m*
i   oro *m*
n   goud *n*
d   Gold *n*

3040 GONAD,   md
 GONIDIA
 An ovary or testis, site of origin eggs or
 spermatozoa.
f glande *f* sexuelle, gonade *f*
e gónada *f*, gonidia *f*
i ghiandola *f* sessuale, gonada *f*
n geslachtsklier, gonade
d Geschlechtsdrüse *f*, Gonade *f*

3041 GOOD GEOMETRY   rt
 A geometry which entails small corrections.
f bonne géométrie *f*
e buena geometría *f*
i buona geometria *f*
n goede geometrie
d günstige Geometrie *f*, gute Geometrie *f*

3042 GOUDSMIT Γ-SUM RULE   ma
 For a given electron configuration, the
 sum of the values corresponding to a
 given value of the magnetic quantum
 number M is independent of the field
 strength H.
f règle *f* de totaux Γ de Goudsmit
e regla *f* de totales Γ de Goudsmit
i regola *f* di somma Γ di Goudsmit
n Γ-somregel *f* van Goudsmit
d Goudsmitscher Γ-Summensatz *m*

3043 GOUDSMIT-UHLENBECK   np
 ASSUMPTION
 The assumption of the existence of nuclear
 spin made by Goudsmit and Uhlenbeck
 in their explanation of the causes of
 multiple atomic spectra.
f postulat *m* de Goudsmit-Uhlenbeck
e postulado *m* de Goudsmit-Uhlenbeck
i postulato *m* di Goudsmit-Uhlenbeck
n veronderstelling van Goudsmit en
 Uhlenbeck
d Annahme *f* von Goudsmit und Uhlenbeck

3044 GRADED VACUUM SPARK GAP   cd
 Component part of the Columbus II
 thermonuclear installation.
f éclateur *m* échelonné à vide
e chispómetro *m* escalonado de vacío
i spinterometro *m* a gradinata a vuoto
n gradatievacuümvonkbrug
d gradierte Vakuumfunkenstrecke *f*

3045 GRADIENT   ge
 The rate at which a variable quantity
 increases or decreases.
f gradient *m*
e gradiente *m*
i gradiente *m*
n gradiënt
d Gradient *m*

3046 GRADIENT COUPLING   np
 Type of postulated coupling between
 nucleons and other particles, in which the
 interaction energy depends explicitly on
 the first order derivatives of the wave

functions with respect to position and time.
f couplage *m* à gradient
e acoplamiento *m* de gradiente
i accoppiamento *m* a gradiente
n gradiëntkoppeling
d Gradientenkupplung *f*

3047 GRADING   mi
 The classification of ores according to
 size.
f triage *m*
e clasificación *f*
i classificazione *f*
n classificeren *n*
d Klassieren *n*

3048 GRAIN, see 1563

3049 GRAIN,   ph
 PHOTOGRAPHIC GRAIN
 A small particle of metallic silver
 remaining in a photographic emulsion after
 developing and fixing.
f grain *m*
e grano *m*
i grano *m*
n korrel
d Korn *n*

3050 GRAIN BOUNDARY   cr
 The surface separating two regions of a
 solid in which the crystal axes are
 differently oriented.
f limite *f* de granulation
e límite *m* de grano,
 superficie *f* intergranular
i contorno *m* di grano
n korrelgrens
d Korngrenze *f*, Kristallitbegrenzung *f*

3051 GRAIN COUNTING   ph
 In the photographic emulsion technique,
 the method of determining the mass of a
 particle by counting the number of grains
 per unit length along its track.
f comptage *m* des grains
e recuento *m* de los granos
i conteggio *m* dei grani
n korreltelling
d Körnerzählung *f*

3052 GRAIN IMPLANTATION GUN   md, ra
 A gun-shaped appliance for implanting
 radioactive grains into the body.
f pistole *f* d'injection de grains radioactifs
e pistola *f* de inyección de granos radiactivos
i pistola *f* d'iniezione di grani radioattivi
n injectiepistool *n* voor radioactieve
 korrels
d Injektionspistole *f* für radioaktive
 Körner

3053 GRAIN NOISE,   ph
 TRACK NOISE
 Spurious scattering of a track in a nuclear
 emulsion due to the finite size of the grains
 in a track.

f trouble *m* de trace
e enturbiamiento *m* de traza
i torbido *m* di traccia
n spoorvertroebeling, spoorvervaging
d Spurtrübung *f*, Spurverwaschung *f*

3054 GRAININESS, ph
PHOTOGRAPHIC GRAININESS
Visible coarseness in a photographic
image under specified conditions, due to
silver grains.
f granulation *f*
e granulación *f*
i granitura *f*, granulosità *f*
n korreligheid
d Körnigkeit *f*

3055 GRAM-ATOM, un
GRAM-ATOMIC WEIGHT
The weight of a gram-atom.
f atome-gramme *m*
e átomo-gramo *m*
i grammo-atomo *m*
n gramatoom *n*
d Grammatom *n*

3056 GRAM-ELEMENT np, un
SPECIFIC ACTIVITY
Total radioactivity of a given isotope per
gram of element.
f activité *f* spécifique par élément-gramme
e actividad *f* específica por gramo de
elemento
i attività *f* specifica per grammo
d'elemento
n specifieke activiteit per gramelement
d spezifische Aktivität *f* je Grammelement

3057 GRAM-EQUIVALENT un
The gram-atomic weight of an element or
formula weight of a radical, divided by
its valence.
f gramme-équivalent *m*
e gramo-equivalente *m*
i grammo-equivalente *m*
n gramequivalent *n*
d Grammäquivalent *n*, Val *n*

3058 GRAM-MOLE, ch, un
GRAM-MOLECULE
That weight of a pure substance which,
when expressed in grams, gives the same
number as its molecular weight.
f mole *m*, molécule-gramme *m*
e molécula-gramo *m*
i grammo-molecola *f*
n grammolecule
d Grammol *n*, Grammolekül *n*, Mol *n*

3059 GRANULAR-GRAPHITE-COOLED rt
REACTOR
A nuclear reactor design in which
granular graphite serves as both moderator and coolant.
f réacteur *m* refroidi et modéré par
graphite granuleux
e reactor *m* enfriado y moderado de grafito
granular

i reattore *m* refrigerato e moderato a
grafite granulare
n reactor met korrelgrafiet als koelmiddei
en moderator
d Graphitkugelnreaktor *m*

3060 GRANULOCYTE, md
POLYMORPHONUCLEAR
LEUCOCYTE
Any cell containing conspicuous granules,
especially a leucocyte containing neutrophil, basophil or eosinophil granules in its
cytoplasm.
f granulocyte *m*,
leucocyte *m* granulé polynucléaire
e granulocito *m*
i granulocito *m*,
leucocito *m* polinucleato
n gekorreld wit bloedlichaampje *n*,
granulocyt
d granulierter Leukozyt *m*,
Granulozyt *m*, Hämozyt *m*

3061 GRANULOCYTOPENIA md
Decrease of number of granulocytes in the
blood.
f granulocytopénie *f*
e granulocitopenia *f*
i granulocitopenia *f*
n granulocytopenie
d Granulozytopenie *f*

3062 GRAPHITE mt
A form of carbon in which the atoms are
hexagonally arranged in planes.
f graphite *m*
e grafito *m*
i grafite *f*
n grafiet *n*
d Graphit *m*

3063 GRAPHITE LATTICE fu, rt
f réseau *m* d'éléments combustibles
modéré au graphite
e celosía *f* de elementos combustibles
moderada con grafito
i reticolo *m* ad elementi combustibili
moderato alla grafite
n met grafiet gemodereerd splijtstof-
rooster *n*
d graphitmoderiertes Brennstoffgitter *n*

3064 GRAPHITE MODERATED rt
REACTOR,
GRAPHITE REACTOR
A nuclear reactor in which the moderator
consists of graphite.
f réacteur *m* modéré au graphite
e reactor *m* moderado con grafito
i reattore *m* moderato alla grafite
n met grafiet gemodereerde reactor
d graphitmoderierter Reaktor *m*

3065 GRAPHITE PEBBLES fu
Balls of graphite about 1 inch in diameter
and used in heterogeneous enriched fuel
reactors as a reflector.
f billes *pl* de graphite

e bolas *pl* de grafito
i ciottoli *pl* di grafite,
   sfere *pl* di grafite
n grafietkogels *pl*
d Graphitkugeln *pl*

3066 GRAPHITE REFLECTOR   rt
f réflecteur *m* en graphite
e reflector *m* en grafito
i riflettore *m* in grafite
n grafietreflector
d Graphitreflektor *m*

3067 GRAVEYARD, see 819

3068 GRAVITATIONAL ENERGY   gp
   The energy inherent to gravitational
   force.
f énergie *f* de la gravitation
e energía *f* gravitacional
i energia *f* gravitazionale
n zwaartekrachtenergie
d Schwerkraftenergie *f*

3069 GRAVITATIONAL FIELD   gp
   A region in which a particle is subject
   to the gravitational force.
f champ *m* de la gravitation
e campo *m* gravitacional
i campo *m* gravitazionale
n zwaartekrachtveld *n*
d Schwerkraftfeld *n*

3070 GRAVITON   np
   Elementary quantum of the gravitational
   field.
f graviton *m*
e gravitón *m*
i gravitone *m*
n graviton *n*
d Graviton *n*

3071 GRAVITY SEGREGATION   ch, is
   Separation of materials by making use
   of their different weights.
f ségrégation *f* par gravité,
   séparation *f* par gravité
e segregación *f* por gravedad,
   separación *f* por gravedad
i segregazione *f* per gravità,
   separazione *f* per gravità
n scheiden *n* door zwaartewerking,
   uitzijgen *n* door zwaartewerking
d Ausseigern *n* durch Schwerewirkung,
   Trennung *f* durch Schwerewirkung

3072 GRAY,   rt
      GREY
   In reactor technology, of a body or medium,
   absorbing a significant part of, but not
   all, the neutrons of some specified energy
   incident on it.
f gris adj
e gris adj
i grigio adj
n grijs adj
d grau adj

3073 GRAY BODY   np
   A body which absorbs a great deal but not
   all of the incident neutrons of a specified
   energy.
f corps *m* gris
e cuerpo *m* gris
i corpo *m* grigio
n grijs lichaam *n*
d grauer Körper *m*

3074 GRAY CAVITY, see 757

3075 GREEN'S FUNCTION   ma
   If the integral operator of a kernel is so
   defined that it is the inverse of a
   differential operator, the kernel is known
   as the Green's function.
f fonction *f* de Green
e función *f* de Green
i funzione *f* di Green
n functie van Green
d Greensche Funktion *f*

3076 GREEN SALT,   ch
      URANIUM TETRAFLUORIDE
   A solid green compound, an intermediate
   product in the production of uranium
   hexachloride gas.
f tétrafluorure *m* d'uranium
e tetrafluoruro *m* de uranio
i tetrafluoruro *m* d'uranio
n uraniumtetrafluoride *n*
d Urantetrafluorid *n*

3077 GRENZ RAY TUBE,   xr
      GRENZ TUBE
   An X-ray tube having a window of low
   absorption permitting the transmission
   of X-rays produced at low voltages.
f tube *m* à rayons limite
e tubo *m* de rayos límite
i tubo *m* di raggi limite
n zachte röntgenbuis
d Grenzstrahlenröhre *f*

3078 GRENZ RAYS   ra, xr
   Electromagnetic rays produced when
   charged particles are decelerated.
f rayons *pl* limite
e rayos *pl* límite
i raggi *pl* limite
n zachte röntgenstralen *pl*
d Grenzstrahlen *pl*

3079 GRID   xr
   An assembly of strips of metal, opaque
   to X-rays, assembled edgewise and inter-
   leaved with material of low absorption.
f grille *f*
e rejilla *f*
i griglia *f*
n raster *n*
d Blende *f*, Rasterblende *f*

3080 GRID, see 2850

3081 GRID IONIZATION CHAMBER, see 2840

3082 GRID RATIO   xr
    1. In radiography, the ratio of the depth of
    the opaque strips of a grid, measured in
    the direction of the primary beam, to the
    spacing between them.
    2. In grid therapy, the ratio of the total
    area of holes to the total area of the grid.
f   rapport *m* de grille
e   proporción *f* reticular
i   rapporto *m* di griglia
n   rasterverhouding
d   Blendenverhältnis *n*

3083 GRID THERAPY,   xr
        RASTER THERAPY
    Radiotherapy in which the body surface
    is irradiated through a protective sheet
    having holes arranged in a regular
    pattern.
f   traitement *m* par grille
e   terapia *f* reticular
i   terapia *f* reticolare
n   rastertherapie
d   Rastertherapie *f*

3084 GRINDING   mi
    Ore treatment in order to facilitate the
    recovery of metal contained.
f   broyage *m*
e   machaqueo *m*, quebrantadura *f*
i   frantumazione *f*, macinazione *f*
n   malen *n*
d   Mahlen *n*, Zerkleinern *n*

3085 GROSS FLUX VARIATION,   rt
        MACROSCOPIC FLUX VARIATION
    In a reactor, the spatial flux variation
    which is determined by plotting the actual
    flux at corresponding points in a lattice of
    cells and drawing a smooth curve through
    these points.
f   variation *f* de flux de grande répartition,
    variation *f* de flux spatiale
e   variación *f* de flujo de gran repartición
i   variazione *f* di flusso di grande
    ripartizione
n   macroscopische fluxvariatie,
    ruimtelijke fluxvariatie
d   makroskopische Flussvariation *f*,
    räumliche Flussvariation *f*

3086 GROSS FLUX DISTRIBUTION,   rt
        MACROSCOPIC FLUX DISTRIBUTION
    Distribution of the neutron flux or of the
    power density in a multiplying system,
    when the system is considered as sub-
    divided in homogeneous regions.
f   distribution *f* de flux de grande répartition,
    distribution *f* de flux spatiale
e   distribución *f* de flujo de gran repartición
i   distribuzione *f* di flusso di grande
    ripartizione
n   macroscopische fluxdistributie,
    ruimtelijke fluxdistributie
d   makroskopische Flussverteilung *f*,
    räumliche Flussverteilung *f*

3087 GROTTHUS-DRAPER LAW   ra
    A physico-chemical law applicable to
    radiobiology which states that absorbed
    radiation alone is effective.
f   loi *f* de Grotthus-Draper
e   ley *f* de Grotthus-Draper
i   legge *f* di Grotthus-Draper
n   wet van Grotthus en Draper
d   Grotthus-Drapersches Gesetz *n*

3088 GROUND CONCENTRATION   sa
    Concentration of a contaminated effluent
    in the atmospheric air near the ground
    surface.
f   concentration *f* au sol
e   concentración *f* al suelo
i   concentrazione *f* al suolo
n   concentratie aan het grondoppervlak
d   Konzentration *f* am Erdboden

3089 GROUND DISPOSAL OF   rw
        EFFLUENT
    A method for the disposal of radioactive
    liquid waste into suitable strata in the
    ground.
f   décharge *f* d'effluent dans la terre
e   descraga *f* de efluente en la tierra
i   scarico *m* d'affluente nella terra
n   lozen *n* van de radioactieve afvoer in
    de aarde
d   Beseitigung *f* der aktiven Ausströmung in
    der Erde

3090 GROUND STATE,   np
        NORMAL ENERGY LEVEL,
        NORMAL STATE
    The state of a quantized system, e.g. a
    nucleus, atom or molecule, which is that
    of lowest energy.
f   état *m* fondamental, niveau *m* normal
e   estado *m* fundamental,
    nivel *m* energético normal
i   livello *m* fondamentale,
    stato *m* fondamentale
n   grondtoestand, nulniveau *n*
d   Grundzustand *m*, Normalzustand *m*

3091 GROUND STATE BETA   np
        DISINTEGRATION
    The total energy released in a beta
    transition between isobars in their ground
    states.
f   énergie *f* de désintégration bêta dans
    l'état fondamental
e   energía *f* de desintegración beta en el
    estado fundamental
i   energia *f* di disintegrazione beta nello
    stato fondamentale
n   bêtadesintegratie-energie in de grond-
    toestand
d   Betazerfallsenergie *f* im Grundzustand

3092 GROUND STATE   np
        DISINTEGRATION ENERGY
    The disintegration energy when all reactant
    and product nuclei are in their ground
    states.

f énergie *f* de désintégration dans l état
fondamental
e energía *f* de desintegración en el estado
fundamental
i energia *f* di disintegrazione nello stato
fondamentale
n desintegratie-energie in de grondtoestand
d Zerfallsenergie *f* im Grundzustand

3093 GROUND SURVEY   mi
General examination of an area, e.g. for
locating ores.
f levé *m* terrestre
e levantamiento *m* terrestre
i rilevamento *m* di terreni
n terreinonderzoek *n*
d Geländeaufnahme *f*

3094 GROUND ZERO (GB),   nw
HYPOCENTER (US),
SURFACE ZERO
The region on the surface of land and
water vertically below or above the
centre(er) of a burst of a nuclear weapon.
f hypocentre *m*
e hipocentro *m*
i ipocentro *m*
n hypocentrum *n*
d Hypozentrum *n*

3095 GROUNDWATER LEVEL,   rw
GROUNDWATER TABLE
f niveau *m* de la nappe d'eau souterraine
e nivel *m* de la napa freática,
nivel *m* freático
i livello *m* dell'acqua sotterranea,
stato *m* dell'acqua sotterranea
n grondwaterstand
d Grundwasserspiegel *m*,
Grundwasserstand *m*

3096 GROUNDWATER STRATIFICATION is
INVESTIGATION
f investigation *f* du niveau de la nappe
d'eau souterraine
e investigación *f* del nivel de la napa
freática
i investigazione *f* del livello dell'acqua
solterranea
n onderzoek *n* van de grondwaterstand
d Untersuchung *f* des Grundwasserspiegels

3097 GROUP, see 2517

3098 GROUP DIFFUSION METHOD,   rt
GROUP METHOD
A theoretical treatment of nuclear
reactors in which it is postulated that the
energy of the neutrons, from the source
energy to the thermal energy, can be
divided into a finite set of energy intervals
or groups.
f méthode *f* de groupes
e método *m* de grupos
i metodo *m* di gruppi
n groepenmethode
d Gruppentheorie *f*

3099 GROUP REMOVAL CROSS   cs
SECTION
The weighted average cross section,
characteristic of an energy group, that will
account for the removal of neutrons from
that group by all processes.
f section *f* efficace d'extraction de groupe
e sección *f* eficaz de remoción de grupo
i sezione *f* d'urto di rimozione di gruppo
n groepdoorsnede voor verwijdering
d Gruppenverlustquerschnitt *m*

3100 GROUP TRANSFER   cs
SCATTERING CROSS SECTION
The weighted average cross section,
characteristic of the energy group structure, that will account for the transfer
of neutrons by scattering from one
specified group to another specified group.
f section *f* efficace de transfert de groupe
par diffusion
e sección *f* eficaz de transferencia de grupo
por dispersión
i sezione *f* d'urto di trasporto di gruppo
per deviazione
n intergroepdoorsnede voor verstrooiing
d Gruppenübergangsquerschnitt *m*

3101 GROWTH,   mg
IRRADIATION GROWTH,
RADIATION GROWTH
Of materials, the change in shape under
zero external stress without significant
change in volume, which may occur during
irradiation or during thermal cycling.
f grandissement *m* par rayonnement
e crecimiento *m* por radiación
i accrescimento *m* di radiazione
n stralingsgroei
d Strahlungswachstum *m*

3102 GROWTH CURVE   np
An activity which grows through the decay
of the parent substance, or as a result of
irradiation, plotted against time.
f courbe *f* de grandissement
e curva *f* de crecimiento
i curva *f* d'accrescimento
n groeikromme,
kromme voor activiteitsstijging
d Anstiegskurve *f*, Wachstumskurve *f*

3103 GROWTH SPIRAL   cr
A structure observed on the surface of
crystals, after growth.
f spirale *f* de croissance
e espiral *m* de crecimiento
i spirale *f* di crescita
n groeispiraal
d Wachstumsspirale *f*

3104 GROWTH STOP   cr
A ledge, one or more lattice spacings high,
on the surface of a crystal, where crystal
growth may take place.
f échelon *m* de croissance,
gradin *m* de croissance
e escalón *m* de crecimiento

i   gradino *m* di crescita
n   groeitrede
d   Wachstumstreppe *f*

3105 GUARD CIRCUIT,   rt, sa
      SAFETY CIRCUIT
      An electronic circuit incorporated in the
      instrumentation of a nuclear reactor.
f   circuit *m* de sécurité
e   circuito *m* de seguridad
i   circuito *m* di sicurezza
n   veiligheidsschakeling
d   Sicherheitsschaltung *f*

3106 GUARD RING   ct
      An auxiliary electrode in a counter tube
      or ionization chamber to control potential
      gradients, reduce insulator leakage
      and/or define the sensitive volume.
f   anneau *m* de garde,
      électrode *f* auxiliaire
e   anillo *m* de guarda, electrodo *m* auxiliar
i   anello *m* di guardia,
      elettrodo *m* ausiliario
n   schutring *m*, waakring *m*
d   Hilfselektrode *f*, Schutzring *m*

3107 GUARD TUBE   ct
      A metal tube around the end of the anode
      wire of a cylindrical ionization chamber
      or proportional counter.
f   tube *m* de garde
e   tubo *m* de guarda
i   tubo *m* di guardia
n   schutbuis, waakbuis
d   Schutzrohr *n*

3108 GUARDING COUNTER,   ct
      SHIELDING COUNTER
      A counter mounted in such a way as to
      cover a measuring counter.
f   compteur *m* de garde
e   contador *m* de guarda
i   contatore *m* di guardia
n   waakteller
d   Schutzzähler *m*

3109 GUIDE FIELD,   pa
      GUIDING FIELD
      The magnetic flux that holds particles in
      a stable orbit during the acceleration
      period in a betatron or synchrotron.
f   champ *m* de guidage,
      champ *m* magnétique guide

e   campo *m* magnético de guía
i   campo *m* magnetico di guida
n   geleidingsveld *n*, leiveld *n*
d   Führungsfeld *n*

3110 GUIDE SHIELD   pp
      Component part of the Ixion thermonuclear
      installation.
f   écran *m* de guidage
e   pantalla *f* de guía
i   schermo *m* di guida
n   leischerm *n*
d   Führungsschild *m*

3111 GUMMITE   mi
      A mineral containing lead, calcium,
      barium and uranium.
f   gummite *f*
e   gumita *f*
i   gummite *f*
n   gummiet *n*
d   Gummit *m*

3112 GUNK   ge
      A term denoting an undesirable nondes-
      cript material usually semi-solid.
f   crasse *f*
e   mugre *f*
i   lordura *f*
n   smurrie
d   Matsch *m*

3113 GURWITSCH RAYS,   ra
      MITOGENIC RADIATION
      Electromagnetic radiation, the dubious
      existence of which was postulated by
      Gurwitsch and supposed to have origin in
      tissue cells as the result of mitotic
      division.
f   rayonnement *m* mitogénétique
e   radiación *f* mitogenética
i   radiazione *f* mitogenetica
n   mitogenetische straling
d   mitogenetische Strahlung *f*

3114 GYROMAGNETIC RATIO   ma
      The ratio of the magnetic moment of a
      system to its angular momentum.
f   rapport *m* gyromagnétique
e   relación *f* giromagnética
i   rapporto *m* giromagnetico
n   gyromagnetische verhouding
d   gyromagnetisches Verhältnis *n*

3115 GYROMAGNETIC RESONANCE,
      see 1599

H

3116 H BAR, see 1851

3117 H BOMB, see 2899

3118 HAFNIUM   ch
Metallic element, symbol Hf, atomic
number 72.
f hafnium *m*
e hafnio *m*
i afnio *m*
n hafnium *n*
d Hafnium *n*

3119 HAHN TECHNIQUE   ra
The measurement of emanating power to
study physical changes in solids.
f méthode *f* de Hahn
e método *m* de Hahn
i metodo *m* di Hahn
n methode van Hahn
d Hahnsches Verfahren *n*

3120 HALF LIFE,   np
HALF-VALUE PERIOD,
RADIOACTIVE HALF LIFE
The time in which the amount of a radioactive nuclide decays to half its initial
value.
f période *f* radioactive
e período *m* radiactivo
i periodo *m* radioattivo
n halveringstijd
d Halbwertzeit *f*

3121 HALF THICKNESS,   np
HALF-VALUE LAYER,
HALF-VALUE THICKNESS,
HVL, HVT
The thickness of the attenuating layer that
reduces the current density to one-half of
its initial value.
f CDA *f*, couche *f* de demi-atténuation
e capa *f* de semiatenuación, CSA *f*,
espesor *m* semirreductor
i spessore *m* di dimezzamento,
strato *m* di semiattenuazione
n halveringsdikte
d Halbwertschicht *f*, Halbwertdicke *f*

3122 HALF TIME OF EXCHANGE   ch
The time required for half the net
realizable exchange of atoms in a chemical
exchange reaction to take place.
f demi-période *f* d'échange
e semiperíodo *m* de intercambio
i semiperiodo *m* di scambio
n uitwisselingshalveringstijd
d Austauschhalbwertzeit *f*

3123 HALF WIDTH   sp
The full width of an energy peak in a
spectrum as measured at half amplitude

from the baseline and expressed either as
an absolute energy or as a percentage of
the average value.
f largeur *f* de moitié
e anchura *f* media
i larghezza *f* di valore medio
n halveringsbreedte
d Halbwertbreite *f*

3124 HALIDE-FREON   sa
LEAK DETECTOR
A leak detector filled with freon, a
halogenated compound.
f détecteur *m* de fuites à remplissage
d'hydrocarbure halogéné
e detector *m* de fugas con relleno
de hidrocarburo halogenado
i rivelatore *m* di fughe con riempimento
d'idrocarburo alogenato
n lekdetector met vulling van gehalogeneerde
koolwaterstof
d Leckdetektor *m* mit Füllung aus
halogeniertem Kohlenwasserstoff

3125 HALL COEFFICIENT,   ma
HALL CONSTANT
The constant of proportionality R in the
relation $E_h = RJ \times H$, where E_h is the
transverse electric field (Hall field), J
is the current density and H is the
magnetic field.
f constante *f* de Hall
e constante *f* de Hall
i costante *f* di Hall
n hallconstante
d Hall-Konstante *f*

3126 HALL EFFECT   ec
The development of a transverse, electric
potential gradient in a current carrying
conductor upon the application of a
magnetic field.
f effet *m* Hall
e efecto *m* Hall
i effetto *m* Hall
n halleffect *n*
d Hall-Effekt *m*

3127 HALL EFFECT   ec, me
MAGNETOMETER
A sub-assembly for the rapid measurement
of usual magnetic field by an electronic
method based on the measurement of the
difference of potential produced in a
semiconductor crystal by the Hall effect.
f magnétomètre *m* à effet Hall
e magnetómetro *m* por efecto Hall
i magnetometro *m* ad effetto Hall
n hallmagnetometer
d Hall-Magnetometer *n*

3128 HALL MOBILITY, see 898

3129 HALL PROBE   pp
A magnetic probe used in studying plasmas
based on the Hall effect.
f sonde f de Hall
e sonda f de Hall
i sonda f di Hall
n hallsonde
d Hall-Sonde f

3130 HALLWACHS EFFECT   ra
The ability of ultraviolet radiation to
discharge a negatively charged body in
a vacuum.
f effet m Hallwachs
e efecto m Hallwachs
i effetto m Hallwachs
n hallwachseffect n
d Hallwachs-Effekt m

3131 HALMATOGENESIS,   md
      MUTATION,
      SALFATORY EVOLUTION,
      TRANSGENATION
Gradual variation towards a definite
change of structure.
f mutation f
e halmatogénesis f, mutación f
i mutazione f
n mutatie
d Mutation f

3132 HALOGEN   ch
A member of the seventh group of elements,
i.e. those lacking just one electron to
make a closed outer shell.
f halogène m
e halógeno m
i alogeno m
n halogeen n
d Halogen n

3133 HALOGEN COUNTER TUBE,   ct
      HALOGEN-QUENCHED
      COUNTER TUBE
A self-quenched counter tube in which
quenching is obtained by the addition of
a halogen, generally bromine, to the rare
gas filling.
f tube m compteur à halogène
e tubo m contador de halógeno
i tubo m contatore ad alogeno
n halogeentelbuis,
   telbuis met halogeenbijvulling
d Halogenzählrohr n,
   Zählrohr n mit Halogenzugabe

3134 HAMMER TRACKS   ph
Tracks produced in nuclear emulsions
which are attributed to the expulsion of a
Li-8 nucleus from a heavier nucleus as
a result of cosmic ray bombardment.
f traces pl dues au noyau Li-8,
   traces pl en forme de marteau
e trazas pl debidas al nucleo de Li-8,
   trazas pl en forma de martillo
i traccie pl dovute al nucleo di Li-8,
   traccie pl in forma di martello

n hamervormige sporen pl,
   sporen pl van de Li-8-kern
d hammerförmige Spuren pl,
   Li-8-Kernspuren pl

3135 HAND AND FOOT MONITOR   me, sa
A radiation monitor designed to measure
the amount of radioactive contamination
on hands and feet.
f moniteur m pour pieds et mains
e monitor m para manos y pies
i avvisatore m per mani e piedi
n handen- en voetenmonitor
d Monitor m zur Prüfung von Händen und
   Füssen

3136 HANSEN LINEAR ACCELERATOR   pa
A linear accelerator that produces very
high energy electrons.
f accélérateur m linéaire de Hansen
e acelerador m lineal de Hansen
i acceleratore m lineare di Hansen
n lineaire versneller volgens Hansen
d Linearbeschleuniger m nach Hansen

3137 HANSON AND MCKIBBEN   ct
      COUNTER,
      LONG COUNTER
A detector for fast neutrons in which a
proportional counter containing BF_3 is
surrounded by a large cylinder of wax.
f compteur m à enveloppe cylindrique en cire
e contador m de envoltura cilíndrica en
   cera
i contatore m ad involucro cilindrico in
   cera
n teller met cylindrische wasomhulling
d Zähler m mit zylindrischer Wachshülle

3138 HAPLOID   md
Having a single complete set of chromosomes, the condition of the gamete nucleus.
f haploïde adj
e haploide adj
i aploide adj
n haploïde adj
d haploid adj

3139 HARD   ra, xr
A qualitative term describing the ability
of radiation to penetrate matter.
f dur adj
e duro adj
i duro adj
n hard adj
d hart adj

3140 HARD COMPONENT OF   cr
      COSMIC RAYS
That portion of cosmic radiation which
penetrates a moderate thickness of an
absorber.
f composante f dure de rayons cosmiques
e componente m duro de rayos cósmicos
i componente m duro di raggi cosmici
n harde component van kosmische stralen
d harte Komponente f von kosmischen
   Strahlen

3141 HARD-CORE GEOMETRY   np
Elementary shape of the configuration of
the containing magnetic field in thermonuclear research.
f géométrie f à noyau dur
e geometría f de núcleo duro
i geometria f a nucleo duro
n opstelling met harde kern
d Anordnung f mit festem Kern

3142 HARD-CORE PINCH DEVICE,   pp
TUBULAR PINCH DEVICE
A discharge device containing a central
solid conductor, whereas the discharge
takes place in an annular space
surrounding the conductor.
f dispositif m à pincement tubulaire
e dispositivo m de estricción tubular
i dispositivo m a contrazione tubolare
n plasmavat n met buisvormige insnoering
d Plasmagefäss n mit röhrenförmiger
Einschnürung

3143 HARD RADIATION,   ra
PENETRATING RADIATION
Ionizing radiation of short wavelength
and high penetration.
f rayonnement m dur
e radiación f dura
i radiazione f dura
n harde straling
d harte Strahlung f

3144 HARD X-RAY   xr
An X-ray having high penetrating power.
f rayon m X dur
e rayo m X duro
i raggio m X duro
n harde röntgenstraal
d harter Röntgenstrahl m

3145 HARDENING   ra
Of radiation, to increase the average
energy of a radiation beam by filtering
out the softer or lower energy components.
f durcissement m
e endurecimiento m
i indurimento m
n harding
d Härtung f

3146 HARDNESS   mg, xr
1. A property defining resistance to
deformation.
2. A term for qualitatively specifying the
penetrating power of X-rays.
f dureté f
e dureza f
i durezza f
n hardheid, hardte
d Härte f

3147 HARTREE METHOD,   ma
SELF-CONSISTENT FIELD METHOD
A method of treating the problem of the
many-eletron atom in which the eigenfunction for the system is represented as

the product of 2 single-electron functions
which must satisfy specified conditions.
f méthode f de Hartree
e método m de Hartree
i metodo m di Hartree
n hartreemethode
d Hartree-Verfahren n

3148 HARTREE UNITS, see 424

3149 HATCHETTOLITE   mi
A mineral containing a decomposed
pyrochlorine compound.
f hatchettolite f
e hatchettolita f
i hatchettolite f
n hatchettoliet n
d Hatchettolit m

3150 HAZARD REPORT,   sa
SAFETY REPORT
Report accompanying a project for a
nuclear plant and containing an analysis
of all hazards inherent to this plant with
instructions how to minimize risks.
f manuel m de sécurité
e guía f de seguridad,
manual m de seguridad
i relazione f sulla sicurezza
n gevarenhandleiding,
veiligheidsvademecum n
d Sicherheitsvorschriften pl

3151 HEAD AMPLIFIER,   ec
PRE-AMPLIFIER
A unit containing the input stage of an
amplifying system and designed to be
mounted integrally with, or close to, the
detector.
f pré-amplificateur m
e preamplificador m
i preamplificatore m
n voorversterker
d Vorverstärker m

3152 HEAD-END TREATMENT   mt, rt
Mechanical and chemical operations
carried out on irradiated fuel elements in
the head section of the processing plant.
f opérations pl initiales,
traitement m d'entrée
e operaciones pl iniciales,
tratamiento m de entrada
i operazioni pl iniziali,
trattamento m di testa
n ingangsbewerkingen pl
d Eingangsbearbeitungen pl

3153 HEAD OF A LIQUID   gp
Pressure expressed as the height of the
liquid column necessary to develop that
pressure at the base.
f hauteur f de pression de liquide
e altitud f de presión de líquido
i altezza f di pressione di liquido
n drukhoogte van een vloeistof
d Druckhöhe f einer Flüssigkeit

3154 HEAD-ON COLLISION, see 943

3155 HEALTH HAZARD,   md, sa
     RADIATION HAZARD
The danger to health arising from
exposure to ionizing radiation.
f risque *m* d'irradiation
e riesgo *m* de irradiación
i rischio *m* d'irradiazione
n stralingsgevaar *n*
d Strahlengefährdung *f*

3156 HEALTH MONITOR,   me, ra
     RADIATION MONITOR
A radiation measuring assembly provided
with devices intended to draw attention
to an event or situation which may result
in harmful consequences.
f moniteur *m* de radioprotection,
   moniteur *m* de rayonnement
e monitor *m* de radiación,
   vigía *f* de radiación
i avvisatore *m* di radiazione,
   segnalatore *m* di radiazione
n stralingsmonitor, stralingsverklikker
d Strahlenüberwachungsgerät *n*,
   Strahlungsmonitor *m*,
   Strahlungswarngerät *n*

3157 HEALTH PHYSICS   np, sa
The branch of physics dealing with
protection against radiation.
f radioprotection *f*,
   science *f* de la protection contre les
   rayonnements ionisants
e física *f* médica y sanitaria,
   ciencia *f* de la protección contra las
   radiaciones ionizantes
i fisica *f* sanitaria
n leer van de stralingsbescherming
d Gesundheitsphysik *f*,
   Strahlenschutzphysik *f*

3158 HEALTH PHYSICS   me, sa
     LOG RATEMETER
Used in association with an ionization
chamber in monitoring applications in the
vicinity of reactors and accelerators.
f ictomètre *m* logarithmique de radioprotection
e impulsímetro *m* de física médica y
   sanitaria
i rateometro *m* logaritmico di fisica
   sanitaria
n logarithmische tempometer voor
   stralingsbescherming
d logarithmischer Leistungsmesser *m* für
   Gesundheitsphysik

3159 HEAT,   gp
     THERMAL ENERGY
Energy possessed by a substance in the
form of kinetic energy of molecular
translation, rotation and vibration.
f chaleur *f*, énergie *f* thermique
e calor *m*, energía *f* térmica
i calore *m*, energia *f* termica

n thermische energie, warmte
d thermische Energie *f*, Wärme *f*

3160 HEAT BALANCE   rt
A heat-energy account drawn up for a
boiler, engine trial, nuclear reactor, etc.
showing how the heat energy supplied is
expended by the plant.
f bilan *m* thermique
e balance *m* térmico
i bilancio *m* termico
n warmtebalans
d Wärmebilanz *f*

3161 HEAT CONDUCTION, see 1297

3162 HEAT CONDUCTIVITY,   gp
     THERMAL CONDUCTIVITY
The process of heat transport through a
substance, excluding heat transfer due to
mass flow in the substance.
f conductivité *f* thermique
e conductividad *f* térmica
i conduttività *f* termica
n warmtegeleidingsvermogen *n*
d Wärmeleitfähigkeit *f*

3163 HEAT ENGINE   gp
An engine in which air or gas is heated
and cooled in a closed cycle.
f moteur *m* à air chaud
e motor *m* de aire caliente,
   motor *m* térmico
i motore *m* ad aria calda
n heetgasmotor, hete-luchtmotor
d Heissluftmotor *m*

3164 HEAT EXCHANGER   ch, rt
An apparatus for transferring heat from
one medium to another.
f échangeur *m* de chaleur
e cambiador *m* de calor,
   intercambiador *m* de calor
i scambiatore *m* di calore
n warmte-uitwisselaar, warmtewisselaar
d Wärmeaustauscher *m*

3165 HEAT EXCHANGER LEAK   rt, sa
     DETECTION PROBE
f sonde *f* détectrice de pertes d'échangeur
   de chaleur
e sonda *f* detectora de pérdidas de
   cambiador de calor
i sonda *f* rivelatrice di perdite di
   scambiatore di calore
n lekzoeker voor warmte-uitwisselaar
d Wärmeaustauscherlecksucher *m*

3166 HEAT FLUSH   is
A method of separating isotopes by using
superfluidity, depending on the fact that
a temperature gradient produces a flow
of helium-4 and not of helium-3.
f coup *m* de chaleur
e golpe *m* de calor
i colpo *m* di calore
n warmtestoot
d Wärmestoss *m*

3167 HEAT OF COMBUSTION, see 1246

3168 HEAT OF EMISSION   ec
The extra amount of energy to be supplied
to an electron emitting surface to keep it
at a constant temperature.
f chaleur f d'émission
e calor m de emisión
i calore m d'emissione
n emissiewarmte
d Emissionswärme f

3169 HEAT OF EVAPORATION,
HEAT OF VAPORIZATION,
see 2386

3170 HEAT OF FORMATION, see 2785

3171 HEAT OF FUSION, see 2902

3172 HEAT OF RADIOACTIVITY   np
The quantity of heat generated by radioactive decay per unit mass per unit time.
f chaleur f de décroissance,
chaleur f de désintégration
e calor m de decaimiento,
calor m de desintegración
i calore m di decadimento,
calore m di disintegrazione
n desintegratiewarmte
d Zerfallswärme f

3173 HEAT OUTPUT OF A REACTOR,   rt
THERMAL OUTPUT OF A REACTOR
The output power of a nuclear reactor
expressed in thermal kW.
f puissance f thermique d'un réacteur
e potencia f térmica de un reactor
i potenza f termica d'un reattore
n thermisch vermogen n van een reactor
d Wärmeleistung f eines Reaktors

3174 HEAT PRODUCTION REACTOR,   rt
HEAT REACTOR,
INDUSTRIAL HEAT REACTOR,
PROCESS HEAT REACTOR
A nuclear reactor designed primarily
to supply heat energy for industrial
purposes.
f réacteur m de production de chaleur
e reactor m de producción de calor
i reattore m di produzione di calore
n reactor voor warmteproduktie
d Reaktor m zur Wärmeerzeugung

3175 HEAT RELEASE, see 135

3176 HEAT TRANSFER,   gp
HEAT TRANSMISSION
The transfer of heat from, e.g., one point
to another.
f transfert m de chaleur,
transmission f de chaleur
e termotransmisión f,
transmisión f de calor
i trasmissione f di calore
n warmteoverdracht
d Wärmeübertragung f

3177 HEAT TRANSFER COEFFICIENT   gp
The rate of flow of heat through a medium
or a system, expressed as the amount of
heat passing through unit area, per unit
time, and per degree temperature
difference.
f coefficient m de transmission de chaleur
e coeficiente m de transmisión de calor
i coefficiente m di trasmissione di calore
n warmteoverdrachtscoëfficiënt
d Wärmeübertragungskoeffizient m

3178 HEAT TRANSFER CYCLE   gp
A closed circuit of heat transfer used, e.g.,
in the hot air engine.
f cycle m de transmission de chaleur
e ciclo m de transmisión de calor
i ciclo m di trasmissione di calore
n warmteoverdrachtskringloop
d Wärmeübertragungskreislauf m

3179 HEAT TRANSFER FLUID   gp
A fluid used as a heat transfer medium in
high temperature processes.
f fluide m évacuateur de chaleur
e flúido m evacuador de calor
i fluido m scambiatore di calore
n warmte overdragend medium n
d Wärmeübertragungsflüssigkeit f

3180 HEAT TRANSFER RIG   cd
The complete arrangement for heat
transfer.
f équipement m de transmission de chaleur
e equipo m de transmisión de calor
i impianto m di trasmissione di calore
n warmteoverdrachtsopstelling
d Wärmeübertragungsaufstellung f

3181 HEAT TREATMENT   mg
A combination of timed heating and cooling
operation performed upon a solid metal
or alloy in order to obtain desired
structure or properties.
f traitement m thermique
e tratamiento m térmico
i trattamento m termico
n warmtebehandeling
d Wärmebehandlung f

3182 HEATING TUBE   cd
A conduit as a component part of a
reactor filled with liquid or gas at a high
temperature.
f tube m de chauffe, tuyau m de chauffe
e tubo m de caldeo, tubo m de calefacción
i tubo m di riscaldamento
n verhittingspijp
d Heizrohr n

3183 HEAVY AGGREGATE   mt, sa
CONCRETE,
HEAVY CONCRETE,
LOADED CONCRETE
A concrete containing an aggregate whose
density is greater than that of gravel, e.g.
barytes, iron ore, steel shot.
f béton m lourd

e hormigón *m* pesado
i calcestruzzo *m* pesante
n verzwaard beton *n*
d Schwerbeton *m*

3184 HEAVY AGGREGATE   rt
       CONCRETE SHIELD,
         HEAVY CONCRETE SHIELD,
         LOADED CONCRETE SHIELD
       A shield made of loaded concrete.
f bouclier *m* en béton lourd
e blindaje *m* de hormigón pesado
i schermo *m* in calcestruzzo pesante
n verzwaard-betonscherm *n*
d Schwerbetonschild *m*

3185 HEAVY ALLOY,   mg
         TUNGSTEN ALLOY
       A protective alloy containing tungsten,
       copper and nickel.
f alliage *m* de tungstène, alliage *m* lourd
e aleación *f* de wolframio,
   aleación *f* pesada
i lega *f* di wolframio, lega *f* pesante
n legering van zware metalen,
   wolfraamlegering
d Schwermetallegierung *f*,
   Wolframlegierung *f*

3186 HEAVY ANODE,   xr
         SOLID ANODE
       An anode of an X-ray tube consisting of
       a solid piece of metal.
f anode *f* massive
e ánodo *m* macizo
i anodo *m* pieno
n massieve anode
d massive Anode *f*, Vollanode *f*

3187 HEAVY ATOM   ch
       An atom of an element with a rather high
       atomic weight.
f atome *m* lourd
e átomo *m* pesado
i atomo *m* pesante
n zwaar atoom *n*
d schweres Atom *n*

3188 HEAVY ATOM METHOD, see 1858

3189 HEAVY COSMIC-RAY TRACKS   ic, ph
       Tracks found in cloud chambers and
       emulsions exposed to cosmic radiation at
       very high altitudes.
f traces *pl* de particules cosmiques lourdes
e trazas *pl* de partículas cósmicas pesadas
i traccie *pl* di particelle cosmiche pesanti
n sporen *pl* van zware kosmische deeltjes
d Spuren *pl* von schweren kosmischen
   Teilchen

3190 HEAVY ELEMENT   ch
       A term often used with reference to the
       actinide elements, especially thorium and
       uranium.
f élément *m* lourd
e elemento *m* pesado

i elemento *m* pesante
n zwaar element *n*
d schweres Element *n*

3191 HEAVY ELEMENT CHEMISTRY   ch
f chimie *f* des éléments lourds
e química *f* de los elementos pesados
i chimica *f* degli elementi pesanti
n chemie van de zware elementen
d Chemie *f* der schweren Elementen

3192 HEAVY FRACTION   ch, is
       A fraction which has a smaller abundance
       of the light isotope.
f fraction *f* lourde
e fracción *f* pesado
i frazione *f* pesante
n zware fractie
d schwere Fraktion *f*

3193 HEAVY HYDRIDE, see 1754

3194 HEAVY HYDROGEN, see 1755

3195 HEAVY HYDROGEN PLANT   ch
f installation *f* d'hydrogène lourd
e equipo *m* de deuterio,
   equipo *m* de hidrógeno pesado
i impianto *m* d'idrogeno pesante
n zware-waterstofinstallatie
d Schwerwasserstoffanlage *f*

3196 HEAVY ICE   ch
       Frozen D_2O, used, e.g., in apparatus for
       generating fast neutrons.
f glace *f* lourde
e hielo *m* pesado
i ghiaccio *m* pesante
n zwaar ijs *n*
d schweres Eis *n*

3197 HEAVY ION   ch, is
       A term used with reference to such ions
       as are formed by carbon-12, carbon-13,
       nitrogen-14, oxygen-16, etc.
f ion *m* lourd
e ión *m* pesado
i ione *m* pesante
n zwaar ion *n*
d schweres Ion *n*

3198 HEAVY ION LINEAR   pa
         ACCELERATOR,
         HILAC
       A linear accelerator to be used for
       chemical transmutation.
f accélérateur *m* linéaire d'ions lourds
e acelerador *m* lineal de iones pesados
i acceleratore *m* lineare d'ioni pesanti
n lineaire zware-ionenversneller
d Schwerionen-Linearbeschleuniger *m*

3199 HEAVY ISOTOPE   is
       An isotope of an element of which the
       weight is higher than of another isotope
       of the same element.
f isotope *m* lourd

e isótopo *m* pesado
i isotopo *m* pesante
n zwaar isotoop *n*
d schweres Isotop *n*

3200 HEAVY/LIGHT WATER HEAT   ma
      EXCHANGER LEAK MONITOR
A monitor designed to detect leakages
between the primary heavy water coolant
circuit and the secondary light water
coolant circuit of a nuclear reactor, by
detecting radioactivity in the secondary
circuit.
f moniteur *m* de pertes entre circuits d'un
   échangeur à eau lourde/eau légère
e vigía *f* de pérdidas para cambiadores
   agua pesada/agua ligera
i monitore *m* di perdite nello scambiatore
   di calore ad acqua pesante/acqua leggera
n monitor voor lekken in een warmtewisselaar met licht en zwaar water
d Dichtheitsmonitor *m* für Schwer/Leicht-
   wasser-Wärmeaustauscher

3201 HEAVY MESON,   cr
      K MESON, KAON
The class of heavier mesons which are
unstable elementary particles, resulting
from the split-up of atomic nuclei, and
are present in cosmic radiation.
f kaon *m*, méson *m* K, méson *m* lourd
e mesón *m* K, mesón *m* pesado
i mesone *m* K, mesone *m* pesante
n K-meson *n*, zwaar meson *n*
d K-Meson *n*, schweres Meson *n*

3202 HEAVY NUCLEUS   np
The nucleus of the heavy elements.
f noyau *m* lourd
e núcleo *m* pesado
i nucleo *m* pesante
n zware kern
d schwerer Kern *m*

3203 HEAVY OXYGEN   ch, is
A denomination for the isotopes of oxygen.
f oxygène *m* lourd
e oxígeno *m* pesado
i ossigeno *m* pesante
n zware zuurstof
d schwerer Sauerstoff *m*

3204 HEAVY PARTICLE   np
A term used in atomic and nuclear physics
to refer to a particle having a mass equal
to or greater than that of a photon.
f particule *f* lourde
e partícula *f* pesada
i particella *f* pesante
n zwaar deeltje *n*
d schweres Teilchen *n*

3205 HEAVY PARTICLE SYNCHROTRON pa
A synchrotron in which the acceleration
of the heavy particles, usually protons, is
realized by means of an electric field of
variable frequency, the particles being

maintained in a stable orbit by a guide
field variable as well.
f synchrotron *m* à particules lourdes
e sincrotrón *m* de partículas pesadas
i sincrotrone *m* a particelle pesanti
n synchrotron *n* met zware deeltjes
d Synchrotron *n* mit schweren Teilchen

3206 HEAVY WATER   rt
1. Deuterium oxide.
2. Water containing an appreciable quantity
of deuterium in the form of D_2O or HDO.
f eau *f* lourde
e agua *f* pesada
i acqua *f* pesante
n zwaar water *n*
d schweres Wasser *n*

3207 HEAVY WATER BOILING   rt
      REACTOR
f réacteur *m* à eau lourde bouillante
e reactor *m* de agua pesada hirviente
i reattore *m* ad acqua pesante in
   ebollizione
n kokend-zwaar-waterreactor
d Schwerwassersiedereaktor *m*

3208 HEAVY WATER CONTENT   me
      METER
A content meter designed for continuously
or discontinuously determining the heavy
water content of a heavy-light water
mixture in a nuclear reactor.
f teneurmètre *m* d'eau lourde
e medidor *m* del porcentaje de agua pesada
i tenorimetro *m* d'acqua pesante
n meter voor het gehalte aan zwaar water
d Gerät *n* zur Bestimmung des
   Schwerwassergehalts

3209 HEAVY WATER COOLED REACTOR,
      see 1756

3210 HEAVY WATER MODERATED
      REACTOR, see 1758

3211 HEAVY WATER PLANT   ch
f installation *f* d'eau lourde
e equipo *m* de agua pesada
i impianto *m* d'acqua pesante
n zwaar-waterinstallatie
d Schwerwasseranlage *f*

3212 HEEL, see 748

3213 HEEL EFFECT   xr
The decrease in intensity at the anode side
of an X-ray beam owing to absorption in
the anode.
f effet *m* de talon
e efecto *m* de talón
i effetto *m* di tallone
n anodeschaduw
d Anodenschatten *m*

3214 HEIGHT OF BURST   nw
The height above the earth's surface at
which a bomb is detonated in the air.

f   altitude *f* d'explosion
e   altitud *f* de explosión
i   altezza *f* d'esplosione
n   ontploffingshoogte
d   Explosionshöhe *f*

3215 HEISENBERG FORCE   np
The nuclear force which changes sign if
both the space co-ordinates and the spin
co-ordinates are interchanged.
f   force *f* d'Heisenberg
e   fuerza *f* de Heisenberg
i   forza *f* di Heisenberg
n   heisenbergkracht
d   Heisenberg-Kraft *f*

3216 HEISENBERG RELATION,   np, qm
     INDETERMINANCY PRINCIPLE
The statement, fundamental in quantum
theory, that no measurement can determine
both the co-ordinate and the momentum
of a particle (or any other pair of
complementary quantities) so accurately
that the product of the errors is less than
Planck's constant.
f   relation *f* d'indétermination d'Heisenberg
e   relación *f* de indeterminación de
   Heisenberg
i   rapporto *m* d'indeterminazione di
   Heisenberg
n   onbepaaldheidsbetrekking van Heisenberg
d   Heisenbergsche Unbestimmtheitsrelation *f*

3217 HEISENBERG REPRESENTATION,   qm
     MATRIX MECHANICS
A mathematical formulation of quantum
mechanics.
f   mécanique *f* de matrices,
   représentation *f* d'Heisenberg
e   mecánica *f* de matrices,
   representación *f* de Heisenberg
i   meccanica *f* di matrici,
   rappresentazione *f* di Heisenberg
n   matrices-rekenwijze,
   opstelling van Heisenberg
d   Heisenberg-Darstellung *f*,
   Matrizenmechanik *f*

3218 HEITLER-LONDON THEORY   ch
     OF COVALENT BONDING
A treatment of the exchange forces between
atoms in which the two interacting
electrons are assumed to be in atomic
orbitals about each of the nuclei, these
orbitals being combined into symmetric
and anti-symmetric functions.
f   théorie *f* de liaisons covalentes
   d'Heitler-London
e   teoría *f* de enlaces covalentes de
   Heitler-London
i   teoria *f* di legami covalenti di
   Heitler-London
n   covalente bindingstheorie van Heitler en
   London
d   kovalente Bindungstheorie *f* von Heitler
   und London

3219 HELIUM   ch
Gaseous element, symbol He, atomic
number 2.
f   hélium *m*
e   helio *m*
i   elio *m*
n   helium *n*
d   Helium *n*

3220 HELIUM COOLING   cl
Cooling method in a reactor whereby
helium is used as the coolant.
f   refroidissement *m* à l'hélium
e   enfriamiento *m* por helio
i   raffreddamento *m* all'elio
n   heliumkoeling
d   Heliumkühlung *f*

3221 HELIUM COUNTER TUBE   ct
A proportional counter tube containing
helium and often intended for measuring
the energy of neutrons, using their
reaction on helium-3.
f   tube *m* compteur à hélium
e   tubo *m* contador de helio
i   tubo *m* contatore ad elio
n   heliumtelbuis
d   heliumgefülltes Zählrohr *n*

3222 HELIUM LEAK DETECTOR,   rt
     LEAK DETECTOR
Apparatus for the spectrometric detection
and location of helium in-leakage into a
vacuum created within an enclosure.
f   détecteur *m* de fuites à l'hélium
e   detector *m* de fugas de helio
i   rivelatore *m* di fughe ad elio
n   heliumlekdetector
d   Heliumleckprüfer *m*, Heliumlecksuchgerät
   n

3223 HELIUM LEAK TEST   te
A test carried out on all welded sources.
f   essai *m* de fuites de l'hélium
e   prueba *f* de fugas del helio
i   prova *f* di fughe dell'elio
n   heliumlekproef
d   Heliumleckprüfung *f*

3224 HELIUM MASS   sp, te
     SPECTROGRAPHIC METHOD
A method for testing the integrity of the
final weld of, e.g., a gamma source.
f   méthode *f* spectrographique de masse
   pour essais de fuites de l'hélium
e   método *m* espectrográfico de masa para
   pruebas de fugas del helio
i   metodo *m* spettrografico di massa per
   prove di fughe dell'elio
n   massaspectrograafmethode voor helium-
   lekproeven
d   Massenspektrographverfahren *n* für
   Heliumleckprüfungen

3225 HELIUM PERMEATION   vt
A phenomenon occurring in extreme
vacuum technique in that the helium of the

atmosphere permeates through glass
walls.
f pénétration *f* d'hélium
e penetración *f* de helio
i attraversamento *m* d'elio
n heliumdoordringing
d Heliumdurchdringung *f*

3226 HELIUM PERMEATION TEST   vt
Method for detecting tightness defects of
a fuel can by using helium permeation.
f détection *f* par ressuage,
essai *m* de pénétration d'hélium
e prueba *f* de penetración de helio
i prova *f* d'attraversamento d'elio
n heliumdoordringingsproef
d Heliumdurchdringungsprüfung *f*

3227 HELIUM PRESSURE TANK   rw
A removable tank for collecting the fission
gases from the accumulator tank.
f récipient *m* d'hélium sous pression
e tanque *m* de helio de presión
i serbatoio *m* d'elio a pressione
n heliumdrukvat *n*
d Heliumdruckgefäss *n*

3228 HELIUM SPECTROMETER   sp
A small mass spectrometer to detect the
presence of helium in a vacuum system.
f spectromètre *m* détecteur d'hélium
e espectrómetro *m* detector de helio
i spettrometro *m* rivelatore d'elio
n heliumdetectiespectrometer,
heliumspectrometer
d Heliumdetektionsspektrometer *n*,
Heliumspektrometer *n*

3229 HELVITE   mi
A mineral containing silicate of beryllium,
iron and manganese.
f helvite *f*
e helvita *f*
i elvite *f*
n helviet *n*
d Helvit *m*

3230 HEMACYTOMETER,
HEMATIMETER,
see 1484

3231 HEMACYTOMETRY,
HEMATIMETRY,
see 677

3232 HEMOGLOBIN   md
The oxygen carrying pigment of the red
blood corpuscles.
f hémoglobine *f*
e hemoglobina *f*
i emoglobina *f*
n hemoglobine *n*
d Hämoglobin *n*

3233 HEPATOLIENOGRAPHY   xr
Radiological examination of the liver and
the spleen after thoratrast injection of
sols of iodine.

f hépatoliénographie *f*
e hepatolienograffa *f*
i epatolienografia *f*
n hepatolienografie
d Hepatolienographie *f*

3234 HEREDITARY DEFECT   md
A defect in the human body caused by
mutation of reproductive cells.
f lésion *f* héréditaire
e defecto *m* hereditario
i difetto *m* ereditario
n erfelijk gebrek *n*
d Erbschaden *m*

3235 HERMETIC SEAL   cd, rt
A seal that prevents passage of air, water,
vapo(u)r and all other gases.
f scellement *m* hermétique
e cierre *m* hermético
i chiusura *f* ermetica
n hermetische afsluiting,
luchtdichte afsluiting
d luftdichter Verschluss *m*

3236 HETEROCHROMATIC X-RAYS,   xr
HETEROGENEOUS X-RAYS
X-Rays of a broad range or considerable
number of frequencies.
f rayons *pl* X hétérogènes
e rayos *pl* X heterogéneos
i raggi *pl* X eterogenei
n heterogene röntgenstralen *pl*
d heterogene Röntgenstrahlen *pl*

3237 HETEROGENEOUS CORE LAY-OUT   rt
A concept previously studied in planning
the DRAGON reactor but not carried out.
f arrangement *m* hétérogène du coeur du
réacteur
e proyecto *m* heterogéneo de la alma del
reactor
i progetto *m* eterogeneo del nocciolo del
reattore
n heterogene kernopstelling
d heterogene Kernanordnung *f*

3238 HETEROGENEOUS MIXTURE   rt
A mixture of fuel and moderator in a
heterogeneous reactor.
f mélange *m* hétérogène
e mezcla *f* heterogénea
i miscela *f* eterogenea
n heterogeen mengsel *n*
d heterogene Mischung *f*

3239 HETEROGENEOUS RADIATION   np, ra
Radiation having several different
frequencies, or a beam of particles of a
variety of energies or containing different
types of particles.
f rayonnement *m* hétérogène
e radiación *f* heterogénea
i radiazione *f* eterogenea
n heterogene straling
d heterogene Strahlung *f*

3240 HETEROGENEOUS REACTOR   rt
A nuclear reactor in which the core
materials are segregated to such an extent
that its neutron characteristics cannot
be accurately described by the assumption
of homogeneous distribution of the
materials throughout the core.
f réacteur *m* hétérogène
e reactor *m* heterogéneo
i reattore *m* eterogeneo
n heterogene reactor
d heterogener Reaktor *m*

3241 HETEROPOLAR BOND   ch
A valence linkage between two atoms,
consisting of a pair of electrons.
f liaison *f* hétéropolaire
e enlace *m* heteropolar
i legame *m* eteropolare
n heteropolaire binding
d heteropolare Bindung *f*

3242 HETEROTOPIC   ch
Having a different atomic number or
charge; the opposite of isotopic.
f hétérotopique adj
e heterotópico adj
i eterotopico adj
n heterotopisch adj
d heterotopisch adj

3243 HETEROZYGOUS   md
Derived from germ cells genetically
unlike.
f hétérozygote adj
e heterozigoto adj
i eterozigoto adj
n heterozygoot adj
d heterozygot adj

3244 HEX,   mt
       URANIUM HEXAFLUORIDE
A gaseous compound used in the
separation of U-235 from natural uranium
by the gaseous diffusion process.
f hexafluorure *m* d'uranium
e hexafluoruro *m* de uranio
i esafluoruro *m* d'uranio
n uraniumhexafluoride *n*
d Uranhexafluorid *n*

3245 HEX RECONVERSION PROCESS,   ch
       URANIUM HEXAFLUORIDE
       RECONVERSION PROCESS
A process whereby uranium hexafluoride
is converted to uranium metal.
f procédé *m* de conversion à hexafluorure
   d'uranium
e procedimiento *m* de conversión con hexafluoruro de uranio
i processo *m* di conversione all'esafluoruro
   d'uranio
n conversieproces *n* met uraniumhexafluoride
d Umwandlungsverfahren *n* mit Uranhexafluorid

3246 HEXAGONAL CLOSE-PACKED   cr
       STRUCTURE
A crystal structure that may be
considered to have the form of a hexagonal
right prism with an atom at each corner,
an atom in the centre(er)s of each of the basal
planes and three atoms located in the
corners of an interior equilateral triangle.
f structure *f* hexagonale compacte
e estructura *f* hexagonal compacta
i struttura *f* esagonale compatta
n hexagonale dichtbezette structuur
d hexagonale dichteste Kugelpackung *f*

3247 HEXAGONAL LATTICE,   fu, rt
       HEXAGONAL PATTERN
f réseau *m* hexagonal
e celosía *f* hexagonal
i reticolo *m* esagonale
n zeshoekig rooster *n*
d sechseckiges Gitter *n*

3248 HEXANE IONIZATION CHAMBER   ic
An ionization chamber in which the
organic gas filling consists of hexane
(C_6H_{14}).
f chambre *f* d'ionisation à hexane
e cámara *f* de ionización con hexano
i camera *f* d'ionizzazione ad esano
n met hexaan gevuld ionisatievat *n*
d hexangefüllte Ionisationskammer *f*

3249 HEXONE,   ch
       METHYL ISOBUTYL KETONE,
       M.I.K.
An organic solvent employed in the
original U.S. process for the solvent
extraction of plutonium from irradiated
uranium.
I méthylisobutylkétone *m*
e metilisobutilcetona *f*
i metilisobutilchetone *m*
n methylisobutylketon *n*
d Methylisobutylketon *n*

3250 HIELMITE,   mi
       HJELMITE
A mineral containing tin, tantalum,
niobium, yttrium, uranium, calcium and
manganese.
f hjelmite *f*
e hielmita *f*
i ielmite *f*
n hjelmiet *n*
d Hjelmit *m*

3251 HIGH-ACTIVITY WASTE   rw
Radioactive waste material from nuclear
power plants of high activity.
f déchets *pl* de haute activité,
   déchets *pl* fortement radioactifs
e desechos *pl* fuertemente radiactivos
i rifiuti *pl* fortemente radioattivi
n zeer radioactief afval *n*
d Abfall *m* hoher Aktivität,
   hochradioaktiver Abfall *m*

3252 HIGH-ACTIVITY WASTE   rw
       CONTAINER
f récipient m de déchets de haute activité
e recipiente m de desechos fuertemente
   radiactivos
i serbatoio m di rifiuti fortemente
   radioattivi
n houder voor zeer radioactief afval
d Sammelbehälter m für Abfall hoher
   Aktivität

3253 HIGH-BACKING-PRESSURE PUMP   vt
f pompe f à vide à contrepression élevée
e bomba f de vacío de alta contrapresión
i pompa f a vuoto ad alta contropressione
n vacuümpomp met hoge tegendruk
d Vakuumpumpe f mit hohem Gegendruck

3254 HIGH-COMPRESSION-RATIO PUMP vt
   A pump backing up the action of a
   four-stage mercury vapo(u)r pump.
f pompe f à rapport volumétrique élevé
e bomba f de alto grado de compresión
i pompa f ad alto rapporto di compressione
n pomp met hoge compressieverhouding
d Pumpe f mit hohem Verdichtungsgrad

3255 HIGH-DENSITY PLASMA   pp
f plasma m de densité élevée
e plasma m de alta densidad
i plasma m d'alta densità
n plasma n met hoge dichtheid
d Plasma n hoher Dichte

3256 HIGH-ENERGY ELECTRON SHEET,
       see 2008

3257 HIGH-ENERGY FISSION, see 2526

3258 HIGH-ENERGY GAMMA   gr
       RADIATION
   Gamma radiation resulting from a nuclear
   reaction in the region of energies from
   about 30 to 50 MeV.
f rayonnement m gamma à grande énergie
e radiación f gamma de alta energía
i radiazione f gamma ad alta energia
n gammastraling met grote energie
d energiereiche Gammastrahlung f

3259 HIGH-ENERGY GAMMA   gr, md, ra
       SOURCE
   A radiation source of which the gamma
   energy is greater than 1 MeV.
f source f gamma à énergie élevée
e fuente f gamma de alta energía
i sorgente f gamma d'alta energia
n gammabron met hoge energie
d Gammaquelle f hoher Energie

3260 HIGH-ENERGY ION INJECTION   pp
   A method used to form the plasma in a
   thermonuclear installation.
f injection f d'ions d'énergie élevée
e inyección f de iones de alta energía
i iniezione f d'ioni ad alta energia
n injectie van hoogenergetische ionen
d Injektion f hochenergetischer Ionen

3261 HIGH-ENERGY LEVEL REACTOR,   rt
       HIGH-POWER REACTOR
   A reactor in which the energy level is in
   the region of 30 to 50 MeV.
f réacteur m à grande puissance
e reactor m de alta potencia
i reattore m ad alta potenza
n reactor met groot vermogen
d Hochleistungsreaktor m

3262 HIGH-ENERGY PARTICLE   ar
   A particle resulting from a nuclear
   reaction in the region of energies from
   about 30 to 50 MeV.
f particule f à grande énergie
e partícula f de alta energía
i particella f ad alta energia
n deeltje n met grote energie
d energiereiches Teilchen n

3263 HIGH-FLUX REACTOR   rt
   A nuclear reactor designed to operate
   with high neutron flux.
f réacteur m à flux élevé
e reactor m de alto flujo
i reattore m ad alto flusso
n hoge-fluxreactor
d Hochflussreaktor m

3264 HIGH-FLUX RESEARCH REACTOR   rt
   A research reactor designed to operate
   with high neutron flux.
f réacteur m de recherche à flux élevé
e reactor m de investigación de alto flujo
i reattore m per ricerca ad alto flusso
n hoge-fluxreactor voor speurwerk
d Hochflussforschungsreaktor m

3265 HIGH-GRADE NUCLEAR FUEL,   fu
       HIGHLY ENRICHED NUCLEAR FUEL
   A nuclear fuel containing a high percentage
   of enriched material.
f combustible m fortement enrichi
e combustible m muy enriquecido
i combustibile m molto arricchito
n in hoge mate verrijkte splijtstof
d hochangereicherter Brennstoff m

3266 HIGH-GRADE ORE   mi
   An ore containing a high percentage of
   metal.
f minerai m à haute teneur
e mineral m de alta graduación,
   mineral m de alta ley
i minerale m ad alto tenore
n erts n met hoog gehalte, rijk erts n
d hochgradiges Erz n,
   hochhaltiges Erz n

3267 HIGH-LEVEL RADIATION   ra
   Radiation with a high-energy level.
f rayonnement m à niveau d énergie élevé
e radiación f de alto nivel de energía
i radiazione f a livello d'energia elevato
n straling met hoog energieniveau
d Strahlung f mit hohem Energieniveau

3268 HIGH-LEVEL SINGLE EXPOSURE,
       see 44

3269 HIGH-LEVEL TRIP,   co
      POWER SETBACK
    The gradual reduction of power achieved
    by continuous insertion of the coarse
    control members at their normal operating
    speed.
  f réduction *f* graduelle de la puissance
  e reducción *f* gradual de la potencia
  i riduzione *f* graduale della potenza
  n geleidelijke vermogensverlaging
  d allmähliche Leistungserniedrigung *f*

3270 HIGH-LEVEL WASTE   rw
    The radioactive liquid waste which at
    present at Hanford is retained indefinitely
    in underground storage tanks.
  f déchets *pl* liquides à radioactivité
    permanente
  e desechos *pl* líquidos de radiactividad
    permanente
  i rifiuti *pl* liquidi a radioattività permanente
  n permanent radioactieve afvalvloeistof
  d permanent radioaktive Abfallflüssigkeit *f*

3271 HIGH-PERFORMANCE RESEARCH   rt
      REACTOR
  f réacteur *m* de recherche à haut rendement
  e reactor *m* de investigaciones de gran
    rendimiento
  i reattore *m* di ricerca ad alto rendimento
  n speurwerkreactor met hoog rendement
  d Forschungsreaktor *m* hoher Leistungs-
    fähigkeit

3272 HIGH-PRESSURE ASSEMBLY   cd
  f ensemble *m* pour haute pression
  e conjunto *m* para alta presión
  i complesso *m* per alta pressione
  n hoge-drukopstelling
  d Hochdruckanordnung *f*

3273 HIGH-PRESSURE CLOUD   ic
      CHAMBER
    A cloud chamber in which the gas is
    maintained at high pressure.
  f chambre *f* à détente à haute pression
  e cámara *f* de expansión de alta presión
  i camera *f* d'ionizzazione ad alta pressione
  n hoge-druknevelvat *n*
  d Hochdrucknebelkammer *f*

3274 HIGH-PRESSURE HEAT   gp
      TRANSFER RIG
  f équipement *m* de transfert de chaleur à
    pression élevée
  e equipo *m* de termotransferencia de alta
    presión
  i attrezzatura *f* di trasmissione di calore
    ad alta pressione
  n warmteoverdrachtsopstelling voor hoge
    druk
  d Hochdruckwärmeübertragungsaufstellung *f*

3275 HIGH-PRESSURE LUBRICANT   cd
    A lubricant used in apparatus in which high
    pressures are used.
  f lubrifiant *m* pour hautes pressions

  e lubricante *m* para altas presiones
  i lubrificante *m* per alte pressioni
  n hoge-druksmeermiddel *n*
  d Hochdruckschmiermittel *n*

3276 HIGH-PRESSURE   ec, ic
      RECOMBINATION
    In a device in which ions formed in a gas
    are attracted towards electrodes by an
    electric field, the recombination of ions
    will take place under high gas pressure
    or low field intensities.
  f recombinaison *f* sous pression élevée
  e recombinación *f* sobre alta presión
  i ricombinazione *f* sotto alta pressione
  n recombinatie onder hoge druk
  d Rekombination *f* unter Hochdruck

3277 HIGH-PRESSURE WATER LOOP   cd
  f circuit *m* fermé à eau sous pression
    élevée
  e circuito *m* cerrado de agua de alta presión
  i circuito *m* chiuso ad acqua ad alta
    pressione
  n gesloten circuit *n* voor water onder hoge
    druk
  d Hochdruckwasserkreislauf *m*

3278 HIGH-RADIATION FLUX   ra
    The time rate of flow of radiant energy of
    high intensity.
  f flux *m* de rayonnement intensif
  e flujo *m* de radiación intensiva
  i flusso *m* di radiazione intensiva
  n intensieve-stralingsflux
  d intensiver Strahlungsfluss *m*

3279 HIGH-RECOMBINATION-RATE   ec
      CONTACT
    A contact between semiconductors or
    between a metal and a semiconductor,
    at which the densities of charge carriers
    are maintained substantially independent
    of current densities.
  f contact *m* à vitesse de recombinaison
    élevée
  e contacto *m* de alta velocidad de
    recombinación
  i contatto *m* ad alta velocità di
    ricombinazione
  n contact *n* met hoge recombinatiesnelheid
  d Kontakt *m* mit hoher Rekombinations-
    geschwindigkeit

3280 HIGH-SPEED CENTRIFUGE,   ch, is
      ULTRACENTRIFUGE
    A centrifuge that is operated at extremely
    high speed and used, e.g. in isotope
    separation.
  f ultracentrifugeur *m*
  e ultracentrífuga *f*
  i ultracentrifuga *f*
  n ultracentrifuge
  d Ultrazentrifuge *f*

3281 HIGH-SPEED ELECTRON   np
      An electron travelling at a high velocity.
f électron *m* rapide
e electrón *m* rápido
i elettrone *m* veloce
n snel elektron *n*
d schnelles Elektron *n*

3282 HIGH-SPEED NEUTRONS, see 2533

3283 HIGH SPOT,   ra
      HOT SPOT
      A small volume so situated with respect
      to one or more sources of radiation that
      the dose therein is significantly above
      the general dose level in the region
      treated.
f point *m* chaud
e punto *m* caliente
i punto *m* caldo
n hete plek,
   plek met grote stralingsdichtheid
d heisse Stelle *f*,
   Stelle *f* hoher Strahlungsdichte

3284 HIGH-TEMPERATURE   mt
      CERAMIC MATERIAL
      Used in building nuclear reactors and able
      to withstand very high temperatures.
f matière *f* céramique résistante aux
   températures élevées
e material *m* cerámico para temperaturas
   muy altas
i materiale *m* ceramico per temperature
   molto elevate
n keramisch materiaal *n* voor zeer hoge
   temperaturen
d keramisches Material *n* für sehr hohe
   Temperaturen

3285 HIGH-TEMPERATURE   rt
      GAS-COOLED REACTOR
      A nuclear reactor in which the cooling gas
      has such a temperature and pressure that
      the reactor may be used directly in a gas
      turbine.
f réacteur *m* à gaz à température élevée
e reactor *m* de gas de alta temperatura
i reattore *m* a gas ad alta temperatura
n met gas gekoelde hoge-temperatuurreactor
d gasgekühlter Hochtemperaturreaktor *m*

3286 HIGH-TEMPERATURE   rt
      HELIUM-COOLED REACTOR
f réacteur *m* à haute température à
   refroidissement à hélium
e reactor *m* de alta temperatura refrigerado
   por helio
i reattore *m* ad elio ad alta temperatura
n met helium gekoelde hoge-temperatuurreactor
d heliumgekühlter Hochtemperaturreaktor *m*

3287 HIGH-TEMPERATURE   mg, mt
      PROCESSING
      A treatment of a material at high tempera-
      tures for improving its properties.

f traitement *m* à température élevée
e tratamiento *m* de alta temperatura
i trattamento *m* ad alta temperatura
n bewerking bij hoge temperatuur
d Hochtemperaturverarbeitung *f*

3288 HIGH-TEMPERATURE   rt
      REACTOR
      A nuclear reactor designed to operate with
      the active section at a high temperature.
f réacteur *m* à température élevée
e reactor *m* de alta temperatura
i reattore *m* ad alta temperatura
n hoge-temperatuurreactor
d Hochtemperaturreaktor *m*

3289 HIGH VACUUM   vt
      A degree of vacuum at which essentially
      no gases or vapo(u)rs are present, so that
      ionization cannot occur.
f vide *m* élevé, vide *m* poussé
e alto vacío *m*
i alto vuoto *m*, vuoto *m* spinto
n hoogvacuüm *n*
d Hochvakuum *n*

3290 HIGH-VACUUM CUT-OFF   vt
      A device which permits the apparatus
      being exhausted to be temporarily
      disconnected from the pump.
f fermeture *f* de vide élevé
e cierre *m* de alto vacío
i chiusura *f* d'alto vuoto
n hoogvacuümafsluiting
d Hochvakuumsperre *f*

3291 HIGH-VACUUM DIFFUSION PUMP   vt
f pompe *f* à diffusion à vide élevé
e bomba *f* de difusión de alto vacío
i pompa *f* a diffusione ad alto vuoto
n hoogvacuümdiffusiepomp
d Hochvakuumdiffusionspumpe *f*

3292 HIGH-VOLTAGE ACCELERATOR   pa
      A device for imparting large amounts of
      kinetic energy to charged particles such
      as electrons, deuterons, etc., by the
      application of high voltages.
f accélérateur *m* à tension élevée
e acelerador *m* de alta tensión
i acceleratore *m* ad alta tensione
n hoogspanningsversneller
d Hochspannungsbeschleuniger *m*

3293 HIGHLY ENRICHED FUEL   fu, rt
      CYCLE
      A new core design possibly adaptable for
      the DRAGON project.
f cycle *m* de combustible très enrichi
e ciclo *m* de combustible de alto
   enriquecimiento
i ciclo *m* di combustibile ad alto
   arricchimento
n in hoge mate verrijkte splijtstofcyclus
d hochangereicherter Brennstoffkreislauf *m*

3294 HIGHLY ENRICHED NUCLEAR FUEL,
     see 3265

3295 HIGHLY RADIOACTIVE,   np
     HOT
f chaud adj, fortement radioactif
e activísimo adj, fuertemente radiactivo
i caldo adj, d'attività elevata,
    fortemente radioattivo
n heet adj, sterk radioactíef
d heiss adj, hochaktiv adj,
    stark radioaktiv

3296 HILAC, see 3198

3297 HISTORADIOGRAPHY   xr
    Radiological examination of tissues.
f historadiographie *f*
e historadiografía *f*
i istoradiografia *f*
n historadiografie
d Historadiographie *f*

3298 HIT THEORY,   ra
     TARGET THEORY
    A theory explaining biological effects of
    radiation on the basis of ionization
    occurring in a very small sensitive region
    within the cell.
f théorie *f* de la cible
e teoría *f* del blanco
i teoria *f* del bersaglio
n treffertheorie
d Treffertheorie *f*

3299 HJELMITE, see 3250

3300 HLOPINITE, see 1075

3301 HODOSCOPE   ct
    An arrangement of radiation counters used
    in cosmic-ray detection.
f hodoscope *m*
e hodoscopio *m*
i odoscopio *m*
n hodoscoop *m*
d Hodoskop *n*

3302 HOKUTOLITE   mi
    Radioactive mixture of Pb and Ba sulphate,
    probably containing Ra, Th and U.
f hokutolite *f*
e hokutolita *f*
i hokutolite *f*
n hokutoliet *n*
d Hokutolit *m*

3303 HOLD-BACK AGENT   ch, is
    The inactive isotope or isotopes of a radio-
    active element or an element of similar
    properties or some reagent which may
    be used to diminish the amount of the
    nuclide coprecipitated or absorbed on a
    particle carrier or absorbent.
f inhibiteur *m* d'entraînement
e agente *m* de retención
i agente *m* di ritenzione

n schutstof
d Rückhaltestoff *m*

3304 HOLD-BACK CARRIER   is
    A carrier used to prevent a particular
    species, e.g. an impurity, from following
    another species in a chemical operation.
f entraîneur *m* de rétention
e portador *m* de retención
i protatore *m* di ritenzione
n retentiedrager
d Rückhalteträger *m*

3305 HOLD-UP,   is
     INVENTORY
    The quantity of process material in a
    plant or any part of a plant.
f charge *f* ouvrable
e carga *f* práctica
i dotazione *f* di materiale,
    impegno *m* di materiale
n materiaalinbrengst, werkinhoud
d Materialeinsatz *m*

3306 HOLE   cr
    A deficiency resulting when an electron is
    removed from an almost full energy band.
f lacune *f*, trou *m*
e hueco *m*, laguna *f*
i buco *m*, vacanza *f* elettronica
n gat *n*
d Elektronenmangelstelle *f*, Loch *n*

3307 HOLE, see 985

3308 HOLE CAPTURE   ec
    The capture of a hole by an impurity
    other than the foreign atom.
f capture *f* de trous
e captura *f* de huecos
i cattura *f* di buche
n gatenvangst
d Löchereinfang *m*

3309 HOLE CONDUCTION, see 1666

3310 HOLLOW ANODE   xr
    An anode of tubular shape and open at
    one end.
f anode *f* cylindrique
e ánodo *m* cilíndrico
i anodo *m* cilindrico
n holle anode, kananode
d Hohlanode *f*

3311 HOLMIUM   ch
    Metallic element, symbol Ho, atomic
    number 67.
f holmium *m*
e holmio *m*
i olmio *m*
n holmium *n*
d Holmium *n*

3312 HOMOGENEOUS IONIZATION   ic
     CHAMBER
    An ionization chamber in which the walls

3532 INTEGRATING DOSE   xr
      RATEMETER
      A dose ratemeter with a measuring system
      that integrates the exposure dose rate with
      respect to time to indicate or record the
      total exposure dose.
f débitmètre *m* à intégration
e medidor *m* de dosis integrador
i rateometro *m* di dose integratore
n integrerende dosistempometer
d integrierender Dosisleistungsmesser *m*

3533 INTEGRATING GAMMA   gr
      DETECTOR·
      A self-contained portable apparatus for
      accurate and reliable indication of
      gamma dose in radiation therapy and
      survey work.
f détecteur *m* gamma intégrateur
e detector *m* gamma integrador
i rivelatore *m* gamma integratore
n integrerende gammadetector
d integrierender Gammadetektor *m*

3534 INTEGRATING INDICATOR   me
      An indicator which sums up the value of
      the quantity indicated with respect to time.
f indicateur *m* intégrateur
e indicador *m* integrador
i indicatore *m* integratore
n integrerende aanwijzer
d integrierender Anzeiger *m*

3535 INTEGRATION IONIZATION   ic
      CHAMBER
      An ionization chamber designed for the
      measurements of the accumulated charge
      caused by individual ionizing events
      occurring during some interval of time.
f chambre *f* d'ionisation à intégration
e cámara *f* de ionización de integración
i camera *f* d'ionizzazione ad integrazione
n ionisatievat *n* met ladingsmeting
d integrierende Ionisationskammer *f*,
      Ionisationskammer *f* mit Ladungsmessung

3536 INTEGRITY OF A WELD   sa, te
      Term used in describing test methods of
      neutron sources in welded containers.
f étanchéité *f* de la soudure
e estanqueidad *f* de la soldadura
i impermeabilità *f* della saldatura
n lasdichtheid
d Dichtigkeit *f* der Schweissstelle

3537 INTENSELY BUNCHED-ION   pa
      SOURCE
      A special form of Van de Graaff machine.
f source *f* d'ions fortement groupés
e fuente *f* de iones de agrupamiento
      intensivo
i sorgente *f* d'ioni ad accumulo intensivo
n ionenbron met intensieve pakketvorming
d Ionenquelle *f* mit intensiver Ballung

3538 INTENSIFYING FACTOR,   xr
      SPEED FACTOR
      The ratio of the exposure time without
      intensifying screens to that when screens
      are used, other conditions being the same.
f facteur *m* d'intensification
e factor *m* de intensificación
i fattore *m* d'intensificazione
n versterkingsfactor
d Verstärkungsfaktor *m*

3539 INTENSIFYING SCREEN   ph, xr
      A layer of suitable material which, when
      placed in close contact with a photo-
      graphic emulsion, adds to the photographic
      effect of the incident radiation.
f écran *m* renforçateur
e pantalla *f* reforzadora
i schermo *m* rinforzatore
n versterkscherm *n*
d Verstärkerschirm *m*

3540 INTENSITOMETER   xr
      A device for determining relative X-ray
      intensities during radiography, in order to
      control exposure time.
f intensitomètre *m*
e intensitómetro *m*
i intensitometro *m*
n intensitometer
d Intensitometer *n*

3541 INTENSITY   ge
      In general, the level or value of any
      quantity, as in a field of ionizing radiation,
      the dose rate, etc.
f intensité *f*
e intensidad *f*
i intensità *f*
n intensiteit, sterkte
d Intensität *f*, Stärke *f*

3542 INTENSITY,   np
      INTENSITY OF RADIATION,
      RADIANT FLUX DENSITY
      At a given place, the energy per unit time
      entering a sphere of unit cross-sectional
      area at that place.
f intensité *f* de rayonnement
e intensidad *f* de radiación
i intensità *f* di radiazione
n stralingsintensiteit
d Strahlenintensität *f*

3543 INTENSITY OF ACTIVATION, see 67

3544 INTENSITY OF RADIOACTIVITY   np
      The number of atoms disintegrating per
      unit time.
f intensité *f* de radioactivité
e intensidad *f* de radiactividad
i intensità *f* di radioattività
n aantal *n* vervalsatomen per tijdeenheid
d Anzahl *f* der zerfallenden Atome je
      Zeiteinheit

3545 INTERACTION   np
Any process in which two particles
approach each other closely enough to
affect each other in any way.
f interaction *f*
e interacción *f*
i interazione *f*
n wisselwerking
d Wechselwirkung *f*

3546 INTERACTION CROSS SECTION   cs
Given by $\sigma = 1/n\,\lambda$, where n is the
number of nuclei per unit volume and
λ is the interaction mean free path.
f section *f* efficace d'interaction
e sección *f* eficaz de interacción
i sezione *f* d'urto d'interazione
n doorsnede voor wisselwerking
d Wirkungsquerschnitt *m* für Wechselwirkung

3547 INTERACTION MEAN FREE PATH   np
The average distance which a given
particle travels before experiencing an
interaction with another particle.
f libre moyen parcours *m*
e camino *m* libre medio,
   trayectoria *f* libre media
i cammino *m* libero medio,
   percorso *m* libero medio
n gemiddelde vrije weglengte
d mittlere freie Weglänge *f*

3548 INTERACTION TIME   np
The transit time across a Debye sphere.
f temps *m* d'interaction
e tiempo *m* de interacción
i tempo *m* d'interazione
n wisselwerkingstijd
d Wechselwirkungszeit *f*

3549 INTERATOMIC FORCE   ..p
f force *f* interatomaire
e fuerza *f* interatómica
i forza *f* interatomare
n interatomaire kracht
d interatomare Kraft *f*

3550 INTERCHANGE   is
The mixing of tracer and added isotopic
carrier such that the two participate to
the same degree in any chemical action,
showing that mixing has occurred in
whatever chemical form the tracer may
have originally been distributed.
f mélange *m* complet
e mezcla *f* completa
i mescolamento *m* completo
n volledige menging
d vollständige Mischung *f*

3551 INTERCHANGE,   md
       TRANSLOCATION
Structural change in which the distal parts
of two chromosomes are exchanged.
f translocation *f*
e translocación *f*
i intercambio *m*

n translocatie
d Translokation *f*

3552 INTERCHANGE INSTABILITY, see 2745

3553 INTERDENDRITIC SEGREGATION,
       see 1418

3554 INTERFACE,   cr
       INTERPLANAR SPACING
The perpendicular distance between a
plane of given Miller indices and a
parallel through the origin of co-ordinates.
f distance *f* réticulaire cristalline,
   écartement *m* des faces, interface *f*
e distancia *f* reticular cristalina,
   espaciamiento *m* de caras,
   separación *f* interplanar
i distanza *f* reticolare cristallina,
   scartamento *m* alla faccie
n netvlakkenafstand
d Netzebenenabstand *m*

3555 INTERFACE REGION   ch, cr
The region of the surface which forms
the boundary between two phases or
systems.
f zone *f* de surfaces limites
e zona *f* interfacial
i zona *f* interfacciale
n grenslaaggebied *n*
d Grenzschichtgebiet *n*

3556 INTERFACE TEMPERATURE   fu
The temperature of the surface of the
fuel element.
f température *f* de surface de l'élément de
   combustible
e temperatura *f* de superficie del elemento
   de combustible
i temperatura *f* di superficie dell'elemento
   di combustibile
n oppervlaktemperatuur van het splijtstofelement
d Oberflächentemperatur *f* des Brennelementes

3557 INTERFERENCE PULSE,   ct
       SPURIOUS PULSE
A pulse caused by an electrical or other
disturbance.
f impulsion *f* parasite
e impulsión *f* espuria
i impulso *m* parassito,
   impulso *m* perturbatore
n storingspuls
d Störimpuls *m*

3558 INTERLOCK   co
A device to prevent activation of a control
unit until a preliminary condition has been
met or to prevent hazardous operations.
f verrouillage *m*
e enclavamiento *m*, intercierre *m*
i dispositivo *m* di blocco
n blokkering, vergrendeling
d Blockierung *f*, Verriegelung *f*

3559 INTERMEDIATE CIRCUIT,   cl
    INTERMEDIATE LOOP
    A circuit between two cooling circuits.
f   circuit *m* intermédiaire
e   circuito *m* intermedio
i   circuito *m* intermedio
n   tussenleiding, verbindingsleiding
d   Verbindungsleitung *f*, Zwischenleitung *f*

3560 INTERMEDIATE COUPLING   ch
    A coupling exhibited by elements such as
    Si, Ge and Sn showing an energy
    separation of the fine structure levels
    which is nearly intermediate between that
    which would result from either j-j or
    L-S coupling.
f   couplage *m* à valeur intermédiaire
e   acoplamiento *m* de valor intermedio
i   accoppiamento *m* a valore intermedio
n   koppeling met middelwaarde
d   Mittelwertkupplung *f*

3561 INTERMEDIATE ENERGY REGION   np
f   zone *f* d'énergie moyenne
e   zona *f* de energía media
i   zona *f* d'energia media
n   gebied *n* van gemiddelde energie
d   Gebiet *n* mittlerer Energie

3562 INTERMEDIATE LEVEL WASTE   rw
    The radioactive liquid waste which usually
    is discharged underground, when the
    long-lived radioisotopes are known to be
    retained in the soil above the water table.
f   déchets *pl* à radioactivité modérée
e   desechos *pl* de radiactividad moderada
i   rifiuti *pl* a radioattività moderata
n   afval met gematigde radioactiviteit
d   Abfall *m* mässiger Radioaktivität

3563 INTERMEDIATE MATERIAL, see 3424

3564 INTERMEDIATE NEUTRONS   np
    Neutrons having a kinetic energy between
    those of epithermal neutrons and fast
    neutrons.
f   neutrons *pl* intermédiaires
e   neutrones *pl* de energía intermedia
i   neutroni *pl* d'energia intermedia
n   middelsnelle neutronen *pl*
d   mittelschnelle Neutronen *pl*

3565 INTERMEDIATE NUCLEUS, see 1268

3566 INTERMEDIATE REACTOR,   rt
    INTERMEDIATE SPECTRUM
    REACTOR
    A nuclear reactor in which fission is
    induced predominantly by intermediate
    neutrons.
f   réacteur *m* à neutrons intermédiaires
e   reactor *m* de neutrones de energía
    intermedia
i   reattore *m* a neutroni d'energia intermedia
n   middelsnelle reactor,
    reactor met middelsnelle neutronen
d   intermediärer Reaktor *m*,
    mittelschneller Reaktor *m*

3567 INTERMEDIATE SPEED   np
    OF NEUTRONS
    The velocity of average value of neutrons.
f   vitesse *f* intermédiaire de neutrons
e   velocidad *f* intermedia de neutrones
i   velocità *f* intermedia di neutroni
n   gemiddelde neutronensnelheid
d   mittlere Neutronengeschwindigkeit *f*

3568 INTERMETALLIC COMPOUND   mg
    A compound consisting of metallic atoms
    only, which are joined by metallic bonds.
f   composé *m* intermétallique
e   compuesto *m* intermetálico
i   composto *m* intermetallico
n   intermetallieke verbinding
d   intermetallische Verbindung *f*

3569 INTERMITTENT-DUTY   ge
    RATING
    The specified output rating of a device when
    operated for specified intervals of time
    other than continuous duty.
f   régime *m* nominal pour service intermittent
e   régimen *m* nominal de trabajo intermitente
i   regime *m* nominale con servizio inter-
    mittente
n   nominale bedrijfsgegevens *pl* bij niet-conti-
    nu bedrijf en gelijkblijvende belasting
d   aussetzender Nennbetrieb *m* mit während
    des Spiels gleichbleibender Belastung

3570 INTERNAL ABSORPTION   ab
    The absorption of radiation within the
    source from which it originates.
f   absorption *f* interne
e   absorción *f* interna
i   assorbimento *m* interno
n   inwendige absorptie
d   innere Absorption *f*

3571 INTERNAL BREEDING RATIO   rt
    The breeding ratio relating only to the
    fissile (fissionable) products in the interior
    of the core of the reactor.
f   rapport *m* de régénération interne
e   relación *f* de regeneración interna
i   rapporto *m* di rigenerazione interna
n   inwendige kweekverhouding
d   inneres Brutverhältnis *n*

3572 INTERNAL CONTAMINATION   md
    Contamination by radioactive substances
    which have penetrated into the organism.
f   contamination *f* interne
e   contaminación *f* interna
i   contaminazione *f* interna
n   inwendige besmetting
d   innerliche Kontamination *f*

3573 INTERNAL CONVERSION   np
    The emission of an electron in the
    de-excitation of a nucleus by direct
    coupling between the excited nucleus and
    an extra nuclear electron, usually one in
    the K, L or M shell.
f   conversion *f* interne
e   conversión *f* interna

i   conversione f interna
n   interne conversie
d   innere Konversion f

3574 INTERNAL CONVERSION   np
      COEFFICIENT,
      INTERNAL CONVERSION FACTOR
      The ratio of the number of internal
      conversion electrons to the number of
      gamma quanta emitted by the atom in the
      de-excitation of a nucleus.
f   facteur m de conversion interne
e   factor m de conversión interna
i   fattore m di conversione interna
n   interne conversiecoëfficiënt,
    interne conversiefactor
d   innerer Konversionskoeffizient m

3575 INTERNAL CONVERSION   np
      ELECTRON
      The electron produced or ejected by
      internal conversion.
f   électron m de conversion interne
e   electrón m de conversión interna
i   elettrone m di conversione interna
n   intern conversie-elektron n
d   inneres Konversionselektron n

3576 INTERNAL CONVERSION RATIO,
      see 1415

3577 INTERNAL ENERGY   np
      The total of the kinetic energy and the
      potential energy of a particle or a system.
f   énergie f interne
e   energía f interna
i   energia f interna
n   inwendige energie
d   innere Energie f

3578 INTERNAL EXPOSURE,   ra, xr
      INTERNAL IRRADIATION
f   exposition f interne,
    irradiation f interne
e   exposición f interna,
    irradiación f interna
i   esposizione f ad irradiazione interna,
    irradiazione f interna
n   inwendige exposie
d   innere Bestrahlung f

3579 INTERNAL PAIR PRODUCTION   np
      The formation of an electron and a posit(r)on
      through de-excitation of an excited nucleus.
f   formation f d'une paire électron-positon
    par conversion interne
e   formación f de un par electrón-positón
    por conversión interna
i   formazione f d'una coppia elettrone-
    positone per conversione interna
n   elektron-positon-paarvorming door
    interne conversie
d   Elektron-Positron-Paarbildung f durch
    innere Konversion

3580 INTERNAL QUENCHING,   ct
      SELF-QUENCHING
      The act of internally terminating a pulse

of ionization current in a Geiger-Müller
counter.
f   autocoupure f, étouffement m
e   autoextinción f
i   autospegnimento m
n   zelfdoving
d   Selbstlöschung f

3581 INTERNAL RADIATION   ra
      That radiation reaching a given point in
      the body or irradiated material which is
      due directly or indirectly to a source of
      radiation inside that body or material.
f   rayonnement m interne
e   radiación f interna
i   radiazione f interna
n   inwendige straling
d   innere Strahlung f

3582 INTERPHASE   gp
      The boundary surface between two phases.
f   couche f limite entre deux phases
e   capa f límite entre dos fases
i   strato m limite tra due fasi
n   fazengrenslaag
d   Phasengrenzschicht f

3583 INTERPLANAR SPACING, see 3554

3584 INTERSECTING STORAGE RINGS   pa
      Storage rings in which the accelerated
      particles are introduced in opposite
      directions and collide at 8 intersection
      points at an angle of 15^0.
f   anneaux pl de stockage à intersection
e   anillos pl de almacenamiento de inter-
    sección
i   anello pl d'immagazzinamento ad
    intersezione
n   elkaar snijdende opslagringen pl
d   einander kreuzende Speicherringe pl

3585 INTERSTICE   np
      A small space within a phase or between
      particles.
f   interstice m
e   intersticio m
i   interstizio m
n   roostertussenruimte
d   Zwischengitterplatz m

3586 INTERSTITIAL APPLIANCE br, md, ra
      An appliance for beta-ray therapy whereby
      the radioactive source is implanted within
      or close to the diseased tissue.
f   appareil m interstitiel
e   aparato m intersticial
i   apparecchio m interstiziale
n   interstitieel apparaat n
d   interstitielles Gerät n

3587 INTERSTITIAL ATOM   cr
      An additional atom or ion placed in between
      the normal sites, causing some deformation
      of the lattice around it.
f   atome m interstitiel
e   átomo m intersticial
i   atomo m interstiziale

n interstitieel atoom *n*
d interstitielles Atom *n*

3588 INTERSTITIAL COMPOUND ch, cr, mg
A compound of a metal or metals and
certain metalloid elements, in which the
metalloid atoms occupy the interstices
between the atoms of the metal lattice.
f composé *m* interstitiel
e compuesto *m* intersticial
i composto *m* interstiziale
n interstitiële verbinding
d interstitielle Verbindung *f*

3589 INTERSTITIAL IRRADIATION   xr
Irradiation of part of the body by radioactive sources introduced into the
tissues.
f irradiation *f* interstitielle
e irradiación *f* intersticial
i irradiazione *f* interstiziale
n interstitiële bestraling
d interstitielle Bestrahlung *f*

3590 INTERSTITIAL TECHNIQUE, see 3408

3591 INTRACAVITARY   br, md, ra
APPLIANCE
An appliance for irradiating lesions within
body cavities.
f appareil *m* intercavitaire
e aparato *m* intercavitario,
aparato *m* para irradiación interna
i apparecchio *m* intercavitale
n endotherapieapparaat *n*
d intrakavitäres Gerät *n*

3592 INTRACAVITARY IRRADIATION   ra
Irradiation of part of the body by
introducing one or more moulds or
applicators into a natural or artifical
body cavity.
f irradiation *f* intercavitaire
e irradiación *f* intercavitaria
i irradiazione *f* intercavitale
n lichaamsholtebestraling
d intrakavitäre Bestrahlung *f*

3593 INTRACAVITARY RÖNTGEN-   xr
THERAPY
A means of irradiating lesions within body
cavities accessible to direct vision or
touch by the use of a short treatment
cone introduced into the cavity exposing
the lesion to direct irradiation.
f röntgenthérapie *f* intercavitaire
e röntgenoterapia *f* intercavitaria
i röntgenterapia *f* intercavitale
n endotherapie, lichaamsholtetherapie
d intrakavitäre Röntgentherapie *f*

3594 INTRAMOLECULAR FORCE   ch
f force *f* intramoléculaire
e fuerza *f* intramolecular
i forza *f* intramolecolare
n intramoleculaire kracht
d intramolekülare Kraft *f*

3595 INTRANUCLEAR FORCES   np
The forces between proton and proton,
proton and neutron and neutron and neutron.
f forces *pl* intranucléaires
e fuerzas *pl* intranucleares
i forze *pl* intranucleari
n intranucleaire krachten *pl*
d intranukleare Kräfte *pl*

3596 INTRINSIC ANGULAR MOMENTUM ma
The angular momentum associated with
axial rotation of an elementary particle.
f moment *m* angulaire intrinsèque
e momento *m* angular intrínseco
i momento *m* angolare intrinseco
n eigen impulsmoment *n*
d Eigendrehimpuls *m*

3597 INTRINSIC COUNTER   ct
EFFICIENCY,
INTRINSIC EFFICIENCY OF A
GAMMA-RAY DETECTOR
The proportion of photons or particles
reaching the sensitive part of a counter
which give rise to counts.
f sensibilité *f* intrinsèque de compteur
e sensibilidad *f* intrínseca de contador
i sensibilità *f* intrinseca di contatore
n intrinsieke tellergevoeligheid
d intrinsike Zählerempfindlichkeit *f*

3598 INTRINSIC ENERGY   gp
A definite quantity of energy possessed
by and inherent in every substance.
f énergie *f* intrinsèque
e energía *f* intrínseca
i energia *f* intrinseca
n eigenenergie, intrinsieke energie
d Eigenenergie *f*

3599 INVENTORY, see 3305

3600 INVERSE ELECTRON CAPTURE   np
A term applied to a hypothetical nuclear
reaction in which a neutrino is captured
by a neutron in a nucleus, with emission
of a negative electron or in which an antineutrino is captured by a nuclear proton,
with emission of a posit(r)on.
f capture *f* inverse d'électron
e captura *f* inversa de electrón
i cattura *f* inversa d'elettrone
n omgekeerde elektronvangst
d umgekehrter Elektroneneinfang *m*

3601 INVERSE NUCLEAR REACTIONS   np
Two nuclear reactions in which the
products of one are the interacting
members of the other.
f réactions *pl* nucléaires inverses
e reacciones *pl* nucleares inversas
i reazioni *pl* nucleari inverse
n omgekeerde kernreacties *pl*
d umgekehrte Kernreaktionen *pl*

3602 INVERSE PHOTOELECTRIC   ec, np
EFFECT
Emission of photons caused by the impact
of electrons.

f   effet *m* photoélectrique inverse
e   efecto *m* fotoeléctrico inverso
i   effetto *m* fotoelettrico inverso
n   omgekeerd fotoelektrisch effect *n*
d   inverser photoelektrischer Effekt *m*

3603 INVERSE SQUARE LAW   ma, ra, xr
The intensity of radiation from a point
source is inversely proportional to the
square of the distance from the source,
if no absorber exists in that region.
f   loi *f* de l'inverse des carrés
e   ley *f* de proporcionalidad a la inversa de
los cuadrados
i   legge *f* dell'inverso dei quadrati
n   kwadratenwet
d   Gesetz *n* der quadratischen Abnahme

3604 INVERSE SUPPRESSOR   ec, xr
A rectifier in the primary circuit of a
transformer used with a self-rectifying
tube with the purpose of lessening inverse
voltage.
f   limitateur *m* d'onde inverse
e   limitador *m* de onda inversa
i   limitatore *m* d'onda inversa
n   tegenspanningsonderdrukker
d   Einrichtung *f* zur Fehlphasenunterdrückung

3605 INVERSE VOLTAGE   ec, xr
The voltage impressed across the X-ray
tube during the half cycle when the anode
is negatively charged.
f   tension *f* inverse
e   tensión *f* inversa
i   tensione *f* inversa
n   tegenspanning
d   Sperrspannung *f*

3606 IODINE   ch
Non-metallic element, symbol I, atomic
number 53.
f   iode *m*
e   yodo *m*
i   iodio *m*
n   jodium *n*
d   Jod *n*

3607 IODINE AIR MONITOR WITH   sa
CONTINUOUS SAMPLING
A continuous air monitor for determining
the presence of iodine in air.
f   moniteur *m* atmosphérique d'iode avec
prélèvement continu
e   monitor *m* atmosférico de yodo con
muestreo continuo
i   monitore *m* atmosferico d'iodio con
prelevamento continuo
n   continue luchtmonitor voor jodium
d   Jod-Luftmonitor *m* mit kontinuierlicher
Probenahme

3608 IODINE AIR MONITOR WITH   sa
DISCONTINUOUS SAMPLING
A discontinuous air monitor for determining the presence of iodine in the
atmosphere.

f   moniteur *m* atmosphérique d'iode avec
prélèvement discontinu
e   monitor *m* atmosférico de yodo con
muestreo discontinuo
i   monitore *m* atmosferico d'iodio con
prelevamento discontinuo
n   discontinue luchtmonitor voor jodium
d   Jod-Luftmonitor *m* mit diskontinuierlicher
Probenahme

3609 IODINE-131,   is, rw
RADIOIODINE
One of the gaseous waste products of
fission.
f   iode-131 *m*
e   yodo-131 *m*
i   iodio-131 *m*
n   jodium-131 *n*
d   Jod-131 *n*

3610 IOFFE COILS   pp
An assembly of magnetic field producing
coils arranged round a magnetic mirror
system to produce a magnetic well.
f   bobines *pl* de Ioffe
e   bobinas *pl* de Ioffe
i   bobine *pl* di Ioffe
n   spoelen *pl* van Ioffe
d   Ioffe-Spulen *pl*

3611 ION   ch, np
A charged atom or molecularly bound
group of atoms or a free electron or other
charged subatomic particle.
f   ion *m*
e   ión *m*
i   ione *m*
n   ion *n*
d   Ion *n*

3612 ION ACCELERATION   pa
The acceleration of ions.
f   accélération *f* d'ions
e   aceleración *f* de iones
i   accelerazione *f* d'ioni
n   ionenversnelling
d   Ionenbeschleunigung *f*

3613 ION ACCELERATOR   pa
An apparatus for accelerating ions.
f   accélérateur *m* d'ions
e   acelerador *m* de iones
i   acceleratore *m* d'ioni
n   ionenversneller
d   Ionenbeschleuniger *m*

3614 ION ACCEPTOR   ch, np
A substance that accepts an ion.
f   accepteur *m* d'ions
e   aceptador *m* de iones
i   accettore *m* d'ioni
n   ionenacceptor
d   Ionenakzeptor *m*

3615 ION ACOUSTIC WAVES   pp
Longitudinal waves which can be propagated in a plasma the electrons of which

have a much higher temperature than the
ions.
f ondes *pl* acoustiques ioniques,
   ondes *pl* pseudosonores
e ondas *pl* acústicas iónicas
i onde *pl* acustiche ioniche
n pseudosone golven *pl*
d pseudoakustische Wellen *pl*

3616 ION AVALANCHE   np
A group of ions freed by cumulative
ionization.
f avalanche *f* ionique
e avalancha *f* iónica
i valanga *f* ionica
n ionenlawine
d Ionenlawine *f* , Trägerlawine *f*

3617 ION BEAM   np
A beam of charged particles which
compare in velocity with those yielded by
radioactive substances, or which exceed
that velocity.
f faisceau *m* d'ions
e haz *m* de iones
i fascio *m* d'ioni
n ionenbundel
d Ionenstrahl *m*

3618 ION BEAM SCANNING   sp
The process of analyzing the mass
spectrum of an ion beam.
f balayage *m* d'un faisceau ionique
e exploración *f* de un haz iónico
i analisi *f* d'un fascio ionico
n massaspectrografische analyse van een
   ionenbundel
d massenspektrographische Analyse *f*
   eines Ionenstrahls

3619 ION CLUSTER   np
The ions close together at the end of the
path of an ionizing particle.
f essaim *m* d'ions, groupe *m* d'ions
e grupo *m* de iones
i grappolo *m* d'ioni, gruppo *m* d'ioni,
   sciame *m* d'ioni
n ionenopeenhoping
d Ionenanhäufung *f*

3620 ION COLLECTION   ic
Achieved by applying a polarizing voltage
between the electrodes of an ionization
chamber.
f collection *f* d'ions
e recolección *f* de iones
i collezione *f* d'ioni
n ionenvergaring
d Ionensammlung *f*

3621 ION COLLECTION CHAMBER   ic
A pulse ionization chamber in which
voltage pulses are due principally to the
collection of positive ions.
f chambre *f* à collection ionique
e cámara *f* de recolección de iones
i camera *f* a collezione d'ioni

n ionisatievat *n* met ionenvergaring
d Ionisationskammer *f* mit Ionensammlung

3622 ION COLLECTION TIME   ic
The time between the quasi-instantaneous
creation of ions by ionizing radiation and
the total collection by the collecting
electrode of the corresponding ions.
f temps *m* de collection d'ions
e tiempo *m* de recolección de iones
i tempo *m* di collezione d'ioni
n ionenvergaringstijd
d Ionensammlungszeit *f*

3623 ION COUNTER   ic
A tubular ionization chamber used for
measuring the ionization of the air.
f compteur *m* d'ions
e contador *m* iónico
i contatore *m* ionico
n ionenteller
d Ionenzähler *m*

3624 ION CYCLOTRON FREQUENCY   pa
The frequency at which an ion of charge q
in esu and of mass m rotates in an orbit
in a steady uniform magnetic field of
magnetic flux density B in emu.
f fréquence *f* de cyclotron d'un ion
e frecuencia *f* de ciclotrón de un ión
i frequenza *f* di ciclotrone d'un ione
n cyclotronfrequentie van een ion
d Ion *n* mit Zyklotronfrequenz

3625 ION CYCLOTRON RESONANCE   pa
      HEATING
A form of magnetic pumping in which the
frequency of the applied magnetic field
is so chosen that it equals the ion
cyclotron frequency.
f chauffage *m* par résonance à la fréquence
   d'un cyclotron à ions
e calentamiento *m* por resonancia de la
   frecuencia de un ciclotrón de iones
i riscaldamento *m* per risonanza alla
   frequenza d'un ciclotrone ad ioni
n resonantieverhitting in een ionencyclotron
d Resonanzerhitzung *f* im Ionenzyklotron

3626 ION DENSITY   np
The number of ion pairs per unit volume.
f densité *f* ionique,
   nombre *m* volumique d'ions,
   nombre *m* volumique de paires d'ions
e densidad *f* iónica
i densità *f* ionica
n ionendichtheid
d Ionendichte *f* , Ionenkonzentration *f*

3627 ION DETECTOR   sp
Apparatus for use in mass spectrometers
which distinguishes between the required
ions and those undesirable ions which
constitute a large part of the background
noise.
f détecteur *m* séparateur d'ions
e detector *m* selectivo de iones

i   rivelatore *m* selettivo d'ioni
n   selectieve ionendetector
d   selektiver Ionendetektor *m*

3628 ION DOSE   ra, xr
A dose produced by X-rays or gamma
radiation and equal to the quotient of the
electric charge of the ions created in a
specified volume directly or indirectly
by irradiation, by the mass of air in this
volume.
f   dose *f* ionique
e   dosis *f* iónica
i   dose *f* ionica
n   ionendosis
d   Ionendosis *f*

3629 ION ENERGY SELECTOR   np, sp
A dispersive device for ion beams, i.e.
one which separates ions having different
energies.
f   séparateur *m* énergétique d'ions
e   separador *m* energético de iones
i   separatore *m* energetico d'ioni
n   ionenenergiescheider
d   Ionenenergiescheider *m*

3630 ION ENGINE   ms
A reaction engine designed for space
travel, in which thrust is produced by a
stream of positive ions obtained as a
result of nuclear fission or fusion.
f   moteur *m* ionique
e   motor *m* iónico
i   motore *m* ionico
n   ionenmotor
d   Ionenmotor *m*

3631 ION EXCHANGE   ch
A chemical process involving the revers-
ible interchange of ions between a solution
and a particular solid material.
f   échange *m* d'ions
e   intercambio *m* de iones
i   scambio *m* d'ioni
n   ionenwisseling
d   Ionenaustausch *m*

3632 ION EXCHANGER RESIN   ma
MONITORING EQUIPMENT
An equipment designed to determine the
uranium contents in various solutions
in connection with ion exchanger resin
processing and which includes an alpha
activity measuring assembly associated
with an automatic sample changer.
f   équipement *m* de contrôle des résines-
échangeuses d'ions
e   radiovigía *f* de intercambio iónico
i   apparecchiatura *f* per controllo di
prodotti resinosi scambiatori d'ioni
n   uitrusting voor gehaltecontrole met
ionenwisselaar
d   Überwachungseinrichtung *f* für
Ionenaustauscher

3633 ION FLOW   np
The flow of a group of ions.

f   flux *m* ionique
e   flujo *m* iónico
i   flusso *m* ionico
n   ionenstroom
d   Ionenfluss *m*

3634 ION GUN,   np
ION SOURCE
A device in which gas ions are produced,
focused, accelerated, and emitted as a
narrow beam.
f   source *f* d'ions
e   fuente *f* de iones
i   sorgente *f* d'ioni
n   ionenbron
d   Ionenquelle *f*

3635 ION IMPLANTATION   ec
Method of introducing impurity atoms into
semiconductors.
f   injection *f* d'ions
e   inyección *f* de iones
i   iniezione *f* d'ioni
n   ioneninbrengst
d   Ioneneinbring *m*

3636 ION IMPLANTATION   ms
A technique used i.a. for studying the
behavio(u)r of gas bubbles in solids.
f   implantation *f* d'ions
e   injerto *m* de iones
i   innestatura *f* d'ioni
n   ionenimplantatie
d   Ionenimplantation *f*

3637 ION-ION RECOMBINATION   np
One of the basic modes for recombination.
f   recombinaison *f* ion-ion
e   recombinación *f* ión-ión
i   ricombinazione *f* ione-ione
n   ion-ion-recombinatie
d   Ion-Ion-Rekombination *f*

3638 ION LIMIT   ic
In a cloud chamber the value of the
expansion ratio when droplets of moisture
will condense on whatever ions are
present after expansion.
f   limite *f* de parcours d'ionisation
e   límite *m* de trayectoria de ionización
i   limite *m* di percorso d'ionizzazione
n   einde *n* van de ionenbaan
d   Ende *n* der Ionenbahn

3639 ION MOBILITY,   is
IONIC MOBILITY,
IONIC TRANSFER
The mobility or transference of different
ions in an electrolytic solution or melt
under the influence of an electric field.
f   mobilité *f* ionique
e   movilidad *f* iónica
i   mobilità *f* ionica
n   ionenbeweeglijkheid
d   Ionenbeweglichkeit *f*

3640 ION MOBILITY   is
      ISOTOPE SEPARATION
      A process based on the difference in
      mobility of different ions in an electrolytic
      solution under the influence of an electric
      field.
f séparation *f* d'isotopes par mobilité
   ionique
e separación *f* de isótopos por movilidad
   iónica
i separazione *f* d'isotopi per mobilità
   ionica
n isotopenscheiding door ionenbeweeglijkheid
d Isotopentrennung *f* durch Ionenbeweglichkeit

3641 ION NUMBER. DENSITY   np
      The number of positive or negative ions,
      divided by volume.
f nombre *m* volumique d'ions
e densidad *f* de iones
i densità *f* d'ioni
n ionendichtheid
d Ionendichte *f*

3642 ION PAIR   np
      A positive ion and a negative ion, usually
      an electron, that have charges of the same
      magnitude, and are formed from a neutral
      atom or molecule by any ionizing
      mechanism.
f paire *f* d'ions
e par *m* de iones
i coppia *f* d'ioni
n ionenpaar *n*
d Ionenpaar *n*

3643 ION-PAIR YIELD,   np
      M/N RATIO,
      YIELD PER ION PAIR
      The quotient of the number of molecules,
      M, of a given kind produced or converted,
      divided by the number N of ion pairs
      from high energy radiation.
f rapport *m* M/N,
   rendement *m* de paires d'ions,
   rendement *m* par paire d'ions
e relación *f* M/N,
   rendimiento *m* de pares de iones,
   rendimiento *m* por par de iones
i rapporto *m* M/N,
   rendimento *m* di coppie d'ioni,
   rendimento *m* per coppia d'ioni
n ionenpaaropbrengst, M/N-verhouding,
   opbrengst per ionenpaar
d Ausbeute *f* je Ionenpaar,
   Ionenpaarausbeute *f*, M/N-Verhältnis *n*

3644 ION PROPULSION   ms
      A method of obtaining propulsion for
      spaceships by expelling ions and electrons
      from a combustion chamber.
f propulsion *f* ionique
e propulsión *f* iónica
i propulsione *f* ionica
n voortstuwing door ionen
d Ionenantrieb *m*

3645 ION TRAJECTORY   np
      The path of one ion in a beam.
f parcours *m* d'un ion
e trayectoria *f* de un ión
i percorso *m* d'un ione
n ionenbaan
d Ionenbahn *f*

3646 ION TRANSFER   md
      The introduction of ions into the tissues
      of the body by the use of electric currents.
f transfert *m* d'ions
e transferencia *f* de iones
i trasporto *m* d'ioni
n iontoforese
d Iontophorese *f*

3647 ION TRANSIT TIME   ec
      Time taken by an ion in moving between
      two specified points.
f temps *m* de transit d'un ion
e tiempo *m* de tránsito de un ión
i tempo *m* di transito d'un ione
n looptijd van een ion
d Ionenlaufzeit *f*

3648 ION YIELD   np
      The number of ion pairs produced per
      incident particle or quantum.
f rendement *m* ionique
e rendimiento *m* iónico
i rendimento *m* ionico
n ionenopbrengst
d Ionenausbeute *f*

3649 IONIC BOND, see 2192

3650 IONIC CENTRIFUGE   ch, is
      An instrument used for the separation of
      isotopes.
f centrifugeur *m* ionique
e centrífuga *f* iónica
i centrifuga *f* ionica
n ionencentrifuge
d Ionenzentrifuge *f*

3651 IONIC CHARGE   np
      The whole charge which an ion carries.
f charge *f* ionique
e carga *f* iónica
i carica *f* ionica
n ionlading
d Ionenladung *f*

3652 IONIC CONDUCTION   np
      The continuous movement of charges
      within a substance due to the displacement
      of ions in a crystal lattice, the movement
      being maintained by a continuous
      contribution of external energy.
f conduction *f* ionique
e conducción *f* iónica
i conduzione *f* ionica
n ionengeleiding
d Ionenleitung *f*

3653 IONIC CRYSTAL   cr
      A crystal which consists effectively of ions

bound together by their electrostatic
attraction.
f   cristal *m* ionique
e   cristal *m* iónico
i   cristallo *m* ionico
n   ionenkristal *n*
d   Ionenkristall *m*

3654 IONIC EQUILIBRIUM   ch
In any ionization, at any particular
temperature and pressure, the conditions
at which the rate of dissociation of
un-ionized molecules, or other particles, to
form ions is equal to the rate of
combination of the ions to form the
un-ionized molecules or other particles so
that activities and concentrations remain
constant as long as the conditions are
unchanged.
f   équilibre *m* ionique
e   equilibrio *m* iónico
i   equilibrio *m* ionico
n   ionenevenwicht *n*
d   Ionengleichgewicht *n*

3655 IONIC MIGRATION METHOD, see 2098

3656 IONIC POTENTIAL   np
The ratio of ionic charge to radius.
f   rapport *m* charge-rayon d'un ion
e   relación *f* carga-radio de un ión
i   rapporto *m* carica-raggio d'un ione
n   invloed van lading op ionstraal,
    verhouding lading-ionstraal
d   Verhältnis *n* Ladung zu Ionenradius

3657 IONIC RADIUS   np
A radius which fixes the dimensions and
structure of ionic crystals.
f   rayon *m* d'ion
e   radio *m* de ión
i   raggio *m* d'ione
n   ionstraal, straal van een ion
d   Ionenradius *m*

3658 IONIC VALENCE, see 2204

3659 IONIUM   ch, is
The common name for 8.0 x 10^4y Th-230,
a member of the uranium series.
f   ionium *m*
e   ionio *m*
i   ionio *m*
n   ionium *n*
d   Ionium *n*

3660 IONIUM AGE,   np
      RADIUM AGE
The age calculated from the numbers of
radium (or ionium) atoms present originally, now and when equilibrium is established with ionium (or uranium).
f   âge *m* de radium
e   edad *f* de radio
i   età *f* di radio
n   radiumleeftijd
d   Radiumalter *n*

3661 IONIZATION   ch
The formation of ions by dividing molecules
or by adding or removing electrons to or
from atoms, molecules or groups of
molecules.
f   ionisation *f*
e   ionización *f*
i   ionizzazione *f*
n   ionisatie
d   Ionisation *f*, Ionisierung *f*

3662 IONIZATION BY COLLISION, see 1231

3663 IONIZATION CHAMBER   ic
A detector which employs an electric
field for the collection at the electrodes,
without gas multiplication, of charges
associated with the ions produced in the
sensitive volume by ionizing radiation.
f   chambre *f* d'ionisation
e   cámara *f* de ionización
i   camera *f* d'ionizzazione
n   ionisatiekamer, ionisatievat *n*
d   Ionisationskammer *f*

3664 IONIZATION CHAMBER   ic, ra
      DOSEMETER
A dosemeter combined with an ionization
chamber.
f   dosimètre *m* à chambre d'ionisation
e   dosímetro *m* de cámara de ionización
i   dosimetro *m* a camera d'ionizzazione
n   dosismeter met ionisatievat
d   Dosismesser *m* mit Ionisationskammer

3665 IONIZATION CHAMBER   ma
      EXPOSURE RATEMETER
A measuring assembly for the exposure,
in which the detector is a built-in or
separate ionization chamber.
f   débitmètre *m* d'exposition à chambre
    d'ionisation
e   intensímetro *m* con cámara de ionización
i   rateometro *m* d'esposizione a camera
    d'ionizzazione
n   exposietempometer met ionisatievat
d   Dosisleistungsmesser *m* mit Ionisations-
    kammer

3666 IONIZATION CHAMBER WITH   ic
      INTERNAL GAS SOURCE
An ionization chamber in which the filling
gas consists in all or in part of the radioactive gas whose activity is to be
measured.
f   chambre *f* d'ionisation à source interne
    gazeuse
e   cámara *f* de ionización de fuente interna
    de gas
i   camera *f* d'ionizzazione a sorgente
    interna di gas
n   ionisatievat *n* met interne gasbron
d   Gasfüll-Ionisationskammer *f*,
    Ionisationskammer *f* mit innerer
    Gasquelle

3667 IONIZATION COUNTER   ct, ic
An ionization chamber which has no
internal amplification by gas multiplication
and which is used for counting ionizing
particles.
f compteur m à ionisation
e contador m de ionización
i contatore m ad ionizzazione
n ionisatieteller·
d ·Ionisationszähler m

3668 IONIZATION CROSS SECTION   cs, np
The probability that a particle or photon
passing through matter will undergo
ionization by collision.
f section f d'ionisation
e sección f de ionización
i sezione f d'ionizzazione
n ionisatiedoorsnede
d Ionisationsquerschnitt m,
   Ionisierungsquerschnitt m

3669 IONIZATION CURRENT, see 2966

3670 IONIZATION DEFECT   np
For a heavy charged particle stopped in
a gas, the difference between the total
kinetic energy of the particle and the
total kinetic energy of the alpha particle
causing the same amount of ionization
in the gas.
f défaut m d'ionisation
e defecto m de ionización
i difetto m d'ionizzazione
n ionisatiedefect n
d Ionisationsdefekt m

3671 IONIZATION DEGREE, see 2808

3672 IONIZATION DENSITY   np
The number of ion pairs divided by
volume at a given moment in an irradiated
substance.
   densité f d'ionisation
e densidad f de ionización
i densità f d'ionizzazione
n ionisatiedichtheid
d Ionisierungsdichte f

3673 IONIZATION ENERGY,   np
       IONIZING ENERGY
The minimum energy needed to ionize an
atom or a molecule being in the
fundamental state.
f énergie f d'ionisation
e energía f de ionización
i energia f d'ionizzazione
n ionisatie-energie
d Ionisationsenergie f

3674 IONIZATION GAGE (US),   me
       IONIZATION GAUGE (GB),
       IONIZATION MANOMETER
A ga(u)ge in which the rate of collection
of positive ions on an electrode is used
as a measure of the residual gas pressure
in the tube.
f manomètre m à ionisation
e manómetro m de ionización
i manometro m ad ionizzazione
n ionisatiemanometer
d Ionisationsmanometer n,
   Ionisierungsmanometer n

3675 IONIZATION PATH,   ic, np
       IONIZATION TRACK
The trail of ion pairs produced by an
ionizing particle in its passage through
matter.
f parcours m d'ionisation
e trayectoria f de ionización
i percorso m d'ionizzazione
n ionisatiebaan
d Ionisationsstrecke f,
   Ionisierungsstrecke f

3676 IONIZATION POTENTIAL   ch, np
For a particular kind of atom, the energy
per unit charge to remove an electron
from the atom to an infinite distance.
f potentiel m d'ionisation
e potencial m de ionización
i potenziale m d'ionizzazione
n ionisatiepotentiaal
d Ionisationspotential n

3677 IONIZATION PRESSURE   ch
An increase in the pressure within a
gaseous discharge tube due to ionization.
f pression f d'ionisation
e presión f de ionización
i pressione f d'ionizzazione
n ionisatiedruk
d Ionisierungsdruck m

3678 IONIZATION RATE   np
The number of pairs of ions with opposite
charges produced per unit volume of gas
in unit time.
f taux m d'ionisation,
   vitesse f d'ionisation
e porcentaje m de ionización,
   velocidad f de ionización
i velocità f d'ionizzazione
n ionisatiesnelheid
d Ionisationsgeschwindigkeit f,
   Ionisierungsgeschwindigkeit f

3679 IONIZATION SPECTROMETER, see 764

3680 IONIZATION TIME   ec
The time interval between the initiation
of conditions for and the establishment of
conduction at some stated value of tube
voltage drop.
f temps m d'ionisation
e tiempo m de ionización
i tempo m d'ionizzazione
n ionisatietijd
d Ionisationszeit f, Ionisierungszeit f

3681 IONIZATION TRACK   ra
The visible manifestation of the path of an
ionizing particle in a cloud chamber,

bubble chamber or in a nuclear emulsion.
f   trace f d'ionisation
e   traza f de ionización
i   traccia f d'ionizzazione
n   ionisatiespoor n
d   Ionisationsspur f

3682 IONIZED ATOM   np
An ion, which is an atom that has acquired
an electric charge by gain or loss of
electrons surrounding its nucleus.
f   atome m ionisé
e   átomo m ionizado
i   atomo m ionizzato
n   geïoniseerd atoom n
d   Atomion n, ionisiertes Atom n

3683 IONIZED-GAS ANEMOMETER   ma
An assembly for the measurement of the
velocity of gas and comprising an
ionizing radiation source included in an
ionization chamber through which the gas
studied is flowing, the value of the velocity
being determined from the current of the
chamber.
f   anémomètre m à ionisation
e   anemómetro m de gas ionizado
i   anemometro m ad ionizzazione
n   ionisatie-anemometer
d   Anemometer n nach der Gasionisierungs-
    methode

3684 IONIZING COLLISION   cr
One of the collision forms when a
cosmic-ray primary enters the atmosphere
and collides with a molecule of air.
f   collision f ionisant
e   colisión f ionizante
i   collisione f ionizzante
n   ioniserende botsing
d   ionisierender Stoss m

3685 IONIZING EVENT   np
A process in which an ion or a group of
ions is produced by interaction of a
single particle or photon with matter.
f   événement m ionisant
e   acontecimiento m ionizante inicial
i   evento m ionizzante
n   ionisatieproces n
d   Ionisierungsereignis n

3686 IONIZING PARTICLE   np
A particle that directly produces ion pairs
in its passage through a substance.
f   particule f ionisante
e   partícula f ionizante
i   particella f ionizzante
n   ioniserend deeltje n
d   ionisierendes Teilchen n

3687 IONIZING RADIATION   ra
Any electromagnetic or particulate
radiation capable of producing ions, directly or indirectly, in its passage through
matter.
f   rayonnement m ionisant

e   radiación f ionizante
i   radiazione f ionizzante
n   ioniserende straling
d   ionisierende Strahlung f

3688 IONIZING RADIATION BACK-   me
       SCATTER SOIL DENSITY METER
A portable density meter designed to
determine soil density by measurement
of the radiation backscattered by the soil.
f   densimètre m de sol à rétrodiffusion de
    rayonnement ionisant
e   densímetro m por retrodispersión por
    radiación ionizante
i   densimetro m del suolo a sparpagliamento
    all'indietro di radiazione ionizzante
n   dichtheidsmeter voor grond berustend op
    terugverstrooiing van ioniserende straling
d   Bodendichtemessgerät n nach der Rück-
    streuungsmethode

3689 IONIZING RADIATION CALCIUM   me
       AND IRON CONTENT IN ORE METER
A content meter designed to determine
continuously calcium and iron content of
ore samples by measurement of the
characteristic X-radiation of these two
metals.
f   teneurmètre m en fer et calcium de
    minerais par rayonnement ionisant
e   valorímetro m de hierro y calcio para
    minerales por radiación ionizante
i   tenorimetro m in ferro e calcio di
    minerali a radiazione ionizzante
n   ijzer-calciumgehaltemeter berustend
    op ioniserende straling
d   Gerät n zur Bestimmung des Kalzium-
    und Eisengehalts in Erzen mittels
    ionisierender Strahlung

3690 IONIZING RADIATION   ma
       DENSITY METER
A measuring assembly that includes an
ionizing radiation source and is designed
to determine either the density of the
material or the average specific gravity
of a heterogeneous mixture, using the
variation, within a defined geometry, of
the absorption or diffusion of radiation.
f   densimètre m par rayonnement ionisant
e   densímetro m por radiación ionizante
i   densimetro m a radiazione ionizzante
n   dichtheidsmeter berustend op ioniserende
    straling
d   Dichtemessgerät n mittels ionisierender
    Strahlung

3691 IONIZING RADIATION   ma
       FOLLOWING LEVEL METER
A level meter including an on-off type
level indicator associated with a servo-
mechanism by means of which the
source-detector set is compelled to
follow the level.
f   ensemble m de mesure de niveau à
    poursuite automatique par rayonnement
    ionisant,

limnimètre *m* à poursuite automatique par
rayonnement ionisant
e indicador *m* móvil de nivel por radiación
ionizante
i livellometro *m* ad inseguimento
automatico a radiazione ionizzante
n volgniveaumeter berustend op ioniserende
straling
d Füllstandsmessgerät *n* mittels
ionisierender Strahlung mit Nachlauf

**3692 IONIZING RADIATION   ma
LEVEL METER**
A measuring assembly that includes an
ionizing radiation source and is designed
for the measurement or indication of the
level in a container of liquid or granular
substances, even when direct access to
that level is not possible.
f limnimètre *m* par rayonnement ionisant
e nivelímetro *m* por radiación ionizante
i livellometro *m* a radiazione ionizzante
n niveaumeter berustend op ioniserende
straling
d Füllstandsmessgerät *n* mittels
ionisierender Strahlung

**3693 IONIZING RADIATION ON-OFF   ma
LEVEL INDICATOR**
An indicator including an ionizing
radiation source and determining the
absence or the presence, on the path
between source and detector, of the
material contained in an enclosure.
f signaleur *m* de dépassement de niveau par
rayonnement ionisant
e indicador *m* de presencia de nivel por
radiación ionizante
i segnalatore *m* di livello a soglia a
radiazione ionizzante
n niveau-indicator berustend op
ioniserende straling
d Grenzfüllstandsanzeigegerät *n* mittels
ionisierender Strahlung

**3694 IONIZING RADIATION   ra
PROXIMITY INDICATOR**
An indicator including an ionizing radiation
source and a radiation detector, designed
to give an estimate of the relative
proximity of two objects, by using the
direct or scattered radiation.
f signaleur *m* de proximité par rayonnement
ionisant
e indicador *m* de proximidad por radiación
ionizante
i segnalatore *m* di prossimità a radiazione
ionizzante
n nabijheidsindicator berustend op
ioniserende straling
d Wegindikator *m* durch ionisierende
Strahlung

**3695 IONIZING RADIATION SOIL   ma
MOISTURE METER,
NEUTRON MOISTURE METER**
A content meter including a fast neutron
source and designed to determine soil

water contents through counting the
neutrons moderated by the hydrogen nuclei
in the water molecules.
f humidimètre *m* de sol par rayonnement
ionisant
e valorímetro *m* de la humedad del suelo
por radiación ionizante
i umidimetro *m* del suolo a radiazione
ionizzante
n vochtigheidsmeter voor grond berustend
op ioniserende straling
d Gerät *n* zur Bestimmung der Boden-
feuchtigkeit mittels ionisierender
Strahlung

**3696 IONIZING RADIATION   ma
STATIC LEVEL METER**
A level meter for continuous measurement
of a level and that includes a fixed ionizing
radiation source and a detector disposed
in such a way that the radiation imparted
to the detector is a function of the level
value.
f ensemble *m* statique de mesure de niveau
par rayonnement ionisant,
statolimnimètre *m* par rayonnement
ionisant
e nivelímetro *m* estático por radiación
ionizante
i livellometro *m* statico a radiazione
ionizzante
n statische niveaumeter berustend op
ioniserende straling
d Füllstandsmessgerät *n* mit feststehender
Strahlungsquelle

**3697 IONIZING RADIATION SULPHUR   ma
CONTENT METER FOR HYDRO-
CARBONS**
A content meter designed to determine the
sulphur content of hydrocarbons by
measurement of residual radiation after
absorption in the hydrocarbon.
f sulfoteneurmètre *m* d'hydrocarbures,
teneurmètre *m* en soufre d'hydrocarbures
e valorímetro *m* de azufre para hidrocarburos por radiación ionizante
i solfotenorimetro *m* per idrocarburi a
radiazione ionizzante
n gehaltemeter voor zwavel in koolwaterstoffen berustend op ioniserende straling
d Gerät *n* zur Bestimmung des Schwefel-
gehalts in Kohlenwasserstoffen mittels
ionisierender Strahlung

**3698 IONIZING RADIATION   ma
THICKNESS METER**
A measuring assembly that includes an
ionizing radiation source and is designed
for non-destructive measurement of the
thickness of a material by means of
ionizing radiation.
f épaisseurmètre *m* par rayonnement
ionisant
e calibrador *m* por radiación ionizante
i spessimetro *m* a radiazione ionizzante
n diktemeter berustend op ioniserende
straling

d Dickenmessgerät *n* mittels ionisierender
   Strahlung

**3699 IONIZING RADIATION TRANS-   ma
       MISSION DENSITY METER**
   A thickness meter including an ionizing
   radiation source and designed to determine
   material thickness by measurement of the
   radiation transmitted through this
   material.
f épaisseurmètre *m* à transmission de
   rayonnement ionisant
e densímetro *m* por transmisión de
   radiación ionizante
i densimetro *m* a trasmissione di radiazione
   ionizzante
n dichtheidsmeter berustend op transmissie
   van ioniserende straling
d Gerät *n* zur Bestimmung der Dichte
   mittels Durchstrahlung

**3700 IONIZING RADIATION TRANS-   ma
       MISSION SOIL DENSITY METER**
   A portable density meter designed to
   determine soil density by measurement
   of the radiation transmitted through the
   soil.
f densimètre *m* de sol à transmission de
   rayonnement ionisant
e densímetro *m* del suelo de transmisión
   por radiación ionizante
i densimetro *m* del suolo a trasmissione
   di radiazione ionizzante
n dichtheidsmeter voor grond berustend
   op transmissie van ioniserende straling
d Gerät *n* zur Bestimmung der Bodendichte
   mittels Durchstrahlung

**3701 IONIZING RADIATION TRANS-   ma
       MISSION THICKNESS METER**
   Thickness meter measuring the radiation
   transmitted through the material.
f densimètre *m* à transmission de rayonnement ionisant
e calibrador *m* por transmisión de radiación
   ionizante
i spessimetro *m* a trasmissione di radiazione ionizzante
n diktemeter berustend op transmissie van
   ioniserende straling
d Gerät *n* zur Bestimmung der Dicke
   mittels Durchstrahlung

3702 IONOGENIC   ch
   Descriptive of materials which develop
   or supply ions.
f ionogène adj
e ionógeno adj
i ionogeno adj
n ionogeen adj
d ionogen adj

3703 IRIDIUM   ch
   Metallic element. symbol Ir. atomic
   number 77.
f iridium *m*
e iridio *m*

i iridio *m*
n iridium *n*
d Iridium *n*

3704 IRINITE   mi
   A thorium variety of loparite, containing
   about 11 % of Th.
f irinite *f*
e irinita *f*
i irinite *f*
n iriniet *n*
d Irinit *m*

3705 IRON   ch
   Metallic element, symbol Fe, atomic
   number 26.
f fer *m*
e hierro *m*
i ferro *m*
n ijzer *n*
d Eisen *n*

3706 IRON ABSORBER   ab
   Absorber used in connection with a
   Mössbauer source.
f absorbant *m* en fer
e absorbedor *m* de hierro
i assorbente *m* in ferro
n ijzerabsorbens *n*
d Eisenabsorbens *n*

**3707 IRRADIATED URANIUM   ma
       REPROCESSING CONTROL ASSEMBLY**
   An assembly designed for the continuous
   monitoring of fission product contents in
   uranium chemical reprocessing solutions.
f ensemble *m* de contrôle du traitement de
   l'uranium irradié
e conjunto *m* regulador del tratamiento de
   uranio irradiado
i complesso *m* di controllo dell'uranio
   irradiato
n opstelling voor de controle bij de opwerking van bestraald uranium
d Überwachungsanlage *f* für die Aufbereitung von bestrahltem Uran

3708 IRRADIATION   ra
   Exposure to ionizing radiation.
f radioexposition *f*
e irradiación *f*
i irradiazione *f*
n bestraling
d Bestrahlung *f*

3709 IRRADIATION CHANNEL, see 2443

**3710 IRRADIATION-CHEMICAL   ch, ms
       SYNTHESIS OF MERCAPTANS**
   A process used to manufacture inter-
   mediates in the production of synthetic
   rubber.
f synthèse *f* chimique de mercaptans par
   irradiation
e síntesis *f* química de mercaptanos por
   irradiación
i sintesi *f* chimica di mercaptani per
   irradiazione

n   synthetische vervaardiging van mercap-
    tanen met behulp van bestraling
d   synthetische Herstellung *f* von Merkap-
    tanen mittels Bestrahlung

3711 IRRADIATION CORROSION   mg
    Corrosion due to the presence of ionizing
    radiation.
f   corrosion *f* par irradiation
e   corrosión *f* por irradiación
i   corrosione *f* sotto irradiazione
n   bestralingscorrosie
d   Bestrahlungskorrosion *f*

3712 IRRADIATION CREEP   mg, ra
    The creep resulting from irradiation.
f   fluage *m* d'irradiation
e   fluencia *f* de irradiación
i   deformazione *f* plastica d'irradiazione,
    scorrimento *m* d'irradiazione
n   bestralingskruip
d   Bestrahlungskriechen *n*

3713 IRRADIATION CYLINDER   ra
    The complete set of apparatus for carrying
    out irradiation, enclosed in a tubular
    container.
f   cylindre *m* d'irradiation
e   cilindro *m* de irradiación
i   astuccio *m* d'irradiazione
n   bestralingsbus
d   Bestrahlungsbuchse *f*

3714 IRRADIATION GROWTH, see 3101

3715 IRRADIATION HARDENING   mg
    Hardening provoked in materials as a
    consequence of irradiation.
f   durcissement *m* par irradiation
e   endurecimiento *m* por irradiación
i   indurimento *m* da irradiazione
n   bestralingsharding
d   Bestrahlungshärtung *f*

3716 IRRADIATION HEAT   ra
    Energy dissipated in the form of heat in a
    material due to irradiation.
f   chaleur *f* d'irradiation,
    chauffage *m* par irradiation
e   calentamiento *m* por irradiación,
    calor *m* de irradiación
i   calore *m* da irradiazione,
    riscaldamento *m* da irradiazione
n   bestralingswarmte
d   Bestrahlungswärme *f*

3717 IRRADIATION LEVEL, see 827

3718 IRRADIATION REACTOR   rt
    A nuclear reactor used primarily as a
    source of nuclear radiation for irradiation
    of materials or for medical purposes.
f   réacteur *m* d'irradiation
e   reactor *m* de irradiación
i   reattore *m* d'irradiazione
n   bestralingsreactor
d   Bestrahlungsreaktor *m*

3719 IRRADIATION TEST, see 567

3720 IRREVERSIBLE PROCESS   ch
    A process which takes place in one
    direction only, and therefore proceeds to
    completion.
f   procédé *m* irréversible
e   procedimiento *m* irreversible
i   processo *m* irreversibile
n   onomkeerbaar proces *n*
d   irreversibles Verfahren *n*

3721 ISHIKAWAITE   mi
    A samarskite containing about 20 % UO_2.
f   ishikawaite *f*
e   ishikawaita *f*
i   ishikawaite *f*
n   ishikawaiet *n*
d   Ishikawait *m*

3722 ISITRON METHOD, see 2199

3723 ISLANDS OF ISOMERISM   ch
    The regions of the periodic table
    (around Z = 50 and Z = 82) where large
    numbers of isomeric nuclides occur.
f   îles *pl* d'isomérie
e   islas *pl* de isomería
i   isole *pl* d'isomeria
n   isomerie-eilanden *pl*
d   Isomerieinsel *pl*

3724 ISOBARIC SPACE,   np
    ISOTOPIC SPACE
    A symbolic space in which certain
    orientations occur.
f   espace *m* isobarique, espace *m* isotopique
e   espacio *m* isobárico, espacio *m* isotópico
i   spazio *m* isobarico, spazio *m* isotopico
n   spinruimte
d   Spinkonfigurationsraum *m*, Spinraum *m*

3725 ISOBARIC SPIN,   is
    ISOSPIN, ISOTOPIC SPIN
    The spin characteristic for the two
    quantum states, the proton and the neutron,
    of one and the same particle.
f   spin *m* isotopique
e   espín *m* isotópico
i   spin *m* isotopico
n   isotopenspin
d   Isospin *m*, Isotopenspin *m*

3726 ISOBARIC SPIN   is
    QUANTUM NUMBER,
    ISOTOPIC SPIN QUANTUM NUMBER,
    ISOTOPIC VARIABLE
    A nuclear quantum number based on the
    view that the proton and the neutron are
    different states of the same elementary
    particle, the nucleon.
f   nombre *m* quantique de spin isotopique
e   número *m* cuántico de espín isotópico
i   numero *m* quantico di spin isotopico
n   isotopenspinquantumgetal *n*
d   Isospinquantenzahl *f*

3727 ISOBARIC TRANSFORMATION,   np
     ISOBARIC TRANSMUTATION
Any nuclear transformation in which the
resulting nucleus is an isobar, i.e. has
essentially the same nuclear mass as the
initial nucleus.
f transmutation *f* isobarique
e transmutación *f* isobárica
i trasmutazione *f* isobarica
n isobare transmutatie
d isobare Transmutation *f*

3728 ISOBARIC TRIAD   np
A group of three neighbo(u)ring isobaric
nuclides.
f triplet *m* isobarique
e tríada *f* isobárica
i triade *f* isobarica
n isobaar trio *n*
d isobares Tripel *n*

3729 ISOBARS,   np
     NUCLEAR ISOBARS
Nuclides having the same mass number
but different atomic numbers.
f nucléides *pl* isobares
e isobaros *pl* nucleares
i isobari *pl* nucleari
n kernisobaren *pl*
d Kernisobaren *pl*

3730 ISOCHRONE CYCLOTRON   pa
A cyclotron in which the guide field is
arranged in such a way that even at the
highest energies, the accelerated ions
can maintain constant their angular
velocities notwithstanding the increase of
their relativistic mass.
f cyclotron *m* isochrone
e ciclotrón *m* isócrono
i ciclotrone *m* isocrono
n isochroon synchrotron *n*
d Isochronsynchrotron *n*

3731 ISOCOUNT CONTOURS (GB),   ct
     ISOPULSE CONTOURS (US)
The curves formed by the intersection of
a series of isocount surface with a
specified surface.
f courbes *pl* d'isocomptage
e curvas *pl* de isocontaje,
   curvas *pl* de isocuenta
i curve *pl* d'isoconteggio
n isopulskrommen *pl*
d Isoimpulskurven *pl*,
   Kurven *pl* gleicher Zählrate

3732 ISOCOUNT SURFACE (GB),   np
     ISOPULSE SURFACE (US)
A surface on which the counting rate is
everywhere the same.
f surface *f* d'isocomptage
e superficie *f* de isocontaje,
   superficie *f* de isocuenta
i superficie *f* d'isoconteggio
n isopulsvlak *n*
d Ebene *f* gleicher Zählrate,
   Isoimpulsebene *f*

3733 ISODIAPHERES   np
Nuclides having the same value of (N-Z),
the difference between the number of
neutrons and protons.
f isodiaphères *pl*
e isodiáferos *pl*
i isodiaferi *pl*
n isodiaferen *pl*
d Isodiaphere *pl*

3734 ISODOSE   ab
Descriptive of a locus at every point of
which the absorbed dose is the same.
f isodose *f*
e isodosis *f*
i isodose *f*
n isodosis
d Isodosis *f*

3735 ISODOSE CHART   ra
Chart showing the distribution of radiation
in a medium by means of lines or surfaces
drawn through points receiving equal
doses.
f carte *f* d'isodoses
e gráfica *f* de isodosis
i tavola *f* d'isodosi
n isodosistabel
d Isodosistafel *f*

3736 ISODOSE CONTOUR,   ra
     ISODOSE CURVE
The curve obtained at the intersection of
a particular isodose surface with a given
plane.
f courbe *f* d'isodose
e curva *f* de isodosis
i curva *f* d'isodose
n isodosiskromme
d Isodosiskurve *f*

3737 ISODOSE EQUALIZING, see 2694

3738 ISODOSE SURFACE   ra
A surface on which the dose received is
everywhere the same.
f surface *f* d'isodose
e superficie *f* de isodosis
i superficie *f* d'isodose
n isodosisvlak *n*
d Isodosisfläche *f*

3739 ISOELECTRONIC   np
Pertaining to similar electronic
arrangements.
f isoélectronique adj
e isoelectrónico adj
i isoelettronico adj
n iso-elektronisch adj
d gleichelektronisch adj, isoelektronisch adj

3740 ISOELECTRONIC SEQUENCE   np
A series of atoms having the same extranuclear electronic configuration.
f atomes *pl* isoélectroniques
e átomos *pl* isoelectrónicos
i atomi *pl* isoelettronici
n iso-elektronische atomen *pl*
d isoelektronische Atome *pl*

3741 ISOMER SEPARATION   ch
   Chemical separation of isomers, made
   possible when the radiation emitted in
   their formation has different effects on
   chemical bonds.
f séparation ƒ d'isomères
e separación ƒ de isómeros
i separazione ƒ d'isomeri
n isomerenscheiding
d Isomerentrennung ƒ

3742 ISOMERIC NUCLEI,   np
        ISOMERS, NUCLEAR ISOMERS
   Nuclides having the same mass number
   and atomic number, but occupying different
   nuclear energy states.
f nucléides pl isomères
e isómeros pl nucleares
i isomeri pl nucleari
n kernisomeren pl
d Kernisomeren pl

3743 ISOMERIC STATE   np
   An excited nuclear state having a mean
   life long enough to be observed.
f état m isomérique
e estado m isomérico
i stato m isomerico
n isomere toestand
d isomerer Zustand m

3744 ISOMERIC TRANSITION   np
   A transition between two isomeric states
   of a nucleus or from an isomeric state to
   the ground state.
f transition ƒ isomérique
e transición ƒ isomérica
i transizione ƒ isomerica
n isomere overgang
d isomerer Übergang m

3745 ISOMERISM   np
   The occurrence of isomers.
f isomérie ƒ
e isomería ƒ, isomerismo m
i isomeria ƒ
n isomerie
d Isomerie ƒ

3746 ISOMORPHIC,   cr
        ISOSTRUCTURAL
   Having the same crystalline form.
f isomorphe adj
e isomorfo adj
i isomorfo adj
n isomorf adj
d isomorph adj

3747 ISOMORPHISM   cr
   Exhibited by substances in the strictest
   sense when they have analogous crystalline
   structures and are mutually soluble in
   the solid state.
f isomorphisme m
e isomorfismo m
i isomorfismo m
n isomorfie
d Homöomorphie ƒ, Isomorphie ƒ

3748 ISORAD MAP   mi
   A map which gives a picture of the isorad
   family.
f plan-compteur m
e mapa ƒ de isorrados
i carta ƒ d'isoradi
n isoradenkaart
d Isoradenplan m

3749 ISORADS   mi
   Curves of equal radioactivity of soils
   obtained by measures carried out
   systematically at the surface, in the quarry
   or in the pit.
f isorades pl
e isorrados pl
i isoradi pl
n isoraden pl
d Isoraden pl

3750 ISOSTERIC MOLECULE   ch
   One or two or more molecules possessing
   essentially the same valence configuration,
   usually the same total number and
   arrangement of valence electrons.
f molécule ƒ isostérique
e molécula ƒ isoestérica
i molecola ƒ isosterica
n isostere molecule
d isosteres Molekül n

3751 ISOTONES   np
   Nuclides having the same neutron numbers
   but different atomic numbers.
f isotones pl
e isótonos pl
i isotoni pl
n isotonen pl
d Isotonen pl

3752 ISOTOPE BALANCE   is
   Material balance applied to a particular
   isotope.
f bilan m isotopique
e balance m isotópico
i bilancio m isotopico
n isotopenbalans
d Isotopenbilanz ƒ

3753 ISOTOPE CASK (US),   is
        ISOTOPE CONTAINER,
        ISOTOPE FLASK (GB)
   A usually lead-lined flask for storing and
   transporting isotopes.
f château m de transport d'isotopes
e recipiente m blindado para isótopos
i bara ƒ d'isotopi
n isotopentransportvat n
d Isotopentransportgefäss n

3754 ISOTOPE CHART   is
   Any of a set of charts in which the
   properties of atomic nuclei are summarized
f diagramme m d'isotopes,
      table ƒ d'isotopes
e diagrama m de isótopos,
      gráfica ƒ de isótopos
i diagramma m d'isotopi,

tavola *f* d'isotopi
n isotopentabel
d Isotopentafel *f*

3755 ISOTOPE CONTAINING me
INSTRUMENT
A device for studying something by shining
on it the invisible radiation from a small
quantity of a suitable radioactive source.
f appareil *m* isotopique
e aparato *m* isotópico
i apparecchio *m* isotopico
n isotopenapparaat *n*
d Isotopenapparat *m*

3756 ISOTOPE HANDLING is
CALCULATOR,
SAFE HANDLING CALCULATOR
A calculator designed to give rapid and
direct solution to handling and shielding
problems associated with the more
commonly used gamma emitters.
f calculateur *m* de manipulation d'isotopes,
disque *m* à calcul pour la manipulation
d'isotopes
e calculador *m* de manipulación de
isótopos,
disco *m* calculador de manipulación de
isótopos
i calcolatore *m* per la manipolazione
d'isotopi,
disco *m* calcolatore da manipolare isotopi
n rekenschijf voor isotopenbehandeling
d Handhabungsberechner *m*,
Rechenscheibe *f* für Isotopenbehandlung

3757 ISOTOPE MILKER, see 2934

3758 ISOTOPE MIXING, ch
MIXING
In separation of isotopes by gaseous
diffusion through barriers, the process,
diffusion, turbulent convection or others
whereby the concentration gradient of the
lighter isotope normal to the diffusion
barrier is kept as small as possible.
f homogénéisation *f* du résidu
e homogenización *f* del residuo
i omogenizzazione *f* del residuo
n homogenisatie van het residu
d Homogenisierung *f* des Rückstands

3759 ISOTOPE MIXTURE, is
MIXTURE OF ISOTOPES
A chemical substance in which a number
of isotopes are present.
f mélange *m* d'isotopes, mélange *m* isotopique
e mezcla *f* de isótopos, mezcla *f* isotópica
i miscela *f* d'isotopi, miscela *f* isotopica
n isotopenmengsel *n*
d Isotopengemisch *n*

3760 ISOTOPE MIXTURE VALUE is
In isotope separation, a measure of the
difficulty of preparing a quantity of an
isotope mixture.
f valeur *f* de mélange isotopique
e valor *m* de mezcla isotópica
i valore *m* di miscela isotopica
n isotopenmengwaarde
d Isotopenmischwert *m*

3761 ISOTOPE PRODUCTION is
CALCULATOR
A calculator used to obtain readily the
specific activity of any radioisotope
produced by reactor irradiation.
f calculateur *m* de la production isotopique
e calculador *m* de la producción isotópica
i calcolatore *m* della produzione d'isotopi
n rekenschijf voor isotopenproduktie
d Isotopenproduktionsberechner *m*

3762 ISOTOPE PRODUCTION is, rt
REACTOR
A nuclear reactor primarily employed to
produce radioisotopes.
f réacteur *m* de production d'isotopes
e reactor *m* de producción de isótopos
i reattore *m* di produzione d'isotopi
n reactor voor isotopenproduktie
d Isotopenerzeugungsreaktor *m*

3763 ISOTOPE PRODUCTION UNIT is
f département *m* de production d'isotopes
e instalación *f* para producir isótopos
i impianto *m* di produzione d'isotopi
n isotopenproduktie-inrichting
d Isotopenproduktionswerkstatt *f*

3764 ISOTOPE SEPARATION is
The field of knowledge and practice
concerned with changing the relative
abundance of isotopes.
f séparation *f* d'isotopes
e separación *f* de isótopos
i separazione *f* d'isotopi
n isotopenscheiding
d Isotopentrennung *f*

3765 ISOTOPE SEPARATION FACTOR is
The ratio of the abundance ratio of two
isotopes after processing to their
abundance ratio before processing.
f facteur *m* de séparation
e factor *m* de separación
i fattore *m* di separazione
n scheidingsfactor
d Trennfaktor *m*

3766 ISOTOPE SEPARATION METHODS is
The methods used to separate isotopes
from isotope containing mixtures.
f méthodes *pl* de séparation d'isotopes
e métodos *pl* de separación de isótopos
i metodi *pl* di separazione d'isotopi
n isotopenscheidingsmethoden *pl*
d Isotopentrennungsverfahren *pl*

3767 ISOTOPE SEPARATION PLANT is
f installation *f* de séparation des isotopes
e equipo *m* de separación de los isótopos
i impianto *m* di separazione degl'isotopi
n isotopenscheidingsinstallatie
d Isotopentrennungsanlage *f*

3768 ISOTOPE SEPARATOR   is
An apparatus for chemically separating the
various isotopes of an element.
f séparateur *m* d'isotopes
e separador *m* de isótopos
i separatore *m* d'isotopi
n isotopenscheider
d Isotopentrenner *m*

3769 ISOTOPE SHIFT   sp
In atomic spectroscopy, the slight
difference in wavelength of a given line
in the spectrum emitted by one isotope
from that of the corresponding line emitted
by another isotope of the same element.
f déplacement *m* isotopique
e desplazamiento *m* isotópico
i spostamento *m* isotopico
n isotopieverschuiving
d Isotopieverschiebung *f*

3770 ISOTOPE SPECIFIC ACTIVITY   is
Total radioactivity of a given isotope
per gram of the radioactive isotope.
f activité *f* spécifique d'isotope
e actividad *f* específica de isótopo
i attività *f* specifica d'isotopo
n specifieke isotopenactiviteit
d spezifische Isotopenaktivität *f*

3771 ISOTOPE THERAPY   md
Radiotherapy by means of radioisotopes.
f thérapie *f* isotopique
e terapia *f* isotópica
i terapia *f* isotopica
n isotopentherapie
d Isotopentherapie *f*

3772 ISOTOPE TRANSPORT,   is
      TRANSPORT
The net rate at which the desired isotope
is carried towards the production end of
the plant.
f transport *m* d'isotope
e transporte *m* de isótopo
i trasporto *m* d'isotopo
n isotopentransport *n*
d Isotopentransport *m*

3773 ISOTOPES   is
Nuclides having the same atomic number
but different mass number.
f isotopes *pl*
e isótopos *pl*
i isotopi *pl*
n isotopen *pl*
d Isotope *pl*

3774 ISOTOPIC   is
Concerning isotopes.
f isotopique adj
e isotópico adj
i isotopico adj
n isotopen-, isotopisch adj
d Isotopen-, isotopisch adj

3775 ISOTOPIC ABUNDANCE   is
The relative number of atoms of a
particular isotope in a mixture of the
isotopes of an element, expressed as a
fraction of all the atoms of the element.
f abondance *f* isotopique,
   teneur *f* isotopique
e abundancia *f* isotópica
i abbondanza *f* isotopica
n abondantie, isotoopgehalte *n*
d Isotopenhäufigkeit *f*

3776 ISOTOPIC ANALYSIS   an, is
Determination of the isotope content.
f analyse *f* isotopique
e análisis *f* isotópica
i analisi *f* isotopica
n isotopenanalyse
d Isotopenanalyse *f*

3777 ISOTOPIC ATOMIC WEIGHT   is
The comparative atomic weight of an
isotope, calculated on the basis of an
atomic weight of 16.000 for the lighter
isotope of oxygen.
f poids *m* atomique isotopique
e peso *m* atómico isotópico
i peso *m* atomico isotopico
n isotopenatoomgewicht *n*
d Isotopenatomgewicht *n*

3778 ISOTOPIC BLOOD   md, ms
      VOLUME MEASUREMENT
Radioactive tracer method for blood
volume measurement.
f mesure *f* isotopique du volume sanguin
e medida *f* isotópica del volumen sanguíneo
i misura *f* isotopica del volume sanguigno
n isotopische bloedvolumemeting
d isotopische Blutvolumenmessung *f*

3779 ISOTOPIC CARRIER   is
A quantity of an element which may be
mixed with radioactive isotopes of that
element given a ponderable quantity, to
facilitate chemical operations.
f porteur *m* d'isotope
e portador *m* de isótopo
i portatore *m* d'isotopo
n isotopendrager
d Isotopenträger *m*

3780 ISOTOPIC COMPOSITION   ch, is
The percentage of the various isotopes
constituting a mixture.
f composition *f* isotopique
e composición *f* isotópica
i composizione *f* isotopica
n isotopische samenstelling
d isotopische Zusammensetzung *f*

3781 ISOTOPIC CONTINUOUS   me
      WEIGHING DEVICE
A device for continuously weighing materials on moving conveyors, measuring the
absorption of radiation by the material
occupying a unit area of the moving belt.

f   appareil *m* isotopique de pesage continu
e   aparato *m* isotópico de peso continuo
i   apparecchio *m* isotopico di pesatura
    continua
n   isotopenapparaat *n* voor continue weging
d   Isotopenapparat *m* für kontinuierliches
    Wägen

3782 ISOTOPIC DATING   is
    The determination of the age of, e.g., wood,
    wine, etc. by fixing the isotope age by
    means of counters.
f   datation *f* isotopique
e   determinación *f* de la edad por isótopos
i   datazione *f* isotopica
n   isotopische ouderdomsbepaling
d   isotopische Datierung *f*

3783 ISOTOPIC DENSITY GAGE (US),   me
      ISOTOPIC DENSITY GAUGE (GB)
    A device operating by measuring the
    amount of radiation that can penetrate a
    fixed thickness of the material under test.
f   densimètre *m* isotopique
e   densímetro *m* isotópico
i   densimetro *m* isotopico
n   isotopenapparaat *n* voor dichtheidsmeting
d   Gerät *n* zur Bestimmung der Dichte
    mittels Isotopen

3784 ISOTOPIC DILUTION   is
    The mixing of a particular nuclide with
    one or more of its isotopes.
f   dilution *f* isotopique
e   dilución *f* isotópica
i   diluizione *f* isotopica
n   isotopenverdunning
d   Isotopenverdünnung *f*

3785 ISOTOPIC DILUTION ANALYSIS   ch
    A method whereby the amount of some
    element in a specimen is found by
    observing how the isotopic composition of
    that element is changed by the addition
    of the known amount of a known alobar.
f   analyse *f* par dilution isotopique
e   análisis *f* por dilución isotópica
i   analisi *f* per diluizione isotopica
n   analyse door isotopenverdunning,
    isotopenverdunningsmethode
d   Isotopenverdünnungsanalyse *f*

3786 ISOTOPIC EFFECT   is
    Small differences that may be detectable in
    the chemical or physical properties of two
    isotopes of their compounds.
f   effet *m* isotopique
e   efecto *m* isotópico
i   effetto *m* isotopico
n   isotopie-effect *n*
d   Isotopieeffekt *m*

3787 ISOTOPIC ENRICHMENT   is
    A process by which the relative abundances
    of the isotopes of a given element are
    altered in a batch, thus producing the
    element enriched in a particular isotope.

f   enrichissement *m* isotopique
e   enriquecimiento *m* isotópico
i   arricchimento *m* isotopico
n   isotopenverrijking
d   Isotopenanreicherung *f*

3788 ISOTOPIC EQUILIBRIUM   is
    The relative abundance of the various
    isotopes of an element as occurring in
    natural material.
f   équilibre *m* isotopique
e   equilibrio *m* isotópico
i   equilibrio *m* isotopico
n   isotopisch evenwicht *n*
d   isotopisches Gleichgewicht *n*

3789 ISOTOPIC EXCHANGE   ch, is
    A process whereby isotopic atoms in
    different valency states or in different
    molecules, or in different sites in the
    same molecules, exchange places.
f   échange *f* isotopique
e   intercambio *m* isotópico
i   scambio *m* isotopico
n   isotopenuitwisseling
d   Isotopenaustausch *m*

3790 ISOTOPIC GENERATOR   ms
    A thermoelectric device powered by heat
    from a radioisotope.
f   générateur *m* isotopique
e   generador *m* isotópico
i   generatore *m* isotopico
n   isotopenstroombron
d   Isotopenstromquelle *f*

3791 ISOTOPIC INCOHERENCE   is
    In the interpretation of neutron
    diffraction the presence of isotopes in
    the diffracting material results in the
    introduction of incoherence.
f   incohérence *f* isotopique
e   incoherencia *f* isotópica
i   incoerenza *f* isotopica
n   isotopische incoherentie
d   isotopische Inkohärenz *f*

3792 ISOTOPIC INDICATOR, see 3443

3793 ISOTOPIC LABORATORY   is
f   laboratoire *m* isotopique
e   laboratorio *m* para isótopos
i   laboratorio *m* per isotopi
n   isotopenlaboratorium *n*
d   Isotopenlaboratorium *n*

3794 ISOTOPIC LEVEL DETECTOR   me, sa
    An apparatus for determining the level in
    a tank containing a liquid which is either
    under pressure or is hermetically sealed
    to prevent contact with the atmosphere.
f   détecteur *m* isotopique de niveau
e   detector *m* isotópico de niveles
i   rivelatore *m* isotopico di livelli
n   isotopische niveaudetector
d   isotopischer Niveaudetektor *m*

3795 ISOTOPIC LEVEL GAGE (US),   me
        ISOTOPIC LEVEL GAUGE (GB)
     An instrument used to determine the level
     of a substance in a closed container.
f limnimètre m isotopique
e nivelímetro m isotópico
i livellometro m isotopico
n isotopenapparaat n voor niveaumeting
d Isotopenfüllstandsmessgerät n

3796 ISOTOPIC MASS   is
     The relative mass referred to the mass
     of a neutral atom of the naturally most
     abundant isotope of oxygen ($_8O^{16}$) taken as
     16,00000.
f masse f isotopique
e masa f isotópica
i massa f isotopica
n isotopenmassa
d Isotopenmasse f

3797 ISOTOPIC MOISTURE GAGE (US),   me
        ISOTOPIC MOISTURE GAUGE (GB)
     An instrument depending on the slowing-
     down effect that hydrogen nuclei have upon
     neutrons having approximately the same
     mass, the neutrons being produced by the
     action of a radioactive source.
f humidimètre m isotopique
e valorímetro m isotópico de la humedad
i umidimetro m isotopico
n isotopenapparaat n voor vochtigheids-
     bestemming
d Isotopenapparat m zur Feuchtigkeits-
     bestimmung

3798 ISOTOPIC NUMBER, see 1782

3799 ISOTOPIC RATE OF EXCHANGE   is
     The velocity of the isotopic exchange
     reaction.
f vitesse f d'échange isotopique
e velocidad f de intercambio isotópico
i velocità f di scambio isotopico
n isotopenuitwisselingssnelheid
d Isotopenaustauschgeschwindigkeit f

3800 ISOTOPIC RATIO, see 26

3801 ISOTOPIC SPACE, see 3724

3802 ISOTOPIC SPIN, see 3725

3803 ISOTOPIC SPIN QUANTUM NUMBER,
        ISOTOPIC VARIABLE, see 3726

3804 ISOTOPIC THICKNESS GAGE (US),   me
        ISOTOPIC THICKNESS GAUGE (GB)
     A device usually measuring the basis
     weight or mass per unit area of the product,
     which must have substantially constant
     composition and therefore density if the
     figure is to be used as a true measure of
     thickness.

f épaisseurmètre m isotopique
e calibrador m isotópico
i spessimetro m isotopico
n isotopendiktemeter
d Isotopendickenmessgerät n

3805 ISOTRON   is
     A device for isotope separation based on
     the electrical sorting of ions.
f isotron m
e isotrón m
i isotrone m
n isotron n
d Isitron n

3806 ISOTROPIC BODY,   ch, gp
        ISOTROPIC MEDIUM
     A medium whose properties are the same
     in whatever direction they are measured.
f substance f isotropique
e substancia f isotrópica
i sostanza f isotropica
n isotrope stof
d isotrope Substanz f

3807 ISOTROPIC GRAPHITE   mt
     Used in nuclear reactors on account of its
     useful properties.
f graphite m isotropique
e grafito m isotrópico
i grafite f isotropica
n isotroop grafiet n
d isotropes Graphit n

3808 ISOTROPIC SCATTERING   np
     Scattering without a preferential direction.
f diffusion f isotropique
e dispersión f isotrópica
i deviazione f isotropica
n isotrope verstrooiing
d isotrope Streuung f

3809 ISOTROPIC SOURCE OF   ra
        RADIATION
     A source which emits equally in all
     directions.
f source f isotropique de rayonnement
e fuente f isotrópica de radiación
i sorgente f isotropica di radiazione
n isotrope stralingsbron
d isotrope Strahlungsquelle f

3810 ITERATED FISSION   np, rt
        EXPECTATION,
        ITERATED FISSION PROBABILITY
     In a critical reactor, the average value,
     after many generations, of the number of
     fissions per generation arising from the
     daughter neutrons of a given neutron.
f espérance f de descendance,
     espérance f de fission itérée,
     probabilité f de fission itérée
e esperanza f de fisión iterada
i probabilità f di fissione iterata,
     speranza f di discendenza
n verwachtingswaarde van herhaalde splijting
d asymptotische Spalterwartung f

3811 IZOD TEST   te
An impact test in which a standard notched
specimen supported at one end as a
cantilever beam is broken by the impact
of a moving pendulum.
f essai *m* d'Izod
e prueba *f* de Izod
i prova *f* d'Izod
n kerfslagproef van Izod
d Kerbschlagversuch *m* von Izod

J

3812   j-j COUPLING   np
A coupling in which the total angular
momenta of the individual particles
interact with one another.
f   couplage *m* j-j
e   acoplamiento *m* j-j
i   accoppiamento *m* j-j
n   j-j-koppeling
d   j-j-Kopplung *f*

3813   JACKET, see 858

3814   JACKETING, see 863

3815   JAVELIN-SHAPED FUEL ROD   fu
Fuel rods used in the fast breeder reactor
at Dounreay.
f   barre *f* combustible en forme de javelin
e   barra *f* combustible en forma de jabalina
i   barra *f* combustibile in forma di
giavellotto
n   speervormig splijtstofelement *n*
d   speerförmiger Brennstoffstab *m*

3816   JAW   cd
Component part of manipulators.
f   mordache *f*, mors *m*
e   mordaza *f*
i   ganascia *f*, griffa *f*
n   grijper
d   Greifer *m*, Klaue *f*

3817   JET PUMP,   vt
LIQUID JET PUMP
A pump in which a jet of high velocity
fluid is used to accelerate another fluid.
f   pompe *f* à jet de liquide
e   bomba *f* de chorro de líquido
i   pompa *f* a getto di liquido
n   vloeistofstraalpomp
d   Flüssigkeitsstrahlpumpe *f*

3818   JET SEPARATION,   is
NOZZLE JET SEPARATION
A method of separating isotopes by using
a unit containing the mixture into which
a jet of gas is introduced.
f   séparation *f* d'isotopes à jets de gaz
e   separación *f* de isótopos por chorros de
gas
i   separazione *f* d'isotopi per getti di gas
n   isotopenscheiding door gasstralen
d   Isotopentrennung *f* durch Gasstrahlen

3819   JIG,   mi
JIG TABLE
An apparatus for sorting and grading ores.

f   crible *m* vibrant
e   criba *f* vibradora
i   crivello *m* oscillante
n   deintoestel *n*, jigmachine, zeeftoestel *n*
d   Rüttelklassierer *m*, Schwingsieb *n*,
Siebsetzmaschine *f*

3820   JOHANNITE   mi
A secondary mineral containing about
50.8 % of U.
f   johannite *f*
e   johannita *f*
i   johannite *f*
n   johanniet *n*
d   Johannit *m*

3821   JOSHI EFFECT   np
In a gas discharge, the effect of light on
the discharge current.
f   effet *m* Joshi
e   efecto *m* Joshi
i   effetto *m* Joshi
n   joshi-effect *n*
d   Joshi-Effekt *m*

3822   JOULE HEATING,   pp
OHMIC HEATING
A method of plasma heating by the Joule
effect resulting from the resistance of
the plasma.
f   chauffage *m* ohmique,
chauffage *m* par effet Joule
e   calentamiento *m* Joule,
calentamiento *m* óhmico
i   riscaldamento *m* omico,
riscaldamento *m* per effetto Joule
n   verwarming door joule-effect
d   Joulesche Erwärmung *f*

3823   JUMP OF ELECTRONS, see 2114

3824   JUNCTION PARTICLE   ct, me
DETECTOR
A detector in which the sensitive area is
the junction region of a few tens of microns
in thickness, in which a strong electrical
field exists.
f   détecteur *m* de particules à jonction
semiconductrice
e   detector *m* de partículas de junta
semiconductora
i   rivelatore *m* di particelle a giunzione
semiconduttrice
n   deeltjesdetector met halfgeleidergrenslaag
d   Teilchendetektor *m* mit Halbleitergrenz-
schicht

K

3825 K CAPTURE, np
K ELECTRON CAPTURE
Electron capture from the K shell by the
nucleus of the atom.
f capture f K
e captura f K
i cattura f K
n K-vangst
d K-Einfang m

3826 K ELECTRON np
An electron having an orbit of such
dimensions that the electron constitutes
part of the first shell of electrons
surrounding the atomic nucleus.
f électron m K
e electrón m K
i elettrone m K
n K-elektron n
d K-Elektron n

3827 K/L RATIO np
The ratio of the number of internal
conversion electrons from the K shell
to the number of internal conversion
electrons from the L shell emitted in the
de-excitation of a nucleus.
f relation f K/L
e relación f K/L
i rapporto m K/L
n K/L-verhouding
d K/L-Verhältnis n

3828 K LINE sp
One of the characteristic lines in the
X-ray spectrum of an atom.
f raie f K
e línea f K
i riga f K
n K-lijn
d K-Linie f

3829 K MESON,
KAON, see 3201

3830 K RADIATION np, ra
The radiations emitted when K electrons
are excited.
f rayonnement m K
e radiación f K
i radiazione f K
n K-straling
d K-Strahlung f

3831 K SHELL np
The innermost of the hypothetical shells
or spherical regions surrounding the
nucleus of an atom, each shell containing
one or more orbital electrons.
f couche f K
e capa f K
i strato m K

n K-schil
d K-Schale f

3832 KAHLERITE mi
A rare secondary mineral containing
about 47 % of U.
f kahlérite f
e kahlerita f
i kahlerite f
n kahleriet n
d Kahlerit m

3833 KARYOCERITE mi
A mineral near melanocerite but containing
more Th.
f caryocérite f
e cariocerita f
i cariocerite f
n karyoceriet n
d Karyozerit m

3834 KARYOKINESIS, md
MITOSIS
In biology, a form of nuclear division in
which the daughter nuclei come to have
the same number and kinds of chromosomes as the parent nucleus, characteristic
of most divisions other than those
involving meiosis.
f caryocinèse f, division f mitotique,
mitose f
e cariocinesis f, mitosis f
i cariocinesi f, divisione f cariocinetica,
divisione f mitotica
n indirecte celdeling, karyokinese, mitose
d indirekte Zellteilung f, Karyokinese f,
Mitose f

3835 KARYOPLASM, see 1606

3836 KASOLITE mi
A mineral containing lead and uranium.
f kasolite f
e kasolita f
i kasolite f
n kasoliet n
d Kasolit m

3837 KATABATIC WIND FALL-OUT,
see 1966

3838 KELOID,
KELOMA, KELOS, see 1036

3839 KERMA ma, np
For indirectly ionizing particles, the sum
of the initial kinetic energies of all
charged particles liberated by indirectly
ionizing particles in an element of matter
divided by the mass of that element.
f kerma m
e kerma m

i   kerma *m*
n   kerma
d   Kerma *m*

3840 KERMA RATE   np
      The increment of kerma per unit time.
f   débit *m* de kerma
e   intensidad *f* kerma
i   intensità *f* kerma
n   kermatempo *n*
d   Kermarate *f*

3841 KERNEL   ma
      A function of two sets of variables used
      to define an integral operator.
f   noyau *m* intégral
e   núcleo *m* integral
i   nucleo *m* integrale
n   integraalkern
d   Integralkern *m*

3842 keV,   un
      KILO-ELECTRONVOLT
      A unit of energy, equal to 10^3eV.
f   keV *m*, kilo-électron-volt *m*
e   keV *m*, kilo-electrón-voltio *m*
i   chilo-elettronevolt *m*, keV *m*
n   keV, kilo-elektronvolt
d   keV *n*, Kilo-Elektronenvolt *n*

3843 KEY SUBSTANCE, see 552

3844 KHLOPINITE, see 1075

3845 KICK SORTER (GB), see 228

3846 KILOCURIE   un
      One thousand curies, symbol kCi.
f   kilocurie *m*
e   kilocurie *m*
i   chilocurie *m*
n   kilocurie
d   Kilocurie *n*

3847 KILOTON   un
      A unit used in specifying the yield of a
      fission bomb, equal to the explosive power
      of 1,000 tons of trinitrotoluene.
f   kilotonne *f*
e   kilotonelada *f*
i   chiloton *m*
n   kiloton
d   Kiloton *n*

3848 KILOTON BOMB   nw
      A nuclear bomb, the explosive power of
      which is equivalent to one thousand tons
      of trinitrotoluene.
f   bombe *f* à force explosive égale à 1000
      tonnes de TNT,
      bombe *f* kilotonne
e   bomba *f* de fuerza explosiva igual a
      1000 toneladas de TNT,
      bomba *f* kilotón
i   bomba *f* a forza esplosiva uguale a
      1000 tonnellate di TNT,
      bomba *f* chiloton

n   kilotonbom
d   Kilotonbombe *f*

3849 KINETIC BLURRING,   xr
      MOVEMENT BLUR
      Blurring of the radiogram by movement
      of the object.
f   flou *m* cinétique, flou *m* de mouvement
e   borrosidad *f* cinemática,
      borrosidad *f* debida al movimiento
i   sfocatura *f* di movimento,
      soffuso *m* di movimento
n   bewegingsonscherpte
d   Bewegungsunschärfe *f*

3850 KINETIC ENERGY,   gp
      Energy possessed by a body because of its
      motion.
f   énergie *f* cinétique
e   energía *f* cinética
i   energia *f* cinetica
n   arbeidsvermogen *n* van beweging,
      kinetische energie
d   Bewegungsenergie *f*, kinetische Energie *f*

3851 KINETIC INSTABILITY,   pp
      MICRO-INSTABILITY
      Instability which cannot be deduced from
      macroscopic plasma equations but refers
      to microscopic equations for plasma
      particle distribution functions.
f   instabilité *f* cinétique,
      micro-instabilité *f*
e   inestabilidad *f* cinética,
      microinestabilidad *f*
i   instabilità *f* cinetica,
      micro-instabilità *f*
n   kinetische instabiliteit,
      micro-instabiliteit
d   kinetische Instabilität *f*,
      Mikroinstabilität *f*

3852 KINETIC MOMENTUM   np, pp
      Momentum of a charged particle in an
      electromagnetic field.
f   moment *m* cinétique
e   momento *m* cinético
i   momento *m* cinetico
n   kinetisch moment *n*
d   kinetisches Moment *n*

3853 KINETIC PRESSURE   pp
      Kinetic energy density due to thermal
      agitation of plasma constituting particles.
f   pression *f* cinétique
e   presión *f* cinética
i   pressione *f* cinetica
n   kinetische druk
d   kinetischer Druck *m*

3854 KINETIC TEMPERATURE   pp
      Temperature of a particle system following
      a Maxwellian distribution law and in
      thermal equilibrium with the medium.
f   température *f* cinétique
e   temperatura *f* cinética
i   temperatura *f* cinetica

n   kinetische temperatuur
d   kinetische Temperatur *f*

3855 KINK INSTABILITY,   pp
      SAUSAGE INSTABILITY
      Plasma instability in which the local
      deformation (kink) tends to grow because
      the lines of force of the self-induced
      confining field are crowded on the concave
      side of the field.
f   instabilité *f* à coques
e   inestabilidad *f* por deformación local de
    crecimiento
i   instabilità *f* per deformazione locale
    d'accrescimento
n   instabiliteit door locale vervormings-
    aanwas
d   Instabilität *f* gegen Knickung

3856 KLEIN-GORDON EQUATION   np
      In nuclear quantum theory, a relativistic
      form of the Schrödinger equation.
f   équation *f* de Klein-Gordon
e   ecuación *f* de Klein-Gordon
i   equazione *f* di Klein-Gordon
n   vergelijking van Klein en Gordon
d   Klein-Gordonsche Gleichung *f*

3857 KLEIN-NISHINA FORMULA   ma, np
      A formula that expresses the cross
      section of an unbound electron for
      scattering of a photon in the Compton
      effect, as a function of the energy of the
      photon.
f   formule *f* de Klein-Nishina
e   fórmula *f* de Klein-Nishina
i   formula *f* di Klein-Nishina
n   klein-nishinaformule
d   Klein-Nishina-Formel *f*

3858 KLEIN PARADOX   np
      Consequences of the Dirac electron
      theory that a particle could penetrate a
      potential barrier of greater than 1 MeV
      by making a transition from a positive
      energy state to a negative energy state.
f   paradoxe *m* de Klein
e   paradoja *f* de Klein
i   paradosso *m* di Klein
n   paradox van Klein
d   Kleinsches Paradox *n*

3859 KNIGHT SHIFT   np
      The change in the magnetic flux density
      or in the frequency for nuclear magnetic
      resonance that results from the magnetic
      field of the oriented conduction electrons.
f   déplacement *m* de Knight
e   desplazamiento *m* de Knight
i   spostamento *m* di Knight
n   verschuiving van Knight
d   Knightsche Verschiebung *f*

3860 KNOCK-ON   np
      The process whereby a particle is set in
      motion on being struck by a fast-moving
      particle or photon.

f   communication *f* d'énergie, percussion *f*
e   comunicación *f* de energía, percusión *f*
i   spostamento *m*
n   aanstoten *n*
d   Anstossen *n*

3861 KNOCK-ON ATOM,   np
      KNOCKED-ON ATOM
      An atom of material which recoils after
      collision with an energetic particle such
      as a neutron, fission fragment, ion or
      atom moving through a solid.
f   atome *m* percuté
e   átomo *m* bombardeado,
    átomo *m* percutido
i   atomo *m* spostato
n   aangestoten atoom *n*
d   angestossenes Atom *n*

3862 KNOCK-ON PROTON,   np
      KNOCKED-ON PROTON,
      RECOIL PROTON
      The nucleus which receives half the
      neutron energy in a collision between a
      fast neutron and a nucleus of a hydrogen
      atom.
f   proton *m* percuté
e   protón *m* bombardeado,
    protón *m* percutido
i   protone *m* spostato
n   aangestoten proton *n*
d   angestossenes Proton *n*

3863 KNOCKING OUT, see 1867

3864 KNUDSEN FLOW, see 2827

3865 KRUSKAL LIMIT   pp
      The limiting value of the current of a
      discharge in plasma physics, given by the
      condition $I < \pi r_o^2 B_2 / L$, where I is the
      current, r_o and L the radius and length
      of the plasma and B_2 the axial stabilizing
      magnetic flux density.
f   limite *f* de Kruskal
e   límite *m* de Kruskal
i   limite *m* di Kruskal
n   kruskalgrens
d   Kruskal-Grenze *f*

3866 KRYPTON   ch
      Gaseous element, symbol Kr, atomic
      number 36.
f   krypton *m*
e   criptón *m*
i   cripton *m*
n   krypton *n*
d   Krypton *n*

3867 KRYPTON-85 SOURCE   ra
      A source in which krypton gas is sealed
      hermetically in a metal tube or in a
      container provided with a metal foil
      window.
f   source *f* à krypton-85

e fuente *f* de criptón-85
i sorgente *f* a cripton-85
n krypton-85-bron
d Krypton-85-Quelle *f*

3868 KRYPTOSCOPE   xr
A small portable fluorscopic screen at one
end of a light-tight hood for use in
undarkened areas.
f cryptoscope *m*
e criptoscopio *m*
i criptoscopio *m*
n kryptoscoop
d Kryptoskop *n*

3869 KURIE PLOT, see 2566

3870 KYMOGRAPH   xr
Mechanical device for radiographic
recording of the motion of an object.
f kymographe *m*
e quimógrafo *m*
i chimografo *m*
n kymograaf
d Kymograph *m*

3871 KYMOGRAPHY   xr
A method of recording in a single radiograph the excursions of moving organs
in the body.
f kymographie *f*
e quimografía *f*
i chimografia *f*
n kymografie
d Kymographie *f*

L

3872 L CAPTURE,   np
      L ELECTRON CAPTURE
Electron capture from the L shell by the
nucleus of the atom.
f capture f L
e captura f L
i cattura f L
n L-vangst
d L-Einfang m

3873 L ELECTRON   np
An electron having an orbit of such
dimensions that the electron constitutes
part of the second shell of electrons
surrounding the atomic nucleus, counting
out from the nucleus.
f électron m L
e electrón m L
i elettrone m L
n L-elektron n
d L-Elektron n

3874 L MESONS,   cr
      LIGHT MESONS
Mu-mesons and pi-mesons are sometimes
referred to as L mesons in order to
distinguish them from the heavier K
mesons.
f mésons pl L, mésons pl légers
e mesones pl L, mesones pl livianos
i mesoni pl L, mesoni pl leggeri
n L-mesonen pl, lichte mesonen pl
d L-Mesonen pl, leichte Mesonen pl

3875 L RADIATION   np, xr
One of a series of X-rays characteristic
of each element, that is emitted when that
element, commonly as the metal, is used
as an anti-cathode in an X-ray tube and
the electrons of its L shell are excited.
f rayonnement m L
e radiación f L
i radiazione f L
n L-straling
d L-Strahlung f

3876 L SHELL   np
The second layer of electrons about the
nucleus of an atom.
f couche f L
e capa f L
i strato m L
n L-schil
d L-Schale f

3877 LABELLED ATOM (GB),   is
      TAGGED ATOM (US)
A tracer atom which can be detected
easily, and which is introduced into a
system to study a process of structure.
f atome m marqué
e átomo m marcado
i atomo m marcato
n gemerkt atoom n
d markiertes Atom n

3878 LABELLED COMPOUND (GB),   ch, is
      TAGGED COMPOUND (US)
A material which contains labelled
molecules.
f composé m marqué
e compuesto m marcado
i composto m marcato
n gemerkte verbinding
d markierte Verbindung f

3879 LABELLED FERTILIZERS   is
Fertilizers containing an isotopic tracer.
f engrais pl minéraux marqués
e abonos pl marcados
i concimi pl marcati
n gemerkte kunstmeststoffen pl
d markierte Kunstdünger pl

3880 LABELLED INSECTICIDES (GB),   is
      TAGGED INSECTICIDES (US)
Insecticides containing an isotopic tracer.
f insecticides pl marquées
e insecticidas pl marcadas
i insetticidi pl marcati
n gemerkte insekticiden pl
d markierte Insektizide pl

3881 LABELLED MOLECULES (GB),   ch, is
      TAGGED MOLECULES (US)
Molecules containing an isotopic tracer.
f molécules pl marquées
e moléculas pl marcadas
i molecole pl marcate
n gemerkte moleculen pl
d markierte Moleküle pl

3882 LABILE,   ge
      UNSTABLE
A term used i.a. in chemistry and
frequently in isotope exchange reactions.
f instable adj, labile adj
e inestable adj, lábil adj
i instabile adj, labile adj
n instabiel adj, labiel adj, onstabiel adj
d instabil adj, labil adj

3883 LABORATORY SYSTEM   ma
A frame of reference attached to the
observer's laboratory and hence usually
at rest relative to the surface of the earth.
f système m de référence du laboratoire
e sistema m de referencia del laboratorio
i sistema m di riferenza del laboratorio
n laboratoriumcoördinatensysteem n
d Laboratoriumbezugssystem n

3884 LABYRINTH SEAL   cd
A seal which prevents fluid from escaping
past a rotating shaft.

f garniture f en labyrinthe,
   scellement m en labyrinthe
e junto m labiríntico,
   sello m labiríntico
i tenuta f a labirinto
n labyrintaf sluiting
d Labyrinthdichtung f,
   Labyrinthverschluss m

3885 LAMB SHIFT   np
   The very small energy difference between
   the $2^2P^1/2$ and $2^2S^1/2$ levels of the
   hydrogen atom, due to the interaction of
   the electron with an electromagnetic
   field.
f déplacement m de Lamb, effet m Bethe
e desplazamiento m de Lamb,
   efecto m Bethe
i spostamento m di Lamb
n bethe-effect n, lambverschuiving
d Bethe-Effekt m, Lambsche Verschiebung f

3886 LAMBDA LIMITING PROCESS   np
   Method of defining a point electron as the
   limiting case in which a time-like vector
   $\lambda\mu$ tends to zero.
f limitation f lambda
e limitación f lambda
i limitazione f lambda
n lambdabegrenzing
d Lambdabegrenzung f

3887 LAMBDA PARTICLE   np
   A hyperon having an extremely short life
   (about 3.7×10^{-10} second) and a mass
   between that of neutrons and deuterons.
f particule f lambda
e partícula f lambda
i particella f lambda
n lambdadeeltje n
d Lambdateilchen n

3888 LAMBERTITE,   mi
      URANOPHANE, URANOTIL
   A rare, uranium containing mineral.
f lambertite f, uranophane m, uranotile f
e lambertita f, uranofano m, uranotila f
i lambertite f, uranofano m, uranotile f
n lambertiet n, uranofaan n, uranotiel n
d Lambertit m, Uranophan m, Uranotil m

3889 LAMINAGRAPH, see 688

3890 LAMINOGRAPHY, see 689

3891 LAMINAR BOUNDARY LAYER,
      see 2599

3892 LAMINAR FLOW,   gp
      VISCOUS FLOW
   In dynamics, a particular type of streamline flow in which fluid in thin parallel
   layers tends to maintain uniform velocity.
f courant m laminaire,
   écoulement m laminaire
e flujo m laminar

i flusso m laminare, flusso m viscoso
n laminaire stroming
d Bandströmung f, laminare Strömung f

3893 LAMINATED SHIELD   sa
   A shield consisting of several homogeneous
   layers, e.g. steel and masonite.
f écran m lamellaire,
   écran m stratifié
e pantalla f laminada
i schermo m stratificato
n gelaagd scherm n
d Mehrlagenschild m

3894 LANCEVIN FUNCTION, see 1424

3895 LANGEVIN ION   np
   An ion moving under the influence of an
   electric field in a gas.
f ion m de Langevin
e ión m de Langevin
i ione m di Langevin
n langevinion n
d Langevinsches Ion n

3896 LANGMUIR DARK SPACE   ec
   A non-luminous region surrounding a
   negatively charged probe inserted in the
   positive column of a glow discharge.
f espace m sombre de Langmuir
e espacio m obscuro de Langmuir
i spazio m oscuro di Langmuir
n donkere ruimte van Langmuir
d Langmuirscher Dunkelraum m

3897 LANGMUIR FREQUENCY, see 2147

3898 LANGMUIR PROBE,   me, pp
      PLASMA FREQUENCY PROBE
   A probe which, when the ionosphere
   plasma is excited with a varying radiofrequency signal, measures plasma resonance frequency, from which measurement
   electron density is calculated.
f sonde f de fréquence de plasma,
   sonde f de Langmuir
e sonda f de frecuencia de plasma,
   sonda f de Langmuir
i sonda f di frequenza di plasma,
   sonda f di Langmuir
n langmuirsonde, plasmafrequentiesonde
d Langmuir-Sonde f,
   Plasmafrequenzsonde f

3899 LANGMUIR WAVES, see 2205

3900 LANTHANIDE CONTRACTION   np
   The sequence of decreasing crystal radii
   of the tripositive rare-earth ions with
   increasing atomic number.
f contraction f de lanthanides
e contracción f de lantánidos
i contrazione f di lantanidi
n lanthanidencontractie
d Lanthanidenkontraktion f

3901 LANTHANIDES   ch
  The rare-earth elements from atomic
  numbers 58 to 71, inclusive.
  f lanthanides *pl*
  e lantánidos *pl*
  i lantanidi *pl*
  n lanthaniden *pl*
  d Lanthanide *pl*

3902 LANTHANUM   ch
  Rare-earth metallic element, symbol La,
  atomic number 57.
  f lanthane *m*
  e lantano *m*
  i lantanio *m*
  n lanthanum *n*
  d Lanthan *n*

3903 LAPLACIEN, see 805

3904 LARMOR FREQUENCY   np
  The angular frequency of precession of
  a charged particle rotating in a magnetic
  field.
  f fréquence *f* de Larmor,
     fréquence *f* gyromagnétique
  e frecuencia *f* de Larmor
  i frequenza *f* di Larmor
  n larmorfrequentie
  d Larmor-Frequenz *f*

3905 LARMOR PRECESSION,   np
      LARMOR THEOREM
  The motion which a charged particle or
  system of charged particles subject to
  a central force directed towards a common
  point experiences when under the
  influence of a small uniform magnetic
  field.
  f précession *f* de Larmor,
     théorème *m* de Larmor
  e precesión *f* de Larmor,
     teorema *m* de Larmor
  i precessione *f* di Larmor,
     teorema *m* di Larmor
  n larmorprecessie, larmortheorema *n*
  d Larmor-Präzession *f*

3906 LARMOR PRECESSION   np
      FREQUENCY
  The frequency of the Larmor precession
  movement.
  f fréquence *f* de précession de Larmor
  e frecuencia *f* de precesión de Larmor
  i frequenza *f* di precessione di Larmor
  n larmorprecessiefrequentie
  d Larmor-Präzessionsfrequenz *f*

3907 LARMOR RADIUS   np
  The radius of the circular path followed
  by a particle having a given charge, mass
  and energy when it moves in a given
  magnetic field.
  f rayon *m* de Larmor
  e radio *m* de Larmor
  i raggio *m* di Larmor
  n larmorstraal
  d Larmor-Radius *m*

3908 LAST MOMENT   rt
      EMERGENCY SHUT-DOWN
  A shut-down system coming into operation
  when all other systems have failed.
  f arrêt *m* d'urgence dernière minute
  e paro *m* de emergencia de última hora
  i arresto *m* d'assoluto periodo
  n uiterste-nooduitschakeling
  d äusserster Notschnellschluss *m*

3909 LATENCY TIME,   ra
      LATENT PERIOD,
      RADIOLOGICAL LATENT PERIOD
  The period between exposure to radiation
  and the onset of a particular effect.
  f temps *m* de latence
  e tiempo *m* de latencia
  i tempo *m* di latenza
  n latente tijd
  d Latenzzeit *f*

3910 LATENT IMAGE FADING   ph
  The total or partial loss of a track by
  a particle having a given charge, mass
  and energy when it moves in a given
  magnetic field.
  f disparition *f* de l'image latente
  e desaparición *f* de la imagen latente
  i scomparto *m* dell'immagine latente
  n verdwijnen *n* van het latente beeld
  d Verschwinden *n* des latenten Bildes

3911 LATENT NEUTRONS   np, rt
  In nuclear reactor theory, a term which
  has been applied to those radioactive
  fission products which give rise to the
  delayed neutrons.
  f neutrons *pl* latents
  e neutrones *pl* latentes
  i neutroni *pl* latenti
  n latente neutronen *pl*
  d latente Neutronen *pl*

3912 LATENT TISSUE INJURY,   md
      LATENT TISSUE LESION,
      RADIOLOGICAL LATENT TISSUE
      INJURY
  Injury which does not become manifest
  until some time after irradiation and
  possibly until some other trauma has
  supervened.
  f lésion *f* latente de tissu
  e lesión *f* latente de tejido
  i lesione *f* latente di tessuto
  n latente weefselbeschadiging
  d latente Gewebeschädigung *f*

3913 LATERAL PROJECTION,   xr
      LATERAL VIEW, PROFILE VIEW
  A radiograph for which the röntgen rays
  traverse a body from side to side.
  f projection *f* latérale, vue *f* de profil,
     vue *f* latérale
  e proyección *f* lateral, vista *f* de perfil,
     vista *f* lateral
  i proiezione *f* laterale, vista *f* di profilo,
     vista *f* laterale
  n zijdelingse projectie

d   Projektion *f* in seitlicher Richtung

3914 LATITUDE EFFECT, see 3015

3915 LATTICE   cr
An orderly arrangement of atoms in a
crystalline material.
f   réseau *m*
e   red *f*
i   reticolo *m*
n   rooster *n*
d   Gitter *n*

3916 LATTICE,   rt
REACTOR LATTICE
An array of fuel and other materials
arranged according to a regular pattern.
f   réseau *m* du réacteur,
réseau *m* multiplicateur
e   celosía *f* del reactor, celosía *f* espacial
i   reticolo *m* del reattore
n   rooster *n* van een reactor
d   Reaktorgitter *n*

3917 LATTICE ANISOTROPY   rt
A property of a multiplying lattice in which
the diffusion coefficient for the neutron
flux density has different values when
measured along the lattice axis or along
a direction perpendicular to that axis.
f   anisotropie *f* de réseau
e   anisotropía *f* de celosía
i   anisotropia *f* di reticolo
n   roosteranisotropie
d   Gitteranisotropie *f*

3918 LATTICE CALCULATION   cr, ma
A method for the evaluation of lattice
sums according to Evjen or Ewald.
f   calcul *m* des réseaux
e   cálculo *m* de las redes
i   calcolo *m* dei reticoli
n   roosterberekening
d   Gitterberechnung *f*

3919 LATTICE CELL, see 933

3920 LATTICE CELL, see 2209

3921 LATTICE CONSTANT,   cr
LATTICE PARAMETER,
LATTICE SPACING
The length of the edges of the unit cell in
a crystal.
f   constante *f* de réseau
e   constante *f* de la red
i   costante *f* del reticolo
n   roosterconstante
d   Gitterkonstante *f*

3922 LATTICE DESIGN   rt
The planning of the assembly of reactive
and moderating material in a reactor.
f   géométrie *f* du réseau
e   diseño *m* de la celosía
i   progetto *m* del reticolo
n   roosterontwerp *n*, roosterplanning
d   Gitterplanung *f*

3923 LATTICE DIMENSIONS   cr
According to the Bragg formula the spacing
of the atomic planes can be deduced from
the X-ray diffraction pattern and a
knowledge of the X-ray wavelength.
f   dimensions *pl* du réseau
e   dimensiones *pl* de la red
i   dimensioni *pl* del reticolo
n   roosterafmetingen *pl*
d   Gitterabmessungen *pl*

3924 LATTICE DISTORTION   cr
Change in the ideal arrangement of a
crystal lattice due to vacancies, interstitial atoms, Frenkel defects or dislocation.
f   distorsion *f* du réseau
e   distorsión *f* de la red
i   distorsione *f* del reticolo
n   roostervervorming
d   Gitterverzerrung *f*

3925 LATTICE ENERGY   cr
The energy required to separate the ions
of a crystal to an infinite distance from
each other.
f   énergie *f* de réseau
e   energía *f* de red
i   energia *f* di reticolo
n   roosterenergie
d   Gitterenergie *f*

3926 LATTICE IMPERFECTION   cr
Deviation from a perfect homogeneous
crystal lattice.
f   défaut *m* du réseau
e   defecto *m* de la red
i   difetto *m* del reticolo
n   roosterfout
d   Gitterfehler *m*

3927 LATTICE PITCH,   fu, rt
LATTICE SPACING
The distance between two fuel elements
or groups of elements, the distance being
filled up with a moderator, and the elements forming part of a lattice.
f   pas *m* du réseau
e   paso *m* de la celosía
i   passo *m* del reticolo
n   elementafstand
d   Elementenabstand *m*

3928 LATTICE PLANE, see 403

3929 LATTICE REACTOR   rt
A heterogeneous reactor in which the
discrete bodies of fuel and moderator are
in the form of long rods or stringers.
f   réacteur *m* à réseau
e   reactor *m* de celosía
i   reattore *m* a reticolo
n   roosterreactor
d   Gitterreaktor *m*

3930 LATTICE STRUCTURE   rt
The form and construction of the lattice in
the reactor.

f   structure *f* du réseau
e   estructura *f* de la celosía
i   struttura *f* del reticolo
n   roosterbouw
d   Gitterbau *m*

3931 LATTICE THEORY,   cr
The theory of the lattice principle in
crystallography.
f   théorie *f* du réseau
e   teoría *f* de la red
i   teoria *f* del reticolo
n   roostertheorie
d   Gittertheorie *f*

3932 LAUE EQUATIONS   cr
A set of three equations which must be
satisfied in order to permit reinforcement
of the contributions scattered from
successive equivalent points along each
of co-ordinate axes.
f   équations *pl* de Laue
e   ecuaciones *pl* de Laue
i   equazioni *pl* di Laue
n   vergelijkingen *pl* van Laue
d   Laue-Gleichungen *pl*

3933 LAUE METHOD   cr
The examination of the diffraction beams
which are produced, for any arbitrary
setting of a crystal, from a white beam of
incident radiation.
f   méthode *f* de Laue
e   método *m* de Laue
i   metodo *m* di Laue
n   methode van Laue
d   Laue-Methode *f*

3934 LAUE PATTERN   cr
The characteristic photographic record
obtained when X-rays from a pinhole or
slit are sent through a single crystal
that diffracts or bends the rays in all
directions.
f   diagramme *m* de Laue
e   diagrama *m* de Laue
i   macchie *pl* di Laue
n   diagram *n* van Laue
d   Laue-Diagramm *n*

3935 LAUNDRY CONTAMINATION   sa
MONITOR
A contamination monitor for laundry.
f   moniteur *m* de contamination pour linge
et tenues de travail
e   monitor *m* de contaminación para ropa
de trabajo
i   monitore *m* di contaminazione da
lavanderia
n   besmettingsmonitor voor wasgoed en
kleding
d   Kontaminationswarngerät *n* für Wäsche
und Arbeitskleidung

3936 LAURITSEN ELECTROSCOPE,   me
QUARTZ FIBRE(ER) ELECTROSCOPE
A rugged yet sensitive electroscope

employing a metallized quartz fibre(er) as
the sensitive element.
f   électroscope *m* à fibre de quartz,
électroscope *m* de Lauritsen
e   electroscopio *m* de fibra de cuarzo,
electroscopio *m* de Lauritsen
i   elettroscopio *m* a fibra di quarzo,
elettroscopio *m* di Lauritsen
n   kwartsdraadelektroscoop,
lauritsenelektroscoop
d   Lauritsen-Elektroskop *n*,
Quartzfadenelektroskop *n*

3937 LAWRENTIUM   ch
Transuranic element, symbol Lw, atomic
number 103.
f   lawrentium *m*
e   lawrentio *m*
i   lawrentio *m*
n   lawrentium *n*
d   Lawrentium *n*

3938 LAYER LATTICE   cr
A solid structure consisting of sheets of
atoms, not necessarily in one plane,
extending throughout the whole crystal and
separated from one another by a distance
which is too large for chemical bonding.
f   réseau *m* à couches atomiques
e   red *f* de capas atómicas
i   reticolo *m* a strati atomici
n   vlakkenrooster *n*
d   Flächengitter *n*

3939 LAYER LINES   cr
More or less horizontal lines formed by
diffraction spots produced by the lattice
planes having the same spacing in the
direction parallel to the axis of rotation.
f   lignes *pl* des plans atomiques
e   líneas *pl* de los planos atómicos
i   linee *pl* dei piani atomici
n   roostervlaklijnen *pl*
d   Gitterflächenlinien *pl*

3940 LAYER OF CHARGE   ec, gp
A simple layer is a sheet of charge of
one sign, such as the surface charge in a
capacitor.
f   couche *f* de charge
e   capa *f* de carga
i   strato *m* di carica
n   ladingslaag
d   Ladungsschicht *f*

3941 LD,   xr
LETHAL DOSE
The dose of radiation required to kill.
f   dose *f* létale
e   dosis *f* letal
i   dose *f* letale
n   dodelijke dosis, letale dosis
d   letale Dosis *f*

3942 LD 50,   ra
MEDIAN LETHAL DOSE
The absorbed dose which will kill, within

a specified time, 50 % of the individuals of
a large population or organisms of a given
species.
f   DL 50, dose f létale 50 %
e   dosis f letal media
i   dose f di sopravvivenza al 50 %,
    dose f letale media
n   LD 50, letale-dosismediaan
d   mittlere Letaldosis f

3943 LD 50 TIME,   ra
      MEDIAN LETHAL TIME, MLT
The time required for the death of 50 %
of the individuals of a large population or
organisms of a given species that has
received a specified absorbed dose.
f   temps m létal 50 %, TL 50
e   período m letal medio
i   periodo m letale medio,
    tempo m di sopravvivenza al 50 %
n   letale-tijdmediaan, LT 50
d   mittlere Letalzeit f

3944 LEACH RESISTANCE   mt, rw
Property of glass studied when using this
material for storing radioactive waste.
f   résistance f aux lessives
e   resistencia f a las lejías
i   resistenza f alle liscive
n   loogvastheid
d   Laugenbeständigkeit f

3945 LEACHING,   ch
      LIXIVIATION
Chemical process used in treating uranium
and other ores.
f   lessivage m, lixiviation f
e   lixiviación f
i   liscivazione f
n   uitloging
d   Auslaugen n

3946 LEAD   ch
Metallic element, symbol Pb, atomic
number 82.
f   plomb m
e   plomo m
i   piombo m
n   lood n
d   Blei n

3947 LEAD AGE   np
A determination of the age of a mineral
by measuring the amount of radiogenic
lead which it contains.
f   âge m de plomb
e   edad f de plomo
i   età f di piombo
n   loodleeftijd
d   Bleialter n

3948 LEAD BRICK   rt, sa
Building stone of reactor walls, etc.
f   brique f plombifère
e   ladrillo m plombífero
i   mattone m al piombo
n   loodhoudende steen
d   bleihaltiger Baustein m

3949 LEAD CASTLE   ra, sa
An enclosure of lead generally designed
to shield a radiation detecting device
against ambient radiation.
f   château m de plomb
e   recipiente m de plomo de guarda
i   castello m di piombo
n   loden mantel, loodkasteel n
d   Bleiabsorptionskammer f, Bleikammer f

3950 LEAD DOOR   sa
Door consisting of outer sheets of stainless
or mild steel plate with poured lead
between.
f   porte f à couche intérieure de plomb
e   puerta f con capa interior de plomo
i   porta f a strato interior di piombo
n   deur met inwendige loodlaag
d   Tür f mit Bleieinlage

3951 LEAD EQUIVALENT,   sa
      PROTECTIVE VALUE
The thickness of metallic lead which
affords the same protection as a given
material under the same conditons of
irradiation.
f   équivalent m de plomb
e   equivalente m de plomo
i   equivalente m in piombo
n   loodequivalent n
d   Bleiäquivalent n, Bleigleichwert m

3952 LEAD GLASS WINDOW   sa, xr
f   fenêtre f de verre au plomb
e   ventana f de vidrio plomado
i   finestra f in vetro al piombo
n   loodglasvenster n
d   Bleiglasfenster n

3953 LEAD IMPREGNATED APRON,
      LEAD-RUBBER APRON,
      see 684

3954 LEAD-LOADED   xr
      INSULATING SLEEVE
Cylindrical sleeve around the neck of an
X-ray tube of plastics material loaded
with lead.
f   manchon m en résine synthétique au
    plomb
e   manguito m de resina sintética al plomo
i   manicotto m di resina sintética al piombo
n   lood-kunststofhuls
d   Blei-Kunststoffmuffe f

3955 LEAD PROTECTION   sa
Protection from ionizing radiation
afforded by the use of metallic lead.
f   protection f en plomb
e   protección f por plomo
i   protezione f in piombo
n   loodhoudende bescherming
d   Bleischutz m

3956 LEAD RUBBER   mt, sa
Rubber containing a high proportion of lead.
f   caoutchouc m plombeux
e   goma f plombífera

i   gomma f piombifera
n   loodrubber
d   Bleigummi m

3957   LEAD RUBBER GLOVES,   sa
          PROTECTIVE GLOVES
       Gloves made from lead rubber as a
       protecting material.
f   gants pl protecteurs
e   guantes pl protectores
i   guanti pl di protezione
n   beschermingshandschoenen pl,
       loodrubberhandschoenen pl
d   Schutzhandschuhe pl

3958   LEAD SCREEN   sa
       A protective screen used in radiation and
       consisting of or containing lead.
f   écran m en plomb
e   blindaje m de plomo
i   schermo m in piombo
n   loodscherm n
d   Bleischirm m

3959   LEAD SLEEVE   xr
       Cylindrical shield around the neck of an
       X-ray tube consisting of lead.
f   manchon m en plomb
e   manguito m de plomo
i   manicotto m di piombo
n   loodhuls
d   Bleimuffe f

3960   LEAK DETECTION ASSEMBLY,   sa
          LEAK DETECTION SYSTEM
f   ensemble m de détection de fuites
e   conjunto m de detección de fugas
i   complesso m di rivelazione di fughe
n   lekdetectieopstelling
d   Lecksuchanordnung f

3961   LEAK DETECTOR, see 3222

3962   LEAK DETECTORS,   vt
          LEAK HUNTERS
       Devices used to locate leaks in
       high-vacuum systems.
f   détecteurs pl de fuites
e   detectores pl de fugas
i   rivelatori pl di fughe
n   lekzoekers pl
d   Lecksucher pl

3963   LEAK RATE   te
       In testing welded sources, this rate should
       be smaller than 10^{-8} standard cc/sec.
f   débit m de fuites, vitesse f de fuites
e   intensidad f de fugas,
       velocidad f de fugas
i   intensità f di fughe, velocità f di fughe
n   lektempo n
d   Leckrate f

3964   LEAK TESTING   sa, te
       The testing of neutron sources, etc., for
       surface contamination.
f   essai m de fuites

e   ensayo m de fugas
i   collaudo m di fughe
n   lekproef
d   Leckprüfung f

3965   LEAKAGE (SHIELDING)   ra
       Escape of radiation through a shield,
       especially by way of holes or cracks
       through the shield.
f   fuite f de rayonnement
e   fuga f de radiación
i   fuga f di radiazione
n   lek
d   Durchlassstrahlung f

3966   LEAKAGE (REACTOR THEORY),   rt
          NEUTRON LEAKAGE
       The net loss of neutrons from a region
       of a reactor by escape across the
       boundaries of the region.
f   fuite f de neutrons
e   fuga f de neutrones
i   fuga f di neutroni
n   lek, neutronenlek
d   Neutronenausfluss m, Neutronenverlust m

3967   LEAKAGE CONTROL, see 1307

3968   LEAKAGE PEAK, see 2932

3969   LEAKAGE PROBABILITY   gr
          FOR GAMMA SOURCES
       The probability that a gamma ray emitted
       inside a sample having an absorption
       coefficient μ will be absorbed inside the
       sample.
f   probabilité f de fuite pour sources gamma
e   probabilidad f de fuga para fuentes gamma
i   probabilità f di fuga per sorgentes gamma
n   ontkomwaarschijnlijkheid voor gamma-
       bronnen
d   Entkommwahrscheinlichkeit f für
       Gammaquellen

3970   LEAKAGE RADIATION   xr
       All radiation emitted from within a
       röntgen tube housing or a teletherapy
       housing except the useful beam of röntgen
       rays or gamma rays.
f   rayonnement m de fuite
e   radiación f de fuga
i   radiazione f dispersa
n   lekstraling
d   Leckstrahlung f

3971   LEAKAGE SPECTRUM,   np, rt, sp
          NEUTRON LEAKAGE SPECTRUM
       The energy distribution of neutrons
       leaving the reactor.
f   spectre m de fuite de neutrons
e   espectro m de fuga de neutrones
i   spettro m di fuga di neutroni
n   neutronenlekspectrum n, verliesspectrum n
d   Neutronenausflussspektrum n

3972   LEAKPROOF,   ge
          LEAKTIGHT
f   étanche adj

e   estanco adj
i   a tenuta
n   lekdicht adj
d   leckdicht adj

3973 LEAKTIGHTNESS   ge
f   étanchéité f
e   estanqueidad f
i   ermeticità f
n   lekdichtheid
d   Leckdichtheit f

3974 LEAPFROG CASCADE   is
A cascade in which the enriched fraction
is fed to the second or subsequent stage
(or the depleted fraction to the
next-but-one preceding stage or an
earlier stage).
f   cascade f à saut
e   cascada f de salto
i   cascata f a sbalzo
n   sprongcascade
d   Sprungkaskade f

3975 LENARD CATHODE-RAY TUBE,   ec
LENARD TUBE
A cathode-ray tube in which the target is
replaced by a window thin enough to
allow the exit of cathode rays.
f   tube m de Lenard
e   tubo m de Lenard
i   tubo m di Lenard
n   lenardbuis
d   Lenard-Röhre f

3976 LENARD RAYS   ra
Cathode rays which have passed outside
the discharge tube.
f   rayons pl de Lenard
e   rayos pl de Lenard
i   raggi pl di Lenard
n   lenardstralen pl
d   Lenard-Strahlen pl

3977 LENGTH OF PLATEAU,   ct
PLATEAU LENGTH
The range of applied voltage over which
the plateau of a radiation counter tube
extends.
f   longueur f du palier, longueur f du plateau
e   longitud f de la meseta,
longitud f del plato
i   lunghezza f del ripiano
n   plateaulengte
d   Plateaulänge f

3978 LENGTH OF WORKING CYCLE,
see 2001

3979 LENS   ec
A arrangement of electrodes used to
produce an electric field that serves to
focus electrons into a beam.
f   lentille f
e   lente f
i   lente f
n   lens
d   Linse f

3980 LEPTON   np
A particle of small mass; specifically
an electron, a positron, a neutrino or
an anti-neutrino.
f   lepton m
e   leptón m
i   leptone m
n   lepton n
d   Lepton n

3981 LEPTON FIELD, see 2142

3982 LEPTON NUMBER   np
The number of leptons minus the number
of anti-leptons in a system.
f   nombre m de leptons
e   número m de leptones
i   numero m di leptoni
n   leptongetal n
d   Leptonzahl f

3983 LET,   ra
LINEAR ENERGY TRANSFER
The average energy locally imparted to a
medium by a charged particle of
specified energy per unit distance
traversed.
f   TLE, transfert m linéique d'énergie
e   transferencia f lineal de energía
i   trasferimento m lineare d'energia
n   energieoverdracht per lengte,
lineïeke energieoverdracht
d   lineare Energieübertragung f

3984 LETHAL DOSE, see 3941

3985 LETHAL MUTATION   md
Mutation leading to the death of the
offspring at any stage.
f   mutation f létale
e   mutación f letal
i   mutamento m letale, mutazione f letale
n   dodelijke mutatie, letale mutatie
d   letale Mutation f

3986 LETHARGY,   np
NEUTRON LETHARGY
The lethargy of a neutron is, to within an
additive constant, the negative of the
neutral logarithm of the energy of the
neutron.
f   léthargie f
e   letargia f
i   letargia f
n   lethargie
d   Lethargie f

3987 LEUCEMIA,   md
LEUKEMIA, LEUKOCYTHEMIA
A disease in which there is often a great
overproduction of white blood cells, or
a relative overproduction of immature
white cells, and great enlargement of the
spleen.
f   leucémie f
e   leucémia f, leucocitemia f
i   leucemia f
n   leukemie
d   Leukämie f, Weissblütigkeit f

3988 LEUCOCYTE, md
  LEUKOCYTE, WHITE CORPUSCLE,
  WHITE GLOBULE
  A colo(u)rless blood corpuscle.
f globule *m* blanc, leucocyte *m*
e corpúsculo *m* blanco, glóbulo *m* blanco,
  leucocito *m*
i globulo *m* bianco, leucocito *m*
n leukocyt *n*, wit bloedlichaampje *n*
d Leukozyt *m*, weisses Blutkörperchen *n*

3989 LEUCOCYTOSIS, md
  LEUKOCYTOSIS
  An increase in the number of leucocytes
  in the blood.
f leucocytose *f*
e leucocitosis *f*
i leucocitosi *f*
n leukocytose
d Leukozytose *f*

3990 LEUCOGRAM, md
  LEUKOGRAM
  The picture of the distribution of different
  white corpuscles per cubic millimetre(er)
  of blood.
f leucogramme *m*
e leucograma *m*
i leucogramma *m*
n wit bloedbeeld *n*
d Leukogramm *n*, weisses Blutbild *n*

3991 LEVEL ge
  The level of a quantity is its magnitude,
  especially when considered in relation to
  an arbitrary reference value.
f niveau *m*
e nivel *m*
i livello *m*
n niveau *n*, peil *n*
d Niveau *n*, Pegel *m*

3992 LEVEL DEMAND SIGNAL co
  In an automatic control system, the
  reference signal which sets the power level
  at which the nuclear reactor should
  operate.
f signal *m* de niveau de puissance de consigne
e señal *f* de nivel de potencia de consigna
i segnale *m* d'impostazione di livello
  di potenza
n instelvermogensniveausignaal *n*
d Solleistungsniveausignal *n*

3993 LEVEL DENSITY, np
  NUCLEAR LEVEL DENSITY
  The number of energy levels of a nucleus
  per unit energy interval at a particular
  energy.
f densité *f* de niveau nucléaire
e densidad *f* de nivel nuclear
i densità *f* di livello nucleare
n kernniveaudichtheid
d Kernniveaudichte *f*

3994 LEVEL DISTRIBUTION np
  The distribution of the energy levels.

f distribution *f* des niveaux d'énergie
e distribución *f* de los niveles de energía
i distribuzione *f* dei livelli d'energia
n energieniveauverdeling
d Energieniveauverteilung *f*

3995 LEVEL SPACING, np
  NEUTRON LEVEL SPACING
  The average separation between nuclear
  energy levels at a particular excitation
  energy. It is the reciprocal of the level
  density.
f écartement *m* de niveau
e espacio *m* entre niveles
i spazio *m* tra livelli
n niveauafstand
d Niveauabstand *m*

3996 LEVEL WIDTH, see 2277

3997 LID TANK SHIELDING FACILITY rt, te
  A facility at the Oak Ridge Laboratory
  used in testing the shielding properties
  of materials.
f installation *f* d'essai de blindage en
  tanque accessible
e equipo *m* de prueba de blindaje en
  tanque acesible
i impianto *m* di prova di schermatura in
  serbatoio accessibile
n afschermingsbeproevingsinstallatie in
  van deur voorziene tank
d Abschirmungsprüfungsanlage *f* in mit
  einer Tür versehenem Behälter

3998 LIEBIGITE, mi
  URANOTHALLITE
  A calcium uranocarbonate containing
  about 33.6 % U.
f uranothallite *f*
e uranotalita *f*
i uranotallite *f*
n uranothalliet *n*
d Uranothallit *m*

3999 LIENOGRAPHY xr
  Radiological examination of the spleen.
f liénographie *f*
e lienografía *f*
i lienografia *f*
n lienografie
d Lienographie *f*

4000 LIFE LINE, np
  SPACE-TIME PATH OF A
  CLASSICAL PARTICLE
  A curve drawn in space-time to represent
  the position of a particle as a function
  of time.
f parcours *m* espace-temps d'une particule
  classique
e trayectoria *f* espacio-tiempo de una
  partícula clásica
i percorso *m* spazio-tempo d'una
  particella classica
n ruimte-tijd-baan van een klassiek deeltje
d Raum-Zeit-Bahn *f* eines klassischen
  Teilchens

4001 LIFE OF A GEIGER-MÜLLER   ct
      COUNTER TUBE
      A value expressed in the number of dis-
      charges produced before its characteris-
      tics become unacceptable because of
      deterioration due to normal processes
      involved in its operation.
f durée *f* de vie d'un tube compteur
   Geiger-Müller
e vida *f* de un tubo contador Geiger-Müller
i periodo *m* d'efficienza d'un tubo
   contaore Geiger-Müller
n levensduur van een geiger-müllertelbuis
d Lebensdauer *f* eines Geiger-Müller-
   Zählrohrs

4002 LIGHT ATOM   np
      An atom of an element with an atomic
      weight below a certain value.
f atome *m* léger
e átomo *m* liviano
i atomo *m* leggero
n licht atoom *n*
d leichtes Atom *n*

4003 LIGHT-BEAM LOCALIZING   xr
      DEVICE,
      LIGHT-BEAM POINTER
      A device used to direct the incident beam
      of X-rays upon the desired area of the
      surface of an object.
f localisateur *m* lumineux
e localizador *m* luminoso
i localizzatore *m* luminoso
n lichtvizier *n*
d Lichtvisier *n*

4004 LIGHT ELEMENT   ch
      An element with an atomic weight below
      a certain value.
f élément *m* léger
e elemento *m* liviano
i elemento *m* leggero
n licht element *n*
d leichtes Element *n*

4005 LIGHT ELEMENT CHEMISTRY   ch
f chimie *f* des éléments légers
e química *f* de los elementos livianos
i chimica *f* degli elementi leggeri
n chemie van de lichte elementen
d Chemie *f* der leichten Elementen

4006 LIGHT FRACTION   ch, is
      A fraction which has a greater abundance
      of the light isotope.
f fraction *f* légère
e fracción *f* ligera
i frazione *f* leggera
n lichte fractie
d leichte Fraktion *f*

4007 LIGHT GUIDE,   ec
      LIGHT LINE, LIGHT PIPE
      Optical system sometimes placed between
      a scintillator and a photomultiplier, which
      reduces photon losses by means of
      transmission and internal reflection.
f conduit *m* de lumière
e conducto *m* de luz, tubo *m* luminoso
i condotto *m* di luce
n lichtpijp
d Lichtrohr *n*

4008 LIGHT MESONS, see 3874

4009 LIGHT NUCLEUS   np
      The nucleus of the light elements.
f noyau *m* léger
e núcleo *m* liviano
i nucleo *m* leggero
n lichte kern
d leichter Kern *m*

4010 LIGHT QUANTUM,   np
      PHOTON
      A photon of visible light.
f photon *m*, quantum *m* de lumière
e cuanto *m* de luz, fotón *m*
i fotone *m*, quanto *m* di luce
n foton *n*, lichtquant
d Lichtquant *m*, Photon *n*

4011 LIGHT WATER   ge
      A term used to differentiate ordinary
      water from heavy water.
f eau *f* légère
e agua *f* liviana
i acqua *f* leggera
n licht water *n*
d leichtes Wasser *n*

4012 LIGHT WATER REACTOR,   rt
      WATER MODERATED REACTOR,
      WATER REACTOR
f réacteur *m* à eau légère
e reactor *m* de agua liviana
i reattore *m* ad acqua leggera
n licht-waterreactor, waterreactor
d Leichtwasserreaktor *m*, Wasserreaktor *m*

4013 LIMEN,   ge
      THRESHOLD
      The least value of a current, voltage or
      other quantity that produces the minimum
      detectable response.
f seuil *m*
e umbral *m*
i soglia *f*
n drempel
d Schwelle *f*

4014 LIMIT   ge
      The upper or lower bound of a physical
      quantity.
f limite *f*
e límite *m*
i limite *m*
n grens
d Grenze *f*

4015 LIMITATION OF MOBILITY,   cr, np
      MOBILITY LIMITATION
      The limitation of mobility of electrons in
      crystals by scattering due to thermal
      vibration of the lattice or by impurities.

f limitation *f* de mobilité
e limitación *f* de movilidad
i limitazione *f* di mobilità
n beweeglijkheidsbegrenzing
d Beweglichkeitsbegrenzung *f*

4016 LIMITED PROPORTIONALITY ct
Said of a counter when the gas amplification depends on the number of ions
produced in the initial ionizing event, and
also on the voltage.
f proportionalité *f* limitée
e proporcionalidad *f* limitada
i proporzionalità *f* limitata
n begrensde proportionaliteit,
beperkte evenredigheid
d begrenzte Proportionalität *f*

4017 LIMITED PROPORTIONALITY ct
REGION,
REGION OF LIMITED
PROPORTIONALITY
In counting tubes, that part of the
characteristic curve of pulse versus
voltage in which the gas amplification
depends on the number of ions produced
in the initial ionizing event, and also on the
voltage.
f zone *f* de proportionalité limitée
e zona *f* de proporcionalidad limitada
i zona *f* di proporzionalità limitata
n gebied *n* van beperkte evenredigheid
d Bereich *m* begrenzter Proportionalität

4018 LIMITED SAFE rt
Term used to qualify the safety of a
container by reference to specific
isotopic ratios, e.g. with uranium to
concentration limits or density
restrictions.
f de sûreté limitée
e de seguridad limitada
i di sicurezza limitata
n begrensd veilig
d begrenzt sicher

4019 LIMITED SECTOR sp
SPECTROMETER
A magnetic spectrometer in which the
magnetic field operates over considerably
less than 180° of the circular trajectory
of the charged particles.
f spectromètre *m* à secteur limité
e espectrómetro *m* de sector limitado
i spettrometro *m* a settore limitato
n sectorspectrometer
d Sektorspektrometer *n*

4020 LIMITER ec
A circuit for preventing the amplitude of
an electrical signal from exceeding a
predetermined value.
f écrêteur *m*
e limitador *m*, truncador *m*
i circuito *m* limitatore, limitatore *m*
n begrenzer
d Begrenzer *m*

4021 LINAC, pa
LINEAR ACCELERATOR
An accelerator in which a number of
electrodes are so arranged that, when a
potential difference is applied at an
appropriate frequency, the particles
passing through them receive successive
increments of energy.
f accélérateur *m* linéaire
e acelerador *m* lineal
i acceleratore *m* lineare
n lineaire versneller
d Linearbeschleuniger *m*

4022 LINDEMANN ELECTROMETER, me
QUARTZ FIBRE(ER) ELECTROMETER
An electrometer using a metallized quartz
fibre(er) mounted on and perpendicular to a
quartz torsion fibre(er), in such a way that
the former fibre(er) is positioned in a
system of electrodes.
f électromètre *m* à fibre de quartz,
électromètre *m* de Lindemann
e electrómetro *m* de fibra de cuarzo,
electrómetro *m* de Lindemann
i elettrometro *m* a fibra di quarzo,
elettrometro *m* di Lindemann
n kwartsdraadelektrometer,
lindemannelektrometer
d Lindemann-Elektrometer *n*,
Quarzfadenelektrometer *n*

4023 LINDEMANN GLASS xr
Glass of low X-ray absorption containing
lithium, boron and beryllium.
f verre *m* de Lindemann
e vidrio *m* de Lindemann
i vetro *m* di Lindemann
n lindemannglas *n*
d Lindemann-Glas *n*

4024 LINE FOCUS xr
An elongated rectangular focus so placed
that an effectively square source of
X-rays can be achieved.
f foyer *m* linéaire
e foco *m* lineal
i fuoco *m* lineare
n lijnfocus
d Strichfokus *m*

4025 LINE FOCUS TUBE xr
A röntgen tube in which a rectangular focus
spot on the surface of the anode is
projected as a square spot in the direction
of the central ray.
f tube *m* à foyer linéaire
e tubo *m* de foco lineal
i tubo *m* con fuoco lineare
n lijnfocusbuis
d Strichfokusröhre *f*

4026 LINE FOCUSING, see 343

4027 LINE OF MAGNETIC FORCE (GB), ec
MAGNETIC LINE OF FORCE (US)
A line drawn in a magnetic field such that

its direction at every point is the direction
of the magnetic force at that point.
f   ligne f de force
e   línea f de fuerza
i   linea f di forza
n   krachtlijn
d   Kraftlinie f

4028 LINE SPECTRUM   sp
A spectrum in which the radiation is
concentrated in narrow regions called
lines.
f   spectre m de raies
e   espectro m de líneas, espectro m de rayas
i   spettro m di righe
n   lijnenspectrum n
d   Linienspektrum n

4029 LINE WIDTH   sp
A measure in the spread of wavelength
of radiation that normally is characterized
by a single wavelength of energy value.
f   largeur f de raies spectrales
e   anchura f de líneas espectrales
i   larghezza f di righe spettrali
n   lijnbreedte
d   Linienbreite f

4030 LINEAR ABSORPTION   np
COEFFICIENT
The fractional decrease in intensity per
unit distance traversed.
f   coefficient m d'absorption linéaire
e   coeficiente m de absorción lineal
i   coefficiente m d'assorbimento lineare
n   lineaire absorptiecoëfficiënt
d   linearer Absorptionskoeffizient m

4031 LINEAR ACCELERATOR INJECTOR rt
Component part of the first stage of a
Nimrod reactor.
f   injecteur m d'accélérateur linéaire
e   inyector m de acelerador lineal
i   iniettore m d'acceleratore lineare
n   injector van een lineaire versneller
d   Injektor m eines Linearbeschleunigers

4032 LINEAR ACTIVITY,   gr
LINEAR INTENSITY
The activity per unit length of an
elongated gamma ray source.
f   activité f linéique
e   actividad f lineal,
actividad f por unidad de longitud
i   attività f lineare,
attività f per unità di lunghezza
n   lineïeke activiteit
d   Aktivität f je Längeneinheit

4033 LINEAR ATTENUATION   ma
COEFFICIENT
$\frac{dJ}{dx} = \mu \, J$, where J is the current density
of particles or quanta parallel to the
x-direction.
f   coefficient m d'atténuation linéique
e   coeficiente m de atenuación lineal

i   coefficiente m d'attenuazione lineare
n   lineïeke verzwakkingscoëfficiënt,
verzwakkingscoëfficiënt per lengte
d   linearer Schwächungskoeffizient m

4034 LINEAR CONTROL   con
ELECTROMECHANISM
An electromechanism designed to perform
a linear motion of one or several nuclear
reactor control elements.
f   électromécanisme m de commande linéaire
e   electromecanismo m de regulación lineal
i   elettromeccanismo m di comando lineare
n   elektromechanisme n voor rechtlijnig
bewegende regelelementen
d   elektromechanischer Hubantrieb m für
Steuerstäbe

4035 LINEAR DELAY UNIT   ec
A basic function unit comprising an
electronic circuit which, in response to an
input pulse and after a given time interval,
delivers a substantially identical output
pulse.
f   élément m retard linéaire
e   elemento m retardo lineal
i   elemento m ritardo lineare
n   lineaire vertraagschakel
d   lineares Verzögerungsgerät n

4036 LINEAR DIRECT   ec
CURRENT AMPLIFIER
An amplifier whose output quantity is a
linear function of the input quantity, even
when the frequency of the input quantity
approaches zero.
f   amplificateur m linéaire pour courant
continu
e   amplificador m lineal de corriente continua
i   amplificatore m lineare per corrente
continua
n   lineaire gelijkstroomversterker
d   linearer Gleichstromverstärker m

4037 LINEAR ELECTRON   ec, pa
ACCELERATOR
An evacuated metal tube in which electrons
are accelerated through a series of small
gaps so arranged and spaced that, at a
specific excitation frequency, the stream
of electrons on passing through successive
gaps gains additional energy from the
electric field in each gap.
f   accélérateur m linéaire d'électrons
e   acelerador m lineal de electrones
i   acceleratore m lineare d'elettroni
n   lineaire elektronenversneller
d   Linearelektronenbeschleuniger m

4038 LINEAR ENERGY TRANSFER, see 3983

4039 LINEAR EXTRAPOLATION   np
DISTANCE
In the one-group model of neutron transport,
the distance beyond the boundary of a
medium to a point at which the tangent to
the asymptotic neutron flux density at
the boundary goes to zero.

f   longueur f linéaire d'extrapolation
e   distancia f lineal de extrapolación
i   lunghezza f lineare d'estrapolazione
n   lineaire extrapolatieafstand
d   lineare Extrapolationslänge f,
    lineare Extrapolationsstrecke f

**4040 LINEAR GAMMA   me
        ABSORPTIOMETER**
f   absorptiomètre m linéaire à rayonnement
    gamma
e   absorciómetro m lineal de radiación
    gamma
i   assorbimetro m lineare a radiazione
    gamma
n   lineaire absorptiemeter met gamma-
    straling
d   Linearabsorptionsmesser m mit Gamma-
    strahlung

4041 LINEAR ION DENSITY   ic
    Number of ions of one kind existing per
    unit length.
f   nombre m linéique d'ions
e   densidad f de iones por unidad de
    longitud
i   densità f d'ioni per unità di lunghezza
n   ionendichtheid per lengte-eenheid
d   Ionendichte f je Längeneinheit

4042 LINEAR IONIZATION   ic
    The number of elementary charges of one
    sign produced over a small length of the
    path of an ionizing particle divided by
    that length.
f   ionisation f linéique
e   ionización f por longitud
i   ionizzazione f per lunghezza
n   ionisatie per lengte, lineïeke ionisatie
d   Ionisation f je Länge

**4043 LINEAR MEASURING   me
        ASSEMBLY**
    A measuring assembly in which the
    amplitude of the output signal is a linear
    function of the input signal.
f   ensemble m de mesure linéaire
e   conjunto m de medida lineal
i   complesso m di misura lineare
n   lineaire stralingsmeetopstelling
d   linear messende Anordnung f

4044 LINEAR PINCH DEVICE   pp
    Component part of the Columbus thermo-
    nuclear experimental rig.
f   dispositif m de striction linéaire
e   dispositivo m de estricción lineal
i   dispositivo m di contrazione lineare
n   lineair insnoeringsapparaat n
d   linearer Einschnürungsapparat m

**4045 LINEAR POWER MEASURING   me
        ASSEMBLY BASED ON GAMMA
        RADIATION**
    A power measuring assembly in which the
    output signal is a linear function of the
    input signal.

f   ensemble m de mesure linéaire de puissance
    par mesure du rayonnement gamma
e   conjunto m de medida lineal de potencia
    por la radiación gamma
i   complesso m di misura lineare di potenza
    per mezzo della misura della densità di
    flusso gamma
n   opstelling voor lineaire vermogensmeting
    door middel van de gammastraling
d   Anordnung f zur linearen Leistungs-
    messung mittels Gammastrahlung

**4046 LINEAR POWER MEASURING   me
        ASSEMBLY BASED ON THE
        NEUTRON FLUX DENSITY**
    A power measuring assembly in which the
    output signal is a linear function of the
    input signal.
f   ensemble m de mesure linéaire de
    puissance par mesure de la densité de
    flux neutronique
e   conjunto m de medida lineal de potencia
    por densidad de flujo neutrónico
i   complesso m di misura lineare di potenza
    per mezzo della densità del flusso
    neutronico
n   opstelling voor lineaire vermogensmeting
    door middel van de neutronenfluxdichtheid
d   Anordnung f zur linearer Leistungs-
    messung mittels Neutronenstrahlung

4047 LINEAR PULSE AMPLIFIER   ec
    A pulse amplifier which, within the limit
    of its normal operating characteristics,
    delivers an output pulse of amplitude
    proportional to that of the input pulse.
f   amplificateur m linéaire d'impulsions
e   amplificador m lineal de impulsos
i   amplificatore m lineare d'impulsi
n   lineaire pulsversterker
d   linearer Impulsverstärker m

4048 LINEAR RANGE   np
    The range expressed in units of length.
f   portée f linéaire
e   alcance m lineal
i   portata f lineare
n   lineaire dracht
d   lineare Reichweite f

4049 LINEAR RATEMETER   ct, ec, me
    An electronic sub-assembly which gives a
    continuous indication proportional to the
    average counting rate over a predeter-
    mined time interval.
f   ictomètre m linéaire
e   impulsímetro m lineal
i   rateometro m lineare di conteggio
n   lineaire telsnelheidsmeter,
    lineaire teltempometer
d   lineares Zählratenmessgerät n

**4050 LINEAR SPEED,   gp
        LINEAR VELOCITY**
    A vector quantity which denotes at once
    the time rate and the direction of a linear
    motion.

f vitesse f linéaire
e velocidad f lineal
i velocità f lineare
n lineaire snelheid
d lineare Geschwindigkeit f,
  Lineargeschwindigkeit f

4051 LINEAR STOPPING POWER np
Of a material for charged particles, the
average energy lost by a charged particle
of specific energy in traversing a small
path length divided by that small path
length.
f pouvoir m d'arrêt linéique
e poder m de frenado por unidad de longitud
i potere m di rallentamento per unità di
  lunghezza
n lineïek stoppend vermogen n,
  stoppend vermogen n per lengte
d Bremsvermögen n je Längeneinheit

4052 LINGERING PERIOD np
The length of time which an electron
remains in its orbit of highest excitation
before jumping to a lower orbit, and
radiating the difference in energy.
f période f d'attardement
e perfodo m de permanencia
i periodo m di permanenza nello stato
  eccitato
n verblijftijd
d Verweilzeit f

4053 LIP APPLICATOR md, ra
An applicator which can be applied to the
lip of a patient.
f localisateur m labial
e localizador m labial
i localizzatore m labiale
n lipconus
d Lippenlokalisator m

4054 LIQUID COOLING cl
f refroidissement m à liquide
e enfriamiento m por líquido
i raffreddamento m a liquido
n vloeistofkoeling
d Flüssigkeitskühlung f

4055 LIQUID COUNTER ct
A counter for measuring the radioactivity
of a liquid.
f compteur m de la radioactivité de liquides
e contador m de la radiactividad de líquidos
i contatore m della radioattività di liquidi
n teller voor vloeistofmetingen
d Flüssigkeitszähler m

4056 LIQUID COUNTER TUBE ct
A counter tube intended to measure the
activity of a liquid.
f tube m compteur à jupe
e tubo m contador para líquidos
i tubo m contatore per liquidi
n vloeistoftelbuis
d Flüssigkeitszählrohr n

4057 LIQUID DROP MODEL np
A model in which the atomic nucleus is
imagined to behave much like a drop of
liquid.
f modèle m de la goutte liquide
e modelo m de la gota líquida
i modello m della goccia liquida
n druppelmodel n
d Tropfenmodell n

4058 LIQUID EMULSION ph
Nuclear photographic emulsion manufactured in gel form so that, after heating, it
can be poured to the desired shape and at
the desired time.
f émulsion f liquéfiable
e emulsión f licuable
i emulsione f a liquefazione
n vervloeibare emulsie
d verflüssigbare Emulsion f

4059 LIQUID-FILLED sa
  SHIELDING WINDOW
f fenêtre f de blindage à remplissage
  liquide
e ventana f de blindaje con llenado líquido
i finestra f di schermatura a riempimento
  liquido
n schermvenster n met vloeistofvulling
d Schirmfenster n mit Flüssigkeitsfüllung

4060 LIQUID FILM PLUTONIUM ma
  CONTENT METER
A content meter designed to determine the
content of irradiated uranium reprocessing
solutions by the measurement of specific
alpha activity.
f teneurmètre m en plutonium à film
  liquide
e plutoniómetro m de película líquida
i plutoniotenorimetro m a pellicola liquida
n gehaltemeter voor plutonium met
  vloeistoffilm
d Flüssigkeitsfilm-Gehaltmessgerät n
  für Plutonium

4061 LIQUID-FLOW COUNTER ct
A counter provided with an internal tube for
measuring the radioactivity of a flowing
liquid.
f compteur m à fluide traversant
e contador m de flúido de paso
i contatore m a fluido traversante
n vloeistofdoorstroomtelbuis
d Flüssigkeitsdurchflusszählrohr n

4062 LIQUID HYDROGEN LOOP cd
f circuit f fermé à hydrogène liquide
e circuito m cerrado de hidrógeno líquido
i circuito m chiuso ad idrogeno liquido
n gesloten circuit n voor vloeibare waterstof
d Kreislauf m mit flüssigem Wasserstoff

4063 LIQUID JET PUMP, see 3817

4064 LIQUID LEVEL METER ma
An assembly for measuring the level of a

liquid associated with the operation of a
nuclear reactor.
f limnimètre *m*
e nivelímetro *m*
i livellometro *m*
n niveaumeter
d Füllstandsmessgerät *n* für Flüssigkeiten

4065 LIQUID- LIQUID EXTRACTION   ch
A process in which two immiscible liquids
are brought into contact to effect a
redistribution of solutes between them.
f extraction *f* par partage liquide-liquide
e extracción *f* por repartición líquido-
líquido
i estrazione *f* per ripartizione liquido-
liquido
n vloeistof-vloeistof-extractie
d Flüssig-Flüssig-Extraktion *f*

4066 LIQUID- LIQUID   ch
EXTRACTION PLANT
f installation *f* d'extraction par partage
liquide-liquide
e equipo *m* de extracción por repartición
líquido-líquido
i impianto *m* d'estrazione per ripartizione
liquido-liquido
n vloeistof-vloeistof-extractie-installatie
d Flüssig-Flüssig-Extraktionsanlage *f*

4067 LIQUID-METAL COOLANT   cl
f réfrigérant *m* métallique liquide
e refrigerante *m* metálico líquido
i refrigerante *m* metallico liquido
n koelmiddel *n* uit vloeibaar metaal
d Flüssigmetallkühlmittel *n*

4068 LIQUID-METAL COOLED   rt
REACTOR
A nuclear reactor in which the coolant is
a liquid metal or alloy.
f réacteur *m* à réfrigérant liquide métallique
e reactor *m* de refrigerante líquido metálico
i reattore *m* a refrigerante liquido
metallico
n door vloeibaar metaal gekoelde reactor
d flüssigmetallgekühlter Reaktor *m*

4069 LIQUID-METAL FUEL   fu
A fuel consisting of a fissile (fissionable)
material, e.g. uranium dissolved in a liquid
metal, as, e.g., sodium or bismuth.
f combustible *m* métallique liquide
e combustible *m* metálico líquido
i combustibile *m* metallico liquido
n vloeibare metallieke splijtstof
d Flüssigmetallbrennstoff *m*

4070 LIQUID-METAL FUEL REACTOR   rt
A nuclear reactor employing fissile
(fissionable) material dissolved in a
liquid metal as the fuel.
f réacteur *m* à combustible métallique
liquide
e reactor *m* de combustible metálico
líquido

i reattore *m* con combustibile metallico
liquido
n reactor met vloeibare metallieke splijtstof
d Flüssigmetallreaktor *m*,
Reaktor *m* mit Flüssigmetallbrennstoff

4071 LIQUID-METAL PUMP   cl
f pompe *f* de métal liquide
e bomba *f* de metal líquido
i pompa *f* di metallo liquido
n vloeibaar-metaalpomp
d Flüssigmetallpumpe *f*

4072 LIQUID MODERATOR   con
LEVEL METER
A level meter designed to measure the
level of a liquid moderator in a nuclear
reactor.
f limnimètre *m* d'un modérateur liquide
e nivelímetro *m* de moderador líquido
i livellometro *m* d'un moderatore liquido
n niveaumeter voor de moderator
d Füllstandsmessgerät *n* für flüssige
Moderatoren

4073 LIQUID SCINTILLATOR   ct
A scintillator consisting of a liquid in
which has been dissolved some organic
material that causes it to behave like a
phosphor.
f scintillateur *m* liquide
e centelleador *m* líquido
i scintillatore *m* liquido
n vloeibare scintillator, vloeistofscintillator
d flüssiger Szintillator *m*,
Flüssigkeitsszintillator *m*

4074 LIQUID SCINTILLATOR COUNTER   ct
A scintillation counter which employs
a liquid as the scintillator.
f compteur *m* à scintillateur liquide
e contador *m* de centelleador líquido
i contatore *m* a scintillatore liquido
n teller met vloeibare scintillator
d Zähler *m* mit flüssigem Szintillator

4075 LIQUID SCINTILLATOR   ra
DETECTOR
A scintillation detector in which the
scintillator is a liquid.
f détecteur *m* à scintillateur liquide
e detector *m* de centelleador líquido
i rivelatore *m* a scintillatore liquido
n detector met vloeibare scintillator,
scintillatiedetector met vloeistof
d Detektor *m* mit flüssigem Szintillator

4076 LIQUID SCINTILLATOR PULSE   ma
COUNTING ASSEMBLY
A low background pulse counting assembly
designed to measure the activity of a
solution of a low energy beta emitter or
of a solution of low specific activity, mixed
with a liquid scintillator.
f ensemble *m* de mesure à scintillateur
liquide
e conjunto *m* contador de centelleador
líquido

i   complesso *m* di misura a scintillatore
    liquido
n   telopstelling met vloeibare scintillator
d   Zählanordnung *f* mit flüssigem Szintillator

4077 LIQUID SHUT-DOWN SYSTEM   rt
    A shut-down system supplementing the
    moderator draining system in order to
    realize a faster shut-down.
f   système *m* d'arrêt à liquide
e   sistema *m* de paro con líquido
i   sistema *m* d'arresto a liquido
n   uitschakeling door middel van vloeistof
d   Flüssigabschaltung *f*

4078 LIQUID TARGET   ra
    The target material in a radiation source
    is made of liquid metal which forms part
    of a liquid stream in order to remove the
    heat produced.
f   cible *f* métallique liquide
e   blanco *m* metálico líquido
i   bersaglio *m* metallico liquido
n   treflaag uit vloeibaar metaal
d   Flüssigmetallauffänger *m*,
    Flüssigmetalltreffschicht *f*

4079 LIQUID-WALL   ic
    IONIZATION CHAMBER
    An ionization chamber designed to measure
    the alpha or beta activity in a liquid so
    situated that its surface constitutes the
    wall of the chamber.
f   chambre *f* d'ionisation à paroi liquide
e   cámara *f* de ionización de pared liquida
i   camera *f* d'ionizzazione a parete liquida
n   ionisatievat *n* met vloeistofwand
d   Flüssigwändekammer *f*,
    Ionisationskammer *f* mit Flüssigkeitswand

4080 LIQUID WASTE   rw
f   déchets *pl* liquides
e   desechos *pl* líquidos
i   rifiuti *pl* liquidi
n   vloeibare afval
d   flüssiger Abfall *m*

4081 LITHIUM   ch
    Metallic element, symbol Li, atomic
    number 8.
f   lithium *m*
e   litio *m*
i   litio *m*
n   lithium *n*
d   Lithium *n*

4082 LITHIUM DEUTERIDE   ch, mt
f   deutérure *m* de lithium
e   deuteruro *m* de litio
i   deuteruro *m* di litio
n   lithiumdeuteride *n*
d   Lithiumdeuterid *n*

4083 LIVE REFLECTOR   rt
    A neutron reflector which is so designed
    as to be removable.
f   réflecteur *m* amovible

e   reflector *m* amovible
i   riflettore *m* removibile
n   verplaatsbare reflector
d   abnehmbarer Reflektor *m*

4084 LIVING ORGANISM   md
    An organism in which all the vital organs
    are in a perfect state.
f   organisme *m* vivant
e   organismo *m* vivo
i   organismo *m* vivo
n   gezond organisme *n*
d   gesunder Organismus *m*

4085 LIVING TISSUE   md
    Tissue which is completely intact.
f   tissu *m* vivant
e   tejido *m* vivo
i   tessuto *m* vivo
n   gezond weefsel *n*
d   gesundes Gewebe *n*

4086 LIXIVIATION, see 3945

4087 L/M RATIO   np
    The ratio of the number of integral
    conversion electrons from the L shell to
    the number from the M shell, emitted in
    the de-excitation of a nucleus.
f   relation *f* L/M
e   relación *f* L/M
i   rapporto *m* L/M
n   L/M-verhouding
d   L/M-Verhältnis *n*

4088 LMTD,   gp
    LOGARITHMIC MEAN
    TEMPERATURE DIFFERENCE
    The logarithmic mean of the terminal
    temperature differences in a heat
    exchanger.
f   différence *f* moyenne logarithmique des
    températures
e   diferencia *f* media logarítmica de las
    temperaturas
i   differenza *f* media logaritmica delle
    temperature
n   logarithmisch gemiddelde *ñ* van het
    temperatuurverschil
d   logarithmischer Mittelwert *m* der
    Temperaturunterschiede

4089 LOAD, see 1000

4090 LOAD (TO), see 1001

4091 LOAD FACTOR   rt
    The total time in a year in which, e.g., a
    nuclear reactor is in full power.
f   facteur *m* de charge
e   factor *m* de carga
i   fattore *m* di carico
n   werkelijk belaste tijd
d   Nutzlastzeit *f*

4092 LOAD HOIST, see 1006

4093 LOAD HOLE,
LOAD TUBE, see 1007

4094 LOADED CONCRETE, see 3183

4095 LOADED CONCRETE SHIELD, see 3184

4096 LOADING   ge
The force on an object or structure or
element of a structure.
f charge f
e carga f
i carica f
n belasting
d Belastung f

4097 LOADING   ph
Incorporating in a nuclear emulsion
substances like deuterium, boron or
uranium.
f imprégnation f
e impregnación f
i impregnazione f
n beladen n
d Imprägnierung f

4098 LOADING   rt
Introduction of substances in a nuclear
reactor to be irradiated for experimental
or industrial purposes.
f enfournement m
e implantación f
i introduzione f
n invoeren n
d Einführen n

4099 LOADING,
CHARGING, see 1027

4100 LOADING FACE, see 1028

4101 LOADING MACHINE, see 1029

4102 LOADING POINT   ch
In two-phase countercurrent flow in a
column the flow rate below the flooding
point at which either phase begins to
seriously impede the flow of the other.
f vitesse f critique du courant
e velocidad f crítica del flujo
i velocità f critica del flusso
n kritieke stroomsnelheid
d kritische Strömungsgeschwindigkeit f

4103 LOADING PROCEDURE, see 1030

4104 LOCAL DOSE RATE   xr
The exposure dose rate in occupied space,
with a patient, phantom or other object
present in the useful beam.
f débit m de dose ambiant
e variación f con el tiempo de la dosis
ambiente
i emissione f di dose ambientale
n plaatselijke doseringssnelheid
d lokale Dosisleistung f

4105 LOCAL FALL-OUT   nw
Fall-out from the large particles in the
fireball of the bomb.
f retombées pl locales
e depósito m local, poso m local
i ricaduta f locale
n plaatselijke neerslag
d örtlicher Ausfall m,
örtlicher Niederschlag m

4106 LOCALIZATION   xr
The determination of the position of an
object in the body by radiographic means.
f localisation f
e localización f
i localizzazione f
n plaatsbepaling
d Lagebestimmung f, Lokalisierung f

4107 LOCALIZER, see 314

4108 LODE,   mi
VEIN
A fissure in the country rock, filled with
mineral or later deposition.
f filon m, veine f
e filón m, vena f
i filone m, vena f
n ader
d Ader f

4109 LOG DECODER,   cd
LOG DECODING BOARD
Component part of the 400 channel pulse
height analyzer.
f décodeur m logarithmique des contenus
des canaux
e descodificador m logarítmico de los
contenidos de los canales
i decodificatore m logaritmico dei
contenuti dei canali
n logarithmisch decodeerapparaat n van de
kanaalinhoud
d logarithmischer Dekodierapparat m des
Kanalinhalts

4110 LOGARITHMIC COUNTING-RATE   ct
METER
A counting-rate meter which gives an
indication proportional to the logarithm
of the count rate.
f débitmètre m logarithmique de coups
e contador m logarítmico de velocidad de
recuento
i contatore m logaritmico di velocità di
conteggio
n logarithmische telsnelheidsmeter
d logarithmischer Zählgeschwindigkeits-
messer m

4111 LOGARITHMIC DIRECT   ec
CURRENT AMPLIFIER
An amplifier in which the output signal has
a logarithmic relation to the input signal.

f   amplificateur *m* logarithmique pour courant
    continu
e   amplificador *m* logarítmico de corriente
    continua
i   amplificatore *m* logaritmico per corrente
    continua
n   logarithmische gelijkstroomversterker
d   logarithmischer Gleichstromverstärker *m*

4112   LOGARITHMIC ENERGY   np
       DECREMENT
       In reactor physics, the mean value of the
       change in the natural logarithm of the
       neutron energy in a single collision.
f   décrément *m* logarithmique d'énergie
e   decremento *m* logarítmico de energía
i   decremento *m* logaritmico d'energia
n   logarithmisch energiedecrement *n*
d   logarithmisches Energiedekrement *n*

4113   LOGARITHMIC MEASURING   me
       ASSEMBLY
       A measuring assembly in which the
       amplitude of the output signal is a
       logarithmic function of the input signal.
f   ensemble *m* de mesure logarithmique
e   conjunto *m* de medida logarítmica
i   complesso *m* di misura logaritmica
n   logarithmische meetopstelling
d   logarithmisch messende Anordnung *f*

4114   LOGARITHMIC POWER   me
       MEASURING ASSEMBLY BASED
       ON GAMMA RADIATION
       A power measuring assembly in which the
       output signal is a logarithmic function of
       the input signal.
f   ensemble *m* de mesure logarithmique de
    puissance par mesure du rayonnement
    gamma
e   conjunto *m* de medida logarítmica de
    potencia por la radiación gamma
i   complesso *m* di misura logaritmica di
    potenza per mezzo della misura della
    densità di flusso gamma
n   opstelling voor logarithmische vermogens-
    meting door middel van de gammastraling
d   Anordnung *f* zur logarithmischen
    Leistungsmessung mittels Gammastrahlung

4115   LOGARITHMIC POWER   me
       MEASURING ASSEMBLY BASED ON
       THE NEUTRON FLUX DENSITY
       A power measuring assembly in which the
       output signal is a logarithmic function of
       the input signal.
f   ensemble *m* de mesure logarithmique de
    puissance par mesure de la densité de
    flux neutronique
e   conjunto *m* de medida logarítmica de
    potencia por densidad de flujo neutrónico
i   complesso *m* di misura logaritmica di
    potenza per mezzo della densità del
    flusso neutronico
n   opstelling voor logarithmische vermogens-
    meting door middel van de neutronen-
    fluxdichtheid

d   Anordnung *f* zur logarithmischen Leistungs-
    messung mittels Neutronenstrahlung

4116   LOGARITHMIC PULSE   ec
       AMPLIFIER
       A pulse amplifier which, within the limits
       of its normal operating characteristics,
       delivers an output pulse of an amplitude
       proportional to the logarithm of the
       input pulse amplitude.
f   amplificateur *m* logarithmique d'impulsions
e   amplificador *m* logarítmico de impulsos
i   amplificatore *m* logaritmico d'impulsi
n   logaritmische pulsversterker
d   logarithmischer Impulsverstärker *m*

4117   LOGARITHMIC RATEMETER ct, ec, me
       An electronic sub-assembly which gives a
       continuous indication proportional to the
       logarithm of the average counting rate over
       a predetermined time interval.
f   ictomètre *m* logarithmique
e   impulsímetro *m* logarítmico
i   rateometro *m* logaritmico di conteggio
n   logarithmische telsnelheidsmeter,
    logarithmische teltempometer
d   logarithmisches Zählratenmessgerät *n*

4118   LOGIC SIGNAL   ec
       A digital signal of which the symbols
       correspond to the rules of a particular
       algebra.
f   signal *m* logique
e   señal *f* lógica
i   segnale *m* logico
n   logisch signaal *n*
d   logisches Signal *n*

4119   LONE ELECTRON   ec, np
       An electron that is alone on an energy
       level.
f   électron *m* célibataire
e   electrón *m* solitario
i   elettrone *m* solitario
n   één elektron *n* op één niveau,
    eenzaam elektron *n*
d   Einzelelektron *n*

4120   LONG COUNTER, see 3137

4121   LONG-LIVED FISSION PRODUCT   np
       A fission product with a comparatively
       long lifetime.
f   produit *m* de fission à longue période
e   producto *m* de fisión de vida larga
i   prodotto *m* di fissione di lunga durata
n   langlevend splijtprodukt *n*
d   langlebiges Spaltprodukt *n*

4122   LONG-LIVED RADIATION   ra
       Radiation with a comparatively long time
       of duration.
f   rayonnement *m* à longue période
e   radiación *f* de vida larga
i   radiazione *f* di lunga durata
n   langlevende straling
d   langlebige Strahlung *f*

4123 LONG RANGE ALPHA   ar
      PARTICLES
      Alpha particles producing lines in an
      alpha particle spectrum due to groups
      that have very low intensities relative to
      those for the main groups.
f particules *pl* alpha à portée longue
e partículas *pl* alfa de alcance largo
i particelle *pl* alfa a portata lunga
n alfadeeltjes *pl* met verre dracht
d weitreichende Alphateilchen *pl*

4124 LONG TERM   rt
      REACTIVITY CHANGES
f variations *pl* de réactivité à longue
   échéance
e variaciones *pl* seculares de reactividad
i variazioni *pl* di reattività a lungo
   termine
n reactiviteitsveranderingen *pl* op lange
   termijn
d Langzeitreaktivitätsänderungen *pl*

4125 LONGITUDE EFFECT   cr
      The variation in cosmic ray intensity
      along the equator as a function of longitude.
f effet *m* de longitude
e efecto *m* de longitud
i effetto *m* di longitudine
n lengte-effect *n*
d Effekt *m* der geographischen Länge

4126 LONGITUDINAL FINNING   cd, cl
      Method of arranging cooling fins to
      improve the heat characteristics of
      nuclear reactors.
f à ailettes longitudinales
e con aletas longitudinales
i ad alette longitudinali
n in de lengterichting van koelribben
   voorzien
d Längsberippung *f*,
   mit Längsrippen versehen

4127 LOOP, see 2444

4128 LOOSELY SPACED LATTICE,   fu, rt
      WIDELY SPACED LATTICE
      A fuel lattice with a large pitch.
f réseau *m* à pas large
e celosía *f* de paso ancha
i reticolo *m* lasco
n rooster *n* met grote elementen afstand
d Gitter *n* mit grössem Elementenabstand

4129 LORDOTIC PROJECTION,   xr
      LORDOTIC VIEW
      A radiograph of the chest made with the
      film behind the patient and the röntgen
      rays going obliquely upward.
f projection *f* lordotique,
   vue *f* lordotique
e proyección *f* lordótica, vista *f* lordótica
i proiezione *f* lordotica, vista *f* lordotica
n lordotische projectie
d Projektion *f* in schräger Richtung

4130 LORENTZ DISSOCIATION   np
      Dissociation of molecular ions by the
      Lorentz ionization process.
f dissociation *f* de Lorentz
e disociación *f* de Lorentz
i dissociazione *f* di Lorentz
n lorentzdissociatie
d Lorentz-Dissoziation *f*

4131 LORENTZ GAS   np
      An idealized model of a fully ionized gas
      in which all electron-electron interactions
      are neglected and the ions are assumed
      to be stationary.
f gaz *m* d'électrons de Lorentz
e gas *m* de electrones de Lorentz
i gas *m* d'elettroni di Lorentz
n lorentzgas *n*
d Lorentzgas *n*

4132 LORENTZ IONIZATION   np
      Ionization of neutral atoms obtained by
      projecting them at high velocity into a
      magnetic field.
f ionisation *f* de Lorentz
e ionización *f* de Lorentz
i ionizzazione *f* di Lorentz
n lorentzionisatie
d Lorentz-Ionisation *f*

4133 LOSCHMIDT NUMBER, see 481

4134 LOSS CONE   pp
      In a magnetic mirror, a cone having the
      same symmetry axis as the mirror and
      a top angle θ_m defined by

      $$\sin \theta_m = \frac{1}{\sqrt{R}}$$, where R is the mirror
      ratio.
f cône *m* de perte
e cono *m* de pérdida
i cono *m* di perdita
n verlieskegel
d Verlustkonus *m*

4135 LOSS OF COOLANT ACCIDENT   cl
      Escape of cooling liquid in a reactor due
      to a failure in the cooling circuit.
f incident *m* dangereux de perte de réfrigérant
e incidente *m* peligroso de pérdida de
   refrigerante
i incidente *m* pericoloso da perdita di
   refrigerante
n gevaarlijk koelmiddelverlies *n*
d gefährlicher Kühlmittelverlust *m*

4136 LOSS OF FLOW ACCIDENT   cl
      An accident due to flow reduction in the
      cooling circuit.
f incident *m* dangereux de réduction de flux
e incidente *m* peligroso de reducción de
   flujo
i incidente *m* pericoloso da perdita di
   circolazione
n gevaarlijke doorstromingsreductie
d gefährliche Durchströmungsreduktion *f*

4137 LOT-SORTING CONVEYOR   ma
      GRADING EQUIPMENT
      An equipment which includes one or
      several detectors whose thresholds are
      determined according to the weight and are
      sited along the path of the mined products
      to control the grading of the lots in
      several categories of activity/weight.
f   équipement *m* d'estimation et de triage sur
      bande par lot
e   equipo *m* selector clasificador de lotes
i   apparecchiatura *f* per stima e cernita
      di lotti
n   uitrusting voor sortering van ertsen aan
      de band
d   Klassiereinrichtung *f* bei der Sortierung
      der Ladungen am Förderband

4138 LOVOZERITE   mi
      A complex silicate of titanium and zirco-
      nium containing a small amount of
      thorium.
f   lovozérite *f*
e   lovozerita *f*
i   lovozerite *f*
n   lovozeriet *n*
d   Lovozerit *m*

4139 LOW-BACKGROUND PULSE   ma
      COUNTING ASSEMBLY
      A pulse counting assembly to measure
      very low activities and designed so as to
      reduce the background of the assembly by
      means of shielding, of anticoincidence
      arrays of detectors to eliminate the effect
      of cosmic rays, and/or by any other
      appropriate means.
f   ensemble *m* de mesure à faible mouvement
      propre
e   conjunto *m* contador por impulsos para
      débiles actividades
i   complesso *m* di misura a basso fondo
n   telopstelling met klein nuleffect
d   Zählanordnung *f* zur Messung niedriger
      Aktivitäten

4140 LOW-DENSITY PLASMA   pp
f   plasma *m* de basse densité
e   plasma *m* de baja densidad
i   plasma *m* di bassa densità
n   plasma *n* met lage dichtheid
d   Plasma *n* mit niedriger Dichte

4141 LOW-ENERGY GAMMA   gr, md, ra
      SOURCE
      A radiation source of which the gamma
      energy is smaller than 1 MeV.
f   source *f* gamma à basse énergie
e   fuente *f* gamma de baja energía
i   sorgente *f* gamma a bassa energia
n   gammabron met lage energie
d   Gammaquelle *f* niedriger Energie

4142 LOW-ENERGY NEUTRON   np
      A neutron resulting from a reaction in
      which the energy level is below 30 MeV.
f   neutron *m* à faible énergie

e   neutrón *m* de baja energía
i   neutrone *m* di bassa energia
n   neutron *n* met geringe energie
d   Neutron *n* niedriger Energie

4143 LOW-ENERGY PARTICLE   np
      A particle resulting from a reaction in
      which the energy level is below 30 MeV.
f   particule *f* à faible énergie
e   partícula *f* de baja energía
i   particella *f* di bassa energia
n   deeltje *n* met lage energie
d   Teilchen *n* niedriger Energie

4144 LOW-ENRICHED FUEL CYCLE   rt
      The fuel cycle used in the high
      temperature reactor project DRAGON.
f   cycle *m* de combustible légèrement enrichi
e   ciclo *m* de combustible de bajo enrique-
      cimiento
i   ciclo *m* di combustibile di basso
      arricchimento
n   in geringe mate verrijkte splijtstofcyclus
d   schwachangereicherter Spaltstoffzyklus *m*

4145 LOW-FLUX REACTOR   rt
      A nuclear reactor having a relatively low
      neutron flux, i.e. a reactor which operates
      at a low power density.
f   réacteur *m* à flux bas
e   reactor *m* de bajo flujo
i   reattore *m* a basso flusso
n   lage-fluxreactor
d   Niederflussreaktor *m*

4146 LOW-FLUX RESEARCH REACTOR   rt
      A research reactor designed to operate
      with low neutron flux.
f   réacteur *m* de recherche à flux bas
e   reactor *m* de investigación de bajo flujo
i   reattore *m* per ricerca a basso flusso
n   lage-fluxreactor voor speurwerk
d   Niederflussforschungsreaktor *m*

4147 LOW-GRADE ORE   mi
      Ore containing small quantities of a
      mineral.
f   minerai *m* de qualité inférieure
e   mineral *m* de calidad inferior
i   minerale *m* di cattiva qualità
n   arm erts *n*
d   armes Ertz *n*, geringwertiges Erz *n*

4148 LOW-LEVEL BETA   br, ct
      COUNTING ASSEMBLY
      A device for measuring beta activities as
      low as a few tenths of a count per minute.
f   ensemble *m* de mesure de faibles activités
      bêta
e   conjunto *m* de medida de bajas actividades
      beta
i   complesso *m* di misura di basse attività
      beta
n   opstelling voor het meten van lage
      bêta-activiteit
d   Anordnung *f* zum Messen niedriger
      Betaaktivität

4149 LOW-LEVEL WASTE   rw
The radioactive liquid waste which is
usually separations process cooling water
or steam condensate and is discharged to
the ground surface.
f   déchets pl de faible radioactivité
e   desechos pl de baja radiactividad
i   rifiuti pl a bassa radioattività
n   afval van geringe activiteit
d   Abfall m niedriger Aktivität

4150 LOW-POWER RANGE,   rt
      SOURCE RANGE
Range of operation in which the reactor
is subcritical and its power depends upon
its multiplying the neutron flux from an
added source.
f   domaine m des sources
e   régimen m de fuentes
i   campo m di sorgenti
n   bronnengebied n, subkritiek gebied n,
    subkritisch gebied n
d   Quellenbereich m

4151 LOW-PRESSURE CLOUD   ic
      CHAMBER
A cloud chamber in which the gas is
maintained at low pressure.
f   chambre f à détente à basse pression
e   cámara f de expansión de baja presión
i   camera f d'ionizzazione a bassa
    pressione
n   lage-druknevelvat n
d   Niederdrucknebelkammer f

4152 LOW-TEMPERATURE REACTOR   rt
A nuclear reactor designed to operate at
relatively low temperature, i.e. at such
a temperature that it is impossible to
generate mechanical power with good
efficiency.
f   réacteur m à température basse
e   reactor m de baja temperatura
i   reattore m a bassa temperatura
n   lage-temperatuurreactor
d   Niedertemperaturreaktor m

4153 LOW VACUUM   vt
A degree of vacuum at which so much gas
or vapo(u)r is still present that ionization
can occur in an electron tube or valve.
f   vide m peu poussé
e   bajo vacío m
i   basso vuoto m
n   laag vacuüm n
d   Feinvakuum n, niedriges Vakuum n

4154 LOWER TRIP LEVEL   rt, sa
f   seuil m inférieur d'arrêt
e   umbral m inferior de paro
i   soglia f inferiore di scatto
n   onderuitschakeldrempel
d   Unterabschaltschwelle f

4155 L-S COUPLING,   np
      RUSSEL-SAUNDERS COUPLING
The interaction between the resultant
orbital angular momentum of two or more

particles and their resultant intrinsic
angular momentum.
f   couplage m L-S
e   acoplamiento m L-S
i   accoppiamento m L-S
n   L-S-koppeling
d   L-S-Kopplung f

4156 LUMINESCENCE   ra
A phenomenon in which the absorption of
primary radiation by a substance gives
rise to the emission of light, characteristic of the substance.
f   luminescence f
e   luminiscencia f
i   luminescenza f
n   luminescentie
d   Lumineszenz f

4157 LUMINESCENCE THRESHOLD,   ra
      THRESHOLD OF LUMINESCENCE
The lowest frequency of radiation that is
capable of exciting a luminescent material.
f   seuil m de luminescence
e   umbral m de luminiscencia
i   soglia f di luminescenza
n   luminescentiedrempel
d   Lumineszenzschwelle f

4158 LUMINESCENT CENTER (US),   ra
      LUMINESCENT CENTRE (GB)
An atom or group of atoms which, when
suitably excited, can produce luminescence.
f   centre m luminogène
e   centro m luminógeno
i   centro m luminogeno
n   luminescentiecentrum n
d   Lumineszenzzentrum n

4159 LUMINESCENT MATERIAL,   ch
      PHOSPHOR
Any substance which can store energy and
release it later in the form of light.
f   substance f luminescente
e   fósfor m, substancia f luminiscente
i   sostanza f luminescente
n   fosfor, luminescerende stof
d   Leuchtstoff m,
    lumineszierende Substanz f, Phosphor m

4160 LUMINOPHORE, see 2187

4161 LUMINOUS EFFICIENCY
      OF A SCINTILLATOR, see 2261

4162 LUMP   fu
f   bloc m
e   bloque m, cartucho m, trozo m
i   blocco m, bloccolo m
n   blok n, klompje m
d   Block m, Brocken m, Klumpen m

4163 LUMP-SORTING GRADING   mi
      EQUIPMENT
An equipment which includes one or
several activity measuring assemblies
with adjustable threshold, and whose
detectors are arranged along the path of

the mined products to control ejectors for
the sorting of previously calibrated lumps
into several categories of activity.
f équipement *m* de triage caillou par caillou
e equipo *m* selector de terrones
i apparecchiatura *f* per cernita di pietre
n uitrusting voor sortering van ertsblokken
d Klassiereinrichtung *f* bei der Stück-
   sortierung

4164 LUNGS   md
The critical organ for most insoluble
radioactive dusts and vapo(u)rs from the
atmosphere.
f poumons *pl*
e pulmones *pl*
i polmoni *pl*
n longen *pl*
d Lungen *pl*

4165 LUTECIUM   ch
Rare earth metallic element, symbol Lu,
atomic number 71.
f lutétium *m*
e lutecio *m*
i lutecio *m*
n lutetium *n*
d Lutetium *n*

4166 LYMPH   md
An almost colo(u)rless fluid circulating
in the lymphatic vessels of vertebrates.
f lymphe *f*
e linfa *f*
i linfa *f*
n lymf, lymfe
d Lymphe *f*

4167 LYMPH GLAND,   md
     LYMPH NODE
An aggregation of connective tissue
crowded with lymphocytes and surrounded
with a fibrous capsule.
f ganglion *m* lymphatique
e ganglio *m* linfático,
   glándula *f* linfática
i ganglio *m* linfatico
n lymfklier
d Lymphdrüse *f*, Lymphknoten *m*

4168 LYMPHOCYTE   md
A type of leucocyte characterized by a
single sharply defined nucleus and scanty
cytoplasm.

f lymphocyte *m*
e linfocito *m*
i linfocito *m*
n lymfocyt *n*
d Lymphozyt *m*, Lymphzelle *f*

4169 LYMPHOGRAPHY   xr
Radiological examination of the lymph
gland.
f lymphographie *f*
e linfografía *f*
i linfografia *f*
n lymfografie
d Lymphographie *f*

4170 LYMPHOPENIA   md
Decrease in the proportion of lymphocytes
in the blood.
f lymphopénie *f*
e linfopenía *f*
i linfopenia *f*
n lymfopenie
d Lymphopenie *f*

4171 LYNDOCHITE   mi
A calcium-thorium-euxenite containing
uranium.
f lyndochite *f*
e lindoquita *f*
i lindochite *f*
n lyndochiet *n*
d Lyndochit *m*

4172 LYOTROPIC SERIES   ch
A listing of ions in sequence of their ion
exchange possibility.
f série *f* lyotropique
e serie *f* liotrópica
i serie *f* liotropica
n lyotropische reeks
d lyotropische Reihe *f*

4173 LYSHOLM GRID,   xr
     STATIONARY GRID
An anti-diffusion grid which does not
move during exposure.
f grille *f* de Lysholm, grille *f* fixe
e rejilla *f* de Lysholm, rejilla *f* fija
i griglia *f* di Lysholm, griglia *f* fissa
n lysholmraster *n*, vast raster *n*
d feststehende Streustrahlenblende *f*

M

4174 M CAPTURE, np
M ELECTRON CAPTURE
A mode of electron capture similar to K electron capture.
f capture *f* M
e captura *f* M
i cattura *f* M
n M-vangst
d M-Einfang *m*

4175 M ELECTRON np
An electron characterized by having a principal quantum number of value 3.
f électron *m* M
e electrón *m* M
i elettrone *m* M
n M-elektron *n*
d M-Elektron *n*

4176 M LINE sp
One of the lines in the M series of X-rays that are characteristic of the various elements and are produced by excitation of the electrons of the M shell.
f raie *f* M
e línea *f* M
i riga *f* M
n M-lijn
d M-Linie *f*

4177 M RADIATION ra, xr
One of a series of X-rays characteristic of each element that is excited when that element, commonly as a metal, is used as an anticathode in an X-ray tube, so as to excite the electron of the M shell.
f rayonnement *m* M
e radiación *f* M
i radiazione *f* M
n M-straling
d M-Strahlung *f*

4178 M SHELL np
The collection of all those electrons in an atom that are characterized by the principal quantum number 3.
f couche *f* M
e capa *f* M
i strato *m* M
n M-schil
d M-Schale *f*

4179 M THEORY, pp
MOTOR THEORY
The theory of imploding current sheaths in pinch devices developed by M. Rosenbluth and by Russian research workers.
f théorie *f* des gaines de courant à implosion
e teoría *f* de las vainas de corriente de implosión
i teoria *f* delle guaine di corrente ad implosione
n theorie van imploderende stroomomhullingen
d Theorie *f* der implodierenden Stromhüllen

4180 MACH CONE, nw
MACH FRONT, MACH STEM
The shock front formed by the fusion of the incident and reflected shock fronts from an explosion.
f front *m* de Mach
e frente *f* de Mach
i fronte *f* di Mach
n machkegel
d Machscher Kegel *m*

4181 MACH NUMBER nw
The ratio of the speed of an object to the speed of sound in the undisturbed medium in which the object is travel(l)ing.
f nombre *m* de Mach
e número *m* de Mach
i numero *m* di Mach
n machgetal *n*
d Mach-Zahl *f*

4182 MACH REGION nw
The region on the surface at which the Mach stem has formed as the result of a particular explosion in the air.
f région *f* de Mach
e región *f* de Mach
i regione *f* di Mach
n machgebied *n*
d Mach-Gebiet *n*

4183 MACH WAVE nw
The shock wave set up by an object travel(l)ing with a Mach number greater than unity.
f onde *f* de Mach
e onda *f* de Mach
i onda *f* di Mach
n machgolf
d Mach-Welle *f*

4184 MACKINTOSHITE mi
A mineral containing thorium, uranium, cesium, lanthanum and yttrium.
f mackintoshite *f*
e mackintoshita *f*
i mackintoshite *f*
n mackintoshiet *n*
d Mackintoshit *m*

4185 MACROMOLECULE ch
A term sometimes applied to a crystalline solid, such as diamond, but also to molecules of large molecular weight.
f macromolécule *f*
e macromolécula *f*
i macromolecola *f*
n macromolecule
d Makromolekül *n*

4186 MACROPARTICLES np
Name given to the large-size particles.
f macroparticules *pl*
e macropartículas *pl*
i macroparticelle *pl*
n macrodeeltjes *pl*
d Makroteilchen *pl*

4187 MACROPOROSITY mt
The occurrence of large holes between groups of crystals of graphite.
f macroporosité *f*
e macroporosidad *f*
i macroporosità *f*
n macroporeusheid
d Makroporosität *f*

4188 MACROSCOPIC CROSS SECTION, see 1542

4189 MACROSCOPIC FLUX DISTRIBUTION, see 3086

4190 MACROSCOPIC FLUX VARIATION, see 3085

4191 MACROSCOPIC PROPERTY rt
A nuclear reactor property that can be treated independently of other factors.
f propriété *f* macroscopique
e propiedad *f* macroscópica
i proprietà *f* macroscopica
n macroscopische eigenschap
d makroskopisches Kennzeichen *n*

4192 MACROSCOPIC REMOVAL np CROSS SECTION
Conventional cross section introduced in the multigroup model.
f section *f* macroscopique de déplacement
e sección *f* macroscópica de remoción
i sezione *f* d'urto macroscopica di rimozione
n macroscopische verplaatsingsdoorsnede
d makroskopischer Ausscheidquerschnitt *m*

4193 MACROSCOPIC STATE, np MACROSTATE
A state in which the individuality of the electron does not play a role.
f état *m* macroscopique
e estado *m* macroscópico
i stato *m* macroscopico
n macroscopische toestand
d makroskopischer Zustand *m*, Makrozustand *m*

4194 MAGIC NUCLEI np
Nuclei with n protons and/or m neutrons, where n and m are magic numbers.
f noyaux *pl* magiques
e núcleos *pl* mágicos
i nuclei *pl* magici
n magische kernen *pl*
d magische Kerne *pl*

4195 MAGIC NUMBERS np
Numbers 2, 8, 20, 28, 50, 82 and 126;

nuclides possessing these numbers of neutrons or of protons have exceptional stability.
f nombres *pl* magiques
e números *pl* mágicos
i numeri *pl* magici
n magische getallen *pl*
d magische Zahlen *pl*

4196 MAGNESIOTHERMY mg
Manufacturing process of uranium by reducing a compound, usually a fluoride, by magnesium.
f magnésiothermie *f*
e magnesiotermia *f*
i magnesiotermia *f*
n magnesiothermie
d Magnesiothermie *f*

4197 MAGNESIUM ch
Metallic element, symbol Mg, atomic number 12.
f magnésium *m*
e magnesio *m*
i magnesio *m*
n magnesium *n*
d Magnesium *n*

4198 MAGNETIC ANALYZER, see 239

4199 MAGNETIC BARRIERS, pp MAGNETIC MIRRORS
The regions of high field strength in adiabatic containment.
f miroirs *pl* magnétiques
e espejos *pl* magnéticos
i specchi *pl* magnetici
n magnetische spiegels *pl*
d magnetische Spiegel *pl*

4200 MAGNETIC BOTTLE, see 120

4201 MAGNETIC DIPOLE, np MAGNETIC DOUBLET
An elementary dipole associated with nuclear particles.
f dipôle *m* magnétique
e dipolo *m* magnético
i dipolo *m* magnetico
n magnetische dipool
d magnetischer Dipol *m*

4202 MAGNETIC DIPOLE DENSITY, ec MAGNETIC MOMENT DENSITY
The volume density of magnetic moment.
f intensité *f* du moment magnétique
e intensidad *f* del momento magnético
i intensità *f* del momento magnetico
n intensiteit van het magnetisch moment
d Intensität *f* des magnetischen Momentes

4203 MAGNETIC DIPOLE MOMENT, np MAGNETIC MOMENT
Associated with the intrinsic spin of a particle and with the orbital motion of a particle in a system.
f moment *m* magnétique
e momento *m* magnético

i   momento *m* magnetico
n   magnetisch moment *n*
d   magnetisches Moment *n*

4204 MAGNETIC FIELD   ec, gp
The space in the neighbo(u)rhood of an
electric current, or of a permanent magnet,
throughout which the forces due to the
current or magnet can be detected.
f   champ *m* magnétique au sens qualitatif
e   campo *m* magnético
i   campo *m* magnetico
n   magneetveld *n*, magnetisch veld *n*
d   Magnetfeld *n*, magnetisches Feld *n*

4205 MAGNETIC FIELD COIL   pp
Component part of any thermonuclear
installation.
f   bobine *f* de champ magnétique
e   bobina *f* de campo magnético
i   bobina *f* di campo magnetico
n   magneetveldspoel
d   Magnetfeldspule *f*

4206 MAGNETIC FIELD INDEX, see 2588

4207 MAGNETIC FIELD STRENGTH,   ec
MAGNETIC INTENSITY,
MAGNETIZING FORCE
An axial vector quantity which, together
with magnetic induction, specifies a
magnetic field at any point in space.
f   champ *m* magnétique au sens quantitatif
e   intensidad *f* de campo magnético
i   forza *f* magnetica,
intensità *f* di campo magnetico
n   magnetische veldsterkte
d   magnetische Erregung *f*,
magnetische Feldstärke *f*

4208 MAGNETIC FLUX   ec
Flux of the magnetic induction.
f   flux *m* magnétique
e   flujo *m* magnético
i   flusso *m* magnetico
n   magnetische flux
d   magnetischer Fluss *m*

4209 MAGNETIC FLUX DENSITY,
MAGNETIC INDUCTION,
see 2747

4210 MAGNETIC FOCUSING   pa
Focusing of an electron or particle beam
by means of a magnetic field.
f   focalisation *f* magnétique
e   enfoque *m* magnético
i   focalizzazione *f* magnetica
n   magnetische focussering
d   magnetische Fokussierung *f*

4211 MAGNETIC GATE   ec
A gate circuit employed in magnetic
amplifiers.
f   gâchette *f* magnétique,
porte *f* magnétique
e   puerta *f* magnética

i   porta *f* magnetica
n   magnetische poort
d   magnetische Torschaltung *f*,
magnetisches Gatter *n*

4212 MAGNETIC LATITUDE EFFECT,
see 3015

4213 MAGNETIC LENS   ec
An apparatus which produces a distribution
of magnetic fields such that they have a
focusing effect on a beam of charged
particles.
f   lentille *f* magnétique
e   lente *m* magnético
i   lente *f* magnetica
n   magnetische lens
d   magnetische Linse *f*

4214 MAGNETIC LINE OF FORCE (US),
see 4027

4215 MAGNETIC MOMENT OF AN   ma
ATOM OR NUCLEUS
The maximum expectation value of the
component of the electromagnetic moment
in direction of the magnetic field.
f   moment *m* magnétique d'un atome ou d'un
noyau
e   momento *m* magnético de un átomo o de
un núcleo
i   momento *m* magnetico d'un atomo o d'un
nucleo
n   magnetisch moment *n* van een atoom of
een kern
d   magnetisches Moment *n* eines Atoms oder
eines Kerns

4216 MAGNETIC MOMENT OF AN   np
ORBITAL ELECTRON
An integral multiple of the magnetic
moment of a Bohr magneton.
f   moment *m* magnétique d'un électron
satellite
e   momento *m* magnético de un electrón
planetario
i   momento *m* magnetico d'un elettrone
orbitale
n   magnetisch moment *n* van een schaalelektron
d   magnetisches Moment *n* eines Hüllen-
elektrons

4217 MAGNETIC PLASMOID   pp
A discrete piece of plasma produced by
discharges in magnetic fields.
f   plasmoïde *m* produit par décharge en
champ magnétique
e   plasmoide *m* producido por descarga en
campo magnético
i   plasmoide *m* prodotto per scarica in
campo magnetico
n   door ontlading in magneetveld gevormd
plasmoïde *n*
d   durch Entladung in Magnetfeld gebildetes
Plasmoid *n*

4218 MAGNETIC PRESSURE pp
Term used to describe the plasma containment force by the magnetic field.
f   pression f magnétique
e   presión f magnética
i   pressione f magnetica
n   magnetische druk
d   magnetischer Druck m

4219 MAGNETIC PUMPING pp
A method for heating or moving a plasma
by varying the magnetic field.
f   pompage m magnétique,
    variation f du champ magnétique
e   bombeo m magnético,
    variación f del campo magnético
i   pompaggio m magnetico,
    variazione f del campo magnetico
n   magnetisch pompen n,
    variatie van het magnetisch veld
d   Änderung f des Magnetfeldes,
    magnetisches Pumpen n

4220 MAGNETIC QUANTUM np, qm
    NUMBER
A quantum number that determines the
component of the angular momentum
vector of an atomic electron or group of
electrons along the externally applied
magnetic field.
f   nombre m quantique magnétique
e   número m cuántico magnético
i   numero m quantico magnetico
n   magnetisch quantumgetal n
d   magnetische Quantenzahl f

4221 MAGNETIC RESONANCE gp
When a substance containing nuclear or
electronic spin or orbital magnetic
moments is placed in a strong magnetic
field, its energy levels are split into two
or more levels, depending on the orientation of the magnetic moment of the field.
f   résonance f magnétique
e   resonancia f magnética
i   risonanza f magnetica
n   magnetische resonantie
d   magnetische Resonanz f

4222 MAGNETIC RESONANCE sp
    SPECTROMETER
A spectrometer using the principle of
nuclear magnetic resonance for chemical
analysis purposes.
f   spectromètre m à résonance magnétique
e   espectrómetro m de resonancia magnética
i   spettrometro m di risonanza magnetica
n   magnetische resonantiespectrometer
d   magnetisches Resonanzspektrometer n

4223 MAGNETIC RESONANCE sp
    SPECTRUM
A spectrum produced by varying the radiofrequency electromagnetic field that is
superimposed on a steady or slowly
varying magnetic field about which the
atoms of a material precess, to make

molecules change their magnetic quantum
numbers as they absorb or emit quanta of
radiowaves.
f   spectre m de résonance magnétique
e   espectro m de resonancia magnética
i   spettro m di risonanza magnetica
n   magnetisch resonantiespectrum n
d   magnetisches Resonanzspektrum n

4224 MAGNETIC RIGIDITY, np
    MOMENTUM-CHARGE RATIO
A measure of the momentum of a particle
equal to the product of the magnetic intensity perpendicular to the path of the particle
and the resultant radius of the curvature of
the path of the particle.
f   rigidité f magnétique
e   rigidez f magnética
i   rigidità f magnetica
n   magnetische stijfheid
d   magnetische Steifigkeit f

4225 MAGNETIC SEPARATOR mi
An apparatus used, e.g., in ore treatment
to separate magnetic from non-magnetic
materials.
f   séparateur m magnétique
e   separador m magnético
i   separatore m magnetico
n   magnetische separator,
    magnetische sorteerinrichting
d   magnetische Trenn- und Sortiervor-
    richtung f

4226 MAGNETIC SPECTROGRAPH sp
An instrument which utilizes the curvature
of the path of the particle in a magnetic
field to obtain the velocity spectra of
charged particles in radioactive decay.
f   spectrographe m magnétique
e   espectrógrafo m magnético
i   spettrografo m magnetico
n   magnetische spectrograaf,
    snelheidsspectrograaf
d   Geschwindigkeitsspektrograph m

4227 MAGNETIC SPECTROMETER sp
A spectrometer using a magnetic field for
measuring the magnetic rigidity of charged
particles.
f   spectromètre m magnétique
e   espectrómetro m magnético
i   spettrometro m magnetico
n   magnetische spectrometer
d   magnetisches Spektrometer n

4228 MAGNETIC WELL pp
A form of magnetic field used to contain
hot plasma. The magnetic field strength
increases in all directions away from the
centre(er) so that plasma there cannot
readily spill out.
f   puits m magnétique
e   pozo m magnético
i   pozzo m magnetico
n   magnetische put
d   magnetischer Topf m

4229 MAGNETOHYDRODYNAMIC CONVERSION pp
Magnetohydrodynamic technique of fluids tending to allow the directed kinetic energy of the plasma particles, moving perpendicularly to a magnetic field, to be converted into electrical energy.
f conversion *f* magnétodynamique
e conversión *f* magnetodinámica
i conversione *f* magnetodinamica
n magnetohydrodynamische conversie
d magnetohydrodynamische Umwandlung *f*

4230 MAGNETOHYDRODYNAMIC INSTABILITY pp
Instability which may be deduced from the equations of ideal magnetohydrodynamics, in which the plasma and the magnetic lines of force are brought to macroscopic movement of the assembly.
f instabilité *f* magnétodynamique
e inestabilidad *f* magnetodinámica
i instabilità *f* magnetodinamica
n magnetohydrodynamische instabiliteit
d magnetohydrodynamische Instabilität *f*

4231 MAGNETOHYDRODYNAMIC WAVES pp
Waves which can be propagated in a fluid conductor, e.g. a plasma in the presence of a magnetic field.
f ondes *pl* magnétodynamiques
e ondas *pl* magnetodinámicas
i onde *pl* magnetodinamiche
n magnetohydrodynamische golven *pl*
d magnetohydrodynamische Wellen *pl*

4232 MAGNETON np
A unit in which the magnetic moment of atomic particles may be measured.
f magnéton *m*
e magnetón *m*
i magnetone *m*
n magneton *n*
d Magneton *n*

4233 MAGNOX fu, mg
Magnesium alloy used in cladding fuel elements.
f magnox *m*
e magnox *m*
i magnox *m*
n magnox
d Magnox *m*

4234 MAIN QUANTUM NUMBER, see 2633

4235 MAJORANA FORCE np
That nuclear force which changes sign if in the wave function of the two particles their space co-ordinates are interchanged.
f force *f* de Majorana
e fuerza *f* de Majorana
i forza *f* di Majorana
n majoranakracht
d Majorana-Kraft *f*

4236 MAJORANA PARTICLE np
Neutrino which is identical with the anti-neutrino so that double beta decay could occur with the emission and absorption of a neutrino rather than the emission of two neutrinos.
f particule *f* de Majorana
e partícula *f* de Majorana
i particella *f* di Majorana
n majoranadeeltje *n*
d Majorana-Teilchen *n*

4237 MAKE-UP ge
An operation necessary to re-instate the initial condition of a facility or device.
f apport *m*, renivellement *m*
e repuesto *m*
i reintegrazione *f*, rimpiazzo *m*
n bijvullen *n*
d Auffüllung *f*, Nachfüllung *f*

4238 MAKE-UP SHIELDING rt, sa
Shielding used when the normal shielding slabs have been removed during refuelling
f blindage *m* de replace
e blindaje *m* de reemplazo
i schermatura *f* di rimpiazzo
n vervangingsscherm *n*
d Ersatzschild *m*

4239 MALIGNANT GROWTH, md
MALIGNANT NEOPLASM
Said of a neoplasm when the growth invades the tissue of the host or spreads to distant parts or both.
f croissance *f* maligne, néoplasma *m* malin
e crecimiento *m* maligno,
neoplasma *m* maligno
i neoformazione *f* maligna,
neoplasma *m* maligno
n kwaadaardig neoplasma *n*,
kwaadaardige nieuwvorming
d bösartige Neubildung *f*,
malignes Neoplasma *n*

4240 MALIGNANT TUMOR md
A tumor capable of metastasising.
f tumeur *f* maligne
e tumor *m* maligno
i tumore *m* maligno
n kwaadaardig gezwel *n*
d bösartige Geschwulst *f*,
bösartiger Tumor *m*

4241 MAMMOGRAPHY xr
The radiological examination of the breasts with or without the injection of a contrast medium.
f mammographie *f*, radiographie *f* du sein
e mamografía *f*, radiografía *f* de mama
i mammografia *f*
n mammografie
d Mammographie *f*, Maxographie *f*

4242 MANGANESE ch
Metallic element, symbol Mn, atomic number 25.

f   manganèse *m*
e   manganeso *m*
i   manganese *m*
n   mangaan *n*
d   Mangan *n*

4243 MANHATTAN PROJECT   nw
A project of the War department (USA)
lasting from August 1942 to August 1946 in
which the nuclear bomb was developed.
f   projet *m* Manhattan
e   proyecto *m* Manhattan
i   progetto *m* Manhattan
n   manhattanproject *n*
d   Manhattan-Projekt *n*

4244 MANIFOLD OF ELECTRONIC   np
STATES
The sum or totality of the electronic
terms of an atom or molecule.
f   totalité *f* des états électroniques
e   totalidad *f* de los estados electrónicos
i   totalità *f* degli stati elettronici
n   totaal *n* van de elektronische toestanden
d   Summe *f* der Elektronenzustände

4245 MANIPULATOR   rt
A hand-operated or -controlled device
for remotely operating tongs or other tools
over or through a shielding wall.
f   télémanipulateur *m*
e   manipulador *m*
i   manipolatore *m*
n   manipulator
d   Ferngreifer *m*, Manipulator *m*

4246 MANY-BODY FORCES   np
An interaction between two particles that
becomes modified when a third particle
is present.
f   forces *pl* entre plusieurs corps
e   fuerzas *pl* entre múltiples cuerpos
i   forze *pl* tra multipli corpi
n   veellichamenkrachten *pl*
d   Mehrkörperkräfte *pl*

4247 MANY-GROUP MODEL,   np, rt
MULTIGROUP MODEL
A model which divides the neutron
population into a finite number of energy
groups with each group being assigned a
single effective energy.
f   modèle *m* à plusieurs groupes
e   modelo *m* de grupos múltiples
i   modello *m* a più gruppi
n   veelgroepsmodel *n*
d   Mehrgruppenmodell *n*

4248 MARGINAL STABILITY   pp
In a series of states dependent in a
continuous manner on a parameter, the
state corresponding to the limit values of
that parameter relative to the stable
states.
f   stabilité *f* marginale
e   estabilidad *f* marginal
i   stabilità *f* marginale

n   grenswaardenstabiliteit
d   Grenzwertestabilität *f*

4249 MARX EFFECT   ra
The reduction in the energy of a photoelectric emission by the simultaneous
incidence of radiation of lower frequency
than that producing the emission.
f   effet *m* de Marx
e   efecto *m* de Marx
i   effetto *m* di Marx
n   marxeffect *n*
d   Marx-Effekt *m*

4250 mAs,   un
MILLIAMPÈRE-SECOND
A unit commonly used to measure the
product of average röntgen-ray tube
current and exposure time.
f   mAs *f*, milliampère-seconde *f*
e   mAs *m*, miliampere-segundo *m*,
miliamperio-segundo *m*
i   mAs *m*, milliampere-secondo *m*
n   milliampère-seconde
d   mAs *f*, Milliampere-Sekunde *f*

4251 MASK   ra
A device made of sheet lead or lead rubber,
used to restrict the area irradiated.
f   masque *m*
e   máscara *f*
i   maschera *f*
n   masker *n*
d   Maske *f*

4252 MASS   gp
The physical measure of the principal
inertia property of a body.
f   masse *f*
e   masa *f*
i   massa *f*
n   massa
d   Masse *f*

4253 MASS ABSORPTION   ab, np
COEFFICIENT
The linear absorption coefficient, divided
by the density of the absorber in grams
per cubic cm.
f   coefficient *m* d'absorption massique
e   coeficiente *m* de absorción de masa
i   coefficiente *m* d'assorbimento di massa
n   massieke absorptiecoëfficiënt
d   Massenabsorptionskoeffizient *m*

4254 MASS ABUNDANCE   is
MASS CONCENTRATION
The abundance calculated in terms of
weights of isotopes, rather than the number
of atoms.
f   concentration *f* en pourcentages de poids
e   concentración *f* en porcentajes de pesos
i   concentrazione *f* in procentuali di pesi
n   concentratie in gewichtsprocenten
d   Konzentration *f* in Gewichtsprozenten

4255 MASS ASSIGNMENT   np
The determination of the mass of a radioactive species in using several nuclear
bombardments.
f détermination f du nombre de masse
e determinación f del número de masa
i determinazione f del numero di massa
n massagetalbepaling, massawaardebepaling
d Massenwertbestimmung f,
   Massenzahlbestimmung f

4256 MASS ATTENUATION   ma
      COEFFICIENT
The linear attenuation coefficient divided
by the mass density of the substance.
f coefficient m d'atténuation massique
e coeficiente m de atenuación de masa
i coefficiente m d'attenuazione di massa
n verzwakkingscoëfficiënt per massadichtheid
d Massenschwächungskoeffizient m

4257 MASS BALANCE,   ch
      MATERIAL BALANCE
The right relation between feed, product
and waste.
f bilan m matière
e balance m de materia
i bilancio m di materia
n materiaalbalans
d Materialbilanz f, Mengenbilanz f

4258 MASS COEFFICIENT OF   np
      REACTIVITY
The partial derivative of reactivity with
respect to the mass of a given substance
in a specified location.
f coefficient m massique de réactivité
e coeficiente m de masa de reactividad
i coefficiente m di massa di reattività
n massacoëfficiënt van de reactiviteit
d Massenkoeffizient m der Reaktivität

4259 MASS CONVERSION FACTOR,
      MASS-ENERGY CONVERSION
      FORMULA, see 392

4260 MASS DECREMENT   np
The number obtained by deducting the mass
number of a nuclide from its mass
measured on the physical scale of atomic
weights.
f décrément m de masse
e decremento m de masa
i decremento m di massa
n massadecrement n
d Massendekrement n

4261 MASS DEFECT   np
The difference between the sum of the
masses of the nucleons constituting a
nucleus and the mass of that nucleus.
f . défaut m de masse
e defecto m de masa
i difetto m di massa
n massatekort n
d Massendefekt m

4262 MASS EFFECT,   np
      PACKING EFFECT
The difference between the observed mass
of a nucleus and that calculated by adding
the masses of the constituent elementary
particles.
f effet m de masse, effet m de tassement
e efecto m de empaquetamiento,
   efecto m de masa
i effetto m d'impaccamento,
   effetto m di massa
n massa-effect n, pakkingseffect n
d Masseneffekt m, Packungseffekt m

4263 MASS ENERGY ABSORPTION   np
      COEFFICIENT
For indirectly ionizing particles in a
medium of mass density
$$\mu_{en}/\rho = \frac{\mu K}{\rho}(1-G)$$ where $\mu K/\rho$ is the
mass energy transfer coefficient and G
is the proportion of the energy of secondary
charged particles that is lost to bremsstrahlung in the material.
f coefficient m d'absorption d'énergie
   massique
e coeficiente m de absorción de energía
   másica
i coefficiente m d'assorbimento d'energia
   di massa
n energieabsorptiecoëfficiënt per massadichtheid
d Massenabsorptionskoeffizient m

4264 MASS-ENERGY EQUIVALENCE gp, ma
The equivalence of a quantity m and a
quantity of energy E, when the two
quantities are related by the energy-mass
relation $E = mc^2$.
f équivalence f masse-énergie
e equivalencia f masa-energía
i equivalenza f massa-energia
n gelijkwaardigheid van massa en energie
d Masse-Energie-Äquivalenz f,
   Massenäquivalent m der Energie

4265 MASS-ENERGY RELATION, see 2054

4266 MASS-ENERGY TOTAL   gp
The sum of the mass and the energy of
any substance or particle.
f total m masse-énergie
e suma f de la masa y de la energía
i somma f della massa e dell'energia
n som van massa en energie
d Summe f von Masse und Energie

4267 MASS ENERGY TRANSFER   np
      COEFFICIENT
For a beam of indirectly ionizing particles
or quanta in a medium of mass density
$$\mu\frac{K}{\rho} = \frac{1}{E\rho}\frac{dE_K}{dx},$$ where E is the sum of the
energies, exclusive of rest energies of the
indirectly ionizing particles incident on a
layer of thickness dx in a time interval,
and dE_K is the sum of the initial kinetic

energies of decharged particles liberated in the layer during this time interval.
f coefficient m de transfert d'énergie massique
e coeficiente m de transferencia de energía másica
i coefficiente m di trasporto d'energia di massa
n energieoverdrachtscoëfficiënt per massadichtheid
d Massenaustauschzahl f

4268 MASS EXCESS   np
$\Delta = M_a - Am_u$ where M_a is the mass of the atom, A the nuclear number and m_u the unified atomic mass constant.
f excès m de masse
e exceso m de masa
i eccesso m di massa
n massaoverschot n
d Massenüberschuss m

4269 MASS FLOW   gp
The mass of fluid flowing past or through a particular reference plane, per unit of time.
f débit-masse m
e flujo m de masa
i flusso m di massa
n doorstroming
d Durchfluss m, Massendurchsatz m, Mengenstrom m

4270 MASS FORMULA   np
An equation giving the atomic mass of a nuclide as a function of its atomic number and mass number.
f formule f massique
e fórmula f de masa
i formula f di massa
n massaformule
d Massenformel f

4271 MASS NUMBER,   np
       NUCLEON NUMBER
Number of nucleons in a nucleus.
f nombre m de masse, nombre m de nucléons
e número m de masa, número m de nucleones
i numero m di massa, numero m di nucleoni
n massagetal n, nucleongetal n
d Massenzahl f, Nukleonenzahl f

4272 MASS OF THE ELECTRON   np
A quantity analogous to mass, which characterizes the inertia effects of the electron in an electric or magnetic field, and which is conditioned by its charge.
f masse f de l'électron
e masa f del electrón
i massa f dell'elettrone
n massa van het elektron
d Elektronenmasse f, Masse f des Elektrons

4273 MASS OF THE INTERACTING   np
       NUCLEI
f masse f des noyaux initiaux
e masa f de los núcleos iniciales

i massa f dei nuclei agenti
n massa van de reagerende kernen
d Masse f der Ausgangskerne

4274 MASS OF THE PRODUCED   np
       NUCLEI
f masse f des noyaux produits
e masa f de los núcleos producidos
i massa f dei nuclei prodotti
n massa van de geproduceerde kernen
d Masse f der erzeugten Kerne

4275 MASS OF THE UNIVERSE   gp
A quantity of order $M = c^3 T^3$, where c is the density of matter in the universe and T is the age.
f masse f de l'univers
e masa f del universo
i massa f dell'universo
n massa van het heelal
d Masse f des Weltalls

4276 MASS RANGE   np
Expressed in units of surface density, also the mass per unit area of a layer of thickness equal to the linear range.
f portée f massique
e alcance m de masa
i portata f di massa
n massieke dracht
d Massenreichweite f, Reichweite f in g/cm^2

4277 MASS SEPARATION   rt
Adjusting the spacing between two or more bodies of fissile (fissionable) material or parts of the core of a nuclear reactor to control the rate of fission taking place in them.
f séparation f en masses partielles sous-critiques
e separación f en masas parciales subcríticas
i separazione f in masse parziali sottocritiche
n scheiding in subkritische deelmassa's
d Zerlegung f in unterkritische Teilmassen

4278 MASS SPECTROGRAPH   me
A device for analyzing a substance in terms of the ratios of charge q to mass m of its components.
f spectrographe m de masse(s)
e espectrógrafo m de masa
i spettrografo m di massa
n massaspectrograaf
d Massenspektrograph m

4279 MASS-SPECTROGRAPHIC METHOD OF ISOTOPE SEPARATION, see 2086

4280 MASS SPECTROMETER   me
A device similar to the mass spectrograph but so designed that the beam constituents of a given mass-to-charge ratio are focused on an electrode and detected or measured electrically.

f spectromètre *m* de masse(s)
e espectrómetro *m* de masa
i spettrometro *m* di massa
n massaspectrometer
d Massenspektrometer *n*

4281 MASS SPECTRUM sp
A spectrum showing the distribution in mass or in mass-to-charge ratio of ionized atoms, molecules or molecular fragments.
f spectre *m* de masse
e espectro *m* de masa
i spettro *m* di massa
n massaspectrum *n*
d Massenspektrum *n*

4282 MASS STOPPING POWER np
The linear stopping power divided by the mass density of the substance.
f pouvoir *m* d'arrêt massique
e poder *m* de frenado másico
i potere *m* di rallentamento per densità di massa
n stoppend vermogen *n* per massadichtheid
d Bremsvermögen *n* je Massendichte

4283 MASS SYNCHROMETER me
An instrument which may be used for the absolute determination of atomic masses and for mass analysis.
f synchromètre *m* de masse(s)
e sincrómetro *m* de masa
i sincrometro *m* di massa
n massasynchrometer
d Massensynchromesser *m*

4284 MASS TRANSFER ch, gp
The chemical or physical dissolution of a solid in a fluid with re-deposition elsewhere.
f transfert *m* de masse
e transferencia *f* de masa
i trasferimento *m* di massa
n stofuitwisseling
d Stoffaustausch *m*

4285 MASS TRANSFER COEFFICIENT ch
The rate of transfer of mass across unit area of phase contact under unit driving force.
f coefficient *m* de transfert de masse
e coeficiente *m* de transferencia de masa
i coefficiente *m* di trasferimento di massa
n stofuitwisselingsgetal *n*
d Austauschzahl *f*, Stoffaustauschzahl *f*

4286 MASS TRANSFER RIG, mt, rt
MATERIAL TRANSFER RIG
The complete set of apparatus for transporting materials in or near a reactor.
f ensemble *m* de transfert de matériaux
e conjunto *m* de transporte de materiales
i complesso *m* di trasporto di materiali
n materiaaltransportopstelling
d Materialtransportanordnung *f*

4287 MASS VELOCITY ch
Mass rate of flow per unit cross-sectional area.
f densité *f* de courant massique
e densidad *f* de corriente de masa, velocidad *f* másica
i densità *f* di corrente di massa, velocità *f* di massa
n massastroomdichtheid
d Mengenflussdichte *f*, Produkt *n* aus Geschwindigkeit und Dichte

4288 MASS YIELD CURVE np
The plot of the chain yields as a function of the isobaric mass.
f courbe *f* de rendement massique
e curva *f* de rendimiento másico
i curva *f* di rendimento di massa
n massarendementkromme
d Massenausbeutekurve *f*

4289 MASSEY FORMULA ec, ma
A formula which gives the probability of secondary electron emission by an excited atom approaching a metallic surface.
f formule *f* de Massey
e fórmula *f* de Massey
i formula *f* di Massey
n masseyformule
d Massey-Formel *f*

4290 MASTER SLAVE MANIPULATOR rt
Mechanical hand used to handle active materials.
f robot *m*, télémanipulateur *m* asservi
e piloto-operador *m*
i manipolatore *m* asservito
n kunsthand, op afstand bediende manipulator
d magische Hände *pl*, Parallelmanipulator *m*

4291 MASUYITE mi
A rare secondary mineral containing about 74 % of U.
f masuyite *f*
e masuita *f*
i masuite *f*
n masuyiet *n*
d Masuyit *m*

4292 MATERIAL BALANCE, see 4257

4293 MATERIAL BUCKLING np
A parameter providing a measure of the multiplying properties of a medium as a function of the materials and their disposition.
f laplacien *m* matière
e laplaciano *m* materia
i parametro *m* fisico di criticità
n materiële bolling, materiële welving
d materielle Flussdichtewölbung *f*

4294 MATERIAL ECONOMY rt
The efficiency with which material, in particular fissile (fissionable) or fertile material, is used.

f   éconcmie f des matériaux
e   economía f de los materiales
i   grado m d'utilizzazione dei materiali
n   materiaaleconomie
d   Materialökonomie f

4295 MATERIAL INVENTORY   rt
      The total quantity of a given material
      present in a specified installation in a
      given moment.
f   inventaire m de matériel
e   inventario m de material
i   materiale m di dotazione,
      materiale m d'impegno
n   materiaalinventaris
d   Materialeinsatz m,
      materielle Ausstattung f

4296 MATERIAL PARTICLE, see 1316

4297 MATERIALIZATION   np
      The transformation of a photon into an
      electron-posit(r)on pair
f   matérialisation f
e   materialización f
i   materializzazione f
n   materialisatie
d   Materialisation f

4298 MATERIALS PROCESSING REACTOR rt
      A nuclear reactor employed primarily to
      improve the properties of materials.
f   réacteur m de traitement de matériaux
e   reactor m de tratamiento de materiales
i   reattore m di trattamento di materiali
n   reactor voor materiaalbewerking
d   Materialbearbeitungsreaktor m

4299 MATERIALS TESTING REACTOR,   rt
      MTR
      A nuclear reactor used for testing
      materials and reactor components in
      intense radiation fields.
f   réacteur m d'essai de matériaux
e   reactor m de ensayo de materiales
i   reattore m per prova di materiali
n   reactor voor materiaalonderzoek
d   Materialprüfreaktor m

4300 MATRIX MECHANICS, see 3217

4301 MATTERHORN PROJECT   np, pp
      A classified project carried on by
      Princeton University to investigate the
      controlled use of thermonuclear reactors.
f   projet m Matterhorn
e   proyecto m Matterhorn
i   progetto m Matterhorn
n   matterhornproject n
d   Matterhorn-Projekt n

4302 MAXIMUM CREDIBLE   rt, sa
      ACCIDENT
      The worst accident in a reactor or nuclear
      energy installation that, by agreement,
      need be taken into account in devising
      protective measures.

f   accident m maximal prévisible
e   accidente m máximo previsible
i   incidente m massimo verosimile
n   ergst denkbaar ongeluk n
d   grösster anzunehmender Unfall m

4303 MAXIMUM ENERGY TRANSFER   np
      The maximum energy that can be trans-
      ferred in an elastic collision by a photon
      to a particle is given by

$$E_{max} = \left(\frac{2}{2 + m_o c^2\,hv}\right) hv$$ where m_o is the

      rest mass of the particle, c is the velocity
      of light, h is Planck's constant and v is the
      frequency of the photon.
f   transfert m maximal d'énergie
e   transferencia f máxima de energía
i   trasferimento m massimo d'energia
n   maximale energieoverdracht
d   maximale Energieübertragung f

4304 MAXIMUM PERMISSIBLE   ra
      ACCUMULATED DOSE,
      MAXIMUM PERMISSIBLE
      CUMULATIVE DOSE,
      MAXIMUM PERMISSIBLE
      INTEGRATED DOSE
      The maximum permissible dose for a long
      period exposure of personnel to ionizing
      radiation.
f   dose f accumulée maximale admissible
e   dosis f acumulada máxima permisible
i   dose f massima ammissibile per
      esposizione professionale
n   maximaal toegestane geaccumuleerde
      dosis
d   höchstzugelassene akkumulierte Dosis f

4305 MAXIMUM PERMISSIBLE   md, ra
      BODY BURDEN
      The total quantity of radiation in a body
      which, when allowance has been made for
      the distribution, will cause a dose rate of
      300 m rem per week in a critical organ.
f   charge f corporelle maximale admissible,
      quantité f maximale admissible dans
      l'organisme
e   cantidad f máxima permisible en el cuerpo
i   quantità f massima ammissibile nel corpo
n   maximaal toegestane lichaamsbelasting
d   höchstzugelassene Körperbelastung f

4306 MAXIMUM PERMISSIBLE   ra
      BONE DOSE
      The maximum dose that can safely be
      delivered to a bone.
f   dose f maximale admissible pour os
e   dosis f máxima permisible para huesos
i   dose f massima ammissibile per ossa
n   maximaal toegestane botdosis
d   höchstzugelassene Knochendosis f

4307 MAXIMUM PERMISSIBLE   md
      CONCENTRATION,
      MPC
      The recommended upper limit for the
      concentration of a specified radioactive

substance in any material liable to enter the human body.
- f CMA,
 concentration *f* maximale admissible
- e concentración *f* máxima permisible
- i concentrazione *f* massima ammissibile
- n maximaal toegestane concentratie
- d höchstzugelassene Konzentration *f*

4308 MAXIMUM PERMISSIBLE DOSE, ra
 MPD
The recommended upper limit for the dose which may be received during a specified period by a person exposed to ionizing radiation.
- f DMA, dose *f* maximale admissible
- e dosis *f* máxima permisible
- i dose *f* massima ammissibile
- n maximaal toegestane dosis
- d höchstzugelassene Dosis *f*

4309 MAXIMUM PERMISSIBLE ra
 DOSE EQUIVALENT,
 MPDE
The largest dose equivalent received within a specified period which is permitted by a regulatory committee on the assumption that there is no appreciable probability of somatic or genetic injury.
- f EDMA,
 équivalent *m* de dose maximale admissible
- e equivalente *m* de dosis máxima permisible
- i equivalente *m* di dose massima ammissibile
- n maximaal toegestaan dosisequivalent *n*
- d höchstzugelassene Äquivalentdosis *f*

4310 MAXIMUM PERMISSIBLE ra
 DOSE RATE
The dose rate which, if constant during a specified period, would give rise to the maximum permissible dose for that period.
- f taux *m* de DMA,
 taux *m* de dose maximale admissible
- e intensidad *f* de dosis máxima permisible
- i intensità *f* di dose massima ammissibile
- n maximaal toegestane doseringssnelheid
- d höchstzugelassene Dosisleistung *f*,
 HZD-Leistung *f*

4311 MAXIMUM PERMISSIBLE FLUX ra
That flux of radiation which, if constant during a specified period, would give rise to the maximum permissible dose in that period.
- f flux *m* maximal admissible
- e flujo *m* máximo permisible
- i flusso *m* massimo ammissibile
- n maximaal toegestane flux
- d höchstzugelassener Flux *m*

4312 MAXIMUM PERMISSIBLE LEVEL ra
A term used to refer loosely to the maximum permissible concentration, dose, rate or flux.
- f niveau *m* maximal admissible
- e nivel *m* máximo permisible
- i livello *m* massimo ammissibile

- n maximaal toegestaan niveau *n*
- d höchstzugelassenes Niveau *n*

4313 MAXIMUM RANGE np
For a group of ionizing particles the greatest distance in a specified direction at which their ionization can be detected.
- f parcours *m* maximal
- e trayectoria *f* máxima
- i percorso *m* massimo
- n maximale baanlengte
- d maximale Bahnlänge *f*

4314 MAXIMUM RESIDUAL me
 SYSTEMATIC ERRORS
Errors estimated from the correction factors applied to the observed measurements, from assumptions made for the decay scheme and radiation characteristics of the nuclide, and from knowledge of the details of the techniques used.
- f erreurs *pl* systématiques maximales résiduelles
- e errores *pl* sistemáticos máximos residuales
- i errori *pl* sistematici massimi residui
- n maximale systematische restfouten *pl*
- d höchstzugelassene systematische Restfehler *pl*

4315 MAXIMUM VALENCE, ch
 POSITIVE VALENCE
The highest valence shown by an element in any of its compounds.
- f valence *f* maximale, valence *f* positive
- e valencia *f* máxima, valencia *f* positiva
- i valenza *f* massima, valenza *f* positiva
- n maximale valentie, positieve valentie
- d maximale Valenz *f*, positive Valenz *f*

4316 MAXWELL-BOLTZMANN qm
 CLASSICAL STATISTICS,
 MAXWELL-BOLTZMANN
 STATISTICS
Study of the probabilities of the macroscopic states of a system of non-quantized states.
- f statistique *f* classique de Maxwell-Boltzmann,
 statistique *f* de Maxwell-Boltzmann
- e estadística *f* clásica de Maxwell-Boltzmann,
 estadística *f* de Maxwell-Boltzmann
- i statistica *f* classica di Maxwell-Boltzmann,
 statistica *f* di Maxwell-Boltzmann
- n klassieke statistiek van Maxwell en Boltzmann,
 statistiek van Maxwell en Boltzmann
- d klassische Maxwell-Boltzmannsche Statistik *f*,
 Maxwell-Boltzmannsche Statistik *f*

4317 MAXWELL-BOLTZMANN gp
 DISTRIBUTION,
 MAXWELLIAN DISTRIBUTION
The velocity distribution, as computed in the kinetic theory of gases, of the

molecules of a gas in thermal equilibrium.
f   distribution f de Maxwell-Boltzmann,
    distribution f maxwellienne
e   distribución f de Maxwell
i   distribuzione f di Maxwell
n   snelheidsverdeling van Maxwell,
    verdeling van Maxwell
d   Maxwellsche Geschwindigkeitsverteilung f

4318 MAXWELL-BOLTZMANN   gp
      QUANTUM STATISTICS
      Quantum statistics derived from the
      classical statistics of Maxwell-Boltzmann
      taking into account the indiscernability of
      the particles.
f   statistique f quantique de Maxwell-
    Boltzmann
e   estadística f cuántica de Maxwell-
    Boltzmann
i   statistica f quantica di Maxwell-
    Boltzmann
n   quantenstatistiek van Maxwell en
    Boltzmann
d   Maxwell-Boltzmannsche Quantenstatistik f

4319 MAXWELL SPECTRUM   sp
      A spectrum representing the distribution
      of particles behaving in accordance with
      the Maxwell-Boltzmann distribution law.
f   spectre m de Maxwell
e   espectro m de Maxwell
i   spettro m di Maxwell
n   maxwellspectrum n
d   Maxwell-Spektrum n

4320 MAXWELLIAN CROSS SECTION   cs
      Thermal cross section corresponding to
      the maxwellian distribution of the neutron
      energies.
f   section f efficace maxwellien
e   sección f eficaz de Maxwell
i   sezione f d'urto di Maxwell
n   maxwelldoorsnede
d   Maxwell-Wirkungsquerschnitt m

4321 MAZE COUNTER TUBE, see 2472

4322 mCi δ,   un
      MILLICURIE DESTROYED
      The amount of radiation emitted by a
      specimen of a radioactive nuclide during
      the time that its activity falls by one
      millicurie.
f   mCi δ m, millicurie m détruit
e   mCi δ m, milicurie m destruido
i   mCi δ m, millicurie m distrutto
n   millicurie-vernietigd
d   Millicurie-détruit n

4323 mCih,   un
      MILLICURIE HOUR
      Unit of number of disintegration.
f   mCih m, millicurie-heure m
e   mCih m, milicurie-hora m
i   mCih m, millicurie-ora m
n   mCih n, millicurie-uur n
d   mCih f, Millicurie-Stunde f

4324 MEAN CURRENT   ma
      MEASURING ASSEMBLY
      A measuring assembly which uses the
      current output of its radiation detector(s)
      as the basis of operation.
f   ensemble m de mesure à courant
e   conjunto m medidor por corriente
i   complesso m di misura a corrente
n   meetopstelling met stroomaflezing
d   Anordnung f zur Messung des
    Detektorstromes

4325 MEAN FREE IONIZING PATH   np
      Mean free path for an ionizing collision.
f   libre moyen parcours m d'ionisation
e   trayectoria f libre media de ionización
i   percorso m libero medio d'ionizzazione
n   gemiddelde vrije ionisatieweglengte
d   mittlere freie Ionisierungsweglänge f

4326 MEAN FREE PATH   np
      The average distance a particle travels
      between successive collisions with the
      other particles of an ensemble.
f   libre moyen parcours m
e   camino m libre medio,
    trayectoria f libre media
i   cammino m libero medio,
    percorso m libero medio
n   gemiddelde vrije weglengte
d   mittlere freie Weglänge f

4327 MEAN FREE TIME   np
      The average time between two collisions.
f   libre moyen temps m
e   tiempo m libre medio
i   tempo m libero medio
n   gemiddelde vrije tijd
d   mittlere freie Zeit f

4328 MEAN IONIZATION ENERGY,   np
      MEAN IONIZING ENERGY
      The average energy lost by an ionizing
      particle in producing an ion pair in a gas.
f   énergie f moyenne d'ionisation,
    perte f moyenne d'énergie par pairs d'ions
e   energía f media de ionización
i   energia f media d'ionizzazione
n   gemiddelde ionisatie-energie
d   mittlere Ionisationsenergie f

4329 MEAN LIFE, see 478

4330 MEAN LIFE OF AN ATOMIC STATE,
      see 479

4331 MEAN LINEAR RANGE,   np, ra
      MEAN RANGE
      The average distance that a particle
      penetrates a given substance under
      specified conditions.
f   portée f moyenne
e   alcance m medio, promedio m de recorrido
i   portata f media
n   gemiddelde dracht
d   mittlere Reichweite f

4332 MEAN MASS RANGE np, ra
The mean linear range multiplied by the
mass density of the substance.
f portée *f* moyenne en masse
e alcance *m* medio en masa,
promedio *m* de recorrido en masa
i portata *f* media in massa
n gereduceerde dracht
d Reichweite *f* in g/cm^3

4333 MEAN NEUTRON LIFETIME np
The average time for which a neutron
survives in a given medium.
f vie *f* moyenne neutronique
e vida *f* media neutrónica
i vita *f* media neutronica
n gemiddelde neutronlevensduur
d mittlere Neutronenlebensdauer *f*

4334 MEAN RADIOACTIVITY me, ra
METER,
MEAN VALUE METER
An apparatus for measuring the mean
value of radioactivity.
f activimètre *m* des valeurs moyennes
e activímetro *m* de valores medios
i attivimetro *m* di valori medi
n gemiddelde-activiteitsmeter
d mittlerer Aktivitätsmesser *m*

4335 MEAN REFUELLING RATE fu, rt
f vitesse *f* de rechargement moyenne
e velocidad *f* de relleno media
i velocità *f* di ricarica media
n gemiddeld opladingstempo *n*
d mittlere Aufladungsrate *f*

4336 MEAN SQUARE LENGTH np
OF MODERATION
The average distance travel(l)ed by a
thermal neutron from formation to capture.
f carré *m* moyen du parcours de modération
e cuadrado *m* medio de la longitud de
moderación
i quadrato *m* medio della lunghezza di
moderazione
n gemiddelde kwadratische remlengte
d mittlere quadratische Bremslänge *f*

4337 MEAN SURVIVAL TIME mc
The arithmetic mean of the individual
survival times of a given group of persons,
animals or organisms.
f durée *f* de survie moyenne
e tiempo *m* de supervivencia
i tempo *m* di sopravvivenza medio
n gemiddelde overlevingstijd
d mittlere Überlebenszeit *f*

4338 MEAN TEMPERATURE gp
DIFFERENCE,
MTD
The average temperature difference acting
to cause heat exchange.
f différence *f* moyenne de températures
e diferencia *f* media de temperaturas
i differenza *f* media di temperature

n gemiddeld temperatuurverschil *n*
d mittlerer Temperaturunterschied *m*

4339 MEAN VELOCITY gp
The average value of a number of different
velocities in a system.
f vitesse *f* moyenne
e velocidad *f* media
i velocità *f* media
n gemiddelde snelheid
d mittlere Geschwindigkeit *f*

4340 MEASURING ASSEMBLY, me
RADIATION MEASURING ASSEMBLY,
RADIATION METER
An assembly including one or several
radiation detectors and associated
sub-assemblies or basic function units
used to measure quantities connected with
ionizing radiation.
f ensemble *m* de mesure de rayonnement,
radiamètre *m*
e conjunto *m* medidor de radiación,
radiámetro *m*
i complesso *m* di misura di radiazione,
radiametro *m*
n stralingsmeetopstelling, stralingsmeter
d Strahlungsmessanordnung *f*,
Strahlungsmessgerät *n*

4341 MEASURING ASSEMBLY FOR ma
DETERMINATION OF WHOLE-BODY
GAMMA ACTIVITY
An assembly which measures the total
gamma radiation, including bremsstrahlung
emitted by the body, and uses one or
several scintillators heavily shielded
against natural ambient radiation.
f ensemble *m* de mesure de l'activité
gamma globale du corps
e conjunto *m* medidor de la actividad gamma
global del cuerpo
i complesso *m* di misura dell'attività
gamma globale del corpo
n meetopstelling voor de in een lichaam
aanwezige gamma-activiteit
d Anordnung *f* zur Bestimmung der
Ganzkörpergammaaktivität

4342 MEASURING ASSEMBLY me
UTILIZING IONIZING RADIATION
An assembly including one or more
radiation detectors and associated
sub-assemblies or basic function units
and designed to measure physical
quantities by utilizing ionizing radiation.
f ensemble *m* de mesure par rayonnement
ionisant
e conjunto *m* de medida por radiación
ionizante
i complesso *m* di misura per mezzo di
radiazione ionizzante
n meetopstelling met behulp van ioniserende
straling
d Messanordnung *f* mit ionisierender
Strahlung als Messmittel

4343 MECHANICAL ARM   cd, rt
An electromechanical robot device used
for the remote handling of radioisotopes.
f bras *m* mécanique
e brazo *m* mecánico
i braccio *m* meccanico
n mechanische arm
d mechanischer Arm *m*

4344 MECHANICAL BOND   rt
A bond less intimate than a metallurgical
bond.
f liaison *f* mécanique
e ligazón *f* mecánica
i contatto *m* meccanico,
legamento *m* meccanico
n mechanische binding
d mechanische Verbindung *f*

4345 MECHANICAL ELECTRO-   cd, cl
MAGNETIC PUMP
Electromagnetic pump in which the liquid
metal is agitated by Foucault currents
induced in the liquid by the magnetic field
of an electromagnet rotating at constant
velocity.
f pompe *f* électromagnétique mécanique
e bomba *f* electromagnética mecánica
i pompa *f* elettromagnetica meccanica
n mechanische elektromagnetische pomp
d mechanische elektromagnetische Pumpe *f*

4346 MECHANICAL FORCE   gp
The force exerted by mechanical means.
f force *f* mécanique
e fuerza *f* mecánica
i forza *f* meccanica
n mechanische kracht
d mechanische Kraft *f*

4347 MECHANICAL MASS   np
The part of the mass of a particle which
is supposed to be an intrinsic property
of the particle.
f masse *f* mécanique
e masa *f* mecánica
i massa *f* meccanica
n mechanische massa
d mechanische Masse *f*

4348 MECHANICAL REGISTER,   ct
MESSAGE REGISTER
An electromechanical device for
registering counts.
f enregistreur *m* mécanique,
numéroteur *m* électromécanique
e registrador *m* mecánico
i registratore *m* meccanico
n mechanisch register *n*,
mechanische schrijver
d mechanischer Schreiber *m*,
mechanisches Zählwerk *n*

4349 MECHANICAL STRESS   gp
A mechanical force acting on a unit area in
a solid, as in the theory of elasticity.
f effort *m* mécanique

e esfuerzo *m* mecánico
i sollecitazione *f* meccanica
n mechanische belasting,
mechanische spanning
d mechanische Beanspruchung *f*

4350 MEDIAN LETHAL DOSE, see 3942

4351 MEDIAN LETHAL TIME, see 3943

4352 MEDIASTINOGRAPHY   xr
Radiological examination of the
mediastinal space by substernal injection
of abrodil.
f médiastinographie *f*
e radiografía *f* de mediastino
i mediastinografia *f*
n mediastinografie
d Mediastinographie *f*

4353 MEDICAL ACTIVITY METER,   ma
MEDICAL RADIOACTIVITY METER
A (radio)activity meter designed to
localize, by means of appropriate probes,
the tissues having fixed radionuclides.
f activimètre *m* médical,
radioactivimètre *m* médical
e activímetro *m* médico,
radiactivímetro *m* médico
i attivimetro *m* medico,
radioattivimetro *m* medico
n medische activiteitsmeter
d Aktivitätsmessgerät *n* für medische
Zwecke

4354 MEDICAL RADIOLOGY (GB),   ra, xr
RADIOLOGY (US)
The science of the application of X-rays,
gamma rays or other penetrating ionizing
radiation.
f radiologie *f*, radiologie *f* médicale
e radiología *f* médica
i radiologia *f*, radiologia *f* medica
n radiologie
d medizinische Radiologie *f*, Radiologie *f*

4355 MEDULLARY MEMBRANE, see 2249

4356 MEGA-ELECTRONVOLT,   un
MeV,
MILLION ELECTRONVOLTS (US)
A common unit in nuclear science, equal
to 10^6 eV. Symbol MeV.
f méga-électron-volt *m*, MeV *m*
e mega-electrón-voltio *m*, MeV *m*
i mega-elettronevolt *m*, MeV *m*
n mega-elektronvolt, MeV
d Mega-Elektronenvolt *n*, MeV *n*

4357 MEGATON BOMB   nw
A nuclear bomb, the explosive power of
which is equivalent to one million tons of
trinitrotoluene.
f bombe *f* à force explosive égale à
1,000,000 tonnes de TNT,
bombe *f* mégatonne

e bomba *f* de fuerza explosiva igual a
1,000,000 toneladas de TNT,
bomba *f* megatón
i bomba *f* a forza esplosiva uguale a
1,000,000 tonnellate di TNT,
bomba *f* megaton
n megatonbom
d Megatonbombe *f*

4358 MEGAWATT-DAY PER TON un
A unit used for expressing the burnup
of fuel in a reactor, specially the number
of megawatt-days of heat output per
metric ton of fuel.
f mégawatt-jour *m* par tonne
e megawatt-día *m* per tonelada
i megawatt-giorno *m* per tonnellata
n megawatt-dag per ton
d Megawatt-Tag *m* je Tonne

4359 MEIOSIS md
In biology, nuclear division in which the
members of each pair of chromosomes
separate and form different nuclei thus
reducing the number of chromosomes by
half.
f méiose *f*
e meiosis *f*
i meiosi *f*
n reductiedeling
d Reduktionsteilung *f*

4360 MELANOCERITE mi
A borosilicate of the Ce and Y metals,
containing a small percentage of Th.
f mélanocérite *f*
e melanocerita *f*
i melanocerite *f*
n melanoceriet *n*
d Melanozerit *m*

4361 MELT-OUT rt
A condition that the solid fuel in the reactor
rises in temperature so much that it melts.
f fusion *f*
e derretimiento *m*
i fusione *f*
n smelten *n*
d Schmelzen *n*

4362 MEMBRANE (GB), see 542

4363 MENDELEEVITE mi
A titanium betafite containing about
13.7 % of U.
f mendélévite *f*
e mendeleevita *f*
i mendelevite *f*
n mendelejeviet *n*
d Mendelejevit *m*

4364 MENDELEVIUM ch
Transuranic radioactive element, symbol
Mv, atomic number 101.
f mendélévium *m*
e mendelevio *m*
i mendelevio *m*

n mendelevium *n*
d Mendelevium *n*

4365 MERCURY ch
Liquid metallic element, symbol Hg,
atomic number 80.
f mercure *m*
e mercurio *m*
i mercurio *m*
n kwik *n*
d Quecksilber *n*

4366 MERCURY-DIFFUSION PUMP vt
A diffusion pump which uses mercury as
its working fluid.
f pompe *f* à diffusion à mercure
e bomba *f* de difusión de mercurio
i pompa *f* a diffusione a mercurio
n kwikdiffusiepomp
d Quecksilberdiffusionspumpe *f*

4367 MESIC ATOM, np
MESONIC ATOM
The atom formed when a negative meson is
captured by the potential field of a nucleus
and exists for a short time in one of its
quantum mechanically allowed energy
levels.
f atome *m* mésonique
e átomo *m* mesónico
i atomo *m* mesonico
n mesonisch atoom *n*
d mesonisches Atom *n*

4368 MESON np
Any elementary particle with a rest mass
between those of an electron and a proton.
f méson *m*
e mesón *m*
i mesone *m*
n meson *n*
d Meson *n*

4369 MESON FIELD cr
A field in which the mesons are considered
as carriers of energy quanta.
f champ *m* mésonique
e campo *m* mesónico
i campo *m* mesonico
n mesonenveld *n*
d Kernfeld *n*, Mesonenfeld *n*

4370 MESON FIELD THEORY, cr, np
MESON THEORY OF NUCLEAR
FORCES
A theory of nuclear forces in which the
meson field acts as a basis of the exchange
between a proton and a neutron.
f théorie *f* du champ mésonique,
théorie *f* mésonique des forces
nucléaires
e teoría *f* del campo mesónico,
teoría *f* mesónica de las fuerzas
nucleares
i teoria *f* del campo mesonico,
teoria *f* mesonica delle forze nucleari
n mesonentheorie van de kernkrachten
d Mesonentheorie *f* der Kernkräfte

4371 MESOTHORIUM-I   ch
The common name for 6.74 Ra-228, a
member of the thorium series, symbol
MsTh$_1$.
f mésothorium-I m
e mesotorio-I m
i mesotorio-I m
n mesothorium-I n
d Mesothorium-I n

4372 MESOTHORIUM-II   ch
The common name for 6.13 Ac-228,
a member of the thorium series, symbol
MsTh$_2$.
f mésothorium-II m
e mesotorio-II m
i mesotorio-II m
n mesothorium-II n
d Mesothorium-II n

4373 MESSAGE REGISTER, see 4348

4374 METAL FOIL DETECTOR, see 2769

4375 METAL FOIL WINDOW   br, md, ra
Component part of radiation sources.
f fenêtre f en feuille métallique
e ventana f de hoja metálica
i finestra f a foglia metallica
n metaalfoelievenster n
d Metallfoliefenster n

4376 METAL SCREEN   xr
An intensifying screen of metal, usually
lead, which emits electrons instead of
light under the influence of X-rays or
other ionizing radiation.
f écran m renforçateur métallique
e pantalla f intensificadora metálica
i schermo m rinforzatore metallico
n metalen versterkscherm n
d metallischer Verstärkerschirm m

4377 METAL-SHEATHED   cd, me
      THERMOCOUPLE
f couple m thermoélectrique à gaine
   métallique
e par m termoeléctrico de vaina metálica
i coppia f termoelettrica a guaina
   metallica
n thermo-element n met metaalomhulling
d Thermoelement n mit Metallhülle

4378 METALLIC BOND   np
A special type of bond existing in metals,
in which the valence electrons of the
constituent atoms are free to move in the
periodic lattice.
f liaison f métallique
e enlace m metálico
i legame m metallico
n metaalbinding
d metallische Bindung f

4379 METALLIC FOIL,   rt
      FOIL
Thin sheet of metal used in light-water
moderated reactors for measuring

thermal neutron flux distribution.
f feuille f
e hoja f
i foglia f
n foelie
d Folie f

4380 METALLIC-FUELLED   rt
      GAS-COOLED REACTOR
f réacteur m à combustible métallique à
   refroidissement par gaz
e reactor m de combustible metálico enfriado
   al gas
i reattore m a combustibile metallico con
   raffreddamento a gas
n met gas gekoelde reactor met metallieke
   brandstof
d gasgekühlter Reaktor m mit metallischem
   Brennstoff

4381 METALLOGRAPHY   mg
The science dealing with the constitution
and structure of metals and alloys.
f métallographie f
e metalografía f
i metallografia f
n metaalkunde, metallografie
d Metallographie f

4382 METALLURGICAL BOND, see 1830

4383 METALLURGICAL NUCLEUS,   mg
      NUCLEUS
In metallurgy, a small cluster of atoms of
a new and more stable phase formed
within a less stable phase as a preliminary
to its conversion to the stable variety.
f germe m
e germen m
i germe m
n kiem
d Keim m

4384 METALLURGICAL PLANT   mg
An installation where metallurgical
processes and tests are carried out.
f installation f métallurgique
e equipo m metalúrgico
i impianto m metallurgico
n metallurgisch bedrijf n
d Metallhütte f

4385 METALLURGY   mg
The science dealing with the processing of
metals, from their recovery from ores
to their purification, alloying and fabrication into industrial articles.
f métallurgie f
e metalurgia f
i metallurgia f
n metallurgie
d Metallurgie f

4386 METAMICT CRYSTALS   cr
Crystals of which the space lattice
arrangement of their atoms or molecules
has been disturbed without alteration of
their chemical composition.

f   cristaux *pl* métamictes
e   cristales *pl* metamictos
i   cristalli *pl* metamitti
n   metamicte kristallen *pl*
d   metamikte Kristalle *pl*

4387 METAMICT STATE   np
The amorphous condition brought about in
certain originally crystalline materials
containing uranium or thorium as a result
of bombardment by recoil nuclei and alpha
particles produced in the radioactive decay
of those elements and their daughter
products.
f   état *m* métamicte
e   estado *m* metamicto
i   stato *m* metamitto
n   metamicte toestand
d   metamikter Zustand *m*

4388 METAPHASE   md
A stage in nuclear division in which the
divided chromosomes lie in a plane at
right angles to the plane of the division
spindle and midway between its poles.
f   métaphase *f*
e   metafase *f*
i   metafase *f*
n   metafaze
d   Metaphase *f*

4389 METASTABLE   qm
Of a state of a system, capable of undergoing a quantum transition to a state of
lower energy, but having a relatively long
lifetime as compared with the most rapid
quantum transitions of similar systems.
f   métastable adj
e   metaestable adj
i   metastabile adj
n   metastabiel adj
d   metastabil adj

4390 METASTABLE ATOMIC STATE   np
An excited atomic energy level wherein
the atom cannot give up its energy in the
form of radiation by an allowed transition
but must ultimately return to the normal
state by some other process.
f   état *m* atomique métastable
e   estado *m* atómico metaestable
i   stato *m* atomico metastabile
n   metastabiele energietoestand van een
     atoom
d   metastabiler Energiezustand *m* eines Atoms

4391 METASTABLE ATOMS   np
One of the causes of secondary ionization.
f   atomes *pl* métastables
e   átomos *pl* metaestables
i   atomi *pl* metastabili
n   metastabiele atomen *pl*
d   ·metastabile Atome *pl*

4392 METASTABLE EQUILIBRIUM   gp
A condition of pseudo-equilibrium in which
the free energy of a system is at a minimum

with respect to infinitesimal changes, but
not with respect to major changes.
f   équilibre *m* métastable
e   equilibrio *m* metaestable
i   equilibrio *m* metastabile
n   metastabiel evenwicht *n*
d   metastabiles Gleichgewicht *n*

4393 METASTABLE NUCLEI   np
Nuclei in excited states that have
measurable lifetime.
f   noyaux *pl* métastables
e   núcleos *pl* metaestables
i   nuclei *pl* metastabili
n   metastabiele kernen *pl*
d   metastabile Kerne *pl*

4394 METASTABLE STATE   np
An excited state from which all possible
quantum transitions to lower states are
forbidden transitions by the appropriate
selection rules.
f   état *m* métastable
e   estado *m* metaestable
i   stato *m* metastabile
n   metastabiele toestand
d   metastabiler Zustand *m*

4395 METASTASIC ELECTRON,   np
       METASTATIC ELECTRON
An electron that moves from one atom to
another, or from one shell to another in a
given atom, or to the nucleus of an atom.
f   électron *m* métastatique
e   electrón *m* metastático
i   elettrone *m* metastatico
n   metastatisch elektron *n*
d   metastatisches Elektron *n*

4396 METASTASIS   md, np
1. A fundamental change in the position
or orbit of a particle.
2. Growth in the body of malignant neo-
plastic cells at a distance from the original
or parent cancer.
f   métastase *f*
e   metástasis *f*
i   metastase *f*
n   metastase
d   Metastase *f*

4397 METATORBERNITE   mi
A common secondary mineral, containing
about 50.8 % of U.
f   métatorbernite *f*
e   metatorbernita *f*
i   metatorbernite *f*
n   metatorberniet *n*
d   Metatorbernit *m*

4398 META–URANOCIRCITE,   mi
       URANOCIRCITE
Belongs to the meta series of hydrates of
the metatorbernite group.
f   méta-uranocircite *f*
e   metauranocircita *f*
i   metauranocircite *f*

n   meta-uranoci rkiet n
d   Meta-Uranozirkit m

4399   METAZEUNERITE   mi
A common secondary mineral, containing
about 46.4 % of U.
f   métazeunérite f
e   metazeunerita f
i   metazeunerite·f
n   metazeuneriet n
d   Metazeunerit m

4400   METHANE INHIBITOR   mt
An inhibitor added to the coolant gas to
prevent reaction between graphite and the
carbon dioxide coolant.
f   inhibiteur m à méthane
e   inhibidor m de metano
i   inibitore m a metano
n   methaaninhibitor
d   Methaninhibitor m

4401   METHANE IONIZATION   ic
CHAMBER
An ionization chamber in which the organic
gas filling consists of methane (CH_4).
f   chambre f d'ionisation à méthane
e   cámara f de ionización con metano
i   camera f d'ionizzazione a metano
n   met methaan gevuld ionisatievat n
d   methangefüllte Ionisationskammer f

4402   METHYL ISOBUTYL KETONE,
M.I.K., see 3249

4403   mgh,   un
MILLIGRAM-HOUR
The product of the mass in milligrams
of radium by the time in hours.
f   mgh m, milligramme-heure m
e   mgh m, miligramo-hora m
i   mgh m, milligrammo-ora m
n   milligram-uur n
d   mgh f, Milligrammstunde f

4404   MICHEL PARAMETER   br
A parameter in the expression of the
momentum spectrum in the beta decay of
a mu meson.
f   paramètre m de Michel
e   parámetro m de Michel
i   parametro m di Michel
n   michelparameter
d   Michel-Parameter m

4405   MICROAMMETER-RATEMETER me, pa
Device for measuring the intensity and the
counting of charges at the target of a
particle accelerator.
f   mesureur-ictomètre m de charges
e   impulsímetro m medidor de cargas
i   rateometro m misuratore di cariche
n   tempo-ladingsmeter
d   Rate-Ladungsmesser m

4406   MICROANALYSIS,   an
MICROCHEMICAL ANALYSIS
Chemical analysis conducted with very

small quantities of samples and reagents
by use of small scale equipment.
f   microanalyse f
e   microanálisis f
i   microanalisi f
n   microanalyse
d   Mikroanalyse f

4407   MICROBEAM   ra
A beam of radiation usually consisting of
fast charged particles or ultraviolet
radiation, collimated to dimensions less
than those of the individual cells of the
specimen.
f   microfaisceau m
e   microhaz m
i   microfascio m
n   microbundel
d   Mikrobündel n

4408   MICROBIOLOGICAL   an
CONTAMINATION
A contamination of radioactive compounds
occurring readily in dilute solutions kept
for long periods and subject to occasional
withdrawal of samples.
f   contamination f microbiologique
e   contaminación f microbiológica
i   contaminazione f microbiologica
n   microbiologische besmetting
d   mikrobiologische Kontamination f

4409   MICROCHEMISTRY,   ch
TRACE CHEMISTRY
Chemistry which involves the use of the
microscope.
f   microchimie f
e   microquímica f
i   microchimica f
n   microchemie
d   Mikrochemie f

4410   MICROCURIE   un
One millionth of a curie, symbol μ Ci.
f   microcurie m
e   microcurie m
i   microcurie m
n   microcurie
d   Mikrocurie n

4411   MICRO-INSTABILITY, see 3851

4412   MICRO-IRRADIATION   ra
The irradiation of a biological specimen
with a microbeam.
f   microradioexposition f
e   microirradiación f
i   microirradiazione f
n   microbestraling
d   Mikrobestrahlung f

4413   MICROLITE   mi
An ore containing about 10 % of U and
0.2 % of Th.
f   microlite f
e   microlita f
i   microlite f
n   microliet n
d   Mikrolit m

4414 MICROMANIPULATOR   co
A device in reactor control by means of
which fine adjustments of apparatus may
be realized.
f   micromanipulateur *m*
e   micromanipulador *m*
i   micromanipolatore *m*
n   fijnregelmanipulator, micromanipulator
d   Feinregelmanipulator *m*,
    Mikromanipulator *m*

4415 MICROPOROSITY   mt
The appearance of small spaces between
graphite crystal due to irregular packing.
f   microporosité *f*
e   microporosidad *f*
i   microporosità *f*
n   microporeusheid
d   Mikroporosität *f*

4416 MICRORADIOGRAPHY   xr
Radiography of small objects or fine
structure with a view to subsequent great
optical enlargement of the radiograph.
f   microradiographie *f*
e   microrradiografía *f*
i   microradiografia *f*
n   microradiografie
d   Mikroradiographie *f*

4417 MICRORADIOMETER   me
A radiometer used for measuring weak
radiant power, in which a thermopile is
supported on and connected directly to
the moving coil of a galvanometer.
f   microradiamètre *m*
e   microrradiámetro *m*
i   microradiametro *m*
n   microstralingsmeter
d   Mikrostrahlungsmesser *m*

4418 MICROSCOPIC CONCENTRATION,   ch
      TRACE CONCENTRATION
A concentration of a substance below the
usual limits of chemical detection.
f   concentration *f* microscopique
e   concentración *f* microscópica
i   concentrazione *f* microscopica
n   microscopische concentratie,
    spoortjesconcentratie
d   mikroskopische Konzentration *f*,
    Spurenkonzentration *f*

4419 MICROSCOPIC CROSS SECTION   cs
The cross section per target nucleus,
atom or molecule.
f   section *f* efficace microscopique
e   sección *f* eficaz microscópica
i   sezione *f* d'urto microscopica
n   microscopische doorsnede
d   mikroskopischer Wirkungsquerschnitt *m*

4420 MICROSCOPIC SEGREGATION,
      see 1418

4421 MICROSCOPIC STATE,   np
      MICROSTATE
The state in which it is supposed that the
electrons of a system are individualized
and that the position and the velocity of
each electron are known.
f   état *m* microscopique
e   estado *m* microscópico
i   stato *m* microscopico
n   microscopische toestand
d   mikroskopischer Zustand *m*,
    Mikrozustand *m*

4422 MICROTRON   pa
An accelerator operating on a modified
cyclotron principle.
f   microtron *m*
e   microtrón *m*
i   microtrone *m*
n   microtron *n*
d   Mikrotron *n*

4423 MICROWAVE ABSORPTION   sp
      SPECTRUM,
      MICROWAVE SPECTRUM
That portion of the spectrum of a molecule
which lies in the so-called microwave
region of frequencies.
f   spectre *m* micro-ondes
e   espectro *m* microondas
i   spettro *m* microonde
n   microgolvenspectrum *n*
d   Mikrowellenspektrum *n*

4424 MICROWAVE SPECTROSCOPY   sp
Measurement of the absorption or emission,
by atomic or molecular systems, of the
electromagnetic radiation of wavelengths
in the range of 0.1 mm to 10 cm.
f   spectroscopie *f* à micro-ondes
e   espectroscopia *f* de microondas
i   spettroscopia *f* a microonde
n   microgolfspectroscopie
d   Mikrowellenspektroskopie *f*

4425 MICTURITION UROGRAPHY   xr
Radiological examination of the
micturition organs.
f   urographie *f* de miction
e   urografía *f* de micción
i   urografia *f* di minzione
n   mictie-urografie
d   Miktionsurographie *f*

4426 MIDDLE FRACTION,   ch, is
      RABBIT FRACTION
The flow recycled in a rabbit stage.
f   fraction *f* de recyclage
e   fracción *f* de recirculación
i   frazione *f* riciclata
n   teruggevoerde fractie
d   rückgeführte Fraktion *f*

4427 MIGRATION AREA   np
The sum of the slowing down area from
fission energy to thermal energy and the
diffusion area for thermal neutrons.

f aire *f* de migration
e área *f* de migración
i area *f* di migrazione
n migratieoppervlak *n*
d Wanderfläche *f*

4428 MIGRATION LENGTH   np
The square root of the migration area.
f longueur *f* de migration
e longitud *f* de migración
i lunghezza *f* di migrazione
n migratielengte
d Wanderlänge *f*

4429 MIGRATION OF FISSION   rt
PRODUCTS
The unwanted travel of fission products
in or near a nuclear reactor.
f migration *f* de produits de fission
e migración *f* de productos de fisión
i migrazione *f* di prodotti di fissione
n zwerven *n* van splijtingsprodukten
d Spaltproduktwanderung *f*

4430 MILKER, see 2934

4431 MILKINESS,   ph
TURBIDITY
Unsharpness due to radiation scattered by
a photographic emulsion.
f aspect *m* laiteux, flou *m* d'émulsion
e borrosidad *f* de emulsión
i sfumatore *m* d'emulsione
n verstrooiingsonscherpte
d milchiges Aussehen *n*, Emulsionstrübung *f*

4432 MILLER INDICES   cr
In crystallography, three numbers by which
a crystal face may be delineated, symbols
h, k and l.
f indices *pl* de Miller
e índices *pl* de Miller
i indici *pl* di Miller
n millerindexen *pl*, millerindices *pl*
d Miller-Indizes *pl*

4433 MILLIAMPÈRE-SECOND, see 4250

4434 MILLICURIE   un
One thousanth of a curie. Symbol mCi.
f millicurie *m*
e milicurie *m*
i millicurie *m*
n millicurie
d Millicurie *n*

4435 MILLICURIE DESTROYED, see 4322

4436 MILLICURIE HOUR, see 4323

4437 MILLIGOAT DOSE   ra
The dose making no allowance for
self-shielding and to be expected in a
small sphere of flesh.
f dose *f* sans autoblindage
e dosis *f* sin autoblindaje
i dose *f* senza autoschermatura

n dosis zonder zelfafscherming
d Dosis *f* ohne Selbstabschirmung

4438 MILLIGRAM-HOUR, see 4403

4439 MILLIKAN COSMIC-RAY   cr, me
METER
A device consisting of an ionization
chamber and a quartz fibre(er) electroscope
used to record the intensity of cosmic
radiation.
f appareil *m* de mesure de rayons cosmiques
de Millikan
e aparato *m* de medida de rayos cósmicos
de Millikan
i apparecchio *m* di misura di raggi cosmici
di Millikan
n millikanmeter voor kosmische stralen
d Millikan-Messgerät *n* für kosmische
Strahlen

4440 MILLIKAN METHOD,   np
MILLIKAN OIL DROP EXPERIMENT
A method of measuring the charge of an
electron.
f méthode *f* de Millikan
e método *m* de Millikan
i metodo *m* di Millikan
n millikanproef, proef van Millikan
d Öltröpfchenmethode *f* nach Millikan

4441 MILLIMASS UNIT,   un
mMU
$1/16000$ of the mass of atom ^{16}O.
f unité *f* millimasse
e milésima *f* de unidad de masa,
unidad *f* milimasa
i unità *f* di millimassa
n eenduizendste massaeenheid
d tausendstel Masseneinheit *f*, TME *f*

4442 MILLING   mi
In mining engineering, pulverization of the
ore and removing valueless material and
harmful constituents from the ore.
f trituration *f*
e molido *m*
i macinazione *f*
n fijnmalen *n*
d Feinmahlen *n*

4443 MILLINILE   un
A unit for indicating a change in
reactivity of 10^{-5}.
f P.C.M. *m*
e milinile *m*
i millinile *m*
n millinile
d Millinile *n*

4444 MILLION ELECTRONVOLTS (US),
see 4356

4445 MILLIRÖNTGEN   un
One thousandth of a röntgen.
f milliröntgen *m*
e miliröntgen *m*

i milliröntgen *m*
n milliröntgen
d Milliröntgen *n*

4446 MINE   mi
Any place where minerals and ores,
metals or precious stones are obtained
by appropriate means.
f mine *f*
e mina *f*
i miniera *f*
n mijn
d Bergwerk *n*, Grube *f*

4447 MINERAL   mi
A body processed by processes of
inorganic nature.
f minéral *m*
e mineral *m*
i minerale *m*
n delfstof, mineraal *n*
d Gestein *n*, Mineral *n*

4448 MINERAL CONCENTRATE,   mi
ORE CONCENTRATE
An ore treated in such a way that its
content of valuable material is
considerably enlarged.
f concentré *m* de minerai
e concentrado *m* de mineral
i concentrato *m* di minerale
n ertsconcentraat *n*
d Erzkonzentrat *n*

4449 MINERAL CONCENTRATION,   mi
ORE CONCENTRATION
The amount of mineral present in an ore
containing the mineral.
f concentration *f* minérale
e concent:ación *f* mineral
i concentrazione *f* minerale
n ertsgehalte *n*
d Erzgehalt *m*

4450 MINERALOGY   mi
The scientific study of minerals.
f minéralogie *f*
e mineralogía *f*
i mineralogia *f*
n mineralogie
d Mineralogie *f*

4451 MINIMUM B CONFIGURATION   pp
Magnetic configuration in which the field
intensity has a minimum value in the region
where it is desired to contain the plasma
and increases in all directions from that
region.
f configuration *f* à champ minimal
e configuración *f* de campo mínimo
i configurazione *f* a campo minimo
n configuratie bij minimaal veld
d· Konfiguration *f* bei Minimalfeld.

4452 MINIMUM CREEP RATE   mt
A rate needed to keep pace with the rate
of differential shrinkage.

f vitesse *f* minimale de fluence
e velocidad *f* mínima de fluencia
i velocità *f* minima di scorrimento
n minimale kruipsnelheid
d minimale Kriechgeschwindigkeit *f*

4453 MINIMUM IONIZATION   np
The smallest possible value of the specific
ionization that a charged particle can
produce in passing through a particular
substance.
f ionisation *f* minimale
e ionización *f* mínima
i ionizzazione *f* minima
n minimale ionisatie
d minimale Ionisation *f*,
minimale Ionisierung *f*

4454 MINIMUM NUMBER OF   ch
THEORETICAL PLATES,
MINIMUM NUMBER OF
THEORETICAL STAGES
The number of plates (stages) below which
it is not possible to obtain the desired
separation of a mixture, as, for example,
in a continuous distillation.
f nombre *m* minimal de plaques théoriques
e número *m* mínimo de placas teóricas
i numero *m* minimo di piatti teorici
n theoretisch minimaal schotelaantal *n*
d theoretische Mindestbodenzahl *f*

4455 MINIMUM REFLUX RATIO   is
In isotope separation the reflux ratio
below which it is not possible to obtain
the desired separation, even if an infinite
number of stages were to be used.
f taux *m* minimal de reflux
e relación *f* mínima de reflujo
i rapporto *m* minimo di riflusso
n minimale terugloopverhouding
d Mindestrücklaufverhältnis *n*

4456 MINIMUM WAVELENGTH, see 753

4457 MINING ENGINEERING   mi
That branch of engineering chiefly
concerned with the sinking and equipment
of mine shafts and workings and all
operations incidental to the winning and
preparation of minerals.
f exploitation *f* minière
e explotación *f* minera, minería *f*
i industria *f* mineraria,
coltivazione *f* mineraria
n mijnbouw, mijnexploitatie
d Bergbau *m*

4458 MINOMETER, see 1948

4459 MIRROR GEOMETRY   pp
Elementary shape of the configuration of
the containing magnetic field in thermonuclear research.
f géométrie *f* à miroirs
e geometría *f* de espejos
i geometria *f* a specchi

n   spiegelgeometrie, spiegelopstelling
d   Spiegelanordnung f, Spiegelgeometrie f

4460 MIRROR MACHINE, see 120

4461 MIRROR NUCLEI   np
Pairs of nuclei, each member of which
would be transformed into the other by
exchanging all neutrons for protons and
vice versa.
f   noyaux pl miroirs
e   núcleos pl especularmente simétricos
i   nuclei pl speculari
n   spiegelkernen pl
d   Spiegelkerne pl

4462 MIRROR NUCLIDES   np
Pairs of nuclides possessing mirror nuclei.
f   nucléides pl miroirs
e   núclidos pl especularmente simétricos
i   nuclidi pl speculari
n   spiegelnucliden pl
d   Spiegelnuklide pl

4463 MIRROR RATIO   pp
In plasma technique, in a magnetic mirror
configuration, the ratio between the
magnetic field intensities on the direction
axis at the point with the highest and at
the point with the lowest intensity.
f   rapport m de miroir
e   relación f de espejo
i   rapporto m di specchio
n   spiegelverhouding
d   Spiegelverhältnis n

4464 MITIGATION OF HAZARDS   sa
Measure taken to reduce the danger of
contamination of human and animal
foodstuffs.
f   limitation f de dangers
e   limitación f de riesgos
i   riduzione f di rischi
n   gevaarbegrenzing, gevaarvermindering
d   Gefahrenbegrenzung f

4465 MITOGENIC RADIATION, see 3113

4466 MITOSIS, see 3834

4467 MITOTIC   md
Relating to an indirect or karyokinetic
division.
f   mitotique adj
e   mitótico adj
i   mitotico adj
n   mitotisch adj
d   mitotisch adj

4468 MIXED CRYSTALS,   gp
        SOLID SOLUTION
A single phase solid containing more than
one component.
f   cristaux pl mixtes, solution f solide
e   cristales pl mezclados, solución f sólida
i   cristalli m misti, soluzione f solida
n   mengkristallen pl, vaste oplossing
d   feste Lösung f, Mischkristalle pl

4469 MIXED SPECTRUM REACTOR   rt
A nuclear reactor in which fission is
induced by thermal neutrons in one region
of the core and by fast neutrons in another
region of the core.
f   réacteur m à spectre mixte,
    réacteur m à zone thermique et à zone
    rapide
e   reactor m de zona térmica y zona rápida
i   reattore m a zona termica ed a zona
    veloce
n   reactor met thermische en snelle
    neutronen
d   Reaktor m mit thermischen und schnellen
    Neutronen

4470 MIXER SETTLER   ch, is
A form of stage-wise contactor for solvent
extraction in which the two phases are
mixed in one chamber, and, passing to a
second, are allowed to settle and leave
through separate ports.
f   mélangeur-clarificateur m
e   mezclador-asentador m
i   mescolatore-chiarificatore m
n   meng- en neerslagapparaat n
d   Misch- und Klärgerät n

4471 MIXING   ch
The bringing together of different
materials in order to obtain a homogeneous
or heterogeneous aggregate.
f   mélange m
e   mezclado m
i   mescolatura f
n   menging
d   Mischung f

4472 MIXING, see 3758

4473 MIXING EFFICIENCY   is
A measure of the effectiveness of the
promoted mixing process in terms of the
stage separation factor which would be
obtained under conditions of perfect
mixing.
f   rendement m de homogénéisation
e   rendimiento m de homogenización
i   rendimento m d'omogenizzazione
n   homogeniseringsrendement n
d   Homogenisierungsausbeute f,
    Mischungsgrad m

4474 MIXTURE   ch
A heterogeneous or homogeneous
aggregation of different materials.
f   mélange m
e   mezcla f
i   miscela f
n   mengsel n
d   Gemisch n

4475 MIXTURE OF ISOTOPES, see 3759

4476 MLT, see 3943

4477 M/N RATIO, see 3643

4478 MOBILE REACTOR,   rt
      PROPULSION REACTOR
   A nuclear power reactor designed to
   propel aircraft, surface ships, sub-
   marines, etc.
f réacteur *m* de propulsion,
   réacteur *m* mobile
e reactor *m* de propulsión, reactor *m* móvil
i reattore *m* di propulsione,
   reattore *m* mobile
n mobiele reactor, voortstuwingsreactor
d Antriebsreaktor *m*, beweglicher Reaktor *m*

4479 MOBILITY   np, pa
   In a medium, the average drift velocity
   imparted to a charged particle by an
   electric field, divided by the field strength.
f mobilité *f*
e movilidad *f*
i mobilità *f*
n beweeglijkheid
d Beweglichkeit *f*

4480 MOBILITY LIMITATION, see 4015

4481 MOBILITY OF A CHARGED   np
      PARTICLE
   The quotient of the velocity in the
   direction of an electric field by the
   intensity of this field.
f mobilité *f* d'un porteur électrisé
e movilidad *f* de una partícula cargada
i mobilità *f* d'una particella caricata
n beweeglijkheid van een geladen deeltje
d Beweglichkeit *f* eines geladenen Teilchens

4482 MOCK-UP REACTOR,   rt
      REACTOR SIMULATOR
   A device used to study the behavio(u)r of
   a given reactor design.
f maquette *m* de réacteur
e simulador *m* de reactor
i modello *m* di reattore
n reactormodel *n*
d Nachbildungsreaktor *m*, Reaktormodell *n*

4483 MODERATED REACTOR   rt
f réacteur *m* à modération,
   réacteur *m* modéré
e reactor *m* de moderación,
   reactor *m* moderado
i reattore *m* a moderazione,
   reattore *m* moderato
n gemodereerde reactor
d moderierter Reaktor *m*

4484 MODERATION   np, rt
   The process by which neutron energy is
   reduced through scattering collisions
   without appreciable capture.
f modération *f*
e moderación *f*
i moderazione *f*
n afremming, moderatie
d Bremsung *f*, Moderierung *f*

4485 MODERATION RATIO,   rt
      MODERATOR MERIT
   A measure of the efficiency of a moderator.
f rapport *m* de modération
e relación *f* de moderación
i rapporto *m* di moderazione
n moderatieverhouding, remverhouding
d Bremsverhältnis *n*,
   Moderierungsverhältnis *n*

4486 MODERATOR   np, rt
   A material used to reduce, by scattering
   collisions and without appreciable capture,
   the kinetic energy of neutrons.
f modérateur *m*, ralentisseur *m*
e moderador *m*
i moderatore *m*
n moderator, remstof
d Moderator *m*

4487 MODERATOR CONTROL   co, rt
   Control of a nuclear reactor by adjustment
   of the properties, position or quantity of
   the moderator in such a way as to change
   the reactivity.
f commande *f* par le modérateur
e regulación *f* por el moderador
i regolazione *f* col moderatore
n regeling met moderator
d Moderatortrimmung *f*

4488 MODERATOR COOLANT   cl, rt
   A substance acting simultaneously as a
   coolant and a moderator.
f modérateur *m* réfrigérant
e moderador *m* refrigerante
i moderatore *m* refrigerante
n koelende moderator, koelende remstof
d kühlender Moderator *m*

4489 MODERATOR COOLING SYSTEM   cl
   A cooling system used in a reactor plant
   when the cooling of the moderator is
   treated as a separate problem from the
   cooling of the fuel system.
f système *m* indépendant de refroidissement
   du modérateur
e sistema *m* independiente de enfriamiento
   del moderador
i sistema *m* indipendente di raffreddamento
   del moderatore
n onafhankelijk moderatorkoelsysteem *n*
d unabhängiges Moderatorkühlsystem *n*

4490 MODERATOR DUMPING   rt, sa
      SAFETY MECHANISM
   A safety mechanism actuated by the discharge of the moderator.
f mécanisme *m* de sécurité par vidange du
   modérateur
e mecanismo *m* de seguridad por descarga
   del moderador
i meccanismo *m* di sicurezza mediante
   scarico del moderatore
n veiligheidsmechanisme *n* berustend op
   lozing van de moderator
d Sicherheitsvorrichtung *f* mittels Schnell-
   ablass des Moderators

4491 MODERATOR LATTICE rt
A lattice formed by the moderator elements only.
f réseau *m* modérateur
e celosía *f* moderadora
i reticolo *m* moderatore
n moderatorrooster *n*, remstofrooster *n*
d Moderatorgitter *n*

4492 MODULUS OF ELASTICITY, gp
YOUNG'S MODULUS
The ratio of stress to strain within the elastic range.
f module *m* d'élasticité
e módulo *m* de elasticidad
i modulo *m* d'elasticità
n elasticiteitsmodulus
d Elastizitätsmodul *m*

4493 MODULUS OF RIGIDITY, see 1172

4494 MODULUS OF RUPTURE, gp
RUPTURE MODULUS
A measure of the ultimate strength of the breaking load per unit area of a specimen as determined from a torsion or move commonly from a loading test.
f module *m* de rupture
e módulo *m* de ruptura
i modulo *m* di rottura
n breukmodulus
d Bruchmodul *m*

4495 MOISTURE LOSS ch
In the drying of solids, the evaporated quantity of moisture.
f perte *f* d'humidité
e pérdida *f* de humedad
i perdita *f* d'umidità
n vochtverlies *n*
d Feuchtigkeitsverlust *m*

4496 MOLE FRACTION ch
The number of moles of one component in a phase or system containing more than one component, divided by the total number of moles of all the components. As used in isotope separation, synonymous with abundance.
f fraction *f* molaire
e fracción *f* molar
i frazione *f* molare
n molenbreuk
d Molenbruch *m*

4497 MOLECULAR ABUNDANCE sp
In mass spectroscopy, the fraction of molecules, in a mixture of molecules, which has a given total mass number.
f abondance *f* moléculaire
e abundancia *f* molecular
i abbondanza *f* molecolare
n moleculaire abondantie
d molekulare Häufigkeit *f*

4498 MOLECULAR ABUNDANCE RATIO is
Ratio of the molar fraction of a given isotope to the sum of the mole fractions of the other isotopes.
f richesse *f*
e relación *f* de abundancia molecular
i rapporto *m* d'abbondanza molecolare
n moleculaire abondantieverhouding
d molekulares Häufigkeitsverhältnis *n*

4499 MOLECULAR BEAM np
Neutral molecules or atoms travel(l)ing through a vacuum in a narrow beam at thermal velocities.
f faisceau *m* moléculaire
e haz *m* molecular
i fascio *m* molecolare
n moleculaire bundel
d Molekularstrahlen *pl*

4500 MOLECULAR BOND ch
A valence linkage between two atoms consisting of a pair of electrons, both of which have been furnished by one of the atoms.
f liaison *f* moléculaire
e enlace *m* molecular
i legame *m* molecolare
n moleculebinding
d Molekülbindung *f*

4501 MOLECULAR COLLISION ch, ra
Collision between molecules which may be perfectly elastic or inelastic, resulting in chemical reaction, radiation, etc.
f choc *m* moléculaire
e choque *m* molecular
i urto *m* molecolare
n moleculaire botsing
d molekularer Stosz *m*

4502 MOLECULAR DEUTERIUM IONS pp
Ions fired into a carbon arc mounted in a magnetic field with mirror configuration.
f ions *pl* moléculaires de deutérium
e iones *pl* moleculares de deuterio
i ioni *pl* molecolari di deuterio
n moleculaire deuteriumionen *pl*
d molekulare Deuteriumionen *pl*

4503 MOLECULAR DIFFUSION ch
A process whereby one substance gradually impenetrates another substance as a result of the continuous thermal motion of the individual molecules.
f diffusion *f* moléculaire
e difusión *f* molecular
i diffusione *f* molecolare
n moleculaire diffusie
d Molekulardiffusion *f*

4504 MOLECULAR DISTILLATION ch, is
A distillation separation based on the difference between the rates at which molecules of different masses evaporate.
f distillation *f* moléculaire
e destilación *f* molecular
i distillazione *f* molecolare
n filmverdamping, moleculaire destillatie

d Kurzwegdestillation f,
   Molekulardestillation f

4505 MOLECULAR DISTILLATION is
   METHOD
   The separation of isotopes by means of
   molecular distillation.
f méthode f de distillation moléculaire
e método m de destilación molecular
i metodo m di distillazione molecolare
n moleculaire-destillatiemethode
d Molekulardestillationsverfahren n

4506 MOLECULAR EFFUSION ch, is
   The passage of a gas through an orifice
   small compared with the mean free path
   of gas molecules.
f effusion f moléculaire
e efusión f molecular
i effusione f molecolare
n moleculaire uitstroming
d Molekularausströmung f

4507 MOLECULAR EXCITATION ch
   The process of putting an atom or a
   molecule into a condition in which the
   total energy of its interior or mechanism
   is greater than it is in the normal or
   ground state.
f excitation f moléculaire
e excitación f molecular
i eccitazione f molecolare
n molecule-aanslag
d Molekülanregung f

4508 MOLECULAR FLOW gp
   The type of flow occurring when the mean
   free path of gas molecules is large in
   comparison with the dimensions of the
   channel in which the gas is flowing.
f écoulement m moléculaire
e flujo m molecular
i flusso m molecolare
n moleculaire stroming
d Molekularströmung f

4509 MOLECULAR FREE PATH, ch, ra
   MOLECULAR MEAN FREE PATH
   The average free path or distance
   travel(l)ed by a molecule between collisions
   in a gas or in a collision.
f libre moyen parcours m moléculaire
e trayectoria f libre media molecular
i percorso m libero medio molecolare
n gemiddelde vrije weglengte van een
   molecule
d mittlere freie Weglänge f eines Moleküls

4510 MOLECULAR HEAT ch
   The heat capacity of a substance.
f chaleur f moléculaire
e calor m molecular
i calore m molecolare
n moleculewarmte, molwarmte
d Molekülwärme f, Molwärme f

4511 MOLECULAR ION np
   A molecule having a positive or negative
   charge through loss or gain of an electron.
f ion m moléculaire
e ión m molecular
i ione m molecolare
n moleculair ion n
d Molekularion n

4512 MOLECULAR ION BREAKUP np
   The process of dissociating a molecular
   ion into its atomic components, used in the
   injection of high energy deuterium ions
   into a magnetic field with the ultimate
   goal of fusion power.
f décomposition f d'ion moléculaire
e descomposición f de ión molecular
i decomposizione f d'ione molecolare
n ontleding van een moleculair ion
d Zersetzung f eines Molekularions

4513 MOLECULAR MASS ch
   The mass of a molecule.
f masse f moléculaire
e masa f molecular
i massa f molecolare
n moleculemassa
d Molekülmasse f

4514 MOLECULAR NUMBER ch
   The sum of the atomic numbers of the
   atoms in a molecule.
f nombre m moléculaire
e número m molecular
i numero m molecolare
n moleculegetal n
d Molekülzahl f

4515 MOLECULAR ORBITAL np
   The orbital or wave function of an
   electron in motion within a molecule and
   subject to the fields of other electrons or
   nuclei.
f orbite f moléculaire
e órbita f molecular
i orbita f molecolare
n moleculaire baan
d molekulare Bahn f

4516 MOLECULAR SPECTRUM sp
   An array of bands instead of distinct lines,
   but arranged, like lines, in groups and
   series.
f spectre m de molécule
e espectro m de molécula
i spettro m di molecola
n moleculespectrum n
d Molekülspektrum n

4517 MOLECULAR STOPPING POWER np
   The stopping power closely equal to the
   sum of the atomic stopping powers of the
   constitutional atoms.
f pouvoir m d'arrêt moléculaire
e poder m de frenado molecular
i potere m di rallentamento molecolare

n moleculair stoppend vermogen *n*,
   stoppend vermogen *n* per molecule
d molekulares Bremsvermögen *n*

4518 MOLECULAR STRUCTURE   ch
   The internal structure of the molecule.
f structure *f* de la molécule
e estructura *f* de la molécula
i struttura *f* della molecola
n moleculestructuur
d Molekularstruktur *f*, Molekülstruktur *f*

4519 MOLECULAR VACUUM PUMP   vt
   A pump consisting of a drum which rotates
   at high speed within a cylindrical casing.
f pompe *f* moléculaire
e bomba *f* molecular
i pompa *f* molecolare
n moleculaire pomp
d Molekularpumpe *f*

4520 MOLECULAR WEIGHT   ch
   The weight of a molecule measured by
   taking as a unit one sixteenth of the weight
   of an atom of oxygen.
f poids *m* moléculaire
e peso *m* molecular
i peso *m* molecolare
n moleculegewicht *n*
d Molekulargewicht *n*, Molgewicht *n*

4521 MOLECULE   ch
   The smallest unit which still has the
   properties of the compound, theoretically
   separated by mechanical means.
f molécule *f*
e molécula *f*
i molecola *f*
n molecule, molekuul
d Molekül *n*

4522 MOLECULE CONVERSION YIELD   np
   The number of molecules produced or
   converted per 100 eV of energy absorbed.
f rendement *m* de conversion moléculaire
e rendimiento *m* de conversión molecular
i rendimento *m* di conversione molecolare
n moleculair conversierendement *n*
d Molekularkonversionsausbeute *f*

4523 MÖLLER SCATTERING   np
   The first theory on scattering in which the
   theoretical matrix element for the
   scattering of electrons by electrons was
   derived.
f diffusion *f* de Möller
e dispersión *f* de Möller
i deviazione *f* di Möller
n möllerverstrooiing
d Möller-Streuung *f*

4524 MOLTEN BISMUTH COOLING   cl
   A process whereby molten bismuth is used
   as the cooling agent in a nuclear reactor.
f refroidissement *m* par bismut fondu
e enfriamiento *m* por bismuto fundido
i raffreddamento *m* per bismuto fuso

n koeling door gesmolten bismut
d Kühlung *f* mit geschmolzenem Wismut

4525 MOLTEN SALT REACTOR   rt
   A nuclear reactor which makes use of a
   fluid fuel system consisting of a mixed
   salt solution of fissile (fissionable) and
   fertile materials which solution is
   circulated from the reactor through an
   external heat exchanger.
f réacteur *m* à sels fondus
e reactor *m* de sal fundida
i reattore *m* a sali fusi
n gesmolten-zoutreactor
d Salzschmelzenreaktor *m*

4526 MOLTING, see 1745

4527 MOLYBDENUM   ch
   Metallic element, symbol Mo, atomic
   number 42.
f molybdène *m*
e molibdeno *m*
i molibdeno *m*
n molybdeen *n*
d Molybdän *n*

4528 MOLYBDENUM WASHER   cd, rt
   Designed to separate the breeder ends from
   the enriched uranium elements.
f rondelle *f* en molybdène
e arandela *f* de molibdeno
i rondella *f* in molibdeno
n molybdeenonderlegplaatje *n*
d Molybdänunterlegscheibe *f*

4529 MOMENT OF INERTIA   gp
   A measure of the ease with which a mass
   can be rotated.
f moment *m* d'inertie
e momento *m* de inercia
i momento *m* d'inerzia
n traagheidsmoment *n*
d Trägheitsmoment *n*

4530 MOMENTS METHOD   ma
   A method used in gamma shielding
   calculations, developed by Spencer and
   Fano, involving simplifying assumptions
   which permit numerical approximate
   solutions of the Boltzmann transport
   equation.
f méthode *f* des moments
e método *m* de los momentos
i metodo *m* dei momenti
n momentenmethode
d Momentenmethode *f*

4531 MOMENTUM   gp
   The quantity of motion in a moving body,
   being always proportional to the mass
   multiplied into the velocity.
f impulsion *f*, quantité *f* de mouvement
e cantidad *f* de ímpetu, impulsión *f*
i impulso *m*, quantità *f* di moto
n bewegingsgrootheid, impuls
d Bewegungsgrösse *f*, Impuls *m*

4532 MOMENTUM-CHARGE RATIO,
see 4224

4533 MONAZITE,   mi
    TURNERITE
A mineral phosphate of cerium metals,
usually containing thorium also as
$Th_3(PO_4)$.
f   monazite f, turnerite f
e   monazita f, turnerita f
i   monazite f, turnerite f
n   monaziet n, turneriet n
d   Monazit m, Turnerit m

4534 MONITOR   sa
A radiation detector designed to meet
the special needs of monitoring.
f   moniteur m
e   monitor m, vigía f
i   avvisatore m, monitore m, segnalatore m
n   monitor
d   Monitor m

4535 MONITOR IONIZATION   ic, xr
    CHAMBER
An ionization chamber mounted in an
X-ray beam and connected to a continuously reading instrument to serve as an
indicator of X-ray output.
f   chambre f d'ionisation de contrôle
e   cámara f de ionización de control
i   camera f d'ionizzazione di controllo
n   controle-ionisatievat n
d   Kontrollionisationskammer f

4536 MONITORING   sa
Periodic or continuous determination of
the amount of ionizing radiation or radioactive contamination present.
f   contrôle f de l'activité, surveillance f
e   reconocimiento m, vigilancia f
i   controllo m di radiazione ionizzante,
    controllo m di radioattività
n   activiteitsbewaking
d   Aktivitätskontrolle f

4537 MONOCHROMATIC RADIATION   ra
Radiation having one frequency or one
wavelength.
f   rayonnement m monochromatique
e   radiación f monocromática
i   radiazione f monocromatica
n   monochromatische straling
d   einfarbige Strahlung f,
    monochromatische Strahlung f

4538 MONOCHROMATIC X-RAYS, see 3317

4539 MONOCRYSTAL MINERAL   ra
    SCINTILLATOR DETECTOR
A scintillation detector in which the
scintillator is a monocrystal mineral.
f   détecteur m à scintillateur minéral
    monocristallin
e   detector m de centelleador mineral
    monocristalino
i   rivelatore m a scintillatore a mono-
    cristallo minerale

n   scintillatiedetector met mineraal enkel-
    kristal
d   Detektor m mit anorganischem Ein-
    kristallszintillator

4540 MONOCRYSTAL ORGANIC   ra
    SCINTILLATOR DETECTOR
A scintillation detector in which the
scintillator is a monocrystal organic
substance.
f   détecteur m à scintillateur organique
    monocristallin
e   detector m de centelleador orgánico
    monocristalino
i   rivelatore m a scintillatore a monocristallo
    organico
n   scintillatiedetector met organisch
    enkelkristal
d   Detektor m mit organischem Ein-
    kristallszintillator

4541 MONOCYTE, see 3349

4542 MONO-ENERGETIC RADIATION   ro
Particle radiation of a given type in which
all particles have the same energy.
f   rayonnement m monoénergétique
e   radiación f monoenergética
i   radiazione f monoenergetica
n   mono-energetische straling
d   monoenergetische Strahlung f

4543 MONOLAYER,   ch
    MONOMOLECULAR LAYER,
    UNILAYER
A film on a solid or liquid surface one
molecule thick.
f   couche f monomoléculaire
e   capa f monomolecular
i   strato m monomolecolare
n   monomoleculaire laag
d   monomolekulare Schicht f

4544 MONOPOLE   np
A particle with an isolated magnetic charge
f   monopôle m
e   monopolo m
i   monopolo m
n   monopool
d   Monopol m

4545 MONTE CARLO METHOD   ma
A method of solving physical problems by
a series of statistical experiments
performed by applying mathematical
operations to random numbers.
f   méthode f Monte Carlo
e   método m Monte Carlo
i   metodo m Monte Carlo
n   Monte Carlo-methode
d   Monte-Carlo-Verfahren n

4546 MORSE EQUATION   ma, np
An equation relating the potential energy
of a diatomic molecule to the internuclear
distance.
f   équation f de Morse
e   ecuación f de Morse

i   equazione *f* di Morse
n   vergelijking van Morse
d   Morse-Gleichung *f*

4547   MOSAIC   ph
A photomicrograph of a track in an
emulsion, prepared from a number of
consecutive fields of view and recon-
structed as though the track lay in one
plane.
f   mosaïque *f*
e   mosaico *m*
i   mosaico *m*
n   mozaïek *n*
d   Mosaik *n*

4548   MOSAIC STRUCTURE   cr
A crystal is said to have this structure
when its concept consists of mosaic
blocks, misaligned with respect to one
another by a few minutes of arc.
f   structure *f* mosaïque
e   estructura *f* mosaica
i   struttura *f* a mosaico
n   mozaïekstructuur
d   Mosaiktextur *f*

4549   MOSELEY LAW   np
A statement of the relation between the
atomic number of an element and the
frequency of its characteristic radiation.
f   loi *f* de Moseley
e   ley *f* de Moseley
i   legge *f* di Moseley
n   wet van Moseley
d   Moseleysches Gesetz *n*

4550   MÖSSBAUER EFFECT   ra
The emission of gamma photons without
recoil of the source atoms from a source
used for preparing certain isotopes.
f   effet *m* Mössbauer
e   efecto *m* Mössbauer
i   effetto *m* Mössbauer
n   mössbauereffect *n*
d   Mössbauer-Effekt *m*

4551   MÖSSBAUER SOURCE   ra
A source utilizing the Mössbauer effect.
f   source *f* de Mössbauer
e   fuente *f* de Mössbauer
i   sorgente *f* di Mössbauer
n   mössbauerbron
d   Mössbauer-Quelle *f*

4552   MOTOR THEORY, see 4179

4553   MOTT SCATTERING FORMULA   np
Gives the differential cross section for
Coulomb scattering of fast nuclei of the
same kind, taking into account the special
properties, which follow from quantum
theory, of collisions between identical
particles.
f   formule *f* de diffusion de Mott
e   fórmula *f* de dispersión de Mott
i   formula *f* di deviazione di Mott

n   verstrooiingsformule van Mott
d   Mottsche Streuformel *f*

4554   MOULAGE,   ra
     RADIUM MOLD
A cast of a surface or cavity holding
radioactive sources for therapy.
f   moulage *m*, moulage *m* de radium
e   molde *m*, molde *m* de radio
i   modellatura *f*
n   moulage, radiummoulage
d   Moulage *f*

4555   MOVEMENT BLUR, see 3849

4556   MOVING BEAM THERAPY,   xr
     MOVING FIELD THERAPY
Radiation therapeutic technique in which
the radiation beam moves during treatment.
f   thérapie *f* par faisceau mobile,
    traitement *m* par champ mobile
e   irradiación *f* con campo móvil,
    terapia *f* por haz móvil
i   terapia *f* con fascio mobile,
    trattamento *m* con campo mobile
n   bewegingsbestraling
d   Bewegungsbestrahlung *f*,
    Therapie *f* mit bewegtem Strahlenbündel

4557   MOVING GRID, see 806

4558   MPC, see 4307

4559   MPD, see 4308

4560   MPDE, see 4309

4561   MTD, see 4338

4562   MTR, see 4299

4563   MU MESON,   np
     MUON
A meson with a mean life 2×10^{-6} s,
decaying into an electron, a neutrino and
an antineutrino.
f   méson *m* mu, muon *m*
e   mesón *m* mu, muón *m*
i   mesone *m* mu, muone *m*
n   mu-meson *n*, muon *n*
d   mu-Meson *n*, Muon *n*, Myon *n*

4564   MUCOSA,   md
     MUCOUS COAT,
     TUNICA MUCOSA
Mucous membrane lining gastrointestinal
tract and air passages.
f   membrane *f* muqueuse, muqueuse *f*,
    tunique *f* muqueuse
e   membrana *f* mucosa, mucosa *f*,
    túnica *f* mucosa
i   mucosa *f*, tunica *f* mucosa
n   mucosa, slijmvlies *n*
d   Mucosa *f*, Schleimhaut *f*, Tunica *f* mucosa

4565 MULTICHANNEL AMPLITUDE   me
  ANALYZER WITH STORAGE
  FUNCTION
  An amplitude analyzer which includes a
  storage function to record the number of
  pulses received per channel.
f analyseur *m* d'amplitude multicanal à
  mémoire
e analizador *m* de amplitud multicanal
  con memoria
i analizzatore *m* d'ampiezza multicanale
  a memoria
n veelkanalige pulshoogteanalysator met
  geheugen
d Vielkanal-Impulshöhenanalysator *m* mit
  Speicherung

4566 MULTICHANNEL PULSE   ec
  AMPLITUDE ANALYZER
  An instrument for determining the
  distribution function of a set of pulses in
  terms of their amplitudes.
f analyseur *m* d'amplitude multicanal
e analizador *m* de amplitud de impulsos
  multical
i analizzatore *m* d'ampiezza multicanale,
  classificatore *m* d'impulsi a canali
  multipli
n veelkanalige pulshoogteanalysator
d Mehrkanal-Amplitudenanalysator *m*,
  Mehrkanal-Impulshöhenanalysator *m*

4567 MULTIFLOW COOLING SYSTEM,   cl
  MULTIPASS COOLING SYSTEM
  The most economical heat exchange is
  obtained by circulating the coolant through
  the core of a reactor in such a way as to
  make most efficient use of the non-
  uniformity of heat generation.
f système *m* de refroidissement à
  plusieurs voies
e sistema *m* de enfriamiento de varias
  trayectorias
i sistema *m* di raffreddamento a più
  percorsi
n meerwegskoelsysteem *n*
d Mehrwegskühlsystem *n*

4568 MULTIGROUP MODEL, see 4247

4569 MULTIGROUP THEORY   np, rt
  A theory of neutron transport in which a
  neutron population is represented by
  several neutron energy groups.
f théorie *f* à plusieurs groupes
e teoría *f* de grupos múltiples
i teoria *f* a più gruppi
n veelgroepstheorie
d Gruppentransporttheorie *f*

4570 MULTIPARAMETER ANALYZER,   me
  MULTIPARAMETER ANALYZING
  ASSEMBLY
  Device for analyzing the simultaneous
  values of several parameters with single
  classification as determined by the result
  of this analysis.

f chaîne *f* d'analyse multiparamètre,
  ensemble *m* d'analyse paramétrique
e cadena *f* analizadora multiparamétrica
i catena *f* analizzatrice multiparametrica
n multiparameteranalysator
d Multiparameteranalysator *m*

4571 MULTIPLE BRANCHING DECAY,   np
  MULTIPLE BRANCHING
  DISINTEGRATION
  The occurrence of more than two modes
  by which a radionuclide can undergo
  radioactive decay.
f désintégration *f* à embranchement
  multiple
e desintegración *f* de bifurcación múltiple
i disintegrazione *f* a ramificazione
  multipla
n desintegratie met meervoudige vertakking
d Mehrfachverzweigungszerfall *m*

4572 MULTIPLE CRIT   rt
  Something more than a critical mass of
  fissile (fissionable) material.
f masse *f* critique excédante
e masa *f* crítica excedente
i massa *f* critica eccedente
n rijkelijk kritieke massa
d reichlich kritische Masse *f*

4573 MULTIPLE DECAY,
  MULTIPLE DISINTEGRATION,
  see 767

4574 MULTIPLE IONIZATION   np
  The loss of an electron by an already
  positively ionized atom (or molecule) or
  the addition of an electron to an already
  negatively ionized atom (or molecule).
f ionisation *f* multiple
e ionización *f* múltiple
i ionizzazione *f* multipla
n meervoudige ionisatie
d Mehrfachionisation *f*

4575 MULTIPLE PRODUCTION   np
  The production of two or more particles
  in a single collision between two nucleons.
f production *f* multiple en collision unique
e producción *f* múltiple en colisión simple
i produzione *f* multipla in collisione
  semplice
n meervoudige produktie bij eenvoudige
  collisie
d Mehrfacherzeugung *f* bei einfacher
  Kollision

4576 MULTIPLE SCATTERING   np
  Any scattering of a particle or photon in
  which the final deviation is the sum of
  many deviations, resulting from many
  individual scattering processes.
f diffusion *f* multiple
e dispersión *f* múltiple
i deviazione *f* multipla
n meervoudige verstrooiing
d Mehrfachstreuung *f*

4577 MULTIPLE TUBE COUNTS   ct
Spurious counts in radiation counter
tubes induced by previous tube counts.
f comptages *pl* parasites
e recuentos *pl* parásitos
i conteggi *pl* parassiti
n parasitaire tellingen *pl*
d Störzählungen *pl*

4578 MULTIPLET   sp
A spectrum line which, when examined
under high resolution, is composed of two,
three, four òr more closely packed fine
lines.
f multiplet *m*
e multiplete *m*
i multipletto *m*
n multiplet *n*
d Linienkomplex *n*, Multiplett *n*

4579 MULTIPLICATION   np
The process by which additional neutrons
are produced by a chain reaction in an
assembly containing fissile (fissionable)
material.
f multiplication *f*
e multiplicación *f*
i moltiplicazione *f*
n vermeerdering, vermenigvuldiging
d Vermehrung *f*

4580 MULTIPLICATION CONSTANT,
see 2034

4581 MULTIPLICATION FACTOR,   np
REPRODUCTION CONSTANT
The ratio of the total number of neutrons
produced during a time interval to the
total number of neutrons lost by absorption
and leakage during the same interval.
f facteur *m* de multiplication
e factor *m* de multiplicación
i fattore *m* di moltiplicazione
n vermenigvuldigungsfactor
d Multiplikationsfaktor *m*,
Vermehrungsfaktor *m*

4582 MULTIPLICITY   ma, np
The number 2S + 1, representing the
number of ways of vectorially coupling the
orbital angular momentum vector L with
the spin angular momentum S of an atom.
f multiplicité *f*
e multiplicidad *f*
i moltiplicità *f*
n meervoudigheid, multipliciteit
d Multiplizität *f*, Vielfachheit *f*

4583 MULTIPLYING   rt
In reactor technology, capable of supporting
a neutron-induced fission chain reaction.
f multiplicateur adj
e multiplicador adj
i moltiplicante adj, moltiplicatore adj
n vermenigvuldigend adj
d multiplizierend adj

4584 MULTIPLYING MEDIUM   np
A medium in which neutron multiplication
occurs.
f milieu *m* multiplicateur
e medio *m* multiplicador
i mezzo *m* moltiplicante,
sistema *m* moltiplicante
n vermenigvuldigend medium *n*
d multiplizierendes Medium *n*

4585 MULTIPOLE MOMENT   np
The electric and magnetic multipole
moments of a system in a given state are
measures of the charge, current and
magnetic distributions in the state, and
determine the interaction of the system
with weak external fields.
f moment *m* multipolaire
e momento *m* multipolar
i momento *m* moltipolare
n multipoolmoment *n*
d Multipolmoment *n*

4586 MULTIPOLE RADIATION   ra
The radiation field in free space may be
expanded into multipole fields, electric
and magnetic, as well as into plane waves.
f rayonnement *m* multipolaire
e radiación *f* multipolar
i radiazione *f* moltipolare
n meerpolige straling, multipoolstraling
d Multipolstrahlung *f*

4587 MULTIPROBE RADIATION   me
METER
A measuring assembly, usually employed
for health physics purposes with a set of
probes which cau be used for various
measurements, an appropriate probe
being chosen according to the phenomenon
to be measured.
f polyradiamètre *m*
e radiámetro *m* múltiple
i poliradiametro *m*
n meersondige stralingsmeter,
stralingsmeter met verschillende meet-
sondes
d Strahlungsmessgerät *n* mit Detektorsatz

4588 MULTIPURPOSE REACTOR   rt
A nuclear reactor so designed as to have
different ways of operation.
f réacteur *m* à plusieurs desseins
e reactor *m* de múltiples aprovechamientos
i reattore *m* a molteplici scopi
n reactor met meervoudig doel
d Mehrzweckreaktor *m*

4589 MULTIREGION REACTOR,   rt
MULTIZONE REACTOR
A nuclear reactor in which the core is
constituted of regions or zones having
different multiplication properties, due to
different enrichment or poisoning of the
fuel.
f réacteur *m* à plusieurs régions

e reactor *m* de múltiples regiones
i reattore *m* a più regioni
n reactor met zonenkern
d Mehrzonenreaktor *m*

4590 MULTISTAGE DIFFUSION UNIT   is
A set of apparatus in an isotope separation
plant of the diffusion type, in which the
process is carried out in a number of
stages.
f ensemble *m* de diffusion à plusieurs
étages
e conjunto *m* de difusión de etapas múltiples
i complesso *m* di diffusione a stadi multipli
n meertrapsdiffusieaggregaat *n*
d mehrstufiges Diffusionsaggregat *n*

4591 MULTISTAGE PROCESS   is
A process of separating isotopes carried
out in a number of stages.
f procédé *m* à plusieurs étages
e procedimiento *m* de etapas múltiples
i processo *m* a stadi multipli
n meertrapsproces *n*
d Mehrstufenverfahren *n*

4592 MULTISTAGE TUBE   xr
A röntgen tube in which the cathode rays
are accelerated by multiple pierced
anodes at progressively higher potentials.
f tube *m* à plusieurs étages
e tubo *m* de varias etapas
i tubo *m* a più elettrodi
n cascadebuis
d Mehrstufenröhre *f*

4593 MUON, see 4563

4594 MUON NUMBER   np
Quantum number introduced to explain
the behavio(u)r of two kinds of neutrinos
towards a muon and an electron.
f nombre *m* muonique
e número *m* muónico
i numero *m* muonico
n muongetal *n*
d Muonzahl *f*

4595 MUONIUM   np
A bound system consisting of a positive
mu meson and an electron.
f muonium *m*
e muonio *m*
i muonio *m*
n muonium *n*
d Muonium *n*

4596 MURAL TYPE RADIATION   ra, sa
BEACON,
WALL TYPE RADIATION BEACON
f balise *f* de rayonnement murale
e baliza *f* de radiación mural
i faro *m* di radiazione murale
n wandstralingsbaken *n*
d Wandstrahlungsbake *f*

4597 MUSHROOM HEAD   nw
The spread-out dust cloud after the latter
has soared some five or six miles in the
upper atmosphere.
f tête *f* de champignon
e cabeza *f* fungiforme
i testa *f* a fungo
n paddestoel
d Rauchpilz *m*, Staubpilz *m*

4598 MUTANT   md
A strain of an organism which differs
from the normal in an inherited character
difference.
f mutant *m*
e mutante *m*
i mutante *m*
n mutant
d Mutant *m*

4599 MUTANT GENE   md
A gene undergoing mutation.
f gène *m* mutant
e gen *m* mutante
i gene *m* mutante
n muterend gen *n*
d mutierendes Gen *n*

4600 MUTATION, see 3131

4601 MUTUAL ATTRACTION   gp
The attractive forces between particles
in a system.
f attraction *f* mutuelle
e atracción *f* mutua
i attrazione *f* mutua
n wederzijdse aantrekking
d gegenseitige Anziehung *f*

4602 MUTUAL DIFFUSION   ch, gp
Diffusion of two substances towards each
other.
f diffusion *f* mutuelle
e difusión *f* mutua
i diffusione *f* mutua
n wederzijdse diffusie
d gegenseitige Diffusion *f*

4603 MUTUAL INTERACTION   gp
The complex of forces between particles
in a system.
f interaction *f* mutuelle
e interacción *f* mutua
i interazione *f* mutua
n wederzijdse wisselwerking
d gegenseitige Wechselwirkung *f*

4604 MUTUAL REPULSION   gp
The repulsive forces between particles in
a system.
f. répulsion *f* mutuelle
e repulsión *f* mutua
i ripulsione *f* mutua
n wederzijdse afstoting
d gegenseitige Abstossung *f*

4605 MYELOGRAPHY   xr   f   myélographie *f*
The radiological examination of the space   e   mielografía *f*
between the theca and the spinal chord,   i   mielografia *f*
following the injection of air or other   n   myeolografie
contrast medium.   d   Myelographie *f*

N

4606 N ELECTRON   np
An electron characterized by having a
principal quantum number of value 4.
f électron *m* N
e electrón *m* N
i elettrone *m* N
n N-elektron *n*
d N-Elektron *n*

4607 n HOUR SAMPLER   ma
An apparatus designed to trap on a fixed
filter the dust contained in known volume
of air passing through the filter in n hours.
f appareil *m* de prélèvement (n heures)
e muestreador *m* (n horas)
i apparecchio *m* di campionatura (n ore)
n n-uur-monsternemer
d n-Stunden-Probenehmer *m*

4608 n HOUR SAMPLING MONITOR   ma
An n-hour sampler equipped with an alarm,
which is tripped if the activity of the
trapped dust on the filter is higher than a
preset value.
f moniteur *m* de contamination d'échantillons
   d'air (n heures)
e vigía *f* horaria de contaminación de
   muestras de aire (n horas)
i monitore *m* di contaminazione di
   campioni d'aria (n ore)
n besmettingsmonitor met luchtmonster-
   neming gedurende n uur
d Luftmonitor *m* mit n Stunden Sammelzeit

4609 N LINE   xr
One of the lines in N series of X-rays which
are characteristic of the various elements
and are produced by excitation of electrons
of the N shell.
f raie *f* N
e línea *f* N
i riga *f* N
n N-lijn
d N-Linie *f*

4610 N RADIATION   ra
One of a series of X-rays due to the
excitation of the N shell.
f rayonnement *m* N
e radiación *f*
i radiazione *f* N
n N-straling
d N-Strahlung *f*

4611 N SHELL   np
The collection of all those electrons in the
atom which have the principal quantum
number 4.
f couche *f* N
e capa *f* N
i strato *m* N
n N-schil
d N-Schale *f*

4612 NAEGITE   mi
A mineral containing zirconium, yttrium,
niobium, tantalum, thorium and uranium.
f naégite *f*
e naegita *f*
i naegite *f*
n naëgiet *n*
d Naegit *m*

4613 NAGATELITE   mi
A phosphation variety of allamite.
f nagatélite *f*
e nagatelita *f*
i nagatelite *f*
n nagateliet *n*
d Nagatelit *m*

4614 NaK   mt
An alloy of sodium and potassium used as
a liquid-metal heat transfer fluid in a
reactor.
f alliage *m* NaK
e aleación *f* NaK
i lega *f* NaK
n natriumkaliumlegering
d Natrium-Kaliumlegierung *f*

4615 NARROW BEAM   ra
A radiation beam from which scattered
radiation is excluded.
f faisceau *m* étroit
e haz *m* estrecho
i fascio *m* stretto
n verstrooiingsvrije bundel
d streustrahlenfreies Bündel *n*

4616 NARROW BEAM ABSORPTION   ab
Absorption measured under conditions in
which scattered radiation is excluded from
the measuring apparatus.
f absorption *f* à champ étroit,
   absorption *f* à faisceau étroit
e absorción *f* de haz estrecho
i assorbimento *m* a fascio stretto
n absorptie bij klein veld,
   verstrooiingsvrije absorptie
d Kleinfeldabsorption *f*,
   streustrahlenfreie Absorption *f*

4617 NARROW BEAM GEOMETRY ma, rt, sa
In radiation shielding calculations, the
conditions assumed when it is considered
that a single scattering event removes the
photon from the beam for good.
f géométrie *f* à champ étroit,
   géométrie *f* à faisceau étroit
e geometría *f* de haz estrecho
i geometria *f* di fascio stretto
n geometrie bij klein veld,
   verstrooiingsvrije geometrie
d Kleinfeldgeometrie *f*,
   streustrahlenfreie Geometrie *f*

4618 NARROW SHOWER   cr
A shower of cosmic-ray particles extending over a rather small area.
f gerbe *f* étroite
e chaparrón *m* estrecho
i sciame *m* stretto
n smalle bui
d schmaler Schauer *m*

4619 NASOPHARYNGEAL   md, ra
APPLICATOR
An applicator which is applied in the nasal
and mouth cavity, containing up to 100 mCi
strontium-90.
f capsule *f* porte-strontium pour la
cavité nasopharyngienne
e aplicador *m* de estronico para la cavidad
nasofaríngea
i applicatore *m* di stronzio per la cavità
nasofaringea
n strontiumcapsule voor neusholte
d Strontiumkapsel *f* für den Nasenrachen-
raum

4620 NASTURAN,   mi
PITCHBLENDE, URANITE
A mineral consisting largely of uranium
oxide; the most important ore of uranium
and radium.
f pechblende *f*, péchurane *f*, uranite *f*
e pechblenda *f*, pechurana *f*, uranita *f*
i pechblenda *f*, uranite *f*
n pekblende, uraniet *n*
d Nasturan *n*, Pechblende *f*, Uranpecherz *n*

4621 NATIVE ROCK, see 1494

4622 NATURAL ABUNDANCE   is
Of a specified isotope of an element, the
isotopic abundance in the element as found
in nature.
f teneur *f* isotopique naturelle
e abundancia *f* isotópica natural
i abbondanza *f* isotopica naturale
n natuurlijke abondantie
d natürliche Isotopenhäufigkeit *f*

4623 NATURAL ACTIVITY,   np
NATURAL RADIOACTIVITY
Radioactivity of naturally occurring
nuclides.
f radioactivité *f* naturelle
e radiactividad *f* natural
i radioattività *f* naturale
n natuurlijke radioactiviteit
d natürliche Radioaktivität *f*

4624 NATURAL BACKGROUND RADIATION,
see 502

4625 NATURAL CIRCULATION   cl, rt
REACTOR
A reactor in which the coolant is made to
circulate without pumping owing to the
different densities of its cold and
reactor-heated portions, i.e. by natural
convection.

f réacteur *m* à circulation naturelle
e reactor *m* de circulación natural
i reattore *m* a circolazione naturale
n reactor met natuurlijke circulatie
d Reaktor *m* mit Naturumlauf

4626 NATURAL ELEMENT   ch
An element as found in nature, i.e. with
its accompanying isotopes.
f élément *m* naturel
e elemento *m* natural
i elemento *m* naturale
n natuurlijk element *n*
d natürliches Element *n*

4627 NATURAL IRON,   ge, mt
ORDINARY IRON
Material used, e.g., in manufacturing metal
foil supports for Mössbauer sources.
f fer *m* naturel, fer *m* ordinaire
e hierro *m* natural
i ferro *m* naturale
n gewoon ijzer *n*, natuurlijk ijzer *n*
d gewöhnliches Eisen *n*, natürliches Eisen *n*

4628 NATURAL LEAK   me
The rate of loss of charge of an electric
measuring instrument, e.g. an electrometer, arising from causes other than
ionization by the radiation to be
measured.
f fuite *f* propre
e fuga *f* propia
i fuga *f* propria
n eigenverlies *n*
d Eigenverlust *m*

4629 NATURAL NEUTRON   np
A neutron generated from a natural source.
f neutron *m* naturel
e neutrón *m* natural
i neutrone *m* naturale
n natuurlijk neutron *n*
d natürliches Neutron *n*

4630 NATURAL RADIOACTIVE   np
ELEMENT,
NATURAL RADIONUCLIDE
Naturally occurring nuclide exhibiting
radioactivity.
f élément *m* radioactif naturel
e elemento *m* radiactivo natural
i elemento *m* radioattivo naturale
n natuurlijk radioactief element *n*
d natürliches radioaktives Element *n*

4631 NATURAL URANIUM   fu, mg
f uranium *m* naturel
e uranio *m* natural
i uranio *m* naturale
n natuurlijk uranium *n*
d natürliches Uran *n*

4632 NATURAL URANIUM REACTOR   rt
A nuclear reactor employing natural
uranium as fuel.
f réacteur *m* à uranium naturel

e reactor *m* de uranio natural
i reattore *m* ad uranio naturale
n reactor met natuurlijk uranium
d Natururanreaktor *m*

4633 NATURAL WIDTH   np
The width of an atomic or nuclear energy
level due to its spontaneous transition
lifetime.
f largeur *f* naturelle
e anchura *f* natural
i larghezza *f* naturale
n natuurlijke breedte
d natürliche Breite *f*

4634 NAVAL REACTOR   rt
A reactor used in the nuclear propulsion
of ships.
f réacteur *m* naval
e reactor *m* naval
i reattore *m* navale
n scheepsreactor
d Schiffsreaktor *m*

4635 NAVAL RESEARCH REACTOR   ms, rt
A water-moderated nuclear reactor of the
swimming pool type with an operating
power level of 100 kilowatts and two
possible core positions.
f réacteur *m* naval de recherche
e reactor *m* naval de investigaciones
i reattore *m* navale di ricerca
n speurwerkscheepsreactor
d Forschungsschiffsreaktor *m*

4636 NECKING DOWN   mg, te
Reducing locally the area of a ductile
tension test specimen in the region of
fracture.
f entaillage *m*
e entalladura *f*
i intaccatura *f*
n inkepen *n*, insnoeren *n*
d Einkerben *n*

4637 NECROSIS   md
Death of a circumscribed portion of tissue.
f nécrose *f*
e necrosis *f*
i necrosi *f*
n necrose, weefselafsterving
d Gewebstod *m*, Nekrose *f*

4638 NEEDLE COUNTER   ct
Counter fitted with a comparatively long
needle of stainless steel.
f compteur *m* à aiguille
e contador *m* con aguja
i contatore *m* ad ago
n naaldteller
d Nadelzähler *m*

4639 NEGATIVE CHARGE   ec, np
A condition of having excess electrons, or
in particular, the total charge represented
by these electrons.
f charge *f* négative

e carga *f* negativa
i carica *f* negativa
n negatieve lading
d negative Ladung *f*

4640 NEGATIVE ELECTRON,
NEGATRON, see 2099

4641 NEGATIVE GLOW   ra
The luminous region in a gas discharge
at moderately low pressure between the
cathode dark space and the Faraday dark
space.
f lueur *f* cathodique
e luminosidad *f* negativa
i bagliore *m* catodico, bagliore *m* negativo
n negatief glimlicht *n*
d negatives Glimmlicht *n*

4642 NEGATIVE ION, see 264

4643 NEGATIVE ION BEAM   ra
f faisceau *m* d'ions négatifs
e haz *m* de iones negativos
i fascio *m* d'ioni negativi
n negatieve ionenbundel
d negativer Ionenstrahl *m*

4644 NEGATIVE PROTON, see 304

4645 NEGATIVE REACTIVITY   rt, sa
Decrease of reactivity which may be
produced by a neutron absorber, e.g. a
control rod, when it is introduced in the
core of the reactor.
f antiréactivité *f*,
réactivité *f* négative
e reactividad *f* negativa
i reattività *f* negativa
n negatieve reactiviteit
d negative Reaktivität *f*

4646 NEGATIVE VALENCE   ch
A electrovalence possessed by an atom
because it has become ionized by addition
of an electron or electrons.
f valence *f* négative
e valencia *f* negativa
i valenza *f* negativa
n negatieve valentie, negatieve waardigheid
d negative Valenz *f*, negative Wertigkeit *f*

4647 NEGATOSCOPE, see 2602

4648 NEODYMIUM   ch
Rare earth metallic element, symbol Nd,
atomic number 60.
f néodyme *m*
e neodimio *m*
i neodimio *m*
n neodymium *n*
d Neodym *n*

4649 NEOFORMATION,   md
NEOPLASM
A new growth of cells which is more or
less unstrained and not governed by the

usual limitation of normal growth.
f néoformation f, néoplasma m
e neoformación f, neoplasma m
i neoformazione f, neoplasma m
n abnormale weefselvorming, nieuwvorming
d Gewebsneubildung f, Neoplasma n, Neubildung f

4650 NEON   ch
Gaseous element, symbol Ne, atomic number 10.
f néon m
e neón m
i neon m
n neon n
d Neon n

4651 NEPTUNIUM   ch
Radioactive element, symbol Np, atomic number 93.
f neptunium m
e neptunio m
i nettunio m
n neptunium n
d Neptunium n

4652 NEPTUNIUM FAMILY,   np
        NEPTUNIUM SERIES
The series of nuclides resulting from the decay of the long-lived synthetic nuclide Np-257.
f famille f du neptunium
e familia f del neptunio
i famiglia f del nettunio
n neptuniumreeks
d Neptuniumreihe f

4653 NET TRANSPORT   is
The rate at which the product isotope is transported through an isotope separating system.
f transport m net
e transporte m neto
i trasporto m netto
n nettodoorloop, nettotransport n
d Nettodurchsatz m, Nettotransport m

4654 NEUTRAL   np
Having the same number of electrons and protons.
f neutre adj
e neutro adj
i neutro adj
n neutraal adj
d neutral adj

4655 NEUTRAL ATOM,   np
        NON-IONIZED ATOM,
        UN-IONIZED ATOM
An atom which has no overall or resultant electric charge.
f atome m neutre
e átomo m neutro
i atomo m neutro
n neutraal atoom n
d neutrales Atom n

4656 NEUTRAL INSTABILITY   np
Plasma instability characterized by an oscillating disturbance which does not change in amplitude with time.
f instabilité f neutre
e inestabilidad f neutra
i instabilità f neutra
n neutrale instabiliteit
d neutrale Instabilität f

4657 NEUTRAL MESON   cr
A meson without an electric charge.
f méson m neutre
e mesón m neutro
i mesone m neutro
n neutraal meson n
d neutrales Meson n

4658 NEUTRAL MOLECULE   ch
In general a molecule without electrical charge; the term is often applied to a system of two ions of opposite but equal charge, in a solvent.
f molécule f neutre
e molécula f neutra
i molecola f neutra
n neutrale molecule
d neutrales Molekül n

4659 NEUTRETTO   np
Mass particle identical with the electron but without charge.
f neutretto m
e neutreto m
i neutretto m
n neutretto n
d Neutretto n

4660 NEUTRINO   np
A particle with no charge and essentially zero rest mass, and spin $1/2$. So far two kinds of neutrino have been established, one associated with the emission of electrons, and one with muons.
f neutrino m
e neutrino m
i neutrino m
n neutrino n
d Neutrino n

4661 NEUTRON   np
An electrically neutral elementary particle with a rest mass approximately equal to one unified atomic mass unit.
f neutron m
e neutrón m
i neutrone m
n neutron n
d Neutron n

4662 NEUTRON ABSORBER   ab, np
A material with which neutrons interact significantly by reactions resulting in their disappearance.
f absorbeur m de neutrons
e absorbedor m de neutrones
i assorbitore m di neutroni

n   absorbens *n*, absorberend materiaal *n*
d   Neutronenabsorbermaterial *n*

4663   NEUTRON ABSORBER   ab, np
An object with which neutrons interact
significantly or predominantly by reactions
resulting in their disappearance without
production of other neutrons.
f   absorbant *m* de neutrons
e   absorbente *m* de neutrones
i   assorbitore *m* di neutroni
n   absorptieplaatje *n*
d   Neutronenabsorber *m*

4664   NEUTRON-ABSORBING   sa
GLASS PLATE
f   plaque *f* de verre à absorption de neutrons
e   placa *f* de vidrio absorbedora de neutrones
i   piastra *f* di vetro ad assorbimento di
neutroni
n   neutronen absorberende glasplaat
d   neutronenabsorbierende Glasplatte *f*

4665   NEUTRON ABSORPTION   ab, np
Nuclear interaction in which the incident
neutron disappears even when one or
more neutrons are subsequently emitted
accompanied by other particles, e.g. in
fission.
f   absorption *f* de neutrons
e   absorción *f* de neutrones
i   assorbimento *m* neutronico
n   neutronenabsorptie
d   Neutronenabsorption *f*

4666   NEUTRON ABSORPTION   cs
CROSS SECTION
The cross section for the neutron
absorption process. It is the difference
between the total cross section and the
scattering cross section.
f   section *f* efficace d'absorption des
neutrons
e   sección *f* eficaz de absorción de los
neutrones
i   sezione *f* d'urto d'assorbimento dei
neutroni
n   absorptiedoorsnede,
doorsnede voor neutronenabsorptie
d   Neutronenabsorptionsquerschnitt *m*

4667   NEUTRON ACTIVATION   np
The interaction of a neutron with matter
by being absorbed in the nucleus of an
atom.
f   activation *f* par neutrons
e   activación *f* por neutrones
i   attivazione *f* da neutroni
n   neutronenactivering
d   Neutronenaktivierung *f*

4668   NEUTRON AGE, see 139

4669   NEUTRON ALBEDO, see 176

4670   NEUTRON ATOMIC MASS   np
The atomic mass of a neutron
$n = 1.008982 \pm 0.000003$ amu.

f   masse *f* du neutron
e   masa *f* del neutrón
i   massa *f* del neutrone
n   massa van het neutron
d   Neutronenmasse *f*

4671   NEUTRON ATTENUATION,   np
NEUTRON BEAM ATTENUATION
Reduction in the intensity of a beam of
neutron on passage through matter.
f   atténuation *f* du faisceau neutronique
e   atenuación *f* del haz neutrónico
i   attenuazione *f* del fascio neutronico
n   verzwakking van de neutronenbundel
d   Neutronenstrahlschwächung *f*

4672   NEUTRON BALANCE,   rt
NEUTRON CONSERVATION
PRINCIPLE
In nuclear reactors, the rate of change of
neutron density per unit volume is equal
to the rate of absorption and the rate of
leakage.
f   bilan *m* neutronique,
principe *m* de conservation de l'équilibre
neutronique
e   balance *m* de neutrones,
principio *m* de conservación del
equilibrio neutrónico
i   bilancio *m* dei neutroni,
principio *m* di conservazione dell'
equilibrio neutronico
n   neutronenbalans,
principe *n*.van het behoud van het
neutronenevenwicht
d   Neutronenbilanz *f*,
Prinzip *n* der Erhaltung des Neutronengleichgewichts

4673   NEUTRON BEAM   np
Stream of neutrons, usually fast neutrons.
f   faisceau *m* neutronique
e   haz *m* neutrónico
i   fascio *m* neutronico
n   neutronenbundel
d   Neutronenstrahl *m*

4674   NEUTRON BINDING ENERGY   np
The energy required to remove a single
neutron from a nucleus.
f   énergie *f* de liaison neutronique
e   energía *f* de enlace neutrónico
i   energia *f* di legame neutronico
n   neutronenbindingsenergie
d   Neutronenbindungsenergie *f*

4675   NEUTRON BOMBARDMENT   np, rt
The bombardment of fissile (fissionable)
material with neutrons.
f   bombardement *m* de neutrons
e   bombardeo *m* con neutrones,
bombardeo *m* neutrónico
i   bombardamento *m* neutronico
n   beschieting met neutronen,
neutronenbombardement *n*
d   Beschiessung *f* mit Neutronen,
Neutronenbeschuss *m*

4676 NEUTRON BOOSTER, see 728

4677 NEUTRON BURST   rt
Sudden multiplication of the number of
neutrons in a reactor due to a fast
uncontrolled divergence.
f bouffée f de neutrons
e explosión f de neutrones
i esplosione f di neutroni
n neutronenuitbarsting
d Neutronenausbruch m

4678 NEUTRON CAPTURE   np
A process whereby a system acquires an
additional neutron.
f capture f de neutrons,
   capture f neutronique
e captura f de neutrones,
   captura f neutrónica
i cattura f di neutroni,
   cattura f neutronica
n neutronenvangst
d Neutroneneinfang m

4679 NEUTRON CAPTURE   np
      CROSS SECTION
f section f efficace de capture neutronique
e sección f eficaz de captura neutrónica
i sezione f d'urto di cattura neutronica
n vangstdoorsnede voor neutronen
d Neutroneneinfangsquerschnitt m

4680 NEUTRON CHAIN REACTION   np
A chain reaction in which a neutron plus
a fissile (fissionable) atom causes a
fission resulting in a number of neutrons
which in turn cause other fissions.
f réaction f neutronique en chaîne
e reacción f neutrónica en cadena
i reazione f neutronica in catena
n neutronenkettingreactie
d Neutronenkettenreaktion f

4681 NEUTRON CHOPPER   np
A device for modulating a beam of neutrons.
f hacheur m de faisceau neutronique
e modulador m de haz neutrónico
i modulatore m di fascio neutronico
n neutronenbundelhakker
d Neutronenstrahlzerhacker m

4682 NEUTRON COLLIMATOR,   np
      NEUTRON HOUWITZER
A collimating apparatus for the production
of a stream of neutrons.
f collimateur m de neutrons
e colimador m de neutrones
i collimatore m di neutroni
n neutronencollimator
d Neutronenkollimator m

4683 NEUTRON COLLISION RADIUS   np
The nuclear radius determined by fast-
neutron transmission experiments.
f rayon m efficace de neutron pour collision
e radio m eficaz de neutrón para colisión
i radio m efficace di neutrone per collisione

n effectieve botsingsradius
d effektiver Kollisionsradius m

4684 NEUTRON CONSERVATION   np
      OF BALANCE
The underlying principle for any theory
of nuclear reactors.
f conservation f de l'équilibre neutronique
e conservación f del equilibrio neutrónico
i conservazione f dell'equilibrio
   neutronico
n behoud n van het neutronenevenwicht
d Erhaltung f des Neutronengleichgewichts

4685 NEUTRON CONVERTER, see 1389

4686 NEUTRON COUNTER   ct
f compteur m de neutrons
e contador m de neutrones
i contatore m di neutroni
n neutronenteller
d Neutronenzähler m

4687 NEUTRON CROSS SECTION   np
Cross section for reactions with neutrons.
f section f efficace de neutrons
e sección f eficaz de neutrones
i sezione f d'urto dei neutroni
n effectieve neutronendoorsnede
d Neutronenwirkungsquerschnitt m

4688 NEUTRON CRYSTALLOGRAPHY   cr
The study of crystals by neutron
diffraction.
f analyse f cristallographique par diffraction
   neutronique.
e análisis f cristalográfica por difracción
   neutrónica
i analisi f cristallografica per diffrazione
   neutronica
n kristalanalyse met behulp van
   neutronendiffractie
d Kristallanalyse f mittels Neutronen-
   beugung

4689 NEUTRON CURRENT,   np
      NEUTRON CURRENT DENSITY
The number of neutrons crossing unit
surface per unit time, in a direction normal
to the surface.
f densité f de courant neutronique
e densidad f de corriente neutrónica
i densità f di corrente neutronica
n neutronenstroomdichtheid
d Neutronenstromdichte f

4690 NEUTRON CURTAIN, see 1582

4691 NEUTRON CYCLE   np, rt
The life cycle of neutrons in a reactor,
beginning with the fission process and
continuing until all neutrons have caused
further fission or have been otherwise
absorbed or have leaked out.
f cycle m neutronique
e ciclo m neutrónico
i ciclo m neutronico

n neutronenkringloop
d Neutronenzyklus *m*

4692 NEUTRON DECAY   np
The spontaneous beta decay of a neutron.
f désintégration *f* de neutron
e desintegración *f* de neutrón
i disintegrazione *f* di neutrone
n neutrondesintegratie, neutronverval *n*
d Neutronenzerfall *m*

4693 NEUTRON DENSITY,   np
        NEUTRON NUMBER DENSITY
The number of free neutrons per unit
volume.
f nombre *m* volumique de neutrons
e densidad *f* neutrónica
i densità *f* neutronica
n neutronendichtheid
d Neutronenzahldichte *f*

4694 NEUTRON DETECTION   np
Carried out, e.g., by making use of the
charged particles produced by the interaction of neutrons with atomic nuclei.
f détection *f* de neutrons
e detección *f* de neutrones
i rivelazione *f* di neutroni
n neutronendetectie
d Neutronendetektion *f*

4695 NEUTRON DIFFRACTION   ms
Diffraction method used to study crystal
structure in a manner very similar to
X-ray diffraction.
f diffraction *f* à l'aide de neutrons
e difracción *f* con neutrones
i diffrazione *f* con neutroni
n diffractie door middel van neutronen
d Beugung *f* mittels Neutronen

4696 NEUTRON DIFFRACTOMETER   me
An instrument used in diffraction analysis
for measuring with a neutron beam the
intensity of the diffracted beams at
different angles.
f diffractomètre *m* neutronique
e difractómetro *m* neutrónico
i diffrattometro *m* neutronico
n neutronendiffractometer
d Neutronenbeugungsmessgerät *n*

4697 NEUTRON DIFFUSION   np
A phenomenon in which neutrons in a
medium tend, through a process of
successive scattering collisions, to migrate
from regions of high concentration to
regions of low concentration.
f diffusion *f* des neutrons
e difusión *f* de los neutrones
i diffusione *f* neutronica
n neutronendiffusie
d Neutronendiffusion *f*

4698 NEUTRON DIFFUSION   np
        COEFFICIENT
The coefficient which, multiplied by the

gradient of the neutron flux in a specific
point, gives the net neutron current in that
point.
f coefficient *m* de diffusion des neutrons
e coeficiente *m* de difusión de los neutrones
i coefficiente *m* di diffusione neutronica
n neutrondiffusiecoëfficiënt
d Neutronendiffusionskoeffizient *m*

4699 NEUTRON ECONOMY   np, rt
Degree to which neutrons are used in
desired ways instead of being lost by
leakage or by useless absorptions.
f économie *f* neutronique
e economía *f* neutrónica
i grado *m* d'utilizzazione dei neutroni
n neutroneneconomie
d Neutronenökonomie *f*

4700 NEUTRON-ELECTRON   np
        INTERACTION
The weak electrostatic force existing
between the neutron and the electron.
f interaction *f* neutron-électron
e interacción *f* neutrón-electrón
i interazione *f* neutrone-elettrone
n neutron-elektron-wisselwerking
d Neutron-Elektron-Wechselwirkung *f*

4701 NEUTRON ENERGY   np
The kinetic energy of a neutron, usually
expressed in electron volt units.
f énergie *f* neutronique
e energía *f* neutrónica
i energia *f* neutronica
n neutronenenergie
d Neutronenenergie *f*

4702 NEUTRON ENERGY DISTRIBUTION np
f distribution *f* de l'énergie neutronique,
    distribution *f* énergétique des neutrons
e distribución *f* de la energía neutrónica
i distribuzione *f* dell'energia neutronica
n energieverdeling van neutronen
d Neutronenenergieverteilung *f*

4703 NEUTRON ENERGY GROUP   np
One of a set of groups consisting of
neutrons having energies within arbitrarily
chosen intervals.
f groupe *m* d'énergie de neutrons,
    groupe *m* de neutrons par énergie
e grupo *m* de neutrones por energía
i gruppo *m* di neutroni per energia
n neutronengroep
d Neutronenenergiegruppe *f*

4704 NEUTRON ESCAPE,   np
        NEUTRON LEAKAGE
The undesired escape of neutrons in a
nuclear reactor.
f fuite *f* de neutrons
e fuga *f* de neutrones,
i fuga *f* di neutroni,
n neutronenlek, neutronenverlies
d Neutronenabfluss *m*, Neutronenverlust *m*

4705 NEUTRON EXCESS, see 1782

4706 NEUTRON FISSION-   me
      SCINTILLATION DETECTOR,
      SCINTILLATOR SLOW NEUTRON
      DETECTOR
      A slow neutron detector containing a
      material capable of producing lumines-
      cence when bombarded by fission
      fragments, which material is mixed with
      a uranium compound.
f compteur *m* à scintillation de neutrons
   lents
e contador *m* de centelleo de neutrones
   lentos
i contatore *m* a scintillazione di
   neutroni lenti
n neutronenscintillatieteller
d Neutronenszintillationszähler *m*

4707 NEUTRON FLUX   np
      The product of the neutron density and
      the average speed.
f flux *m* de neutrons, flux *m* neutronique
e flujo *m* de neutrones, flujo *m* neutrónico
i flusso *m* di neutroni, flusso *m* neutronico
n neutronenflux
d Neutronenfluss *m*

4708 NEUTRON FLUX DENSITY, see 2748

4709 NEUTRON FLUX DENSITY   me, np
      INDICATOR
      An indicator designed to give an estimate
      of neutron flux density.
f signaleur *m* de la densité de flux de
   neutrons
e indicador *m* de la densidad de flujo de
   neutrones
i segnalatore *m* della densità di flusso
   neutronico
n indicator voor neutronenfluxdichtheid
d Neutronenfluenzanzeiger *m*

4710 NEUTRON FLUX DENSITY   me, np
      METER
      A measuring assembly for neutron flux
      density.
f fluxmètre *m* de neutrons
e fluxímetro *m* de neutrones
i flussometro *m* neutronico
n meter voor neutronenfluxdichtheid
d Gerät *n* zur Bestimmung der Neutronenfluenz

4711 NEUTRON FLUX DENSITY   np, me
      MONITOR
      A monitor designed to measure and
      respond to neutron flux density.
f moniteur *m* de la densité de flux de neutrons
e monitor *m* de la densidad de flujo de
   neutrones
i monitore *m* della densità di flusso
   neutronico
n monitor voor neutronenfluxdichtheid
d Warngerät *n* für Neutronenfluenz

4712 NEUTRON FLUX DENSITY   ma
      SCANNING ASSEMBLY
      An assembly designed to chart the neutron

flux density of a nuclear reactor core by
determining the distribution of the activity
induced at different points of a suitable
wire or tape exposed at known locations
in the core.
f ensemble *m* d'exploration de la densité
   de flux de neutrons
e conjunto *m* explorador de la densidad del
   flujo de neutrones
i complesso *m* di misura del profilo della
   densità di flusso neutronico
n aftaster voor de neutronenfluxdichtheid
d Registrieranordnung *f* zur Aufnahme der
   Neutronenfluenzverteilung

4713 NEUTRON FLUX DISTRIBUTION,
      see 2749

4714 NEUTRON FORMATION   np
      BY STRIPPING
      If a high energy deuteron strikes a target
      nucleus in such a way as to graze the edge
      of the latter, the proton in the deuteron
      may be stripped.
f formation *f* de neutron par cassure,
   formation *f* de neutron par stripage
e formación *f* de neutrón por despojo de
   fotón
i formazione *f* di neutrone per
   spogliamento di fotone
n neutronvorming door strippen
d Neutronbildung *f* durch Herausreissen
   eines Photons,
   Neutronbildung *f* durch Photonstrippung

4715 NEUTRON GENERATOR,   rt
      NEUTRON PRODUCER
      A term usually applied to a low energy
      accelerator when used for neutron
      production.
f générateur *m* de neutrons
e generador *m* de neutrones
i generatore *m* di neutroni
n neutronenproducent
d Neutronengenerator *m*

4716 NEUTRON HARDENING,   np, sp
      SPECTRUM HARDENING
      Spectral hardening of neutrons.
f durcissement *m* du spectre des neutrons
e endurecimiento *m* del espectro de los
   neutrones
i indurimento *m* dello spettro dei neutroni
n spectrumverharding van neutronen
d Erhöhung *f* der mittleren Neutronenenergie,
   Härtung *f* des Neutronenspektrums

4717 NEUTRON HOUWITZER, see 4682

4718 NEUTRON-INDUCED REACTION   np
f réaction *f* induite par neutrons
e reacción *f* inducida por neutrones
i reazione *f* indotta per neutroni
n door neutronen geïnduceerde reactie
d neutroneninduzierte Reaktion *f*

4719 NEUTRON IRRADIATION   rt
An undesired effect in the upper regions
of a reactor vessel, prevented by
arranging a shield above the core.
f irradiation f par neutrons
e irradiación f por neutrones
i irradiazione f per neutroni
n bestraling door neutronen
d Bestrahlung f durch Neutronen

4720 NEUTRON LEAKAGE, see 3966

4721 NEUTRON LEAKAGE SPECTRUM,
see 3971

4722 NEUTRON LETHARGY, see 3986

4723 NEUTRON LEVEL SPACING, see 3995

4724 NEUTRON LIFETIME   np
f vie f d'une génération neutronique
e duración f de vida de una generación
neutrónica
i tempo m medio d'una generazione
neutronica,
vita f media d'una generazione neutronica
n levensduur van een neutronengeneratie
d Lebensdauer f einer Neutronengeneration

4725 NEUTRON MAGNETIC MOMENT   np
Equal in magnitude to 1.913 Bohr
magnetons.
f moment m magnétique du neutron
e momento m magnético del neutrón
i momento m magnetico del neutrone
n magnetisch moment n van het neutron
d magnetisches Moment n des Neutrons

4726 NEUTRON MOISTURE METER,
see 3695

4727 NEUTRON MULTIPLICATION   np, rt
The process in which a neutron produces
on the average more than one neutron in
a medium containing fissile (fissionable)
material.
f multiplication f de neutrons
e multiplicación f de neutrones
i moltiplicazione f di neutroni
n vermenigvuldiging van neutronen
d Neutronenmultiplikation f,
Vermehrung f von Neutronen

4728 NEUTRON MULTIPLICATION   np
FACTOR
The ratio of the number of neutrons in
any one generation to the number of
corresponding neutrons in the previous
generation.
f facteur m de multiplication de neutrons
e factor m de multiplicación de neutrones '
i fattore m di moltiplicazione di neutroni
n neutronenvermenigvuldigingsfactor
d Neutronenvermehrungsfaktor m

4729 NEUTRON-NEUTRON REACTION   np
A neutron-produced and neutron-producing
reaction in which the bombarding neutron
is captured by the target nucleus, which
emits two neutrons as a result.
f réaction f neutron-neutron
e reacción f neutrón-neutrón
i reazione f neutrone-neutrone
n neutron-neutron-reactie
d Neutron-Neutron-Reaktion f

4730 NEUTRON NUMBER   np
The number of neutrons in a nucleus.
f nombre m de neutrons
e número m de neutrones
i numero m di neutroni
n neutrongetal n
d Neutronenzahl f

4731 NEUTRON NUMBER DENSITY,
see 4693

4732 NEUTRON OPTICS   ms
The study of optical phenomena, as
diffraction, by means of neutron beams.
f optique f neutronique
e óptica f neutrónica
i ottica f neutronica
n neutronenoptiek
d Neutronenoptik f

4733 NEUTRON OR PARTICLE   np
CURRENT DENSITY,
NUCLEAR CURRENT DENSITY
A vector such that its component along
the normal to a surface equals the net
number of particles crossing that surface
in the positive direction per unit area per
unit time.
f densité f de courant de neutrons ou
particules
e densidad f de corriente de neutrones o
partículas
i densità f di corrente di neutroni o
particelle
n stroomdichtheid van neutronen of
deeltjes
d Stromdichte f von Neutronen oder Teilchen

4734 NEUTRON PERIOD,   np
NEUTRON RATE
The time constant of the neutron flux.
f constante f de temps du flux neutronique,
période f neutronique
e constante f de tiempo del flujo neutrónico,
período m neutrónico
i costante f di tempo del flusso neutronico,
tempo m di divergenza del flusso
neutronico
n tijdconstante van de neutronenflux
d Zeitkonstante f des Neutronenflusses

4735 NEUTRON PHYSICS   np
f physique f neutronique
e física f neutrónica
i fisica f neutronica
n neutronenfysica
d Neutronenphysik f

4736 NEUTRON PILE,   me
       SIMPSON PILE
A stack of thermal neutron detectors at the
interior of a producing and slowing down
block of neutrons (lead foils and paraffin).
f pile f à neutrons, pile f de Simpson
e pila f de Simpson
i pila f di Simpson
n simpsonzuil
d Simpson-Säule f

4737 NEUTRON POPULATION   np
The free neutrons present in a given
medium.
f densité f de neutrons libres
e densidad f de neutrones libres
i popolazione f neutronica
n vrije-neutronendichtheid
d freie Neutronendichte f

4738 NEUTRON-PROTON   np, qm
       EXCHANGE FORCES
Forces postulated in an attempt to explain
nuclear forces in terms of an energy
contribution brought about quantum
mechanically as the result of the exchange
of forces between a proton and a neutron.
f forces pl d'échange neutron-proton
e fuerzas pl de intercambio neutrón-protón
i forze pl d'interscambio neutrone-protone
n uitwisselingskrachten pl tussen neutronen
   en protonen
d Austauschkräfte pl zwischen Neutronen
   und Protonen

4739 NEUTRON-PROTON REACTION   np
A nuclear reaction in which a bombarding
neutron is captured by the target nucleus
and the compound nucleus thus formed
emits a proton.
f réaction f neutron-proton
e reacción f neutrón-protón
i reazione f neutrone-protone
n neutron-proton-reactie
d Neutron-Proton-Reaktion f

4740 NEUTRON RADIATION   np, ra
The emission of neutrons.
f rayonnement m neutronique
e radiación f neutrónica
i radiazione f neutronica
n neutronenstraling
d Neutronenstrahlung f

4741 NEUTRON RADIATIVE CAPTURE   np
The capture of a slow neutron by an
atomic nucleus with the prompt emission
of one or more gamma rays, the total
energy of which is equal to the binding
energy of the neutron in the compound
nucleus.
f capture f de neutrons radiative
e captura f de neutrones radiativa
i cattura f di neutroni radiativa
n radiogene neutronenvangst
d Neutron-Gamma-Strahlungseinfang m,
   radiogener Neutroneneinfang m

4742 NEUTRON RADIOGRAPHY   ms
Radiography by means of neutrons.
f radiographie f neutronique
e radiografía f neutrónica
i radiografia f neutronica
n neutronenradiografie
d Neutronenradiographie f

4743 NEUTRON REFLECTION   np, rt
Neutrons may be reflected by crystalline
materials according to the Bragg law for
the de Broglie wavelength characteristic
of their energy, or they may be totally
reflected by highly polished surfaces of
selected materials at angles smaller than
their critical angle.
f réflexion f de neutrons
e reflexión f de neutrones
i riflessione f dei neutroni
n neutronenreflectie
d Neutronenreflexion f

4744 NEUTRON REFLECTOR,   rt
       REFLECTOR
In reactor technology, a part of a reactor
placed adjacent to the core for the purpose
of returning some of the escaping
neutrons back into the core by scattering
collisions.
f réflecteur m
e reflector m
i riflettore m
n reflector
d Reflektor m

4745 NEUTRON REST MASS   np
Equal to $1.67474 \pm 0.000 \times 10^{-24}$ gram.
f masse f de repos du neutron
e masa f en reposo del neutrón
i massa f di riposo del neutrone
n rustmassa van een neutron
d Ruhmasse f eines Neutrons

4746 NEUTRON SCATTERING,   np
       SCATTERING OF NEUTRONS
The change in direction of a neutron owing
to a collision with another particle or a
system.
f diffusion f de neutrons
e dispersión f de neutrones
i deviazione f di neutroni
n neutronenverstrooiing
d Neutronenstreuung f

4747 NEUTRON SOURCE   np
An apparatus or a material emitting or
capable of emitting neutrons.
f source f de neutrons
e fuente f de neutrones
i sorgente f di neutroni
n neutronenbron
d Neutronenquelle f

4748 NEUTRON SPECTROMETER   sp
An instrument designed to determine the
spectrum of fast neutrons.
f spectromètre m neutronique

e espectrómetro *m* neutrónico
i spettrometro *m* neutronico
n neutronenspectrometer
d Neutronenspektrometer *n*

4749 NEUTRON SPECTRUM   np, sp
The energy distribution of neutrons.
f spectre *m* neutronique
e espectro *m* neutrónico
i spettro *m* neutronico
n neutronenspectrum *n*
d Neutronenspektrum *n*

4750 NEUTRON SPEED   np
The magnitude of the neutron velocity.
f vitesse *f* de neutrons
e velocidad *f* de neutrones
i velocità *f* di neutroni
n neutronensnelheid
d Neutronengeschwindigkeit *f*

4751 NEUTRON TEMPERATURE   np
The energy possessed by neutrons in
thermal equilibrium with their surroundings
may be expressed in terms of temperature
by means of the relation $E = 3 K^{T/2}$,
where E is neutron energy, K is
Boltzmann's constant and·T is the neutron
temperature on the Kelvin scale.
f température *f* neutronique
e temperatura *f* neutrónica
i temperatura *f* neutronica
n neutronentemperatuur
d Neutronentemperatur *f*

4752 NEUTRON THERAPY   md
Irradiation with neutrons for therapeutic
purposes.
f neutronthérapie *f*
e neutronoterapia *f*
i neutronoterapia *f*
n neutronentherapie
d Neutronentherapie *f*

4753 NEUTRON THERMOPILE   ra
A neutron detector in which the hot
junctions of thermocouples are in thermal
contact with a material which heats under
the effect of neutron absorption.
f thermopile *f* à neutrons
e pila *f* termoneutrónica,
   termopila *f* para neutrones
i termopila *f* per neutroni
n thermozuil voor neutronen
d Thermoelementsäule *f* zum Neutronennachweis,
   Thermosäule *f* für Neutronen

4754 NEUTRON TIME OF FLIGHT   np
The time T taken for a neutron to travel
a given distance D, expressed in micro-
seconds $T = \frac{0.0723D \text{ (metre(er))}}{\sqrt{E(\text{MeV})}}$ where E
is the neutron energy.
f temps *m* de vol du neutron
e tiempo *m* de vuelo del neutrón
i tempo *m* di volo del neutrone

n looptijd van een neutron
d Neutronenflugzeit *f*

4755 NEUTRON VELOCITY   me
SELECTOR
Instrument in which neutrons of a
particular velocity or range of velocities
are selected for detection.
f sélecteur *m* de vitesse des neutrons
e selector *m* de velocidad de los neutrones
i selettore *m* di velocità dei neutroni
n snelheidsselector voor neutronen
d Geschwindigkeitsselektor *m* für Neutronen,
   Neutronenmonochromator *m*

4756 NEUTRON WAVELENGTH   ma, np
The wavelength of a de Broglie wave
associated with a neutron.
f longueur *f* d'onde du neutron
e longitud *f* de onda del neutrón
i lunghezza *f* d'onda del neutrone
n neutronengolflengte
d Neutronenwellenlänge *f*

4757 NEUTRON YIELD PER ABSORPTION,
see 2371

4758 NEUTRON YIELD PER FISSION, np
NU FACTOR
The average number of primary fission
neutrons, including delayed neutrons,
emitted per fission.
f facteur *m* nu
e factor *m* nu
i fattore *m* nu,
   resa *f* neutronica per fissione
n neutronenopbrengst per splijting,
   nu-factor
d Neutronenausbeute *f* je Spaltung,
   Nu-Faktor *m*

4759 NEUTRONGRAPHY   ra
Radiography by means of neutron beams.
f neutrongraphie *f*
e neutronografía *f*
i neutronografia *f*
n neutronografie
d Neutronographie *f*

4760 NEUTRONTIGHT   rt, sa
Said of a material when no neutrons can
escape through it or be absorbed by it.
f étanche aux neutrons
e a prueba de fuga de neutrones,
   estanco a los neutrones
i a tenuta di neutroni,
   garantito contro le fughe di neutroni
n neutronendicht adj
d neutronendicht adj

4761 NEUTROPHIL GRANULE   md
A granule accepting or being stained by
neutral dyes.
f granule *m* neutrophile
e gránulo *m* neutrófilo
i granulo *m* neutrofilo
n neutrofiel korreltje *n*
d neutrophiles Körnchen *n*

4762 NEWTONIAN MECHANICS, see 1114

4763 NICKEL   ch
Metallic element, symbol Ni, atomic
number 28.
f nickel *m*
e níquel *m*
i nichel *m*, nichelio *m*
n nikkel *n*
d Nickel *n*

4764 NICOLAYITE,   mi
      THORÓGUMMITE
A thorium silicate ore containing
2.5 - 31.4 % of U and 18.2 - 50.8 % of Th.
f nicolayite *f*, thorogummite *f*
e nicolaita *f*, torogumita *f*
i nicolaite *f*, torogummite *f*
n nicolayiet *n*, thorogummiet *n*
d Nicolayit *m*, Thorogummit *m*

4765 NINE-TENTH PERIOD   np
In radioactivity that period of time that
0.1 (or 10 %) of the radioactive atoms
have decayed.
f période *f* de neuf dixièmes
e período *m* de nueve décimos
i periodo *m* di nove decimi
n negen-tiende periode
d Neunzehntel-Periode *f*

4766 NIOBITE, see 1240

4767 NIOBIUM   ch
Metallic element, symbol Nb, atomic
number 41.
f niobium *m*
e niobio *m*
i niobio *m*
n niobium *n*
d Niob *n*

4768 NITROGEN   ch
Gaseous element, symbol N, atomic
number 7.
f azote *m*, nitrogène *m*
e nitrógeno *m*
i azoto *m*
n stikstof
d Stickstoff *m*

4769 NIVENITE   mi
A mineral containing small quantities
of uranium; a variety of uraninite.
f nivénite *f*
e nivenita *f*
i nivenite *f*
n niveniet *n*
d Nivenit *m*

4770 NO-BOND RESISTANCE, see 3364

4771 NOBELIUM   ch
The tenth synthetic element obtained
by bombarding curium with ions
accelerated by radioactive carbon, symbol
No, atomic number 102.

f nobélium *m*
e nobelio *m*
i nobelio *m*
n nobelium *n*
d Nobelium *n*

4772 NOBLE GASES, see 3467

4773 NOHLITE   mi
Yttrium and other rare earths containing
mineral and about 13.0 % of U.
f nohlite *f*
e nolita *f*
i nolite *f*
n nohliet *r*
d Nohlit *m*

4774 NOISE   ec
Any unwanted disturbance in a system.
f bruit *m*
e ruido *m*
i rumore *m*
n ruis
d Rauschen *n*

4775 NOMINAL ACTIVITY   ra
The activity at which a radiation source
is manufactured for stock.
f activité *f* nominale
e actividad *f* nominal
i attività *f* nominale
n nominale activiteit
d Nennaktivität *f*

4776 NOMINAL LOAD, see 2886

4777 NOMINAL NUCLEAR WEAPON   nw
A nuclear weapon yielding energies of the
order of 10,000 to 20,000 tons of TNT.
f arme *f* nucléaire nominale
e arma *f* nuclear nominal
i arma *f* nucleare nominale
n nominaal kernwapen *n*
d nominelle Kernwaffe *f*

4778 NON-AQUEOUS REPROCESSING,
      see 1980

4779 NON-CENTRAL FORCE   np
A nuclear force depending on other
co-ordinates than the distance of the point
where it effects its influence at another
point of the considered system.
f force *f* non-centrale
e fuerza *f* no central
i forza *f* non centrale
n niet-centrale kracht
d nichtzentrale Kraft *f*

4780 NON-COHERENT, see 3430

4781 NON-CROSSING RULE   np
For an infinitely slow change of inter-
nuclear distance, two electronic states of
the same species cannot cross each other.
f loi *f* de non-entrecroisement
e ley *f* de imposibilidad de cruzamiento

i legge *f* di non incrociamento
n wet *f* van de niet-overkruising
d Gesetz *n* der Nichtüberkreuzung

4782 NON-DEGENERATE GAS   np
Gas formed by a system of particles the
concentration of which is sufficiently low
for the Maxwell-Boltzmann law to apply.
f gaz *m* non-dégénéré
e gas *m* no degenerado
i gas *m* non degenere
n niet-ontaard gas *n*
d nichtentartetes Gas *n*

4783 NON-DESTRUCTIVE   an, ms
ESTIMATION OF HYDROGEN
Carried out by using a plutonium fission
counter to detect neutrons slowed down
by collisions with hydrogen nuclei.
f détermination *f* non-destructive d'hydrogène
e determinación *f* no destructiva de
hidrógeno
i determinazione *f* non distruttiva
d'idrogeno
n niet-destructieve bepaling van waterstof
d zerstörungsfreie Bestimmung *f* von
Wasserstoff

4784 NON-ELASTIC   cs
CROSS SECTION,
NON-ELASTIC INTERACTION
CROSS SECTION
The difference between the total cross
section and the elastic scattering cross
section.
f section *f* efficace non-élastique
d'interaction
e sección *f* eficaz no elástica de interacción
i sezione *f* d'urto anelastica d'interazione
n niet-elastische doorsnede
d Wirkungsquerschnitt *m* für nicht-
elastische Wechselwirkung

4785 NON-EQUILIBRIUM   ma
DIFFUSION EQUATION
The diffusion equation for thermal
neutrons modified to include a source
term which describes the increase in
thermal neutron density due to the
slowing down of fission neutrons.
f équation *f* de diffusion non-équilibrée
e ecuación *f* de difusión no equilibrada
i equazione *f* di diffusione non equilibrata
n onstabiele diffusievergelijking
d instabile Diffusionsgleichung *f*

4786 NON-ESCAPE PROBABILITY,   np
NON-LEAKING PROBABILITY
A measure of the probability that neutrons
will not leak from a finite reactor system,
but will remain in the system until they
are absorbed.
f probabilité *f* de non-fuite,
probabilité *f* de permanence
e probabilidad *f* de permanencia
i probabilità *f* di permanenza

n behoudsfactor, behoudskans
d Erhaltungswarscheinlichkeit *f*,
Leckfaktor *m*,
Verbleibwahrscheinlichkeit *f*

4787 NON-FISSION CAPTURE   np
Term usually applied to the capture of
neutrons by uranium nuclei which does not
result in a nuclear fission of the capturing
nucleus.
f capture *f* sans fission, capture *f* stérile
e captura *f* sin fisión
i cattura *f* senza fissione
n splijtingsloze vangst
d spaltungsloser Einfang *m*

4788 NON-IONIZED ATOM, see 4655

4789 NON-ISOTOPIC CARRIER   is
A carrier in which the added substance
is a different element from the tracer.
f porteur *m* non-isotopique
e portador *m* no isotópico
i portatore *m* non isotopico
n niet-isotopische drager
d nichtisotopischer Träger *m*

4790 NON-LINEARITY   ge, me, ra
The deviation of any functional relationship
from direct proportionality, e.g. in
radiation instruments.
f non-linéarité *f*
e falta *f* de linealidad, no linealidad *f*
i nonlinearità *f*
n niet-lineariteit, niet-rechtlijnig verband *n*
d Nichtlinearität *f*

4791 NON-QUANTIZED SYSTEM, see 1116

4792 NON-RADIOACTIVE ISOTOPE,   is
STABLE ISOTOPE
f isotope *m* non-radioactif
e isótopo *m* no radiactivo
i isotopo *m* non radioattivo
n niet-radioactief isotoop *n*
d nichtradioaktives Isotop *n*

4793 NON-SCREEN FILM   xr
Film with the photographic emulsion
specially designed to be sensitive to
X-rays rather than visible light.
f film *m* sans écran
e película *f* sin pantalla
i pellicola *f* senza schermo
n film zonder scherm
d folienloser Film *m*

4794 NON-SELF-MAINTAINED DISCHARGE,
see 2589

4795 NON-SPILL PIPE COUPLING   cd
A coupling which safeguards against
spillage of radioactive liquids.
f joint *m* de tuyaux hermétique aux liquides
e manguito *m* para tubos sin derramiento de
líquido
i manicotto *m* di tuberia senza perdita di
liquido

n   verliesvrije koppeling
d   verlustfreie Kupplung *f*

4796 NORMAL ATOM   np
An atom which has no overall electric
charge and in which all the electrons
surrounding the nucleus are at their
lowest energy levels.
f   atome *m* normal
e   átomo *m* normal
i   atomo *m* normale
n   normaal atoom *n*
d   normales Atom *n*

4797 NORMAL BAND   np
The lowest band of the energy diagram for
a given substance corresponding to the
normal state in the absence of any energy
supplied by an external source.
f   bande *f* normale
e   banda *f* normal
i   banda *f* normale
n   normale band
d   Normalband *n*

4798 NORMAL ENERGY LEVEL,
NORMAL STATE, see 3090

4799 NORMAL GLOW DISCHARGE   np
The glow discharge characterized by the
fact that the working voltage decreases or
remains constant as the current increases.
f   décharge *f* luminescente normale
e   descarga *f* luminiscente normal
i   scarica *f* luminescente normale
n   normale glimontlading
d   normale Glimmentladung *f*

4800 NORMAL SHUT-DOWN   sm
SAFETY ASSEMBLY
A safety assembly which performs the
nuclear reactor shut-down by means of
reversible mechanism enabling the
insertion in the reactor, in due time and
at correct speed, of the necessary
negative reactivity.
f   ensemble *m* de sécurité normale d'arrêt
e   conjunto *m* de seguridad de paro
i   complesso *m* di sicurezza normale
    d'arresto
n   veiligheidsopstelling voor een normale
    stop
d   Normalschnellschlusseinrichtung *f*

4801 NORMAL VALENCE   ch
The valence that an element exhibits in the
majority of its compounds.
f   valence *f* normale
e   valencia *f* normal
i   valenza *f* normale
n   normale valentie
d   normale Valenz *f*

4802 NORMALIZED PLATEAU SLOPE   ct
A figure of merit for a counter tube, the
percentage change in counting rate divided
by the percentage change in voltage using
the threshold values as a base.

f   pente *f* normalisée de plateau
e   pendiente *f* normalizada de meseta
i   pendenza *f* normalizzata di ripiano
n   genormaliseerde plateauhelling
d   normalisierter Plateauanstieg *m*

4803 NORTH-SOUTH EFFECT   cr
An effect due to the influence of the earth's
magnetic field on the paths of the cosmic
particles.
f   effect *m* nord-sud
e   efecto *m* norte-sur
i   effetto *m* nord-sud
n   noord-zuid-effect *n*
d   Nord-Süd-Effekt *m*

4804 NOTCH   mg, te
The localized reduction in area of a
ductile tension test specimen in the region
of fracture.
f   entaille *m*
e   entalla *f*
i   intaccatura *f*, intaglio *m*
n   inkerving, insnoering, kerf
d   Kerbe *f*

4805 NOTCHED SPECIMEN,   te
NOTCHED TEST BAR
A piece of metal used in a Charpy test.
f   éprouvette *f* entaillée
e   probeta *f* entallada
i   provetta *f* intagliata
n   gekerfde proefstaaf
d   gekerbter Probestab *m*

4806 NOVACEKITE   mi
A rare secondary mineral containing
about 52 % of U.
f   novacékite *f*
e   novacequita *f*
i   novacechite *f*
n   novacekiet *n*
d   Novacekit *m*

4807 NOZZLE JET SEPARATION, see 3818

4808 NU FACTOR, see 4758

4809 NUCLEAR ABSORPTION   np
Absorption of energy by the nucleus of an
atom.
f   absorption *f* nucléaire
e   absorción *f* nuclear
i   assorbimento *m* nucleare
n   kernabsorptie
d   Kernabsorption *f*

4810 NUCLEAR ACTIVITY, see 100

4811 NUCLEAR ACTIVITY CONCENTRATION,
see 103

4812 NUCLEAR ANGULAR PRECESSION
FREQUENCY, see 1596

4813 NUCLEAR ASH,   fu
SPENT FUEL
Reactor fuel material which has lost its

useful content of fissile (fissionable)
material.
f combustible *m* épuisé
e combustible *m* agotado
i combustibile *m* esaurito
n verbruikte splijtstof
d ausgebrannter Brennstoff *m*

4814 NUCLEAR ATOM,   np
        STRIPPED ATOM
An atomic nucleus without surrounding
electrons.
f atome *m* dépouillé, atome *m* nucléaire
e átomo *m* despojado, átomo *m* nuclear
i atomo *m* spogliato, atomo *m* nucleare
n naakt atoom *n*, sterk geïoniseerd atoom *n*
d elektronenberaubtes Atom *n*,
   hochionisiertes Atom *n*, nacktes Atom *n*

4815 NUCLEAR ATTRACTION   np
The force active in the atomic nucleus,
which overcomes the neutral repulsion of
the positive charges of the proton.
f attraction *f* intranucléaire
e atracción *f* intranuclear
i attrazione *f* intranucleare
n intranucleaire aantrekking, kernaantrekking
d intranukleare Anziehung *f*,
   Kernanziehung *f*

4816 NUCLEAR BARRIER, see 541

4817 NUCLEAR BATTERY, see 367

4818 NUCLEAR BINDING ENERGY   np
The energy that would be required to
separate an atom number Z and mass
number A into Z hydrogen atoms and
A–Z neutrons.
f énergie *f* de liaison nucléaire
e energía *f* de enlace nuclear
i energia *f* di legame nucleare
n kernbindingsenergie
d Kernbindungsenergie *f*

4819 NUCLEAR BOHR MAGNETON,   np
        NUCLEAR MAGNETON
Given by $\mu_1 = eh/4\pi M_o c = \mu 1.836$
erg-gauss^{-2}, where M_o is the rest mass
of the proton.
f magnéton *m* nucléaire
e magnetón *m* nuclear
i magnetone *m* nucleare
n kernmagneton *n*
d Kernmagneton *n*

4820 NUCLEAR BOMB, see 1

4821 NUCLEAR BOMBARDMENT   np
Any process in which sub-atomic particles
are directed atomic nuclei at high
velocity.
f bombardement *m* nucléaire
e bombardeo *m* nuclear
i bombardamento *m* nucleare
n kernbeschieting
d Kernbeschiessung *f*, Kernbeschuss *m*

4822 NUCLEAR BREEDER, see 779

4823 NUCLEAR CASCADE   np
A sequence of nuclear reactions which end
after a certain time.
f cascade *f* nucléaire
e cascada *f* nuclear
i cascata *f* nucleare
n kerncascade
d Kernkaskade *f*

4824 NUCLEAR CHAIN REACTION, see 981

4825 NUCLEAR CHARGE   np
Of a nuclide, the electric charge carried
by each nucleus.
f charge *f* nucléaire
e carga *f* nuclear
i carica *f* nucleare
n kernlading
d Kernladung *f*

4826 NUCLEAR CHEMISTRY   ch
The study of nuclei and nuclear reactions,
using methods and concepts analogous to
those of ordinary chemistry.
f chimie *f* nucléaire
e química *f* nuclear
i chimica *f* nucleare
n kernchemie
d Kernchemie *f*

4827 NUCLEAR COLLISION   cr
One of the collision forms when a cosmic-
ray primary enters the atmosphere and
collides with a molecule of air.
f collision *f* nucléaire
e colisión *f* nuclear
i collisione *f* nucleare
n kernbotsing
d Kernstoss *m*

4828 NUCLEAR CONSTANTS   ma, np
The constants peculiar and belonging to
nuclear science.
f constantes *pl* nucléaires
e constantes *pl* nucleares
i costanti *pl* nucleari
n kernconstanten *pl*
d Kernkonstanten *pl*

4829 NUCLEAR CONSTITUTION,   np
        NUCLEAR STRUCTURE
The internal structure of the nucleus.
f structure *f* nucléaire
e estructura *f* nuclear
i struttura *f* nucleare
n kernbouw, kernstructuur
d Aufbau *m* des Atomkerns, Kernbau *m*,
   Kernstruktur *f*

4830 NUCLEAR CROSS SECTION   cs
The cross section of an atomic nucleus for
a particular process.
f section *f* efficace nucléaire
e sección *f* eficaz nuclear
i sezione *f* d'urto nucleare

n   werkzame kerndoorsnede
d   Wirkungsquerschnitt *m* des Kerns

4831 NUCLEAR CURRENT DENSITY,
      see 4733

4832 NUCLEAR DELAY   rt
      In reactors, the time between the
      establishment òf an undesired nuclear
      condition and the start of the corrective
      motion of a control rod.
f   temps *m* de mise en marche de la barre
    de commande
e   tiempo *m* de arranque de la barra de
    regulación
i   tempo *m* d'avviamento della barra di
    regolazione
n   aanlooptijd van de regelstaaf
d   Anlaufzeit *f* des Steuerstabs

4833 NUCLEAR DEPTH CHARGE   nw
      A nuclear bomb which can be set to
      detonate at a given depth below the
      surface of the sea.
f   grenade *f* sous-marine nucléaire
e   carga *f* de profundidad nuclear
i   bomba *f* nucleare di profondità
n   dieptekernbom
d   Wasserkernbombe *f*

4834 NUCLEAR DIAMETER   ma, np
      The diameter of a nucleon.
f   diamètre *m* nucléaire
e   diámetro *m* nuclear
i   diametro *m* nucleare
n   kerndiameter
d   Kerndurchmesser *m*

4835 NUCLEAR DISINTEGRATION   np
      Transformation of a nucleus, possibly a
      compound nucleus, involving a splitting into
      more nuclei or the emission of particles
      or photons.
f   désintégration *f* nucléaire
e   desintegración *f* nuclear
i   disintegrazione *f* nucleare
n   kerndesintegratie
d   Kernumwandlung *f*

4836 NUCLEAR DISINTEGRATION   np
      ENERGY,
      Q VALUE
      The energy evolved, or the negative of the
      energy absorbed, in a nuclear disintegra-
      tion.
f   énergie *f* de désintégration nucléaire
e   energía *f* de desintegración nuclear
i   energia *f* di disintegrazione nucleare
n   desintegratie-energie, kernvervalenergie
d   Kernzerfallsenergie *f*

4837 NUCLEAR DISPERSION   np
      RELATIONS
      In many nuclear reactions the cross
      section for a given effect shows a
      dependence on the wavelength of the
      incident wave that is similar to classical

dispersion formula, with maxima occurring
due to resonance.
f   relations *pl* nucléaires de dispersion
e   relaciones *pl* nucleares de dispersión
i   relazioni *pl* nucleari di dispersione
n   kernbetrekkingen *pl* voor dispersie
d   Kernbeziehungen *pl* für Dispersion

4838 NUCLEAR ELECTRON   np
      An electron which is emitted from the
      nucleus of an atom.
f   électron *m* nucléaire
e   electrón *m* nuclear
i   elettrone *m* nucleare
n   kernelektron *n*
d   Kernelektron *n*

4839 NUCLEAR EMULSION,   ph
      A photographic emulsion specially
      prepared to record the tracks of nuclear
      fragments and other fast particles passing
      through it.
f   émulsion *f* nucléaire
e   emulsión *f* nuclear
i   emulsione *f* nucleare
n   fotografische kernemulsie
d   kernphysikalische Emulsion *f*

4840 NUCLEAR ENERGY, see 381

4841 NUCLEAR ENERGY LEVELS   np
      Difference in the energy of atomic nuclei.
f   niveaux *pl* énergétiques du noyau
e   niveles *pl* energéticos del núcleo
i   livelli *pl* energetici del nucleo
n   kernenergieniveaus *pl*
d   Kernniveaus *pl*, Kernterme *pl*

4842 NUCLEAR ENGINEERING   ge, ms
      Research and development associated
      with construction and practical use of
      nuclear reactors and their products.
f   génie *f* nucléaire, technique *f* nucléaire
e   ingeniería *f* nuclear, técnica *f* nuclear
i   ingegneria *f* nucleare
n   kerntechniek, toegepaste kernwetenschap
d   Kerntechnik *f*, Kernverfahrenstechnik *f*

4843 NUCLEAR EQUATION,   rt
      NUCLEAR REACTION EQUATION,
      REACTION FORMULA
      An equation showing the changes in
      composition of an atomic nucleus during
      a nuclear reaction.
f   formule *f* de la réaction nucléaire
e   fórmula *f* de la reacción nuclear
i   formula *f* della reazione nucleare
n   kernreactieformule, kernreactievergelijking
d   Kernreaktionsformel *f*,
    Kernreaktionsgleichung *f*

4844 NUCLEAR ERA, see 363

4845 NUCLEAR EVAPORATION   np
      Displacements of neutrons due to thermal
      agitation as postulated in the interior of a
      static nuclear model according to Bohr or
      Frenkel.

f   évaporation *f* du noyau
e   evaporación *f* del núcleo
i   evaporazione *f* del nucleo
n   kernverdamping
d   Kernverdampfung *f*

4846 NUCLEAR FIELD   np
The effect of the total of intranuclear
forces.
f   champ *m* nucléaire
e   campo *m* nuclear
i   campo *m* nucleare
n   kernveld *n*
d   Kernfeld *n*

4847 NUCLEAR FISSION   np
The division of a heavy nucleus into two
or more parts with masses of equal order
of magnitude, usually accompanied by the
emission of neutrons, gamma rays and,
rarely, small charged nuclear fragments.
f   fission *f* nucléaire
e   fisión *f* nuclear
i   fissione *f* nucleare
n   kernsplijting
d   Kernspaltung *f*

4848 NUCLEAR FLUID   np
Colloquial term for the hypothetical binding
mass of a nucleus.
f   fluide *m* de liaison nucléaire
e   flúido *m* de enlace nuclear
i   fluido *m* di legame nucleare
n   kernbindingsfluïde *n*
d   Kernbindungsfluid *n*

4849 NUCLEAR FORCES   np
The forces acting between two nucleons
at close quarters, over and above their
electromagnetic interaction.
f   forces *pl* nucléaires
e   fuerzas *pl* nucleares
i   forze *pl* nucleari
n   kernkrachten *pl*
d   Kernkräfte *pl*

4850 NUCLEAR FUEL, see 2848

4851 NUCLEAR FUSION   np, rt
The process in which nuclei undergo
nuclear fusion reactions.
f   fusion *f* nucléaire
e   fusión *f* nuclear
i   fusione *f* nucleare
n   kernfusie, kernversmelting
d   Kernfusion *f*, Kernverschmelzung *f*

4852 NUCLEAR FUSION REACTION, see 2903

4853 NUCLEAR GRADE,   np
    NUCLEAR QUALITY
Said of a material which, taking into
account its chemical and physical properties, can be used for nuclear purposes.
f   qualité *f* nucléaire
e   grado *m* nuclear
i   grado *m* nucleare

n   bruikbaarheid voor kernreacties
d   Brauchbarkeit *f* für Kernreaktionen

4854 NUCLEAR GRAPHITE   mt
A graphite which, taking into account its
chemical and physical properties, can be
used for nuclear purposes.
f   graphite *m* nucléaire
e   grafito *m* nuclear
i   grafite *f* di purezza nucleare
n   voor kernreacties geschikt grafiet *n*
d   für Kernreaktionen geeigneter Graphit *m*

4855 NUCLEAR GYROMAGNETIC   np
    RATIO
The ratio of the magnetic moment of the
nucleus, μ, to the nuclear angular
momentum quantum number I.
f   relation *f* gyromagnétique nucléaire
e   relación *f* giromagnética nuclear
i   rapporto *m* giromagnetico nucleare
n   gyromagnetische verhouding van de kern
d   gyromagnetisches Verhältnis *n* des Kerns

4856 NUCLEAR HAZARD   sa
The possibility of damaging effects by
nuclear reactions.
f   danger *m* nucléaire
e   riesgo *m* nuclear
i   rischio *m* nucleare
n   kernreactiegevaar *n*
d   Kernreaktionsgefahr *f*

4857 NUCLEAR HEAT   rt
Heat derived from nuclear reactions.
f   chaleur *f* de réaction nucléaire
e   calor *m* de reacción nuclear
i   calore *m* di reazione nucleare
n   kernreactiewarmte
d   Kernreaktionswärme *f*

4858 NUCLEAR IMPORTANCE FUNCTION,
    see 3411

4859 NUCLEAR INDUCTION   np
The radio frequency signal induced in a
coil surrounding a specimen in a magnetic
field if the spins of the nuclei in the
specimen are disturbed so as to make them
precess.
f   induction *f* nucléaire
e   inducción *f* nuclear
i   induzione *f* nucleare
n   kerninductie
d   Kerninduktion *f*

4860 NUCLEAR INTERACTION   np
The system of exchange forces in the
nucleus which keep it together.
f   action *f* mutuelle nucléaire,
    interaction *f* nucléaire
e   interacción *f* nuclear
i   interazione *f* nucleare
n   kernwisselwerking
d   Kernwechselwirkung *f*,
    Wechselwirkung *f* zwischen Kernen

4861 NUCLEAR ISOBARS, see 3729

4862 NUCLEAR ISOMERISM   np
The occurrence of nuclear isomers.
f isomérie f nucléaire
e isomería f nuclear,
isomerismo m nuclear
i isomeria f nucleare
n kernisomerie
d Kernisomerie f

4863 NUCLEAR ISOMERS, see 3742

4864 NUCLEAR LEVEL   np
One of the energy values at which a
nucleus can exist for an appreciable time.
f niveau m nucléaire
e nivel m nuclear
i livello m nucleare
n kernniveau n
d Kernniveau n

4865 NUCLEAR LEVEL DENSITY, see 3993

4866 NUCLEAR LIMITATIONS   an
Limitations in activation analysis due to
self-shielding, resonance capture and
fast neutron and gamma ray reactions.
f limitations pl dues à la physique nucléaire
e limitaciones pl debidas a la física nuclear
i limitazione pl dovute alla fisica nucleare
n begrenzingen pl van kernfysische aard
d Beschränkungen pl kernphysikalischer Art

4867 NUCLEAR MAGNETIC ALIGNMENT np
Evidence of nuclear magnetic alignment
has been obtained by studying radioactive
emanation from atoms such as Co-60 and
Co-58 at very low temperatures.
f alignement m magnétique nucléaire
e alineación f magnética nuclear
i allineamento m magnetico nucleare
n magnetische kernuitrichting
d magnetische Kernausrichtung f

4868 NUCLEAR MAGNETIC MOMENT   np
The magnetic moment of an electrically
charged particle possessing angular
momentum.
f moment m magnétique nucléaire
e momento m magnético nuclear
i momento m magnetico nucleare
n magnetisch kernmoment n
d magnetisches Kernmoment n,
magnetisches Moment n des Atomkerns

4869 NUCLEAR MAGNETIC   np
RESONANCE
Resonance encountered in energy transfers
between an r-f magnetic field and a
nucleus placed in a constant magnetic field
that is sufficiently strong to decouple the
nucleus from its orbital electrons.
f résonance f magnétique nucléaire
e resonancia f magnética nuclear
i risonanza f magnetica nucleare
n kernmagnetische resonantie
d kernmagnetische Resonanz f

4870 NUCLEAR MAGNETIC   sp
RESONANCE SPECTROSCOPY,
RADIOFREQUENCY SPECTROSCOPY
Measurement of the absorption or
emission, by atomic or molecular systems,
of electromagnetic radiation in the
frequency range of 1 Mc/s to 500 Mc/s.
f spectroscopie f de la résonance
magnétique nucléaire
e espectroscopia f de la resonancia
magnética nuclear
i spettroscopia f della risonanza
magnetica nucleare
n hoogfrequentiekernspectroscopie
d Hochfrequenzkernspektroskopie f

4871 NUCLEAR MAGNETON, see 4819

4872 NUCLEAR MASS   np
Approximately the sum of the masses of
the protons and neutrons composing the
nucleus.
f masse f nucléaire
e masa f nuclear
i massa f nucleare
n kernmassa
d Kernmasse f

4873 NUCLEAR MATTER   np
Nucleons packed together as densely as
they are in nuclei.
f matière f nucléaire
e materia f nuclear
i materia f nucleare
n kernmaterie
d Kernmaterie f

4874 NUCLEAR MEDICINE   md, ms
The branch of medicine concerned with the
problems resulting from the use of nuclear
energy and nuclear weapons.
f médecine f nucléaire
e medicina f nuclear
i medicina f nucleare
n kerngeneeskunde
d Kernmedizin f, Nuklearmedizin f

4875 NUCLEAR MODEL   np
A description of the atomic nucleus based
on simplifying assumptions, designed to
explain some of the properties of nuclei.
f modèle m nucléaire
e modelo m nuclear
i modello m nucleare
n kernmodel n
d Kernmodell n

4876 NUCLEAR PACKING   np
The concentration of particles within the
nucleus of an atom.
f concentration f des particules
e concentración f de las partículas
i concentrazione f delle particelle
n kerndeeltjesopstelling
d Kernteilchenanordnung f

4877 NUCLEAR PARAMAGNETIC RESONANCE — np

The precession of nuclear spins in a magnetic field under perturbation by an alternating magnetic field of sharply defined frequency.

f résonance *f* paramagnétique nucléaire
e resonancia *f* paramagnética nuclear
i risonanza *f* paramagnetica nucleare
n kernparamagnetische resonantie
d kernparamagnetische Resonanz *f*

4878 NUCLEAR PARAMAGNETISM — np

Paramagnetism associated with nuclear magnetic moments.

f paramagnétisme *m* nucléaire
e paramagnetismo *m* nuclear
i paramagnetismo *m* nucleare
n kernparamagnetisme *n*
d Kernparamagnetismus *m*

4879 NUCLEAR PARTICLE — np

A particle believed to exist as such in the nucleus of atoms or of certain atoms.

f particule *f* nucléaire
e partícula *f* nuclear
i particella *f* nucleare
n kerndeeltje *n*
d Kernteilchen *n*

4880 NUCLEAR PERIODICITY — np

Expression used to describe the systematics of the nuclear build-up.

f périodicité *f* nucléaire
e periodicidad *f* nuclear
i periodicità *f* nucleare
n kernperiodiciteit
d Kernperiodizität *f*

4881 NUCLEAR PHYSICS — np

The science of the atomic nuclei, sub-atomic particles and nuclear reactions.

f physique *f* nucléaire
e física *f* nuclear
i fisica *f* nucleare
n kernfysica
d Kernphysik *f*

4882 NUCLEAR PLATE — ph

The glass plate supporting the nuclear emulsion.

f plaque *f* nucléaire
e placa *f* nuclear
i lastra *f* nucleare
n kernemulsieplaat
d Kernemulsionsplatte *f*, Kernspurplatte *f*

4883 NUCLEAR POISON — np

A substance which, because of its high neutron absorption cross section, can reduce reactivity.

f poison *m* nucléaire
e veneno *m* nuclear
i veleno *m* nucleare
n gif *n*, kerntechnisch gif *n*
d Neutronengift *n*, Reaktorgift *n*

4884 NUCLEAR POLARIZATION — np

Alignment of the spin magnetic moments of atomic nuclei in the same direction, giving a net macroscopic magnetic moment.

f polarisation *f* nucléaire
e polarización *f* nuclear
i polarizzazione *f* nucleare
n kernpolarisatie
d Kernpolarisation *f*

4885 NUCLEAR POTENTIAL — np

The potential energy of some specified particle as a function of its position near a nucleus.

f potentiel *m* nucléaire
e potencial *m* nuclear
i potenziale *m* nucleare
n kernpotentiaal
d Kernpotential *n*

4886 NUCLEAR POTENTIAL ENERGY — np

The average total potential energy of all of the nucleons in a nucleus due to the specifically nuclear forces between them, but excluding the electrostatic potential energy.

f énergie *f* potentielle nucléaire
e energía *f* potencial nuclear
i energia *f* potenziale nucleare
n potentiële energie van de atoomkern
d potentielle Energie *f* des Atomkerns

4887 NUCLEAR POWER, see 405

4888 NUCLEAR POWER PLANT — ms, rt

A plant in which nuclear energy is used for generating power.

f centrale *f* à énergie nucléaire, installation *f* à énergie nucléaire
e central *f* de energía nuclear
i impianto *m* nucleotermoelettrico
n kernenergiecentrale
d Kernenergieanlage *f*, Leistungsreaktoranlage *f*

4889 NUCLEAR-POWER SUBMARINE — ms

f sous-marin *m* à propulsion nucléaire
e submarino *m* de propulsión nuclear
i sottomarino *m* a propulsione nucleare
n onderzeeboot met kernaandrijving
d Reaktor-U-Boot *n*, Unterseeboot *n* mit Kernreaktorantrieb

4890 NUCLEAR POWERED DESALINATION PLANT, NUCLEAR POWERED DESALTING PLANT — ms

f installation *f* de dessalage alimentée par énergie nucléaire
e instalación *f* de desalación alimentada por energía nuclear
i impianto *m* di desalinizzazione alimentato per energia nucleare
n met kernenergie gevoede ontzoutings-installatie
d mit Kernenergie gespeiste Entsalzungs-anlage *f*

4891 NUCLEAR POWERED ROCKET, nw
NUCLEAR ROCKET
A rocket propelled by the reaction to
released nuclear energy.
f fusée *f* nucléaire
e cohete *m* de propulsión nuclear
i razzo *m* a propulsione nucleare
n kernraket, raket met kernaandrijving
d Kernrakete *f*,
Rakete *f* mit nuklearem Treibstoff

4892 NUCLEAR PROJECTILE nw
A projectile, the warhead of which may
contain a nuclear reactor.
f projectile *m* nucléaire
e proyectil *m* nuclear
i proiettile *m* nucleare
n kernprojectiel *n*, tactisch kernwapen *n*
d Kernprojektil *n*, nukleares Geschoss *n*,
taktische Kernwaffe *f*

4893 NUCLEAR PROPERTIES np
The properties of particles resulting
from a nuclear reaction.
f propriétés *pl* nucléaires
e propiedades *pl* nucleares
i proprietà *pl* nucleari
n kerneigenschappen *pl*
d Kerneigenschaften *pl*,
nukleare Eigenschaften *pl*

4894 NUCLEAR PROPULSION ms
Propulsion by means of nuclear energy.
f propulsion *f* nucléaire
e propulsión *f* nuclear
i propulsione *f* nucleare
n kernaandrijving, kernvoortstuwing
d Kernenergieantrieb *m*, Reaktorantrieb *m*

4895 NUCLEAR QUADRIPOLE *∫* ma
MOMENT
Expectation value of the quantity,
$(1/e) \int (3z^2 - r^2)\rho (x, y, z) dxdydz$ in the
quantum state with the nuclear spin in
the field direction; (x, y, z) is the nuclear
charge density, e is the elementary charge.
f moment *m* quadripolaire nucléaire
e momento *m* cuadripolar nuclear
i momento *m* quadripolare nucleare
n quadrupoolmoment *n* van een kern
d Quadrupolmoment *n* eines Kerns

4896 NUCLEAR QUADRIPOLE np
RESONANCE
Similar to nuclear magnetic resonance
except that it involves the interaction of an
inhomogeneous electric field with the
nuclear electric quadrupole moment,
instead of the interaction of a homogeneous
magnetic field with the nuclear dipole
moment.
f résonance *f* quadripolaire nucléaire
e resonancia *f* cuadripolar nuclear
i risonanza *f* quadripolare nucleare
n quadripolaire kernresonantie
d Kernquadrupolresonanz *f*

4897 NUCLEAR RADIATIONS np, ra
The three types of radiation emitted by
naturally radioactive substances.
f rayonnements *pl* nucléaires
e radiaciones *pl* nucleares
i radiazioni *pl* nucleari
n kernstralingen *pl*,
radioactieve stralingen *pl*
d Kernstrahlungen *pl*,
radioaktive Strahlungen *pl*

4898 NUCLEAR RADIUS ma
The average radius of the volume in which
the nuclear matter is included.
f rayon *m* nucléaire
e radio *m* nuclear
i raggio *m* nucleare
n kernstraal
d Kernradius *m*

4899 NUCLEAR REACTION np, rt
Any nuclear disintegration caused by the
impact of a particle or a photon and
resulting in fragments specified as to
character and state of excitation.
f réaction *f* nucléaire
e reacción *f* nuclear
i reazione *f* nucleare
n kernreactie
d Kernreaktion *f*

4900 NUCLEAR REACTION ENERGY, np
Q-VALUE, REACTION ENERGY
In a nuclear reaction, the sum of the
kinetic and radiant energies of the
reactants minus the sum of the kinetic
and radiant energies of the reaction
products.
f énergie *f* de réaction, valeur *f* Q
e energía *f* de reacción
i tonalità *f* termica
n reactie-energie
d Energietönung *f* der Kernreaktion,
Q-Wert *m*

4901 NUCLEAR REACTION EQUATION,
see 4843

4902 NUCLEAR REACTOR, rt
REACTOR
A reactor in which a self-sustaining
nuclear fission chain reaction can be
maintained and controlled.
f réacteur *m*, réacteur *m* nucléaire
e reactor *m*, reactor *m* nuclear
i reattore *m*, reattore *m* nucleare
n kernreactor, reactor
d Kernreaktor *m*, Reaktor *m*

4903 NUCLEAR REARRANGEMENT, np
NUCLEAR TRANSFORMATION
The transition of a nucleus into one with
a different configuration of a more stable
character after a nuclear reaction.
f transformation *f* nucléaire
e transformación *f* nuclear
i trasformazione *f* nucleare

n kernomvorming, kerntransformatie
d Kernumordnung f

4904 NUCLEAR REPULSION .   np
The mutual repulsion of particles inside
the nucleus.
f répulsion f intranucléaire
e repulsión f intranuclear
i ripulsione f intranucleare
n intranucleaire afstoting, kernafstoting
d intranukleare Abstossung f,
Kernabstossung f

4905 NUCLEAR RESEARCH   np
f recherches pl nucléaires
e investigaciones pl nucleares
i ricerche pl nucleari
n kernresearch, kernspeurwerk n
d Kernforschung f

4906 NUCLEAR RESONANCE LEVEL,   np
RESONANCE LEVEL
An excited level of the compound system
capable of being formed in a collision
between two systems.
f niveau m de résonance nucléaire
e nivel m de resonancia nuclear
i livello m di risonanza nucleare
n kernresonantieniveau n
d Kernresonanzniveau n

4907 NUCLEAR SCIENCE   np
That part of science which is concerned
with the study of atomic nuclei and of
phenomena caused by radiation emitted
from atomic nuclei.
f science f nucléaire
e ciencia f nuclear
i scienza f nucleare
n kernwetenschap
d Kernwissenschaft f

4908 NUCLEAR SELECTION RULES,   np
SELECTION RULES
A set of statements that serve to classify
transitions of a given type in terms of the
spin and parity quantum numbers of the
initial and final states of the systems
involved in the transition.
f lois pl de sélection nucléaire,
règles pl d'exception nucléaire
e leyes pl de selección nuclear
i leggi pl di selezione nucleare
n kernselectiewetten pl
d Kernauswahlregeln pl

4909 NUCLEAR SITING POLICY   rt
The policy concerned with the selection
of the proper site for a nuclear station.
f étude f d'élection de domicile d'un
réacteur
e estudio m de ubicación de un reactor
i studio m dei problemi d'ubicazione d'un
reattore,
studio m di selezione di domicilio d'un
reattore
n vestigingsonderzoek n van een reactor
d Baugeländewahlstudium n eines Reaktors

4910 NUCLEAR SPECIES   np
A nuclide of given charge, mass number
and quantum state.
f espèce f nucléaire
e especie f nuclear
i specie f nucleare
n kernsoort, kerntype n
d Kernart f

4911 NUCLEAR SPIN   np
The rotation of the nucleus of the atom.
f spin m nucléaire
e espín m nuclear
i spin m nucleare
n kernspin
d Kernspin m

4912 NUCLEAR SPIN EFFECT   sp
Effect caused by nuclear spin on the
hyperfine structure in a spectrum.
f effet m de spin nucléaire
e efecto m de espín nuclear
i effetto m di spin nucleare
n kernspineffect n
d Kernspineffekt m

4913 NUCLEAR STABILITY   np
Present when the intranuclear forces are
in equilibrium.
f stabilité f nucléaire
e estabilidad f nuclear
i. stabilità f nucleare
n kernstabiliteit
d Kernstabilität f

4914 NUCLEAR STAR,   ic, ph
SIGMA STAR
A typical many-particle nuclear reaction
so called from its appearance in nuclear
emulsion photographs or in the cloud
chamber.
f étoile f nucléaire, étoile f sigma
e estrella f nuclear
i stella f nucleare
n ster in de kernemulsie
d Stern m der Kernspuremulsion,
Zertrümmerungsstern m

4915 NUCLEAR STRUCTURE, see 4829

4916 NUCLEAR SUPERHEATING   rt
Superheating of the vapo(u)r produced in
a reactor by means of nuclear energy.
f surchauffe f nucléaire
e recalentamiento m nuclear
i surriscaldamento m nucleare
n kerntechnische oververhitting
d nukleare Überhitzung f

4917 NUCLEAR SURFACE TENSION,   np
SURFACE ENERGY
A term of binding energy which represents
the loss of binding by the nucleons at the
surface of the nucleus.
f énergie f de surface,
tension f superficielle nucléaire
e energía f de superficie
i energia f di superficie

n oppervlakenergie
d Oberflächenenergie *f*

4918 NUCLEAR TARGET,   ge
         TARGET
         A piece of layer of material, suitably
         supported and cooled for bombardment
         by fast particles.
f cible *f*
e blanco *m*
i bersaglio *m*
n treflaag, trefplaat
d Auffänger *m*, Treffplatte *f*, Treffschicht *f*

4919 NUCLEAR TEMPERATURE   np
         COEFFICIENT,
             TEMPERATURE COEFFICIENT OF
             REACTIVITY
         The temperature coefficient in a nuclear
         reactor determined by the effect on the
         nuclear cross sections.
f coefficient *m* de température de la
     réactivité
e coeficiente *m* de temperatura de la
     reactividad
i coefficiente *m* di temperatura della
     reattività
n temperatuurcoëfficiënt van de reactiviteit
d Temperaturkoeffizient *m* der Reaktivität

4920 NUCLEAR TRANSITION   np
         A change in nuclear configuration.
f transition *f* nucléaire
e transición *f* nuclear
i transizione *f* nucleare
n kernovergang
d Kernübergang *m*

4921 NUCLEAR UNDERGROUND BURST,
         see 422

4922 NUCLEAR UNDERWATER BURST,
         see 423

4923 NUCLEAR WARHEAD   nw
         A warhead that contains fissile (fissionable)
         or fusionable material.
f cône *f* de charge nucléaire
e cabeza *f* de guerra nuclear,
     ojiva *f* de guerra nuclear
i testa *f* esplosiva nucleare
n kernprojectielkop
d Kernsprengkopf *m*

4924 NUCLEAR WEAPONS   nw
         Bombs, missiles and other devices in
         which the explosive power is derived from
         nuclear energy.
f armes *pl* nucléaires
e armas *pl* nucleares
i arme *pl* nucleari
n kernwapens *pl*
d Kernwaffen *pl*

4925 NUCLEATION   cr
         The formation of germ cells in
         crystallization.

f formation *f* de germes de cristaux
e formación *f* de gérmenes de cristales,
     germinación *f*
i formazione *f* di germi cristalli
n kristalkiemvorming,
     vorming van kristallisatiekernen
d Bildung *f* von Kristallisationskernen,
     Keimbildung *f*

4926 NUCLEIC ACID,   ch, md
         NUCLEINIC ACID
         A constituent of the cell nucleus, composed
         of a union between phosphoric acid, ribose
         or desoxyribose and the four bases:
         adenine, guanine, cystosine and uracil or
         thymine.
f acide *f* nucléique
e ácido *m* nucleico
i acido *m* nucleico
n nucleïnezuur *n*
d Nukleinsäure *f*

4927 NUCLEOGENESIS   np
         The hypothetical process or processes
         by which the nuclei formed in nature were
         originally formed, or are still being
         formed.
f nucléogenèse *f*
e nucleogenesis *f*
i nucleogenesi
n kernvorming, nucleogenese
d Atomkernbildung *f*, Nukleogenese *f*

4928 NUCLEON   np
         A proton or a neutron.
f nucléon *m*
e nucleón *m*
i nucleone *m*
n nucleon *n*
d Nukleon *n*

4929 NUCLEON NUMBER, see 4271

4930 NUCLEONICS   np
         The practical applications of nuclear
         science and the techniques associated with
         these applications.
f technique *f* nucléaire
e nucleónica *f*
i tecnica *f* nucleare
n kerntechniek, toegepaste kernwetenschap
d angewandte Kernwissenschaft *f*,
     Kerntechnik *f*

4931 NUCLEOPLASM, see 1606

4932 NUCLEOPROTEIN   md
         Protein conjugated with nucleic acid.
f nucléoprotéine *f*
e nucleoproteína *f*
i nucleoproteina *f*
n nucleoproteïde *n*, nucleoproteïne *n*
d Nukleoproteid *n*, Nukleoprotein *n*

4933 NUCLEOR   np
         The core of a nucleon, as, according to
         some theories, each nucleon is composed
         of a nucleor surrounded by a pion cloud.

f   coeur *m* du noyau
e   corazón *m* del núcleo
i   cuore *m* del nucleo
n   binnenste *n* van de kern, hart *n* van de kern
d   Kerninneres *n*

4934 NUCLEUS   md
Of a cell, a definitely delineated body
within the cell, containing the chromosomes.
f   noyau *m*
e   núcleo *m*
i   nucleo *m*
n   celkern
d   Zellkern *m*

4935 NUCLEUS, see 396   np

4936 NUCLEUS, see 4383   mg

4937 NUCLEUS SCREENING   np
The reduction of the electric field about
a nucleus by the space of the surrounding
electrons.
f   blindage *m* du noyau
e   blindaje *m* del núcleo
i   schermatura *f* del nucleo
n   kernafscherming
d   Kernabschirmung *f*

4938 NUCLEUS SHELL STRUCTURE,   np
    SHELL STRUCTURE OF THE
    NUCLEUS
The arrangement of the quantum states of
nucleons of a given kind in a nucleus in
groups of approximately the same energy.
f   structure *f* quantique du noyau
e   estructura *f* cuántica del núcleo
i   struttura *f* quantica del nucleo
n   schilstructuur van de kern
d   Schalenaufbau *m* des Kerns

4939 NUCLIDE   np
A species of atom characterized by its
mass number, atomic number and nuclear
energy state, provided that the mean life
in that state is long enough to be
observable.
f   nucléide *m*
e   núclido *m*
i   nuclide *m*
n   nuclide *n*
d   Nuklid *n*

4940 NUCLIDE MASS   un
The mass of the neutral atom of a nuclide
in amu.
f   masse *f* atomique d'un nucléide
e   masa *f* atómica de un núclido
i   massa *f* atomica d'un nuclide
n   nuclidemassa
d   Nuklidmasse *f*

4941 NULL VALENCE   ch
According to the electronic conception of
valence, a condition in which an element
has no valence because, in its normal
state, it has a complete outer electronic
shell.

f   valence *f* zéro
e   valencia *f* nula
i   valenza *f* nulla
n   nulvalentie, nulwaardigheid
d   Nullvalenz *f*, Nullwertigkeit *f*

4942 NUMBER OF NEUTRONS   np, rt
    PER FISSION
The average number of neutrons, both
prompt and delayed, emitted per fission.
f   nombre *m* des neutrons par fission
e   número *m* de los neutrones por fisión
i   numero *m* dei neutroni per fissione
n   aantal *n* neutronen per splijting
d   Anzahl *f* Neutronen je Spaltung

4943 NUMBER OF PRODUCED   np, rt
    NEUTRONS PER ABSORBED NEUTRON
The average number of neutrons produced
per neutron absorbed by a fissile
(fissionable) nucleus or in fuel materials.
f   nombre *m* des neutrons produits par
    neutron absorbé
e   número *m* de los neutrones producidos
    por neutrón absorbido
i   numero *m* dei neutroni prodotti per
    neutrone assorbito
n   aantal *n* geproduceerde neutronen per
    geabsorbeerd neutron
d   Anzahl *f* der erzeugten Neutronen je
    absorbiertes Neutron

4944 NUMBER OF THEORETICAL   ch
    PLATES,
    NUMBER OF THEORETICAL STAGES
The number of hypothetical devices for
bringing two streams of material into
such perfect contact that they leave in
equilibrium with each other.
f   nombre *m* de plateaux théoriques
e   número *m* de platos teóricos
i   numero *m* di piatti teorici
n   aantal *n* theoretische schotels
d   Zahl *f* der theoretischen Boden

4945 NUMBER OF TRANSFER UNITS   ch
The total number of units in a distillation
or a diffusion process.
f   nombre *m* d'unités de transfert
e   número *m* de unidades de transferencia
i   numero *m* d'unità di trasporto
n   aantal *n* overgangscellen,
    uitwisselingseenheden *pl*
d   Austauscheinheiten *pl*,
    Zahl *f* der Übertragungseinheiten

4946 NUMERICAL CONSTANT   ma
A constant which is represented by figures
and does not contain letters.
f   constante *f* métrique,
    constante *f* numérique
e   constante *f* numérica
i   costante *f* numerica
n   cijferconstante
d   Zahlenkonstante *f*

4947 NUPAC MONITORING SET   sa   f ensemble *m* moniteur
A warning system that detects undesirable   e conjunto *m* monitor
radiation emanating from a nuclear reactor   i complesso *m* monitore
and associated equipment.   n monitoropstelling
                                             d Monitoraggregat *n*

O

4948 O ELECTRON   np
An electron having an orbit of such
dimensions that the electron constitutes
part of the fifth shell of electrons
surrounding the atomic nucleus.
f électron *m* O
e electrón *m* O
i elettrone *m* O
n O-elektron *n*
d O-Elektron *n*

4949 O MANIPULATOR,   cd, co
        OVERHEAD MANIPULATOR
A heavy-work manipulator.
f télémanipulateur *m* robuste
e manipulador *m* robusto
i manipolatore *m* robusto
n krachtmanipulator
d Kraftmanipulator *m*

4950 O SHELL   np
The fifth layer of electrons about the
nucleus of an atom.
f couche *f* O
e capa *f* O
i strato *m* O
n O-schil
d O-Schale *f*

4951 OBLIQUE ANODE   xr
In an X-ray tube an anode, the surface of
which forms an oblique angle with the
anode support.
f anode *f* inclinée, anode *f* oblique
e ánodo *m* oblícuo
i anodo *m* obliquo
n schuine anode
d Schräganode *f*

4952 OBLIQUE PROJECTION,   xr
        OBLIQUE VIEW
A radiograph for which the röntgen ray
traverses the body obliquely.
f projection *f* oblique, vue *f* oblique
e proyección *f* oblícua, vista *f* oblícua
i proiezione *f* obliqua, vista *f* obliqua
n half-zijdelingse projectie,
    schuine projectie
d Projektion *f* in schräger Richtung

4953 OCCUPANCY FACTOR   xr
A time consideration used in establishing
radiological controls such that radiation
exposures under planned conditions will
be within permissible limits.
f surveillance *f* du temps d'exposition
e vigilancia *f* del tiempo de exposición
i sorveglianza *f* del tempo d'esposizione
n exposietijdbewaking
d Expositionszeitüberwachung *f*

4954 OCCUPATION NUMBER   np
The number of electrons occupying the

separate shells or electric states.
f nombre *m* d'occupation
e número *m* de ocupación
i numero *m* d'occupazione
n bezettingsgetal *n*
d Besetzungszahl *f*

4955 OCCUPATIONAL EXPOSURE   ra
That exposure to ionizing radiation which
is associated with a person's occupation
and takes place during his working hours.
f exposition *f* professionnelle
e exposición *f* profesional
i esposizione *f* professionale
n beroepsexposie
d berufsbedingte Exponierung *f*,
    berufsbedingte Exposition *f*

4956 OCCUPIED AREA,   ra
        OCCUPIED SPACE
Space which may be occupied by personnel
and in which a radiation hazard may exist.
f espace *m* occupé, zone *f* occupée
e espacio *m* ocupado, zona *f* ocupada
i spazio *m* occupato, zona *f* occupata
n bevolkt gevarengebied *n*
d bewohntes Gebiet *n* im Strahlenbereich

4957 ODD-EVEN NUCLEI   np
Nuclei in which there are an odd number
of protons and an even number of neutrons.
f noyaux *pl* impair-pair
e núcleos *pl* par-impar
i nuclei *pl* impari-pari
n oneven-even-kernen *pl*
d ug-Kerne *pl*, ungerade-gerade-Kerne *pl*

4958 ODD-EVEN RULE OF   np
        NUCLEAR STABILITY
A rule concerning the number of protons
Z and neutrons or forming a nucleus,
stating that the stability of the nucleus
depends on the odd or even character of
Z and N.
f loi *f* impair-pair de stabilité nucléaire
e ley *f* de paridad de la estabilidad nuclear
i legge *f* di parità della stabilità nucleare
n oneven-even-regel van de kernstabiliteit
d ungerade-gerade-Regel *f* der Kern-
    stabilität

4959 ODD MOLECULES   ch
A few unusual molecules that have an odd
number of valence electrons.
f molécules *pl* impair
e moléculas *pl* impares
i molecole *pl* dispari
n oneven moleculen *pl*
d ungerade Moleküle *pl*

4960 ODD NUMBER   ma
A number not divisible by two without a
remainder.

f   nombre m impair
e   número m impar
i   numero m dispari
n   oneven getal n
d   ungerade Zahl f

4961 ODD-ODD NUCLEI   np
Nuclei in which there are an odd number
of protons and àlso an odd number of
neutrons.
f   noyaux pl impair-impair
e   núcleos pl impar-impar
i   nuclei pl impari-impari
n   oneven-oneven-kernen pl
d   ungerade-ungerade-Kerne pl, uu-Kerne pl

4962 ODD TERM OF ATOM   ma
A term for which $\in 1_i$, summed over all
the electrons of the atom, is odd.
f   terme m atomique impair
e   término m atómico impar
i   termine m atomico dispari
n   oneven atoomterm
d   ungerader Atomterm m

4963 OEDEMA, see 1589

4964 OFF-LOAD REFUELLING   fu, rt
f   recharge f hors opération
e   reaprivisionamiento m fuera de operación
i   ricarica f fuori esercizio
n   oplading buiten bedrijf
d   Aufladung f ausser Betrieb

4965 OFF-SHORE SEA BED   is
     MOVEMENT
One of the problems to be studied by
using isotopic tracers.
f   mouvement m du fond de la mer au large
e   movimiento m del fondo del mar de
    costafuera
i   movimento m del fondo marino in alto mare
n   beweging van de zeebodem buitengaats
d   Bewegung f des Meeresbodens auf
    offener See

4966 OFFSET,   co
     STEADY STATE DEVIATION
A sustained deviation due to an inherent
characteristic of positioning-controlled
action.
f   écart m permanent
e   desfase f, desviación f permanente
i   scarto m permanente
n   permanente afwijking
d   bleibende Regelabweichung f

4967 OHMIC HEATING, see 3822

4968 OIL-IMMERSED TUBE   xr
An X-ray tube in which the heat generated
is dissipated by immersing the tube in oil.
f   tube m à rayons X à bain d'huile
e   tubo m de rayos X de inmersión en aceite
i   tubo m a raggi X ad immersione in olio
n   in olie gedompelde röntgenbuis
d   Ölröhre f

4969 OIL-VAPOR DIFFUSION   vt
     PUMP (US),
     OIL-VAPOUR DIFFUSION PUMP (GB)
A diffusion pump which uses a low
vapo(u)r-pressure oil as its working fluid.
f   pompe f à diffusion à vapeur d'huile
e   bomba f de difusión de vapor de aceite
i   pompa f a diffusione a vapore d'olio
n   oliedampdiffusiepomp
d   Öldampfdiffusionspumpe f

4970 OLD QUANTUM MECHANICAL
     THEORY, see 695

4971 OMEGATRON   me
An instrument in which ions are caused
to move in spiral paths by a high-frequency
electric field applied at right angles to a
constant magnetic field.
f   omégatron m
e   omegatrón m
i   omegatrone m
n   omegatron n
d   Massenspektrometer n nach dem
    Zyklotronprinzip,
    Omegatron n

4972 ON-LOAD REFUELLING, see 2887

4973 ONCE-THROUGH COOLING   cl
A cooling system in which the cooling
medium is used only once.
f   refroidissement m à passage unique
e   enframiento m de paso único
i   raffreddamento m a passaggio unico
n   enkele doorloopkoeling
d   Kühlung f mit einmaligem Kühlmittel-
    durchlauf

4974 ONCE-THROUGH STEAM BOILER,
     see 572

4975 1/E FLUX   np
Neutron flux of which the density varies
inversely with the neutron energy.
f   flux m en $1/E$
e   flujo m $1/E$
i   flusso m $1/E$
n $1/E$-flux
d $1/E$-Flux m

4976 1/E SPECTRUM   sp
Neutron spectrum corresponding to a
distribution of these neutrons inversely
proportional to their energy.
f   spectre m en $1/E$
e   espectro m $1/E$
i   spettro m $1/E$
n $1/E$-spectrum n
d $1/E$-Spektrum n

4977 ONE-GROUP MODEL,   np
     ONE-GROUP THEORY
A simplified treatment of neutron migration, in which neutrons of all energies are
regarded as having the same energy.
f   modèle m à un groupe d'énergies
e   modelo m de un grupo de energías

i modello *m* ad un gruppo d'energie
n eengroepsmodel *n*
d Eingruppenmodell *n*

4978 ONE-GROUP TREATMENT rt
    REACTOR
A treatment of nuclear theory in which it
is assumed that the production, diffusion
and absorption of neutrons occur at a
single energy, the thermal energy.
f théorie *f* de réacteur à groupe unique
e teoría *f* de reactor de grupo único
i teoria *f* di reattore a gruppo unico
n eengroepsreactortheorie
d Eingruppenreaktortheorie *f*

4979 ONE-PARTICLE MODEL np
A nuclear model in which the properties
considered are attributed to the effect of
one nucleon.
f modèle *m* à particule unique
e modelo *m* de partícula única
i modello *m* a particella unica
n enkeldeeltjesmodel *n*
d Einzelteilchenmodell *n*

4980 ONE-TENTH PERIOD np
In radioactivity that period of time that
0.9 (or 90 %) of the radioactive atoms
have decayed.
f un dixième *m* de période
e un décimo *m* de período
i un decimo *m* di periodo
n één-tiende periode
d Einzehntel-Periode *f*

4981 1/v DETECTOR ra
A neutron detector for which the cross
section of the detection reaction varies
inversely with neutron velocity.
f détecteur *m* en 1/v
e detector *m* en 1/v
i rivelatore *m* 1/v
n 1/v-detector
d 1/v-Detektor *m*

4982 O-P PROCESS, np
    OPPENHEIMER-PHILLIPS PROCESS
A nuclear reaction in which a deuteron
of low energy gives its neutron to a
nucleus without entering it.
f procédé *m* Oppenheimer-Phillips
e procedimiento *m* Oppenheimer-Phillips
i processo *m* Oppenheimer-Phillips
n oppenheimer-phillipsproces *n*
d Oppenheimer-Phillips-Prozess *m*

4983 OPACITY ra
Imperviousness to radiation, especially
to light.
f opacité *f*
e opacidad *f*
i opacità *f*
n ondoorlaatbaarheid, opaciteit
d Opazität *f*, Strahlenundurchlässigkeit *f*,
Undurchlässigkeit *f*

4984 OPAQUE ra
Preventing the passage of radiation or
particles.
f opaque adj
e opaco adj
i opaco adj
n ondoorlatend adj, ondoorschijnend adj,
opaak adj
d opak adj, strahlenundurchlässig adj,
undurchlässig adj

4985 OPEN AIR IONIZATION CHAMBER,
    see 2821

4986 OPEN COOLING SYSTEM cl
f système *m* de refroidissement ouvert
e sistema *m* de enfriamiento abierto
i sistema *m* di raffreddamento aperto
n open koelsysteem *n*
d offenes Kühlsystem *n*

4987 OPEN CYCLE ge, gp
Cycle of operation of a heat engine in
which the power fluid is used only once
and replaced with fresh fluid.
f cycle *m* puvert
e ciclo *m* abierto
i ciclo *m* aperto
n open kringloop, open systeem *n*
d offener Kreis *m*, offener Kreislauf *m*

4988 OPERATING CHARACTERISTIC, ma
    OPERATING LINE
A mathematical or graphical relation,
based on material and energy balance,
which relates the properties of one of two
streams in contact to those of the other
at the same point in the apparatus.
f caractéristique *f* de fonctionnement,
régime *m* d'opération
e curva *f* de trabajo, línea *f* de operación
i curva *f* di lavoro, linea *f* d'operazione
n werklijn
d Arbeitskennlinie *f*, Arbeitskurve *f*

4989 OPERATING RANGE, see 3515

4990 OPHTHALMIC APPLICATOR br, md, ra
An applicator comprising beta emitting
silver foil elements embedded in silver or
plastics cups.
f porte-émetteur-bêta *m* pour buts
ophtalmiques
e aplicador *m* beta para intentos
oftálmicos
i applicatore *m* beta per disegni oftalmici
n bêtacapsule voor oftalmische doeleinden
d Beta-Kapsel *f* für ophthalmische Zwecke

4991 OPTICAL MODEL, see 1143

4992 OPTICAL PUMPING pp
Optical process allowing the pumping of a
system at a specified energy state to a
higher energy state, the energy needed for
the transition being supplied by the photon
absorption of which the wavelengths are
equal to those of light.

f   pompage *m* optique
e   bombeo *m* óptico
i   pompaggio *m* ottico
n   optisch pompen *n*
d   optisches Pumpen *n*

4993 OPTICAL RESONANCE   ra
Luminescence wherein the frequencies or
wavelengths of the exciting and emitted
radiation are the same except for a minor
Doppler shift.
f   résonance *f* optique
e   resonancia *f* óptica
i   risonanza *f* ottica
n   resonantiestraling
d   optische Resonanz *f*

4994 ORANGE OXIDE,   ch, mt
      URANIUM TRIOXIDE
An intermediate product in the refining of
uranium.
f   trioxyde *m* d'uranium
e   trióxido *m* de uranio
i   ossido *m* giallo d'uranio,
    triossido *m* d'uranio
n   uraniumtrioxyde *n*
d   Urantrioxyd *n*

4995 ORANGE PEEL EFFECT   mg
The effect of irradiation on the surface
of uranium fuel elements
f   effet *m* peau d'orange,
    effet *m* pelure d'orange.
e   efecto *m* de corteza de naranja
i   effetto *m* di buccia d'arancio
n   sinaasappelschileffect *n*
d   Apfelsinenschaleneffekt *m*

4996 ORANGITE   mi
A thorium orthosilicate containing mineral.
f   orangite *f*
e   orangita *f*
i   orangite *f*
n   orangiet *n*
d   Orangit *m*

4997 ORBIT   np
The path described by a particle, or by
the centroid of a body, under the influence
of a gravitational or other force field.
f   orbite *f*
e   órbita *f*
i   orbita *f*
n   baan
d   Bahn *f*

4998 ORBIT SHIFT COILS   pa
A set of coils through which a pulse of
current is passed to alter momentarily
the guiding field in such a way as to cause
the orbit radius to increase or decrease,
or the plane of the orbit to rise, lower or
tilt, thereby causing the accelerated
particles to strike a target placed outside
the stable orbit or to enter a deflector for
the production of an external beam.
f   bobines *pl* de déviation,
    enroulements *pl* déflecteurs d'orbite

e   bobinas *pl* de desplazamiento orbital,
    bobinas *pl* desviadoras
i   bobine *pl* di deviazione orbitale
n   baanverschuivingsspoelen *pl*
d   Ablenkspulen *pl*

4999 ORBITAL   np
Pertaining to the extranuclear electrons
of an atom.
f   satellite adj
e   planetario adj
i   orbitale adj
n   schaal-
d   Hüllen-

5000 ORBITAL,
      ATOMIC ORBITAL, see 399

5001 ORBITAL ANGULAR   ma
      MOMENTUM
The angular momentum H associated with
the motion of a particle moving in an
orbit.
f   moment *m* angulaire orbital
e   momento *m* angular orbital
i   momento *m* angolare orbitale
n   baanimpulsmoment *n*
d   Bahndrehimpuls *m*

5002 ORBITAL ELECTRON,   np
      SHELL ELECTRON
An electron remaining with a high degree
of probability in the immediate neighbourhood of a nucleus, where it occupies a
quantized orbital.
f   électron *m* orbital, électron *m* planétaire
e   electrón *m* orbital, electrón *m* planetario
i   elettrone *m* orbitale, elettrone *m* planetario
n   schilelektron *n*
d   Schalenelektron *n*

5003 ORBITAL ELECTRON CAPTURE   np
Electron capture in which the electron
comes from an atomic or molecular
orbital of the atom or molecule containing
the transforming nucleus.
f   capture *f* d'un électron orbital
e   captura *f* de un electrón orbital
i   cattura *f* d'un elettrone orbitale
n   vangst van een schilelektron
d   Einfang *m* eines Schalenelektrons

5004 ORBITAL MOMENT   ma, np
The moment of momentum of an orbital
electron.
f   moment *m* orbital
e   momento *m* orbital
i   momento *m* orbitale
n   baanmoment *n*
d   Bahnimpuls *m*, Bahnmoment *m*

5005 ORBITAL MOTION   np
The motion of a particle in its orbit.
f   mouvement *m* orbital
e   movimiento *m* orbital
i   movimento *m* orbitale
n   baanbeweging
d   Bahnbewegung *f*

5006 ORBITAL QUANTUM NUMBER, see 490

5007 ORDER-DISORDER cr
TRANSFORMATION,
ORDER-DISORDER TRANSITION
The transition from the ordered arrangement of a crystal to a disordered
arrangement.
f transformation *f* ordre-désordre
e transformación *f* orden-desorden
i trasformazione *f* ordine-disordine
n orde-wanorde-omzetting
d Ordnung-Unordnung-Übergang *m*

5008 ORDERED SCATTERING, see 763

5009 ORDERING np
In a system of a large number of particles,
the order, or organization, of the particles
increases as the temperature decreases.
f arrangement *m*
e arreglo *m*
i ordinamento *m*
n ordening
d Ordnung *f*

5010 ORDINARY IRON, see 4627

5011 ORE mi
Any material containing valuable metallic
constituents for the sake of which it is
mined and worked.
f minerai *m*
e mineral *m*
i minerale *m*
n erts *n*
d Erz *n*

5012 ORE BEARING SHALE mi
f schiste *m* métallifère
e esquisto *m* metalífero
i scisto *m* metallifero
n ertshoudend leisteen *n*
d erzhaltiger Schiefer *m*

5013 ORE CONCENTRATE, see 4448

5014 ORE CONCENTRATION, see 4449

5015 ORE CONTENT METER ma
A measuring assembly designed to
determine the metal content of ore by
laboratory measurement on a given number
of specimens of that ore.
f teneurmètre *m* de minerais
e valorímetro *m* de minerales
i tenorimetro *m* di minerali
n gehaltemeter voor ertsen
d Messgerät *n* zur Bestimmung des
Erzgehalts

5016 ORE DEPOSIT mi
A natural occurrence or accumulation of
ore.
f gisement *m* minéral, gîte *m* minéral
e yacimiento *m* mineral
i giacimento *m* di minerale

n ertsafzetting
d Erzlagerstätte *f*

5017 ORE EXTRACTION ch
Act of extracting a desired substance from
the ore.
f extraction *f* du minerai
e extracción *f* del mineral
i estrazione *f* del minerale
n ertsextractie, ertsscheiding
d Erzzerlegung *f*, Extraktion *f* des Erzes

5018 ORE GRADING AND SORTING ma
EQUIPMENT AND ASSEMBLY
An equipment and an assembly using
natural or artificially caused radioactivity for the removal of sterile ore and
the classification of ores in categories
according to metal contents.
f équipement *m* et ensemble *m* utilisé
pour l'estimation et le triage des minerais
e conjunto *m* y equipo *m* selector
clasificador
i apparecchiatura *f* e complesso *m*
utilizzati per la stima e cernita di
minerali
n uitrusting en opstelling voor zuivering
en sortering van ertsen
d Einrichtung *f* und Anordnung *f* zum
Klassieren und Sortieren von Erzen

5019 ORE REFINING mi
To free ore from impurities.
f raffinage *m* du minerai
e refinado *m* de minerales
i depurazione *f* di minerali,
raffinazione *f* di minerali
n ertsraffinage, ertszuivering
d Raffinieren *n* des Erzes

5020 ORGAN md
Any part of an organism performing
some definite function.
f organe *m*
e órgano *m*
i organo *m*
n orgaan *n*
d Organ *n*

5021 ORGANIC COOLED REACTOR rt
A nuclear reactor in which the coolant is
an organic compound.
f réacteur *m* à réfrigérant organique
e reactor *m* de refrigerante orgánico
i reattore *m* a refrigerante organico
n reactor met organisch koelmiddel
d Reaktor *m* mit organischem Kühlmittel

5022 ORGANIC MODERATED REACTOR rt
A nuclear reactor in which the moderator
is an organic compound.
f réacteur *m* à modérateur organique
e reactor *m* de moderador orgánico
i reattore *m* a moderatore organico
n reactor met organische moderator
d Reaktor *m* mit organischem Moderator

5023 ORGANIC MODERATOR   rt
f   modérateur *m* organique
e   moderador *m* orgánico
i   moderatore *m* organico
n   organische moderator, organische remstof
d   organischer Moderator *m*

5024 ORGANIC QUENCHED   ct
         COUNTER TUBE,
         ORGANIC VAPOR COUNTER TUBE
         (US),
         ORGANIC VAPOUR COUNTER
         TUBE (GB)
     A self-quenching counter tube in which
     quenching is obtained by the addition of
     an organic vapo(u)r, e.g. methanol.
f   tube *m* compteur à vapeur organique
e   tubo *m* contador de vapor orgánico
i   tubo *m* contatore a vapore organico
n   organische-damptelbuis,
     telbuis met organische-dampbijvulling
d   Zählrohr *n* mit organischem Löschzusatz,
     Zählrohr *n* mit Zugabe von organischem
     Dampf

5025 ORIENTATION   cr
     The direction in space of the axes of a
     crystal, or in a polycrystalline solid,
     the relation between similar axes of
     different crystals.
f   orientation *f*
e   orientación *f*
i   orientamento *m*
n   oriëntatie, textuur
d   Orientierung *f*, Textur *f*

5026 ORIENTATION,   ch
         BOND DIRECTION, see 716

5027 ORIENTATION EFFECT   ch
     A basis of calculating the atttactive forces
     between molecules from the interaction
     of molecular dipoles due to their
     respective orientation.
f   effet *m* d'orientation
e   efecto *m* de orientación
i   effetto *m* d'orientamento
n   oriënteringseffect *n*
d   Orientierungseffekt *m*

5028 ORIFICING THE COOLANT   cl, rt
     A method of equalizing the non-uniformity
     of the neutron flux and the power level by
     using cooling systems with orifices of
     different sizes.
f   réglage *m* par variation d'orifices
e   regulación *f* por variación de orificios
i   regolazione *f* per variazione d'orifici
n   regeling door variatie van de uitstroom-
     openingen
d   Regelung *f* durch Variation der
     Ausströmungsöffnungen

5029 ORIGINAL FISSION   np
     In a nuclear reactor the first fission
     occurring.
f   fission *f* primordiale
e   fisión *f* original, fisión *f* primordial
i   fissione *f* originale
n   eerste splijting, primaire splijting
d   Erstspaltung *f*, Primärspaltung *f*

5030 ORTHODIAGRAPHY,   xr
         ORTHORADIOSCOPY
     A fluoroscopic technique for locating and
     recording the actual size of an organ,
     especially the heart.
f   orthoradioscopie *f*
e   ortorradioscopia *f*
i   ortoradioscopia *f*
n   orthodiagrafie, orthodiascopie
d   Orthodiagraphie *f*, Orthoröntgenoskopie *f*

5031 ORTHOHELIUM   ch
     A helium in which the two electrons of the
     helium atom have a spin parallel to each
     other.
f   orthohélium *m*
e   ortohelio *m*
i   ortoelio *m*
n   orthohelium *n*
d   Orthohelium *n*

5032 ORTHOHELIUM TERMS   ch, sp
     One group or system of terms in the
     spectrum of helium that is due to atoms in
     which the spins of the two atoms are
     parallel to each other.
f   termes *pl* orthohélium
e   términos *pl* ortohelio
i   termini *pl* ortoelio
n   orthoheliumtermen *pl*
d   Orthoheliumterme *pl*

5033 ORTHOHYDROGEN   ch, gp
     That form of molecular hydrogen in which
     the two nuclear spins in each molecule
     are parallel.
f   orthohydrogène *m*
e   ortohidrógeno *m*
i   ortoidrogeno *m*
n   orthowaterstof
d   Orthowasserstoff *m*

5034 ORTHOPOSITONIUM,   np
         ORTHOPOSITRONIUM
     The state of posit(r)onium in which the
     spins of the electron and posit(r)on are
     parellel.
f   orthopositonium *m*
e   ortopositonio *m*
i   ortopositonio *m*
n   orthopositonium *n*
d   Orthopositronium *n*

5035 ORTHOSCOPIC VIEW   xr
     A three-dimensional view seen in normal
     perspective.
f   projection *f* orthoscopique,
     vue *f* orthoscopique
e   proyección *f* ortoscópica,
     vista *f* ortoscópica
i   proiezione *f* ortoscopica,
     vista *f* ortoscopica
n   orthoscopische projectie
d   orthoskopische Projektion *f*

5036 OSCILLATING CRYSTAL METHOD   cr
A modification of the rotating crystal
method in examining crystals.
f méthode f à cristal oscillant
e método m de cristal oscilador
i metodo m a cristallo oscillante
n methode met oscillerend kristal
d Schwingkristallverfahren n

5037 OSCILLATION   co
A periodic change of the controlled
variable from one value to another.
f oscillation f
e oscilación f
i oscillazione f
n schommeling, trilling
d Schwingung f

5038 OSMIUM   ch
Metallic element, symbol Os, atomic
number 76.
f osmium m
e osmio m
i osmio m
n osmium n
d Osmium n

5039 OSTEOGENIC   md
Derived from or composed of tissue
concerned in growth or repair of bone.
f ostéogénique adj
e osteogénico adj
i osseogenico adj
n osteogeen adj
d osteogen adj

5040 OUTAGE   rt
Common term to indicate the time that a
nuclear reactor is not in operation.
f période f hors service,
  temps m d'inactivité
e período m de interrupción
i periodo m d'inoperosità
n buitenbedrijftijd
d Ausserbetriebszeit f

5041 OUTER BREMSSTRAHLUNG   ra, xr
Bremsstrahlung whereby the energy loss
by radiation far exceeds that by ionization
as a stopping mechanism in matter.
f rayonnement m de freinage externe
e radiación f de frenado exterior
i radiazione f di rallentamento esterno
n uitwendige remstraling
d äussere Bremsstrahlung f

5042 OUTER CAPSULE   md, ra
The capsule surrounding the inner capsule.
f capsule f extérieure
e cápsula f exterior
i capsula f esteriore
n buitencapsule
d Aussenkapsel f

5043 OUTER ORBIT   np
The orbit farthest away from the nucleus.
f orbite f externe

e órbita f exterior
i orbita f esterna
n buitenbaan
d Aussenbahn f

5044 OUTER-SHELL ELECTRON, see 1300

5045 OUTERWORK FUNCTION   np
The energy gap between the crest of the
potential barrier at the surface and the
potential plateau.
f travail m externe
e trabajo m externo
i lavoro m esterno
n totale uittreearbeid
d äussere Austrittsarbeit f

5046 OUTFALL PIPE,   rw
     OUTFALL SEWER
A pipeline to conduct radioactive fluid
waste to the sea.
f égout m d'évacuation
e tubo m de desembucadura
i fogna f di scarico
n afvoerleiding
d Abflussleitung f

5047 OUTPUT   ge
The point where the useful energy of a
device or circuit is delivered.
f sortie f
e salida f
i uscita f
n uitgang
d Ausgang m

5048 OUTPUT   is
In isotope separation the product extracted
from a cascade.
f soutirage m
e descarga f, toma f
i prelievo m
n aftap
d Entnahme f

5049 OUTPUT PULSE   ec
In instrumentation for nuclear reactors,
a pulse produced in a scaler whenever a
prescribed number of input pulses have
been received.
f impulsion f de sortie
e impulsión f de salida
i impulso m d'uscita
n uitgangspuls
d Ausgangsimpuls m

5050 OVERALL COEFFICIENT   gp
     OF HEAT TRANSFER.
The average rate of flow of heat through
a medium or a system, expressed as the
amount of heat passing through unit area,
per unit time and per degree temperature
difference.
f coefficient m total de transmission de
  chaleur
e coeficiente m total de transmisión de calor
i coefficiente m totale di trasmissione di
  calore

n   totale warmteoverdrachtscoëfficiënt
d   Wärmedurchgangszahl *f*

5051 OVERALL EFFICIENCY   ge
      The mean value of the efficiency of any
      system, process, etc.
f   efficacité *f* totale
e   rendimiento *m* total
i   rendimento *m* .totale
n   totaal rendement *n*
d   Gesamtwirkungsgrad *m*

5052 OVERALL ENRICHMENT   ch
         PER STAGE
      The percentage of isotope enrichment in
      a single stage of the separation process.
f   facteur *m* d'enrichissement par étage
e   factor *m* de enriquecimiento por etapa
i   fattore *m* d'arricchimento per stadio
n   totale verrijkingsfactor per trap
d   Gesamtanreicherungsfaktor *m* je Stufe

5053 OVERALL ERROR,   me
         TOTAL ERROR
      The sum of the system error and the
      accidental error of a measuring instrument.
f   erreur *f* totale
e   error *m* total
i   errore *m* totale
n   totale fout
d   Gesamtfehler *m*

5054 OVERALL LOAD FACTOR   rt
f   facteur *m* de charge moyen
e   factor *m* de carga medio
i   fattore *m* di carica medio
n   gemiddelde belastingsfactor
d   mittlerer Belastungsfaktor *m*

5055 OVERALL TREATMENT TIME   xr
      The duration of a course of treatment in
      days or weeks from the first irradiation
      to the last.
f   durée *f* de traitement totale
e   duración *f* de tratamiento total
i   durata *f* di trattamento totale
n   totale behandelingsduur
d   Gesamtbehandlungsdauer *f*

5056 OVERCOMPRESSION   ic
      In the operation of an expansion chamber,
      a technique whereby the required waiting
      time is shortened by raising the pressure
      above its normal value for a short time
      immediately after each expansion.
f   surpression *f*
e   sobrepresión *f*
i   sovrapressione *f*
n   overdruk
d   Überdruck *m*

5057 OVERDOSAGE,   md, xr
         OVERDOSING
f   surdosage *m*
e   superdosificación *f*
i   sovradosaggio *m*
n   overdosering
d   Überdosierung *f*

5058 OVERDOSE   md, xr
f   dose *f* excessive
e   dosis *f* excesiva
i   dose *f* eccessiva
n   te hoge dosis
d   Überdosis *f*

5059 OVERFLUX RELAY   sa
      Sensitive relay of the moving-contact type
      for releasing the safety-rod latches and
      shutting down the reactor.
f   relais *m* de déverrouillage
e   relevador *m* de descerrajadura
i   relè *m* di disserramento
n   ontgrendelrelais *n*
d   Entriegelrelais *n*

5060 OVERHAUSER EFFECT   np
      Effect leading to a dynamic polarization
      of nuclear spins.
f   effet *m* Overhauser
e   efecto *m* Overhauser
i   effetto *m* Overhauser
n   overhausereffect *n*
d   Overhauser-Effekt *m*

5061 OVERHEAD   ch
      The total vapo(u)rs leaving the top of a
      tower or column.
f   vapeur *f* de tête
e   vapor *m* en cabeza
i   vapore *m* di testa
n   topstroom
d   Dampf *m* am Kolonnentopf, Dampfstrom *m*

5062 OVERHEAD MANIPULATOR, see 4949

5063 OVERHEAD PRODUCT   ch
      The distillate resulting from a distillation
      process.
f   distillat *m* de tête, produit *m* de tête
e   destilado *m* de cabeza,
      producto *m* de cabeza
i   distillato *m* di testa, prodotto *m* di testa
n   topdestillaat *n*, topprodukt *n*
d   Kopfdestillat *n*

5064 OVERLOAD INTERLOCK   sa, xr
         X-RAY UNIT
      A unit in which the presetting of voltage,
      current and time are interlinked in such a
      way that if their product exceeds the
      permissible loading of the X-ray tube, the
      latter cannot be energized.
f   dispositif *m* automatique contre la
      surcharge
e   dispositivo *m* automático antisobrecarga
i   dispositivo *m* automatico di sicurezza
      contro il sovraccarico
n   overbelastingsautomaat
d   Überlastungsschutzautomat *m*

5065 OVERMODERATED   rt
      Of a multiplying system, having a
      moderator-to-fuel volume ratio greater
      than that which makes some specified
      reactor parameter an extremum.
f   sur-modéré adj

e sobremoderado adj
i sovramoderato adj
n overgemodereerd adj
d übermoderiert adj

5066 OVERPRESSURE   gp, nw
1. The transient pressure, exceeding the
ambient pressure, manifested in the
shock wave from an explosion.
2. Excessive pressure.
f surpression f
e sobrepresión f
i sovrapressione f
n overdruk
d Überdruck m

5067 OVERRADIATION ALARM   sa
A radiation detector which trips an alarm
when a predetermined level of radioactivity is reached.
f installation f d'alarme de radioactivité
   dangereuse
e equipo m de alarma de radiactividad
   peligrosa
i impianto m d'allarme di radioattività
   pericolosa
n waarschuwingsseintoestel n voor
   gevaarlijke radioactiviteit
d Warnungsgerät n für gefährliche
   Radioaktivität

5068 OVERRIDE PROTECTION   sa
The protection against the danger that the
power of a reactor exceeds a certain
pre-set level.
f protection f contre l'outrepassage de la
   puissance
e protección f contra exceso de potencia
i protezione f contro la sovrappotenza
n bescherming tegen vermogensoverschrijding
d Leistungsüberhöhungsschutz m

5069 OVERVOLTAGE   ct
The difference between the operating
voltage and the Geiger-Müller threshold.
f surtension f
e sobretensión f
i sovratensione f
n overspanning
d Überspannung f

5070 OXIDATION-REDUCTION CYCLE   ch
Reaction involving only the transfer of
electrons from one uncompleted ion to
another in an ionizing event.
f cycle m d'oxydoréduction
e ciclo m de oxidorreducción

i ciclo m d'ossidoriduzione
n redoxcyclus
d Redoxzyklus m

5071 OXIDATION STATE   ch
An atom is said to be in the oxidation state
when it has lost or gained electrons from
its normal state.
f état m d'oxydation
e estado m de oxidación
i stato m d'ossidazione
n oxydatietoestand
d Oxydationszustand m

5072 OXIDATIVE SLAGGING   cg
The removal of fission products from
molten metal.
f scorification f
e escorificación f
i scorificazione f
n verslakken n
d Verschlacken n

5073 OXIDIZER UNIT   ch
The apparatus in which oxidation of
materials used in a reactor takes place.
f dispositif m d'oxydation
e dispositivo m de oxidación
i dispositivo m d'ossidazione
n oxydatietoestel n
d Oxydationsgerät n

5074 OXYGEN   ch
Gaseous element, symbol O, atomic
number 8.
f oxygène m
e oxígeno m
i ossigeno m
n zuurstof
d Sauerstoff m

5075 OXYGEN EFFECT   ra
The increased sensitivity to radiation of
biological material when exposed to the
presence of oxygen.
f effet m d'oxygène
e efecto m de oxígeno
i effetto m d'ossigeno
n zuurstofeffect n
d Sauerstoffeffekt m

5076 OXYPHILE GRANULE   md
A granule accepting, or being stained by,
acid dyestuffs.
f granule m oxyphile
e gránulo m oxífilo
i granulo m ossifilo
n oxyfiel korreltje n
d azidophiles Körnchen n

P

5077 P ELECTRON   np
An electron characterized by having a
principal quantum number of 6.
f électron *m* P
e electrón *m* P
i elettrone *m* P
n P-elektron *n*
d P-Elektron *n*

5078 P SHELL   np
The collection of electrons characterized
by the principal quantum number of 6.
f couche *f* P
e capa *f* P
i strato *m* P
n P-schil
d P-Schale *f*

5079 PACK ROLLING,   mg
      ROLL BONDING
The rolling of two or more metal sheets
arranged in layers, or the rolling of a
metal sheet while sandwiched between two
other sheets.
f colaminage *m*, laminage *m* en paquet
e laminación *f* en paquete
i laminazione *f* a pacco
n pakketwalsen *n*
d Paketwalzen *n*, Walzen *n* im Pack

5080 PACKAGE FILLING MONITOR,   me
      RADIOISOTOPE PACKAGE MONITOR
Monitor used for ascertaining whether
the package contains the prescribed
quantity of material.
f moniteur *m* isotopique de produits emballés
e monitor *m* isotópico de productos
   empaquetados
i monitore *m* isotopico di prodotti imballati
n isotopenmonitor voor verpakte produkten
d Isotopenmonitor *m* für verpackte Produkte

5081 PACKAGE IRRADIATION PLANT   ms
A plant for sterilizing on a large scale
many materials such as plastics, glass,
animal products, medical supplies and
pharmaceuticals, etc.
f installation *f* d'irradiation de produits
   emballés
e equipo *m* de irradiación de productos
   empaquetados
i impianto *m* d'irradiazione di prodotti
   imballati
n installatie voor het bestralen van verpakte
   produkten
d Anlage *f* zum Bestrahlen von verpackten
   Produkten

5082 PACKAGE MONITOR   ms, sa
An instrument for checking that the dose
rate at the surface of packages containing
radioactive materials does not exceed
specific levels.

f moniteur *m* de produits radioactifs
   emballés
e monitor *m* de productos radiactivos
   empaquetados
i monitore *m* di prodotti radioattivi imballati
n monitor voor verpakte radioactieve
   produkten
d Monitor *m* für verpackte radioaktive
   Produkte

5083 PACKAGE REACTOR   rt
A compact transportable power reactor.
f réacteur *m* compact transportable,
   réacteur *m* préfabriqué
e reactor *m* pequeño transportable,
   reactor *m* prefabricado
i reattore *m* compatto trasportabile,
   reattore *m* prefabbricato
n kleine verplaatsbare reactor
d kleiner transportabler Reaktor *m*

5084 PACKAGED COMPONENT   rt
A component constructed in such a way
that it can be mounted as a unit in the
reactor.
f élément *m* de construction prêt à
   fonctionner
e elemento *m* de construcción completo en
   su mismo
i elemento *m* di costruzione autonomo
n bedrijfsklaar onderdeel *n*
d betriebsfertiger Baustein *m*

5085 PACKED COLUMN,   ch
      PACKED TOWER
A vertical column, used in distillation,
absorption, exposition and the like processes, containing packing, e.g. rings,
saddles or crushed rock.
f colonne *f* à garnissage, colonne *f* garnie
e columna *f* atestada, columna *f* de relleno
i colonna *f* a riempimento
n gevulde kolom, kolom met vullichamen
d Füllkörperkolonne *f*, Füllkörperturm *m*

5086 PACKING   ge
The gathering or compression of material
or material particles into a relatively
confined space.
f compression *f*
e compresión *f*
i compressione *f*
n comprimeren *n*
d Ballung *f*, Packung *f*, Zusammenpressen *n*

5087 PACKING EFFECT, see 4262

5088 PACKING FRACTION   ma, np
$f = \triangle_r / A$, where $\triangle_r$ is the relative mass
excess and A the nuclear number.
f fraction *f* de tassement
e fracción *f* de empaquetamiento
i frazione *f* d'impacchettamento

n pakkingsfractie
d Packungsanteil *m*

5089 PAIR ATTENUATION   np, ra
      COEFFICIENT
      That part of the total attenuation
      coefficient that is attributable to pair
      production.
f coefficient *m* d'atténuation de formation
   de paires
e coeficiente *m* de atenuación de formación
   de pares
i coefficiente *m* d'attenuazione di formazione
   di coppie
n verzwakkingscoëfficiënt voor paarvorming
d Schwächungskoeffizient *m* für Paarbildung

5090 PAIR CONVERSION   np
      The conversion of a photon into an electron
      and a positron when the proton traverses
      a strong electric field, such as that
      surrounding a nucleus or an electron.
f conversion *f* d'une paire
e conversión *f* de un par
i conversione *f* d'una coppia
n paarconversie
d Paarkonversion *f*

5091 PAIR CREATION,   np
      PAIR EMISSION, PAIR FORMATION,
      PAIR PRODUCTION
      The conversion of a photon into an
      electron and a posit(r)on, when the proton
      traverses a strong electric field such
      as that surrounding a nucleus or an
      electron.
f production *f* de paires
e formación *f* de pares
i produzione *f* di coppie
n paarvorming
d Paarbildung *f*

5092 PAIR PRODUCTION   ab, np
      ABSORPTION
      The absorption of gamma rays or other
      photons in the process of pair production.
f absorption *f* avec production de paires
e absorción *f* con formación de pares
i assorbimento *m* con produzione di coppie
n absorptie met paarvorming
d Absorption *f* mit Paarbildung

5093 PAIR SPECTROMETER   sp
      An instrument for determining the energy
      of gamma radiation by observing the sum
      of the kinetic energies of the electron and
      posit(r)on in the pair produced by the
      gamma radiation.
f spectromètre *m* aux paires
e espectrómetro *m* por pares
i spettrometro *m* alle coppie
n paarspectrometer
d Paarspektrometer *n*

5094 PAIRED ELECTRON   np
      One of two electrons which are shared by
      two atoms to form a valence bond.

f électron *m* commun
e electrón *m* común
i elettrone *m* comune
n gemeenschappelijk elektron *n*
d geteiltes Elektron *n*

5095 PAIRING ENERGY   np
      The binding energy which arises from the
      difference in binding between an even and
      an odd number of identical nucleons.
f énergie *f* de création d'une paire,
   énergie *f* de parité
e energía *f* de formación de un par
i energia *f* di produzione d'una coppia
n paarvormingsenergie
d Paarbildungsenergie *f*

5096 PALLADIUM   ch
      Metallic element, symbol Pd, atomic
      number 46.
f palladium *m*
e paladio *m*
i palladio *m*
n palladium *n*
d Palladium *n*

5097 PALLADIUM LEAK TEST   sa, rt
      A technique used in vacuum testing and
      leak detection by means of heated
      palladium.
f essai *m* d'étanchéité au palladium
e prueba *f* de estanquecidad con paladio
i prova *f* di tenuta ermetica al palladio
n dichtheidsproef met palladium
d Palladiumleckprüfung *f*

5098 PANETH RULE   ch
      Adsorption of radioelements is promoted
      by formation of an insoluble compound with
      the adsorbing substance, especially when
      the radioelement is present in a negative
      radical.
f règle *f* de Paneth
e regla *f* de Paneth
i regola *f* di Paneth
n regel van Paneth
d Panethsche Regel *f*

5099 PANORAMIC X-RAY MACHINE   xr
      An X-ray machine with a wide-beam angle
      which permits the preparation of
      full-mouth dental surveys with one
      photographic plate.
f appareil *m* à rayons X panoramique
e aparato *m* de rayos X panorámico
i apparecchio *m* per raggi X panoramico
n panoramaröntgenapparaat *n*
d Panoramaröntgenapparat *m*

5100 PARAHELIUM   ch
      Helium composed of atoms in which the
      spins of the two electrons are opposing
      each other.
f parahélium *m*
e parahelio *m*
i paraelio *m*
n parahelium *n*
d Parahelium *n*

395   5114 PAR-

5101 PARAHYDROGEN   ch
      That form of molecular hydrogen in which
      the two nuclear spins in each molecule
      are anti-parallel.
f parahydrogène *m*
e parahidrógeno *m*
i paraidrogeno *m*
n parawaterstof
d Parawasserstoff *m*

5102 PARALLAX PANORAMAGRAM   xr
      An autostereogram, in each vertical strip
      of which the image changes progressively
      from a leftward aspect at one edge to a
      rightward aspect at the other edge.
f panoramagramme *m* parallactique
e panoramagrama *m* paraláctico
i panoramagramma *m* parallattico
n parallactisch panoramagram *n*
d parallaktisches Panoramagramm *n*

5103 PARALLAX STEREOGRAM   xr
      An autostereogram consisting of a
      composite production in which vertical
      strips from a left eye and a right eye
      view are interdigitated and exposed to
      view through a screen of alternate opaque
      and transparent lines.
f stéréogramme *m* parallactique
e estereograma *m* paraláctico
i stereogramma *m* parallattico
n parallactisch stereogram *n*
d parallaktisches Stereogramm *n*

5104 PARALLEL FLOW, see 1170

5105 PARALLEL-PLATE CHAMBER   ic
      An ionization chamber with plane
      parallel electrodes.
f chambre *f* d'ionisation à plaques
   parallèles
e cámara *f* de ionización de placas
   paralelas
i camera *f* d'ionizzazione a lamiere
   parallele
n ionisatievat *n* met parallelle platen
d Parallelplattenionisationskammer *f*

5106 PARALLEL-PLATE COUNTER   ct
      TUBE
      A radiation counter tube using parallel
      plate electrodes.
f tube *m* compteur à plaques parallèles
e tubo *m* contador de placas paralelas
i tubo *m* contatore a lamiere parallele
n telbuis met parallelle platen
d Parallelplattenzählrohr *n*

5107 PARALYSIS CIRCUIT   ct
      A circuit which renders the counting
      system inoperative for a pre-determined
      time after its recorded pulse.
f circuit *m* de blocage,
   circuit *m* de paralysie
e circuito *m* de bloqueo,
   circuito *m* de parálisis
i circuito *m* di bloccaggio,
   circuito *m* di paralisi

n blokkeerschakeling
d Blockierschaltung *f*, Sperrschaltung *f*

5108 PARALYSIS TIME   ct
      Constant and known value imposed on the
      resolving time, usually in order to make
      the correction for resolving time more
      accurate.
f temps *m* de paralysie
e tiempo *m* de parálisis
i tempo *m* di paralisi
n blokkeringstijd
d Blockierungszeit *f*

5109 PARAMAGNETIC ABSORPTION   ab
      Absorption of particles by paramagnetic
      substances.
f absorption *f* paramagnétique
e absorción *f* paramagnética
i assorbimento *m* paramagnetico
n paramagnetische absorptie
d paramagnetische Absorption *f*

5110 PARAMAGNETIC RESONANCE   np
      Resonance observable in a paramagnetic
      material as a peak in the energy absorption
      spectrum at a frequency related to the
      strength of the applied magnetic field and
      to the gyromagnetic ratio.
f résonance *f* paramagnétique
e resonancia *f* paramagnética
i risonanza *f* paramagnetica
n paramagnetische resonantie
d paramagnetische Resonanz *f*

5111 PARAMETER   ge, un
      One of the constants entering into a
      functional equation and corresponding to
      some characteristic property, dimension
      or degree of freedom.
f paramètre *m*
e parámetro *m*
i parametro *m*
n parameter
d Parameter *m*

5112 PARAPOSITONIUM,   np
      PARAPOSITRONIUM
      The state of posit(r)onium in which the
      spins of electron and positron are anti-
      parallel.
f parapositonium *m*
e parapositonio *m*
i parapositonio *m*
n parapositonium *n*
d Parapositronium *n*

5113 PARASITIC CAPTURE   np
      Absorption of neutrons not leading to
      fission or any other desired process.
f capture *f* parasite
e captura *f* parásita
i cattura *f* parassita
n parasitaire vangst
d parasitärer Einfang *m*

5114 PARASITIC COUNT   ct
      A count causing an accidental increase in

the value of the background of a device.
f coup *m* parasite
e impulso *m* de cuenta parásito,
   impulso *m* de recuento parásito
i impulso *m* di conteggio parassito
n parasitaire telpuls
d parasitärer Zählimpuls *m*

5115 PARAVERTEBRAL IRRADIATIONra, xr
Radiological treatment of the surroundings
of the vertebra.
f irradiation *f* paravertébrale
e irradiación *f* paravertebral
i irradiazione *f* paravertebrale
n paravertebrale bestraling
d paravertebrale Bestrahlung *f*

5116 PARENT   np
Of a nuclide, that radioactive nuclide from
which it is formed by decay.
f parent *m*, précurseur *m*, substance *f* père
e ascendiente *m*, substancia *f* padre
i sostanza *f* progenitrice
n moeder, moederstof
d Ausgangsstoff *m*, Mutterstoff *m*

5117 PARENT ATOM   np
The atom containing the original nucleus.
f atome *m* père, père *m* atomique
e átomo *m* original, átomo *m* padre
i atomo *m* padre, atomo *m* progenitore
n moederatoom *n*
d Ausgangsatom *n*, Mutteratom *n*

5118 PARENT ELEMENT   ch, np
An element which produces another element
or isotope, whether by spontaneous radioactive decay or as target element in a
nuclear bombardment.
f élément *m* père
e elemento *m* original, elemento *m* padre
i elemento *m* padre, elemento *m* progenitore
n moederelement *n*
d Ausgangselement *n*, Mutterelement *n*

5119 PARENT MASS PEAK,   sp
PARENT PEAK
The component of a mass spectrum that is
caused by the undissociated molecule.
f raie *f* mère
e línea *f* madre
i riga *f* madre
n uitgangslijn
d Ausgangslinie *f*,
   Bezugslinie *f* im Massenspektrum

5120 PARENT MATERIAL,   is
RADIOACTIVE PRECURSOR
A radioactive isotope that upon disintegration yields a specific nuclide,
the daughter.
f matière *f* originale,
   précurseur *m* radioactif
e materia *f* original
i materia *f* progenitrice
n uitgangsmateriaal *n*
d Ausgangsstoff *m*

5121 PARHELIUM,   sp
PARHELIUM TERMS
One group or system of terms in the
spectrum of helium that is due to atoms in
which the spins of the two electrons are
opposing each other.
f parhélium *m*
e parhelio *m*
i parelio *m*
n parhelium *n*
d Parhelium *n*

5122 PARITY   ma
A symmetry property of a spatial wave
function.
f parité *f*
e paridad *f*
i parità *f*
n pariteit
d Parität *f*

5123 PARSONSITE   mi
A rare secondary mineral, containing
about 26.4 % of U.
f parsonsite *f*
e parsonsita *f*
i parsonsite *f*
n parsonsiet *n*
d Parsonsit *m*

5124 PARTIAL BODY   ra, xr
IRRADIATION
f irradiation *f* partielle du corps
e irradiación *f* parcial del cuerpo
i irradiazione *f* parziale del corpo
n gedeeltelijke lichaamsbestraling
d Teilkörperbestrahlung *f*

5125 PARTIAL CONDENSATION, see 1721

5126 PARTIAL CRIT   rt
Something less than a critical mass of
fissile (fissionable) material.
f masse *f* critique partielle
e masa *f* crítica parcial
i massa *f* critica parziale
n bijna voldoende kritische massa
d nahezu genügende kritische Masse *f*

5127 PARTIAL CROSS SECTION   cs
A cross section for a particular process
among several competing processes.
f section *f* efficace partielle
e sección *f* eficaz parcial
i sezione *f* d'urto parziale
n gedeeltelijke doorsnede
d Teilquerschnitt *m*

5128 PARTIAL DECAY CONSTANT,   np
PARTIAL DISINTEGRATION
CONSTANT
For a radionuclide, the probability per
unit time for the spontaneous decay of a
nucleus by one of two or more modes of
decay.
f constante *f* de désintégration partielle
e constante *f* de desintegración parcial

i   costante *f* di disintegrazione parziale
n   partiële desintegratieconstante,
    partiële vervalconstante
d   partielle Zerfallskonstante *f*

5129 PARTIAL EXPOSURE   ra
    Exposure to radiation of only part of the
    body or material under consideration.
f   exposition *f* partielle
e   exposición *f* parcial
i   esposizione *f* parziale
n   gedeeltelijke exposie
d   Teilexposion *f*, Teilexposition *f*

5130 PARTIAL LEVEL WIDTH,   np
        PARTIAL WIDTH
    A level width assigned to each of several
    different transitions from a given nuclear
    level.
f   largeur *f* de niveau partielle
e   ancho *m* de nivel parcial
i   larghezza *f* di livello parziale
n   deelniveaubreedte
d   Teilniveaubreite *f*

5131 PARTIAL SHIELD,   ra
        SHADOW SHIELD
    A shield which does not surround an
    irradiation source completely.
f   écran *m* partial
e   pantalla *f* parcial
i   schermo *m* parziale
n   deelscherm *n*
d   Teilschirm *m*

5132 PARTIAL WAVE   np
    An incident beam or plane wave of particles
    may be regarded as a coherent super-
    position of a number of partial waves, each
    having a definite angular momentum.
f   onde *f* partielle
e   onda *f* parcial
i   onda *f* parziale
n   deelgolf
d   Teilwelle *f*

5133 PARTIALLY OCCUPIED BAND   np
    An energy band not all the levels of which
    correspond to the energy of each of two
    electrons with opposite spins in a given
    substance in a given state.
f   bande *f* partiellement occupée
e   banda *f* parcialmente ocupada
i   banda *f* parzialmente occupata
n   gedeeltelijk bezette band
d   teilweise besetztes Energieband *n*

5134 PARTICLE   np
    A very small portion of matter and
    smaller than an atom.
f   particule *f*
e   partícula *f*
i   particella *f*
n   deeltje *n*, partikel *n*
d   Partikel *n*, Teilchen *n*

5135 PARTICLE ABSORPTION   np
    A nuclear or atomic interaction, whereby

the incident particle disappears as a free
particle.
f   absorption *f* de particules
e   absorción *f* de partículas
i   assorbimento *m* di particelle
n   deeltjesabsorptie
d   Teilchenabsorption *f*

5136 PARTICLE ACCELERATOR   pa
    Any device for accelerating charged
    particles.
f   accélérateur *m* de particules
e   acelerador *m* de partículas
i   acceleratore *m* di particelle
n   deeltjesversneller
d   Teilchenbeschleuniger *m*

5137 PARTICLE ANALYZER,   sp
        PARTICLE SPECTRUM ANALYZER,
        RADIATION SPECTRUM ANALYZER
    A general term for instruments for the
    measurement of alpha, beta and gamma
    ray spectra.
f   spectromètre *m* de particules
e   espectrómetro *m* de partículas
i   spettrometro *m* di particelle
n   deeltjesspectrometer
d   Teilchenspektrometer *n*

5138 PARTICLE BOOSTER   pa
    An accelerator receiving the accelerated
    particles from the first accelerator and
    injecting them after acceleration in a
    synchrotron.
f   accélérateur *m* intermédiaire
e   acelerador *m* intermedio
i   acceleratore *m* intermedio
n   tussenversneller
d   Zwischenbeschleuniger *m*

5139 PARTICLE CHARGE   np
    The electrical charge of a particle.
f   charge *f* d'une particule
e   carga *f* de una partícula
i   carica *f* d'una particella
n   lading van een deeltje
d   Teilchenladung *f*

5140 PARTICLE COUNTER,   ct
        PARTICLE DETECTOR
    Device for counting or detecting particles
    of atomic magnitude.
f   compteur *m* de particules
e   contador *m* de partículas
i   contatore *m* di particelle
n   deeltjesteller
d   Teilchenzähler *m*

5141 PARTICLE CURRENT DENSITY,
        see 1578

5142 PARTICLE EMISSION, see 1421

5143 PARTICLE FLUENCE   np
    At a given point in space, the number of
    neutrons incident on a small sphere in a
    time interval divided by cross sectional
    area of that sphere and by the time interval.

f fluence f des particules
e fluencia f de las partículas
i fluenza f delle particelle
n deeltjesfluentie
d Teilchenfluenz f

5144 PARTICLE FLUX   np
      For particles of a given energy, the
      product of particle density with speed.
f flux m de particules
e flujo m de partículas
i flusso m di particelle
n deeltjesflux
d Teilchenfluss m

5145 PARTICLE FLUX DENSITY, see 2719

5146 PARTICLE FLUX   me, np
      DENSITY INDICATOR
      An indicator designed to give an
      estimation of particle flux density.
f signaleur m de la densité de flux de
   particules
e indicador m de la densidad de flujo de
   partículas
i segnalatore m della densità di flusso
   di particelle
n indicator voor deeltjesfluxdichtheid
d Teilchenfluenzanzeiger m

5147 PARTICLE FLUX   me, np
      DENSITY METER
      A measuring assembly for particle flux
      density.
f fluxmètre m de particules
e fluxímetro m de partículas
i flussometro m di particelle
n meter voor deeltjesfluxdichtheid
d Gerät n zur Bestimmung des
   Teilchenfluenz

5148 PARTICLE FLUX   me, np
      DENSITY MONITOR
      A monitor designed to measure and
      respond to particle flux density.
f moniteur m de la densité de flux de
   particules
e monitor m de la densidad de flujo de
   partículas
i monitore m della densità di flusso
   di particelle
n moniteur voor deeltjesfluxdichtheid
d Warngerät n für Teilchenfluenz

5149 PARTICLE INJECTION   pa
      Operation which consists in introducing
      particles meant to be accelerated into an
      accelerator.
f injection f de particules
e inyección f de partículas
i iniezione f di particelle
n deeltjesinjectie, inschieten n van deeltjes
d Teilcheneinschuss m, Teilcheninjektion f

5150 PARTICLE RADIATION, see 1422

5151 PARTICLE SIZE   an, me
      ANALYZER
      An analyzer which determines particle
      size from the intensity of the beta
      radiation scattered by a layer of sediment.
f analysateur m des dimensions de
   particules
e analizador m de las dimensiones de
   partículas
i analizzatore m delle dimensioni di
   particelle
n apparaat n voor het bepalen van
   deeltjesafmetingen
d Apparat m zur Teilchengrössenbestimmung

5152 PARTICLE TRACK   ct
      COMPUTER
      An analog(ue) computer enabling the shape
      and path of the particle beam within an
      evacuated tube to be visually displayed
      for any combination of positions and
      settings.
f calculateur m analogique du parcours du
   faisceau
e computadora f analógica de la trayectoria
   del haz
i calcolatore m analogico del percorso del
   fascio
n analogonrekentuig n voor het bepalen van
   de bundelbaan
d Analogrechner m zur Bestimmung der
   Elektronenbahn

5153 PARTICULATE ACTIVITY   np, nw
      The radioactivity present in the air due to
      the presence of radioactive particles.
f activité f de particules,
   activité f particulaire
e actividad f de partículas
i attività f di particelle
n deeltjesactiviteit
d Teilchenaktivität f

5154 PARTICULATE MATTER   np
      Said of a substance consisting of a large
      number of particles, such as smoke, sand,
      powder, etc.
f matière f subdivisée
e materia f particulada
i materia f suddivisa
n fijnverdeelde substantie
d feinunterteilte Substanz f

5155 PARTICULATES   ch
      Solid particles dispersed in a gas stream
      as opposed to the liquid droplets, aerosols
      or vapo(u)rs which it may contain.
f particules pl matérielles
e partículas pl materiales
i particelle pl materiali
n vaste deeltjes pl
d Feststoffteilchen pl

5156 PARTITION COEFFICIENT, see 1916

5157 PASCHEN'S LAW   np
      The law stating that, at a constant

temperature, the breakdown voltage is a
function only of the product of the gas
pressure by the distance between parallel
plane electrodes.
f   loi f de Paschen
e   ley f de Paschen
i   legge f di Paschen
n   wet van Paschen
d   Paschensches Gesetz n

5158 PASTE REACTOR, see 2726

5159 PATH LENGTH   np
The distance measured along the path of
the particle.
f   longueur f du parcours
e   longitud f de la trayectoria
i   lunghezza f del percorso
n   weglengte
d   Weglänge f

5160 PAULI EXCLUSION PRINCIPLE   qm
A law of quantum mechanics stating that
no two identical fermions can exist in the
same quantum state.
f   principe m d'exclusion de Pauli
e   principio m de exclusión de Pauli
i   principio m d'esclusione di Pauli
n   pauliverbod n
d   Ausschliessungsprinzip n, Pauli-Prinzip n

5161 PAULI-FERMI PRINCIPLE   np
Each level of a quantized system can
include one, two or no electrons. If there
are two electrons, they must have spins
in opposite directions.
f   principe m de Pauli-Fermi
e   principio m de Pauli-Fermi
i   principio m di Pauli-Fermi
n   principe n van Pauli en Fermi
d   Pauli-Fermisches Prinzip n

5162 PEAK DOSE   ra
The dose delivered to an irradiated body
or material by external radiation
measured at that depth below the surface
of the material where the absorbed dose
rate is a maximum.
f   dose f maximale
e   dosis f máxima
i   dose f massima
n   maximale dosis
d   Höchstdosis f

5163 PEAK PARTICLE VELOCITY   np
The maximum absolute value of the
instantaneous particle velocity in a
specified time interval.
f   vitesse f maximale de particules
e   velocidad f máxima de partículas
i   velocità f massima di particelle
n   maximale deeltjessnelheid
d   maximale Teilchengeschwindigkeit f

5164 PEAKING FLUX   np, rt
Local increase of the flux due i.a. to the
heterogeneous state of the core.

f   pic m de flux
e   pico m de flujo
i   picco m di flusso
n   fluxpiek
d   Flussspitze f

5165 PEAKING POWER   np, rt
Local increase of the power due i.a. to
the heterogeneous state of the core.
f   pic m de puissance
e   pico m de potencia
i   picco m di potenza
n   vermogenspiek
d   Leistungsspitze f

5166 PEBBLE BED REACTOR   rt
A nuclear reactor in which some or all of
the essential nuclear material is in the
form of a stationary bed of small balls
in contact with each other.
f   réacteur m à lit de boulets
e   reactor m de lecho de bolas
i   reattore m a letto di sfere
n   ballenreactor, kogelbedreactor
d   Kugelbettreaktor m,
    Reaktor m mit Brennstoffkugeln

5167 PEGMATITE   mi
A coarse-grained acidic rock resulting
from the solidification of the last portion
of a molten ro(k mass, usually in dykes
or sills.
f   pegmatite f
e   pegmatita f
i   pegmatite f
n   pegmatiet n
d   Pegmatit m

5168 PELIOSIS,   md
        PURPLE, PURPURA
Large hemorrhagic spots in or under the
skin of mucous tissues.
f   péliose f, purpura m
e   peliosis f, púrpura f
i   peliosi f, porpora f
n   purpura
d   Blutfleckenkrankheit f, Peliosis f,
    Purpura f

5169 PELLICLE   ph
In the photographic emulsion technique,
a thick emulsion which can be used
without a supporting plate.
f   pellicule f à autosupport
e   emulsión f autosoportada
i   emulsione f autoportante
n   zelfdragende emulsie
d   selbsttragende Emulsion f

5170 PELTIER EFFECT   gp
The evolution or absorption of heat at a
junction between two similar conductors,
across which a current passes.
f   effet m Peltier
e   efecto m Peltier
i   effetto m Peltier
n   peltiereffect n
d   Peltier-Effekt m

5171 PELTIER EFFECT   cl, ec
     REFRIGERATOR
     A photomultiplier unit containing a
     cylindrical refrigerator open at both ends.
f   enceinte *f* réfrigérée à effet Peltier
e   envoltura *f* refrigerada de efecto Peltier
i   involucro *m* refrigerato da effetto Peltier
n   koelmantel met peltiereffect
d   Kühlmantel *m* mit Peltier-Effekt

5172 PELVIGRAPHY   xr
     Radiological examination of the pelvis.
f   pelvigraphie *f*
e   pelvigrafía *f*
i   pelvigrafia *f*
n   bekkenfotografie, pelvigrafie
d   Beckenröntgenographie *f*, Pelvigraphie *f*

5173 PELVIMETRY   xr
     The radiological determination of the
     dimensions of the pelvic outlet.
f   pelvimétrie *f*
e   pelvimetría *f*
i   pelvimetria *f*
n   bekkenmeting, pelvimetrie
d   Beckenmessung *f*, Pelvimetrie *f*

5174 PENDULUM   gp
     A body suspended so as to be free to
     swing or oscillate.
f   pendule *m*
e   péndulo *m*
i   pendolo *m*
n   slinger
d   Pendel *n*

5175 PENDULUM THERAPY   xr
     Moving-field therapy in which the source
     of radiation swings to and fro in an arc
     concave to the patient.
f   radiothérapie *f* pendulaire
e   radioterapia *f* pendular
i   radioterapia *f* pendolare
n   slingertherapie
d   Pendeltherapie *f*

5176 PENETRABILITY   ra
     The susceptibility of a substance or an
     object to be traversed by radiation.
f   pénétrabilité *f*
e   penetrabilidad *f*
i   penetrabilità *f*
n   doordringbaarheid
d   Durchdringbarkeit *f*, Durchlässigkeit *f*

5177 PENETRATING CAPACITY,   ra
     PENETRATING POWER,
     RADIATION HARDNESS
     The power of radiations to traverse
     materials or objects.
f   dureté *f* de rayonnement,
     pouvoir *m* de pénétration
e   dureza *f* de radiación,
     poder *m* de penetración
i   durezza *f* di radiazione,
     potere *m* di penetrazione
n   doordringingsvermogen *n*,
     stralingshardheid

d   Durchdringungsvermögen *n*,
     Strahlungshärtegrad *m*

5178 PENETRATING COMPONENT   np, ra
     A component penetrating through the
     potential barrier.
f   composante *f* pénétrante
e   componente *m* penetrante
i   componente *m* penetrante
n   doordringende component
d   durchdringende Komponente *f*

5179 PENETRATING RADIATION, see 3143

5180 PENETRATING SHOWER   cr
     A shower containing particles, mainly
     mu mesons and nucleons, capable of
     penetrating 15 to 20 cm of lead.
f   gerbe *f* pénétrante
e   chaparrón *m* penetrante
i   sciame *m* penetrante
n   doordringende bui
d   durchdringender Schauer *m*

5181 PENETRATION   ra
     The passage of particles or radiation
     through matter or objects.
f   pénétration *f*
e   penetración *f*
i   penetrazione *f*
n   doordringing
d   Durchdringung *f*

5182 PENETRATION FACTOR,   np
     PENETRATION PROBABILITY,
     TRANSMISSION COEFFICIENT
     In a nuclear reaction, the probability that
     an incident particle will pass through the
     barrier of the nucleus.
f   facteur *m* de pénétration,
     probabilité *f* de pénétration
e   factor *m* de penetración,
     probabilidad *f* de penetración
i   fattore *m* di penetrazione,
     probabilità *f* di penetrazione
n   doordringingsfactor, doordringingskans,
     transmissiecoëfficiënt
d   Durchdringungsfaktor *m*,
     Durchdringungswahrscheinlichkeit *f*

5183 PENETRATION POTENTIAL   np
     The potential needed by a particle to
     penetrate the potential barrier.
f   potentiel *m* de pénétration
e   potencial *m* de penetración
i   potenziale *m* di penetrazione
n   doordringingspotentiaal
d   Durchdringungspotential *n*

5184 PENETRATION PROBABILITY,   np
     TUNNEL EFFECT
     The probability that a particle penetrates
     the potential barrier.
f   effet *m* tunnel
e   efecto *m* túnel
i   effetto *m* tunnel
n   tunneleffect *n*
d   Tunneleffekt *m*

5185 PENETROMETER,   me, xr
         QUOTIMETER
   An instrument for measuring the
   penetrating power of a beam of X-rays
   or other penetrating radiation by comparing transmission through various absorbers.
f pénétramètre *m*
e penetrómetro *m*
i penetrametro *m*
n kwaliteitsmeter, penetrometer
d Penetrometer *n*

5186 PENETROMETER SENSITIVITY   me
   The smallest change in thickness which
   can be detected in a radiograph,
   expressed as a percentage of the total
   thickness, the object being assumed to
   be of specific homogeneous material.
f sensibilité *f* de pénétramètre
e sensibilidad *f* de penetrómetro
i sensibilità *f* di penetrametro
n kwaliteitsmetergevoeligheid,
   penetrometergevoeligheid
d Penetrometerempfindlichkeit *f*

5187 PENTA-ETHER   mt
   An organic solvent used for the solvent
   extraction of uranium.
f penta-éther *m*
e penta-éter *m*
i penta-etere *m*
n penta-ether
d Dibutyläther *m* des Tetraäthylenglykols,
   Pentaäther *m*

5188 PENUMBRA,   ra
         REGION OF PARTIAL SHADOW
   A region behind an object in a beam of
   radiation such that if from any point in
   this region straight lines are drawn to
   different points in the source, some pass
   through the object and some do not.
f pénombre *f*
e penumbra *f*
i penombra *f*
n halfschaduw
d Halbschatten *m*

5189 PERCENTAGE DEPTH DOSE   ra
   The ratio, expressed as a percentage, of
   the absorbed dose at any given depth
   within a body to the absorbed dose at some
   reference point of the body along the
   central ray.
f pourcentage *m* de dose en profondeur,
   rendement *m* en profondeur
e porcentaje *m* de dosis en profundidad,
   rendimiento *m* en profundidad
i dose *f* specifica assorbita in profondità,
   rendimento *m* in profondità
n procentuele dieptedosis
d relative Tiefendosis *f*

5190 PERCENTAGE LOSS   np
   The total number of neutrons absorbed or
   lost by leakage, etc.
f pourcentage *m* des pertes

e porcentaje *m* de pérdidas
i perdita *f* percentuale
n totaal neutronenverlies *n*
d Gesamtneutronenverlust *m*

5191 PERCOLATING FILTER, see 641

5192 PERCOLATION RATE   rw
   The rate of transmission of water
   through a porous material.
f vitesse *f* de percolation
e velocidad *f* de percolación
i velocità *f* di percolazione
n sijpelsnelheid, percolatiesnelheid
d Durchseihgeschwindigkeit *f*,
   Perkolationsgeschwindigkeit *f*

5193 PERFECT MOSAIC CRYSTAL, see 3382

5194 PERFORATED PLATE,   ch
         SIEVE PLATE
   In a column, a plate or tray containing
   numerous perforations.
f plateau *m* à trous, plateau *m* perforé
e plato *m* perforado
i piatto *m* forato
n zeefschotel
d Siebboden *m*

5195 PERHAPSATRON   np
   A smaller version of the Zeta apparatus
   for investigating controlled fusion of
   hydrogen.
f perhapsatron *m*
e perhapsatrón *m*
i perapsatrone *m*
n perhapsatron *n*
d Perhapsatron *n*

5196 PERIDUROGRAPHY   xr
   Radiological examination of the peridural
   cavity.
f péridurographie *f*
e peridurografía *f*
i peridurografia *f*
n peridurografie
d Peridurographie *f*

5197 PERIOD   ge
   A variable quantity whose characteristics
   are reproduced at equal intervals of an
   independent variable is said to be periodic,
   and the minimum interval after which the
   same characteristics recur is known as
   the period.
f période *f*
e período *m*
i periodo *m*
n periode
d Periode *f*

5198 PERIOD DEMAND SIGNAL   rt
   In automatic systems for power increase
   of a nuclear reactor, the reference signal
   which stabilizes that period if it is desired
   that the reactor power is increased.
f signal *m* de la constante de temps de
   consigne

e señal f del período de consigna
i segnale m d'impostazione del tempo di divergenza
n signaal n voor de ingestelde.tijdconstante
d Signal n der Sollzeitkonstante

5199 PERIOD METER,   me
    TIME CONSTANT METER
An electronic subassembly which, in association with one or more detectors, is used to indicate the time constant (period) of a nuclear reactor.
f périodemètre m
e medidor m de período, periodímetro m
i indicatore m del tempo di divergenza, unità f per la misura della costante di tempo
n crescentiemeter, tijdconstantemeter
d Periodenmessgerät n, Zeitkonstantemesser m

5200 PERIOD RANGE,   rt
    TIME CONSTANT RANGE
The range of power level within which the reactor time constant rather than reactor power is of primary importance for reactor control.
f domaine m de divergence
e régimen m de divergencia
i campo m del tempo di divergenza, intervallo m del tempo di divergenza
n tijdconstantegebied n
d Periodenbereich m, Zeitkonstantenbereich m

5201 PERIOD SHUT-DOWN,   rt
    PERIOD TRIP
Shut-down of a reactor provoked by a signal corresponding to a too short period.
f arrêt m d'urgence par constante de temps insuffisante
e paro m de emergencia por período demasiado breve
i arresto m rapido di divergenza
n noodstop door te lage tijdconstante
d Notabschaltung f wegen zu kleiner Zeitkonstante

5202 PERIODIC SYSTEM   ch
An arrangement of the elements in a systematic grouping, based on the atomic number of each element.
f système m périodique
e sistema m periódico
i sistema m periodico
n periodiek systeem n
d periodisches System n

5203 PERIODIC TIME   ge
The period when the independent variable is time.
f période f de temps
e período m de tiempo
i periodo m di tempo
n tijdsperiode
d Zeitperiode f

5204 PERIODICITY   gp
The recurrence of an event, phenomenon or characteristic feature at regular intervals of time, distance or other variable.
f périodicité f
e periodicidad f
i periodicità f
n periodiciteit
d Periodizität f

5205 PERIOST,   md
    PERIOSTEUM
The tough fibrous membrane surrounding bone.
f périoste m
e periostio m
i periostio m
n beenvlies n, periostium n
d Beinhaut f, Knochenhaut f, Periost n

5206 PERIPHERAL ARTERIOGRAPHY   xr
f artériographie f périphérique
e arteriografía f periférica
i arteriografia f periferica
n periferische arteriografie
d periphere Arteriographie f

5207 PERIPHERAL ELECTRON, see 1300

5208 PERIPHERAL REGION,   gp
    PERIPHERY
The volume surrounding an object in which the gravitational, magnetic or electric fields of the object produce observable effects.
f périphérie f
e periferia f
i periferia f
n periferie
d Peripherie f

5209 PERISCOPE   me
Instrument sometimes used in nuclear reactors for viewing events inside.
f périscope m
e periscopio m
i periscopio m
n periscoop
d Periskop n

5210 PERMANENT LIGHT SOURCE   is, ms
A light source made by incorporating a radioactive isotope into a phosphor, which is excited by the ionizing radiation and produces light.
f source f permanente de lumière
e fuente f permanente de luz
i sorgente f permanente di luce
n permanente lichtbron
d permanente Lichtquelle f

5211 PERMANENT SET,   mg
    PLASTIC STRAIN
The term used in reference to the permanent distortion of metals under applied stresses which strain the material beyond its elastic limit.

f déformation *f* permanente
e deformación *f* permanente
i deformazione *f* permanente
n permanente vervorming
d Beanspruchung *f* oberhalb der
   Elastizitätsgrenze,
   bleibende Verformung *f*

**5212 PERPENDICULAR AXIAL   xr
        ARC THERAPY**
A special form of moving beam therapy.
f irradiation *f* perpendiculaire axiale
e irradiación *f* con eje perpendicular
i irradiazione *f* con asse perpendicolare
n axiale slingerbestraling
d axiale Pendelbestrahlung *f*

5213 PERSISTENCE, see 134

5214 PERSISTENT RADIATION   np
The radiation emitted by fission products
in slow radioactive decay.
f rayonnement *m* persistant
e radiación *f* persistente
i radiazione *f* persistente
n langdurige uitstraling
d Dauerstrahlung *f*

5215 PERSONAL AIR SAMPLER   ra, sa
An air sampler for toxic, non-toxic and
radioactive dusts, gases, fumes, mists,
etc.
f appareil *m* de prélèvement d'échantillons
   d'air individuel
e muestreador *m* de aire individual
i apparecchio *m* da disporre campioni
   d'aria individuale
n individuële luchtmonsternemer
d individueller Luftprobenehmer *m*

5216 PERSONAL DOSEMETER   ma
Small dosemeter giving an evaluation of
the exposure carried by the carrier.
f dosimètre *m* individuel
e dosímetro *m* particular
i dosimetro *m* individuale
n persoonlijke dosismeter
d Privatdosismesser *m*

5217 PERSONAL EXPOSUREMETER   me
Small exposuremeter measuring the
exposure received by the carrier.
f dosimètre *m* individuel,
   exposimètre *m* individuel
e exposímetro *m* particular
i esposimetro *m* individuale
n persoonlijke exposiemeter
d Privatexponierungsmesser *m*,
   Privatexpositionsmesser *m*

5218 PERSONAL MONITORING   sa
Periodic or continuous determination of
the amount of ionizing radiation or radioactive contamination in or upon a person
or his clothing.
f contrôle *m* individuel
e control *m* de seguridad individual

i controllo *m* individuale
n persoonlijke bewaking
d individuelle Überwachung *f*

5219 PERTURBATION   pp
A small contribution to a physical quantity,
such that the problem into which the
quantity enters can be solved exactly or in
a far simpler manner than otherwise if the
perturbation is neglected.
f perturbation *f*
e perturbación *f*
i perturbazione *f*
n storing
d Störung *f*

5220 PERTURBATION SAMPLE   rt
A piece of material which, when placed in
a reactor, causes a change (perturbation)
in the neutron flux.
f échantillon *m* de perturbation
e muestra *f* de perturbación
i campione *m* di perturbazione
n storingsmonster *n*
d Störungsprobestück *n*

5221 PERTURBATION THEORY   rt
A mathematical method used to determine
the effects of small localized changes
resulting, for example, from localized
poisoning or temperature changes.
f théorie *f* des perturbations
e teoría *f* de las perturbaciones
i teoria *f* delle perturbazioni
n storingstheorie
d Störungstheorie *f*

5222 pH-METER   ma
An assembly for measuring the pH of a
liquid associated with the operation of a
nuclear reactor.
f pH-mètre *m*
e pehachímetro *m*
i pH-metro *m*
n pH-meter
d pH-Messgerät *n*

5223 PHANTOM   ra
A volume of phantom material either large
enough to provide full backscatter or
constructed to simulate some special
shape, such as part of the human body.
f fantôme *m*
e fantasma *m*
i fantasma *m*
n fantoom *n*
d Phantom *n*

5224 PHANTOM MATERIAL   ra
A solid or liquid whose absorbing and
scattering properties for a given radiation
are similar to those of a given biological
material, such as part or whole of the
human body.
f matière *f* phantôme
e materia *f* fantasma
i materia *f* fantasma

n fantoommateriaal *n*
d Phantommaterial *n*

5225 PHANTOSCOPE   an, sp
An analyzer for rapid routine radiation
investigation based on the identification
of radiation by the observation of
characteristic radiation spectra.
f phantoscope *m*
e fantoscopio *m*
i fantoscopio *m*
n fantoscoop
d Phantoskop *n*

5226 PHASE BOUNDARY   ch
The boundary layer between two substances
which are in different phases.
f couche *f* limite de phases
e capa *f* límite de fases
i strato *m* limite di fasi
n fazengrenslaag
d Phasengrenzschicht *f*

5227 PHASE DIAGRAM, see 1317

5228 PHASE FOCUSING   pa
A focusing method to make the particles
revert to their correct place if they have
got ahead or fallen behind and are thus
getting out of phase.
f focalisation *f* de phase
e enfoque *m* de fase, focalizacíon *f* de fase
i focalizzazione *f* di fase
n fazefocussering
d Phasenfokussierung *f*

5229 PHASE INTEGRAL   ma
If p and q are two canonical variables of
the classical mechanics of a periodically
moving system, the integral $\int$ pdq taken
over one period of the movement is
called the phase integral.
f intégrale *f* de phase
e integral *f* de fase
i integrale *m* di fase
n faze-integraal
d Phasenintegral *n*

5230 PHASE RELATIONSHIP,   ch
          PHASE RULE
A law of equilibration between phases
of a chemically homogeneous mixture or
of a pure substance.
f règle *f* des phases
e regla *f* de las fases,
    relación *f* de las fases
i regola *f* delle fasi
n fazenregel
d Phasenregel *f*

5231 PHASE SHIFT   np
In the wave description of scattering, the
change in phase suffered by the wave in
passing through the scattering field, e.g.,
of a nucleus.
f déphasage *m*
e desfase *m*

i sfasamento *m*, spostamento *m* di fase
n fazeverschuiving
d Phasenverschiebung *f*

5232 PHASE SPACE, see 2948

5233 PHASE SPACE CELL   ma, np
An elementary hypervolume in the phase
space.
f cellule *f* de l'espace de phase
e célula *f* del espacio de fase
i cellula *f* dello spazio di fase
n cel, elementair hypervolume *n*
d Phasenraumzelle *f*

5234 PHASE SPACE DISTRIBUTION, see 257

5235 PHASE TRANSITION,   ch
          TRANSITION
A change from one phase to another, or
a change in the number of phases present
in a system.
f transition *f* de phase
e transición *f* de fase
i transizione *f* di fase
n fazeovergang
d Phasenübergang *m*

5236 PHASOTRON, see 2758

5237 PHENACITE   mi
A mineral, a silicate of beryllium.
f phénacite *f*
e fenacita *f*
i fenacite *f*
n fenaciet *n*
d Phenazit *m*

5238 PHLEBOGRAPHY   xr
The radiological examination of veins
following the injection of a contrast medium.
f phlébographie *f*
e flebografía *f*
i flebografia *f*
n flebografie
d Phlebographie *f*

5239 PHOSPHATE GLASS   me, mt
A special glass which fluoresces under
ultraviolet light following gamma
irradiation.
f verre *m* phosphaté
e vidrio *m* al fosfato
i vetro *m* al fosfato
n fosfaatglas *n*
d Phosphatglas *n*

5240 PHOSPHOR, see 4159

5241 PHOSPHORESCENCE   ec, ra
Emission of radiation by a substance as a
result of previous absorption of radiation
of shorter wavelength.
f phosphorescence *f*
e fosforescencia *f*
i fosforescenza *f*
n fosforescentie
d Phosphoreszenz *f*

5242 PHOSPHORUS   ch
Non-metallic element, symbol P, atomic
number 15.
f   phosphore *m*
e   fósforo *m*
i   fosforo *m*
n   fosfor *n*
d   Phosphor *m*

5243 PHOSPHURANYLITE   ch, mi
A secondary mineral containing about
63 % of U.
f   phosphuranylite *f*
e   fosfuranilita *f*
i   fosfuranilite *f*
n   fosfuranyliet *n*
d   Phosphuranylit *m*

5244 PHOTOCATHODE   ec
A cathode producing principally
photoelectric emission.
f   photocathode *f*
e   fotocátodo *m*
i   catodo *m* fotoelettrico, fotocatodo *m*
n   fotokatode
d   Photokatode *f*

5245 PHOTOCATHODE STANDARD   ec
        SENSITIVITY
The quotient of the photocathode current
due to the incident luminous flux emitted
by a non-filtered incandescent source at
the colo(u)r temperature of 2854°K, by
that flux.
f   sensibilité *f* conventionnelle d'une
    photocathode
e   sensibilidad *f* patrón de un fotocátodo
i   sensibilità *f* campione d'un fotocatodo
n   standaardgevoeligheid van een fotokatode
d   Normalempfindlichkeit *f* einer
    Photokatode

5246 PHOTODISINTEGRATION,   np
        PHOTONUCLEAR REACTION
Any nuclear reaction caused by a photon
and resulting in the emission of charged
fragments or neutrons.
f   photodésintégration *f*,
    réaction *f* photonucléaire
e   fotodesintegración *f*,
    reacción *f* fotonuclear
i   fotodisintegrazione *f*,
    reazione *f* fotonucleare
n   fotodesintegratie, fotokernreactie
d   Kernphotoeffekt *m*, Photokernreaktion *f*,
    Photozerfall *m*

5247 PHOTOEFFECT,   ec
        PHOTOELECTRIC EFFECT
The complete absorption of a photon by
an atom with the emission of an orbital
electron.
f   effet *m* photoélectrique
e   efecto *m* fotoeléctrico
i   effetto *m* fotoelettrico
n   foto-effect *n*, foto-elektrisch effect *n*
d   Photoeffekt *m*, lichtelektrischer Effekt *m*,
    photoelektrischer Effekt *m*

5248 PHOTOELECTRIC ABSORPTION   ab
The absorption of photons in the photoelectric effect.
f   absorption *f* photoélectrique
e   absorción *f* fotoeléctrica
i   assorbimento *m* fotoelettrica
n   foto-elektrische absorptie
d   photoelektrische Absorption *f*

5249 PHOTOELECTRIC   np, ra
        ATTENUATION COEFFICIENT
That part of the total attenuation
coefficient that is attributable to the photoelectric effect.
f   coefficient *m* d'atténuation photoélectrique
e   coeficiente *m* de atenuación fotoeléctrico
i   coefficiente *m* d'attenuazione fotoelettrico
n   foto-elektrische verzwakkingscoëfficiënt
d   photoelektrischer Schwächungskoeffizient *m*

5250 PHOTOELECTRIC PEAK   ec
In a gamma radiation spectrum, the part
of the response curve corresponding to the
photoelectric effect region, and denoted
by the energy corresponding to the highest
ordinate, and also by the full width at
half maximum.
f   pic *m* photoélectrique
e   pico *m* fotoeléctrico
i   picco *m* fotoelettrico
n   foto-elektrische piek
d   photoelektrische Spitze *f*

5251 PHOTOELECTRIC THRESHOLD ec, ra
The quantum of energy λV_o that is just
sufficient to release an electron from a
given system in the photoelectric effect.
f   seuil *m* photoélectrique
e   umbral *m* fotoeléctrico
i   soglia *f* fotoelettrica
n   foto-elektrische drempelwaarde
d   Photoschwelle *f*,
    Schwellenenergie *f* für Photoelektronen-
    auslösung

5252 PHOTOELECTRON   ec
An electron liberated through the photoelectric effect.
f   photoélectron *m*
e   fotoelectrón *m*
i   fotoelettrone *m*
n   foto-elektron *n*
d   Photoelektron *n*

5253 PHOTOFISSION   np
A fission caused by a photon.
f   photofission *f*
e   fotofisión *f*
i   fotofissione *f*
n   fotosplijting
d   Photospaltung *f*

5254 PHOTOFLUOROGRAPH,   xr
        PHOTORÖNTGEN UNIT, PR UNIT
A device used in radiography by means of
which the normal sized image of a
fluorescent screen is photographed on a
small film.

f appareil *m* de radiographie
e aparato *m* de radiofotografía
i apparecchio *m* per radiofotografia
n schermbeeldapparaat *n*
d Schirmbildgerät *n*

5255 PHOTOFLUOROGRAPHY, see 2741

5256 PHOTOGRAPHIC DENSITY, see 1715

5257 PHOTOGRAPHIC DOSEMETER   me, ra
A dosemeter using one or more photographic emulsions as an essential element.
f dosimètre *m* photographique
e dosímetro *m* fotográfico
i dosimetro *m* fotografico
n fotografische dosismeter
d photographischer Dosismesser *m*

5258 PHOTOGRAPHIC GRAIN, see 3049

5259 PHOTOGRAPHIC GRAININESS, see 3054

5260 PHOTOGRAPHIC METHOD OF   ph, ra
RADIATION DETECTION
f méthode *f* photographique de détection du
rayonnement
e método *m* fotográfico de detección de
la radiación
i metodo *m* fotografico di rivelamento della
radiazione
n fotografische stralingsdetectiemethode
d photographisches Strahlungsdetektions-
verfahren *n*

5261 PHOTOIONIZATION, see 402

5262 PHOTOMAGNETIC EFFECT   np
Photodisintegration ascribable to the
magnetic vector of the photon.
f effet *m* photomagnétique
e efecto *m* fotomagnético
i effetto *m* fotomagnetico
n fotomagnetisch effect *n*
d photomagnetischer Effekt *m*

5263 PHOTOMESON   np
A meson produced by the interaction of
a photon with a nucleus.
f photoméson *m*
e fotomesón *m*
i fotomesone *m*
n fotomeson *n*
d Photomeson *n*

5264 PHOTOMULTIPLIER   ec
A sensitive detector of light in which the
initial electron current, derived from
photoelectric emission, is amplified by
successive stages of secondary electron
emission.
f photomultiplicateur *m*
e multiplicador *m* fotoeléctrico de electrones
i moltiplicatore *m* fotoelettrico d'elettroni
n fotomultiplicator
d Photoelektronenvervielfacher *m*

5265 PHOTOMULTIPLIER COUNTER,   ct
SCINTILLATION COUNTER
The combination of phosphor, photomultiplier tube and associated circuits for
counting oscillations.
f compteur *m* à scintillation,
ensemble *m* de comptage à scintillation
e contador *m* de centelleo,
contador *m* de escintilación
i contatore *m* a scintillazione
n scintillatieteller
d Szintillationszähler *m*

5266 PHOTOMULTIPLIER TUBE   ec
A vacuum tube containing a photosensitive
layer which serves as the cathode for an
electron multiplier.
f tube *m* photomultiplicateur
e tubo *m* fotomultiplicador
i tubo *m* fotomoltiplicatore
n fotomultiplicatorbuis,
fotovermenigvuldiger
d Photovervielfacherröhre *f*,
Sekundärelektronenvervielfacher *m* mit
Photokatode

5267 PHOTON, see 4010

5268 PHOTON EMISSION CURVE   np
The curve representing the variation with
time of the photon emission rate
corresponding to a single excitation of a
scintillating material.
f courbe *f*, d'émission de photons
e curva *f* de emisión de fotones
i curva *f* d'emissione di fotoni
n kromme voor fotonenemissie
d Kurve *f* für Photonenemission

5269 PHOTON EMISSION   np, ra, sp
SPECTRUM
The relative numbers of optical photons
emitted by a scintillator material per unit
wavelength as a function of the wavelength.
f spectre *m* d'émission de photons
e espectro *m* de emisión de fotones
i spettro *m* d'emissione di fotoni
n fotonenemissiespectrum *n*
d Photonenemissionsspektrum *n*

5270 PHOTON ENERGY   np
The energy of a photon is h_v, where h is
the Planck constant and v is the frequency
associated with the photon.
f énergie *f* du photon
e energía *f* del fotón
i energia *f* del fotone
n fotonenergie
d Photonenergie *f*

5271 PHOTONEUTRON   np
A neutron resulting from photodisintegration
f photoneutron *m*
e fotoneutrón *m*
i fotoneutrone *m*

n fotoneutron *n*
d Photoneutron *n*

5272 PHOTOPROTON   np
A proton resulting from photodisintegration.
f photoproton *m*
e fotoprotón *m*
i fotoprotone *m*
n fotoproton *n*
d Photoproton *n*

5273 PHOTOTIMER   xr
An automatic device containing a
photo-cell or an ionization chamber, for
radiographic control.
f photominuterie *f*
e fotocronómetro *m*
i fotoscatto *m*
n belichtingsautomaat
d Zeitschalter *m*

5274 PHYSICAL MASS UNIT, see 233

5275 PHYSICAL SCALE OF   ch
ATOMIC WEIGHTS
A scale of atomic weights based on
assigning the value of exactly sixteen to
the mass of the most abundant isotope
of oxygen.
f table *f* physique de poids atomiques
e tabla *f* física de pesos atómicos
i tavola *f* fisica di pesi atomici
n atoomgewichten *pl* (fysische schaal)
d Atomgewichte *pl* (physikalische Skala)

5276 PHYSICAL TRACER   is
A tracer attached by purely physical means
to the object being traced.
f traceur *m* physique
e trazador *m* físico
i tracciante *m* fisico
n fysische indicator
d physikalischer Indikator *m*,
physikalischer Tracer *m*

5277 π MESON,   np
PION
A meson with a mean life $< 10^{-15}$s, and
usually decaying into two photons.
f pion *m*
e pión *m*
i pione *m*
n pion *n*
d Pion *n*

5278 PICK-UP,   np
PICK-UP REACTION
A nuclear reaction in which the incident
particle picks up a nucleon from the target
nucleus and proceeds with that nucleon
bound to itself.
f enlèvement *m*, rapt *m*
e extracción *f*, recogida *f*
i distacco *m*
n extractie, schaking
d Herausreissen *n* eines Nukleons

5279 PICK-UP COILS,   pp
ROGOVSKI COILS
A probe used i.a. to study plasmas,
consisting of coils which encircle the
current to be measured.
f bobines *pl* de Rogovski,
ceinture *f* de Rogovski
e bobinas *pl* de Rogovski
i bobine *pl* di Rogovski
n rogovskispoelen *pl*
d Rogovski-Spulen *pl*

5280 PICKET FENCE THERMONUCLEAR
REACTOR, see 1585

5281 PICKLING   ch
The removal of chemical compounds,
usually oxides, from the surface of a
metal by chemical or electrochemical
means.
f décapage *m*
e decapaje *m*
i decapaggio *m*
n beitsen *n*
d Beizen *n*

5282 PIEZOELECTRIC PROBE   me, pp
A probe used for measuring plasma
pressions.
f sonde *f* piézoélectrique
e sonda *f* piezoeléctrica
i sonda *f* piezoelettrica
n piezo-elektrische sonde
d piezoelektrische Sonde *f*

5283 PILBARITE   mi
A mineral containing thorium, uranium,
silicon and lead.
f pilbarite *f*
e pilbarita *f*
i pilbarite *f*
n pilbariet *n*
d Pilbarit *m*

5284 PILE   rt
Name given to the first nuclear reactor,
on account of its shape.
f pile *f*
e pila *f*
i -
n -
d Meiler *m*

5285 PILE EQUATION, see 1515

5286 PILE FACTOR   rt
A term used in connection with the
irradiation of materials in nuclear reactors
and particularly in the production of
radioisotopes to denote the magnitude of
the thermal neutron flux in the
irradiation.
f facteur *m* de pile
e factor *m* de irradiación óptima
i fattore *m* d'irradiazione ottima
n relatieve bestralingsfactor
d Optimalbestrahlungsfaktor *m*

5287 PILE GUN rt
A probe for inserting apparatus into a
nuclear reactor so as to permit
measurement during operation.
f sonde f de réacteur
e sonda f de reactor
i sonda f di reattore
n reactorsonde
d Reaktorsonde f

5288 PILE OSCILLATOR, me
REACTIVITY OSCILLATOR,
REACTOR OSCILLATOR
A device which produces periodic
variations of reactivity by movement of
a sample for the purpose of measuring
reactor properties or nuclear cross
sections.
f oscillateur m de pile
e oscilador m de reactor
i oscillatore m in reattore
n reactoroscillator
d Reaktoroszillator m

5289 PILE-UP ct, ra
In a radiation detector, a condition where
several pulses occur sufficiently close
together to be indistinguishable in time
by the instrument and so produce the same
effect as larger but less numerous
pulses.
f empilement m d'impulsions
e amontonamiento m de impulsos
i ammuchiamento m d'impulsi
n pulsophoping
d Impulsanhäufung f

5290 PILOT PLANT ge
A plant for testing new processes,
applications, etc.
f atelier m d'essais, installation f pilote
e fábrica f experimental,
instalación f de ensayos
i impianto m pilota,
impianto m sperimentale
n proeffabriek
d Versuchsanlage f, Versuchswerk n

5291 PIMPLING fu, rt
Of a canned fuel element, production of
protuberances on the surface of the can
as a result of reactions inside the can.
f gondolage m
e vesiculación f
i rigonfiamento m, vescicatura f
n blaasjesvorming
d Blasenbildung f

5292 PIN AND ARC INDICATOR xr
A beam direction indicator consisting of
a geometrical instrument which indicates
the direction by reference to a known
.surface mark.
f rapporteur m d'angle
e indicador m geométrico
i indicatore m geometrico
n hoekindicator
d Winkelindikator m

5293 P.I.N. SEMICONDUCTOR, see 1253

5294 P.I.N. SEMICONDUCTOR DETECTOR,
see 1254

5295 PINCH, pp
PINCH EFFECT, RHEOSTRICTION
The constriction of a plasma caused by
the magnetic field of a high current in the
plasma itself.
f effet m de pincement, effet m de striction
e efecto m de estricción, efecto m de pinza
i effetto m di contrazione,
effetto m di strizione
n insnoeringseffect n, pincheffect n,
reostrictie, stroominsnoering
d eigenmagnetische Kontraktion f,
Einschnürungseffekt m, Pinch-Effekt m

5296 PIPE, see 1350

5297 PISEKITE mi
An impure gummite variety.
f pisékite f
e pisequita f
i pisechite f
n pisekiet n
d Pisekit m

5298 PITCH rt
In a heterogeneous reactor, the distance
between the centre(er)s of adjacent fuel
channels.
f pas m entre les canaux
e espaciado m de canales
i passo m tra i canali
n kanaalafstand
d Kanalabstand m

5299 PITCHBLENDE, see 4620

5300 PITTINITE mi
An impure gummite variety.
f pittinite f
e pitinita f
i pittinite f
n pittiniet n
d Pittinit m

5301 PITUITARY IRRADIATION, see 3371

5302 PLACENTOGRAPHY xr
The radiological examination of the
placenta with or without a contrast medium
f placentographie f
e placentografía f
i placentografia f
n placentografie
d Plazentographie f

5303 PLAIT POINT mg
The point on the solubility curve of a
ternary solubility diagram at which the
compositions of the conjugate layers
are identical.
f point m de pliage
e punto m de plegado
i punto m di piega

n   plooipunt *n*, vouwpunt *n*
d   Faltpunkt *m*

5304 PLAITING   fu
The total of undulations formed on a fuel
can by thermal cycling.
f   plissée *f*
e   superficie *f* ondulada
i   superficie *f* ondulata
n   gegolfd oppervlak *n*
d   Wellenoberfläche *f*

5305 PLANAR IMPLANT   xr
Implant in a single plane or in two
dimensions.
f   implant *m* en surface
e   injerto *m* bidimensional,
    injerto *m* laminar
i   innesto *m* in superficie
n   vlakke implantatie
d   Oberflächenimplantat *n*

5306 PLANCK LAW   gp
The fundamental law of the quantum
theory.
f   loi *f* de Planck
e   ley *f* de Planck
i   legge *f* di Planck
n   wet van Planck
d   Plancksches Gesetz *n*

5307 PLANCK'S CONSTANT,   np, un
      QUANTUM OF ACTION
A universal constant, the factor which
relates the energy of a photon to its
frequency. Symbol h.
f   constante *f* de Planck
e   constante *f* de Planck
i   costante *f* di Planck
n   constante van Planck
d   Plancksche Konstante *f*

5308 PLANE SOURCE   rt
In reactors, a neutron source of plane
form as, e.g., a slab.
f   source *f* plane
e   fuente *f* plana
i   sorgente *f* piana
n   platte bron
d   flache Quelle *f*

5309 PLANIGRAPH, see 688

5310 PLANIGRAPHY   xr
Tomography of plane thin layers.
f   planigraphie *f*
e   planigrafía *f*
i   planigrafia *f*
n   planigrafie
d   Körperschichtaufnahme *f* mittels
    Planigraph,
    Planigraphie *f*

5311 PLANNED EMERGENCY EXPOSURE,
      see 2224

5312 PLANT PROVING RUN   ge
f   marche *f* d'essai d'une installation

e   marcha *f* de prueba de un equipo
i   marcia *f* di prova d'un impianto
n   proefdraaien *n* van een installatie
d   Versuchsbetrieb *m* einer Anlage

5313 PLASMA   pp
The region in a gas discharge which
contains very nearly equal numbers of
positive ions and electrons, and hence is
nearly neutral.
f   plasma *m*
e   plasma *m*
i   plasma *m*
n   plasma *n*
d   Plasma *n*

5314 PLASMA BALANCE   pp
The condition when ionization balances
diffusion.
f   équilibre *m* de plasma
e   equilibrio *m* de plasma
i   equilibrio *m* di plasma
n   plasma-evenwicht *n*
d   Plasmagleichgewicht *n*

5315 PLASMA CONTAINMENT, see 1327

5316 PLASMA DIAMAGNETISM, see 1772

5317 PLASMA FREQUENCY, see 2147

5318 PLASMA FREQUENCY PROBE,
      see 3898

5319 PLASMA GUN   pp
A device designed to produce plasmoids.
f   canon *m* à plasma
e   cañon *m* de plasma
i   cannone *m* a plasma
n   plasmakanon *n*
d   Plasmakanone *f*

5320 PLASMA INSTABILITY   pp
The tendency of a magnetically confined
plasma to move to the walls of the
discharge tube, owing to the effect of local
deformations of the plasma.
f   instabilité *f* de plasma
e   inestabilidad *f* de plasma
i   instabilità *f* di plasma
n   plasma-instabiliteit
d   Plasmainstabilität *f*

5321 PLASMA OSCILLATIONS   pp
A plasma is capable of supporting various
modes of vibration and propagate several
types of waves.
f   oscillations *pl* du plasma
e   oscilaciones *pl* del plasma
i   oscillazioni *pl* del plasma
n   plasmatrillingen *pl*
d   Plasmaschwingungen *pl*

5322 PLASMA PRESSURE   pp
The pressure of a plasma in the discharge
chamber.
f   pression *f* du plasma
e   presión *f* del plasma

i   pressione *f* del plasma
n   plasmadruk
d   Plasmadruck *m*

5323 PLASMA PROPULSION   pp
A scheme for propulsion of a vehicle in
space by the use of a plasma which is
accelerated up to velocities of the order
of $10^7 cm/sec$.
f   propulsion *f* par plasma
e   propulsión *f* por plasma
i   propulsione *f* per plasma
n   voortstuwing door een plasma
d   Plasmaantrieb *m*

5324 PLASMA PURITY   pp
f   pureté *f* du plasma
e   pureza *f* del plasma
i   purezza *f* del plasma
n   plasmazuiverheid
d   Plasmareinheit *f*

5325 PLASMA TORCH   pp
A torch using a plasma jet at a very
high temperature (higher than $15,000^oC$),
realized by injecting a non-oxidizing gas
in an electric arc which heats and
completely ionizes it.
f   chalumeau *m* à plasma
e   antorcha *f* de plasma
i   fiaccola *f* a plasma
n   plasmafakkel
d   Plasmafackel *f*

5326 PLASMAGENES   md
Postulated units of inheritance located in
the cytoplasma of the cell.
f   plasmagènes *pl*
e   plasmágenes *pl*
i   plasmageni *pl*
n   plasmagenen *pl*
d   Plasmagene *pl*

5327 PLASMOID   pp
A discrete piece of plasma.
f   plasmoïde *m*
e   plasmoide *m*
i   plasmoide *m*
n   plasmoïde *n*
d   Plasmoid *n*

5328 PLASMON   pp
A name given to a hypothetical particle
associated with the various waves which
may exist in a plasma.
f   plasmon *m*
e   plasmón *m*
i   plasmone *m*
n   plasmon *n*
d   Plasmon *n*

5329 PLASTIC-MODERATED REACTOR   rt
f · réacteur *m* à modérateur plastique
e   reactor *m* de moderador plástico
i   reattore *m* a moderatore plastico
n   reactor met kunstofmoderator
d   Reaktor *m* mit Kunststoffmoderator

5330 PLASTIC RANGE   gp
The stress range in which a material will
not fail when subjected to the action of a
force, but will not recover completely, so
that a permanent deformation results.
f   zone *f* plastique
e   zona *f* plástica
i   zona *f* plastica
n   plastisch gebied *n*
d   Plastizitätsgebiet *n*

5331 PLASTIC SCINTILLATOR   ct
A scintillator consisting of a plastic
material produced by dissolving an organic
substance in a monomer and then
polymerizing.
f   scintillateur *m* plastique
e   centelleador *m* plástico
i   scintillatore *m* plastico
n   kunststofscintillator
d   Kunststoffszintillator *m*

5332 PLASTIC SCINTILLATOR   ra
      DETECTOR
A scintillation detector in which the
scintillator contains a plastic substance.
f   détecteur *m* à scintillateur plastique
e   detector *m* de centelleador plástico
i   rivelatore *m* a scintillatore plastico
n   scintillatiedetector met kunststof
d   Detektor *m* mit Kunststoff-Szintillator

5333 PLASTIC STRAIN, see 5211

5334 PLATE,   ch
      TRAY
Component part of a distillation column.
f   plateau *m*
e   piso *m*, plato *m*
i   piatto *m*
n   schotel
d   Boden *m*

5335 PLATE COLUMN,   ch
      PLATE TOWER
A column containing a number of horizontal
plates such as sieve plates or bubble
plates, generally equally placed one above
the other.
f   colonne *f* à plateaux
e   columna *f* de platos
i   colonna *f* a piatti
n   schotelkolom
d   Bodenkolonne *f*

5336 PLATE EFFICIENCY   ch
The ratio of the number of theoretical
plates needed to the number of plates
actually required.
f   rendement *m* des plateaux
e   rendimiento *m* de los platos
i   rendimento *m* dei piatti
n   nuttig effect *n* per schotel,
     schotelrendement *n*
d   Bodenwirkungsgrad *m*

5337 PLATE PENETROMETER,   xd
      STRIP PENETROMETER
      A plate of similar material to the specimen
      under examination having a thickness of
      1 or 2 per cent of the specimen thickness,
      and having holes of different diameters.
f pénétramètre *m* à plaque
e penetrómetro *m* de placa
i penetrametro *m* a piastra
n plaatkwaliteitsmeter
d Plattenhärtemesser *m*

5338 PLATEAU   ct
      That portion of the plateau characteristic
      for which the counting rate is approxi-
      mately independent of voltage.
f palier *m*, plateau *m*
e meseta *f*, plato *m*
i ripiano *m*
n plateau *n*
d Plateau *n*

5339 PLATEAU CHARACTERISTIC,   ct
      PLATEAU CHARACTERISTIC
      CURVE
      Of a Geiger-Müller counter tube, a curve
      showing the counting rate as a function of
      the voltage applied to the counter tube with
      all other parameters constant.
f caractéristique *f* de palier,
   courbe *f* caractéristique de palier
e característica *f* de meseta
i caratteristica *f* di ripiano
n plateaukarakteristiek
d Plateaucharakteristik *f*

5340 PLATEAU LENGTH, see 3977

5341 PLATEAU SLOPE   ct
      The percentage change in count rate for a
      given change, usually 100 volts, in
      applied voltage.
f pente *f* du palier
e pendiente *f* de la meseta
i pendenza *f* del ripiano
n plateauhelling, plateausteilheid
d Plateausteilheit *f*

5342 PLATELET,   md
      THROMBOCYTE
      A small colo(u)rless corpuscle present in
      the blood of all mammals in large
      numbers, believed to play a role in the
      clotting of the blood.
f plaquette *f* sanguine, thrombocyte *m*
e plaqueta *f* sanguinea, trombocito *m*
i piastrina *f*, trombocito *m*
n bloedplaatje *n*, trombocyt
d Blutplättchen *n*, Thrombozyt *m*

5343 PLATELET COUNT,   md
      THROMBOCYTE COUNT
      The number of platelets per cubic
      millimetre(er) of blood.
f numération *f* thrombocytaire
e numeración *f* trombocitaria
i numerazione *f* trombocitaria

n trombocytengetal *n*
d Thrombozytenzahl *f*

5344 PLATINUM   ch
      Metallic element, symbol Pt, atomic
      number 78.
f platine *m*
e platino *m*
i platino *m*
n platina *n*
d Platin *n*

5345 PLENUM   ge
      Descriptive of any enclosed space where
      the pressure of the enclosed atmosphere
      of gas is greater than the ambient or
      outside atmospheric pressure.
f plein *m*
e cámara *f* plena, pleno *m*
i camera *f* di pressione
n drukruimte
d ausgefüllter Raum *m*, Druckkammer *f*

5346 PLEOCHROIC HALO   ar, mi
      A microscopic colo(u)red ring or group
      of concentric rings, found in certain
      minerals, formed by the alpha rays from
      minute radioactive inclusions.
f halo *m* pléochroïque
e halo *m* pleocróico
i alone *m* pleocroico
n veelkleurige halo
d mehrfarbiger Lichthof *m*,
   pleochroitischer Halo *m*

5347 PLEUROGRAPHY   xr
      The radiological examination of the
      pleural cavity following the injection of
      air or other contrast medium.
f pleurographie *f*
e pleurograffa *f*
i pleurografia *f*
n pleurografie
d Pleurographie *f*

5348 PLUG   rt
      In a nuclear reactor, a piece of material
      used to prevent the escape of radiation
      from channels or experimental holes.
f bouchon *m*
e tapón *m*
i tappo *m*
n plug, prop
d Pfropfen *m*, Stopfen *m*

5349 PLUG-IN UNIT,   cd
      SUBCHASSIS
      A component of an assembly of instruments
      and/or electronic parts which can be
      readily inserted as a unit.
f bloc *m* interchangeable, tiroir *m*
e unidad *f* enchufable
i unità *f* intercambiabile
n insteekblok *n*
d Einsteckblock *m*

5350 PLUGGING LOOP   cd, sa
   Component part of reactor instrumentation
   for controlling the plugging of tubes by
   oxidation.
f boucle f détectrice d'obstructions
e bucle f detectora de obstrucciones
i cappio m rivelatore d'intasature
n verstoppingenzoeker
d Stockungssucher m

5351 PLUMBING   cd, rt
   The complex system of pipelines, valves,
   fittings,and ga(u)ges of a nuclear reactor,
   etc.
f tuyauterie f
e plomería f
i impianto m idraulico
n buisleiding
d Rohrleitung f

5352 PLUMBONIOBITE   mi
   A plumbian variety of samarskite.
f plumboniobite f
e plumboniobita f
i plumboniobite f
n plumboniobiet n
d Plumboniobit m

5353 PLUME   ge
   The stream of gas, warm air or smoke
   leaving a stack or chimney.
f panache m
e penacho m
i pennacchio m
n pluim
d Fahne f

5354 PLUME,   nw
   COLUMN, see 1241

5355 PLURAL PRODUCTION   np
   Occurring when a fast nucleon produces
   particles in several successive collisions
   with nucleons in one nucleus.
f production f multiple en collisions
   successives
e producción f múltiple en colisiones
   sucesivos
i produzione f multipla in collisioni
   consecutivi
n veelvoudige produktie bij opeenvolgende
   collisies
d Mehrfacherzeugung f bei aufeinander-
   folgenden Kollisionen

5356 PLURAL SCATTERING   np
   Any scattering of a particle or photon in
   which the final displacement is the vector
   sum of a small number of displacements.
f diffusion f multiple
e dispersión f múltiple
i deviazione f multipla
n veelvoudige verstrooiing
d Vielfachstreuung f

5357 PLUTONIUM   ch
   Radioactive element, symbol Pu, atomic
   number 94.
f plutonium m

e plutonio m
i plutonio m
n plutonium n
d Plutonium n

5358 PLUTONIUM AEROSOL   ma
   MONITOR
   A monitor designed for the continuous
   measurement of atmospheric contamination
   due to plutonium aerosols, by taking into
   account parasitic alpha emitters.
f moniteur m pour aérosols de plutonium.
e vigía f de aerosoles de plutonio
i monitore m per aerosoli di plutonio
n monitor voor plutoniumaerosolen
d Überwachungsgerät n für Plutoniumaerosole

5359 PLUTONIUM CYCLE   rt
   Cycle based on the transformation of
   U-238 into U-239.
f cycle m du plutonium
e ciclo m del plutonio
i ciclo m del plutonio
n plutoniumcyclus
d Plutoniumzyklus m

5360 PLUTONIUM ENRICHED FUEL   fu
f combustible m en plutonium enrichi
e combustible m de plutonio enriquecido
i combustibile m in plutonio arricchito
n splijtstof uit verrijkt plutonium
d Brennstoff m aus angereichertem
   Plutonium

5361 PLUTONIUM PRODUCTION   rt
   REACTOR
   A nuclear reactor whose primary purpose
   is to produce plutonium.
f réacteur m de production du plutonium
e reactor m de producción del plutonio
i reattore m di produzione del plutonio
n reactor voor plutoniumproduktie
d Plutoniumproduktionsreaktor m

5362 PLUTONIUM REACTOR   rt
   A nuclear reactor in which plutonium is
   the fuel.
f réacteur m au plutonium
e reactor m de plutonio
i reattore m al plutonio
n plutoniumreactor
d Plutoniumreaktor m

5363 PNEUMORÖNTGENOGRAPHY   xr
   Contrast photography, in which the
   contrast medium is a gas.
f pneumoröntgenographie f
e neumoröntgenografía f
i pneumoröntgenografia f
n pneumoröntgenografie
d Pneumoröntgenographie f

5364 POCKET ALARM DOSEMETER   me, ra
   An instrument to warn the bearer in the
   event of ambient radioactivity going beyond
   a preset level.
f signaleur m individuel de poche
e dosímetro m de alarma de bolsillo

i   dosimetro *m* d'allarme tascabile
n   waarschuwende zakdosismeter
d   warnender Taschendosismesser *m*

5365 POCKET BATTERY MONITOR   me
Radiation counter of small dimensions and
operated by batteries.
f   moniteur *m* portatif à piles
e   monitor *m* de bolsillo con pilas
i   monitore *m* tascabile con pile
n   zakmonitor met batterij
d   Taschenmonitor *m* mit Batterie

5366 POCKET CHAMBER,   ic
POCKET IONIZATION CHAMBER
A small-sized ionization chamber used
for monitoring radiation exposure of
personnel.
f   chambre *f* d'ionisation de poche
e   cámara *f* de ionización de bolsillo
i   camera *f* d'ionizzazione tascabile
n   zakionisatievat *n*
d   Taschenionisationskammer *f*

5367 POCKET DOSEMETER, see 2791

5368 POCKET EXPOSUREMETER, see 2792

5369 POD BOILER   rt
A boiler which is accommodated in
vertical, lined and insulated holes within
the thickness of the pressure vessel wall.
f   chaudière *f* encastrée
e   caldera *f* montada en vaina
i   caldaia *f* incassata
n   in wandholte ingelaten stoomketel
d   in Wandhöhle eingebauter Dampfkessel *m*

5370 PODBIELNIAK CONTACTOR   ch
A solvent extraction unit in which the
heavy phase is fed to the centre(er) of a
rapidly moving spiral and is removed
from the periphery; the lighter phase is
fed in the opposite direction.
f   spirale *f* contactrice de Podbielniak
e   espiral *f* contactora de Podbielniak
i   spirale *f* contattrice di Podbielniak
n   contactspiraal van Podbielniak
d   Kontaktspirale *f* nach Podbielniak

5371 POINT COUNTER TUBE   ct
A counter tube, using gas amplification,
in which the central electrode is a point
or a small sphere.
f   tube *m* compteur à pointe
e   tubo *m* contador de punta
i   tubo *m* contatore a punta
n   punttelbuis, telbuis met puntelektrode
d   Spitzenzählrohr *n*

5372 POINT LATTICE   cr
A regular arrow of points in space, as for
example the sites of atoms in a crystal.
f   réseau *m* de points
e   red *f* espacial de puntos
i   reticolo *m* di punti
n   puntrooster *n*
d   Punktgitter *n*

5373 POINT SOURCE   np
No finite source of radiation is a point
source but any source viewed from a
distance sufficiently great compared with the
linear size of the source may be considered
as a point source.
f   source *f* ponctuelle
e   fuente *f* puntiforme
i   sorgente *f* di punta
n   puntvormige bron
d   punktförmige Quelle *f*, Punktquelle *f*

5374 POISON   rt
Any non-fissile (non-fissionable) element
in a reactor with appreciable absorption
cross section.
f   poison *m*
e   veneno *m*
i   veleno *m*
n   gif *n*
d   Gift *n*

5375 POISON LIMIT   rt
Poison concentration in such a manner that
the reactor cannot become critical.
f   limite *f* d'empoisonnement
e   límite *m* de envenenamiento
i   limite *m* d'avvelenamento
n   vergiftigingsgrens
d   Vergiftungsgrenze *f*

5376 POISON RANGE   rt
Concentration gap of the poisons in which
the reactor can still operate.
f   intervalle *m* d'empoisonnement
e   intervalo *m* de envenenamiento
i   intervallo *m* d'avvelenamento
n   vergiftigingsmarge
d   Vergiftungsspielraum *m*

5377 POISONING   rt
The phenomenon of poison formation in a
reactor.
f   empoisonnement *m*
e   envenenamiento *m*
i   avvelenamento *m*
n   vergiftiging
d   Vergiftung *f*

5378 POISONING COMPUTER,   ma, rt
XENON POISONING COMPUTER
An analog(ue) computer which calculates the
level of xenon poisoning in a nuclear
reactor from the continuously measured
neutron flux.
f   calculatrice *f* analogique de l'effet
d'empoisonnement
e   calculadora *f* de envenenamiento xenón,
xenómetro *m*
i   calcolatore *m* analogico per effetti
d'avvelenamento
n   analogonrekenmachine voor vergiftigingseffecten,
rekenmachine voor de xenonvergiftiging
d   Analogrechner *m* für Vergiftungseffekte,
Xenonvergiftungsrechner *m*

5379 POISONING OF A REACTOR,			rt
	REACTOR POISONING
The ratio of the number of thermal
neutrons absorbed by poison to the number
of those absorbed in fuel.
f	empoisonnement *m* du réacteur
e	envenenamiento *m* del reactor
i	avvelenamento *m* del reattore
n	reactorvergiftiging
d	Reaktorvergiftung *f*

5380 POISONING PREDICTOR,			ma, rt
	XENON POISONING PREDICTOR
An analog(ue) computer used in conjunction
with a poisoning computer to predict the
behavio(u)r of xenon poisoning following a
given change in operation.
f	prédicteur *m* d'empoisonnement xénon
e	predictor *m* de envenenamiento xenón
i	calcolatore *m* analogico per calcolo
	preliminare d'avvelenamento,
	previsore *m* dell'avvelenamento xenon
n	analogonrekenmachine voor voorafgaande
	berekening van xenonvergiftiging
d	Analogrechner *m* für Xenonvergiftungs-
	voraussage

5381 POLARIZATION					gp
The process of bringing about a partial
separation of electrical charges of
opposite signs in a body by the superposition of an external field.
f	polarisation *f*
e	polarización *f*
i	polarizzazione *f*
n	polarisatie
d	Polarisation *f*

5382 POLARIZED BEAM				ra
Of corpuscular radiation, a beam in which
the individual particles have non-zero spin
and in which the distribution of the
values of the spin components varies with
the direction in which these components
are measured.
f	faisceau *m* polarisé
e	haz *m* polarizado
i	fascio *m* polarizzato
n	gepolariseerde bundel
d	polarisierter Strahl *m*

5383 POLARIZED IONIC BOND,			ch
	SEMICOVALENT BOND
A valence bond between atoms which is
neither wholly electrostatic (ionic),
nor chemical (covalent), in nature, but has
properties of both types.
f	liaison *f* ionique polarisée
e	enlace *m* iónico polarizado
i	legame *m* ionico polarizzato
n	gepolariseerde ionbinding
d	polarisierte Ionenbindung *f*

5384 POLARIZING VOLTAGE, see 1215

5385 POLONIUM						ch
Metallic element, symbol Po, atomic
number 84.

f	polonium *m*
e	polonio *m*
i	polonio *m*
n	polonium *n*
d	Polonium *n*

5386 POLYCRASE,
	POLYCRASITE, see 2379

5387 POLYCRYSTAL MINERAL			ra
	SCINTILLATOR DETECTOR
A scintillation detector in which the
scintillator is a polycrystal mineral.
f	détecteur *m* à scintillateur minéral
	polycristallin
e	detector *m* de centelleador mineral
	policristalino
i	rivelatore *m* a scintillatore a policristallo
	minerale
n	polykristallijne scintillatiedetector
d	Detektor *m* mit polykristallinem
	anorganischem Szintillator

5388 POLYCRYSTALLINE				cr
Said of a crystal body when made up of a
number of small crystals in a mass.
f	polycristallin adj
e	policristalino adj
i	policristallino adj
n	polykristallijn adj
d	polykristallin adj

5389 POLYCYTHEMIA					md
A disease characterized by an overproduction of erythrocytes.
f	polycythémie *f*
e	policitemia *f*
i	policitemia *f*
n	polycytemie
d	Polyzythämie *f*

5390 POLYELECTRONS				np
A probably existing group of electrons and
positrons with temporary stability.
f	polyélectrons *pl*
e	polielectrones *pl*
i	polielettroni *pl*
n	polyelektronen *pl*
d	Polyelektronen *pl*

5391 POLYENERGETIC RADIATION			ra
Particle radiation of a given type in which
the emitted particles have different
energies.
f	rayonnement *m* polyénergétique
e	radiación *f* polienergética
i	radiazione *f* polienergetica
n	polyenergetische straling
d	polyenergetische Strahlung *f*

5392 POLYMORPHONUCLEAR LEUCOCYTE,
	see 3060

5393 PONDERABLE AMOUNT,			ge
	PONDERABLE QUANTITY
Any amount of material obtained in a
specified process.
f	quantité *f* pondérable

e cantidad *f* ponderable
i quantità *f* ponderabile
n weegbare hoeveelheid
d wägbare Menge *f*

5394 PONDERATOR   pa
A term suggested for a device used to
produce high-energy particles, when the
speed of the particles becomes so great,
approaching that of light, than an increase
of energy results in an appreciable
increase in mass.
f pondérateur *m*
e ponderador *m*
i ponderatore *m*
n ponderator
d Ponderator *m*

5395 POOL REACTOR,   rt
    SWIMMING POOL REACTOR
A nuclear reactor whose fuel elements
are immersed in a pool of water which
serves as moderator, coolant and
biological shield.
f réacteur *m* piscine
e reactor *m* de alberca, reactor *m* de piscina
i reattore *m* a piscina
n bassinreactor
d Schwimmbadreaktor *m*

5396 POOR GEOMETRY, see 516

5397 POROSITY GENERATION   mt
An undesirable effect due to crystal
shape changes at high dose.
f génération *f* de porosité
e generación *f* de porosidad
i generazione *f* di porosità
n poreusheidontwikkeling
d Porositätsentwicklung *f*

5398 POROUS DIFFUSION, see 2985

5399 POROUS FUEL PARTICLE KERNEL fu
The core of the fuel particles used in the
DRAGON reactor.
f noyau *m* poreux d'une particule de
   combustible
e núcleo *m* poroso de una partícula de
   combustible
i nucleo *m* poroso d'una particella di
   combustibile
n poreuse kern van een splijtstofdeeltje
d poröser Kern *m* eines Brennstoffteilchens

5400 POROUS REACTOR   rt
A nuclear reactor composed of porous
material, or of an aggregate of small
particles with coolant or fluid fuel
flowing through the pores.
f réacteur *m* poreux
e reactor *m* poroso
i reattore *m* poroso
n poreuse reactor
d poröser Reaktor *m*

5401 PORTABLE GAMMA BACK-   ma
    SCATTER THICKNESS METER
A portable thickness meter, which is a
type of gamma backscatter thickness
ga(u)ge, and which is frequently used for
the measurement of pipe wall thickness.
f épaisseurmètre *m* portatif à rétrodiffusion
   gamma
e calibrador *m* portátil por retrodispersión
   gamma
i spessimetro *m* portatile a sparpagliamento
   gamma dell'indietro
n draagbare diktemeter berustend op gamma-
   terugverstrooiing
d tragbares Dickenmessgerät *n* nach der
   Gammarückstreumethode

5402 PORTABLE IONIZATION   ic
    CHAMBER
f chambre *f* d'ionisation portative
e cámara *f* de ionización portátil
i camera *f* d'ionizzazione portatile
n draagbaar ionisatievat *n*
d tragbare Ionisationskammer *f*

5403 PORTABLE LINEAR   ct
    RATEMETER
f ictomètre *m* linéaire portatif
e impulsímetro *m* lineal portátil
i rateometro *m* lineare portatile
n draagbare lineaire tempometer
d tragbarer linearer Leistungsmesser *m*

5404 PORTABLE LOGARITHMIC   me
    SCINTILLATOR EXPOSURE
    RATEMETER
f débitmètre *m* d'exposition logarithmique
   portatif à scintillateur
e dosímetro *m* logarítmico portátil de
   centelleador
i rateometro *m* d'esposione logaritmico
   tascabile a scintillatore
n draagbare logarithmische exposietempometer met scintillator
d tragbarer logarithmischer Dosisleistungsmesser *m* mit Szintillator.

5405 PORTABLE PROSPECTING   ma
    RADIATION METER
A portable prospecting assembly containing
its own power source and designed to
measure particle or photon flux density.
f radiamètre *m* de prospection portatif
e radiámetro *m* portátil de exploración
i radiametro *m* portatile di prospezione
n draagbare prospectiestralingsmeter
d tragbares Lagerstättensuchgerät *n*

5406 PORTABLE SUCTION FAN   co, sa
Used to collect and accumulate aerosols
on built-in filters.
f aspirateur *m* étanche portatif
e aspirador *m* de polvo portátil
i aspirapolvere *m* portatile
n draagbare stofzuiger
d tragbarer Staubsauger *m*

5407 POSITION MEASURING   con
      ASSEMBLY
A measuring assembly designed for the
indication of the instantaneous position of
control elements in a nuclear reactor.
f ensemble *m* de mesure de position
e conjunto *m* medidor de posición
i posiziometro *m*
n opstelling voor plaatsbepaling
d Stellungsmessanordnung *f*

5408 POSITIVE CHARGE   np
The condition existing in a body having
fewer electrons than normal.
f charge *f* positive
e carga *f* positiva
i carica *f* positiva
n positieve lading
d positive Ladung *f*

5409 POSITIVE ION, see 926

5410 POSITIVE ION BEAM   ra
f faisceau *m* d'ions positifs
e haz *m* de iones positivos
i fascio *m* d'ioni positivi
n positieve ionenbundel
d positiver Ionenstrahl *m*

5411 POSITIVE ION RAYS, see 860

5412 POSITIVE NUCLEUS   kp
The nucleus of an atom which always
carries a positive charge of magnitude
depending upon the particular atomic
species.
f noyau *m* positif
e núcleo *m* positivo
i nucleo *m* positivo
n positieve kern
d positiver Kern *m*

5413 POSITIVE VALENCE, see 4315

5414 POSITON,   np
      POSITRON
A positive electron.
f positon *m*
e positón *m*
i positone *m*
n positon *n*
d Positron *n*

5415 POSITON DECAY,
      POSITON DISINTEGRATION,
      POSITRON DECAY,
      POSITRON DISINTEGRATION,
      see 588

5416 POSITON EMISSION CONDITIONS,   np
      POSITRON EMISSION CONDITIONS
A conditional relationship for the emission
of posit(r)ons from nuclei.
f conditions *pl* d'émission pour les positons
e condiciones *pl* de emisión para los
   positones
i condizioni *pl* d'emissione per i positoni
n emissievoorwaarden *pl* voor positonen
d Emissionsbedingungen *pl* für Positronen

5417 POSITONIUM,   np
      POSITRONIUM
A quasi-stable system consisting of a
posit(r)on and an electron bound together.
f positonium *m*
e positonio *m*
i positonio *m*
n positonium *n*
d Positronium *n*

5418 POSTERIOR-ANTERIOR VIEW, see 284

5419 POSTERIOR PROJECTION, see 283

5420 POSTIRRADIATION   md, xr
Irradiation after an operation.
f postirradiation *f*
e postirradiación *f*
i postirradiazione *f*
n nabestraling
d Nachbestrahlung *f*

5421 POSTIRRADIATION ANNEALING   mg
A method for removal of the fast neutron
effect.
f recuit *m* après l'irradiation
e recocido *m* después de la irradiación
i ricottura *f* dopo l'irradiazione
n uitgloeien *n* na bestraling
d Ausglühen *n* nach Bestrahlung

5422 POSTIRRADIATION EXAMINATION   ra
Examination of materials after irradiation
has taken place.
f examen *m* après l'irradiation
e examen *m* después de la irradiación
i esame *m* dopo l'irradiazione
n onderzoek *n* na de bestraling
d Prüfung *f* nach der Bestrahlung

5423 POTASSIUM   ch
Metallic element, symbol K, atomic
number 19.
f potassium *m*
e potasio *m*
i potassio *m*
n kalium *n*
d Kalium *n*

5424 POTENTIAL BARRIER, see 541

5425 POTENTIAL CURVE,   ge
      POTENTIAL-ENERGY CURVE
In an energy diagram, a curve representing
the variation of potential energy as a
function of a parameter characterizing
the degree of freedom of a particle.
f courbe *f* d'énergie potentielle,
   courbe *f* de potentiel
e curva *f* de energía potencial,
   curva *f* de potencial
i curva *f* d'energia potenziale,
   curva *f* di potenziale
n potentiaalkromme
d Potentialkurve *f*

5426 POTENTIAL DIAGRAM   me
Of an electron-optical system, a diagram
showing the equipotential curves in a plane

of symmetry of an electron-optical system.
f diagramme *m* du potentiel
e diagrama *m* del potencial
i diagramma *m* del potenziale
n equipotentiaallijnendiagram *n*,
   potentiaaldiagram *n*
d Potentialbild *n*, Potentialschaubild *n*

5427 POTENTIAL GRADIENT ma
At a point, the potential difference per
unit length measured along a conductor
or through a dielectric.
f gradient *m* de potentiel
e gradiente *m* de potencial
i gradiente *m* di potenziale
n potentiaalgradiënt
d Potentialgradient *m*

5428 POTENTIAL JUMP ge
A sudden change in the value of the
potential.
f saut *m* de potentiel
e salto *m* de potencial
i discontinuità *f* di potenziale
n potentiaalsprong
d Potentialsprung *m*

5429 POTENTIAL PLATEAU np
The lowest level of the free electrons.
f palier *m* de potentiel
e meseta *f* de potencial
i ripiano *m* di potenziale
n potentiaalplateau *n*
d Potentialplateau *n*

5430 POTENTIAL SCATTERING np
The reflexion of the incident wave at the
nuclear surface.
f diffusion *f* potentielle
e dispersión *f* potencial
i deviazione *f* potenziale
n potentiaalverstrooiing
d Potentialstreuung *f*

5431 POTENTIAL TROUGH, np
   POTENTIAL WELL
The region of an energy level diagram
delimited by the sides of two neighbouring
potential hills.
f puits *m* de potentiel
e pozo *m* de potencial
i pozzo *m* di potenziale
n potentiaalkuil
d Potentialmulde *f*, Potentialsenke *f*,
   Potentialtopf *m*

5432 POTENTIOMETRIC POSITION con
   MEASURING ASSEMBLY
A position measuring assembly in which
the signal is transmitted in the form of an
electrical potential difference which is a
function of the position of the wiper of a
potentiometer connected to the controlled
element.
f ensemble *m* de mesure de position à
   potentiomètre
e conjunto *m* medidor de posición con
   potenciómetro

i posiziometro *m* potenziometrico
n opstelling voor plaatsbepaling met een
   potentiometer
d Stellungsmessanordnung *f* mit
   Potentiometergeber

5433 POTTER-BUCKY GRID, see 806

5434 POWDER METALLURGY mg
The art of producing powder and of
utilizing metal powders for the production
of massive materials and shaped objects.
f métallurgie *f* des poudres
e pulvimetalurgia *f*
i metallurgia *f* delle polveri
n poedermetallurgie
d Pulvermetallurgie *f*,
   Sintermetallurgie *f*

5435 POWDER METHOD, see 1629

5436 POWDER PATTERN cr
An X-ray diffraction pattern obtained
from a sample consisting of many small
crystals oriented at random.
f cristallogramme *m* à poudre de cristal
e cristalograma *m* de polvo de cristal
i cristallogramma *m* a polvere di
   cristallo
n poederdiagram *n*
d Pulveraufnahme *f*, Pulverdiagramm *n*

5437 POWER BREEDER, rt
   POWER BREEDER REACTOR
A nuclear reactor designed to produce
both useful power and fuel.
f réacteur *m* surrégénérateur de puissance
e reactor *m* regenerador de potencia
i reattore *m* rigeneratore a potenza
n energieleverende kweekreactor,
   vermogenskweekreactor
d Energiebrutreaktor *m*,
   Leistungsbrutreaktor *m*

5438 POWER COEFFICIENT rt
Of a nuclear reactor, the derivative of
reactivity with respect to power.
f coefficient *m* de puissance
e coeficiente *m* de potencia
i coefficiente *m* di potenza
n vermogenscoëfficiënt
d Leistungskoeffizient *m*

5439 POWER CONTROL co
Variation of the power of a reactor by
modifying the reactivity.
f réglage *m* de puissance
e regulación *f* de potencia
i regolazione *f* di potenza
n vermogensregeling
d Leistungsregelung *f*

5440 POWER CONTROL MEMBER rt
A control member used for adjusting the
reactivity of a nuclear reactor or for
modifying its flux distribution.
f élément *m* de commande de puissance
e elemento *m* de regulación de potencia

i elemento *m* di regolazione di potenza
n vermogensregelelement *n*
d Leistungssteuerelement *n*

5441 POWER CONTROL ROD   co
A rod for controlling the power level of a
nuclear reactor.
f barre *f* de commande de la puissance
e barra *f* de regolación de la potencia
i barra *f* di regolazione della potenza
n vermogensregelstaaf
d Leistungsregelstab *m*

5442 POWER DENSITY   rt
The power generated per unit volume of a
reactor core.
f puissance *f* volumique
e densidad *f* de potencia
i densità *f* di potenza
n vermogensdichtheid, volumiek vermogen *n*
d Leistungsdichte *f*

5443 POWER EXCURSION,   rt
      REACTOR EXCURSION
Very rapid increase of reactor power
above the normal operating level.
f excursion *f* de puissance du réacteur
e excursión *f* de potencia del reactor
i escursione *f* di potenza del reattore
n uitschieter van het reactorvermogen
d Leistungsexkursion *f*,
   Reaktorexkursion *f*

5444 POWER LEVEL   rt
The power production of a nuclear reactor
in watts, equal to the product of total
number of nuclear fissions per second,
the energy in ergs released per nuclear
fission, and the number of watts per ergs
per second.
f niveau *m* de puissance
e nivel *m* de potencia
i livello *m* di potenza
n vermogensniveau *n*
d Leistungshöhe *f*, Leistungspegel *m*

5445 POWER LEVEL DETECTOR,   me, rt
      POWER LEVEL INDICATOR
Assembly usually containing a slow neutron
detector followed by amplifiers and
indicating and/or recording instruments,
designed to measure continually the
power level of a reactor based on the
neutron flux.
f détecteur *m* du niveau de puissance
e detector *m* del nivel de potencia
i rivelatore *m* del livello di potenza
n vermogensniveaudetector
d Leistungsniveaudetektor *m*

5446 POWER LEVEL SAFETY SYSTEM rt
Designed for the monitoring of reactor
operating power levels.
f appareil *m* de contrôle de sécurité à
   haut flux
e aparato *m* de control de seguridad de
   alto flujo

i apparecchio *m* di controllo di sicurezza ad
   alto flusso
n hoge-fluxveiligheidscontrole-apparaat *n*
d Hochflussicherheitskontrollgerät *n*

5447 POWER MEASURING   ma
      ASSEMBLY BASED ON ACTIVATION
A measuring assembly designed to
determine the thermal power of a nuclear
reactor by measuring the activation of an
appropriate material.
f ensemble *m* de mesure de puissance par
   activation
e conjunto *m* medidor de potencia por
   activación
i complesso *m* di misura di potenza ad
   attivazione
n opstelling voor vermogensmeting door
   activering
d Anordnung *f* zur Leistungsmessung mittels
   Aktivierung

5448 POWER MEASURING   ma
      ASSEMBLY BASED ON GAMMA
      RADIATION
A measuring assembly designed to
determine the thermal power of a nuclear
reactor by measuring the gamma flux
density at a specified point or points.
f ensemble *m* de mesure de puissance au
   moyen de rayonnement gamma
e conjunto *m* medidor de potencia por
   radiación gamma
i complesso *m* di misura di potenza
   tramite radiazione gamma
n opstelling voor vermogensmeting door
   middel van gammastraling
d Anordnung *f* zur Leistungsmessung
   mittels Gammastrahlung

5449 POWER MEASURING   ma
      ASSEMBLY BASED ON NEUTRON
      FLUX DENSITY
A measuring assembly designed to determine the thermal power of a nuclear
reactor by measuring the neutron flux
density at a specified point or points.
f ensemble *m* de mesure de puissance au
   moyen de la densité de flux neutronique
e conjunto *m* medidor de potencia por
   densidad del flujo neutrónico
i complesso *m* di misura di potenza
   tramite flusso neutronico
n opstelling voor vermogensmeting door
   middel van de neutronenfluxdichtheid
d Anordnung *f* zur Leistungsmessung
   mittels Neutronenstrahlung

5450 POWER OVERRIDE   rt
f outrepassage *m* de puissance
e exceso *m* de potencia
i eccesso *m* di potenza, sovrappotenza *f*
n vermogensoverschrijding
d Leistungsüberhöhung *f*

5451 POWER PER UNIT AREA   rt
Power generated in a reactor core per
unit area.

f   puissance *f* par unité de surface
e   potencia *f* por unidad de superficie
i   potenza *f* per unità di superficie
n   vermogen *n* per eenheid van oppervlak
d   Leistung *f* je Flächendichte der Leistung

5452 POWER RANGE   rt
      The range of power level.
f   domaine *m* de puissance
e   régimen *m* de potencia
i   campo *m* di potenza,
    intervallo *m* di potenza
n   vermogensgebied *n*
d   Leistungsbereich *m*

5453 POWER RATING,   ge
      RATED OUTPUT POWER
      The power available at the output
      terminals of a device when the device is
      operated according to manufacturer's
      specifications.
f   puissance *f* de sortie
e   potencia *f* de salida
i   potenza *f* d'uscita
n   uitgangsvermogen *n*
d   Ausgangsleistung *f*

5454 POWER RATIO   ge
      The ratio of the power output to the
      power input of a device.
f   gain *m*
e   amplificación *f*
i   amplificazione *f*
n   versterking
d   Leistungsverhältnis *n*, Verstärkung *f*

5455 POWER REACTOR   rt
      A nuclear reactor whose primary purpose
      is to produce power.
f   réacteur *m* de puissance
e   reactor *m* de energía,
    reactor *m* de potencia
i   reattore *m* di potenza
n   energiereactor
d   Energiereaktor *m*, Leistungsreaktor *m*

5456 POWER SETBACK, see 3269

5457 POWER SUPPLY UNIT   ge
      Basic function unit designed to supply
      electrical power to an assembly or
      sub-assembly.
f   élément *m* d'alimentation
e   unidad *f* de alimentación
i   elemento *m* d'alimentazione
n   voedingselement *n*
d   Stromversorgungsteil *m*

5458 PR UNIT, see 5254

5459 PRACTICAL RANGE,   np
      REAL RANGE
      The rectilinear range between the ends
      of the path of an ionizing particle.
f   portée *f* réelle
e   alcance *m* real
i   portata *f* reale

n   werkelijke dracht
d   reelle Reichweite *f*,
    tatsächliche Reichweite *f*

5460 PRASEODYMIUM   ch
      Rare earth metallic element, symbol Pr,
      atomic number 59.
f   praséodyme *m*
e   praseodimio *m*
i   praseodimio *m*
n   praseodymium *n*
d   Praseodym *n*

5461 PRE-AMPLIFIER, see 3151

5462 PRECIPITATION HARDENING, see 142

5463 PRECIPITATION UNIT,   rt
      WIRE MACHINE
      In reactor instrumentation, an electrostatic
      device in which the solid daughters of the
      rare-gas fission products are collected on
      a wire and subsequently monitored for
      radioactivity.
f   capteur *m* de dépôt radioactif
e   captador *m* de depósito radiactivo
i   ricettore *m* di deposito radioattivo
n   neerslagvanger
d   Niederschlagsammler *m*

5464 PRECURSOR   np
      Of a nuclide, any radioactive nuclide in a
      decay chain.
f   précurseur *m*
e   precursor *m*
i   progenitore *m*
n   moedernuclide *n*
d   Mutternuklid *n*

5465 PREDETONATION   nw
      A detonation of, e.g., a nuclear bomb prior
      to the scheduled moment.
f   détonation *f* prématurée
e   detonación *f* prematura
i   detonazione *f* prematura
n   voortijdige ontploffing
d   vorzeitige Explosion *f*

5466 PREDISSOCIATION   ch
      The loss of energy by a complex molecule
      due to dissociation before being able to
      emit radiation.
f   prédissociation *f*
e   predisociación *f*
i   predissociazione *f*
n   predissociatie
d   Prädissoziation *f*

5467 PREFERENTIAL RECOMBINATION np
      Recombination which takes place
      immediately after the ion pair is formed.
f   recombinaison *f* préférentielle
e   recombinación *f* preferente
i   ricombinazione *f* preferenziale
n   voorkeursrecombinatie
d   bevorzugte Rekombination *f*

5468 PREFERRED ORIENTATION .   cr
When an ideal isotropic solid is plastically
deformed, or plastically deformed and
annealed, a partial rearrangement of the
crystals frequently occurs so that similar
axes display a common preferred
orientation.
f orientation f préférée
e orientación f preferente
i orientamento m preferito
n voorkeurtextuur
d bevorzugte Orientierung f,
   bevorzugte Textur f

5469 PREFORMED PRECIPITATE   ch
A precipitate used for co-separation of
tracer which has been formed before
admixture with the tracer species to be
absorbed.
f précipité m préformé
e precipitado m preformado
i precipitato m preformato
n aanwezige neerslag
d vorgefällter Niederschlag m

5470 PRE-IONIZATION, see 447

5471 PRE-IRRADIATION   md, xr
Irradiation of a patient before an
operation.
f pré-irradiation f
e preirradiación f
i preirradiazione f
n voorbestraling
d Vorbestrahlung f

5472 PRESET APPARATUS   xr
X-Ray apparatus in which the values of
such variables as voltage, current and
time can be accurately set before the
exposure is made.
f appareil m à pré-ajustage
e aparato m de ajuste previo,
   aparato m preajustado
i apparecchio m a·predisposizione
n apparaat n met voorinstelling,
   vooraf ingesteld apparaat n
d Apparat m mit Voreinstellung

5473 PRESSURE BELL   rt
Component part of a reactor vessel into
which the gas circulators discharge.
f dôme m à pression
e casquete m de presión
i duomo m a pressione
n drukkoepel
d Druckhaube f

5474 PRESSURE COEFFICIENT   rt
In a nuclear reactor the derivative of the
reactivity with regard to the pressure of
the fluids present in the core.
f coefficient m de pression
e coeficiente m de presión
i coefficiente m di pressione
n drukcoëfficiënt
d Druckkoeffizient m

5475 PRESSURE DROP   cl
Loss in pressure of the coolant caused
by friction in the channels, pipes and
fittings and to changes in the cross-
sectional area.
f chute f de pression
e caída f de presión
i caduta f di pressione
n drukverlies n
d Druckabfall m, Druckgefälle n

5476 PRESSURE FLUSH   is
In superfluidity separation making use of
the fact that a surface film of superfluid
helium transmits helium-4 only.
f coup m de pression
e golpe m de presión
i colpo m di pressione
n drukstoot
d Druckstoss m

5477 PRESSURE FRONT,   nw
      SHOCK FRONT
The fairly sharp boundary between the
pressure disturbance created by an
explosion and the ambient atmosphere.
f front m de choc
e frente m de choque, frente m de presión
i fronte f di pressione
n stootfront n
d Stossfront f

5478 PRESSURE METER   ma
An assembly for measuring the pressure
at a point in a nuclear reactor circuit.
f manomètre m
e manómetro m
i manometro m
n manometer
d Druckmessgerät n

5479 PRESSURE SUPPRESSION (GB),   rt, sa
      VAPOR SUPPRESSION (US)
A safety system that can be incorporated
in the design of structures housing
water reactors.
f enlèvement m de la pression
e supresión f de la presión
i soppressione f della pressione
n drukwegneming
d Druckbeseitigung f

5480 PRESSURE TUBE   cd
A metal tube used in reactor constructions
and able to withstand high pressures.
f tube m de force
e tubo m resistente a la presión
i tubo m in pressione
n drukvaste buis
d druckfestes Rohr n

5481 PRESSURE TUBE REACTOR   rt
A nuclear reactor whose fuel assemblies
and coolant are confined in tubes that
withstand the pressure of the coolant.
f réacteur m à tubes de force
e reactor m de tubos de presión

i reattore *m* a tubi in pressione
n drukbuizenreactor
d Druckröhrenreaktor *m*

5482 PRESSURE VESSEL, see 1416

5483 PRESSURE VESSEL REACTOR rt
A nuclear reactor in which the pressure
of the coolant is supported by an envelope
outside the core.
f réacteur *m* à récipient sous pression
e reactor *m* de recipiente sobre presión
i reattore *m* a recipiente in pressione
n drukvatreactor
d Druckgefässreaktor *m*

5484 PRESSURIZED CASING ge
A casing that will withstand high inward
and outward pressures.
f gaine *f* de force,
   gaine *f* résistante à la pression
e envoltura *f* presurizada,
   envoltura *f* resistente a la presión
i involucro *m* pressurizzato,
   involucro *m* resistente alla pressione
n drukvaste omhulling
d druckfeste Hülle *f*

5485 PRESSURIZED GAS FLOW ic
   IONIZATION CHAMBER
An ionization chamber for monitoring gases
under pressure.
f chambre *f* d'ionisation à circulation
   de gaz pressurisé
e cámara *f* de ionización de circulación
   de gas presurizado
i camera *f* d'ionizzazione a circolazione
   di gas sotto pressione
n ionisatievat *n* met gasdoorstroming
   onder druk
d Ionisationskammer *f* mit Druckgas-
   strömung

5486 PRESSURIZED HEAVY rt
   WATER REACTOR
A nuclear reactor in which the coolant and
the moderator consist of heavy water
under pressure.
f réacteur *m* à eau lourde sous pression
e reactor *m* de agua pesada presurizada
i reattore *m* ad acqua pesante in pressione
n zwaar-waterreactor onder druk
d Schwerwasserdruckreaktor *m*

5487 PRESSURIZED REACTOR rt
A nuclear reactor whose primary coolant
is maintained under such a pressure that
no bulk boiling occurs.
f réacteur *m* pressurisé
e reactor *m* de refrigerante presurizado
i reattore *m* a refrigerante in pressione
n drukreactor
d Druckreaktor *m*

5488 PRESSURIZED WATER REACTOR rt
A nuclear reactor in which pressurized
water serves both as moderator and
coolant.

f réacteur *m* à eau sous pression
e reactor *m* de agua presurizada
i reattore *m* ad acqua in pressione
n drukwaterreactor
d Druckwasserreaktor *m*

5489 PRESTRESSED CONCRETE mt
Concrete used in the construction of
nuclear power plants.
f béton *m* précontraint
e hormigón *m* pretensado
i calcestruzzo *m* precompresso
n voorgespannen beton *n*
d Spannbeton *m*, vorgespannter Beton *m*

5490 PRESTRESSED CONCRETE rt
   PRESSURE VESSEL
f caisson *m* en béton précontraint
e recipiente *m* de presión en hormigón
   pretensado
i recipiente *m* di pressione in calcestruzzo
   precompresso
n drukvat *n* uit voorgespannen beton
d Druckgefäss *n* aus Spannbeton

5491 PRIMARY COOLANT cl
A coolant used to remove heat from a
primary source, such as a reactor core
or a breeding blanket.
f fluide *m* de refroidissement primaire
e refrigerante *m* primario
i fluido *m* termovettore primario,
   refrigerante *m* primario
n primair koelmiddel *n*
d Primärkühlmittel *n*

5492 PRIMARY COOLANT CIRCUIT cl
A system for circulating a coolant used
to remove heat from a primary heat
source, such as a reactor core or a
breeding blanket.
f circuit *m* de refroidissement primaire
e circuito *m* de enfriamiento primario
i circuito *m* di raffreddamento primario
n primair koelsysteem *n*
d Primärkühlkreislauf *m*

5493 PRIMARY COSMIC RAYS cr
Probably consists of atomic nuclei, some
of which may have energies of
10^{10}-10^{15}eV.
f rayons *pl* cosmiques primaires
e rayos *pl* cósmicos primarios
i raggi *pl* cosmici primari
n primaire kosmische stralen *pl*
d primäre kosmische Strahlen *pl*

5494 PRIMARY CREEP, see 3487

5495 PRIMARY ELECTRON np
An electron released from an atom by
internal forces.
f électron *m* incident, électron *m* primaire
e electrón *m* incidente, electrón *m* primario
i elettrone *m* iniziale, elettrone *m* primario
n primair elektron *n*
d Primärelektron *n*

5496 PRIMARY FILTER   cd, xr
A sheet of material, usually metal, placed
in a beam of radiation to absorb, as far as
possible, the less penetrating components.
f filtre *m* primaire
e filtro *m* primario
i filtro *m* primario
n primair filter *n*
d Primärfilter *m*

5497 PRIMARY FISSION YIELD, see 1855

5498 PRIMARY ION   np
That ion which has the greater energy
after a collision between two ions.
f ion *m* primaire
e ión *m* primario
i ione *m* primario
n primair ion *n*
d primäres Ion *n*

5499 PRIMARY ION PAIR   np
An ion pair produced directly by the
causative radiation.
f paire *f* d'ions primaire
e par *m* de iones primario
i coppia *f* d'ioni primaria
n primair ionenpaar *n*
d primäres Ionenpaar *n*

5500 PRIMARY IONIZATION   ct, np
1. In collision theory:
The ionization produced by the primary
particles.
2. In counter tubes:
The total ionization produced by incident
radiation without gas amplification.
f ionisation *f* primaire
e ionización *f* primaria
i ionizzazione *f* primaria
n primaire ionisatie
d Primärionisation *f*

5501 PRIMARY IONIZING EVENT, see 3491

5502 PRIMARY NATURAL   np
      RADIONUCLIDES,
         PRIMARY RADIONUCLIDES
Natural radionuclides having lifetimes
exceeding several hundred million years.
f radionucléides *pl* naturels primaires
e radionúclidos *pl* naturales primarios
i radionuclidi *pl* naturali primari
n primaire natuurlijke nucliden *pl*
d primäre natürliche Nuklide *pl*

5503 PRIMARY PROTECTIVE BARRIER   xr·
Material used to absorb the useful beam
of radiation sufficiently to prevent
exposure exceeding the maximum permissible dose.
f écran *m* primaire de radioprotection
e barrera *f* primaria de radioprotección
i schermo *m* primario di radioprotezione
n primair veiligheidsscherm *n*,
   primaire beschermingswand
d Primärstrahlenschutzwand *f*

5504 PRIMARY QUANTUM   np, qm
      NUMBER
In quantum mechanics theory most
commonly used to identify the electronic
orbital of the electrons.
f nombre *m* quantique principal
e número *m* cuántico principal
i numero *m* quantico principale
n hoofdquantumgetal *n*
d Hauptquantenzahl *f*

5505 PRIMARY RADIATION   ra
Radiation direct from the source.
f rayonnement *m* primaire
e radiación *f* primaria
i radiazione *f* primaria
n primaire straling
d primäre Strahlung *f*

5506 PRIMARY REACTOR, see 1985

5507 PRIMARY SEPARATION PROCESS   ch
The solvent extraction process by which
plutonium is separated from irradiated
uranium and fission products.
f procédé *m* de séparation primaire
e procedimiento *m* de separación primaria
i processo *m* di separazione primaria
n primair scheidingsproces *n*
d Primärtrennungsverfahren *n*

5508 PRIMARY SPECIFIC IONIZATION   np
The number of ion clusters produced per
unit track length.
f ionisation *f* primaire spécifique
e ionización *f* primaria específica
i ionizzazione *f* primaria specifica
n specifieke primaire ionisatie
d spezifische Primärionisation *f*

5509 PRIMARY X-RAYS   xr
X-Rays emanating directly from the source.
f rayons *pl* X primaires
e rayos *pl* X primarios
i raggi *pl* X primari
n primaire röntgenstralen *pl*
d primäre Röntgenstrahlen *pl*

5510 PRINCIPLE OF CONTINUITY, see 1341

5511 PRINCIPLE OF PHASE STABILITY, pa
      PRINCIPLE OF SELF-PHASING
In a particle accelerator the principle that
the rotation of a charged particle is
automatically synchronized with the
changing frequency of the accelerating
potential.
f principe *m* d'autophasage
e principio *m* de las partículas estabilizadas
   en fase
i. principio *m* delle particelle stabilizzate
   in fase
n principe *n* van de automatische faze-
   stabilisatie
d Prinzip *n* der Autophasierung,
   Prinzip *n* der selbständigen Phasen-
   stabilisierung

5512 PRO-ACTINIDES   pl
A name proposed for elements of the last
row of the periodic system in oxidation
state -5.
f pro-actinides *pl*
e proactínidos *pl*
i proattinidi *pl*
n proactiniden *pl*
d Proaktinide *pl*

5513 PROBABILITY OF COLLISION   np
The probability that an electron collides
with an atom or molecule when moving
through a distance of one centimetre(er).
f probabilité *f* de choc
e probabilidad *f* de choque
i probabilità *f* d'urto
n botsingskans
d Stosswahrscheinlichkeit *f*

5514 PROBABILITY OF IONIZATION   np
The ratio of the number of collisions
followed by ionization to the total number
of collisions in a gas during a specified
time.
f probabilité *f* d'ionisation
e probabilidad *f* de ionización
i probabilità *f* d'ionizzazione
n ionisatiekans
d Ionisationswahrscheinlichkeit *f*

5515 PROBABLE ERROR   me
The amount of error that is most likely
to occur during a measurement.
f erreur *f* probable
e error *m* probable
i errore *m* probabile
n waarschijnlijke fout
d wahrscheinlicher Fehler *m*

5516 PROBE UNIT   ct
A form of head amplifier having a
low-impedance output and very suitable
for use with a Geiger-Müller counter.
f sonde *f* à amplificateur
e sonda *f* con amplificador
i sonda *f* ad amplificatore
n sonde met versterker
d Sonde *f* mit Verstärker

5517 PROBE UNIT   ra
A radiation detector, with or without
amplifier, of shape suitable for inserting
into cavities.
f sonde *f* de détection
e sonda *f* de detección
i sonda *f* di rivelazione
n detectorsonde
d Detektorsonde *f*

5518 PROCESS HEAT REACTOR, see 3174

5519 PRODUCTION FACTOR,   np, rt
THERMAL FISSION FACTOR
In nuclear reactors, the average number
of fast fission neutrons emitted upon
capture of one thermal neutron in the fuel.

f facteur *m* de fission thermique
e factor *m* de fisión térmica
i fattore *m* di fissione termica
n thermische splijtingsfactor
d thermischer Spaltfaktor *m*

5520 PRODUCTION RATE, see 1504

5521 PRODUCTION REACTOR   rt
A reactor whose primary purpose is to
produce fissile (fissionable) materials to
perform irradiation on an industrial scale.
f réacteur *m* de production
e reactor *m* de producción
i reattore *m* di produzione
n produktiereactor
d Produktionsreaktor *m*

5522 PROFILE VIEW, see 3913

5523 PROGRAMMED ACTION   sm
SAFETY ASSEMBLY
A safety assembly which controls a
decrease of power of a nuclear reactor
according to a program(me) down to a
value which is not necessarily zero.
f ensemble *m* de sécurité programmé
e conjunto *m* de seguridad de acción
programada
i complesso *m* di sicurezza a programma
n veiligheidsopstelling met
geprogrammeerde werking
d von Grenzwertprogrammen abhängige
Schutzeinrichtung *f*

5524 PROHIBITED AREA   sa
The area that can be entered only on
special permit.
f zone *f* interdite
e zona *f* prohibida
i zona *f* proibita
n verboden gebied *n*
d verbotenes Gebiet *n*

5525 PROJECTED FOCAL SPOT, see 2031

5526 PROMETHIUM   ch
Radioactive element, symbol Pm, atomic
number 61.
f prométhium *m*
e prometio *m*
i prometio *m*, promezio *m*
n promethium *n*
d Promethium *n*

5527 PROMETHIUM LUMINOUS   mt
COMPOUND
A luminous compound containing
promethium-147, bound in small highly
insoluble particles.
f composé *m* lumineux à prométhium
e compuesto *m* luminoso de prometio
i composto *m* luminoso a promezio
n promethiumhoudende lichtgevende
verbinding
d promethiumhaltiger Leuchtstoff *m*

5528 PROMOTED MIXING   is
In separation by gaseous diffusion through
membranes, the processes (diffusion,
convection, or other) whereby the
concentration gradient normal to the
diffusion membrane is kept as small as
possible.
f homogénéisation *f*
e homogenización *f*
i omogenizzazione *f*
n homogenisatie
d Homogenisierung *f*

5529 PROMPT CRITICAL   rt
Fulfilling the condition that a nuclear
chain reacting medium is critical
utilizing prompt neutrons only.
f critique instantané
e crítico con neutrones instantáneos
i critico istantaneo
n prompt kritiek, prompt kritisch
d prompt-kritisch

5530 PROMPT CRITICALITY   rt
Criticality condition due to prompt
neutrons only.
f criticité *f* instantanée
e criticidad *f* instantánea
i criticità *f* istantanea
n prompte kriticiteit
d prompte Kritizität *f*

5531 PROMPT DOSE   ra
The dose due to prompt fission gammas
and neutrons from an atomic detonation.
f dose *f* instantanée
e dosis *f* inmediata, dosis *f* instantánea
i dose *f* istantanea
n prompte dosis
d prompte Dosis *f*

5532 PROMPT DROP OF REACTIVITY   rt
A decrease in reactivity caused by a
sudden variation of the reactivity, due to
prompt neutrons.
f réponse *f* négative instantanée
e diminución *f* instantánea de la reactividad
i risposta *f* negativa istantanea
n prompte reactiviteitsverlaging
d prompte Reaktivitätsabnahme *f*

5533 PROMPT GAMMA RADIATION,
see 2688

5534 PROMPT JUMP OF REACTIVITY   rt
An increase in reactivity caused by a
sudden variation of the reactivity, caused
by prompt neutrons.
f réponse *f* positive instantanée
e aumento *m* instantáneo de la reactividad
i risposta *f* positiva istantanea
n prompte reactiviteitsstijging
d . prompte Reaktivitätszunahme *f*

5535 PROMPT NEUTRON FRACTION   np
The ratio of the mean number of prompt
neutrons per fission to the mean total
number of neutrons per fission.

f fraction *f* de neutrons instantanés
e fracción *f* de neutrones instantáneos
i frazione *f* di neutroni istantanei
n fractie prompte neutronen
d Anteil *m* der prompten Neutronen

5536 PROMPT NEUTRONS   np
Neutrons accompanying the fission process
without measurable delay.
f neutrons *pl* instantanés
e neutrones *pl* instantáneos
i neutroni *pl* istantanei
n prompte neutronen *pl*
d prompte Neutronen *pl*

5537 PROMPT RADIATION   ra
Radiation emitted within a time too short
for measurement, including gamma rays,
characteristic X-rays, conversion and
Auger electrons, prompt neutrons and
annihilation radiation.
f rayonnement *m* instantanée
e radiación *f* instantánea
i radiazione *f* istantanea
n prompte straling
d prompte Strahlung *f*

5538 PROMPT REACTIVITY   rt
Reactivity due solely to prompt neutrons.
f réactivité *f* instantanée
e reactividad *f* instantánea
i reattività *f* istantanea
n prompte reactiviteit
d prompte Reaktivität *f*

5539 PROPANE IONIZATION CHAMBER   ic
An ionization chamber in which the organic
gas filling consists of propane (C_3H_8).
f chambre *f* d'ionisation à propane
e cámara *f* de ionización con propano
i camera *f* d'ionizzazione a propano
n met propaan gevuld ionisatievat *n*
d propangefüllte Ionisationskammer *f*

5540 PROPHASE   md
The first stages of mitosis, during which
the chromosomes shorten and thicken
preparatory to forming a metaphase plate.
f prophase *f*
e profase *f*
i profase *f*
n profaze, voorstadium *n*
d Prophase *f*

5541 PROPORTIONAL ACTION,   co
PROPORTIONAL-POSITION ACTION
That action in which there is a continuous
linear relation between the value of a
controlled variable and the position of a
final control element.
f action *f* proportionnelle
e acción *f* proporcional
i azione *f* proporzionale
n proportionele werking
d proportionales Verhalten *n*

5542 PROPORTIONAL BAND   co
In a proportional-position controller the
range of values of the controlled variable
which corresponds to the full operating
range of the final control element.
f bande *f* proportionnelle
e banda *f* proporcional propia
i banda *f* proporzionale intrinseca
n proportionaliteitsgebied *n*
d innerer Proportionalbereich *m*

5543 PROPORTIONAL COUNTER TUBE   ct
A radiation counter tube or chamber
operated in the proportional region.
f tube *m* compteur proportionnel
e tubo *m* contador proporcional
i tubo *m* contatore proporzionale
n proportionele telbuis
d Proportionalzählrohr *n*

5544 PROPORTIONAL IONIZATION   ic
CHAMBER
An ionization chamber in which the initial
ionization current is amplified by electron
multiplication in a region of high electric
field strength, as it is in a proportional
counter.
f chambre *f* d'ionisation proportionnelle
e cámara *f* de ionización proporcional
i camera *f* d'ionizzazione proporzionale
n proportioneel ionisatievat *n*
d Proportionalionisationskammer *f*

5545 PROPORTIONAL REGION   ct
The range of operating voltage for a
counter tube or ionization chamber in
which the gas amplification is greater than
one and independent of the primary
ionization.
f région *f* de proportionnalité
e región *f* de proporcionalidad
i dominio *m* di proporzionalità
n proportionaliteitsgebied *n*
d Proportionalbereich *m*

5546 PROPULSION   ge
The act or process of propelling.
f propulsion *f*
e propulsión *f*
i propulsione *f*
n aandrijving, voortstuwing
d Antrieb *m*

5547 PROPULSION REACTOR, see 4478

5548 PROSPECTING   mi
To explore a region for minerals.
f prospection *f*
e exploración *f*
i prospezione *f*
n prospectie
d Lagerstättensuche *f*, Schürfung *f*

5549 PROSPECTING AUDIO-   ma
INDICATOR
A portable prospecting radiation indicator
that contains its own power supply and

provides an audible indication of photon
flux density.
f audio-signaleur *m* de prospection,
radiophone *m* de prospection
e indicador *m* auditivo de exploración
i audiosegnalatore *m* per prospezione
n audio-indicator voor prospectie
d Lagerstättensuchgerät *n* mit akustischer
Anzeige

5550 PROSPECTING RADIATION   me
METER WITH G.M. COUNTER TUBE
A portable prospecting assembly containing
its own power source and designed to
measure, by means of one or several
G.M. counter tubes, particle or photon flux
density.
f radiamètre *m* de prospection à tube
compteur Geiger-Müller
e radiámetro *m* de exploración con contador
G.M.
i radiametro *m* di prospezione a tubo
contatore G.M.
n prospectiestralingsmeter met geigertelbuis,
stralingsmeter voor prospectie met geigertelbuis
d Lagerstättensuchgerät *n* mit G.M.Zählrohr

5551 PROTACTINIDES   np
The name sometimes used for the elements
of atomic number 89 to 103 when they are
in the oxidation state +5.
f protactinides *pl*
e protactínidos *pl*
i protattinidi *pl*
n protactiniden *pl*
d Protaktinide *pl*

5552 PROTACTINIUM   ch
Radioactive element, symbol Pa, atomic
number 91.
f protactinium *m*
e protactinio *m*
i protattinio *m*
n protactinium *n*
d Protaktinium *n*

5553 PROTECTION   sa
Against radiation: this term covers all
provisions designed to reduce exposure
to external radiation or to lessen the
ingestion or inhalation of radioactive
materials.
f protection *f*
e protección *f*
i protezione *f*
n bescherming
d Schutz *m*

5554 PROTECTION FACTOR   nw
The composite factor, including distance,
shielding materials and type of radiation,
describes how much a given shelter will
reduce the dose rate from that existing
outside.
f facteur *m* de protection

e factor m de protección
i fattore m di protezione
n beschermingsfactor
d Schutzfaktor m

5555 PROTECTION SURVEY,   sa
       RADIATION SURVEY
    Evaluation of the radiation hazards
    incidental to the production, use or
    existence of radioactive materials or
    other sources of radiation under a specific
    set of conditions.
f contrôle m de protection,
    contrôle m de rayonnements
e control m de protección,
    control m de radiaciones
i controllo m di protezione,
    controllo m di radiazioni
n stralingsgevaarcontrole,
    stralingsverkenning
d Strahlenschutzüberwachung f,
    Strahlungsgefahrkontrolle f

5556 PROTECTIVE BARRIER   xr
    Material used to absorb ionizing radiation
    for protection purposes.
f écran m protecteur
e barrera f protectora
i schermo m protettivo
n beschermingswand, veiligheidsscherm n
d Schutzwand f

5557 PROTECTIVE CLOTHES, see 2842

5558 PROTECTIVE COATING   fu
    A coating, e.g. on fuel elements, for
    reasons of protection.
f barrière f protectrice,
    couche f protectrice
e capa f protectora
i strato m protettivo
n beschermende laag
d Schutzschicht f

5559 PROTECTIVE GLOVES, see 3957

5560 PROTECTIVE LEAD GLASS   mt, sa
    Glass containing a high proportion of lead
    components.
f verre m au plomb protecteur
e vidrio m al plomo protector
i vetro m piombifero protettivo
n veiligheidsloodglas n
d Bleischutzglas n

5561 PROTECTIVE MATERIAL   mt, sa
f matière f protectrice
e material m protector
i materiale m protettivo
n beschermend materiaal n
d Schutzstoff m

5562 PROTECTIVE SCREEN   sa
f écran m protecteur
e pantalla f protectora
i schermo m protettivo
n scherm n
d Schirm m

5563 PROTECTIVE VALUE, see 3951

5564 PROTIUM   is
    The isotope of hydrogen having a mass
    number of 1.
f protium m
e protio m
i protio m
n protium n
d Protium n

5565 PROTON   np
    1. A positively charged elementary particle
    of mass number 1 and charge equal in
    magnitude to the electronic charge e.
    2. The nucleus of an atom of hydrogen of
    mass number 1.
f proton m
e protón m
i protone m
n proton n
d Proton n

5566 PROTON ATOMIC MASS   np
    The atomic mass of the proton
    $H^+ = 1.007593 \pm 0.000003$ amu.
f masse f du proton
e masa f del protón
i massa f del protone
n massa van het proton
d Protonenmasse f

5567 PROTON BINDING ENERGY   np
    The energy required to remove a single
    proton from a nucleus.
f énergie f de liaison d un proton
e energía f de enlace de un protón
i energia f di legame d un protone
n protonenbindingsenergie
d Protonenbindungsenergie f

5568 PROTON MAGNETIC MOMENT   ma, np
    The magnetic moment of a proton
    $\mu_0 = (1.41045 \pm 0.00009) \times 10^{-23}$ erg
    gauss^{-1}.
f moment m magnétique du proton
e momento m magnético del protón
i momento m magnetico del protone
n magnetisch moment n van een proton
d magnetisches Moment n eines Protons

5569 PROTON MAGNETIC RESONANCE,   np
       PROTON RESONANCE
    Magnetic resonance using the magnetic
    moment of the proton.
f résonance f magnétique du proton
e resonancia f magnética del protón
i risonanza f magnetica del protone
n magnetische resonantie van een proton
d magnetische Resonanz f eines Protons

5570 PROTON MICROSCOPE,   me
       PROTON SCATTERING MICROSCOPE
    A microscope similar to the electron
    microscope but using protons instead of
    electrons as the charged particles.

f   microscope m protonique
e   microscopio m protónico
i   microscopio m protonico
n   protonenmicroscoop
d   Protonenmikroskop n

5571 PROTON NUMBER, see 397

5572 PROTON-PROTON   np
      CHAIN REACTION
      A series of thermonuclear reactions
      initiated by a reaction of two protons
      that provides the energy of some stars.
f   réaction f en chaîne proton-proton
e   cadena f de reacción protón-protón
i   catena f di reazione protone-protone
n   proton-proton-kettingreactie
d   Proton-Proton-Kettenreaktion f

5573 PROTON-SYNCHROTRON, see 1438

5574 PROTON-TO-ELECTRON-MASS   np
      RATIO
      The ratio of the proton mass H^+, to the
      electron rest mass m_e
f   rapport m des masses proton/électron
e   relación f de las masas protón/electrón
i   rapporto m delle masse protone/elettrone
n   proton-elektron-massaverhouding
d   Proton-Elektron-Massenverhältnis n

5575 PROTOTYPE REACTOR   rt
      A nuclear reactor that is the first of a
      series of the same design.
f   réacteur m prototype
e   reactor m prototipo
i   reattore m prototipo
n   prototypereactor
d   Prototypreaktor m

5576 PROTRACTION, see 1944

5577 PSEUDO-IMAGE   ra, xr
      The shadow cast by an irradiated object
      of smaller width than the source, in the
      region beyond the point of convergence
      of the umbra.
f   pseudo-image f
e   seudoimagen f
i   pseudoimmagine f
n   pseudobeeld n
d   Pseudobild n

5578 PSEUDOSCALAR COUPLING   np
      Type of interaction energy postulated
      between a π meson and a nucleon which
      consists of the product of the pseudo-
      scalar field and a bilinear pseudoscalar
      function of the nucleus wave functions.
f   couplage m pseudoscalaire
e   acoplamiento m seudoescalar
i   accoppiamento m pseudoscalare
n   pseudoscalaire koppeling
d   pseudoskalare Kupplung f

5579 PSEUDOSCALAR PARTICLE   np
      A particle, such as a π meson, which may
      be described by a pseudoscalar field.
f   particule f pseudoscalaire
e   partícula f seudoescalar
i   particella f pseudoscalare
n   pseudoscalair deeltje n
d   pseudoskalares Teilchen n

5580 PSEUDOSCALAR QUANTITY   ma
      A scalar quantity which changes sign in
      the transition from a right-handed to a
      left-handed co-ordinate system.
f   grandeur f pseudoscalaire
e   magnitud f seudoescalar
i   grandezza f pseudoscalare
n   pseudoscalaire grootheid, pseudoscalar
d   Pseudoskalar f, pseudoskalare Grösse f

5581 PSEUDOSCOPIC VIEW   xr
      A reversed stereoscopic view in which
      near objects appear to be distant and vice
      versa.
f   projection f pseudoscopique,
    vue f pseudoscopique
e   proyección f seudoscópica,
    vista f seudoscópica
i   proiezione f pseudoscopica,
    vista f pseudoscopica
n   pseudoscopische projectie
d   pseudoskopische Projektion f

5582 PSEUDOVECTOR, see 487

5583 PSEUDOVECTOR COUPLING   np
      Type of interaction energy postulated
      between a nuclear and other particles in
      which a bilinear pseudovector function of
      the nucleon wavefunctions appears.
f   couplage m pseudovectoriel
e   acoplamiento m seudovectorial
i   accoppiamento m pseudovettoriale
n   pseudovectoriële koppeling
d   pseudovektorielle Kupplung f

5584 PULP GRADING EQUIPMENT   ma
      An equipment designed to grade the uranium
      content of a pulped ore and the weight of
      dry materials contained in this ore by
      measurement of activity flow-rate and
      density.
f   équipement m de contrôle du minerai
    sous forme de pulpe
e   equipo m clasificador de pulpa
i   apparecchiatura f per controllo di
    minerali sotto forma di polpa
n   uitrusting voor gehaltecontrole van erts
    in pulpvorm
d   Klassiereinrichtung f bei der
    Schwimmaufbereitung

5585 PULSE, see 3413

5586 PULSE AMPLIFIER   ec
      An electronic amplifier which, within the
      limits of its normal operating characteris-
      tics, delivers a single output pulse for
      each input pulse.
f   amplificateur m d'impulsions
e   amplificador m de impulsos

i amplificatore *m* d'impulsi
n pulsversterker
d Impulsverstärker *m*

5587 PULSE AMPLITUDE DISCRIMINATOR,
      PULSE HEIGHT DISCRIMINATOR,
      see 231

5588 PULSE AMPLITUDE·   ec
      DISCRIMINATOR UNIT
      A basic function unit containing a pulse
      amplitude discriminator.
f élément *m* discriminateur d'amplitude
e unidad *f* discriminadora de amplitud
i elemento *m* discriminatore d'ampiezza
n pulshoogtediscriminator
d Impulshöhendiskriminator *m*

5589 PULSE AMPLITUDE SELECTOR,   ec
      PULSE HEIGHT SELECTOR
      An electronic circuit which permits only
      those voltage pulses that have amplitudes
      between predetermined levels to be passed
      to the succeeding circuits.
f sélecteur *m* d'amplitude
e selector *m* de amplitude
i selettore *m* d'ampiezza
n pulshoogtekiezer
d Impulshöhenwähler *m*

5590 PULSE AMPLITUDE   ct
      SELECTOR UNIT
      A basic function unit containing a pulse
      amplitude selector.
f élément *m* sélecteur d'amplitude
e unidad *f* selectora de amplitud
i elemento *m* selettore d'ampiezza
n pulshoogtekiezer
d Impulshöhendifferentialdiskriminator *m*

5591 PULSE AMPLITUDE-TO-TIME   ec, ma
      CONVERTER
      An electronic sub-assembly, designed to
      provide, according to the type of apparatus
      1. An output pulse the duration of which is
      proportional to the amplitude of the input
      pulse.
      2. Two output pulses, one delayed with
      respect to the other by a time interval
      that is proportional to the amplitude of the
      input pulse.
f convertisseur *m* amplitude-temps
e convertidor *m* amplitud-tiempo
i convertitore *m* ampiezza-tempo
n amplitude-tijd-omzetter
d Amplituden-Zeit-Wandler *m*

5592 PULSE COUNTING AND/OR   ma
      COUNTING RATE ASSEMBLY
      A measuring assembly to determine the
      number (pulse counting assembly) and/or
      counting rate (pulse counting ratemeter
      assembly) of the electrical output pulses
      of its radiation detector(s).
f ensemble *m* de mesure à impulsions
e conjunto *m* contador per impulsos o
   conjunto medidor de la intensidad de
   emisión

i complesso *m* di misura ad impulsi
n opstelling met pulstelling
d Anordnung *f* zur Impulszählung und/oder
   Zählratenmessung

5593 PULSE COUNTING RATE   ma
      ASSEMBLY
      A pulse counting assembly which includes
      a detector whose output pulses are applied
      to a ratemeter usually linear or
      logarithmic.
f ensemble *m* de mesure de taux de comptage
e conjunto *m* contador diferencial
i complesso *m* di misura del rateo di
   conteggio
n opstelling voor teltempometing,
   teltempo-opstelling
d Strahlungsmessgerät *n* mit
   Zählratenanzeige

5594 PULSE GENERATOR, see 3414

5595 PULSE HEIGHT ANALYZER (US),
      see 228

5596 PULSE IONIZATION CHAMBER,
      see 1488

5597 PULSE OVERLAPPING,   ct
      PULSE PILE-UP
      In pulse circuits, the probable overlapping
      of two or more pulses to produce single
      pulses of amplitude different from normal.
f superposition *f* d'impulsions
e superposición *f* de impulsos
i sovrapposizione *f* d'impulsi
n pulsoverlapping
d Impulsüberlappung *f*

5598 PULSE RISE TIME   ct
      The time interval required for the function
      presenting the pulse to increase from 10 %
      to 90 % of its peak value.
f temps *m* de montée d'une impulsion
e tiempo *m* de crecimiento de un impulso
i tempo *m* di salita d'un impulso
n stijgtijd van een puls
d Anstiegzeit *f* eines Impulses

5599 PULSE SHAPE DISCRIMINATOR   ec
      A circuit designed to produce different
      output signals for different shapes of the
      input pulses in order to distinguish between
      pulses of different origin.
f discriminateur *m* de forme
e discriminador *m* de forma
i discriminatore *m* di forma
n pulsvormdiscriminator
d Impulsformdiskriminator *m*

5600 PULSE SHAPE   ec
      DISCRIMINATOR UNIT
      A basic function unit containing a pulse
      shape discriminator.
f élément *m* discriminateur de forme
e unidad *f* discriminadora de forma
i elemento *m* discriminatore di forma
n pulsvormdiscriminator
d Impulsformdiskriminator *m*

5601 PULSED COLUMN   ch
A packed column or plate column to which
more intimate mixing of the phases is
achieved by imparting a pulsing movement
to the liquid contents of the column.
f colonne *f* pulsée
e columna *f* pulsada
i colonna *f* pulsata
n gepulsde kolom
d gepulste Kolonne *f*, gepulste Säule *f*

5602 PULSED NEUTRON EXPERIMENT *   np
Experiment in which the decay of neutron
bursts is studied.
f expérience *f* pulsée
e experimento *m* pulsado
i sperimento *m* pulsato
n gepulsde-neutronenexperiment *n*
d gepulste-Neutronenexperiment *n*

5603 PULSED NEUTRON SOURCE   np
Apparatus used for producing short
bursts of neutrons.
f source *f* pulsée de neutrons
e fuente *f* pulsada de neutrones
i sorgente *f* pulsata di neutroni
n gepulsde neutronenbron
d gepulste Neutronenquelle *f*

5604 PULSED NEUTRONS   np
Neutrons produced in bursts by pulsed
neutron sources.
f neutrons *pl* pulsés
e neutrones *pl* pulsados
i neutroni *pl* pulsati
n gepulsde neutronen *pl*
d gepulste Neutronen *pl*

5605 PULSED REACTOR   rt
A nuclear reactor designed to produce
intense bursts of neutrons for short
intervals of time.
f réacteur *m* pulsé
e reactor *m* pulsado
i reattore *m* pulsato
n gepulsde reactor
d gepulster Reaktor *m*

5606 PUMP MIXER SETTLER   ch, is
A mixer settler in which pumps are used
both as the mixers and to move the phases
from stage to stage.
f mélangeur-clarificateur *m* à pompes
e mezclador-asentador *m* de bombas
i mescolatore-chiarificatore *m* a pompe
n meng-neerslagapparaat *n* met circulatie-
pompen
d Misch- und Klärgerät *n* mit Umlauf-
pumpen

5607 PURE SUBSTANCE   ch
A substance having well defined properties
which can be represented by numerical
constants, whatever the size of the sample.
f substance *f* pure
e substancia *f* pura
i sostanza *f* pura
n zuivere stof
d Reinstoff *m*

5608 PURGING   cg
The scouring of nuclear reactor fuel
elements and process piping by the
mildly abrasive action of diatomaceous
earth.
f rodage *m*
e alisadura *f*
i smerigliatura *f*
n afschuren *n*
d Scheuern *n*

5609 PURPLE,
PURPURA, see 5168

5610 PYELOGRAPHY   xr
Radiological examination of the upper
urinary tract following the injection of
a contrast medium.
f pyélographie *f*
e pielografía *f*
i pielografia *f*
n pyelografie
d Pyelographie *f*

5611 PYROCHEMISTRY   ch
Chemistry of, e.g., irradiated fuel elements
at high temperatures.
f pyrochimie *f*
e piroquímica *f*
i pirochimica *f*
n pyrochemie
d Pyrochemie *f*

5612 PYROCHLORE   mi
Mineral containing small quantities of
cerium, thorium, uranium and titanium.
f pyrochlore *m*
e pirocloro *m*
i pirocloro *m*
n pyrochloor *n*
d Pyrochlor *n*

5613 PYROLYTIC CARBON-SILICON   mt
CARBIDE MIXTURE
Materials used in coating fuel pellets.
f mélange *m* de carbone pyrolitique et
carbure de silicium
e mezcla *f* de carbón pirolítico y carburo
de silicio
i miscela *f* di carbonio pirolitico a carburo
di silicio
n mengsel *n* van pyrolitische koolstof en
siliciumcarbide
d Gemisch *n* von pyrolitischem Kohlenstoff
und Siliziumkarbid

5614 PYROMETALLURGICAL
REPROCESSING, see 1980

5615 PYROTRON   np, rt
A machine that uses magnetic mirrors in
a long straight tube to reflect charged
particles and prevent end leaks.
f pyrotron *m*
e pirotrón *m*
i pirotrone *m*
n pyrotron *n*
d Pyrotron *n*

5616 Q ELECTRON						np
An electron having an orbit of such
dimensions that the electron constitutes
part of the seventh shell of electrons
surrounding the atomic nucleus.
f électron *m* Q
e electrón *m* Q
i elettrone *m* Q
n Q-elektron *n*
d Q-Elektron *n*

5617 Q SHELL						np
The seventh layer of electrons in motion
about the nucleus of an atom.
f couche *f* Q
e capa *f* Q
i strato *m* Q
n Q-schil
d Q-Schale *f*

5618 Q VALUE,
NUCLEAR DISINTEGRATION ENERGY,
see 4836

5619 Q VALUE,
NUCLEAR REACTION ENERGY,
see 4900

5620 QUADRIVALENT,					ch
TETRAVALENT
Said of an element the atom of which has
four electrons in its shell.
f tétravalent adj
e tetravalente adj
i tetravalente adj
n vierwaardig adj
d vierwertig adj

5621 QUADRUPLET					sp
A spectrum line which, when examined
under high resolution, is composed of
four closely-packed, fine lines.
f quadruplet *m*
e cuadrupleto *m*
i quadrupletto *m*
n quadruplet *m*
d Quadruplett *n*

5622 QUADRUPOLAR FIELD				ec, pp
A magnetic field which may be generated
by four rectilinear currents of alternating
sense distributed over the periphery of a
chamber.
f champ *m* quadripolaire
e campo *m* cuadripolar
i campo *m* quadripolare
n vierpolig veld *n*
d Vierpolfeld *n*

5623 QUADRUPOLE BROADENING			sp
In spectroscopy, broadening of spectral
lines caused by interaction of the orbital

electrons with a nucleus possessing a
quadrupole field.
f élargissement *m* quadripolaire
e ensanchamiento *m* cuadripolar
i allargamento *m* quadripolare
n quadrupoolverbreding
d Quadrupolverbreiterung *f*

5624 QUADRUPOLE ELECTRICAL			ra
MOMENT,
QUADRUPOLE MOMENT
When the radiation field due to a set of
moving electric or magnetic charges is
expanded in a series of powers of the
product of the charges times space
coordinates, the sum of the quadrative
terms is the quadrupole moment.
f moment *m* quadripolaire
e momento *m* cuadripolar
i momento *m* quadripolare
n quadrupoolmoment *n*
d Quadrupolmoment *n*

5625 QUADRUPOLE LENS				pa
An electrostatic or magnetic lens based
on the strong focusing principle and used
for focusing beams of charged particles
obtained from accelerators.
f lentille *f* quadripolaire
e lente *f* cuadripolar
i lente *f* quadripolare
n vierpolige lens
d vierpolige Linse *f*

5626 QUADRUPOLE RADIATION			ra
Radiation emitted by a quadrupole.
f rayonnement *m* de quadrupole
e radiación *f* de cuadrupolo
i radiazione *f* di quadrupolo
n quadrupoolstraling
d Quadrupolstrahlung *f*

5627 QUALITY						ra
Of ionizing radiation, a relative energy
assessment based on penetrating power.
f qualité *f*
e cualidad *f*
i qualità *f*
n kwaliteit
d Qualität *f*

5628 QUALITY FACTOR				ra
A linear-energy-transfer-dependent
factor by which absorbed dose is multiplied in obtaining dose equivalent in order
to express on a common scale, for all
ionizing radiations, the irradiation
incurred by exposed persons.
f facteur *m* de qualité
e factor *m* de cualidad
i fattore *m* di qualità
n kwaliteitsfactor
d Bewertungsfaktor *m*

5629 QUANTA CURRENT DENSITY, see 1579

5630 QUANTITY OF RADIATION   ra
Of specified radiation incident on a
specified medium, loosely used to imply
absorbed dose or, less usually, exposure
dose.
f   quantité f de rayonnement
e   cuantidad f de radiación
i   quantità f di radiazione
n   stralingskwantiteit
d   Strahlungsquantität f

5631 QUANTIZATION   qm
1. The existence of discrete values only,
according to the quantum theory, for some
physical quantity.
2. The mathematical procedure for
computing those values.
f   quantification f
e   cuantificación f, cuantización f
i   quantizzazione f
n   quantisatie
d   Quantelung f, Quantisierung f

5632 QUANTIZATION DISTORTION,   qm
     QUANTIZATION NOISE
Inherent distortion introduced in process
of quantization.
f   distorsion f de quantification
e   distorsión f de cuantización
i   distorsione f di quantizzazione
n   quantisatievervorming
d   Quantelungsverzerrung f

5633 QUANTIZATION LEVEL   qm
One of the subrange values obtained by
quantization.
f   niveau m de quantification
e   nivel m de cuantización
i   livello m di quantizzazione
n   quantisatieniveau n
d   Quantelungsniveau n

5634 QUANTIZE (TO)   qm
To restrict a variable to a discrete number
of possible values.
f   quantifier v
e   cuantificar v, cuantizar v
i   quantizzare v
n   quantiseren v
d   quanteln v

5635 QUANTIZED FIELD THEORY   qm
A theory in which electromagnetic fields
and the fields of matter are represented
by mathematical operators that describe
the elementary processes of creation and
annihilation of particles or photons.
f   théorie f du champ quantifié
e   teoría f del campo cuantizado
i   teoria f del campo quantizzato
n   theorie van het gequantiseerde veld
d   Theorie f des gequantelten Feldes

5636 QUANTIZED SYSTEM   qm
A system of particles the energies of which
can have discrete values only.

f   système m quantifié
e   sistema m cuantificado
i   sistema m quantizzato
n   gequantiseerd systeem n
d   gequanteltes System n

5637 QUANTUM, see 2284

5638 QUANTUM CONDITION   ma, np
The mathematical condition that must be
satisfied for any given quantum state of
an atom or other system.
f   condition f quantique
e   condición f cuántica
i   condizione f quantica
n   quantumvoorwaarde
d   Quantenbedingung f

5639 QUANTUM EFFICIENCY,   qm
     QUANTUM YIELD
The average number of electrons photo-
electrically emitted from the photocathode
per incident photon of a given wavelength.
f   rendement m quantique
e   rendimiento m cuántico
i   rendimento m quantico
n   quantumopbrengst
d   Quantenausbeute f, Quantenertrag m

5640 QUANTUM ELECTRODYNAMICS   qm
Quantized field theory of the interaction
between electrons, positrons and radiations
based on the quantized form of the
Maxwell equations and the Dirac electron
theory.
f   électrodynamique f quantique
e   electrodinámica f cuántica
i   elettrodinamica f quantica
n   quantumelektrodynamica
d   Quantenelektrodynamik f

5641 QUANTUM EMISSION   qm
The emission of a quant, e.g. a photon.
f   émission f quantique
e   emisión f cuántica
i   emissione f quantica
n   quantumemissie
d   Quantenemission f

5642 QUANTUM JUMP,   qm
     QUANTUM TRANSITION
An abrupt readjustment which is
accompanied by the emission or absorption
of a quantum of radiant energy.
f   saut m quantique
e   salto m cuántico
i   salto m quantico
n   quantumsprong
d   Quantensprung m, Quantenübergang m

5643 QUANTUM LIMIT, see 753

5644 QUANTUM MECHANICAL SYSTEM, qm
     QUANTUM MECHANICS
A branch of science dealing with the
description of atomic and molecular properties, based on the dualistic relation
between matter and wave radiation.

f   mécanique f quantique
e   mecánica f cuántica
i   meccanica f quantica
n   quantummechanica
d   Quantenmechanik f

5645 QUANTUM NUMBER   qm
A number assigned to one of the various
values of a quantized quantity in its
discrete range.
f   nombre m quantique
e   número m cuántico
i   numero m quantico
n   quantumgetal n
d   Quantenzahl f

5646 QUANTUM OF ACTION, see 5307

5647 QUANTUM POSTULATE   qm
The postulate that a system emitting
monochromatic radiation of frequency
changes its energy always by whole
multiples of hv, where h is Planck's
constant.
f   postulat m du quantum
e   postulado m del cuanto
i   postulato m del quanto
n   quantumpostulaat n
d   Quantumpostulat n

5648 QUANTUM STATE   qm
One of the states in which an atom may
exist permanently or momentarily.
f   niveau m d'énergie
e   nivel m de energía
i   livello m d'energia
n   quantige toestand
d   Energieniveau n, Quantenzustand m

5649 QUANTUM STATISTICS   qm
Statistics of the distribution of particles
of a given type among the various possible
energy corresponding to a particular
quantum.
f   statistique f quantique
e   estadística f cuántica
i   statistica f quantica
n   quantumstatistiek
d   Quantenstatistik f

5650 QUANTUM THEORY   qm
Conception that energy is produced and
radiated only in multiples of minimum
quantities (quanta), the quantum being
related to the frequency of the radiation.
f · théorie f des quanta
e   teoría f de los cuantos
i   teoria f dei quanti
n   quantumtheorie
d   Quantentheorie f

5651 QUANTUM VOLTAGE   qm
The voltage through which an electron
must be accelerated to acquire the energy
corresponding to a particular quantum.
f   tension f quantique
e   tensión f cuántica

i   tensione f quantica
n   quantumspanning
d   Quantenspannung f

5652 QUARTZ FIBER DOSEMETER   me, ra
(US),
     QUARTZ FIBRE DOSEMETER (GB)
A dosemeter using the Lauritzen
electroscope principle.
f   dosimètre m à fibre de quartz
e   dosímetro m de fibra de cuarzo
i   dosimetro m a fibra di quarzo
n   kwartsdraaddosismeter
d   Quarzfadendosismesser m

5653 QUARTZ FIBER ELECTROMETER,
see 4022

5654 QUARTZ FIBER ELECTROSCOPE,
see 3936

5655 QUARTZ SPECTROGRAPH   me, sp
A spectrograph the optical system of which
is made of quartz.
f   spectrographe m à quartz
e   espectrógrafo m de cuarzo
i   spettrografo m a quarzo
n   kwartsspectrograaf
d   Quarzspektrograph m

5656 QUASI-STATIONARY   gp, np
ENERGY LEVEL
It is assumed that at any moment a field
will have the same energy level which
would belong to a stationary field for
charges at the same moment. This
condition is called the quasi-stationary
energy level.
f   niveau m quasistationnaire
e   nivel m cuasiestacionario
i   livello m quasistazionario
n   quasistationnair niveau n
d   quasistationäres Niveau n

5657 QUATERNARY FISSION   np
The hypothetical break-up of a nucleus
into four fragments.
f   fission f quaternaire
e   fisión f cuaternaria
i   fissione f quaternaria
n   kern-in-vierensplijting
d   Kernspaltung f in vier Bruchstücke

5658 QUENCHING   ct
The process of inhibiting continuous or
multiple discharges in a counter tube
which uses gas amplification.
f   coupage m, extinction f
e   extinción f
i   estinzione f, spegnimento m
n   doving
d   Löschung f

5659 QUENCHING CIRCUIT   ct
A circuit which diminishes, suppresses
or reverses the voltage applied to a counter
tube in order to inhibit multiple

discharges from a single ionizing event.
f   circuit *m* coupeur, circuit *m* d'extinction
e   circuito *m* de extinción
i   circuito *m* d'estinzione,
    circuito *m* di spegnimento
n   doofschakeling
d   Löschkreis *m*

5660 QUENCHING GAS   ct
One of the components of the mixed
gas-filling of a Geiger-Müller counter
tube which is intended to ensure
self-quenching of the discharge.
f   gaz *m* de coupage
e   gas *m* de extinción
i   gas *m* d'estinzione, gas *m* di spegnimento
n   doofgas *n*
d   Löschgas *n*

5661 QUENCHING OF ORBITAL   np
      ANGULAR MOMENTUM
An electric field may be so strong that it
causes rapid precession of the orbit of
an electron moving about an atom,
averaging the associated magnetic moment
to zero.
f   extinction *f* du spin
e   extinción *f* del espín
i   estinzione *f* dello spin
n   spindoving
d   Spinlöschung *f*

5662 QUENCHING RESISTOR   ct
A resistor in the circuit of a counter able
to quench it.
f   résistance *f* de coupure,
    résistance *f* d'extinction
e   resistencia *f* de extinción
i   resistenza *f* d'estinzione,
    resistenza *f* di spegnimento
n   doofweerstand
d   Löschwiderstand *m*

5663 QUOTIMETER, see 5185

R

5664 R VALUE   rt
   The percentage decrease in density, for
   1 per cent burn-up, resulting from swelling.
f valeur *f* R
e valor *m* R
i valore *m* R
n R-waarde
d R-Wert *m*

5665 RABBIT,   rt
   SHUTTLE
   A small sample container propelled
   pneumatically through a tube leading from
   the laboratory to a location in a nuclear
   reactor or other device where irradiation
   can take place, and designed to provide
   short irradiation times and, particularly,
   short transit times to the laboratory.
f furet *m*
e objeto *m* buscado, tubo *m* de ensamble
i cartuccia *f* portacampione, spola *f*
n schietkoker
d Rohrpostkapsel *f*

5666 RABBIT (GB),   is
   SINGLE STAGE RECYCLE (US)
   A modification of the simple gaseous
   diffusion stage wherein the diffused gas
   is separated into two fractions, one of
   which is advanced to the next stage,
   whilst the other is re-circulated within
   the original stage.
f recyclage *m* à étage unique
e recirculación *f* de etapa única
i recircolazione *f* a stadio unico
n kringloop van Hertz
d Einstufenrückführung *f*,
   Kreislauf *m* nach Hertz

5667 RABBIT FRACTION, see 4426

5668 RABBIT HOLE   rt
f tubo *m* pneumatique
e tubo *m* neumático
i passaggio *m* per impianto pneumatico
n buizenpostkanaal *n*, schietkokerkanaal *n*
d Rohrpostkanal *m*

5669 RABBITTITE   mi
   A rare secondary mineral containing about
   31 % of U.
f rabbittite *f*
e rabitita *f*
i rabbittite *f*
n rabbittiet *n*
d Rabbittit *m*

5670 RABI METHOD   sp
   A method for determining nuclear
   moments by radiofrequency spectroscopy.
f méthode *f* de Rabi

e método *m* de Rabi
i metodo *m* di Rabi
n rabimethode
d Rabi-Verfahren *n*

5671 RACETRACK   ec, pp
   In thermonuclear experiments, a
   continuous discharge tube, oval in plan
   like a racetrack.
f tube *m* électronique de forme ovale
e tubo *m* electrónico de forma oval
i tubo *m* elettronico di forma ovale
n ovale ontladingsbuis
d Ovalentladungsröhre *f*

5672 RACETRACK,   ic
   TRACK
   A multiple-unit electromagnetic separator
   in which a common magnetic field serves
   a large number of separators, i.e.
   evacuated chambers each containing an ion
   source and a collector.
f piste *f*
e pista *f*, trayectoria *f* cerrada magnética
i traccia *f*
n renbaan
d Rennbahn *f*

5673 RAD ·   xr
   A special unit of absorbed dose equal to
   100 ergs/gram.
f rad *m*
e rad *m*
i rad *m*
n rad
d Rad n

5674 RADIAC INSTRUMENTS   nw
   A contraction of radioactivity, detection,
   identification and computation, used to
   describe various radiation monitoring and
   detection instruments designed for defence.
f instruments *pl* moniteurs
e instrumentos *pl* monitores,
   instrumentos *pl* radiac
i strumenti *pl* monitori
n monitorinstrumenten *pl*,
   radiac-instrumenten *pl*
d Monitorinstrumente *pl*,
   Radiac-Instrumente *pl*

5675 RADIAL DIFFUSION   rt
   COEFFICIENT
   Flux density diffusion coefficient in the
   direction perpendicular to the axis of an
   anisotropic multiplying lattice.
f coefficient *m* de diffusion radial
e coeficiente *m* de difusión radial
i coefficiente *m* di diffusione radiale
n radiale diffusiecoëfficiënt
d radialer Diffusionskoeffizient *m*

5676 RADIAL DISTRIBUTION METHOD   an
A statistical method for the analysis of
data obtained by measuring the intensity
of X-ray diffraction at various angles.
f méthode f de la distribution radiale
e método m de la distribución radial
i metodo m della distribuzione radiale
n radiale verdelingsmethode
d radiales Verteilungsverfahren n

5677 RADIAL FORCE   gp
The force issuing along the radius.
f force f radiale
e fuerza f radial
i forza f radiale
n radiale kracht
d radiale Kraft f

5678 RADIAL QUANTUM NUMBER   qm
In the Bohr theory of the atom, the quantum
number that characterizes the momentum
of an electron in the direction of the
radius vector.
f nombre m quantique radial
e número m cuántico radial
i numero m quantico radiale
n radiaal quantumgetal n
d radiale Quantenzahl f

5679 RADIAL REARRANGEMENT   fu, rt
OF FUEL,
RADIAL SHUFFLING OF FUEL
f altération f radiale de l'arrangement du
combustible
e rearreglo m radial del combustible
i alterazione f radiale della disposizione
del combustibile
n radiale wijziging van de splijtstofopstelling
d radiale Abänderung f der Brennstoffanordnung

5680 RADIANT ENERGY   ra
Energy emitted or transmitted by radiation.
f énergie f de rayonnement
e energía f de radiación
i energia f di radiazione
n stralingsenergie
d Strahlungsenergie f

5681 RADIANT ENERGY DENSITY   ra
Radiant energy per unit volume, expressed in such units as ergs/cm^2.
f énergie f rayonnante par unité de volume,
énergie f rayonnante volumique
e energía f de radiación por unidad de
volumen
i energia f di radiazione per unità di
volume
n stralingsenergie per volume-eenheid
d Strahlungsenergie f je Volumeneinheit

5682 RADIANT FLUX,   ra
RADIANT POWER
Power emitted, transferred or received
in the form of radiation.
f flux m énergétique
e flujo m energético

i flusso m d'energia radiante
n stralingsstroom
d Energiefluss m, Strahlungsfluss m

5683 RADIANT FLUX DENSITY, see 3542

5684 RADIATING ATOM   np, ra
An atom which is emitting radiation
during the transition of one or more of its
electrons from higher to lower energy
states.
f atome m rayonnant
e átomo m radiante
i atomo m radiante
n stralend atoom n
d strahlendes Atom n

5685 RADIATION   ra
The emission and propagation of energy
through space or through a material
medium in the form of waves.
f rayonnement m
e radiación f
i radiazione f
n straling
d Strahlung f

5686 RADIATION ANALYZING   ma
ASSEMBLY
A measuring assembly designed to analyze
the output signals from its radiation
detector(s) as a function of a given
parameter.
f ensemble m d'analyse de rayonnement
e conjunto m analizador de radiación
i complesso m d'analisi di radiazione
n stralingsanalyseopstelling
d Anordnung f zur Strahlungsanalyse

5687 RADIATION BEACON   ra, sa
Generating an audible or visual signal upon
ambient radioactivity rising beyond a
preset level.
f balise f d'alarme, balise f de rayonnement
e baliza f de radiación
i faro m di radiazione
n stralingsbaken n
d Strahlungsbake f

5688 RADIATION BURN   md, ra, xr
A burn caused by over-exposure to radiant
energy.
f brûlure f par rayonnement
e quemadura f por radiación
i bruciatura f per radiazione
n röntgenverbranding, stralingsverbranding
d Röntgenverbrennung f,
Strahlungsverbrennung f

5689 RADIATION CHANNEL,   ra
RADIATION HARDWARE
The whole set of apparatus, connecting
cables and electronic tubes allowing to
transfer to the control desk of a nuclear
reactor the signals enabling the desk
people to consider the intensity of the
radiation in the zone concerned.

f chaîne *f* de contrôle de rayonnement
e equipo *m* de control de radiación
i apparecchiatura *f* di controllo di radiazione
n stralingscontroleapparatuur
d Strahlungskontrollgeräte *pl*

5690 RADIATION CHARGE METER, see 1011

5691 RADIATION CHEMICAL YIELD ch, np
The number of molecules transformed per photon absorbed.
f rendement *m* radiochimique
e rendimiento *m* radioquímico
i rendimento *m* radiochimico
n radiochemisch rendement *n*
d radiochemische Ausbeute *f*

5692 RADIATION CHEMISTRY ch, ra
The study of the chemical effects of radiation on matter.
f chimie *f* du rayonnement
e química *f* de la radiación
i chimica *f* della radiazione
n stralingschemie
d Strahlenchemie *f*, Strahlungschemie *f*

5693 RADIATION CONE xr
The cone formed by the emitted X-rays towards the irradiated object.
f cône *m* de rayons
e cono *m* de rayos
i cono *m* di raggi
n stralenkegel
d Strahlenkegel *m*

5694 RADIATION CONTRAST xr
Contrast in a radiograph, usually expressed in terms of density difference.
f contraste *m* de rayonnement
e contraste *m* de radiación
i contrasto *m* di radiazione
n stralingscontrast *n*
d Strahlungskontrast *m*

5695 RADIATION COOLING cl
Cooling system in which the heat is transmitted into the ambiance by radiation.
f refroidissement *m* par rayonnement
e enfriamiento *m* por radiación
i raffreddamento *m* per radiazione
n stralingskoeling
d Strahlungskühlung *f*

5696 RADIATION COUNTER ct
Radiation measuring device comprising a radiation detector in which individual ionizing events cause electrical pulses, and the associated equipment for processing and counting the pulses.
f ensemble *m* de comptage
e contador *m* de radiación
i contatore *m* di radiazione
n stralingssteller
d Strahlungszähler *m*

5697 RADIATION COUNTER EFFICIENCY ct
A measure of the probability that a count will be recorded when a specified radiation is incident on a reactor in a specified manner.
f rendement *m* d'ensemble de comptage
e rendimiento *m* de contador de radiación
i rendimento *m* di contatore di radiazione
n stralingstellerrendement *n*
d Strahlungszählerausbeute *f*

5698 RADIATION DAMAGE, see 713

5699 RADIATION DAMPING np, ra
The damping due to loss of energy through radiation.
f amortissement *m* par rayonnement
e amortiguación *f* por radiación
i smorzamento *m* per radiazione
n stralingsdemping
d Strahlungsdämpfung *f*

5700 RADIATION DANGER ZONE sa
A zone within which the maximum permissible dose rate of concentration is exceeded.
f zone *f* de rayonnement dangereux
e zona *f* de radiación peligrosa
i zona *f* di radiazione pericolosa
n gevaarlijke stralenzone
d strahlengefährdete Zone *f*

5701 RADIATION DECOMPOSITION, RADIOLYSIS ch
The chemical decomposition of materials by ionizing radiation.
f décomposition *f* par rayonnement, radiolyse *f*
e descomposición *f* por radiación, radiólisis *f*
i decomposizione *f* per radiazione, radiolisi *f*
n ontleding door straling, radiolyse
d Radiolyse *f*, Zerlegung *f* durch Strahlung

5702 RADIATION DETECTION, see 1747

5703 RADIATION DETECTOR me, ra
Any device whereby radiation is made to produce some physical effect suitable for observation and measurement.
f détecteur *m* de rayonnement
e detector *m* de radiación
i rivelatore *m* di radiazione
n stralingsdetector
d Strahlungsdetektor *m*

5704 RADIATION DOSEMETER, see 1947

5705 RADIATION EFFECT md
The effect of radiation on living organisms, etc.
f effet *m* de rayonnement
e efecto *m* de radiación
i effetto *m* di radiazione
n stralingseffect *n*, werking van de straling
d Strahlungswirkung *f*

5706 RADIATION ENERGY   ra
The individual energy of the particles or
photons constituting the radiation.
f énergie f d'un rayonnement
e energía f de una radiación
i energia f d'una radiazione
n stralingsenergie
d Strahlungsenergie f

5707 RADIATION EXCITATION   np, ra
Excitation of a gas under the influence
of electromagnetic radiation.
f excitation f par rayonnement
e excitación f por radiación
i eccitazione f per radiazione
n aanslag door straling
d Anregung f durch elektromagnetische
Strahlung

5708 RADIATION EXPOSURE INDICATOR,
see 2464

5709 RADIATION FIELD   ra
Region through which energy is radiated.
f champ m de rayonnement
e campo m de radiación
i campo m di radiazione
n stralingsveld n
d Strahlungsfeld n

5710 RADIATION FLUX   ra
The radiation power, or energy per unit
time, passing through a surface.
f flux m de rayonnement
e flujo m de radiación
i flusso m di radiazione
n stralingsflux
d Strahlungsfluss m

5711 RADIATION GENETICS   md
The study of the effect of radiations on the
genetic organs.
f génétique f des rayonnements
e genética f de las radiaciones
i genetica f delle radiazioni
n stralengenetica
d Strahlungsgenetik f

5712 RADIATION GROWTH, see 3101

5713 RADIATION HARDNESS, see 5177

5714 RADIATION HAZARD, see 3155

5715 RADIATION HYGIENE   md
The art of keeping health in the presence
of radiation hazard.
f hygiène f du rayonnement
e higiene f contra irradiación,
higiene f de radiación
i igiene f delle radiazioni
n stralingshygiëne
d Strahlungshygiene f

5716 RADIATION INDICATOR   me, ra
An assembly for evaluating quickly but
often coarsely a quantity in connection
with ionizing radiation.

f signaleur m de rayonnement
e indicador m de radiación
i indicatore m di radiazione
n stralingsaanduider, stralingsverkenner
d Strahlungsanzeiger m,
Strahlungsindikator m

5717 RADIATION INDUCED   md
GENETIC EFFECT
The effect of radiation on the genetic
organs of living beings.
f effet m génétique induit par rayonnement
e efecto m genético inducido por radiación
i effetto m genetico indotto per radiazione
n door straling veroorzaakt genetisch
effect n
d genetischer Strahlungseffekt m

5718 RADIATION INJURY,   md
RADIOLESION
The medical term to describe localized
injurious effect due to exposure to
ionizing radiation without adequate
protection.
f radiolésion f
e radiolesión f
i danno m biologico da irradiazione
n stralingsletsel n
d Strahlenschaden m

5719 RADIATION IONIZATION,   np, ra
The ionization of the atoms or molecules
of a gas or vapo(u)r by the action of
electromagnetic radiation such as:
ultraviolet radiation, X-rays, gamma
rays, etc.
f ionisation f par rayonnement
e ionización f por radiación
i ionizzazione f per radiazione
n ionisatie door straling,
stralingsionisatie
d Strahlungsionisation f

5720 RADIATION LENGTH   np, ra
The mean path length required for the
reduction, by the factor $1/e$, of the energy
of relativistic charged particles as they
pass through matter.
f longueur f de rayonnement
e longitud f de radiación
i lunghezza f di radiazione
n stralingsdracht
d Strahlungslänge f

5721 RADIATION LEVEL   ra
The intensity of radiation at the point in
question.
f niveau m de rayonnement
e nivel m de radiación
i livello m di radiazione
n stralingsniveau n
d Strahlungsniveau m

5722 RADIATION LOSS,   ra
RADIATIVE LOSS
The energy loss occurring when a charged
particle passes through matter.
f perte f par rayonnement

e pérdida f por radiación
i perdita f per radiazione
n stralingsverlies n
d Strahlungsverlust m

5723 RADIATION MAZE   sa
A thick-walled corridor or duct designed
to reduce the passage of ionizing radiation
from one end to the other.
f chicane f, labyrinthe m de rayonnement
e laberinto m de radiación
i labirinto m di radiazione
n stralingsdoolhof, stralingssluis
d Strahlenschleuse f

5724 RADIATION MEASUREMENT   me, ra
The measurement by means of appropriate
apparatus of the effect and the intensity
of radiation.
f mesure f du rayonnement
e medida f de la radiación
i misura f della radiazione
n stralingsmeting
d Strahlungsmessung f

5725 RADIATION MEASURING ASSEMBLY,
RADIATION METER, see 4340

5726 RADIATION MONITOR, see 3156

5727 RADIATION MULTI-PARAMETER   ma
ANALYZING ASSEMBLY
A measuring assembly intended to record
simultaneously the information delivered
by its radiation detectors, and so analyze
this information as a function of several
parameters in order to establish
correlations.
f ensemble m d'analyse multiparamétrique
e conjunto m analizador multiparamétrico
de radiación
i complesso m d'analisi multiparametrica
di radiazione
n meervoudige stralingsanalysator,
opstelling voor meervoudige analyse
d Anordnung f zur Mehrparameteranalyse

5728 RADIATION OUTPUT   ra
From a sealed source, the exposure dose
rate, expressed in röntgens per hour in
air 1 metre(er) from the source in line with
the major axis of the source, as far as
possible in conditions of freedom from
scattered radiation.
f puissance f de sortie d'une source
e potencia f de salida de una fuente
i potenza f d'uscita d'una sorgente
n uitgangsvermogen n van de straling
d Strahlungsausgangsleistung f

5729 RADIATION   ch, ms
POLYMERIZATION
Polymerization of organic materials by
means of radiation.
f polymérisation f par rayonnement
e polimerización f por radiación
i polimerizzazione f per radiazione

n stralingspolymerisatie
d Strahlungspolymerisation f

5730 RADIATION POTENTIAL   ra
The potential difference in volts,
necessary to cause an atom to emit
radiation of a frequency characteristic
of the material.
f potentiel m de rayonnement
e potencial m de radiación
i potenziale m di radiazione
n stralingspotentiaal
d Strahlungspotential n

5731 RADIATION PRESSURE   ra
The pressure exerted by electromagnetic
radiation on a surface exposed to it.
f pression f de rayonnement
e presión f de radiación
i pressione f di radiazione
n stralingsdruk
d Strahlungsdruck m

5732 RADIATION PURITY,   ra
RADIOACTIVE PURITY
For a radioactive sample, the activity
of a specified radionuclide expressed as
per cent of the total activity of the sample.
f pureté f radioactive
e pureza f radiactiva
i purezza f radioattiva
n radioactieve zuiverheid
d radioaktive Reinheit f

5733 RADIATION RESISTANCE,   mc, ra
RADIORESISTANCE,
RESISTANCE TO RADIATION
Relative resistance of cells, tissues,
organs or organisms to the action of
radiation.
f radiorésistance f,
résistance f au rayonnement
e radiorresistencia f,
resistencia f a la radiación
i radioresistenza f,
resistenza f alla radiazione
n resistentie tegen straling
d Strahlenfestigkeit f, Strahlungsresistenz f

5734 RADIATION SELF-   is
DECOMPOSITION
Decomposition by self-radiation.
f auto-décomposition f par rayonnement
e autodescomposición f por radiación
i autodecomposizione f per radiazione
n zelfontleding door straling
d Selbstzersetzung f durch Strahlung

5735 RADIATION SHADOW,   nw
SHADOW
A surface shielded from ionizing radiation
used in ascertaining the centre(er) of a
nuclear explosion.
f ombre f de rayonnement
e sombra f de radiación
i ombra f di radiazione
n stralingsschaduw
d Strahlungsschatten m

5736 RADIATION SHIELD   ra, sa
A shield or wall of lead or other material
that effectively absorbs nuclear radiation.
f écran *m* absorbant le rayonnement
e pantalla *f* absorbente la radiación
i schermo *m* assorbente la radiazione
n straling absorberend scherm *n*
d Strahlenschild *m*

5737 RADIATION SICKNESS   md
A self-limited syndrome characterized
by nausea, vomiting, diarrhoea and physical
depression following exposure to
appreciable doses of ionizing radiation.
f maladie *f* d'irradiation,
   maladie *f* de rayonnement,
   radiotoxémie *f*
e enfermedad *f* radiativa, radiotoxemia *f*
i malattia *f* d'irradiazione
n bestralingskater, stralenkater
d Strahlenkrankheit *f*

5738 RADIATION SOURCE   ra
An apparatus or a material emitting or
capable of emitting ionizing radiation.
f source *f* de rayonnement
e fuente *f* de radiación
i sorgente *f* di radiazione
n stralingsbron
d Strahlungsquelle *f*

5739 RADIATION SPECTRUM, see 2289

5740 RADIATION SPECTRUM ANALYZER,
       see 5137

5741 RADIATION STABILITY   ra
The stability of materials towards high
energy radiation.
f stabilité *f* sous rayonnement
e estabilidad *f* contra radiación
i stabilità *f* contro radiazione
n stralingsvastheid
d Strahlungsbeständigkeit *f*

5742 RADIATION STERILIZATION,   ra
       STERILIZATION BY IRRADIATION
Sterilization by means of irradiation.
f stérilisation *f* par irradiation
e esterilización *f* por irradiación
i sterilizzazione *f* per irradiazione
n sterilisatie door bestraling
d Sterilisierung *f* durch Bestrahlung,
   Strahlungssterilisierung *f*

5743 RADIATION SURVEY, see 5555

5744 RADIATION SYNDROME   xr
A complex of symptoms caused by
radiation.
f syndrome *m* d'irradiation,
   syndrome *m* de rayonnement
e síndrome *m* por irradiación,
   síndrome *m* por radiación
i sindrome *m* da irradiazione,
   sindrome *m* da radiazione
n bestralingssyndroom *n*, stralensyndroom *n*

d Bestrahlungssyndrom *n*, Strahlensyndrom *n*

5745 RADIATION THERAPY   ra, xr
Treatment of disease with any type of
ionizing radiation.
f thérapie *f* par rayonnement
e terapia *f* por radiación
i terapia *f* per radiazione
n radiotherapie, stralingstherapie
d Strahlentherapie *f*

5746 RADIATION TRAP   sa
A system for absorbing radiation so that
the scattered radiation is reduced to an
acceptable level.
f piège *m* de rayonnement
e trampa *f* de radiación
i trappola *f* di radiazione
n stralingsval
d Strahlenfalle *f*

5747 RADIATION ULCER   md
f ulcère *m* par rayonnement
e úlcera *f* por radiación
i ulcera *f* da radiazione
n stralingsulcus
d Strahlenulkus *m*

5748 RADIATION WINDOW   ra
A window that is transparent to alpha,
beta, gamma and/or X-rays while
protecting from foreign matter the item
that it covers.
f fenêtre *f* transparente au rayonnement
e ventana *f* transparente a la radiación
i finestra *f* trasparente alla radiazione
n stralendoorlatend venster *n*
d strahlendurchlässiges Fenster *n*

5749 RADIATIONLESS TRANSITION   np
A transition taking place between two
energy states of a system in which the
necessary energy is taken from it or given
up to it by direct interaction with another
system or particle rather than by
absorption or emission of electromagnetic
radiation.
f transition *f* sans rayonnement
e transición *f* sin radiación
i transizione *f* senza radiazione
n stralingsloze overgang
d strahlungsfreier Übergang *m*

5750 RADIATIVE CAPTURE   np
Capture of a particle by a nucleus followed
by immediate emission of gamma
radiation.
f capture *f* radiative
e captura *f* radiativa
i cattura *f* radiativa
n stralingsvangst, vangst met n, γ reactie,
   vangst met p, γ reactie
d Strahlungseinfang *m*

5751 RADIATIVE CAPTURE   np
       CROSS SECTION
The cross section of radiative capture.

f section *f* efficace de capture radiative
e sección *f* eficaz de captura radiativa
i sezione *f* d'urto di cattura radiativa
n stralingsvangstdoorsnede
d Wirkungsquerschnitt *m* für Strahlungseinfang

5752 RADIATIVE COLLISION   np
A collision between two charged particles
in which part of the kinetic energy is
converted directly into electromagnetic
radiation.
f collision *f* radiative
e colisión *f* radiativa
i collisione *f* radiativa
n stralingsbotsing, stralingscollisie
d Strahlungskollision *f*

5753 RADIATIVE CORRECTION   ma, np
Difference between the theoretical values
of some property of a dynamical system
as computed from the quantized field
theory of the system and from the
corresponding unquantized field theory.
f correction *f* radiative
e corrección *f* radiativa
i correzione *f* radiativa
n stralingscorrectie
d Strahlungskorrektion *f*

5754 RADIATIVE INELASTIC   cs
SCATTERING CROSS SECTION
The cross section for the radiative
inelastic scattering process.
f section *f* efficace de diffusion inélastique
   radiative
e sección *f* eficaz de difusión inelástica
   radiativa
i sezione *f* d'urto di deviazione
   anelastica radiativa
n doorsnede voor stralende inelastische
   verstrooiing
d Wirkungsquerschnitt *m* für unelastische
   Streuung mit Strahlungsemission

5755 RADIATIVE LOSS, see 5722

5756 RADIATIVE RECOMBINATION   np
Recombination carried out by means of
radiation.
f recombinaison *f* par rayonnement
e recombinación *f* por radiación
i ricombinazione *f* per radiazione
n stralingsrecombinatie
d Strahlungsrekombination *f*

5757 RADIATIVE TRANSITION   np, qm
A transition between two energy eigen-
states of a quantum mechanical system
accompanied by the emission of electromagnetic radiation.
f transition *f* radiative
e transición *f* radiativa
i transizione *f* radiativa
n stralingsovergang
d Strahlungsübergang *m*

5758 RADIATOR   ra
A body which emits energy quanta or
certain material particles.
f élément *m* rayonnant
e radiador *m*
i radiatore *m*
n straler
d Strahler *m*

5759 RADIATOR TUBE   xr
An air-cooled X-ray tube in which a finned
radiator is fitted to the external end of the
anode stem.
f tube *m* à rayons X avec ailettes de
   refroidissement
e tubo *m* de rayos X con aletas de enfriamiento
i tubo *m* a raggi X con alette di raffreddamento
n röntgenbuis met koelribben
d Röntgenröhre *f* mit Kühlrippen

5760 RADIOACTINIUM   ch
The common name for 18.6d Th-227, a
member of the actinium series.
f radioactinium *m*
e radiactinio *m*
i radioattinio *m*
n radioactinium *n*
d Radioaktinium *n*

5761 RADIOACTIVATION ANALYSIS, see 63

5762 RADIOACTIVE   np
Possessing or pertaining to radioactivity.
f radioactif adj
e radiactivo adj
i radioattivo adj
n radioactief adj
d radioaktiv adj

5763 RADIOACTIVE AGE   np
Of a geological or archaeological object,
the time, estimated from measurement of
the isotopic composition, during which the
content of a radioactive species within that
object has remained unchanged except for
radioactive decay.
f âge *m* radioactif
e edad *f* radiactiva
i età *f* radioattiva
n radioactieve leeftijd
d radioaktives Alter *n*

5764 RADIOACTIVE BATTERY, see 367

5765 RADIOACTIVE BY-PRODUCT, see 72

5766 RADIOACTIVE CARBON, see 73

5767 RADIOACTIVE CHAIN,
RADIOACTIVE SERIES, see 1639

5768 RADIOACTIVE CONCENTRATION   np
Of a radioactive material, the activity per
unit mass, volume or mole of material.

f   concentration *f* radioactive
e   concentración *f* radiactiva
i   concentrazione *f* radioattiva
n   radioactieve concentratie
d   radioaktive Konzentration *f*

5769 RADIOACTIVE CONTAMINANT,
      see 1331

5770 RADIOACTIVE CONTAMINATION,
      see 1334

5771 RADIOACTIVE DECAY LAW   np

The exponential law $N = N_o e^{-\lambda t}$, which

governs the decrease with time of the
number of atoms of a radioactive species,
provided the number is large.
f   loi *f* de décroissance radioactive
e   ley *f* de decaimiento radiactivo
i   legge *f* di decadimento radioattivo
n   wet van radioactief verval
d   Gesetz *n* des radioaktiven Zerfalls

5772 RADIOACTIVE DECONTAMINATION,
      see 1653

5773 RADIOACTIVE DEPOSIT   mi
An ore deposit containing radioactive
material.
f   gisement *m* radioactif
e   yacimiento *m* radiactivo
i   giacimento *m* radioattivo
n   radioactieve afzetting
d   radioaktive Lagerstätte *f*

5774 RADIOACTIVE DEPOSIT,   np
      ACTIVE DEPOSIT, see 75

5775 RADIOACTIVE DISPLACEMENT ch, np
Change of place of a nuclide in the periodic
system of the elements as a result of its
disintegration.
f   déplacement *m* radioactif
e   desplazamiento *m* radiactivo
i   spostamento *m* radioattivo
n   radioactieve verschuiving
d   radioaktive Verschiebung *f*

5776 RADIOACTIVE DISPLACEMENT   np
      LAW,
      SODDY-FAJANS LAW
f   loi *f* de déplacement radioactif
e   ley *f* de desplazamiento radiactivo
i   legge *f* di spostamento radioattivo
n   wet van de radioactieve verschuiving,
    wet van Soddy en Fajans
d   radioaktives Verschiebungsgesetz *n*

5777 RADIOACTIVE DRY FALL-OUT   rw
The fraction of radioactive fall-out,
retained by the earth's surface, in the
absence of atmospheric precipitations
as rain, snow, etc.
f   dépôt *m* radioactif sec
e   depósito *m* radiactivo seco
i   ricaduta *f* radioattiva secca
n   droge radioactieve neerslag
d   trockener radioaktiver Ausfall *m*

5778 RADIOACTIVE DUST, see 76

5779 RADIOACTIVE EFFLUENT, see 77

5780 RADIOACTIVE EFFLUENT DISPOSAL,
      see 78

5781 RADIOACTIVE EFFLUENT DRAIN
      PIPE, see 79

5782 RADIOACTIVE EFFLUENT PLANT
      AREA, see 80

5783 RADIOACTIVE ELEMENT, see 81

5784 RADIOACTIVE EMANATION,   np
      RADIOACTIVE GAS
A gas created in a nuclear reactor.
f   gaz *m* radioactif
e   gas *m* radiactivo
i   gas *m* radioattivo
n   radioactief gas *n*
d   radioaktives Gas *n*

5785 RADIOACTIVE EQUILIBRIUM   np
A condition which may occur in the course
of the decay of a radioactive parent
having shorter-lived descendants in which
the ratio of a descendant is independent
of time.
f   équilibre *m* radioactif
e   equilibrio *m* radiactivo
i   equilibrio *m* radioattivo
n   radioactief evenwicht *n*
d   radioaktives Gleichgewicht *n*

5786 RADIOACTIVE FALL-OUT, see 82

5787 RADIOACTIVE FAMILY,   np
      RADIOACTIVE SERIES,
      TRANSFORMATION SERIES
A number of radioactive nuclides, each
except the first being the daughter product
of the previous one; the final member, the
end product, although stable, is included
in the family.
f   famille *f* radioactive
e   familia *f* radiactiva
i   famiglia *f* radioattiva
n   radioactieve reeks
d   radioaktive Serie *f*, Zerfallsreihe *f*

5788 RADIOACTIVE FISSION PRODUCT,
      see 83

5789 RADIOACTIVE GO-DEVIL   ms
Mechanical pipe-cleaner containing a
radioactive substance for locating its
position in case of being jammed.
f   râcleur *m* radioactif
e   rascador *m* radiactivo
i   scovolo *m* radioattivo
n   radioactieve buiskrabber
d   radioaktiver Molch *m*,
    radioaktiver Rohrkratzer *m*

5790 RADIOACTIVE GRAIN   br, md, ra
Radioactive material of small dimensions
used for interstitial applications.

f grain *m* radioactif
e grano *m* radiactivo
i grano *m* radioattivo
n radioactieve korrel
d radioaktives Korn *n*

5791 RADIOACTIVE HAIRPIN br, md, ra
A radioactive wire shaped like a hairpin.
f boucle *f* métallique radioactive
e lazo *m* metálico radiactivo
i laccio *m* metallico radioattivo
n radioactieve draadlus
d radioaktive Drahtschleife *f*

5792 RADIOACTIVE HALF LIFE, see 3120

5793 RADIOACTIVE HEAT, np
RADIOGENIC HEAT
The heat produced within the earth by the
disintegration of radioactive nuclides.
f chaleur *f* radiogénique
e calor *m* radiogénico
i calore *m* radiogenico
n radiogene warmte
d radiogene Wärme *f*

5794 RADIOACTIVE INCINERATOR, see 97

5795 RADIOACTIVE ISOTOPE, is
RADIOISOTOPE
A radioactive isotope of a specified
element.
f radio-isotope *m*
e isótopo *m* radiactivo, radioisótopo *m*
i radioisotopo *m*
n radio-isotoop *n*
d radioaktives Isotop *n*, Radioisotop *n*

5796 RADIOACTIVE LOGGING me
Underground radiation measuring method
for locating radioactive minerals, by means
of a detecting probe which is lowered at
various depths in a bore-hole.
f carottage *m* radioactif, radiocarottage *m*
e radiosondeo *m*
i radiocarottaggio *m*
n activiteitsmeting in boorgaten
d Bohrlochvermessung *f*

5797 RADIOACTIVE MATERIAL, see 88

5798 RADIOACTIVE NUCLEUS np
A nucleus, as end product of a nuclear
chain reaction, may have been made
radioactive artificially.
f noyau *m* radioactif
e núcleo *m* radiactivo
i nucleo *m* radioattivo
n radioactieve kern
d radioaktiver Kern *m*

5799 RADIOACTIVE NUCLIDE np
A radioactive minute highly positively
charged central portion of an atom.
f nucléide *m* radioactif
e núclido *m* radiactivo
i nuclido *m* radioattivo

n radioactief nuclide *n*
d radioaktives Nuklid *n*

5800 RADIOACTIVE PERIOD, see 478

5801 RADIOACTIVE POISON, see 89

5802 RADIOACTIVE PRECURSOR, see 5120

5803 RADIOACTIVE PRODUCT, see 90

5804 RADIOACTIVE PURITY, see 5732

5805 RADIOACTIVE RELATIONSHIP np
The relation between the parent substance,
the daughter product and the end product.
f filiation *f* radioactive
e filiación *f* radiactiva
i membri *pl* d'una serie radioattiva
n radioactieve verwantschap
d radioaktive Verwandschaft *f*

5806 RADIOACTIVE SAMPLING
EQUIPMENT, see 91

5807 RADIOACTIVE SOURCE md, np
Any quantity of radioactive material which
is intended for use as a source of ionizing
radiation.
f source *f* radioactive
e fuente *f* radiactiva
i sorgente *f* radioattiva
n radioactieve bron
d radioaktive Quelle *f*

5808 RADIOACTIVE STANDARD, np
RADIOACTIVITY STANDARD,
REFERENCE SOURCE
A specimen of a radioactive nuclide the
activity or quantity of which in the
specimen is known, having been determined
at a specified time to a specified accuracy
by a recognized standardizing organization.
f étalon *m* radioactif
e muestra *f* radiactiva,
patrón *m* de radiactividad
i campione *m* radioattivo
n radioactief standaardpreparaat *n*
d radioaktives Standardpräparat *n*

5809 RADIOACTIVE TRACER md
A material, recognizable by its radioactivity, which is introduced into a system
in order to trace the behavio(u)r of some
component of that system.
f traceur *m* radioactif
e trazador *m* radiactivo
i tracciante *m* radioattivo
n radioactieve indicator
d radioaktiver Tracer *m*

5810 RADIOACTIVE TRANSFORMATION np
The spontaneous transformation with a
measurable lifetime of a nuclide into one
or more different nuclides.
f transformation *f* radioactive,
transition *f* radioactive

e transformación *f* radiactiva,
   transición *f* radiactiva
i transizione *f* radioattiva,
   trasformazione *f* radioattiva
n radioactieve omzetting,
   radioactieve overgang
d radioaktive Umwandlung *f*,
   radioaktiver Übergang *m*

5811 RADIOACTIVE WASTE, see 95

5812 RADIOACTIVE WASTE HANDLING BAY,
      see 96

5813 RADIOACTIVE WATER, see 98

5814 RADIOACTIVE WATER HOMING, see 99

5815 RADIOACTIVE WIRE   md, ra
      Radioactive material in wire form used
      for interstitial and intracavitary
      application.
f fil *m* radioactif
e hilo *m* radiactivo
i filo *m* radioattivo
n radioactieve draad
d radioaktiver Faden *m*

5816 RADIOACTIVITY   np
      The property of certain nuclides of
      spontaneously emitting particles or
      gamma radiation or of emitting X-radiation
      following orbital electron capture or of
      undergoing spontaneous fission.
f radioactivité *f*
e radiactividad *f*
i radioattività *f*
n radioactiviteit
d Radioaktivität *f*

5817 RADIOACTIVITY METER, see 106

5818 RADIOACTIVITY SIMULATOR   me
      Used for the calibration of high-energy
      gamma dose ratemeters.
f simulateur *m* de radioactivité
e simulador *m* de radiactividad
i simulatore *m* di radioattività
n analogiemodel *n* voor radioactiviteit
d Radioaktivitätsanalogiemodell *n*

5819 RADIOAUTOGRAPH, see 460

5820 RADIOBIOLOGICAL ACTION   md
      The effect of radiation upon living matter.
f effet *m* radiobiologique
e efecto *m* radiobiológico
i effetto *m* radiobiologico
n biologisch stralingseffect *n*,
   radiobiologische werking
d radiobiologische Wirkung *f*,
   strahlenbiologische Wirkung *f*

5821 RADIOBIOLOGICAL   bi
      SENSITIVE VOLUME
      The total of the most sensitive points in
      the production of a radiolesion.

f volume *m* sensible radiobiologique
e volumen *m* sensible radiobiológico
i volume *m* sensibile radiobiologico
n radiobiologisch gevoelig volume *n*
d radiobiologisch empfindliches Volumen *n*

5822 RADIOBIOLOGY   mc
      That branch of science which deals with
      the effects of radiation on biological
      systems and the study of the behavio(u)r
      of radioactive material in living matter.
f radiobiologie *f*
e radiobiología *f*
i radiobiologia *f*
n radiobiologie, stralingsbiologie
d Radiobiologie *f*, Strahlenbiologie *f*

5823 RADIOCARBON AGE,
      RADIOCARBON DATING, see 885

5824 RADIOCARDIOGRAPHY   md
      Measuring the passage of radioactive
      blood through the chambers of the heart
      with specially constructed Geiger-Müller
      counter tubes.
f radiocardiographie *f*
e radiocardiografía *f*
i radiocardiografia *f*
n radiocardiografie
d Radiokardiographie *f*

5825 RADIOCARTOGRAPH, see 687

5826 RADIOCESIUM, see 844

5827 RADIOCHEMICAL ANALYSIS   an
      The determination of the absolute
      disintegration rate of a radionuclide in a
      mixture based on the counting rate of a
      sample that has been separated and
      purified, in measured yield, by
      appropriate chemical procedures.
f analyse *f* radiochimique
e análisis *f* radioquímica
i analisi *f* radiochimica
n radiochemische analyse
d radiochemische Analyse *f*

5828 RADIOCHEMICAL PURITY   ch, is
      Of a radioactive material, consisting
      primarily of a given radioisotope in a
      stated chemical form, the proportion of
      that radioisotope that is in the stated
      chemical form.
f pureté *f* radiochimique
e pureza *f* radioquímica
i purità *f* radiochimica
n radiochemische zuiverheid
d radiochemische Reinheit *f*

5829 RADIOCHEMISTRY   ch
      The production of radioactive nuclides
      and their chemical compounds by
      chemically processing irradiated materials
      or naturally radioactive materials; their
      use in elucidating ordinary chemical
      problems and the study of the special
      techniques involved.

f   radiochimie *f*
e   radioquímica *f*
i   radiochimica *f*
n   radiochemie
d   Radiochemie *f*

5830 RADIOCHROMATOGRAPH   ma
A measuring assembly designed to draw a
representative curve of the activity of
different organic components of a mixture,
labelled with radionuclides and deposited
by a chromatograph method, on a paper
strip moving in front of a radiation
detector.
f   radiochromatographe *m*
e   radiocromatógrafo *m*
i   radiocromatografo *m*
n   radiochromatograaf
d   Radiochromatograph *m*

5831 RADIOCOLLOID   np
Radioactive material in a truly or
apparently colloidal condition.
f   radiocolloïde *m*
e   radiocoloide *m*
i   radiocolloide *m*
n   radioactief colloïde *n*
d   Radiokolloid *n*

5832 RADIOCRYSTALLOGRAPHY, see 1551

5833 RADIODE,   ra
     RADIUM CAPSULE, RADIUM CELL
A sealed radium container in the form of
a thin-walled tube, usually of metal,
normally loaded into other containers,
such as tubes or needles.
f   capsule *f* à radium
e   cápsula *f* de radio
i   capsula *f* a radio
n   radiumcapsule
d   Radiumkapsel *f*

5834 RADIODERMATITIS, see 53

5835 RADIODIAGNOSIS (US), see 1768

5836 RADIOECOLOGY   np
Branch of ecology which studies the
relations of living organisms with
radiations or those radioelements which
pollute a medium.
f   radioécologie *f*
e   radioecología *f*
i   radioecologia *f*
n   radio-ecologie
d   Radio-Ökologie *f*

5837 RADIOELEMENT, see 81

5838 RADIOFREQUENCY PLASMOID   pp
A discrete piece of plasma produced in
low-pressure electrodeless discharges.
f   plasmoïde *m* produit par décharges à
haute fréquence
e   plasmoide *m* producido por descargas
de alta frecuencia

i   plasmoide *m* prodotto per scariche ad alta
frequenza
n   door hoogfrequente ontlading gevormd
plasmoïde *n*
d   durch Hochfrequenzentladung gebildetes
Plasmoid *n*

5839 RADIOFREQUENCY SPECTROSCOPY,
see 4870

5840 RADIOGENIC   np
Resulting from radioactive decay.
f   radiogénique adj
e   radiogénico adj
i   radiogenico adj
n   radiogeen adj
d   durch Radiozerfall entstanden, radiogen adj

5841 RADIOGENIC HEAT, see 5793

5842 RADIOGRAPH,   ra, xr
     RÖNTGENOGRAM
A photographic image produced by a beam
of penetrating radiation after passing
through an object.
f   radiogramme *m*
e   radiograma *m*
i   radiogramma *m*
n   röntgenfoto
d   Radiogramm *n*

5843 RADIOGRAPHIC PUTTY   xr
A blocking medium used in radiography to
reduce the effect of scattered radiation
and to shield portions of the X-ray film
that would otherwise be overexposed.
f   substance *f* antidiffuseuse
e   substancia *f* antidispergente
i   sostanza *f* antideviatrice
n   antiverstrooiingsmiddel *n*
d   Antistreuungsmittel *n*

5844 RADIOGRAPHIC STEREOMETRY,   xr
     STEREOMETRIC LOCALIZATION
The process of finding the position and
dimensions of details within an object
by measurements made in radiographs
taken from different directions.
f   localisation *f* stéréométrique,
stéréométrie *f* radiographique
e   estereometría *f* radiográfica,
localización *f* estereométrica
i   localizzazione *f* stereometrica,
stereometria *f* radiografica
n   radiografische stereometrie,
stereometrische plaatsbepaling
d   radiographische Stereometrie,
stereometrische Lagebestimmung *f*

5845 RADIOGRAPHY   ra, xr
Art or the act of producing radiographs.
f   radiographie *f*
e   radiografía *f*
i   radiografia *f*
n   radiografie
d   Radiographie *f*

5846 RADIOIODINE, see 3609

5847 RADIOISOTOPE, see 5795

5848 RADIOISOTOPE CONCENTRATION   is
The concentration of a radioisotope in
active material, expressed as the ratio of
the number of radioisotope atoms to the
total number of atoms of the same number
present.
f concentration *f* radioisotopique
e concentración *f* radioisotópica
i concentrazione *f* radioisotopica
n radio-isotopische concentratie
d Radioisotopenkonzentration *f*

5849 RADIOISOTOPE PACKAGE MONITOR,
see 5080

5850 RADIOISOTOPE THERAPY, see 1575

5851 RADIOISOTOPIC GENERATOR, see 367

5852 RADIOISOTOPIC PURITY   is
Of a radioactive material, consisting
primarily of a given radioisotope, the
proportion of the total activity that is
attributable to the radioisotope stated.
f pureté *f* radioisotopique
e pureza *f* radioisotópica
i purità *f* radioisotopica
n radio-isotopische zuiverheid
d radioisotopische Reinheit *f*

5853 RADIOLESION, see 5718

5854 RADIOLOGICAL DEFENCE   nw
Defence against the effects of radioactivity
from atomic weapons, including detection
and measurement of radioactivity, and
decontamination of areas and equipment.
f défense *f* contre l' emploi tactique
des moyens radioactifs,
défense *f* radiologique
e defensa *f* radiológica
i difesa *f* contro la guerra radioattiva
n verdediging tegen radioactieve oorlogvoering
d Abwehr *f* eines radiologischen Krieges

5855 RADIOLOGICAL FILTER   cd, xr
A filter utilized to absorb part of the rays
emitted.
f filtre *m* radiologique
e filtro *m* radiológico
i filtro *m* radiologico
n röntgenfilter *n*
d Röntgenfilter *m*

5856 RADIOLOGICAL LATENT PERIOD,
see 3909

5857 RADIOLOGICAL LATENT TISSUE
INJURY, see 3912

5858 RADIOLOGICAL PHYSICS   gp, xr
The physics applied to and belonging to
radiology.

f physique *f* radiologique
e física *f* radiológica
i fisica *f* radiologica
n fysica van de radiologie,
fysische radiologie
d physikalische Radiologie *f*,
Strahlenphysik *f*

5859 RADIOLOGICAL WARFARE   nw
Warfare involving weapons that produce
radioactivity, such as atomic bombs and
shells.
f guerre *f* radiologique
e guerra *f* radiológica
i guerra *f* radioattiva
n radioactieve oorlogvoering
d radiologischer Krieg *m*

5860 RADIOLOGY (US), see 4354

5861 RADIOLUCENT,   ra, xr
RADIOPARENT (US),
RADIOTRANSPARENT (GB)
Permitting passage of X-rays or other
forms of radiation.
f radiotransparent adj
e radiotransparente adj
i radiotrasparente adj
n doorlatend adj voor straling
d strahlendurchlässig adj

5862 RADIOLUMINESCENCE   ra
Light emission caused by radiation from
radioactive materials.
f radioluminescence *f*
e radioluminiscencia *f*
i radioluminescenza *f*
n radioluminescentie
d Radiolumineszenz *f*

5863 RADIOLYSIS, see 5701

5864 RADIOLYSIS OF SOLVENTS   ch, ra
Damage to solvents caused by radiation.
f décomposition *f* radiolytique des solvants,
radiolyse *f* des solvants
e descomposición *f* radiolítica de los
solventes,
radiólisis *f* de los solventes
i decomposizione *f* radiolitica dei solventi,
radiolisi *f* dei solventi
n radiolyse van oplosmiddelen,
radiolytische ontleding van oplosmiddelen
d Radiolyse *f* der Lösungsmittel,
radiolytische Zerlegung *f* von Lösungs-
mitteln

5865 RADIOLYTIC ATTACK   mt
A damaging effect on graphite in reactor
usage.
f attaque *f* radiolytique
e ataque *m* radiolítico
i attacco *m* radiolitico
n radiolytische aantasting
d radiolytischer Angriff *m*

5866 RADIOLYTIC OXIDATION   rt
Oxidation due to radiation decomposition.

f   oxydation *f* radiolytique
e   oxidación *f* radiolítica
i   ossidazione *f* radiolitica
n   radiolytische oxydatie
d   radiolytische Oxydierung *f*

5867 RADIOLYTIC WEIGHT LOSS   rt
Loss of weight due to radiation decomposition.
f   perte *f* de poids par radiolyse
e   pérdida *f* de peso por radiolisis
i   perdita *f* di peso per radiolisi
n   gewichtsverlies *n* door radiolyse
d   Gewichtsverlust *m* durch Radiolyse

5868 RADIOMETALLOGRAPHY   mg
Examination of the crystalline structure
and other characteristics of metals and
alloys with X-ray equipment.
f   radiométallographie *f*
e   radiometalografía *f*
i   radiometallografia *f*
n   röntgenmetallografie
d   Röntgenmetallographie *f*

5869 RADIOMETRIC ANALYSIS   an, ch
A method of quantitative chemical analysis
for a radioactive component based on a
measurement of its disintegration rate.
f   analyse *f* radiométrique
e   análisis *f* radiométrica
i   analisi *f* radiometrica
n   radiometrische analyse
d   radiometrische Analyse *f*

5870 RADIOMETRIC BORE-HOLE
LOGGING ASSEMBLY, see 735

5871 RADIOMETRIC MAP   mi
A detailed geological map showing the
distribution of radioactivity.
f   carte *f* radiométrique
e   mapa *f* radiométrica
i   carta *f* radiometrica
n   radiometrische landkaart
d   radiometrische Landkarte *f*

5872 RADIOMETRIC PROSPECTING   mi
Prospection technique based on measuring
the natural activities of areas.
f   prospection *f* radiométrique
e   exploración *f* radiométrica
i   prospezione *f* radiometrica
n   radiometrische prospectie
d   radiometrische Schürfung *f*

5873 RADIOMETRY   me
f   radiométrie *f*
e   radiometría *f*
i   radiometria *f*
n   radiometrie, stralingsmeting
d   Radiometrie *f*, Strahlungsmessung *f*

5874 RADIONUCLIDE   np
A radioactive nuclide.
f   radionucléide *m*
e   radionúclido *m*

i   radionuclide *m*
n   radionuclide *n*
d   Radionuklid *n*

5875 RADIO-OPAQUE (GB),   ra
RADIOPAQUE (US)
Not appreciably penetrable by X-rays or
other forms of radiation.
f   opaque adj aux rayonnements
e   opaco adj a las radiaciones
i   opaco adj alle radiazioni
n   niet-doorlatend adj voor straling
d   strahlenundurchlässig adj

5876 RADIOPHOTOLUMINESCENCE   ra
Luminescence exhibited by certain
minerals as a result of irradiation with
beta and gamma rays, followed by
exposure to light.
f   radiophotoluminescence *f*
e   radiofotoluminiscencia *f*
i   radiofotoluminescenza *f*
n   radiofotoluminescentie
d   Radiophotolumineszenz *f*

5877 RADIOPROSPECTING ASSEMBLY   ma
An assembly designed for radiometric
prospecting based on the detection of
natural or artificially caused ionizing
radiation.
f   ensemble *m* de radioprospection
e   conjunto *m* de exploración radiométrica
i   complesso *m* di radioprospezione
n   opstelling voor radioprospectie
d   Anordnung *f* zur Lagerstättensuche

5878 RADIORESISTANCE, see 5733

5879 RADIOSCOPE   me
An electroscope used to measure the
quantity of radioactive material.
f   radioscope *m*
e   radioscopio *m*
i   radioscopio *m*
n   radioscoop
d   Radioskop *n*

5880 RADIOSCOPY, see 2744

5881 RADIOSENSITIVE   ra
Sensitive to radiation.
f   radiosensible adj
e   radiosensible adj
i   radiosensibile adj
n   stralingsgevoelig adj
d   strahlungsempfindlich adj

5882 RADIOSENSITIVITY   ra
Relative susceptibility of cells, tissues,
organs or organisms to the injurious
action of radiation.
f   radiosensibilité *f*
e   radiosensibilidad *f*
i   radiosensibilità *f*
n   stralingsgevoeligheid
d   Strahlungsempfindlichkeit *f*

5883 RADIOSTRONTIUM,   ch, is
         STRONTIUM-90
      A radioisotope of strontium used in beta
      and bremsstrahlung sources.
  f strontium-90 *m*
  e estroncio-90 *m*
  i stronzio-90 *m*
  n strontium-90 *n*
  d Strontium-90 *n*

5884 RADIOTHERAPY   md
      The use of ionizing radiation, except
      ultraviolet, for medical treatment.
  f radiothérapie *f*
  e radioterapia *f*
  i radioterapia *f*
  n stralingstherapie
  d Strahlentherapie *f*

5885 RADIOTHERMOLUMINESCENCE   ra
      Luminescence exhibited by certain
      minerals as a result of irradiation with
      beta and gamma rays followed by heating.
  f radiothermoluminescence *f*
  e radiotermoluminiscencia *f*
  i radiotermoluminescenza *f*
  n radiothermoluminescentie
  d Radiothermolumineszenz *f*

5886 RADIOTHORIUM   ch
      The common name for 1.90 Th-228, a
      member of the thorium series.
  f radiothorium *m*
  e radiotorio *m*
  i radiotorio *m*
  n radiothorium *n*
  d Radiothorium *n*

5887 RADIOTOXICITY   md
      Toxicity connected with the radiations of
      a radioactive element present in the
      organism.
  f radiotoxicité *f*
  e radiotoxicidad *f*
  i radiotossicità *f*
  n stralingsgiftigheid
  d Strahlungsgiftigkeit *f*

5888 RADIOTRANSPARENT (GB), see 5861

5889 RADIOTROPISM   ra
      Turning or bending of a plant or other
      organism in response to some form of
      radiation.
  f radiotropisme *m*
  e radiotropismo *m*
  i radiotropismo *m*
  n radiotropie, radiotropisme *n*
  d Radiotropismus *m*

5890 RADIUM   ch
      Radioactive element, symbol Ra, atomic
      number 88.
  f radium *m*
  e radio *m*
  i radio *m*
  n radium *n*
  d Radium *n*

5891 RADIUM AGE, see 3660

5892 RADIUM CAPSULE,
         RADIUM CELL, see 5833

5893 RADIUM CONTAINER   ra
      A container for a radium capsule.
  f récipient *m* de radium
  e recipiente *m* de radio
  i recipiente *m* di radio
  n radiumhouder
  d Radiumbehälter *m*

5894 RADIUM CONTENT   ra
      The quantity of radium in any radium
      container expressed in terms of the mass
      of radium element present.
  f contenu *m* de radium
  e contenido *m* de radio
  i contenuto *m* di radio
  n radiumgehalte *n*
  d Radiumgehalt *m*

5895 RADIUM EQUIVALENT   md, ra
      The number of milligram(me)s of
      radium-226 in the form of a point source
      screened by 0.5 mm platinum which gives
      the same exposure rate at a distance of
      25 cm in air.
  f équivalent *m* de radium
  e equivalente *m* de radio
  i equivalente *m* di radio
  n radiumequivalent *n*
  d Radiumäquivalent *n*

5896 RADIUM MOLD, see 4554

5897 RADIUM NEEDLE   ra
      A radium container in the form of a needle.
  f aiguille *f* de radium
  e aguja *f* de radio
  i ago *m* di radio
  n radiumnaald
  d Radiumnadel *f*

5898 RADIUM PACK   ra
      An applicator holding radium sources on
      the outside of the body.
  f applicateur *m* de radium
  e compresa *f* de radio
  i radiumsopporto *m*
  n radiumapplicator
  d Radiumpackung *f*

5899 RADIUM PLAQUE   ra
      A radium container in which the radium is
      distributed over the surface.
  f plaque *f* radiofère
  e placa *f* radiofero
  i placca *f* radiofero
  n radiumplak
  d Flachträger *m*, Plattenträger *m*,
      Radiumplakete *f*

5900 RADIUM SEED   ra
      A permanent implant.
  f semence *f* à radium
  e sencilla *f* de radio

i implantazione *f* di radio
n radiumzaadje *n*
d Dauerimplantat *n*, Radiumkapillare *f*

5901 RADIUM THERAPY md, ra
The use of radium and radon in the
treatment of disease.
f radiumthérapie *f*
e terapia *f* por radio
i radiumterapia *f*
n radiumtherapie
d Radiumtherapie *f*

5902 RADIUM TUBE ra
A radium container in the form of a
blunt-ended tube.
f tube *m* à radium
e tubo *m* de radio
i tubo *m* a radio
n radiumbuisje *n*
d Radiumröhrchen *n*

5903 RADIUS PARAMETER ma, np
The effective radius of a nucleus divided
by the cube root of its mass number A.
f paramètre *m* du rayon
e parámetro *m* del radio
i parametro *m* del raggio
n radiusparameter, straalparameter
d Radiusparameter *m*, Strahlparameter *m*

5904 RADON ra
Radioactive gaseous element, symbol Rn,
atomic number 86.
f radon *m*
e radón *m*
i radon *m*
n radon *n*
d Radon *n*

5905 RADON CONTAINER ra
A sealed container for holding radon.
f tube *m* à radon
e tubo *m* de radón
i tubo *m* a radon
n radonbuisje *n*
d Radonröhrchen *n*

5906 RADON CONTENT ra
The quantity of radon in any radon
container, normally expressed in
millicuries.
f contenu *m* de radon
e contenido *m* de radón
i contenuto *m* di radon
n radongehalte *n*
d Radongehalt *m*

5907 RADON CONTENT METER FOR ma
HEALTH PHYSICS PURPOSES
An assembly designed for health physics
to measure the content of radon and its
daughters in the atmosphere.
f émanomètre *m* de radioprotection
e radonómetro *m* para radioprotección
i radontenorimetro *m* per fisica sanitaria
n radonmeter voor stralingsbescherming

d Gerät *n* zur Bestimmung des Radongehalts
für Strahlenschutzzwecke

5908 RADON CONTENT METER FOR ma
PROSPECTING PURPOSES
An assembly used in prospecting to
measure the counting rate and/or the
number of counts corresponding to the
alpha particle emission rate from radon
and its daughters in the air sample.
f émanomètre *m* de prospection
e radonómetro *m* para uso minero
i radontenorimetro *m* per prospezione
n radonmeter voor prospectie
d Gerät *n* zur Bestimmung des Radongehalts
für Lagerstättensuche

5909 RADON EFFECT ra
Variation of radon velocity caused by
underground rock movement.
f effet *m* radon
e efecto *m* radón
i effetto *m* radon
n radoneffect *n*
d Radoneffekt *m*

5910 RADON EFFECT me, sa
SEISMIC PREDICTOR
A device for predicting earthquakes by
measuring the change in velocity of
ascending radon.
f prédicteur *m* de tremblements de terre
par effet radon
e predictor *m* de terremotos por efecto radón
i previsore *m* di terremoti per effetto radon
n aardbevingvoorspeller door middel van
radoneffect
d Erdbebenvoraussager *m* mittels Radoneffekt

5911 RAFFINATE ch
Any refined product obtained by fractional
distillation.
f raffinat *m*
e refinado *m*
i raffinato *m*
n raffinaat *n*
d Raffinat *n*

5912 RAFFINATE LAYER ch
The liquid layer in a solvent extraction
system from which the required solute has
been extracted.
f couche *f* de raffinat
e capa *f* de refinado
i strato *m* di raffinato
n raffinaatlaag
d Raffinatschicht *f*

5913 RAIN-OUT, nw
WASH-OUT
Fall-out deposited on earth by rain.
f retombées *pl* entraînées par la pluie
e depósito *m* arrastrado por la lluvia
i ricaduta *f* provocata dalle precipitazioni
n door regen meegevoerde neerslag
d durch Regen mitgeführter Ausfall *m*

5914 RAMSAUER EFFECT   ec, ic
Increase of the mean free path of free
electrons of low energy, e.g. in argon.
f   effet *m* Ramsauer
e   efecto *m* Ramsauer
i   effetto *m* Ramsauer
n   ramsauereffect *n*
d   Ramsauer-Effekt *m*

5915 RANDITE   mi
A mineral containing calcium uranyl
carbonate.
f   randite *f*
e   randita *f*
i   randite *f*
n   randiet *n*
d   Randit *m*

5916 RANDOM COINCIDENCE, see 40

5917 RANDOM ERROR, see 42

5918 RANDOM EVENTS   np
Events whose occurrence in no way affects
the occurrence, non-occurrence, or any
other characteristics of any future event.
f   événements *pl* aléatoires
e   sucesos *pl* de azar
i   eventi *pl* accidentali
n   toevallige gebeurtenissen *pl*
d   Zufallsereignisse *pl*

5919 RANDOM VARIABLE, see 984

5920 RANDOM VELOCITY   np
The velocity at or near the peak of the
distribution curve and equivalent to the
mean energy of electron temperature.
f   vitesse *f* complexe
e   velocidad *f* irregular
i   velocità *f* errata
n   ongerichte snelheid
d   ungerichtete Geschwindigkeit *f*

5921 RANDOM WALK   np
The path followed by a particle as it
makes random scattering collisions in a
medium.
f   parcours *m* erratique
e   trayectoria *f* irregular
i   percorso *m* errato
n   zwerftocht
d   Irrfahrt *f*

5922 RANGE, see 2285

5923 RANGE-ENERGY RELATION   ma, np
The relation, usually expressed in the form
of a graph, between the range of particles
of a given type and initial kinetic energy
and the energy.
f   relation *f* portée-énergie
e   relación *f* alcance-energía
i   rapporto *m* portata-energia
n   dracht-energie-betrekking
d   Reichweite-Energie-Beziehung *f*

5924 RANGE OF NUCLEAR FORCES   np
Used to indicate r_o when the curve of a
square well potential represents the
potential between two nucleons.
f   portée *f* des forces nucléaires
e   alcance *m* de las fuerzas nucleares
i   portata *f* delle forze nucleari
n   dracht van de kernkrachten
d   Reichweite *f* der Kernkräfte

5925 RANGE STRAGGLING   np
The variation in the range of particles
having the same initial energy.
f   dispersion *f* statistique de la portée,
    fluctuation *f* de portée
e   dispersión *f* estadística de alcance
i   dispersione *f* statistica di portata
n   drachtspreiding
d   Reichweitenstreuung *f*

5926 RARE GASES, see 3467

5927 RAREFACTION   nw
In an explosion, a condition existing at the
centre(er) of the explosion, in which the
pressure, after a rise induced by the
explosion, drops below that which existed
prior to the explosion.
f   diminution *f* de la pression
e   diminución *f* de la presión
i   diminuzione *f* della pressione
n   drukdaling
d   Druckminderung *f*

5928 RAREFACTION WAVE,   nw
         SUCTION WAVE
A pressure wave or rush of air or water
induced by rarefaction.
f   onde *f* de succion
e   onda *f* de succión
i   onda *f* aspirante
n   zuiggolf
d   Saugwelle *f*

5929 RASTER THERAPY, see 3083

5930 RATCHETTING   rt
Progressive relative movement, e.g. of can
and fuel, resulting from thermal cycling,
due to differences in thermal expansion.
f   mouvement *m* de long en-large,
    rochetage *m*
e   movimiento *m* de vaivén
i   corrugamento *m*
n   ongewenste verschuiving
d   unerwünschte Verschiebung *f*

5931 RATE ACTION (US), see 1739

5932 RATE OF EXCHANGE   np
The rate of a reaction in which atoms of
the same element in two different molecular species or in two different sites in the
same molecular species exchange places.
f   taux *m* d'échange, vitesse *f* d'échange
e   tasa *f* de intercambio,
    velocidad de intercambio

i tasso *m* di scambio, velocità *f* di scambio
n uitwisselingstempo *n*
d Austauschrate *f*

5933 RATE OF INFLOW gp
The rate at which a quantity enters into a
volume.
f vitesse *f* de débit à l' entrée,
   volume *m* d' apport
e velocidad *f* de influjo
i velocità *f* d' afflusso
n instroomsnelheid
d Einströmungsgeschwindigkeit *f*

5934 RATED CURRENT ge
The designated limit in rms amperes
or direct current amperes which an
electrical device will carry continuously
without exceeding the limit of observable
temperature rise.
f courant *m* nominal
e corriente *f* nominal
i corrente *f* nominale
n nominale stroom
d Nennstrom *m*

5935 RATED FREQUENCY ge
The frequency at which a device is
designed to operate.
f fréquence *f* nominale
e frecuencia *f* nominal
i frequenza *f* nominale
n nominale frequentie
d Nennfrequenz *f*

5936 RATED IMPULSE-WITHSTAND ge
      VOLTAGE
An assigned crest value of a specified
impulse voltage wave which a device must
withstand without flashover or other
electrical failure.
f tension *f* non-disruptive nominale
   d' impulsion
e tensión *f* no disruptiva nominal de
   impulso
i tensione *f* non disruttiva nominale
   d' impulso
n nominale doorslagvastheid
d Nennimpulsaushaltespannung *f*

5937 RATED OUTPUT POWER, see 5453

5938 RATED VOLTAGE ge
The voltage at which an electrical device
is designed to operate under usual service
conditions.
f tension *f* nominale
e tensión *f* nominal
i tensione *f* nominale
n nominale spanning
d Nennspannung *f*

5939 RATED WATT CONSUMPTION, ge
      WATTAGE RATING
A rating expressing the maximum power
that a device can safely handle continuously.
f puissance *f* absorbée normale

e consumo *m* nominal en vatios
i consumo *m* nominale di corrente
n nominaal wattverbruik *n*
d Nennaufnahme *f*

5940 RATEMETER, see 1493

5941 RATEMETER DISCRIMINATOR ct
An apparatus which unites in a single unit
the three functions of discrimination,
shaping and ratemetering.
f ictomètre *m* discriminateur
e impulsímetro *m* discriminador
i rateometro *m* discriminatore
n discriminerende tempometer
d Diskriminatorleistungsmesser *m*

5942 RATING ge
Of a machine, device or equipment, a
designated limit of operating characteristics, based on specified conditions.
f régime *m* nominal
e régimen *m* nominal atribuido
i regime *m* nominale
n nominale bedrijfsgegevens *pl*
d Nennbetrieb *m*

5943 RATING CHART xr
For X-ray tubes, a set of curves giving the
relation between the maximum permissible
tube current and the time of exposure for
the high tension indicated in every curve.
f courbes *pl* de charge
e curvas *pl* de carga
i curve *pl* di carico
n belastingskrommen *pl*
d Belastungsdiagramm *n*

5944 RATIO ge
The value obtained when one quantity is
divided by another of the same kind, to
indicate their relative proportions.
f rapport *m*, taux *m*
e relación *f*
i rapporto *m*
n verhouding
d Verhältnis *n*

5945 RATIO CONTROL co
Control of the rate of change of the
independent variable in an automatic
control system.
f commande *f* de proportion
e regulación *f* de relación
i controllo *m* di rapporto
n verhoudingsregeling
d Verhältnisregelung *f*

5946 RATIO OF ACTIVITY DENSITIES,
      see 1286

5947 RAUVITE mi
A mineral containing calcium, uranium
and vanadium.
f rauvite *f*
e rauvita *f*
i rauvite *f*

n rauviet *n*
d Rauvit *m*

5948 RAY   ra
1. In particle propagation, the direction of
propagation.
2. Frequently, the moving particles or
photons themselves.
f rayon *m*
e rayo *m*
i raggio *m*
n straal
d Strahl *m*

5949 RAY DIVERGENCE   ra
The spreading or widening with distance
of beams of particles.
f divergence *f* de rayons
e divergencia *f* de rayos
i divergenza *f* di raggi
n bundeldivergentie, straalspreiding
d Strahlendivergenz *f*

5950 RAYLEIGH DISTILLATION   ch
A distillation wherein the composition of
the residue changes continuously during
the course of the distillation.
f distillation *f* de Rayleigh
e destilación *f* de Rayleigh
i distillazione *f* di Rayleigh
n rayleighdestillatie
d Rayleighsche Destillation *f*

5951 RBE,   xr
RELATIVE BIOLOGICAL
EFFECTIVENESS
For a particular living organism or part
of an organism, the ratio of the absorbed
dose of a reference radiation that
produces a specified biological effect
to the absorbed dose of the radiation of
interest that produces the same biological
effect.
f EBR, efficacité *f* biologique relative
e EBR, efectividad *f* biológica relativa
i EBR, efficacia *f* biologica relativa
n relatieve biologische efficiëntie
d RBW, relative biologische Wirksamkeit *f*

5952 REACTION   ge
An action wherein one or more substances
are changed into one or more new
substances.
f réaction *f*
e reacción *f*
i reazione *f*
n reactie
d Reaktion *f*

5953 REACTION CHANNEL, see 986

5954 REACTION ENERGY, see 4900

5955 REACTION FORMULA, see 4843

5956 REACTION INHIBITION, see 3482

5957 REACTION RATE   rt
The rate at which fission takes place in a
nuclear reactor, commonly expressed as
the number of nuclei undergoing fission in
unit time.
f vitesse *f* de réaction
e velocidad *f* de reacción
i velocità *f* di reazione
n reactiesnelheid, reactietempo *n*
d Reaktionsgeschwindigkeit *f*,
Reaktionsrate *f*

5958 REACTIVITY   ma, np
A parameter giving deviation from
criticalitv of a nuclear chain reacting
medium such that positive values correspond to a supercritical state and negative
values to a subcritical state.
f réactivité *f*
e reactividad *f*
i reattività *f*
n reactiviteit
d Reaktivität *f*

5959 REACTIVITY COEFFICIENT,   ma
REACTIVITY TEMPERATURE
COEFFICIENT
The partial derivative of reactivity with
respect to some specified parameter.
f coefficient *m* de réactivité
e coeficiente *m* de reactividad
i coefficiente *m* di reattività
n reactiviteitscoëfficiënt
d Reaktivitätskoeffizient *m*

5960 REACTIVITY EXCESS, see 2398

5961 REACTIVITY INCREMENT   rt
The increase in reactivity in a nuclear
reactor.
f augmentation *f* de la réactivité
e aumento *m* de la reactividad
i aumento *m* della reattività
n reactiviteitsstijging
d Reaktivitätsanstieg *m*

5962 REACTIVITY METER   ec, me
An electronic sub-assembly which,
connected to one or more detectors, gives
an indication of the reactivity of a nuclear
reactor.
f réactimètre *m*
e medidor *m* de reactividad, reactímetro *m*
i misuratore *m* di reattività,
unità *f* per la misura della reattività
n reactiviteitsmeter
d Gerät *n* zur Messung der Reaktivität,
Reaktivitätsmesser *m*

5963 REACTIVITY NOISE   rt
In a nuclear reactor operating at a steady
mean power, the variation in neutron flux
caused by changes in reactivity arising
from non-nuclear phenomena such as
mechanical vibration, surface waves and
bubbling in liquid moderator, etc.
f bruit *m* de réactivité

e ruido *m* de reactividad
i rumore *m* di reattività
n reactiviteitsruis
d Reaktivitätsrauschen *n*

5964 REACTIVITY OSCILLATOR,
REACTOR OSCILLATOR, see 5288

5965 REACTIVITY POWER   rt
COEFFICIENT
The change of reactivity per unit change
of reactor thermal power when other
variables are not independently changed.
f coefficient *m* de puissance de réactivité
e coeficiente *m* de potencia de reactividad
i coefficiente *m* di potenza di reattività
n vermogenscoëfficiënt van de reactiviteit
d Leistungskoeffizient *m* der Reaktivität

5966 REACTIVITY RATE   rt
The partial derivative of reactivity with
respect to time.
f taux *m* de variation de réactivité
e grado *m* de variación de reactividad
i tasso *m* di variazione di reattività
n reactiviteitstempo *n*
d Reaktivitätsrate *f*

5967 REACTIVITY UNITS   un
The units used or proposed in nuclear
physics and engineering.
f unités *pl* de réactivité
e unidades *pl* de reactividad
i unità *pl* di reattività
n reactiviteitseenheden *pl*
d Reaktivitätseinheiten *pl*

5968 REACTOR, see 4902

5969 REACTOR AUXILIARY SYSTEMS   rt
Systems required for proper operation of
a nuclear reactor which are not part of
the reactor itself.
f appareillage *m* auxiliaire du réacteur
e equipo *m* auxiliar del reactor
i apparecchiatura *f* ausiliaria del reattore
n reactortoebehoren *n*
d Reaktorzusatzgeräte *pl*

5970 REACTOR CELL, see 933

5971 REACTOR CHEMISTRY   ch, rt
f chimie *f* du réacteur
e química *f* del reactor
i chimica *f* del reattore
n reactorchemie
d Reaktorchemie *f*

5972 REACTOR CONTAINMENT   rt
The prevention of release, even under the
conditions of a reactor accident, of
unacceptable quantities of radioactive
material beyond a controlled zone.
f retenue *f* d'un réacteur
e contenimiento *m* de un reactor
i contenimento *m* d'un reattore
n beheersing van een reactor
d Beherrschung *f* eines Reaktors

5973 REACTOR CONTAINMENT,
CONTAINMENT VESSEL, see 1330

5974 REACTOR CONTROL   co, rt
The intentional variation of the reaction rate
in a nuclear reactor, or the adjustment of
reactivity to maintain steady-state
operation.
f commande *f* d'un réacteur
e regulación *f* de un reactor
i regolazione *f* d'un reattore
n reactorregeling
d Reaktorsteuerung *f*

5975 REACTOR CORE, see 1414

5976 REACTOR CROSS SECTION   rt
The average effective cross section
corresponding to the real energy distribution of the neutrons in a nuclear reactor.
f section *f* efficace de réacteur
e sección *f* eficaz de reactor
i sezione *f* d'urto di reattore
n werkzame reactordoorsnede
d Reaktorwirkungsquerschnitt *m*

5977 REACTOR DESIGN   rt
The complete preparation of lay-out, etc.,
of a nuclear reactor plant.
f dessein *m* du réacteur,
projet *m* du réacteur
e diseño *m* del reactor,
proyecto *m* del reactor
i disegno *m* del reattore,
progetto *m* del reattore
n reactorontwerp *n*, reactorplanning
d Reaktorplanung *f*, Reaktorprojektierung *f*

5978 REACTOR DOME,   rt
REACTOR SPHERE
A ball-shaped outer housing of a nuclear
reactor.
f dôme *m* du réacteur
e cúpula *f* del reactor
i cupola *f* del reattore
n reactorkoepel
d Reaktorkuppel *f*

5979 REACTOR EVOLUTION   rt
Reactivity changes in a nuclear reactor
provoked by changes occurring in the
composition of the fuel material during the
irradiation period.
f évolution *f* du réacteur
e comportamiento *m* del reactor
i andamento *m* del reattore
n reactorgedrag *n*
d Verhalten *n* des Reaktors

5980 REACTOR EXCURSION, see 5443

5981 REACTOR FAMILY,
REACTOR SYSTEM, see 2519

5982 REACTOR LATTICE, see 3916

5983 REACTOR LOOP   rt
In a reactor, a piping system through which
a fluid may flow as a part of reactor
operation or for experimental purposes.
f boucle f de réacteur
e circuito m de reactor
i circuito m di reattore
n reactorkringloop, reactorlus
d Reaktorversuchskreislauf m

5984 REACTOR NOISE   rt
In a nuclear reactor operating at a steady
mean power, the variation in neutron flux
caused by changes in neutron population
within the reactor arising from statistical
fluctuations of the fission process.
f bruit m de réacteur
e ruido m de reactor
i rumore m di reattore
n reactorruis
d Reaktorrauschen n

5985 REACTOR PERIOD,   rt
      REACTOR TIME CONSTANT
The time required for the neutron flux
density in a reactor to change by a factor
of e (2·718) when the flux density is
rising or falling exponentially.
f constante f de temps d'un réacteur
e perfodo m del reactor
i tempo m di divergenza del reattore
n tijdconstante van een reactor
d Reaktorperiode f,
   Zeitkonstante f eines Reaktors.

5986 REACTOR POISONING, see 5379

5987 REACTOR SAFETY FUSE   rt, sa
A self-contained device designed to
respond to excessive temperature or flux
density in a nuclear reactor and to act
to reduce the reaction rate to a safe level.
f fusible m de sécurité d'un réacteur
e fusible m de seguridad de un reactor
i fusibile m di sicurezza d'un reattore
n reactorveiligheid
d Reaktorschutzsicherung f

5988 REACTOR SHIMMING,   rt
      SHIMMING
Coarse regulation of a reactor carried out
to correct the reactivity variations of large
amplitude spread over a long period.
f compensation f
e compensación f
i compensazione f
n grofregeling
d Trimmen n

5989 REACTOR SIMULATOR, see 4482

5990 REACTOR SPECTRUM, see 2681

5991 REACTOR THEORY   rt
The physical laws laid down for constructing a nuclear reactor.
f théorie f du réacteur

e teoría f del reactor
i teoria f del reattore
n reactortheorie
d Reaktortheorie f

5992 REACTOR TRIP   rt, sa
The rapid reduction of reactor power to a
negligible level, achieved by the full
insertion of the coarse control members
and the fine control members at their
emergency speed.
f arrêt m rapide du réacteur
e paro m rápido del reactor
i scatto m del reattore
n snelle reactoruitschakeling
d Schnellabschaltung f

5993 REACTOR VAULT   rt
A concrete biologically shielding
department housing the core of the nuclear
reactor and the primary cooling circuits.
f voûte f du réacteur
e bóveda f del reactor
i volta f del reattore
n reactorgewelf n
d Reaktorgewölbe n

5994 REACTOR VESSEL, see 1416

5995 REAL ABSORPTION COEFFICIENT,
      see 14

5996 REAL RANGE, see 5459

5997 REARRANGEMENT OF FUEL,   rt
      SHUFFLING OF FUEL
f altération f de l'arrangement du combustible
e rearreglo m del combustible
i alterazione f della disposizione del
   combustibile
n wijziging van de splijtstofopstelling
d Abänderung f der Brennstoffanordnung

5998 REBOILER   ch
The heat transfer equipment which
generates vapo(u)r at the bottom of a
continuous distillation unit.
f bouilleur m, rebouilleur m
e evaporador m
i bollitore m
n verdamper
d Verdampfer m

5999 RECALESCENCE   mg
The liberation of heat when a metal is
cooled through the lower critical point.
f récalescence f
e recalescencia f
i recalescenza f
n recalescentie
d Rekaleszenz f

6000 RECALESCENT POINT   mg
The temperature at which there is a
sudden liberation of heat as a heated
metal is cooled.

f   point *m* de récalescence
e   punto *m* de recalescencia
i   punto *m* di recalescenza
n   recalescentiepunt *n*
d   Rekaleszenzpunkt *m*

6001 RECESSIVE CHARACTER   md
In genetics, of a pair of contrasted
characteristics, the one which will not
appear in the hybrid from cross-breeding
homozygous parents unlike with respect
to this characteristic.
f   caractère *m* recessif
e   carácter *m* recesivo
i   carattere *m* recessivo
n   recessief karakter *n*
d   rezessiver Charakter *m*

6002 RECIPROCAL LATTICE   cr
A mathematical device much used in the
interpretation of diffraction problems in
three-dimensional structures.
f   réseau *m* réciproque
e   red *f* recíproca
i   reticolo *m* reciproco
n   reciprook rooster *n*
d   reziprokes Gitter *n*

6003 RECIPROCAL VELOCITY REGION   np
The energy region in which the capture
cross section for neutrons by a given
element is inversely proportional to the
neutron velocity.
f   zone *f* de vitesse réciproque
e   zona *f* de velocidad recíproca
i   zona *f* di velocità reciproca
n   reciprook snelheidsgebied *n*
d   reziproker Geschwindigkeitsbereich *m*

6004 RECIPROCATING COLUMN   ch
A plate column used in solvent extraction
in which the plates are mounted on an
oscillating axial rod.
f   colonne *f* à plateaux vibrante
e   columna *f* de platos oscilante
i   colonna *f* di piatti oscillante
n   trillende schotelkolom
d   vibrierende Bodenkolonne *f*

6005 RECIPROCATING GRID, see 806

6006 RECOIL   np
The motion of an atom because of the
emission of an alpha particle, etc., or a
quantum of radiation.
f   recul *m*
e   rechazo *m*, retroceso *m*
i   rinculo *m*
n   terugslag, terugstoot
d   Rückstoss *m*

6007 RECOIL ATOM   np
An atom which is suddenly deflected or
reversed in its path according to the
principle of conservation of momentum
owing to the recoil action from the
emission of a particle or photon.

f   atome *m* de recul
e   átomo *m* de rechazo, átomo *m* de retroceso
i   atomo *m* di rinculo
n   terugslagatoom *n*, terugstootatoom *n*
d   Rückstossatom *n*

6008 RECOIL ELECTRON   np
An electron set in motion by a collision.
f   électron *m* de recul
e   electrón *m* de rechazo,
    electrón *m* de retroceso
i   elettrone *m* di rinculo
n   terugslagelektron *n*, terugstootelektron *n*
d   Rückstosselektron *n*

6009 RECOIL NUCLEUS   np
A nucleus that recoils as a result of a
collision with a nuclear particle or as in
radioactivity, as the result of the ejection
of a particle from it.
f   noyau *m* de recul
e   núcleo *m* de rechazo, núcleo *m* de retroceso
i   nucleo *m* di rinculo
n   terugslagkern, terugstootkern
d   Rückstosskern *m*

6010 RECOIL PARTICLE   np
A particle that has been set into motion by
collision or by a process involving the
ejection of another particle.
f   particule *f* de recul
e   partícula *f* de rechazo,
    partícula *f* de retroceso
i   particella *f* di rinculo
n   terugslagdeeltje *n*, terugstootdeeltje *n*
d   Rückstossteilchen *n*

6011 RECOIL PARTICLE COUNTER   ct
TUBE
A counter tube in which ionization in the
gas filling is produced by recoil particles
resulting from the collision of fast
neutrons with nuclei of light atoms.
f   tube *m* compteur à particules de recul
e   tubo *m* contador de partículas de rechazo
i   tubo *m* contatore a particelle di rinculo
n   terugstoottelbuis
d   Rückstossteilchenzählrohr *n*

6012 RECOIL PROTON, see 3862

6013 RECOIL PROTON   ct
COUNTER TUBE
A counter tube in which ionization in the
gas filling is produced by recoil protons
resulting from the collision of fast
neutrons with nuclei of light atoms.
f   tube *m* compteur à protons de recul
e   tubo *m* contador de protones de rechazo,
    tubo *m* contador de protones de retroceso
i   tubo *m* contatore a protoni di rinculo
n   terugstoottelbuis
d   Rückstosszählrohr *n*

6014 RECOIL PROTON COUNTER   ct, ma
TUBE FAST NEUTRON FLUXMETER
An assembly designed to measure fast

neutron flux density in which the detector is a recoil proton counter tube.
f fluxmètre *m* de neutrons rapides à tube compteur à protons de recul
e flujómetro *m* de neutrones rápidos de contador por rechazo
i flussometro *m* di neutroni veloci a tubo contatore a protoni di rinculo
n fluxdichtheidsmeter voor snelle neutronen met telbuis voor terugslagprotonen
d Gerät *n* zur Messung der Fluenz schneller Neutronen mit Rückstossprotonenzählrohr

6015 RECOIL PROTON   ic
  IONIZATION CHAMBER
An ionization chamber in which ionization in the gas filling is produced by recoil particles resulting from the collision of fast neutrons with nuclei of light atoms.
f chambre *f* d'ionisation à protons de recul
e cámara *f* de ionización de protones de rechazo,
 cámara *f* de ionización de protones de retroceso
i camera *f* d'ionizzazione a protoni di rinculo
n terugstootionisatievat *n*
d Rückstossionisationskammer *f*

6016 RECOIL RADIATION   ra
Radiation emitted during nuclear disintegration in such a way that there is an observable recoil of the nucleus.
f rayonnement *m* de recul
e radiación *f* de rechazo,
 radiación *f* de retroceso
i radiazione *f* di rinculo
n terugslagstraling, terugstootstraling
d Rückstossstrahlung *f*

6017 RECOMBINATION   np
The return of an ionized atom or molecule to its electrically neutral state, by the gain of an electron in the case of positive ions and by the loss of an electron for negative ions.
f recombinaison *f*
e recombinación *f*
i ricombinazione *f*
n recombinatie
d Rekombination *f*

6018 RECOMBINATION COEFFICIENT   np
In an ionized gas, the ratio of the time rate of recombination of ions to the product of the positive-ion density and negative-ion density.
f coefficient *m* de recombinaison
e coeficiente *m* de recombinación
i coefficiente *m* di ricombinazione
n recombinatiecoëfficiënt
d Rekombinationskoeffizient *m*

6019 RECOMBINATION VELOCITY   ec
The quotient of the normal component of the electron (hole) current density at the surface by the excess electron (hole) charge density at the surface.

f vitesse *f* de recombinaison
e velocidad *f* de recombinación
i velocità *f* di ricombinazione
n recombinatiesnelheid
d Rekombinationsgeschwindigkeit *f*

6020 RECOMBINER   ch
A heated catalyst used to recombine the liberated oxygen of heavy water and deuterium.
f catalyseur *m* recombinateur
e catalizador *m* recombinador
i catalizzatore *m* ricombinatore
n herenigende katalysator
d wiedervereinigender Katalysator *m*

6021 RECORDING UNIT   ct
A basic function unit comprising an electromechanical device for counting electrical pulses.
f élément *m* numéroteur
e unidad *f* registradora
i elemento *m* numeratore
n registreereenheid
d Zählwerk *n*

6022 RECOVERY   md, xr
In radiology, the return towards normal of a particular cell, tissue or organism after radiation injury.
f guérison *f*, restauration *f*
e curación *f*, restablecimiento *m*
i guarigione *f*, ristabilimento *m*
n genezing, herstel *n*
d Genesung *f*, Heilung *f*

6023 RECOVERY,   ch
  EXTRACTION, see 2486

6024 RECOVERY,   mg
  RELIEVING
In metallurgy, the term denotes the removal of residual stresses, usually those that are a result of work-hardening.
f détente *f*
e remoción *f* de tensiones
i distensione *f*
n ontlating, ontspanning
d Entspannung *f*

6025 RECOVERY RATE   md, xr
The rate at which recovery takes place following after radiation injury.
f vitesse *f* de restauration
e velocidad *f* de restablecimiento
i velocità *f* di ristabilimento
n herstelsnelheid
d Genesungsgeschwindigkeit *f*

6026 RECOVERY TIME,   co
  CORRECTION TIME, see 1423

6027 RECOVERY TIME,   ct
  COUNTER RECOVERY TIME, see 1468

6028 RECRYSTALLIZATION   cr
A process which occurs upon annealing cold-worked metals in which the original

stressed and distorted grains are replaced
by stress-free grains.
f recristallisation f
e recristalización f
i ricristallizzazione f
n rekristallisatie
d Rekristallisation f

6029 RECTANGULAR WELL,   ma, np
   SPHERICAL WELL, SQUARE WELL
   A potential well which assumes a potential
   which is constant and negative inside a
   certain radius, zero outside.
f puits m rectangulaire
e pozo m rectangular
i pozzo m rettangolare
n rechthoekige put
d rechteckiger Topf m

6030 RECTIFICATION   ch
   Any fractional distillation carried out in
   a distillation column.
f rectification f
e rectificación f
i rettificazione f
n rectificeren n
d Rektifizierung f

6031 RECTIFIER   ch
   That section of a cascade between the
   feed point and product withdrawal point.
f rectificateur m
e rectificador m
i rettificatore m
n rectificeerapparaat n
d Rektifizierapparat m

6032 RECTILINEAR MANIPULATOR,
   see 1502

6033 RECTILINEAR MOTION   np
   OF A PARTICLE
   The motion of a particle in a straight line.
f mouvement m rectilinéaire de la particule
e movimiento m rectilineal de la partícula
i moto m rettilineare della particella
n rechtlijnige deeltjesbeweging
d geradlinige Teilchenbewegung f

6034 RECYCLED FUEL   fu
   Fuel for a nuclear reactor that has been
   reprocessed.
f combustible m recyclé
e combustible m reciclado,
   combustible m recirculado
i combustibile m recircolato
n opgewerkt splijtmateriaal n
d aufgearbeiteter Brennstoff m

6035 RECYCLING   rt
   The returning of a material or a stream
   to an operation for further processing.
f recyclage m
e reciclaje m, recirculación f
i rimessa f in ciclo
n terugvoering
d Rückführung f

6036 RED CORPUSCLE, see 2365

6037 RED COUNT   md
   The number of red corpuscles per cubic
   millimetre(er) of blood.
f numération f érythrocytaire
e numeración f eritrocitaria,
   recuento m eritrocitario
i conteggio m eritrocitario,
   numerazione f eritrocitaria
n aantal n rode bloedlichaampjes,
   erythrocytenaantal n
d Erythrozytenzahl f

6038 REDOX PROCESSES   ch
   Chemical processes which involve both
   reduction and oxidation steps. An
   example is the solvent extraction process.
f procédés pl d'oxydo-réduction,
   procédés pl rédox
e procedimientos pl redox,
   procedimientos pl reducción-oxidación
i processi pl redox,
   processi pl riduzione-ossidazione
n redoxprocessen pl,
   reductie-oxydatie-processen pl
d Redoxverfahren pl

6039 REDUCED COOLANT FLOW   cl
f flux m réduit du réfrigérant
e flujo m reducido del refrigerante
i flusso m ridotto del refrigerante
n gereduceerde koelmiddelstroom
d reduzierte Kühlmittelströmung f

6040 REDUCED MASS   np
   The kinetic energy of a system of two
   particles can be written as the sum of the
   kinetic energies of the mass centre(er) and
   of the relative motion; the mass constant
   in the latter is the reduced mass.
f masse f réduite
e masa f reducida
i massa f ridotta
n gereduceerde massa
d reduzierte Masse f

6041 REDUCTION COEFFICIENT, see 2021

6042 REDUCTION IN BULK   rw
   Chemical process carried out for reducing
   the volume of, e.g., radioactive waste by
   evaporation or incineration.
f réduction f massique
e reducción f másica
i riduzione f di volume
n massareductie
d Mengenreduktion f

6043 RE-ENRICHMENT   fu
   Uranium hexafluoride reclaimed from spent
   fuel is converted into new fuel element
   and re-enriched in U-235 by gaseous
   diffusion.
f réenrichissement m
e reenriquecimiento m
i riarricchimento m

herverrijking
Neuanreicherung f, Wideranreicherung f

6044 RE-ENRICHMENT PLANT   fu
f installation f de réenrichissement
e equipo m de reenriquecimiento
i impianto m di riarricchimento
n herverrijkingsinstallatie
d Neuanreicherungsanlage f

6045 RE-ENTRANT GAS COOLING   cl
The cool gas is not all fed directly to the
fuel channels, but instead about 50 % is
ducted to the top of the core and then
passed through inter-brick spaces
throughout the core to the bottom.
f refroidissement m par gaz ré-entrant
e enfriamiento m por gas reentrante
i raffreddamento m per gas rientrante
n koeling met terugleiding van het koelgas
d Rückleitungskühlung f

6046 RE-ENTRANT IONIZATION CHAMBER,
see 2798

6047 REFERENCE   ge
Any value, level or magnitude of a
quantity against which other values of the
quantity are lined.
f référence f
e referencia f
i riferimento m
n betrekking, referentie, vergelijking,
verwijzing
d Bezug m, Vergleich m

6048 REFERENCE FRAME,
REFERENCE SYSTEM, see 2814

6049 REFERENCE INPUT   co
VARIABLE (US),
REFERENCE VARIABLE (GB)
The value of the input quantity which
precedes the controlled condition.
f grandeur f de référence
e magnitud f de referencia,
magnitud f piloto
i grandezza f di riferimento
n referentiegrootheid, vergelijkingsgrootheid
d Bezugsgrösse f

6050 REFERENCE SOURCE   ra
A radiation source with similar radiation
characteristics and of such a form that the
absorption and scattering of radiation is
similar to that of an unknown source to be
measured.
f source f de référence
e fuente f de referencia
i sorgente f di riferimento
n referentiebron, vergelijkingsbron
d Bezugsquelle f, Vergleichsquelle f

6051 REFERENCE SOURCE,   np
RADIOACTIVE STANDARD, see 5808

6052 REFLECTANCE OF A   np
NUCLEAR BARRIER,
REFLECTION PROBABILITY
In the scattering of particles by the
potential barrier of a nucleus, the
probability that a particle will be
reflected rather than transmitted through
the barrier.
f pouvoir m réflecteur de la barrière de
potentiel,
probabilité f de réflexion
e probabilidad f de reflexión,
reflectancia f de la barrera de potencial
i probabilità f di riflessione,
riflettanza f della barriera di potenziale
n reflectievermogen n van een potentiaalstoep, reflectiewaarschijnlijkheid
d Reflexionsvermögen n einer Potentialschwelle,
Reflexionswahrscheinlichkeit f

6053 REFLECTED PRESSURE   nw
The total pressure which results
instantaneously at the surface when a
shock wave travel(l)ing in one medium
strikes another medium.
f pression f réfléchie
e presión f reflejada
i pressione f riflessa
n gereflecteerde druk
d Reflexdruck m

6054 REFLECTED REACTOR   rt
A nuclear reactor containing a reflector.
f réacteur m à réflecteur
e reactor m con reflector
i reattore m con riflettore
n reactor met reflector
d Reaktor m mit Reflektor

6055 REFLECTED SHOCK WAVE   nw
A shock wave resulting from an explosion,
especially from the explosion of an
airburst bomb, which is reflected from a
surface or object.
f onde f de choc réfléchie
e onda f de choque reflejada
i onda f d'urto riflessa
n teruglopende stootgolf
d zurücklaufende Stosswelle f

6056 REFLECTION EFFECT   np
A possible cause of a deficiency of low
energy electrons.
f effet m de réflexion
e efecto m de reflexión
i effetto m di riflessione
n reflectie-effect n
d Reflexionseffekt m

6057 REFLECTION FACTOR (GB),   ec
REPELLENCE FACTOR (US)
The ratio of electrons reflected to
electrons entering a reflector space, as in
a reflex klystron.
f facteur m de réflexion
e factor m de reflexión

i   fattore m di riflessione
n   reflectiefactor
d   Reflexionsfaktor m

6058 REFLECTION TARGET   xr
A target so arranged that the useful X-ray
beam emerges from the surface on which
the electron stream is incident.
f   ecible f réfléchissante
e   blanco m reflector
i   bersaglio m riflettente
n   reflecterende trefplaat
d   reflektierende Treffplatte f,
    reflektierender Auffänger m

6059 REFLECTOR, see 4744

6060 REFLECTOR CONTROL   co, rt
Control of nuclear reactor by adjustment
of the properties, position or quantity
of the reflector in such a way as to change
the reactivity.
f   commande f par réflecteur
e   regulación f por reflector
i   regolazione f con riflettore
n   regeling met reflector
d   Reflektorsteuerung f

6061 REFLECTOR ECONOMY,   rt
      REFLECTOR SAVING
The reduction which can be made, without
changing reactivity, in a specified
dimension of the core of a bare reactor
when a given reflector is added.
f   économie f due au réflecteur
e   economía f por uso del reflector
i   risparmio m per riflettore
n   reflectorwinst
d   Reflektorersparnis n

6062 REFLECTOR TANK   rt
The support and container for the reflector
of a reactor.
f   récipient m du réflecteur liquide
e   recipiente m del reflector líquido
i   recipiente m del riflettore liquido
n   reflectorvloeistoftank
d   Reflektorflüssigkeitbehälter m

6063 REFLUX   ch, rw
The countercurrent recycle of a portion of
an effluent.
f   reflux m
e   reflujo m
i   riflusso m
n   terugloop, terugstroom
d   Rücklauf m, Rückstrom m

6064 REFLUX RATIO   ch
Quantity of reflux employed per unit quantity of overhead product removed or the
ratio of the rates of flow of the backward
to the forward flowing streams in a
countercurrent system.
f   taux m de reflux
e   relación f de reflujo
i   rapporto m di riflusso

n   terugloopverhouding
d   Rücklaufverhältnis n

6065 REFRIGERANT   cl
A substance which is suitable as the
working medium of a cycle of operations
wherein refrigeration is accomplished.
f   réfrigérant m
e   refrigerante m
i   refrigerante m
n   koelmiddel n
d   Kühlmittel n

6066 REFUELLING   rt
The extraction of depleted elements and
the insertion of new ones.
f   rechargement m
e   reaprovisionamiento m, relleno m,
    reposición f
i   ricarica f
n   oplading
d   Aufladung f

6067 REGENERATED FUEL   fu
Spent fuel material which has undergone
a reconditioning treatment.
f   combustible m régénéré
e   combustible m regenerado
i   combustibile m rigenerato
n   opgewerkte splijtstof
d   aufgearbeiteter Brennstoff m

6068 REGENERATION,   ch, mg
      REPROCESSING
The restoration or purification of a used
material to a usable condition or
composition.
f   régénération f
e   regeneración f
i   rigenerazione f
n   opwerken n
d   Aufarbeitung f

6069 REGENERATION LOSS,   fu
      REPROCESSING LOSS
Loss of fissile (fissionable) fertile or other
valuable materials in reprocessing
operations.
f   pertes pl de régénération
e   pérdidas pl de regeneración
i   perdite pl di rigenerazione
n   opwerkverliezen pl
d   Aufarbeitungsverluste pl

6070 REGENERATIVE PROCESS   md
The process by which damaged cells are
replaced by new ones of the same type.
f   processus m de régénération
e   proceso m de regeneración
i   processo m di rigenerazione
n   vernieuwingsproces n
d   Erneuerungsprozess m

6071 REGENERATIVE REACTOR, see 1390

6072 REGION, see 2285

6073 REGION OF LIMITED
      PROPORTIONALITY, see 4017

6074 REGION OF PARTIAL SHADOW,
      see 5188

6075 REGMOGRAPHY   xr
      A new technique of X-ray cinematography
      in which within a very short exposure
      time (1/1000 of a second and less) 4-6
      pictures per second are realized.
      f règmographie f
      e regmografía f
      i regmografia f
      n regmografie
      d Regmographie f

6076 REGULATED STAY AREA   sa
      The area in which those persons who have
      to carry out operations subject to
      radiation have to obey special regulations.
      f zone f à séjour réglementé
      e zona f de residencia reglamentada
      i zona f di soggiorno secondo il regolamento
      n aan verblijfsvoorschriften onderworpen
         gebied n
      d Gebiet n mit Aufenthaltsbestimmungen

6077 REGULATED WORK AREA   ra, sa
      Area in which only those persons are
      admitted who have to carry out operations
      subject to radiation.
      f zone f à conditions de travail
         réglementées
      e zona f de accesibilidad limitada
      i zona f d'accessibilità limitata
      n beperkt toegankelijk gebied n
      d beschränkt zugängliches Gebiet n

6078 REGULATING, see 2616

6079 REGULATING MEMBER, see 2618

6080 REGULATING ROD, see 2619

6081 REGULATION   co
      The maintaining of a variable of a circuit
      or device at essentially a constant level.
      f réglage m
      e regulación f
      i regolazione f
      n regeling
      d Regelung f

6082 REIGNITION, see 1469

6083 RELATIVE ABUNDANCE, see 26

6084 RELATIVE APERTURE   pa
      The ratio of the minimum vertical or
      horizontal clearance for particle passage
      to the particle orbit radius in the accele-
      rating chamber of an accelerator.
      f ouverture f relative
      e abertura f relativa
      i apertura f relativa
      n relatieve apertuur
      d relative Apertur f

6085 RELATIVE ATOMIC MASS   np
      The relation A_r of the atomic mass of a
      nuclide to the unified constant of atomic
      mass $A_r \dfrac{M_a}{m\mu}$
      f masse f atomique relative
      e masa f atómica relativa
      i massa f atomica relativa
      n relatieve atoommassa
      d relative Atommasse f

6086 RELATIVE BIOLOGICAL
      EFFECTIVENESS, see 5951

6087 RELATIVE CALIBRATION   co
      A calibration showing the relation between
      the control rod position and the reactivity
      change in units, e.g. effective inches.
      f jaugeage m relatif
      e calibración f relativa
      i taratura f relativa
      n relatieve ijking
      d relative Eichung f

6088 RELATIVE CONVERSION   np, rt
      RATIO
      The instantaneous conversion ratio in a
      reactor, relative to the instantaneous
      conversion ratio in fuel of the same
      composition in some specified neutron
      spectrum.
      f rapport m relatif de conversion
      e relación f relativa de conversión
      i rapporto m relativo di conversione
      n relatieve conversieverhouding
      d relatives Konversionsverhältnis n

6089 RELATIVE IMPORTANCE   np
      For neutrons of type A relative to neutrons
      of type B, the average number of neutrons
      with velocity and position B which must be
      added to a critical system to keep the
      chain reaction rate constant after removal
      of a neutron with velocity and position A.
      f importance f relative
      e importancia f relativa
      i importanza f relativa
      n relatief gewicht n
      d relativer Einfluss m

6090 RELATIVE ISOTOPIĆ ABUNDANCE,
      see 2809

6091 RELATIVE MASS DEFECT   np
      $B_r = B/m_u$, where B is the mass defect
      and m_u the unified atomic mass constant.
      f défaut m de masse relatif
      e defecto m de masa relativo
      i difetto m di massa relativo
      n relatief massatekort n
      d relativer Massendefekt m

6092 RELATIVE MASS EXCESS   np
      $\triangle_r = A/m_u$ where $\triangle$ is the mass excess,
      A the nuclear number and m_u the unified
      atomic mass constant.

f   excès *m* de masse relatif
e   exceso *m* de masa relativo
i   eccesso *m* di massa relativo
n   relatief massaoverschot *n*
d   relativer Massenüberschuss *m*

6093 RELATIVE PLATEAU SLOPE   ct
The relative change of counting rate within the plateau region for a given change of the applied voltage.
f   pente *f* relative de palier
e   inclinación *f* relativa de meseta
i   pendenza *f* relativa di ripiano
n   relatieve plateauhelling
d   relative Plateauneigung *f*

6094 RELATIVE SPECIFIC   np
IONIZATION
The specific ionization for a particle of a given medium, relative either to that for the same particle and energy in a standard medium or the same particle and medium at a specified energy.
f   ionisation *f* spécifique relative
e   ionización *f* específica relativa
i   ionizzazione *f* specifica relativa
n   relatieve specifieke ionisatie
d   relative spezifische Ionisation *f*

6095 RELATIVE STOPPING POWER   np
The ratio of the stopping power of a given substance to that of a standard substance.
f   pouvoir *m* d'arrêt relatif
e   poder *m* de frenado relativo
i   potere *m* di rallentamento relativo
n   relatief stoppend vermogen *n*
d   Bezugsbremsvermögen *n*

6096 RELATIVE VOLATILITY   ch, gp
The quotient obtained when the ratio of the concentrations of any pair of substances in the vapo(u)r is divided by the ratio of the same pair of substances in the liquid with which the vapo(u)r is in equilibrium.
f   volatilité *f* relative
e   volatilidad *f* relativa
i   volatilità *f* relativa
n   relatieve vluchtigheid
d   relative Flüchtigkeit *f*

6097 RELATIVISTIC KINETIC ENERGY   np
The kinetic energy of a relativistic particle
is given by $T = m_0c^2 \left[\dfrac{1}{\sqrt{1-v^2c^2}} - 1 \right]$ where m_0
is the rest mass of the particle, v is the velocity of the particle and c is the speed of light.
f   énergie *f* cinétique relativiste
e   energía *f* cinética relativista
i   energia *f* cinetica relativista
n   relativistische kinetische energie
d   relativistische kinetische Energie *f*

6098 RELATIVISTIC MASS   np
The mass of a particle moving at a velocity exceeding about one-tenth the velocity of light.

f   masse *f* relativiste
e   masa *f* relativista
i   massa *f* relativista
n   relativistische massa
d   relativistische Masse *f*

6099 RELATIVISTIC MASS EQUATION   ma
The equation for the relativistic mass of a particle or body having a given rest mass and velocity.
f   équation *f* de masse relativiste
e   ecuación *f* de masa relativista
i   equazione *f* di massa relativista
n   relativistische massavergelijking
d   relativistische Massengleichung *f*

6100 RELATIVISTIC PARTICLE   np
A particle whose velocity is so large that its mass in motion is significantly greater than its rest mass.
f   particule *f* relativiste
e   partícula *f* relativista
i   particella *f* relativista
n   relativistisch deeltje *n*
d   relativistisches Teilchen *n*

6101 RELATIVISTIC TRACK   ic, ph
The track, in a photographic emulsion or in a bubble chamber, of a particle moving with a speed near that of light.
f   trace *f* relativiste
e   traza *f* relativista
i   traccia *f* relativista
n   relativistisch spoor *n*
d   relativistische Spur *f*

6102 RELATIVISTIC VELOCITY   np
A particle velocity sufficiently large that mass and other properties of the particles are significantly different from the at-rest values.
f   vitesse *f* relativiste
e   velocidad *f* relativista
i   velocità *f* relativista
n   relativistische snelheid
d   relativistische Geschwindigkeit *f*

6103 RELATIVITY   gp
A principle that postulates the equivalence of the description of the universe, in terms of physical laws, by various observers or for various frames of reference.
f   relativité *f*
e   relatividad *f*
i   relatività *f*
n   relativiteit
d   Relativität *f*

6104 RELAXATION   gp
A term applied to the process whereby a physical system reaches equilibrium or a steady state after a sudden change in conditions.
f   relaxation *f*
e   relajación *f*
i   attenuazione *f*, rilassamento *m*
n   relaxatie
d   Relaxation *f*

6105 RELAXATION DISTANCE,   ma
      RELAXATION LENGTH
   For a physical quantity which decreases
   exponentially with distance, the distance
   over which the quantity drops by a factor e.
   (e=2·71828)
f longueur f de relaxation
e longitud f de relajación
i lunghezza f d'attenuazione
n relaxatielengte
d Relaxationslänge f

6106 RELAXATION METHODS   ma
   A mathematical method for solving a
   complicated set of simultaneous linear
   equations, which may represent a
   differential equation.
f méthodes m de relaxation
e métodos pl de relajación
i metodi pl di rilassamento
n relaxatiemethoden pl
d Relaxationsverfahren pl

6107 RELAXATION TIME   ma
   For a physical quantity which decreases
   exponentially with time, the time in which
   the quantity drops by a factor e.
   (e=2·71828)
f temps m de relaxation
e tiempo m de relajación
i tempo m d'attenuazione
n relaxatietijd
d Relaxationszeit f

6108 RELEASE OF ENERGY, see 2286

6109 RELEASED ENERGY   ge, pp
   The energy set free by a reaction.
f énergie f libérée
e energía f liberada
i energia f liberata, energia f rilasciata
n vrijgemaakte energie
d freigesetzte Energie f

6110 RELIABILITY   ge
   The probability that a device will perform
   its purpose adequately for the period of
   time intended under the operating conditions
   encountered.
f sécurité f de fonctionnement,
   sécurité f de service
e seguridad f de funcionamiento,
   seguridad f de servicio
i sicurezza f d'esercizio,
   sicurezza f di funzionamento
n bedrijfszekerheid
d Betriebssicherheit f,
   Betriebszuverlässigkeit f

6111 RELIABILITY INDEX   ge
   A quantitive figure of merit related to the
   reliability of a piece of equipment, such
   as the number of failures per 1,000
   operations or the number of failures in a
   specified number of operating hours.
f indice m de sécurité de fonctionnement
e índice m de seguridad de funcionamiento

i indice m di sicurezza d'esercizio
n betrouwbaarheidsgetal n
d Zuverlässigkeitszahl f

6112 RELIABILITY TEST   te
   A test designed specifically to evaluate the
   level and uniformity and reliability of
   equipment under various environmental
   conditions.
f essai m de sécurité de fonctionnement
e prueba f de seguridad de funcionamiento
i prova f di sicurezza d'esercizio
n bedrijfszekerheidsproef
d Betriebssicherheitsprüfung f

6113 RELIEVING, see 6024

6114 REM,   un, xr
      RÖNTGEN EQUIVALENT MAN
   The special unit of dose equivalent.
f rem m
e rem m
i rem m
n rem
d Rem n

6115 REMOTE AREA   sa
      MONITORING SYSTEM
f télécontrôle m de contamination du terrain
e telecontrol m de contaminación de área
i controllo m a distanza di contaminazione
   di zona
n afstandscontrole van gebiedsbesmetting
d Fernüberwachung f von Gebiets-
   kontamination

6116 REMOTE CONTROL   co
   A control carried out from a distance.
f télécommande f
e telerregulación f
i teleregolazione f
n afstandsregeling, verreregeling
d Fernsteuerung f

6117 REMOTE HANDLING,   co
      REMOTE MANIPULATION
   The handling of the remote control
   elements.
f télémanipulation f
e telemanipulación f
i telemanipolazione f
n afstandsbediening, verrebediening
d Fernbedienung f

6118 REMOTE HANDLING   cd, rt
      EQUIPMENT,
      REMOTE MANIPULATING
      EQUIPMENT
   In a nuclear reactor, those components
   which can be handled from a safe position.
f ensemble m de télémanipulation
e conjunto m de telemanipulación
i complesso m di telemanipolazione
n afstandsbedieningsapparatuur,
   verrebedieningsopstelling
d Fernbedienungsgeräte pl

6119 REMOTE INDICATOR   me
An indicator located at a distance from the
data gathering sensing element.
f téléindicateur *m*
e teleindicador *m*
i teleindicatore *m*
n verreaanwijzer
d Fernanzeiger *m*

6120 REMOTE LIQUID SAMPLING   ch
f échantillonnage *m* de liquides à distance
e muestreo *m* de líquidos a distancia
i prelevamento *m* di campioni liquidi a
   distanza
n vloeistofmonsterneming op afstand
d Fernprobenahme *f* von Flüssigkeiten

6121 REMOTE METERING,   me
      TELEMETERING
Transmitting the readings of instruments
to a remote location by appropriate means.
f télémesure *f*
e telemedida *f*
i telemisura *f*
n telemeting, verremeting
d Fernmessung *f*

6122 REMOVABLE GRAPHITE BLOCKS   rt
The construction accepted for the core
of the DRAGON reactor.
f blocs *pl* en graphites transportables
e bloques *pl* de grafito amovibles
i blocchi *pl* di grafite amovibili
n verplaatsbare grafietblokken *pl*
d demontierbare Graphitblöcke *pl*

6123 REMOVAL   np
The phenomenon that a particle is
withdrawn from an assembly or from an
energy level, without an absorption
follow-up.
f déplacement *m*
e remoción *f*
i rimozione *f*
n verplaatsing
d Ausscheidung *f*, Versetzung *f*

6124 REMOVAL CROSS SECTION, see 2040

6125 RENARDITE   mi
A mineral containing lead, uranium and
phosphorus.
f rénardite *f*
e renardita *f*
i renardite *f*
n renardiet *n*
d Renardit *m*

6126 RENORMALIZATION   np
Adding to the mechanical mass of a
particle its extra mass due to self-inter-
action, to give a sum equal to the
measured mass.
f rénormalisation *f* de masse
e renormalización *f* de masa
i rinormalizzazione *f* di massa
n massarenormering
d Massenrenormierung *f*

6127 REP,   un, xr
      RÖNTGEN EQUIVALENT PHYSICAL
A unit of absorbed dose equal to the
absorbed dose in water which has
received an exposure dose of one röntgen.
f rep *m*
e rep *m*
i rep *m*
n rep
d Rep *n*

6128 REPELLENCE FACTOR (US), see 6057

6129 REPELLING OF ELECTRONS   np
A phenomenon observed in cathode-ray
beams of high density and resulting in an
increase in beam diameter.
f répulsion *f* d'électrons
e repulsión *f* de electrones
i ripulsione *f* d'elettroni
n afstoten *n* van elektronen,
   elektronenafstoting
d Elektronenabstossung *f*

6130 REPLACEABLE BOILER UNIT   cd, rt
f ensemble *m* de chaudière démontable
e conjunto *m* de caldera recambiable
i complesso *m* di caldaia sostituibile
n uitwisselbare stoomketelopstelling
d austauschbare Dampfkesselanordnung *f*

6131 REPROCESSING, see 6068

6132 REPROCESSING LOSS, see 6069

6133 REPRODUCTION CONSTANT, see 4581

6134 REPULSIVE FORCE, see 2780

6135 REPULSIVE POTENTIAL   np
The force between two atoms keeping them
apart at short distances is due to the
overlapping of the electron clouds,
especially of the closed shells of electrons.
f potentiel *m* de répulsion
e potencial *m* de repulsión
i potenziale *m* di ripulsione
n afstotingspotentiaal
d Abstossungspotential *n*

6136 RERADIATION   ra
The scattering of incident radiation.
f rerayonnement *m*
e reradiación *f*
i riradiazione *f*
n heruitstraling
d Wiederausstrahlung *f*

6137 RESEARCH REACTOR   rt
A nuclear reactor of any power level used
primarily as a research tool for basic or
applied research.
f réacteur *m* de recherche
e reactor *m* de investigación
i reattore *m* per ricerca
n speurwerkreactor
d Forschungsreaktor *m*

6138 RESERVOIR   ge
A receptacle specially constructed to store
a large supply of liquid material.
f réservoir *m*
e depósito *m*
i cisterna *f*, serbatoio *m*
n reservoir *n*
d Speicher *m*, Zisterne *f*

6139 RESET TIME   ct
The time required in a decade scaler
tube to move the electron beam back to
the starting position.
f durée *f* de retour
e tiempo *m* de reposición
i tempo *m* di ritorno
n terugslagtijd
d Rückstellzeit *f*

6140 RESIDUAL ACTIVITY,   np
        RESIDUAL RADIOACTIVITY
Radioactivity remaining in a substance or
system at a specified time after a period of
decay.
f activité *f* résiduelle
e actividad *f* residual
i attività *f* residua
n restactiviteit
d Restaktivität *f*

6141 RESIDUAL CURRENT   ic
The current which continues to be produced
by an ionizing chamber no longer submitted to external radiation, and which is
due either to activation of the component
materials of the chamber or to their
contamination.
f courant *m* résiduel
e corriente *f* residual
i corrente *f* residua
n reststroom
d Reststrom *m*

6142 RESIDUAL ERROR   ma
In an experiment, the sum of the random
errors and the uncorrected systematic
errors.
f erreur *f* résiduelle
e error *m* residual
i errore *m* residuo
n restfout
d Restfehler *m*

6143 RESIDUAL IONIZATION   np
Ionization of air or other gas in a closed
chamber, not accounted for by recognizable
neighbo(u)ring agencies.
f ionisation *f* résiduelle
e ionización *f* residual
i ionizzazione *f* residua
n restionisatie
d Restionisation *f*

6144 RESIDUAL NUCLEAR   nw, ra
        RADIATION,
        RESIDUAL RADIATION
Nuclear radiation emitted by radioactive
material deposited after an atomic burst.
f rayonnement *m* résiduel
e radiación *f* residual
i radiazione *f* residua
n reststraling
d Reststrahlen *pl*, Reststrahlung *f*

6145 RESIDUAL NUCLEUS   np
The heavy nucleus which is the end
product of a nuclear transformation.
f noyau *m* résiduel
e núcleo *m* residual
i nucleo *m* residuo
n restkern
d Restkern *m*

6146 RESIDUAL RANGE   np
The distance over which the particle can
still produce ionization after having lost
already some of its energy in passing
through matter.
f portée *f* résiduelle
e alcance *m* residual
i portata *f* residua
n restdracht
d Restreichweite *f*

6147 RESIN WOOL   mt
Material used in filtering air.
f laine *f* de résine
e lana *f* de resina
i lana *f* di resina
n kunstharswol
d Kunstharzwolle *f*

6148 RESISTANCE TO RADIATION, see 5733

6149 RESISTIVE INSTABILITY   pp
Instability resulting from the macroscopic
equations of a plasma with a finite
electrical conductivity.
f instabilité *f* résistive
e inestabilidad *f* resistiva
i instabilità *f* resistiva
n weerstandsinstabiliteit
d Widerstandsinstabilität *f*

6150 RESOLUTION,   me, sp
        RESOLVING POWER
The ratio of the mean value of two
quantities to the smallest difference
between them which a spectrometer is
able to distinguish.
f pouvoir *m* de résolution
e poder *m* de resolución
i potere *m* di risolvenza
n scheidend vermogen *n*
d Auflösungsvermögen *n*

6151 RESOLUTION TIME,
        RESOLVING TIME, see 1470

6152 RESOLUTION TIME CORRECTION,   ct
        RESOLVING TIME CORRECTION
Correction to the observed counting rate
to allow for the probability of the occurrence of events within the resolution time.

f   correction ƒ du temps de résolution
e   corrección ƒ del tiempo de resolución
i   correzione ƒ del tempo di risolvenza
n   correctie voor de scheidingstijd,
    scheidingstijdcorrectie
d   Auflösungszeitkorrektion ƒ

6153 RESONANCE   np
      In nuclear processes, the increased
      probability that a nuclear process will
      occur if the energy of the incoming
      particle or photon is in the vicinity of a
      particular value.
f   résonance ƒ
e   resonancia ƒ
i   risonanza ƒ
n   resonantie
d   Resonanz ƒ

6154 RESONANCE ABSORPTION ENERGY,
      see 2282

6155 RESONANCE ABSORPTION   np
      OF NEUTRONS
      Neutron absorption in the resonance energy
      range.
f   absorption ƒ de neutrons par résonance
e   absorción ƒ de neutrones por resonancia
i   assorbimento m di neutroni per risonanza
n   resonantieabsorptie van neutronen
d   Resonanzabsorption ƒ von Neutronen

6156 RESONANCE ABSORPTION   sp, te
      SPECTRUM,
      RESONANCE SPECTRUM
      A spectrum resulting from the excitation
      of an atom from its ground state to an
      excited state by the absorption of
      radiation, and the subsequent emission of
      light during the return of the atom to the
      ground state.
f   spectre m de résonance
e   espectro m de resonancia
i   spettro m di risonanza
n   resonantiespectrum n
d   Resonanzspektrum n

6157 RESONANCE CAPTURE   np
      The capture of an incident particle into
      a resonance level of the resulting
      compound nucleus.
f   capture ƒ de résonance
e   captura ƒ de resonancia
i   cattura ƒ di risonanza
n   resonantievangst
d   Resonanzeinfang m

6158 RESONANCE CAPTURE   np
      OF NEUTRONS
      Radiative capture of neutrons in the
      resonance energy range.
f   capture ƒ de résonance de neutrons
e   captura ƒ de resonancia de neutrones
i   cattura ƒ di risonanza di neutroni
n   resonantievangst van neutronen
d   Resonanzeinfang m von Neutronen

6159 RESONANCE CONCEPT   np
      OF NUCLEAR REACTION
      Nuclei exist in discretely spaced energy
      levels, which may be observed through the
      increase in reaction cross sections at the
      energies corresponding to the levels.
f   concept m de résonance de la réaction
    nucléaire
e   concepto m de resonancia de la reacción
    nuclear
i   concetto m di risonanza della reazione
    nucleare
n   resonantieopvatting van de kernreactie
d   Resonanzdeutung ƒ der Kernreaktion

6160 RESONANCE CROSS SECTION, see 2027

6161 RESONANCE ENERGY   np
      The kinetic energy of a particle that will
      be captured or scattered preferentially
      because of the presence of an appropriate
      level of resonance in the compound
      nucleus formed by the incident particle and
      the target nucleus.
f   énergie ƒ de résonance
e   energía ƒ de resonancia
i   energia ƒ di risonanza
n   resonantie-energie
d   Resonanzenergie ƒ

6162 RESONANCE ESCAPE   np
      PROBABILITY
      In an infinite medium, the probability that
      a neutron which is being slowed down from
      fission energy will reach thermal energy
      rather than suffer resonance capture.
f   probabilité ƒ d'échappement de résonance,
    facteur m antitrappe
e   probabilidad ƒ de escape a la resonancia
i   fattore m di trasparenza alla risonanza,
    probabilità ƒ di fuga alla risonanza
n   resonantieontsnappingskans
d   Bremsnutzung ƒ

6163 RESONANCE FISSION   np
      The fission of a nucleon provoked by a
      neutron with an energy equal to that of a
      fission resonance band of that nucleon.
f   fission ƒ de résonance
e   fisión ƒ de resonancia
i   fissione ƒ di risonanza
n   resonantiesplijting
d   Resonanzspaltung ƒ

6164 RESONANCE FLUORESCENCE   ra
      The emission of radiation by a gas or
      vapo(u)r at the same frequency as the
      exciting radiation.
f   fluorescence ƒ de résonance
e   fluorescencia ƒ de resonancia
i   fluorescenza ƒ di risonanza
n   resonantiefluorescentie
d   Resonanzfluoreszenz ƒ

6165 RESONANCE HEATING   pp
      Of a plasma, heating by motion of the ions

excited at or near the frequency of one of their natural motions, e.g. the ion cyclotron frequency.
f   chauffage m par résonance
e   calentamiento m por resonancia
i   riscaldamento m per risonanza
n   resonantieverhitting
d   Resonanzheizung f

6166   RESONANCE INTEGRAL   ma, np
The integral over all or some specified portion of the resonance energy range of the product of the absorption cross section of a nuclide and the reciprocal of the neutron energy.
f   intégrale f de résonance
e   integral f de resonancia
i   integrale m di risonanza
n   resonantie-integráal
d   Resonanzintegral n

6167   RESONANCE LEVEL, see 4906

6168   RESONANCE NEUTRON FLUX   np
The flux, or rate of flow through a unit area per unit time, of neutrons of resonance energies.
f   flux m neutronique de résonance
e   flujo m de neutrónico de resonancia
i   flusso m neutronico di risonanza
n   neutronenflux voor resonantie
d   Resonanzneutronenfluss m

6169   RESONANCE NEUTRONS   np
Neutrons having kinetic energy in the resonance energy range.
f   neutrons pl de résonance
e   neutrones pl de resonancia
i   neutroni pl di risonanza
n   resonantieneutronen pl
d   Resonanzneutronen pl

6170   RESONANCE PEAK   np
A peak in the cross section curve occurring at certain resonance energies.
f   pic m de résonance
e   cresta f de resonancia,
    pico m de resonancia
i   cresta f di risonanza,
    picco m di risonanza
n   resonantiepiek
d   Resonanzspitze f

6171   RESONANCE PENETRATION   np
The penetration of a nucleus by a charged particle whose energy corresponds to one of the energy levels in the nucleus.
f   pénétration f de résonance
e   penetración f de resonancia
i   penetrazione f di risonanza
n   doordringing bij resonantie
d   Resonanzdurchgang m

6172   RESONANCE RADIATION   np, ra
The emission of radiation by a gas or vapo(u)r as a result of excitation of atoms to higher energy levels by incident photons at the resonance frequency of the gas or vapo(u)r.
f   rayonnement m de résonance
e   radiación f de resonancia
i   radiazione f di risonanza
n   resonantiestraling
d   Resonanzstrahlung f

6173   RESONANCE REGION   np
The region in which neutrons are captured by resonance, their energy varying from 6 to 10 eV.
f   région f des énergies de résonance
e   zona f de resonancia
i   zona f di risonanza
n   resonantiegebied n
d   Resonanzgebiet n

6174   RESONANCE SCATTERING   np
The scattering caused by the incident wave penetrating the surface of the nucleus and interacting with its interior.
f   diffusion f résonante
e   dispersión f resonadora
i   deviazione f risonatrice
n   resonantieverstrooiing
d   Resonanzstreuung f

6175   RESONANCE SPECTRAL LINE   sp
One of the spectral lines emitted when an electron undergoes transfer to a lower state in a given atom.
f   raie f spectrale de résonance
e   línea f espectral de resonancia
i   riga f spettrale di risonanza
n   resonantiespectrumlijn
d   Resonanzspektrallinie f

6176   REST ENERGY,   np
       REST MASS ENERGY
The energy of a particle in a system in which it appears to be at rest.
f   énergie f au repos
e   energía f en reposo
i   energia f in riposo
n   rustenergie
d   Ruheenergie f

6177   REST MASS   np
The mass of a particle at rest when moving with a velocity low compared with that of light.
f   masse f au repos
e   masa f en reposo
i   massa f in riposo
n   rustmassa
d   Ruhemasse f, Ruhmasse f

6178   RESTORING FORCE   me
The force needed to bring the indicating component part of a measuring instrument back to its original position.
f   force f de rappel
e   fuerza f de restitución
i   forza f di ripristina
n   terugstelkracht
d   Rückstellkraft f

6179 RESTORING TORQUE me
The torque which tends to bring the moving
element back to the mechanical zero of the
instrument.
f couple *m* antagoniste
e par *m* antagonista
i coppia *f* antagonista
n terugdrijvend koppel *n*
d Richtmoment *n*, Rückstellmoment *n*

6180 RESTRAINT TANK rt
In reactor technology, the vat enclosing and
supporting the reactor core.
f récipient *m* enveloppant
e recipiente *m* envolvente
i serbatoio *m* circondante
n omsluitingsvat *n*
d Umschliessungsgefäss *n*

6181 RESTRICTED AREA, see 71

6182 RETENTION, np
RETENTION FRACTION
Of atoms undergoing a nuclear transformation, that fraction which remains in, or
reverts to, its initial chemical condition,
and is thus retained in the specimen on
subsequent chemical separation.
f rétention *f*
e retención *f*
i ritenzione *f*
n niet-afscheidbaar deel *n*
d Retention *f*, Rückstand *m*

6183 RETENTION BASIN rw
The concrete pool or steel tank which
holds reactor effluent or separations plant
cooling water until it is released to the
river or ground.
f récipient *m* de dépôt
e estanque *m* de retención
i serbatoio *m* di deposito
n bewaartank, opslagvat *n*
d Aufbewahrungsgefäss *n*

6184 RETENTION COEFFICIENT, see 3484

6185 RETICULAR DENSITY cr
The number of points per unit area of a
network, as in that of a plane in a crystal
lattice.
f densité *f* réticulaire
e densidad *f* reticular
i densità *f* reticolare
n roosterdichtheid
d Gitterdichte *f*

6186 RETICULAR STRUCTURE cr
A crystal structure containing a network.
f structure *f* réticulaire
e estructura *f* reticular
i struttura *f* reticolare
n roosterstructuur
d Gitterstruktur *f*

6187 RETURN SIGNAL, see 2547

6188 REVERSE ISOTOPE is
DILUTION ANALYSIS
A process used for determining the radiochemical purity of a labelled (tagged)
compound.
f analyse *f* inverse d'isotopes par dilution
e análisis *f* inversa de isótopos por dilución
i analisi *f* inversa d'isotopi per diluizione
n omgekeerde oplossingsanalyse van
isotopen
d umgekehrte Lösungsanalyse *f* von Isotopen

6189 REVERSIBLE PROCESS ch
Any cycle of operations in which the
different operations can be performed
reversely with a reversal of their effects.
f procédé *m* réversible
e procedimiento *m* reversible
i processo *m* reversibile
n omkeerbaar proces *n*
d reversibeles Verfahren *n*

6190 REVERSIBLE SCALER ct
A scaler with a single output, which for
each incoming signal pulse adds one to its
contents, or subtracts one from its
contents, according to an auxiliary control.
f échelle *f* de comptage reversible
e escala *f* de recuento reversible
i unità *f* per conteggio reversibile
n omkeerbare pulsteller
d Vor-Rückwärtszähler *m*

6191 RHENIUM ch
Metallic element, symbol Re, atomic
number 75.
f rhénium *m*
e renio *m*
i renio *m*
n rhenium *n*
d Rhenium *n*

6192 RHEOSTRICTION, see 5295

6193 RHEOTRON pa
A modified type of betatron.
f rhéotron *m*
e reotrón *m*
i reotrone *m*
n rheotron *n*
d Rheotron *n*

6194 RHM, gr, xr
RÖNTGEN PER HOUR AT 1 METRE
A unit of quantity of a gamma ray source,
under specified conditions of shielding, such
that, at a distance of 1 metre in air, its
gamma rays produce an exposure dose rate
of one röntgen per hour.
f r/h à 1m, röntgen *m* par heure à 1 mètre
e r/h a 1m, röntgen *m* por hora a 1 metro
i r/h a 1m, röntgen *m* all'ora ad 1 metro
n röntgenuur *n* op een meter afstand
d Dosiskonstante *f*,
Röntgen *n* je Stunde in 1m

6195 RHODIUM   ch
Metallic element, symbol Rh, atomic
number 45.
f rhodium m
e rodio m
i rodio m
n rhodium n
d Rhodium n

6196 RICHARDSON-DUSHMAN   ma
     EQUATION
The equation relating current density
with temperature, surface state and work
function in thermionic emission.
f équation f de Richardson-Dushman
e ecuación f de Richardson-Dushman
i equazione f di Richardson-Dushman
n vergelijking van Richardson en Dushman
d Richardson-Dushmansche Gleichung f

6197 RICHARDSON PLOT   ma
By plotting the logarithm of the thermionic
current per square Kelvin degree, i.e.
$\log I/T^2$, against the reciprocal of
absolute temperature, a straight line is
obtained whose slope is a measure of the
activation energy involved.
f caractéristique f de Richardson
e característica f de Richardson
i caratteristica f di Richardson
n richardsonkarakteristiek
d Richardson-Kennlinie f

6198 RICHETITE   mi
An ore containing lead and uranium.
f richetite f
e richetita f
i richetite f
n richetiet n
d Richetit m

6199 RIG   ge
A composite device.
f équipement m
e equipo m
i astuccio m
n opstelling
d Aufstellung f

6200 RING COUNTER   ct
A loop of interconnected bistable elements
such that one and only one is in a specified
state at any given time and such that, as
input signals are counted, the position of
the one specified state moves in an ordered
sequence around the loop.
f compteur m annulaire
e contador m anular
i contatore m anulare
n ringteller
d Ringzähler m

6201 RING FOCUSING   sp
In particle spectrometers, the lens-like
action of a suitably designed field whereby
particles which diverge at first will come
together again on a circle.

f focalisation f annulaire
e enfoque m anular, focalización f anular
i focalizzazione f anulare
n ringfocussering
d Ringfokussierung f

6202 RING SCALER,   ct
     RING SCALING CIRCUIT
A multistable scaling circuit consisting
of any number of stages equal to the
desired scaling factor, arranged in a
circle so that a special state is present
in one stage, and each input pulse causes
this state to transfer one unit around the
circle.
f échelle f en anneau
e circuito m anular de escalímetro
i circuito m numeratore d'impulsi anulare
n ringdeelschakeling
d Untersetzer m in Ringschaltung

6203 RIOMETER   me
Apparatus for detecting the arrival near
earth of ionized materials or ionizing
radiation originated in the sun, by
measuring the absorption in the atmosphere
of the cosmic radio noise.
f riomètre m
e riómetro m
i riometro m
n riometer
d Riometer n

6204 ROD   fu
A relatively long and slender body of
material used in or in conjunction with
a nuclear reactor.
f barre f
e barra f
i barra f
n staaf
d Stab m

6205 ROD ASSEMBLY   rt
The complete set of uranium rods in a
reactor.
f ensemble m de barres
e conjunto m de barras
i complesso m di barre
n staafopstelling
d Stabanordnung f

6206 ROD BANK   sa
A number of control or safety rods
firmly grouped together.
f groupe m de barres
e grupo m de barras
i gruppo m di barre
n stavengroep
d Stäbengruppe f

6207 ROD DROP EXPERIMENT   cl, te
Experiment for measuring the control
effected by the free fall of a rod in the
system.
f expérience f de barre descendante
e experimento m de barra bajante

i prova *f* a caduta d'una barra
n staafvalproef
d Stabfallprobe *f*

6208 ROD LATTICE   rt
A lattice in a nuclear reactor composed
of rods.
f réseau *m* de barres
e celosía *f* de barras
i reticolo *m* di barre
n staafrooster *n*
d Stabgitter *n*

6209 ROD POSITION INDICATOR   cl, sa
f indicateur *m* de la position d'une barre
e indicador *m* de la posición de una barra
i indicatore *m* della posizione d'una barra
n staafstandindicator
d Stabstellenzeiger *m*

6210 ROGERSITE   mi
A decomposition product of samarskite.
f rogersite *f*
e rogersita *f*
i rogersite *f*
n rogersiet *n*
d Rogersit *m*

6211 ROGOVSKI COILS, see 5279

6212 ROLL BONDING, see 5079

6213 ROLL SWAGING,   mg
SWAGING
Swaging method used in manufacturing fuel
plates.
f étampage *m* rotatif
e estampado *m* rotatorio
i martellatura *f* rotante
n roterend hameren *n*
d rotierend Hämmern *n*

6214 RÖNTGEN   xr
The unit of exposure of X- or gamma
radiation.
f röntgen *m*
e röntgen *m*
i röntgen *m*
n röntgen
d Röntgen *n*

6215 RÖNTGEN APPARATUS,   xr
X-RAY APPARATUS
The assembly of electrical devices used
to produce röntgen rays.
f appareil *m* à rayons X
e aparato *m* de rayos X
i apparecchio *m* per raggi X
n röntgenapparaat *n*
d Röntgenapparat *m*

6216 RÖNTGEN DACTYLOGRAM   xr
A radiograph of the hands.
f dactylogramme *m* radiologique
e röntgenodactilograma *m*
i röntgenodattilogramma *m*
n röntgendactylogram *n*
d Röntgendaktylogramm *n*

6217 RÖNTGEN EQUIVALENT MAN, see 6114

6218 RÖNTGEN EQUIVALENT PHYSICAL,
see 6127

6219 RÖNTGEN PER HOUR AT 1 METRE(ER),
see 6194

6220 RÖNTGEN RAYS,   xr
X-RAYS
Electromagnetic radiation of wavelength
less than about 100 angstroms.
f rayons *pl* X
e rayos *pl* X
i raggi *pl* X
n röntgenstralen *pl*
d Röntgenstrahlen *pl*

6221 RÖNTGENCINEMATOGRAPHY, see 1102

6222 RÖNTGENOGRAM, see 5842

6223 RÖNTGENOGRAPHY   ra, xr
Radiography by means of röntgen rays.
f röntgenographie *f*
e röntgenograffa *f*
i röntgenografia *f*
n röntgenfotografie
d Röntgenographie *f*

6224 RÖNTGENOLOGY   xr
That part of radiology which pertains to
X-rays.
f röntgenblogie *f*
e röntgenología *f*
i röntgenologia *f*
n röntgenologie
d Röntgenologie *f*

6225 RÖNTGENOSCOPE, see 2742

6226 RÖNTGENOSCOPY, see 2744

6227 RÖNTGENTHERAPY,   xr
X-RAY THERAPY
The treatment of disease by X-rays.
f röntgenthérapie *f*,
thérapie *f* par rayons X
e röntgenoterapia *f*,
terapia *f* por rayos X
i röntgenterapia *f*, terapia *f* con raggi X
n röntgentherapie
d Röntgentherapie *f*

6228 ROSENBLUM DETECTOR,   ra
SPARK DETECTOR
A radiation detector in which the passage,
between electrodes, of a strongly ionizing
particle produces a spark accompanied
by a voltage pulse of a measurable amplitude.
f détecteur *m* à étincelle,
détecteur *m* Rosenblum
e detector *m* de chispa,
detector *m* Rosenblum
i rivelatore *m* a scintilla,
rivelatore *m* Rosenblum
n detector van Rosenblum, vonkdetector

d Funkendetektor *m*,
   Rosenblum-Detektor *m*

6229 ROTARY COLUMN   ch
A column containing a rapidly rotating rod
as a mixing implement.
f colonne *f* à rotation
e columna *f* de rotación
i colonna *f* a rotazione
n draaiingskolom
d Drehungskolonne *f*, Drehungsturm *m*

6230 ROTATING-ANODE TUBE   xr
An X-ray tube in which the anode rotates
continuously.
f tube *m* à anode rotative
e tubo *m* de ánodo rotativo
i tubo *m* ad anodo rotativo
n buis met draaiende anode
d Drehanodenröhre *f*

6231 ROTATING-CRYSTAL METHOD   cr
The examination of the diffracted beams
from a single crystal irradiated by monochromatic radiation, the crystal being
usually rotated about an axis perpendicular
to the incident beam.
f méthode *f* à cristal rotatoire
e método *m* de cristal rotatorio
i metodo *m* a cristallo rotativo
n methode met draaiend kristal
d Drehkristallverfahren *n*

6232 ROTATING FROZEN SEAL   cd
A seal formed by allowing liquid metal to
leak along the rotating shaft to an annular
space around the shaft. By cooling this the
metal is frozen except for a thin film
around the rotating shaft.
f scellement *m* axial congélé rotatoire
e cerrado *m* axial congelado rotatorio
i tenuta *f* ermetica congelata rotativa
n draaibare bevroren asafsluiting
d erfrorener drehbarer Achsenverschluss *m*

6233 ROTATING SHIELD   cd, rt
A shield in a nuclear reactor rotating
around the core and used during refuelling
operations.
f écran *m* rotatif
e pantalla *f* rotativa
i schermo *m* rotativo
n draaibaar scherm *n*
d Drehschild *m*

6234 ROTATION PHOTOGRAPH   cr, ph
The photographic record of the diffracted
beams produced when a slender beam of
X-rays is directed on a rotating single
crystal.
f photographie *f* de cristal rotatoire
e fotografía *f* de cristal rotatorio
i fotografia *f* di cristallo rotativo
n opname van draaiend kristal
d Drehkristallaufnahme *f*

6235 ROTATION THERAPY   xr
Radiation therapy during which either the

patient revolves in front of the source of
radiation or the source revolves around the
patient.
f cyclothérapie *f*
e cicloterapia *f*
i terapia *f* rotatoria
n rotatiebestraling
d Therapie *f* mit Rotationsbestrahlung

6236 ROTATIONAL CONTROL   con
      ELECTROMECHANISM
An electromechanism designed to perform
a rotational motion of one or several
nuclear reactor control elements.
f électromécanisme *m* rotatif de commande
e electromecanismo *m* rotativo de regulación
i elettromeccanismo *m* rotativo di comando
n elektromechanisme *n* voor draaiende
   regelelementen
d elektromechanischer Drehantrieb *m*
   für Steuerstäbe

6237 ROTATIONAL FINE SPECTRUM   sp
The fine structure of the spectrum supposed
to be due to the rotation of the nucleus in
conformity with quantum conditions.
f structure *f* fine due à la rotation
e estructura *f* fina debida a la rotación
i struttura *f* fina dovuta alla rotazione
n draaiingsfijnstructuur
d Drehungsfeinstruktur *f*

6238 ROTATIONAL QUANTUM NUMBER   qm
A quantum number determining the total
angular quantum number of a particle,
exclusive of spin.
f nombre *m* quantique relatif à une impulsion
   de rotation
e número *m* cuántico relativo a un impulso
   de rotación
i numero *m* quantico relativo ad un impulso
   di rotazione
n quantumgetal *n* van het impulsmoment
d Drehimpulsquantenzahl *f*,
   Rotationsquantenzahl *f*

6239 ROTATIONAL TRANSFORM   pp
A twist of a magnetic field so that the lines
of force do not close upon themselves.
f transformation *f* rotationnelle
e transformación *f* rotacional
i trasformazione *f* rotazionale
n rotatietransformatie
d Rotationstransformation *f*

6240 ROUGH VACUUM   vt
A vacuum with pressure between 760 and 1
Torr.
f vide *m* grossier
e vacío *m* aproximado
i vuoto *m* grossolano
n ruw vacuüm *n*
d Grobvakuum *n*

6241 ROUTINE REPLACEMENT OF   fu
      FUEL ELEMENTS
In a nuclear reactor, the normal replacement of depleted fuel elements by new ones.
f substitution *f* courante du combustible

e   substitución *f* normal del combustible
i   sostituzione *f* d'abitudine del combustibile
n   routine-verwisseling van splijtstof-
    elementen
d   routinemässiger Brennstoffwechsel *m*

6242 RUBBER BOOT SEAL   cd
A type of seal using a rubber tube used
for enclosing joints, such as hinges, in a
mechanism, which require lubrication.
f   scellement *m* à tube de caoutchouc
e   cerrada *f* de tubo de goma
i   tenuta *f* ermetica a tubo di gomma
n   rubberbuisafsluiting
d   Gummirohrverschluss *m*.

6243 RUBIDIUM   ch
Metallic element, symbol Rb, atomic
number 37.
f   rubidium *m*
e   rubidio *m*
i   rubidio *m*
n   rubidium *n*
d   Rubidium *n*

6244 RULE OF MAXIMUM   np
       MULTIPLICITY
An empirical statement that the energy of
interaction between the electrons in any
atom is at a minimum when their initial
spin is greatest.
f   loi *f* de la multiplicité maximale
e   ley *f* de la multiplicidad máxima
i   legge *f* della moltiplicità massima
n   wet van de maximale meervoudigheid
d   Regel *f* der maximalen Vermehrung

6245 RUNAWAY   rt
An increase in power or reactivity of a
nuclear reactor that cannot be controlled
by the normal control system although it
might possibly be terminated safely by
the emergency shutdown system.
f   emballement *m*
e   embalamiento *m*
i   perdita *f* di controllo
n   uit de hand lopen
d   Durchgehen *n*

6246 RUNAWAY ELECTRONS   np
A term referring to a class of electrons
in an ionized gas under the influence
of an externally applied electric field.
f   électrons *pl* de fuite, électrons *pl* découplés
e   electrones *pl* de fuga
i   elettroni *pl* di fuga
n   ontsnappingselektronen *pl*
d   Entkommungselektronen *pl*

6247 RUNAWAY REACTION,   np, rt
       UNCONTROLLABLE REACTION
A thermonuclear reaction which can be
initiated but not controlled or stopped.

f   réaction *f* d'emballement,
    réaction *f* incontrôlée
e   reacción *f* irrefrenable,
    reacción *f* no controlada
i   reazione *f* incontrollata,
    reazione *f* non controllata
n   onbeheerste reactie
d   unkontrollierte Reaktion *f*

6248 RUPTURE MODULUS, see 4494

6249 RUSSEL-SAUNDERS COUPLING,
       see 4155

6250 RUTHENIUM   ch
Metallic element, symbol Ru, atomic
number 44.
f   ruthénium *m*
e   rutenio *m*
i   rutenio *m*
n   rhutenium *n*
d   Rhutenium *n*

6251 RUTHERFORD ATOM   np
A model of the atom which visualizes a
concentrated nucleus carrying a positive
charge and representing the major portion
of the atom's mass.
f   atome *m* de Rutherford
e   átomo *m* de Rutherford
i   atomo *m* di Rutherford
n   atoommodel *n* volgens Rutherford
d   Rutherfordsches Atommodell *n*

6252 RUTHERFORD DISPERSION   np
       FORMULA,
       RUTHERFORD SCATTERING
       FORMULA
The classical expression for the effective
cross section, about the nucleus of an atom,
which an alpha particle must enter in order
to be scattered into a solid angle at a
deviation from the initial direction.
f   formule *f* de diffusion de Rutherford
e   fórmula *f* de dispersión de Rutherford
i   formula *f* di deviazione di Rutherford
n   verstrooiingsformule van Rutherford
d   Rutherfordsche Streuungsformel *f*

6253 RUTHERFORDINE, see 1777

6254 RYDBERG CONSTANT,   sp
       RYDBERG WAVE NUMBER
A constant which appears in the expression
for wave numbers of various atomic
spectra.
f   constante *f* de Rydberg
e   constante *f* de Rydberg
i   costante *f* di Rydberg
n   constante van Rydberg, rydbergconstante
d   Rydberg-Konstante *f*

S

6255 S ELECTRON np
An electron having an orbital angular
momentum quantum number of zero.
f électron *m* S`
e electrón *m* S
i elettrone *m* S
n S-elektron *n*
d S-Elektron *n*

6256 S EVENT cr, np
A phenomenon interpretable as the decay
of a K meson or a hyperon while at rest.
f événement *m* S
e acontecimiento *m* S
i evento *m* S
n S-proces *n*
d S-Ereignis *n*

6257 S STATE np
A state of zero orbital angular momentum.
f état *m* S
e estado *m* S
i stato *m* S
n S-toestand
d S-Zustand *m*

6258 S VALUE rt
The percentage increase in volume, for
1 per cent burn-up, resulting from
swelling.
f valeur *f* S
e valor *m* S
i valore *m* S
n S-waarde
d S-Wert *m*

6259 SABUGALITE mi
A rare secondary mineral containing about
54 % of U.
f sabugalite *f*
e sabugalita *f*
i sabugalite *f*
n sabugaliet *n*
d Sabugalit *m*

6260 SADDLE POINT np
In the liquid drop model of fission, on a
three-dimensional plot of potential energy
against degrees of nuclear distortion, the
point which shows the minimum energy a
nucleus must possess in order to become
sufficiently distorted to undergo fission.
f col *m*, point *m* de minimum
e punto *m* de silla
i punto *m* di sella
n zadelpunt *n*
d Sattelpunkt *m*

6261 SAFE CONCENTRATION rt
For a fissile (fissionable) substance in
solution in an infinite medium, the product
of the critical concentration by an
appropriate safety coefficient.

f concentration *f* sûre
e concentración *f* segura
i concentrazione *f* sicura
n veilige concentratie
d sichere Konzentration *f*

6262 SAFE DISTANCE, sa
         SAFETY DISTANCE
The distance from a radiation source
beyond which the safety conditions issued
by the competent staff are satisfied.
f distance *f* de sécurité
e distancia *f* de seguridad
i distanza *f* di sicurezza
n veilige afstand
d sicherer Abstand *m*

6263 SAFE GEOMETRY rt
A geometry in which a system containing
fissile (fissionable) material is incapable
of supporting a self-sustaining nuclear
chain reaction.
f géométrie *f* sûre
e geometría *f* segura
i geometria *f* sicura
n veilige geometrie
d sichere Geometrie *f*

6264 SAFE HANDLING CALCULATOR,
         see 3756

6265 SAFE MASS rt
The product of the critical mass by an
appropriate safety coefficient.
f masse *f* sûre
e masa *f* segura
i massa *f* sicura
n veilige massa
d sichere Masse *f*

6266. SAFETY ASSEMBLY sa
An assembly designed to receive information
from various assemblies measuring the
conditions of a nuclear reactor and able to
initiate automatic action on one or several
safety members in order to ensure
integrity of the nuclear reactor.
f ensemble *m* de sécurité
e conjunto *m* de seguridad
i complesso *m* di sicurezza
n veiligheidsopstelling
d Sicherheitsanordnung *f*

6267 SAFETY BANK, cl, sa
         SAFETY ROD BANK
A set of firmly connected safety rods.
f groupe *m* de barres de sécurité
e grupo *m* de barras de seguridad
i gruppo *m* di barre di sicurezza
n samenstel *n* van veiligheidsstaven
d Aggregat *n* von Sicherheitsstäben

6268 SAFETY CHANNEL AMPLIFIER ec
Used in association with uncompensated
ionization chambers or current output
fission chambers in the control of reactor
level alarms.
f amplificateur *m* statique de sécurité
e amplificador *m* estático de seguridad
i amplificatore *m* statico di sicurezza
n statische veiligheidsversterker
d statischer Sicherheitsverstärker *m*

6269 SAFETY CIRCUIT, see 3105

6270 SAFETY ELEMENT,
SAFETY MEMBER, see 2227

6271 SAFETY INJECTION SYSTEM sa
A system for injecting cooling liquid when
the normal cooling liquid is leaking out
or has leaked out.
f système *m* d'injection de sécurité
e sistema *m* de inyección de seguridad
i impianto *m* d'allagamento di sicurezza
n noodkoelmiddelinjectiesysteem *n*
d Notkühlmitteleinspritzverfahren *n*

6272 SAFETY MECHANISM sm
A mechanism designed to initiate or
accomplish a rapid decrease in reactivity,
for example by movement of safety
elements.
f mécanisme *m* de sécurité
e mecanismo *m* de seguridad
i meccanismo *m* di sicurezza
n veiligheidsmechanisme *n*
d Sicherheitsvorrichtung *f*

6273 SAFETY REGULATIONS sa
The total of measures taken for the safety
of personnel and apparatus in a nuclear
plant.
f instructions *pl* de sécurité
e instrucciones *pl* de seguridad
i prescrizioni *pl* di sicurezza
n veiligheidsvoorschriften *pl*
d Sicherheitsvorschriften *pl*

6274 SAFETY REPORT, see 3150

6275 SAFETY ROD, rt, sa
SCRAM ROD, SHUT-DOWN ROD
A rod controlling a large amount of
reactivity and capable of bringing the
reactor below critical in a very short
time.
f barre *f* de sécurité
e barra *f* de seguridad
i barra *f* di sicurezza
n veiligheidsstaaf
d Abschaltstab *m*, Notstab *m*,
Sicherheitsstab *m*

6276 SAFETY SIGNAL sa
A signal given when a specific operation
may be carried out.
f signal *m* d'autorisation
e señal *f* de seguir adelante
i segnale *m* di consenso

n vrijgeefsignaal *n*
d Freigabesignal *n*

6277 SAFETY SWITCH, rt, sa
SCRAM BUTTON,
SHUT-DOWN SWITCH
A button which, in addition to dropping the
rods, resets the safety circuit after it
has been tripped.
f commutateur *m* de sécurité
e interruptor *m* de seguridad
i interruttore *m* di sicurezza
n veiligheidsschakelaar
d Notschalter *m*, Schnellschlussschalter *m*

6278 SAGITTA METHOD ph
A method of determining the multiple
scattering of a track in an emulsion by
measuring at regular intervals the
distance between the track and a straight
line.
f méthode *f* de la flêche
e método *m* de la sagita
i metodo *m* della freccia
n pijlmethode
d Pfeilverfahren *n*

6279 SALEITE mi
Uranium ore belonging to the phosphate
range.
f saléite *f*
e saleita *f*
i saleite *f*
n saleïet *n*
d Saleit *m*

6280 SALFATORY EVOLUTION, see 3131

6281 SALPINGOGRAPHY, see 3374

6282 SALT SCREEN, see 2743

6283 SALTING-OUT ch
Decrease of solubility inherent to a salt
when another soluble salt is added to the
solution.
f relargage *m*
e precipitación *f* por salazón
i precipitazione *f* per addizione di sale
n uitzouten *n*
d Aussalzen *n*

6284 SALTING-OUT AGENT ch
f agent *m* relargant, relargant *m*
e agente *m* de precipitación por adición de sal
i agente *m* di precipitazione per addizione
di sale
n uitzoutingsagens *n*
d Aussalzungsstoff *m*

6285 SAMARIUM ch
Rare earth element, symbol Sm, atomic
number 62.
f samarium *m*
e samario *m*
i samario *m*
n samarium *n*
d Samarium *n*

6286 SAMARIUM POISONING   rt
Poisoning of the reactor by the isotope
samarium-149.
f empoisonnement *m* samarium
e envenenamiento *m* de samario
i avvelenamento *m* da samario
n samariumvergiftiging
d Samariumvergiftung *f*

6287 SAMARSKITE   mi
A mineral containing lead, niobium,
tantalum and uranium.
f samarskite *f.*
e samarskita *f*
i samarskite *f*
n samarskiet *n*
d Samarskit *m*

6288 SAMIRESITE   mi
A uranium ore of the oxide class.
f samirésite *f*
e samiresita *f*
i samiresite *f*
n samiresiet *n*
d Samiresit *m*

6289 SAMPLED DATA CONTROL,   co
SAMPLING CONTROL
A control in which the difference between
the independent variable and the controlled
variable is measured and correction made
only at intermittent intervals.
f commande *f* à échantillonnage
e mando *m* por muestreo,
regulación *f* por muestreo
i controllo *m* a segnali campionati
n aftastregeling
d Abtastregelung *f*

6290 SAMPLING   te
Selecting a small statistically determined
portion of the total group under consideration for tests used to infer the value of
one or several characteristics of the
entire group.
f échantillonnage *m*
e muestreo *m*
i prelevamento *m* di campioni
n monsterneming
d Probenahme *f*

6291 SAND FILTER   rw
f filtre *m* à sable
e filtro *m* de arena
i filtro *m* a sabbia, filtro *m* a terra
n zandfilter *n*
d Sandfilter *m*

6292 SAND FILTRATION   rw
The use of sand filters in the treatment
of radioactive waste.
f filtration *f* par sable
e filtración *f* por arena
i filtrazione *f* per terra
n zandfiltratie
d Sandfiltrierung *f*

6293 SANDWICH IRRADIATION   ra
The irradiation of tissues from opposite
sides.
f irradiation *f* en sandwich
e irradiación *f* en sandwich
i irradiazione *f* in sandwich
n sandwichbestraling
d Sandwich-Bestrahlung *f*

6294 SANDWICHING   ph
In the photographic emulsion technique,
a system composed of layers of emulsion
between which are interposed thin layers
of another material, in which some process
is to be studied.
f sandwich *n*
e método *m* sandwich
i metodo *m* sandwich
n sandwichopstelling
d Sandwich-Anordnung *f*

6295 SARCOMA   md
Malignant neoplasma composed of cells
initiating the appearance of the supportive
and lymphatic tissues.
f sarcome *m*
e sarcoma *m*
i sarcoma *m*
n kwaadaardig bindweefselgezwel *n*,
sarcoom *n*
d bösartige Bindegewebsgeschwulst *f*,
Sarkom *n*

6296 SARGENT CURVES,   br
SARGENT DIAGRAM
A plot of log λ versus log E for beta-
emitters where λ is the disintegration
constant and E is the upper energy limit
in the beta spectrum.
f courbes *pl* de Sargent,
diagramme *m* de Sargent
e curvas *pl* de Sargent,
diagrama *m* de Sargent
i curve *pl* di Sargent,
diagramma *m* di Sargent
n sargentdiagram *n*, sargentkrommen *pl*
d Sargent-Diagramm *n*, Sargent-Kurven *pl*

6297 SATELLITE PULSE   ct
A pulse following a scintillation pulse
after 0.2 to 2 μ sec. and showing a ragged
bump predominantly caused by ion feedback
between cathode and first dynode.
f impulsion *f* satellite
e impulso *m* satélite
i impulso *m* satellite
n satellietpuls
d Satellitimpuls *m*

6298 SATURATED ACTIVITY,   np, rt
SATURATED RADIOACTIVITY
The maximum activity obtained by
activation in a definite flux in a nuclear
reactor.
f activité *f* à saturation
e actividad *f* de saturación

i attività *f* a saturazione
n verzadigingsactiviteit
d Sättigungsaktivität *f*

6299 SATURATION												ic
Of an ionization chamber, the condition
reacted when the potential difference
between the electrodes is such that
practically all the ions formed by the
radiation are collected and no ions are
produced by collision.
f saturation *f*
e saturación *f*
i saturazione *f*
n verzadiging
d Sättigung *f*

6300 SATURATION ACTIVITY									is
The amount of an active isotope present in
a given sample after irradiation for an
infinite time, when the rate of decay
becomes equal to the rate of production.
f activité *f* de saturation
e actividad *f* de saturación
i attività *f* di saturazione
n verzadigingsactiviteit
d Sättigungsaktivität *f*

6301 SATURATION CURRENT									ic
Of a current ionization chamber, the
ionization current obtained when the applied
voltage is sufficiently high for substantially all the ions to be collected before
gas multiplication occurs.
f courant *m* de saturation
e corriente *f* de saturación
i corrente *f* di saturazione
n verzadigingsstroom
d Sättigungsstrom *m*

6302 SATURATION CURVE									ic
Of a current ionization chamber, a curve
characteristic of the variation of output
current against the voltage applied,
permitting the determination of saturation
current and voltage.
f courbe *f* de saturation
e curva *f* de saturación
i curva *f* di saturazione
n verzadigingskromme
d Sättigungskurve *f*

6303 SATURATION POTENTIAL,							ic
         SATURATION VOLTAGE
The minimum voltage necessary to obtain
saturation current in a current ionization
chamber.
f tension *f* de saturation
e tensión *f* de saturación
i tensione *f* di saturazione
n verzadigingsspanning
d Sättigungsspannung *f*

6304 SATURN												rt
A thermonuclear device of the magnetic
mirror design which has an equatorial
ring plasma source.

f appareil *m* saturne
e aparato *m* saturno
i apparecchio *m* saturno
n saturnustoestel *n*
d Saturn-Apparat *m*

6305 SAUSAGE INSTABILITY, see 3855

6306 SCALAR FORCE											np
A nuclear force depending on distance
only.
f force *f* scalaire
e fuerza *f* escalar
i forza *f* scalare
n scalaire kracht
d Skalarkraft *f*

6307 SCALAR FUNCTION										ma
A scalar quantity that has a definite value
for each value of some other scalar
quantity.
f fonction *f* scalaire
e función *f* escalar
i funzione *f* scalare
n scalaire functie
d Skalarfunktion *f*

6308 SCALAR QUANTITY										ma
A quantity that has only magnitude, such as
resistance, time or temperature but not
direction.
f grandeur *f* scalaire, scalaire *f*
e cuantidad *f* escalar, magnitud *f* escalar
i grandezza *f* scalare, quantità *f* scalare
n scalaire grootheid, scalar
d Skalar *m*, Skalargrösse *f*

6309 SCALE OF EIGHT CIRCUIT								ct
A scaling circuit with a scaling factor 8.
f circuit *m* d'échelle de huit
e circuito *m* desmultiplicador de relación
   8:1
i circuito *m* numeratore a rapporto 8:1
n achtdeler
d Achteruntersetzerschaltung *f*

6310 SCALE OF TEN CIRCUIT									ct
A scaling circuit with factor 10.
f circuit *m* d'échelle de dix
e circuito *m* desmultiplicador de relación
   10:1
i circuito *m* numeratore a rapporto 10:1
n tiendeler
d dekadische Untersetzerschaltung,
   Zehneruntersetzerschaltung *f*

6311 SCALE OF TWO CIRCUIT									ct
A scaling circuit with factor two.
f circuit *m* d'échelle de deux
e circuito *m* desmultiplicador de relación
   2:1
i circuito *m* numeratore a rapporto 2:1
n tweedeler
d Zweieruntersetzerschaltung *f*

6312 SCALER												ct
An apparatus for counting electrical pulses

and containing one or more scaling circuits.
f   échelle f de comptage
e   escala f de recuento
i   unità f per conteggio
n   pulsteller
d   Zählgerät n

6313 SCALING   ct
Counting pulses with a scaler when the
pulses occur too fast for direct counting
by conventional means.
f   démultiplication f d'impulsions
e   desmultiplicación f de impulsos
i   demoltiplicazione f d'impulsi
n   pulsdeling
d   Impulsteilung f

6314 SCALING CIRCUIT   ct
A device that produces an output pulse
whenever a prescribed number of pulses
have been received.
f   circuit m d'échelle,
    circuit m démultiplicateur
e   circuito m desmultiplicador
i   circuito m numeratore d'impulsi
n   deelschakeling
d   Untersetzerschaltung f, Zählschaltung f

6315 SCALING FACTOR,   ct
      SCALING RATIO
A factor, the value of which is the number
of pulses required at the input of a
scaling circuit in order to produce an
output pulse.
f   facteur m d'échelle
e   factor m de contador de impulsos,
    factor m de escalfmetro
i   fattore m di numeratore d'impulsi
n   deelfactor
d   Untersetzerfaktor m, Zählfaktor m

6316 SCALING UNIT   ct
A basic function unit containing a scaling
circuit.
f   élément m démultiplicateur
e   unidad f escala
i   elemento m demoltiplicatore
n   deeleenheid
d   Zählstufe f

6317 SCALPING   mg
The removal of the surface layers of metal
ingots, billets or slabs by machining or
other means prior to finishing.
f   enlèvement m de la croûte superficielle
e   descrostación f
i   eliminazione f della crosta superficiale
n   scalperen n,
    verwijderen n van de oppervlakhuid
d   Beseitigung f der Oberflächenkruste

6318 SCALPING   mi
A screening operation for crushed ore, in
which 85 to 95 percent of the material
passes through the screen.
f   tamisage m préliminaire
e   cribado m preliminar

i   vagliatura f grossa
n   grofzeven n
d   Grobsieben n

6319 SCANDIUM   ch
Metallic element, symbol Sc, atomic
number 21.
f   scandium m
e   escandio m
i   scandio m
n   scandium n
d   Skandium n

6320 SCANNING   me
The sequential measurement of some
quantity at a number of positions on an
area or in a volume.
f   balayage m, exploration f
e   exploración f
i   scansione f
n   aftasten n, onderzoeken n
d   Abtasten n, Untersuchen n

6321 SCANNING GAMMA   sp
      SPECTROMETER
Used for pulse height analysis of detector
outputs.
f   spectromètre m gamma à canal mobile
e   espectrómetro m gamma de canal móvil
i   spettrometro m gamma a canale mobile
n   gammaspectrometer met aftastkanaal
d   Gammaspektrometer n mit Abtastkanal

6322 SCANNING GEAR   rt, te
The mechanism used to collect gas
samples from the fuel channels for any
volatile fission product and so detect the
presence of a punctured fuel element can.
f   appareil m d'échantillonnage
e   aparato m de muestreo
i   apparecchio m per prelevamentore di
    campioni
n   gasmonstersteker
d   Gasprobenahmegerät n

6323 SCATTER ABSORPTION   np
      COEFFICIENT
The linear, mass, atomic or electronic
absorption coefficient that represents the
decrease in intensity of the beam through
all scattering collisions, but not through
any absorptive process.
f   coefficient m d'absorption de diffusion
e   coeficiente m de absorción de dispersión
i   coefficiente m d'assorbimento di
    deviazione
n   verstrooiingsabsorptiecoëfficiënt
d   Streuungsabsorptionskoeffizient m

6324 SCATTER UNSHARPNESS   np
Unsharpness due to radiation scattered by
an irradiated object, intensifying screen
or fluorescent screen.
f   flou m dû à la diffusion
e   borrosidad f debida a la dispersión
i   sfocato m dovuto alla deviazione
n   onscherpte door verstrooiing
d   Unschärfe f durch Streuung

6325 SCATTERED BEAM ra
A beam that during its passage through a
substance has been deviated in direction.
f faisceau *m* diffusé
e haz *m* dispersado
i fascio *m* deviato
n verstrooide bundel
d gestreuter Strahl *m*, Streustrahl *m*

6326 SCATTERED NEUTRON np
A neutron which during its passage through
a substance has been deviated in direction.
f neutron *m* diffusé
e neutrón *m* dispersado
i neutrone *m* deviato
n verstrooid neutron *n*
d gestreutes Neutron *n*, Streuneutron *n*

6327 SCATTERED PARTICLE np
A particle which, during its passage
through a substance, has been deviated in
direction.
f particule *f* diffusée
e partícola *f* dispersada
i particella *f* deviata
n verstrooid deeltje *n*
d Streuteilchen *n*

6328 SCATTERED RADIATION np, ra
Radiation which, during its passage through
a substance, has been deviated in direction.
f rayonnement *m* diffusé
e radiación *f* dispersada
i radiazione *f* deviata
n verstrooide straling
d Streustrahlung *f*

6329 SCATTERED RADIATION me
PROXIMITY INDICATOR
A proximity indicator using scattered
ionizing radiation.
f signaleur *m* de proximité par rayonnement
ionisante diffusé
e indicador *m* de proximidad por radiación
ionizante dispersada
i segnalatore *m* di prossimità a radiazione
ionizzante a deviazione
n nabijheidsindicator berustend op
verstrooide ioniserende straling
d Wegindikator *m* durch ionisierende
Streustrahlung

6330 SCATTERED WAVE np
A wave which during its passage through
a substance has been deviated in direction.
f onde *f* diffusée
e onda *f* dispersada
i onda *f* deviata
n verstrooide golf
d gestreute Welle *f*, Streuwelle *f*

6331 SCATTERED X-RAYS xr
X-Rays which, during their passage through
a substance, have been deviated in direction
and modified by an increase in wavelength.
f rayons *pl* X diffusés
e rayos *pl* X dispersados

i raggi *pl* X deviati
n verstrooide röntgenstralen *pl*
d Streuröntgenstrahlen *pl*

6332 SCATTERING np
A process in which a change in direction
or energy of an incident particle is caused
by a collision with a particle or a system
of particles.
f diffusion *f*
e dispersión *f*
i deviazione *f*
n verstrooiing
d Streuung *f*

6333 SCATTERING AMPLITUDE np
A quantity closely related to the intensity
of scattering of a wave by a central force
field such as that of a nucleus, in nuclear
potential scattering.
f amplitude *f* de diffusion
e amplitud *f* de dispersión
i ampiezza *f* di deviazione
n verstrooiingsamplitude
d Streuungsamplitude *f*

6334 SCATTERING ANGLE np
The angle between the initial and final
lines of motion of a scattered particle.
f angle *m* de diffusion
e ángulo *m* de dispersión
i angolo *m* di deviazione
n verstrooiingshoe
d Streuwinkel *m*

6335 SCATTERING ATTENUATION np, ra
COEFFICIENT
That part of the total attenuation
coefficient that is attributable to scattering.
f coefficient *m* d' atténuation de diffusion
e coeficiente *m* de atenuación de dispersión
i coefficiente *m* d'attenuazione di deviazione
n verzwakkingscoëfficiënt voor verstrooiing
d Schwächungskoeffizient *m* für Streuung

6336 SCATTERING CENTER (US), np
SCATTERING CENTRE (GB)
Any microscopic or submicroscopic
particle, molecule, atom or nucleus which
causes scattering of photon or particulate
radiation.
f centre *m* de diffusion
e centro *m* de dispersión
i centro *m* di deviazione
n verstrooiingscentrum *n*
d Streuzentrum *n*

6337 SCATTERING COEFFICIENT np
A proportionality constant which
corresponds exactly to absorption
coefficient, except that the process involved
is scattering rather than absorption.
f coefficient *m* de diffusion
e coeficiente *m* de dispersión
i coefficiente *m* di deviazione
n verstrooiingscoëfficiënt
d Streukoeffizient *m*

6338 SCATTERING CROSS SECTION,
      see 1115

6339 SCATTERING FREQUENCY   np
      A frequency present when no forward and
      no backward scattering is counted.
f fréquence f de diffusion
e frecuencia f de dispersión
i frequenza f di deviazione
n verstrooiingsfrequentie
d Streufrequenz f

6340 SCATTERING LENGTH, see 2564

6341 SCATTERING LOSS   np
      The portion of the transmission loss that
      is due to scattering within the medium.
f perte f de diffusion
e pérdida f de dispersión
i perdita f di deviazione
n verstrooiingsverlies n
d Streuverlust m

6342 SCATTERING MEAN FREE PATH   np
      The average scattering length.
f libre parcours m moyen de diffusion
e trayectoria f libre media de dispersión
i percorso m libero medio di deviazione
n gemiddelde verstrooiingsweglengte
d mittlere Streuweglänge f

6343 SCATTERING OF NEUTRONS, see 4746

6344 SCATTERING PROBABILITY   np
      When electrons or atoms collide with
      atoms, the resulting scattered current per
      unit incident current per unit path length,
      per unit pressure at 0^oC, per unit solid
      angle in the direction θ.
f probabilité f de diffusion
e probabilidad f de dispersión
i probabilità f di deviazione
n verstrooiingskans,
   verstrooiingswaarschijnlijkheid
d Streuungswahrscheinlichkeit f

6345 SCAVENGER   ch
      An unselective precipitate to remove from
      solution by adsorption or precipitation a
      large fraction of one or more unwanted
      radionuclides.
f coprécipitant m, entraîneur m
e barredor m
i coprecipitante m
n uitwasmiddel n
d reinigendes Fällungsmittel n

6346 SCAVENGING   ch
      The use of an unselective precipitate to
      remove from solution by adsorption or
      co-precipitation a large fraction of one
      or more unwanted radionuclides.
f coprécipitation f, entraînement m
e depuración f
i coprecipitazione f
n reinigen n, uitwassen n
d Reinigungsfällung f

6347 SCHEIBEL COLUMN,   ch
      STIRRING VANE COLUMN
      A solvent extractor intermediate between
      the mixer settler and the packed column.
f colonne f à ailettes mélangeuses
e columna f de aletas mezcladoras
i colonna f ad alette mescolatrici
n kolom met mengvleugels
d Kolonne f mit Mischflügeln

6348 SCHMIDT LIMITS,   np
      SCHMIDT LINES
      Two lines in the plot of nuclear magnetic
      moment against nuclear spin that show the
      relationship to be expected according to
      the one-particle model.
f lignes pl de Schmidt
e líneas pl de Schmidt
i linee pl di Schmidt
n schmidtlijnen pl
d Schmidt-Linien pl

6349 SCHOEPITE   mi
      A mineral containing uranium sesquioxide.
f schoepite f
e schoepita f
i schoepite f
n schoepiet n
d Schoepit m

6350 SCHOTTKY DEFECT,   cr, np
      VACANCY
      An unoccupied atomic site in the structure
      of a crystal from which an atom is missing.
f défaut m de Schottky, lacune f
e defecto m de Schottky, hueco m de celosía
i difetto m di Schottky, lacuna f di reticolo
n roostergat n, vacature
d Fehlstelle f, Gitterloch n, Leerstelle f

6351 SCHRÖDINGER EQUATION,   ma
      SCHRÖDINGER WAVE EQUATION
      An equation descriptive of the wave motion
      of elementary particles.
f équation f de Schrödinger
e ecuación f de Schrödinger
i equazione f di Schrödinger
n golfvergelijking van Schrödinger
d Schrödinger-Gleichung f

6352 SCHRÖDINGER FUNCTION,   ma
      SCHRÖDINGER WAVE FUNCTION
      A function X of the coordinates that
      determines the state of a system and
      satisfies the Schrödinger wave equation.
f fonction f de Schrödinger
e función f de Schrödinger
i funzione f di Schrödinger
n golffunctie van Schrödinger
d Schrödinger-Funktion f

6353 SCHROECKINGERITE   mi
      A mineral containing sodium, potassium,
      sulphur and uranium.
f schroeckingérite f
e schroeckingerita f
i schroeckingerite f

n schroeckingeriet *n*
d Schroeckingerit *m*

6354 SCINTIGRAM,   md
       SCINTISCAN
    Diagram obtained by means of scintigraphy.
f scintigramme *m*
e escintigrama *m*
i scintigramma *m*
n scintigram *n*
d Szintigramm *n*

6355 SCINTIGRAPHY   md
    A method of visualizing a critical organ
    after introduction of a labelled compound
    in the organism, by scanning the region
    under examination with a scintillation
    detector.
f scintigraphie *f*
e escintigraffa *f*
i scintigrafia *f*
n scintigrafie
d Szintigraphie *f*

6356 SCINTILLATING MATERIAL   ct
    Any substance constituting an appropriate
    medium for the detection of radiation by
    means of scintillation.
f matière *f* scintillante
e material *m* de centelleo,
    material *m* escintilante
i materiale *m* di scintillazione
n scintillerend materiaal *n*
d szintillierendes Material *n*

6357 SCINTILLATION   ct, ra
    A flash of light produced in a phosphor by
    an ionizing event.
f scintillation *f*
e centelleo *m*, escintilación *f*
i scintillazione *f*
n scintillatie
d Szintillation *f*

6358 SCINTILLATION COUNTER, see 5265

6359 SCINTILLATION-COUNTER   ct
       CAESIUM RESOLUTION (GB),
       SCINTILLATION-COUNTER
       CESIUM RESOLUTION (US)
    The scintillation counter energy resolution
    for the gamma-ray or conversion electron
    from c(a)esium -137.
f résolution *f* de compteur à scintillation
    pour Cs-137
e resolución *f* de contador de centelleo
    por Cs-137
i potere *m* di risolvenza di contatore a
    scintillazione per Cs-137
n scheidend vermogen *n* van een
    scintillatieteller voor Cs-137
d Szintillationszählerauflösungsvermögen
    ·*n* für Cs-137

6360 SCINTILLATION-COUNTER HEAD   ct
    The combination of scintillators and photo-
    tubes or photocells to produce electric

pulses in response to ionizing radiation.
f tête *f* de compteur à scintillation
e cabeza *f* de contador de centelleo
i testina *f* di contatore a scintillazione
n scintillatietellerkop
d Szintillationszählerkopf *m*

6361 SCINTILLATION-COUNTER   ct
       TIME DISCRIMINATION
    The smallest interval of time that can
    separate two events individually
    discernible by a given scintillation counter.
f intervalle *m* minimal de temps entre
    deux impulsions
e intervalo *m* mínimo de tiempo entre dos
    impulsos
i intervallo *m* minimo di tempo tra due
    impulsi
n kleinste tijdinterval *n* tussen twee tellen
d kleinstes Zeitintervall *n* für zwei Impulse

6362 SCINTILLATION COUNTING   te
       TEST FOR RADON
    The appliance is immersed in a solution of
    a phosphor in an organic liquid under
    vacuum and the leakage of radon is
    measured by liquid scintillation counting.
f essai *m* de fuites de radon à scintillateur
    liquide
e prueba *f* de fugas de radón con centelleador
    líquido
i prova *f* di fughe di radon a scintillatore
    liquido
n radonlekproef met vloeistofscintillator
d Radonleckprüfung *f* mit Flüssigkeits-
    szintillator

6363 SCINTILLATION CRYSTALS   cr, ct
    Materials used to produce the scintillations
    in scintillation counters.
f cristaux *pl* scintillateurs
e cristales *pl* centelleadores,
    cristales *pl* escintiladores
i cristalli *pl* scintillatori
n scintillatiekristallen *pl*
d Szintillatorkristalle *pl*

6364 SCINTILLATION DECAY TIME   ct
    The time required for the rate of the
    emission after a single excitation to
    decrease from 90 % to 10 % of the maximum
    value.
f temps *m* de décroissance d'une scintillation
e tiempo *m* de decaimiento de un centelleo
i tempo *m* di decadimento d'una scintillazione
n daaltijd van een scintillatie
d Abklingzeit *f* einer Szintillation

6365 SCINTILLATION DETECTOR   ct
    A radiation detector using a medium in
    which a burst of luminescence radiation is
    produced along the path of an ionizing
    particle.
f détecteur *m* à scintillation
e detector *m* de centelleo
i rivelatore *m* a scintillazione
n scintillatiedetector
d Szintillationsdetektor *m*

6366 SCINTILLATION DURATION,   ct
      SCINTILLATION TIME
      The time interval from the emission of the
      first optical photon of a scintillation until
      90 % of the optical photons of the
      scintillation have been emitted.
   f durée f de scintillation
   e duración f de centelleo
   i durata f di scintillazione
   n scintillatietijd
   d Szintillationszeit f

6367 SCINTILLATION PROBE   te
      An apparatus used i.a. in testing
      polonium-210 sources by a wipe test.
   f sonde f à scintillation
   e sonda f de centelleo
   i sonda f a scintillazione
   n scintillatiesonde
   d Szintillationssonde f

6368 SCINTILLATION RISE TIME   ct
      The time required for the rate of
      emission of optical photons of a scintilla-
      tion to increase from 10 % to 90 % of its
      maximum value.
   f temps m de montée d'une scintillation
   e tiempo m de aumento de un centelleo
   i tempo m di salita d'una scintillazione
   n stijgtijd van een scintillatie
   d Anstiegzeit f einer Szintillation

6369 SCINTILLATION   ct, sp
      SPECTROMETER
      A scintillation counter adapted to the study
      of energy distribution.
   f spectromètre m à scintillation
   e espectrómetro m de centelleo
   i spettrometro m a scintillazione
   n scintillatiespectrometer
   d Szintillationsspektrometer n

6370 SCINTILLATOR   ct
      A special kind of phosphor which gives a
      brief flash of light when a fast charged
      particle passes through it.
   f scintillateur m
   e centelleador m, escintilador m
   i scintillatore m
   n scintillator
   d Szintillator m

6371 SCINTILLATOR CONVERSION   ct
      EFFICIENCY
      The ratio of the optical photon energy
      emitted by a scintillator to the incident
      energy of a particle or photon of ionizing
      radiation.
   f rendement m de conversion d'un scintilla-
      teur
   e rendimiento m de conversión de un
      centelleador
   i rendimento m di conversione d'uno
      scintillatore
   n conversierendement n van een scintillator
   d Umwandlungsgrad m eines Szintillators

6372 SCINTILLATOR EXPOSURE   ma
      RATEMETER
      A measuring assembly for the exposure
      rate, in which the detector is a scintillator.
   f débitmètre m d'exposition à scintillateur
   e dosímetro m de centelleador
   i rateometro m d'esposizione a scintillatore
   n exposietempometer met scintillator
   d Dosisleistungsmesser m mit Szintillator

6373 SCINTILLATOR FAST   ct, ma
      NEUTRON FLUXMETER
      An assembly designed to measure fast
      neutron flux density, in which the detector
      is a scintillator.
   f fluxmètre m de neutrons rapides à
      scintillateur
   e flujómetro m de neutrones rápidos de
      centelleador
   i flussometro m di neutroni veloci a
      scintillatore
   n fluxdichtheidsmeter voor snelle neutronen
      met scintillator
   d Gerät n zur Messung der Fluenz schneller
      Neutronen mit Szintillator

6374 SCINTILLATOR PHOTON   ct
      DISTRIBUTION
      The curve giving the number of optical
      photons produced by a scintillator when it
      totally absorbs the energy of a mono-
      energetic particle of photon of ionizing
      radiation.
   f courbe f de répartition des photons d'un
      scintillateur
   e curva f de distribución de los fotones
      de un centelleador
   i curva f di distribuzione dei fotoni d'uno
      scintillatore
   n fotonenverdelingskromme van een
      scintillator
   d Photonenverteilungskurve f eines
      Szintillators

6375 SCINTILLATOR PROSPECTING   me
      RADIATION METER
      A portable prospecting radiation meter
      containing its own power supply, and
      designed to measure particle or photon flux
      density by means of a scintillator.
   f radiamètre m de prospection à
      scintillateur,
      scintillomètre m
   e radiámetro m de exploración de centelleador
   i radiametro m di prospezione a scintillatore
   n prospectiestralingsmeter met scintillator
   d Lagerstättensuchgerät n mit Szintillations-
      zähler

6376 SCINTILLATOR SLOW NEUTRON
      DETECTOR, see 4706

6377 SCINTILLATOR TOTAL   ra
      CONVERSION EFFICIENCY
      The ratio of the optical photon energy
      produced to the energy of a particle or

photon of ionizing radiation that is totally absorbed in the scintillator material.
f rendement *m* de la conversion totale d'un scintillateur
e rendimiento *m* de la conversión total de un centelleador
i rendimento *m* della conversione totale d'uno scintillatore
n rendement *n* van de totale conversie van een scintillator
d Gesamtumwandlungsgrad *m* eines Szintillators

6378 SCINTISCANNING ct
The use of scintillation counters to scan distribution of gamma emitting tracers in a human body.
f scintigraphie *f*
e scintigrafía *f*
i scintigrafia *f*
n scintigrafie
d Szintigraphie *f*

6379 SCRAM, see 2226

6380 SCRAM BUTTON, see 6277

6381 SCRAM ROD, see 6275

6382 SCREEN, ge
SHIELD
Material so disposed in relation to a field as to reduce its penetration into an assigned region.
f blindage *m*, matériel *m* de blindage
e blindaje *m*, material *m* de blindaje
i medio *m* di schermaggio, schermatura *f*
n afscherming, afschermmiddel *n*
d Abschirmstoff *m*, Abschirmung *f*

6383 SCREENAGE, is, np
SELF-SCREENAGE
The filtration afforded by a container of radioactive material.
f autoblindage *m*
e autoblindaje *m*
i autoschermatura *f*
n eigen afscherming
d Eigenabschirmung *f*

6384 SCREENED HANDLE is, sa
A handle for manipulating isotope preparations screened from the source by a plastics perforated disk.
f poignée *f* blindée
e manivela *f* de blindaje
i manubrio *m* schermato
n afgeschermd handvat *n*
d abgeschirmter Griff *m*

6385 SCREENING CONSTANT, np
SCREENING NUMBER
The atomic number minus the apparent atomic number that is effective for a given process.
f constante *f* d'effet d'écran, nombre *m* d'effet d'écran
e constante *f* de efecto de pantalla, número *m* de efecto de pantalla
i costante *f* d'effetto di schermo, numero *m* d'effetto di schermo
n afschermingsconstante, afschermingsgetal *n*
d Abschirmungskonstante *f*, Abschirmungszahl *f*

6386 SCREENING OF NUCLEUS, see 2157

6387 SCREW DISLOCATION cr
A dislocation whose Burgers vector is parallel to the line of dislocation.
f dislocation *f* en vie
e dislocación *f* en tornillo
i dislocazione *f* in vite
n schroefdislocatie
d Schraubenversetzung *f*

6388 SCRUBBING (US), ch
STRIPPING (GB)
In chemical engineering, the removal of a component from a solid or liquid by contact with a gas or vapo(u)r.
f lavage *m*
e lavado *m*
i lavatura *f*
n uitwassen *n*
d Abtreibung *f*, Abtrieb *m*

6389 SCYLLA PINCH, pp
THETA-PINCH
Pinch effect in the case of a longitudinal magnetic field and an azimuthal current.
f striction *f* azimuthale, striction *f* orthogonale
e estricción *f* theta
i contrazione *f* teta
n thêta-insnoering
d Theta-Einschnürung *f*

6390 SEA DISPOSAL OF rw
RADIOACTIVE WASTE
f dépôt *m* des déchets radioactifs en mer
e depósito *m* de desechos radiactivos en el mar
i deposito *m* di rifiuti radioattivi nel mare
n lozing van radioactief afval in zee
d Abführung *f* von radioaktivem Abfall ins Meer

6391 SEAL vt
A vacuum tight connection between glass and glass or glass and metal.
f scellement *m*
e cierre *m* hermético
i chiusura *f* ermetica
n versmelting
d Verschmelzung *f*

6392 SEALED GLASS AMPOULE is, md
f ampoule *f* en verre scellée
e ampolla *f* de vidrio sellada
i ampolla *f* in vetro sigillata
n afgesmolten glasampul
d abgeschmolzene Glasampulle *f*

6393 SEALED-IN FUEL ELEMENT (GB), see 2240

6394 SEALED SOURCE   ra
A radiation source sealed in a container or
having a bonded cover, where the container
or cover has sufficient mechanical strength
to prevent contact with and dispersion of
the radioactive material under the
conditions of use and wear for which it
was designed.
f source f scellée
e fuente f sellada
i sorgente f sigillata
n gesloten bron
d geschlossenes radioaktives Präparat n,
   umschlossener radioaktiver Stoff m,
   umschlossener radioaktiver Strahler m

6395 SEALING MEDIUM   cd
A substance used to seal off various
parts of a nuclear reactor.
f substance f d'étoupage,
   substance f de scellement
e substancia f obturadora
i ermetico m, sostanza f di tenuta
n afdichtmassa, afsluitmiddel n,
   dichtingsmateriaal n
d Abdichtmasse f, Dichtungsmasse f

6396 SECOND IONIZATION POTENTIAL np
Pertains to the removal of the most
loosely bound electron from an atom from
which one electron has already been
removed.
f deuxième potentiel m d'ionisation
e segundo potencial m de ionización
i secondo potenziale m d'ionizzazione
n tweede ionisatiepotentiaal
d zweites Ionisierungspotential n

6397 SECOND QUANTIZATION   qm
Process whereby a classical field is
analyzed as an ensemble of particles.
f deuxième quantification f
e segunda cuantificación f
i seconda quantizzazione f
n tweede quantisatie
d zweite Quantelung f, zweite Quantisierung f

6398 SECOND RADIATION CONSTANT   ma
A constant, c_2, equal to the product of

Planck's constant, h, and the velocity of
light, c, divided by Boltzmann's constant
k.
f deuxième constante f de rayonnement
e segunda constante f de radiación
i seconda costante f di radiazione
n tweede stralingsconstante
d zweite Strahlungskonstante f

6399 SECOND STATE CREEP,   mg
      SECONDARY CREEP
The creep that becomes constant a certain
time after the initial or primary creep.
f fluage m secondaire
e fluencia f secundaria
i scorrimento m secondario
n secondaire kruip
d sekundäres Kriechen n

6400 SECOND TOWNSEND   ic
      COEFFICIENT
The number of secondary electrons
escaping from the cathode of an ionization
chamber per positive ion produced in the
gas.
f deuxième coefficient m de Townsend
e segundo coeficiente m de Townsend
i secondo coefficiente m di Townsend
n tweede townsendcoëfficiënt
d zweiter Townsend-Koeffizient m

6401 SECOND TOWNSEND   np
      DISCHARGE
A semi-self-maintained discharge in which
the additional ions are due to the secondary
electrons emitted by the cathode.
f deuxième décharge f de Townsend
e segunda descarga f de Townsend
i seconda scarica f di Townsend
n tweede townsendontlading
d zweite Townsend-Entladung f

6402 SECONDARY COOLANT   cl
A coolant used to remove heat from the
primary coolant circuit.
f fluide m de refroidissement secondaire
e refrigerante m secundario
i fluido m termovettore secondario,
   refrigerante m secondario
n secondair koelmiddel n
d Sekundärkühlmittel n

6403 SECONDARY COOLANT CIRCUIT   cl
A system for circulating a coolant used
to remove heat from the primary coolant
circuit.
f circuit m de refroidissement secondaire
e circuito m de enfriamiento secundario
i circuito m di raffreddamento secondario
n secondair koelsysteem n
d Sekundärkühlkreislauf m

6404 SECONDARY COSMIC RAYS   cr
Rays produced when primary cosmic rays
interact with nuclei and electrons, e.g., in
the earth's atmosphere.
f rayons pl cosmiques secondaires
e rayos pl cósmicos secundarios
i raggi pl cosmici secondari
n secondaire kosmische stralen pl
d sekundäre kosmische Strahlen pl

6405 SECONDARY ELECTRON   ec, np
An electron ejected from an atom, molecule
or surface as a result of a charged
particle or photon.
f électron m secondaire
e electrón m secundario
i elettrone m secondario
n secondair elektron n
d Sekundärelektron n

6406 SECONDARY EMISSION   ec, np
The ejection of electrons from a solid or
liquid as a result of charged-particle
impact.

f émission f secondaire
e emisión f secundaria
i emissione f secondaria
n secondaire emissie
d Sekundäremission f

6407 SECONDARY EMISSION RATIO ec
The average number of electrons emitted
from a surface per incident primary
electron.
f taux m d'émission secondaire
e relación f de emisión secundaria
i rapporto m d'emissione secondaria
n secondaire-emissieverhouding
d Sekundäremissionsgrad m

6408 SECONDARY EXTINCTION cr
Occurs with imperfect crystals and arises
from the underlying of mosaic blocks by
identically oriented blocks nearer to the
surface of the crystal.
f extinction f secondaire
e extinción f secundaria
i estinzione f secondaria
n secondaire doving
d sekundäre Löschung f

6409 SECONDARY FILTER cd, xr
A sheet of material of low atomic number
relative to that of the primary filter,
placed in the filtered beam of radiation
to remove characteristic radiation
produced in the primary filter.
f filtre m secondaire
e filtro m secundario
i filtro m secondario
n secondair filter n
d Sekundärfilter m

6410 SECONDARY HEAT EXCHANGER cl
The heat exchanger following the main
one in a chemical process.
f échangeur m de chaleur secondaire
e cambiador m de calor secundario
i scambiatore m di calore secondario
n secondaire warmte-uitwisselaar
d sekundärer Wärmeaustauscher m

6411 SECONDARY ION np
That ion which has the smaller energy
after a collision between two ions.
f ion m secondaire
e ión m secundario
i ione m secondario
n secondair ion n
d Sekundärion n

6412 SECONDARY IONIZATION np
The ionization including that due to
delta rays.
f ionisation f secondaire
e ionización f secundaria
i ionizzazione f secondaria
n secondaire ionisatie
d Sekundärionisation f

6413 SECONDARY NATURAL np
RADIONUCLIDES,
SECONDARY RADIONUCLIDES
Natural radionuclides having short lifetimes in a geological sense; they are decay
products of primary natural radionuclides.
f radionucléides pl naturels secondaires
e radionúclidos pl naturales secundarios
i radionuclidi pl naturali secondari
n secondaire natuurlijke radionucliden pl
d sekundäre natürliche Radionuklide pl

6414 SECONDARY NEUTRON np
A neutron produced by the interaction with
matter of a radiation regarded as primary.
f neutron m secondaire
e neutrón m secundario
i neutrone m secondario
n secondair neutron n
d sekundäres Neutron n

6415 SECONDARY NUCLEAR ms
AUXILIARY POWER PROJECT,
SNAP PROJECT
A project for the auxiliary use of nuclear
power for satellites and space vehicles.
f projet m SNAP
e proyecto m SNAP
i progetto m SNAP
n SNAP-project n
d SNAP-Projekt n

6416 SECONDARY PARAMETER me
An additional rating or characteristic
needed to evaluate the operation of a part
beyond its normal limits, such as its
temperature coefficient.
f paramètre m secondaire
e parámetro m secundario
i parametro m secondario
n secondaire parameter
d Sekundärparameter m

6417 SECONDARY PROTECTION sa, xr
BARRIER
Barrier that reduces the stray radiation
to the maximum permissible dose rate.
f écran m secondaire de radioprotection
e barrera f secundaria de radioprotección
i schermo m secondario di radioprotezione
n secondair veiligheidsscherm n,
secondaire beschermingswand
d Sekundärstrahlenschutzwand f

6418 SECONDARY QUANTUM NUMBER,
see 490

6419 SECONDARY RADIATION ra
Particles or photons produced by the
interaction with matter of a radiation
regarded as primary.
f rayonnement m secondaire
e radiación f secundaria
i radiazione f secondaria
n secondaire straling
d Sekundärstrahlung f

6420 SECONDARY X-RAYS   xr
   The X-rays emitted by any matter
   irradiated by primary X-rays.
f rayons *pl* X secondaires
e rayos *pl* X secundarios
i raggi *pl* X secondari
n secondaire röntgenstralen *pl*
d sekundäre Röntgenstrahlen *pl*

6421 SECTOR FOCUSED CYCLOTRON   pa
   A cyclotron in which the focusing is
   realized by azimuthal variations of the
   guide field due to the sectorizing of the
   pole faces.
f cyclotron *m* à focalisation par secteurs
e ciclotrón *m* de enfoque por sectores
i ciclotrone *m* a focalizzazione per settori
n cyclotron *n* met sectorfocussering
d Zyklotron *n* mit Sektorfokussierung

6422 SECULAR EQUILIBRIUM,   np
   SECULAR RADIOACTIVE
   EQUILIBRIUM
   A radioactive equilibrium in which the
   half-life of the parent is long compared
   with the time of the experiment.
f équilibre *m* radioactif séculaire
e equilibrio *m* radiactivo secular
i equilibrio *m* radioattivo secolare
n langdurig radioactief evenwicht *n*
d dauerndes radioaktives Gleichgewicht *n*

6423 SEDIMENT,   ch
   SETTLING
   The separated particles from a suspension
   by means of gravity or of another force.
f sédiment *m*
e sedimento *m*
i sedimento *m*
n afzetting, neerslag, sediment *n*
d Ablagerung *f*, Niederschlag *m*, Sediment *n*

6424 SEDIMENTATION,   ch
   SETTLING
   The separation of suspended solid particles
   from a liquid by the action of gravity or of
   another force.
f sédimentation *f*
e sedimentación *f*
   sedimentazione *f*
n afzetten *n*, sedimentatie
d Absitzen *n*, Sedimentation *f*

6425 SEED,   rt
   SPIKE
   A fuel assembly containing more fissile
   material than the surrounding fuel.
f semence *f*
e madre *f*, semilla *f*
i pacchetto *m* d'elementi combustibili
   arricchiti specialmente,
   seme *m*
n speciaal verrijkt splijtstofpakket *n*
d Saatelement *n*

6426 SEED CORE   rt
   A reactor core having local regions

(seeds) of fuel of an enrichment which is
much higher than the enrichment of the
remainder of the core.
f coeur *m* à germes
e núcleo *m* con combustibles enriquecidos
   localmente
i nocciolo *m* a combustibili arricchiti
   localmente
n kern met plaatselijk verrijkte splijtstof
d Kern *m* mit örtlich angereichertem
   Brennstoff

6427 SEED CORE REACTOR   rt
   A nuclear reactor having a core containing
   local regions (seeds) of enriched fuel,
   distributed in a lattice of fuel of lower
   enrichment or of fertile material.
f réacteur *m* à coeur à germes
e reactor *m* de combustibles enriquecidos
   localmente
i reattore *m* a combustibili arricchiti
   localmente
n reactor met plaatselijk verrijkte splijtstof
d Reaktor *m* mit örtlich angereichertem
   Brennstoff

6428 SEEDING,   rt
   SPIKING
   The replacement of some fuel elements in
   the core of a reactor by others containing
   more fissile material than normal, so as
   to increase the reactivity of the reactor.
f ensemencement *m*, intensification *f* locale,
   spiking *m*
e intensificación *f* local
i intensificazione *f* locale
n plaatselijke intensivering, spikkelen *n*
d örtliche Intensivierung *f*, Spicken *n*

6429 SEEPAGE PIT   rw
   Hole or depression at the ground surface
   into which solutions are admitted for
   percolation into the ground.
f puits *m* de percolation
e pozo *m* de percolación
i pozzo *m* di percolazione
n doorsijpelput
d Durchsickergrube *f*

6430 SEFSTRÖMITE   mi
   A mixture of ilmenite with minor amounts
   of radioactive minerals.
f sefströmite *f*
e sefströmita *f*
i sefströmite *f*
n sefströmiet *n*
d Sefströmit *m*

6431 SEGREGATION   mg
   A variation in the concentration of
   alloying elements or constituents of an
   alloy.
f ségrégation *f*
e segregación *f*
i segregazione *f*
n segregeren *n*, zijgen *n*, zijgeren *n*
d Seigern *n*

6432 SEISMIC METHOD   mi
A geophysical method used to study rock
structures by recording the reflections
of shock waves from controlled explosions.
f méthode *f* séismique, méthode *f* sismique
e reconocimiento *m* sísmico
i metodo *m* sismico
n opsporing van mineralen door middel van
  seismometers
d Sprengseismik *f*

6433 SEISMIC SITE, see 2010

6434 SELECTION RULES, see 4908

6435 SELECTIVE ABSORBER   ab, md
A concentrated particular substance in
an organ or specified tissue which absorbs
radiation to a high degree.
f absorbant *m* sélectif, absorbeur *m* sélectif
e absorbedor *m* selectivo,
  absorbente *m* selectivo
i assorbente *m* selettivo,
  assorbitore *m* selettivo
n selectief absorbens *n*,
  selectief absorptiemiddel *n*
d selektives Absorbens *n*,
  selektives Absorptionsmittel *n*

6436 SELECTIVE ABSORPTION   ab
Absorption of radiation at some function
of frequency.
f absorption *f* sélective
e absorción *f* selectiva
i assorbimento *m* selettivo
n selectieve absorptie
d selektive Absorption *f*

6437 SELECTIVE ISOTOPE   is, mc
   ABSORPTION,
   SELECTIVE LOCALIZATION,
   SPECIFIC ABSORPTION
The concentration of a particular substance,
which may be isotopically labelled, in an
organ or specified tissue to a greater level
than would be accounted for by the normal
diffusion process.
f affinité *f* différentielle,
  localisation *f* sélective
e afinidad *f* diferencial,
  localización *f* selectiva
i affinità *f* differenziale,
  localizzazione *f* selettiva
n selectieve localisatie, voorkeursafzetting
d bevorzugte Ablagerung *f*,
  selektive Speicherung *f*

6438 SELENIUM   ch
Non-metallic element, symbol Se, atomic
number 34.
f sélénium *m*
e selenio *m*
i .selenio *m*
n seleen *n*, selenium *n*
d Selen *n*

6439 SELF-ABSORPTION   ab
The absorption of radiation by the emitting
body.

f auto-absorption *f*
e autoabsorción *f*
i autoassorbimento *m*
n zelfabsorptie
d Selbstabsorption *f*

6440 SELF-CHARGE   np
Extra contribution to the electric charge
of a charged particle due to vacuum
polarization arising from the field produced
by the original charge.
f charge *f* propre
e carga *f* propia
i carica *f* propria
n eigenlading
d Eigenladung *f*

6441 SELF-CONSISTENT FIELD   pp
In a plasma, the sum of electric and
magnetic fields satisfying the Maxwell
equations in which the charge and current
densities are determined by the particle
distribution.
f champ *m* électromagnétique autoconsistant
e campo *m* electromagnético autoconsistente
i campo *m* elettromagnetico autoconsistente
n zelfonderhoudend elektromagnetisch veld *n*
d elektromagnetisches Eigenfeld *n*

6442 SELF-CONSISTENT FIELD METHOD,
      see 3147

6443 SELF-ENERGY   np
Of a particle, the contribution to its
energy, according to a given theoretical
model, which comes from its interaction
with its own field.
f énergie *f* propre
e energía *f* propia
i energie *f* propria
n eigenenergie
d Eigenenergie *f*

6444 SELF-FILTRATION, see 3480

6445 SELF-FRACTIONATING OIL PUMP   vt
An oil diffusion pump which incorporates
a fractionating device which provides for
the segregation of the more volatile
constituents of the oil into the region where
they will do no harm.
f pompe *f* à diffusion à huile à autofraction
e bomba *f* de difusión de aceite auto-
  fraccionadora
i pompa *f* a diffusione ad olio auto-
  frazionatrice
n zelffractionerende oliediffusiepomp
d selbstfraktionierende Öldiffusionspumpe *f*

6446 SELF-IRRADIATION   is
A non-desired phenomenon occurring in
encapsulated labelled (tagged) compounds
during storage, the result being decomposition.
f autorayonnement *m*
e autorradiación *f*
i autoradiazione *f*
n zelfstraling
d Selbststrahlung *f*

6447 SELF-LIMITING CHAIN REACTION   rt
A chain reaction in which the moderator
automatically keeps the reaction within
pre-set limits.
f réaction f nucléaire en chaîne à
   automodération
e reacción f nuclear en cadena de
   automoderación
i reazione f nucleare in catena ad auto-
   moderazione
n zelfremmende kettingreactie
d selbstbremsende Kettenreaktion f

6448 SELF-MAINTAINING DISCHARGE   np
Ionic conduction in a gas caused by the
application of an electric field sufficient
to cause creation of the necessary supply
of ions by collisions between electrons and
molecules.
f décharge f autonome
e descarga f autónoma
i scarica f autosostenuta
n zelfstandige ontlading
d selbständige Entladung f

6449 SELF-MAINTAINING NUCLEAR   rt
     CHAIN REACTION,
        SELF-SUPPORTING NUCLEAR
        CHAIN REACTION,
        SELF-SUSTAINING NUCLEAR
        CHAIN REACTION
A nuclear reaction which goes on once
started without any further influence from
outside.
f réaction f en chaîne auto-entretenue
e reacción f nuclear en cadena autónoma
i reazione f nucleare a catena auto-
   alimentata
n zichzelfonderhoudende kettingreactie
d sich selbsterhaltende Kettenreaktion f

6450 SELF-MULTIPLYING   rt
     CHAIN REACTION
A chain reaction in which the number of
neutrons generated is multiplied automatically without interference from outside.
f réaction f en chaîne automultiplicatrice
e reacción f en cadena automultiplicada
i reazione f in catena automoltiplicata
n zelfvermeerderende kettingreactie
d selbstvermehrende Kettenreaktion f

6451 SELF-PROTECTED TUBE   xr
An X-ray tube the construction of which
includes protection against excessive
emission of radiation outside the useful
beam.
f tube m à autoprotection
e tubo m autoprotegido
i tubo m ad autoprotezione
n zelfbeschermende buis
d Röntgenröhre f mit Primärstrahlungs-
   abschirmung

6452 SELF-QUENCHED   ct
     COUNTER TUBE
A Geiger-Müller counter tube which is

quenched by means of a suitable component
in the counting gas.
f tube m compteur autocoupeur
e tubo m contador autoextinctor,
   tubo m contador autointerruptor
i tubo m contatore ad autospegnimento,
   tubo m contatore autoestintore
n zelfdovende telbuis
d selbstlöschendes Zählrohr n

6453 SELF-QUENCHING, see 3580

6454 SELF-RADIATION, see 998

6455 SELF-REGULATION   rt
An inherent tendency under certain
conditions of a reactor to operate at a
constant power level because of the effect
on reactivity of a change in power level.
f autorégulation f
e autorregulación f
i autoregolazione f
n zelfregeling
d Selbstregelung f

6456 SELF-SCATTERING   np, ra
Scattering of radiation by the material
that emits the radiation.
f autodiffusion f
e autodispersión f
i autodeviazione f
n eigenverstrooiing, zelfverstrooiing
d Eigenstreuung f, Selbststreuung f

6457 SELF-SCREENAGE, see 6383

6458 SELF-SEALING COUPLING   cd
f couplage m à autoscellement
e acoplamiento m autosellante
i accoppiamento m ad autochiusura ermetica
n zelfdichtende koppeling
d selbstdichtende Kopplung f

6459 SELF-SHIELDING   ab, ra, sa
Shielding of the inner parts of a body
through absorption of radiation in its outer
parts.
f autoprotection f
e autoblindaje m, autoprotección f
i autoschermatura f
n zelfafscherming
d Selbstabschirmung f

6460 SELF-SHIELDING FACTOR   ab, ra, sa
The factor by which the value of a
radiation quantity inside an irradiated body
is reduced by self-shielding.
f facteur m d'autoprotection
e factor m de autoblindaje,
   factor m de autoprotección
i fattore m d'autoschermatura
n zelfafschermingsfactor
d Selbstabschirmfaktor m

6461 SEMICIRCULAR PATH   pa
f parcours m semicirculaire
e trayectoria f semi-circular

i   percorso *m* semicircolare
n   halfcirkelvormige baan
d   halbkreisförmige Bahn *f*

6462 SEMICONDUCTOR DETECTOR   ra
    A radiation detector using either ionization
    or the creation of structural faults, in a
    semiconductive medium.
f   détecteur *m* à semiconducteur,
    détecteur *m* à jonction
e   detector *m* de semiconductor
i   rivelatore *m* a semiconduttore
n   halfgeleiderdetector
d   Halbleiterdetektor *m*

6463 SEMICONDUCTOR   ct, ra
    DETECTOR DOSEMETER
    A dosemeter combined with a semi-
    conductor detector.
f   dosimètre *m* à détecteur semiconducteur
e   dosímetro *m* de detector semiconductor
i   dosimetro *m* a rivelatore semiconduttore
n   dosismeter met halfgeleiderdetektor
d   Dosismesser *m* mit Halbleiterdetektor

6464 SEMICOVALENT BOND, see 5383

6465 SEMI-EMPIRICAL   ma
    MASS FORMULA
    A mass formula based on the liquid-drop
    model of the nucleus.
f   formule *f* semi-empirique de masse
e   fórmula *f* semiempírica de masa
i   formula *f* semiempirica di massa
n   semi-empirische massaformule
d   halbempirische Massenformel *f*

6466 SENGIERITE   mi
    A rare secondary mineral containing about
    43 % of U.
f   sengiérite *f*
e   sengierita *f*
i   sengierite *f*
n   sengeriet *n*
d   Sengierit *m*

6467 SENSING ELEMENT,   co, me
    SENSOR
    An element that detects a change in a
    selective physical quantity and converts
    that change into a useful signal for a
    measuring, recording or control system.
f   capteur *m*
e   captador *m*, órgano *m* sensor
i   elemento *m* sensibile
n   gevoelig element *n*
d   Messfühler *m*

6468 SENSITIVE   me
    Of an instrument, capable of measuring
    or indicating very small quantities of the
    measurement.
f   sensible adj
e   sensible adj
i   sensibile adj
n   gevoelig adj
d   empfindlich adj

6469 SENSITIVE GLASS   mi
    A glass which discolours or darkens by
    irradiation.
f   verre *m* sensible au rayonnement
e   vidrio *m* sensible a la radiación
i   vetro *m* sensibile alla radiazione
n   stralingsgevoelig glas *n*
d   strahlungsempfindliches Glas *n*

6470 SENSITIVE LINING   ic
    The substance applied as a lining inside
    an ionization chamber and which reacts
    with neutrons to produce charged ionizing
    particles.
f   dépôt *m* sensible
e   capa *f* sensitiva
i   strato *m* sensitivo
n   gevoelige laag
d   empfindliche Schicht *f*

6471 SENSITIVE LINING IONIZATION
    CHAMBER, see 3508

6472 SENSITIVE REGION,   md
    SENSITIVE VOLUME
    In radiology, part of a cell particularly
    sensitive to radiation.
f   part *m* sensible, région *f* sensible
e   región *f* sensible
i   regione *f* sensibile
n   gevoelig deel *n*
d   empfindlicher Bereich *m*

6473 SENSITIVE TIME   ic
    The duration of supersaturation required
    for track formation following expansion in
    a cloud chamber.
f   durée *f* de sensibilité,
    temps *m* de sensibilité
e   duración *f* de sensibilidad,
    tiempo *m* de sensibilidad
i   durata *f* di sensibilità,
    tempo *m* di sensibilità
n   gevoelige tijd
d   empfindliche Zeit *f*, wirksame Zeit *f*

6474 SENSITIVE VOLUME   ra
    That part of a radiation detector from which
    an output signal could originate.
f   volume *m* sensible, volume *m* utile
e   volumen *m* sensible
i   volume *m* sensibile
n   gevoelige ruimte
d   empfindliches Volumen *n*, Zählvolumen *n*

6475 SENSITIVITY   me
    The response of a measuring device divided
    by the value of the quantity measured.
f   sensibilité *f*
e   sensibilidad *f*
i   sensibilità *f*
n   gevoeligheid
d   Empfindlichkeit *f*

6476 SENSITIZED DECOMPOSITION   ch
    A chemical decomposition that is brought
    about by the presence of another substance

which absorbs the exciting radiation.
f décomposition *f* sensibilisée
e descomposición *f* sensibilizada
i decomposizione *f* sensibilizzata
n gesensibiliseerde ontleding
d sensibilisierte Zersetzung *f*

6477 SEPARATED ORBIT CYCLOTRON   pa
A cyclotron-like accelerator, in which the
particles remain in the neighbo(u)rhood of
a helical orbit, the distance between the
helices facilitating the beam extraction.
f cyclotron *m* à orbites séparées
e ciclotrón *m* de órbitas separadas
i ciclotrone *m* ad orbite separate
n cyclotron *n* met gescheiden banen
d Zyklotron *n* mit getrennten Bahnen

6478 SEPARATING UNIT   ch, is
A subdivisional part of an isotope separating plant.
f groupe *m* de séparation
e grupo *m* de separación
i gruppo *m* di separazione
n scheidingsgroep
d Trenngruppe *f*

6479 SEPARATION   ch
The process of dividing a mixture into its
component parts.
f séparation *f*
e separación *f*
i separazione *f*
n scheiding
d Trennung *f*

6480 SEPARATION   is
The separation factor minus one, often
expressed as a percentage and only used
when the separation factor is small.
f coefficient *m* de séparation
e coeficiente *m* de separación
i coefficiente *m* di separazione
n scheidingscoëfficiënt
d Trennungskoeffizient *m*

6481 SEPARATION BY MEANS OF IONS IN
LIQUIDS, see 2080

6482 SEPARATION COLUMN,   ch, is
SEPARATION TOWER
Part of a chemical plant where a separation
process is carried out.
f colonne *f* de séparation
e columna *f* de separación
i colonna *f* di separazione
n scheidingskolom
d Trennsäule *f*

6483 SEPARATION EFFICIENCY   is
The quotient of the difference in isotopic
abundance at the entrance and at the exit
of a separative element.
f rendement *m* de séparation
e rendimiento *m* de separación
i rendimento *m* di separazione
n scheidingsrendement *n*
d Trennungsausbeute *f*

6484 SEPARATION ENERGY, see 638

6485 SEPARATION FACTOR   ch, is
Of a system, the ratio of the abundance ratio
of material taken from the product end of
the system to the abundance ratio of the
material emerging from the waste end.
f facteur *m* de séparation
e factor *m* de separación
i fattore *m* di separazione
n scheidingsfactor
d Trennfaktor *m*

6486 SEPARATION FACTOR, see 2304 fu, rt

6487 SEPARATION PLANT   ch, is
A chemical plant in which isotopes are
separated.
f installation *f* de séparation
e equipo *m* de separación
i impianto *m* di separazione
n scheidingsinstallatie
d Trennanlage *f*

6488 SEPARATION POTENTIAL   ch
Separative work content per mole.
f potentiel *m* de séparation
e potencial *m* de separación
i potenziale *m* di separazione
n scheidingspotentiaal
d Trennpotential *n*

6489 SEPARATION PROCESS   ch, is
A process for the separation of the
constituents of a mixture.
f procédé *m* de séparation
e procedimiento *m* de separación
i processo *m* di separazione
n scheidingsproces *n*
d Trennverfahren *n*

6490 SEPARATION TUBE,   is
THERMAL DIFFUSION TUBE
A tube used in isotope separation by
thermal diffusion.
f colonne *f* de diffusion thermique
e columna *f* de difusión térmica
i colonna *f* di diffusione termica
n thermische diffusiekolom
d Trennrohr *n*

6491 SEPARATIVE DUTY,   ch, is
SEPARATIVE EFFORT,
SEPARATIVE POWER
A measure of the separation job imposed
on any isotope separation process.
f pouvoir *m* de séparation
e poder *m* de separación
i potere *m* di separazione
n scheidend vermogen *n*
d Trennvermögen *n*

6492 SEPARATIVE EFFICIENCY   is
Of a cascade, the ratio of the change in
separative work content produced by a
cascade in unit time to the integrated
separative power of its elements.
f efficacité *f* de séparation

e eficacia f de separación
i efficacia f di separazione
n nuttig scheidingseffect n,
   scheidingsrendement n
d Trenngüte f, Trennungsnutzeffekt m

6493 SEPARATIVE ELEMENT   ch
Any separative unit which may be
considered to be the basic element of a
large apparatus containing many similar
units.
f élément m séparateur
e elemento m separador
i elemento m separatore
n scheidingselement n
d Trennelement n

6494 SEPARATIVE WORK,   ch, is
      SEPARATIVE WORK CONTENT
In isotope separation, a measure of the
difficulty of producing from a supply of
material of a given abundance a product
of some other abundance.
f travail m de séparation
e trabajo m de separación
i lavoro m di separazione
n scheidingsarbeid
d Trennungsarbeit f

6495 SEQUENCING-MIXING UNIT, see 2520

6496 SEQUESTERING AGENT, see 1264

6497 SERBER FORCE   np
A nuclear force which is a mixture of an
ordinary and an exchange force of the
Majorana type.
f force f de Serber
e fuerza f de Serber
i forza f di Serber
n serberkracht
d Serber-Kraft f

6498 SERIAL RADIOGRAPHY   xr
A technique for making a number of
radiographs of the same object in
succession.
f radiographie f en séries
e radiografía f en serie
i radiografia f in serie
n serie-opname, serieradiografie
d Reihenaufnahme f, Serienradiographie f

6499 SERIES DECAY,
      SERIES DISINTEGRATION, see 978

6500 SERVOCONTROL MECHANISM,   co
      SERVOMECHANISM
An automatic control system which uses
an external source of energy to control an
output position in prescribed relationship
to an input variable.
f servomécanisme m
e servomecanismo m
i servomeccanismo m
n servomechanisme n
d Servomechanismus m

6501 SERVOSYSTEM   co
An automatic control system for maintaining
a process condition at or near a predetermined value by activation of an
element.
f système m asservi
e servosistema m
i servosistema m
n servosysteem n
d Servosystem n

6502 SET POINT,
      SET VALUE (GB), see 3441

6503 SETTING TIME   co
The time needed to arrange, e.g., an
instrument before use.
f temps m d'affichage
e tiempo m de ajuste, tiempo m de calado
i tempo m di predisposizione
n insteltijd
d Einstellzeit f

6504 SETTLING,
      SEDIMENT, see 6423

6505 SETTLING,
      SEDIMENTATION, see 6424

6506 SETTLING TIME, see 1423

6507 SEX CHROMOSOME   md
A chromosome which has no similar
homologue in the heterozygous sex and
which is closely bound up with sex
determination.
f chromosome m sexuel
e cromosoma m sexual
i cromosoma m sessuale
n geslachtschromosoom n
d Geschlechtschromosom n

6508 SEX LINKAGE   md
The inheritance of certain characteristics
which are determined by genes located
in the sex chromosomes.
f transmission f des caractères sexuels
e herencia f lijada al sexo,
   vinculación f sexual
i eredità f ligata al sesso
n aan het geslacht gebonden erfelijke
   eigenschappen pl
d geschlechtsverbundene Vererbung f

6509 SEXUAL CELL, see 2915

6510 SHADOW, see 5735

6511 SHADOW SCATTERING, see 1801

6512 SHADOW SHIELD, see 5131

6513 SHADOWING, see 1366

6514 SHAKING   ch, mi
A term generally used in chemical
technology and mineralogy.

f   secouement *m*
e   sacudidura *f*, sacudimiento *m*
i   scolimento *m*, scossa *f*
n   schudden *n*
d   Schütteln *n*

6515 SHAPING UNIT   ec
A basic function unit comprising an
electronic circuit designed to provide an
output pulse of predetermined characteristics in response to an input pulse having a
different shape.
f   élément *m* de mise en forme
e   unidad *f* de impulso modulado
i   elemento *m* formatore
n   pulsvormer
d   Impulsformer *m*

6516 SHARED ELECTRONS   np
Electrons forming a covalent bond between
atoms.
f   électrons *pl* pártagés
e   electrones *pl* compartidos
i   elettroni *pl* comuni
n   gemeenschappelijke elektronen *pl*
d   anteilige Elektronen *pl*,
    gemeinsame Elektronen *pl*

6517 SHARPITE   mi
A mineral containing about 68.5 % of U.
f   sharpite *f*
e   sharpita *f*
i   sharpite *f*
n   sharpiet *n*
d   Sharpit *m*

6518 SHEAR STRESS,   gp
SHEARING STRESS
The stress which accompanies shear in an
elastic body.
f   effort *m* de cisaillement
e   tensión *f* de cizallamiento
i   forza *f* di taglio
n   schuifspanning
d   Schubspannung *f*

6519 SHEATH, see 858

6520 SHEATHED PUMP, see 862

6521 SHEDDING, see 1745

6522 SHELF LIFE   is
The life during which a radioactive
compound may be stored on a shelf.
f   vie *f* en magasin
e   vida *f* en almacén
i   vita *f* in magazzino
n   opslaglevensduur
d   Lagerungslebensdauer *f*

6523 SHELL   np
A group of electrons, supposed to form
part of the outer structure of an atom, and
having a common energy level.
f   couche *f*
e   capa *f*

i   strato *m*
n   schil
d   Schale *f*

6524 SHELL ELECTRON, see 5002

6525 SHELL MODEL   np
A nuclear model in which shell structure
is postulated.
f   modèle *m* des couches
e   modelo *m* de capas,
    modelo *m* estratiforme
i   modello *m* di strati
n   schilmodel *n*
d   Schalenmodell *n*

6526 SHELL STRUCTURE OF THE ATOM,
see 413

6527 SHELL STRUCTURE OF THE
NUCLEUS, see 4938

6528 SHIELD   ra, sa
A body of material intended to reduce the
intensity of radiation entering a region.
f   blindage *m*, bouclier *m*, écran *m*
e   blindaje *m*, pantalla *f*
i   schermo *m*
n   scherm *n*
d   Abschirmung *f*, Schirm *m*

6529 SHIELD,   ge
SCREEN, see 6382

6530 SHIELD TEST POOL FACILITY,
see 812

6531 SHIELD WALL   rt
In reactor technology, a wall surrounding
the reactor core and its restraint tank.
f   paroi *f* de blindage
e   pared *f* de blindaje
i   parete *f* di schermaggio
n   afschermwand
d   Abschirmwand *f*

6532 SHIELDED BOX   rt, sa
A box similar to a glove box but suitable
for work on gamma-active materials.
f   enceinte *f* blindée contre rayons gamma
e   caja *f* blindada contra rayos gamma
i   scatola *f* schermata contro raggi gamma
n   tegen gammastralen afgeschermde kast
d   gegen Gammastrahlen abgeschirmter
    Kasten *m*

6533 SHIELDED NUCLIDE   np
An active nuclide which cannot be formed
by decay of a precursor because the
nuclide from which it would have to be
formed is stable.
f   nucléide *m* blindé
e   núclido *m* blindado
i   nuclido *m* schermato
n   afgeschermd nuclide *n*
d   abgeschirmtes Nuklid *n*

6534 SHIELDED X-RAY TUBE,   xr
    SHOCK-PROOF TUBE
An X-ray tube enclosed in a grounded
metal container except for a small window
through which X-rays emerge.
f   tube *m* à rayons X à gaine métallique
e   tubo *m* de rayos X de vaina metálica
i   tubo *m* a raggi X a guaina metallica
n   afgeschermde röntgenbuis
d   abgeschirmte Röntgenröhre *f*

6535 SHIELDING   ge
The arrangement of shields provided for
any particular circumstances or
establishment.
f   blindage *m*
e   apantallamiento *m*, blindaje *m*
i   schermatura *f*
n   afscherming
d   Abschirmung *f*

6536 SHIELDING BARRIER, see 2992

6537 SHIELDING CALCULATIONS   ma, sa
Calculation of the proper thickness and
configuration of a shield for the reduction
of radiation levels to a desired value.
f   calculs *pl* de blindage
e   cálculos *pl* de blindaje
i   calcoli *pl* di schermatura
n   afschermingsberekeningen *pl*
d   Abschirmungsberechnungen *pl*

6538 SHIELDING COUNTER, see 3108

6539 SHIELDING POND,   sa
    SHIELDING POOL
A tank fitted so that highly radioactive
material can be manipulated under water
by an operator standing above the tank,
shielded from the radiation by a depth
of water.
f   réservoir *m* à blindage d'eau
e   tanque *m* de blindaje de agua
i   serbatoio *m* a schermatura d'acqua
n   tank met waterafscherming
d   wasserabgeschirmter Tank *m*

6540 SHIELDING SLAB   rt
f   plaque *f* de blindage
e   placa *f* de blindaje
i   piastra *f* di schermatura
n   afschermplaat
d   Abschirmplatte *f*

6541 SHIELDING WINDOWS   sa
Window for shielded boxes or cells.
f   fenêtres *pl* blindées, hublots *pl*
e   ventanas *pl* blindadas
i   finestre *pl* schermanti
n   afschermvensters *pl*
d   Abschirmfenster *pl*

6542 SHIM ELEMENT,   co
    SHIM MEMBER
A control member used to compensate for
long-term reactivity and flux density
distribution effects in a reactor.

f   élément *m* de compensation
e   grupo *m* de barras de corrección
i   gruppo *m* di barre di regolazione a lungo
    termine
n   stellichaam *n*
d   Trimmelement *n*

6543 SHIM MECHANISM   co, sa
A device by which a control rod intervenes
after a certain delay with respect to the
fine control rod and so allows the last one
to operate always with efficiency.
f   mécanisme *m* de compensation
e   mecanismo *m* de compensación
i   meccanismo *m* di ripresa
n   compensatiemechanisme *n*
d   Trimmechanismus *m*

6544 SHIM ROD   co
A coarse regulating member which ensures
compensation of large amplitude reactivity
variations over a long period.
f   barre *f* de compensation
e   barra *f* de corrección
i   barra *f* di regolazione a lungo termine
n   compensatiestaaf
d   Trimmstab *m*

6545 SHIM ROD BANK   cl, sa
A set of firmly connected shim rods.
f   groupe *m* de barres de compensation
e   grupo *m* de barras de corrección
i   gruppo *m* di barre di regolazione a lungo
    termine
n   samenstel *n* van compensatiestaven
d   Aggregat *n* von Trimmstäben

6546 SHIMMING   co, pa
The adjustment of a magnetic field to
achieve desired characteristics by means
of thin spacers, shims of soft iron or
compensating coils.
f   ajustage *m* précis de champ
e   ajuste *m* exacto del campo
i   aggiustaggio *m* esatto del campo
n   fijninstelling van het veld
d   Feldfeinkorrektion *f*

6547 SHIMMING, see 5988   rt

6548 SHINKOLOBWITE, see 1074

6549 SHOCK COIL   pp
Component part of the Ixion thermonuclear
installation.
f   bobine *f* de choc
e   bobina *f* de choque
i   bobina *f* d'urto
n   schokspoel
d   Stossspule *f*

6550 SHOCK FRONT, see 5477

6551 SHOCK HEATING   pp
Of a plasma, heating produced by the
passage of a shock wave.
f   chauffage *m* par onde de choc
e   calentamiento *m* por onda de choque

i   riscaldamento *m* per onda d'urto
n   verhitting door stootgolf
d   Stosswellenheizung *f*

6552 SHOCK WAVE   nw
A wave in which an abrupt, finite change
takes place in pressure and particle
velocity.
f   onde *f* de choc
e   onda *f* de choque
i   onda *f* d'urto
n   schokgolf, stootgolf
d   Stosswelle *f*

6553 SHOCKPROOF   ec, ge, xr
A term applied to those components of
a high voltage circuit which are entirely
surrounded by earthed metal enclosures.
f   antichoc adj
e   protegido adj contra choques
i   antiscosso adj, protetto adj contro scosse
n   hoogspanningsveilig adj
d   hochspannungssicher adj

6554 SHOCKPROOF TUBE HOUSING   xr
A component of, e.g., a nuclear shower.
grounded metal external surface, used to
eliminate danger of electric shock to
personnel.
f   gaine *f* de tube antichoc
e   envuelta *f* de tubo contra choques
eléctricos
i   guaina *f* per tubo antiscossa
n   hoogspanningsveilige omhulling
d   hochspannungssichere Hülle *f*

6555 SHORT FOCAL DISTANCE THERAPY,
see 1322

6556 SHORT-LIVED   is, np
Said of a radioactive element or an
isotope when the lifetime is very short.
f   à vie courte, de vie courte,
période *f* courte de vie
e   de vida corta
i   di vita corta
n   kortlevend adj, met korte levensduur
d   kurzlebig adj

6557 SHORT-LIVED ISOTOPE   is
Said of an isotope of which the activity
disappears within a comparatively short
time.
f   isotope *m* de vie courte
e   isótopo *m* de vida corta
i   isotopo *m* di vita corta
n   kortlevend isotoop *n*
d   kurzlebiges Isotop *n*

6558 SHORT-LIVED   is, np
RADIOACTIVE SUBSTANCE
f   substance *f* radioactive à courte vie
e   substancia *f* radiactiva de vida corta
i   sostanza *f* radioattiva di vita corta
n   kortlevende radioactieve stof
d   kurzlebige radioaktive Substanz *f*

6559 SHORT PERIOD OF RISE   rt
Said of a reactor when only a short time is
necessary to bring it to normal operation.
f   période *f* courte de montée en puissance
e   perfodo *m* corto de arranque
i   periodo *m* corto d'avviamento
n   korte aanlooptijd
d   kurze Anlaufzeit *f*

6560 SHOWER   np
A large amount of charged particles.
f   gerbe *f*
e   chaparrón *m*
i   sciame *m*
n   bui
d   Schauer *m*

6561 SHOWER PARTICLE   np
A component of, e.g., a nuclear shower.
f   particule *f* de gerbe
e   partícula *f* de chaparrón
i   particella *f* di sciame
n   buideeltje *n*
d   Schauerteilchen *n*

6562 SHOWER UNIT   np
The mean path length required for the
reduction, by the factor $1/2$, of the energy
of relativistic charged particles as they
pass through matter.
f   parcours *m* de gerbe
e   trayectoria *f* de chaparrón
i   percorso *m* medio d'uno sciame
n   buiweglengte
d   Schauerweglänge *f*

6563 SHRINKAGE CAVITY, see 1350

6564 SHRINKAGE FACTOR   ph
In the photographic emulsion method, the
ratio of the thickness of the unprocessed
emulsion to that after processing.
f   facteur *m* de contraction
e   factor *m* de contracción
i   fattore *m* di contrazione
n   krimpfactor
d   Schrumpffaktor *m*

6565 SHUFFLING OF FUEL, see 5997

6566 SHUT-DOWN   rt
The deliberate reduction of reactor power
to a negligible level by causing the reactor
to become subcritical and maintaining it
subcritical.
f   arrêt *m*, fermeture *f*
e   paro *m*
i   arresto *m*
n   uitschakeling
d   Abschaltung *f*

6567 SHUT-DOWN AMPLIFIER,   ec, rt
TRIP AMPLIFIER
A radiation instrument incorporating an
amplifier, which initiates the shut-down
of a reactor when the power exceeds a
predetermined safe level.

f amplificateur *m* du signal d'arrêt,
  amplificateur *m* incorporé dans l'appareil
  d'arrêt
e amplificador *m* de la señal de paro,
  amplificador *m* en el aparato de paro
i amplificatore *m* del segnale d'arresto,
  amplificatore *m* nell'apparecchio d'arresto
n in het uitschakelapparaat ingebouwde
  versterker,
  uitschakelsignaalversterker
d Abschaltsignalverstärker *m*,
  Verstärker *m* im Abschaltapparat

6568 SHUT-DOWN POWER, see 137

6569 SHUT-DOWN PROCEDURE   rt, sa
  Sequence of operations used for shutting
  down a nuclear reactor.
f mesures *pl* d'arrêt
e operaciones *pl* de paro
i operazioni *pl* d'arresto
n uitschakelingsmaatregelen *pl*
d Abschaltverfahren *n*

6570 SHUT-DOWN REACTIVITY   rt
  Reactivity of a reactor after a shut-down
  with the normal means.
f réactivité *f* à l'arrêt
e reactividad *f* de paro
i reattività *f* all'arresto
n reactiviteit bij uitschakeling
d Abschaltreaktivität *f*

6571 SHUT-DOWN ROD, see 6275

6572 SHUT-DOWN SWITCH, see 6277

6573 SHUT-OFF MEMBER   rt
  That control member which is used for
  routine shut-down.
f élément *m* d'arrêt
e elemento *m* de paro
i elemento *m* d'arresto
n uitschakelelement *n*
d Abschaltelement *n*

6574 SHUTTER   ra
  A movable plate of absorbing material
  used to cover a window or beam hole when
  radiation is not desired.
f obturateur *m*
e obturador *m*
i otturatore *m*
n sluiter
d Verschluss *m*

6575 SHUTTLE, see 5665

6576 SIALOGRAPHY   xr
  The radiological examination of the
  salivary ducts and alveoli following the
  injection of a contrast medium.
f sialographie *f*
e sialografía *f*
i sialografia *f*
n sialografie
d Sialographie *f*

6577 SIDE REFLECTED REACTOR   rt
  A reflected reactor in which the reflector
  is only arranged at the lateral surface.
f réacteur *m* à réflecteur latéral
e reactor *m* de reflector lateral
i reattore *m* a riflettore laterale
n reactor met zijreflector
d Reaktor *m* mit Seitenreflektor

6578 SIDE STREAM,   ch
       SLIP STREAM
  A stream withdrawn from some point along
  the height of the column.
f coupe *f* latérale
e corte *m* lateral
i taglio *m* laterale
n zijdelingse aftap
d Seitenschnitt *m*

6579 SIDE-THRUST EFFECT   pp
  The result of the force developed on a
  charged particle moving in a magnetic
  field.
f effet *m* de déplacement latéral
e efecto *m* de desplazamiento lateral
i effetto *m* di spostamento laterale
n effect *n* van zijwaartse verplaatsing
d Effekt *m* der lateralen Verschiebung,
  seitliche Ablenkung *f*

6580 SIEVE PLATE, see 5194

6581 SIEVE PLATE IRRADIATION   md, xr
f irradiation *f* à tamis
e irradiación *f* con tamiz
i irradiazione *f* a vaglio
n zeefbestraling
d Siebbestrahlung *f*

6582 SIEVERT CHAMBER   ic
  An ionization chamber of very small size.
f chambre *f* d'ionisation de Sievert
e cámara *f* de ionización de Sievert
i camera *f* d'ionizzazione di Sievert
n ionisatievat *n* van Sievert
d Sievert-Ionisationskammer *f*

6583 SIGMA AMPLIFIER   ec
  The source of the signal for operating the
  safety circuit.
f amplificateur *m* sigma
e amplificador *m* sigma
i amplificatore *m* sigma
n sigmaversterker
d Sigmaverstärker *m*

6584 SIGMA MESON   np, ph
  Any meson which, at the end of its path,
  is observed to cause a star.
f méson *m* sigma
e mesón *m* sigma
i mesone *m* sigma
n sigmameson *n*
d Sigmameson *n*

6585 SIGMA PARTICLE   np
  A hyperon having an extremely short life,

about 10^{-10} second, and a mass between
that of neutrons and deuterons and either
a positive or negative charge.
f   particule f sigma
e   partícula f sigma
i   particella f sigma
n   sigmadeeltje n
d   Sigmateilchen n

6586 SIGMA PILE, see 1827

6587 SIGMA STAR, see 4914

6588 SIGMATRON   pa, xr
A cyclotron and betatron operating in
tandem to produce billion-volt X-rays.
f   sigmatron m
e   sigmatrón m
i   sigmatrone m
n   sigmatron n
d   Sigmatron n

6589 SIGMOID CURVE   ra
S-shaped curve, often characteristic of
a dose-effect curve in radiobiological
studies.
f   courbe f sigmoïde
e   curva f sigmoide
i   curva f sigmoide
n   S-kromme
d   S-Kurve f

6590 SILICOFLUORIDE PROCESS   ch
A process for producing beryllium by
mixing the powdered mineral with sodium
ferricfluoride and/or sodium silicofluoride.
f   procédé m au fluorure de silicium
e   procedimiento m al fluoruro de silicio
i   processo m al fluoruro di silicio
n   siliciumfluorideproces n
d   Siliziumfluoridverfahren n

6591 SILICON   ch
Non-metallic element, symbol Si, atomic
number 14.
f   silicium m
e   silicio m
i   silicio m
n   silicium n
d   Silizium n

6592 SILICON CARBIDE   mt
A suggested canning material.
f   carbure m de silicium
e   carburo m de silicio
i   carburo m di silicio
n   siliciumcarbide n
d   Siliziumkarbid n

6593 SILT MOVEMENT   ms
An object for studying by means of radio-
active tracers.
f   mouvement m des alluvions
e   movimiento m de las sedimentaciones
    fluviales
i   movimento m dell'alluvione
n   slibbeweging
d   Anschwemmungsbewegung f

6594 SILVER   ch
Metallic element, symbol Ag, atomic
number 47.
f   argent m
e   plata f
i   argento m
n   zilver n
d   Silber n

6595 SILVER MATRIX   ar, md, ra
Carrier of the radioisotope in alpha foils.
f   matrice f en argent
e   matriz f en plata
i   matrice f in argento
n   zilvermatrijs
d   Silbermatrize f

6596 SILVER REACTOR   ch
A process vessel using silver nitrate to
remove iodine vapo(u)rs from a gas stream.
f   piège d'iode
e   trampa f de yodo
i   trappola f d'iodio
n   jodiumvanger
d   Silberjodidbildner m

6597 SIMPLE CASCADE   is
A cascade in which the enriched fraction
is fed to the succeeding stage, and the
depleted fraction to the preceding stage.
f   cascade f simple
e   cascada f simple
i   cascata f semplice
n   eenvoudige cascade
d   einfache Kaskade f

6598 SIMPLE DISTILLATION   ch
A distillation in which the vapo(u)r evolved
from the boiling liquid is removed as fast
as it is formed without the return of the
condensed vapo(u)r to the boiling liquid.
f   distillation f globalement équilibrée,
    distillation f simple
e   destilación f simple
i   distillazione f semplice
n   eenvoudige destillatie
d   einfache Destillation f

6599 SIMPLE PROCESS   gp, is
The simple physical process, such as a
passage through a membrane, upon which
an isotope separation is based.
f   procédé m unitaire
e   procedimiento m de una sola etapa,
    procedimiento m simple
i   processo m di separazione semplice
n   elementair proces n, enkelvoudig proces n
d   einfaches Verfahren n, Einzelprozess m

6600 SIMPLE PROCESS FACTOR   ch, is
The separation factor obtained by a simple
process.
f   facteur m d'enrichissement unitaire,
    facteur m de séparation unitaire
e   factor m de procedimiento de una sola
    etapa
i   fattore m di processo di separazione
    semplice

n elementaire scheidingsfactor,
   factor van enkelvoudig proces
d Einzelprozessfaktor *m*,
   elementarer Trennfaktor *m*

6601 SIMPSON PILE, see 4736

6602 SIMULATION   ge
      The representation of physical systems
      and phenomena by computers, models or
      other equipment.
f simulation *f*
e simulación *f*
i simulazione *f*
n nabootsing
d Nachbildung *f*

6603 SIMULATOR   ge
      A computer or other piece of equipment
      that simulates a desired system or
      condition and shows the effects of various
      applied charges, such as a reactor
      simulator.
f simulateur *m*
e simulador *m*
i simulatore *m*
n analogiemodel *n*, nabootsing, simulator
d Analogiemodell *n*, Simulator *m*

6604 SIMULTANEOUS MULTISECTION   xr
      LAMINAGRAPHY
f tomographie *f* multisection simultanée
e tomografía *f* simultánea
i tomografia *f* simultanea
n simultaantomografie
d Simultan-Schichtverfahren *n*

6605 SINGLE CHANNEL PULSE   ct, ec
      AMPLITUDE SELECTOR UNIT
      A pulse amplitude selecting unit in which
      the amplitude range scanned by a single
      channel may be continuously shifted
      through the spectrum.
f élément *m* sélecteur d'amplitude à canal
   mobile
e unidad *f* selectora de amplitud con canal
   móvil
i elemento *m* selettore d'ampiezza a
   canale mobile
n eenkanalige pulshoogtekiezer
d Einkanalimpulshöhendiskriminator *m*

6606 SINGLE CHANNEL PULSE HEIGHT
      ANALYZER, see 1789

6607 SINGLE-COATED FILM,   ph, xr
      SINGLE EMULSION FILM
f film *m* à émulsion unique,
   film *m* à simple couche
e película *f* de capa única,
   película *f* de emulsión única
i pellicola *f* con emulsione unica,
   pellicola *f* con strato semplice
n eenlaagsfilm, enkel begoten film
d Einschichtfilm *m*,
   einseitig begossener Film *m*

6608 SINGLE CRYSTAL   cr
      A macroscopic specimen of a solid in which
      all parts have the same crystallographic
      orientation.
f monocristal *m*
e monocristal *m*
i monocristallo *m*
n enkelkristal *n*
d Einkristall *m*

6609 SINGLE CRYSTAL CAMERA   cr
      An X-ray camera in which use is made of
      a single crystal.
f chambre *f* à monocristal
e cámara *f* de monocristal
i camera *f* a monocristallo
n enkelkristalcamera
d Einkristallkammer *f*

6610 SINGLE DIFFUSION STAGE   ch, is
      A single stage in an isotope separating
      process based on the diffusion principle.
f étage *m* unique de diffusion
e etapa *f* única de difusión
i stadio *m* unico di diffusione
n enkelvoudige diffusietrap
d Einzeldiffusionsstufe *f*

6611 SINGLE LINE SOURCE   ra, sp
      A Mössbauer source producing a single
      spectral line.
f source *f* à raie unique
e fuente *f* de línea única
i sorgente *f* a riga unica
n enkellijnbron
d Einzellinienquelle *f*

6612 SINGLE PARTICLE NUCLEAR MODEL,
      see 3451

6613 SINGLE ROD LATTICE   rt
f réseau *m* à barre combustible unique
e celosía *f* de una barra combustible
i reticolo *m* a barra combustibile semplice
n enkelstaafrooster *n*
d Einstabgitter *n*

6614 SINGLE SCATTERING   np
      The deflection of a particle from its
      original path to one encounter with a
      single scattering centre(er) in the material
      traversed.
f diffusion *f* unique
e dispersión *f* única
i deviazione *f* unica
n enkelvoudige verstrooiing
d Einfachstreuung *f*

6615 SINGLE-SLIT KYMOGRAPHY   xr
f kymographie *f* avec obturateur à fente
   unique
e quimografía *f* de una sola ranura
i chimografia *f* a fessura unica
n kymografie met enkele spleet
d Einschlitzkymographie *f*

6616 SINGLE-STAGE RECYCLE (US),
      see 5666

6617 SINGLE TURN COIL   pp
      A component part of the magnetic assembly
      around a plasma.
f bobine *f* à spire continue
e bobina *f* de espira continua
i bobina *f* a spira unica
n doorlopende spoelwinding,
   uit één winding bestaande spoel
d Einzelspule *f*

6618 SINGLET   sp
      A spectrum line which, even when
      examined under high resolution, is still
      a single line.
f singulet *m*
e singulete *m*
i singuletto *m*
n singlet
d Singulett *n*

6619 SINOGRAPHY   xr
      The radiological examination of the large
      intracranial venous sinuses.
f radiographie *f* des sinus
e radiografía *f* de los senos
i radiografia *f* dei seni
n sinografie
d Nebenhöhlenradiographie *f*

6620 SINTERING   mg
      The process of bonding metal or other
      powders by heating to form a strong
      cohesive body.
f frittage *m*
e sinterización *f*
i sinterizzazione *f*
n sinteren *n*
d Sintern *n*

6621 SITE   rt
      In reactor technology, the place chosen
      for building.
f site *m*, terrain *m* à bâtir
e lugar *m*, situación *f*
i luogo *m*, sito *m*
n bouwplaats, ligging
d Baustelle *f*, Lage *f*

6622 SITE MONITORING   sa
      Continuous or periodic measurement of
      the local dose rate.
f surveillance *f* locale
e vigilancia *f* local
i controllo *m* locale, sorveglianza *f* locale
n plaatselijke bewaking
d lokale Kontrolle *f*, lokale Überwachung *f*

6623 SITING   rt
f situation *f*
e ubicación *f*
i ubicazione *f*
n vestiging
d Niederlassung *f*

6624 SITING CRITERIA   rt
f critères *pl* de situation
e criterios *pl* de ubicación
i criteri *pl* d'ubicazione
n vestigingseisen *pl*
d Niederlassungsbedingungen *pl*

6625 SKIAGRAPH,   xr
      SKIOGRAPH
      An apparatus which measures the intensity
      of X-rays.
f skiographe *m*
e esquiágrafo *m*
i schiografo *m*
n skiograaf
d Skiograph *m*

6626 SKIN DOSE   ra
      The dose of radiation, which may be
      absorbed dose or exposure, received at or
      by the skin from all directions by a single
      radiation exposure or a series of
      exposures.
f dose *f* à la peau
e dosis *f* en la piel
i dose *f* alla pelle
n huiddosis
d Hautdosis *f*

6627 SKIN TOLERANCE DOSE   ra
      The maximum dose that can be safely
      delivered to the skin.
f dose *f* maximale admissible à la peau
e dosis *f* máxima permisible en la piel
i dose *f* massima ammissibile alla pelle
n maximaal toelaatbare huiddosis
d höchstzugelassene Hautdosis *f*

6628 SKIP DISTANCE, see 1120

6629 SKLODOWSKITE, see 1074

6630 SKY SHINE (US), see 165

6631 SKY SHINE SHIELD (US), see 166

6632 SLAB DISSOLVER   ch, is
      A vessel for dissolving highly enriched
      materials.
f récipient *m* de solution pour plaques
   enrichies
e recipiente *m* de solución para placas
   enriquecidas
i recipiente *m* di soluzione per piastre
   arricchite
n oplosvat *n* voor verrijkte platen
d Lösungsgefäss *n* für angereicherte Platten

6633 SLANT DISTANCE   nw
      The distance from a given location,
      usually on the earth's surface, to the point
      at which the explosions occurred.
f distance *f* oblique, distance *f* réelle
e distancia *f* oblicua, distancia *f* real
i distanza *f* obliqua, distanza *f* reale
n hoekafstand
d Winkelentfernung *f*

6634 SLATER METHOD   ma
      A method for treating the problem of the

many-electron atom, involving anti-
symmetrical functions, which yields the
relative values of coulomb and exchange
energy.
f méthode f de Slater
e método m de Slater
i metodo m di Slater
n slatermethode
d Slatersches Verfahren n

6635 SLIP   cr
A process of plastic deformation of the
metals by the simple shear displacement
of one part of a crystal relative to another.
f glissement m
e deslizamiento m
i scorrimento m
n glijden n
d Gleiten n

6636 SLIP BANDS,   cr
      SLIP LINES
Lines formed on the surface of plastically
deformed single crystals defining planes
in which shear displacement has taken
place.
f lignes pl de glissement
e líneas pl de deslizamiento
i linee pl di scorrimento
n glijlijnen pl
d Gleitlinien pl

6637 SLIP PLANE   cr
An atomic plane of a crystal along which
slip may be supposed to have taken place
in order to create an edge dislocation.
f plan m de glissement
e plano m de deslizamiento
i piano m di scorrimento
n glijvlak n
d Gleitebene f

6638 SLIP STREAM, see 6578

6639 SLIT   cr, me
The long narrow opening by which radiation
enters or leaves certain diffraction or
other optical instruments.
f fente f
e hendidura f, lumbrera f
i fenditura f, fessura f
n spleet
d Spalt m

6640 SLIT SOURCE   ra
A source of radiation in which the emission
opening is slit-shaped.
f source f à fente
e fuente f hendida
i sorgente f a fessura
n spleetbron
d spaltförmige Quelle f

6641 SLOAN-LAWRENCE ALTERNATING pa
      CURRENT LINEAR ACCELERATOR
A device for accelerated charged nuclear
particles in a straight line by means of

successive multiple application of radiofrequency voltages.
f accélérateur m linéaire de Sloan-Lawrence
e acelerador m lineal de Sloan-Lawrence
i acceleratore m lineare di Sloan-Lawrence
n lineaire versneller van Sloan en Lawrence
d Linearbeschleuniger m nach Sloan und
  Lawrence

6642 SLOW CHOPPER   sp
Used in time-of-flight apparatus in which
a burst of neutrons is produced by a
mechanical chopper over a long flight path.
f hacheur m lent
e modulador m lento
i modulatore m lento
n langzame hakker
d langsamer Zerhacker m

6643 SLOW-FAST REACTOR   rt
An Argonne National Laboratory study
project of a two-region nuclear reactor.
f réacteur m combiné lent et rapide
e reactor m combinado lento y rápido
i reattore m combinato lento e veloce
n reactor met langzaam en snel gebied
d Doppelreaktor m mit langsamen und
  schnellen Neutronen

6644 SLOW NEUTRON CAPTURE   np
f capture f de neutrons lents
e captura f de neutrones lentos
i cattura f di neutroni lenti
n vangst van langzame neutronen
d Einfang m langsamer Neutronen

6645 SLOW NEUTRON FISSION   np
Nuclear fission produced by slow neutrons.
f fission f par neutrons lents
e fisión f por neutrones lentos
i fissione f per neutroni lenti
n splijting met langzame neutronen
d Spaltung f mit langsamen Neutronen

6646 SLOW NEUTRON FLUXMETER   ma
An assembly to measure slow neutron flux
density, in which the detector is a boron
counter tube.
f fluxmètre m de neutrons lents
e flujómetro m de neutrones lentos
i flussometro m di neutroni lenti
n fluxdichtheidsmeter voor langzame
  neutronen
d Gerät n zur Bestimmung der Fluenz
  langsamer Neutronen

6647 SLOW NEUTRONS   np
Neutrons of kinetic energy less than some
specified value.
f neutrons pl lents
e neutrones pl lentos
i neutroni pl lenti
n langzame neutronen pl
d langsame Neutronen pl

6648 SLOW REACTION   np
In nuclear physics, a term applied to a

nuclear reaction which takes place within 10^{-10} second.
f réaction f lente
e reacción f lenta
i reazione f lenta
n langzame reactie
d langsame Reaktion f

6649 SLOW REACTOR, rt
    THERMAL REACTOR
A nuclear reactor in which fission is produced predominantly by thermal neutrons.
f réacteur m à neutrons thermiques
e reactor m de neutrones térmicos
i reattore m termico
n reactor met thermische neutronen, thermische reactor
d thermischer Reaktor m

6650 SLOWING DOWN np
Decrease in energy of a nuclear particle.
f ralentissement m
e retardación f
i rallentamento m
n afremming, verlangzaming
d Bremsung f, Verlangsamung f

6651 SLOWING-DOWN AREA np
In an infinite homogeneous medium, one-sixth of the mean square vector distance between the neutron source and the point where the neutrons reach a given average energy.
f aire f de ralentissement
e área f de retardación
i area f di rallentamento
n afremoppervlak n
d Bremsfläche f

6652 SLOWING-DOWN np
    CROSS SECTION
The cross section for the slowing-down process.
f section f efficace de ralentissement
e sección f de retardación
i sezione f d'urto di rallentamento
n afremdoorsnede
d Bremsquerschnitt m

6653 SLOWING-DOWN DENSITY np
The number of neutrons per unit volume and unit time which, in slowing down, pass a given energy volume.
f densité f de ralentissement
e densidad f de retardación
i densità f di rallentamento
n afremdichtheid
d Bremsdichte f

6654 SLOWING-DOWN KERNEL ma
The probability that a neutron will go from one position to another while slowing down through a specified energy range.
f noyau m de l'intégrale de ralentissement
e núcleo m de retardación
i nucleo m di rallentamento

n integraalkern voor afremming
d Bremskern m

6655 SLOWING-DOWN LENGTH np
The square root of the slowing-down area.
f longueur f de ralentissement
e longitud f de retardación
i lunghezza f di rallentamento
n afremlengte
d Bremslänge f

6656 SLOWING-DOWN POWER np
For a given medium, the product of the average logarithmic energy decrement and the macroscopic neutron scattering cross section.
f pouvoir m de ralentissement
e poder m de retardación
i potere m di rallentamento
n afremvermogen n
d Bremsvermögen n

6657 SLOWING-DOWN TIME np
The average time needed by a neutron in a given medium to pass from one energy state to another one.
f temps m de ralentissement
e tiempo m de retardación
i tempo m di rallentamento
n afremtijd
d Bremszeit f

6658 SLUDGE rw
A wet solid settling out from a liquid.
f schlamm m de clarification, sédiment m fangeux
e sedimento m fangoso
i sedimento m fangoso
n brijig sediment n
d Klärschlamm m

6659 SLUG, see 2878 fu

6660 SLUG, ge
    STEAM VOID
In heat transfer, the pocket of vapo(u)r which is formed in tubes when the temperature of their heated surface, usually the tube surface, and of the bulk fluid are both above the boiling point of the fluid.
f bulle f de vapeur
e borbollón m de vapor
i bolla f di vapore
n dampbel
d Dampfblase f

6661 SLURRY rw
A watery mud or any substance resembling it.
f schlamm m, suspension f
e papilla f, suspensión f aguada
i melma f, sospensione f acquosa
n brij, natte suspensie
d nasse Suspension f, Schlamm m

6662 SLURRY REACTOR, see 2727

6663 SLURRYING   ch, fu, mi
Used as a method of transporting ore
during the leaching stages of uranium
extraction and as a means of transporting
spent ion exchange material of uranium
reprocessing methods.
f sédimentation *f*, traitement *m* du schlamm
e sedimentación *f*,
  tratamiento *m* de la papilla
i lavorazione *f* della melma,
  sedimentazione *f*
n brijverwerking
d Aufschlämmung *f*, Schlämmung *f*

6664 SMEAR TEST,   ra, sa
     WIPE TEST
A method of estimating the loose, i.e.
easily removed, radioactive contamination
from a surface by rubbing that surface
with a given material and examining the
material for collected radioactive
contamination.
f essai *m* de frottement
e prueba *f* de frotamiento
i prova *f* di strofinamento
n wrijfproef
d Reibeprüfung *f*, Wischtest *m*

6665 SNAP PROJECT, see 6415

6666 SODDITE,   mi
     SODDYITE
A secondary mineral containing about
71 % of U.
f soddyite *f*
e sodita *f*
i soddita *f*
n soddyiet *n*
d Soddyit *m*

6667 SODDY-FAJANS LAW, see 5776

6668 SODIUM   ch
Metallic element, symbol Na, atomic
number 11.
f sodium *m*
e sodio *m*
i sodio *m*
n natrium *n*
d Natrium *n*

6669 SODIUM COOLED REACTOR,   rt
     SODIUM REACTOR
A nuclear reactor in which the cooling
agent consists of liquid sodium.
f réacteur *m* refroidi au sodium
e reactor *m* refrigerado al sodio
i reattore *m* raffreddato al sodio
n reactor met natriumkoeling
d natriumgekühlter Reaktor *m*

6670 SODIUM-GRAPHITE REACTOR   rt
A nuclear reactor in which the coolant
consists of sodium and the moderator
of graphite.
f réacteur *m* sodium-graphite
e reactor *m* sodio-grafito

i reattore *m* sodio-grafite
n natrium-grafietreactor
d Natrium-Graphitreaktor *m*

6671 SODIUM METAPHOSPHATE   ch, cl
Useful addition to solutions for
decontaminating protective clothing in
active laundries.
f métaphosphate *m* de sodium
e metafosfato *m* de sodio
i metafosfato *m* di sodio
n natriummetafosfaat *n*
d Natriummetaphosphat *n*

6672 SODIUM PHOTON COUNTER   ct
A photon counter in which the sensitive
substance contains pure sodium.
f compteur *m* de photons à sodium
e contador *m* de fotones a sodio
i contatore *m* di fotoni a sodio
n natriumfotonenteller
d Natriumphotonenzähler *m*

6673 SODIUM URANATE   ch
f uranate *m* de sodium
e uranato *m* de sodio
i uranato *m* di sodio
n natriumuranaat *n*
d Natriumuranat *n*

6674 SODIUM-VOID COEFFICIENT   rt
In a fast neutron reactor, the reactivity
coefficient due to the presence of sodium
in the primary cooling circuit.
f coefficient *m* de sodium
e coeficiente *m* de sodio
i coefficiente *m* di sodio
n natriumcoëfficiënt
d Natriumkoeffizient *m*

6675 SOFT   rq, xr
Said of a radiation of low penetration.
f mou adj
e blando adj
i molle adj
n zacht adj
d weich adj

6676 SOFT COMPONENT   ct
That portion of cosmic radiation which is
absorbed in a moderate thickness of an
absorber.
f composante *f* molle
e componente *m* blando
i componente *m* molle
n zachte component
d weiche Komponente *f*

6677 SOFT RADIATION   ra
Ionizing radiation of relatively long wavelength and low penetrating power.
f rayonnement *m* mou
e radiación *f* blanda
i radiazione *f* molle
n zachte straling
d weiche Strahlung *f*

6678 SOFT SHOWER, see 908

6679 SOFT X-RAY xr
An X-ray having a comparatively long
wavelength and poor penetrating power.
f rayon *m* X mou
e rayo *m* X blando
i raggio *m* X molle
n zachte röntgenstraal
d weicher Röntgenstrahl *m*

6680 SOIL MOISTURE PROBE me
A probe using neutron scattering
technique for soil moisture measurement.
f sonde *f* d' humidimètre du sol
e sonda *f* de valorímetro de la humedad
del suelo
i sonda *f* d'umidimetro del suolo
n bodemvochtigheidssonde,
grondvochtigheidssonde
d Bodenfeuchtigkeitssonde *f*

6681 SOL-GEL PROCESS fu
Method for manufacturing ceramic fuels.
f procédé *m* sol-gel
e procedimiento *m* sol-gel
i processo *m* sol-gel
n sol-gel-proces *n*
d Sol-Gel-Verfahren *n*

6682 SOLAR ENERGY np
f énergie *f* solaire
e energía *f* solar
i energia *f* solare
n zonne-energie
d Sonnenenergie *f*

6683 SOLAR STREAM, cr
SOLAR WIND
Corpuscular emission of low energy produced by the sun, and composed mostly
of photons and electrons moving at a speed
of 400-500 km/s and of which the intensity
rises considerably during sun eruptions.
f vent *m* solaire
e emisión *f* solar de partículas
i emissione *f* solare di particelle
n deeltjesstroom van de zon
d solarer Teilchenstrom *m*

6684 SOLDERING ge
Joining of metals, using lower melting
non-ferrous metals as a solder.
f soudage *m*, soudure *f*
e soldadura *f*
i saldatura *f*
n solderen *n*
d Löten *n*

6685 SOLID ANGLE gp
The ratio of the area of the surface of the
portion of a sphere enclosed by the angle,
to the square of the radius of the sphere.
f angle *m* solide
e ángulo *m* sólido
i angolo *m* solido
n ruimtehoek
d Raumwinkel *m*

6686 SOLID ANODE, see 3186

6687 SOLID PHASE, ch, gp
SOLID STATE
A state of aggregation in which the
substance possesses both definite volume
and definite shape.
f état *m* solide, phase *f* solide
e estado *m* sólido, fase *f* sólida
i fase *f* solida, stato *m* solido
n vaste faze, vaste toestand
d feste Phase *f*, fester Zustand *m*

6688 SOLID SOLUTION, see 4468

6689 SOLID-STATE PHYSICS gp
That branch of physics which deals with
structure and properties of solids,
especially semiconductors.
f physique *f* des matières solides
e física *f* de las substancias sólidas
i fisica *f* delle sostanze solide
n physica van de vaste stoffen
d Festkörperphysik *f*

6690 SOLIDIFICATION, see 2833

6691 SOLIDIFYING UNIT, see 2832

6692 SOLUBILITY ch
Quality or state of being soluble.
f solubilité *f*
e solubilidad *f*
i solubilità *f*
n oplosbaarheid
d Löslichkeit *f*

6693 SOLUBILITY ma
Capability of being solved or explained.
f solubilité *f*
e solubilidad *f*
i risolvibilità *f*, solubilità *f*
n oplosbaarheid
d Lösbarkeit *f*

6694 SOLUBLE POISON rt
A nuclear poison used in the form of a
soluble compound.
f poison *m* soluble
e veneno *m* en solución
i veleno *m* in soluzione
n oplossingsgif *n*
d Lösungsgift *n*

6695 SOLUTION ma
The action or process of solving.
f solution *f*
e resolución *f*, solución *f*
i soluzione *f*
n oplossing
d Auflösung *f*

6696 SOLUTION, ch
DISSOLUTION, see 1904

6697 SOLUTION CHEMISTRY ch
The chemistry of substances in dissolved
state.
f chimie *f* des solutions
e química *f* de las soluciones
i chimica *f* delle soluzioni

n oplossingschemie
d Lösungschemie *f*

6698 SOLUTION TYPE REACTOR, see 317

6699 SOLUTION TYPE REACTOR TANK,
see 1417

6700 SOLVATE ch
Complex formed between an inorganic
compound and an organic solvent in
solvent extraction processes.
f solvate *m*
e solvato *m*
i solvato *m*
n solvaat *n*
d Solvat *n*

6701 SOLVENT ch
Usually a liquid which dissolves another
compound to form a homogeneous
one-phase liquid mixture.
f solvant *m*
e solvente *m*
i solvente *m*
n oplosmiddel *n*
d Lösungsmittel *n*

6702 SOLVENT EXTRACTION ch, is
The extraction of that component of a
solution which is present in excess.
f extraction *f* par solvants
e extracción *f* por solventes
i estrazione *f* per solventi
n oplossingsextractie
d Lösungsmittelextraktion *f*,
Solventextraktion *f*

6703 SOMATIC CELLS md
Body cells, usually having two sets of
chromosomes, as opposed to germ cells.
f cellules *pl* somatiques
e células *pl* somáticas
i cellule *pl* somatiche
n lichaamscellen *pl*, somatische cellen *pl*
d somatische Zellen *pl*

6704 SOMATIC EFFECT OF md
RADIATION
The effect limited to the exposed individual,
as distinguished from genetic effects.
f effet *m* somatique du rayonnement
e efecto *m* somático de la radiación
i effetto *m* somatico della radiazione
n somatisch stralingseffect *n*
d somatischer Strahlungseffekt *m*

6705 SOMATIC NUMBER md
The number of somatic cells in the body.
f nombre *m* somatique
e número *m* somático
i numero *m* somatico
n somatisch getal *n*
l somatische Zahl *f*

6706 SORET EFFECT, ch
THERMAL DIFFUSION
The phenomenon by which a temperature

gradient in a mixture of two or more fluids
tends to establish a concentration gradient.
f diffusion *f* thermique, effet *m* Soret
e difusión *f* térmica, efecto *m* Soret
i diffusione *f* termica, effetto *m* Soret
n soreteffect *n*, thermische diffusie
d Soret-Phänomen *n*, Thermodiffusion *f*

6707 SORPTION ch
A general term used in chemistry for
the processes of absorption, adsorption
and persorption.
f sorption *f*
e sorción *f*
i sorbimento *m*
n sorptie
d Sorption *f*

6708 SOURCE ACTIVITY, see 108

6709 SOURCE COMPARTMENT ra
Compartment containing the irradiated
cobalt-60 rods for industrial processes.
f compartiment *m* de source
e compartimiento *m* de fuente
i compartimento *m* di sorgente
n bronafdeling
d Quellenabteilung *f*

6710 SOURCE CONTAINER ra
The receptable for a sealed source, either
unmounted or in its source holder.
f récipient *m* de source
e recipiente *m* de fuente
i serbatoio *m* di sorgente
n bronvat *n*
d Quellenbehälter *m*

6711 SOURCE HOLDER ra
A mechanical support for a sealed source.
f support *m* de source
e soporte *m* de fuente
i sopporto *m* di sorgente
n bronhouder
d Quellenhaltevorrichtung *f*

6712 SOURCE INTERLOCK rt
In a reactor, a safety device for
inactivating the reactive space.
f verrouillage *m* de la source de neutrons
e enclavamiento *m* de la fuente de neutrones
i bloccaggio *m* della sorgente
n bronvergrendeling
d Quellenverriegelung *f*

6713 SOURCE OF ENERGY ge
f source *f* d'énergie
e fuente *f* de energía
i sorgente *f* d'energia
n energiebron
d Energiequelle *f*

6714 SOURCE OF RADIATION ra
The device from which radiation takes
place.
f source *f* de rayonnement
e fuente *f* de radiación
i sorgente *f* di radiazione

n stralingsbron
d Strahlungsquelle *f*

6715 SOURCE RANGE, see 4150

6716 SOURCE REACTOR   rt
A nuclear reactor specially designed to
supply a stable flux of neutrons having a
well-determined energy spectrum,
principally for conducting exponential or
shielding experiments or for calibrating
detectors.
f réacteur *m* source
e reactor *m* fuente
i reattore *m* sorgente
n bronreactor
d Quellenreaktor *m*

6717 SOURCE STRENGTH   ra
The strength of a radiation source may be
defined either by specifying its radiation
output or by stating the radioactivity of
its contents.
f intensité *f* de source
e intensidad *f* de fuente
i intensità *f* di sorgente
n bronsterkte
d Quellenstärke *f*

6718 SOURCE TO FILM DISTANCE, see 2762

6719 SPACE CHARGE   ec, np
The electric charge carried by a cloud or
stream of electrons or ions in a vacuum or
a region of low gas pressure, when the
charge is sufficient to produce local
changes in the potential distribution.
f charge *f* spatiale
e carga *f* espacial
i carica *f* spaziale
n ruimtelading
d Raumladung *f*

6720 SPACE-CHARGE BARRIER   ec, np
A potential barrier due to the space charge
effect.
f barrière *f* de charge spatiale
e barrera *f* de carga espacial
i barriera *f* di carica spaziale
n ruimteladingsbarrière
d Raumladungsbarriere *f*

6721 SPACE-CHARGE   ec, np
      COMPENSATION
Measures to counteract the space charge
effect.
f compensation *f* de la charge spatiale
e compensación *f* de la carga espacial
i compensazione *f* della carica spaziale
n ruimteladingscompensatie
d Raumladungskompensation *f*

6722 SPACE-CHARGE DENSITY   ec, np
The net electric charge per unit volume.
f charge *f* spatiale volumique
e carga *f* espacial volúmica
i carica *f* spaziale volumica

n volumieke ruimtelading
d Raumladung *f* je Volumeneinheit

6723 SPACE-CHARGE DISTORTION   ec, np
An undesired effect caused by space charge.
f distorsion *f* par charge spatiale
e distorsión *f* por carga espacial
i distorsione *f* per carica spaziale
n ruimteladingsvervorming
d Raumladungsverzerrung *f*

6724 SPACE-CHARGE EFFECT   ec, np
The effect of the electric charge existing
in the space between and adjacent to
electrodes due to the presence of
electrons and/or ions.
f effet *m* de charge spatiale
e efecto *m* de carga espacial
i effetto *m* di carica spaziale
n ruimteladingseffect *n*
d Raumladewirkung *f*, Raumladungseffekt *m*

6725 SPACE-CHARGED   is
      LIMITED OPERATION
A condition of operation of an electromagnetic separator under which optimum
focusing of the ion beam is lost due to
space charge.
f fonctionnement *m* limité par la charge
   spatiale
e funcionamiento *m* limitado por la carga
   espacial
i funzionamento *m* limitato dalla carica
   spaziale
n door ruimtelading begrensde isotopenscheiding
d durch Raumladung begrenzte
   Isotopentrennung *f*

6726 SPACE GROUP   cr
A mutually consistent infinite group of
symmetry elements by which are
determined the relative positions in which
atoms or sets of atoms can occur in a
unit cell.
f groupe *m* spatial
e grupo *m* espacial
i gruppo *m* spaziale
n ruimtegroep
d Raumgruppe *f*

6727 SPACE GROUP EXTINCTIONS   cr
The total absence of Bragg reflection from
certain sets of crystal planes, caused by
the symmetry of the space group.
f extinctions *pl* de groupes spatiaux
e extinciones *pl* de grupos espaciales
i estinzioni *pl* di gruppi spaziali
n ruimtegroependovingen *pl*
d Raumgruppenlöschungen *pl*

6728 SPACE LATTICE, see 771

6729 SPACE-TIME   gp
A space of four dimensions which specify
the space and time co-ordinates of an
event.

f   continuum *m* espace-temps,
    monde-espace-temps
e   mundo-espacio-tiempo
i   monde-spazio-tempo
n   ruimte-tijd-wereld
d   Raum-Zeit-Welt

6730 SPACE-TIME PATH OF A CLASSICAL
      PARTICLE, see 4000

6731 SPALLATION   np
      A nuclear reaction in which the energy of
      the incident particle or photon is so high
      that several nuclei are emitted from the
      target nucleus, which is thus reduced both
      in mass number and atomic number by
      several units.
f   spallation *f*
e   espalación *f*
i   spallazione *f*
n   spallatie
d   Spallation *f*

6732 SPALLATION FRAGMENT   np
      One of more than two or three particles
      ejected during a high energy bombardment.
f   fragment *m* de spallation
e   fragmento *m* de espalación
i   frammento *m* di spallazione
n   spallatiebrokstuk *n*
d   Spallationsbruchstück *n*

6733 SPALLING, see 2426

6734 SPARGE PIPE   cl, sa
      A pipe used for distribution of emergency
      cooling water.
f   tube *m* d'arrosage
e   tubo *m* rociador
i   tubo *m* d'annaffiatòio
n   sproeibuis
d   Spritzrohr *n*

6735 SPARGING, see 804

6736 SPARK CHAMBER   me
      A track chamber in which the paths of
      ionizing particles are indicated by a
      succession of sparks.
f   chambre *f* à étincelles
e   cámara *f* de chispas
i   camera *f* a scintille
n   vonkenvat *n*
d   Funkenkammer *f*

6737 SPARK DETECTOR, see 6228

6738 SPECIFIC ABSORPTION, see 6437

6739 SPECIFIC ACTIVITY,   np
      SPECIFIC RADIOACTIVITY
      The activity of a radioactive nuclide
      divided by the total mass of the nuclide
      present in the sample.
f   activité *f* massique
e   actividad *f* específica
i   attività *f* specifica

n   massieke activiteit, specifieke activiteit
d   spezifische Aktivität *f*

6740 SPECIFIC BINDING ENERGY   np
      The binding energy per particle.
f   énergie *f* de liaison spécifique
e   energía *f* de enlace específica
i   energia *f* di legame specifica
n   specifieke bindingsenergie
d   spezifische Bindungsenergie *f*

6741 SPECIFIC BURN-UP, see 2866

6742 SPECIFIC CHARGE, see 1010

6743 SPECIFIC ELECTRONIC CHARGE,
      see 2108

6744 SPECIFIC ENERGY   gp
      Internal energy per unit mass.
f   énergie *f* spécifique
e   energía *f* específica
i   energia *f* specifica
n   soortelijke energie
d   spezifische Energie *f*

6745 SPECIFIC GAMMA-RAY   gr
      CONSTANT,
      SPECIFIC GAMMA-RAY EMISSION
      For a nuclide emitting gamma radiation,
      the product of exposure rate at a given
      distance from a point source of that
      nuclide and the square of that distance
      divided by the activity of the source.
f   constante *f* spécifique de rayonnement
    gamma
e   constante *f* específica de radiación gamma
i   costante *f* specifica di radiazione gamma
n   specifieke gammastralingsconstante
d   spezifische Gammastrahlenkonstante *f*

6746 SPECIFIC GRAVITY,   gp
      SPECIFIC WEIGHT, UNIT WEIGHT
      The ratio between the density of a
      substance at a given temperature and the
      density of some substance associated with
      it.
f   poids *m* spécifique
e   peso *m* específico
i   peso *m* specifico
n   soortelijk gewicht *n*
d   Dichte *f*, spezifisches Gewicht *n*

6747 SPECIFIC GRAVITY   ch
      CONCENTRATION
      Concentration of chemical deposits by
      making use of the difference in specific
      gravity.
f   concentration *f* par gravité
e   concentración *f* por gravedad
i   concentrazione *f* per gravitá
n   graviteitsconcentratie
d   gravitative Konzentration *f*,
    Konzentration *f* im Schwerfeld

6748 SPECIFIC HEAT   gp
      The quantity of heat required to raise the

temperature of unit mass of a substance
by one degree of temperature.
f chaleur *f* spécifique
e calor *m* específico
i calore *m* specifico
n soortelijke warmte
d spezifische Wärme *f*

6749 SPECIFIC IONIZATION   np
The number of ion pairs formed per unit
distance along the track of an ion passing
through matter.
f ionisation *f* linéique,
   ionisation *f* spécifique
e ionización *f* específica
i ionizzazione *f* specifica
n ionisatie per eenheid van weglengte,
   soortelijke ionisatie
d spezifische Ionisation *f*

6750 SPECIFIC IONIZATION   np
      COEFFICIENT
The average number of pairs of ions with
opposite charges that electrons with a
specified kinetic energy produce in a gas
at a specified pressure and temperature
over a unit distance.
f coefficient *m* spécifique d'ionisation
e coeficiente *m* específico de ionización
i coefficiente *m* specifico d'ionizzazione
n differentiële ionisatiecoëfficiënt
d differentieller Ionisationskoeffizient *m*

6751 SPECIFIC POWER   rt
The power produced per unit mass of fuel
in a reactor.
f puissance *f* massique
e potencia *f* específica
i potenza *f* specifica
n massiek vermogen *n*, specifiek vermogen *n*
d spezifische Leistung *f*

6752 SPECIFIC STRENGTH   gr, np
The gamma-ray strength per unit volume
of a radioactive source.
f intensité *f* spécifique
e intensidad *f* específica
i intensità *f* specifica
n gammaquanten *pl* per cm^3,
   specifieke intensiteit,
d Gammaquanten *pl*/cm^3,
   spezifische Stärke *f*

6753 SPECTRAL DISTRIBUTION   sp
The radiant power emitted by a source as
a function of wavelength and frequency.
f distribution *f* spectrale
e distribución *f* espectral
i distribuzione *f* spettrale
n spectrale verdeling
d Spektralverteilung *f*

6754 SPECTRAL HARDENING   sp
The shifting towards the higher energy
bands of the energy spectrum of thermal
neutrons, due to the partial capture of the
medium with which they are in equilibrium.

f durcissement *m* du spectre
e endurecimiento *m* del espectro
i indurimento *m* dello spettro
n spectrumverharding
d spektrale Härtung *f*

6755 SPECTRAL LINE   sp
One of the lines together forming the
spectrum.
f raie *f* spectrale
e línea *f* espectral
i riga *f* spettrale
n spectraallijn
d Spektrallinie *f*

6756 SPECTRAL QUANTUM YIELD   qm, sp
The average number of electrons photo-
electrically emitted from a photocathode
per incident photon of a given wavelength.
f rendement *m* quantique spectral
e rendimiento *m* cuántico espectral
i rendimento *m* quantico spettrale
n spectrale quantumopbrengst
d spektrale Quantenausbeute *f*

6757 SPECTRAL RESPONSE   ec
      CHARACTERISTIC,
      SPECTRAL RESPONSE CURVE
Of a photocathode the curve expressing the
relationship between quantum conversion
efficiency and the wavelength of the
incident radiation.
f caractéristique *f* spectrale,
   courbe *f* de réponse spectrale
e característica *f* de respuesta espectral
i caratteristica *f* spettrale
n spectrale gevoeligheidskarakteristiek,
   spectrumkarakteristiek
d spektrale Verteilungscharakteristik *f*

6758 SPECTRAL SHIFT CONTROL   co, rt
A special type of moderator control.
f commande *f* par dérive spectrale
e regulación *f* por desplazamiento espectral
i regolazione *f* con spostamento spettrale
n regeling met spectrumverschuiving
d Spektralsteuerung *f*

6759 SPECTRAL SHIFT REACTOR   rt
A nuclear reactor in which, for control or
other purposes, the neutron spectrum may
be adjusted by varying the properties or
amount of moderator.
f réacteur *m* à dérive spectrale
e reactor *m* de desplazamiento espectral
i reattore *m* a spostamento spettrale
n reactor met spectrumverschuiving
d Reaktor *m* mit Spektralverschiebung

6760 SPECTROGRAM   sp
A record produced by a spectrograph.
f spectrogramme *m*
e espectrograma *m*
i spettrogramma *m*
n spectrogram *n*
d Spektrogramm *n*

6761 SPECTROGRAPH   me, sp
   An instrument used to produce a record
   of a spectrum.
   f   spectrographe *m*
   e   espectrógrafo *m*
   i   spettrografo *m*
   n   spectrograaf
   d   Spektrograph *m*

6762 SPECTROGRAPHY   sp
   The recording of a spectrum.
   f   spectrographie *f*
   e   espectrografía *f*
   i   spettrografia *f*
   n   spectrografie
   d   Spektrographie *f*

6763 SPECTROMETER, see 2288

6764 SPECTROSCOPE   sp
   An instrument that spreads individual
   wavelengths in a radiation to permit
   observation of the resulting spectrum.
   f   spectroscope *m*
   e   espectroscopio *m*
   i   spettroscopio *m*
   n   spectroscoop
   d   Spektroskop *n*

6765 SPECTROSCOPY   sp
   That branch of physical science that
   deals with the measurement and analysis
   of spectra.
   f   spectroscopie *f*
   e   espectroscopia *f*
   i   spettroscopia *f*
   n   spectroscopie
   d   Spektroskopie *f*

6766 SPECTRUM   sp
   A visual display, a photographic record
   or a plot of the distribution of the intensity
   of radiation of a given kind as a function
   of its wavelength, energy, frequency,
   momentum, mass or any related quantity.
   f   spectre *m*
   e   espectro *m*
   i   spettro *m*
   n   spectrum *n*
   d   Spektrum *n*

6767 SPECTRUM HARDENING, see 4716

6768 SPECTRUM STABILIZER   ec, sp
   A sub-assembly designed to be associated
   with a radiation spectrometer for
   compensating the drift of the detectors and
   the electronic circuit in order to stabilize
   the abscissa of a defined peak.
   f   stabilisateur *m* de spectre
   e   estabilizador *m* de espectro
   i   stabilizzatore *m* di spettro
   n   spectrumstabilisator
   d   Impulshöhenspektrometerstabilisierung *f*

6769 SPEED FACTOR, see 3538

6770 SPENT FUEL, see 4813

6771 SPENT FUEL   fu
   HANDLING SYSTEM
   Equipment used in nuclear reactor
   technology to remove spent fuel from a
   reactor and package this highly radioactive material for transport to a
   processing facility.
   f   traitement *m* de combustible épuisé
   e   tratamiento *m* de combustible agotado
   i   trattamento *m* di combustibile esaurito
   n   behandeling van verbruikte splijtstof
   d   Behandlung *f* von ausgebranntem
      Brennstoff

6772 SPENT FUEL STORAGE   fu
   The storage of spent fuel in a fuel cooling
   installation.
   f   emmagasinement *m* de combustible épuisé
   e   almacenamiento *m* de combustible agotado
   i   immagazzinamento *m* di combustibile
      esaurito
   n   opslaan *n* van verbruikte splijtstof
   d   Speicherung *f* von ausgebranntem Brennstoff

6773 SPERMATOGENESIS   md
   The successive divisions which transform
   the spermatogonium into the mature
   spermatozoon.
   f   spermatogenèse *f*
   e   espermatogénesis *f*
   i   spermatogenesi *f*
   n   spermatogenese
   d   Spermatogenese *f*

6774 SPEVACK PROCESS (US), see 3358

6775 SPHERICAL CONTAINMENT VESSEL,
      see 1328

6776 SPHERICAL REACTOR   rt
   A nuclear reactor in which the core is
   approximately sphere-shaped.
   f   réacteur *m* sphérique
   e   reactor *m* esférico
   i   reattore *m* sferico
   n   bolreactor
   d   Kugelreaktor *m*

6777 SPHERICAL WELL, see 6029

6778 SPIKE, see 6425

6779 SPIKING   is
   The act of adding an isotopic tracer.
   f   addition *f* d'indicateur
   e   adición *f* de indicador
   i   addizione *f* di tracciante
   n   indicatortoevoeging
   d   Tracerzugabe *f*

6780 SPIKING,   rt
      SEEDING, see 6428

6781 SPILL   rw
   The accidental release of radioactive
   liquids.
   f   perte *f* accidentelle
   e   pérdida *f* accidental
   i   perdita *f* accidentale

n   weglekken *n*
d   Leckverlust *m*

6782 SPIN   np
Of an elementary particle, its angular
momentum in the absence of orbital
motion.
f   spin *m*
e   espín *m*, spin *m*
i   spin *m*
n   spin
d   Spin *m*

6783 SPIN ANGULAR MOMENTUM   ma
The intrinsic angular momentum of an
elementary particle.
f   moment *m* angulaire du spin
e   momento *m* angular del espín
i   momento *m* angolare dello spin
n   spinimpulsmoment *n*
d   Spindrehimpuls *m*

6784 SPIN-DEPENDENT FORCE   np
Force between two particles which depend
on their relative spin orientations and
possibly on their spin directions relative
to the line joining the particles.
f   force *f* dépendant du spin
e   fuerza *f* dependiente del espín
i   forza *f* dipendente dello spin
n   van de spin afhankelijke kracht
d   spinabhängige Kraft *f*

6785 SPIN EFFECT   np
The influence of the spin on the normal
chemical properties of the molecules,
especially with the light nucleon species.
f   effet *m* de spin
e   efecto *m* de espín
i   effetto *m* di spin
n   spineffect *n*
d   Spineffekt *m*

6786 SPIN-LATTICE RELAXATION   np
The relaxation process by which the electrons in a solid return to an equilibrium
distribution after a disturbance that
places more electron spins in states of
high energy than would be in such states
according to the Boltzmann factor.
f   relaxation *f* spin-réseau
e   relajación *f* espín-red
i   rilassamento *m* spin-reticolo
n   spin-rooster-relaxatie
d   Spin-Gitter-Relaxation *f*

6787 SPIN MAGNETIC MOMENT   np
The magnetic moment of a magneton
connected with the spin impuls moment
of an electron.
f   moment *m* magnétique du spin
e   momento *m* magnético del espín
i   momento *m* magnetico dello spin
n   magnetisch spinmoment *n*
d   magnetisches Spinmoment *n*

6788 SPIN MAGNETIC RESONANCE   np
The magnetic moment associated with the
intrinsic spin of a particle.

f   résonance *f* magnétique du spin
e   resonancia *f* magnética del espín
i   risonanza *f* magnetica dello spin
n   magnetische spinresonantie
d   magnetische Spinresonanz *f*

6789 SPIN-ORBIT COUPLING   np
The interaction between intrinsic and
orbital angular momentum of a particle.
f   couplage *m* spin-orbite
e   acoplamiento *m* espín-órbita
i   accoppiamento *m* spin-orbita
n   spin-baan-koppeling
d   Spin-Bahn-Kopplung *f*

6790 SPIN ORIENTATION   np
The direction of rotation of the spin.
f   orientation *f* du spin
e   orientación *f* del espín
i   orientamento *m* dello spin
n   spinorientatie, spinrichting
d   Spinorientierung *f*, Spinrichtung *f*

6791 SPIN QUANTUM NUMBER   np
An integral term in the expression for the
contribution to the total angular
momentum of the electron that is due to
the rotation of the electron on its own axis.
f   nombre *m* quantique de spin
e   número *m* cuántico de espín
i   numero *m* quantico di spin
n   spinquantumgetal *n*
d   Spinquantenzahl *f*

6792 SPIN STATE   gp, qm
A system is said to be in a definite spin
state when the quantum mechanical wave
function describing the system is an
eigenfunction of the various spin operators
corresponding to the square of the total
spin angular momentum and the
component(s) of the spin angular momentum
being used to designate the system.
f   état *m* de spin
e   estado *m* de espín
i   stato *m* di spin
n   spintoestand
d   Spinzustand *m*

6793 SPINDLE-SHAPED PLASMOID   pp
A discrete piece of plasma shaped like
a spindle.
f   plasmoïde *m* fuselé
e   plasmoide *m* husiforme
i   plasmoide *m* fusiforme
n   spilvormig plasmoïde *n*
d   spindelförmiges Plasmoid *n*

6794 SPINNING ELECTRON   np
An electron rotating around its axis.
f   électron *m* tournant
e   electrón *m* giratorio
i   elettrone *m* rotante
n   draaiend elektron *n*
d   Elektron *n* mit Eigendrehimpuls

6795 SPINNING TOP   xr
A disk of dense material having an
eccentric aperture which, when spun above

a film, enables a record to be made of
variations of the X-ray output during the
exposure.
f toupie *f* en plomb
e peonza *f* de plomo
i trottola *f* di piombo
n tolletje *n*
d Bleihandkreisel *m*

6796 SPINTHARISCOPE me
An instrument in which scintillations are
visually observed through a microscope.
f spinthariscope *m*
e espintariscopio *m*
i spintariscopio *m*
n spinthariscoop
d Spinthariskop *n*

6797 SPIRAL RIDGED ACCELERATOR pa
A cyclotron or synchrotron which uses
spiral ridges on the surface of the magnet
pole tips to improve beam focusing and
enable radial redesign of the accelerator.
f accélérateur *m* à pôles magnétiques à
rainures en hélices
e acelerador *m* de polos magnéticos con
ranuras en hélices
i acceleratore *m* a poli magnetici con
scanalatura elicoidale
n versneller met van spiraalgroeven
voorziene magneetpolen
d Beschleuniger *m* mit von Spiralrillen
versehenen Magnetpolen

6798 SPLEEN md
A large gland-like organ one of whose
functions is to disintegrate the red blood
corpuscles and release hemoglobin.
f rate *f*
e bazo *m*
i milza *f*, splene *m*
n milt
d Milz *f*

6799 SPLIT-FLOW REACTOR rt
A nuclear reactor in which the coolant
enters at the centre(er) of the core and
flows outward at both ends or vice-versa.
f réacteur *m* à réfrigérant divisé
e reactor *m* de refrigerante dividido
i reattore *m* a refrigerante diviso
n reactor met centraal naar binnen geleid
koelmiddel,
reactor met gedeelde koelmiddelstroom
d Reaktor *m* mit geteiltem Kühlmittelfluss

6800 SPLIT LINE SOURCE ra, sp
A Mössbauer source producing a split
spectral line.
f source *f* à raie spectrale subdivisée
e fuente *f* de línea espectral subdividida
i sorgente *f* a riga spettrale suddivisa
n bron met gespleten spectrumlijn
d Quelle *f* mit aufgespalteter Spektrallinie

6801 SPLITTING, see 1786

6802 SPLITTING RATIO ch, is
In isotope separation, the ratio between the
quantity of enriched product leaving a
separation element and the quantity entering
the element.
f coefficient *m* de partage
e coeficiente *m* de partición
i coefficiente *m* di partizione
n delingscoëfficiënt
d Teilungskoeffizient *m*

6803 SPONGE METAL, see 653

6804 SPONTANEOUS DECAY, np
SPONTANEOUS DISINTEGRATION,
SPONTANEOUS TRANSMUTATION
The change of radioactive elements into
other radioactive elements, which form
a radioactive series or family, the end of
which is a stable isotope.
f désintégration *f* spontanée
e desintegración *f* espontánea
i disintegrazione *f* spontanea
n spontaan verval *n*
d spontaner Zerfall *m*

6805 SPONTANEOUS FISSION np
Nuclear fission which occurs without the
addition of particles or energy to the
nucleus.
f fission *f* spontanée
e fisión *f* espontánea
i fissione *f* spontanea
n spontane splijting
d spontane Spaltung *f*

6806 SPONTANEOUS NUCLEAR np
REACTION
The spontaneous decay of a nucleus into
other nuclei and elementary particles.
f réaction *f* nucléaire spontanée
e reacción *f* nuclear espontánea
i reazione *f* nucleare spontanea
n spontane kernreactie
d spontane Kernreaktion *f*

6807 SPONTANEOUS TRANSFORMATION gp
A transformation occurring by virtue of
inherent properties or energy as
distinguished from processes carried out
by deliberate application of external force.
f transformation *f* spontanée
e transformación *f* espontánea
i trasformazione *f* spontanea
n spontane omzetting
d spontane Umwandlung *f*

6808 SPRAY POINTS pa
A row of sharp points charged to a high
D.C. potential, the purpose of which is to
charge and discharge the conveyer belt in
a Van de Graaff generator.
f peignes *pl*, points *pl* de décharge
e puntas *pl* de descarga,
puntas *pl* de inducción
i punte *pl* di scarica
n sproeipunten *pl*
d Sprühstellen *pl*

6809 SPRAY RADIATION   ra
        TREATMENT,
          TOTAL BODY RADIATION,
          WHOLE BODY EXPOSURE
    A radiation therapeutic technique whereby
    the entire body is irradiated.
 f irradiation *f* globale, irradiation *f* totale
 e irradiación *f* total
 i irradiazione *f* corporale totale
 n totale bestraling
 d Ganzkörperbestrahlung *f*

6810 SPRINKLING FILTER, see 641

6811 SPURIOUS COUNTS   ct
    Counts caused by any agency other than
    the passage into or through a radiation
    detector of photons or particles to which
    it is sensitive.
 f coups *pl* parasites
 e impulsos *pl* espurios,
      recuentos *pl* accidentales
 i impulsi *pl* falsi
 n valse telpulsen *pl*
 d Fehlstösse *pl*, unechte Zählimpulse *pl*,
      unechte Zählstösse *pl*

6812 SPURIOUS PULSE, see 3557

6813 SQUARE CASCADE   is
    A cascade in which the turnover is the same
    in each stage.
 f cascade *f* carrée, cascade *f* constante
 e cascada *f* constante, cascada *f* cuadrada
 i cascata *f* costante, cascata *f* quadrata
 n rechthoekige cascade
 d rechteckige Kaskade *f*

6814 SQUARE WELL, see 6029

6815 SQUARED-OFF CASCADE   is
    An approximation of an ideal cascade and
    comprising sections in each of which all
    the stages have the same turnover.
 f cascade *f* arrondie
 e cascada *f* descuadrada
 i cascata *f* squadrata
 n afgeronde cascade
 d abgerundete Kaskade *f*

6816 STABILITY   ge
    As applied to a reactor, a tendency to hold
    a steady power level without action by the
    control system.
 f stabilité *f*
 e estabilidad *f*
 i stabilità *f*
 n stabiliteit
 d Stabilität *f*

6817 STABILITY PARAMETER   np

    Parameter equal to the relation $\dfrac{Z^2}{A}$ where

    Z is the atomic number and A the mass
    number of a given element.
 f paramètre *m* de stabilité
 e parámetro *m* de estabilidad
 i parametro *m* di stabilità
 n stabiliteitsparameter
 d Stabilitätsparameter *m*

6818 STABILIZED GLASS   ra
    Glass which darkens less than ordinary
    glass when it is irradiated.
 f verre *m* non-décolorant
 e vidrio *m* no descolorante
 i vetro *m* non decolorante
 n niet-verkleurend glas *n*
 d entfärbungsfreies Glas *n*

6819 STABILIZED PINCH   pp
    A controlled thermonuclear device in the
    shape of a cylinder or torus which uses
    the self-constricting pinch effect for
    confinement of a plasma and an axial
    magnetic field for stabilization of this
    plasma-field configuration against
    hydrodynamic instabilities.
 f striction *f* stabilisée
 e estricción *f* estabilizada
 i contrazione *f* stabilizzata
 n gestabiliseerde insnoering
 d stabilisierte Einschnürung *f*

6820 STABILIZED POWER   ge
        SUPPLY UNIT
    An electric power supply unit the voltage
    (current) of which remains, as a function
    of time, within two specified limits when
    operating at its nominal condition.
 f élément *m* d'alimentation stabilisée
 e unidad *f* de alimentación estable
 i elemento *m* alimentatore stabilizzato
 n gestabiliseerde voedingseenheid
 d stabilisierter Netzteil *m*

6821 STABILIZER   xr
    A device for maintaining constant tube
    voltage or current.
 f stabilisateur *m*
 e estabilizador *m*
 i stabilizzatore *m*
 n stabilisator
 d Stabilisator *m*

6822 STABILIZING MAGNETIC FIELD   pp
    In the thermonuclear program, a magnetic
    field which is added to an experimental
    device to increase the stability of the
    plasma-field configuration.
 f champ *m* magnétique stabilisateur
 e campo *m* magnético estabilizador
 i campo *m* magnetico stabilizzatore
 n stabiliserend magnetisch veld *n*
 d stabilisierendes Magnetfeld *n*

6823 STABLE   ge
    Of an atomic or nuclear system, incapable
    of spontaneous changes.
 f stable adj
 e estable adj
 i stabile adj
 n stabiel adj
 d stabil adj

6824 STABLE ISOTOPE, see 4792

6825 STABLE NUCLIDE   ch, np
A nuclide showing no radioactivity.
f nucléide *f* stable
e núclido *m* estable
i nuclido *m* stabile
n stabiel nuclide *n*
d stabiles Nuklid *n*

6826 STABLE ORBIT, see 2344

6827 STABLE REACTOR PERIOD   rt
The reactor period after a lapse of
sufficient time to permit the contribution
of transient terms to damp out.
f constante *f* de temps stable du réacteur
e perfodo *m* estable del reactor
i tempo *m* di divergenza stabile del reattore
n stabiele tijdconstante van de reactor
d stabile Zeitkonstante *f* des Reaktors

6828 STABLE STATE,   ge
STEADY STATE
A system incapable of spontaneous
changes.
f état *m* stable
e estado *m* estable
i stato *m* stabile
n stabiele toestand
d stabiler Zustand *m*

6829 STABLE TRACER ISOTOPE   is
A stable isotope used as a tracer.
f isotope *m* traceur stable
e isótopo *m* trazador estable
i isotope *m* tracciante stabile
n stabiel speurderisotoop *n*
d stabiles Indikatorisotop *n*

6830 STACK   ph
In the photographic·emulsion technique,
a system composed of a number of
pellicles, one on top of another, which
are exposed together for the study of
penetrating particles.
f empilement *m*, pile *f* de pellicules
e pila *f* de películas
i pila *f* di pellicole
n emulsiepakket *n*
d Emulsionspaket *n*, Schichtenstapel *m*

6831 STACK GASES   rw
Gaseous waste products released to the
atmosphere through chimney stacks.
f fumées *pl*, gaz *pl* de combustion
e gases *pl* de combustión, humos *pl*
i fumi *pl*, gas *pl* di combustione
n rookgassen *pl*, schoorsteengassen *pl*
d Rauchgase *pl*

6832 STAGE   is
A stage of a cascade consists of all
separative elements operating in parallel
on material of the same concentration.
f étage *m*
e etapa *f*
i stadio *m*
n trap
d Stufe *f*

6833 STAGE CASCADE, see 901

6834 STAGE CIRCULATION,   is
STAGE MASS FLOW, TURNOVER
The sum of flows entering a stage from
other stages in a cascade.
f circulation *f* totale
e circulación *f* total
i circolazione *f* totale
n totale omloop
d Gesamtumlauf *m*

6835 STAGE EFFICIENCY   ch, is
A form of contactor used, e.g., in solvent
extraction stage given by the change in
concentration across stage expressed as
a percentage of the change across a
theoretical stage.
f rendement *m* d'étage
e rendimiento *m* de etapa
i rendimento *m* di stadio
n traprendement *n*
d Stufenausbeute *f*

6836 STAGE HOLD-UP   is
The hold-up in a single stage.
f charge *f* d'étage
e carga *f* de etapa
i carica *f* di stadio
n trapinhoud
d Stufeneinsatz *m*

6837 STAGE NOISE   ph
In the photographic emulsion technique,
the imperfection of the mechanical stage
of a microscope which results in the
motion not being strictly rectilinear.
f bruit *m* de tablette
e ruido *m* de platina
i rumore *m* di piatto
n tafelruis
d Kreuztischrauschen *n*

6838 STAGE SEPARATION FACTOR,
see 2042

6839 STAGE-WISE CONTACTOR   ch, is
A form of contactor used,e.g.,in solvent
extraction.
f appareil *m* de contact par étage
e aparato *m* de contacto por etapa
i apparecchio *m* di contatto per stadio
n contactapparaat *n* per trap
d Kontaktapparat *m* je Stufe

6840 STAGNATION POINT   gp
A point at which moving fluid comes
entirely at rest.
f point *m* de stagnation
e punto *m* neutro
i punto *m* di stagnamento
n stuwpunt *n*
d Staupunkt *m*

6841 STAINLESS STEEL   ge, mt
Material used for general purposes and in
particular for manufacturing Mössbauer
sources.

f   acier *m* inoxydable,
    acier *m* résistant à la corrosion
e   acero *m* corrosiorresistente,
    acero *m* inoxidable
i   acciaio *m* inossidabile,
    acciaio *m* resistente alla corrosione
n   corrosievast staal *n*, roestvast staal *n*
d   nichtrostender Stahl *m*, rostfreier Stahl *m*

6842 STANDARD ABSORBER   ab, te
    Material used in testing Mössbauer
    sources.
f   absorbant *m* étalon
e   absorbedor *m* patrón
i   assorbente *m* campione
n   standaardabsorbens *n*
d   Normalabsorbens *n*

6843 STANDARD ABSORPTION CURVES   ra
    Graphs showing the amount of radiation
    of a beam of röntgen rays transmitted
    by increasing thicknesses of absorbing
    material, usually with specified
    röntgen-ray tube voltages of constant
    potential.
f   courbes *pl* d'atténuation de référence
e   curvas *pl* de atenuación de referencia
i   curve *pl* normali d'attenuazione
n   standaardabsorptiekrommen *pl*
d   Standardabsorptionskurven *pl*

6844 STANDARD DEVIATION   ma
    In statistical analysis a measure of the
    deviation of the individual values of a
    series from their mean value.
f   déviation *f* normale
e   desviación *f* normal
i   deviazione *f* normale
n   normale afwijking
d   normale Abweichung *f*

6845 STANDARD ERROR   me
    The standard deviation of the mean
    experimental result.
f   erreur *f* étalon
e   error *m* patrón
i   errore *m* campione
n   standaardfout
d   Normalfehler *m*

6846 STANDARD IONIZATION   ic
        CHAMBER
    Ionization chamber for the absolute
    measurement of exposure (doses).
f   chambre *f* d'ionisation étalon,
    chambre *f* étalon
e   cámara *f* de ionización patrón
i   camera *f* d'ionizzazione campione
n   standaardionisatievat *n*
d   Normalionisationskammer *f*

6847 STANDARD MAN   ra
    In the recommendations of the International
    Commission on Radiological Protection
    (ICRP) the averaged characteristics of
    the adult human body.
f   homme *m* standard

e   hombre *m* patrón
i   uomo *m* campione
n   standaardmens
d   Normalmensch *m*

6848 STANDING-OFF DOSE   ra
    That maximum dose of radiation which
    personnel are allowed to receive as an
    occupational exposure before they are
    moved to work not involving exposure to
    radiation.
f   dose *f* de changement de poste,
    dose *f* maximale admissible profession-
    nelle
e   dosis *f* de traslado,
    dosis *f* máxima permisible profesional
i   dose *f* di trasferimento,
    dose *f* massima ammissibile
    professionale
n   maximaal toelaatbare beroepsdosis,
    overplaatsingsdosis
d   höchstzugelassene berufsbedingte Dosis *f*,
    Versetzungsdosis *f*

6849 STANDPIPE   rt
    An open vertical pipe connected to a
    pipe line, to ensure that the pressure head
    at that point cannot exceed the length of
    that pipe.
f   tube *m* ascendant ouvert
e   tubo *m* vertical abierto
i   serbatoio *m* piezometrico
n   standpijp
d   Standrohr *n*

6850 STAR   ic, ph
    A group of tracks due to ionizing particles
    originating at a common point, e.g. in a
    nuclear emulsion or in a cloud chamber,
    so named because of its appearance.
f   étoile *f*
e   estrella *f*
i   stella *f*
n   ster
d   Stern *m*

6851 START-UP   rt
f   démarrage *m*
e   arranque *m*, puesta *f* en marcha
i   avviamento *m*
n   start
d   Start *m*

6852 START-UP ACCIDENT   rt
    An accident occurring when starting a
    reactor before the period meters have
    started to function.
f   accident *m* au démarrage
e   accidente *m* a la puesta en marcha
i   incidente *m* all'avviamento
n   onregelmatigheid bij het starten
d   Startzwischenfall *m*,
    Unfall *m* beim Anfahren des Reaktors

6853 START-UP PROCEDURE   rt
    A specified procedure for bringing a
    reactor into operation.

f   méthode f de démarrage
e   método m de puesta en marcha
i   norme pl d'avviamento
n   op gang brengen n,
    starten n en op niveau brengen n
d   Startverfahren n

6854 START-UP TIME   is
The time during which a separation plant
is operated without withdrawing product
and whilst the required isotope concentration is being reached at the output.
f   temps m de démarrage
e   tiempo m de puesta en marcha
i   tempo m d'avviamento
n   starttijd
d   Startzeit f

6855 STARTING TIME, see 2346

6856 STARTING VOLTAGE   ct
For a counter tube, the minimum voltage
that must be applied to obtain counts with
the particular circuit with which it is
associated.
f   tension f d'amorçage
e   tensión f de cebado
i   tensione f d'accensione
n   ontsteekspanning
d   Einsatzspannung f, Zündspannung f

6857 STATIC ELIMINATOR   sa
f   éliminateur m d'électrostatique
e   eliminador m de electrostática
i   eliminatore m d'elettrostatica
n   opheffer van statische elektriciteit
d   Antistatikgerät n,
    Gerät n zur Beseitigung elektrostatischer
    Aufladungen

6858 STATIC REFLECTOR, see 2684

6859 STATIC VALUE   ge
The slope of the line from the origin to the
operating point on the appropriate
characteristic curve.
f   valeur f statique
e   valor m estático
i   valore m statico
n   statische waarde
d   statischer Nert m

6860 STATIONARY-ANODE TUBE   xr
An X-ray tube having a stationary anode.
f   tube m à anode fixe
e   tubo m de ánodo estacionario
i   tubo m ad anodo stazionario
n   buis met stilstaande anode
d   Festanodenröhre f

6861 STATIONARY GRID, see 4173

6862 STATIONARY STATE   gp, qm
One of the discrete energy states in which
a quantized particle or system may
exist, according to the quantum theory.
f   état m stationnaire

e   estado m estacionario
i   stato m stazionario
n   stationaire toestand
d   stationärer Zustand m

6863 STATISTICAL ANALYSIS   an
Applies to the analysis of a combination of
data obtained from individual events or
from separate entries or events.
f   analyse f statistique
e   análisis f estadística
i   analisi f statistica
n   statistische analyse
d   statistische Analyse f

6864 STATISTICAL COUNTER TIME-LAG,
see 1461

6865 STATISTICAL COUNTING ERROR,
see 1486

6866 STATISTICAL ERROR   me
The random error involved in any
measurement.
f   erreur f statistique
e   error m estadístico
i   errore m statistico
n   statistische fout
d   statistischer Fehler m

6867 STATISTICAL FLUCTUATION,   np
STATISTICAL STRAGGLING
That variation in range, ionization or
direction which is due to fluctuations in the
distance between collisions in the stopping
medium and in the energy loss and
deflection angle per collision.
f   dispersion f statistique,
    fluctuation f statistique
e   dispersión f estadística,
    fluctuación f estadística
i   dispersione f statistica,
    fluttuazione f statistica
n   statistische fluctuatie,
    statistische strooiing
d   statistische Streuung f

6868 STATISTICAL METHOD   is
A method of separating isotopes depending
on small differences in the average
behavio(u)r of the molecules of the
different isotopes.
f   méthode f statistique
e   método m estadístico
i   metodo m statistico
n   statistische methode
d   statistisches Verfahren n

6869 STATISTICAL MODEL   np
A nuclear model in which the subsequent
emission of particles is often described
by invoking the method of statistical
mechanics.
f   modèle m statistique
e   modelo m estadístico
i   modello m statistico
n   statistisch model n
d   statistisches Modell n

6870 STATISTICAL UNCERTAINTY   ma, me
General term for the estimated amount by
which the observed or estimated value of
a quantity may depart from the true value.
f incertitude f statistique
e incertidumbre f estadística
i incertezza f statistica
n statistische onzekerheid
d statistische Unsicherheit f

6871 STATISTICAL WEIGHT   ma
The number of microscopic states
contained in a macroscopic state.
f poids m statistique
e peso m estadístico
i peso m statistico
n statistisch gewicht n
d statistisches Gewicht n

6872 STATISTICAL WEIGHT,   rt
      WEIGHTING FACTOR,
      WEIGHTING FUNCTION
In a nuclear reactor, a measure on the
effect on reactivity of inserting a small
absorber at some given point.
f fonction f de pondération
e función f de ponderación
i funzione f di peso
n gewichtsfunctie
d Einflussfunktion f

6873 STATOR PUMP   cl
A type of pump having no moving parts and
using the magnetic induction principle.
f pompe f à induction magnétique
e bomba f de inducción magnética
i pompa f ad induzione magnetica
n magnetische inductiepomp
d magnetische Induktionspumpe f

6874 STEADY STATE   co
State reached, e.g., in a measuring or
control system under steady conditions.
f état m permanent, régime m établi
e régimen m estacionario,
   régimen m permanente
i regime m d'equilibrio,
   regime m stazionario
n blijvende toestand
d Beharrungszustand m

6875 STEADY STATE,   ge
      STABLE STATE, see 6828

6876 STEADY STATE COSMOLOGY, see 1343

6877 STEADY STATE DEVIATION, see 4966

6878 STEADY STATE ERROR   me
An error which remains after initial
transients or fluctuations have been
damped out.
f erreur f permanente
e error m permanente
i errore m permanente
n blijvende fout
d permanenter Fehler m

6879 STEAM-COOLED REACTOR,   rt
      SUPERHEATED STEAM REACTOR
A nuclear reactor cooled by superheated
steam.
f réacteur m à vapeur surchauffée
e reactor m de vapor recalentado
i reattore m a vapore surriscaldato
n met oververhitte stoom gekoelde reactor
d heissdampfgekühlter Reaktor m

6880 STEAM CYCLE   rt
f cycle m de vapeur
e ciclo m de vapor
i ciclo m di vapore
n stoomkringloop
d Dampfkreislauf m

6881 STEAM DISTILLATION   ch
A distillation in which vaporization of the
volatile constituents is effected at a
temperature lower than the normal boiling
point by the introduction of steam directly
into charge.
f distillation f à la vapeur d'eau,
   entraînement m à la vapeur d'eau
e destilación f por arrastre de vapor de agua
i distillazione f in corrente di vapore
   d'acqua
n destillatie in waterdampstroom
d Wasserdampfdestillation f

6882 STEAM GENERATING REACTOR rt
A nuclear reactor in which the core
generates superheated steam.
f réacteur m générateur de vapeur
e reactor m generador de vapor
i reattore m generatore di vapore
n stoomontwikkelende reactor
d dampferzeugender Reaktor m

6883 STEAM SURGE DRUMS   rt, sa
In a boiling water reactor, vessels
inserted in the steam outlet conduit of the
reactor, made to suppress the sudden
pressure variations in the reactor vessel.
f récipient pl de compensation à vapeur
e recipientes pl de compensación de vapor
i corpi pl cilindrici di vapore per
   compensazione
n compensatiestoomvaten pl,
   drukvereffeningsvaten pl
d Kompensationsdampfgefässe pl

6884 STEAM VOID, see 6660

6885 STEEL LINING   rt
In a reactor vessel made from prestressed
concrete a steel wall lining the interior of
the vessel to safeguard the vessel against
inflow of coolant.
f peau f d'étanchéité
e revestimiento m de acero
i rivestimento m in acciaio
n staalvoering
d Stahlfutter n

6886 STEENSTRUPINE   mi
Complex silicate of rare earths, containing
about 6.2 % of Th.
f steenstrupine *f*
e steenstrupina *f*
i steenstrupine *f*
n steenstrupine *n*
d Steenstrupin *n*

6887 STELLAR ENERGY   gp
The energy originated by the stars.
f énergie *f* stellaire
e energía *f* estelar
i energia *f* stellare
n sterre-energie
d Sternenergie *f*

6888 STELLAR GENERATOR,   pp
STELLARATOR
A device in which ionized gas is confined
in an endless tube by means of a very
strong externally-applied magnetic field.
f stellarator *m*
e stelarator *m*
i stellaratore *m*
n stellarator
d Stellarator *m*

6889 STELLARATOR GEOMETRY   pp
Elementary shape of the configuration of
the containing magnetic field in thermonuclear research.
f géométrie *f* du stellarateur
e geometría *f* del stelarator
i geometria *f* dello stellaratore
n stellaratorgeometrie
d Stellaratorgeometrie *f*

6890 STEM RADIATION   xr
Röntgen rays given off from parts of the
anode other than the focus.
f rayonnement *m* extrafocal
e radiación *f* extrafocal
i radiazione *f* estrafocale
n steelstraling
d ausserfokale Strahlung *f*

6891 STEP   cd
An off-set in the side of a hole through a
shield or in the corresponding plug or
other closure so that when the hole is
closed the mating steps will form a
zig-zag joint.
f marche *f*
e escalón *m*
·i scalino *m*
n trede
d Absatz *m*

6892 STEP-BY-STEP EXCITATION   np
The successive transitions of an atom or
a molecule by stages to higher levels of
excitation.
f excitation *f* par degrés
e excitación *f* por pasos
i eccitazione *f* graduale
n stapsgewijze aanslag
d stufenweise Anregung *f*

6893 STEP PENETROMETER,   me, xr
STEP-WEDGE PENETROMETER
A penetrometer of similar material to the
specimen under observation, having steps
ranging usually from 1 to 5 per cent of the
specimen thickness.
f pénétramètre *m* comparatif
e penetrómetro *m* de cuña graduata
i penetrametro *m* a cuneo a gradinata
n getrapte kwaliteitsmeter
d Stufenpenetrometer *n*

6894 STEP TIME   ct
The time needed in a decade scaler tube
to move the electron beam from one
position to the next position.
f durée *f* du pas
e duración *f* del paso
i durata *f* del passo
n staptijd
d Schrittzeit *f*

6895 STEP-WEDGE   xr
A block of material in the form of a
series of steps used to compare the
radiographic effects of X-rays under
various conditions.
f coin *m* à degrés, coin *m* gradué
e cuña *f* escalonada, cuña *f* graduada
i cuneo *m* a gradinata
n getrapte wig
d Stufenkeil *m*

6896 STEPPED LABYRINTH SEAL   cd
A form of labyrinth seal in which baffles
take the form of closely opposed steps on
stator and rotor, usually set so that only
axial clearance need be closely defined.
f garniture *f* en labyrinthe échelonnée
e junto *m* labiríntico escalonado
i tenuta *f* a labirinto a gradinata
n getrapte labyrintafsluiting
d versetzter Labyrinthverschluss *m*

6897 STEREOFLUOROSCOPY,   xr
STEREORADIOSCOPY
The use of the fluoroscope for
stereoscopy.
f stéréoradioscopie *f*
e estereorradioscopia *f*
i stereoradioscopia *f*
n stereoscopisch doorlichten *n*
d Stereodurchleuchtung *f*

6898 STEREOGRAM   ph, xr
Any type of stereoscopic radiograph,
photograph, or picture.
f stéréogramme *m*
e estereograma *m*
i stereogramma *m*
n stereogram *n*
d Stereogramm *n*

6899 STEREOGRAPHY,   xr
STEREORADIOGRAPHY
The production of a pair of radiographs of
an object from two slightly different
angles, in order to be viewed stereoscopic-
ally.

f stéréographie *f*, stéréoradiographie *f*
e estereografía *f*, estereorradiografía *f*
i stereografia *f*, stereoradiografia *f*
n radiostereografie, stereografie
d Stereographie *f*, Stereoröntgenographie *f*

6900 STEREOMETRIC LOCALIZATION,
       see 5844

6901 STEREOSCOPY   ph, xr
       The three-dimensional visual effect
       resulting from binocular vision.
f stéréoscopie *f*
e estereoscopia *f*
i stereoscopia *f*
n stereoscopie
d Stereoskopie *f*

6902 STERILITY   md
       Temporary or permanent inability to breed.
f stérilité *f*
e esterilidad *f*
i sterilità *f*
n onvruchtbaarheid, steriliteit
d Sterilität *f*, Unfruchtbarkeit *f*

6903 STERILIZATION BY IRRADIATION,
       see 5742

6904 STICKING PROBABILITY   np
       In a nuclear reaction, the probability that
       a particle which has reached the surface
       of a nucleus will be absorbed by it and
       form a compound nucleus.
f probabilité *f* d'adhérence
e probabilidad *f* de adherencia
i probabilità *f* d'aderenza
n vasthoudwaarschijnlijkheid
d Haftwahrscheinlichkeit *f*

6905 STIFFNESS   gp
       The ability of a system to resist a
       prescribed deviation.
f rigidité *f*
e rigidez *f*
i rigidezza *f*
n buigvastheid, stijfheid
d Biegungsfestigkeit *f*, Steifigkeit *f*

6906 STIGMATIC FOCUSING   sp
       In particle spectrometers, the lens-like
       action of a suitably designed field whereby
       particles which diverge at first will come
       together again in a point.
f focalisation *f* stigmatique
e enfoque *m* estigmático,
       focalización *f* estigmática
i focalizzazione *f* stigmatica
n stigmatische focussering
d stigmatische Fokussierung *f*

6907 STILL   ch
       Any apparatus in which a substance is
       vaporized and the vapo(u)rs are condensed.
f appareil *m* de distillation
e aparato *m* de destilación
i apparecchio *m* di distillazione

n destilleerapparaat *n*
d Destillierapparat *m*

6908 STIRRING, see 146

6909 STIRRING VANE COLUMN, see 6347

6910 STOCHASTIC VARIABLE, see 984

6911 STOKES'LAW   np, ra
       The wavelength of luminescence excited
       by radiation is always greater than that of
       the exciting radiation.
f loi *f* de Stokes
e ley *f* de Stokes
i legge *f* di Stokes
n stokeswet
d Stokessches Gesetz *n*

6912 STOP-START UNIT   ct
       A basic function unit designed for the
       automatic stopping and starting of counting,
       with preselection of a count number or
       counting time.
f élément *m* arrêt-marche
e unidad *f* de arranque y paro
i elemento *m* d'inizio ed arresto
n stop-start-schakel
d Start-Stopp-Einheit *f*

6913 STOPPING   np
       The loss of kinetic energy of an ionizing
       particle as a result of energy losses along
       its path through matter.
f arrêt *m*
e frenado *m*
i rallentamento *m*
n stoppen *n*
d Abbremsung *f*, Bremsung *f*

6914 STOPPING CROSS SECTION, see 416

6915 STOPPING EQUIVALENT   np
       For a given thickness of a substance, that
       thickness of a standard substance that
       produces the same energy loss.
f équivalent *m* d'arrêt
e equivalente *m* de frenado
i equivalente *m* di rallentamento
n stoppend dikte-equivalent *n*
d Bremsäquivalent *n*

6916 STOPPING NUMBER   np
       The dimensionless factor B in the
       Bethe-Bloch formula for the deviation of
       the linear stopping power of a material for
       a fast charged particle.
f nombre *m* d'arrêt
e número *m* de frenado
i numero *m* di rallentamento
n stopgetal *n*
d Bremszahl *f*

6917 STOPPING POTENTIAL   np
       That potential required to bring an
       emitted particle to rest.
f potentiel *m* d'arrêt

e potencial *m* de frenado
i potenziale *m* di rallentamento
n stoppende potentiaal
d Bremspotential *n*

6918 STOPPING POWER np
A measure of the effect of a substance
upon the kinetic energy of a charged
particle passing through it.
f pouvoir *m* d'arrêt
e poder *m* de frenado
i potere *m* di rallentamento
n stoppend vermogen *n*
d Bremsvermögen *n*

6919 STORAGE ge
The keeping of material in any depository.
f stockage *m*
e almacenamiento *m*
i immagazzinamento *m*
n bewaring, opslag
d Aufbewahrung *f*, Speicherung *f*

6920 STORAGE CAVERN rw
Hollow underground space for storing
radioactive waste.
f caverne *f* d'emmagasinage
e caverna *f* de depósito
i caverna *f* di deposito
n opslaghol *n*
d Aufbewahrungsgrube *f*

6921 STORAGE CONTAINER ra
A container for a sealed source for use
at an unknown future moment.
f récipient *m* de stockage
e recipiente *m* de almacenamiento
i recipiente *m* di deposito
n opslagvat *n*
d Lagerbehälter *m*

6922 STORAGE RING pa
Toroidal chamber used in some synchro-
trons to keep the particles in circulation
in a stable orbit until they are used.
f anneau *m* de stockage
e anillo *m* de almacenamiento
i anello *m* d'immagazzinamento
n opslagring
d Speicherring *m*

6923 STORAGE RING SYNCHROTRON pa
f synchrotron *m* à anneau de stockage
e sincrotrón *m* de anillo de almacenamiento
i sincrotrone *m* ad anello d'immagazzinamento
n synchrotron *n* met opslagring
d Synchrotron *n* mit Speicherring

6924 STORAGE SPACE ge
f dépôt *m*, magasin *m*
e almacén *m*, depósito *m*
i deposito *m*, magazzino *m*
n opslagruimte
d Aufbewahrungsraum *m*

6925 STORAGE TANK, see 47

6926 STORED ENERGY ra, rt
1. In an irradiated material, the energy
stored as the result of disordering or
excitation of atoms.
2. In a graphite reactor, the energy stored
as a result of the Wigner effect.
f énergie *f* emmagasinée
e energía *f* almacenada
i energia *f* immagazzinata
n achtergebleven energie,
opgezamelde energie
d gespeicherte Energie *f*

6927 STRAGGLING np
The fluctuation of a property such as range
or energy of charged particles, caused by
the random character of the energy losses
on passing through matter.
f dispersion *f* statistique, fluctuation *f*
e dispersión *f* estadística, fluctuación *f*
i dispersione *f* statistica, fluttuazione *f*
n fluctuatie, spreiding, statistische dispersie
d Schwankung *f*, statistische Streuung *f*

6928 STRAGGLING PARAMETER np
Equal approximately to 1 or 2 per cent of
the range for radioactive particles.
f paramètre *m* de dispersion,
paramètre *m* de fluctuation
e parámetro *m* de dispersión,
parámetro *m* de fluctuación
i parametro *m* di dispersione,
parametro *m* di fluttuazione
n fluctuatieparameter, strooiingsparameter
d Streuungsparameter *m*

6929 STRAIN mg
The deformation produced in a solid as
a result of stress.
f déformation *f* sous charge
e deformación *f* sobre carga
i deformazione *f* sotto carico
n vervorming door belasting
d Formänderung *f* durch Spannungskraft,
überelastische Drehung *f*

6930 STRAIN HARDENING, mg
WORK HARDENING
An increase in hardness and strength
caused by plastic deformation.
f écrouissage *m*
e endurecimiento *m* por deformación en frío
i incrudimento *m* per deformazione al freddo
n koudversteviging
d Kaltverfestigung *f*

6931 STRANGE PARTICLES np
Elementary particles which possess a
strangeness effect different from zero;
these are at present K-mesons and
hyperons.
f particules *pl* étranges
e partículas *pl* extrañas
i particelle *pl* estranee
n vreemde deeltjes *pl*
d Fremdteilchen *pl*

6932 STRANGENESS np
A quantum number used at present to
account for the fact that certain transformations between elementary particles
happen much more slowly than expected.
f étrangeté f
e extrañeza f
i singolarità f
n vreemdheid
d Seltsamkeit f

6933 STRANGENESS NUMBER np
A quantum number describing some
property of fundamental particles.
f nombre m d'étrangeté
e número m de extrañeza
i numero m di singolarità
n vreemdheidsgetal n
d Seltsamkeitszahl f

6934 STRATIFICATION mi
The natural layering of materials built
up because of sequential deposition, etc.
f stratification f
e estratificación f
i stratificazione f
n gelaagdheid, stratificatie
d Schichtung f

6935 STRATIGRAPH, see 688

6936 STRATIGRAPHIC AGE gp
The relative age based on the position of
a specimen in the sequence of geologic
strata.
f âge m stratigraphique
e edad f estratigráfica
i età f stratigrafica
n stratigrafische leeftijd
d stratigraphisches Alter n

6937 STRATIGRAPHY, see 689

6938 STRATOSPHERE gp
The portion of the earth's atmosphere
above the tropopause.
f stratosphère f
e estratosfera f
i stratosfera f
n stratosfeer
d Stratosphäre f

6939 STRATOSPHERIC FALL-OUT nw
Fall-out from the particles which were
taken up into the stratosphere and then
deposited in the course of some years all
over the globe.
f retombées pl stratosphériques
e depósito m estratosférico,
poso m estratosférico
i ricaduta f stratosferica
n stratosfeerneerslag
d Stratosphärenausfall m,
Stratosphärenniederschlag m

6940 STRAY NEUTRON np
A neutron not serving any useful purpose.
f neutron m vagabond

e neutrón m disperso
i neutrone m disperso
n verstrooid neutron n, zwerfneutron n
d Streuneutron n,
vagabondierendes Neutron n

6941 STRAY RADIATION ra
Radiation not serving any useful purpose.
f rayonnement m parasite
e radiación f dispersa
i radiazione f dispersa
n strooistraling, verstrooide straling
d Streustrahlung f

6942 STREAMING, see 995

6943 STREAMING FACTOR np
The attenuation per unit length of
equivalent homogeneous material divided
by the attenuation per unit length of the
material which contains voids.
f facteur m d'inhomogénéité
e factor m de encanalamiento
i fattore m di scanalatore
n kanaalverliesfactor
d Kanaleffektfaktor m, Kanalverlustfaktor m

6944 STREAMING POTENTIAL is
A difference of electrical potential between
a porous diaphragm, or other permeable
solid, and a liquid which is passing through
it.
f potentiel m d'écoulement
e potencial m de flujo
i potenziale m di flusso
n stromingspotentiaal
d Strömungspotential n

6945 STREAMLINE FLOW gp
Fluid flow such that the local velocity
of the fluid is steady both in direction and
magnitude.
f flux m à vitesse uniforme
e flujo m de velocidad uniforme
i flusso m di velocità uniforme
n gelijkmatige stroming
d gleichmässige Strömung f

6946 STRENGTH FUNCTION np
Equal to the average neutron width of the
resonances divided by the spacing between
the resonances.
f fonction f densité
e función f densidad
i funzione f densità
n dichtheidsfunctie
d Dichtefunktion f

6947 STRESS CORROSION mg
Chemical corrosion, such as that of reactor
pressure vessels, that is accelerated by
stress concentrations, either built in, or
resulting from a load.
f corrosion f par l'état latent d'efforts
e corrosión f por el estado latente de
esfuerzos,
tensocorrosión f
i corrosione f per lo stato latente di sforzi,

corrosione *f* sotto sforzi
n corrosie door latente spanningen
d Korrosion *f* durch latente Spannungen

6948 STRESS TO RUPTURE TEST   te
A type of high temperature test in which
the stress required to cause rupture in a
stated time is determined.
f essai *m* d'effort de rupture
e prueba *f* de esfuerzo de ruptura
i prova *f* di sforzo di rottura
n breukbelastingsproef
d Bruchbelastungsprüfung *f*

6949 STRINGER, see 816

6950 STRIP, see 1582

6951 STRIP (TO), see 1724

6952 STRIP PENETROMETER, see 5337

6953 STRIPPED ATOM, see 4814

6954 STRIPPER   ch, is
The section of a cascade between the feed
point and the waste withdrawal point.
f section *f* d'épuisement,
section *f* de séparation
e sección *f* de extracción,
sección *f* de separación
i sezione *f* d'esaurimento,
sezione *f* di separazione
n afscheidingssectie
d Abscheider *m*, Abstreifer *m*

6955 STRIPPING   np
A nuclear reaction in which a constituent
nucleon of the bombarding nucleus is
captured by the struck nucleus, without
a compound nucleus having been formed.
f cassure *f* en vol, stripage *m*
e rebote *m*
i sfioramento *m*
n strippen *n*
d Stripping *n*

6956 STRIPPING (US),
BACKWASHING (GB), see 515

6957 STRIPPING (GB),
SCRUBBING (US), see 6388

6958 STRIPPING CASCADE, see 514

6959 STRIPPING FILM,   ph
STRIPPING-OFF EMULSION
In a photographic emulsion technique, a
thick emulsion which can easily be
stripped off its glass base.
f émulsion *f* à détacher,
émulsion *f* pelliculable
e emulsión *f* despegable
i emulsione *f* staccabile
n aftrekbare emulsie
d Abziehemulsion *f*

6960 STRONG-COUPLING MODEL   np
A nuclear model in which the incident
particle is assumed to interact so
strongly with the nucleons it encounters
that it is absorbed in the nucleus.
f modèle *m* à couplage fort
e modelo *m* de acoplamiento fuerte
i modello *m* ad accoppiamento forte
n model *n* met sterke koppeling
d Modell *n* mit starker Kopplung

6961 STRONG FOCUSING, see 212

6962 STRONG FOCUSING ACCELERATOR,
see 211

6963 STRONTIUM   ch
Metallic element, symbol Sr, atomic
number 38.
f strontium *m*
e estroncio *m*
i stronzio *m*
n strontium *n*
d Strontium *n*

6964 STRONTIUM-90, see 5883

6965 STRONTIUM TITANATE   ch
An insoluble and stable strontium compound
used as a source of strontium-90.
f titanate *m* de strontium
e titanato *m* de estroncio
i titanato *m* di stronzio
n strontiumtitanaat *n*
d Strontiumtitanat *n*

6966 STRONTIUM UNIT (GB),   un
SUNSHINE UNIT (US)
A unit of the concentration of strontium-90
in a given organic medium, e.g. soil,
milk, bone, relative to the concentration
of calcium in the same medium.
f unité *f* de strontium
e unidad *f* de estroncio
i unità *f* di stronzio
n strontiumeenheid
d Strontium-Einheit *f*

6967 STRUCK PARTICLE, see 709

6968 STRUCTURE FACTOR   cr
The amplitude, measured at unit distance,
of the wave scattered in any chosen
direction by a unit cell, for an incident
plane wave of unit amplitude.
f facteur *m* structurel
e factor *m* estructural
i fattore *m* strutturale
n structuurfactor
d Strukturfaktor *m*

6969 STUDTITE   mi
A hydrated carbonate of uranium and lead.
f studtite *f*
e studtita *f*
i studtite *f*

n   studtiet *n*
d   Studtit *m*

6970 SUBASSEMBLY   ge
      A removable part of an assembly which
      effects a partial function.
f   sous-ensemble *m*
e   subconjunto *m*
i   unità *f*
n   deelopstelling, subsamenstel *n*
d   Teilanordnung *f*

6971 SUBATOMIC   np
      Pertaining to particles smaller than
      atoms, such as electrons, protons and
      neutrons.
f   subatomique adj
e   subatómico adj
i   subatomico adj
n   subatomisch adj
d   subatomisch adj

6972 SUBATOMIC PARTICLE   np
      Particle yielded by processes or reactions
      in which atoms undergo disintegration.
f   particule *f* subatomique
e   partícula *f* subatómica
i   particella *f* subatomica
n   subatomisch deeltje *n*
d   subatomisches Teilchen *n*

6973 SUBCADMIUM NEUTRONS   np
      Neutrons of kinetic energy less than the
      cadmium cut-off frequency.
f   neutrons *pl* subcadmiques
e   neutrones *pl* subcádmicos
i   neutroni *pl* subcadmici
n   subcadmiumneutronen *pl*
d   Subkadiumneutronen *pl*

6974 SUBCHASSIS, see 5349

6975 SUBCRITICAL   rt
      Having an effective multiplication constant
      less than unity, so that the nuclear chain
      reaction is not self-sustaining.
f   sous-critique adj
e   subcrítico adj
i   sottocritico adj
n   onderkritiek adj, onderkritisch adj
d   unterkritisch adj

6976 SUBCRITICAL ASSEMBLY   rt
      An assembly of fissile material and
      moderator which is subcritical.
f   assemblage *m* sous-critique
e   ensamblaje *m* subcrítico
i   struttura *f* sottocritica
n   onderkritieke opstelling,
    onderkritische opstelling
d   unterkritische Anordnung *f*

6977 SUBCRITICAL MULTIPLICATION   np
      The ratio of the total number of neutrons
      resulting from fission and a source which
      exist in equilibrium in a subcritical
      assembly to the total number of neutrons

which would exist in the assembly due to
the source in the absence of fission.
f   multiplication *f* sous-critique
e   multiplicación *f* subcrítica
i   moltiplicazione *f* sottocritica
n   onderkritieke vermenigvuldiging,
    onderkritische vermenigvuldiging
d   unterkritische Vermehrung *f*

6978 SUBDIVIDED CAPACITOR (US),
      see 865

6979 SUBJECT CONTRAST   ra, xr
      Contrast arising from variation in
      radiation opacity within an irradiated
      object.
f   contraste *m* de l'objet
e   contraste *m* del sujeto,
    substancia *f* de contraste
i   contrasto *m* del soggetto,
    diradazione *f*
n   objectcontrast *n*
d   Objektkontrast *m*

6980 SUBJECT CONTRAST RANGE   xr
      The range of thickness or radiation opacity
      of material in an object which is to be
      examined.
f   latitude *f* du contraste de l'objet
e   latitud *f* del contraste del sujeto
i   latitudine *f* del contrasto del soggetto
n   omvang van het objectcontrast
d   Objektkontrastumfang *m*

6981 SUBJECTIVE CONTRAST   xr
      Qualitative contrast in a radiograph or
      fluorescent screen reproduction as esti-
      mated visually.
f   contraste *m* subjectif
e   contraste *m* subjetivo
i   contrasto *m* soggettivo
n   subjectief contrast *n*
d   subjektiver Kontrast *m*

6982 SUBLETHAL LEVEL   is
      The level below which an organism, being
      injected with an active substance, is able
      to survive.
f   niveau *m* sous-létal
e   nivel *m* subletal
i   livello *m* sottoletale
n   subletaal niveau *n*
d   subletales Niveau *n*

6983 SUBLEVEL   rt
      Tha part of a nuclear reactor building
      below ground level.
f   sous-niveau *m*
e   subnivel *m*
i   sottolivello *m*
n   ondergrondse bouw
d   Unterniveau *n*

6984 SUBMARINE REACTOR   rt
      A nuclear reactor used for the production
      of power to be used in propelling a
      submarine.

f réacteur *m* de sous-marin
e reactor *m* para submarino
i reattore *m* per sottomarino
n onderzeebootreactor
d U-Bootreaktor *m*

6985 SUBSURFACE BURST   nw
The detonation of a nuclear weapon
beneath the surface of the ground or
beneath the surface of water.
f explosion *f* sous surface
e explosión *f* subsuperficial
i esplosione *f* sotto superficie
n explosie onder de oppervlakte
d unteroberflächliche Explosion *f*

6986 SUCCESSIVE GAMMA RAYS, see 902

6987 SUCCESSIVE   me, rt
SUBSTITUTION METHOD
Method for measuring material buckling
of a lattice consisting in replacing in a
reactor with known characteristics an
increasing number of fuel elements by the
elements of the lattice to be examined.
f méthode *f* de substitution progressive
e método *m* de substitución progresiva
i metodo *m* di sostituzione progressiva
n methode van de geleidelijke verwisseling
d fortschreitendes Austauschverfahren *n*

6988 SUCTION WAVE, see 5928

6989 SULFATE PROCESS (US),   ch
SULPHATE PROCESS (GB)
A process for producing beryllium in which
the ore is fused with alkali or alkaline
earth carbonates or just fused by itself.
The frit produced is treated with heated
sulphuric acid.
f procédé *m* du sulfate
e procedimiento *m* al sulfato
i processo *m* al solfato
n sulfaatproces *n*
d Sulfatverfahren *n*

6990 SULFUR (US),   ch
SULPHUR (GB)
Nonmetallic element, symbol S, atomic
number 16.
f soufre *m*
e azufre *m*
i zolfo *m*
n zwavel
d Schwefel *m*

6991 SULFUR DISK (US),   np
SULPHUR DISK (GB)
Used as activation detectors for fast
neutrons.
f disque *m* de soufre
e disco *m* de azufre
i disco *m* di zolfo
n zwavelschijf
d Schwefelscheibe *f*

6992 SULFUR NEUTRON FLUX (US),   np
SULPHUR NEUTRON FLUX (GB)
Neutron flux measured by using sulphur

disks or pellets.
f flux *m* neutronique au soufre
e flujo *m* neutrónico al azufre
i flusso *m* neutronico allo zolfo
n met behulp van zwavel bepaalde neutronenflux
d mittels Schwefel bestimmter Neutronenfluss *m*

6993 SULFUR PELLET (US),   mt
SULPHUR PELLET (GB)
Used as activation detectors for fast
neutrons.
f comprimé *m* de soufre
e comprimido *m* de azufre
i pastiglia *f* di zolfo
n zwavelpastille
d Schwefelpastille *f*

6994 SUNSHINE UNIT (US), see 6966

6995 SUPER CHOPPER   np
Device used in time of flight method,
chopping the neutron beam into pulses.
f hacheur *m* ultrarapide
e moderador *m* ultrarrápido
i moderatore *m* ultraveloce
n ultrasnelle hakker
d ultraschneller Zerhacker *m*

6996 SUPERCONDUCTING MAGNET   an, ra
Magnet used in studying radiation induced
damage in materials.
f aimant *m* à supraconductibilité
e imán *m* de supraconductibilidad
i magnete *m* a superconduttività
n suprageleidingsmagneet
d Supraleitfähigkeitsmagnet *m*

6997 SUPERCRITICAL   rt
Having an effective multiplication
constant greater than one, so that the
nuclear chain reaction is not merely
self-sustaining but increases.
f surcritique adj
e supercrítico adj
i ipercritico adj
n overkritiek adj, overkritisch adj
d überkritisch adj

6998 SUPERFICIAL DENSITY, see 320

6999 SUPERFICIAL RÖNTGENTHERAPY, xr
SURFACE RÖNTGENTHERAPY
Röntgen ray treatment directed to lesions
on the surface of the body, usually with
low energy radiation.
f röntgenthérapie *f* superficielle
e radioterapia *f* superficial,
röntgenoterapia *f* superficial
i röntgenterapia *f* superficiale
n röntgenoppervlaktherapie
d Oberflächenröntgentherapie *f*

7000 SUPERFICIAL VELOCITY   ch
The flow rate through a tower or other
process equipment based on the overall
cross section, neglecting the cross
sectional area blocked by internal obstacles.

f vitesse *f* superficielle
e velocidad *f* superficial
i velocità *f* superficiale
n oppervlaksnelheid
d Oberflächengeschwindigkeit *f*

7001 SUPERFLUIDITY METHOD   is
The separation of isotopes by using
helium-3 and helium-4 superfluidity.
f méthode *f* de suprafluidité
e método *m* de superfluidez
i metodo *m* di superfluidità
n supervloeibaarheidsmethode
d Superflüssigverfahren *n*

7002 SUPERHEATED STEAM REACTOR,
see 6879

7003 SUPERLATTICE   cr
A lattice in which the atoms of one element
occupy regular positions in the lattice of
another element without forming a
compound.
f réseau *m* superposé
e red *f* superpuesta
i reticolo *m* sovrapposto
n gesuperponeerd rooster *n*, suprastructuur
d übergeordnetes Teilgitter *n*,
Überstruktur *f*

7004 SUPERNATE   ch
The liquor remaining after a mixture of
sludge and liquid has settled.
f couche *f* surnageante
e capa *f* sobrenadante
i liquido *m* sovrastante,
strato *m* sovrastante
n bovenlaag
d Überschicht *f*,
überstehende Flüssigkeit *f*

7005 SUPERSATURATED   ch
Beyond saturation.
f sursaturé adj
e supersaturado adj
i soprasaturo adj
n oververzadigd adj
d übersättigt adj

7006 SUPPLY VOLTAGE   ge
The potential of the power source under
operating conditions.
f tension *f* d'alimentation
e tensión *f* de alimentación
i tensione *f* d'alimentazione
n voedingsspanning
d Speisespannung *f*

7007 SUPPORTING GRAPHITE   fu, rt
SLEEVE
Annular support for the channels
containing the fuel pins of a nuclear
reactor.
f manchon *m* de support en graphite
e manguito *m* de soporte en grafito
i manicotto *m* di sopporto in grafite
n grafietsteunmanchet
d Graphittragmanschette *f*

7008 SURFACE APPLICATOR   md, ra
An applicator which is applied to the
surface of the part to be treated.
f capsule *f* superficielle
e aplicador *m* superficial
i applicatore *m* superficiale
n oppervlakcapsule
d Oberflächenkapsel *f*

7009 SURFACE BARRIER DIODE   cd, ec
Component part of a junction particle
detector.
f diode *f* à barrière de surface
e dfodo *m* de barrera superficial
i diodo *m* a barriera superficiale
n oppervlakkeerlaagdiode
d Oberflächensperrschichtdiode *f*

7010 SURFACE BARRIER   ra
SEMICONDUCTOR DETECTOR
A semiconductor detector utilizing a
junction due to a surface inversion layer.
f détecteur *m* semiconducteur à barrière
de surface
e detector *m* semiconductor de barrera
superficial
i rivelatore *m* semiconduttore a barriera
superficiale
n halfgeleiderdetektor met keerlaag
d Oberflächensperrschichtdetektor *m*

7011 SURFACE BURST, see 418

7012 SURFACE CONTAMINATION   sa, te
The deposit of radioactive materials on
any surface.
f contamination *f* surfacique
e contaminación *f* superficial
i contaminazione *f* di superficie
n oppervlakbesmetting
d Oberflächenkontamination *f*

7013 SURFACE CONTAMINATION   me
INDICATOR
An indicator designed to give an estimate
of the activity per unit surface associated
with the contamination of the examined
object.
f signaleur *m* de contamination surfacique
e indicador *m* de contaminación superficial
i segnalatore *m* di contaminazione di
superficie
n oppervlakbesmettingsindicator
d Oberflächenkontaminationsanzeiger *m*

7014 SURFACE CONTAMINATION   me
METER
A measuring assembly for determining the
activity per unit surface, associated with
the contamination of an object.
f contaminamètre *m* surfacique
e medidor *m* de contaminación superficial
i contaminametro *m* di superficie
n oppervlakbesmettingsmeter
d Gerät *n* zur Bestimmung der
Oberflächenkontamination

7015 SURFACE CONTAMINATION   sa
      MONITOR
      A monitor designed to measure the
      activity per unit surface associated with
      the contamination of the examined object
      and to give a warning when it exceeds a
      predetermined value.
f   moniteur *m* de contamination surfacique
e   monitor *m* de contaminación superficial
i   monitore *m* di contaminazione di superficie
n   oppervlakbesmettingsmonitor
d   Warngerät *n* für Oberflächenkontamination

7016 SURFACE DENSITY   gp
      The quantity per unit area of anything
      distributed over a surface.
f   densité *f* de surface,
      épaisseur *f* surfacique
e   densidad *f* superficial
i   densità *f* superficiale
n   oppervlaktedichtheid
d   Oberflächendichte *f*

7017 SURFACE DOSE, see 2584

7018 SURFACE DOSE RATE   ra
      A dose rate in tissue equivalent material
      measured for surface applicators either
      by an extrapolation chamber or by using a
      scintillation probe to compare the dose
      rate from the appliance with that from a
      similar one calibrated by an extrapolation
      chamber.
f   taux *m* de dose superficielle
e   intensidad *f* de dosis superficial
i   intensità *f* di dose superficiale
n   oppervlaktedoseringssnelheid
d   Oberflächendosisleistung *f*

7019 SURFACE ENERGY, see 4916

7020 SURFACE ION DENSITY   ic
      Number of ions of one kind per unit area.
f   nombre *m* surfacique d'ions
e   densidad *f* de iones por unidad de
      superficie
i   densità *f* d'ioni per unità di superficie
n   ionendichtheid per oppervlakte-eenheid
d   Ionendichte *f* je Oberflächeneinheit

7021 SURFACE IRRADIATION   ra
      Irradiation of a part of the body by
      applying a mold or applicator loaded with
      radioactive material to the surface of the
      body.
f   irradiation *f* superficielle
e   irradiación *f* superficial
i   irradiazione *f* superficiale
n   oppervlakbestraling
d   Oberflächenbestrahlung *f*

7022 SURFACE PHOTOELECTRIC   np
      EFFECT
      The surface phenomena associated with
      the release of a bound electron from a
      material surface by an incident photon
      whose energy is absorbed in the action.

f   effet *m* photoélectrique superficiel
e   efecto *m* fotoeléctrico superficial
i   effetto *m* fotoelettrico superficiale
n   foto-elektrisch oppervlakte-effect *n*
d   lichtelektrischer Oberflächeneffekt *m*

7023 SURFACE RÖNTGENTHERAPY,
      see 6999

7024 SURFACE TENSION   gp
      A property possessed by liquid surfaces
      whereby they appear to be covered by a
      thin elastic membrane in a state of tension.
f   tension *f* superficielle
e   tensión *f* superficial
i   tensione *f* superficiale
n   oppervlakspanning
d   Oberflächenspannung *f*

7025 SURFACE TENSION EFFECT   np
      A correction applied to computations of
      nuclear attractive energy.
f   effet *m* de tension superficielle
e   efecto *m* de tensión superficial
i   effetto *m* di tensione superficiale
n   oppervlakspanningseffect *n*
d   Oberflächenspannungseffekt *m*

7026 SURFACE THERAPY TUBE   xr
      X-Ray tube used in superficial röntgen
      therapy.
f   tube *m* pour traitement en surface
e   tubo *m* de röntgenoterapia superficial
i   tubo *m* di röntgenterapia superficiale
n   röntgenoppervlaktherapiebuis
d   Oberflächentherapieröhre *f*

7027 SURFACE WRINKLING,   rt
      WRINKLING
      Of uranium, surface roughening produced
      by the elevation or depression of
      neighbo(u)ring grains or group of grains
      of dissimilar orientation.
f   craquellement *m*
e   arrugamiento *m*
i   corrugamento *m*
n   rimpeling
d   Aufrauhung *f*, Runzelbildung *f*

7028 SURFACE ZERO, see 3094

7029 SURVEY   sa
      The complete set of measures to safeguard
      a site or a building against noxious radio-
      activity.
f   contrôle *m*, surveillance *f*
e   control *m*, vigilancia *f*
i   controllo *m*, sorveglianza *f*
n   bewaking, controle
d   Kontrolle *f*, Überwachung *f*

7030 SURVEY INSTRUMENT   me, sa
      A portable instrument used for detecting
      and measuring radiation.
f   appareil *m* de surveillance,
      instrument *m* de contrôle
e   instrumento *m* de control

i strumento *m* di sorveglianza
n mobiele stralingsmeter
d Überwachungsgerät *n*

7031 SURVEYED AREA   co, sa
f terrain *m* surveillé
e zona *f* vigilada
i zona *f* sorvegliata
n bewaakt gebied *n*
d überwachtes Gebiet *n*

7032 SURVIVAL CURVE   ra
1. The relation between the number or
percentage of organisme surviving at a
given time and the dose of radiation.
2. The relation between the percentage of
individuals surviving at different time
intervals after a particular dose of
radiation and the length of the intervals.
f courbe *f* de survie
e curva *f* de supervivencia
i curva *f* di sopravvivenza
n overlevingskromme
d Überlebenskurve *f*

7033 SUSCEPTIBILITY   ec
The ratio of the polarization in a dielectric
to the electric intensity responsible for it.
f susceptibilité *f*
e susceptibilidad *f*
i suscettibilità *f*
n susceptibiliteit
d Suszeptibilität *f*

7034 SUSPENSION REACTOR   rt
A nuclear reactor which makes use of a
fluid fuel system in which the particle
size of the fissile (fissionable) or fertile
material is so small that it forms a
suspension under operating conditions.
f réacteur *m* à combustible en suspension
e reactor *m* de combustible en suspensión
i reattore *m* a sospensione
n suspensiereactor
d Suspensionsreaktor *m*

7035 SUSTAINED REACTION   np, rt
A reaction which is maintained without
any interruption.
f réaction *f* entretenue
e reacción *f* persistente
i reazione *f* sostenuta
n ononderbroken reactie
d ununterbrochene Reaktion *f*

7036 SWAGING, see 6213

7037 SWARM OF PARTICLES   np
A comparatively small number of particles
keeping together.
f essaim *m* de particules
e énjambre *m* de partículas
i sciame *m* di particelle
n deeltjeszwerm
d Teilchenschwarm *m*

7038 SWARTZITE   mi
A rare secondary mineral containing about
33 % of U.

f swartzite *f*
e swartzita *f*
i swartzite *f*
n swartziet *n*
d Swartzit *m*

7039 SWELLING   fu
Of fissile (fissionable) materials, a change
in volume without, necessarily, any change
in shape, which may occur during
irradiation.
f gonflement *m*
e hinchamiento *m*
i rigonfiamento *m*
n zwelling
d Schwellung *f*

7040 SWIMMING POOL REACTOR, see 5395

7041 SYMBOLIC REPRESENTATION   ma
The use of symbols and notations.
f représentation *f* symbolique
e representación *f* simbólica
i rappresentazione *f* simbolica
n symbolische voorstelling
d symbolische Darstellung *f*

7042 SYMMETRY ENERGY, see 346

7043 SYNCHROCYCLOTRON, see 2758

7044 SYNCHRONOUS TIMER   cd, ra
A timer which is operated by a synchronous
motor.
f minuterie *f* synchrone
e cronómetro *m* sincrónico
i cronometro *m* sincronico
n synchrone tijdschakelaar
d Synchronzeitschalter *m*

7045 SYNCHRONOUS TRANSMITTER   con
POSITION MEASURING ASSEMBLY
A position measuring assembly in which
the signal is transmitted by the voltages
induced in the stator (or rotor) windings
of a synchronous transmitter, the rotor
(or stator) of which has an angular position
related to the controlled element.
f ensemble *m* de mesure de position à
transmetteurs synchrones
e conjunto *m* medidor de posición por
transmisores síncronos
i posiziometro *m* a trasmettitori sincroni
n opstelling voor plaatsbepaling met behulp
van synchrone overdracht
d Stellungsmessanordnung *f* mit Drehfeld-
geber

7046 SYNCHROPHASOTRON   pa
A 10,000 MeV proton accelerator.
f synchrophasotron *m*
e sincrofasotrón *m*
i sincrofasotrone *m*
n synchrofasotron *n*
d Synchrophasotron *n*

7047 SYNCHROTRON   pa
A device for accelerating particles in a
circular orbit in an increasing magnetic
field by means of an alternating electric

field applied in synchronism with the
orbital motion.

f synchrotron *m*
e sincrotrón *m*
i sincrotrone *m*
n synchrotron *n*
d Synchrotron *n*

7048 SYNCHROTRON CAPTURE
       EFFICIENCY, see 875

7049 SYNCHROTRON OSCILLATIONS   pa
Forced oscillations of the particles in a
synchrotron, due to the action of the
accelerating electric field.

f oscillations *pl* synchrotron
e oscilaciones *pl* sincrotrón
i oscillazioni *pl* sincrotrone
n synchrotrontrillingen *pl*
d Synchrotronschwingungen *pl*

7050 SYNDROME   md
The complex of symptoms associated with
any disease.

f syndrome *m*
e síndrome *m*
i sindrome *m*
n syndroom *n*
d Syndrom *n*

7051 SYSTEM OF PARTICLES   np
A collective name for an arrangement or
configuration of particles.

f système *m* de particules
e sistema *m* de partículas
i sistema *m* di particelle
n deeltjessysteem *n*
d Teilchensystem *n*

7052 SYSTEMATIC ERRORS   ge
Errors that have an orderly character and
can be corrected by calibration.

f erreurs *pl* systématiques
e errores *pl* sistemáticos
i errori *pl* sistematici
n systematische fouten *pl*
d systematische Fehler *pl*

7053 SYSTEMIC REACTION   md
Generalized reaction of the whole body,
rather than localized reaction.

f réaction *f* générale
e reacción *f* general
i reazione *f* generale
n algemene reactie
d Gesamtverhalten *n*

7054 SZILARD-CHALMERS PROCESS   np
A chemical change caused by a nuclear
transformation in which there is no change
in atomic number.

f procédé *m* de Szilard-Chalmers
e procedimiento *m* de Szilard-Chalmers
i processo *m* di Szilard-Chalmers
n proces *n* van Szilard en Chalmers
d Szilard-Chalmers-Verfahren *n*

T

7055　TABLE OF ISOTOPES　　　　　is
A table containing in a condensed form the isotopes known at the present time, together with some of the nuclear data associated with them.
f　table f d'isotopes
e　tabla f de isótopos
i　tabella f d'isotopi
n　isotopentabel
d　Isotopentabelle f, Isotopenverzeichnis n

7056　TAGGED ATOM (US), see 3877

7057　TAGGED COMPOUND, see 3878

7058　TAGGED INSECTICIDES (US), see 3880

7059　TAGGED MOLECULES, see 3881

7060　TAIL-END TREATMENT　　　mt, rt
Mechanical and chemical operations carried out on irradiated fuel elements in the tail section of the processing plant.
f　opérations pl terminales, traitement m de sortie
e　operaciones pl terminales, tratamiento m de salida
i　operazioni pl terminali, trattamento m di coda
n　uitgangsbewerkingen pl
d　Ausgangsbearbeitungen pl

7061　TAILINGS　　　　　　　　　mi
The rejected portion of an ore.
f　queue f, refus pl de broyage
e　colas pl de primera flotación, productos pl estériles
i　scarti pl di miniera
n　ertsafval n, wasrest
d　Erzabfall m, Grubenklein n, Wascherze pl

7062　TAILINGS,　　　　　　　　ch
BOTTOMS, see 748

7063　TAMM-DANCOFF METHOD　　np
Technique for approximating the wave function of a system of interacting particles.
f　méthode f de Tamm-Dancoff
e　método m de Tamm-Dancoff
i　metodo m di Tamm-Dancoff
n　methode van Tamm en Dancoff
d　Tamm-Dancoffsches Verfahren n

7064　TAMPER　　　　　　　　　nw
The reflector on an atomic bomb, made of heavy material, which also helps to delay expansion of the fissile (fissionable) material on explosion.
f　réflecteur m de bombe
e　reflector m de bomba
i　riflettore m di bomba
n　bomreflector
d　Bombenreflektor m, Tamper m

7065　TANDEM GENERATOR　　　　pa
A doubled Van de Graaff machine.
f　accélérateur m tandem, générateur m de Van de Graaff en série
e　generador m de Van de Graaff en serie
i　generatore m di Van de Graaff in serie
n　vandegraaffseriegenerator
d　Van-de-Graaff-Seriengenerator m

7066　TANGENTIAL PROJECTION,　　xr
TANGENTIAL VIEW
A radiograph for which the central ray is tangential to the surface.
f　projection f tangentielle, vue f tangentielle
e　proyección f tangencial, vista f tangencial
i　proiezione f tangenziale, vista f tangenziale
n　tangentiële projectie
d　Projektion f in tangentialer Richtung

7067　TANK FARM　　　　　　　rw
A place where waste-containing tanks from reactor buildings are temporarily collected.
f　dépôt m de cuves, entrepôt m de tanques
e　depósito m de recipientes, patio m de tanques
i　deposito m di serbatoi
n　tankbergplaats
d　Tankaufbewahrungsstelle f, Tanklager n

7068　TANK TYPE REACTOR　　　rt
A nuclear reactor in which the reactor core is enclosed by a core pressure vessel.
f　réacteur m chaudière
e　reactor m de tanque
i　reattore m a cisterna
n　tankreactor
d　Tankreaktor m

7069　TANTALUM　　　　　　　ch
Metallic element, symbol Ta, atomic number 73.
f　tantale m
e　tantalio m
i　tantalo m
n　tantaal n, tantalium n
d　Tantal n

7070　TAPERED CASCADE　　　　is
A stage cascade in which feed is introduced part-way along it and the separated material passing up the plant decreases in volume as unwanted material is removed.
f　cascade f à volume d'étage croissant et décroissant
e　cascada f ahusada
i　cascata f rastremata
n　cascade met toe- en afnemende trappen
d　Kaskade f mit ab- und zuhnehmenden Stufen

7071 TARED FILTER an, ch
A filter weighed in such a way that the
weight of material subsequently collected
on it may be estimated.
f filtre *m* taré
e filtro *m* tarado
i filtro *m* tarato
n getarreerd filter *n*
d tarierter Filter *m*

7072 TARGET, see 4918

7073 TARGET AREA ra
An area exposed to a beam of particles
or radiation and usable for the arrangement of instruments or materials to be
irradiated for studying the phenomena
produced by the particles or the radiation.
f aire *f* de la cible
e área *f* del blanco
i area-bersaglio *m*
n trefoppervlak *n*
d Treffoberfläche *f*

7074 TARGET DEUTERON np
A deuteron emitted from a target.
f deutéron *m* de cible
e deuterón *m* de blanco
i deuterone *m* di bersaglio
n treflaagdeuteron *n*, trefplaatdeuteron *n*
d Auffängerdeuteron *n*, Treffplattendeuteron *n*

7075 TARGET MATERIAL np
Material subjected to bombardment by
nuclear particles, etc.
f matière *f* de cible
e material *m* de blanco
i materiale *m* di bersaglio
n treflaagmateriaal *n*, trefplaatmateriaal *n*
d Auffängermaterial *n*, Treffplattenmaterial *n*

7076 TARGET NUCLEUS np
The initially stationary atom or nucleus in
a nuclear reaction.
f noyau-cible *m*
e núcleo-blanco *m*
i nucleo-bersaglio *m*
n trefkern
d Auffängerkern *n*

7077 TARGET PARTICLE, see 709

7078 TARGET PREPARATION np
f manufacture *f* de la cible
e fabricación *f* del blanco
i fabbricazione *f* del bersaglio
n treflaagvervaardiging,
trefplaatvervaardiging
d Auffängeranfertigung *f*,
Treffplattenanfertigung *f*

7079 TARGET THEORY, see 3298

7080 TAU MESON cr
A meson which is seen to disintegrate
into three pi mesons.
f méson *m* tau

e mesón *m* tau
i mesone *m* tau
n tau-meson *n*
d Tau-Meson *n*

7081 TAUTOMERIC TRANSFORMATION gp
When a compound exists as a mixture of
two isomers in equilibrium, the two forms
are inconvertible.
f transformation *f* tautomère
e transformación *f* tautómera
i trasformazione *f* tautomera
n tautomere overgang
d tautomerer Übergang *m*

7082 TBP, ch, mt
TRIBUTYL PHOSPHATE
Important complexing solvent used in the
purification of uranium and thorium.
f TBP *m*, phosphate *m* de tributyle,
tributylphosphate *m*
e fosfato *m* de tributilo, tributilofosfato *m*
i fosfato *m* di tributile, tributilfosfato *m*
n tributylfosfaat *n*
d TBP *n*, Tributylphosphat *n*

7083 TBP PROCESS, ch
THOREX PROCESS (US),
TRIBUTYL PHOSPHATE
EXTRACTION PROCESS (GB)
A process for the recovery of thorium and
U-233 from fission products by chemical
means.
f procédé *m* d'extraction au phosphate de
tributyle
e procedimiento *m* de extracción al fosfato
de tributilo
i processo *m* d'estrazione al fosfato di
tributile
n tributylfosfaatextractieproces *n*
d Tributylphosphatextraktionsverfahren *n*

7084 TECHNETIUM ch
Radioactive element, symbol Tc, atomic
number 43.
f technétium *m*
e tecnecio *m*
i tecnezio *m*
n technetium *n*
d Technetium *n*

7085 TELECURIETHERAPY, ra
TELETHERAPY
Treatment of disease with radioactive
source at a distance from the body.
f télécuriethérapie *f*, téléthérapie *f*
e telecurieterapia *f*, teleterapia *f*
i telecurieterapia *f*, teleterapia *f*
n telecurietherapie
d Telecurietherapie *f*

7086 TELECURIETHERAPY UNIT, ra
TELETHERAPY UNIT
Apparatus containing a large quantity of
radiactive material which is designed to
emit a beam of gamma rays suitable for
external radiation at a distance from the
body.

f   appareil *m* de télécuriethérapie
e   aparato *m* de telecurieterapia
i   apparecchio *m* di telecurieterapia
n   telecurietherapieapparaat *n*
d   Telecurietherapieapparat *m*

7087 TELE-IRRADIATION   ra, xd
f   télé-irradiation *f*
e   teleirradiación *f*
i   teleirradiazione *f*
n   afstandsbestraling, verrebestraling
d   Abstandsbestrahlung *f*, Fernbestrahlung *f*

7088 TELEMETERING, see 6121

7089 TELERADIUM THERAPY   ra
Teletherapy with radium.
f   téléradiumthérapie *f*
e   telerradioterapia *f*
i   teleradioterapia *f*
n   teleradiumtherapie
d   Teleradiumtherapie *f*

7090 TELLURIUM   ch
Non-metallic element, symbol Te, atomic
number 52.
f   tellurem *m*
e   telurio *m*
i   tellurio *m*
n   tellurium *n*, telluur *n*
d   Tellur *n*

7091 TEMPERATURE COEFFICIENT   ge
The amount of change in the value of a
performance characteristic per degree
change in temperature.
f   coefficient *m* de température
e   coeficiente *m* de temperatura
i   coefficiente *m* di temperatura
n   temperatuurcoëfficiënt
d   Temperaturbeiwert *m*,
    Temperaturkoeffizient *m*

7092 TEMPERATURE COEFFICIENT OF
      REACTIVITY, see 4919

7093 TEMPERATURE CYCLE   ph
A method of processing thick photographic
emulsion to ensure uniform development.
f   cycle *m* de température
e   ciclo *m* de temperatura
i   ciclo *m* di temperatura
n   diffusieproces *n* bij lage temperatuur
d   Tieftemperaturdiffusionsverfahren *n*

7094 TEMPERATURE FACTOR   cr
The factor which expresses the reduction
of intensity of a reflection due to the
thermal vibrations of the atoms in a crystal.
f   facteur *m* de température
e   factor *m* de temperatura
i   fattore *m* di temperatura
n   temperatuurfactor
d   Debyescher Wärmefaktor *m*,
    Temperaturfaktor *m*

7095 TEMPERATURE GRADIENT   gp
The maximum rate of decrease of

temperature with distance, e.g. in the
atmosphere.
f   gradient *m* de température
e   gradiente *m* de temperatura
i   gradiente *m* di temperatura
n   temperatuurgradiënt
d   Temperaturgradient *m*

7096 TEMPERATURE METER   ma
An assembly for measuring the temperature
at a point in a nuclear reactor.
f   thermomètre *m*
e   termómetro *m*
i   termometro *m*
n   temperatuurmeter, thermometer
d   Temperaturmessgerät *n*

7097 TENSILE STRENGTH   mg
The resistance offered by a material to
tensile stresses, as measured by the
tensile force per unit cross-sectional
area required to break it.
f   résistance *f* à la traction
e   resistencia *f* a la tracción
i   resistenza *f* alla trazione
n   trekvastheid
d   Zugfestigkeit *f*

7098 TENSOR FORCE   np
A nuclear force depending on the angle
between spin direction and the line
connecting the two nucleons.
f   force *f* nucléaire tensorielle
e   fuerza *f* tensorial
i   forza *f* tensoriale
n   tensorkracht
d   Tensorkraft *f*

7099 TENTH-THICKNESS VALUE,   ra
      TENTH-VALUE THICKNESS, T.T.V.
The thickness of a specified substance
which, when introduced into the path of a
given beam of radiation, reduces the effect
of the beam to one-tenth.
f   couche *f* d'atténuation au dixième
e   espesor *m* de valor decimal
i   spessore *m* di riduzione a 1/10
n   één-tiende-waarde-dikte
d   Zehntelwertdicke *f*

7100 TERBIUM   ch
Rare earth metallic element, symbol Tb,
atomic number 65.
f   terbium *m*
e   terbio *m*
i   terbio *m*
n   terbium *n*
d   Terbium *n*

7101 TERM DIAGRAM   ma
An energy-level diagram in which the energy
levels are represented by the wave number
of the photon which would be emitted if
the system dropped from the given state to
the ground state.
f   diagramme *m* de termes
e   diagrama *m* de términos
i   diagramma *m* di termini

n termschema *n*
d Termschema *n*

7102 TERM DISPLACEMENT   ma
The displacement of the members of a
compound quantity.
f déplacement *m* de termes
e desplazamiento *m* de los términos
i spostamento *m* dei termini
n termverschuiving
d Termverschiebung *f*

7103 TERNARY FISSION,   np
TRIPARTITION
The very rare break-up of a heavy nucleus
into three fragments of comparable mass.
f fission *f* ternaire, tripartition *f*
e fisión *f* ternaria
i fissione *f* ternaria
n kern-in-drieën-splijting, tripartitie
d Kernspaltung *f* in drei Bruchstücke

7104 TERPHENYL   mt
A material belonging to the class of
organic compounds known as the
polyphenyls.
f terphényle *m*
e terfenilo *m*
i terfenile *m*
n terfenyl *n*
d Terphenyl *n*

7105 TERTIARY CREEP,   mg
THIRD STATE CREEP
The rapid deformation which may occur
after the secondary creep.
f fluage *m* tertiaire
e fluencia *f* terciaria
i scorrimento *m* terziario
n tertiaire kruip
d tertiäres Kriechen *n*

7106 TEST HOLE, see 2443

7107 TEST PARTICLE   pp
Additional exterior particle introduced into
a plasma of which the interaction with the
plasma is studied.
f particule *f* témoin
e partícula *f* de investigación
i particella *f* di ricerca
n speurdeeltje *n*
d Forschungsteilchen *n*

7108 TEST REACTOR   rt
A nuclear reactor specially designed to
test the behavio(u)r of materials and
components under the neutron and gamma
fluxes and temperature conditions of an
operating reactor.
f réacteur *m* d'essai
e reactor *m* de prueba
i reattore *m* di prova
n beproevingsreactor, onderzoekreactor
d Prüfreaktor *m*, Testreaktor *m*

7109 TEST REPORT   te
A report issued when a radiation source

has been the subject of a routine measurement in comparison with a laboratory
standard.
f rapport *m* d'essais
e informe *m* de pruebas
i rapporto *m* di prove
n beproevingsrapport *n*
d Prüfungsschein *m*

7110 TEST RIG, see 2439

7111 TETRAVALENT, see 5620

7112 THALLIUM   ch
Metallic element, symbol Tl, atomic
number 81.
f thallium *m*
e talio *m*
i tallio *m*
n thallium *n*
d Thallium *n*

7113 THEORETICAL PLATE,   ch
THEORETICAL STAGE
A hypothetical device for bringing two
streams of materials into such perfect
contact that they leave in equilibrium
with each other.
f plateau *m* théorique
e plato *m* teórico
i piatto *m* teorico
n theoretische schotel
d theoretischer Boden *m*

7114 THEORY OF CHARGE TRANSFER,   np
THEORY OF CHARGE TRANSPORT
A theory used i.a. to understand the
influence of the ion current in photoelectric
processes.
f théorie *f* du transfert de la charge
e teoría *f* de la transferencia de la carga
i teoria *f* del trasferimento della carica
n ladingsoverdrachttheorie
d Ladungsübertragungstheorie *f*

7115 THERAPEUTIC DOSE   md
The dose administered in therapy treatment.
f dose *f* de thérapie
e dosis *f* de terapia
i dose *f* di terapia
n therapiedosis
d Therapiedosis *f*

7116 THERAPEUTIC RADIOLOGY   rq, xr
That part of radiology which deals with the
treatment of diseases by radiations.
f radiothérapie *f*
e radioterapia *f*
i radioterapia *f*
n therapeutische radiologie
d Strahlentherapie *f*

7117 THERAPEUTIC TYPE   xr
PROTECTIVE TUBE HOUSING
f gaine *f* protectrice de tube à usage médical
e envuelta *f* protectora de tubo terapeútico
i guaina *f* protettiva per tubo d'uso medico

n   omhulling voor therapiebuis
d   Therapieröhrenschutzgehäuse *n*

7118   THERAPY TUBE   xr
An X-ray tube designed for use in X-ray
therapy.
f   tube *m* de thérapie
e   tubo *m* de terapia
i   tubo *m* di terapia
n   therapiebuis
d   Therapieröhre *f*

7119   THERMAL AGITATION   ec, np
Random movements of the free electrons
in a conductor.
f   agitation *f* thermique
e   agitación *f* térmica
i   agitazione *f* termica
n   thermische beweging
d   thermische Bewegung *f*

7120   THERMAL AGITATION VOLTAGE   ec
The voltage produced by thermal agitation.
f   tension *f* d'agitation thermique
e   tensión *f* de agitación térmica
i   tensione *f* d'agitazione termica
n   thermische bewegingsspanning
d   thermische Bewegungsspannung *f*

7121   THERMAL ANALYSIS   mg
A method of determining the temperature
of phase transformations by measuring
the discontinuities in the slopes of the
temperature-time curves obtained upon
heating or cooling.
f   analyse *f* thermique
e   análisis *f* térmica
i   analisi *f* termica
n   thermische analyse
d   Thermoanalyse *f*

7122   THERMAL BOND   fu
The contact between the cladding and the
nuclear fuel.
f   liaison *f* thermique
e   ligazón *f* térmica
i   contatto *m* termico
n   thermische binding
d   thermische Verbindung *f*

7123   THERMAL BREEDER   rt
A reactor of the breeding type in which
thermal neutrons are generated.
f   surrégénérateur *m* à neutrons thermiques
e   reactor *m* regenerador de neutrones
    térmicos
i   reattore *m* rigeneratore a neutroni
    termici
n   thermische kweekreactor
d   thermischer Brüter *m*,
    thermischer Brutreaktor *m*

7124   THERMAL CAPTURE   np
Capture of thermal neutrons.
f   capture *f* thermique
e   captura *f* térmica
i   cattura *f* termica

n   thermische-neutronenvangst
d   Thermoneutroneneinfang *m*

7125   THERMAL CHEMICAL REACTION   rt
A reaction to be avoided by means of
partial cooling of the graphite in an
advanced gas-cooled reactor.
f   réaction *f* chimique thermique
e   reacción *f* química térmica
i   reazione *f* chimica termica
n   thermochemische reactie
d   thermochemische Reaktion *f*

7126   THERMAL COLUMN   rt
A large body of moderator material, located
adjacent to or inside the reactor, whose
purpose is to provide thermal neutrons for
experiments.
f   colonne *f* thermique
e   columna *f* térmica
i   colonna *f* termica
n   thermische kolom
d   thermische Säule *f*

7127   THERMAL CONDUCTIVITY, see 3162

7128   THERMAL CROSS SECTION   cs
The cross section for interaction by
thermal neutrons.
f   section *f* efficace thermique
e   sección *f* eficaz térmica
i   sezione *f* d'urto termica
n   thermische doorsnede
d   thermischer Wirkungsquerschnitt *m*

7129   THERMAL CYCLE   ra
In reactor technology, a cycle of operations
in which heat is transferred from one part
of a system to another, or converted to or
from another form of energy by changes in
the temperature of a medium.
f   cycle *m* thermique
e   ciclo *m* térmico
i   ciclo *m* termico
n   warmtekringloop
d   Wärmekreislauf *m*

7130   THERMAL CYCLING   mt
A test carried out on materials, particularly on metallic uranium and its alloys,
which consists in varying periodically the
temperature of the test piece between two
given temperatures.
f   cyclage *m* thermique,
    essai *m* à cycle thermique
e   prueba *f* de ciclo térmico,
    tratamiento *m* térmico alternativo
i   trattamento *m* termico alternato
n   thermische proef met temperatuurwisseling
d   Temperaturwechselbeanspruchung *f*,
    Temperaturwechselprüfung *f*

7131   THERMAL DIFFUSION, see 6706

7132   THERMAL DIFFUSION METHOD   is
The separation of isotopes by means of
thermal diffusion.

f   méthode *f* de diffusion thermique
e   método *m* de difusión térmica
i   metodo *m* di diffusione termica
n   thermische-diffusiemethode
d   Thermodiffusionsverfahren *n*

7133 THERMAL DIFFUSION PLANT   ch, is
A plant in which the thermal diffusion
separation of isotopes is carried out.
f   installation *f* de diffusion thermique
e   instalación *f* de difusión térmica
i   impianto *m* di diffusione termica
n   installatie voor thermische diffusie
d   Thermodiffusionsanlage *f*

7134 THERMAL DIFFUSION TUBE, see 6490

7135 THERMAL ENERGY, see 3159

7136 THERMAL ENERGY REGION,   np
THERMAL RANGE
The energy region of the thermal neutrons.
f   domaine *m* d'énergie thermique
e   zona *f* de energía térmica
i   zona *f* d'energia termica
n   thermisch energiegebied *n*
d   thermisches Energiegebiet *n*

7137 THERMAL ENERGY YIELD,   nw
THERMAL YIELD
The part of the total energy yield of the
nuclear explosion which is radiated as
thermal energy.
f   rendement *m* thermique énergétique
e   rendimiento *m* térmico energético
i   rendimento *m* termico energetico
n   warmte-energieopbrengst
d   Wärmeenergieausbeute *f*

7138 THERMAL EQUILIBRIUM   ge
Considered to be reached when a two-to-one
change in the test time interval does not
produce a change due to thermal effects in
the parameter being measured that is
greater than the required accuracy of the
measurement.
f   équilibre *m* thermique
e   equilibrio *m* térmico
i   equilibrio *m* termico
n   thermisch evenwicht *n*, warmte-evenwicht *n*
d   Wärmegleichgewicht *n*

7139 THERMAL EXCITATION   np
The acquisition of excess energy by atoms
or molecules by collision processes
with other particles.
f   excitation *f* thermique
e   excitación *f* térmica
i   eccitazione *f* termica
n   thermische aanslag
d   thermische Anregung *f*

7140 THERMAL FISSION,   np
THERMAL NEUTRON FISSION
Nuclear fission produced by thermal
neutrons.
f   fission *f* provoquée par des neutrons
thermiques,

fission *f* thermique
e   fisión *f* por neutrones térmicos,
fisión *f* térmica
i   fissione *f* termica
n   splijting door thermische neutronen,
thermische splijting
d   Spaltung *f* durch thermische Neutronen,
thermische Spaltung *f*

7141 THERMAL FISSION FACTOR, see 5519

7142 THERMAL INELASTIC   cs
SCATTERING CROSS SECTION
The cross section for the thermal inelastic
scattering process.
f   section *f* efficace de diffusion inélastique
thermique
e   sección *f* eficaz de dispersión inelástica
térmica
i   sezione *f* d'urto per deviazione anelastica
termica
n   doorsnede voor thermische inelastische
verstrooiing
d   Wirkungsquerschnitt *m* für unelastische
Streuung thermischer Neutronen

7143 THERMAL INERTIA   rt
The reciprocal of thermal response.
f   inertie *f* thermique
e   inercia *f* térmica
i   inerzia *f* termica
n   thermische traagheid
d   Wärmeträgheit *f*

7144 THERMAL INSTABILITY   rt
A positive temperature coefficient in a
nuclear reactor, especially in a component
having low heat capacity.
f   instabilité *f* thermique
e   inestabilidad *f* térmica
i   instabilità *f* termica
n   thermische instabiliteit
d   thermische Instabilität *f*

7145 THERMAL INSULATION   pp
In the presence of a large magnetic field,
the thermal conductivity of a plasma
transverse to the field can be cut down.
This effect is used to provide thermal
insulation.
f   isolation *f* thermique
e   aislamiento *m* térmico
i   isolamento *m* termico
n   thermische insluiting, warmte-isolatie
d   Wärmeeinschliessung *f*, Wärmeisolation *f*

7146 THERMAL IONIZATION   cs
The ionization of atoms or molecules by
heat, as in a flame.
f   ionisation *f* thermique
e   ionización *f* térmica
i   ionizzazione *f* termica
n   thermische ionisatie
d   thermische Ionisation *f*

7147 THERMAL LEAKAGE FACTOR   rt
In one-group theory of a nuclear reactor,
the number of thermal neutrons lost from

a reactor core divided by the number of
thermal neutrons produced in the core.
f facteur *m* de fuite de neutrons thermiques
e factor *m* de fuga de neutrones térmicos
i fattore *m* di fuga di neutroni termici
n lekfactor voor thermische neutronen
d thermischer Entkommfaktor *m*

7148 THERMAL MEDIUM   rt
A medium containing a fissile (fissionable)
substance and in which the slowing-down
nuclei are present in sufficient numbers to
cause, in absence of a reflector, predominance of fissions induced by thermal
neutrons.
f milieu *m* thermique
e medio *m* de neutrones térmicos
i mezzo *m* di neutroni termici
n thermische-neutronenmilieu *n*
d Milieu *n* thermischer Neutronen

7149 THERMAL MOTION   cr
Effect caused by motion of atoms and
resulting in disturbing the regularity of a
crystal lattice.
f mouvement *m* thermique
e movimiento *m* térmico
i movimento *m* termico
n warmtebeweging
d Wärmebewegung *f*

7150 THERMAL NEUTRON BEAM   np
Used for studying structural defects in
materials on an atomic scale.
f faisceau *m* de neutrons thermiques
e haz *m* de neutrones térmicos
i fascio *m* di neutroni termici
n thermische-neutronenbundel
d thermischer Neutronenstrahl *m*

7151 THERMAL NEUTRON   np
CHAIN REACTION
A chain reaction in which nuclear fission
is produced by thermal neutrons.
f réaction *f* en chaîne à neutrons
thermiques
e reacción *f* en cadena de neutrones
térmicos
i reazione *f* in catena a neutroni termici
n kettingreactie met thermische neutronen
d Kettenreaktion *f* mit thermischen
Neutronen

7152 THERMAL NEUTRONS   np
Neutrons essentially in thermal equilibrium
with the medium in which they exist.
f neutrons *pl* thermiques
e neutrones *pl* térmicos
i neutroni *pl* termici
n thermische neutronen *pl*
d thermische Neutronen *pl*

7153 THERMAL OUTPUT OF A REACTOR,
see 3173

7154 THERMAL POWER   ma
MEASURING ASSEMBLY
An assembly including sub-assemblies for

measuring the temperature and the
flow-rates of the cooling fluid(s) associated
with a computer, and designed to determine
the thermal power of a nuclear reactor.
f ensemble *m* de mesure de la puissance
thermique
e conjunto *m* medidor de la potencia térmica
i complesso *m* di misura della potenza
termica
n meetopstelling voor thermisch vermogen
d Wärmeleistungsmessanordnung *f*

7155 THERMAL POWER PLANT   rt
A power plant in which energy is produced
by the fission of thermal neutrons.
f centrale *f* thermique
e central *f* térmica
i centrale *f* termica
n thermische centrale
d Thermokraftwerk *n*

7156 THERMAL PROTECTION   rt
Protection of inert surroundings from the
thermal effects of radiation and particles.
f protection *f* thermique
e protección *f* térmica
i protezione *f* termica
n warmtebescherming
d Wärmeschutz *m*

7157 THERMAL RADIATION   gp
1. Radiation emitted by bodies which are
not at zero temperature.
2. Electromagnetic radiation which causes
heating of materials it strikes.
f rayonnement *m* thermique
e radiación *f* térmica
i radiazione *f* termica
n warmtestraling
d Wärmestrahlung *f*

7158 THERMAL RADIATION BURN, see 2687

7159 THERMAL RANGE, see 7136

7160 THERMAL REACTOR, see 6649

7161 THERMAL RESPONSE   rt
The rate of temperature rise in a reactor
running at its rated power if no heat is
withdrawn by cooling.
f réponse *f* thermique,
taux *m* d'augmentation de la température
e proporción *f* de ascenso de la temperatura,
reacción *f* térmica
i proporzione *f* d'ascensione della
temperatura,
reazione *f* termica
n thermische reactie
d Temperaturanstiegrate *f*, Wärmereaktion *f*

7162 THERMAL SHIELD   ra, sa
A shield intended to reduce heat generation
by ionizing radiation in, and heat transfer
to, exterior regions.
f bouclier *m* thermique
e blindaje *m* térmico
i schermo *m* termico

n thermisch scherm *n*
d thermische Abschirmung *f*,
   thermischer Schild *m*

7163 THERMAL SHOCK   rt
The shock in a component resulting from
an instantaneous change of temperature
in a reactor.
f choc *m* thermique
e choque *m* térmico
i urto *m* termico
n warmteschok
d Wärmestoss *m*

7164 THERMAL SHOCK SHIELD   rt, sa
Thin laminations put up between the
coolant surface and the structure of a
reactor to filter out severe gradients.
f écran *m* de choc thermique
e pantalla *f* contra choque térmico
i schermo *m* contro urto termico
n warmtestootschild *n*
d Wärmestossschild *n*

7165 THERMAL SIPHON   cd
A closed loop containing fluid, a vertical
member of which is kept at a different
temperature from that of another vertical
member.
f siphon *m* thermique
e sifón *m* térmico
i sifone *m* termico
n thermosifon
d Thermosiphon *m*

7166 THERMAL SLEEVE   rt
Heat insulating devices for the boiler
tubes in a nuclear reactor.
f manchette *f* calorifuge
e manguito *m* aislante de calor
i manicotto *m* adiatermico
n warmte-isolerende mof
d wärmeisolierende Muffe *f*

7167 THERMAL SPIKE   np
The momentary zone of high temperature
produced in a solid or liquid along the
track of a fission fragment or other high
energy particle.
f pointe *f* thermique
e pico *m* térmico
i picco *m* termico
n thermische piek, thermische uitschieter
d thermischer Störungsbereich *m*

7168 THERMAL STABILITY   gp
The tendency to remain in a given state
when subjected to heat.
f stabilité *f* thermique
e estabilidad *f* térmica
i stabilità *f* termica
n warmtebestendigheid
d Wärmebeständigkeit *f*

7169 THERMAL STRESS   mg, rt
Caused by the resistance of a structural
element in a reactor due to temperature
differences.

f tension *f* thermique
e tensión *f* térmica
i tensione *f* termica
n warmtespanning
d Wärmespannung *f*

7170 THERMAL UTILIZATION   np, rt
     FACTOR
In an infinite medium, the ratio of the
number of thermal neutrons absorbed in a
fissile (fissionable) nuclide or in nuclear
fuel, as specified, to the total number of
thermal neutrons absorbed.
f facteur *m* d'utilisation thermique
e factor *m* de utilización térmica
i fattore *m* d'utilizzazione termica
n thermische nutsfactor
d thermischer Nutzungsfaktor *m*

7171 THERMAL YIELD, see 7137

7172 THERMALIZATION   np
Establishment of thermal equilibrium
between neutrons and their surroundings.
f thermalisation *f*
e termalización *f*
i termalizzazione *f*
n thermalisatie
d Abbremsen *n* der Neutronen auf thermische
   Geschwindigkeit,
   Thermalisierung *f*

7173 THERMALIZATION RANGE   np
The energy range in which the neutrons
in a specified medium are considered to
become thermal neutrons.
f zone *f* de thermalisation
e zona *f* de termalización
i intervallo *m* di termalizzazione
n thermalisatiegebied *n*
d Thermalisierungsgebiet *n*

7174 THERMION   ec
An electron or ion liberated by thermionic
emission.
f thermion *m*
e termión *m*
i termione *m*
n thermion *n*
d Thermion *n*

7175 THERMIONIC EMISSION   ec
Electron or ion emission from a solid or
liquid as a result of heat.
f émission *f* thermoélectronique,
   émission *f* thermoionique
e emisión *f* termoiónica
i emissione *f* termionica
n thermionische emissie
d thermionische Emission *f*

7176 THERMOCHEMISTRY   ch
That branch of chemistry which treats of
the relation between chemical action and
heat.
f thermochimie *f*
e termoquímica *f*
i termochimica *f*

n   thermochemie
d   Thermochemie *f*, Wärmechemie *f*

7177   THERMOLUMINESCENCE   ra
       Luminescence produced in a material by
       moderate heat.
f   thermoluminescence *f*
e   termoluminiscencia *f*
i   termoluminescenza *f*
n   thermoluminescentie
d   Thermolumineszenz *f*

7178   THERMOLUMINESCENCE   me, ra
       DOSEMETER
       A device using the thermoluminescent
       properties of lithium fluoride to measure
       radiation dose.
f   dosimètre *m* thermoluminescent
e   dosímetro *m* termoluminiscente
i   dosimetro *m* termoluminescente
n   thermoluminescentiedosismeter
d   Thermolumineszenzdosismesser *m*

7179   THERMONUCLEAR   np, rt
       Pertaining to nuclear reaction caused by
       intense heat.
f   thermonucléaire adj
e   termonuclear adj
i   termonucleare adj
n   thermonucleair adj
d   thermonuklear adj

7180   THERMONUCLEAR APPARATUS   pp, rt
       An apparatus in which plasma may be
       contained and heated to high temperatures
       such that thermonuclear reactions might
       be expected to take place within the plasma.
f   appareil *m* thermonucléaire
e   aparato *m* termonuclear
i   apparecchio *m* termonucleare
n   thermonucleair apparaat *n*
d   thermonukleares Gerät *n*

7181   THERMONUCLEAR BOMB   nw
       A bomb in which the energy is mainly
       derived from fusion or partly from fission
       and partly from fusion.
f   bombe *f* thermonucléaire
e   bomba *f* termonuclear
i   bomba *f* termonucleare
n   thermonucleaire bom
d   thermonukleare Bombe *f*

7182   THERMONUCLEAR CONDITIONS   np
       Conditions necessary to obtain the
       realization of a plasma well-contained and
       of such a temperature and density to
       release by fusion reactions a very high
       amount of energy.
f   conditions *pl* thermonucléaires
e   condiciones *pl* termonucleares
i   condizioni *pl* termonucleari
n   thermonucleaire voorwaarden *pl*
d   thermonukleare Bedingungen *pl*

7183   THERMONUCLEAR ENERGY   rt
       Fusion energy in which the acceleration is
       caused solely by thermal agitation.

f   énergie *f* thermonucléaire
e   energía *f* termonuclear
i   energia *f* termonucleare
n   thermonucleaire energie
d   thermonukleare Energie *f*

7184   THERMONUCLEAR REACTION,
       see 2903

7185   THETA-PINCH, see 6389

7186   THICK SOURCE   ab, np
       A radioactive source in which self-absorp-
       tion or scattering is important.
f   source *f* épaisse
e   fuente *f* densa
i   sorgente *f* densa
n   dikke bron
d   dicke Quelle *f*

7187   THICK TARGET   np
       A target of such thickness that there is
       appreciable energy loss or absorption of
       the incident particles or photons traversing
       it.
f   cible *f* épaisse
e   blanco *m* grueso
i   bersaglio *m* grosso
n   dikke trefplaat
d   dicke Treffplatte *f*

7188   THICK-WALL   ic
       IONIZATION CHAMBER
       An ionization chamber whose walls are
       thick enough to allow electronic equilibrium
       to be built up and also to ensure that the
       knock-on particles entering the sensitive
       volume arise from the wall material.
f   chambre *f* d'ionisation à paroi épaisse
e   cámara *f* de ionización de pared gruesa
i   camera *f* d'ionizzazione a parete grossa
n   dikwandig ionisatievat *n*
d   dickwandige Ionisationskammer *f*

7189   THICKNESS,   ml, np
       THICKNESS DENSITY
       A method of expressing the range of mono-
       energetic charged particles in a given
       medium.
f   épaisseur *f*
e   espesor *m*
i   spessore *m*
n   dikte
d   Dicke *f*

7190   THICKNESS GAGING (US),   me
       THICKNESS GAUGING (GB)
       In nuclear terminology, the determination
       of the thickness of a material by means of,
       e.g., low energy gamma radiation.
f   mesure *f* d'épaisseur
e   medida *f* de espesor
i   misura *f* di spessore
n   diktemeting
d   Dickenmessung *f*

7191   THIMBLE   rt
       A tube closed at one end and designed to

enclose the control rods, experimental
devices, etc. to be introduced into the
reactor.
f chaussette f
e estuche m
i astuccio m, fodero m
n koker
d Büchse f

7192 THIMBLE IONIZATION CHAMBER ic
A small cylindrical or spherical ionization
chamber, usually with walls of organic
material.
f chambre f dé,
   chambre f d'ionisation à dé à coudre
e cámara f de ionización de dedal
i camera f d'ionizzazione a ditale
n vingerhoedionisatievat n
d Fingerhutionisationskammer f,
   Kleinstionisationskammer f

7193 THIN-DOWN, see 1678

7194 THIN SOURCE ab, np
A radioactive source in which self-absorption or scattering is not important.
f source f mince
e fuente f delgada
i sorgente f sottile
n dunne bron
d dünne Quelle f

7195 THIN TARGET np
A target of such small thickness that there
is negligible energy loss or absorption
of the incident particles or photons
traversing it.
f cible f mince
e blanco m delgado
i bersaglio m sottile
n dunne trefplaat
d dünne Treffplatte f

7196 THIN-WALL COUNTER TUBE ct
Counter tube in which the envelope is of
such low absorption as to permit the
detection of radiation of low penetrating
power.
f tube m compteur à paroi mince
e tubo m contador de pared delgada
i tubo m contatore a parete sottile
n dunwandige telbuis
d dünnwandiges Zählrohr n

**7197 THIN-WALL IONIZATION ic
CHAMBER**
An ionization chamber with one or all of
its walls so thin that they do not interact
appreciably with the incident radiation.
f chambre f d'ionisation à paroi mince
e cámara f de ionización de pared delgada
i camera f d'ionizzazione a parete sottile
n .dunwandig ionisatievat n
d dünnwandige Ionisationskammer f

7198 THIN-WINDOW COUNTER TUBE ct
A counter tube in which a portion of the

envelope is of such low absorption as to
permit the detection of radiation of low
penetrating power.
f tube m compteur à fenêtre mince
e tubo m contador de ventana delgada
i tubo m contatore a finestra sottile
n telbuis met dun venster
d Zählrohr n mit dünnem Fenster

7199 THIRD STATE CREEP, see 7105

**7200 THOMAS-FERMI ma
DIFFERENTIAL EQUATION**
An equation which occurs in studying the
electron distribution in an atom.
$$y'\sqrt{x} = y^{3/2}$$
f équation f différentielle de Thomas-Fermi
e ecuación f diferencial de Thomas-Fermi
i equazione f differenziale di Thomas-Fermi
n differentiaalvergelijking van Thomas
   en Fermi
d Differentialgleichung f von Thomas und
   Fermi

7201 THOMAS-FERMI MODEL ma
A method for the calculation of atomic
energy levels based on a statistical
treatment of the assembly of electrons.
f modèle m de Thomas-Fermi
e modelo m de Thomas-Fermi
i modello m di Thomas-Fermi
n model n volgens Thomas en Fermi
d Modell n nach Thomas und Fermi

7202 THOMSON ATOM np
An early model of an atom in which the
positive charge was considered to be
distributed continuously throughout the
volume of a sphere, and in which negative
electrons were imbedded.
f atome m de Thomson
e átomo m de Thomson
i atomo m di Thomson
n thomsonatoom n
d Thomson-Atom n

7203 THOMSON CROSS SECTION, see 1115

7204 THOMSON PARABOLA METHOD np
The method of investigating the charge-to-
mass ratio of positive ions in which the
ions are acted upon by electric and
magnetic fields applied in the same
direction normal to the path of the ions.
f méthode f de parabole de Thomson
e método m de parábola de Thomson
i metodo m di parabola di Thomson
n paraboolmethode van Thomson
d Parabelmethode f nach Thomson

7205 THOMSON SCATTERING np
The scattering of electromagnetic
radiation by electrons.
f diffusion f de Thomson
e dispersión f de Thomson
i deviazione f di Thomson
n klassieke verstrooiing, thomsonverstrooiing
d Thomson-Streuung f

7206 THORAEUS FILTER   cd, xr
A primary filter of tin, with a secondary
filter of copper to absorb the characteristic radiation of the tin, and a tertiary
filter of alumin(i)um to absorb the
characteristic radiation of copper.
f filtre *m* de Thoraeus
e filtro *m* de Thoraeus
i filtro *m* di Thoraeus
n thoraeusfilter *n*
d Thoraeus-Filter *m*

7207 THOREX·PROCESS (US), see 7083

7208 THORIA,   ch
        THORIUM DIOXIDE
A thorium compound.
f bioxyde *m* de thorium, thorine *f*
e dióxido *m* de torio, toria *f*
i biossido *m* di torio
n thoriumdioxyde *n*
d Thorerde *f*, Thoriumdioxyd *n*

7209 THORIANITE   ch
A mineral consisting largely of thorium
oxides, with oxides of cerium and uranium.
f thorianite *f*
e torianita *f*
i torianite *f*
n thorianiet *n*
d Thorianit *m*

7210 THORIDES   ch
The name used for the elements of atomic
numbers 89 to 103 when they are in the
oxidation state +4.
f thorides *pl*
e torides *pl*
i toridi *pl*
n thoriden *pl*
d Thoride *pl*

7211 THORITE   ch
A thorium silicate, containing about 10 %
of U and 25-63 % of Th.
f thorite *f*
e torita *f*
i torite *f*
n thoriet *n*
d Thorit *m*

7212 THORIUM   ch
Metallic element, symbol Th, atomic
number 90.
f thorium *m*
e torio *m*
i torio *m*
n thorium *n*
d Thor *n*, Thorium *n*

7213 THORIUM CONTENT METER   ma
A content meter designed to determine the
thorium content of an ore sample by means
of a method based on the difference between
the half lives of radon and thoron.
f teneurmètre *m* en thorium
e torímetro *m*

i toriotenorimetro *m*
n gehaltemeter voor thorium
d Gerät *n* zur Bestimmung des Thorium-
   gehalts

7214 THORIUM CONTENT METER   ma
        BY BETA-ALPHA-QUASICOINCIDENCE
A content meter designed to determine
the thorium content of a complex ore
sample (at least 5 % thorium) by means of
a method based on quasicoincidence beta,
alpha (thorium C and C').
f teneurmètre *m* en thorium par pseudo-
   coïncidence bêta-alpha
e torímetro *m* por seudocoincidencia
   beta-alfa
i toriotenorimetro *m* a pseudocoincidenza
   beta-alfa
n gehaltemeter voor thorium berustend op
   de pseudocoïncidentie bêta-alfa
d Gerät *n* zur Bestimmung des Thorium-
   gehalts mittels Beta-Alphakoinzidenz

7215 THORIUM CYCLE   fu, mt
A cycle based on the transformation of
thorium-232 into uranium-233.
f cycle *m* du thorium
e ciclo *m* del torio
i ciclo *m* del torio
n thoriumcyclus
d Thoriumzyklus *m*

7216 THORIUM DECAY SERIES,   np
        THORIUM SERIES
The series of nuclides resulting from the
decay of Th-232.
f famille *f* du thorium
e familia *f* del torio
i famiglia *f* del torio
n thoriumreeks
d Thoriumreihe *f*

7217 THORIUM FISSION   np
A nuclear process in which thorium is
used as the fissile (fissionable) material.
f fission *f* de thorium
e fisión *f* de torio
i fissione *f* di torio
n thoriumsplijting
d Thoriumspaltung *f*

7218 THORIUM NITRATE   ch
The most common water-soluble salt of
thorium and used in solvent extraction.
f nitrate *m* de thorium
e nitrato *m* de torio
i nitrato *m* di torio
n thoriumnitraat *n*
d Thoriumnitrat *n*

7219 THORIUM ORE   mi
f minerai *m* de thorium
e mineral *m* de torio
i minerale *m* di torio
n thoriumerts *n*
d Thoriumerz *n*

7220   THORIUM REACTOR   rt
A nuclear breeder reactor with thorium
in the breeding blanket.
f   réacteur *m* à thorium
e   reactor *m* de torio
i   reattore *m* a torio
n   thoriumreaktor
d   Thoriumreaktor *m*

7221   THOROGUMMITE, see 4764

7222   THORON   ch
The common name for 54.5 sEm-220,
a member of the thorium series, an isotope
of radon.
f   émanation *f* du thorium, thoron *m*
e   torón *m*
i   toron *m*
n   thoron *n*
d   Radonisotop-220 *n*, Thoron *n*

7223   THOROTUNGSTITE   mi
A mineral containing tungsten, thorium,
cerium and zirconium.
f   thorotungstite *f*
e   torotungstita *f*
i   torotungstite *f*
n   thorotungstiet *n*
d   Thorotungstit *m*

7224   THREE-F BOMB, see 2657

7225   THRESHOLD, see 4013

7226   THRESHOLD DETECTOR   me, ra
Activation detector insensitive to neutrons
with an energy lower than a specified
value.
f   détecteur *m* à seuil
e   detector *m* de umbral
i   rivelatore *m* a soglia
n   drempeldetector
d   Schwellendetektor *m*

7227   THRESHOLD DOSE   ra
The minimum absorbed dose that will
produce a specified effect.
f   dose *f* seuil
e   dosis *f* umbral
i   dose *f* soglia
n   drempeldosis
d   Schwellenwertdosis *f*

7228   THRESHOLD ENERGY, see 2677

7229   THRESHOLD FREQUENCY   np, ra
The minimum frequency of radiation for
a given surface for the emission of
electrons.
f   fréquence *f* de seuil
e   frecuencia *f* de umbral
i   frequenza *f* di soglia
n   drempelfrequentie
d   Schwellenfrequenz *f*

7230   THRESHOLD OF LUMINESCENCE,
see 4157

7231   THRESHOLD POTENTIAL, see 312

7232   THRESHOLD SENSITIVITY   co, me
The smallest amount of a quantity that can
be detected by a measuring instrument or
automatic control system.
f   sensibilité *f* de seuil
e   sensibilidad *f* de umbral
i   sensibilità *f* di soglia
n   drempelgevoeligheid
d   Schwellenempfindlichkeit *f*

7233   THRESHOLD TO RESPONSE   ct
TO PULSES
The minimum amplitude of a pulse
required for a given circuit of the system
associated with the detector to perform its
function in response to that pulse.
f   seuil *m* de réponse aux impulsions
e   umbral *m* de respuesta a los impulsos
i   soglia *f* di risposta agl'impulsi
n   drempelwaarde voor pulsen
d   Ansprechschwelle *f* für Impulse

7234   THRESHOLD VALUE   co
The minimum input which produces a
corrective action in the power element of
an automatic controller.
f   valeur *f* de seuil
e   valor *m* de umbral
i   valore *m* di soglia
n   drempelwaarde
d   Schwellenwert *m*

7235   THRESHOLD VOLTAGE   ct
The lowest voltage at which all pulses
produced in the counter by any ionizing
event are of the same size, regardless of
the size of the primary ionizing event.
f   tension *f* de seuil
e   tensión *f* de umbral
i   tensione *f* di soglia
n   drempelspanning
d   Schwellenspannung *f*

7236   THRESHOLD WAVELENGTH   np
The wavelength of the incident radiant
energy above which there is no photo-
emissive effect.
f   seuil *m* de longueur d'onde photo-
électronique
e   umbral *m* de longitud de onda fotoelectrónica
i   soglia *f* di lunghezza d'onda fotoelettronica
n   drempelgolflengte
d   obere Grenzwellenlänge *f*

7237   THROMBOCYTE, see 5342

7238   THROMBOCYTE COUNT, see 5343

7239   THUCHOLITE   mi
A uranium mineral with a U_3O_8 content
between 2 and 8 %.
f   thucholite *f*
e   tucholita *f*
i   tucholite *f*
n   thucholiet *n*
d   Thucholit *m*

7240 THULIUM ch
Rare earth metallic element, symbol Tm, atomic number 69.
f thulium *m*
e tulio *m*
i tulio *m*
n thulium *n*
d Thulium *n*

7241 THYROID, md
THYROID GLAND
The largest of the cartilages of the larynx.
f glande *f* thyroïde, thyroïde *f*
e glándula *f* tiroides, tiroides *m*
i glandola tiroidea, tiroide *f*
n schildklier
d Schilddrüse *f*

7242 THYROID SEEKER, is, md
THYROID SEEKING ELEMENT
Any element which in vivo is incorporated in the thyroid, in preference to other tissue.
f substance *f* thyroïdiphile
e substancia *f* tiroidifila
i sostanza *f* tiroidifila
n schildklierzoeker
d Schilddrüsensucher *m*

7243 TIGHT BINDING APPROXIMATION ma
One of two alternative approaches to the problem of calculating the energy of an electron in a solid.
f approximation *f* d'une liaison serrée
e aproximación *f* de enlace cerrado
i approssimazione *f* di legame fisso
n benadering bij vaste binding
d Näherung *f* mit fester Bindung

7244 TIME-AMPLITUDE ec, ma
CONVERTER
An electronic sub-assembly designed to provide a voltage pulse with an amplitude proportional to the time interval between two signals.
f convertisseur *m* temps-amplitude
e convertidor *m* tiempo-amplitud
i convertitore *m* tempo-ampiezza
n tijd-amplitude-omzetter
d Zeit-Amplituden-Wandler *m*

7245 TIME-BASE UNIT ec, ge
A basic function unit comprising an electronic circuit which provides pulses of a given shape in a rigorously periodic way.
f élément *m* base de temps
e unidad *f* de impulso periódico
i elemento *m* base dei tempi
n tijdbasiseenheid
d Zeitbasis *f*

7246 TIME CONSTANT ge
The time required for an electrical quantity to rise to 63.2 per cent of its final value or to fall to 36.8 per cent of its initial value.

f constante *f* de temps
e constante *f* de tiempo
i costante *f* di tempo
n tijdconstante
d Zeitkonstante *f*

7247 TIME CONSTANT METER, see 5199

7248 TIME CONSTANT RANGE, see 5200

7249 TIME CONVERTER ec
An electronic sub-assembly designed to supply an output signal which is the conversion of a digital signal of the time interval between two input signals.
f convertisseur *m* de temps
e convertidor *m* de tiempo
i convertitore *m* di tempo
n tijdomzetter
d Zeitumwandler *m*

7250 TIME DEPENDENCE gp
f en dépendance du temps,
en fonction de temps
e en función de tiempo
i in funzione di tempo
n tijdsafhankelijkheid
d Zeitabhängigkeit *f*

7251 TIME DEPENDENT rt
NUCLEAR REACTION
f réaction *f* nucléaire en fonction de temps
e reacción *f* nuclear en función de tiempo
i reazione *f* nucleare in funzione di tempo
n tijdsafhankelijke kernreactie
d zeitabhängige Kernreaktion *f*

7252 TIME-INTERVAL COUNTER ct
An electronic counter used to measure a time interval by counting the number of pulses received from a radiofrequency generator in that time interval.
f compteur *m* électronique d'intervalle de temps
e contador *m* electrónico de intervalo de tiempo
i contatore *m* elettronico d'intervallo di tempo
n elektronische tijdsintervalmeter
d elektronischer Intervallmesser *m*

7253 TIME LAG ct
The time between the occurrence of the primary ionizing event and the occurrence of the count.
f retard *m* d'amorçage
e retardo *m* de cebado
i ritardo *m* d'accensione
n ontsteekvertraging
d Zündverzögerung *f*

7254 TIME-OF-FLIGHT ge
The time required for a nuclear particle to travel from the source to a detector.
f temps *m* de vol
e tiempo *m* de vuelo
i tempo *m* di volo

n looptijd
d Flugzeit *f*, Laufzeit *f*

7255 TIME-OF-FLIGHT ANALYZER ra
An apparatus designed to analyse the
distribution in velocities of the particles
in a beam according to the different
times-of-flight over a given flight path.
f analyseur *m* de temps de vol
e analizador *m* de tiempo de vuelo
i analizzatore *m* di tempo di volo
n looptijdanalysator
d Laufzeitanalysator *m*, Flugzeitanalysator *m*

7256 TIME-OF-FLIGHT ma
ANALYZING ASSEMBLY
A measuring assembly designed to analyze
the output from its radiation detector(s)
as a function of their time distribution.
f ensemble *m* d'analyse de temps de vol
e conjunto *m* analizador del recorrido
i complesso *m* d'analisi di tempo di volo
n opstelling voor looptijdanalyse
d Flugzeitanalyseanordnung *f*

7257 TIME-OF-FLIGHT METHOD np
Method for measuring cross sections as
a function of energy in nuclear reactors.
f méthode *f* de temps de vol
e método *m* de tiempo de vuelo
i metodo *m* di tempo di volo
n looptijdmethode
d Flugzeitmethode *f*, Laufzeitmethode *f*

7258 TIME-OF-FLIGHT me, sp
NEUTRON SPECTROMETER
An analyzing assembly designed to
determine the energy spectrum of the
neutrons in a beam, by measuring the
times-of-flight over a given flight path.
f spectromètre *m* de neutrons à temps
de vol
e espectrómetro *m* de neutrones de tiempo
de vuelo
i spettrometro *m* di neutroni a tempo di volo
n looptijdspectrometer voor neutronen
d Flugzeitneutronenspektrometer *n*,
Laufzeitneutronenspektrometer *n*

7259 TIME OF LIBERATION ec, np
The time necessary to liberate an electron
from an emitting surface.
f temps *m* de libération
e tiempo *m* de liberación
i tempo *m* di liberazione
n vrijmakingstijd
d Auslösezeit *f*, Freiwerdezeit *f*

7260 TIME REVERSAL np
A principle of nuclear physics concerned
with the postulation that if time is reversed,
the sequence of operations that have
occurred will occur again, but in the
reverse order.
f inversion *f* de temps
e inversión *f* de tiempo
i inversione *f* di tempo

n tijdinversie
d Zeitinversion *f*

7261 TIME SORTER np, sp
An apparatus for sorting pulses according
to the time at which they occur in relation
to some standard time.
f sélecteur *m* d'impulsions en dépendance
du temps
e selector *m* de impulsos cronodependiente,
selector *m* de impulsos dependiente del
tiempo
i selettore *m* d'impulsi in funzione del tempo
n tijdsafhankelijke pulsselector
d zeitabhängiger Pulswähler *m*

7262 TIME-VARIABLE FIELD, ec
VARIABLE FIELD
A variable scalar (or vector) field is a field
in which the scalar (or vector) at any point
changes during the time under
consideration.
f champ *m* variable
e campo *m* variable
i campo *m* variabile
n veranderlijk veld *n*
d veränderliches Feld *n*

7263 TIMER, me
TIMING UNIT
A device which controls the operations of
different parts of an experimental set up,
switching on and off the equipment and
taking readings at different times.
f chronorégulateur *m*
e cronoregulador *m*
i cronoregolatore *m*
n regelende tijdschakelaar
d reglender Zeitschalter *m*

7264 TIMER SUBCHASSIS ma
A subchassis which, thanks to the
programming of the operation, makes it
possible to extend the use of the scalers.
f tiroir *m* programmateur
e subchasis *m* programador
i sottocomplesso *m* programmatore
n programmerend subchassis *n*
d programmierender Bauteil *m*

7265 TIN ch
Metallic element, symbol Sn, atomic
number 50.
f étain *m*
e estaño *m*
i stagno *m*
n tin *n*
d Zinn *n*

7266 TISSUE md
The fundamental structure of which animal
and plant organs are composed.
f tissu *m*
e tejido *m*
i tessuto *m*
n weefsel *n*
d Gewebe *n*

7267 TISSUE DOSE   ra
   The dose received by a given irradiated
   tissue within the body.
f   dose *f* au tissu, dose *f* tissulaire
e   dosis *f* en el tejido, dosis *f* hística
i   dose *f* al tessuto
n   weefseldosis
d   Gewebedosis *f*

7268 TISSUE-EQUIVALENT   ic
   IONIZATION CHAMBER
   Ionization chamber in which the material
   of the walls, electrodes and gas are so
   chosen as to produce ionization essentially
   equivalent to that characteristic of the
   tissue under consideration.
f   chambre *f* d'ionisation équivalente au tissu
e   cámara *f* de ionización equivalente al
   tejido
i   camera *f* d'ionizzazione equivalente al
   tessuto
n   aan weefsel equivalent ionisatievat *n*
d   gewebeäquivalente Ionisationskammer *f*

7269 TISSUE-EQUIVALENT   ic, ra
   IONIZATION CHAMBER DOSE
   RATEMETER
   A dose ratemeter combined with a tissue-
   equivalent ionization chamber.
f   ictomètre *n* à chambre d'ionisation
   équivalente au tissu
e   impulsímetro *m* en cámara de ionización
   equivalente al tejido
i   rateometro *m* a camera d'ionizzazione
   equivalente al tessuto
n   dosistempometer met aan weefsel
   equivalent ionisatievat
d   Dosisleistungsmesser *m* mit gewebeäquivalenter Ionisationskammer

7270 TISSUE-EQUIVALENT   ic, ra
   IONIZATION CHAMBER DOSEMETER
   A dosemeter combined with a tissue-
   equivalent ionization chamber.
f   dosimètre *m* à chambre d'ionisation
   équivalente au tissu
e   dosímetro *m* de cámara de ionización
   equivalente al tejido
i   dosimetro *m* a camera d'ionizzazione
   equivalente al tessuto
n   dosismeter met aan weefsel equivalent
   ionisatievat
d   Dosismesser *m* mit gewebeäquivalenter
   Ionisationskammer

7271 TISSUE-EQUIVALENT   ra
   MATERIAL
   A solid or liquid whose absorbing and
   scattering properties for a given radiation
   are similar to those of a given biological
   material, such as part or whole of the
   human body.
f   substance *f* équivalente au tissu
e   substancia *f* equivalente al tejido
i   sostanza *f* equivalente al tessuto
n   aan weefsel equivalent materiaal *n*
d   gewebeäquivalenter Stoff *m*

7272 TITANIUM   ch
   Metallic element, symbol Ti, atomic
   number 22.
f   titane *m*
e   titanio *m*
i   titanio *m*
n   titaan *n*, titanium *n*
d   Titan *n*

7273 TNT EQUIVALENT,   nw
   YIELD
   A measure of the energy released in the
   explosion of a nuclear bomb, expressed
   in tons of TNT.
f   équivalent *m* de TNT
e   equivalente *m* de TNT
i   equivalente *m* di TNT
n   TNT-equivalent *n*
d   TNT-Äquivalent *n*

7274 TODDITE   mi
   A mineral containing uranium; probably
   a uranoan variety of columbite.
f   toddite *f*
e   toddita *f*
i   toddite *f*
n   toddiet *n*
d   Toddit *m*

7275 TOLERANCE   ge
   A permissible deviation from a specified
   value.
f   tolérance *f*
e   tolerancia *f*
i   tolleranza *f*
n   tolerantie
d   Toleranz *f*

7276 TOMOGRAPH, see 688

7277 TOMOGRAPHY, see 689

7278 TONGS   cd, co
   Component parts of remote handling
   apparatus.
f   pinces *pl*, tenailles *pl*
e   garras *pl*, tenazas *pl*
i   pinze *pl*, tenaglie *pl*
n   grijpers *pl*, tangen *pl*
d   Greifer *pl*, Zangen *pl*

7279 TOP AND BOTTOM   rt
   REFLECTED REACTOR
   A reflected reactor in which the reflector
   is only arranged at the top and bottom
   surfaces.
f   réacteur *m* à réflecteur supérieur et
   inférieur
e   reactor *m* de reflector superior y inferior
i   reattore *m* a riflettore superiore ed
   inferiore
n   reactor met boven- en onderreflector
d   Reaktor *m* mit Oben- und Untenreflektor

7280 TORBERNITE, see 1410

7281 TOROID, see 1930 ·

7282 TOROIDAL PINCH DEVICE pp
A pinch device used in the perhapsatron.
f dispositif *m* de striction toroïdale
e dispositivo *m* de extricción toroidal
i dispositivo *m* di contrazione toroidale
n toroïdaal insnoeringsapparaat *n*
d toroidaler Einschnürungsapparat *m*

7283 TOROIDAL PLASMOID pp
A discrete piece of plasma like a toroid.
f plasmoïde *m* toroïdal
e plasmoide *m* toroidal
i plasmoide *m* toroidale
n toroïdplasmoïde *n*
d Toroidplasmoid *n*

7284 TORQUE, gp
TORSIONAL FORCE
The moment of tangential effort.
f moment *m* de torsion
e par *m* de rotación, par *m* de torsión
i momento *m* torcente
n torsiemoment *n*
d Drehmoment *n*

7285 TOTAL ABSORPTION ab, np
COEFFICIENT
The linear, mass, atomic or electronic
absorption coefficient that represents a
decrease in intensity of the beam through
all single absorptive or scattering
collisions.
f coefficient *m* d'absorption totale
e coeficiente *m* de absorción total
i coefficiente *m* d'assorbimento totale
n totale absorptiecoëfficient
d Gesamtabsorptionskoeffizient *m*

7286 TOTAL ANGULAR MOMENTUM, np, qm
TOTAL QUANTUM NUMBER
The number which gives the resultant of
the magnetic field engendered by the
electron due to its orbital movement and
due to its revolving on its own axis.
f nombre *m* quantique interne
e número *m* cuántico interno
i numero *m* quantico interno
n inwendig quantumgetal *n*,
totaal quantumgetal *n*
d äquatoriale Quantenzahl *f*,
magnetische Quantenzahl *f*

7287 TOTAL BODY RADIATION, see 6809

7288 TOTAL CROSS SECTION, cs
TOTAL MICROSCOPIC CROSS
SECTION
The sum of all cross sections corresponding to the various reactions or processes
between incident particle and target
particle.
f section *f* efficace totale
e sección *f* eficaz total
i sezione *f* d'urto totale
n totale doorsnede
d totaler Wirkungsquerschnitt *m*

7289 TOTAL EFFECTIVE COLLISION cs
CROSS SECTION
The sum of the effective collision cross
sections of atoms or molecules in a unit
volume of a gas.
f section *f* spécifique de choc
e sección *f* específica de choque
i sezione *f* specifica d'urto
n totale werkzame botsingsdoorsnede
d totaler Wirkungsquerschnitt *m* für Stoss

7290 TOTAL ELECTRON np
BINDING ENERGY
The energy required to remove all the
electrons of an atom to infinite distance
from the nucleus and from each other,
leaving only the bare nucleus.
f énergie *f* totale de liaison électronique
e energía *f* total de enlace electrónico
i energia *f* totale di legame elettronico
n totale elektronenbindingsenergie
d totale Elektronenbindungsenergie *f*

7291 TOTAL ERROR, see 5053

7292 TOTAL FIELD DOSE, ra, xr
TOTAL GIVEN DOSE,
TOTAL SURFACE DOSE
The sum of the field doses given to a
single specified area during a whole course
of treatment.
f dose *f* globale en surface
e dosis *f* total de superficie
i dose *f* globale di superficie
n totale oppervlakdosis
d Gesamtoberflächendosis *f*

7293 TOTAL FILTER xr
Filter made up of inherent and added filters.
f filtre *m* total
e filtro *m* total
i filtro *m* totale
n totaal filter *n*
d Gesamtfilter *m*

7294 TOTAL IONIZATION np
The total number of elementary charges
of one sign produced by an ionizing particle
along its entire path.
f ionisation *f* totale
e ionización *f* total
i ionizzazione *f* totale
n totale ionisatie
d Gesamtionisation *f*

7295 TOTAL MACROSCOPIC cs
CROSS SECTION
The sum of total cross sections for all
atoms in a given volume divided by that
volume.
f section *f* efficace macroscopique totale,
section *f* efficace volumique totale
e sección *f* eficaz macroscópica total
i sezione *f* d'urto macroscopica totale
n totale macroscopische doorsnede
d gesamter makroskopischer Querschnitt *m*

7296 TOTAL MEAN FREE PATH np
The mean distance a particle travels
before colliding.
f libre parcours *m* moyen total
e camino *m* libre medio total,
trayectoria *f* libre media total
i cammino *m* libero medio totale,
percorso *m* libero medio totale
n totale gemiddelde vrije weglengte
d totale mittlere freie Weglänge *f*

7297 TOTAL NEUTRON np
SOURCE DENSITY
The rate of production of neutrons divided
by volume.
f densité *f* totale d'une source de neutrons
e densidad *f* total de una fuente de
neutrones
i densità *f* totale d'una sorgente di
neutroni
n totale brondichtheid van neutronen
d totale Quellendichte *f* für Neutronen

7298 TOTAL NUCLEAR BINDING np
ENERGY
The energy required to break up a nucleus
into its constituent nucleons.
f énergie *f* totale de liaison nucléaire
e energía *f* total de enlace nuclear
i energia *f* totale di legame nucleare
n totale kernbindingsenergie
d Gesamtkernbindungsenergie *f*

7299 TOTAL SPECIFIC ec, ra
IONIZATION
Specific ionization including ionization due
to delta rays.
f ionisation *f* spécifique totale
e ionización *f* específica total
i ionizzazione *f* specifica totale
n totale specifieke ionisatie
d totale spezifische Ionisation *f*

7300 TOTAL THERMAL POWER rt
The total of the energies dissipated in
unit time by all the fissions produced in
a nuclear reactor.
f puissance *f* thermique totale
e potencia *f* térmica total
i potenza *f* termica totale
n totaal thermisch vermogen *n*
d Gesamtwärmeleistung *f*

7301 TOTAL TRANSITION np
PROBABILITY
The total of the various transition
probabilities issued from the same initial
state.
f probabilité *f* de transition totale
e probabilidad *f* de transición total
i probabilità *f* di transizione totale
n totale overgangskans
d Gesamtübergangswahrscheinlichkeit *f*

7302 TOTAL WHITE COUNT, see 1795

7303 TOTAL WIDTH, see 2277

7304 TOUSCHEK EFFECT pa
The effect provoking the life reduction
of a beam of electrons in a storage ring.
f effet *m* Touschek
e efecto *m* Touschek
i effetto *m* Touschek
n touschekeffect *n*
d Touschek-Effekt *m*

7305 TOWER, see 1242

7306 TOWNSEND AVALANCHE ct
A term used in counter technology to
describe a process which is essentially
a cascade multiplication of ions.
f avalanche *f* de Townsend
e avalancha *f* de Townsend
i valanga *f* di Townsend
n townsendlawine
d Townsend-Lawine *f*

7307 TOWNSEND CHARACTERISTIC np
The current-voltage characteristic curve
for a phototube at constant illumination and
at voltages below that at which a glow
discharge occurs.
f caractéristique *f* de Townsend
e característica *f* de Townsend
i caratteristica *f* di Townsend
n townsendkarakteristiek
d Townsendsche Kennlinie *f*

7308 TOWNSEND COEFFICIENT np
The number of ionizing collisions per
centimetre(er) of path in the direction of
the applied electric field.
f coefficient *m* de Townsend
e coeficiente *m* de Townsend
i coefficiente *m* di Townsend
n townsendcoëfficiënt
d Townsend-Koeffizient *m*

7309 TOWNSEND DISCHARGE, see 2589

7310 TOWNSEND THEORY ct
The theory describing the formation of an
electron avalanche in a counter.
f théorie *f* de Townsend
e teoría *f* de Townsend
i teoria *f* di Townsend
n townsendtheorie
d Townsendsche Theorie *f*

7311 TRACE an, ch
A small quantity of material measurable
only by special techniques.
f trace *f*
e traza *f*
i traccia *f*
n spoor *n*
d Spur *f*

7312 TRACE CHEMISTRY, see 4409

7313 TRACE CONCENTRATION, see 4418

7314 TRACER, see 3443

7315 TRACER CHEMISTRY   ch, is
The use of isotopic tracers in chemical
studies.
f   chimie *f* des indicateurs isotopiques
e   química *f* de los trazadores isotópicos
i   chimica *f* dei traccianti isotopici
n   tracerchemie
d   Indikatorenchemie *f*, Tracerchemie *f*

7316 TRACER COMPOUND   ch, is
A compound which by its ease of detection
enables a reaction or process to be
studied conveniently.
f   composé *m* traceur
e   compuesto *m* trazador
i   composto *m* tracciante
n   tracerverbinding
d   Indikatorverbindung *f*, Tracerverbindung *f*

7317 TRACER STUDIES   ch, is
A technique for, e.g., studying the role of
an element in a biological, chemical or
physical process.
f   études *pl* avec traceurs
e   estudios *pl* con trazadores
i   studi *pl* con traccianti
n   onderzoek *n* met radioactieve indicators,
    traceronderzoek *n*
d   angewandte Isotopenforschung *f*,
    Indikatoruntersuchungen *pl*

7318 TRACK   ic, np
Visual manifestation of the path of an
ionizing particle in a cloud chamber or
nuclear emulsion.
f   trace *f*
e   traza *f*
i   traccia *f*
n   spoor *n*
d   Bahnspur *f*

7319 TRACK,   ic
       RACE-TRACK, see 5672

7320 TRACK CHAMBER   ic
A chamber which makes the path of
ionizing particles visible.
f   chambre *f* à trace
e   cámara *f* de traza, cámara *f* trazadora
i   camera *f* a traccia
n   sporenvat *n*
d   Bahnspurkammer *f*, Spurenkammer *f*

7321 TRACK DISTORTION   ph
In the photographic emulsion technique,
the deviation of a track from a straight
line caused by deformation of the emulsion
during processing or by other artefacts.
f   distorsion *f* de trace
e   distorsión *f* de traza
i   distorsione *f* di traccia
n   spoorvervorming
d   Spurverzerrung *f*

7322 TRACK NOISE, see 3053

7323 TRACK TAPER   ph
The characteristic appearance of the ends

of tracks of multiply charged particles
due to the reduction in the range of delta
rays as the velocity of the primary particle
decreases.
f   conicité *f* de trace, réduction *f* de trace
e   conicidad *f* de traza, reducción *f* de traza
i   conicità *f* di traccia, riduzione *f* di traccia
n   spoorreductie, spoortoespitsing
d   Spurkonizität *f*, Spurreduktion *f*

7324 TRAINING REACTOR   rt
A nuclear reactor operated primarily for
training in reactor operation and
instructing in reactor behavio(u)r.
f   réacteur *m* d'entraînement
e   reactor *m* de adiestramiento
i   reattore *m* d'addestramento
n   opleidingsreactor
d   Ausbildungsreaktor *m*

7325 TRAJECTORY   np
A path in space which a particle or
system traverses.
f   parcours *m*
e   camino *m*, trayectoria *f*
i   cammino *m*, percorso *m*
n   baan
d   Bahn *f*

7326 TRANSCURIUM ELEMENTS   ch
Elements of which the atomic number is
higher than 96 and which are placed above
curium in the periodic system.
f   transcuriens *pl*
e   elementos *pl* transcúricos
i   elementi *pl* trascurici
n   transcuriumelementen *pl*
d   Transcuriumelemente *pl*

7327 TRANSFER CANAL   rt
In a swimming-pool reactor a water-filled
canal which may be connected to the basin
and at the other side to a water reservoir.
f   canal *m* de transfert
e   canal *m* de transferencia
i   canale *m* di trasferimento
n   verbindingskanaal *n*
d   Verbindungskanal *m*

7328 TRANSFER FUNCTION   ma
A relationship between one system
variable and another that enables the
second variable to be determined from the
first.
f   fonction *f* de transfert, transmittance *f*
e   función *f* de transferencia, transmitancia *f*
i   funzione *f* di trasferimento, trasmittenza *f*
n   overdrachtsfunctie
d   Übertragungsfunktion *f*

7329 TRANSFER FUNCTION METER   ma
An assembly for determining the transfer
function of a nuclear reactor by measuring
the variation of the neutron flux density
caused by a modulation of the reactivity.
f   transféromètre *m*
e   transferómetro *m*

i   trasferimetro m
n   meetopstelling voor de overdrachtsfunctie
d   Gerät n zur Bestimmung der
    Übertragungsfunktion

7330 TRANSFER PORT   rt, sa
    The hole or compartment for inserting or
    removing equipment or materials into or
    from a glove box.
f   rond m de transport, trou m de transport
e   orificio m de transporte
i   orificio m di trasporto
n   doorgeefopening
d   Durchlassöffnung f

7331 TRANSFER UNIT,   ch, is
    TRANSPORT UNIT
    One of the component parts in a chemical
    installation in which transfer of the
    treated substances takes place in a
    number of stages.
f   élément m de transport,
    unité f de transfert
e   elemento m de transporte,
    unidad f de transferencia
i   elemento m di trasporto,
    unità f di trasporto
n   overgangscel
d   Austauscheinheit f, Übergangszelle f,
    Übertragungseinheit f

7332 TRANSFORMATION   np
    In nuclear physics the change of one
    nuclide into another.
f   transformation f
e   transformación f
i   trasformazione f
n   transformatie
d   Umwandlung f

7333 TRANSFORMATION SERIES, see 5787

7334 TRANSGENATION, see 3131

7335 TRANSIENT   ge
    A term applied to short-time phenomena
    which take place in a system owing to a
    sudden change of conditions.
f   phénomène m transitoire
e   fenómeno m transitorio
i   fenomeno m transitorio
n   kortstondig verschijnsel n,
    voorbijgaand verschijnsel n
d   flüchtiger Vorgang m,
    schnellvergehender Vorgang m

7336 TRANSIENT EQUILIBRIUM,   np
    TRANSIENT RADIOACTIVE
    EQUILIBRIUM
    A radioactive equilibrium in which the
    half-life of the parent is short compared
    with the time of the experiment and in which
    a decline of the parent activity is
    observable.
f   équilibre m radioactif transitoire
e   equilibrio m radiactivo momentáneo
i   equilibrio m radioattivo momentaneo

n   kortstondig radioactief evenwicht n
d   laufendes radioaktives Gleichgewicht n,
    Übergangsgleichgewicht n

7337 TRANSIT ANGLE,   np
    TRANSIT PHASE ANGLE
    The product of angular frequency and the
    time taken for an electron to traverse a
    given path.
f   angle m de transit
e   ángulo m de tránsito
i   angolo m di transito
n   looptijdhoek
d   Laufzeitwinkel m

7338 TRANSIT BOX   rt, sa
    A box with inside and outside sliding doors
    attached to a glove box for introducing
    components or materials.
f   boîte f de passage
e   caja f de pasaje
i   cassa f di passaggio
n   doorschuifdoos
d   Durchschiebekasten m

7339 TRANSIT DOSE   nw
    The gamma dose recieved after nuclear
    detonation due to direct irradiation by the
    passing fall-out cloud, radiation from the
    base surge as it passes or envelopes the
    dosed object, or radiation from
    contaminated water as a ship passes
    through it.
f   dose f de passage
e   dosis f de tránsito
i   dose f di passaggio
n   voorbijgaansdosis
d   Vorbeigehensdosis f

7340 TRANSIT TIME   np
    The time taken by a charged particle in
    moving between two specified points.
f   durée f de parcours, temps m de transit
e   duración f de recorrido,
    tiempo m de tránsito
i   durata f di propagazione,
    tempo m di transito
n   looptijd
d   Laufzeit f

7341 TRANSIT TIME SPREAD   ct, ec
    A property of a photomultiplier tube
    which is shown by a lengthening of
    scintillation pulses.
f   fluctuation f du temps de transit
e   irregularidad f del tiempo de tránsito
i   erraticità f del tempo di transito
n   looptijdspreiding
d   Laufzeitstreuung f

7342 TRANSITION   qm
    The process whereby a quantum mechanical
    system changes from one energy eigenstate
    to another.
f   transition f
e   transición f
i   transizione f

n overgang
d Übergang *m*

7343 TRANSITION,   ch
       PHASE TRANSITION, see 5235

7344 TRANSITION CURVE, see 2011

7345 TRANSITION EFFECT   np, ra
       A change in the intensity of the secondary
       radiation associated with a beam of
       primary radiation as the latter passes from
       a vacuum into a material medium or from
       one medium into another.
f effet *m* de transition
e efecto *m* de transición
i effetto *m* di transizione
n overgangseffect *n*
d Übergangseffekt *m*

7346 TRANSITION ENERGY   pa
       In a proton synchrotron, the energy at
       which phase focusing changes sign.
f énergie *f* de transition
e energía *f* de transición
i energia *f* di transizione
n overgangsenergie
d Übergangsenergie *f*

7347 TRANSITION MULTIPOLE   np
       MOMENTS
       Multiple moments which determine
       radiative transitions between two states
       and therefore depend on both states.
f moments *pl* multipôles de transition
e momentos *pl* multipolos de transición
i momenti *pl* multipoli di transizione
n multipoolmomenten *pl* bij overgang
d Übergangsmultipolmomente *pl*

7348 TRANSITION PROBABILITY   gp
       The probability that a system in state i
       will undergo a transition to state f.
f probabilité *f* de transition
e probabilidad *f* de transición
i probabilità *f* di transizione
n overgangskans
d Übergangswahrscheinlichkeit *f*

7349 TRANSITIONAL LEUCOCYTE, see 3349

7350 TRANSLATIONAL MOTION,   pa
       TRANSLATIONAL MOVEMENT
       A movement in which all particles move
       with the same velocity and acceleration in
       parallel paths.
f mouvement *m* de translation
e movimiento *m* de translación
i movimento *m* di traslazione
n translatiebeweging
d Translationsbewegung *f*

7351 TRANSLOCATION, see 3551

7352 TRANSMISSION COEFFICIENT,
       see 5182

7353 TRANSMISSION CURVE, see 17

7354 TRANSMISSION EXPERIMENT   ra
       A way of studying the interaction of
       radiation with a material by measuring the
       transmitted beam rather than the scattered
       radiation or any effect upon the material,
       e.g., induced radioactivity.
f expérience *f* de transmission
e prueba *f* de transmisión
i prova *f* di trasmissione
n doorlaatproef
d Durchgangsversuch *m*

7355 TRANSMISSION METHOD OF   np
       CROSS SECTION DETERMINATION
       The determination of absorption and
       scattering cross section by means of the
       measurement of the loss of intensity of
       the incident beam in passing through the
       substance.
f détermination *f* de la section efficace par
   la méthode de transmission
e determinación *f* de la sección eficaz por
   el método de transmisión
i determinazione *f* della sezione d'urto per
   il metodo di trasmissione
n werkzame-doorsnedebepaling met behulp
   van de transmissiemethode
d Wirkungsquerschnittbestimmung *f* mittels
   Transmissionsverfahren

7356 TRANSMUTATION   ch
       The change of one element into another.
f transmutation *f*
e transmutación *f*
i trasmutazione *f*
n transmutatie
d Transmutation *f*

7357 TRANSMUTATION EQUATION   ma, np
       An equation for the change of one atom into
       another, which differs from it in nuclear
       charge, mass or stability.
f équation *f* de transmutation
e ecuación *f* de transmutación
i equazione *f* di trasmutazione
n transmutatievergelijking
d Transmutationsgleichung *f*

7358 TRANSPARENT   ra
       Permitting the passage of radiation or
       particles.
f transparent adj
e transparente adj
i trasparente adj
n doorlatend adj, transparant adj
d durchlässig adj, transparent adj

7359 TRANSPLUTONIUM ELEMENTS   ch
       Elements of which the atomic number is
       higher than 94 and which are placed above
       plutonium in the periodic system.
f transplutoniens *pl*
e elementos *pl* transplutónicos
i elementi *pl* transplutonici
n transplutoniumelementen *pl*
d Transplutoniumelemente *pl*

7360 TRANSPORT, see 3772

7361 TRANSPORT CROSS SECTION   cs
The total cross section less the product
of the scattering cross section and the
average cosine of the scattering angle in
the laboratory system.
f   section *f* efficace de transport
e   sección *f* eficaz de transporte
i   sezione *f* d'urto di trasporto
n   transportdoorsnede
d   Transportquerschnitt *m*

7362 TRANSPORT EQUATION   np
The Boltzmann equation, especially applied
to the distribution of neutrons in a reactor.
f   équation *f* du transport
e   ecuación *f* del transporte
i   equazione *f* del trasporto
n   transportvergelijking
d   Transportgleichung *f*

7363 TRANSPORT KERNEL   ma, np
The Green's function that appears in the
integral equation form of the transport
equation.
f   noyau *m* intégral de transport
e   núcleo *m* integral de transporte
i   nucleo *m* integrale di trasporto
n   integraalkern voor transport
d   Transportintegralkern *m*

7364 TRANSPORT MEAN   np
      FREE PATH
The reciprocal of the macroscopic transport cross section.
f   libre parcours *m* moyen de transport
e   trayectoria *f* libre media de transporte
i   percorso *m* libero medio di trasporto
n   gemiddelde vrije weglengte voor transport,
    transportweglengte
d   Transportweglänge *f*

7365 TRANSPORT THEORY   rt
In reactor technology, a theory for the
treatment of neutron and gamma migration
in a medium, based on the linear Boltzmann
transport equation.
f   théorie *f* du transport
e   teoría *f* del transporte
i   teoria *f* del trasporto
n   transporttheorie
d   Transporttheorie *f*

7366 TRANSPORT UNIT, see 7331

7367 TRANSPORTABLE REACTOR   rt
A nuclear power reactor designed to be
transported to a location and fixed there
permanently for the generation of local
electric power.
f   réacteur *m* transportable
e   reactor *m* transportable
i   reattore *m* trasportabile
n   verplaatsbare reactor
d   transportabler Reaktor *m*

7368 TRANSURANIC ELEMENTS   ch
Elements with atomic number higher than
that of uranium, i.e. above 92.

f   transuraniens *pl*
e   elementos *pl* transuránicos
i   elementi *pl* transuranici
n   transuranen *pl*
d   Transurane *pl*

7369 TRANSVERSE FIELD   cc
An electric field not containing a
magnetic-field component in the direction
of propagation of electric magnetic waves.
f   champ *m* transversal
e   campo *m* transversal
i   campo *m* trasversale
n   dwarsveld *n*, transversaal veld *n*
d   Querfeld *n*, Transversalfeld *n*

7370 TRANSVERSE FINNING, see 1105

7371 TRANSVERSE-FOCUSING   np
      ELECTRIC FIELD
A focusing electric field not containing a
magnetic field component in the direction
of propagation of particles.
f   champ *m* électrique focalisateur transversal
e   campo *m* eléctrico transversal de enfoque
i   campo *m* elettrico trasversale di
    focalizzazione
n   elektrisch transversaal focusseringsveld *n*
d   elektrisches Querfokussierungsfeld *n*

7372 TRANSVERSELY-FINNED FUEL
      ELEMENT, see 1106

7373 TRAPPING CENTER (US), see 2165

7374 TRAUMA,   md
      TRAUMATISM
Violent injury of an organism from an
external source.
f   traumatisme *m*
e   traumatismo *m*
i   traumatismo *m*
n   trauma
d   Trauma *n*

7375 TRAVEL(L)ING WAVE   pa
      LINEAR ACCELERATOR
A linear accelerator in which particles
are accelerated by the electric component
of a travel(l)ing wave field set up in a
wave-guide.
f   accélérateur *m* linéaire à ondes progres-
    sives
e   acelerador *m* lineal de ondas progresivas
i   acceleratore *m* lineare ad onde progressive
n   lineaire versneller met lopende golven
d   Linearbeschleuniger *m* mit fortschreitenden
    Wellen,
    Linearbeschleuniger *m* mit Wanderwellen

7376 TRAY, see 5334

7377 TREATMENT CONE,
      CONE OF RADIATION, see 1305

7378 TREATMENT CONE,
      APPLICATOR, see 314

7379 TRENCH   rw
Open ditch into which radioactive waste,
solid or liquid, is placed and then filled
with dirt to prevent movement of radioactive contamination by wind erosion.
f fossée *f*
e foso *m*
i fosso *m*
n sloot
d Graben *m*

7380. TRIANGULAR LATTICE   fu, rt
A reactor lattice of triangular shape.
f réseau *m* triangulaire
e celosía *f* triangular
i reticolo *m* triangolare
n driehoekig rooster *n*
d Dreieckgitter *n*

7381 TRIBOLUMINESCENCE   np
Luminescence generated by friction.
f triboluminescence *f*
e triboluminiscencia *f*
i tribolum inescenza *f*
n triboluminescentie
d Tribolumineszenz *f*

7382 TRIBUTYL PHOSPHATE, see 7082

7383 TRIBUTYL PHOSPHATE
EXTRACTION PROCESS (GB), see 7083

7384 TRICKLING FILTER, see 641

7385 TRIGLYCOL DICHLORIDE   ch, mt
The first organic solvent used in the
separation for the extraction of plutonium
from uranium.
f dichlorure *m* de glycol de triéthylène
e dicloruro *m* de glicolo de trietileno
i dicloruro *m* di glicolo di trietilene
n triglycolbichloride *n*
d Triglykolbichlorid *n*

7386 TRILAURYLAMINE   ch
Substance used for purifying plutonium
extracts of irradiated fuel.
f trilaurylamine *f*
e trilaurilo-ámina *f*
i trilauriloamina *f*
n trilaurylamine *n*
d Trilaurylamin *n*

7387 TRINITY BOMB, see 174

7388 TRIP   rt, sa
A reduction in reactor power initiated by
any of the safety circuits of the reactor.
f arrêt *m* lent
e paro *m* lento
i scatto *m* lento
n langzame uitschakeling
d langsame Abschaltung *f*

7389 TRIP AMPLIFIER, see 6567

7390 TRIP LEVEL   rt, sa
The power level in a reactor at which the

shut-down system is going into operation.
f seuil *m* d'arrêt
e umbral *m* de paro
i soglia *f* d'arresto, soglia *f* di scatto
n uitschakeldrempel
d Abschaltschwelle *f*

7391 TRIP MARGIN   rt
Power interval determined by the difference
in power corresponding to the upper
threshold of activating a safety circuit and
the operating power of a reactor.
f intervalle *m* de puissance
e intervalo *m* de potencia
i franco di potenza
n vermogensmarge
d Leistungsspielraum *m*

7392 TRIPARTITION, see 7103

7393 TRIPLE POINT   nw
The intersection of incident, reflected and
fused shock fronts accompanying an air
burst.
f triple point *m*
e triple punto *m*
i triplo punto *m*
n driehoekspunt *n*
d Dreieckspunkt *m*

7394 TRIPLET   sp
A spectrum line which, when examined
under high resolution, is composed of three
closely-packed, fine lines.
f triplet *m*
e triplete *m*
i tripletto *m*
n triplet *n*
d Triplett *n*

7395 TRITIATED WATER   sa
Water containing a certain percentage of
tritium.
f eau *f* tritiée
e agua *f* tritiada
i acqua *f* tritiata
n tritiumhoudend water *n*
d tritiumhaltiges Wasser *n*

7396 TRITIATION   is
A method of tritium labelling (tagging) by
means of reductions with tritiated metal
hydrides.
f tritiation *f*
e tritiación *f*
i tritiazione *f*
n tritiatie
d Tritiation *f*

7397 TRITIUM   ch, is
The isotope of hydrogen having a mass
number of 3.
f tritium *m*
e tritio *m*
i tritio *m*
n tritium *n*
d Tritium *n*

7398 TRITIUM AIR MONITOR   sa
An air monitor for determining the
presence of tritium in the atmosphere.
f moniteur *m* atmosphérique de tritium
e monitor *m* atmosférico de tritio
i monitore *m* atmosferico di tritio
n luchtmonitor voor tritium
d Tritiumluftmonitor *m*

7399 TRITIUM LABILIZATION   is
A form of becoming unstable which can
occur under certain biological conditions.
f instabilisation *f* de tritium
e labilización *f* de tritio
i instabilizzazione *f* di tritio
n labiel worden *n* van tritium
d Tritiuminstabilisierung *f*

7400 TRITIUM LUMINOUS COMPOUND   mt
A self-luminous compound prepared by
incorporating tritium into a stable
transparent polymer which is used to coat
the phosphor particles.
f composé *m* lumineux à tritium
e compuesto *m* luminoso con tritio
i composto *m* luminoso a tritio
n tritiumhoudende lichtgevende verbinding
d tritiumhaltiger Leuchtstoff *m*

7401 TRITIUM SURFACE   me
      CONTAMINATION METER
A surface contamination meter for tritium.
f contaminamètre *m* surfacique pour tritium.
e medidor *m* de contaminación superficial
   para tritio
i contaminametro *m* di superficie per tritio
n oppervlakbesmettingsmeter voor tritium
d Gerät *n* zur Bestimmung der Tritium-
   oberflächenkontamination

7402 TRITIUM TARGET   ra
A convenient source of monoenergetic
neutrons consisting of absorbed tritium in
a thin layer of titanium or zirconium
deposited on suitable backing metals.
f cible *f* en tritium
e blanco *m* en tritio
i bersaglio *m* in tritio
n tritiumtreflaag, tritiumtrefplaat
d Tritiumauffänger *m*, Tritiumtreffschicht *f*

7403 TRITIUM/TRITIUM OXIDE IN   me, sa
      AIR MONITOR
A monitor for detecting the presence of
tritium oxide in air after elimination of the
tritium.
f moniteur *m* atmosphérique d' oxyde de
   tritium
e monitor *m* atmosférico de óxido de tritio
i monitore *m* atmosferico d' ossido di tritio
n tritiumoxydemonitor in lucht
d Tritiumoxydmonitor *m* in Luft

7404 TRITON   np
The nucleus of an atom of tritium.
f triton *m*
e tritón *m*

i tritone *m*
n triton *n*
d Triton *n*

7405 TRITONITE   mi
A borosilicate of cerium and other
elements, containing about 7.5 % of Th.
f tritonite *f*
e tritonita *f*
i tritonite *f*
n tritoniet *n*
d Tritonit *m*

7406 TROCHOIDAL MASS ANALYZER   sp
A mass spectrometer in which the ion
beams traverse trochoidal paths in
mutually perpendicular electric and
magnetic fields.
f spectromètre *m* de masse(s) à parcours
   trochoïde
e espectrómetro *m* de masa de trayectoria
   trocoide
i spettrometro *m* di massa a percorso
   trocoide
n trochoïdale massaspectrograaf
d Trochoidenmassenspektrometer *n*,
   Zykloidenmassenspektrometer *n*

7407 TROCHOTRON   ec
A multi-electrode tube in which the electron
beam follows a trochoidal path under the
influence of crossed electric and magnetic
fields.
f trochotron *m*
e trocotrón *m*
i trocotrone *m*
n trochotron *n*
d Trochotron *n*

7408 TROEGERITE   mi
A rare secondary mineral containing about
55 % of U.
f troegerite *f*
e troegerita *f*
i troegerite *f*
n troegeriet *n*
d Trögerit *m*

7409 TROPOPAUSE   gp
The layer in the atmosphere between the
troposphere and the stratosphere.
f tropopause *f*
e tropopausa *f*
i tropopausa *f*
n tropopause
d Tropopause *f*, Tropopausenschicht *f*

7410 TROPOSPHERE   gp
That portion of the earth's atmosphere
extending from the surface up to about
10 km.
f troposphère *f*
e troposfera *f*
i troposfera *f*
n troposfeer
d Troposphäre *f*

7411 TROPOSPHERIC AIR   is
      MOVEMENT
      One of the objects to be studied by using
      isotopic tracers.
f mouvement *m* d'air troposphérique
e movimiento *m* de aire troposférica
i movimento *m* d'aria troposferica
n troposferische luchtbeweging
d troposphärische Luftbewegung *f*

7412 TROPOSPHERIC FALL-OUT   nw
      Fall-out from the fine particles of debris.
f retombées *pl* troposphériques
e depósito *m* troposférico,
   poso *m* troposférico
i ricaduta *f* troposferica
n troposfeerneerslag
d Troposphärenausfall *m*,
   Troposphärenniederschlag *m*

7413 TRUE ABSORPTION   ab, np
      That part of the total absorption which
      arises from the complete or partial loss
      of energy of photons of the primary beam.
f absorption *f* réelle
e absorción *f* real
i assorbimento *m* reale
n werkelijke absorptie
d absolute Absorption *f*, echte Absorption *f*

7414 TRUE ABSORPTION   ab, np
      COEFFICIENT
      The linear, mass, atomic or electronic
      absorption coefficient that represents the
      decrease in intensity of the beam through
      absorptive collisions only.
f coefficient *m* d'absorption réelle
e coeficiente *m* de absorción real
i coefficiente *m* d'assorbimento reale
n werkelijke absorptiecoëfficiënt
d absoluter Absorptionskoeffizient *m*

7415 TRUE ACTIVITY CONSTANT   ra
      OF A SOURCE
      The activity constant of a source given in
      disintegrations per minute with an
      accuracy of $\pm$ 5 % maximum overall error.
f constante *f* d'activité réelle d'une source
e constante *f* de actividad real de una fuente
i costante *f* d'attività reale d'una sorgente
n werkelijke activiteitsconstante van een bron
d wirkliche Aktivitätskonstante *f* einer Quelle

7416 TRUE COINCIDENCE   ct
      A coincidence due to the detection of only
      one particle or photon, or two or more
      particles or photons of common origin.
f coïncidence *f* vraie
e coincidencia *f* verdadera
i coincidenza *f* vera
n ware coïncidentie
d absolute Koinzidenz *f*, echte Koinzidenz *f*

7417 TSCHEFFKINITE, see 1071

7418 T.T.V., see 7099

7419 TUBE CONVEYER   np, rt
      A passage in a nuclear reactor for a rabbit
      or pile oscillator.
f tube *m* d'entrée
e tubo *m* de entrada
i tubo *m* d'ingresso
n toevoerbuis
d Zuführungsrohr *n*

7420 TUBE COUNT   ct
      A terminated discharge produced by an
      ionizing event.
f coup *m*, impulsion *f* de comptage
e colpo *m*, impulso *m* de recuento
i impulso *m* di conteggio
n buistel
d Zählstoss *m*

7421 TUBE FILTER   xr
      A filter which can be attached to an
      X-ray tube.
f filtre *m* de gaine
e filtro *m* de envuelta
i filtro *m* di guaina
n buisomhullingsfilter *n*
d Röhrengehäusefilter *m*

7422 TUBE FOCUS,   xr
      X-RAY FOCAL SPOT
      The part of the target of an X-ray tube
      which is struck by the main electron
      stream.
f foyer *m* de tube
e foco *m* de tubo
i fuoco *m* di tubo
n buisfocus
d Röhrenfokus *m*

7423 TUBE STAND   xr
      A support for holding X-ray tubes in
      position.
f support *m* de tube
e soporte *m* del tubo
i sopporto *m* del tubo
n buisstatief *n*
d Röhrenstativ *n*

7424 TUBULAR PINCH DEVICE, see 3142

7425 TUMOR   md
      An abnormal or morbid swelling or
      enlargement in any part of the body of an
      animal or plant.
f tumeur *f*
e tumor *m*
i tumore *m*
n gezwel *n*, tumor
d Geschwulst *f*, Tumor *m*

7426 TUMOR DOSE   md, xr
      The depth dose at the tumo(u)r under
      treatment.
f dose *f* tumorale
e dosis *f* tumoral

i dose *f* tumorale
n tumordosis
d Tumordosis *f*

7427 TUNGSTEN,   ch
      WOLFRAM
    Metallic element, symbol W, atomic
    number 74.
f tungstène *m*
e tungsteno *m*, volframio *m*
i tungsteno *m*, volframio *m*
n wolfraam *n*, wolfram *n*
d Wolfram *n*

7428 TUNGSTEN ALLOY, see 3185

7429 TUNGSTEN CARBIDE   mg
    Very hard compound of tungsten and carbon.
f carbure *m* de tungstène
e carburo *m* de tungsteno
i carburo *m* di tungsteno
n wolfraamcarbide *n*, wolframcarbide *n*
d Wolframkarbid *n*

7430 TUNGSTEN DISK   xr
    Component part of X-ray tube targets.
f disque *m* en tungstène
e disco *m* de tungsteno
i disco *m* a tungsteno
n wolfraampastille
d Wolframpastille *f*, Wolframscheibe *f*

7431 TUNICA MUCOSA, see 4564

7432 TUNNEL EFFECT, see 5184

7433 TURBIDITY, see 4431

7434 TURBULENT FLOW   gp
    Fluid flow in which the local velocity of the
    fluid medium varies erratically with time
    in both direction and magnitude.
f courant *m* turbulent
e corriente *f* turbulenta
i corrente *f* turbolenta
n turbulente stroming, wervelstroming
d Wirbelströmung *f*

7435 TURBULENT HEATING   pp
    Heating method of a plasma in which the
    ordered movements of the particles
    created by exterior sources are converted
    into disordered movements by micro-
    instability excitation.
f chauffage *m* turbulent
e calentamiento *m* turbulento
i riscaldamento *m* turbolento
n wervelverhitting
d Wirbelheizung *f*

7436 TURNERITE, see 4533

7437 TURNOVER   mc
    In radiobiology, the continued biological
    renewal of a given substance, without
    overall change in net concentration.
f cycle *m* métabolique, renouvellement *m*

e ciclo *m* metabólico, renovación *f*
i ciclo *m* metabolico, rinnovo *m*
n vernieuwing
d Umwandlung *f*

7438 TURNOVER,   is
    STAGE CIRCULATION, see 6834

7439 TURNOVER RATE   mc
    In radiobiology, during the steady state of
    turnover, the amount of substance turned
    over in unit time.
f vitesse *f* de renouvellement,
    vitesse *f* du cycle métabolique
e velocidad *f* de renovación,
    velocidad *f* del ciclo metabólico
i velocità *f* del ciclo metabolico,
    velocità *f* di rinnovo
n vernieuwingstempo *n*
d Umwandlungsrate *f*

7440 TURNOVER RATE CONSTANT   md
f constante *f* de renouvellement
e constante *f* de renovación
i costante *f* di rinnovo
n vernieuwingsconstant e
d Umwandlungskonstante *f*

7441 TURNOVER TIME   is
    In isotope separation, the ratio of stage
    hold-up to turnover.
f temps *m* d'écoulement
e tiempo *m* de circulación
i tempo *m* di circolazione
n omlooptijd, omzettijd
d Umlaufzeit *f*

7442 TURNOVER TIME   mc
    In radiobiology, during a steady state of
    turnover, the time required to turn over
    the amount of a given substance present in
    the tissue.
f temps *m* de renouvellement,
    temps *m* du cycle métabolique
e tiempo *m* de renovación,
    tiempo *m* del ciclo metabólico
i tempo *m* del ciclo metabolico,
    tempo *m* di rinnovo
n vernieuwingstijd
d Umwandlungszeit *f*

7443 TWINNING   cr
    A process in which a region in a crystal
    assumes an orientation which is symmetric-
    ally related to the basis orientation of the
    crystal.
f maclage *m*
e maclado *m*
i formazione *f* di cristalli geminati
n tweelingkristalvorming
d Zwillingbildung *f*

7444 TWO-BEAM INSTABILITY   pp
    A plasma instability which may occur when
    two particle beams intersect.
f instabilité *f* par faisceau entrecroisés
e inestabilidad *f* por haces secantes
i instabilità *f* da intersezione di fasci

n   instabiliteit door elkaar kruisende bundels
d   Instabilität *f* durch einanderkreuzende
    Strahlen

7445 TWO-GROUP MODEL   np
A model in which the neutrons are divided
in two groups of energy, fast and slow.
f   modèle *m* à deux groupes d'énergie
e   modelo *m* de dos grupos de energía
i   modello *m* a due gruppi d'energia
n   tweegroepenmodel *n*
d   Zweigruppenmodell *n*

7446 TWO-GROUP THEORY   rt
A treatment of nuclear reactor theory
in which it is assumed that the neutrons
are at two energies, fast and thermal.
f   théorie *f* à deux groupes
e   teoría *f* de dos grupos
i   teoria *f* a due gruppi
n   tweegroepentheorie
d   Zweigruppentheorie *f*

7447 TWO-PHASE SYSTEM   ch
A system in which a substance is present
in two different phases.
f   système *m* à deux phases
e   sistema *m* de dos fases
i   sistema *m* a due fasi
n   tweefazensysteem *n*
d   Zweiphasensystem *n*

7448 2 π PULSE COUNTING   ma
      ASSEMBLY
A pulse-counting assembly which includes
a 2 π radiation detector.
f   ensemble *m* de mesure à impulsions 2 π
e   conjunto *m* contador por impulsos 2 π
i   complesso *m* di misura ad impulsi 2 π
n   2 π -telopstelling
d   2 π -Impulszählanordnung *f*

7449 2 π IONIZATION CHAMBER   ic
An ionization chamber with which the
radiation emitted by a radioactive source
may be detected within a solid angle of
- 2 π steradians.
f   chambre *f* d'ionisation 2 π

e   cámara *f* de ionización 2 π
i   camera *f* d'ionizzazione 2 π
n   2 π -ionisatievat *n*
d   2 π -Ionisationskammer *f*

7450 TWO-REGION REACTOR,   rt
      TWO-ZONE REACTOR
A nuclear reactor in which the core
consists of two regions.
f   réacteur *m* à deux régions
e   reactor *m* de dos regiones
i   reattore *m* a due regioni
n   tweezonenreactor
d   Zweizonenreaktor *m*

7451 TWO-STATE PROCESS   np
A process used in obtaining photo-ionization
in pure gases.
f   processus *m* à deux étages
e   proceso *m* de dos etapas
i   processo *m* a due stadi
n   proces *n* in twee etappes
d   Zweistufenverfahren *n*

7452 2200 METER PER SECOND FLUX
      DENSITY, see 1376

7453 TWO-W CONCEPT   nw
The concept that the explosion of a weapon
of energy yield W on the earth's surface
produces blast phenomena identical with
those produced by a weapon of twice the
yield, i.e. two-W, burst in free air, i.e.
away from any reflecting surface.
f   projet *m* de deux W
e   proyecto *m* de dos W
i   progetto *m* di due W
n   twee-W-concept *n*
d   zwei-W-Konzept *n*

7454 TYUYAMUNITE   mi
A natural hydrated vanadate of calcium and
uranium.
f   tyuyamunite *f*
e   tiuiamunita *f*
i   tiuiamunite *f*
n   tyuyamuniet *n*
d   Tyuyamunit *m*

U

7455 ULTRACENTRIFUGE, see 3280

7456 ULTRAFAST PINCH   pp
Experiment existing in a linear discharge
between electrodes in which the current
rise is so rapid that there is no time for
particle collisions during the collapse
period and one has what is called a
collision-free shock.
f   striction *f* ultrarapide
e   estricción *f* ultrarrápida
i   contrazione *f* ultraveloce
n   ultrasnelle insnoering
d   ultraschnelle Einschnürung *f*

7457 ULTRAHIGH VACUUM, see 2500

7458 ULTRASONIC CLEANING   mt
The cleaning of the outside of an active
source by means of ultrasonic oscillations.
f   nettoyage *m* ultrasonore
e   limpieza *f* ultrasónica
i   pulitura *f* ultrasonica
n   ultrasone reiniging
d   Ultraschallreinigung *f*

7459 ULTRASONIC LEVEL GAGE (US),   me
         ULTRASONIC LEVEL METER (GB)
An instrument for close control of
uranium recovery processes.
f   limnimètre *m* ultrasonore
e   nivelímetro *m* ultrasónico
i   livellometro *m* ultrasonico
n   ultrasone niveaumeter
d   Ultraschallfüllstandsmessgerät *n*

7460 UMBRA   ra
A region behind an object in a beam of
radiation such that a straight line drawn
from any point in this region to any point
in the source passes through the object.
f   ombre *f*
e   sombra *f*
i   ombra *f*
n   schaduw
d   Schatten *m*

7461 UMKLAPP PROCESS, see 2701

7462 UMOHOITE   mi
A rare secondary molybdenum and
uranium containing mineral.
f   umohoïte *f*
e   umohoíta *f*
i   umoite *f*
n   umohoïet *n*
d   Umohoit *m*

7463 UNCANNED FUEL ELEMENT   fu
Ceramic fuel element in a thermal
breeder reactor with a U-233/Th cycle.
f   élément *m* combustible nu

e   elemento *m* combustible desnudo
i   elemento *m* combustibile nudo
n   naakt splijtstofelement *n*
d   nacktes Brennelement *n*

7464 UNCHARGED PARTICLE   np
A particle that has no electrical charge.
f   particule *f* neutre
e   partícula *f* neutra
i   particella *f* neutra
n   neutraal deeltje *n*
d   neutrales Teilchen *n*,
    ungeladenes Teilchen *n*

7465 UNCONTROLLABLE REACTION,
         see 6247

7466 UNDERGROUND BURST, see 422

7467 UNDERMODERATED   rt
Of a multiplying system, having a
moderator to fuel volume ratio less than
that which makes some specific reactor
parameter an extremum.
f   sous-modéré adj
e   submoderado adj
i   sottomoderato adj
n   ondergemodereerd adj
d   untermoderiert adj

7468 UNDERWATER BURST, see 423

7469 UNIFIED ATOMIC MASS CONSTANT,
         see 391

7470 UNIFIED MODEL   np
A nuclear model in which features of the
individual particle model and the collective
model are combined.
f   modèle *m* unifié
e   modelo *m* unificado
i   modello *m* unificato
n   combinatiemodel *n*
d   Kombinationsmodell *n*

7471 UNIFORM FIELD   ec
An electric field having uniform properties
over its whole area.
f   champ *m* uniforme
e   campo *m* uniforme
i   campo *m* uniforme
n   gelijkmatig veld *n*, uniform veld *n*
d   einheitliches Feld *n*, uniformes Feld *n*

7472 UNILAYER, see 4543

7473 UN-IONIZED ATOM, see 4655

7474 UNIPOLAR ARC   ec, pp
An electric arc between a metallic surface
and a plasma situated near that surface.
f   arc *m* unipolaire

e arco *m* unipolar
i arco *m* unipolare
n eenpolige boog
d Einpolbogen *m*

7475 UNIT un
A quantity adopted as a standard of measurement.
f unité *f*
e unidad *f*
i unità *f*
n eenheid
d Einheit *f*

7476 UNIT AREA, ge
UNIT SURFACE
The area of a square centimetre(er).
f unité *f* de surface
e unidad *f* de superficie
i unità *f* di superficie
n eenheid van oppervlakte
d Einheit *f* von Oberfläche

7477 UNIT CELL, see 2209

7478 UNIT CELL PARAMETERS cr
The lengths of the edges of the unit cell, together with the integral axes.
f paramètres *pl* des cellules élémentaires
e parámetros *pl* de la células elementales
i parametri *pl* delle celle elementari
n parameters *pl* van de elementaire cellen
d Elementarzellenparameter *pl*

7479 UNIT CHARGE ge
The electric charge that will exert a repelling force of 1 dyne on an equal and like charge 1 cm away in a vacuum, assuming that each charge is concentrated at a point.
f unité *f* de charge
e unidad *f* de carga
i unità *f* di carica
n eenheidslading
d Einheitsladung *f*

7480 UNIT LENGTH un
The length of one centimetre(er).
f unité *f* de longueur
e unidad *f* de longitud
i unità *f* di lunghezza
n eenheid van lengte
d Einheit *f* von Länge

7481 UNIT PROCESS cl
A chemical process in which all reactions take place in a single apparatus.
f procédé *m* unitaire
e procedimiento *m* unitario
i processo *m* unitario
n eenheidsproces *n*
d Einheitsverfahren *n*

7482 UNIT TIME ge
The 1/86400th part of a solar day.
f unité *f* de temps
e unidad *f* de tiempo

i unità *f* di tempo
n eenheid van tijd
d Einheit *f* von Zeit

7483 UNIT VOLUME ge
The volume of a cubic centimetre(er).
f unité *f* de volume
e unidad *f* de volumen
i unità *f* di volume
n eenheid van volume
d Einheit *f* von Volumen

7484 UNIT WEIGHT, see 6746

7485 UNIVERSAL INSTABILITY pp
Instability of a non-uniform plasma in a magnetic field, connected with the existence of a pressure gradient.
f instabilité *f* universelle
e inestabilidad *f* universal
i instabilità *f* universale
n universele instabiliteit
d Universalinstabilität *f*

7486 UNIVERSE ge
The totality of space surrounding and including the earth.
f univers *m*
e universo *m*
i universo *m*
n heelal *n*
d Universum *n*, Weltall *n*

7487 UNLOADING fu, rt
The removal of spent fuel from the reactor.
f déchargement *m*
e descarga *f*
i scaricamento *m*
n ontlading
d Entladung *f*

7488 UNLOADING FACE fu, rt
That side of the reactor where unloading takes place.
f front *m* de décharge
e frente *f* de descarga
i fronte *f* di scaricamento
n ontlaadzijde
d Entladefläche *f*, Entladeseite *f*

7489 UNLOADING MACHINE fu
A machine for removing fuel from a nuclear reactor.
f appareil *m* de décharge
e máquina *f* de descarga
i macchina *f* di scaricamento
n ontlaadapparaat *n*
d Entladegerät *n*, Entlademaschine *f*

7490 UNMODIFIED SCATTER np, ra
Radiation scattered without change of photon energy.
f diffusion *f* non-modifiée
e dispersión *f* no modificada
i deviazione *f* non modificata
n ongewijzigde verstrooiing
d Streuung *f* ohne Energieänderung

7491 UNSEALED SOURCE   ra
Any radioactive source which is not
a sealed source.
f source *f* non-scellée
e fuente *f* no sellada
i sorgente *f* non sigillata
n niet-gesloten bron
d offener radioaktiver Stoff *m*

7492 UNSHARPNESS, see 681

7493 UNSTABLE, see 3882

7494 UPPER TRIP LEVEL   rt, sa
f seuil *m* supérieure d'arrêt
e umbral *m* superior de paro
i soglia *f* superiore d'arresto,
   soglia *f* superiore di scatto
n bovenuitschakeldrempel
d Oberabschaltschwelle *f*

7495 UPSCATTERING   np
A scattering process in which a neutron
gains energy.
f diffusion *f* accélératrice
e dispersión *f* aceleradora
i deviazione *f* acceleratrice
n opstrooiing
d Aufwärtsstreuung *f*

7496 UPTAKE, see 3519

7497 UPTAKE FACTOR   is, md
The fraction of the quantity of a radioactive isotope which is absorbed by the
body or goes to the critical organ.
f facteur *m* d'apport
e factor *m* de aportación
i fattore *m* d'assorbimento
n opneemfactor
d Aufnahmefaktor *m*

7498 UPTAKE RATE   is, md
The velocity at which a radioactive isotope
is absorbed by the body.
f vitesse *f* d'apport
e velocidad *f* de aportación
i velocità *f* d'assorbimento
n opneemsnelheid
d Aufnahmegeschwindigkeit *f*

7499 UPWIND FALL-OUT, see 234

7500 URACONITE   mi
A mineral containing zippeite and
uranopilite.
f uraconite *f*
e uraconita *f*
i uraconite *f*
n uraconiet *n*
d Uraconit *m*

7501 URANIDES   ch
The name used for the elements of atomic
numbers 89 to 103 when they are in the
oxidation state +6.
f uranides *pl*

e uránidos *pl*
i uranidi *pl*
n uraniden *pl*
d Uranide *pl*

7502 URANINITE   mi
A natural oxide of uranium.
f uraninite *f*
e uraninita *f*
i uraninite *f*
n uraniniet *n*
d Uraninit *m*

7503 URANITE, see 4620

7504 URANIUM   ch
Metallic element, symbol U, atomic
number 97.
f uranium *m*
e uranio *m*
i uranio *m*
n uraan *n*, uranium *n*
d Uran *n*

7505 URANIUM AGE   mi, np
The age of a mineral as calculated from the
numbers of ionium atoms present originally,
now, and when equilibrium is established
with uranium.
f âge *m* d'uranium
e edad *f* de uranio
i età *f* d'uranio
n uraniumleeftijd
d Uranalter *n*

7506 URANIUM CHALCOGENIDE   ch
f chalcogénure *m* d'uranium
e calcogenuro *m* de uranio
i calcogenuro *m* d'uranio
n uraniumchalcogenide *n*
d Uranchalkogenid *n*

7507 URANIUM CONCENTRATE   ch, mi
A product resulting from physical and
chemical treatment of uranium ores.
f concentré *m* uranifère
e concentrado *m* uranífero
i concentrato *m* uranifero
n uraniumconcentraat *n*
d Urankonzentrat *n*

7508 URANIUM CONTENT   mi
The percentage of uranium in an ore.
f teneur *f* en uranium
e contenido *m* de uranio
i contenuto *m* d'uranio
n uraniumgehalte *n*
d Urangehalt *m*

7509 URANIUM CONTENT METER BY   ma
      BETA AND GAMMA RADIOACTIVITY
A content meter designed to determine the
uranium content of an ore sample by means
of the measurement of beta and gamma
activity of this sample.
f teneurmètre *m* en uranium par radioactivité bêta et gamma

e uranió metro *m* por actividad beta-gamma
i uraniotenorimetro *m* per radioattività beta e gamma
n gehaltemeter voor uranium
d Gerät *n* zur Bestimmung des Uranium-gehalts

7510 URANIUM CREEP, see 1440

7511 URANIUM DEPOSITS mi
The occurrence of uranium ores in the earth's crust.
f gisements *pl* d'uranium
e yacimientos *pl* de uranio
i giacimenti *pl* d'uranio
n uraniumafzettingen *pl*
d Uranlagerstätten *pl*

7512 URANIUM ENRICHED FUEL fu
f combustible *m* en uranium enrichi
e combustible *m* de uranio enriquecido
i combustibile *m* in uranio arricchito
n splijtstof uit verrijkt uranium
d Brennstoff *m* aus angereichertem Uran

7513 URANIUM-GRAPHITE LATTICE rt
A lattice in a reactor composed of uranium and graphite rods, slugs or slabs.
f réseau *m* uranium-graphite
e celosía *f* uranio-grafito
i reticolo *m* uranio-grafite
n uranium-grafietrooster *n*
d Uran-Graphitgitter *n*

7514 URANIUM HALIDE ch
A binary halogen compound of uranium.
f halogénure *m* d'uranium
e halogenuro *m* de uranio
i alogenuro *m* d'uranio
n uraniumhalogenide *n*
d Uranhalogenid *n*

7515 URANIUM HEXAFLUORIDE, see 3244

7516 URANIUM HEXAFLUORIDE
RECONVERSION PROCESS, see 3245

7517 URANIUM ORE mi
f minerai *m* d'uranium
e mineral *m* de uranio
i minerale *m* d'uranio
n uraniumerts *n*
d Uranerz *n*

7518 URANIUM OXIDES ch
f oxydes *pl* d'uranium
e óxidos *pl* de uranio
i ossidi *pl* d'uranio
n uraniumoxyden *pl*
d Uranoxyde *pl*

7519 URANIUM OXOSALTS ch
Salts of the oxo-acids of uranium.
f sels *pl* des oxo-acides d'uranium
e sales *pl* de los oxoácidos de uranio
i sali *pl* degli ossoacidi d'uranio
n oxozouten *pl* van uranium
d Oxosalze *pl* des Urans

7520 URANIUM-RADIUM SERIES np
Radioactive decay series starting with U-238 and ending with the inactive lead isotope with mass number 206.
f famille *f* uranium-radium
e familia *f* uranio-radio
i famiglia *f* uranio-radio
n uranium-radiumreeks
d Uran-Radiumreihe *f*

7521 URANIUM REACTOR rt
A nuclear reactor in which the principal fuel is uranium.
f réacteur *m* à uranium
e reactor *m* de uranio
i reattore *m* ad uranio
n uraniumreactor
d Uranreaktor *m*

7522 URANIUM RECONNAISSANCE mi
A survey of a specified area for the detection of uranium ores.
f exploration *f* d'uranium
e exploración *f* de uranio
i ricognizione *f* per uranio
n uraanspeurtocht
d Uranrekognoszierung *f*

7523 URANIUM SERIES np
The series of nuclides resulting from the decay of uranium-238.
f famille *f* de l'uranium
e familia *f* del uranio
i famiglia *f* dell'uranio
n uraniumreeks
d Uranreihe *f*

7524 URANIUM TETRAFLUORIDE, see 3076

7525 URANIUM-THORIUM REACTOR rt
A nuclear reactor in which the fuel consists of U-233 or U-235 and thorium.
f réacteur *m* à uranium-thorium
e reactor *m* de uranio-torio
i reattore *m* ad uranio-torio
n uranium-thoriumreactor
d Uran-Thoriumreaktor *m*

7526 URANIUM TRIOXIDE, see 4994

7527 URANIUM-235, see 55

7528 URANOCHALCITE mi
An ill-defined uranium sulph(f)ate of doubtful validity, probably a variety of sklodowskite.
f uranochalcite *f*
e uranocalcita *f*
i uranocalcite *f*
n uranochalciet *n*
d Uranochalzit *m*

7529 URANOCIRCITE, see 4398

7530 URANOLEPIDITE, mi
VANDENBRANDEITE
A rare secondary mineral, containing about 60 % of U.
f vandenbrandéite *f*

e   vandenbrandeita *f*
i   vandenbrandeite *f*
n   vandenbrandeïet *n*
d   Vandenbrandeit *m*

7531 URANOPHANE,
       URANOTIL, see 3888

7532 URANOPILITE   mi
      A secondary mineral containing about 68 %
      of U.
f   uranopilite *f*
e   uranopilita *f*
i   uranopilite *f*
n   uranopiliet *n*
d   Uranopilit *m*

7533 URANOSPATHITE   mi
      A rare secondary mineral containing
      about 46 % of U.
f   uranospathite *f*
e   uranospatita *f*
i   uranospatite *f*
n   uranospathiet *n*
d   Uranospathit *m*

7534 URANOSPHAERITE   mi
      An alteration product of uraninite,
      containing about 43 % of U.
f   uranosphérite *f*
e   uranosferita *f*
i   uranosferite *f*
n   uranosferiet *n*
d   Uranosphärit *m*

7535 URANOSPINITE   mi
      A rare secondary mineral containing about
      46 % of U.
f   uranospinite *f*
e   uranospinita *f*
i   uranospinite *f*
n   uranospiniet *n*
d   Uranospinit *m*

7536 URANOTHALLITE, see 3998

7537 URANOTHORITE   mi
      Uranoan variety of thorite.
f   uranothorite *f*
e   uranotorita *f*
i   uranotorite *f*
n   uranothoriet *n*
d   Uranothorit *m*

7538 URANYL   ch
      Name given to the radical UO_2^{++} of a
      number of important uranium
      compounds.
f   uranyle *m*
e   uranilo *m*
i   uranile *m*
n   uranyl-
d   Uranyl-

7539 URANYL FLUORIDE   ch
      A white solid substance, when pure,
      prepared by the reaction of hydrofluoric
      acid with a uranium oxide.

f   fluorure *m* d'uranyle
e   fluoruro *m* de uranilo
i   fluoruro *m* d'uranile
n   uranylfluoride *n*
d   Uranylfluorid *n*

7540 URANYL NITRATE   ch
      Uranium compound, an important inter-
      mediate product in various chemonuclear
      processes.
f   nitrate *m* d'uranyle
e   nitrato *m* de uranilo
i   nitrato *m* d'uranile
n   uranylnitraat *n*
d   Uranylnitrat *n*

7541 URANYL SULFATE (US),   ch
       URANYL SULPHATE (GB)
      Uranium compound used in light water
      moderated reactors.
f   sulfate *m* d'uranyle
e   sulfato *m* de uranilo
i   solfato *m* d'uranile
n   uranylsulfaat *n*
d   Uranylsulfat *n*

7542 URETHROGRAPHY   xr
      The radiological examination of the lumen
      of the urethra following the injection of
      a contrast medium.
f   uréthrographie *f*
e   uretrografía *f*
i   uretrografia *f*
n   urethrografie
d   Urethrographie *f*

7543 UROGRAPHY   xd
      The radiological examination of the
      urinary tract.
f   urographie *f*
e   urografía *f*
i   urografia *f*
n   urografie
d   Urographie *f*

7544 USEFUL BEAM   ra, xd
      That part of the primary radiation from
      a röntgen tube or enclosed radioactive
      source that comes out of the source and its
      housing through the aperture, diaphragm
      or cone.
f   faisceau *m* utile
e   haz *m* útil
i   fascio *m* utile
n   nuttige bundel
d   Nutzstrahl *m*

7545 USEFUL CAPTURE, see 2648

7546 USEFUL NEUTRONS   np
      Neutrons the capture of which provokes
      a fission.
f   neutrons *pl* utiles
e   neutrones *pl* útiles
i   neutroni *pl* utili
n   nuttige neutronen *pl*
d   nützliche Neutrone *pl*

7547 USEFUL POWER   ge
    The relation between applied energy to the
    resultant energy.
f puissance *f* utile
e potencia *f* útil
i potenza *f* utile
n nuttig vermogen *n*
d Nutzleistung *f*

7548 USEFUL THERMAL POWER   rt
    That part of the total thermal power which
    can be delivered at such a temperature
    that it can be utilized as an energy source.
f puissance *f* thermique utilisable

e potencia *f* térmica útil
i potenza *f* termica utile
n nuttig warmtevermogen *n*
d Nutzwärmeleistung *f*

7549 U.V. EXCITATION FLUORIMETER,
    see 1340

7550 UVANITE   mi
    A uranium and vanadium containing mineral.
f uvanite *f*
e uvanita *f*
i uvanite *f*
n uvaniet *n*
d Uvanit *m*

V

7551 V EVENT   cr
A phenomenon interpretable as the
in-flight decay of a K meson or hyperon.
f événement *m* V
e acontecimiento *m* V
i evento *m* V
n V-proces *n*
d V-Ereignis *n*.

7552 V PARTICLE   np
A neutral meson or hyperon decaying in
a cloud chamber, the tracks of its charged
decay products forming a broken line like
the letter V.
f particule *f* à trace en forme de V,
   particule *f* V
e partícula *f* de´traza en forma de V
i particella *f* a traccia in forma di V
n deeltje *n* met V-spoor
d Teilchen *n* mit V-Spur

7553 VACANCY, see 6350

7554 VACUUM   ge, rt
An enclosed space from which practically
all air has been removed.
f vide *m*
e vacío *m*
i vuoto *m*
n luchtledige ruimte, vacuüm *n*
d Luftleere *f*, Vakuum *n*

7555 VACUUM ELECTRON   ec, np
In the Dirac electron theory, an electron
in one of the negative energy states which
are supposed to be all filled for the case
of a vacuum.
f électron *m* de vacuum
e electrón *m* de vacío
i elettrone *m* di vuoto
n vacuümelektron *n*
d Vakuumelektron *n*

7556 VACUUM FOREPUMP   vt
A vacuum pump capable of lowering the
pressure down to about 0.001 mm of
mercury.
f pompe *f* à vide préliminaire
e bomba *f* de vacío preliminar
i pompa *f* di vuoto preliminare
n voorvacuümpomp
d Vorvakuumpumpe *f*

7557 VACUUM GAGE (US),   me
       VACUUM GAUGE (GB)
A device that indicates the absolute gas
pressure in a vacuum system.
f manomètre *m* à vide
e manómetro *m* de vacío
i manometro *m* a vuoto
n vacuümmanometer
d Vakuummanometer *n*

7558 VACUUM LEAK DETECTOR   me, vt
An instrument, usually a mass spectro-
meter, used to detect and locate leaks in
a high-vacuum system.
f détecteur *m* de fuites dans un système
   à vide
e detector *m* de fugas en un sistema de
   vacío
i rivelatore *m* di fughe in un sistema a vuoto
n lekzoeker in een vacuümsysteem
d Lecksucher *m* im Vakuumsystem

7559 VACUUM METER WITH   ma
       ALPHA EMITTER
A vacuum meter for the measurement of
low pressures and comprising an alpha
radiation source included in an ionization
chamber communicating with the enclosure,
the vacuum of which is to be determined.
f manomètre *m* de vide à émetteur alpha
e vacuómetro *m* por emisión alfa
i manometro *m* di vuoto a radiazione alfa
n vacuümmeter met alfastralen
d Vakuummessgerät *n* mit Alphastrahlen

7560 VACUUM POLARIZATION   np
Process by which an electromagnetic field
generates virtual electron-posit(r)on pairs
which modify the charge and current
distribution which produced the original
electromagnetic field.
f polarisation *f* du vide
e polarización *f* del vacío
i polarizzazione *f* del vuoto
n vacuümpolarisatie
d Polarisation *f* des Vakuums,
   Vakuumpolarisation *f*

7561 VACUUM PUMP   vt
A device to remove gases and vapo(u)rs
from a closed receptacle.
f pompe *f* à vide
e bomba *f* de vacío
i pompa *f* a vuoto
n vacuümpomp
d Vakuumpumpe *f*

7562 VACUUM RESERVOIR   vt
A large vessel inserted between the fore-
pump and high vacuum pump in a vacuum
system.
f réservoir *m* à vide
e recipiente *m* de vacío
i serbatoio *m* a vuoto
n vacuümreservoir *n*
d Vakuumzwischengefäss *n*

7563 VACUUM SEAL   cd, vt
An airtight junction.
f scellement *m* à vide
e cierre *m* hermético
i chiusura *f* ermetica

n   vacuümverbinding
d   Vakuumverbindung *f*

7564   VACUUM SPECTROGRAPH   sp
A spectrograph operating in a vacuum to
avoid absorption of long wavelength
radiations by the air.
f   spectrographe *m* à vide
e   espectrógrafo *m* de vacío
i   spettrografo *m* a vuoto
n   vacuümspectrograaf
d   Vakuumspektrograph *m*

7565   VACUUM VOLATILIZATION   fu
A process used in treating irradiated fuel.
f   volatilisation *f* sous vide
e   volatilización *f* en el vacío
i   volatilizzazione *f* sotto vuoto
n   vervluchtiging onder vacuüm
d   Vakuumverflüchtigung *f*

7566   VALENCE,   ch
   VALENCY
The property of an atom or radical to
combine with another atom or radical in
definite proportions, or a number
representing the proportion in which a
given atom or radical combines.
f   valence *f*
e   valencia *f*
i   valenza *f*
n   valentie, waardigheid
d   Valenz *f*, Wertigkeit *f*

7567   VALENCE ANGLES   ch
The angles between the successive valence
bonds of an atom.
f   angles *pl* de valence
e   ángulos *pl* de valencia
i   angoli *pl* di valenza
n   valentiehoeken *pl*
d   Valenzwinkel *pl*

7568   VALENCE BAND,   cr, sp
   VALENCY BAND
The range of energy states in the spectrum
of a solid crystal in which lie the energies
of the valence electron which bind the
crystal together.
f   bande *f* de valence
e   banda *f* de valencia
i   banda *f* di valenza
n   valentieband, waardigheidsband
d   Valenzband *n*, Wertigkeitsband *n*

7569   VALENCE BOND,   ch
   VALENCY BOND
The bond formed between electrons of two
or more atoms.
f   liaison *f* de valence
e   enlace *m* de valencia
i   legame *m* di valenza
n   valentiebinding, waardigheidsbinding
d   Valenzbindung *f*, Wertigkeitsbindung *f*

7570   VALENCE CRYSTAL   cr
A crystal bound together by covalent bonds.

f   cristal *m* de valence
e   cristal *m* de valencia
i   cristallo *m* di valenza
n   valentiekristal *n*
d   Valenzkristall *m*

7571   VALENCE ELECTRON, see 1300

7572   VALENCE NUMBER   ch
A number assigned to an atom or ion that
is equal to its valence, preceded by a plus
or minus sign to indicate whether the ion
is positive or negative, or whether the
atom, in reaching the state of oxidation
under consideration, has lost or gained
electrons from its normal state.
f   nombre *m* de valence
e   número *m* de valencia
i   numero *m* di valenza
n   valentiegetal *n*, waardigheidsgetal *n*
d   Valenzzahl *f*, Wertigkeitszahl *f*

7573   VALENCE SHELL,   np
   VALENCY SHELL
The group of electrons constituting the
outer electronic shell of an atom.
f   couche *f* de valence
e   capa *f* de valencia
i   corteccia *f* di valenza
n   valentieschil, waardigheidsschil
d   Valenzschale *f*, Wertigkeitsschale *f*

7574   VALUE   ge
The magnitude of a quantity.
f   valeur *f*
e   valor *m*
i   valore *m*
n   waarde
d   Wert *m*

7575   VALUE FUNCTION   is
The function V defining the molar
separation potential with relation to the
natural isotopic abundance.
f   fonction *f* de valeur
e   función *f* de valor
i   funzione *f* di valore
n   waardefunctie
d   Wertfunktion *f*

7576   VALUE OF AN   is
   ISOTOPIC MIXTURE
In isotopic separation, a measure of the
difficulty of preparing a quantity of an
isotopic mixture.
f   valeur *f* de mélange isotopique
e   valor *m* de mezcla isotópica
i   valore *m* di miscela isotopica
n   waarde van isotopisch mengsel
d   Wert *m* einer Isotopenmischung

7577   VAN ALLEN RADIATION BELTS   ra
The name applied to the two toroidal belts
of radiation which encircle the earth at
high altitude, composed of charged particles
(almost entirely electrons and protons)
which are trapped in the earth's magnetic
field.

f   ceintures *pl* de Van Allen
e   cinturones *pl* de Van Allen
i   cinture *pl* di Van Allen
n   stralingsgordels *pl* van Van Allen
d   Van-Allen-Gürtel *pl*

7578 VAN DE GRAAFF ACCELERATOR,
      see 2197

7579 VANADIUM   ch
      Metallic element, symbol V, atomic
      number 23.
f   vanadium *m* ·
e   vanadio *m*
i   vanadio *m*
n   vanadium *n*
d   Vanadium *n*

7580 VANDENBRANDEITE, see 7530

7581 VANDENDRIESSCHEITE   mi
      A rare secondary mineral containing
      about 60 % of U.
f   vandendriesschéite *f*
e   vandendriesscheita *f*
i   vandendriesscheite *f*
n   vandendriesscheïet *n*
d   Vandendriesscheit *m*

7582 VANOXITE   mi
      A mineral containing vanadium and
      uranium.
f   vanoxite *f*
e   vanoxita *f*
i   vanossite *f*
n   vanoxiet *n*
d   Vanoxit *m*

7583 VAPOR CHROMATOGRAPHY (US),
      VAPOUR CHROMATOGRAPHY (GB),
      see 2961

7584 VAPOR JET PUMP (US),   vt
      VAPOUR JET PUMP (GB)
      A pump particularly suitable for obtaining
      ultra-high vacua.
f   pompe *f* à jet de vapeur
e   bomba *f* de chorro de vapor
i   pompa *f* a getto di vapore
n   dampstraalpomp
d   Dampfstrahlpumpe *f*

7585 VAPOR SUPPRESSION (US), see 5479

7586 VAPORIZATION, see 2385

7587 VAPORIZATION HEAT, see 2386

7588 VARIABLE   ge, ma
      A quantity that may assume a number of
      distinct values.
f   variable *f*
e   variable *f*
i   variabile *f*
n   variabele, veranderlijke
d   Grösse *f*, Variable *f*

7589 VARIABLE ENERGY   pa
      CYCLOTRON
      A new type of cyclotron used for studying
      radiation induced damage.
f   cyclotron *m* à puissance variable
e   ciclotrón *m* de potencia variable
i   ciclotrone *m* a potenza variabile
n   cyclotron *n* met veranderlijk vermogen
d   Zyklotron *n* mit variabler Leistung

7590 VARIABLE FIELD, see 7262

7591 VARYING DUTY   ge
      A type of duty in which the amount of
      load, and length of time the load is applied,
      are subject to considerable variation.
f   service *m* interrompu à charge variable
e   servicio *m* variable
i   erogazione *f* variabile
n   intermitterend bedrijf *n* met veranderlijke
    belasting
d   aussetzender Betrieb *m* mit veränderlicher
    Belastung

7592 VARYING MODERATOR HEIGHT   co
      A method for controlling reactivity and
      hence reactor power.
f   hauteur *f* de modérateur variable
e   altitud *f* de moderador variable
i   altezza *f* di moderatore variabile
n   veranderlijke moderatorhoogte
d   veränderliche Moderatorhöhe *f*

7593 VASOGRAPHY   xr
      Radiological examination of the vessels.
f   vasographie *f*
e   vasografía *f*
i   vasografia *f*
n   vasografie
d   Vasographie *f*

7594 VECTOR,   ma
      VECTOR QUANTITY
      A quantity that has direction as well as
      numerical value.
f   grandeur *f* vectorielle, vecteur *m*
e   magnitud *f* vectorial, vector *m*
i   grandezza *f* vettoriale, vettore *m*
n   vector, vectoriële grootheid
d   Vektor *m*, vektorielle Grösse *f*

7595 VECTOR FLUX, see 258

7596 VECTOR MODEL OF ATOM   np
      A model in which special rules set up for
      the vector addition have been considered.
f   modèle *m* vectoriel d'un atome
e   modelo *m* vectorial de un átomo
i   modello *m* vettoriale d'un atomo
n   vectormodel *n* van een atoom
d   Vektormodell *n* eines Atoms

7597 VEHICLE-BORNE SCINTILLATOR   ma
      PROSPECTING RADIATION METER
      A vehicle-borne scintillator prospecting
      radiation meter including in most cases a
      recorder the mechanism of which is

driven by the movement of the bearing
vehicle.
f   radiamètre *m* de prospection porté à
    scintillateur
e   radiámetro *m* de exploración móvil con
    centelleador
i   radiametro *m* di prospezione auto-
    trasportato a scintillatore
n   prospectiestralingsmeter met scintillator
    op voertuig
d   fahrzeuggebundenes Lagerstättensuch-
    gerät *n* mit Szintillationszähler

7598   VEHICLE-BORNE SCINTILLATOR   ma
       PROSPECTING RADIATION METER
       WITH AMPLITUDE ANALYZER
       A vehicle-borne selective scintillator
       prospecting radiation meter using an
       amplitude analyzer, in which an
       identification line permits the determina-
       tion of the nature of the radioactive element
       detected.
f   radiamètre *m* de prospection porté à
    analyseur d'amplitude à scintillateur
e   radiámetro *m* portátil de exploración con
    analizador de amplitud y con centelleador
i   radiametro *m* di prospezione autotrasporta-
    to ad analizzatore d'ampiezza ed a
    scintillatore
n   prospectiestralingsmeter met amplitude-
    analysator en scintillator op voertuig
d   fahrzeuggebundenes Lagerstättensuch-
    gerät *n* mit Impulshöhenanalysator und
    Szintillator.

7599   VEHICLE-BORNE SELECTIVE   ma
       SCINTILLATOR PROSPECTING
       RADIATION METER
       A vehicle-borne scintillator prospecting
       radiation meter distinguishing between
       radiation due to thorium, uranium or other
       radioactive elements by means of selective
       measurement of photon flux density
       obtained through the use of amplitude
       discriminators.
f   radiamètre *m* sélectif de prospection porté
    à scintillateur
e   radiámetro *m* selectivo móvil de
    exploración con centelleador
i   radiametro *m* selettivo di prospezione
    autotrasportato a scintillatore
n   selectieve prospectiestralingsmeter met
    scintillator op voertuig
d   fahrzeuggebundenes selektives Lager-
    stättensuchgerät *n* mit Szintillator

7600   VEIN, see 4108

7601   VELOCITY DISTRIBUTION,   np
       VELOCITY SORTING
       Any process of selecting electrons
       according to their velocities.
f   distribution *f* de vitesse,
    répartition *f* de vitesse
e   distribución *f* de velocidad
i   distribuzione *f* di velocità
n   snelheidsverdeling
d   Geschwindigkeitsverteilung *f*

7602   VELOCITY SELECTOR   np
       A device which mechanically selects from
       a beam of neutrons with a continuous
       velocity distribution just those neutrons
       within a narrow range of velocities.
f   sélecteur *m* de vitesse
e   selector *m* de velocidad
i   selettore *m* di velocità
n   snelheidskiezer
d   Geschwindigkeitswähler *m*

7603   VENOGRAPHY   xr
       Radiological examination of the veins.
f   vénographie *f*
e   venografía *f*
i   venografia *f*
n   venografie
d   Venographie *f*

7604   VENTRICULOGRAPHY   xr
       The radiological examination of the intra-
       cranial ventricles following direct
       introduction of air.
f   ventriculographie *f*
e   ventriculografía *f*
i   ventriculografia *f*
n   ventriculografie
d   Ventrikulographie *f*

7605   VERTICAL ACCELERATOR   pa
       RELATIVE APERTURE
       The ratio of the minimum vertical clearance
       for particle passage in the accelerating
       chamber to the particle orbit radius.
f   ouverture *f* relative verticale d'un
    accélérateur
e   abertura *f* relativa vertical de un
    acelerador
i   apertura *f* relativa verticale d'un
    acceleratore
n   relatieve verticale apertuur van een
    deeltjesversneller
d   relative vertikale Apertur *f* eines
    Teilchenbeschleunigers

7606   VESICULOGRAPHY   xr
       The radiological examination of the seminal
       vesicles, vas deferens and ejaculatory duct
       following the injection of a contrast medium.
f   vésiculographie *f*
e   vesiculografía *f*
i   vesiculografia *f*
n   vesiculografie
d   Vesikulographie *f*

7607   VIBRATING CAPACITOR   ec
       A capacitor in which the capacitance
       varies periodically so that an alternating
       e.m.f. is produced which is proportional
       to the charge on the insulated electrode.
f   condensateur *m* vibrant
e   condensador *m* vibrante
i   condensatore *m* vibrante
n   trilcondensator
d   Schwingkondensator *m*

7608   VIBRATING CONTACTOR,
       VIBRATOR, see 1081

7609 VIBRATING REED ELECTROMETER,
see 2002

7610 VIETINGHOFITE						mi
An iron containing variety of samarskite.
f vietinghofite *f*
e vietinghofita *f*
i vietinghofite *f*
n vietinghofiet *n*
d Vietinghofit *m*

7611 VIEWING SCREEN, see 2602

7612 VIEWING SYSTEM					co, rt
System used for optical control of
processes in a nuclear reactor.
f système *m* de contrôle optique
e sistema *m* de control óptico
i sistema *m* di controllo ottico
n optisch controlesysteem *n*
d optisches Überwachungssystem *n*

7613 VIRGIN NEUTRON FLUX				np
A flux of neutrons that have suffered no
collisions and therefore have lost none of
the energy with which they were born.
f flux *m* de neutrons vierges
e flujo *m* de neutrones vírgenes
i flusso *m* di neutroni vergini
n maagdelijke neutronenflux
d jungfräulicher Neutronenfluss *m*

7614 VIRGIN NEUTRONS					np
Neutrons from any source, before they
make a collision.
f neutrons *pl* vierges
e neutrones *pl* vírgenes
i neutroni *pl* vergini
n maagdelijke neutronen *pl*
d jungfräuliche Neutronen *pl*

7615 VIRTUAL LEVEL,				ma, np
VIRTUAL STATE
1. A nuclear state that is so broad that
its existence has to be deduced indirectly.
2. A nuclear state which appears as an
intermediate state when the transition
between two other states of that nucleus
is computed.
f niveau *m* virtuel
e nivel *m* virtual
i livello *m* virtuale
n virtueel niveau *n*, virtuele toestand
d quasistatisches Niveau *n*,
virtuelles Niveau *n*

7616 VIRTUAL PARTICLE				np
A particle emitted and absorbed in a
virtual process.
f particule *f* virtuelle
e partícula *f* virtual
i particella *f* virtuale
n virtueel deeltje *n*
d virtuelles Teilchen *n*

7617 VIRTUAL PHOTON					np
A photon emitted and absorbed in a virtual
process.

f photon *m* virtuel
e fotón *m* virtual
i fotone *m* virtuale
n virtueel foton *n*
d virtuelles Photon *n*

7618 VIRTUAL PROCESS					np
Process which may be pictured as the
emission of a particle or quantum followed
so quickly by its absorption or further
interaction that the energy and momentum
of the particle in this intermediate state
are ill defined.
f processus *m* virtuel
e proceso *m* virtual
i processo *m* virtuale
n virtueel proces *n*
d virtueller Prozess *m*

7619 VIRTUAL QUANTUM					ma
In second and higher order perturbation
theory, a matrix element connecting an
initial state with a final state, involves
intermediate states in which energy is not
conserved.
f quantum *m* virtuel
e cuanto *m* virtual
i quanto *m* virtuale
n virtueel quantum *n*
d virtueller Quant *m*

7620 VIRTUAL REACTOR, see 3390

7621 VIRTUAL SOURCE, see 3391

7622 VIRTUAL STATE					np
An intermediate state between an initial
and a final state.
f état *m* virtuel
e estado *m* virtual
i stato *m* virtuale
n virtuele toestand
d virtueller Zustand *m*

7623 VISCOSITY						gp
Resistance of a fluid to the relative
motion of its particles.
f viscosité *f*
e viscosidad *f*
i viscosità *f*
n viscositeit
d Viskosität *f*

7624 VISCOUS FLOW, see 3892

7625 VISIBILITY DISTANCE				nw
The horizontal distance at which a large
dark object can just be seen against the
horizon sky in daylight.
f distance *f* de visibilité
e distancia *f* de visibilidad
i distanza *f* di visibilità
n zicht *n*, zichtbare afstand
d Sicht *f*, Sichtbereich *m*

7626 VISUAL RANGE, see 2497

7627 VLASOV EQUATION   pp
A kinetic equation applicable only to
plasmas.
f équation f de Vlasov
e ecuación f de Vlasov
i equazione f di Vlasov
n vlasovvergelijking
d Vlasovsche Gleichung f

7628 VOGLIANITE   mi
A hydrous calcium and uranium sulphate
of doubtful validity.
f voglianite f
e voglianita f
i voglianite f
n voglianiet n
d Voglianit m

7629 VOGLITE   mi
A mineral containing calcium, uranium
and copper.
f voglite f
e voglita f
i voglite f
n vogliet n
d Voglit m

7630 VOID   rt
An empty space in a reactor core.
f vide m
e oquedad f, vacío m
i vuoto m
n caviteit, leegte
d Leerraum m, Leerstelle f

7631 VOID COEFFICIENT OF   rt
REACTIVITY
The partial derivative of reactivity with
respect to the volume fraction of voids
in a specified location.
f coefficient m de vide de réactivité
e coeficiente m de vacío de reactividad
i coefficiente m di vuoto di reattività
n caviteitscoëfficiënt van de reactiviteit
d Dampfblasenkoeffizient m der Reaktivität,
   Leerraumkoeffizient m der Reaktivität

7632 VOLATILE FISSION PRODUCT,
see 2658

7633 VOLATILITY   gp
f volatilité f
e volatilidad f
i volatilità f
n vluchtigheid
d Flüchtigkeit f

7634 VOLBORTHITE   mi
A secondary mineral containing about
3 % of U.
f volborthite f
e volbortita f
i volbortite f
n volborthiet n
d Volborthit m

7635 VOLCANIC ROCK, see 3385

7636 VOLTAGE MULTIPLIER RECTIFIER,
see 1169

7637 VOLTAGE PULSE   ct
Change in voltage in the central electrode
system of a radiation counter.
f impulsion f de tension
e impulso m de tensión
i impulso m di tensione
n spanningspuls, spanningsstoot
d Spannungsimpuls m

7638 VOLTAGE RATING   ge
The maximum sustained voltage that can
safely be applied to or taken from an
electrical or electronic device without
risking the possibility of a breakdown.
f tension f nominale
e tensión f nominal
i tensione f nominale
n nominale spanning
d Nennspannung f

7639 VOLUME DOSE   ra
The product of absorbed dose and the
volume of the absorbing mass.
f dose f absorbée dans le volume,
   dose f volume
e dosis f absorbida en el volumen,
   dosis f por volumen
i dose f assorbita nel volume
n dosis in een volume, volumedosis
d Volumdosis f

7640 VOLUME EFFECT   is
An isotopic effect for the heavy atoms which
exceeds the effect of the mutual motion.
f effet m de volume
e efecto m de volumen
i effetto m di volume
n volume-effect n
d Volumeneffekt m

7641 VOLUME ENERGY   ch
That binding energy which is due to the
saturated exchange force between the
nucleons and which is proportional to the
number of nucleons in the nucleus.
f énergie f de volume
e energía f de volumen
i energia f di volume
n volume-energie
d Volumenenergie f

7642 VOLUME IMPLANT   xr
Implant in tissue in three dimensions.
f implant m en volume
e esteroinjerto m, injerto m tridimensional
i innesto m in volume
n volume-implantatie
d Volumenimplantat n

7643 VOLUME ION DENSITY   ic
Number of ions of one kind per unit volume.
f nombre m volumique d'ions
e densidad f de iones por unidad de volumen
i densità f d'ioni per unità di volume

n ionendichtheid per volume-eenheid
d Ionendichte f je Volumeneinheit

7644 VOLUME IONIZATION,   np
         VOLUME IONIZATION DENSITY
Average ionization density in a given
volume irrespective of the specific
ionization of the ionizing particles.
f densité f volumétrique d'ionisation,
   ionisation f volumique
e densidad f volumétrica de ionización,
   ionización f de volumen
i densità f volumetrica d'ionizzazione,
   ionizzazione f di volume
n ruimtelijke ionisatiedichtheid
d räumliche Ionisationsdichte f,
   Volumionisation f

7645 VOLUME RECOMBINATION   ct, ic
Takes place between positive and negative
ions at low energies throughout the
volume of an ionization chamber or counter.
f recombinaison f de volume
e recombinación f de volumen
i ricombinazione f di volume
n volumerecombinatie
d Volumenrekombination f

7646 VOLUME RECOMBINATION RATE   ec
The time rate at which free electrons and
holes recombine within the volume of a
semiconductor.

f vitesse f de recombinaison volumique
e velocidad f de recombinación de volumen
i velocità f di ricombinazione di volume
n volumerecombinatiesnelheid
d Volumenrekombinationsgeschwindigkeit f

7647 VORTEX FINDER,   cd
         VORTEX TUBE
Component part of a vortex separator.
f tube m turbulent
e tubo m turbulento
i tubo m turbolento
n wervelbuis
d Wirbelrohr n

7648 VORTEX SEPARATOR   ch
A device for freeing entrained gases from
a moving liquid. Used to separate radiolytic gases in aqueous homogeneous
reactors.
f séparateur m turbulent
e separador m turbulento
i separatore m turbolento
n wervelscheider
d Wirbeltrenner m

W

7649 W VALUE   np
The amount of energy which a nuclear
particle loses on the average to form one
ion pair.
f valeur *f* W
e valor *m* W
i valore *m* W
n W-waarde
d W-Wert *m*

7650 WALL ABSORPTION   br, gr, np
The decrease in beta-ray or gamma-ray
output resulting from absorption in the
wall of the container.
f absorption *f* aux parois
e absorción *f* en las paredes
i assorbimento *m* alle pareti
n wandabsorptie
d Wandabsorption *f*

7651 WALL EFFECT   ic
The contribution to the ionization in an
ionization chamber by electrons liberated
from the walls.
f effet *m* de paroi
e efecto *m* de pared
i effetto *m* di parete
n wandeffect *n*
d Wandeffekt *m*

7652 WALL-LESS IONIZATION   ic
     CHAMBER
An ionization chamber in which the
sensitive volume is not defined by walls,
but by the lines of force of an electrical
field determined by the form and
arrangement of the electrodes, and the
potential difference between the electrodes.
f chambre *f* d'ionisation sans paroi
e cámara *f* de ionización sin pared
i camera *f* d'ionizzazione senza parete
n wandloos ionisatievat *n*
d wandlose Ionisationskammer *f*

7653 WALL RECOMBINATION   np
One of the basic modes of recombination.
f recombinaison *f* à la paroi
e recombinación *f* de pared
i ricombinazione *f* alla parete
n wandrecombinatie
d Wandrekombination *f*

7654 WALL SCATTERING   ct
In a counter tube, the scattering into the
wall of the tube.
f diffusion *f* dans la paroi
e dispersión *f* en la pared
i deviazione *f* nella parete
n verstrooiing in de wand
d Streuung *f* in die Wand

7655 WALL TYPE RADIATION BEACON,
     see 4596

7656 WALPURGITE   mi
A mineral containing bismuth, uranium and
arsenic.
f walpurgite *f*
e walpurgita *f*
i walpurgite *f*
n walpurgiet *n*
d Walpurgit *m*

7657 WARHEAD   nw
The part of a shell containing the
explosive, chemical, radioactive or other
charge.
f cône *m* de charge
e cabeza *f* cargada, cono *m* de carga
i testa *f* di carica, testa *f* esplosiva
n granaatkop
d Granaatkopf *m*

7658 WARNING ASSEMBLY   sm
An assembly giving visual and possibly
audible indication of the existence, even
temporary, of abnormal conditions not
likely to be of immediate serious
consequences in a nuclear reactor.
f ensemble *m* d'avertissement
e conjunto *m* de alarma
i complesso *m* di segnalazione
n waarschuwingsopstelling
d Warnanlage *f*

7659 WARNING LABEL   is, sa
A label fastened to or affixed to an
isotope container.
f étiquette *f* d'alarme
e etiqueta *f* de alarma
i etichetta *f* d'allarme
n waarschuwingsetiket *n*
d Warnungszettel *m*

7660 WARNING LIGHT   rt, sa
Light in the electrical circuits of reactors
to warn the personnel of an abnormal
condition.
f lumière *f* d'avertissement
e lámpara *f* de alarma, luz *f* indicadora
i lampadina *f* spia, spia *f* luminosa
n waarschuwingslamp, waarschuwingslicht *n*
d Warnlampe *f*, Warnlicht *n*

7661 WARNING SHIELD   is, sa
A standardized shield which should
accompany any transport of radioactive
materials.
f plaque *f* avertissante
e placa *f* de aviso
i piastra *f* avvisatrice
n waarschuwingsembleem *n*
d Warnungsschild *m*

7662 WASH-OUT, see 5913

7663 WASTE   ch, is
Depleted material withdrawn from the
stripper of a plant.
f matière *f* usée, rejet *m*
e material *m* usado
i materiale *m* esaurito
n afgewerkt materiaal *n*
d ausgenütztes Material *n*

7664 WASTE   rt
Discarded radioactive material.
f déchets *pl* radioactifs
e desechos *pl* radiactivos
i rifiuti *pl* radioattivi
n radioactieve afval
d radioaktiver Abfall *m*

7665 WASTE DISPOSAL   rw
f élimination *f* des déchets
e eliminación *f* de los desechos
i eliminazione *f* dei rifiuti,
   sistemazione *f* dei rifiuti,
   smaltimento *m* dei rifiuti
n afvalverwijdering
d Abfallbeseitigung *f*

7666 WASTE RECOVERY   rw
The treatment of waste in order to obtain
useful material.
f récupération *f* des déchets
e recuperación *f* de los desechos
i ricupero *m* dei rifiuti
n afvalregeneratie, afvalverwerking
d Abfallaufarbeitung *f*, Abfallverwertung *f*

7667 WASTE TREATMENT   rw
f traitement *m* des déchets
e tratamiento *m* de los desechos
i trattamento *m* dei rifiuti
n afvalbehandeling, afvalverwerking
d Abfallverwertung *f*

7668 WASTE WATERS   nw
Waters leaving a nuclear installation with ·
an activity lower than the authorized level.
f eaux *pl* résiduaires
e agua *f* residual
i acqua *f* di fogna
n afvalwater *n*
d Abwasser *n*

7669 WATER ACTIVITY METER,   ma
      WATER RADIOACTIVITY METER
A (radio)activity meter designed for the
continuous measurement of the activity of
water by measuring the activity of the
aerosol obtained by reducing the water to
a fine spray.
f activimètre *m* de l'eau,
   radioactivimètre *m* de l'eau
e activímetro *m* por agua,
   radiactivímetro *m* por agua
i attivimetro *m* per acqua,
   radioattivimetro *m* per acqua
n wateractiviteitsmeter
d Gerät *n* zur Bestimmung der Radioaktivität in Wasser

7670 WATER BOILER, see 3316

7671 WATER BOILING REACTOR, see 700

7672 WATER COLUMN, see 1241

7673 WATER LOGGING   fu
The penetration of water from the primary
cooling circuit into the interior of a fuel
element.
f pénétration *f* d'eau
e penetración *f* de agua
i penetrazione *f* d'acqua
n indringen *n* van water
d Eindringen *n* von Wasser

7674 WATER MODERATED REACTOR,
      WATER REACTOR, see 4012

7675 WATER MONITOR   sa
Any device for detecting and measuring
waterborne radioactivity for warning and
control purposes.
f moniteur *m* d'eau
e monitor *m* para agua
i avvisatore *m* per acqua,
   monitore *m* per acqua
n watermonitor
d Wassermonitor *m*

7676 WATER MONITORING   sa
Periodic or continuous determination of
the amount of ionizing radiation or radioactive contamination in water.
f contrôle *f* de l'eau
e control *m* del agua
i controllo *m* dell'acqua
n waterbewaking
d Wasserüberwachung *f*

7677 WATT'S FORMULA   np
The distribution in the energies of neutrons
from fission can be approximated by the
expression $N(E)dE = e^{-E} \sin h\sqrt{2EdE}$,
where N(E) is the number of neutrons
in the energy interval E to E + dE.
f formule *f* de Watt
e fórmula *f* de Watt
i formula *f* di Watt
n formule van Watt
d Wattsche Formel *f*

7678 WATTAGE RATING, see 5939

7679 WAVE DIFFRACTION   gp
The change of direction of a wave inciding
at a certain angle in a medium.
f diffraction *f* d'onde
e difracción *f* de onda
i diffrazione *f* d'onda
n golfbreking
d Wellenbrechung *f*

7680 WAVE EQUATION   gp
The partial differential equation of wave
motion.
f équation *f* d'onde

e ecuación f de onda
i equazione f d'onda
n golfvergelijking
d Wellengleichung f

7681 WAVE FUNCTION   gp
The solution of a differential or partly
differential equation for wave propagation
through a medium.
f fonction f d'onde
e función f de onda
i funzione f d'onda
n golffunctie
d Wellenfunktion f

7682 WAVE MECHANICS   gp
A general physical theory which ascribes
wave characteristics to the fundamental
entities of atomic structure.
f mécanique f ondulatoire
e mecánica f ondulatoria
i meccanica f ondulatoria
n golfmechanica
d Wellenmechanik f

7683 WEAK INTERACTION   cr, np
The interaction that is responsible for
beta decay and the decay of some mesons
and hyperons.
f interaction f faible
e interacción f débil
i interazione f debole
n zwakke wisselwerking
d schwache Wechselwirkung f

7684 WEDGE FILTER   ra
A filter so constructed that its thickness
varies continuously or in steps from one
end to another.
f filtre m en coin
e filtro m de cuña
i filtro m a cuneo
n wigfilter n
d Keilfilter m

7685 WEDGE SPECTROGRAPH   sp
A spectrograph in which the density of the
radiation passing through the entrance slit
is varied by moving an optical wedge.
f spectrographe m à coin
e espectrógrafo m de cuña
i spettrografo m a cuneo
n wigspectrograaf
d Keilspektrograph m

7686 WEIGHTING FACTOR,
WEIGHTING FUNCTION, see 6872

7687 WEISS MAGNETON   np, un
A unit of magnetic moment, equal to
1.87×10^{-21}erg oersted, about one-fifth
of a Bohr magneton.
f magnéton m de Weiss
e magnetón m de Weiss
i magnetone m di Weiss
n magneton n van Weiss
d Magneton n von Weiss

7688 WEISSENBERG METHOD   cr
An experimental technique for the X-ray
analysis of crystal structure.
f méthode f de Weissenberg
e método m de Weissenberg
i metodo m di Weissenberg
n methode van Weissenberg
d Weissenberg-Methode f

7689 WEIZSÄCKER-WILLIAMS METHOD   ra
Method of computing bremsstrahlung
emitted in the collision of two particles
with relative kinetic energy large compared
with their rest energy.
f méthode f de Weizsäcker-Williams
e método m de Weizsäcker-Williams
i metodo m di Weizsäcker-Williams
n methode van Weizsäcker en Williams
d Methode f von Weizsäcker und Williams

7690 WELD DECAY   mg
Probably due to chromium carbide
precipitation.
f attaque f de la soudure
e ataque m de la soldadura
i attacco m della saldatura
n lasplaatsaantasting
d Schweissstellenzerstörung f

7691 WELDED SEAL   cd
f scellement m soudé
e cierre m hermético soldado
i chiusura f ermetica saldata
n gelaste afdichting
d Schweissabdichtung f

7692 WELDED SOURCE   np, rt
A neutron source encapsulated in a welded
container.
f source f soudée
e fuente f soldada
i sorgente f saldata
n gelaste bron
d geschweisste Quelle f

7693 WELDING   mg
A process of joining metals by the
application of heat or pressure or both.
f soudure f
e soldadura f
i saldatura f
n lassen n
d Schweissen n

7694 WELL COUNTER   ct
A radiation counter having a heavy tubular
shield closed at one end, in which the
radiation detector and the radioactive
sample are inserted to reduce the effect
of background radiation.
f compteur m à canal pour échantillons
e contador m de canal de muestras
i contatore m a canali di campioni
n telbuis met monsterkanaal
d Zählrohr n mit Probenkanal

7695 WELL LOGGING   ms
A technique of investigation of bore holes,

especially oil wells, by nuclear measurements.
f radiocarottage *m*
e radiosondeo *m*
i radiocarotaggio *m*
n activiteitsmetingen *pl* in boorgaten
d radiometrische Bohrlochvermessung *f*

7696 WELL-MODERATED   rt
Of a multiplying system, having a moder-
ator-to-fuel ratio such that the lower-
energy part of the neutron spectrum can be
approximated by a Harwell distribution
and such that the greater part of the
neutron population falls within the
distribution.
f bien modéré
e bien moderado
i ben moderato
n goed gemodereerd
d gutmoderiert, richtig moderiert

7697 WELL-TYPE   ic
IONIZATION CHAMBER
An ionization chamber intended mainly
for the measurement of activity of gamma
emitting sources, having a central
cylindrical well in which these sources are
put.
f chambre *f* d'ionisation à puits
e cámara *f* de ionización de pozo,
   cámara *f* de ionización hueca
i camera *f* d'ionizzazione a pozzo
n ionisatievat *n* met put
d Schachtkammer *f*, Topfionisationskammer
   f

7698 WELL-TYPE   ra
SCINTILLATOR DETECTOR
A scintillation detector in which the
scintillator is shaped like a well.
f détecteur *m* à scintillateur à puits
e detector *m* de centelleador de pozo
i rivelatore *m* a scintillatore a pozzo
n scintillatiedetector met put
d Detektor *m* mit Bohrlochszintillator

7699 WELL-TYPE SODIUM   cr
IODIDE CRYSTAL
f cristal *m* d'iodure de sodium type puits
e cristal *m* de yoduro de sodio tipo pozo
i cristallo *m* d'ioduro di sodio tipo pozzo
n putvormig natriumjodidekristal *n*
d Bohrloch-Natriumjodidkristall *m*

7700 WESTCOTT CROSS SECTION, see 2047

7701 WET CRITICALITY   rt
Criticality obtained with a liquid coolant.
f criticité *f* en présence de réfrigérant
e criticidad *f* con refrigerante
i criticità *f* con refrigerante
n criticiteit bij vloeistofkoeling
d Kritizität *f* bei Flüssigkühlung *f*

7702 WET REPROCESSING, see 318

7703 WET-STEAM COOLED   rt
REACTOR
A nuclear reactor in which the coolant is
a mixture of water and steam.
f réacteur *m* refroidi par vapeur humide
e reactor *m* enfriado por vapor húmedo
i reattore *m* refrigerato a vapore umido
n met vochtige stoom gekoelde reactor
d nassdampfgekühlter Reaktor *m*

7704 WHITE CORPUSCLE,
WHITE GLOBULE, see 3988

7705 WHITE COUNT   md
The number of white corpuscles per
cubic millimetre(er) of blood.
f numération *f* leucocytaire
e numeración *f* leucocitaria
i numerazione *f* leucocitaria
n telling van witte bloedlichaampjes
d Leukozytenzählung *f*

7706 WHOLE-BODY EXPOSURE, see 6809

7707 WHOLE-BODY RADIATION   ma
METER
A measuring assembly for determination
of the whole gamma activity of the human
body.
f anthroporadiamètre *m*
e radiámetro *m* corporal
i radiametro *m* biologico
n anthroporadiometer,
   stralingsmeter voor in het lichaam
   aanwezige activiteit
d Ganzkörperzähler *m*

7708 WHOLE-BODY RADIATION   ma
METER WITH AMPLITUDE
ANALYZER
A measuring assembly including a whole-
body radiation meter and an amplitude
analyzer, designed to identify radionuclides
present in the human body and to evaluate
their respective activities.
f anthroporadiamètre *m* à analyseur
   d'amplitude
e radiámetro *m* corporal con analizador de
   amplitud
i radiametro *m* biologico ad analizzatore
   d'ampiezza
n anthroporadiometer met amplitudeanalysator,
   stralingsmeter voor in het lichaam aanwezige activiteit met amplitudeanalysator
d Ganzkörperzähler *m* mit Impulshöhenanalysator

7709 WIDELY SPACED LATTICE, see 4128

7710 WIDMANSTÄTTEN STRUCTURE   mg
A structure in which a geometrical
metallographic pattern is produced by the
generation of a new phase within the body
of the parent phase.
f structure *f* de Widmanstätten

e estructura *f* de Widmanstätten
i struttura *f* di Widmanstätten
n widmanstättenstructuur
d Widmanstättensches Gefüge *n*

7711 WIGNER ENERGY   np
Energy stored as a result of the Wigner
effect.
f énergie *f* Wigner
e energía *f* Wigner
i energia *f* Wigner
n wignerenergie
d Wigner-Energie *f*

7712 WIGNER EFFECT, seê 1868

7713 WIGNER FORCE   np
A nuclear force unaffected by interchanges.
f force *f* Wigner
e fuerza *f* Wigner
i forza *f* Wigner
n wignerkracht
d Wigner-Kraft *f*

7714 WIGNER GAP   np, rt
Space left open between two graphite
blocks in a nuclear reactor to allow
Wigner growth.
f espace *m* de Wigner
e espacio *m* de Wigner
i spazio *m* di Wigner
n wignerafstand
d Wigner-Abstand *m*

7715 WIGNER GROWTH   rt
A change in physical dimensions due to the
Wigner effect, which may be important in
reactors using graphite as moderator or
reflector.
f croissance *f* par effet Wigner,
   expansion *f* Wigner
e crecimiento *m* por efecto Wigner
i crescita *f* per effetto Wigner
n groei door wignereffect
d Wachstum *m* durch Wigner-Effekt

7716 WIGNER NUCLEI   np
A special case of mirror nuclei, comprising
pairs of odd-mass-number isobars for
which the atomic number and the neutron
number differ by one.
f noyaux *pl* de Wigner
e núcleos *pl* de Wigner
i nuclei *pl* di Wigner
n wignerkernen *pl*
d Wigner-Kerne *pl*

7717 WIGNER NUCLIDES   np
Pairs of nuclides possessing Wigner nuclei.
f nucléides *pl* de Wigner
e núclidos *pl* de Wigner
i nuclidi *pl* di Wigner
n wignernucliden *pl*
d Wigner-Nuklide *pl*

7718 WIGNER RELEASE   rt
An operation on a nuclear reactor to bring

about the release, under controlled
conditions, of the Wigner energy in the
graphite moderator.
f libération *f* de l'énergie Wigner,
   relâchement *m* Wigner
e desprendimiento *m* Wigner
i rilascio *m* Wigner
n wignerontspanning
d Wigner-Entspannung *f*

7719 WIGNER THEOREM   qm
A prediction from quantum theory, stating
that in a collision of the second kind,
angular momentum of electron spin is
conserved.
f théorème *m* de Wigner
e teorema *m* de Wigner
i teorema *m* di Wigner
n wignertheorema *n*
d Wignersches Theorem *n*

7720 WIIKITE   mi
A uranium containing mineral of the
samarskite group.
f wiikite *f*
e wiikita *f*
i wiikite *f*
n wiikiet *n*
d Wiikit *m*

7721 WILKIN'S EFFECT   fu
The phenomenon whereby the small
absorption and possibly the large mean
free path of the end caps of short fuel
element slugs results in a higher flux at the
ends of the slugs than at the middle.
f effet *m* de Wilkin
e efecto *m* de Wilkin
i effetto *m* di Wilkin
n effect *n* van Wilkin
d Wilkin-Effekt *m*

7722 WILSON CHAMBER, see 1138

7723 WINDOW   ct, ra
An aperture for the passage of particles
or radiation.
f fenêtre *f*
e ventana *f*
i finestra *f*
n venster *n*
d Fenster *n*

7724 WINDOW   np
An energy range of relatively high
transparency in the total neutron cross
section of a material.
f zone *f* d'énergie transparente
e zona *f* de energía transparente
i zona *f* d'energia trasparente
n transparant energiegebied *n*
d transparentes Energiegebiet *n*

7725 WINDOW,   ec
      CHANNEL WIDTH, see 993

7726 WINDOW AMPLIFIER, see 994

7727 WINDOW COUNTER TUBE   ct
A counter tube in which a portion of the
envelope is of such low absorption as to
permit the detection of radiation of low
penetrating power.
f tube *m* compteur à fenêtre
e tubo *m* contador con ventana
i tubo *m* contatore a finestra
n telbuis met venster, venstertelbuis
d Fensterzählrohr *n*

7728 WINDOW SCATTERING   ct
In a counter tube, the scattering into the
window of the tube.
f diffusion *f* dans la fenêtre
e dispersión *f* en la ventana
i deviazione *f* nella finestra
n verstrooiing in het venster
d Streuung *f* in das Fenster

7729 WINDOWLESS PHOTOMULTIPLIER ec
A photomultiplier in which no material is
interposed between the source of photons
and the target used as a photocathode.
f photomultiplicateur *m* sans fenêtre
e fotomultiplicador *m* sin ventana
i fotomoltiplicatore *m* senza finestra
n vensterloze fotomultiplicatorbuis
d fensterlose Photovervielfacherröhre *f*

7730 WINEBOTTLE POTENTIAL   np
A potential of nuclear forces characterized
by a low central elevation in the bottom part
of the curve.
f potentiel *m* à fond de bouteille
e potencial *m* de fondo de botella
i potenziale *m* a fondo di bottiglia
n wijnflespotentiaal
d Weinflaschenpotential *n*

7731 WIPE TEST, see 6664

7732 WIRE MACHINE, see 5463

7733 WIRE PENETROMETER   xr
A penetrometer incorporating a series of
wires graded in diameter and usually of
similar material to that under consideration.
f pénétramètre *m* à fils
e penetrómetro *m* de hilos
i penetrametro *m* a fili
n draadkwaliteitsmeter
d Drahthärtemesser *m*

7734 WOBBLE SEAL   cd
A flexible seal through the centre(er) of
which passes a lever.
f scellement *m* flexible
e cierre *m* hermético flexible
i chiusura *f* ermetica flessibile
n flexibele afdichting
d flexible Abdichtung *f*

7735 WOLFRAM, see 7427

7736 WOOD-PLASTICS MATERIAL   ms
IRRADIATION
Used in production of wood-plastics
composites by means of gamma radiation
from cobalt-60.
f irradiation *f* de bois plastique
e irradiación *f* de plástico de madera
i irradiazione *f* di legno plastico
n bestraling van kunststofhout
d Bestrahlung *f* von Kunststoffholz

7737 WORK FUNCTION, see 2100

7738 WORK HARDENING, see 6930

7739 WRINKLING, see 7027

7740 X-RAY ANALYSIS   cr, xr
      Determination of the internal structure of
      crystalline solids by means of the
      diffraction pattern.
 f   analyse f par rayons X
 e   análisis f por rayos X
 i   analisi f per raggi X
 n   röntgenanalyse
 d   Röntgenanalyse f

7741 X-RAY APPARATUS, see 6215

7742 X-RAY CAMERA,   an, xr
      X-RAY DIFFRACTION CAMERA
      An apparatus for obtaining a photographic
      record of the diffraction beams produced
      when a crystalline specimen is irradiated
      in a beam of X-rays.
 f   chambre f à rayons X
 e   cámara f de rayos X
 i   camera f a raggi X
 n   röntgencamera
 d   Röntgenkamera f

7743 X-RAY COVERAGE   xr
      The area covered by the X-rays emitted.
 f   champ m d'irradiation
 e   campo m de irradiación
 i   campo m d'irradiazione
 n   bestraald veld n
 d   Röntgenstrahlenfeld n

7744 X-RAY CRYSTALLOGRAPHY   cr, xr
      The study of the arrangement of the atoms
      in a crystal by use of röntgen rays.
 f   cristallographie f à rayons X
 e   cristalografía f con rayos X
 i   cristallografia f a raggi X
 n   röntgenkristallografie
 d   Röntgenkristallographie f

7745 X-RAY DERMATITIS, see 53

7746 X-RAY DIFFRACTION, see 1800

7747 X-RAY DIFFRACTOMETER   cr, xr
      An instrument used in X-ray analysis to
      measure the intensities of the diffracted
      beams at different angles.
 f   diffractomètre m à rayons X
 e   difractómetro m de rayos X
 i   diffrattometro m a raggi X
 n   röntgendiffractometer
 d   Röntgenbeugungsgerät n

7748 X-RAY EXCITATION   ma
      FLUORIMETER,
      X-RAY FLUORESCENCE CONTENT
      METER
      A content meter for the determination of
      one or several elements in liquid or solid
      samples, by measurement of X-ray
      fluorescence excited by X-rays.
 f   fluorimètre m à excitation X,
      teneurmètre m à fluorescence X
 e   fluorímetro m por excitación X,
      valorímetro m por fluorescencia X
 i   tenorimetro m a fluorescenza X
 n   gehaltemeter berustend op röntgen-
      fluorescentie
 d   Röntgenfluoreszenzmesseinrichtung f

7749 X-RAY FILM   ph, xr
      A film base coated with an emulsion
      designed for use with X-rays.
 f   radiofilm m
 e   película f radiográfica
 i   pellicola f radiografica
 n   röntgenfilm
 d   Röntgenfilm m

7750 X-RAY FLUORESCENCE   ma
      THICKNESS METER
      A thickness meter including a source of
      ionizing radiation and designed to
      determine material thickness by measure-
      ment of X-ray fluorescence excited in the
      material itself or in the supporting
      material.
 f   épaisseurmètre m à fluorescence X
 e   calibrador m por fluorescencia X
 i   spessimetro m a fluorescenza X
 n   diktemeter berustend op röntgenfluorescen-
      tie
 d   Röntgenfluoreszenzdickenmessgerät n

7751 X-RAY FOCAL SPOT, see 7422

7752 X-RAY GONIOMETER   cr, xr
      An instrument that determines the positions
      of the electric axes of a quartz crystal by
      reflecting X-rays from the atomic planes
      of the crystal.
 f   goniomètre m à rayons X
 e   goniómetro m de rayos X
 i   goniometro m a raggi X
 n   röntgengoniometer
 d   Röntgengoniometer n

7753 X-RAY HARDNESS   xr
      The penetrating power of X-rays, which is
      an invers° function to the wavelength.
 f   dureté f de rayons X
 e   dureza f de rayos X
 i   durezza f di raggi X
 n   röntgenstralenhardheid
 d   Röntgenstrahlenhärte f

7754 X-RAY INDUCED MUTATION   ms, xr
      Mutation achieved in plant breeding by
      irradiation with X-rays.
 f   mutation f induits par rayons X
 e   mutación f inducida por rayos X
 i   mutazione f indotta per raggi X
 n   door röntgenstralen veroorzaakte mutatie
 d   durch Röntgenstrahlen verursachte
      Mutation f

7755 X-RAY OUTPUT   xr
The quantity of X-rays emitted in relation
to the energy supplied to the anode.
f rendement *m* des rayons X
e rendimiento *m* de rayos X
i rendimento *m* di raggi X
n stralenrendement *n*
d Strahlenausbeute *f*

7756 X-RAY OUTPUT AT FULL RATINGS xr
The maximum output of an X-ray tube at
full load.
f rendement *m* optimal
e rendimiento *m* óptimo
i rendimento *m* ottimo
n optimaal rendement *n*
d optimale Ausbeute *f*

7757 X-RAY SPECTROGRAM   cr, sp, xr
A record of an X-ray diffraction pattern.
f spectrogramma *m* à rayons X
e espectrograma *m* de rayos X
i spettrogramma *m* a raggi X
n röntgenspectrogram *n*
d Röntgenspektrogramm *n*

7758 X-RAY SPECTROGRAPH   cr, sp, xr
An X-ray spectrometer equipped with
photographic or other recording apparatus.
f spectrographe *m* à rayons X
e espectrógrafo *m* de rayos X
i spettrografo *m* a raggi X
n röntgenspectrograaf
d Röntgenspektrograph *m*

7759 X-RAY SPECTROMETER   sp, xr
An instrument for determining the energy
distribution of X-rays.
f spectromètre *m* à rayons X
e espectrómetro *m* de rayos X
i spettrometro *m* a raggi X
n röntgenspectrometer
d Röntgenspektrometer *n*

7760 X-RAY SPECTRUM   sp
When cathode-rays fall upon a specimen
of some element the resulting X-rays
consist of a continuous spectrum upon
which are superimposed certain groups of
much sharper lines characteristic of the
element.
f spectre *m* de rayons X
e espectro *m* de rayos X
i spettro *m* di raggi X
n röntgenspectrum *n*
d Röntgenspektrum *n*

7761 X-RAY STRUCTURE   cr
The atomic or ionic structure of
substances determined by X-ray diffraction
patterns.
f structure *f* aux rayons X
e estructura *f* a los rayos X
i struttura *f* ai raggi X
n röntgenstructuur
d Röntgenstruktur *f*

7762 X-RAY THERAPY, see 6227

7763 X-RAY THERAPY LOCALIZER,
see 314

7764 X-RAY TUBE   xr
A vacuum tube designed for the production
of X-rays.
f tube *m* à rayons X
e tubo *m* de rayos X
i tubo *m* a raggi X
n röntgenbuis
d Röntgenröhre *f*

7765 X-RAYS, see 6220

7766 XENON   ch
Gaseous element, symbol Xe, atomic
number 54.
f xénon *m*
e xenón *m*
i xenon *m*
n xenon *n*
d Xenon *n*

7767 XENON BUILD-UP   rt
AFTER SHUT-DOWN
Temporary increase of xenon poisoning
occurring in thermal reactors during the
first hours after a shut-down.
f surempoisonnement *m* xénon
e surenvenenamiento *m* xenón
i suravvelenamento *m* xenon
n verhoogde xenonvergiftiging
d überhöhte Xenonvergiftung *f*

7768 XENON EFFECT,   rt
XENON POISONING
The reduction in reactivity caused by
neutron capture in Xe-135, a fission product
which is a nuclear poison.
f effet *m* xénon, empoisonnement *m* xénon
e envenenamiento *m* xenón
i avvelenamento *m* xenon
n xenoneffect *n*, xenonvergiftiging
d Xenonvergiftung *f*

7769 XENON INSTABILITY,   rt
XENON OSCILLATION
Oscillation in the power level in localized
parts of a large reactor, due to the
dependence of the poisoning on the thermal
flux density.
f instabilité *f* xénon
e inestabilidad *f* xenón
i instabilità *f* xenon
n xenoninstabiliteit
d Xenoninstabilität *f*

7770 XENON OVERRIDE   rt
That part of the excess reactivity provided
in a reactor to enable it to start up even
when the xenon poisoning is at its maximum
after shut-down.
f réserve *f* de réactivité par xénon
e exceso *m* de xenón
i riserva *f* di reattività per xenon
n xenoncompensatie
d Xenonaktivitätsreserve ·*f*

7771 XENON-PLUS-TEMPERATURE			rt
An undesired effect tending to instabilize
the axial flux shape.
f température *f* excessive par xénon
e temperatura *f* excesiva por xenón
i temperatura *f* eccessiva per xenon
n door xenon veroorzaakte te hoge
   temperatuur
d durch Xenon verursachte zu hohe
   Temperatur *f*

7772 XENON POISONING COMPUTER,
      see 5378

7773 XENON POISONING PREDICTOR,
      see 5380

7774 XENOTIME						mi
An yttrium phosphate containing about
3.6 % of U and 2.2 % of Th.
f xénotime *m*
e xenotima *f*
i xenotime *f*
n xenotiem *n*
d Xenotim *m*

7775 X-Y DISPLAY UNIT					me
A device for the visual display of storage
channel contents by a cathode ray tube.
f unité *f* de visualisation X-Y
e unidad *f* de presentación visual X-Y
i unità *f* di presentazione visuale X-Y
n X-Y-zichtbeeldapparaat *n*
d X-Y-Sichtgerät *n*

Y

7776 YAMACHUCHULITE,   mi
      YAMACHUTULITE
A variety of zirconia containing thorium.
f yamachuchulite *f*
e yamachuchulita *f*
i yamachuchulite *f*
n yamachuchuliet *n*
d Yamachuchulit *m*

7777 YIELD,   gp, nw
      ENERGY YIELD, see 2294

7778 YIELD,
      TNT EQUIVALENT, see 7273

7779 YIELD PER ION PAIR, see 3643

7780 YLEM   np
Proposed name of a hypothetical substance
of density about $10^{13}g/cm^3$, consisting
chiefly of neutrons, out of which all nuclei
may have been formed.
f plasma *m* primordial, ylem *m*
e plasma *m* primordial
i plasma *m* primordiale
n oerstof
d Urplasma *n*

7781 YOUNG'S MODULUS, see 4492

7782 YTTERBIUM   ch
Rare earth metallic element, symbol Yb,
atomic number 70.
f ytterbium *m*
e iterbio *m*
i itterbio *m*
n ytterbium *n*
d Ytterbium *n*

7783 YTTRIALITE   mi
A thorium-yttrium silicate containing a
small amount of Th.
f yttrialite *f*
e itrialita *f*
i ittrialite *f*
n yttrialiet *n*
d Yttrialit *m*

7784 YTTRIUM   ch
Rare earth metallic element, symbol Y,
atomic number 39.
f yttrium *m*
e itrio *m*
i ittrio *m*
n yttrium *n*
d Yttrium *n*

7785 YTTRIUM-90 ROD   br, md, ra
A rod consisting of pressed and sintered
yttrium oxide.
f barre *f* en yttrium-90
e barra *f* en itrio-90
i barra *f* in ittrio-90
n yttrium-90-staaf
d Yttrium-90-Stab *m*

7786 YTTROCRASITE   mi
A mineral similar to delorenzite and
containing calcium, lead, thorium and
uranium.
f yttrocrasite *f*
e itrocrasita *f*
i ittrocrasite *f*
n yttrocrasiet *n*
d Yttrokrasit *m*

7787 YTTROGUMMITE   mi
A decomposition product of cleveite.
f yttrogummite *f*
e itrogumita *f*
i ittrogummite *f*
n yttrogummiet *n*
d Yttrogummit *m*

7788 YTTROTANTALITE   mi
A mineral containing yttrium, tantalum,
calcium and iron.
f yttrotantalite *f*
e itrotantalita *f*
i ittrotantalite *f*
n yttrotantaliet *n*
d Yttrotantalit *m*

7789 YUKAWA KERNEL, see 1821

7790 YUKAWA POTENTIAL   ma, np
A potential function of the form
$$V = -\frac{V_0}{r}e^{-r/b},$$ where V_0 and b are
constants, and r the distance from the
nucleus.
f potentiel *m* de Yukawa
e potencial *m* de Yukawa
i potenziale *m* di Yukawa
n yukawapotentiaal
d Yukawa-Potential *n*

7791 YUKAWA WELL   ma, np
A potential well calculated from Yukawa's
meson theory.
f puits *m* de Yukawa
e pozo *m* de Yukawa
i pozzo *m* di Yukawa
n yukawakuil
d Yukawa-Topf *m*

Z

7792 ZERO-ENERGY LEVEL   np
The energy level of a system of particles
at absolute zero.
f niveau *m* d'énergie zéro
e nivel *m* de energía cero
i livello *m* d'energia zero
n nulenergieniveau *n*
d Nullenergieniveau *n*

7793 ZERO-ENERGY REACTOR,   rt
      ZERO-POWER REACTOR
A reactor designed to be used at such a
low power that no cooling system is needed.
f réacteur *m* de puissance zéro
e reactor *m* de potencia cero
i reattore *m* di potenza zero
n nulenergiereactor
d Nullenergiereaktor *m*

7794 ZERO-FIELD EMISSION CURRENT,
      see 2587

7795 ZERO-POINT ENERGY   np
The total energy of a system of particles
at absolute zero temperature.
f énergie *f* au zéro absolu
e energía *f* al cero absoluto
i energia *f* allo zero assoluto
n nulpuntenergie
d Nullpunktenergie *f*

7796 ZINC   ch
Metallic element, symbol Zn, atomic
number 30.
f zinc *m*
e cinc *m*, zinc *m*
i zinco *m*
n zink *n*
d Zink *n*

7797 ZINC BROMIDE   ch, mt
Chemical compound used in solution as a
stop to low-level radioactivity.
f bromure *m* de zinc
e bromuro *m* de zinc
i bromuro *m* di zinco
n zinkbromide *n*
d Zinkbromid *n*

7798 ZIPPEITE   mi
A mineral containing uranium and sulphur.
f zippéite *f*
e zipeita *f*
i zippeite *f*
n zippeïet *n*
d Zippeit *m*

7799 ZIRCEX PROCESS, see 263

7800 ZIRCONIUM   ch
Metallic element, symbol Zr, atomic
number 40.
f zirconium *m*
e zirconio *m*
i zirconio *m*
n zirkonium *n*, zirkoon *n*
d Zirkon *n*

7801 ZIRCONIUM HYDRIDE   rt
      MODERATED REACTOR
f réacteur *m* modéré par hydrure de
   zirconium
e reactor *m* moderado por hidruro de
   zirconio
i reattore *m* moderato per idruro di
   zirconio
n reactor met zirconiumhydridemoderator
d Reaktor *m* mit Zirkonhydridmoderator

7802 ZIRKELITE   mi
A mineral containing calcium, iron,
zirconium, titanium and thorium.
f zirkélite *f*
e zirquelita *f*
i zirchelite *f*
n zirkeliet *n*
d Zirkelit *m*

7803 ZONE AXIS   cr
The axis through the centre(er) of a crystal
which is parallel to the edge of the zone.
f axe *m* zonal
e eje *m* de zona
i asse *m* di zona
n zone-as
d Zonenachse *f*

7804 ZONE OF A CRYSTAL   cr
A set of faces of a crystal meeting in a
series of edges, all of which are parallel.
f zone *f* de cristal
e zona *f* de cristal
i zona *f* di cristallo
n kristalzone
d Gittergerade *f*, Kristallzone *f*

7805 ZWITTERION, see 226

7806 ZYGOTE   md
The cell resulting from the union of two
gametes.
f zygote *m*
e cigoto *m*
i cigoto *m*
n zygote
d Zygote *f*

FRANÇAIS

altération axiale de l'arrangement du combustible 486
- de dimensions 1839
- de l'arrangement du
combustible 5997
- radiale de l'arrangement du
combustible 5679
alternance de multiplicités 213
altitude d'explosion 3214
- effective de cheminée 2044
alumine 216
aluminium 217
amas de recul 144
amassement 145
ambiance 2314
américium 221
amniographie 223
amorçage parasitaire 1469
amortissement par rayonnement
5699
ampangabéite 225
amplificateur 227
- à fenêtre 994
- à hacheur 1082
- à impulsion de courant 1581
- de charge 1016
- de coïncidence 1184
- de compteur 1453
- de flux 1929
- d'électromètre 2095
- d'impulsions 5586
- du rendement neutronique 728
- du signal d'arrêt 6567
- incorporé dans l'appareil
d'arrêt 6567
- linéaire à commutation
automatique de sensibilité 457
- linéaire à seuil 634
- linéaire d'impulsions 4047
- linéaire pour courant continu
4036
- logarithmique d'impulsions
4116
- logarithmique pour courant
continu 4111
- sigma 6583
- statique de sécurité 6268
amplification due au gaz 2958
- due au gaz du tube compteur
1460
amplitude de diffusion 6333
ampoule 232
- à rupture de pointe 776
- en verre évacuée 2381
- en verre scellée 6392
- évacuée scellée 2382
anaérobique 235
analysateur d'amplitude à canal
mobile 1789
- des dimensions de particules
5151
analyse cristallographique par
diffraction neutronique 4688
- de cristaux par diffraction 1797
- de Feather 2542
- fluorimétrique 2737
- inverse d'isotopes par dilution
6188
- isotopique 3776
- par absorption 12

analyse par activation 63
- par dilution 1836
- par dilution isotopique 3785
- par radiocristallographie
directe 1858
- par rayons X 7740
- radiochimique 5827
- radiométrique 5869
- statistique 6863
- thermique 7121
analyseur d'amplitude 228
- d'amplitude multicanal 4566
- d'amplitude multicanal à
mémoire 4565
- de temps de vol 7255
- magnétique 239
anaphase 240
anaphorèse 241
andersonite 243
androgénie 244
anémie 245
anémomètre à ionisation 3683
angiocardiographie 246
angiographie 248
- cérébrale 961
angiopneumocardiographie 247
angle de Bragg 755
- de déviation 249, 1669
- de diffusion 6334
- de réflexion 251
- de transit 7337
- d'incidence 250
angles de valence 7567
angle solide 6685
anio 264
anisotropie de réseau 3917
anneau de garde 3106
- de stockage 6922
- en uranium 778
anneaux de stockage à
intersection 3584
annerodite 268
annihilation 269
anode 277, 288
- à cible insérée 3324
- cylindrique 3310
- inclinée 4951
- massive 3186
- oblique 4951
anse chromatique 1090
anthracène 285
anthroporadiamètre 7707
- à analyseur d'amplitude 7708
anthroporadiocartographe 687
antibaryon 286
anticathode 288
antichoc 6553
anticoïncidence 289
antilepton 298
antimatière 299
antimoine 300
antineutrino 301
antineutron 302
antiparticule 303
antiproton 304
antiréactivité 4645
aortographie 308
aplatissement du flux 2750

appareil 310
- à différence 1778
- à pré-ajustage 5472
- à rayons X 6215, 7741
- à rayons X panoramique 5099
- circulateur de gaz 2963
- d'échantillonnage 6322
- de chargement 1029
- de contact par étage 6839
- de contrôle de sécurité à
haut flux 5446
- de décharge 7489
- de distillation 6907
- de mesure de la contamination
1337
- de mesure de rayons
cosmiques de Millikan 4439
- de mesure du temps de
doublement 1961
- de mesure indicateur 3442
- de prélèvement 4607
- de prélèvement d'aérosols
atmosphériques 164
- de radiographie 5254
- de radioscopie 2742
- de surveillance 7030
- de télécuriethérapie 7086
- d'impact annulaire 274
- d'irradiation bêta extra-
corporéel du sang 612
- individuel de prélèvement
d'échantillons d'air 5215
- intercavitaire 3591
- interstitiel 3586
- isotopique 3755
- isotopique de pesage continu
3781
appareillage auxiliaire 242
- auxiliaire du réacteur 5969
- de commande 1358
- de contrôle 1358
- de réglage 1358
- expérimental 2439
appareil saturne 6304
appareils de sécurité
auxiliaires 467
appareil thermonucléaire 7180
appauvrissement 1731
applicateur 313
- de radium 5898
application d'ailettes 2627
apport 4237
- à un organe 3519
approche sous-critique 315
approximation adiabatique 115
- de Born 736
- de diffusion 1807
- d'une liaison serrée 7243
arc unipolaire 7474
argent 6594
argon 323
arme à fission nucléaire 2678
- à fusion nucléaire 2905
- à implosion 3410
- nucléaire nominale 4777
armes nucléaires 4924
arrachement d'un électron 2118
arrangement 5009
- hétérogène du coeur du
réacteur 3237

gaz inertes 3467
- nobles 3467
- non-dégénéré 4782
- radioactif 5784
gène 3004
- mutant 4599
générateur de neutrons 4715
- de Van de Graaff en série 7065
- d'impulsions 3414
- en cascade 903
- horloge 128
- isotopique 3790
- Van de Graaff 2197
génération de porosité 5397
générations de noyaux 3010
génétique des rayonnements 5711
génie nucléaire 4842
géochronologie 3012
géométrie 3021
- à champ étroit 4617
- à champ large 792
- à faisceau étroit 4617
- à faisceau large 792
- à miroire 4459
- à noyau dur 3141
- de comptage 1487
- du réseau 3922
- du stellarateur 6889
- fausse 516
- favorable 2541
- sûre 6263
- toujours sûre 219
géométriquement sûr 3020
gerbe 6560
- cosmique 1435
- d'Auger 440
- électrophotonique 908
- en cascade 908
- étroite 4618
- explosive 831
- extensive 440
- pénétrante 5180
germanium 3027
germe 4383
GeV 632
giga-électron-volt 632
gilpinite 3031
gisement 1734
- d'origine secondaire 1736
- minéral 5016
- radioactif 5773
gisements d'uranium 7511
gîte minéral 5016
glace lourde 3196
glande sexuelle 3040
- thyroïde 7241
glandulographie 3034
glissement 6635
globule blanc 3988
- rouge 2365
glucine 576
gonade 3040
gondolage 5291
gonflement 7039
goniomètre à rayons X 7752
gradient 3045
- de potentiel 5427
- de température 7095
gradin de croissance 3104

grain 1563, 3049
- radioactif 5790
gramme-équivalent 3057
grandeur de référence 6049
- indépendante 3439
- pseudoscalaire 5580
- réglée 1371
- scalaire 6308
grandeurs conjuguées 1310
grandeur vectorielle 7594
grandissement par rayonnement 3101
granulation 3054
granule basophile 553
- neutrophile 4761
- oxyphile 5076
granulocyte 3060
granulocytopénie 3061
graphe de fluence 2712
graphite 3062
- boraté 733
- imperméable 3406
- isotropique 3807
- nucléaire 4854
grappe d'éléments combustibles 816
graviton 3070
grenade sous-marine nucléaire 4833
grille 3079
- antidiffusante 296
- d'Akerlund 173
- de Lysholm 4173
- fine 2622
- fixe 4173
- mobile 806
- oscillante 806
grippage 2837
gris 3072
groupe 2517
- de barres 6206
- de barres de compensation 6545
- de barres de sécurité 6267
- d'énergie de neutrons 4703
- de neutrons par énergie 4703
- de séparation 6478
- d'étages 906
- d'ions 3619
groupement d'ions 815
groupe spatial 6726
guérison 6022
guerre radiologique 5859
gummite 3111

hacheur 1081
- de faisceau neutronique 4681
- lent 6642
- rapide 2522
- ultrarapide 6995
hafnium 3118
halogène 3132
halogénure d'uranium 7514
halo pléochroïque 5346
haploïde 3138
hatchettolite 3149
hauteur de barrière 544
- de modérateur variable 7592
- de pression de liquide 3153
hélium 3219

helvite 3229
hématimètre 1484
hémoglobine 3232
hépatoliénographie 3233
hérédité protoplasmique 1607
héritage protoplasmique 1607
hétérotopique 3242
hétérozygote 3243
hexafluorure d'uranium 3244
historadiographie 3297
hjelmite 3250
hodoscope 3301
hokutolite 3302
holmium 3311
homme standard 6847
homogénéisation 5528
- du résidu 3758
- isotopique 2336
homozygote 3322
hublots 6541
humidimètre de sol par rayonnement ionisant 3695
- isotopique 3797
humidité critique 1521
- d'équilibre 2347
huttonite 3347
hydrochloruration anhydrique de combustibles zirconifères 263
hydroeuxénite 225
hydrogéné 3361
hydrogène 3351
- lourd 1755
hydrure lourd 1754
hygiène du rayonnement 5715
hyperconjugaison 3364
hyperfragment 3366
hypernoyau 3367
hypéron 3368
hypocentre 3094
hypothèse de la création continue 1343
hystérésis 3373
- de tube compteur 1474
hystérosalpingographie 3374

ianthinite 3375
ictomètre 1493
- à chambre d'ionisation équivalente au tissu 7269
- discriminateur 5941
- linéaire 4049
- linéaire à différence 1780
- linéaire portatif 5403
- logarithmique 4117
- logarithmique de radio-protection 3158
îles d'isomérie 3723
impact d'électrons 2134
implant 3407
implantation 3407
- d'ions 3636
implant en surface 5305
- en volume 7642
implosion 3409
importance relative 6089
imprégnation 4097
impulsion 3413, 4531
- angulaire 259
- de comptage 7420
- de courant d'obscurité 1616

moment orbital 5004
- quadratique 2795
- quadripolaire 5624
- quadripolaire nucléaire 4895
moments multipolaires de transition 7347
monazite 4533
monde-espace-temps 6729
moniteur 4534
- atmosphérique 150
- atmosphérique C-14 975
- atmosphérique de tritium 7398
- atmosphérique d'iode avec prélèvement continu 3607
- atmosphérique d'iode avec prélèvement discontinu 3608
- atmosphérique d'oxyde de tritium 7403
- atmosphérique en continu 1342
- d'activité globale du fluide de refroidissement 1396
- d'eau 7675
- de cendres de charbon 1153
- de contamination alpha pour les mains 195
- de contamination bêta pour les mains 599
- de contamination d'échantillons d'air 4608
- de contamination pour linge et tenues de travail 3935
- de contamination surfacique 7015
- de contamination surfacique des sols 2709
- de criticité 1532
- de denrées 2772
- d'effluent 2052
- de la densité de flux d'électrons 2131
- de la densité de flux de neutrons 4711
- de la densité de flux de particules 5148
- de la densité de flux gamma 2929
- de pertes entre circuits d'un échangeur à eau lourde/eau légère 3200
- de poussière radioactive transportée par l'air 1999
- de produits radioactifs emballés 5082
- de radioprotection 3156
- de rayonnement 3156
- de rupture de gaine 2511
- de rupture de gaine à collection électrostatique 2193
- de rupture de gaine à effet Cerenkov 964
- de rupture de gaine à séparation des produits de fission 2666
- de rupture de gaine par détection de neutrons retardés 1698
- de terrain 321
- de vivres 2772
- d'habits 1137

moniteur de fond 500
- gamma de criticité 1533
- isotopique de produits emballés 5080
- neutronique de criticité 1534
- portatif à piles 5365
- portique bêta-gamma 596
- pour aérosols de plutonium 5358
- pour pieds et mains 3135
monocristal 6608
monocyte 3349
monopôle 4544
montecharge 1006
mordache 3816
mors 3816
mosaïque 4547
moteur à air chaud 3163
- ionique 3630
mou 6675
moulage 4554
- de radium 4554
mouvement brownien 798
- d'air troposphérique 7411
- de long en large 5930
- des alluvions 6593
- de translation 7350
- du fond de la mer au large 4965
- du fond de la mer côtier 1155
- orbital 5005
- propre d'un appareil 501
- rectilinéaire de la particule 6033
- thermique 7149
- transversal 1537
moyen de blocage 674
- de contraste 1351
multiplet 4578
multiplicateur 4583
- d'électrons 2141
- de neutrons 728
multiplication 4579
- de charge 1013
- d'électrons 2140
- de neutrons 4727
- due au gaz 2974
- sous-critique 6977
multiplicité 4582
muon 4563
muonium 4595
muqueuse 4564
mutant 4598
mutation 3131
- chromosomatique 1091
- de gènes 3005
- induite par rayons X 7754
- létale 3985
myélographie 4605

naégite 4612
nagatélite 4613
nécrose 4637
négaton 2099
négatoscope 2602
néodyme 4648
néoformation 4649
néon 4650
néoplasma 4649
- bénin 570

néoplasma malin 4239
neptunium 4651
nettoyage ultrasonore 7458
neutralisation de charge 1014
neutre 4654
neutretto 4659
neutrino 4660
neutron 4661
- à faible énergie 4142
- diffusé 6326
neutrongraphie 4759
neutron libre 2828
- naturel 4629
neutrons de fission 2662
- de résonance 6169
neutron secondaire 6414
neutrons épicadmiques 2319
- épithermiques 2332
- froids 1205
- instantanés 5536
- intermédiaires 3564
- latents 3911
- lents 6647
- pulsés 5604
- rapides 2533
- retardés 1701
- subcadmiques 6973
- thermiques 7152
- utiles 7546
- vierges 7614
neutronthérapie 4752
neutron vagabond 6940
nickel 4763
nicolayite 4764
nid à poussière 2000
niobite 1240
niobium 4767
nitrate de thorium 7218
- d'uranyle 7540
nitrogène 4768
niveau 3991
- accepteur 37
- caractéristique de Fermi 2556
- de donneur 1928
- de Fermi 2555
- de flux 2754
- de la nappe d'eau souterraine 3095
- d'énergie 2275, 5648
- d'énergie d'impureté 3419
- d'énergie zéro 7792
- de puissance 5444
- de quantification 5633
- de rayonnement 5721
- de résonance nucléaire 4906
- d'excitation 2420
- d'irradiation du combustible 2866
- maximal admissible 4312
- normal 3090
- nucléaire 4864
- quasistationnaire 5656
- sous-létal 6982
- virtuel 7615
niveaux d'énergie atomique 382
- énergétiques du noyau 4841
nivénite 4769
nobélium 4771
nohlite 4773

panache 5353
panne 775
panoramagramme parallactique 5102
paradoxe de Klein 3858
parahélium 5100
parahydrogène 5101
paramagnétisme nucléaire 4878
paramètre 5111
- de choc 3398
- de dispersion 6928
- de fluctuation 6928
- de Michel 4404
- de ralentissement 480
- de stabilité 6817
- du rayon 5903
paramètres atomiques 400
- de diffusion 1824
- des cellules élémentaires 7478
paramètre secondaire 6416
parapositonium 5112
parcours 7325
- de diffusion 2564
- de gerbe 6562
- de réaction 1986
- d'ionisation 3675
- d'un électron 2161
- d'un ion 3645
- erratique 5921
- espace-temps d'une particule classique 4000, 6730
- maximal 4313
- semicirculaire 6461
- total 3529
parent 5116
parhélium 5121
parité 5122
paroi de blindage 6531
parsonsite 5123
particule 5134
- à faible énergie 4143
- à grande énergie 3262
- alpha 197
- à trace en forme de V 7552
- bêta 602
- bombardée 709
- cascade 907
- chargée 1022
- combustible enrobée 1157
- combustible enrobée à rétention de produits de fission 2665
- Compton 1277
- constituante 1316
- consumée 1319
- de désintégration 1645
- de gerbe 6561
- de Majorana 4236
- de recul 6010
- diffusée 6327
- élémentaire 2212
- émise 2058
- en équilibre 2343
- fondamentale 2212
- incidente 710
- ionisante 3686
- lambda 3887
- liée 750
- lourde 3204
- neutre 7464

particule nucléaire 4879
- primordiale 3492
- projectile 710
- pseudoscalaire 5579
- relativiste 6100
particules alpha à portée longue 4123
- alpha retardées 1691
- de recul à la fission 2672
- directement ionisantes 1861
- étranges 6931
- identiques 3383
particule sigma 6585
particules indirectement ionisantes 3447
- matérielles 5155
particule subatomique 6972
- témoin 7107
- virtuelle 7616
part sensible 6472
pas du réseau 3927
pas entre les canaux 5298
passeur automatique d'échantillons 458
pastille 1247
P.C.M. 4443
peau d'étanchéité 6885
pechblende 4620
péchurane 4620
pegmatite 5167
peignes 6808
pelage 1041
péliose 5168
pellicule à autosupport 5169
pelvigraphie 5172
pelvimétrie 5173
pendule 5174
pénétrabilité 5176
pénétramètre 5185
- à fils 7733
- à plaque 5337
- comparatif 6893
pénétration 5181
- d'eau 7673
- de résonance 6171
- d'hélium 3225
pénombre 5188
penta-éther 5187
pente du palier 5341
- normalisée de plateau 4802
- relative de palier 6093
percussion 3860
père atomique 5117
perhapsatron 5195
péridurographie 5196
période 5197
- biologique 645
- comparative 1248
- courte de montée en puissance 6559
- courte de vie 6556
- d'attardement 4052
- de mise en train 2346
- de neuf dixièmes 4765
- de temps 5203
- hors service 5040
périodemètre 5199
période neutronique 4734
- radioactive 3120
- résultante 2032

périodicité 5204
- nucléaire 4880
périoste 5205
- interne 2249
périphérie 5208
périscope 5209
perles à cobalt-60 1166
persistance 134
perte accidentelle 6781
- de comptage 1489
- de coude 569
- de diffusion 6341
- d'énergie 1678
- d'énergie par atome 2278
- d'énergie par paire d'ions formée 2279
- de poids par radiolyse 5867
- d'humidité 4495
- moyenne d'énergie par pairs d'ions 4328
- par effet de couronne 1419
- par enlargement 2295
- par rayonnement 5722
pertes de régénération 6069
- par coïncidence 1190
perturbation 5219
petit foyer 2621
phantoscope 5225
phase gazeuse 2987
- solide 6860
phénacite 5237
phénomène d'échange 1005
- transitoire 7335
phlébographie 5238
pH-mètre 5222
phosphate de tributyle 7082
phosphore 5242
phosphorescence 5241
phosphuranylite 5243
photocathode 5244
photodésintégration 5246
photoélectron 5252
photofission 5253
photographie de cristal rotatoire 6234
photoionisation 402
photoméson 5263
photominuterie 5273
photomultiplicateur 5264
- sans fenêtre 7729
photon 4010
- d'annihilation 272
photoneutron 5271
photon virtuel 7617
photoproton 5272
physique atomique 359
- des matières solides 6689
- des particules fondamentales 2896
- neutronique 4735
- nucléaire 4881
- radiologique 5858
pic d'échappement 2932
- de déplacement de fragments de fission 2676
- de flux 5164
- de fuite 2932
- de puissance 5165
- de résonance 6170
- photoélectrique 5250

ségrégation par gravité 3071
sélecteur d'amplitude 5589
- d'anticoïncidences 292
- de coïncidences 1193
- de coïncidences rapides 2523
- de coïncidences rapides à
  quatre voies 2793
- de vitesse 7602
- de vitesse des neutrons 4755
- d'impulsions en dépendance
  du temps 7261
sélection de signal 2993
sélénium 6438
sels des oxo-acides d'uranium
  7519
semence 6425
- à radium 5900
semiconducteur compensé 1253
- P.I.N. compensé 1253
sengiérite 6466
sensibilité 6475
- au défaut d'épaisseur 2698
- aux contrastes 2698
- conventionnelle d'une
  photocathode 5245
- de pénétramètre 5186
- de précision 3435
- de seuil 7232
- du compteur 1457
- intrinsèque de compteur 3597
sensible 6468
séparateur de gaz 2980
- d'isotopes 3768
- électromagnétique d'isotopes
  2084
- électrostatique 2200
- énergétique d'ions 3629
- magnétique 4225
- répétiteur d'isotope 2934
- turbulent 7648
séparation 6479
- centrifuge 951
- chimique 1065
- d'impulsions 2993
- d'isomères 3741
- d'isotopes 3764
- d'isotopes à jets de gaz 3818
- d'isotopes par mobilité
  ionique 3640
- électrolytique 2080
- électrostatique 2202
- en masses partielles
  sous-critiques 4277
- par gravité 3071
série de Balmer 522
- lyotropique 4172
service interrompu à charge
  variable 7591
servomécanisme 6500
seuil 4013
seuil cadmium effectif 2024
- d'arrêt 7390
- de Geiger-Müller 3002
- de longueur d'onde photo-
  électronique 7236
- de luminescence 4157
- d'énergie 1588, 2677
- de réponse aux impulsions
  7233
- inférieur d'arrêt 4154

seuil photoélectrique 5251
- supérieure d'arrêt 7494
sharpite 6517
sialographie 6576
sigmatron 6588
signal analogique 237
- binaire 637
- d'autorisation 6276
- de la constante de temps
  de consigne 5198
- de niveau de puissance de
  consigne 3992
- de réaction 2547
signaleur atmosphérique 148
- de contamination alpha 188
- de contamination bêta 586
- de contamination gamma 2922
- de contamination surfacique
  7013
- de contamination surfacique
  des sols 2707
- de dépassement de niveau par
  rayonnement ionisant 3693
- de la densité de flux
  d'électrons 2129
- de la densité de flux de
  neutrons 4709
- de la densité de flux de
  particules 5146
- de la densité de flux gamma
  2927
- de pertes du circuit de
  refroidissement 1395
- de pertes du fluide de
  refroidissement 1397
- de proximité par rayonnement
  ionisant 3694
- de proximité par rayonnement ionisant diffusé 6329
- de proximité par rayonnement ionisant direct 1857
- de rayonnement 5716
- de rupture de gaine 2510
- d'exposition 2464
- individuel de poche 5364
signal logique 4118
- numérique 1832
silicium 6591
simulateur 6603
- de radioactivité 5818
simulation 6602
singulet 1017, 6618
siphon thermique 7165
site 6621
- résistante aux tremblements
  de terre 2010
situation 6623
skiographe 6625
sklodowskite 1074
soddyite 6666
sodium 6668
solidification 2833
solubilité 6692, 6693
solution 1904, 6695, 6696
- de combustible 2879
- de combustible irradié 1906
- sans fumée 2892
- solide 4468
solvant 6701
solvate 6700

sommier 2850
sonde à amplificateur 5516
- à scintillation 6367
- bêta intégrale 3523
- de détection 5517
- de fréquence de plasma 3898
- de Hall 3129
- de Langmuir 3898
- de réacteur 5287
- détectrice de pertes
  d'échangeur de chaleur 3165
- d'humidimètre du sol 6680
- piézoélectrique 5282
sorption 6707
sortie 5047
soudage 6684
soudure 6684, 7693
soufre 6990
soupape de réserve 1993
source active 93
- à fente 6640
- à introduction guidée 136
- à krypton-85 3867
- à raie spectrale subdivisée
  6800
- à raie unique 6611
- bêta céramique 956
- bêta ponctuelle 603
- de Mössbauer 4551
- d'énergie 6713
- de neutrons 4747
- de neutrons plane infinie 3472
- de rayonnement 5738, 6714
- de rayonnement gamma 2942
- de rayonnement pour cliniques
  1127
- de référence 6050
- d'ions 3634
- d'ions fortement groupés
  3537
- discoïde 1884
- épaisse 7186
- gamma à basse énergie 4141
- gamma à énergie élevée 3259
- isotropique de rayonnement
  3809
- mince 7194
- non-scellée 7491
- permanente de lumière 5210
- plane 5308
- ponctuelle 5373
- pulsée de neutrons 5603
- radioactive 5807
- scellée 6394
- soudée 7692
- virtuelle 3391
sous-critique 6975
sous-ensemble 6970
sous-marin à propulsion
  nucléaire 4889
sous-modéré 7467
sous-niveau 6983
sous-produit 836
- radioactif 72
soutirage 5048
spallation 6731
spectre 6766
- alpha 200
- atomique 414
- bêta 616

spectre continu 1347
- continu de rayons X 1348
- d'absorption 22
- d'arc 319
- de bandes 527
- de fission 2675
- de fuite de neutrons 3971
- de fuites de rayons gamma 2933
- de masse 4281
- de Maxwell 4319
- d'émission 2232
- d'émission de photons 5269
- de molécule 4516
- de raies 4028
- de rayons X 7760
- de résonance 6156
- de résonance magnétique 4223
- du rayonnement gamma 2946
- du transfert de charge 1020
- énergique 2289
- énergétique de fission 2655
- en 1/E 4976
- micro-ondes 4423
- neutronique 4749
spectres électroniques de bandes 2173
spectrogramme 6760
- à rayons X 7757
spectrographe 6761
- à coin 7685
- à cristal 1559
- à quartz 5655
- à rayons X 7758
- à vide 7564
- de masse(s) 4278
- de masse(s) à double focalisation 1956
- de masse(s) de Dempster 1709
- magnétique 4226
spectrographie 6762
spectromètre 2288
- à cristal 764
- à rayons alpha 206
- à rayons bêta 615
- à rayons gamma anti-Compton 294
- à rayons X 7759
- à résonance magnétique 4222
- à scintillation 6369
- à secteur limité 4019
- automatique à deux faisceaux 452
- aux paires 5093
- de masse(s) 4280
- de masse(s) à parcours trochoïde 7406
- de neutrons à temps de vol 7258
- de particules 5137
- détecteur d'hélium 3228
- du rayonnement gamma 2945
- gamma à canal mobile 6321
- magnétique 4227
- mécanique de neutrons 1083
- neutronique 4748
spectroscope 6764
spectroscopie 6765
- à micro-ondes 4424

spectroscopie au rayonnement gamma 2949
- de la résonance magnétique nucléaire 4870
spermatogenèse 6773
sphère de Debye 1630
spiking 6428
spin 6782
- de voie d'entrée 2309
- de voie de réaction 992
- de voie de sortie 2427
- d'un électron 2158
- isotopique 3725
- nucléaire 4911
spinthariscope 6796
spirale contactrice de Podbielnak 5370
- de croissance 3103
spot lumineux 2760
- lumineux effectif 2031
stabilisateur 6821
- de spectre 6768
stabilité 6816
- atmosphérique 354
- marginale 4248
- nucléaire 4913
- sous rayonnement 5741
- thermique 7168
stable 6823
station de traitement d'effluents radioactifs 80
statistique classique de Maxwell-Boltzmann 4316
- de Bose-Einstein 745
- de Fermi 2569
- de Fermi-Dirac 2560
- de Maxwell-Boltzmann 4316
- quantique 5649
- quantique de Maxwell-Boltzmann 4318
statolimnimètre par rayonnement ionisant 3696
steenstrupine 6886
stellarator 6888
stéréogramme 6898
- parallactique 5103
stéréographie 6899
stéréométrie radiographique 5844
stéréoradiographie 6899
stéréoradioscopie 6897
stéréoscopie 6901
stérilisation par irradiation 5742
stérilité 6902
stockage 6919
stratification 6934
stratigraphe 688
stratigraphie 689
stratosphère 6938
striction azimuthale 6389
- dynamique 2004
- orthogonale 6389
- stabilisée 6819
- ultrarapide 7456
stripage 6955
strontium 6963
strontium-90 5883
structure atomique 417
- aux rayons X 7761

structure cristalline 1561
- de la bande d'énergie 2259
- de la molécule 4518
- de l'atome en couches 413
- de Widmanstätten 7710
- du réseau 3930
- électronique 2183
- extranucléaire 2495
- fine 2623
- fine due à la rotation 6237
- hexagonale compacte 3246
- hyperfine 3365
- mosaïque 4548
- nucléaire 4829
- quantique de l'atome 413
- quantique du noyau 4938
- réticulaire 6186
studtite 6969
stylo-dosimètre 2791
stylo-exposimètre 2792
subatomique 6971
substance adsorbée 124
- antidiffuseuse 5843
- d'addition au réfrigant 1393
- d'aplatissement 2696
- de base 552
- de scellement 6395
- d'étoupage 6395
- équivalente au tissu 7271
- fluorescente 2732
- isotropique 3806
- luminescente 4159
- nucléaire 900
- ostéophile 727
- père 5116
- pure 5607
- radioactive à courte vie 6558
- thyroïdiphile 7242
substitution courante du combustible 6241
- par collision 1236
succession de désintégration 1649
sulfate d'uranyle 7541
sulfoteneurmètre d'hydrocarbures 3697
superposition d'impulsions 5597
supervision d'air 162
- de terrain 322
support de cône 1304
- de source 6711
- de source gamma 2943
- de tube 7423
supports des dés 1661
supposition de congruence 341
surchauffe nucléaire 4916
surcritique 6997
surdosage 5057
surempoisonnement xénon 7767
sûreté absolue, à 2512
- limitée, de 4018
surface de Fermi 2570
- d'isocomptage 3732
- d'isodose 3738
- efficace de cible 2046
- limite 751
sur-modéré 5065
surpression 5056, 5066
- en air libre 2822

valeur de consigne 3441
- de mélange isotopique 3760, 7576
- de seuil 7234
- effective 2048
- ft 1248, 2847
- instantanée 110
- momentanée 110
- prescrite 1743
- Q 4900
- R 5664
- S 6258
- statique 6859
- W 7649
vanadium 7579
vandenbrandéite 7530
vandendriesschéite 7581
vanoxite 7582
vapeur de tête 5061
- radioactive 1333
vaporisation 2385
variable 7588
- d'activation 58
- dépendante 1720
variables conjuguées 1310
variable stochastique 984
variation de flux de grande répartition 3085
- de flux spatiale 3085
- du champ magnétique 4219
variations de réactivité à longue échéance 4124
vasographie 7593
vecteur 7594
- axial 487
- de Burgers 818
veine 4108
vénographie 7603
ventriculographie 7604
vent solaire 6683
verre au plomb protecteur 5560
- de flint 2700
- de Lindemann 4023
- dosimètre 3035
- non-décolorant 6818
- phosphaté 5239
- sensible au rayonnement 6469
verrouillage 3538
- de la source de neutrons 6712
vésiculographie 7606
vêtements de protection contre la contamination 295
- de sûreté 2842
vide 7554, 7630
- élévé 3289
- grossier 6240
- peu poussé 4153
- poussé 3289
- ultra-élevé 2500
vie courte, à 6556
- courte, de 6556
- de combustible 2867
- du compteur 1462
- d'une génération neutronique 4724
- en magasin 6522
vieillissement 143
vie moyenne 478
- moyenne d'un état atomique 479

vie moyenne neutronique 4333
vietinghofite 7610
viscosité 7623
vitesse au point d'engorgement 2706
- complexe 5920
- critique 1530
- critique du courant 4102
- d'Alfven 178
- d'apport 7498
- d'attaque 430
- d'échange 5932
- d'échange isotopique 3799
- de débit à l'entrée 5933
- de fuites 3963
- d'élimination 1124
- de neutrons 4750
- de particules momentanée 3512
- de percolation 5192
- de réaction 5957
- de rechargement moyenne 4335
- de recombinaison 6019
- de recombinaison volumique 7646
- de renouvellement 7439
- de restauration 6025
- différentielle de guérison 1792
- d'ionisation 3678
- du cycle métabolique 7439
- effective de particule 2035
- intermédiaire de neutrons 3567
- linéaire 4050
- maximale de particules 5163
- minimale de fluence 4452
- moyenne 4339
- moyenne de migration 1972
- moyenne de pénétration 1972
- quadratique 2802
- relativiste 6102
vitesses généralisées 3008
vitesse spécifique de désionisation 1685
- superficielle 7000
voglianite 7628
voglite 7629
voie de fission 2651
- d'entrée 2309
- de réaction 986
- de sortie 2427
voile 2766
volatilisation sous vide 7565
volatilité 7633
volatilité relative 6096
volborthite 7634
volume atomique 425
- d'apport 5933
- sensible 6474
- sensible radiobiologique 5821
- utile 6474
voûte du réacteur 5993
vue antéropostérieure 283
- axiale 484
- de profil 3813
- latérale 3913
- lordotique 4129
- oblique 4952
- orthoscopique 5035

vue postéro-antérieure 284
- pseudoscopique 5581
- tangentielle 7066

walpurgite 7656
wiikite 7720

xénon 7766
xénotime 7774

yamachuchulite 7776
ylem 7780
ytterbium 7782
yttrialite 7783
yttrium 7784
yttrocrasite 7786
yttrogummite 7787
yttrotantalite 7780

zinc 7796
zippéite 7798
zirconium 7800
zirkélite 7802
zone à conditions de travail non-réglementées 3425
- à conditions de travail réglementées 6077
- active 92
- à séjour réglementé 6076
- contrôlée 71
- dangereuse surveillée 1373
- d'aplatissement 2693
- de cristal 7804
- de déplacements 1899
- d'énergie moyenne 3561
- d'énergie transparente 7724
- de non-activité 1120
- de proportionalité limitée 4017
- de rayonnement dangereux 5700
- de surfaces limites 3555
- de thermalisation 7173·
- de vitesse réciproque 6003
- d'exclusion 2424
- inactive 3425
- interdite 5524
- morte 1624
- occupée 4956
- plastique 5330
zygote 7806

ESPAÑOL

aplicador beta para intentos
  oftálmicos 4990
- de estroncio para la cavidad
  nasofaríngea 4619
- superficial 7008
aportación 3519
aproximación adiabática 115
- de Born 736
- de difusión 1807
- de enlace cerrado 7243
aquebradización 2220
arandela de molibdeno 4528
arco unipolar 7474
área de difusión 1808
- de fuente activa 94
- del blanco 7073
- de migración 4427
- de moderación 6651
- de retardación 6651
argón 323
arma de fusión nuclear 2678,
  2905
- de implosión 3410
- nuclear nominal 4777
armas nucleares 4924
arranque 6851
- inicial 3495
arrastre 2307
- de gas 2967
arreglo 5009
arrugamiento 7027
arsénico 325
arteriografía 326
- periférica 5206
artrografía 327
ascendiente 5116
asíntota 347
asintótico 348
aspirador de polvo portátil 5406
astatino 342
ataque de la soldadura 7690
- químico 2372
- radiolítico 5865
atenuación 431
- atmosférica de la radiación
  352
- de haz ancho 791
- del haz neutrónico 4671
- geométrica 3016
atenuador 435
- de exposición 2463
atmósfera 350
átomo 356
- bombardeado 3861
- de Bohr 690
- de Bohr-Sommerfeld 694
- de de Broglie 1621
- de impureza 1052
- de rechazo 6007
- de retroceso 6007
- de Rutherford 6251
- despojado 4814
- de Thomson 7202
- donador de quelato 1034
- excitado 2417
- físil 2641
átomo-gramo 3055
átomo hidrogenoide 3357
- intersticial 3587
- ionizado 3682

átomo liviano 4002
- marcado 3877
- mesónico 4367
- neutro 4655
- normal 4796
- nuclear 4814
- original 5117
- padre 5117
- percutido 3861
- pesado 3187
- radiante 5684
átomos isoelectrónicos 3740
- metaestables 4391
atracción del electrón 2102
- intranuclear 4815
- mutua 4601
aumento de actividad 101
- de la reactividad 5961
- instantáneo de la
  reactividad 5534
- permisible presumido de la
  temperatura 1742
autoabsorción 6439
autoblindaje 6383, 6459
autoclave 444
autodescomposición por
  radiación 5734
autodispersión 6456
autoestereograma 463
autoextinción 3580, 6453
autoionización 447
autoprotección 6459
- , de 2512
autorradiación 6446
autorradiografía 461
autorradiograma 460
autorradiólisis 462
autorregulación 6455
autunita 464
avalancha 468
- de Townsend 7306
- electrónica 2103
- iónica 3616
avería 775
ayuda filtrante 2607
azeótropo 488
azufre 6990

baileita 558
bajo vacío 4153
balance de materia 4257
- de neutrones 4672
- energético 2257
- isotópico 3752
- térmico 3160
baliza de radiación 5687
- de radiación de suelo 2710
- de radiación mural 4596
banda 523
- de absorción de un material
  de centelleo 13
- de conducción 1298
- de emisión 2231
- de energía 672, 2258
- de excitación 2409
- de medida 921
- de valencia 7568
- llena 2598
- normal 4797
- ocupada 2598

banda parcialmente ocupada
  5133
- permitida 184
- proporcional propia 5542
- reservada 2774
bandas de deformación 1672
banda vacía 2237
bario 530
barión 546
barn 538
barodifusión 539
barra 6204
- de combustible 2876
- combustible en forma de
  jabalina 3815
- de corrección 6544
- de regulación 10
- de regulación de la potencia
  5441
- de regulación en cadmio 842
- de regulación fina 2619
- de regulación giratoria 1354
- de seguridad 6275
- en itrio-90 7785
barredor 6345
barrera centrífuga 947
- de carga espacial 6720
- de centro elevado 942
- de Coulomb 1441
- de difusión 1809
- de potencial 541
- primaria de radioprotección
  5503
- protectora 5556
- secundaria de radioprotección
  6417
basetita 554
bazo 6798
becquerelita 566
berilímetro 579
berilio 578
beriliosis 577
berilo 575
berquelio 573
beta 583
betafita 625
betaterapia 618
betatópico 626
betatrón 628
beta-uranófano 623
bevatrón 633
bien moderado 7696
bifurcación 766
billietita 635
bineutrón 640
biofísica 650
bióxido de uranio 797
bismuto 654
blanco 4918
- de gas 2982
- delgado 7195
- en deuterio 1759
- en tritio 7402
- grueso 7187
- infinitamente grueso 3476
- metálico líquido 4078
- reflector 6058
blando 6675
blindaje 6382, 6528, 6535
- biológico 648

blindaje de hormigón 1290
- de hormigón pesado 3184
- del núcleo 4937
- de plomo 3958
- de resemplazo 4238
- en boral 732
- por uranio agotado 1729
- térmico 7162
blomstrandita 625
bloque 4162
- de muestro 853
bloqueo 673
- de combustible 2878
bloques de grafito amovibles
  6122
bobina de campo magnético 4205
- de choque 6549
- de espira continua 6617
- en circuito criógeno 1550
bobinas de compensación 1255
- de desplazamiento orbital 4998
- de Ioffe 3610
- de Rogovski 5279
- desviadoras 4998
bodenbenderita 682
bola de fuego 520
bolas de grafito 3065
bolita de cesio-137 845
- en cobalto 1164
bolus 705
bomba atómica 1
- centrífuga 950
- de Alamogordo 174
- de alto grado de compresión
  3254
- de circulación del
  refrigerante 1399
- de corriente conductiva 1301
- de chorro de líquido 3817
- de chorro de vapor 7584
- de difusión 1826
- de difusión de aceite auto-
  fraccionadora 6445
- de difusión de aceite enfriada
  por aire 153
- de difusión de alto vacío 3291
- de difusión de mercurio 4366
- de difusión de vapor de
  aceite 4969
- de fuerza explosiva igual a
  1000 toneladas de TNT 3848
- de fuerza explosiva igual a
  1.000.000 toneladas de TNT
  4357
- de fusión 2899
- de hidrógeno 2899
- de inducción 3456
- de inducción lineal con
  disposición plana de los
  devanados 2692
- de inducción lineal con espacio
  anular 275
- de inducción magnética 6873
- de metal líquido 4071
- de rotor hermético 862
- de vacío 7561
- de vacío de alta contrapresión
  3253
- de vacío preliminar 7556

bomba de vacío preliminar
  elevado 2781
- de vaina de cobalto 1161
- de vapor de mercurio de
  cuatro etapas 2800
- electromagnética 2089
- electromagnética mecánica
  4345
- gamma 2943
- giratoria de aceite de Gaede
  2910
- kilotón 3848
- limpia 1118
- megatón 4357
- molecular 4519
- nuclear 1
bombardear 708
bombardeo 712
- con neutrones 4675
- neutrónico 4675
- nuclear 4821
bombardeos cruzados 1535
bomba sucia 1863
- termonuclear 7181
bombeo 1593
- magnético 4219
- óptico 4992
boral 731
borazol 734
borbollón de vapor 6660
boro 737
borrosidad 681
- cinemática 3849
- debida a la dispersión 6324
- debida al movimiento 3849
- de emulsión 4431
bosón 746
botella magnética 120
bóveda del reactor 5993
branerita 770
brazo mecánico 4343
brevio 786
bröggerita 793
bromo 794
bromuro de zinc 7797
broncografía 796
bucle detectora de
  obstrucciones 5350
buena geometría 3041
burbujeo 804

cabeza cargada 7657
- de banda 525
- de contador de centelleo 6360
- de guerra nuclear 4923
- fungiforme 4597
cabina de carga 1006
cadena 977
- analizadora multiparamétrica
  4570
- de desintegraciones 1639
- de fisión 2649
- de reacción protón-protón
  5572
cadmio 838
caída 2513
- de presión 5475
- de presión de entrada 2310
- de presión de salida
  2430

caja blindada contra rayos
  gamma 6532
- con guantes 3037
- de combustible 2851
- de condensadores 865
- de diversión 1920
- de filtro 2610
- de pasaje 7338
- manipuladora 3037
cajita 912
calandria 846
calcio 850
calciosamarskita 847
calciotermia 848
calciotorita 849
calcogenuro de uranio 7506
calculadora de envenenamiento
  xenón 5378
- de la temperatura en la vaina
  1109
calculador de la producción
  isotópica 3761
- de manipulación de isótopos
  3756
cálculo de las redes 3918
cálculos de blindaje 6537
caldera de paso único 572
- montada en vaina 5369
calentamiento adiabático 118
- Joule 3822
- óhmico 3822
- por irradiación 3716
- por la resonancia del
  ciclotrón 1600
- por onda de choque 6551
- por radiación gamma 2931
- por resonancia 6165
- por resonancia de la
  frecuencia de un ciclotrón
  de iones 3625
- turbulento 7435
calibración absoluta 5
- de una barra de regulación
  1361
- relativa 6087
calibrador 2911
- isotópico 3804
- por fluorescencia X 7750
- por radiación ionizante 3698
- per retrodispersión 511
- por retrodispersión beta 585
- por retrodispersión gamma
  2918
- portátil por retrodispersión
  gamma 5401
- por transmisión de radiación
  ionizante 3701
calibre beta de absorción 584
californio 854
calor 3159
- atómico 387
- de combustión 1246
- de decaimiento 3172
- de desintegración 1643, 3172
- de emisión 3168
- de evaporación 2386
- de fisión 2660
- de formación 2785
- de fusión 2902
- de irradiación 3716

conjunto regulador del tratamiento de uranio irradiado
3707
- y equipo selector clasificador
5018
cono 1303
- de carga 7657
- de compresión 1271
- de pérdida 4134
- de rayos 1305, 5693
conservación de la energía 1311
- de la masa 1312
- del equilibrio neutrónico 4684
- del impulso 1313
constante de acoplamiento 1496
- de actividad real de una
fuente 7415
- de Boltzmann 701
- de desintegración 1640
- de desintegración parcial 5128
- de efecto de pantalla 6385
- de espectro de bandas 528
- de estructura fina 2624
- de Fermi 2557
- de Hall 3125
- de la red 3921
- de masa atómica 391
- de Planck 5307
- de renovación 7440
- de Rydberg 6254
- de tiempo 7246
- de tiempo del flujo
neutrónico 4734
- de tiempo del recortador 1129
- específica de radiación
gamma 6745
- numérica 4946
constantes de fuerzas de enlace
nuclear 2777
- nucleares 4828
constante unificada de masa
atómica 391
constitución química 1043
constreñimiento 1318
consumo 825
- específico 2866
- nominal en vatios 5939
contacto de alta velocidad de
recombinación 3279
contador 1452
- a gas 2965
- alfa 189
- anular 6200
- Cerenkov 962
- cilíndrico 1601
- con aguja 4638
- de anticoincidencias 291
- de canal de muestras 7694
- de centelleador líquido 4074
- de centelleo 5265
- de centelleo de neutrones
lentos 4706
- de coincidencias 1189
- de cristales 1552
- de emisión fría 2586
- de envoltura cilíndrica en cera
3137
- de escintilación 5265
- de flúido 2720
- de flúido de paso 4061

contador de fotones a sodio
6672
- de guarda 3108
- de inmersión 3392
- de ionización 3667
- de la radiactividad de
líquidos 4055
- de neutrones 4686
- de partículas 5140
- de radiación 5696
- de tiempo electrónico 2185
- diferencial de la numeración
leucocitaria 676
- direccional 1860
- dirigido 1860
- electrónico 2177
- electrónico de intervalo de
tiempo 7252
- gaseoso 2968
- de iónico 3623
- logarítmico de velocidad de
recuento 4110
- 4π 2797
contaminación atmosférica 163
- cruzada 1536
- del agua costera 1156
- del ambiente 2315
- interna 3572
- interna excepcional planeada
2225
- interna no planeada 46
- microbiológica 4408
- radiactiva 1334
- superficial 7012
contaminado 1332
contaminante radiactivo 1331
contaminantes aerotransportados
168
contenido del cuerpo 685
- de radio 5894
- de radón 5906
- de uranio 7508
- eficaz de radio 2036
- equivalente de radio 2357
- libre 2823
- real de radio estimado 2368
contenimiento de un reactor
5972
contracción cromosomática 1092
- de lantánidos 3900
contracorriente 1479
contraste de imagen 3387
- del sujeto 6979
- de radiación 5694
- subjetivo 6981
control 7029
- aéreo del aire 131
- de adulteración de semi-
conductores 1931
- de contaminación 1338
- de la concentración 1285
- del agua 7676
- de protección 5555
- de radiaciones 5555
- de seguridad individual 5218
convección 1374
conversión 1380
- de un par 5090
- interna 3573
- magnetodinámica 4229

convertidor amplitud-tiempo
5591
- analogo-numérico 238
- de neutrones 1389
- de tiempo 7249
- numérico de tiempo 1833
- tiempo-amplitud 7244
coordenada 1407
coprecipitación 1411
coproducto 836
corpúsculo 1420
- blanco 3988
coracita 1412
corazón del núcleo 4933
cordilita 1413
corrección de coincidencia 1187
- de la coincidencia accidental
41
- de la geometría 3022
- del momento magnético
anómalo 279
- del tiempo de resolución 6152
- para el tiempo muerto 1626
- radiativa 5753
correlación angular 254
- angular beta-gamma 595
- angular beta-neutrino 601
- angular beta-núcleo de
rechazo 620
- direccional de rayos gamma
sucesives 1859
corriente 2711
- aniónica 266
- catiónica 927
- de conducción 1299
- de convección 1375
- de electrones 2126
- de emisión en campo nulo
2587
- de ionización 2966
- de obscuridad 1615
- de saturación 6301
- electrónica 2115
- incompresible 3434
- nominal 5934
- residual 6141
- turbulenta 7434
corrosión 1426
- de rozamiento 2837
- por el estado latente de
esfuerzos 6947
- por irradiación 3711
corte 1587
- lateral 6578
corvusita 1430
cosmotrón 1438
covalencia 1498
crecimiento 807
- benigno 570
- de cristales 1554
- maligno 4239
- por efecto Wigner 7715
- por radiación 3101
cresta de resonancia 3030,
6170
cría 780
cribado preliminar 6318
criba vibradora 3819
criptón 3866
criptoscopio 3868

detector de radiación 5703
- de radiación fluorescente 2734
- de semiconductor 6462
- de umbral 7226
- en 1/v 1752, 4981
detectores de fugas 3962
- de radiación gamma 2941
detector gamma integrador 3533
- isotópico de niveles 3794
- por activación 66
- Rosenblum 6228
- selectivo de iones 3627
- semiconductor de barrera superficial 7010
- semiconductor por junta de difusión 1804
- semiconductor tipo P.I.N. 1254
deterioro por radiación 713
determinación de la edad por isótopos 3782
- de la fecha por radiacarbón 885
- de la sección eficaz por el método de activación 68
- de la sección eficaz por el método de transmisión 7355
- del número de masa 4255
- no destructiva de hidrógeno 4783
detonación prematura 5465
deuterio 1755
deuterón 1760
- de blanco 7074
deuteruro 1754
- de litio 4082
dewindtita 1766
diafragma 1774
- de rejilla fina 2622
diagnóstico radiológico 1767
diagrama de desintegración 1648
- de fases 1317
- de Fermi 2566
- de Feynman 2580
- de isótopos 3754
- de Kurie 2566
- de Laue 3934
- del potencial 5426
- de Sargent 6296
- de términos 7101
- energético 2276
dialmaita 1922
diamagnetismo de las partículas del plasma 1772
diámetro atómico 377
- eficaz de una barra de regulación 2037
- equivalente de una barra de regulación 2353
- nuclear 4834
dibutoxidimetiléter 1776
dicloruro de glicolo de trietileno 7385
dideriquita 1777
difenilo 651
diferencia de contracción 1794
- media de temperaturas 4338
- media logarítmica de las temperaturas 4088

difracción 1796
- con electrones 2119
- con neutrones 4695
- de onda 7679
- de rayos X 1800
- electrónica 2119
difractómetro 1799
- de rayos X 7747
- neutrónico 4696
difusato 1802
difusión 1806
- ambipolar 220
- de los neutrones 4697
- del plasma a través de un campo magnético 1823
- gaseosa 2985
- hacia atrás 513
- interatómica 378
- inversa de los electrones 492
- molecular 4503
- mutua 4602
- térmica 6706
difusor 1805
dilución 1835
- isotópica 3784
diluente 1834
dimensiones críticas 1514
- de la red 3923
diminución de la presión 5927
- instantánea de la reactividad 5532
di-neutrón 1841
díodo de barrera superficial 7009
dióxido de torio 7208
dipolo 1846
- magnético 4201
di-protón 1849
dirección de la valencia 716
disco calculador de manipulación de isótopos 3756
- de azufre 6991
- de tungsteno 7430
- en cobalto 1163
discontinuidad de absorción 18
discriminador 1869
- de amplitud de impulsos 231
- de forma 5599
- integral 3524
diseño de la celosía 3922
- del reactor 5977
difusión turbulenta 2017
disipación de energía 1901
dislocación 1885
- de un átomo 1867
- en esquina 2020
- en tornillo 6387
disminución de contraste 1840
- de Forbush 2776
disociación de Lorentz 4130
disolución 1904, 6696
disparador de cobalto 1162
dispersión 1892, 6332
- aceleradora 7495
- coherente 1180
- Compton 1274
- Compton doble 1953
- de Bragg 763
- de Coulomb 1449
- de electrones 2156

dispersión de Möller 4523
- de neutrones 4746
- de Thomson 7205
- difrangente 1801
- difusa 1803
- elástica 2061
- electrón-electrón 2123
- en desorden 1889
- en el gas 2976
- en la pared 7654
- en la ventana 7728
- estadística 6867, 6927
- estadística de alcance 5925
- hacia adelante 2788
- hacia atrás 513
- incoherente 3432
- inelástica 3461
- instrumental 3516
- isotrópica 3808
- múltiple 4576, 5356
- no midificada 7490
- potencial 5430
- resonadora 6174
- única 6614
dispositivo automático antisobrecarga 5064
- de congelación 2832
- de estricción lineal 4044
- de estricción toroidal 7282
- de estricción tubular 3142
- de irradiación convergente 1378
- de oxidación 5073
- de seguridad 2227
- de solidificación 2832
- homopolar 3533
- para radiografía por secciones 688
disprosio 2007
distancia de extrapolación 443
- de los átomos en la molécula 379
- de peligro 1612
- de seguridad 6262
- de visibilidad 7625
- focal 2759
- foco-película 2762
- foco-piel 2763
- lineal de extrapolación 4039
- oblicua 6633
- real 6633
- reticular cristalina 3554
distorsión de cuantización 5632
- de la red 3924
- de traza 7321
- por carga espacial 6723
distribución angular 257
- binomial de Bernoulli 574
- de Fermi 2561
- de flujo de gran repartición 3086
- de flujo de neutrones 2749
- de la dosis 1915
- de la energía 2269
- de la energía de los neutrones en el reactor 2681
- de la energía neutrónica 4702
- de la velocidad de los electrones 2167
- de los niveles de energía 3994

distribución de Maxwell 4317
- de velocidad 7601
- electrónica 2120
- espectral 6753
- fina 2620
distribuidor de refrigerante 2863
diuranato de amonio 222
divergencia 1917
- de rayos 5949
divergente 1918
división cromosomática 1094
- de la cromatida 1085
doblete 1844, 1959
dólar 1924
donador 1927
- de electrones 2121
dosa máxima permisible para huesos 4306
dosificación 1936
dosimetría 1949
dosímetro 1947
- con contador G.M. 3000
- con placa de vidrio fluorescente 3035
- de alarma 175
- de alarma de bolsillo 5364
- de bolsillo 2791
- de cámara de ionización 3664
- de cámara de ionización equivalente al tejido 7270
- de centelleador 6372
- de condensador 866
- de detector semiconductor 646 6463
- de electrómetro 2096
- de exposición 2466
- de fibra de cuarzo 5652
- de sulfato ferroso 2575
- fotográfico 5257
- logarítmico portátil de centelleador 5404
- particular 5216
- químico 1044
- termoluminiscente 7178
dosis 1937
- absorbida 7
- absorbida acumulada 1567
- absorbida en el volumen 7639
- absorbida integral 3521
- absorbida por unidad de tiempo 8
- acumulada máxima permisible 4304
- atmosférica 156
- de acumulación 808
- de depilación 1723
- de emergencia 2223
- de eritema 2364
- de exposición del ambiente 3393
- de isótopo 1943
- depositada 1335
- de salida 2428
- de superficie 2584
- de terapia 7115
- de tránsito 7339
- de traslado 6848
- efectiva 2029
- en el tejido 7267
- en la piel 6626
- en profundidad 1738

dosis excesiva 5058
- hística 7267
- inicial 3488
- inmediata 5531
- instantánea 5531
- integral 3525
- iónica 3628
- letal 3941
- letal media 3942
- máxima 5162
- máxima permisible 4308
- máxima permisible en la piel 6627
- máxima permisible en los huesos 725
- máxima permisible profesional 6848
- normal profesional 551
- por volumen 7639
- prolongada 1944
- sin autoblindaje 4437
- total de superficie 7292
- tumoral 7426
- umbral 7227
droogmansita 1974
dumontita 1989
duración de aceleración 28
- de centelleo 6366
- de confinamiento 1329
- de desactivación 1405
- de enfriamiento 1405
- de escintilación 1996
- de funcionamiento 2001
- de generación 3009
- del paso 6894
- de recorrido 7340
- de sensibilidad 6473
- de tratamiento total 5055
- de vida de combustible 2867
- de vida de una generación neutrónica 4724
dureza 3146
- de radiación 5177
- de rayos X 7753
duro 3139

EBR 5951
ecología 2015
economía del combustible 2860
- de los materiales 4294
- neutrónica 4699
- por uso del reflector 6061
ecuación crítica 1515
- de Bethe-Salpeter 630
- de Boltzmann 703
- de continuidad 1341
- de de Broglie 1622
- de difusión 1819
- de difusión no equilibrada 4785
- de Einstein 2054
- de estado 2335
- de Eucken 2376
- de Fermi 140
- de Fokker-Planck 2770
- de Klein-Gordon 3856
- de la edad 140
- de la edad sin captura 141
- del transporte 7362
- de masa relativista 6099

ecuación de Morse 4546
- de onda 7680
- de Richardson-Dushman 6196
- de Schrödinger 6351
- de transmutación 7357
- de transporte de Boltzmann 704
- de Vlasov 7627
- diferencial de Thomas-Fermi 7200
ecuaciones de Laue 3932
ecuación fotoeléctrica de Einstein 2055
- general de difusión 3006
edad 139
- absoluta 4
- de calcio 851
- de Fermi 139
- de plomo 3947
- de radio 3660
- de uranio 7505
- estratigráfica 6936
- geológica 3013
- ósea 723
- química 1037
- radiactiva 5763
edema 1589, 4963
efantinita 2317
efectividad biológica relativa 5951
efecto Auger 438
- Bethe 3885
- biológico de la radiación 644
- centrífugo 948
- Cerenkov 963
- Chadwick-Goldhaber 976
- Compton 1274
- Cottrell 1440
- de altitud 215
- de carga espacial 6724
- de corteza de naranja 4995
- de densidad 1716
- de descarga 1891
- de desplazamiento 1787
- de desplazamiento lateral 6579
- de dilución 1837
- de empaquetamiento 4262
- de encanalamiento 995
- de enlace químico 1039
- de espín 6785
- de espín nuclear 4912
- de estricción 5295
- de extremidad 2244
- de fisión rápida 2524
- de latitud 3015
- de longitud 4125
- de Marx 4249
- de masa 4262
- de orientación 5027
- de oxígeno 5075
- de pantalla del núcleo 2157
- de pared 7651
- de pinza 5295
- de radiación 5705
- de reflexión 6056
- de salida 1891
- de talón 3213
- de tensión superficial 7025
- de transición 7345

efecto de volumen 7640
- de Wilkin 7721
- Doppler 1934
- este-oeste 2012
- fotoeléctrico 5247
- fotoeléctrico atómico 402
- fotoeléctrico inverso 3602
- fotoeléctrico superficial 7022
- fotomagnético 5262
- genético de la radiación 3011
- genético inducido por
  radiación 5717
- geomagnético 3015
- Hall 3126
- Hallwachs 3130
- isotópico 3786
- Joshi 3821
- Mössbauer 4550
- norte-sur 4803
- Overhauser 5060
- Peltier 5170
- radiobiológico 5820
- radón 5909
- Ramsauer 5914
efectos electrocinéticos 2077
efecto somático de la radiación
  6704
- Soret 6706
- Touschek 7304
- túnel 5184
- Wigner 1868
eficacia de separación 6492
- de una barra de regulación
  1367
efluente 2051
- radiactivo 77
efusión molecular 4506
einsteinio 2056
eje de zona 7803
elaboración química 1057
elasticidad 2065
electrización de un gas 2074
electrodesintegración 2076
electrodinámica cuántica 5640
electrodo 2075
- auxiliar 3106
- colector 1214
- en D 1660
- hueco semicilíndrico 1660
- ionizante del aire 160
electroforesis 2189
electromecanismo de regulación
  lineal 4034
- rotativo de regulación 6236
electrómetro 2094
- de fibra de cuarzo 4022
- de lengüeta vibrante 2002
- de Lindemann 4022
electromigración de contracorriente 1478
electrón 2099
- Auger 439
- Compton 1275
- común 5094
- d 1608
- de conducción 1300
- de conversión 1383
- de conversión interna 3575
- de desintegración beta 589
- del átomo 380

electrón de rebote 1275
- de rechazo 6008
- de retroceso 6008
- de vacío 7555
- de valencia 721, 1300
electronegatividad 2170
electrones compartidos 6516
- cósmicos de choque 1434
- cromóforos 1089
- de desintegración de la
  radiación cósmica 1433
- de fuga 6246
- equivalentes 2355
- retrodispersos 509
electrón extranuclear 2493
- giratorio 6794
electrónica 2186
electrón incidente 5495
- inicial 3496
- interno 3501
- K 3826
- L 3873
- libre 2824
- ligado 749
- M 4175
- metastático 4395
- N 4606
- nuclear 4838
- O 4948
electronógeno 2187
electrón óptico 1300
- orbital 5002
- P 5077
- periférico 1300
- planetario 5002
- primario 5495
- Q 5616
- rápido 3281
- S 6255
- secundario 6405
- solitario 4119
electrón-voltio 2188
electroquimógrafo 2078
electroscopio 2190
- de fibra de cuarzo 3936
- de Lauritsen 3936
- para detección de rayos beta
  610
elemento 2207
- activo 81
- combustible 2861
- combustible con aletas
  transversales 1106
- combustible de dispersión
  1894
- combustible de ligazón
  metalúrgica 720
- combustible desnudo 7463
- combustible encamisado 2240
- de construcción completo en
  su mismo 5084
- defectuoso 834
- de impureza 3418
- de paro 6573
- de paro para emergencia
  2227
- de regulación 1357
- de regulación de potencia 5440
- de regulación fina 2618
- de regulación gruesa 1154

elemento de sobrerreactividad
  729
- de transporte 7331
- electronegativo 2169
- fértil 2577
- funcional 549
- liviano 4004
- natural 4626
- original 5118
- padre 5118
- pesado 3190
- portador 895
- puerta 2994
- químico 1045
- radiactivo 81
- radiactivo artificial 329
- radiactivo natural 4630
- retardo lineal 4035
- separador 6493
elementos transcúricos 7326
- transplutónicos 7359
- transuránicos 7368
elevador al aire comprimido 171
eliasita 2213
eliminación de los desechos 7665
- de los desechos radiactivos
  líquidos 78
eliminador de electrostática
  6857
ellsworthita 2214
elutriación 2215
emanación 2217
embalamiento 6245
emisión alfa 187
- corpuscular 1421
- corpuscular asociada 339
- cuántica 5641
- de campo 446
- de partículas 1421
- de rayos beta 611
- electrónica 2124
- frfa 446
- frfa de un contador 1458
- garantizada 971
- iónica de campo 2590
- secundaria 6406
- solar de partículas 6683
- termoiónica 7175
emisor 2234
- alfa 193
- beta 593
- beta-gamma 597
- de neutrones retrasados 1697
- gamma 2924
- emulsión autosoportada 5169
- despegable 6959
- licuable 4058
- nuclear 4839
enalita 2239
encapsulación 2241
encefalografía 2243
encefalograma 2242
encendido parásito 1469
enclavamiento 3558
- de la fuente de neutrones 671 2
endostio 2249
endotelio 2250
endotérmico 2247
endurecimiento 3145
- del espectro 6754

endurecimiento del espectro
de los neutrones 4716
- por deformación en frío 6930
- por envejecimiento 142
- por irradiación 3715
energía 2254
- al cero absoluto 7795
- almacenada 6926
- beta máxima 600
- cinética 3850
- cinética de rotación 260
- cinética relativista 6097
- comunicada a la materia 2274
- de absorción por resonancia
2282
- de activación 67
- de activación para difusión
1818
- de asimetría 346
- de corte 1588
- de deformación 1673
- de desintegración 1642
- de desintegración alfa 192
- de desintegración beta 590
- de desintegración beta en el
estado fundamental 3091, 3092
- de desintegración nuclear 4836
- de dislocación 2281
- de enlace 638
- de enlace de un partícula alfa
198
- de enlace de un protón 5567
- de enlace específica 6740
- de enlace media por nucleón
471
- de enlace neutrónico 4674
- de enlace nuclear 4818
- de equilibrio 2340
- de excitación 2411
- de Fermi 2562
- de fisión 2654
- de formación de un par 5095
- de fusión 2900
- de intercambio 2404
- de ionización 3673
- del borde de banda 524
- del fotón 5270
- de movilidad 1970
- de radiación 5680
- de radiación por unidad de
volumen 5681
- de reacción 4900
- de red 3925
- de resonancia 6161
- de separación 638
- de superficie 4917
- de transición 7346
- de umbral 2677
- de una radiación 5706
- de volumen 7641
- efectiva 2030
- en reposo 6176
- específica 6744
- estelar 6887
- geotérmica 3025
- gravitacional 3068
- inicial de neutrones 3489
- interna 3577
- intrínseca 3598
- liberada 6109

energía media de excitación 476
- media de ionización 4328
- media por par de iones
formado 475
- neutrónica 4701
- nuclear 381, 405
- por unidad de masa 2283
- potencial nuclear 4886
- propia 6443
- química 1046
- solar 6682
- térmica 3159
- termonuclear 7183
- total de enlace electrónico
7290
- total de enlace nuclear 7298
- Wigner 7712
enfermedad radiativa 5737
enfoque 2764
- anular 6201
- astigmático 343
- de acelerador 34
- de fase 5228
- de gradiente alterno 212
- estigmático 6906
- magnético 4210
enfriador 1392
enfriamiento 1401, 1402
- de paso único 4973
- de seguridad 2222
- por aire 154
- por bismuto fundido 4524
- por gas reentrante 6045
- por helio 3220
- por líquido 4054
- por radiación 5695
- posterior 133
- rápido en la fase gamma 2937
- rápido en la región de
temperatura beta 606
enjambre de partículas 7037
enlace atómico 370
- covalente 1499
- cruzado 1539
- de hidrógeno 3353
- de valencia 7569
- heteropolar 3241
- iónico 2192
- iónico polarizado 5383
- metálico 4378
- molecular 4500
- químico 1040
ennegrecimiento de la emulsión
2605
enriquecer 2296
enriquecimiento 2303
- isotópico 3787
ensamblaje subcrítico 6976
ensanchamiento cuadripolar
5623
- de colisión 1226
- de choque 1226
- Doppler 1933
ensayo 334
- químico 1038
ensamblaje combustible 2849
entalla 4804
entalladura 4636
entrada 2991
- de bloqueo 3481

entrada de corriente
equivalente de obscuridad
2352
entrelazamiento 1539
enturbiamiento de traza 3053
envainadura 863
envejecimiento 143
envenenamiento 5377
- de boro 742
- del reactor 5379
- de samario 6286
- xenón 7768
envoltura 659
- de seguridad 1330
- de seguridad esférica 1328
- fértil 660
- hermética 2983
- presurizada 5484
- refrigerada de efecto
Peltier 5171
- resistente a la presión 5484
envolvente 2313
envuelta del tubo de protección
total 2890
- de tubo contra choques
eléctricos 6554
- protectora de tubo de
diagnóstico 1770
- protectora de tubo terapéutico
7117
eosinófilo 2316
epicadmio 2318
epicentro 2321
epidurografía 2322
epifaringografía 2326
epifisis 2327
epilantinita 2323
epitelioma 2329
epitérmico 2330
equilibrio de partículas
cargadas 1024
- de plasma 5314
- electrónico 2125
- iónico 3654
- isotópico 3788
- metaestable 4392
- radiactivo 5785
- radiactivo momentáneo 7336
- radiactivo secular 6422-
- térmico 7138
equipo 2348, 6199
- auxiliar 242
- auxiliar del reactor 5969
- clasificador de pulpa 5584
- de agua pesada 3211
- de alarma de radiactividad
peligrosa 5067
- de control de radiación 5689
- de descontaminación 1658
- de deuterio 3195
- de enfriamiento del
combustible 1406
- de ensayo 335
- de extracción por repartición
líquido-líquido 4066
- de hidrógeno pesado 3195
- de irradiación continua de
cereales 1344
- de irradiación de productos
empaquetados 5081

flujo de radiación intensiva 3278
- de 2200 metros por segundo 1376
- de velocidad uniforme 6945
- electrónico 2127
- energético 5682
- gamma 2925
- iónico 3633
- laminar 3892
- magnético 4208
- máximo permisible 4311
flujómetro 2715
- de neutrones lentos 6646
- de neutrones rápidos con tubo contador 1473
- de neutrones rápidos de centelleador 6373
- de neutrones rápidos de contador por rechazo 6014
- electromagnético 2082
flujo molecular 4508
- neutrónico 4707
- neutrónico al azufre 6992
- neutrónico de resonancia 6168
- neutrónico excesivo 2401
- paralelo 1170
- reducido del refrigerante 6039
- térmico molecular 2827
- 1/E 4975
- vectorial 258
flúor 2740
fluorescencia 2729
- de choque 3396
- de resonancia 6164
fluorimetría 2738
fluorímetro 2736
- por excitación ultravioleta 1340
- por excitación X 7748
fluoruro de boro 740
- de uranilo 7539
fluxímetro de electrones 2130
- de neutrones 4710
- de partículas 5147
- gamma 2928
focalización 2764
- anular 6201
- de fase 5228
- estigmática 6906
foco 2761
- de tubo 7422
- fino 2621
- lineal 4024
- óptico 2031
fondo de radiación 495
- de un aparato 501
formación de gérmenes de cristales 4925
- de neutrón por despojo de fotón 4714
- de pares 5091
- de un par electrón-positón por conversión interna 3579
formanita 2784
fórmula actividad-masa 105
- de Breit-Wigner 783
- de dispersión de Mott 4553
- de dispersión de Rutherford 6252
- de Feather 2543

fórmula de Geiger 2998
- de Klein-Nishina 3857
- de la reacción nuclear 4843
- de los cuatro factores 2794
- de masa 4270
- de Massey 4289
- de Watt 7677
- electrónica 2180
- empírica de masa 2236
- leucocitaria 1795
- semiempírica de masa 6465
fosa de hormigón 1289
fosfato de tributilo 7082
fósfor 4159
fosforescencia 5241
fósforo 5242
fosfuranilita 5243
foso 7379
fotactímetro 2600
- anular 2604
fotocátodo 5244
fotocronómetro 5273
fotodesintegración 5246
fotoelectrón 5252
fotofisión 5253
fotografía de cristal rotatorio 6234
fotoionización 402
fotomesón 5263
fotomultiplicador sin ventana 7729
fotón 4010
- de aniquilación 272
fotoneutrón 5271
fotón virtual 7617
fotoprotón 5272
fourmarierita 2804
fracción 2805
- agotada 1725
fraccionamiento de la dosis 1942
fracción de bifurcación 768
- de combustión nuclear 826
- de consumo 826
- de conversión 1381
- de empaquetamiento 5088
- de enlace 639
- de neutrones instantáneos 5535
- de neutrones retrasados 1699
- de recirculación 4426
- eficaz de neutrones retrasados 2028
- enriquecida 2297
- ligera 4006
- molar 4496
- pesado 3192
fractura de fatiga 2537
- de fragilidad 788
fragmentación del núcleo 2813
fragmento de espalación 6732
- rápido 2528
fragmentos de fisión 2656
francevilita 2815
francio 2816
frecuencia angular de ciclotrón 1595
- angular de precesión nuclear 1596
- atómica 386
- de ciclotrón 1597

frecuencia de ciclotrón de un electrón 2116
- de ciclotrón de un ión 3624
- de colisiones 1229
- de choques 1229
- de difusión 1820
- de dispersión 6339
- de excitación 2412
- de Langmuir 2147
- de Larmor 3904
- de plasma de los electrones 2147
- de precesión de Larmor 3906
- de umbral 7229
- nominal 5935
- patrón de haz atómico 369
freialita 2838
frenado 6913
frente de cargo 1028
- de choque 5477
- de descarga 7488
- de Mach 4180
- de presión 5477
fricción dinámica 2006
frío 1195
fritzscheita 2841
fuente activa 93
- beta cerámica 956
- beta puntual 603
- de criptón-85 3867
- de energía 6713
- de introducción guiada 136
- de iones 3634
- de iones de agrupamiento intensivo 3537
- delgada 7194
- de línea espectral subdividida 6800
- de línea única 6611
- de Mössbauer 4551
- de neutrones 4747
- de neutrones plana infinita 3472
- densa 7186
- de radiación 5738, 6714
- de radiación gamma 2942
- de radiación para clínicas 1127
- de referencia 6050
- discoidal 1884
- gamma de alta energía 3259
- gamma de basa energía 4141
- hendida 6640
- isotrópica de radiación 3809
- no sellada 7491
- permanente de luz 5210
- plana 5308
- pulsada de neutrones 5603
- puntiforme 5373
- radiactiva 5807
- sellada 6394
- soldada 7692
- virtual 3391
fuertemente radiactivo 3295
fuerza ascensional 817
- central 944
- centrífuga 949
- centrípeta 955
- de aniquilación 270
- de arrastre 1967

lábil 3882
labilización de tritio 7399
laboratorio de baja radiactividad 1203
- para isótopos 3793
- radiactivo 3335
ladrillo plombífero 3948
laguna 3306
lambertita 3888
lámina beta 614
laminación en paquete 5079
lámpara de alarma 7660
lana de amianto 333
- de resina 6147
lantánidos 3901
lantano 3902
laplaciano 805
- geométrico 3017
- materia 4293
latitud del contraste del sujeto 6980
lavadero de decontaminación 85
lavado 6388
- cáustico 930
- de gas 2979
lawrentio 3937
lazo 2444
- activo 87
- interior 3422
- metálico radiactivo 5791
lector de cargas 1026
lente 3979
- quadripolar 5625
- magnético 4213
leptón 3980
lesión latente de tejido 3912
letargia 3986
leucemia 3987
leucocitemia 3987
leucocito 3988
leucocitosus 3989
leucograma 3990
levantamiento geológico de terrenos 3014
- terrestre 3093
levigación 2215
ley adiabática de Ehrenfest 2053
- de Barlow 537
- de Bragg 760
- de Child-Langmuir 1072
- de decaimiento radiactivo 5771
- de desintegración 1644
- de desplazamiento radiactivo 5776
- de distribución de Doerner-Hoskins 1923
- de Duane y Hunt 1987
- de Fick 2582
- de Grotthus-Draper 3087
- de imposibilidad de cruzamiento 4781
- de la multiplicidad máxima 6244
- de Moseley 4549
- de paridad de la estabilidad nuclear 4958
- de Paschen 5157
- de Planck 5306

ley de proporcionalidad a la inversa de los cuadrados 3603
- de raíz cúbica 1564
- de Stokes 6911
leyes de cálculo de la onda expansiva 667
- de desplazamiento 1898
- de Hund 3345
- de selección de Fermi 2568
- de selección de Gamow-Teller 2955
- de selección nuclear 4908
ley exponencial de atenuación 2456
liberación condicional 1296
- de energía 2286
- de gas de fisión 2659
lienografía 3999
ligazón 714
- mecánica 4344
- metalúrgica 1830
ligazon térmica 7122
limitación de la señal 505
- de movilidad 4015
- de riesgos 4464
- debides a la física nuclear 4866
- lambda 3886
limitador 4020
- de onda inversa 3604
límite 4014
- cuántico 753
- de captura para cadmio 840
- de envenenamiento 5375
- de grano 3050
- de Kruskal 3865
- de resistencia a la fatiga 2251
- de trayectoria de ionización 3638
límites de secado 1979
límite extrapolado 2496
limpieza ultrasónica 7458
limpio 1117
lindoquita 4171
línea auxiliar 466
- de dislocación 1887
- de fuerza 4027
- de operación 4988
- espectral 6755
- espectral de resonancia 6175
- K 3828
- M 4176
- madre 5119
- N 4609
líneas anti-Stokes 307
- de deslizamiento 6636
- de los planos atómicos 3939
- de Schmidt 6348
linfa 4166
linfocito 4168
linfografía 4169
linfopenia 4170
litio 4018
lixiviación 3945
localización 4106
- estereométrica 5844
- selectiva 6437
localizador 314, 1303
- de compresión 1271
- labial 4053

localizador luminoso 4003
longitud de absorción 21
- de Debye 1628
- de difusión 1822
- de la meseta 3977
- de la trayectoria 5159
- del plato 3977
- de migración 4428
- de moderación 6655
- de onda crítica de absorción 1510
- de onda de Compton 1281
- de onda de de Broglie 1623
- de onda del neutrón 4756
- de onda efectiva 2049
- de onda electrónica 2168
- de onda límite 753
- de radiación 5720
- de relajación 6105
- de relajación eficaz 2039
- de retardación 6655
lote 555
lovozerita 4138
lubricante para altas presiones 3275
lugar 6621
lumbrera 6639
- doble 1957
luminiscencia 4156
luminosidad negativa 4641
lutecio 4165
luz indicadora 7660

mackintoshita 4184
maclado 7443
macromolécula 4185
macropartículas 4186
- en suspensión en el aire 169
macroporosidad 4187
machaqueo 1549, 3084
madre 6425
magnesio 4197
magnesiotermia 4196
magnetodinámica ideal 3378
magnetómetro por efecto Hall 3127
magnetón 4232
- de Bohr 691
- de Weiss 7687
- nuclear 4819
magnitud de referencia 6049
- escalar 6308
magnitudes conjugadas 1310
magnitud independiente 3439
- piloto 6049
- regulada 1371
- seudoescalar 5580
- vectorial 7594
magnox 4233
mala geometría 516
mamografía 4241
mancha focal 2760
- focal efectiva 2031
mandil de caucho al plomo 684
mando 450
- por muestreo 6289
manejo del combustible 2864
manganeso 4242
manguito aislante de calor 7166
- de plomo 3959

manguito de protección 730
- de resina sintética al plomo 3954
- de soporte en grafito 7007
- para tubos sin derramiento de líquido 4795
manipulador 4245
- con articulación de rótula 519
- para usos generales 3007
- rectilíneo 1502
- robusto 4949
manivela de blindaje 6384
manómetro 5478
- de ionización 3674
- de vacío 7557
manual de seguridad 3150
mapa de isorrados 3748
- radiométrica 5871
máquina de descarga 7489
marcha de prueba de un equipo 5312
mAs 4250
masa 4252
- atómica 390
- atómica de un núclido 4940
- atómica relativa 6085
- crítica 1520
- crítica específica 1507
- crítica excedente 4572
- crítica parcial 5126
- de inercia 3469
- del electrón 2101, 4272
- del electrón en reposo 2155
- del neutrón 4670
- de los núcleos iniciales 4273
- de los núcleos producidos 4274
- del protón 5566
- del universo 4275
- efectiva 2033
- electromagnética 2085
- en reposo 6177
- en reposo del neutrón 4745
- isotópica 3796
- mecánica 4347
- molecular 4513
- nuclear 4872
- reducida 6040
- reducida del electrón 2153
- relativista 6098
- segura 6265
máscara 4251
masuita 4291
materia amorfa 224
- cristalina 1562
- descente 2513
- equivalente al aire 159
- explosiva 2450
- extrañea 1547
- fantasma 5224
- fértil 2578
material agotado 1727
- cerámico para temperaturas muy altas 3284
- de blanco 7075
- de blindaje 6382
- de centelleo 6356
- de coproducto 837
- de ligazón 715
- enriquecido 2299
- escintilante 6353

material fértil 662
- físil 2643
materialización 4297
material protector 5561
- radiactivo 88
- usado 7663
materia nuclear 4873
- original 5120
- particulada 5154
matriz en plata 6595
mCi 4322
mCih 4323
mecánica clásica 1114
- cuántica 5644
- de matrices 3217
- newtoniana 1114
- ondulatoria 7682
mecanismo de compensación 6543
- de la barra de regulación 1362
- de regulación 1356
- de seguridad 6272
- de seguridad acelerado 332
- de seguridad por descarga del moderador 4490
- de seguridad por inyección de veneno líquido 2723
- electromagnético de seguridad 2091
medicina nuclear 4874
medida de espesor 7190
- de la radiación 5724
- isotópica del volumen sanguíneo 3778
medidor carbono /hidrógeno para hidrocarburos 886
- de actividad de C-14 y tritio 889
- de contaminación de suelos 2708
- de contaminaciones atmosféricas 149
- de contaminación superficial para tritio 7401
- de contaminación superficial 7014
- de dosis 1946
- de dosis integrador 3532
- del porcentaje de agua pesada 3208
- del tiempo de doblado 1961
- de período 5199
- de reactividad 5962
medio de bloqueo 674
- de contraste 1351
- de dispersión Compton 1278
- de fisión rápida 2529
- de neutrones térmicos 7148
- multiplicador 4584
médula ósea 724
mega-electrón-voltio 4356, 4444
megawatt-día per tonelada 4358
meiosis 4359
melanocerita 4360
membrana 542
- mucosa 4564
mendeleevita 4363
mendelevio 4364

mercurio 4365
meseta 5338
- de contador 1466
- de potencial 5429
mesón 4368
mesones L 3874
- livianos 3874
mesón K 3201
- mu 4563
- neutro 4657
- pesado 3201
- sigma 6584
- tau 7080
mesotorio-I 4371
mesotorio-II 4372
metaestable 4389
metafase 4388
metafosfato de sodio 6671
metal esponjoso 653
metalografía 4381
metalurgia 4385
- extractiva 2492
metástasis 4396
metatorbernita 4397
metauranocircita 4398
metazeunerita 4399
metilisobutilcetona 3249
método analítico por dispersión atómica anómala 278
- catalítico de exposición al tritio 917
- Debye-Scherrer 1629
- de cambio químico 1049
- de Carlson 893
- de centrífuga 953
- de cristal oscilador 5036
- de cristal rotatorio 6231
- de destilación 1911
- de destilación molecular 4505
- de determinación de la relación carga-masa de Dempster 1710
- de difusión gaseosa 2986
- de difusión térmica 7132
- de emanación 2218
- de equilibrio 2342
- de fluorinación 2739
- de grupos 3098
- de Hahn 3119
- de Hartree 3147
- de hidrogenación 3360
- de la distribución radial 5676
- de la sagita 6278
- de Laue 3933
- de los momentos 4530
- de Millikan 4440
- de parábola de Thomson 7204
- de puesta en marcha 6853
- de Rabi 5670
- de separación de isótopos por espectrografía de masa 2086
- de Slater 6634
- de substición progresiva 6987
- de superfluides 7001
- de Tamm-Dancoff 7063
- de tiempo de vuelo 7257
- de Weissenberg 7688
- de Weizsäcker-Williams 7689
- electrolítico 2079
- electromagnético 2086

método electrostático 2199
- espectrográfico de masa para
  pruebas de fugas del helio
  3224
- estadístico 6868
- exponencial 2457
- fotográfico de detección de
  la radiación 5260
- húmedo de reprocesamiento
  318
- Monte Carlo 4545
- sandwich 6294
métodos de relajación 6106
- de separación de isótopos 3766
método sin contacto 1659
MeV 4356, 4444
mezcla 4474
- completa 3550
- de carbón pirolítico y carburo
  de silicio 5613
- de isótopos 3759
mezclado 669, 4471
mezclador-asentador 4470
- de bombas 5606
mezclador de papilla barítica
  534
- formulador en serie 2520
mezcla heterogénea 3238
- isotópica 3759
mgh 4403
microanálisis 4406
microcurie 4410
microhaz 4407
microinestabilidad 3851
microirradiación 4412
microlita 4413
micromanipulador 4414
microporosidad 4415
microquímica 4409
microrradiámetro 4417
microrradiografía 4416
microscopio electrónico 2138
- protónico 5570
microsegregación 1418, 3553
microsonda de Castaing 2150
microtrón 4422
mielografía 4605
migración atómica 394
- de productos de fisión 4429
- iónica de contracorriente 1478
milésima de unidad de masa
  4441
miliampere-segundo 4250
miliamperio-segundo 4250
milicurie 4434
- destruido 4322
milicurie-hora 4323
miligrafía 2689
miligramo-hora 4403
milinile 4443
miliröntgen 4445
mina 4446
mineral 4447, 5011
- complejo 1548
- de alta graduación 3266
- de alta ley 3266
- de calidad inferior 4147
- de torio 7219
- de uranio 7517
mineralogía 4450

mineral tosco 1548
minería 4457
mitosis 3834
mitótico 4467
modelo colectivo 1218
- de acoplamiento fuerte 6960
- de capas 6525
- de dos grupos de energía 7445
- de grupos múltiples 4247
- de la edad de Fermi 2552
- de la estructura in partículas
  alfa 199
- de la gota líquida 4057
- del electrón de Abraham 3
- del núcleo de partículas
  independientes 3438
- del núcleo de partículas
  individuales 3451
- de partícula única 4979
- de retardamiento continuo
  1346
- de Thomas-Fermi 7201
- de un grupo de energías 4977
- estadístico 6869
- estratiforme 6525
- nuclear 4875
- nuclear de evaporación 2387
- óptico del núcleo 1143
- unificado 7470
- vectorial de un átomo 7596
moderación 4484
moderador 4486
- hidrogenado 3362
- orgánico 5023
- refrigerante 4488
- ultrarrápido 6995
modulador 1081
- de haz neutrónico 4681
- lento 6642
- rápido 2522
módulo de elasticidad 4492
- de elasticidad al corte 1172
- de ruptura 4494
- digital de memoria de 4096
  canales 2801
molde 4554
- de radio 4554
molécula 4521
- activada 60
molécula-gramo 3058
molécula homonuclear 3319
- isoestérica 3750
- neutra 4658
moléculas impares 4959
- marcadas 3881
molibdeno 4527
molido 4442
momento angular 261
- angular del espín 6783
- angular intrínseco 3596
- angular orbital 5001
- cinético 3852
- cuadrático 2795
- cuadripolar nuclear 4895
- cuadripolar 5624
- de enlace 717
- de inercia 4529
- de inercia geométrica 259
- dipolo 1848
- dipolo inducido 3446

momento magnético 4203
- magnético anómalo 280
- magnético del espín 6787
- magnético del neutrón 4725
- magnético del protón 5568
- magnético de un átomo·o de
  un núcleo 4215
- magnético de un electrón
  planetario 4216
- magnético nuclear 4868
- multipolar 4585
- orbital 5004
momentos multipolos de
  transición 7347
monazita 4533
monitor 4534
- atmosférico 150
- atmosférico C-14 975
- atmosférico de óxido de tritio
  7403
- atmosférico de tritio 7398
- atmosférico de yodo con
  muestreo continuo 3607
- atmosférico de yodo con
  muestreo discontinuo 3608
- beta-gamma de acceso a
  puerta 596
- de actividad global del
  refrigerante 1396
- de alimentos 2772
- de área 321
- de bolsillo con pilas 5365
- de cenizas de carbón 1153
- de contaminación para ropa
  de trabajo 3935
- de contaminación superficial
  7015
- de contaminación superficial
  de suelos 2709
- de criticidad 1532
- de efluente 2052
- de la densidad de flujo de
  electrones 2131
- de la densidad de flujo de
  neutrones 4711
- de la densidad de flujo de
  partículas 5148
- de la densidad de flujo gamma
  2929
- del fondo 500
- de polvo y gas radiactivo 592
- de productos radiactivos
  empaquetados 5082
- de radiación 3156
- de radiactividad aeroportada
  1999
- de víveres 2772
- gamma de criticidad 1533
- isotópico de productos
  empaquetados 5080
- neutrónico de criticidad 1534
- para agua 7675
- para manos y pies 3135
- para ropas 1137
monocito 3349
monocristal 6608
monopolo 4544
montacargas 1006
mordaza 3816
mosaico 4547

número cuántico del momento
  angular 262
- cuántico interior 3499
- cuántico interno 7286
- cuántico magnético 4220
- cuántico principal 2633, 5504
- cuántico radial 5678
- cuántico relativo a un
  impulso de rotación 6238
- cuántico secundario 490
- de Avogadro 481
- de bariones 547
- de colisiones 1232
- de choques 1232
- de efecto de pantalla 6385
- de excitación 2413
- de extrañeza 6933
- de frenado 6916
- de glóbulos sanguineos 677
- de leptones 3982
- de los neutrones por fisión
  4942
- de los neutrones producidos
  por neutrón absorbido 4943
- de Mach 4181
- de masa 4271
- de neutrones 4730
- de nucleones 4271
- de ocupación 4954
- de platos teóricos 4944
- de protones 397
- de unidades de transferencia
  4945
- de valencia 7572
- entero 3520
- impar 4960
- mínimo de placas teóricas
  4454
- molecular 4514
- muónico 4594
- par 2391
números mágicos 4195
número somático 6705

objeto buscado 5665
obturador 6574
octete electrónico 2143
ojiva de guerra nuclear 4923
omegatrón 4971
onda asociada 340
- de Alfven 179
- de choque 6552
- de choque reflejada 6055
- de Mach 4183
- de succión 5928
- dispersada 6330
- expansiva 668
- parcial 5132
ondas acústicas iónicas 3615
- de Langmuir 2205
- electrostáticas 2205
- magnetodinámicas 4231
opacidad 4983
opaco 4984
- a las radiaciones 5875
operaciones de paro 6569
- iniciales 3152
- terminales 7060
operador de campo 2594
óptica neutrónica 4732

oquedad 7630
orangita 4996
órbita 4997
- de Bohr 692
- de enlace 718
- de equilibrio 2344
- de expansión 2436
- del átomo 398
- electrónica común 722
- electrónica no común 287
- estable 2344
- exterior 5043
- molecular 4515
ordinograma 2712
organismo vivo 4084
órgano 5020
- crítico 1522
- de mando final 2614
- sensor 6467
orientación 5025
- del espín 6790
- preferente 5468
orificio de transporte 7330
- para guantes 3038
oro 3039
ortohelio 5031
ortohidrógeno 5033
ortopositonio 5034
ortorradioscopia 5030
oscilación 5037
oscilaciones de betatrón 629
- del plasma 5321
- sincrotrón 7049
oscilador de reactor 5288
osmió 5038
osteogénico 5039
oxidación radiolítica 5866
óxido de berilio 576
- negro de uranio 658
óxidos de uranio 7518
oxígeno 5074
- pesado 3203

paladio 5096
panoramagrama paraláctico
  5102
pantalla 6528
- absorbente la radiación 5736
- beta 621
- biológica de hormigón 1288
- contra choque térmico 7164
- contra la radiación indirecta
  166
- de guía 3110
- fluorescente 2735
- homogénea 3315
- intensificadora metálica 4376
- laminada 3893
- parcial 5131
- protectora 5562
- radioscópica 2743
- reforzadora 3539
- rotativa 6233
papilla 6661
paradoja de Klein 3858
parahelio 5100
parahidrógeno 5101
paramagnetismo nuclear 4878
parámetro 5111

parámetro de choque 3398
- de dispersión 6928
- de estabilidad 6817
- de fluctuación 6928
- del radio 5903
- de Michel 4404
parámetros atómicos 400
- de difusión 1824
- de la células elementales
  7478
parámetro segundario 6416
par antagonista 6179
parapositonio 5112
par de electrones 2145
- de iones 3642
- de iones primario 5499
- de rotación 7284
- de torsión 7284
pared de blindaje 6531
par electrón-positón 2149
parhelio 5121
paridad 5122
paro 6566
- de emergencia 2226
- de emergencia de última hora
  3908
- de emergencia por período
  demasiado breve 5201
- lento 7388
- por flujo neutrónico
  excesivo 2397
- por temperatura excesiva
  2400
- rápido del reactor 5992
parsonsita 5123
par termoeléctrico de vaina
  metálica 4377
partícula 5134
- alfa 197
- beta 602
- bombardeada 709
- cargada 1022
- cascada 907
- combustible recubierta 1157
- combustible recubierta de
  retención de productos de
  fisión 2665
- Compton 1277
- constitutiva 1316
- consumida 1319
- de alta energía 3262
- de baja energía 4143
- de chaparrón 6561
- de desintegración 1645
- de investigación 7107
- de Majorana 4236
- de rechazo 6010
- de retroceso 6010
- de traza en forma de V 7552
- dispersada 6327
- elemental 2212
- emitida 2058
- en equilibrio 2343
- expulsada 2058
- fundamental 2212
- incidente 710
- ionizante 3686
- lambda 3887
- ligada 750
- neutra 7464

polonio 5385
polvareda 1997
polvo radiactivo 76
ponderador 5394
porcentaje atómico 358
- de dosis en profundidad 5189
- de ionización 3678
- de pérdidas 5190
portador de isótopo 3779
- de retención 3304
portadores de carga 1023
- electrizados 1023
portador no isotópico 4789
positón 5414
positonio 5417
poso estratosférico 6939
- local 4105
- radiactivo 82
- troposférico 7412
postirradiación 5420
postrefrigeración 133
postulado de Goudsmit-Uhlenbeck 3043
- del cuanto 5647
potasio 5423
potencia calorífica 856
- de diseño 1741
- de la capa fértil 663
- de salida 5453
- de salida de una fuente 5728
- específica 2874, 6751
potencial central 945
- constante equivalente 2350
- crítico 1525
- crítico de ionización 1519
- de aparición 312
- de Coulomb 1448
- de deformación 1674
- de desionización 1684
- de excitación 2415
- de extracción 2489
- de Fermi 2567
- de flujo 6944
- de fondo de botella 7730
- de frenado 6917
- de ionización 3676
- de penetración 5183
- de radiación 5730
- de repulsión 6135
- de separación 6488
- de Yukawa 7790
- medio de excitación 477
- nuclear 4885
potencia nominal térmica de un reactor 3510
- por unidad de superficie 5451
- presumida 1741
- remanente 137
- teórica 1741
- térmica de un reactor 3173
- térmica total 7300
- térmica útil 7548
- útil 7547
pozo de Gauss 2997
- de percolación 6429
- de potencial 5431
- de Yukawa 7791
- exponencial 2459
- magnético 4228
- rectangular 6029

praseodimio 5460
preamplificador 3151
precesión de Larmor 3905
precipitación por salazón 6283
precipitado preformado 5469
precipitador electrostático 2200
precisión de medida 48
precursor 5464
- de neutrones retrasados 1700
predictor de envenenamiento productos de fisión 2668
- de envenenamiento xenón 5380
- de terremotos por efecto radón 5910
predisociación 5466
preirradiación 5471
prensado en caliente 3337
preparación 1969
presión cinética 3853
- de arrastre 1968
- de ionización 3677
- del plasma 5322
- de radiación 5731
- dinámica 2005
- magnética 4218
- reflejada 6053
primera constante de radiación 2634
- descarga de Townsend 2636
primer coeficiente de Townsend 2635
- potencial de ionización 2632
principio de Bragg-Gray 759
- de complementariedad 1260
- de conservación del equilibrio neutrónico 4672
- de constitución 810
- de correspondencia 1425
- de exclusión de Pauli 5160
- de Franck-Condon 2817
- de las partículas estabilizadas en fase 5511
- de Pauli-Fermi 5161
- de repartición de Boltzmann 702
proactínidos 5512
probabilidad de adherencia 6904
- de choque 5513
- de dispersión 6344
- de escape a la resonancia 6162
- de excitación 2416
- de fisión 2664
- de fuga para fuentes gamma 3969
- de ionización 5514
- de penetración 5182
- de permanencia 4786
- de reflexión 6052
- de transición 7348
- de transición total 7301
probeta de prueba al choque 3402
- entallada 4805
problema de la selección de los materiales 1251
procedimiento al fluoruro de silicio 6590
- al sulfato 6989

procedimiento con ácido dialquilfosfórico 1613
- continuo 1345
- de cambio con dos temperaturas 1986
- de cargo 1030
- de conversión con hexafluoruro de uranio 3245
- de difusión en barrera 543
- de electro-migración 2098
- de enriquecimiento 2306
- de envenenamiento por productos de fisión 2669
- de etapas múltiples 4591
- de extracción al fosfato de tributilo 7083
- de intercambio de dos temperaturas al hidrógeno sulfurado 3358
- de lotes 557
- de migración iónica 2098
- de partidas 557
- de producción de agua pesada de Clusius y Starke 1145
- de recortado y disolución 1080
- de separación 6489
- de separación con fosfato de bismuto 655
- de separación primaria 5507
- de Szilard-Chalmers 7054
- de tandas 557
- de una sola etapa 6599
- endotérmico 2248
- exotérmico 2432
- extranuclear 2494
- irreversible 3720
- Oppenheimer-Phillips 4982
- químico 1056
- reversible 6189
- seco 1981
- simple 6599
- sol-gel 6681
procedimientos redox 6038
- reducción-oxidación 6038
procedimiento unitario 7481
proceso de dos etapas 7451
- de fustigación 2701
- de regeneración 6070
- virtual 7618
producción múltiple en colisiones sucesivos 5355
- múltiple en colisión simple 4575
producto de bifurcación 765
- de cabeza 5063
- de desintegración 1646
- de desintegración radiactivo 90
- de fisión de vida larga 4121
- de fisión radiactivo 83
- final 2245
- intermedio 3424
productos de fisión 2667
- estériles 7061
producto terminal 2245
profase 5540
progreso de la fisión 2651
prolongación de barra de regulación 1363

prolongación de combustible de una barra de regulación 1364
promedio de recorrido 4331
- de recorrido en masa 4332
prometio 5526
propiedades atómicas 406
- nucleares 4893
propiedad macroscópica 4191
- química 1059
proporcionalidad limitada 4016
proporción de ascenso de la temperatura 7161
- reticular 3082
propulsión 5546
- iónica 3644
- nuclear 4894
- por plasma 5323
protactínidos 5551
protactinio 5552
protección 5553
- biológica 647
- contra exceso de potencia 5068
- por plomo 3955
- térmica 7156
protector ocular de contacto 2502
protegido contra choques 6553
protio 5564
protón 5565
- bombardeado 3862
- negativo 304
- percutido 3862
protón-sincrotrón 1438
protoplasma 1606
proyección anterior 284
- axial 484
- dorsal 283
- frontal 284
- hiperscópica 3369
- hiposcópica 3372
- lateral 3913
- lordótica 4129
- oblícua 4952
- ortoscópica 5035
- posterior 283
- seudoscópica 5581
- tangencial 7066
proyectil nuclear 4892
proyecto de dos W 7453
- del reactor 5977
- heterogéneo de la alma del reactor 3237
- Manhattan 4243
- Matterhorn 4301
- SNAP 6415
prueba 2438
- al choque 3401
- antes de y después de la irradiación 567
- de burbujas 803
- de ciclo térmico 7130
- de esfuerzo de ruptura 6948
- de estanquecidad con paladio 5097
- de fallas, a 2512
- de fatiga 2253
- de frotamiento 6664
- de fuga de neutrones, a 4760
- de fugas 3964

prueba de fugas del helio 3223
- de fugas de radón con centelleador líquido 6362
- de impacto de Charpy 1031
- de inmersión 3394
- de Izod 3811
- de penetración de helio 3226
- de seguridad de funcionamiento 6112
- de transmisión 7354
- de volumen 814
- en no activo 1209
- en reactor 3423
- para más parámetros 3526
puerta 2991
- con capa interior de plomo 3950
- de blindaje 2992
- magnética 4211
puesta en marcha 6851
pulmones 4164
pulvimetalurgía 5434
puntas de descarga 6808
- de inducción 6808
punto caliente 3283
- crítico 1524
- de calefacción 823
- de decalescencia 1634
- de ebullición 698
- de imagen 3389
- de inflexión 3477
- de inundación 2705
- de plegado 5303
- de rayo emergente 2230
- de recalescencia 6000
- de ruptura 773
- de silla 6260
- de verificación 1032
- neutro 6840
pureza del plasma 5324
- química 1062
- radiactiva 5732
- radioisotópica 5852
- radioquímica 5828
púrpura 5168

quebrantadura 3084
quelación 1035
quelato 1033
queloide 1036
queloma 1036
quelos 1036
quemadura por radiación 5688
- por radiación térmica 2687
química de la radiación 5692
- de las soluciones 6697
- de los átomos muy excitados 3328
- de los elementos livianos 4005
- de los elementos pesados 3191
- de los trazadores isotópicos 7315
- del reactor 5971
- nuclear 4826
quimionuclear 1069
quimografía 3871
- de una sola ranura 6615
quimógrafo 3870
quistamina 1604

rabitita 5669
rad 5673
radiación 4610, 5685
- alfa 203
- beta 607
- blanda 6677
- característica 998
- ciclotrón 1598
- coherente 1179
- corpuscular 1422
- cósmica 1432
- de alto nivel de energía 3267
- de aniquilación 273
- de Cerenkov 963
- de cuadrupolo 5626
- de enfrenamiento 785
- de enfrenamiento interna 3497
- de fluoerescencia 998
- de frenado 785
- de frenado exterior 5041
- de frenado interna 3497
- de fuga 3970
- del fondo natural 502
- de partículas 1422
- de rechazo 6016
- de resonancia 6172
- de retroceso 6016
- de vida larga 4122
- directamente ionizante 1862
- dispersa 6941
- dispersada 6328
- dura 3143
- electromagnética 2090
radiaciones nucleares 4897
radiación externa 2478
- extrafocal 6890
radiación gamma 2938
- gamma de alta energía 3258
- gamma de captura 876
- gamma inicial 3490
- gamma inmediata 2688
- gamma instantánea 2688
- heterogénea 3239
- homogénea 3313
- incoherente 3431
- indirecta 165
- indirectamente ionizante 3448
- inicial 3493
- instantánea 5537
- interna 3581
- ionizante 3687
- K 3830
- L 3875
- M 4177
- mitogenética 3113
- monocromática 4537
- monoenergética 4542
- multipolar 4586
- neutrónica 4740
- persistente 5214
- polienergética 5391
- primaria 5505
- residual 6144
- retrodispersa 510
- secundaria 6419
- térmica 7157
radiactinio 5760
radiactividad 5816
- artificial 330
- atmosférica 353

radiactividad beta 608
- en el aire 170
- inducida 330
- natural 4623
radiactivímetro médico 4353
- para gases 2957
- por agua 7669
radiactivo 5762
radiador 5758
radiámetro 4340
- contador para exploración
  con tubo contador
  Geiger-Müller 1490
- corporal 7707
- corporal con analizador de
  amplitud 7708
- de exploración con contador
  G.M. 5550
- de exploración de centelleador
  6375
- de exploración móvil con
  centelleador 7597
- múltiple 4587
- para berilio 582
- portátil de exploración 5405
- portátil de exploración con
  analizador de amplitud y con
  centelleador 7598
- selectivo móvil de exploración
  selectiva con centelleador
  7599
- valorador de carga 1324
radical libre 2829
radio 5890
- atómico 407
radiobiología 5822
radiocardiografía 5824
radiocartógrafo corporal 687
radiocinematografía 1102
radio clásico de electrón 1113
radiocoloide 5831
radiocristalografía 1551
radiocromatógrafo 5830
- en fase gaseosa 2988
radio de aplanamiento 2697
- de Bohr 693
- de ión 3657
- de la barrera de Coulomb 1442
- de Larmor 3907
- del electrón 2152
radiodermatitis 53
radioecología 5836
radio eficaz de neutrón para
  colisión 4683
- electrostático 2201
radiofotografía 2741
radiofotoluminiscencia 5876
radiogénico 5840
radiografía 5845
- de contraste 1352
- de los senos 6619
- de mama 4241
- de mediastino 4352
- electrónica 2151
- en serie 6498
- industrial 3458
- neutrónica 4742
- por secciones 689
radiograma 5842
- de contacto 1321

radioisótopo 5795
radiolesión 5718
radiólisis 5701
- de los solventes 5864
radiología médica 4354
radioluminiscencia 5862
radiometalografía 5868
radiometría 5873
radiómetro de condensador 869
radio nuclear 4898
radionúclido 5874
radionúclidos naturales
  extintos 2480
- naturales inducidos 3452
- naturales primarios 5502
- naturales secundarios 6413
radioquímica 5829
radiorresistencia 5733
radioscopia 2744
radioscopio 5879
radiosensibilidad 5882
radiosensible 5881
radiosondeo 5796, 7695
radioterapia 5884, 7116
- convergente 1377
- de contacto 1322
- pendular 5175
- profunda 1662
- superficial 6999
radiotermoluminiscencia 5885
radiotorio 5886
radiotoxemia 5737
radiotoxicidad 5887
radiotransparente 5861
radiotropismo 5889
radiovigía de intercambio
  iónico 3632
radón 5904
radonómetro para radio-
  protección 5907
- para uso minero 5908
randita 5915
rango de frecuencias 2836
ranura de aceleración 31
- eliminadora del movimiento
  vaivén 305
rascador radiactivo 5789
rauvita 5947
rayo 5948
- central 946
- delta 1705
rayos alfa 207
- anti-Stokes 307
- Becquerel 565
- beta 619
- canales 860
- catódicos 923
- cósmicos 1437
- cósmicos primarios 5493
- cósmicos secundarios 6404
- de Cerenkov 965
- de enfrenamiento 784
- de frenado 784
- de Lenard 3976
- gamma 2947
- límite 3078
- X 6220
- X característicos 999
- X continuos 1349
- X dispersados 6331

rayos X heterogéneos 3236
- X homogéneos 3317
- X primarios 5509
- X secundarios 6420
rayo X blando 6679
- X duro 3144
razón de equivalente de dosis
  1941
reacción 5952
- alfa-neutrón 196
- alfa-protón 201
- con intercambio de energía
  2407
- controlable 1369
- convergente 1379
- de cambio catalítico 916
- de cambio químico 1050
- de captura 877
- de colisión 1235
- de choque 1235
- de fusión nuclear 2903
- deuterón-alfa 1761
- deuterón-neutrón 1762
- deuterón-protón 1763
- divergente 1919
- en cadena automultiplicada
  6450
- en cadena de fisión 2650
- en cadena de neutrones
  térmicos 7151
reacciones nucleares inversas
  3601
reacción explosiva 2451
- fotonuclear 5246
- general 7053
- inducida por neutrones 4718
- irrefrenable 6247
- lenta 6648
- neutrónica en cadena 4680
- neutrón-neutrón 4729
- neutrón-protón 4739
- no controlada 6247
- nuclear 4899
- nuclear con catalizador
  mesónico 919
- nuclear de cadena 981
- nuclear en cadena autónoma
  6449
- nuclear en cadena de automoderación 6447
- nuclear en función de
  tiempo 7251
- nuclear espontánea 6806
- nuclear exotérmica 2434
- persistente 7035
- química 1063
- química térmica 7125
- rápida 2534
- térmica 7161
- termonuclear 2903
reactímetro 5962
reactividad 5958
- de paro 6570
- excedente 2398
- fría 1207
- inherente 811
- instantánea 5538
- limpia 1121
- negativa 4645

sección atómica 375
- de extracción 6954
- de ionización 3668
- de retardación 6652
- de separación 6954
- diferencial para dispersión 1793
- eficaz 1541
- eficaz de absorción 16
- eficaz de absorción de los neutrones 4666
- eficaz de activación 64
- eficaz de captura 874, 2025
- eficaz de captura neutrónica 4679
- eficaz de captura radiativa 5751
- eficaz de choque 2026
- eficaz de difusión inelástica radiativa 5754
- eficaz de dispersión 1115
- eficaz de dispersión coherente 1181
- eficaz de dispersión elástica 2062
- eficaz de dispersión incoherente 3433
- eficaz de dispersión inelástica 3462
- eficaz de fisión 2653
- eficaz de interacción 3546
- eficaz de Maxwell 4320
- eficaz de moderación 6652
- eficaz de neutrones 4687
- eficaz de reactor 5976
- eficaz de remoción 2040
- eficaz de remoción de grupo 3099
- eficaz de resonancia 2027
- eficaz de transferencia de grupo por dispersión 3100
- eficaz de transporte 7361
- eficaz diferencial 1788
- eficaz macroscópica 1542
- eficaz macroscópica total 7295
- eficaz media 473
- eficaz media Doppler 1932
- eficaz microscópica 4419
- eficaz no elástica de interacción 4784
- eficaz nuclear 4830
- eficaz para difusión 1816
- eficaz para rayos gamma 2923
- eficaz parcial 5127
- eficaz térmica 7128
- eficaz térmica efectiva 2047
- eficaz total 7288
- electrónica 2178
- específica de choque 7289
- geométrica 3018
- macroscópica de remoción 4192
secesión 774
sedimentación 6424, 6663
sedimento 6423
- fangoso 6658
sefström ita 6430
segregación 6431

segregación interdentrítica 1418, 3553
- por gravedad 3071
segunda constante de radiación 6398
- cuantificación 6397
- descarga de Townsend 6401
segundo coeficiente de Townsend 6400
- potencial de ionización 6396
seguridad de funcionamiento 6110
- de servicio 6110
- limitada, de 4018
selección de señal 2993
selector de amplitud 5589
- de anticoincidencias 292
- de coincidencias 1193
- de coincidencias rápidas 2523
- de coincidencias rápidas de cuatro canales 2793
- de impulsos cronodependiente 7261
- de impulsos dependiente del tiempo 7261
- de velocidad 7602
- de velocidad de los neutrones 4755
selenio 6438
sello labiríntico 3884
semiconductor compensado 1253
- tipo P.I.N. compensado 1253
semilla 6425
semiperíodo de intercambio 3122
sencilla de radio 5900
sensibilidad 6475
- del contador 1457
- de penetrómetro 5186
- de precisión 3435
- de umbral 7232
- intrínseca de contador 3597
- para defecto de espesor 2698
- para los contrastes 2698
- patrón de un fotocátodo 5245
sensible 6468
sengierita 6466
señal análoga 237
- binaria 182
- del período de consigna 5198
- de nivel de potencia de consigna 3992
- de retorno 2547
- de seguir adelante 6276
- digital 1832
- lógica 4118
- numérica 1832
separación 6479
- centrífuga 951
- de impulsiones 2993
- de isómeros 3741
- de isótopos 3764
- de isótopos par movilidad iónica 3640
- de isótopos por chorros de gas 3818
- electrolítica 2080
- electrostática 2202
- energética entre dos bandas 2273

separación en masas parciales subcríticas 4277
- interplanar 3554
- por gravedad 3071
- química 1065
separador de gas 2980
- de isótopos 3768
- electromagnético de isótopos 2084
- energético de iones 3629
- magnético 4225
- repetidor de isótopo 2934
- turbulent 7648
serie de Balmer 522
- de desintegraciones 1639
- liotrópica 4172
servicio variable 7591
servomecanismo 6500
servosistema 6501
seudoimagen 5577
sharpita 6517
sialografía 6576
sifón térmico 7165
sigmatrón 6588
silicio 6591
simetría de las cargas 1018
simulación 6602
simulador 6603
- de radiactividad 5818
- de reactor 4482
sincrociclotrón 2758
sincrofasotrón 7046
sincrómetro de masa 4283
sincrotrón 7047
- de anillo de almacenamiento 6923
- de electrones 2160
- de partículas pesadas 3205
síndrome 7050
- por irradiación 5744
- por radiación 5744
singulete 1017, 6618
sin portador 897
sinterización 6620
síntesis de Fourier 2803
- química de mercaptanos por irradiación 3710
sistema astrón 345
- clásico 1116
- cuantificado 5636
- de circulación externa 2473
- de control óptico 7612
- de coordenadas 1408
- de dos fases 7447
- de enfriamiento abierto 4986
- de enfriamiento cerrado 1134
- de enfriamiento de varias trayectorias 4567
- de inyección de seguridad 6271
- del centro de masa 940
- de llenado de un tubo contador 1459
- de regulación automática 449
- de paro con líquido 4077
- de partículas 7051
- de reactor de ciclo cerrado 1135
- de referencia 2814

teoría de la edad de Fermi 2553
- de la red 3931
- de las cascadas de la radiación cósmica 909
- de las perturbaciones 5221
- de las vainas de corriente de implosión 4179
- de la transferencia de la carga 7114
- del blanco 3298
- del campo cuantizado 5635
- del campo mesónico 4370
- de los cuantos 5650
- del reactor 5991
- del transporte 7365
- de reactor de grupo único 4978
- de relajación espín-celosía de Casimir y Du Pré 910
- de separación capilar 870
- de Townsend 7310
- electrónica de Drude en metales 1975
- electrónica de la valencia 2184
- mesónica de las fuerzas nucleares 4370
terapia isotópica 3771
- pendular excéntrica 2013
- por haz móvil 4556
- por isótopos de cobalto 1160
- por isótopos radiactivos 1575
- por radiación 5745
- por radio 5901
- por rayos X 6227
- reticular 3083
terbio 7100
terfenilo 7104
termalización 7172
término atómico impar 4962
- par del átomo 2393
términos ortohelio 5032
termión 7174
termoluminiscencia 7177
termómetro 7096
- de la vaina del combustible 2854
- del canal de refrigeración 990
termonuclear 7179
termopila al boro 744
- para neutrones 4753
termoquímica 7176
termotransmisión 3176
tetrafluoruro de uranio 3076
tetravalente 5620
tiempo de ajuste 6503
- de arranque de la barra de regulación 4832
- de aumento de un centelleo 6368
- de calado 6503
- de circulación 7441
- de colección 1216
- de colisión 1237
- de crecimiento de un impulso 5598
- de decaimiento de un centelleo 6364
- de difusión 1829
- de doblado 1960
- de establecimiento 2787

tiempo de fluctuación 2717
- de interacción 3548
- de ionización 3680
- de latencia 3909
- del ciclo metabólico 7442
- de liberación 7259
- de parada 1964
- de parálisis 5108
- de puesta en marcha 2346, 6854
- de recolección de electrones 2111
- de recolección de iones 3622
- de recuperación 1423
- de regulación 1423
- de relajación 6107
- de renovación 7442
- de reposición 6139
- de resolución 1470
- de resolución de coincidencia 1192
- de restitución 1468
- de retardación 6657
- de secado con velocidad constante 1315
- de sensibilidad 6473
- de supervivencia 4337
- de tránsito 7340
- de tránsito de un electrón 2163
- de tránsito de un ión 3647
- de vida de la energía del plasma 2280
- de vuelo 7254
- de vuelo del neutrón 4754
- libre medio 4327
- medio de difusión 474
- muerto 1625
- muerto de un contador 1456
- muerto de un tubo contador Geiger-Müller 1627
tiroides 7241
titanato de estroncio 6965
titanio 7272
tiuiamunita 7454
toddita 7274
tolerancia 7275
toma 5048
tomografía 689
- simultánea 6604
tomógrafo 688
torbernita 1410
toria 7208
torianita 7209
torides 7210
torímetro 7213
- por seudocoincidencia beta-alfa 7214
torio 7212
torita 7211
torogumita 4764
torón 7222
torotungstita 7223
torre 1242
- de enfriamiento 1404
- de extracción 2490
torta de filtro 2608
totalidad de los estados electrónicos 4244
trabajo de extracción 2100

trabajo de salida 2100
- de separación 6494
- externo 5045
- interno 3502
traje protector 2842
trampa 2165
- de flujo 2755
- de haz 564
- de radiación 5746
- de yodo 6596
transferencia de calor de ebullición 697
- de carga 1019
- de electrones 2162
- de energía 2290
- de iones 3646
- de masa 4284
- lineal de energía 3983
- máxima de energía 4303
transferómetro 7329
transformación 7332
- espontánea 6807
- nuclear 4903
- orden-desorden 5007
- radiactiva 5810
- rotacional 6239
- tautómera 7081
transición 7342
- de Auger 441
- de fase 5235
- electrónica 2114, 2164
- isomérica 3744
- libre-libre 2825
- nuclear 4920
- permitida 185
- por conversión 1388
- prohibida 2775
- radiativa 5757
- radiactiva 1540, 5810
- sin radiación 5749
translocación 3551
- de cromosomas 1097
transmisión de calor 3176
transmitancia 7328
- atmosférica 355
transmutación 7356
- artificial de elementos 331
- atómica 421
- beta 604
- isobárica 3727
transparente 7358
transporte de isótopo 3772
- neto 4653
traqueteo 1101
traspaso de ritmo 820
tratamiento de alta temperatura 3287
- de combustible 2873
- de combustible agotado 6771
- de entrada 3152
- de la papilla 6663
- de los desechos 7667
- de salida 7060
- químico 1057
- térmico 3181
- térmico alternativo 7130
traumatismo 7374
trayectoria 7325
- cerrada magnética 5672
- de chaparrón 6562
- de dispersión 2564

unidad discriminadora de carga 1003
- discriminadora de forma 5600
- electrostática 2203
- enchufable 5349
- escala 6316
unidades de Hartree 424
- de reactividad 5967
unidad funcional básica 550
- milimasa 4441
- registradora 6021
- registradora electromecánica 2093
- selectora 2994
- selectora de amplitud con canal móvil 6605
- selectora de amplitudes 5590
- selectora de coincidencias 1194
- selectora de coincidencias diferidas 1694
- selectora por anticoincidencias 293
universo 7486
uraconita 7500
uranato de sodio 6673
uránidos 7501
uranilo 7538
uraninita 7502
uranio 7504
uranio-235 55
uranio agotado 1728
- alfa 208
- beta 622
- enriquecido 2301
- gamma 2951
uraniómetro por actividad beta-gamma 7509
- por fluorescencia 1340
uranio natural 4631
uranita 4620
uranocalcita 7528
uranofano 3888
uranopilita 7532
uranosferita 7534
uranospatita 7533
uranospinita 7535
uranotalita 3998
uranotila 3888
uranotorita 7537
uretrografía 7542
urografía 7543
- de micción 4435
uvanita 7550

vacío 7554, 7630
- aproximado 6240
- ultraelevado 2500
vacuómetro por emisión alfa 7559
vaina 858
valencia 7566
- anómalo 281
- electrónica 2204
- electrostática 2204
- iónica 2204
- libre 2830
- máxima 4315
- negativa 4646
- normal 4801

valencia nula 4941
- positiva 4315
valor 7574
- beta 624
- de consigna 3441
- de mezcla isotópica 3760, 7576
- de referencia 3441
- de umbral 7234
- efectivo 110, 2048
- estático 6859
- ft 1248, 2847
valorímetro 1339
- de azufre para hidrocarburos por radiación ionizante 3697
- de cenizas de carbón 1152
- de hierro y calcio para minerales por radiación ionizante 3689
- de la humedad del suelo por radiación ionizante 3695
- de minerales 5015
- isotópico de la humedad 3797
- por fluorescencia X 7748
valor instantáneo 110
- prescrito 1743
- R 5664
- S 6258
- W 7649
válvula de alivio contra explosiones 297
- de cátodo frío 1199
- de electrómetro 2097
- de seguridad 1993
vanadio 7579
vandenbrandeita 7530
vandendriesscheita 7581
vanoxita 7582
vapor en cabeza 5061
vaporización 2385
vapor radiactivo 1333
variable 7588
- de activación 58
- dependiente 1720
- estocástica 984
variables conjugada 1310
variación con el tiempo de la dosis ambiente 4104
- de flujo de gran repartición 3085
- del campo magnético 4219
variaciones seculares de reactividad 4124
varilla de combustible 2876
vasografía 7593
vector 7594
- axial 487
- de Burgers 818
velo 2766
velocidad absoluta de desintegración 6
- al punto de inundación 2706
- crítica 1530
- crítica del flujo 4102
- cuadrática 2802
- de Alfven 178
- de aportación 7498
- de ataque 430
- de cuenta 1491
- de desintegración 1647
- de eliminación 1124

velocidad de fisión 2671
- de formación de un par 1504
- de fugas 3963
- de influjo 5933
- de intercambio 5932
- de intercambio isotópiço 3799
- de ionización 3678
- del ciclo metabólico 7439
- de neutrones 4750
- de partículas momentánea 3512
- de percolación 5192
- de reacción 5957
- de recombinación 6019
- de recombinación de volumen 7646
- de recuento de fondo natural 496
- de relleno media 4335
- de renovación 7439
- de restablecimiento 6025
- diferencial de curación 1792
- eficaz de partícula 2035
velocidades generalizadas 3008
velocidad específica de desionización 1685
- intermedia de neutrones 3567
- irregular 5920
- lineal 4050
- másica 4287
- máxima de partículas 5163
- media 4339
- media de migración 1972
- media de penetración 1972
- mínima de fluencia 4452
- relativista 6102
- superficial 7000
vena 4108
veneno 5374
- de compensación 824
- en solución 6694
- nuclear 4883
- radiactivo 89
venenos de fisión 2663
venografía 7603
ventana 993, 7723
- de blindaje con llenado líquido 4059
- de hoja metálica 4375
- de vidrio plomado 3952
ventanas blindadas 6541
ventana transparente a la radiación 5748
ventriculografía 7604
verruga 670
vertedero 1990
vesiculación 5291
vesiculografía 7606
vestimentas de seguridad 2842
vida comparativa 1248
- corta, de 6556
- del contador 1462
- de un tubo contador Geiger-Müller 4001
- en almacén 6522
- media 478
- media de un estado atómico 479
- media neutrónica 4333
vidrio al fosfato 5239

ITALIANO

abbondanza 25
- cosmica 1431
- isotopica 3775
- isotopica naturale 4622
- isotopica relativa 26, 2809
- molecolare 4497
abernatite 2
abukumalite 24
acceleratore 32
- ad alta tensione 3292
- a focalizzazione a gradiente
  alternato 211
- a gradiente alternato e campo
  fisso 2683
- a poli magnetici con scanalatura elicoidale 6797
- a tensione continua 1314
- Cockcroft-Walton 1168
- di collisioni 1220
- d'ioni 3613
- di particelle 5136
- di particelle atomiche 362
- elettrostatico 2191, 2197
- elettrostatico a trasportatore
  isolato 3518
- intermedio 5138
- lineare 4021
- lineare ad onde progressive
  7375
- lineare d'elettroni 4037
- lineare di Hansen 3136
- lineare d'ioni pesanti 3198
- lineare di Sloan-Lawrence
  6641
accelerazione 30
- angolare 253
- centripeta 954
- d'ioni 3612
- media 469
accelerometro elettronico 2172
accensione parassitaria 1469
accessorio 39
accettore 36
- d'ioni 3614
acciaio al boro 743
- inossidabile 6841
- resistente alla corrosione
  6841
accoppiamento 1495
- ad autochiusura ermetica 6458
- a gradiente 3046
- a valore intermedio 3560
- elettronico 2113
- j-j 3812
- L-S 4155
- pseudoscalare 5578
- pseudovettoriale 5583
- spin-orbita 6789
accrescimento da radiazione
  3101
accumulo 807
- d'attività 101
- d'ioni 815
acido etilendiaminatetraacetico
  2374
- nucleico 4926
acqua arricchita 2302
- attivata 61
- di fogna 7668
- impoverita 1730

acqua leggera 4011
- libera 2831
- pesante 3206
- radioattiva 98
- tritiata 7395
acquisto d'energia 2266
actinon 54
adamite 112
additività 114
addizione di materiali
  protettivi 1060
- di tracciante 6779
adione 121
adsorbente 125
adsorbimento 126
aerosole 132
affinità differenziale 6437
affluente 2051
afnio 3118
agente di precipitazione per
  addizione di sale 6284
- di ritenzione 3303
- estrattore 2484
- sequestrante 1264
aggiuntatura 807
aggiustaggio esatto del campo
  6546
aggregato di difetti 1147
aggregazione 145
agitazione 146
- termica 7119
ago di combustibile 2870
- di radio 5897
albedo per neutroni 176
aldanite 177
alette longitudinali, ad 4126
- trasversali, ad 1105
alfa 186
alfatopico 209
alfatrone 210
alimentazione 2546
- isolata 2703
allanite 182
allargamento di collisione 1226
- Doppler 1933
- d'urto 1226
- quadripolare 5623
allineamento magnetico nucleare
  4867
- per acqua radioattiva 99
allobari 183
allumina 216
alluminio 217
alogeno 3132
alogenuro d'uranio 7514
alone pleocroico 5346
alterazione della disposizione
  del combustibile 5997
- radiale della disposizione del
  combustibile 486, 5679
alternazione di molteplicità 213
altezza d'esplosione 3214
- di barriera 544
- di moderatore variabile 7592
- di pressione di liquido 3153
- effettiva di fumaiuolo 2044
alto vuoto 3289
ambiente 2314
americio 221
ammassamento 145

ammuchiamento d'impulsi 5289
amniografia 223
ampangabeite 225
ampiezza di deviazione 6333
amplificatore 227
- ad impulso di corrente 1581
- a finestra 994
- a modulatore 1082
- del rendimento neutronico 728
- del segnale d'arresto 6567
- di carica 1016
- di coincidenza 1184
- di contatore 1453
- di flusso 1929
- d'impulsi 5586
- elettrometrico 2095
- lineare a soglia 634
- lineare di commutazione
  automatica di sensibilità 457
- lineare d'impulsi 4047
- lineare per corrente continua
  4036
- logaritmico d'impulsi 4116
- logaritmico per corrente
  continua 4111
- nell'apparecchio d'arresto
  6567
- sigma 6583
- statico di sicurezza 6268
amplificazione 5454
- dovuta al gas del tubo
  contatore 1460
- mediante ionizzazione d'un
  gas 2958
ampolla 232
- a rottura di punta 776
- in vetro sigillata 6392
- in vetro vuotata 2381
- vuotata sigillata 2382
anaerobico 235
anafase 240
anaforesi 241
analisi cristallografica 1551
- cristallografica per
  diffrazione neutronica 4688
- del cristalli per diffrazione
  1797
- di Feather 2542
- d'un fascio ionico 3618
- fluorimetrica 2737
- inversa d'isotopi per
  diluizione 6188
- isotopica 3776
- per assorbimento 12
- per attivazione 63
- per diluizione 1836
- per diluizione isotopica 3785
- per radiocristallografia
  diretta 1858
- per raggi X 7740
- radiochimica 5827
- radiometrica 5869
- statistica 6863
- termica 7121
analizzatore d'ampiezza 228
- d'ampiezza a canale mobile
  1789
- d'ampiezza multicanale 4566
- d'ampiezza multicanale a
  memoria 4565

camera d'ionizzazione a propàno 5539
- d'ionizzazione a protoni di rinculo 6015
- d'ionizzazione a riempimento d'argon 324
- d'ionizzazione a sorgente interna di gas 3666
- d'ionizzazione campione 6846
- d'ionizzazione compensata 1252
- d'ionizzazione con parete equivalente all'aria 158
- d'ionizzazione di Compton 1276
- d'ionizzazione di controllo 4535
- d'ionizzazione differenziale 1779
- d'ionizzazione di Sievert 6582
- d'ionizzazione 2 π 7448
- d'ionizzazione equivalente al tessuto 7268
- d'ionizzazione omogenea 3312
- d'ionizzazione portatile 5402
- d'ionizzazione proporzionale 5544
- d'ionizzazione 4 π 2798
- d'ionizzazione senza parete 7652
- d'ionizzazione tascabile 5366
- di pressione 5345
- di scarica 1866
- di trasferimento di combustibile 2880
- di Wilson 1138
cammino 7325
- libero medio 3547, 4326
- libero medio totale 7296
campanella di gorgogliamento 799
campione di perturbazione 5220
- radioattivo 5808
campo atomico 384
- chiarificatore 1125
- critico d'un tubo contatore 1518
- d'attività 109
- del contatore 1467
- del tempo di divergenza 5200
- di confinamento 1326
- di Coulomb 1446
- di forza effettivo 2038
- di frequenza 2836
- di funzionamento 3515
- d'insensibilità 1624
- di potenza 5452
- di radiazione 5709
- d'irradiazione 7743
- di sorgente 4150
- di tempo d'esposizione 2467
- elastico 2060
- elettrico 2069
- elettrico trasversale di focalizzazione 7371
- elettromagnetico 2081
- elettromagnetico auto-consistente 6441
- elettrone-neutrino 2142
- elettrostatico 2194
- esteriore 2476

campo gravitazionale 3069
- leptone 2142
- magnetico 4204
- magnetico di guida 3109
- magnetico gelato 2844
- magnetico stabilizzatore 6822
- mesonico 4369
- nucleare 4846
- quadripolare 5622
- senza forza 2778
- trasversale 7369
- uniforme 7471
- variabile 7262
canale 859, 985
- di combustibile 2853
- d'irradiazione 561, 2443
- di trasferimento 7327
- gastrointestinale 180
- passante 561
cancro 861
cannone a plasma 5319
- elettronico 2133
capacità di scambio 2402
cappio 2444
- attivo 87
- interiore 3422
- rivelatore d'intasature 5350
capsula 871
- a radiazione gamma 2940
- a radio 5833
- esteriore 5042
- interiore 3498
carattere dominante 1926
- recessivo 6001
caratteristica della velocità di conteggio 1492
- di Richardson 6197
- di ripiano 5339
- di Townsend 7307
- spettrale 6757
carbonato uranilico di calcio 852
carbone attivo 59
carbonio 879
carbonio-14 73, 884
carbonio radioattivo 73, 884
carburano 890
carburo di boro 738
- di silicio 6592
- di tungsteno 7429
carcinoma 861
carcinomatosi 891
carcinosi 891
carica 1000, 4096
- atomica 371
- atomica efficace 2022
- dell'elettrone 1015
- di materiale 555
- di materiale libera 2826
- di stadio 6836
- d'una particella 5139
- ionica 3651
- media 472
caricamento 1027
carica multipletto 1012
- negativa 4639
- nucleare 4825
- positiva 5408
- propria 6440
caricare 1001

carica spaziale 6719
- spaziale volumica 6722
- specifica dell'elettrone 2108
- specifica di una particella 1010
caricatore di dosimetro 1948
caricatore-indicatore 1026
caricatore-lettore 1026
carico corporale 685
- nominale 2886
cariocerite 3833
cariocinesi 3834
carioplasma 900
carnotite 894
carta d'isoradi 3748
- radiometrica 5871
cartuccio ad assorbimento 899
casamatta 2872
cascata 901
- a sbalzo 3974
- costante 6813
- di diffusione 1810
- di riestrazione 514
- di stadi 901
- di stadi di separazione 906
- gamma 902
- ideale 3376
- nucleare 4823
- quadrata 6813
- rastremata 7070
- semplice 6597
- squadrata 6815
cassa a trafori 1506
- di combustibile 2851
- di condensatori 865
- di diversione 1920
- di filtro 2610
- di passaggio 7338
cassetta 912
castello 913
- di piombo 3949
cataforesi 920
catalizzatore 914
- ricombinatore 6020
catena 977
- analizzatrice multipara-metrica 4570
- di decadimenti 1639
- di fissione 2649
- di reazione protone-protone 5572
catione 926
catodo 922
- fotoelettrico 5244
- freddo 1197
- incandescente 3329
cattura 873
- a fissione 2648
- al caldo 3343
- al freddo 1210
- d'elettrone
- di buche 3308
- di neutroni 4678
- di neutroni lenti 6644
- di neutroni radiativa 4741
- di risonanza 6157
- di risonanza di neutroni 6158
- dissociativa 1902
- di un elettrone orbitale 5003
- entropica 2311

contrazione cromosomatica 1092
- di lantanidi 3900
- dinamica 2004
- stabilizzata 6819
- teta 6389
- ultraveloce 7456
controcorrente 1479
controllo 7029
- a distanza di contaminazione di zona 6115
- aereo dell'aria 131
- a segnali campionati 6289
- della concentrazione 1285
- dell'acqua 7676
- di contaminazione 1338
- di drogaggio di semiconduttori 1931
- di protezione 5555
- di radiazione ionizzante 4536
- di radiazione ionizzante d'area 322
- di radiazione ionizzante in aria 162
- di radiazioni 5555
- di radioattività 4536
- di radioattività d'area 322
- di radioattività dell'escrezione 2425
- di radioattività in aria 162
- di rapporto 5945
- individuale 5218
- locale 6622
conversione 1380
- d'una coppia 5090
- interna 3573
- magnetodinamica 4229
convertitore ampiezza-tempo 5591
- analogo-numerico 238
- di neutroni 1389
- di tempo 7249
- numerico di tempo 1833
- tempo-ampiezza 7244
convezione 1374
coordinata 1407
copertura 659
coppia antagonista 6179
- d'elettroni 2145
- d'ioni 3642
- d'ioni primaria 5499
- elettrone-positone 2149
- termoelettrica a guaina metallica 4377
coprecipitante 6345
coprecipitazione 1411, 6346
coracite 1412
cordilite 1413
corpi cilindrici di vapore per compensazione 6883
corpo grigio 3073
- nero 657
corpuscolo 1420
correlazione angolare 254
- angolare beta-gamma 595
- angolare beta-neutrino 601
- angolare beta-nucleo di rinculo 620
- direzionale di raggi gamma successivi 1859
corrente 2711

corrente anionica 266
- cationica 927
- d'emissione in campo nullo 2587
- di conduzione 1299
- di convezione 1375
- di lancio 2587
- d'ionizzazione 2966
- di saturazione 6301
- d'oscurità 1615
- elettronica 2115, 2126
- incompressibile 3434
- nominale 5934
- residua 6141
- turbolenta 7434
correzione della coincidenza accidentale 41
- della geometria 3022
- del momento magnetico anomalo 279
- del tempo di risolvenza 6152
- di coincidenza 1187
- per il tempo morto 1626
- radiativa 5753
corrosione 1426
- di frizione 2837
- per lo stato latente di sforzi 6947
- sotto irradiazione 3711
- sotto sforzi 6947
corrugamento 5930, 7027
corteccia di valenza 7573
corvusite 1430
cosmotrone 1438
costante d'accoppiamento 1496
- d'attività reale d'una sorgente 7415
- d'effetto di schermo 6385
- del reticolo 3921
- di Boltzmann 701
- di decadimento 1640
- di disintegrazione parziale 5128
- di Fermi 2557
- di Hall 3125
- di massa atomica 391
- di Planck 5307
- di rinnovo 7440
- di Rydberg 6254
- di spettro di bande 528
- di struttura fina 2624
- di tempo 7246
- di tempo del flusso neutronico 4734
- di tempo del limitatore 1129
- numerica 4946
- specifica di radiazione gamma 6745
- unificata di massa atomica 391
costanti di forze di legame nucleare 2777
- nucleari 4828
costituzione chimica 1043
- dell'atomo 417
costrizione 1318
covalenza 1498
creazione di difetto 1890
crescita di cristalli 1554
- per effetto Wigner 7715

cresta di risonanza 3030, 6170
cricca da fragilità 787
cricche sullo spigolo 2019
cripton 3866
criptoscopio 3868
crisoberillio 1100
cristalli metamitti 4386
- misti 4468
- scintillatori 6363
cristallito 1563
cristallo d'ioduro di sodio tipo pozzo 7699
- di valenza 7570
cristallografia a raggi X 7744
cristallogramma a polvere di cristallo 5436
cristallo imperfetto 3404
- ionico 3653
- mosaico perfetto 3382
- perfetto 3377
criteri di lesione 1610
- d'ubicazione 6624
criterio di degenerescenza 1508
criticità 1531
- con refrigerante 7701
- istantanea 5530
- ritardata 1696
- senza refrigerante 1977
critico 1509
- istantaneo 5529
- ritardato 1695
crivello oscillante 3819
cromatide 1084
cromatina 900
cromatografia 1087
- a gas 2961
cromo 1088
cromosoma 1090
- sessuale 6507
cromosomi omologhi 3318
cronometro ad impulsi 3416
- atomico 372
- sincronico 7044
cronoregolatore 7263
cronotrone 1099
cuneo a gradinata 6895
cuore del nucleo 4933
cupola 3323
- d'acqua 1925
- del reattore 5978
- di concentrazione 1283
- di focalizzazione 1283
cuproautunite 1571
cuprosklodowskite 1572
cuprozippeite 1573
curie 1574
- equivalente 2351
curieterapia 1575
curio 1577
curite 1576
curva caratteristica del contatore 1454
- d'accrescimento 3102
- d'altezza 214
- d'assorbimento 17
- d'attenuazione 17
- d'attivazione 65
- d'attività 104
- d'emissione di fotoni 5268

diagramma di Feynman 2580
- di Kurie 2566
- di Sargent 6296
- d'isotopi 3754
- di termini 7101
- energetico 2276
dialmaite 1922
diamagnetismo delle particelle
del plasma 1772
diametro atomico 377
- efficace d'una barra di
regolazione 2037
- nucleare 4834
dibutossidimetiletere 1776
dicloruro di glicolo di trietilene
7385
diderichite 1777
difenile 651
difesa contro la guerra
radioattiva 5854
difetto 1665
- del reticolo 3926
- di cristallo 1555
- di Frenkel 2834
- di massa 4261
- di massa relativo 6091
- d'ionizzazione 3670
- di Schottky 6350
- ereditario 3234
differenza di contrazione 1794
- media di temperature 4338
- media logaritmica delle
temperature 4088
diffrattometro 1799
- a raggi X 7747
- neutronico 4696
diffrazione 1796
- con elettroni 2119
- con neutroni 4695
- di raggi X 1800
- d'onda 7679
- elettronica 2119
diffusato 1802
diffusione 1806
- all'indietro 513
- ambipolare 220
- del plasma attraverso un
campo magnetico 1823
- gassosa 2985
- in mezzo poroso 2985
- interatomica 378
- inversa degli elettroni 492
- molecolare 4503
- mutua 4602
- neutronica 4697
- termica 6706
- turbolenta 2017
diffusore 1805
diluente 1834
diluizione 1835
- isotopica 3784
dimensioni critiche 1514
- del reticolo 3923
diminuzione della pressione
5927
- di contrasto 1840
- di Forbush 2776
di-neutrone 1841
diodo a barriera superficiale
7009

dipendenza angolare della
deviazione 255
- dell'energia 2265
dipolo 1846
- magnetico 4201
di-protone 1849
diradazione 6979
direzione della valenza 716
disco a tungsteno 7430
- calcolatore da manipolare
isotopi 3756
- di zolfo 6991
- in cobalto 1163
discontinuità di potenziale 5428
discriminatore 1869
- d'ampiezza 231
- di forma 5599
- integrale 3524
diseccitazione 1664
disegno del reattore 5977
disintegrazione 1638
-: alfa 191
- a ramificazione multipla 4571
- beta 587
- beta doppia 1950
- delle particelle elementari
1878
- di neutrone 4692
- esponenziale 2454
- in catena 978
- nucleare 4835
- nucleare artificiale 328
- nucleare esotermica 2433
- nucleare indotta 3453
- positonica 588
- spontanea 6804
dislocazione 1885
- a scalino 2020
- di un atomo 1867
- in vite 6387
dispersione 1892
- all'indietro 513
- statistica 6867, 6927
- statistica di portata 5925
- strumentale 3516
dispositivo a contrazione
tubolare 3142
- automatico di sicurezza
contro il sovraccarico 5064
- di blocco 3558
- di congelazione 2832
- di contrazione lineare 4044
- di contrazione toroidale 7282
- d'irradiazione convergente
1378
- di sicurezza 2227
- di solidificazione 2832
- d'ossidazione 5073
- omopolare 3321
- per radiografia a sezione 688
disposizione atomica 365
- dei rifiuti radioattivi liquidi
78
disprosio 2007
disseccamento 1982
dissipazione d'energia 1901
dissociazione di Lorentz 4130
dissoluzione 1904, 6696
dissorbimento 1744
- di campo 2583

distacco 5278
- d'un elettrone 2118
distanza degli atomi nella
molecola 379
- di pericolo 1612
- di sicurezza 6262
- di visibilità 7625
- focale 2759
- fuoco-pelle 2763
- fuoco-pellicola 2762
- obliqua 6633
- reale 6633
- reticolare cristallina 3554
distensione 6024
distillato 1908
- di testa 5063
distillazione 1909
- a controcorrente 1477
- azeotropica 489
- di Rayleigh 5950
- discontinua 556
- estrattiva 2491
- frazionata 2807
- in corrente di vapore
d'acqua 6881
- in equilibrio 2339
- molecolare 4504
- semplice 6598
distorsione del reticolo 3924
- di quantizzazione 5632
- di traccia 7321
- per carica spaziale 6723
distributore di refrigerante
2863
distribuzione angolare 257
- binomiale di Bernoulli 574
- dei livelli d'energia 3994
- della dose 1915
- della velocità degli elettroni
2167
- dell'energia 2269
- dell'energia dei neutroni
nel reattore 2681
- dell'energia neutronica 4702
- di Fermi 2561
- di flusso di grande ripartizione 3086
- di flusso neutronico 2749
- di Maxwell 4317
- di velocità 7601
- elettronica 2120
- fina 2620
- spettrale 6753
diuranato d'ammonio 222
divergente 1918
divergenza 1917
- di raggi 5949
divisione cariocinetica 3834
- cromosomatica 1094
- del cromatide 1085
- mitotica 3834
dollaro 1924
dominio di proporzionalità 5545
donatore 1927
- d'elettroni 2121
doppietto 1844, 1959
dosaggio 1936
dose 1937
- aggiunta 808
- alla pelle 6626

elemento d'avviamento 729
- demoltiplicatore 6316
- di controllo finale 2614
- di costruzione autonomo 5084
- difettoso 834
- d'impurezza 3418
- d'inizio ed arresto 6912
- di regolazione 1357
- di regolazione approssimata 1154
- di regolazione di potenza 5440
- di regolazione di precisione 2618
- di ritardo 1690
- discriminatore d'ampiezza 5588
- discriminatore d'ampiezza a corrente continua 1852
- discriminatore di carica 1003
- discriminatore di forma 5600
- di trasporto 7331
- elettronegativo 2169
- fertile 2577
- formatore 6515
- funzionale 549, 550
- leggero 4004
- naturale 4626
- numeratore 6021
- numeratore elettromeccanico 2093
- padre 5118
- pesante 3190
- porta 2994
- portatore 895
- progenitore 5118
- radioattivo 81
- radioattivo artificiale 329
- radioattivo naturale 4630
- ritardo lineare 4035
- selettore d'ampiezza a canale mobile 6605
- selettore d'ampiezze 5590
- selettore di coincidenze 1194
- selettore di coincidenze ritardate 1694
- selettore per anticoincidenze 293
- sensibile 6467
- separatore 6493
elettrizzazione di un gas 2074
elettrochimografo 2078
elettrodinamica quantica 5640
elettrodisintegrazione 2076
elettrodo 2075
- ausiliario 3106
- collettore 1214
- in D 1660
- ionizzante dell'aria 160
elettroforesi 2189
elettromeccanismo di comando lineare 4034
- rotativo di comando 6236
elettrometro 2094
- a fibra di quarzo 4022
- a lamina vibrante 2002
- di Lindemann 4022
elettromigrazione a contro-corrente 1478
elettrone 2099
- Auger 439

elettrone Compton 1275
- comune 5094
- d 1608
- dell'atomo 380
- di conduzione 1300
- di conversione 1383
- di conversione interna 3575
- di disintegrazione beta 589
- di rimbalzo 1275
- di rinculo 6008
- di ripulsione 1275
- di valenza 721, 1300
- di vuoto 7555
- estranucleare 2493
elettronegatività 2170
elettrone iniziale 3496, 5495
- interno 3501
- K 3826
- L 3873
- legato 749
- libero 2824
- M 4175
- metastatico 4395
- N 4606
- nucleare 4838
- O 4948
- orbitale 5002
- ottico 1300
- P 5077
- periferico 1300
- planetario 5002
- primario 5495
- Q 5616
- rotante 6794
- S 6255
- secondario 6405
- solitario 4119
- veloce 3281
elettronevolt 2188
elettronica 2186
elettroni comuni 6516
- cosmici d'urto 1434
- cromofori 1089
- di disintegrazione della radiazione cosmica 1433
- di fuga 6246
- equivalenti 2355
- retrodeviati 509
elettronogeno 2187
elettroscopio 2190
- a fibra di quarzo 3936
- di Lauritsen 3936
- per rivelazione di raggi beta 610
eliasite 2213
eliminatore d'elettrostatica 6857
eliminazione dei rifiuti 7665
- della crosta superficiale 6317
elio 3219
ellsworthite 2214
elutriazione 2215
elvite 3229
emanazione 2217
ematimetro 1484
emettitore 2234
- alfa 193
- beta 593
- beta-gamma 597
- di neutroni ritardati 1697

emettitore gamma 2924
emissione alfa 187
- corpuscolare 1421
- corpuscolare associata 339
- di campo 446
- di dose ambientale 4104
- di particelle 1421
- di raggi beta 611
- elettronica 2124
- fredda 446
- fredda d'un contatore 1458
- garantita 971
- ionica di campo 2590
- quantica 5641
- secondaria 6406
- solare di particelle 6683
- termionica 7175
emoglobina 3232
emulsione a liquefazione 4058
- autoportante 5169
- nucleare 4839
- staccabile 6959
enalite 2239
encefalografia 2243
encefalogramma 2242
endostio 2249
endotelio 2250
endotermico 2247
energia 2254
- allo zero assoluto 7795
- beta massima 600
- chimica 1046
- cinetica 3850
- cinetica di rotazione 260
- cinetica relativista 6097
- comunicata alla materia 2274
- d'asimmetria 346
- d'assorbimento per risonanza 2282
- d'attivazione 67
- d'attivazione per diffusione 1818
- d'eccitazione 2411
- del fotone 5270
- del limite di banda 524
- d'equilibrio 2340
- di deformazione 1673
- di disintegrazione 1642
- di disintegrazione alfa 192
- di disintegrazione beta 590
- di disintegrazione beta nello stato fondamentale 3091
- di disintegrazione nello stato fondamentale 3092
- di disintegrazione nucleare 4836
- di dislocazione 2281
- di Fermi 2562
- di fissione 2654
- di fusione 2900
- di legame 638
- di legame di una particella alfa 198
- di legame di un protone 5567
- di legame media per nucleone 471
- di legame neutronico 4674
- di legame nucleare 4818
- di legame specifica 6740
- di mobilità 1970

monitore della densità di
flusso di particelle 5148
- della densità di flusso
gamma 2929
- della densità di flusso
neutronico 4711
- di cenere di carbone 1153
- di contaminazione alfa per le
mani 195
- di contaminazione beta per
le mani 599
- di contaminazione da
lavanderia 3935
- di contaminazione di campioni
d'aria 4608
- di contaminazione di
superficie 7015
- di contaminazione per
pavimenti 2709
- di criticità 1532
- di perdite nello scambiatore
di calore ad acqua pesante/
acqua leggera 3200
- di polvere e gas radioattivo
592
- di prodotti alimentari 2772
- di prodotti radioattivi
imballati 5082
- di rottura di guaine 2511
- di rottura di guaine ad effetto
Cerenkov 964
- di rottura di guaine a neutroni
ritardati 1698
- di rottura di guaine a
raccolta elettrostatica 2193
- di rottura di guaine a
separazione dei prodotti di
fissione 2666
- gamma di criticità 1533
- isotopico di prodotti imballati
5080
- neutronico di criticità 1534
- per acqua 7675
- per aerosoli di plutonio 5358
- tascabile con pile 5365
monocito 3349
monocristallo 6608
monopolo 4544
montacarichi 1006
mosaico 4547
motore ad aria calda 3163
- ionico 3630
moto rettilineare della
particella 6033
movimento ascensionale 2757
- browniano 798
- d'aria troposferica 7411
- del fondo marino costiero
1155
- del fondo marino in alto
mare 4965
- dell'alluvione 6593
- di traslazione 7350
- orbitale 5005
- termico 7149
- trasversale 1537
mucosa 4564
multipletto 4578
muone 4563
muonio 4595

mutamento letale 3985
mutante 4598
mutazione 3131
- cromosomatica 1091
- di geni 3005
- indotta per raggi X 7754
- letale 3985

naegite 4612
nagatelite 4613
necrosi 4637
negatone 2099
negatoscopio 2602
neodimio 4648
neoformazione 4649
- benigna 570
- maligna 4239
neon 4650
neoplasma 4649
- benigno 570
- maligno 4239
nero 656
nettunio 4651
neutralizzazione di carica 1014
neutretto 4659
neutrino 4660
neutro 4654
neutrone 4661
- deviato 6326
- di bassa energia 4142
- disperso 6940
- libero 2828
- naturale 4629
- secondario 6414
neutroni d'energia intermedia
3564
- di fissione 2662
- di risonanza 6169
- epicadmici 2319
- epitermici 2332
- freddi 1205
- istantanei 5536
- latenti 3911
- lenti 6647
- pulsati 5604
- ritardati 1701
- subcadmici 6973
- termici 7152
- utili 7546
- veloci 2533
- vergini 7614
neutronografia 4759
neutronoterapia 4752
nichel 4763
nichelio 4763
nicolaite 4764
nido a polvere 2000
niobio 4767
niobite 1240
nitrato di torio 7218
- d'uranile 7540
nivenite 4769
nobelio 4771
nocciolo a combustibili
arricchiti localmente 6426
- del reattore 1414
- di bomba 706
nolite 4773
nonlinearità 4790
non radioattivo 1195

norme d'avviamento 6853
novacechite 4806
nube da bomba atomica 373
nuclei di Wigner 7716
- equivalenti 2356
- impari-impari 4961
- impari-pari 4957
- magici 4194
- metastabili 4393
- speculari 4461
nucleo 396, 4934
- atomico 396
nucleo-bersaglio 7076
nucleo caricato 1021
- composto 1268
- di diffusione di Yukawa 1821
- di rallentamento 6654
- di rinculo 6009
- di spostamento 1897
- eccitato 2421
- fissile 2645
nucleogenesi 4927
nucleo integrale 3841
- integrale di trasporto 7363
- leggero 4009
nucleone 4928
- d'evaporazione 2388
nucleo pari-dispari 2392
- pari-pari 2390
- pesante 3202
- poroso di una particella di
combustibile 5399
- positivo 5412
nucleoproteina 4932
- radioattivo 5798
- residuo 6145
nuclide 4939
nuclidi betatopici 627
- di Wigner 7717
- speculari 4462
nuclido radioattivo 5799
- schermato 6533
- stabile 6825
numeratore di pellicola 2603
- elettromeccanico 2092
numerazione di globuli
sanguigni 678
- eritrocitaria 6037
- leucocitaria 7705
- trombocitaria 5343
numeri magici 4195
numero atomico 397
- atomico effettivo 2023
- d'Avogadro 481
- d'eccitazione 2413
- d'effetto di schermo 6385
- dei neutroni per fissione 4942
- dei neutroni prodotti per
neutrone assorbito 4943
- di barioni 547
- di collisioni 1232
- di globuli sanguigni 677
- di leptoni 3982
- di Mach 4181
- di massa 4271
- di neutroni 4730
- di nucleoni 4271
- di piatti teorici 4944
- di protoni 397
- di rallentamento 6916

rapporto d'indeterminazione di Heisenberg 3216
- di prove 7109
- di ramificazione 769
- di resistenza 2252
- di riflusso 6064
- di rigenerazione esterna 2471
- di rigenerazione interna 3571
- di specchio 4463
- dose-effetto 1939
- giromagnetico 3114
- giromagnetico nucleare 4855
- iniziale di conversione 3486
- integrale di conversione 1387
- integrale di rigenerazione 782
- isotopico di carbonio 887
- K / L 3827
- L / M 4087
- massa-energia 2054
- minimo di riflusso 4455
- M / N 3643
- portata-energia 5923
- relativo di conversione 6088
rappresentazione di Heisenberg 3217
- simbolica 7041
rateometro a camera d'ionizzazione equivalente al tessuto 7269
- d'esposizione 2466
- d'esposizione a camera d'ionizzazione 3665
- d'esposizione a scintillatore 6372
- d'esposizione a tubo contatore Geiger-Müller 3000
- d'esposizione beta-gamma a camera d'ionizzazione BBHF 598
- d'esposizione gamma a risposta lineare 491
- d'esposione logaritmico tascabile a scintillatore 5404
- di conteggio 1493
- di conteggio lineare a differenza 1780
- di dose 1946
- di dose integratore 3532
- discriminatore 5941
- lineare di conteggio 4059
- lineare portatile 5403
- logaritmico di conteggio 4117
- logaritmico di fisica sanitaria 3158
- misuratore di cariche 4405
rauvite 5947
razzo a propulsione nucleare 4891
reattività 5958
- all'arresto 6570
- a sistema freddo 1207
- a sistema vergine 1121
- intrinseca 811
- istantanea 5538
- negativa 4645
- ritardata 1702
reattore 4902
- a bassa temperatura 4152
- a basso flusso 4145
- a berillio 580

reattore a ciclo diretto 1854
- a ciclo indiretto 3445
- a circolazione del combustibile 1103
- a circolazione naturale 4625
- a cisterna 7068
- a combustibile e moderatore ceramico 959
- a combustibile fluidizzato 2727
- a combustibile metallico con raffreddamento a gas 4380
- a combustibili arricchiti localmente 6427
- a conversione senza importanza 829
- ad acqua bollente 700
- ad acqua bollente a raffreddamento diretto e scambiatore di calore incorporato 1853
- ad acqua bollente a raffreddamento indiretto e scambiatore di calore incorporato 3444
- ad acqua in pressione 5488
- ad acqua in pressione e veleno di compensazione 828
- ad acqua leggera 4012
- ad acqua pesante in ebollizione 3207
- ad acqua pesante in pressione 5486
- ad alta potenza 3261
- ad alta temperatura 3288
- ad alto flusso 3263
- ad elio ad alta temperatura 3286
- a doppio ciclo 1984
- a due regioni 7450
- a duplice scopo 1985
- ad uranio 7521
- ad uranio arricchito 2300
- ad uranio naturale 4632
- ad uranio-torio 7525
- a fasci 563
- a flusso concentrato 2756
- a fusione 2904
- a fusione controllata 1372
- a gas ad alta temperatura 3285
- a gas di tipo perfezionato 129
- a gas-grafite 2973
- a letto di sfere 5166
- a letto fluidizzato 2725
- al plutonio 5362
- a moderatore organico 5022
- a moderatore plastico 5329
- a moderazione 4483
- a molteplici scopi 4588
- a neutroni d'energia intermedia 3566
- a pasta combustibile 2726
- a piastra infinita 3474
- a piscina 5395
- a piscina con nocciolo visibile 316
- a più regioni 4589
- a raffreddamento ad aria e moderato per grafite 152

reattore a raffreddamento a gas con combustibile ceramico 958
- a raffreddamento per polvere 1998
- a recipiente in pressione 5483
- a refrigerante diviso 6'. 99
- a refrigerante in pressione 5487
- a refrigerante liquido metallico 4068
- a refrigerante organico 5021
- a reticolo 3929
- a riflettore completo 2891
- a riflettore interno 3500
- a riflettore laterale 6577
- a riflettore superiore ed inferiore 7279
- arricchito 2300
- a scambiatore di calore integrato 3527
- a sali fusi 2897
- a sali fusi 4525
- a schermatura differenti 813
- a sospensione 7034
- a spostamento spettrale 6759
- a torio 7220
- a tubi in pressione 5481
- a vapore surriscaldato 6879
- a zona termica ed a zona veloce 4469
- biomedicale 649
- caldo 3338
- caldo e virgine 3334
- chimiconucleare 1058
- cilindrico 1602
- cilindrico finito 2626
- combinato lento e veloce 6643
- compatto trasportabile 5083
- con combustibile in metallico liquido 4069
- con raffreddamento ad acido carbonico 882
- con riflettore 6054
- convertitore 1390
- critico 1526
- cubico 1566
- d'addestramento 7324
- di dimostrazione 1708
- di potenza 5455
- di potenza zero 7793
- di produzione 5521
- di produzione d'elettricità 2073
- di produzione del materiale fissile 2644
- di produzione del plutonio 5361
- di produzione di calore 3174
- di produzione d'isotopi 3762
- di propulsione 4478
- di prova 7108
- di radiochimica 1058
- di ricerca ad alto rendimento 3271
- d'irradiazione 3718
- d'irradiazione di prodotti alimentari 2628
- di trattamento di materiali 4299

rendimento di stadio 6835
- di un rivelatore 1750
- d'omogenizzazione 4473
- energetico 2293, 2294
- energetico di conversione
  d'uno scintillatore 2261
- globale di catena 982
- in profondità 5189
- ionico 3648
- ottimo 7756
- per coppia d'ioni 3643
- per stadio 2810
- quantico 5639
- quantico di conversione 1386
- quantico spettrale 6756
- radiochimico 5691
- termico energetico 7137
- totale 5051
renio 6191
reotrone 6193
resa di fissione 2679
- di fissione cumulativa 1569
- diretta di fissione 1855
- energetica specifica d'un
  combustibili nucleare 827
- neutronica per assorbimento
  2371
- neutronica per fissione 4758
residui di distillazione 748
resistenza alla corrosione 1428
- alla radiazione 5733
- alla trazione 7097
- alle liscive 3944
- all'urto 3400
- d'estinzione 5662
- di spegnimento 5662
- d'oscurità 1617
reticolazione 1539
reticolo 3915
- a barra combustibile semplice
  6613
- ad elementi combustibili
  moderato alla grafite 3063
- a fascio 1150
- a strati atomici 3938
- attivo 84
- cristallino 771
- cubico 1565
- cubico a corpo centrato 686
- del reattore 3916
- di barre 6208
- di diffrazione 1798
- di punti 5372
- esagonale 3247
- infinito 3470
- lasco 4128
- moderatore 4491
- reciproco 6002
- sovrapposto 7003
- triangolare 7380
- uranio-grafite 7513
retrodeviazione 506
rettificatore 6031
rettificazione 6030
r/h a 1m 6194
riarricchimento 6043
ricaduta 82, 2513
- locale 4105
- provocata dalle precipitazioni
  5913

ricaduta radioattiva 82
- radioattiva di vento
  anabatico 234
- radioattiva di vento catabatico
  1966
- radioattiva di vento
  discendente 1966
- radioattiva secca 5777
- stratosferica 6939
- troposferica 7412
ricarica 6066
- fuori esercizio 4964
- in esercizio 2887
ricchezza isotopica 26
ricerche nucleari 4905
ricettore di deposito radioattivo
  5463
richetite 6198
ricognizione per uranio 7522
ricombinatore catalitico 918
ricombinazione 6017
- alla parete 7653
- colonnare 1244
- di volume 7645
- elettrone-ione 2136
- elettrone-ione alla parete
  2137
- iniziale 3494
- ione-ione 3637
- per radiazione 5756
- preferenziale 5467
- sotto alta pressione 3276
ricopertura 863
ricostituzione dello spessore
- dello strato d'emulsione 1740
ricottura 267
- dopo l'irradiazione 5421
ricristallizzazione 6028
ricupero dei rifiuti 7666
riduzione automatica di potenza
  455
- di rischi 4464
- di traccia 7323
- di volume 6042
- graduale della potenza 3269
riestrazione 515
- differenziale 1786
riferimento 6047
rifiuti a bassa radioattività 4149
- a radioattività moderata 3562
- chimici 1068
- fortemente radioattivi 3251
- liquidi 4080
- liquidi a radioattività
  permanente 3270
- radioattivi 95
- radioattivi 7664
riflessione dei neutroni 4743
- di Bragg 761
- integrata 3530
- integrata di raggi X 3531
riflettanza della barriera di
  potenziale 6052
riflettore 4744
- di bomba 7064
- di fondo 747
- fisso 2684
- in grafite 3066
- removibile 4083
riflusso 6063

rifornimento di combustibile
  2868
rifrazione atomica 409
riga K 3828
- M 4176
- madre 5119
- N 4609
- spettrale 6755
- spettrale di risonanza 6175
rigenerazione 780, 6068
- chimica 1064
righe anti-Stokes 307
rigidezza 6905
rigidità dielettrica 2071
- magnetica 4224
rigonfiamento 5291, 7039
rilaseio d'energia 2286
- Wigner 7718
rilassamento 6104
- spin-reticolo 6786
rilevamento di terreni 3093
- geologico di terreni 3014
rilievo 670
rimbalzo 3395
- in gruppo 144
rimessa in ciclo 6035
rimozione 6123
- della pressione 1737
rimpiazzo 4237
rinculo 6006
rinnovo 7437
rinormalizzazione di massa
  6126
riometro 6203
ripiano 5338
- di contatore 1466
- di potenziale 5429
ripresa 1204
ripulsione d'elettroni 6129
- intranucleare 4904
- mutua 4604
riradiazione 6136
riscaldamento adiabatico 118
- da irradiazione 3716
- omico 3822
- per effetto Joule 3822
- per la risonanza del
  ciclotrone 1600
- per onda d'urto 6551
- per radiazione gamma 2931
- per risonanza 6165
- per risonanza alla frequenza
  di un ciclotrone ad ioni 3625
- turbolento 7435
rischi di contaminazione 1336
rischio d'irradiazione 3155
- nucleare 4856
riserva di reattività per xenon
  7770
risoluzione in energia 2287
risolvibilità 6693
risonanza 6153
- dello spin d'un elettrone 2159
- di Fermi 2565
- giromagnetica 1599
- magnetica 4221
- magnetica dello spin 6788
- magnetica del protone 5569
- magnetica nucleare 4869
- ottica 4993
- paramagnetica 5110

tempo di transito di un ione 3647
- di vita dell'energia del
  plasma 2280
- di volo 7254
- di volo del neutrone 4754
- d'urto 1237
- effettivo di dimezzamento
  2032
- libero medio 4327
- medio di diffusione 474
- medio di una generazione
  neutronica 3009, 4724
- morto 1625
- morto d'un contatore 1456
- morto di un tubo contatore
  Geiger-Müller 1627
tenaglie 7278
tenorimetro 1339
- a fluorescenza 1340
- a fluorescenza X 7748
- d'acqua pesante 3208
- di minerali 5015
- in cenere di carbone 1152
- in ferro e calcio di minerali
  a radiazione ionizzante 3689
tensione d'accelerazione 711
- d'accensione 6856
- d'agitazione termica 7120
- d'alimentazione 7006
- dell'elettrodo collettore 1215
- di funzionamento del contatore
  1463
- di saturazione 6303
- di soglia 7235
- disruttiva 1900
- inversa 3605
- nominale 5938, 7638
- non disruttiva nominale
  d'impulso 5936
- quantica 5651
- superficiale 7024
- termica 7169
tenuta, a 3972
- a labirinto 3884
- a labirinto a gradinata 6896
- di neutroni, a 4760
- ermetica assiale congelata
  2845
- ermetica a tubo di gomma 6242
- ermetica congelata rotativa
  6232
- ermetica statica congelata
  2846
teorema di Larmor 3905
- di Wigner 7719
teoria a due gruppi 7446
- atomica 420
- classica degli elettroni di
  Dirac 1850
- dei quanti 5650
- dei quanti di Bohr-Sommerfeld
  695
- d'Eddington 2016
- del bersaglio 3298
- del campo mesonico 4370
- del campo quantizzato 5635
- della diffusione 1828
- delle cascate della radiazione
  cosmica 909

teoria delle guaine di corrente
  ad implosione 4179
- delle perturbazioni 5221
- dell'età di Fermi 2553
- del reattore 5991
- del reticolo 3931
- del trasferimento della
  carica 7114
- del trasporto 7365
- di Bohr-Wheeler 696
- di disintegrazione beta 591
- di disintegrazione beta di
  Fermi 2554
- di Fermi d'accelerazione di
  raggi cosmici 2572
- di Feynman dei positoni 2581
- di legami covalenti di
  Heitler-London 3218
- di più gruppi 4569
- di reattore a gruppo unico
  4978
- di rilasciamento spin-reticolo
  di Casimir e Du Pré 910
- di separazione capillare 870
- di Townsend 7310
- elettronica della valenza 2184
- elettronica di Drude in
  metalli 1975
- mesonica delle forze nucleari
  4370
terapia con fascio mobile 4553
- con radioisotopi 1575
- con raggi X 6227
- isotopica 3771
- pendolare eccentrica 2013
- per isotopi di cobalto 1160
- per radiazione 5745
- reticolare 3083
- rotatoria 6235
terbio 7100
terfenile 7104
termalizzazione 7172
termine atomico dispari 4962
- pari dell'atomo 2393
termini ortoelio 5032
termione 7174
termochimica 7176
termoluminescenza 7177
termometro 7096
- per canali di raffreddamento
  990
- per guaine 2854
termonucleare 7179
termopila al boro 744
- per neutroni 4753
tessuto 7266
- per filtro 2609
- vivo 4085
testa a fungo 4597
- di banda 525
- di carica 7657
- esplosiva 7657
- esplosiva nucleare 4923
testina di contatore a
  scintillazione 6360
tetrafluoruro d'uranio 3076
tetravalente 5620
tiroide 7241
tiuiamunite 7454
titanato di stronzio 6965

titanio 7272
toddite 7274
tolleranza 7275
tomografia 689
- simultanea 6604
tomografo 688
tonalità termica 4900
torbernite 1410
torbido di traccia 3053
torianite 7209
toridi 7210
torio 7212
toriotenorimetro 7213
- a pseudocoincidenza
  beta-alfa 7214
torite 7211
torogummite 4764
toron 7222
torotungstite 7223
torre 1242
- d'estrazione 2490
- di raffreddamento 1402
torta di filtro 2608
totalità degli stati elettronici
  4244
traballamento 1101
traccia 5672, 7311, 7318
- di nebbia 1141
- d'ionizzazione 3681
- finta 2516
- nebulosa 1141
tracciante chimico 1053
- fisico 5276
- isotopico 3443
- radioattivo 5809
traccia relativista 6101
tracce di particelle cosmiche
  pesanti 3189
- di pesce 2638
- dovute al nucleo di Li-8 3134
- in forma di martello 3134
transizione 7342
- d'Auger 441
- di fase 5235
- elettronica 2114, 2164
- isomerica 3744
- libero-libero 2825
- nucleare 4920
- per conversione 1388
- permessa 185
- proibita 2775
- radiativa 1540, 5757, 5810
- senza radiazione 5749
trappola 2165
- a flusso 2755
- di fascio 564
- d'iodio 6596
- di radiazione 5746
trascinamento 2307
- di gas 2967
trascinatore 895
trasferimento d'elettroni 2162
- d'energia 2290
- di calore d'ebullizione 697
- di carica 1019
- di massa 4284
- lineare d'energia 3983
- massimo d'energia 4303
trasferimetro 7329
trasformazione 7332

trasformazione nucleare 4903
- ordine-disordine 5007
- radioattiva 5810
- rotazionale 6239
- spontanea 6807
- tautomera 7081
traslocazione di cromosomi 1097
trasmettenza atmosferica 355
trasmissione di calore 3176
trasmittenza 7328
trasmutazione 7356
- artificiale d'elementi 331
- atomica 421
- beta 604
- isobarica 3727
trasparente 7358
trasporto d'ioni 3646
- d'isotopo 3772
- netto 4653
trattamento ad alta temperatura 3287
- chimico 1057, 1064
- con campo mobile 4556
- dei rifiuti 7667
- del combustibile irradiato 2875
- del combustibile irradiato per via umida 318
- di caoda 7060
- di combustibile 2873
- di combustibile esaurito 6771
- di testa 3152
- pirometallurgico del combustibile irradiato 1980
- termico 3181
- termico alternato 7130
traumatismo 7374
triade isobarica 3728
triboluminescenza 7381
tributilfosfato 7082
trifluoruro di bromo 795
- di cloro 1077
trilauriloamina 7386
trincea schermata 864
triossido d'uranio 4994
tripletto 7394
triplo punto 7393
tritiazione 7396
tritio 7397
tritone 7404
tritonite 7405
triturazione 1549
trocotrone 7407
troegerite 7408
trombocito 5342
tronco dell'atomo 374
tropopausa 7409
troposfera 7410
trottola di piombo 6795
tubatura di gas 2975
tubo acceleratore 29
- ad anodo rotativo 6230
- ad anodo stazionario 6860
- ad autoprotezione 6451
- a diagnostica 1769
- a più elettrodi 4592
- a radio 5902
- a radon 5905
- a raggi catodici 925

tubo a raggi X 7764
- a raggi X a catodo incandescente 3330
- a raggi X ad immersione in olio 4968
- a raggi X a gas 2984
- a raggi X a guaina metallica 6534
- a raggi X con alette di raffreddamento 5759
- con doppio fuoco 1955
- con fuoco lineare 4025
- contatore 1472
- contatore a base decimale 1198
- contatore a campana 568
- contatore a catodo esterno 2472
- contatore a corrente di gas 2970
- contatore ad alogeno 3133
- contatore ad autospegnimento 6452
- contatore ad elio 3221
- contatore ad immersione 1842
- contatore a finestra 7727
- contatore a finestra frontale 2246
- contatore a finestra sottile 7198
- contatore a fissione 2652
- contatore a lamiere parallele 5106
- contatore al boro 739
- contatore alfa 190
- contatore a parete sottile 7196
- contatore a particelle di rinculo 6011
- contatore a protoni di rinculo 6013
- contatore a punta 5371
- contatore a punta irradiata 1025
- contatore a sorgente interna di gas 1475
- contatore a spegnimento esterno 2479
- contatore autoestintore 6452
- contatore a vapore organico 5024
- contatore con atmosfera gassosa 2969
- contatore di Maze 2472
- contatore di neutroni veloci 2530
- contatore Geiger-Müller 2999
- contatore per liquidi 4056
- contatore piano 2691
- contatore proporzionale 5543
- Coolidge 1400
- d'annaffiatoio 6734
- d'esperimento 2440
- di Chaoul 997
- di combustibile 2881
- digestivo 180
- di guardia 3107
- di Lenard 3975
- d'ingresso 7419
- d'inserzione 3507
- di propagazione 1973

tubo di raggi limite 3077
- di riscaldamento 3182
- di röntgenterapia superficiale 7026
- di scappamento 680
- di terapia 7118
- elettronico di forma ovale 5671
- fotomoltiplicatore 5266
- in pressione 5480
- limitatore del campo 2596
- per röntgenterapia profonda 1663
- turbolent 7647
tucholite 7239
tulio 7240
tumore 7425
- benigno 571
- maligno 4240
tungsteno 7427
tunica mucosa 4564
turnerite 4533

ubicazione 6623
u.e.s. 2203
ulcera da radiazione 5747
ultracentrifuga 3280
umidimetro del suolo a radiazione ionizzante 3695
umidimetro isotopico 3797
umidità critica 1521
- d'equilibrio 2347
umoite 7462
un decimo del periodo 4980
unità 6970, 7475
- automatica per conteggio 459
- d'energia 2292
- di carica 7479
- di cascata collettrice 905
- di Hartree 424
- di lunghezza 7480
- di massa atomica 233
- di millimassa 4441
- di peso atomico 428
- di presentazione visuale X-Y 7775
- di reattività 5967
- di stronzio 6966
- di superficie 7476
- di tempo 7482
- di trasporto 7331
- di volume 7483
- elettrostatica 2203
- intercambiabile 5349
- per conteggio 6312
- per conteggio a differenza 1783
- per conteggio reversibile 6190
- per la misura della costante di tempo 5199
- per la misura della reattività 5962
universo 7486
uomo campione 6847
uraconite 7500
uranato di sodio 6673
uranidi 7501
uranile 7538
uraninite 7502

NEDERLANDS

afstandsbedieningsapparatuur
6118
afstandsbestraling 7087
afstandscollisie 1907
afstandscontrole van gebiedsbesmetting 6115
afstandsregeling 6116
afstandswisselwerking 56
afstoten van elektronen 6129
afstotingskracht 2780
afstotingspotentiaal 6135
aftap 1587, 5048
aftasten 6320
- voor de neutronenfluxdichtheid
4712
aftastregeling 6289
aftrekbare emulsie 6959
afvalbehandeling 7667
afval met gematigde radioactiviteit 3562
afvalregeneratie 7666
afval van geringe activiteit 4149
afvalverwerking 7666, 7667
afvalverwijdering 7665
afvalwater 7668
afvlakkingsgebied 2693
afvlakkingsmiddel 2696
afvlakkingsstraal 2697
afvoer 2051
afvoerkist 1920
afvoerleiding 5046
- voor radioactief afvalwater 79
afvoermonitor 2052
afvoerstroom 2051
afvoerstroommonitor 2052
afwijking 1764
afzetten 6424
afzetting 1734, 6423
- uit het mesozoïsche tijdperk
1736
aggregaatterugslag 144
akerlundraster 173
Alamogordobom 174
albedo voor neutronen 176
aldaniet 177
alfa 186
alfa-activiteit 204
alfadeeltje 197
alfadeeltjes met verre dracht
4123
alfadesintegratie 191
alfadesintegratie-energie 192
alfa-emissie 187
alfafactor 205
alfakernmodel 199
alfamonitor voor handbesmetting
195
alfa-neutron-reactie 196
alfa-proton-reactie 201
alfaspectrometer 206
alfaspectrum 200
alfastralen 207
- emitterende foelie 194
alfastraler 193
alfastraling 203
alfatelbuis 190
alfateller 189
alfatoop 209
alfatron 210
alfa-uranium 208

alfaverval 191
alfvengolf 179
alfvensnelheid 178
algemene diffusievergelijking
3006
- reactie 7053
allaniet 182
allobaren 183
aluinaarde 216
alumina 216
aluminium 217
aluminiumoxyde 216
aluminium-siliciumlegering 218
ambipolaire diffusie 220
americium 221
amfoteer ion 226
ammoniumdiuranaat 222
amniografie 223
amorfe stof 224
ampangabeïet 225
amplitude-tijd-omzetter 5591
ampul 232
ampulla 232
ampul met afbreekbare punt 776
anaëroob 235
anafaze 240
anaforese 241
analogiemodel 6603
- voor radioactiviteit 5818
analogonrekenmachine voor
vergiftigingseffecten 5378,
7772
- voor voorafgaande berekening
van xenonvergiftiging 5380,
7773
analogonrekentuig 236
- voor het bepalen van de
bundelbaan 5152
analoog-digitaalomzetter 238
analoog signaal 237
analyse 334
- door isotopenverdunning 3785
andersoniet 243
androgenesis 244
anemie 245
angiocardiografie 246
angiografie 248
angiopneumocardiografie 247
anion 264
anionenuitwisseling 265
annerodiet 268
annihilatie 269
annihilatiefoton 272
annihilatiegammaquantum 271
annihilatiekracht 270
annihilatiestraling 273
anode 277, 288
- met ingelaten trefplaat 3324
anodeschaduw 3213
anomaal magnetisch moment 280
anomale valentie 281
anoxiecel 282
anthraceen 285
anthroporadiometer 7707
- met amplitudeanalysator 7708
antibaryon 286
anticoïncidentie 289
anticoïncidentiekiezer 292
anticoïncidentieschakel 293
anticoïncidentieschakeling 290

anticoïncidentieteller 291
anticomptonspectrometer 294
antideeltje 303
antikatode 288
antilepton 298
antimaterie 299
antimoon 300
antineutrino 301
antineutron 302
antiproton 304
antistokeslijnen 307
antiverstrooiingsmiddel 5843
aortografie 308
apparaat 310
- met voorinstelling 5472
- voor convergentiebestraling
1378
- voor het bepalen van deeltjesafmetingen 5151
- voor snedeopname 688
arbeidscyclus 2001
arbeidsvermogen van beweging
3850
argon 323
arm erts 4147
arseen 325
arsenicum 325
arteriografie 326
arthrografie 327
asbestwol 333
assenverhoudingen 485
astaat 342
astatische actie 2702
astatium 342
astigmatische focussering 343
astonregel 344
astronsysteem 345
asymmetrie-energie 346
asymptoot 347
asymptotisch 348
asymptotische neutronenfluxdichtheid 349
atmosfeer 350
atmosfeerbewaking 162
atmosferische radioactiviteit
353
- stabiliteit 354
- stralingsverzwakking 352
atomaire aanslagfunctie 383
- absorptiecoëfficiënt 361
- massaeenheid 233
- moderatieverhouding 395
- verstrooiingscoëfficiënt 411
- verstrooiingsfactor 412
- verzwakkingscoëfficiënt 366
- wisselwerking 388
atomair stoppend vermogen 416
atoom 356
atoomafstand in de molecule 379
atoombaan 398
atoombatterij 367
atoombinding 370
atoombom 1
atoombomwolk 373
atoombundel 368
atoomdichtheid 376
atoomdislocatie 1867
atoomdoorsnede 375, 377
atoomeigenschappen 406
atoomelektron 380

mozaïekstructuur 4548
M-schil 4178
M-straling 4177
mucosa 4564
multiparameteranalysator 4570
multiplet 4578
multipliciteit 4582
multipoolmoment 4585
multipoolmomenten bij overgang 7347
multipoolstraling 4586
mu-meson 4563
muon 4563
muongetal 4594
muonium 4595
mutant 4598
mutatie 3131
- , door röntgenstralen veroorzaakte 7754
muterend gen 4599
M-vangst 4174
myeolografie 4605

naakt atoom 4814
naakte reflector 529
naakt splijtstofelement 7463
naaldteller 4638
nabestraling 5420
nabije botsing 1131
nabijheidsindicator berustend op directe ioniserende straling 1857
- berustend op ioniserende straling 3694
- berustend op verstrooide ioniserende straling 6329
nabootsing 6602, 6603
nadeelfactor 1865
nadelig schaduweffect 1366
naëgiet 4612
nagateliet 4613
nakoeling 133
nakomende alfadeeltjes 1691
- neutronen 1701
nalichten 134
napuls 138
natrium 6668
natriumcoëfficiënt 6674
natriumfotonenteller 6672
natrium-grafietreactor 6670
natriumkaliumlegering 4614
natriummetafosfaat 6671
natriumuranaat 6673
natte opwerking van splijtstof 318
- suspensie 6661
natuurlijke abondantie 4622
- achtergrondbestraling 499
- achtergrondstraling 502
- breedte 4633
natuurlijk element 4626
natuurlijke radioactiviteit 4623
natuurlijk ijzer 4627
- neutron 4629
- radioactief element 4630
- uranium 4631
nawarmte 135
necrose 4637
neerslag 6423
- , door regen meegevoerde 5913

neerslagdosis 1335
neerslagvanger 5463
neervallend materiaal 2513
negatief glimlicht 4641
- ion 264
- proton 304
negatieve ionenbundel 4643
- ionenstroom 266
- lading 4639
- reactiviteit 4645
- valentie 4646
- waardigheid 4646
negaton 2099
negatoscoop 2602
negen-tiende periode 4765
N-elektron 4606
neodymium 4648
neon 4650
neoplasma 4649
neper per uur 3483
neptunium 4651
neptuniumreeks 4652
netspanningsschommeling 2716
netspanningsvariatie 2716
nettodoorloop 4653
nettotransport 4653
netvlakkenafstand 3554
netvorming 1539
neutraal 4654
- atoom 4655
- deeltje 7464
- meson 4657
neutrale instabiliteit 4656
- molecule 4658
neutretto 4659
neutrino 4660
neutrofiel korreltje 4761
neutron 4661
neutrondesintegratie 4692
neutron-elektron-wisselwerking 4700
neutronen absorberende glasplaat 4664
neutronenabsorptie 4665
neutronenactivering 4667
neutronenbalans 4627
neutronenbindingsenergie 4674
neutronenbombardement 4675
neutronenbron 4747
neutronenbundel 4673
neutronenbundelhakker 4681
neutronencollimator 4682
neutronendetectie 4694
neutronendicht 4760
neutronendichtheid 4693
neutronendiffractometer 4696
neutronendiffusie 4697
neutrondiffusiecoëfficiënt 4698
neutroneneconomie 4699
neutronenergie 4701
neutronenflux 4707
- , met behulp van zwavel bepaalde 6992
neutronenfluxdichtheid 2748
neutronenfluxverdeling 2749
neutronenflux voor resonantie 6168
neutronenfysica 4735
neutronengolflengte 4756
neutronengroep 4703

neutronenkettingreactie 4680
neutronenkringloop 4691
neutronenlek 3966, 4704
neutronenlekspectrum 3971
neutronenomzetter 1389
neutronenopbrengst per absorptie 2371
- per splijting 4758
neutronenoptiek 4732
neutronenoverschot 1782
neutronenproducent 4715
neutronenradiografie 4742
neutronenreflectie 4743
neutronenrendementversterker 728
neutronenscintillatieteller 4706
neutronensnelheid 4750
neutronenspectrometer 4748
neutronenspectrum 4749
neutronenstraling 4740
neutronenstroomdichtheid 4689
neutronenteller 4686
neutronentemperatuur 4751
neutronentherapie 4752
neutronenuitbarsting 4677
neutronenvangst 4678
neutronenverlies 4704
neutronenvermenigvuldigingsfactor 4728
neutronenverstrooiing 4746
neutrongetal 4730
neutronkriticiteitsmonitor 1534
neutron met geringe energie 4142
neutron-neutron-reactie 4729
neutronografie 4759
neutron-proton-reactie 4739
neutronverval 4692
neutronvorming door strippen 4714
nevelspoor 1141
nevelvat 1138
- , door teller gestuurd 1455
nevenquantumgetal 490
newtonmechanica 1114
nicolayiet 4764
niet-actief laboratorium 1203
niet-actieve ruimte 1196
- zone 1120
niet-afscheidbaar deel 6182
niet-centrale kracht 4779
niet-destructieve bepaling van waterstof 4783
niet-doorlatend voor straling 5875
niet-elastische doorsnede 4784
niet-gebonden water 2831
niet-gemeenschappelijke elektronenbaan 287
niet-gesloten bron 7491
niet-gequantiseerd systeem 1116
niet-isotopische drager 4789
niet-lineariteit 4790
niet-ontaard gas 4782
niet-radioactief 1195
- isotoop 4792
niet-rechtlijnig verband 4790
niet-verkleurend glas 6818
nieuwvorming 4649

radioactief water 98
radioactieve afval 7664
- afvoer 77
- afvoerstroom 77
- afzetting 5773
- besmetting 1334
- bron 5807
- buiskrabber 5789
- concentratie 5768
- draad 5815
- draadlus 5791
- indicator 5809
- kern 5798
- koolstof 73, 884
- korrel 5790
- leeftijd 5763
- neerslag 75, 82
- omzetting 5810
- oorlogvoering 5859
- overgang 5810
- reeks 5787
- smetstof 1331
- spanningsbron 367
- stof- en gasmonitor 592
- stralingen 4897
- stijgwindneerslag 234
- valwindneerslag 1966
- verschuiving 5775
- verwantschap 5805
- zuiverheid 5732
radioactinium 5760
radioactiviteit 5816
- , in de lucht aanwezige 170
radioactiviteitsmeter 106
radiobiologie 5822
radiobiologische werking 5820
radiobiologisch gevoelig volume 5821
radiocardiografie 5824
radiocartograaf voor delen van het lichaam 687
radiochemie 5829
radiochemische analyse 5827
- zuiverheid 5828
radiochemisch rendement 5691
radiochromatograaf 5830
- voor de gasfaze 2988
radiodermatitis 53
radio-ecologie 5836
radiofotoluminescentie 5876
radiogeen 5840
radiogene neutronenvangst 4741
- warmte 5793, 5841
radiografie 5845
radiografische stereometrie 5844
radio-isotoop 5795
radio-isotopische concentratie 5848
- zuiverheid 5852
radiologie 4354
radiologiediagnostiek 1767
radioluminescentie 5862
radiolyse 5701
- van oplosmiddelen 5864
radiolytische aantasting 5865
- ontleding van oplosmiddelen 5864
- oxydatie 5866
radiometrie 5873

radiometrische analyse 5869
- landkaart 5871
- prospectie 5872
radionuclide 5874
radioscoop 5879
radiostereografie 6899
radiotherapie 5745
radiothermoluminescentie 5885
radiothorium 5886
radiotropie 5889
radiotropisme 5889
radium 5890
radiumapplicator 5898
radiumbuisje 5902
radiumcapsule 5833
radiumequivalent 5895
radiumgehalte 5894
radiumgehaltecertificaat 970
radiumhouder 5893
radiumleeftijd 3660
radiummoulage 4554
radiumnaald 5897
radiumplak 5899
radiumtherapie 5901
radiumzaadje 5900
radiusparameter 5903
radon 5904
radonbuisje 5905
radoneffect 5909
radongehalte 5906
radonlekproef met vloeistof-scintillator 6362
radonmeter voor prospectie 5908
- voor stralingsbescherming 5907
raffinaat 5911
raffinaatlaag 5912
raket met kernaandrijving 4891
ramsauereffect 5914
randiet 5915
randscheuren 2019
raster 3079
rastertherapie 3083
rasterverhouding 3082
rauviet 5947
rayleighdestillatie 5950
reactie 5952
- , door neutronen geïnduceerde 4718
reactiebaan 986
reactiebaanspin 992
reactie-energie 4900
reactie met energieuitwisseling 2407
reactiesnelheid 5957
reactietempo 5957
reactiviteit 5958
- bij uitschakeling 6570
reactiviteitscoëfficiënt 5959
reactiviteitseenheden 5967
reactiviteitsmeter 5962
reactiviteitsruis 5963
reactiviteitsstijging 5961
reactiviteitstempo 5966
reactiviteitsveranderingen op lange termijn 4124
reactor 4902
- , door poedersuspensie gekoelde 1998

reactor, door vloeibaar metaal gekoelde 4068
- , met beryllium gemodereerde 580
- , met gas gekoelde 2964
- , met grafiet gemodereerde 3064
- , met koolzuur gekoelde 882
- , met lucht gekoelde en met grafiet gemodereerde 152
- , met metallieke brandstof, met gas gekoelde 4380
- , met oververhitte stoom gekoelde 6879
- , met vochtige stoom gekoelde 7703
- , met zwaar water gekoelde 1756
- , met zwaar water gemodereerde 1758
reactorcel 933
reactorchemie 5971
reactorfamilie 2519
reactorgedrag 5979
reactorgewelf 5993
reactorkern 1414
reactorkoepel 5978
reactorkringloop 5983
reactorlus 5983
reactor met berylliumoxyde-moderator 581
- met boven- en onderreflector 7279
- met centraal naar binnen geleid koelmiddel 6799
- met circulerende splijtstof 1103
- met directe kringloop 1854
- met gaskoeling 2964
- met geconcentreerde neutronenflux 2756
- met gedeelde koelmiddel-stroom 6799
- met gefluïdiseerd bed 2725
- met gefluïdiseerde splijtstof 2727
- met geïntegreerde warmte-uitwisselaar 3527
- met gesmolten zouten 2897
- met groot vermogen 3261
- met indirecte kringloop 3445
- met inwendige reflector 3500
- met korrelgrafiet als koel-middel en moderator 3059
- met kunststofmoderator 5329
- met langzaam en snel gebied 6643
- met meervoudig doel 4588
- met middelsnelle neutronen 3566
- met natriumkoeling 6669
- met natuurlijke circulatie 4625
- met natuurlijk uranium 4632
- met nevelkoeling 2767
- met oneindige plaat 3474
- met organische moderator 5022
- met organisch koelmiddel 5021

DEUTSCH

D

nichtisotopischer Träger 4789
Nichtlinearität 4790
nichtquantisiertes System 1116
nichtradioaktiv 1195
nichtradioaktives Isotop 4792
nichtrostender Stahl 6841
nichtvorgesehene äussere nicht-
  zugelassene Exposition 45
- Exposition 43
- innere Kontamination 46
- nichtzugelassene Exposition 44
nichtzentrale Kraft 4779
Nickel 4763
Nicolayit 4764
Niederdrucknebelkammer 4151
Niederflussforschungsreaktor
  4146
Niederflussreaktor 4145
Niederlassung 6623
Niederlassungsbedingungen 6624
Niederschlag 6423
Niederschlagsammler 5463
Niederschlagsdosis 1335
Niedertemperaturreaktor 4152
niedriges Vakuum 4153
Niob 4767
Niobit 1240
Niveau 3991
Niveauabstand 3995
Niveauaustausch 1544
Nivellierungsfilter 2695
Nivenit 4769
N-Linie 4609
Nobelium 4771
Nohlit 4773
nominelle Kernwaffe 4777
Nonaktivitätsprüfung 1209
Nord-Süd-Effekt 4803
Normalabsorbens 6842
Normalband 4797
normale Abweichung 6844
- Glimmentladung 4799
Normalempfindlichkeit einer
  Photokatode 5245
normales Atom 4796
normale Valenz 4801
Normalfehler 6845
Normalionisationskammer 6846
normalisierter Plateauanstieg
  4802
Normalmensch 6847
Normalschnellschlusseinrichtung
  4800
Normalzustand 3090
Notabschaltung 2226
- wegen zu kleiner Zeitkonstante
  5201
Notkühlmitteleinspritzverfahren
  6271
Notkühlung 2222
Notschalter 6277
Not-Schnellschlusseinrichtung
  2228
Notschutzgeräte 467
Notstab 6275
Notstandäquivalentdosis 2223
Novacekit 4806
N-Schale 4611
N-Strahlung 4610

n-Stunden-Probenehmer 4607
Nu-Faktor 4758
nukleare Eigenschaften 4893
nukleares Geschoss 4892
nukleare Überhitzung 4916
Nuklearmedizin 4874
Nukleinsäure 4926
Nukleogenese 4927
Nukleon 4928
Nukleonenzahl 4271
Nukleoproteid 4932
Nukleoprotein 4932
Nuklid 4939
Nuklidmasse 4940
Nulleffektanteil 503
Nulleffekt eines Apparats 501
Nulleffektimpulse 497
Nulleffektmonitor 500
Nullenergieniveau 7792
Nullenergiereaktor 7793
Nullpunktenergie 7795
Nullvalenz 4941
Nullwertigkeit 4941
numerisches Signal 1832
Nutzlastzeit 4091
Nutzleistung 7547
nützliche Neutrone 7546
Nutzstrahl 7544
Nutzwärmeleistung 7548

Oberabschaltschwelle 7494
obere Grenzwellenlänge 7236
Oberflächenbestrahlung 7021
Oberflächendichte 320, 7016
Oberflächendosis 2584
Oberflächendosisleistung 7018
Oberflächenenergie 4917
Oberflächengeschwindigkeit
  7000
Oberflächenimplantat 5305
Oberflächenkapsel 7008
Oberflächenkernexplosion 418
Oberflächenkontamination 7012
Oberflächenkontaminations-
  anzeiger 7013
Oberflächenröntgentherapie
  6999
Oberflächenspannung 7024
Oberflächenspannungseffekt
  7025
Oberflächensperrschicht-
  detektor 7010
Oberflächensperrschichtdiode
  7009
Oberflächentemperatur des
  Brennelementes 3556
Oberflächentherapieröhre 7026
Objektkontrast 6979
Objektkontrastumfang 6980
Ödem 1589, 4963
O-Elektron 4948
offene Luft 2819
- Luftionisationskammer 2821
offener Kreis 4987
- Kreislauf 4987
- radioaktiver Stoff 7491
offenes Kühlsystem 4986
Öffnungsimpuls 2238
Ökologie 2015
Öldampfdiffusionspumpe 4969

Ölröhre 4968
Öltröpfchenmethode nach
  Millikan 4440
Omegatron 4971
opak 4984
Opazität 4983
Oppenheimer-Phillips-Prozess
  4982
Optimalbestrahlungsfaktor 5286
optimale Ausbeute 7756
optische Kupplung 1497
- Resonanz 4993
optisches Kernmodell 1143
- Pumpen 4992
- Überwachungssystem 7612
Orangit 4996
Ordnung 5009
Ordnung-Unordnung-Übergang
  5007
Organ 5020
organischer Moderator 5023
Orientierung 5025
Orientierungseffekt 5027
Orthodiagraphie 5030
Orthohelium 5031
Orthoheliumterme 5032
Orthopositronium 5034
Orthoröntgenoskopie 5030
orthoskopische Projektion 5035
Orthowasserstoff 5033
örtliche Intensivierung 6428
örtlicher Ausfall 4105
- Niederschlag 4105
ortsfester Reflektor 2684
O-Schale 4950
Osmium 5038
osteogen 5039
Ost-West-Effekt 2012
Ovalentladungsröhre 5671
Overhauser-Effekt 5060
Oxosalze des Urans 7519
Oxydationsgerät 5073
Oxydationszustand 5071

Paarbildung 5091
Paarbildungsenergie 5095
Paarbildungsgrad 1504
Paarbildungsrate 1504
Paarkonversion 5090
Paarspektrometer 5093
Packung 5086
Packungsanteil 5088
Packungseffekt 4262
Paketwalzen 5079
Palladium 5096
Palladiumleckprüfung 5097
Panethsche Regel 5098
Panne 775
Panoramaröntgenapparat 5099
Parabelmethode nach Thomson
  7204
Parahelium 5100
parallaktisches Panorama-
  gramm 102
- Stereogramm 5103
Parallelmanipulator 4290
Parallelplattenionisations-
  kammer 5105
Parallelplattenzählrohr
  5106

Tauchzählrohr 1842
Tau-Meson 7080
tausendstel Masseneinheit 4441
tautomerer Übergang 7081
TBP 7082
Technetium 7084
Teigreaktor 2726
Teilanordnung 6970
Teilchen 5134
Teilchenabsorption 5135
Teilchenaktivität 5153
Teilchenbeschleuniger 5136
Teilchendetektor mit Halbleiter-
   grenzschicht 3824
Teilcheneinschuss 5149
Teilchenfluenz 5143
Teilchenfluenzanzeiger 5146
Teilchenfluss 5144
Teilchenflussdichte 2719
Teilchen in Gleichgewicht 2343
Teilcheninjektion 5149
Teilchenladung 5139
Teilchen mit V-Spur 7552
- niedriger Energie 4143
Teilchenschwarm 7037
Teilchenspektrometer 5137
Teilchenstromdichte 1578
Teilchensystem 7051
Teilchenzähler 5140
Teilexposion 5129
Teilexposition 5129
Teilkörperbestrahlung 5124
Teilniveaubreite 5130
Teilquerschnitt 5127
Teilschirm 5131
Teilungskoeffizient 6802
teilweise besetztes Energieband
   5133
Teilwelle 5132
Telecurietherapie 7085
Telecurietherapieapparat 7086
Teleradiumtherapie 7089
Tellur 7090
Temperatur, durch Xenon ver-
   ursachte zu hohe 7771
Temperaturanstiegrate 7161
Temperaturbeiwert 7091
Temperaturfaktor 7094
Temperaturgradient 7095
Temperaturkoeffizient 7091
- der Reaktivität 4919
Temperaturmessgerät 7096
Temperaturwechselbean-
   spruchung 7130
Temperaturwechselprüfung
   7130
Tensorkraft 7098
Terbium 7100
Termschema 2276, 7101
Termverschiebung 7102
Terphenyl 7104
tertiäres Kriechen 7105
Testreaktor 7108
Textur 5025
Thallium 7112
theoretische Mindestbodenzahl
   4454
theoretischer Boden 7113
Theorie der implodierenden
   Stromhüllen 4179

Theorie des gequantelten Feldes
   5635
Therapiedosis 7115
Therapie mit bewegtem
   Strahlenbündel 4556
- mit Radioisotopen 1575
- mit Rotationsbestrahlung 6235
Therapieröhre 7118
Therapieröhrenschutzgehäuse
   7117
Thermalisierung 7172
Thermalisierungsgebiet 7173
Thermion 7174
thermionische Emission 7175
thermische Abschirmung 7162
- Anregung 7139
- Bewegung 7119
- Bewegungsspannung 7120
- Energie 3159
- Instabilität 7144
- Ionisation 7146
- Molekularströmung 2827
- Neutronen 7152
thermischer Brüter 7123
- Brutreaktor 7123
- Entkommfaktor 7147
- Neutronenstrahl 7150
- Nutzungsfaktor 7170
- Reaktor 6649
- Schild 7162
- Spaltfaktor 5519
- Störungsbereich 7167
- Wirkungsquerschnitt 7128
thermische Säule 7126
thermisches Energiegebiet
   7136
thermische Spaltung 7140
thermische Verbindung 7122
thermisch spaltbar durch
   lansame Neutronen 2639
Thermoanalyse 7121
Thermochemie 7176
thermochemische Reaktion 7125
Thermodiffusion 6706
Thermodiffusionsanlage 7133
Thermodiffusionsverfahren 7132
Thermoelement mit Metall-
   hülle 4377
Thermoelementsäule zum
   Neutronennachweis 4753
Thermokraftwerk 7155
Thermolumineszenz 7177
Thermolumineszenzdosis-
   messer 7178
Thermoneutroneneinfang 7124
thermonuklear 7179
thermonukleare Bedingungen
   7182
- Bombe 7181
- Energie 7183
thermonuklearer Cusp-Reaktor
   1585
thermonukleare Reaktion 2903
thermonukleares Gerät 7180
Thermosäule für Neutronen
   4753
Thermosiphon 7165
Theta-Einschnürung 6389
Thomson-Atom 7202
Thomson-Streuung 7205

Thor 7212
Thoraeus-Filter 7206
Thorerde 7208
Thorianit 7209
Thoride 7210
Thorit 7211
Thorium 7212
Thoriumdioxyd 7208
Thoriumerz 7219
Thoriumnitrat 7218
Thoriumreaktor 7220
Thoriumreihe 7216
Thoriumspaltung 7217
Thoriumzyklus 7215
Thorogummit 4764
Thoron 7222
Thorotungstit 7223
Thrombozyt 5342
Thrombozytenzahl 5343
Thucholit 7239
Thulium 7240
Tiefendosis 1738
Tiefentherapieröhre 1663
Tieftemperaturdiffusions-
   verfahren 7093
Titan 7272
TME 4441
TNT-Äquivalent 7273
Tochternuklid 1619
Tochterprodukt 1619
Toddit 7274
Toleranz 7275
Tomograph 688
Tonerde 216
Topfionisationskammer 7697
Torbernit 1410
toroidaler Einschnürungs-
   apparat 7282
Toroidplasmoid 7283
Torschalter 2994
Torschaltung 2991
Torstufe 2994
totale Elektronenbindungs-
   energie 7290
- mittlere freie Weglänge 7296
- Quellendichte für Neutronen
   7297
totaler Wirkungsquerschnitt
   7288
- Wirkungsquerschnitt für
   Stoss 7289
totale spezifische Ionisation
   7299
Totzeit 1625
- eines Geiger-Müller-
   Zählrohrs 1627
Totzeitkorrektion 1626
Totzone 1624
Touschek-Effekt 7304
Townsend-Entladung 2589
Townsend-Koeffizient 7308
Townsend-Lawine 7306
Townsendsche Kennlinie 7307
- Theorie 7310
Tracer 3443
Tracerchemie 7315
Tracerverbindung 7316
Tracerzugabe 6779
tragbare Ionisationskammer
   5402

THE CORRESPONDENCE OF EDWARD HINCKS

The Correspondence of

EDWARD HINCKS

EDITED BY
KEVIN J. CATHCART

VOLUME II
(1850–1856)

UNIVERSITY COLLEGE DUBLIN PRESS

PREAS CHOLÁISTE OLLSCOILE
BHAILE ÁTHA CLIATH

First published 2008 by
UNIVERSITY COLLEGE DUBLIN PRESS
Newman House
86 St Stephen's Green, Dublin 2, Ireland
www.ucdpress.ie

Notes © Kevin J. Cathcart 2008

ISBN 978–1–904558–71–2

British Library Cataloguing in Publication Data
A catalogue record for this title is available
from the British Library

Typeset in Ireland in Haarlemmer by
Elaine Burberry, Bantry, Co. Cork
Text design by Lyn Davies
Printed in England by
CPI Antony Rowe
This book has been printed on acid-free paper

for Declan and Debbie, Kieran and Maeve

CONTENTS

LIST OF ABBREVIATIONS

BIF   Bibliothèque de l'Institut de France

BL   British Library

BM/ANE   British Museum, Department of the Ancient Near East

BM/CA/LB   British Museum, Central Archives, Letter Book

BM/CA/OP   British Museum, Central Archives, Original Papers

Bod. Lib.   Bodleian Library

GIO/H   Griffith Institute, Oxford: Hincks Correspondence

NStuUBG   Niedersächsische Staats- und Universitätsbibliothek, Göttingen

PRONI   Public Record Office of Northern Ireland

StJCLC   St John's College Library, Cambridge

YU/BL   Yale University: Beinecke Rare Books and Manuscripts Library

PREFACE

The letters in this volume cover a period in Edward Hincks's life when his ecclesiastical and academic hopes and aspirations were raised and then dashed on several occasions, so that he was often driven into periods of despondency. Nevertheless, Hincks continued to make remarkable discoveries in his studies of cuneiform texts. Henry Creswicke Rawlinson, his rival in this field, who claimed for himself discoveries that were made by Hincks, was also determined to limit his access to the collection of cuneiform tablets at the British Museum. Unfortunately, few voices were raised in protest against Rawlinson's unacceptable behaviour. In spite of this sometimes gloomy picture of Hincks's life, the correspondence is full of fascinating details of intellectual life in the nineteenth century.

Kind permission to publish the letters in this volume has been granted by the Trustees of the British Museum; the Master and Fellows of St John's College, Cambridge; the Bodleian Library, Oxford; the Management of the Griffith Institute, Oxford; La Commission des bibliothèques et archives de l'Institut de France, Paris, and its President, Madame Hélène Carrère d'Encausse; the Niedersächsische Staats- und Universitätsbibliothek, Göttingen; the Campbell Allen family; the Deputy Keeper of the Public Record Office of Northern Ireland; and the Beinecke Library at Yale University.

Once again I owe words of thanks to friends and colleagues who have generously given me their time and support. Stephanie Dalley (Oxford) read a large part of my manuscript and offered many useful suggestions. I thank her warmly. I am greatly indebted to Anne Mouron (Oxford) who meticulously proofread all my transcriptions of the letters in French. Verena Lepper (Berlin) kindly looked at all the Egyptian hieroglyphs in the letters, ensuring accuracy in this edition. It was always delightful to work in the Bibliothèque de l'Institut de France in Paris and I wish to express a special word of thanks to Fabienne Queyroux for all her help. I had the pleasure of co-operating with Larry J. Schaaf (director) and Kelley Wilder (assistant editor) in the Correspondence of William Henry Fox Talbot project at the University of Glasgow (the project is now hosted by De Montfort University, Leicester). When Talbot's correspondence was moved from Lacock Abbey to the British Library, John Falconer was very helpful in making it available to me. Antoni Üçerler S.J. prepared the cuneiform signs for this volume with admirable patience and skill. Other scholars have generously given me expert advice, including Peter T. Daniels, Graham I. Davies, Mark Geller, Anselm Hagedorn, John F. Healey, Gerard J. Hughes, S.J., Geoffrey Khan, Jaromir Malék, Carl Phillips, Karen Radner, G. Rex Smith, and Annette Zgoll.

It has been a pleasure to work with Barbara Mennell, the Executive Editor at University College Dublin Press, and her assistant editor, Noelle Moran, who guided this book expertly through the press. Elaine Burberry met the challenges of typesetting with aplomb once again. Finally, I dedicate this volume to my sons and daughters-in-law with warm affection.

KEVIN J. CATHCART
Campion Hall, Oxford
January 2008

1850–1856

1850

FROM GEORGE CECIL RENOUARD[1]

7 FEBRUARY 1850

Swanscombe, Dartford, Kent

F. 7. Feb[ry]. 1850

My dear Sir,

I should have long since wished you a happy new year, had not your letter of the 29th. Dec[r]. 1849 been destined to delay.[2] Imprimis,[3] you directed it to me at Tunbridge Wells (i.e. I suppose you did, for it came to me without its cover) not knowing where I was to be found then: and as I had left that place on the 26th. of Dec[r]. without giving my direction to the Post Master, he sent the letter to the Rev[d]. G. Renaud of Bayford House near Hertford, who very obligingly finding out my direction in the Clergy List, forwarded it to me hither where it reached me on the 5th. of Janr[y].

I am waiting with much impatience to see your forth coming paper in the Transactions of the R.I.A.[4] In the mean time I have seen Major Rawlinson & was much pleased to find that his results & yours approximate very closely, except in names of kings & gods.[5] He says that they used words or names of the same import for each other & frequently epithets instead of names; the history & mythology therefore are very difficult to unravel.

The king by whom the obelisk was erected, is in his opinion, Sardanapalus the elder;[6] but I suspect your reading of Dhidhanka (Sesonchis) will prove to be the right one. The chronological difference *alone*, is in its favour. His communication to the R.A.S. was given, I believe, pretty fully in the Athenaeum & Lit. Gaz. but I hardly had time to look at either.[7]

I have been looking again at Rich's brick inscriptions & see how nearly identical they are.[8] From them I apprehend equivalent characters may be ascertained.

I have for translation a paper by Grotefend[9] on some parallel phrases in bricks & in inscriptions; but as I am not sure that I have always found the texts to which he refers, & his paper will be scarcely intelligible without them, I have not yet prepared it for publication. He considers no. 30 (*nu*) in your table of Babyl. ch. of 14th. D. 46[10] (I have none of the later tables at hand, having lent your last papers to Major R.) as the Chaldee & Hebrew word *nāsî'*, being compounded of *na* (28 or 31) a [...][11] which I suppose he thinks equivalent to 𐎟, or (62) or (73) & *u/i* (4) (the position of the wedges being altered, the largest & lowest being placed horizontally the others in the same descending line vertically). ✳ he takes, I believe, for

3

n & an abbreviation of *nasé'*. I must reserve the remainder for another letter as the post hour is come. I have two papers of yours respecting M^r. Ellis's phoneticks which I will soon return.[12] I am going to Town on Monday, but shall not be able to hear Major R. on Saturday, as the hour is too late for me to return that day.

    Believe me, dear Sir,
    faithfully yours
    G. C. Renouard

N.B. K. is the birthplace of Hans Sloane.[13]

MS: GIO/H 456

1   George Cecil Renouard (1780–1867) was a Church of England clergyman and scholar. See *Oxford Dictionary of National Biography* (Oxford, 2004) (hereafter *ODNB*). He was one of Hincks's most regular correspondents.

2   In his diary for 28 December 1849 Hincks wrote: 'Letter from Mr. Renouard with inscription on Lord Aberdeen's stone, at which I worked all day.' The letter has not survived. The entry for 31 December says: 'Began a series of slips for each character in the Aberdeen inscription, in nine varieties, of which I have impressions or copies purporting to be facsimiles.' The entries from Hincks's diary are taken from E. F. Davidson, *Edward Hincks: A Selection from his Correspondence with a Memoir* (London, 1933).

3   In the first place; first (rare usage).

4   Hincks's paper 'On the Khorsabad Inscriptions', which was read at a meeting of the Royal Irish Academy on 25 June 1849, was in the press. It was published in March 1850 in *Transactions of the Royal Irish Academy* 22 (1850), polite literature, pp. 3–72, and included an extensive appendix (pp. 56–65) and *addenda et corrigenda* (pp. 65–72). An entry in his diary for 17 January reads: 'In bed this morning I made an important rectification of my views on the Sanscrit character of the cuneatic syllabary; it caused me to be late rising, and to begin re-writing my appendix.'

5   Henry Creswicke Rawlinson (1810–95), a key contributor in cuneiform decipherment.

6   H. C. Rawlinson, 'On the Inscriptions of Assyria and Babylonia', *Journal of the Royal Asiatic Society* 12 (1850), p. 430. On p. 425 he suggests Assar-adan-apal as a possible Assyrian form of the name Sardanapalus.

7   See the reports of Rawlinson's lectures in *The Athenaeum*, no. 1161 (26 Jan. 1850), pp. 104–5; no. 1166 (2 Mar. 1850), pp. 234–6; *Literary Gazette*, no. 1723 (26 Jan. 1850), pp. 63–4; no. 1727 (23 Feb. 1850), pp. 145–6; no. 1728 (2 Mar. 1850), pp. 164–5. The lectures of 19 January and 16 February were published as 'On the Inscriptions of Assyria and Babylonia', *Journal of the Royal Asiatic Society* 12 (1850), pp. 401–83; and separately as *A Commentary on the Cuneiform Inscriptions of Babylonia and Assyria; including Readings of the Inscription of the Nimrud Obelisk, and a Brief Notice of the Ancient Kings of Nineveh and Babylon* (London, 1850).

8   Claudius James Rich (1787–1821) travelled and collected antiquities in the Near East.

9   See Renouard's letter of 7 March 1850. Georg Friedrich Grotefend (1775–1853) was a German philologist and schoolteacher, who specialised in Latin and Italian. His greatest achievement was the first partial decipherment of Old Persian cuneiform. See M. Pope, *The Story of Decipherment* (London, 1975), pp. 99–102.

10   'On the Three Kinds of Persepolitan Writing, and on the Babylonian Lapidary Characters', *Transactions of the Royal Irish Academy* 21 (1847), polite literature, p. 245.

11   The writing is illegible here.

12   In 1849 Renouard expressed his disapproval of the views found in A. J. Ellis, *The Essentials of Phonetics* (London, 1848) and *The Fonetic Nuz* (London, 1849). He was surprised to find that Hincks admired Ellis's method. See Renouard's letter of 23 November 1849.

13   Sir Hans Sloane (1660–1753), medical doctor and collector, was born in Killyleagh, Co. Down. His collection consisted of more than 80,000 items and it was bought by the British Government in 1753. The British Museum was established by act of parliament on 7 June 1753, six months after Sloane's death. A significant portion of the collection became the core of the foundation of the Natural History Museum in London.

FROM EDWARD CLIBBORN[1]

9 FEBRUARY 1850

Sat. 9[th]. Febry 1850

Dear Sir,

I have paid the drawing of the pillar a visit & shown the tracin[g]s to M[r]. Ball, who is out of health spirits today.[2] He has referred me to Buffon vol. XI. p. 211,[3] where I find the camel with two humps called the Bactrian, by Aristotle, Hist. anim. lib. II, cap. I., & the camel with one hump is the dromadary, the Arabian – see Aris. Hist. anim. lib II. cap. I.[4] The last is the quick camel or ship of the desert, that supposed to be alluded to by Job. See Calmet.[5]

It appears odd to me that two camels are represented in the two pictures & both nearly if not altogether the same. Could not the one to the left be the dromadary, the one to the right the camel? The artist not having marked distinctly the difference in the humps. Both animals are the same species – they are mere varieties – so D[r]. Almond has just said.

Though the outlines are nearly alike, there is a certain mark Ω on the shoulders of the animals in the outline so marked, which is not on those of the others. This mark is found on the shoulders of a great number of animals, horses, bulls, sheep, &c. apparently belonging to the king. It is used exactly as the crux ansata[6] is on things, horses &c. of the Egyptians, but they put it on the horses hind quarter or thy.[7] In the Egypto-Etruscan things found representing as it were the flight of the Etruscans all the horses have the crux ansata in their hind quarters done so.

As D[r]. Osburn has been lately paying great attention to Aristotle's Nat. His., I will drop him a line & let you hear the result.[8]

Most truly yours
Edw[d] Clibborn

MS: GIO/H 115

1   Edward Clibborn was assistant secretary and assistant librarian at the Royal Irish Academy for nearly forty years.

2   Robert Ball (1802–57), Irish naturalist, was a member of the Royal Irish Academy and Director of the Museum in Trinity College, Dublin. Clibborn consulted him about the camels which are found in the Black Obelisk of Shalmaneser III. The tracings of the panels are preserved with Clibborn's letter in the Griffith Institute. See MS: GIO/H 116. Excellent photographs of the Black Obelisk can be found in J. B. Pritchard, *The Ancient Near East in Pictures Relating to the Old Testament* (2nd edn; Princeton, 1969), figures 351–5 (pp. 120–2).

3   G. L. L. de Buffon, *Histoire naturelle, générale, et particulière, avec la description du cabinet du Roi*, vol. II (Paris, 1754), pp. 211–83.

4   Aristotle, *Historia animalium*, II, 1: 499a, 13–30 (Bekker edition).

5   See A. Calmet, *Calmet's Dictionary of the Bible by the Late Mr. Charles Taylor with the Fragments Incorporated* (11th edn; London, 1849), pp. 229–30. There is no reference to Job in the entry on 'Camel' in

Calmet, *An Historical, Critical, Geographical, Chronological, and Etymological Dictionary of the Hebrew Bible in Three Volumes*, vol. 1 (London, 1732), pp. 350–1.

6   *Crux ansata* was a term used for the Egyptian ankh sign on account of its cross-like appearance.

7   An obsolete spelling of 'thigh'.

8   William Osburn (born 1793) was an English antiquarian, whose works were mainly devoted to the relations between Egypt and the Bible. See W. R. Dawson and E. P. Uphill, *Who Was Who in Egyptology*, ed. Bierbrier (3rd edn; London, 1995).

FROM SAMUEL BIRCH[1]
16 FEBRUARY 1850

BM
16. Feby. 1850

My dear Sir,

I have received a small parcel of books addressed to you from M. de Rougé of Paris, and will forward them by any means you will advise me.[2] One is a reprint of a curious paper of his on a tablet at Leyden which *proves* that the Enentefs immediately preceded the XII dynasty.[3] He has not been able to read the text but I think I can perfectly & have sent a paper on it to the R. S. of Literature which will be read next meeting.[4]

Today Major Rawlinson has read a second paper at the Asiatic Society on the inscriptions at Khorsabad giving a very masterly precis of the results of his interpretations.[5] He also reads the king of Egypt who is there mentioned Biari & thinks that he is Pehar as he has been called.[6] Be that as it may I am convinced that the monarchs of the XXI Tanite dynasty were tributaries from their assuming the prenomen of 'High priest of Amen' not the old 'Ra' title. I arrived at this result almost 2 years ago & sent a paper to the Paris Review but it was suppressed owing to Letronne's death.[7] I am now preparing a preçis of the external foreign relations of Egypt and intend D. V. to publish a corpus of the prisoners & foreign nations mentioned in the texts. I should feel obliged if you could let me know the date which induced you to think the 2nd Sallier Papyrus contained an account of a campaign in the Syrian desert. I have long received an inscription from Egypt engraved on a pair of gold bracelets, which confirms my theory.

There is a loose sheet of 4 lithographed pages which has come with M. de R.'s papers commencing Essai sur une stèle funéraire de la collection Passalacqua No. 1393.[8] Let me know if it is not in your parcel when you open it as I will in that case send it by post to you.

Believe me
Yours very sincerely
S. Birch

1   Samuel Birch (1813–85) was an assistant in the British Museum, 1836–44; assistant keeper, Department of Antiquities, 1844–61; keeper of the Oriental, British, and Medieval Antiquities, 1861–6; keeper of Oriental Antiquities, 1866–85.

2   Emmanuel de Rougé (1811–72), a French Egyptologist, became curator of the Egyptian collection at the Louvre in 1849 and Professor of Egyptology at the Collège de France in 1860.

3   E. de Rougé, 'Lettre à M. Leemans sur une stèle égyptienne du Musée de Leyde', *Revue archéologique* 6/2 (1850), pp. 557–75.

4   Birch's paper, entitled 'On the Eleventh Dynasty of Egyptian Kings' was read on 28 February 1850. See *Proceedings of the Royal Society of Literature* 1/20 (1849–50), p. 302; and the report in *Literary Gazette*, no. 1730 (16 Mar. 1850), pp. 200–1.

5   See Renouard's letter of 7 February, n. 7.

6   According to the report, Rawlinson read *Bi-ar-ha = Pe-hur*.

7   Jean Antoine Letronne (1787–1848), French archaeologist and classical scholar.

8   *Essai sur une stèle funéraire de la collection Passalacqua (No. 1393) appartenant au Musée royal de Berlin* (Published privately; Berlin, 1849) = E. de Rougé, *Œuvres diverses*, vol. 1 (Bibliothèque égyptologique 21; Paris, 1907), pp. 331–4. Note that the reprint has 'No. 1353' in the title, which is an error.

FROM SAMUEL SHARPE[1]
28 FEBRUARY 1850

Clements Lane
28 Feb. 1850

Dear Sir,

Many thanks for your welcome letter. I do not read a word of the Assyrian; but reasoning from history say that Major Rawlinson's kings are those mentioned in the Bible. But that is a book which orthodox and unorthodox alike neglect, as far as history is concerned. I will in a few days get one of my children to copy for you my Table of Assyrian History & send it to you. Your paper from the Asiatic Soc^y I will send for with thanks.[2]

I will venture to tell even a D. D. out of the Bible that

1)   Sennacherib's name was also written Jareb, or in the LXX Jarim.[3]

2)   Shalmaneser's name was Shalman.[4]

3)   These Assyrians professed that they were descended from Babylon; as do Major Rawlinson's kings.

4)   Asser was as often spelt Nasser. Tiglath-pul-asser or Tiglath-pul-nasser.[5]

I wish you every success; but the public always take up with the man who first produces any thing *complete*, not who is *first* in the field. So go on with your task, and may you prosper. But I hope you do not desert Egypt. John Kenrick will publish very shortly on Early Egypt & the East.[6] Wilkinson is publishing in Rome on Egyptian Orders of Architecture.[7] I hope you read Bartlett's Forty Days in the

Desert.[8] His tracing the route of the Israelites is very judicious. I have not seen your Lecture in the Literary Gazette but shall look for it.[9]

Yours very truly
Samuel Sharpe

MS: GIO/H 495

1 Samuel Sharpe (1799–1881) was an Egyptologist and biblical scholar. He was a Unitarian and knew Hincks's brother William, who became minister at the Stamford Street Unitarian Chapel in London in 1839.
2 'On the Inscriptions at Van', *Journal of the Royal Asiatic Society* 9 (1848), pp. 387–449.
3 Hebrew *mlk yrb* in Hosea 5: 13 and 10: 6 probably means 'the great king' and can be compared with the Akkadian *šarru rabû*, 'great king', which was a foremost title of Assyrian kings. See J. Barr, *Comparative Philology and the Text of the Old Testament* (Oxford, 1968), p. 123 who refers to the original proposed by Godfrey Rolles Driver. However, Driver and Barr are wrong in claiming that the Septuagint supports the Hebrew text. Ιαρ(ε)ιβ is found in Origen, but not in the Septuagint which has Ιαρ(ε)ιμ. On this point Sharpe is right.
4 In the Old Testament at Hosea 10: 14, Shalman is the person who ransacked Beth-arbel. We do not know who Shalman was. Some scholars interpret the name as a form of Shalmaneser, the name of several Assyrian kings, and others have identified it as the name of a Moabite king Salmanu who is thought to have paid tribute to Tiglath-Pileser III. But see Karen Radner, 'Der Gott Salmānu ("Šulmānu") und seine Beziehung zur Stadt Dūr-Katlimmu', *Die Welt des Orients* 29 (1998), pp. 33–51, esp. pp. 34–5 and nn. 7 and 14.
5 The Akkadian form of the name Tiglath-Pileser is *Tukulti-apil-ešarra*. See A. R. Millard, 'Assyrian Royal Names in Biblical Hebrew', *Journal of Semitic Studies* 21 (1976), p. 7. Sharpe's remarks are based on a wrong interpretation of the late form of the name *tlgt pln(')sr* found in Chronicles (e.g. 1 Chronicles 5: 26).
6 J. Kenrick, *Ancient Egypt under the Pharaohs*, 2 vols (London, 1850).
7 J. G. Wilkinson, *The Architecture of Ancient Egypt: in which the Columns are Arranged in Orders and the Temples Classified, with Remarks on the Early Progress of Architecture, etc.*, 2 vols (London, 1850).
8 W. H. Bartlett, *Forty Days in the Desert on the Track of the Israelites or, A Journey from Cairo, by Wady Feiran, to Mount Sinai and Petra* (London, 1848). Several editions were published within a few years.
9 See the lengthy summary of Hincks's lecture entitled 'The Recently Discovered Assyrian Inscriptions, and the Mode of Deciphering Them', delivered on 30 January 1850 to the Belfast Natural History and Philosophical Society, *Literary Gazette*, no. 1725 (9 Feb. 1850), pp. 110–11. The entry for 30 January in Hincks's diary records that he lectured 'to a very large audience – Mr. Porter spoke in praise, as did Professor McDonald, an acquaintance of Rawlinson'.

TO THE EDITOR OF THE LITERARY GAZETTE
6 MARCH 1850

Killyleagh, C° Down
6th. March

Sir,

My attention has been recently directed to articles in the *Revue Archéologique* for August, 1848, and December, 1849, in which M. de Rougé, the very learned and talented Honorary Keeper of the Egyptian Museum at Paris, has put forward what he considers a *proof* that the *Sésourtasen** dynasty, as he calls them, preceded the *Sevekhotep* dynasty.[1] To the uninitiated public this may appear a matter of little or no consequence, but there are many readers of the *Literary Gazette* who are well aware that it involves a question of vital interest. If this alleged proof be a real one, we must add at least a thousand years to the interval between Manes and Cambyses, turning back the former to a period anterior to that at which even the Septuagintal copies of Genesis place the general deluge. As M. de Rougé has, in the article which I have last mentioned, announced a valuable discovery in relation to the dynasty of the *Entews*, whom he has shown, in the most satisfactory manner, to be the immediate predecessors of the *Sésourtasens* in Thebes (as indeed I had formerly conjectured them to be), I am apprehensive that his authority will induce many persons to receive as a fact what he says that he has proved. I trust, therefore, that you will allow me to point out the defect in M. de Rougé's chain of reasoning, which absolutely nullifies his alleged proof.

The chain of reasoning, as stated by M. de Rougé, is this: 'In an inscription copied at Semné by M. Durand, which is of the reign of *Sevekhotep I*, mention is made of a deceased king, whose praenomen is *Ra-scha-kaou*. This praenomen was borne by two kings, *Sésourtasen III*, and *Nowrehotep II*, the predecessor of *Sevekhotep IV*. The king spoken of must have been the former, *because the latter reigned subsequently to the date of the inscription*.'[2] The whole strength of the argument obviously rests on this last assertion; and this again rests on Lepsius's arrangement of the fragments of the Turin papyrus. Now, in a paper published in the last volume of the Transactions of the Royal Society of Literature, I have *demonstrated* by measurement of the fragments (of which facsimiles have been published by Lepsius himself) that this arrangement is an *impossible* one; and that the fragments numbered 76–80 must have followed, instead of preceding, those numbered 81–83.[3] This being admitted, the king called *Nowrehotep the Second*, would be the *First* of that name who is known to us, and would have preceded by several reigns the king under whom the inscription was cut. I cannot conclude without noticing the importance of watching and correcting any mistake that may be made in archaeology *at once*. Had I refrained from pointing out the gross blunder which Lepsius committed in respect to these fragments until I saw what inference could be drawn from it, persons would say that I found fault with his arrangement only to avoid admitting a conclusion which I could see no other mode of evading. The fact, however, is, that I corrected Lepsius's arrangement in May, 1846, while the inscription which makes that correction of such immense importance was not brought forward till August, 1848.

I am, &c.
Edw. Hincks

9

* I adopt M. de Rougé's spelling of Egyptian names though I by no means admit its entire correctness.

Source: Literary Gazette, no. 1729 (9 Mar. 1850), p. 183

1   E. de Rougé, 'Inscription des rochers de Semné', *Revue archéologique* 5/1 (1848), pp. 311–14; 'Lettre à M. Leemans sur une stèle égyptienne du Musée de Leyde', *Revue archéologique* 6/2 (1850), pp. 557–75.
2   See, for example, *Revue archéologique* 6/2 (1850), pp. 560–1.
3   'On the Portion of the Turin Book of Kings which Follows that Corresponding to the Twelfth Dynasty of Manetho', *Transactions of the Royal Society of Literature*, 2nd ser., 3 (1850), pp. 139–50, esp. pp. 144–5. The article is dated 5 May 1846 and was read on 28 May 1846. Hincks is referring to R. Lepsius, *Auswahl der Wichtigsten Urkunden des ägyptischen Alterthums* (Leipzig, 1842), plates III–VII.

FROM GEORGE CECIL RENOUARD
7 MARCH 1850

Swanscombe, Dartford
Th. 7 March 1850

My dear Sir,

I am very much obliged to you for the specimen of your paper which I am looking for with great eagerness. It puzzles & startles me (to use a favorite word of Father Newman's – him of the Oratory, I mean[1]), but not a whit more than Major Rawlinson's account of the almost numberless ways in which these pristine penmen (if such they were) expressed the names of Gods & Heroes. However, you are, I am sure, on the right scent, & time, patience & penetration such as yours will go very far if it do not complete the solution of this intricate problem.

Various interruptions have prevented me from answering (I fear) your last letter. I have in the interval not only seen Major R. who fully appreciates your powers & whose readings will, I can scarcely doubt, approximate closely to yours, tho' he says that want of materials have often led you into error; but [I] have had a letter from Grotefend. He has got impressions on paper of almost all the cuneatic inscriptions extant & has acquired, as might be expected, a surprising facility in reading off such words & phrases as he can interpret. You have probably seen a summary of his[2] *lectures* to the Asiatic Society, which from the day of the week when that Society meets (Sat.) I could not attend; but I had the pleasure of hearing him read a large portion of the second in which he very ably & candidly summed up the pros & cons of his chronological system. I cannot but think he goes too far back and other objections have occurred to me, but till his views are more developed, it is too soon to criticise them.

The difficulty of rendering D^r. Grotefend's letter intelligible without giving the words to which he refers, in the cuneatic character prevented me from translating it when I wished to do so: & I was therefore not a little rejoiced to learn that supposing his former copy of the 'Inschrift eines Thongefässes mit ninivitischer Keilschrift' sent to Herr D^r. Lee on 25th. of last Sept^r, was lost, he had caused 'eine verbessert Tafel' to be lithographed, in which he had given up what he discovered to be erroneous, & communicated an amended paper on it to the R. S. of Göttingen, copies of which he hoped soon to receive.[3] You have therefore been no loser by my delay; & will be glad to hear that he has discovered his errors.

A part of my difficulties respecting the Babylonian characters arose from my inability to get a copy of M^r. Fischer's Index to the Great Brick.[4] The plate on which he etched it is lost & I believe no more impressions of it remain at the S. H.[5] It was never on sale: but about ten days ago M^r. Norris very kindly lent me his copy which I hope to trace; but in such matters, I am an awkward novice.[6]

Rawlinson (whose eye speaks volumes) has either given or will give on Sat. next a third lecture at the R.A.S. which I wish I could hear.[7] It will, doubtless, be much crowded. He has also twice done much the same thing at the Antiquaries Soc. He has got from Bab. I believe some earthen paterae cover'd internally with inscriptions in ink, very legible, & at first sight [in] the square ordinary Hebrew character, but Norris tells me they are in Pahlaví,[8] different he suspected from Anquetil's Pahlaví which, as he pointed out, is clearly an artificial language such as that of the Desátír.[9] I had no opportunity of enquiring whether Major R. has looked at the Van Inscript. There, I imagine, you stand alone.

A n°. of Galignani was sent to me lately containing M. Loewenstern's claim to the original discovery of the Egyptian-like system of the cuneatics.[10] He certainly threw out a hint of such a notion at the end of his Élémens constitutifs published in 1847.[11]

As the Post hour is come I must only add that I shall read M^r. Ellis with great attention as I find you patronise him & am glad to be able to keep his tract for which pray receive the thanks of

my dear Sir,
Your much obliged & faithful
G. C. Renouard

Major Rawl. is at 39 S^t. James's Street, Piccadilly

MS: GIO/H 457

---

1   John Henry Newman (1801–90), theologian and cardinal, and one of the great religious figures of the nineteenth century, became a Catholic in 1845 and a few years later was appointed superior of the

Birmingham Oratory. A search for Newman's use of 'startle' in his writings at the website of the 'Newman Reader' will reveal how very observant Renouard was. See http://www.newmanreader.org/index.html. Hincks, too, read Newman's writings and an entry in his diary for 1850 reads: 'Read "Lyra Apostolica". If Newman be the author of the poems signed δ, I should regret his fall more than I have ever done yet.'

2   Renouard expresses himself carelessly here. He means Rawlinson's, not Grotefend's lectures.

3   See G. F. Grotefend, 'Bemerkungen zur Inschrift eines Thongefässes mit ninivitischer Keilschrift', *Abhandlungen der königlichen Gesellschaft der Wissenschaften zu Göttingen* 4 (1850), historisch-philologische Klasse, pp. 175–93; 'Keil-Inschriften aus der Gegend von Niniveh, nebst einem persischen Siegel', *Zeitschrift für die Kunde des Morgenlandes* 7 (1850), pp. 63–70 + 1 lithograph. John Lee (formerly Fiott) (1783–1866) was an English antiquary and patron of science, and one of Hincks's correspondents.

4   *A Collection of All the Characters Simple and Compound with their Modifications, which Appear in the Inscription on a Stone Found among the Ruins of Ancient Babylon Sent, in the Year 1801, as a Present to Sir Hugh Inglis Bart. by Harford Jones Esqr., then the Honorable the East India Company's Resident in Bagdad; and, now Deposited in the Company's Library in Leadenhall Street, London. Collected Etched and Published June the 1st 1807 by Thos. Fisher.* Peter Daniels, who has seen the only extant copy in a private collection, describes it as follows: 'It is a single sheet with an image a little smaller than 11 x 14 inches. 287 cuneiform characters are neatly arranged by shape in a 10 x 29 cell grid.' See P. T. Daniels, 'Edward Hincks's Decipherment of Mesopotamian Cuneiform', in K. J. Cathcart (ed.), *The Edward Hincks Bicentenary Lectures* (Dublin, 1994), p. 55, n. 12.

5   S. H. = Society House (Royal Asiatic Society).

6   Edwin Norris (1795–1872), linguist and Assyriologist, was an officer of the Royal Asiatic Society from 1836 until the end of his life. He worked as a translator in the Foreign Office from 1846 to 1866.

7   Rawlinson's lecture had already been given. See *Literary Gazette*, no. 1728 (2 Mar. 1850), pp. 164–5.

8   Many earthenware bowls with incantations inscribed on the inside have been found in Iraq. The inscriptions are mostly in Jewish Aramaic, but some are in Mandaic, Syriac and Pahlavi. See J. B. Segal, with a contribution by Erica C. D. Hunter, *Catalogue of the Aramaic and Mandaic Incantation Bowls in the British Museum* (London, 2000). Hunter discusses the manufacture of the bowls on pp. 163–8.

9   Abraham Hyacinthe Anquetil du Perron (1731–1805) was a French orientalist who translated the Avesta. Renouard referred to the *Desatir* of Mulla Furuz in his letter of 30 January 1847. See *Desâtîr or Sacred Writings of the Ancient Persian Prophets*, published by Mulla Firuz bin Kaus (Bombay, 1818).

10   *Galignani's Messenger* was an English-language newspaper founded in 1814 and published in Paris until 1884. Renouard is referring to an article on 'Babylonian and Assyrian Inscriptions' in the issue for 28 February 1850.

11   I. Löwenstern, *Exposé des éléments constitutifs du système de la troisième écriture cunéiforme de Persépolis* (Paris and Leipzig, 1847).

FROM JOSEPH SAMS[1]

12 MARCH 1850

Londn. 56, Grt. Queen St

12 / 3mo. 1850

J. Sams presents kind remembrance to his much respected D[r]. Hincks, & is very sorry that he has mislaid the address the D[r]. kindly handed, where small parcel could be enclosed. If he could again favour with a line just indicating the place, J. S. wd. be much obliged. After leaving Ireland he took a pretty long round, & staid longer at various places than he expected, as at Liverpool, Manchester,

Lincoln, Nottingham, &c. He has, however, been arrived in this Metropolis some weeks, & much regrets his inability, thro' the circumstances referred to, to forward to D^r. H. the small parcel. He hopes sincerely the D^r. himself, his respected consort, & amiable daughters, continue all favoured with good health. He thought perhaps M^{rs}. H. might be interested with a Specimen of the Fine Linen of Egypt, & therefore makes free to enclose a specimen or two (with a note to verify), of which to request she will please accept, & which J. S. will be obliged to the D^r. to please to hand, with kind respects to herself & daughters. And he waits the favour of a line with the address requested, tho' very sorry to give so much trouble, thro' the accident of having mislaid the one before kindly handed.

Since writing the above, & before it got off, the address has come to hand, & J. S. delivered at Longmans 12 Engravings of some of his antiquities from ancient Egypt, which he thought wd. be interesting, & which they promised to enclose in the booksellers' parcel for Belfast, & of which he requests the D^r. will please accept – a small mark of sincere esteem, & regard.

In the outer part of the roll, he also made free to enclose 6 engravings, of which to request the D^r. wd. he so obliging as, when convenient, to hand to M^r. Getty, of the ballast office, Belfast.[2] J. S. fears in his haste, on leaving Londn., for his residence in the North of Engld., he omitted to put the name on these outer 6, & wd. beg the D^r.'s kindness just to write it. May he also beg the favor of a line, to inform of the safe arrival, & saying also, how the D^r. himself & all the respected family, are.

Darlington, co. of Durham,
4 mo. (Apl.) 6–1850.[3]

[Enclosure with two pieces of linen:]

A note to verify the Specimen of Linen from Ancient Egypt, is on the other side.

Specimen of the Fine Linen of Egypt, & a Specimen of stronger Linen from ancient Egypt. These Specimens are of very considerable interest, from their extreme antiquity, as they may go back, not improbably, not only to the time of Moses, & the Israelites, but perhaps even anterior to that remote period. Brought from that parent of art, & science, the land of Egypt, during a late extensive journey in the East, & presented to,

M^{rs}. (D^r.) Hincks,
by J. Sams.
London 28/ 2mo. (Feby.) 1850

MSS: GIO/H 487, 488

1   Joseph Sams (1784–1860) was an eccentric bookseller and dealer in antiquities. He travelled through Europe and the Near East. See *ODNB*.

2   Edmund Getty (1799–1857) was born in Belfast. He was educated at the Royal Belfast Academical Institution. He became Ballast Master of the Belfast Ballast Board and later Secretary of the Belfast Harbour Board. He and Thomas Dix Hincks were among the founders of the Belfast Natural History Society. See *Dictionary of Ulster Biography* (Belfast, 1993).

3   The letter was delayed for a month. For convenience I have placed it according to the date at the top of the letter.

FROM EDWIN NORRIS
18 MARCH 1850

R. As. Society
18th March, 1850

My dear Sir,

I have just received the copy of the paper on the Khorsabad inscriptions, for which allow me to return you my best thanks.[1] Rawlinson has also received his copy and I left him reading it: he will get to the end before he leaves, but *I* must put it off until to-morrow, as I am closely occupied with the concluding revise of his readings before our Society, which will appear in our journal, and I hope the numbers will be out & ready for delivery this week; though I need not tell you how doubtful all our determinations are when we have to deal with printing Babylonian.

I hope Mr Gill has received the post office order;[2] it was delayed 4 or 5 days from the forgetfullness of the clerk who received it. The numbers shall be distributed as you desire, as soon as they arrive. Mr Sharpe has already twice sent for his copy.[3]

I am
My dear Sir
Yours very sincerely,
Edwin Norris

Read on 25 June 1849

MS: GIO/H308

1   'On the Khorsabad Inscriptions', *Transactions of the Royal Irish Academy* 22 (1850), polite literature, pp. 3–72.

2   Michael Henry Gill was printer to Dublin University (Trinity College) and the Royal Irish Academy. His offices were at 50 Upper Sackville Street, Dublin.

3   Samuel Sharpe, the Egyptologist.

TO SAMUEL BIRCH
19 MARCH 1850

Killyleagh C° Down
19th. March 1850

My dear Sir,

You will I hope have a copy of my paper by the post which follows that by which I now write.[1] I saw an account of your paper at the R.S.L. in the Literary Gazette that came yesterday.[2] Perhaps you saw a letter of mine on the same subject in the preceding Gazette (of the 9th.).[3] I wrote it on receipt of the parcel from Rougé. Is there any *other* proof of a dynasty between the 12th. & 18th. than that which I set aside?

I cannot now lay my hand on your last letter. I believe there was something in it that I should have replied to; but this publication has thrown me into great confusion.

Believe me
Yours very truly
Edw. Hincks

MS: BM/ANE/Corr. 2699

1   'On the Khorsabad Inscriptions', *Transactions of the Royal Irish Academy* 22 (1850), polite literature, pp. 3–72.
2   Birch's paper 'On the Eleventh Dynasty of Egyptian Kings' was reported in *Literary Gazette*, no. 1730 (16 Mar. 1850), pp. 200–1.
3   *Literary Gazette*, no. 1729 (9 Mar. 1850), p. 183.

FROM WILLIAM FRANCIS AINSWORTH[1]
21 MARCH 1850

Thames Villa
Hammersmith
21. March 1850

My dear Sir,

I shall peruse with great interest the memoir which you have been kind enough to forward to me.[2] Your note I will read before the society.[3] I certainly do not think that Calah, Resen & Nineveh can have been in the small territory supposed by Maj. Rawlinson. At the same time I do not think I have made my argument as clear as I would have wished in the Syro-Egyptian paper.[4] I never made it more so in the last numbers of the new Monthly Magazine[5] & we discussed this matter vivâ voce at the S.E.S. last meeting, Maj. Rawlinson insisting upon his view of the matter and stating that the name Calah had been deciphered at Nimrod. We shall always be truly happy to receive any communications you may forward us & we hope to print more than we have heretofore but our funds are still exceedingly limited.

Believe me my dear Sir with great respect for your numerous learned labours.

Yours sincerely
William Francis Ainsworth

The Society's direction is 71 Mortimer St. Cavendish Square.

MS: GIO/H3

1   William Francis Ainsworth (1807–96), geographer and geologist, travelled widely in the Near East. He was surgeon and geologist to the expedition to the Euphrates in 1835. He published *Researches in Assyria, Babylonia and Chaldaea, Forming Part of the Labour of the Euphrates Expedition* (London, 1838); *Travels and Researches in Asia Minor, Mesopotamia, Chaldea, and Armenia*, 2 vols (London, 1842); and *Travels in the Track of Ten Thousand Greeks* (London, 1844).
2   'On the Khorsabad Inscriptions', *Transactions of the Royal Irish Academy* 22 (1850), polite literature, pp. 3–72.
3   A note opposing Rawlinson's identification of Nimrud and Calah was read at a meeting of the Syro-Egyptian Society on 9 April. See *Literary Gazette*, no. 1735 (20 Apr. 1850), p. 280. But Rawlinson was right in this.
4   W. F. Ainsworth, 'Remarks on the Topography of Nineveh', *Original Papers Read before the Syro-Egyptian Society of London*, vol. 1/2 (London, 1850), pp. 15–26, esp. pp. 25–6, where there is a reference to Rawlinson's lecture at the Royal Asiatic Society on 12 January 1850.
5   See [W. F. Ainsworth] 'Layard's Assyrian Researches': a review article on A. H. Layard, *Nineveh and Its Remains*, 2 vols (London, 1849), in *New Monthly Magazine* 85 (1849), pp. 240–51, esp. p. 241; and 'The City of Nimrod', *New Monthly Magazine* 88 (1850), pp. 326–31.

FROM CHRISTIAN CARL JOSIAS VON BUNSEN[1]
28 MARCH 1850

9 Carltonhouse Terrace
London
March 28th. 1850

Chevalier Bunsen presents his respects to Dr. Hincks. He has received Dr. Hincks' elaborate dissertation on the Khorsabad inscriptions and avails himself of the opportunity to express his sincerest thanks for a present so valuable in the study of deciphering those important linguistic documents.[2]

Chevalier Bunsen shall be most happy to receive the copies of the same works inscribed to Professors Lepsius and Bopp and to forward them by the courier of the Legation to Berlin.[3]

Chevalier Bunsen has given orders to transmit the second part of Lepsius' Chronology to the bookseller at Belfast, as it is wished by Dr. Hincks.[4]

MS: GIO/H 99

1   Christian Carl Josias von Bunsen (1791–1860), diplomat and scholar, was Prussian Ambassador to the Court of St James from 1842 to 1854, having previously been ambassador to the Holy See in Rome and to Switzerland.
2   'On the Khorsabad Inscriptions', *Transactions of the Royal Irish Academy* 22 (1850), polite literature, pp. 3–72.
3   The Egyptologist Richard Lepsius (1810–84) was professor at Berlin from 1846; Franz Bopp (1791–1867) was Professor of Sanskrit and Comparative German Literature at Berlin from 1821 to 1851.
4   Only the first part of R. Lepsius, *Die Chronologie der Ägypter* (Berlin, 1849) was published.

⁓⧉⧉⧉⁓

FROM GEORGE CECIL RENOUARD
16 APRIL 1850

Swanscombe, Dartford
T. 16. April, 1850

My dear Sir,

Your letter of the 2ᵈ. instᵗ. would have received a much earlier answer had not many impediments intervened. Some visits to London were among the number, & in one of them, I procured Dennis's book which gave me the Etruscan alphabet to which you referred.[1]

I also found it fully discussed in Müller's able work, written some years before that alphabet was discovered.[2] I had mislaid & lost sight of Lepsius's Inscr. Umbricae et Oscae where I should have seen the Etruscan letters accompanied by valuable discussions.[3] When L. published his book, in 1841, he could not have heard of the patera seen & copied by Mʳ. Dennis; for I find by then in the publications of the 'Società Archeologica di Roma', in the Bulletins for Jan. & Feb. 1846 (p. 7.), that Father Giampietro Secchi gives it as the 'alfabeto etrusco di Bomarzo da mè scoperto l'anno scorso nella tazzetta borghesiana': so that it was not noticed till 1844 (Father Secchi's address was delivered 19 Dec. 1845).[4] This alphabet is valuable on many accᵗˢ. but particularly, as confirming Müller's views & containing exactly the number, & nearly the same order of letters as were given by him. Knowing nothing of Mʳ. Dennis's book, I had a long hunt for it & *lit* upon

17

M. Dennis's Einleitung in die Bücherkunde, Wien, 1795–6, a valuable work which I supposed you had consulted at Dublin.[5] It contains, I believe, an Etruscan alphabet.

Of Loewenstern, I spoke in my last letter.[6] His pretensions = *nil*. I believe the want of Assyrian types & the many required have retarded the appearance of Major R.'s last paper.

You say 'if the Babylonian name of Darius be *Darayyachir*' &c. Your value of *cha*, I fear, is very questionable. It is, I believe, first announced in n. ‡ p. 62 of your 'Khorsabad Inscriptions',[7] where a misprint 'see note in p. 42' for 'p. 46', has given me a great deal of trouble. It is there that you refer to your 'Hieroglyphic Alph.' (1847) p. 75, that you intimate that κι or κυ, *ci* or *cy*, in the time of Galen was probably pronounced *chi* (as in Italian & by the Ionian Greeks, at the present day). κι in Coptic, you remark, is almost always *qi* & there can be little doubt that *q* was *ch*. I should say there can be little doubt that *q* was *g* (perhaps the Arab *q*); for it is used almost promiscuously for *d̲* which is either *j* or *zh*. The Copts are said now to pronounce it as the latter which, you know, is the sound commonly given to *jim* in Morocco. You say p. 75, 'it was unfortunately by Frenchmen that the equivalents of the Coptic letters were taken down when it was yet a living language'.[8] Where you met with that account I know not;[9] but it seems to have mislead you; & had you seen Étienne Quatremère's 'Recherches sur la langue de l'Égypte (Paris, 1808)', you would there have found a complete history of the transition from the ancient Egyptian to the modern Coptic & of the study of the latter esp. in Europe.[10] [. . .] & Raimondi at Rome were the first who studied & had intentions of printing Coptic books.[11] Salmasius got a gram. & lex. from *Rome*:[12] but Pietro della Valle, the learned & instructive Roman traveller, was the first who drew up a complete Coptic grammar which was printed by Kircher (a German) in 1643, in his Ling. Aegypt. Restituta.[13] So you see Italians & Germans – not Frenchmen were the first *Copticians*. The Coptic *q* moreover was, you know, derived from a hieroglyphic & not from a Greek letter. In Egypt the *jim* is still vulgarly pronounced *g* as in get, got, girl & *qa* has nearly the same sound, tho' doubtless deeper in the throat. At Damascus it forms a sort of hiatus, or catch as in gasping. But is it not more probable that your Babylonian syllable was *ṣir* or *tsir*? Had such a change, as you suppose, occurred in the second century, I think traces of it would be found in the writings of the ancient grammarians. *Gh*, *ch*, & *j*, represent sounds to which they probably were strangers, but my paper tells me I must bid you farewell & am,

dear Sir,
truly yours
G. C. Renouard

P.S. It can scarcely be doubted that Salmasius's conjecture Δαριαυην for Δαρινκην (in Strabo p. 785[14]), is right; so nearly do αυ & ηκ approach to each other, for ΑΥ & ΗΚ approach still nearer than the cursive letters. Unfortunately Kramer's (the only good edition) does not yet go so far as to include this passage.

MS: GIO/H 459

1   G. Dennis, *The Cities and Cemeteries of Etruria*, vol. 1 (London, 1848), p. 225.

2   Renouard obviously consulted K. O. Müller, *Die Etrusker* (Breslau, 1828). A new edition by W. Deeke was published in Stuttgart, 1877.

3   R. Lepsius, *Inscriptiones umbricae et oscae* (Leipzig, 1841).

4   G. Secchi, 'Tesoretto di etruschi arredi funebri in oro posseduti al signor cav. Giampietro Campana', *Bullettino dell'instituto di corrispondenza archeologica*, nos 1–2 (Jan.–Feb. 1846), pp. 3–16 (published 15 March 1846). It was published separately as *Descrizione d'alquanti Etruschi arredi in oro posseduti dal sig. cav. Giampietro Campana e interpretazione d'una epigrafe etrusca sopra una fibula con alfabeto etrusco della tazzetta borghesiana scoperto nell'anno CDDCCCXLV* (Rome, 1846).

5   J. N. C. M. Dennis, *Einleitung in die Bücherkunde*, 2 vols (Vienna, 1795–6).

6   See Renouard's letter of 7 March 1850.

7   'On the Khorsabad Inscriptions', *Transactions of the Royal Irish Academy* 22 (1850), polite literature, pp. 62–3. The footnote begins on p. 62 but the relevant sentence is on p. 63: 'I feel myself compelled to value the two characters under consideration as *chi* and *chir*; and accordingly to read the Babylonian name of Darius as *Darayyachir!*' We now know that the Babylonian form is *Dariyamus*, the Old Persian *Darayavaus*.

8   Hincks wrote: 'A doubt, however, may exist whether, in the age of Galen, the Greek syllable κυ commenced with the sound of our K. We know that, in Coptic transcriptions of Greek words, κ followed by ι, or a vowel of similar power, was usually written *q*... Now there can be little doubt that the power of *q* was *ch*. The Coptic is no longer a living language; and, when it was so, it was unfortunately by Frenchmen that the equivalents of its letters were taken down, who, we know, cannot pronounce *ch* or *j*, but substitute for them *sh* and *zh*.' See Hincks, 'An Attempt to Ascertain the Number, Names, and Powers, of the Letters of the Hieroglyphic, or Ancient Egyptian Alphabet; Grounded on the Establishment of a New Principle in the Use of Phonetic Characters', *Transactions of the Royal Irish Academy* 21 (1847), polite literature, pp. 203–4. Renouard refers to one of the separately printed copies which had a different pagination.

9   Reading 'not' for MS: 'it'.

10   E. M. Quatremère, *Recherches critiques et historiques sur la langue et littérature de l'Égypte* (Paris, 1808).

11   The first name is illegible. Giovanni Battista Raimondi (1540–1610) was Director of the Medici Press in Rome.

12   Claudius Salmasius (1588–1653), French classical scholar.

13   In 1626 Pietro della Valle (1586–1652) brought back manuscripts from the Near East, including two Coptic-Arabic dictionaries. Athanasius Kircher (1602–80), the learned German Jesuit and the founder of Coptic studies, made use of Pietro della Valle's Coptic MSS for his *Lingua Aegyptiaca restituta* (Rome, 1643). See R. Solé and Dominique Valbelle, *The Rosetta Stone: The Story of the Decoding of Hieroglyphics* (London, 2002).

14   See Strabo 16.4.27. Renouard is referring to Isaac Casaubon's edition, *Strabonis rerum geographicarum libri XVII* (Paris, 1620), p. 785. In his letter of 21 November 1851, Renouard lamented that only two volumes of G. Kramer's edition, *Strabonis Geographica* (Berlin, 1844; 1847) had been published. However, vol. 3 appeared in 1852.

FROM JAMES WHATMAN BOSANQUET[1]
23 MAY 1850

Claysmore near Enfield
23rd. May 1850

Dear Sir,

I have just been reading with much interest Major Rawlinson's Commentary on the Cuneiform Inscriptions, which of course you have already seen.[2] In it, I observe that he refers frequently to a publication of yours on the inscriptions at Khorsabad.[3] I have been unable to learn where your work is published, and I have taken the liberty, therefore, of writing to you to ask the name of your publisher. Will you also have the goodness to say, where I may have the pleasure of sending a copy of a work on chronology, published by me in 1848, in which I have suggested material alterations in the arrangement of the chronology of the Kings of Babylon & Nineveh, and which it appears to me may possibly throw light upon the question between you and Major Rawlinson, arising out of the inscriptions.[4]

In your remarks on the inscriptions at Van, I observe you say, that the name *Niladan* is clearly legible in the inscriptions, as the son of Assar-adan son of the king who built Koyunjik.[5] The inference from which is unavoidable, that Assar-adan is the same as Assaradinus of the Canon of Ptolemy, whose next successor but one was Kiniladinus, and whose reign is immoveably fixed as commencing in the year B.C. 680. Are you still of the same opinion? If so, then must Assar-adan or Assaradinus of the Canon, be the Esarhaddon son of Sennacherib of Scripture, and the Asordanius son of Sennacherib of Polyhistor, who he says was placed on the throne of Babylon shortly before Sennacherib threatened Jerusalem, because there is no other name in the Canon which can possibly be supposed to represent the son of Sennacherib who reigned at Babylon. The result of which is, that the reign of Hezekiah must be lowered so as to bring his fifteenth year on a level with the fall of Sennacherib and rise of Esarhaddon, that is, to the year B.C. 680.

I have already pointed out in my work, that the years B.C. 680 & 681, counted upwards from the birth of Christ, were in regular series, one Sabbatical, the other a Jubilee, and this I arrive at on the assumption that Christ was born in the Sabbatical year ending the period of 'seven weeks, and three score & two weeks', i.e. at the end of one Jubilee and sixty two Sabbatical weeks. Dan. IX and these two successive fallow years are undoubtedly referred to by Isaiah in the words 'Ye shall eat this year that which groweth of itself, and in the second that which springeth of the same', which words were spoken in the 14th. year of Hezekiah.[6] I have also shown that Demetrius, an Alexandrine Jew, quoted by Clemens Alexandrinus, expressly fixes the date of Sennacherib's invasion to the same year

B.C. 680. Now these conclusions appear to me to be confirmed by the inscriptions, if indeed it can be shown that Assar-adan, son of the builder of Koyunjik, was father of Chiniladinus.

Major Rawlinson reads the titles of the two predecessors of Assar-adan, as Bel-adonim-shu and Arkotsin. Is it possible, that Bel-adonim is the same as Baladan, and father of 'Merodach Baladan, son of Baladan', who sent messengers to Hezekiah and that Acices who preceded him at Babylon is Arkotsin? I see in my table they immediately precede Assaradinus. Again may Biarku, the Egyptian king whose name appears on the monuments be read Tiarku, or Tearchos, i.e. Tirakah King of Ethiopia?

If the reign of Hezekiah is lowered about 30 years, the fall of Jerusalem and the 19th. of Nebuchadnezzar must also be lowered to the same extent, leaving a gap of 30 years between the reign of Nabopolassar which is astronomically fixed, and the reign of Nebuchadnezzar concerning which the copies of the Canon differ. This interval I conceive must be filled by the period of invasion of the Scythians, mentioned by Herodotus, and for which there is no place in the generally received chronology. Thus the principal objection of Major Rawlinson to your conclusion, that the three kings of Khorsabad & Koyunjik are Shalmeneser, Sennacherib, & Esarhadon, viz that there is no space left for the acts of certain mighty kings who reigned between Assar-adan & Nebuchadnezzar, is removed: for according to this arrangement there is a space of more than 100 years between the two.[7]

Believe me to be
Yours very truly
J. W. Bosanquet

MS: GIO/H 51

1   James Whatman Bosanquet (1804–77), banker, chronologist and biblical historian.
2   See Renouard's letter of 7 February 1850, n. 7.
3   'On the Khorsabad Inscriptions', *Transactions of the Royal Irish Academy* 22 (1850), polite literature, pp. 3–72.
4   *Chronology of the Times of Daniel, Ezra, and Nehemiah* (London, 1848). There were frequent exchanges between Hincks and Bosanquet on the subject of chronology.
5   See Hincks, 'On the Inscriptions at Van', *Journal of the Royal Asiatic Society* 9 (1848), pp. 439–40.
6   Isaiah 37: 30.
7   The merits of the arguments put forward by Bosanquet, Hincks and Rawlinson with regard to Biblical chronology cannot be dealt with here. However, note Rawlinson's remark: 'After reading, indeed, and carefully considering all Dr Hincks's arguments, I remain as incredulous as ever of the identity of the Koyunjik king with the Sennacherib of Scripture.' See 'On the Inscriptions of Assyria and Babylonia', *Journal of the Royal Asiatic Society* 12 (1850), p. 454 n.

~~~

WILLIAM DESBOROUGH COOLEY TO THE ATHENAEUM[1]
MAY 1850

[May 1850]

[Dear Sir,]

A recently published volume of the Transactions of the Royal Irish Academy (vol. xxii, part 2), contains a paper on Babylonian Writing, by the Rev. D[r] Edw. Hincks, which is peculiarly interesting at the present moment.[2] It is a sequel to those communications for which the Academy awarded D[r] Hincks the Conyngham gold medal in 1848.[3]

Among the chief literary triumphs of the present age may be justly reckoned its linguistic discoveries, achieved by deciphering and interpreting monumental inscriptions and various remains of ancient writings, all of inestimable value, however imperfect they may be, as tending to establish on a solid basis and to complete our knowledge of the early history and civilization of our species. The hieroglyphics of Egypt and the cuneiform inscriptions, Median, Assyrian, and Babylonian, are now at length revealed and expounded to us, in the nineteenth century, by the learning and sagacity of a few, among whom Dr Hincks holds a foremost place. Indeed, in sound scientific method, boldness of conjecture without rashness, and felicity in seizing on such points as admit of being chronologically determined, he is unrivalled.

To show the importance of these discoveries, I need only refer to the Behístún inscription, copied and explained in a masterly manner by Major Rawlinson, and which throws so steady a light on the history of Persia five-and-twenty centuries ago. And if there be any one inclined to doubt their reality, let him only weigh attentively the remarkable fact, that two highly gifted and accomplished men, Major Rawlinson and D[r] Hincks, unknown to each other and wide asunder, the one on the banks of the Euphrates, the other on the shores of Lough Strangford,[4] both applying themselves to the study of the cuneiform inscriptions, have arrived in general at precisely the same conclusions – have found the same alphabet, the same grammatical forms, and the same terms. The differences between them – for in some points they differ – are not such as to invalidate the conclusions in which they concur, but serve rather as proofs of their respective independence and originality. They form also obvious marks by which lookers-on may readily estimate the progress of investigation; and this brings me to the point which I have immediately in view, namely, the unquestionable priority of D[r] Hincks in the discovery of the ideographic element, which is now admitted to be of frequent occurrence in the Babylonian inscriptions. The existence of this element was already recognized by him in a paper on the Van inscriptions, published in 1848 (Journal of the Royal

Asiatic Society, vol. ix), but he developed his views more completely, and entered into details, in the paper above alluded to, and which was read before the Royal Irish Academy on the 25th of June, 1849. These views have been since adopted in a great measure by Major Rawlinson, who now reads, for example, Assar-adan-pal (Sardanapalus), where he previously read Ninus, thus approximating to Dr Hincks, by whom the same name has been always read Asshurhadin.[5]

It must not be supposed that these remarks are intended to detract from the well-merited reputation of Major Rawlinson; their object is merely to vindicate the merits of a comparatively recluse student, who, with the great disadvantage of pos-sessing but a small supply of texts to study, has nevertheless laboured with signal success – thanks to his extensive learning and great analytical powers – in solving some of the most difficult literary problems which have ever claimed the attention of the learned. The student who, toiling in the fields of literature, is fortunate enough to reclaim something from the waste, must like the husbandman, look well after his landmarks. In proof of this, it will be sufficient to mention that Chevalier Isidore de Löwenstern has this year published a work in which he claims to have discovered the existence of an ideographic element in Babylonian writing, and illustrated his meaning by the very examples which were communicated to him by Dr Hincks (in the paper on the Van inscriptions) nearly two years ago.[6] Discoveries which are worth claiming ought at least to be fairly recorded. In the bye-ways of learning, injustice is easily and often done by mere suppression of the truth.

Suum cuique

Source: The Athenaeum, no. 1178 (25 May 1850), p. 555

1   William Desborough Cooley (1795–1883), Irish geographer and controversialist, was a student at Trinity College, Dublin from 1811 to 1816. He would have met Hincks there. Thomas Kibble Hervie (1799–1859), poet and journalist, was editor of *The Athenaeum* from 1846 to 1853. The previous letter suggests that Charles Wentworth Dilke, the proprietor, had a strong say about the content of the journal.
2   'On the Khorsabad Inscriptions', *Transactions of the Royal Irish Academy* 22 (1850), polite literature, pp. 3–72.
3   It was reported in the minutes of a meeting of the Council of the Royal Irish Academy on Monday, 3 April 1848 'that this Committee do recommend to the Council the Essays "Upon the Number Names and Powers of Letters of the Hieroglyphic, or Ancient Egyptian Alphabet" and the series of essays on the Cuneiform Writing by Dr Hincks as eminently deserving of a Cunningham medal.' After a ballot, it was 'resolved that medals be awarded to Mr. O'Donovan and Dr. Hincks'. See the Academy's Council Minutes, no. 7 (May 1845–Mar. 1849), pp. 283–4. The series of papers on cuneiform writing included 'On the First and Second Kinds of Persepolitan Writing', *Transactions of the Royal Irish Academy* 21 (1846–8), polite literature, pp. 114–31; 'On the Three Kinds of Persepolitan Writing, and on the Babylonian Lapidary Characters', pp. 233–48; 'On the Third Persepolitan Writing, and on the Mode of Expressing Numerals in Cuneatic Characters', pp. 249–56; 'On the Inscriptions of Van', *Journal of the Royal Asiatic Society* 9 (1848), pp. 387–449.
4   The town of Killyleagh, where Hincks's parish church and house were situated, was close to the shore of Strangford Lough.
5   See Rawlinson, 'On the Inscriptions of Assyria and Babylonia', *Journal of the Royal Asiatic Society* 12 (1850), p. 421.
6   See Isidore Löwenstern's angry reply in his letter of 23 August 1850 below.

FROM WILLIAM DESBOROUGH COOLEY
27 MAY 1850

33 King St, Bloomsbury
27 May 1850

Dear Sir,

I hope you have received a copy of the last Athenaeum which I ordered to be sent to you. It contains the long-promised paragraph in vindication of your rights as a discoverer.[1] On my first application to Mr Dilke (propr^r. of the Athenaeum), he gave me to understand that he had no space to spare & should be obliged to defer for two months perhaps, any new matter not of transient interest or urgent nature.[2] However, after issuing a few double N^os., he felt relieved, & hinting to me that brevity was of much importance, he inserted my letter without any demur. The letter in question will, I trust, attain the desired end, namely, of calling attention to your investigations, as fully as a more elaborate performance, for which indeed in the present state of my health – repeated attacks of neuralgia have deprived me almost wholly of the sense of hearing – I have scarcely the requisite powers of application. Besides, several of the papers which you have at different times sent to me, I lent, through a friend, to Dean Milman who has never returned them.[3] And here I may mention, that in the evidence which I gave before the Commissioners in the British Museum, printed last year but only recently made public & which evidence, as you will have perceived, if you be a constant reader of the Athenaeum, has attracted much notice,[4] I illustrated the necessity of rendering the Transactions of Learned Societies as accessible as possible, by the fact that you, the most scientific & successful investigator of the cuneatic inscriptions, were not once mentioned in a (then) recent paper (by Milman) in the Quarterly Review.[5] That was an instance of flagrant ignorance on the part of Milman, who affected nevertheless to excuse his omission of your name, by the difficulty of understanding those Irish!

With respect to the Edinburgh Review, the chief practical difficulty lies in the great number of its literary supporters, in consequence of which each contributor is strictly confined to one departm^t.[6] But I shall not lose sight of this matter, & hope to see you ere long in full enjoyment of your honours as a discoverer, of which indeed you cannot, in the long run, be deprived.

Believe me dear Sir
Yours very sincerely
W. D. Cooley

I have no doubt that any communications from you will be *gladly* though perhaps not *thankfully* received by the Editor of the Athenaeum.

MS: GIO/H 123

1   Cooley's letter to *The Athenaeum*, no. 1178 (25 May 1850), p. 555, follows this letter.
2   Charles Wentworth Dilke (1789–1864) was a proprietor of *The Athenaeum* from 1830 to the end of his life and editor from 1830 to 1846.
3   Henry Hart Milman (1791–1868), historian and Dean of St Paul's, London.
4   In 1847 commissioners were appointed to inquire into the constitution and government of the British Museum. The commission issued its report on 28 March 1850. See *Report of the Commissioners Appointed to Enquire into the Constitution and Government of the British Museum* (London, 1850).
5   See H. H. Milman, 'Persian and Assyrian Inscriptions': a review article on H. C. Rawlinson, *The Persian Cuneiform Inscriptions at Behistun* (London, 1846–7); C. Lassen and N. L. Westergaard, *Über die Keilinschriften der ersten und zweiten Gattung* (Bonn, 1845); N. L. Westergaard, *On the Deciphering of the Second Achaemenian or Median Species of Arrow-head Writing* (Copenhagen, 1844); A. Holtzmann, *Beiträge zur Erklärung der persischen Keilinschriften* (Karlsruhe, 1845); F. Hitzig, *Die Grabinschrift des Darius in Nakshi Rustam erläutert* (Zurich, 1847); J. Mohl, *Lettres de M. Botta sur les découvertes à Khorsabad* (Paris, 1845), in *Quarterly Review* 79 (Dec. 1846–Mar. 1847), pp. 413–49. In a later review Milman referred to Hincks's work only once. See his review of A. H. Layard, *Nineveh and Its Remains* (London, 1848), in *Quarterly Review* 84 (Dec. 1848), pp. 106–53, esp. 142–3.
6   See 'The Edinburgh Review, 1802–1900', in W. E. Houghton (ed.), *Wellesley Index to Victorian Periodicals, 1824–1900*, vol. 1 (Toronto, 1966), pp. 416–29.

FROM THOMAS ROMNEY ROBINSON[1]
JUNE 1850

June 1850

Dear Hincks,

Stevelly[2] thinks a few lines from me might be of use among your testimonials, so I send you the enclosed if you think it is worth using. I have scarcely any friends in Parliament. The only one I could reckon on is Sir Robert Inglis, and if you let me know when the application is likely to be made, I will write to him to back it.[3]

If I had the power as I had the will, I would send you off with a staff to grub, copy, draw, and decypher, as long as you could find anything at Nineveh, Babel or Elymais. But Lord John I fancy cares more for the *present* than the *past*: still I hope that the *present* will put the screw on him in this case.[4]

Yours ever,
T. R. Robinson

Source: Davidson, *Edward Hincks*, p. 47

1   Thomas Romney Robinson (1793–1882), Irish astronomer and physicist. See *ODNB*.
2   John Stevelly was born in Cork in 1794 or 1795, son of George Stevelly. He was educated at Trinity College, Dublin: B.A. 1817, M.A. 1827, LL.B., LL.D. 1844. From 1823 he was Professor of Natural Philosophy at the Royal Belfast Academical Institution and later at Queen's College, Belfast. He died in 1868.
3   Hincks's diary for 4 May 1850 says: 'Determined to try for the pension'. Some of the June correspondence is concerned with Hincks's application and his supporters were: Sir Robert Henry Inglis (1786–1855), M.P. for Oxford University; Robert James Tennent (1803–86), M.P. for Belfast; Thomas Thornely, M.P. (1781–1862); George Alexander Hamilton (1802–71), M.P. for Dublin University; Joseph Napier (1804–82), judge, who was attorney-general for Ireland in 1852 in Lord Derby's first administration; James Charles Chatterton (1794–1874), M.P. for Cork. Hincks's diary for 9 August says: 'This day my memorial was presented to Lord J. Russell by Mr. Tennent and Mr. Thornely – Mr. Hamilton, Sir R. Inglis, Mr. Napier, and Col. Chatterton having signed with them a recommendatory letter.' The application was not successful and Hincks was very despondent.
4   Lord John Russell was prime minister from 1846 to 1852.

FROM JAMES WHATMAN BOSANQUET
1 JUNE 1850

Claysmore Enfield
Middlesex
1ˢᵗ. June 1850

Dear Sir,

It was only the day before yesterday that I received your kind present of your publication on the inscriptions at Khorsabad.[1] This evening I have had the pleasure of forwarding to you by the post a copy of my work on Chronology.[2] I observe that some of our dates nearly coincide: you having deducted 30 years from the reign of Manasseh, while I have lowered the reign of Nebuchadnezzar by 33 years, leaving space for the Scythian invasion. I do not insist upon my arrangement of the Persian chronology, but rather throw it out as a suggestion, but I am satisfied that the reigns of Cyrus and Darius Hystaspes overlapped, and that there is something wrong here in the arrangement of the received system of chronology. Your date of the capture of Samaria by the king of Assyria is exactly that which I have deduced from Demetrius quoted by Clemens Alexandrinus. I have scarcely had opportunity yet to digest your learned observations on the inscriptions.

Believe me
Sir
Yours very truly
J. W. Bosanquet

MS: GIO/H 52

1   'On the Khorsabad Inscriptions', *Transactions of the Royal Irish Academy* 22 (1850), polite literature, pp. 3–72.
2   *Chronology of the Times of Daniel, Ezra, and Nehemiah* (London, 1848).

EDWARD SABINE[1] TO THOMAS DIX HINCKS[2]
11 JUNE 1850

Woolwich
June 11th

My dear Sir,

If your correspondent has reference to the £1000 in the estimates of this year to be expended according to the recommendation of the Royal Society, he in error, I apprehend, in supposing it applicable to the promotion of *literary* & scientific efforts. The Royal Society was originally chartered for the promotion of *natural* knowledge, and still bears that title, and cannot of course recommend beyond its own sphere.

I should think that the appropriate channel through which Dr. Hincks might seek the aid of public money to assist him in his researches is the annual parliamentary grant of annuities amounting to £1200; and that his application might not unreasonably be made through the Royal Society of Literature, but I am quite uninformed as to the customs of that Society.

With our very kind regards,
believe me,
very sincerely yours,
Edward Sabine

Source: Davidson, *Edward Hincks*, p. 45

1   Edward Sabine (1788–1883), army officer, physicist, ornithologist and explorer, was born in Dublin. He was a general secretary in the British Association for the Advancement of Science for twenty years and its president in 1852. He was president of the Royal Society from 1861 to 1871. See *ODNB*.
2   Thomas Dix Hincks (1767–1857), Edward Hincks's father.

HENRY HOLLAND[1] TO THOMAS DIX HINCKS
12 JUNE 1850

25 Brook Street
London
June 12th, 1850

My dear Sir,

I hasten to answer your letter and to assure you of my earnest desire to serve Dr. Hincks in the object to which it relates, in any way in which I may be able to do so. I am obliged to express my fear that this present high tide of economy is not the most favourable condition of things for the prosecution of his claim on public aid; nor do I feel assured that the Ministers have any fund in their power, except the occasional vacancies on the pension list, without coming to Parliament through the Estimates. But I will endeavour to ascertain what may be done and as I often see Lord J. Russell and Lord Lansdowne, I will find some opportunity to put the subject before them.[2]

If your son should come to London I hope he will call upon me, when, in addition to the pleasure of seeing him, I may obtain more explicit knowledge of his views, and the channels through which they may best be forwarded.

Believe me ever,
My dear Sir
Yours faithfully,
H. Holland

Source: Davidson, *Edward Hincks*, pp. 45–6

1   Henry Holland (1788–1873), physician, was born in Knutsford, Cheshire. Through the Holland family connections, he and Edward Hincks and Elizabeth Gaskell were cousins.
2   Henry Petty Fitzmaurice, third marquess of Lansdowne (1780–1863), was lord president of the council and leader of the House of Lords in Lord John Russell's first government, 1846–52.

⚜

TO JOHN LEE[1]
17 JUNE 1850

Killyleagh C° Down
17th. June 1850

My dear Sir,

I received your letter of the 14th. this morning. I feel much obliged by your kind invitation of doing what you can to promote my views. Lord John has so many applicants for his patronage that I have not much hope of his being interested in my favour. I feel so confident that, as to the *main* question at issue between Major Rawlinson & me, I am right and he completely wrong, that I should regret being forced to quit the field. I am curious to see Grotefend's paper.[2] M[r]. H. Greer Belfast is my bookseller; and Longmans are his London correspondent; but as a parcel left there would not reach me till three weeks hence, I think it best to send stamps that it may come by post. I presume it is under a pound weight. I have seen nothing of M[r]. Nash's.[3]

Believe me
My dear Sir,
Your very faithful & obliged servant
Edw. Hincks

MS: BL Add. 47491A, f. 219

1   John Lee (formerly Fiott) (1783–1866) was an English antiquary.
2   Perhaps G. F. Grotefend, 'Keil-Inschriften aus der Gegend von Niniveh, nebst einem persischen Siegel', *Zeitschrift für die Kunde des Morgenlandes* 7 (1850), pp. 63–70 + 1 lithograph. See Renouard's letter of 7 March 1850, n. 3.
3   David William Nash (1809–76), barrister and naturalist. He was a member of the Syro-Egyptian Society and published 'On the Antiquity of the Egyptian Calendar', *Original Papers Read before the Syro-Egyptian Society of London*, vol. 1/2 (London, 1850), pp. 29–57.

FROM HENRY HOLLAND
20 JUNE 1850

25 Brook Street
June 20th, Thursday, 1850

My dear Sir,

In thanking you for the very remarkable paper you have sent me, let me express the admiration with which I regard the zeal and ability you have given to this research.

I should be happy in believing that the means might be amply afforded you for further prosecuting it. But I dare not at the moment feel very sanguine on the subject, seeing the many circumstances which may throw contrariety in the way, and the difficulty of engaging Ministers to think for an instant on a matter of this kind, when so heavily embarrassed on every side by difficulties at home and abroad. Should you deem it worth while to come to London, I hope to see you, to talk over any possible access to the object desired, but none of those you mention seem very feasible. Have you had any communication with Major Rawlinson? He frequently comes to breakfast with me, and when next this happens I will speak to him on the subject. He is a man for whom I have high esteem.[1]

I every year take a rapid run for a few weeks in the autumn, as a relief to my labours during the rest of the year. In my excursion last year I got within 400 miles of Nimrud, following through Armenia the steps of Layard, who had gone over the ground some ten days before.[2]

I am interrupted here and must hastily finish.

Believe me, my dear Sir,
faithfully Yours,
H. Holland

Source: Davidson, *Edward Hincks,* p. 46

1   Holland could scarcely have made a more annoying remark to Hincks. Exactly a month later, however, Hincks was in London and the entry in his diary for 20 July says: 'Called on Mr. Norris, and then with a letter from him on Major Rawlinson, with whom I had some talk.' He also breakfasted with Henry Holland. At the end of July he travelled to Edinburgh for the meeting of the British Association for the Advancement of Science. According to his diary entry for 2 August, he 'heard a paper on Sardinian and Sicilian languages – had some talk in committee room with Tattam and Rawlinson'.

2   Austen Henry Layard (1817–94) was appointed attaché at Constantinople in April 1849. From October 1849 till April 1851 he conducted excavations at Kuyunjik.

FROM THE 6TH DUKE OF MANCHESTER (GEORGE MONTAGU)[1]
29 JUNE 1850

Tandragee June 29th 1850

Dear Dr Hincks,

I should be very glad to hear that you had received assistance from the Government for the prosecution of your discoveries in hieroglyphic and cuneiform writing. When I find that such adversaries as Bunsen & Rawlinson, when writing against your discoveries in these two lines, admit your learning, ingenuity and great sagacity, I feel the more confident that the cause of science & literature would be greatly benefited by your being put into a position to pursue uninterruptedly the task of deciphering these ancient records which you have so successfully commenced.

Believe me
Very truly yours
Manchester

MS: GIO/H 275

---

1   George Montagu, 6th Duke of Manchester (1799–1855), was an orangeman and strongly evangelical.

ISIDORE LÖWENSTERN[1] TO THE ATHENAEUM
23 AUGUST 1850

Paris, August 23

[Dear Sir,]

After an absence of some months from Paris, I find in your journal two articles, to which you will oblige me by admitting the following reply. In your number for the 25th May, I read an article, without any other indication of authorship than the device SUUM CUIQUE, in which the discovery of an ideographic element in Babylonian writing is ascribed to the Rev. Dr. Edw. Hincks.[2] I must protest against this assertion in favour of the learned Director of the Museum of the Louvre, M. Adrien de Longpérier, to whom the merit of this important discovery belongs. He published it nearly three years ago in the 'Revue archéologique' (Oct. 1847, p. 504).[3]

I would not have dwelt on the singular coincidence of the alleged discovery by Mr. Hincks in December 1847 with that of M. de Longpérier two months sooner –

31

a coincidence which no doubt has been quite fortuitous – if the writer of the article had not mentioned my name coupled with the assertion of my having this year published a work in which I am said to claim the discovery of the existence of an ideographic element in Babylonian writing, and to illustrate my meaning by the very examples which were communicated to me by Dr. Hincks in the paper on the Van inscriptions nearly two years ago.

Having quoted M. de Longpérier as the discoverer of the ideographic element in my 'Note sur une table généalogique des rois de Babylone dans Ker-Porter' (Revue archéol. Oct. 1849, p. 417),[4] and referred to this statement in my publication last alluded to, 'Remarques sur la deuxième écriture cunéiforme de Persépolis' (Revue archéol. Févr. 1850, note 1, p. 711),[5] I think that I cannot be considered as making any other pretension to the discovery of the ideographic element in cuneiform writing than that of considering the knowledge of this fact as a consequence of Champollion le Jeune's system of the hieroglyphics of ancient Egypt; a system the existence of which I have discovered in the cuneiform writings of Assyria and Babylonia,* and introduced into this study, and to which the ideographic element due to M. de Longpérier's sagacity, forms the most interesting complement.

In your paper of the 6th July I find in the report of the Syro-Egyptian Society, that a member of that learned body, Mr. D. W. Nash, is said 'to endeavour to show that the so-called Median inscriptions were conceived not in a Tartar dialect, as Major Rawlinson supposed, but in a Semitic tongue, the language of the population of Western Asia prior to the supremacy of the Arian immigrants. This language, though not the modern Pehlevi, is its ancient representative, and the language, not as M. Löwenstern supposes, of merely the Southern Elymaeans, but of the great substratum of the population of Persia and Media'.[6] Though much satisfied to see that Mr. Nash has given so clear an account of the results which I have lately obtained and published in the above-mentioned treatise on the second cuneiform writing of Persepolis ('Rev. arch.' Févr. 1850), I have yet to object that my name in the above report is mentioned only for a special remark. I therefore request the following change in the drawing up of the article: 'M. Löwenstern has endeavoured to show in his work the results for which Mr. Nash, probably unwillingly on his part, has been quoted; and *vice versâ* Mr. Nash proposes an amendment to them, in adopting the language specified by M. Löwenstern as having belonged to the great substratum of the population of Persia and Media, contrary to M. Löwenstern's supposition, who, in conformity to Scripture, considers Madai and its Arian language as ancient in its dwelling-places as Elam and its Semitic tongue in the abodes which he has assigned to them prior to the invasion of the Japhetic Paras.'

I am, &c.
Chevalier Isidore Löwenstern

* NOTE C, *Exposé des éléments constitutifs du système de la troisième écriture cunéiforme de Persépolis*, Paris, 1847.

Source: *The Athenaeum*, no. 1193 (7 Sept. 1850), p. 953

1   Isidore Löwenstern was an orientalist who lived in France.
2   See William Desborough Cooley's letter of May 1850 above. It was published in *The Athenaeum*, no. 1178 (25 May 1850), p. 555.
3   A. de Longpérier, 'Lettre à M. Isidore Löwenstern sur les inscriptions cunéiformes de l'Assyrie', dated 20 September 1847, *Revue archéologique* 4/2 (1848), pp. 501–7. Adrien de Longpérier (1816–82), a French archaeologist, was Director of the Louvre in Paris from 1847.
4   'Note sur une table généalogique des rois de Babylone dans Ker-Porter', *Revue archéologique* 6/2 (1850), pp. 417–20.
5   'Remarques sur la deuxième écriture cunéiforme de Persépolis', *Revue archéologique* 6/2 (1850), pp. 687–728, here p. 711, n. 1.
6   Report of the Syro-Egyptian Society, *The Athenaeum*, no. 1184 (6 July 1850), pp. 713–14.

FROM CHARLES WILLIAM WALL[1]
29 AUGUST 1850

Trin: Coll: Augt. 29 – 50

My dear Hincks,

Many thanks for your letter and inclosure which I received yesterday on my return from England where I passed the last four weeks for the benefit of a little ramble and change of air. I am glad to see you also have taken the recreation of a trip to Edinburgh, where you had the opportunity of meeting some of the more remarkable savans, Rawlinson in particular.[2] I have read the paper to which you refer in the last Athenaeum and feel no little disgust at the egotism and unfairness he betrays in virtually ascribing to himself what really belongs to old Grotefend, the discovery of the alphabetic nature of the Persian species of cuneiform writing; for the manner in which he describes the finding out of the alphabet employed in this writing applies not at all to Grotefend but only to himself.[3] As for the paper you have inclosed to me, I must candidly confess to you, it appears to me no better than mere moonshine and I have very little doubt that, if as much Chinese were laid before you as you have samples of Assyrian cuneiform writing, and if you applied the same industry and ingenuity to the investigation, you could coin as plausible fragments of a language from one set of materials as from the other.[4] Where I think your labours have been really useful is in ascertaining the use of the secondary set of consonants in the Persian cuneiform alphabet, for which I have taken care to shift the credit from the

Major to you in a work which will probably be overlooked and neglected as long as the present rage lasts for deciphering undecipherable records, but which will in time I trust receive due attention.[5]

> Yours dear Hincks
> very truly
> Chas Wm Wall

MS: GIO/H 542

1   Charles William Wall (1780–1862) was Professor of Oriental Languages at Trinity College, Dublin from 1825 to 1849.

2   A meeting of the British Association for the Advancement of Science was held at Edinburgh for a week from 31 July. At this meeting Hincks read a paper 'On the Language and Mode of Writing of the Ancient Assyrians'. See the *Report of the Twentieth Meeting of the British Association for the Advancement of Science; held at Edinburgh in July and August 1850* (1851), p. 140 + 1 plate. The plate illustrates well how Hincks used the grammatical characteristics of Semitic languages to elucidate the writing system. See Daniels, 'Edward Hincks's Decipherment of Mesopotamian Cuneiform', in Cathcart (ed.), *The Edward Hincks Bicentenary Lectures*, pp. 43–6.

3   See *The Athenaeum*, no. 1191 (24 Aug. 1850), pp. 908–9.

4   Wall must have received a copy of Hincks's paper 'On the Khorsabad Inscriptions', *Transactions of the Royal Irish Academy* 22 (1850), polite literature, pp. 3–72. He was prone to making dismissive and some-times ridiculous remarks about ancient Egyptian hieroglyphic writing and Mesopotamian cuneiform. At Trinity College, Dublin, he was a member of the board and for a time he blocked the preparation and publication of a catalogue of the Egyptian papyri in the Trinity College Library by Hincks. He regarded the study of Egyptian hieroglyphic texts with contempt. See the remarks in his letter of 23 March 1843 (vol. 1).

5   'On the Different Kinds of Cuneiform Writing in the Triple Inscriptions of the Persians, and on the Language Transmitted through the First Kind', *Transactions of the Royal Irish Academy* 21 (1848), polite literature, pp. 257–314. Wall praises Hincks's insights into the Persian vowel system and he also mentions Rawlinson's contribution. In a footnote Wall is fulsome in his praise of Hincks: 'Though I venture to differ upon some points with the Rev. Dr. Edward Hincks, late Fellow of Trinity College, Dublin, yet, I must say, I consider no author superior to him, and very few his equals, in ingenuity combined with learning' (p. 273).

TO THE EDITOR OF THE ATHENAEUM
14 SEPTEMBER 1850

> Killyleagh C° Down
> 14th. September 1850

[Sir,]

In your paper of the 7th. inst. there is a letter from Chevalier Isidore Löwenstern which contains a charge, or at least an insinuation, against me of having pub-lished as my own a discovery of M. Longpérier.[1] You will, I hope, do me the justice to publish this vindication of myself.

I have never seen M. Longpérier's paper in the 'Revue Archéologique' of October 1847,[2] nor have I seen M. Löwenstern's paper in the same journal of October 1849;[3] but I have his paper in the 'Revue' of December 1849, in which he evidently claims as *his own* a discovery which, however unimportant it may be, was in fact *mine*.[4] After mentioning the first word in the second Persepolitan inscription on the portal at Persepolis, which signifies 'god', and which M. Westergaard read *anap*, he proceeds: '*I* have recognized that this word ought to be separated into two parts ... The first of these signs is ideographic ... and forms the determinative which precedes every name of Divinity, at the same time that it serves in Assyrian to express the noun "god", both isolated and in the plural. The other two signs *nap* are then to be read *nebo* or *nepo*.'[5] He says again: 'I read the name of Ormazd (West. *Aurázda*) separating the first sign, which is ideographic and the determinative of god, *Ou.rᵃ.z.dᵃ*. Compare with *a.u.r.mᵃ.z.d.a* of the first and (equally taking away the ideographic sign) *ha.u.r.m.az.d.a* of the third writing (3).'[6] The reference is to '*Exposé*', p. 26, indicating a former work of the author, published in 1847.[7] One might expect to find there the reading of the name of Ormazd which he here gives; but on turning to that work I find it transcribed by *a.h.u.r.m.a.z.d.a*; the character which he *now* calls an ideographic sign being *then* read *a*, and the following character which he now reads *ha* being then read *h*. Of course I do not blame M. Löwenstern for correcting his former error; but I do blame him for referring to his former work for what it did not contain, instead of giving to the real author of this discovery the credit to which he is entitled. In my paper 'On the Three Kinds of Persepolitan Writing and on the Babylonian Lapidary Characters', which was read before the Royal Irish Academy in December 1846, I say of the initial character in the second Persepolitan name of Ormazd: 'Besides having a phonetic value, it is used as a non-phonetic initial before the name of Ormazd, as the corresponding Babylonian character is. This name is . . . . . , which I now read *O.ra.wash.ta*'; Transactions of the Royal Irish Academy, vol. xxi, part ii, p. 241.[8] In that paper I did not recognize the ideographic use of this character in the Median word for 'god', which I then read *n'.na.p'.pi*; but in my paper on the Van inscriptions, read a year after, before the Royal Asiatic Society, this is clearly stated, the word being there read *nab* or *nabbi*.[9]

In the letter in the *Athenaeum* of the 25ᵗʰ. of May, the expression used by the kind friend who wrote it, 'discovery of the ideographic element',[10] is not what I should have used myself, and I may take this opportunity of correcting it. To talk of the discovery of an ideographic element being announced by M. Longpérier in 1847, as M. Löwenstern does, is erroneous not merely on account of *my* previous announcement in 1846, but for other reasons. In my paper of December 1846, already quoted, I speak of the use of the determinative prefix to names of countries in the Babylonian inscription at Nakshi Rustam having been communicated to me by Mʳ. Norris.[11] It must have been observed by the person who first copied that inscription, namely, M. Westergaard; and the existence of ideographic characters in the shorter inscriptions at Persepolis was recognized long before this by Director Grotefend. M. Löwenstern, indeed, in his *Exposé* of 1847, denies the existence of any ideographic element in the cuneatic writing;[12] but I am not aware of any other person having made a similar assertion. The existence of both an ideographic and a phonetic element in the Babylonian writing seems to have been admitted from the very first; but the extent to which the ideographic element was used and its different modifications have not been recognized till lately. What I have claimed in my last paper as my discovery, was, 'the existence of ideographic characters *with various uses*', as there explained, *'and the consequent possibility of a character being read in two or more ways according as it was used as a phonograph or an ideograph'*.[13] To illustrate this by the character already referred to, which begins the name of Ormazd. 1. Its *phonetic* value is *an*. It is used with this value when it occurs in ordinary Babylonian words, as *anna*, 'me', an affix to verbs, *annut*, 'that', &c.; and in foreign proper names, as Zar*an*ga. This value I was the first to assign to it; Westergaard and Löwenstern took it for a vowel. 2. Its value as an independent ideograph is 'god', and with the plural sign, 'gods'. This was, I believe, discovered by Grotefend. 3. At the commencement of some proper names of gods and some kindred words, it is a *non-phonetic determinative*, the name being phonetically complete without it. In my paper of December 1846 I explained it as such, and I was the first to do so; but the existence of *other* determinative prefixes was previously known. 4. At the commencement of other proper names of gods and kindred words, it is a part of a *compound ideograph*. Thus, when followed by a character of which the phonetic value is *ac*, it is not to be read *anac* or *ac*, but *Nabu*. I now believe that in this name the elements are both used as ideographs, the latter denoting some epithet which, with the generic character for God, was one way of indicating the god Nebo. Formerly I thought that the first character was to be read thus, and that the second was an arbitrary addition which might or might not be sounded. Major Rawlinson explains the compound in a different manner. However this may be, it is quite certain that it is to be read *Nabu*, constituting the first two syllables of the name of the celebrated king of Babylon; and it is certain also that if it be not a compound ideograph, there are many such to be met with in the inscriptions, as, for instance, that which begins the majority of the Assyrian inscriptions, composed of the ideographs for 'house' and 'great', and signifying 'a palace'. I claim to have discovered the existence of this class of compounds, and also to have first read the group *an.ac* as *Nabu*, in the royal names on the Babylonian bricks. 5. The character for god occurs as an *ideographic element* in Semitic proper names, which were significant in the language; and when it so occurs it is not to be read by its phonetic value *an*, but by the Assyrian word corresponding to its ideographic value, which was *il*, or occasionally *assur*. Thus, in the most ancient form of the name of Babylon, this character, which occupies the second place, was pronounced *il*, the name being a significant one, and denoting 'the gate of god'. The character which precedes it was read by me *bab* in my Van paper of 1847, but I then supposed it to signify 'a province'. Major Rawlinson corrected this to 'gate', which is at the same time the known meaning of the Persian word that corresponds to it at Persepolis, and the meaning of the word itself in the Arabic language.[14] The possibility of the same character being read in different manners according as it was used as a phonograph or an ideograph, is what my friend particularly alluded to in the *Athenaeum* of the 25th May as a discovery in which my priority is unquestioned. To the statement there made, so far as this use of the ideographic element is concerned, M. Löwenstern has given no denial. What he says on the other uses of ideographs is, I have shown, altogether incorrect.

I am, &c.
Edw. Hincks

Source: *The Athenaeum*, no. 1195 (21 Sept. 1850), pp. 999–1000

1   See Löwenstern's letter of 23 August 1850 above. It appeared in *The Athenaeum*, no. 1193 (7 Sept. 1850), p. 953.
2   A. de Longpérier, 'Lettre à M. Isidore Löwenstern sur les inscriptions cunéiformes de l'Assyrie', *Revue archéologique* 4/2 (1848), pp. 501–7. The letter is dated 20 September 1847.
3   I. Löwenstern, 'Note sur une table généalogique des rois de Babylone dans Ker-Porter', *Revue archéologique* 6/2 (1850), pp. 417–20.
4   Löwenstern, 'Remarques sur la deuxième écriture cunéiforme de Persépolis', *Revue archéologique* 6/2 (1850), pp. 687–728.
5   Ibid., pp. 710–11.
6   Ibid., p. 710.
7   *Exposé des éléments constitutifs du système de la troisième écriture cunéiforme de Persépolis* (Paris, 1847), p. 26.
8   'On the Three Kinds of Persepolitan Writing, and on the Babylonian Lapidary Characters', *Transactions of the Royal Irish Academy* 21 (1847), polite literature, p. 241. Hincks modified the citation for the letter.
9   'On the Inscriptions at Van', *Journal of the Royal Asiatic Society* 9 (1848), p. 406.
10 See William Desborough Cooley's letter dated May 1850 to *The Athenaeum* above.
11 Hincks, 'On the Three Kinds of Persepolitan Writing, and on the Babylonian Lapidary Characters', *Transactions of the Royal Irish Academy* 21 (1847), polite literature, p. 243, n.*.
12 Löwenstern, *Exposé des éléments constitutifs du système de la troisième écriture cunéiforme de Persépolis* (Paris, 1847).
13 'On the Khorsabad Inscriptions', *Transactions of the Royal Irish Academy* 22 (1850), polite literature, p. 65, and reference there to §§ 9–23.
14 See Rawlinson, 'On the Inscriptions of Assyria and Babylonia', *Journal of the Royal Asiatic Society* 12 (1850), pp. 478–9, n. 2.

TO SAMUEL BIRCH
7 OCTOBER 1850

Killyleagh 7th. Oct[r]. 1850

My dear Sir,

On this day month, I received your paper on the Egyptian Calendar, for which I am much obliged to you, & which I read with great interest.[1] If your views as to the meaning of *mer* be correct, & it really signify 'a river', it would make a great change in the data on which all of us have hitherto depended in fixing the origin of the Egyptian Calendar. I cannot, however, admit this without very strong evidence. I think the meaning is a sheet of water & that Champollion was right in applying it to the inundation. So the cognate words *vari* Sanskrit 'water' *mare* a sea or extent of water & *var-ât* or *mar-ât* the Assyrian 'Sea' (in the plural) – 'the waters'. It is used too, as determinative of *yam* 'the sea' & other waters which do not appear to be *rivers*. I, however, speak drollingly, throwing out these objections for you to consider.

I delayed writing till I should be able to mention the time when the mummy of the period immediately following the 12th. dynasty would be opened. It is to come off on the 17th. inst.[2] I am not sanguine as to any thing being found which may throw light on the disputed chronological question but it is possible there may be such. Could you write us a few lines (for I presume you cannot come, tho' we should be most happy to see you) as to the characteristics of mummies of the 18th. dynasty if there be any such? Are any mummies known to be of that dynasty or of an earlier one? I know you have King Enteos in the Museum; but are there any others? I believe you have only bones of Menheris & that he was not embalmed.

Excuse this misplacement of the writing[3] &

Believe me
Yours very truly
Edw. Hincks

MS: BM/ANE/Corr. 2700

1   S. Birch, 'Observations on an Egyptian Calendar, of the Reign of Philip Ardaeus', *Archaeological Journal* 7 (1850), pp. 111–20.
2   See the following letter.
3   Hincks wrote on pages 2 and 4 of the folded sheet instead of 1 and 2 or 1 and 3.

⌒⌒⌒

TO SAMUEL BIRCH
17 OCTOBER 1850

Belfast 17th. Oct/50

My dear Sir,

I write a few lines before I return home to thank you for your letter & to say that the mummy was opened this day. We were surprised to find that it was in pieces. Tho' wrapped up in the usual bandages (on some of which were stamps identifying it with that which belonged to the coffin), it does not appear to have been *embalmed*. Some of the bones were bare; others had a bituminous substance attached to them which was generally thought to be the flesh converted into this substance by eremacausis[1] – no bitumen or any similar substance having been used. This will be investigated in the interval between this & Wednesday next when there is to be a public lecture on the mummy or rather a number of statements made to the audience. I am to speak on the hieroglyphical & chronological departments.[2] As the body was that of a lady of rank, it may be safely inferred that the art of embalming was not known yet. May she not have died of some disease that prevented the process being applied to her case? If embalming was not then in use, we have then a limit to the age of the coffin; & as it must have been done 60 years after the termination of the 12th. dynasty when it was made, embalming must have been subsequent to this. I have not yet received Gliddon's book with your letter, but have written for it and hope it will be here before Wednesday.[3] Is there any *mummy* known to be of the 18th. dynasty?

I remain
My dear Sir,
Yours very truly
Edw. Hincks

MS: BM/ANE/Corr. 2701

1   Eremacausis: slow burning.
2   See the published account: *Special Meeting of the Natural History and Philosophical Society of Belfast, Held in the Music-Hall, on Wednesday, the 23rd of October, 1850, Relative to Two Mummies Transmitted from Thebes, by Sir James Emerson Tennent, and Unrolled in the Museum, on the 17th and 18th of the Above Month* (Belfast, 1850) (From *The Northern Whig*, 24 October 1850). The statement by Hincks, including an explanation of the hieroglyphs, is on pp. 10–18.
3   See S. Birch, 'Letter to Mr. George R. Gliddon, on Various Archaeological Criteria for Determining the Relative Epochs of Mummies', dated 23 December 1848, in G. R. Gliddon, *Otia Aegyptiaca: Discourses on Egyptian Archaeology and Hieroglyphical Literature* (London, 1849), pp. 78–87. The letter is appended to Gliddon's lecture 7, one of 'Three Discourses on the Art of Mummification among the Egyptians: its Origin, Nature, and Development'. George Robbins Gliddon (1809–57) was formerly U.S. Consul in Egypt.

TO SAMUEL BIRCH
19 OCTOBER 1850

Killyleagh – 19th. Oct 1850

My dear Sir,

I wrote you a few hurried lines from Belfast respecting the mummy which was unrolled on Thursday. I now write to say that on closer examination the theory which attributed the bituminous substance to eremacausis has been abandoned, there being evident marks of bitumen outside the skin. The substance will be analysed carefully. I have not yet got Gliddon's book & fear it will not reach me till Monday evening – or perhaps not till Wednesday morning;[1] but I saw in Belfast yesterday a paper of D^r. Carpenter's in the Philosophical Transactions from which I infer that the fragmentary state of the body & almost total disappearance of the flesh is not unusual.[2] Yesterday we opened a second mummy which was apparently of much later age, but which presented the same appearances.

I observed that in this the bull Apis on which the mummy was represented going to the tomb had a globe on his head – not on the older one. The mummy was on this carried with its head foremost but in the older one with its feet foremost. The Scarabaeus, too, was placed over the representation in the later one but not in the older. This remarkable representation occurring in both, the differences in it attracted attention. The later case contained many strange representations which I do not recollect to have seen before, tho' I dont doubt their occurring elsewhere. Among others there was a bird with a ram's head on which was the crown 𓏰 as well as I can recollect it. Also the deceased (as I have always considered it, tho' some call him the god Mani) holding up, not the Sun dish as usually, but a barge containing the Sun & other deities [?].

You would very much oblige me if you can let me know the meaning of th[e] title, & the name of the father, *by return of post* (if possible by the afternoon post, which would reach me several hours sooner than the night one) to '7. Murray's Terrace Belfast'.[3] I would then give them on your authority at a lecture which I am to read on Wednesday in Belfast. I incline to read the father's name *ca-nûn-ati*, supposing the second character to be that for a river & the third a determinative – but I doubt. The second character seemed very like the character for *sotp* wanting the ⌣ at the bottom. It occurs only twice.

The 26th. chapter of the Todtenbuch is written on the older mummy case. I mean to take my copy of Lepsius to collate it. On going over the inscription at Belfast & the copy in Lepsius on my return, I observed no differences; but of course I might have overlooked many.

There was an inscription inside the case on three gilt laths – the figures of Amset &c were found on similar laths (gilt also), only 3 were to be found, Hapi having been, I presume, taken away by the Arabs who certainly opened the case.

Believe me
My dear Sir,
Yours very truly
Edw. Hincks

I can add nothing to what you say of *vakh*; it evidently implies a beautiful soul. *Acar*, which accompanies it, is certainly *instructed*; in response to what was to be said by way of passport at different stages of the journey in Hades.

MS: BM/ANE/Corr. 2702

1   Hincks is referring to the letter from Birch to Gliddon published in Gliddon, *Otia Aegyptiaca*, pp. 78–87. For details see the previous letter, n. 2. In his diary for Thursday 24 October Hincks wrote: 'Gliddon's book came, read it over. He praises me to a greater extent than anyone who has yet mentioned me, and censures me less when he differs.'
2   The name is not clearly written but Hinck's may be referring to W. B. Carpenter, 'On the Mutual Relations of the Vital and Physical Forces', *Philosophical Transactions of the Royal Society of London* 140 (1850), pp. 727–57.
3   I have omitted some lines containing poorly drawn hieroglyphs. Birch's letter, dated 21 October 1850 and sent to the address of Hincks's father, Thomas Dix Hincks, was received only after the lecture had been delivered. It is preserved among the Hincks papers (MS GIO/H31) and includes the following reply to Hincks's query: 'The father's name I regard as Atai. Only the previous group *cha nen* is a title, meaning of "like kind" or "order", i.e. the father held the same office as the son.'

1851

ฯ๛๛

Killyleagh 12th. April 1851

My dear Sir,

I saw in the last Athenaeum a notice of an intended publication of a copy of the Turin papyrus – with the writing on the back.[1] This will be most valuable with a view to the arrangement of the fragments & ought to decide the question whether Lepsius or I is right.[2] I feel very confident that *I* shall be found to be so.

Strange enough, I had a letter from Sir G. Wilkinson this week offering to lend me a copy of his work on Egyptian architecture.[3] He took no notice of his copy of the papyrus in his letter; but I mentioned it in my reply, having heard a rumour of what was in the Athenaeum. I have since seen the paragraph itself.

I have made out a list of the variations between my arrangement & Lepsius's, which I promised to send to Sir G. if he wishes for it.[4]

Some arrangement of the fragments must be decided on & the value of the publication will in great measure depend on its being a good one. It would be impossible, or at all events extremely difficult, to carry the view backwards & forwards so as to read in connexion the two pieces of fragments that are separated from one another while others that are not connected with them intervene. I shall be anxious to know what is done in this matter. Perhaps I shall hear from Sir G. next week.

I send along with this a letter of thanks for the Cuneiform Inscriptions, which is of an official character.[5] Allow me also to thank you personally for your good offices in the purchase &

believe me to remain
Yours most truly
Edw. Hincks

MS: BM/ANE/Corr. 2703

1   See *The Athenaeum*, no. 1223 (5 Apr. 1851), p. 383 ('Our Weekly Gossip'). The work appeared under the title J. Gardner Wilkinson, *The Fragments of the Hieratic Papyrus at Turin: Containing the Names of Egyptian Kings, with the Hieratic Inscription at the Back*, 2 vols (London, 1851). The second volume contains facsimiles.
2   R. Lepsius, *Auswahl der wichtigsten Urkunden des ägyptischen Alterthums* (Leipzig, 1842); E. Hincks, 'On the Portion of the Turin Book of Kings which Corresponds to the Sixth Dynasty of Manetho', *Transactions of*

the Royal Society of Literature, 2nd ser., 3 (1850), pp. 128–38 (dated 7 Mar. 1846); 'On the Portion of the Turin Book of Kings which Follows that Corresponding to the Twelfth Dynasty of Manetho', *Transactions of the Royal Society of Literature*, 2nd ser., 3 (1850), pp. 139–50 (dated 5 May 1846); 'On the Portion of the Turin Book of Kings, which Corresponds to the First Five Dynasties of Manetho', *Proceedings of the Royal Society of Literature* 1 (1848), pp. 275–7 (read 9 Nov. 1848). The substance of these papers by Hincks was published again in J. Gardner Wilkinson, *The Fragments of the Hieratic Papyrus at Turin: Containing the Names of Egyptian Kings, with the Hieratic Inscription at the Back*, vol. 1 (London, 1851), pp. 47–60. For a recent study of the fragments, see J. Malék, 'The Original Version of the Royal Canon of Turin', *Journal of Egyptian Archaeology* 68 (1982), pp. 93–106.

3   John Gardner Wilkinson (1797–1875) was the author of *The Architecture of Ancient Egypt*, 2 vols (London, 1850).

4   See Hincks, 'Observations on the Turin Papyrus', in J. Gardner Wilkinson, *The Fragments of the Hieratic Papyrus at Turin: Containing the Names of Egyptian Kings, with the Hieratic Inscription at the Back*, vol. 1 (London, 1851), pp. 47–60.

5   Hincks is referring to the British Museum publication *Inscriptions in the Cuneiform Character from Assyrian Monuments Discovered by A. H. Layard* (London, 1851).

FROM JOHN GARDNER WILKINSON
13 APRIL 1851

Aldermaston
Newbury
13 April 1851

Dear Sir,

I am going to Town tomorrow & will order the copy of the 'Architecture' to be sent as you direct.[1]

I am much obliged by your kind offer to send me some lists of corrections suggested by you for the Turin papyrus, which if you think fit I shall have much pleasure in mentioning as your corrections in the text that I intend to publish with the plates. I only wish I could show you the copy of the back part in order to enable you to verify them. My object in doing this (as well as my architecture) is that it shall be published in a complete & correct form by us, before it is done by foreigners who are apt to say that we do nothing & I naturally wish it to be useful. I know how much attention you have bestowed on this curious record & any remarks of yours will be highly interesting and important. I should also be very glad that you should have an opportunity of studying it before it is published; please therefore let me know if you are likely to be in London this Spring. If not I might possibly be able to send you a copy before it is published.

The fragments as you know were arranged by Seyffarth,[2] & Lepsius has only published them as they were left by him, in the form in which they are put up at the Turin Museum.[3] But Lepsius ought to have given a copy of the back to enable others to benefit by his copy as well as himself – for I doubt not that he copied the back also. I ought to mention to you that I do not publish this at my own expense but that we have formed ourselves into a committee & I am one of the subscribers like everybody else. The Members of the Committee as well as the subscribers pay each £2 for one copy & I am sure that it would be highly gratifying to the other Members of the Committee as well as myself if you would allow your name to be added to our Committee.[4]

My address is always 33 York Street Portman Square.

Believe me
Yours very faithfully
Gardner Wilkinson

MS: GIO/H547

1   J. G. Wilkinson, *The Architecture of Ancient Egypt*, 2 vols (London, 1850).
2   G. Seyffarth, *Remarks upon an Egyptian History, in Egyptian Characters, in the Royal Museum at Turin* (London, 1828).
3   See R. Lepsius, *Auswahl der Wichtigsten Urkunden des ägyptischen Alterthums* (Leipzig, 1842).
4   See Wilkinson, *The Fragments of the Hieratic Papyrus at Turin*, vol. 1, p. vi, for the list of members, including Hincks.

FROM SAMUEL BIRCH
16 APRIL 1851

BM
16. April, 1851

My dear Sir,

I have duly laid your letter before the proper authorities, and hope you will find the inscriptions useful to you. I suppose that you have heard of the 'Bank Notes' of Nebuchadnezzar.[1] They are unfortunately in bad condition but are undergoing a process which will I hope preserve them effectively. I am myself at work on the Turin Canon, but the recent statements of M. Champollion Figeac in the Revue archéologique embroil the matter worse than ever.[2] I suppose that you are aware that Lepsius has published it as it appears at Turin after the reconstruction of the fragments by M. Seyffarth, and that it is not his arrangement of the pieces. Considering Seyffarth's knowledge it is honestly done by him, but it appears that at the time Champollion made his copy many pieces now arranged as in undoubted continuity were in small separate fragments.[3]

Believe me
Yours very sincerely
S. Birch

MS: GIO/H34

1   In a letter to Layard dated 24 February 1851, Rawlinson described tablets brought back by William Kennett Loftus as a kind of 'Govt. bank notes, being in fact the regular currency of the country and payable in gold or silver at the Royal Treasury' (MS: BL Add.38980, ff. 27–8). See M. Trolle Larsen, *The Conquest of Assyria: Excavations in an Antique Land 1840–1860* (London, 1996), p. 287: 'Recorded private economic transactions such as loans and the purchase of land.'

2   J. J. Champollion-Figeac, 'De la table manuelle des rois et des dynasties d'Égypte ou papyrus royal de Turin, de ses fragments originaux de ses copies et de ses interprétations', *Revue Archéologique* 7/2 (1851), pp. 397–407, 461–72, 589–99, 653–5.

3   I have omitted some lines which contain an unclear reference to one of Hincks's publications.

FROM JOHN GARDNER WILKINSON
13 JUNE 1851

33 York Street
Portman Square
13 June 1851

Dear Sir,

I enclose a rough copy of the Turin Papyrus giving you the inscription on the back also. Some of the fractures require still to be put in by the lithographers. Wherever the junction of the pieces has been determined by an examination of the fibres & the correspondence of the parts I have marked lines across the fissure thus: ++++++ These may be looked upon as accurate junctions of the parts.

The pieces are merely copied in the order in which they were placed by Dr Seyffarth & it is this same arrangement or rather temporary juxtaposition of the different pieces which Lepsius has given. I do not know if Lepsius has given any arrangement of his own, but I see that Champollion-Figeac talks of Lepsius' arrangement of the fragments of the Turin Papyrus. I have not seen it – & I fancy he thinks the plate given by Lepsius represents his arrangement of the pieces, whereas everybody knows it is that of Dr Seyffarth.

Your position of Nitocris is fully borne out by the inscription at the back – as you will see. If you have any remarks that you like me to insert as coming from you I shall be happy in doing so.[1] I have had very little to alter or add to Lepsius' copy which I have gone over very carefully while examining the original at Turin.

Believe me, Dear Sir
Yours very truly
Gardner Wilkinson

MS: GIO/H 548

1   J. Gardner Wilkinson, *The Fragments of the Hieratic Papyrus at Turin*, vol. 1, pp. 47–60.

~~~

TO RICHARD SAINTHILL[1]
9 JULY 1851

Killyleagh
9th. July, 1851

Dear Sir,

I this morning received your letter of the 7th. with sulphur casts of a cylinder and coin of Simon Bar-Cochab. I was not before aware of the existence of any coins of his, not having much knowledge as to any description of medals. The legend about the bunch of grapes is clearly *šmʿwn*. All the letters but the last are perfect. This is 'Simeon', the proper name of 'Bar-Cochab', *br-kwkb* 'son of the star'.[2] On the other side we have *lhrwt yrwšlm*, the first word may mean 'for the restoring of', the other is 'Jerusalem'. I am not satisfied as to the first word, the third letter may not be a *r*, and if it be, the word is not correct Hebrew for what I take it to be.[3] Perhaps Eckhel[4] or Bayer[5] explains it.

I am by no means surprised at this character being in use as the national character of the Jews in the time of Hadrian. Gesenius has shewn that the square Hebrew characters now used by the Jews are derived from the Palmyrene; and, if I recollect right, he thinks that they were not invented till the third or fourth century. The tradition of their having been brought from Babylon is rejected by every orientalist of character at the present day; and, as you have justly observed, the fact of the Babylonians having used arrow-headed characters would refute it. Had they adopted Babylonian characters they would have used arrow-headed ones.

I was not aware of your having published against [this] opinion in 1829.[6] *Then* it was pretty generally received, but it has since fallen into disrepute.

The cylinder, I am sorry to say, met with an accident. I attempted to take an impression of the inscription, with a view to see the characters in their proper position, and the heat caused it to break in the middle. The original was evidently used as a *seal*, and so are several other cylinders, but the majority have the characters in their proper positions (not reversed as here). Though I know the values of all the characters, I cannot conjecture the interpretation of the sentence which they compose.[7]

Do you wish for the casts back? If you do, write me a word and I will send them.

Believe me
dear Sir
Yours very truly
Edw. Hincks

Source: Sainthill, *An Olla Podrida*, vol. 2, pp. 25–6[8]

1   Richard Sainthill (1787–1869), wine-merchant, antiquarian and numismatist, moved to Cork from Topsham, Devon in 1801. He was a patron of the Irish artist Daniel Maclise, who did portaits of him. He published, or rather had printed for private circulation, *An Olla Podrida; or, Scraps, Numismatic, Antiquarian, and Literary,* 2 vols (London, 1844–53).

2   See L. Mildenberg, *The Coinage of the Bar Kokhba War* (Aarau, 1984); *idem, Vestigia Leonis: Studien zur antik Numismatik Israels, Palästinas und der östlichen Mittelmeerwelt,* eds U. Hübner and E. A. Knauf (Freiburg Schweiz and Göttingen, 1998), pp. 161–241.

3   The reading *lhrwt yrwšlm* is certain; it occurs on dozens of coins and means 'for the freedom of Jerusalem'.

4   J. Eckhel, *Doctrina numorum veterum,* vol. 3 (Vienna, 1794), pp. 471–7.

5   F. Pérez Bayer, *De numis hebraeo-samaritanis* (Valencia, 1781), pp. 237–8; supp. XIII–XIV; see also pp. 100–4, nn. 11 and 13; table facing p. 171.

6   '1829' is a mistake for '1819'. The reference here is to a letter that Sainthill published in the newspaper *Morning Post,* in which he commented on a medal found at Cork in October 1818. The brass medal came into the possession of George Corlett, who was told that the medal had been found by a young girl watching her father, who was working in a potato garden at the rear of a house in an area known as Friar's Walk in the city of Cork. On one side of the medal there is a figure of the head of Christ and on the other side a Hebrew inscription. There is an engraving of the medal and various scholarly comments on it, including some by Hincks, in T. R. England (ed.), *A Short Memoir of an Antique Medal, Bearing on One Side the Representation of the Head of Christ, and on the Other a Curious Hebrew Inscription, lately Found at Friar's Walk, near the City of Cork, in Ireland; Containing Some Letters and Observations of Different Men of Learning on the Subject; with Divers Translations of the Characters thereon* (London, 1819). In one of the comments, the Rev. William McHale of Killeshandra, Co. Cavan, dismisses the comments by Sainthill in the *Morning Post,* describing him as ignorant and presumptuous (pp. 40–1). After all those years, Sainthill felt bound to reply to the late Rev. McHale's comments. Obviously Hincks did not recall Sainthill's contribution in 1818, even though he had published comments on the medal in the same pamphlet as McHale. See Hincks's letter of 10 December 1818 in vol. 1, pp. 21–2.

7   The language of the inscription on the seal was probably Sumerian which had not yet been deciphered. Hincks felt he could read the individual signs but make no sense of the text as a whole.

8   Hincks's letter is found in Sainthill, 'The Use of the Samaritan Language by the Jews until the Reign of Hadrian, Deduced from the Coins of Judea'. A Letter to J. B. Bergne, *An Olla Podrida,* vol. 2 (London, 1853), pp. 25–40 (see pp. 25–6). Hincks's letter is not in the original article that was published in *Numismatic Chronicle* 14 (1851), pp. 89–104.

FROM JOHN ABRAHAM RUSSELL[1]

AUGUST 1851

August 1851

My dear Hincks,

I chanced to meet Mr. Layard in London a short time since at a dinner party, and I took the opportunity of introducing your name in connexion with the subject of his pursuits, and as he seemed to be under some misapprehension as to the cause of your relinquishing your researches (the value of which he fully admitted) I took the liberty of setting him right, by mentioning what you said in your last letter to me.[2] I said you had good reason to complain of want of due help and encouragement, and I endeavoured to impress on him the importance of this being represented in high quarters. I afterwards met Col. Rawlinson, who heard from Mr. Layard the subject of our conversations and he seemed to take it up very warmly. Both seem to appreciate fully the value of your cooperation, or your labours singly directed to the same subject. Mr. Layard told me some notice was once taken of a letter of yours to him in some periodical, which notice was not written by him or by his authority, and in a *spirit* and in *language* which he did not approve of, and he permitted me to mention this. Both he and Col. R. expressed their desire that you would be kind enough to communicate with them, and suggest in what way they might be of use in removing any difficulties in the way of your resuming your valuable labours in a department in which so few are qualified by their habits, their learning and their ingenuity to work out such results as you have already done. I return home to-morrow or next day and shall be glad to hear from you.

Believe me,
Yours most faithfully,
J. A. Russell

Source: Davidson, *Edward Hincks*, p. 48

1   John Abraham Russell was born in Limerick in 1792. He entered Trinity College, Dublin in 1809 and took his B.A. in 1815, when he was also ordained. On 23 August 1831 he married Frances Thomasina Story. He was Archdeacon of Clogher from 1826 till his death on 29 April 1865. See J. B. Leslie, *Clogher Clergy and Parishes* (Enniskillen, 1929), pp. 47–8. Russell published *Remains of the Late Rev. Charles Wolfe, A.B., Curate of Donoughmore, Diocese of Armagh, with a Brief Memoir of his Life* (6th edn; London, 1836). Charles Wolfe (1791–1823) was the author of the well-known poem 'The Burial of Sir John Moore after Corunna'. His brother John Charles succeeded Russell as Archdeacon of Clogher. Hincks, Russell and Wolfe were contemporaries at Trinity College, Dublin.
2   Layard arrived in London on 1 July 1851 after a long journey from the Near East (he left Mosul on 28 April). He stayed with his cousin, Lady Charlotte Guest, at her London house. See G. Waterfield, *Layard of Nineveh* (London, 1963), p. 227.

FROM JOHN ABRAHAM RUSSELL
22 AUGUST 1851

22nd. August 1851
[My dear Hincks,]
    I received your very interesting letter and have written to Mr. Layard to remind him and Col. Rawlinson of our conversation at Miss Coutts' on 2 occasions.[1] I have made full use of your letter in such way as I thought most judicious, and as Lord Clarendon is now in London I suggested that the matter should be brought

under the notice of some other member of the Govt.[2] I cannot bring myself to think that you are 'to be quietly laid on the shelf', or 'that it is useless to complain'. It will give me great pleasure, if, in my humble way, I may be in any way instrumental in setting you free, or rather in inducing *others* to lend a helping hand to set you free to follow out your valuable labors. I have every hope that Col. R. and Layard will act on my letter and report its contents in some influential quarter, as I seemed to myself to have excited much interest in their minds in what I said of you in conversation – all of which they seemed most ready to admit, and much of which they were much more competent than I am, to appreciate.

I am not sure whether you might not with advantage apply to Lord C. *immediately*, and refer to those gentlemen as ready and most competent to bear testimony to the value of your labors in the same field of inquiry which is now exciting so much interest in the learned world.

[Believe me,
Yours most faithfully,
J. A. Russell]

Source: Davidson, *Edward Hincks*, pp. 48–9

1   Angela Georgina Burdett-Coutts (1814–1906), philanthropist, was described as 'the richest heiress in England'. The parties at her home attracted many people from society. Among her many friends were the Duke of Wellington and Charles Dickens. She corresponded with Austen Henry Layard.
2   George William Frederick Villiers (1800–70) was the fourth earl of Clarendon. He was Viceroy of Ireland from 1847 to 1852.

TO AUSTEN HENRY LAYARD[1]
22 AUGUST 1851

Killyleagh C° Down
22^d. Augt. 1851

My dear Sir,

I had a letter some days ago from the Archdeacon of Clogher[2] in which he mentions that he had seen you and that you were kind enough to wish me to communicate with you as to what could be done to remove the difficulties under which I labour as to carry on my researches. I wrote last week to the Archdeacon, but I could not then suggest any thing definite. Indeed, when I wrote I was rather desponding.

This day's post has brought the word that the Deanery of Armagh is vacant. It is the only deanery in Ireland, having an income attached to it, which can be held along with a living. The income is small; but would, after defraying the expenses connected with the duties, leave me enough to pay a curate & to spend some little on books & travelling.

I have written to Lord Clarendon applying for it. There will be many candidates. Indeed I fear I am late in the field, as the Dean died on the 14th.[3] An effort will of course be made to direct the income to the Ecclesiastical Commissioners – but it is possible that I may succeed; and in order to give me a chance, a good word from any who take an interest for me would be of use.

I think if you would just at this time endeavour to impress on those in high places (Lord Clarendon as well as Lord J. Russell) the importance of enabling me to prosecute my studies, it might be of use.

From what the Archd^n. wrote to me I cannot doubt your friendly intention.

Believe me to remain
My dear Sir
Yours very faithfully
Edw. Hincks

MS: BL Add. 38980, ff. 104–5

1   This letter marks the beginning of frequent and important correspondence between Hincks and Layard from 1851 to 1854.

2   John Abraham Russell; see his letters dated August 1851 and 22 August 1851.

3   Edward Gustavus Hudson died at Glenville, Co. Cork, on 14 August 1851. He was Dean of Armagh from 1841. See W. E. C. Fleming, *Armagh Clergy, 1800–2000* (Dublin, 2001), p. 63. Hincks's application was not successful and Brabazon William Disney (1797–1874) was appointed. See Fleming, *Armagh Clergy 1800–2000*, p. 63.

FROM THOMAS THORNELY[1]
26 AUGUST 1851

26th. August 1851

I have written to Lord John Russell, referring to the application made on your behalf last year in which I joined Mr. Tennent[2] and other Irish and English members, and then I spoke of the vacancy in the Deanery of Armagh, giving the information contained in your letter and stating that if you were appointed to this, it would precisely enable you to pursue the studies alluded to, in our application of last year. I have written as fully as I thought there would be any use in doing. I presume Lord John Russell will refer my letter to the Lord Lieutenant.

I am not sufficiently acquainted with Lord Clarendon to authorise me to write to him direct, and his brother my colleague is absent somewhere on the Continent.[3]

Let me earnestly recommend you to get Mr. Tennent and any other Irish members to write, for I conceive their influence in Irish appointments must be much greater than mine.

[Thomas Thornely]

Source: Davidson, *Edward Hincks*, p. 49

1   Thomas Thornely was born on 1 April 1781, son of William Thornely of Liverpool. He was a merchant at Liverpool until 1835, when he became Liberal M.P. for Wolverhampton. He held this seat until 1859 and was chairman of the public petitions committee from 1844 to 1853. He died in Liverpool on 4 May 1862.
2   Robert James Tennent (1803–86) was M.P. for Belfast from 1847 to 1852.
3   Charles Pelham Villiers, M.P. (1802–98) was Lord Clarendon's brother. He formed a successful constituency partnership with Thomas Thornely in Wolverhampton.

FROM AUSTEN HENRY LAYARD
1 SEPTEMBER 1851

1 Paragon Parade
Cheltenham
Septr. 1/51

My dear Sir,

I found your letter in London on my return from a visit to Wales. I lost no time in seeing Col^l. Rawlinson, who fortunately happened to be still in London, and in consulting with him on what we could do to further your views. At this time no one who would be of any use is in London. L^d. John Russell is in Scotland, and with Lord Clarendon neither Rawlinson nor myself have any acquaintance. Col^l. Rawlinson is leaving town and promised me that he would take every opportunity of bringing your case and your wishes before such persons as might have any influence and would be likely to assist. I need scarcely add that I will not fail to do the same thing, tho' I fear my interest is nothing, and my acquaintance amongst the great far more circumscribed than that of the Colonel. It would indeed give me great pleasure to learn that you are prosecuting research which you have so sucessfully commenced – the materials are now most abundant. I hope to be able to publish a collection of inscriptions far exceeding in size that already in your hands, & there is, amongst the inscribed tablets recently brought to this country, work for years. The general interest is much increased by recent discoveries, and the public at large are more than usually desirous of information on the subject. I should think, therefore, that any appointment which would enable you to continue your investigations, would be very well received indeed, & would redound to the credit & popularity of the Government making it. I shall not fail to express this conviction to all who may be concerned in the matter.

Believe me,
My dear Sir,
Yours very truly,
A. H. Layard

MS: GIO/H 213

TO THE EDITOR OF THE ATHENAEUM
8 SEPTEMBER 1851

Killyleagh C° Down
8ᵗʰ. Sepᵗ. 1851

[Sir,]

I perceive that the very interesting letter from Col. Rawlinson,[1] which you recently published has led to some remarks from a correspondent who signs himself 'J. G.'.[2] I trust I may be permitted to offer some additional remarks on the same subject.

I quite agree with 'J. G.' that the destruction of Sennacherib's army before Jerusalem closely followed his success recorded in the inscription; and that the reason of its not being noticed on the bull is, that the Assyrian monarchs recorded only their successes. I cannot, however, agree with him, that there is any mistake in the Biblical chronology of the conquest of Samaria, the interval between which and Sennacherib's success against Hezekiah is limited to five years. It appears to me that Col. Rawlinson is mistaken in supposing that the conquest of Samaria occurred in the first year of Sargin.

In the first place, I am not disposed to admit that the name which the Colonel reads *Samarina* is the Samaria of Scripture.[3] The initial character of this name is that which begins the names of *Saparda* and *Thattagus* in the inscription on the tomb of Darius, and which forms the second syllable of the name of Persia (*Parsa*) on the portal at Persepolis. From etymological considerations, I have valued it *tsâ*.[4] The next two syllables are written indifferently *mir-i* and *mi-ri*. I am disposed to read the whole name *Ir-tsâmirina*, 'yr smryn; though it is *possible* that the initial character may be a mere determinative. I have been for a very long time undecided as to the city alluded to. Sometimes, I have leaned to *Simyra* (see Ges. *Thes.* 1173[5]); at other times, I have been strongly disposed to consider *tsomerin* as the plural of the Assyrian equivalent of the Hebrew *tomer*, 'a palm-tree', and to identify the city with Tadmor, or Palmyra; but I could never bring myself to think that it was the Hebrew *Shimron*, the former part of which would naturally be expressed by three Assyrian characters totally different from those used.

Besides, I cannot agree with Col. Rawlinson in supposing *Bith-Khumria* to be *the same as Tsamirina*. They appear to me to be clearly different places, though not far distant from each other; and agreeing as I do with Col. Rawlinson that the former is *Beth-'Omri*, or Samaria, I must seek the latter elsewhere. Now, it is from *Tsamirina*, and not from *Bith-Khumria*, that the deportation was made which Col. Rawlinson identifies with the captivity of the Ten Tribes.

But, secondly, if the identity of these places be conceded, I see no proof that this deportation took place in the *first* year of *Sargin*.[6] The inscriptions in which alone it is mentioned do not appear to be in chronological order; so that, though the defeat of the *Negas*, or sovereign, of Susiana and the deportation of the people of *Tsamarin* are placed first, they may not have been first in order of time. What relates to the Egyptians, which immediately follows, occurred in two different years, the second and the seventh, as appears from the inscription in the form of Annals.

The defeat of *Khanun*, king of Gaza (*Khadzithi*, or *Khajithi*, in the genitive, the theme of which would be *Khadzith*, 'zt = 'zh, aided by the *Tartan*, or general, of the Egyptians, at Raphia (on the frontiers of Egypt, where Antiochus was defeated by Ptolemy, 218 B.C.) was in the second year of Sargin;[7] the tribute of the king of Egypt, whose name I think to be Pehor (Bocchoris), and not the title Pharaoh, was not paid till the seventh.

I observe that Col. Rawlinson has been puzzled by the title *Tartan*, the second significant character in which is in the principal copy of the inscription written *lib* in place of *tâ*. This, however, is an error of either the sculptor or the copyist, which I have corrected by means of the other copies. This reading is of great importance, as the name *Tartan* occurs in 2 Kings xviii, 17 (along with *Rab-saris*, chief eunuch, and *Rab-shakeh*, chief butler, which are, like it, names of office), and again Isaiah xx, 1, where it has been supposed that the same person is alluded to. *Tartan* is 'the general'; and in all probability different generals commanded on these two occasions.[8] The same word is used to express 'general' on the Nimrud obelisk. The first character is a homophone of that which here expresses *tar*, and the two others are precisely the same.

I must also express my doubts whether the Sargon and Shalmanezer of Scripture were the same king. In my paper on the Khorsabad inscriptions, I considered Shalmanezer to be a son of the Khorsabad king, an elder brother of Sennacherib.[9]

Col. Rawlinson considers Shalmanezer to be the reading of a title given to *Sargin* at Khorsabad. He has not, however, pointed out the title to which he alludes; and I do not, as yet, see any reason to alter the opinion which I formerly expressed.

While, however, I thus express my dissent from what Col. Rawlinson has stated about the mention of the captivity of Israel in the Khorsabad inscription, thinking that the Assyrian record of this event remains to be discovered, I have no doubt at all of the correctness of what he has stated concerning the account of Sennacherib's war with Hezekiah; and I heartily congratulate him on his having made so important a discovery.[10]

I am, &c.

Edw. Hincks.

Source: The Athenaeum, no. 1246 (13 Sept. 1851), p. 977

1   *The Athenaeum*, no. 1243 (23 Aug. 1851), pp. 902–3.

2   *The Athenaeum*, no. 1245 (6 Sept. 1851), pp. 951–2. Another letter from the same person appeared in *The Athenaeum*, no. 1246 (13 Sept. 1851), p. 977.

3   Rawlinson was right and in due course Hincks accepted this.

4   Hincks's use of *ts* for ṣ and later, *ch* for ṣ, and *s* for š, was utterly confusing. See his letter of 4 May 1853 to Henry Ellis.

5   W. Gesenius, *Thesaurus philologicus criticus linguae hebraeae et chaldaeae Veteris Testamenti*, vol. 3 (Leipzig, 1842), p. 1173.

6   It is generally agreed among modern scholars that Shalmaneser V captured Samaria and Sargon II deported many of the inhabitants.

7   In the Annals of Sargon II (721–705 B.C.) we read that 'Hanno, king of Gaza and also Sib'e the turtan of Egypt set out from Rapihu against me to deliver a decisive battle. I defeated them'. See J. B. Pritchard (ed.), *Ancient Near Eastern Texts Relating to the Old Testament* (3rd edn; Princeton, 1969), p. 285. In 217 B.C. Ptolemy IV of Egypt beat Antiochus III, king of the Seleucid empire.

8   See Rawlinson, 'On the Inscriptions of Assyria and Babylonia', *Journal of the Royal Asiatic Society* 12 (1850), p. 451: 'the Sargon of Isaiah, who sent his general, Tartan, against Ashdod'. Notice Hincks's superior philological skill. For a discussion of the Akkadian loanword *tartan*, 'field marshal', in Biblical Hebrew, see P. V. Mankowski, *Akkadian Loanwords in Biblical Hebrew* (Harvard Semitic Studies 47; Winona Lake IND, 2000), pp. 151–2 and reference there to F. Delitzsch, *Assyrisches Handwörterbuch* (Leipzig, 1896), p. 716. In Neo-Assyrian texts, the sequence of officers and officials after the king's name is: *šarru* ('king'), *turtānu* ('field marshal'), *rab šaqê* ('chief cupbearer'), *nāgir ekalli* ('palace herald'). See *The Assyrian Dictionary*, vol. 18 (Chicago, 2006), pp. 489–90.

9   'On the Khorsabad Inscriptions', *Transactions of the Royal Irish Academy* 22 (1850), polite literature, p. 51. Rawlinson's view was erroneous of course.

10   Notice Hincks's graciousness in recognizing the correctness and importance of this discovery, which Rawlinson announced in *The Athenaeum*, no. 1243 (23 Aug. 1851), p. 903. But see also Austen Henry Layard's sharp remark in his letter of 8 January 1852 below: 'Had not Col. Rawlinson *by an accident* obtained possession of my papers he would not have been able to publish his last discovery relating to Hezekiah.'

꙳꙳

FROM SAMUEL DAVIDSON[1]
8 SEPTEMBER 1851

Independent College
Manchester
Sept[r]. 8. 1851

My dear Sir,

As I am now engaged in rewriting my Biblical Criticism and bringing it up to the present state of knowledge, in various ways,[2] my attention has been directed to the ancient Hebrew language and the investigations of Kopp, which most scholars in Germany immediately followed.[3] Gesenius, however, hesitated to the last. With this view I have been looking over all the papers of yours which I have, as reprinted from the transactions of the R. I. Academy, but I can find nothing in them exactly to my purpose. Presuming that you have Gesenius on the Phenician language, for I see you refer to it, I should like to have your opinion on several points; & by references to G.'s work, perhaps I can make my meaning intelligible.[4]

I believe it is now settled that the old Hebrew character, i.e. that on the Maccabean coins, is substantially the same as the Samaritan. It is also settled that this ancient Hebrew sprung from the Phenician. Now I should like to know the relation of the Phenician character to the characters on the Babylonian bricks which Grotefend, I believe, was the first to give. Was that Babylonian character the oldest of all – the mother of the Phenician. Kopp thought so, and derived the Hebrew from it by the usual processes & changes of time. Gesenius however is disposed to think the inscriptions on the bricks *Old Persian*, & therefore *the daughter* of the Phenician, not the mother. I do not know whether the writing on those old Babylonian bricks has been yet deciphered (Gesen. p. 77).[5]

It lies too low down in point of time for your researches to inquire about the origin of the square Hebrew character, or why you suppose it to be called *the Assyrian*, if indeed that be the right version of the word applied to it. Of course Ezra cannot have been the author of the change, nor I believe can it have been so late as the third century after Christ. Still there are difficulties in the way of placing it about the time of Christ which I did not see at one time, & difficulties in supposing that it arose gradually in the process of time out of the old Hebrew or Samaritan. How did the Jews come to adopt that *Aramaean Egyptian* kind of writing, as Gesenius terms it.[6] Perhaps your inquiries have not been directed to these latter points. They must have been turned however to the Babylonian bricks & the Phenician, as well as the relation between them; & perhaps the characters on those bricks have been deciphered since Gesenius wrote. I should like very much to have your opinion on the subject, as well as to know generally whether your researches have already thrown any new light on the Hebrew writing, or are likely to do it. I am aware that they confirm & illustrate *Hebrew history*.[7]

I am yours very sincerely
Saml. Davidson

MS: GIO/H 152

1   Samuel Davidson (1806–98) held the chair of biblical criticism in the Belfast Academical Institution from 1835 to 1841 and was a close colleague of Thomas Dix Hincks, Edward Hincks's father. In his autobiography he refers to Thomas Dix Hincks as 'one of my best friends'. See *The Autobiography and Diary of Samuel Davidson*, ed. Anne J. Davidson (Edinburgh, 1899), p. 16. There is an informative account of Davidson's career in J. Rogerson, *Old Testament Criticism in the Nineteenth Century: England and Germany* (London, 1984), pp. 197–208.

2   S. Davidson, *A Treatise on Biblical Criticism*, vol. 1. *The Old Testament* (Edinburgh, 1852).

3   U. F. Kopp, *Bilder und Schriften der Vorzeit*, 2 vols (Mannheim, 1819–21).

4   W. Gesenius, *Scripturae linguaeque phoeniciae* (Leipzig, 1837).

5   Ibid., p. 77 ('De antiquissima Persarum scriptura').

6   Ibid., pp. 59–61 ('De scriptura aramaea aegyptica').

7   Davidson does not refer to any of Hincks's research and publications in his book.

⁓⁓⁓

FROM JOHN GARDNER WILKINSON
13 SEPTEMBER 1851

Malvern Wells
13th Sepr 1851

Dear Sir,

I beg to acknowledge the receipt of your remarks on the papyrus which have just reached me safely.[1] I will attend to your wishes when they are printed & will send you the proof. As yet I have not had the plates sent me from the lithographers, & I am not quite ready myself with the text, having been so fully engaged with other things.

I have not yet read your 'remarks', as I lose no time in writing to acknowledge their safe arrival, but they appear so full that it will not be necessary for me to say much on the subject. Do you know any body who would be willing to become a subscriber to our publication?

I have given in the text Manetho's lists of kings to the 19th dynasty, & the 'Theban kings' of Eratosthenes, &c. in order to bring together the most useful authorities for reference. Thanks to your valuable contribution, ample means will be afforded for those who wish to study the papyrus. It will give me great plea-sure to insert it in your name and acknowledge your kindness in drawing it up.

I remain
Dear Sir
Yours very truly
Gardner Wilkinson

MS: GIO/H551

1   'Observations on the Turin Papyrus', in J. Gardner Wilkinson, *The Fragments of the Hieratic Papyrus at Turin: Containing the Names of Egyptian Kings, with the Hieratic Inscription at the Back*, vol. 1, pp. 47–60.

FROM GEORGE CECIL RENOUARD
18 SEPTEMBER 1851

Swanscombe, Dartford
18th. Septr. 1851

My dear Sir,
     It is so long since I last had the pleasure of writing to you, that I fear you must suppose Killyleagh & its excellent pastor have quite escaped from my memory: that, however, is not the case; &, I trust, never will be. The want of any thing to communicate, was, I believe, the main cause of my silence; however it would be useless to trouble you with any further apology for my delinquency, I shall therefore immediately turn to the circumstance which made me resolve to write this letter. I read, a few days ago, a letter from yourself, printed in the Athenaeum, which told me that you still are alive to the Ninevite inscriptions,[1] & that con-sequently I might safely mention some recent papers respecting them which well deserve attention, & which you will be glad to hear of, if they have hitherto escaped your notice.
     The papers to which I particularly allude, are 1st. the commencement of a revision by Prof^r. Oppert (a young German established in France and just appointed to join a literary mission to Mesopotamia[2]) of Col. Rawlinson's text & version of the record found at Bagistane (now Behistún, or rather Bahistán) & 2dly. remarks by Hofrath & Hofmeister *Holtzmann* on Westergaard & De Saulcy's papers respecting the inscriptions published by Botta & Layard, together with Rawlinson's interpretation of parts of them. 1. Oppert, who published a tract of considerable value on the sounds of the ancient Persian language, some years ago

(Das Lautsystem des Altpersischen 8vo, Berlin 1847 56 pp.), has sent his observations on the Behistún inscription to the Journal Asiatique, in which they have appeared in the N^{os}. for Feb. March, Apr. May & June (1851) successively.[3] 2. Holtzmann, whose blackguard attack on Lassen, you perhaps remember,[4] read a long paper of unconnected observations on the inscriptions found at Nineveh, to the German Oriental Society, at their general meeting, in Oct. 1850. This paper was so highly approved of, that he was at last persuaded to print it, though, as he adds, far from being yet prepared to attempt a continued version of any one inscription, or even to assert positively the truth of his own conjectures. This paper is printed in the 2^d. N° of the Zeitschrift der Deutschen Morgenlaendischen Gesellschaft, for 1851; p. 145–178.[5]

As both, or either of these papers may have been seen by you, I shall merely make some short remarks on the results deducible from them, with some statements as to the impression, I believe them to have made elsewhere, reserving any further details till I know whether they would be agreeable to you, as supplying what has not yet fallen under your notice.

1. Oppert, who lives in a provincial town (Laval, in the Dep. of Mayenne) has not access to libraries like those at Paris, & probably scarcely knows your writings by name; but in his first work on the sounds of the ancient Persian letters, he proposed nearly the same pronunciation of the final *yas* as yourself, & has only modified his method in his last papers. (I refer to your remarks on 'The First and Second Kinds of Persepolitan writing', 9 June 1846[6].) He supplies some of the lacunae in the Records of Darius plausibly, if not certainly; &, as far as I can judge, justifies most of his departures from Rawlinson's text & version. He also gives, with much appearance of correctness, many Persian names mentioned by Greek writers. His work, however, is far from finished; & as he has just been appointed to join another expedition to Mesopotamia, it is possible, it may remain unfinished.

2. Holtzmann was unwilling to publish his remarks as[7] they are necessarily unconnected with each other, but such was the impression produced by them when read, that he was pressed by almost all present not to withhold them from the world at large. The result of them is, I believe, much the same as that obtained by Col. Rawlinson, several months ago, as he then told me that he was convinced all the cuneatic languages were *fundamentally* the same, but that the Babylonian approached nearly to the Chaldee of Daniel. As I think it very possible you may not have seen so modern a portion of the Zeitschrift, I shall venture to transcribe a part of his concluding paragraph.

'Da eine vollständige und sichere Entzifferung der medischen Schrift mit dem jetzigen Material, unmöglich erreicht werden kann, so glaube ich mich auf die mitgetheilten Betrachtungen beschränken zu dürfen, und hoffe durch dieselben hinlänglich erwiesen zu haben, dass die Sprache der Inscriften der zweiten

Art eine arische, und zwar der persichen Familie angehörige sei, jedoch mit Beimischung semitischer Elemente.' s. 178.[8]

In the preceding pages, he combats successfully Col. R.'s notion of a Turkish (or, as he calls it, Scythic) element,[9] & shows that it consists for the most part, of Chaldee, or Aramaean words engrafted on the Arian grammar & idiom; so that it stands nearly in the same relation to the contemporary Persian, as the Pahlaví (now said to be the Húzvárish or Khúzí [Susian of the Greeks]) to the Pársí.

Of the latter, a grand grammar has been lately published by Spiegel, a Bavarian orientalist.[10]

If these works are already known to you, my dear Sir, you will, I am sure, excuse the trouble now given with a good intention by

your much obliged & faithful
Geo. C. Renouard

I hope your family, whose kindness & hospitality I shall never forget, are well, & will accept my best remembrances & compliments. I hear you were in London two or three months ago.

Swanscombe, Friday, 19 Sept. 1851

P.S. Your writings seemed little known in France. I fear you do not send copies to the *Société Asiatique* & the *Académie des Inscriptions*. I saw a beautiful map & heard an interesting report of his journey round the *Dead Sea* by M. de Saulcy at a meeting of the Academy.[11]

MS: GIO/H 460

1   See Hincks's letter of 8 September 1851 above.

2   Julius Oppert was born in Hamburg on 9 July 1825 of Jewish parents. He studied law at Heidelberg and various oriental languages at Bonn, Berlin and Kiel. He moved to France in 1847 where he was in a position to hold a professorship. He died on 21 August 1905. See W. Muss-Arnoldt, 'The Works of Jules Oppert', *Beiträge für Assyriologie* 2 (1894), pp. 523–56; C. Bezold, 'Julius Oppert', *Zeitschrift für Assyriologie* 19 (1905), pp. 169–73. Oppert left Paris on 1 October 1851 and sailed on the ship *Hellespont* from Marseille on 9 October. He arrived in Mosul on 1 March 1852 and in Baghdad on 27 May 1852. See *Expédition scientifique en Mésopotamie exécutée par ordre du gouvernement de 1851 à 1854 par MM. F. Fresnel, F. Thomas et J. Oppert*, 2 vols (Paris, 1859–63).

3   J. Oppert, 'Mémoire sur les inscriptions des Achéménides, conçues dans l'idiome des anciens Perses', *Journal Asiatique*, 4th ser., 17 (1851), pp. 255–96, 378–430, 534–67. The series of articles continued in *Journal Asiatique*, 4th ser., 18 (1851), pp. 56–83, 322–66, 553–84; 19 (1852), pp. 140–215.

4   See Renouard's letter of 21 January 1847 in vol. 1, p. 172.

5   A. Holtzmann, 'Über die zweite Art der achämenidischen Keilschrift', *Zeitschrift der deutschen morgenländischen Gesellschaft* 5 (1851), pp. 145–78.

6   'On the First and Second Kinds of Persepolitan Writing', *Transactions of the Royal Irish Academy* 21 (1846), polite literature, pp. 114–31.

7   Renouard wrote 'are' in error. Probably he meant to write 'as'.

8   Holtzmann, 'Über die zweite Art der achämenidischen Keilschrift', *Zeitschrift der deutschen morgenländischen Gesellschaft* 5 (1851), p. 178. Translation: 'Since a complete and reliable decipherment of the
Median script can by no means be attained with the material now available, I believe I can restrict myself
to the observations which I communicated; and I hope to have thus shown sufficiently that the language of
the inscriptions is of the second kind, Aryan and belonging to the Persian family, but with an admixture
of Semitic elements.'

9   Holtzmann refers to Rawlinson, 'On the Inscriptions of Assyria and Babylonia', *Journal of the Royal
Asiatic Society* 12 (1850), pp. 401–83. In his publications Rawlinson used both the terms 'Median' and
'Scythic' for Elamite, which is very confusing.

10   F. Spiegel, *Grammatik der Pârsisprache nebst Sprachproben* (Leipzig, 1851).

11   F. de Saulcy, *Voyage autour de la mer morte et dans les terres bibliques: exécuté de décembre 1850 à avril
1851*, 2 vols (Paris, 1853); Eng. tr.: E. *Narrative of a Journey round the Dead Sea and in the Bible Lands in 1850
and 1851*, tr. E. de Warren, 2 vols (London, 1853).

FROM SAMUEL DAVIDSON
19 SEPTEMBER 1851

[Independent College
Manchester
Sept\". 19. 1851]

[My dear Sir,]

I am obliged by your two letters which, but for other things, I would have
answered sooner. I should like very much to see a cast of the medal you refer to. Is
it possible to get one, and how? And then I should like to know whether it be
really an authentic one, how it was obtained, and when. Is it newly discovered, &
how was it procured for the Museum. Could you give me the letters forming the
inscription on it, or tell me where I could get them.[1]

The coin throws new light on the change of the Hebrew characters, and goes
to corroborate Kopp's view, who refers the change to the third or fourth
century.[2] I confess however that the difficulties against that late period are to me
insuperable. In consequence of those difficulties Hupfeld modified Kopp's view
reducing the time of change to the first century of the Xtian aera.[3] And now the
coin comes in against even this idea, although all scholars adopted Hupfeld's
opinion down to Herbst, in 1840, who objects both to it and to Kopp's.[4] Although
therefore the coin throws light on the general subject, yet it only makes a general
conclusion on it the more embarrassing. I confess I cannot at all see any clear way
of getting at the truth. As soon as I get a particular account of the coin, I shall
write to Hupfeld about it, & put the case before him for reexamination.[5]

I should like much that you would see the inscriptions on the Babylonian tiles
to which Kopp traced up the Hebrew character, & which he thought the oldest of
all.[6] Gesenius *conjectured* that it was nothing but *Old Persian* which Kopp

mistook. If the characters be cuneiform would that not be an argument against deriving the Hebrew ancient character from it?

I have not seen the Report of the British Association for 1850, but will try to borrow it here.[7] Rawlinson's paper I saw in the Athenaeum.[8] I suppose yours will appear on Saturday.[9]

If you have any views which you think throw light on the Hebrew character, or the nature of the Hebrew language, I should be glad if you would communicate them to the public or to me, as I will be attending to this subject for some time. Just at present however, I am at the Septuagint, but will return to the point which the coin touches. But I should like to be more certain about it. If you want any other things out of Gesenius copied, let me know, and I will give them as accurately as I can.

I am yours sincerely
Sam¹. Davidson

Tyrus[10]
ΙΕΡΑΣ

All the coins of *Tyre* given have a head on one side, and letters on the other side, mostly three peculiar ones, along with others (but the latter are in a different place) which are Greek. The above has only three peculiar letters at the place marked. [lṣr]

Sidon

I will make this larger for the purpose of showing better what is on it.

EIT
ΣΙΔΩΝΟ[Σ]
ΘΕΑΣ
[lṣdn(m)]

More coins of Tyre than of Sidon are given by Gesenius.

MS: GIO/H 153

1   Hincks may have mentioned the drawing of a seal inscription, which he received in a letter from Edward Clibborn on 18 April 1849 (see vol. 1, pp. 276–8). The seal, which is probably Phoenician, was sold to the British Museum by a Miss Walsh of Dublin. The inscription on the seal can be translated as follows: 'Belonging to Abd-El'ab, son of Shabeath, servant of Mititti, son of Zidqa.' Sidqa and his son Mitinti II were kings of Ashkelon and are mentioned in Assyrian inscriptions. See the bibliography in n. 1 to Clibborn's letter.
2   For Kopp's views, see his *Bilder und Schriften der Vorzeit*, vol. 2 (Mannheim, 1821), part 4, 'Entwickelung der semitischen Schriftens'. There is no further information about this coin mentioned by Davidson.
3   H. Hupfeld, 'Kritische Beleuchtung einiger dunteln und misverstanden Stellen der alttestamentlichen Textgeschichte', *Theologische Studien und Kritiken* 3 (1830), pp. 247–301.
4   J. G. Herbst, *Historisch-Kritische Einleitung in die Heilige Schriften des Alten Testaments. 1. Allgemeine Einleitung* (Karlsruhe and Freiburg, 1840), pp. 59–63.

5   Hermann Hupfeld (1796–1866), orientalist and biblical commentator, succeeded Wilhelm Gesenius as Professor of Theology at the University of Halle. See further O. Kaiser, *Zwischen Reaktion und Revolution: Hermann Hupfeld (1796–1866) – ein deutsches Professorenleben* (Göttingen, 2005).

6   See the summary of Kopp's position in *Bilder und Schriften der Vorzeit*, vol. 2, pp. 177–8.

7   Hincks, 'On the Language and Mode of Writing of the Ancient Assyrians', *Report of the Twentieth Meeting of the British Association for the Advancement of Science; Held at Edinburgh in July and August 1850* (1851), p. 140 + 1 plate.

8   Rawlinson, *The Athenaeum*, no. 1243 (23 Aug. 1851), pp. 902–3.

9   See Hincks's letter of 8 September to *The Athenaeum* above. Davidson had not noticed its publication on 13 September.

10   On a separate page Davidson made rough drawings of these two coins from Tyre and Sidon. He copied them from Gesenius, *Scripturae linguaeque phoeniciae*, plate 34, B and S; and pp. 99 and 265. See also J. Swinton, 'Observations upon Two Antient Etruscan Coins, never before Illustrated or Explained': A Letter to T. Birch, *Philosophical Transactions* 54 (1764), pp. 99–106, esp. plate II (b), 99, VI and VIII.

⚜

TO AUSTEN HENRY LAYARD

27 SEPTEMBER 1851

Killyleagh C° Down

27th. Septr. 1851

My dear Sir,

I have delayed acknowledging the receipt of, and thanking you for, your very friendly letter of the 1st. until I could tell you the result of the application made in my favour. It has, I am sorry to say, proved unsuccessful; and though my friends in this neighbourhood encourage me to hope for something at a future period, I am myself far from sanguine. There is nothing now vacant that would suit me & nothing likely to be vacant in *Ireland*; as for England, preferment in the church *there* is *never* given to an Irishman.

Under these circumstances, I can do but little in the way of study. I cannot devote to it above ten or twelve hours on an average in a week; and cannot leave home *atall*. If, however, any division of labour could be struck out by which I could examine some of the many documents that have come over *here*, I would do what I could. I am very curious about the Assyrian annals of Sennacherib.[1] I cannot think that Merodach Baladan was the king of Babylon whom he subdued in the first year of his reign.[2]

I hope this letter will not go astray; but at this season of the year every one is moving about, but those who are like myself bound down to the one spot.

Believe me
My dear Sir
Your faithful & obliged
Edw. Hincks

MS: BL Add. 38980, f. 126

1   Hincks is referring to the Annals of Sennacherib found in the Taylor Prism. Colonel Taylor's prism has been in the British Museum since Rawlinson purchased it in June 1855 for £250. Its number is BM 91032 = 55–10–3,1. Colonel Robert Taylor was the East India Company resident at Baghdad from 1828 to 1843. Apparently he knew Arabic well and helped Rawlinson to learn it.

2   See, however, J. A. Brinkman, 'Merodach-Baladan II', *Studies Presented to A. Leo Oppenheim* (Chicago, 1964), pp. 6–53; M. Cogan and H. Tadmor, *II Kings* (The Anchor Bible 11; Garden City, N.Y., 1988), pp. 259–61.

FROM GEORGE CECIL RENOUARD
9 OCTOBER 1851

Swanscombe, Dartford
Th. 9th. Octr. 1851

My dear Sir,

On the 25th. ult., the day after I received your letter, finding from it that you had not seen Holtzman[n's] paper on the Ninevite inscriptions, which I was sure would be read with pleasure by you, I sent the 2^d. Heft of the 5th. vol. of the Zeitschrift der Deutschen Morgenländischen Gesellschaft[1] by post to you at Killeyleigh.[2] It ought therefore to have reached you on the 27th. or 28th. but from your silence I am persuaded that it was lost in the P.O. As the enclosure was very thin, it might easily be worn out in so long a journey, so that the loss is probably ascribable to my own want of caution in not securing it better. As soon as I hear from you that my apprehensions are justified, I will endeavour to replace it; but as there can be little doubt that the cover was torn off in its way to Ireland, an enquiry at the Dublin P.O. might possibly recover the missing pamphlet: or at Belfast through which I presume it passes on its way to Killyleagh, something might possibly be heard of it.

I am sorry to give you so much trouble but it would be premature for me to write to Dublin before I know that my packets miscarried.

I hoped to have seen Col. Rawlinson, when I was in Town lately, but he was in the country. I wished to recommend to his notice a very estimable Frenchman (whom he must already know well by reputation), lately dispatched by the French Gov^t. to Baghdad, accompanied by M. Oppert (whose improved text & version of the Behistún inscription, I have already mentioned to you[3]), for the purpose of discovering & copying ancient monuments. They are accompanied by a very skilful artist.[4] Fresnel, with whom I became acquainted at Paris, is, as I dare say you recollect, the principal promoter of discoveries of the ancient Arabian monuments on which M^r. Forster has made such amusing speculations.[5]

While at Paris, I was so much preoccupied with other matters, & so much afraid of overwalking, that I never entered the Musée at Tuileries tho' living just opposite to it & passing by it every day. I brought back, as, I think, I told you, Ibn Khaldún's Hist. of the Berbers lately printed at Algiers, the 1ˢᵗ. vol. all there on sale, & have since received the 2ᵈ. De Slane's version & notes will, I believe, soon follow.⁶ I have not been so lucky as to find Mʳ. Norris at the R.A.S. since his return from a visit in the country, or I should be able to give you some accᵗ. of what R. is doing. I am sorry for more reasons than one, to hear it reported that he is engaged in a new ed. of Herodotus.⁷ He must be at his post again (as I hear) in Decʳ. so that we may expect his paper on the Assyrian or Median part of the Behistún *Title* soon.⁸

Believe me,
My dear Sir,
very faithfully yours
Geo. C. Renouard

P.S. This letter will not set off till the 10ᵗʰ. Octʳ.

MS: GIO/H 461

1   A. Holtzmann, 'Über die zweite Art der achämenidischen Keilschrift', *Zeitschrift der deutschen morgenländischen Gesellschaft* 5 (1851), pp. 145–78.
2   Renouard's spelling 'Killeyleigh' contrasts with the normal spelling which he uses at the end of the paragraph.
3   See Renouard's letter of 18 September 1851, esp. nn. 2–3.
4   Félix Thomas (1815–75), French artist.
5   Fulgence Fresnel (1795–1855) studied Arabic and Persian at Paris and Rome. He entered the diplomatic service as a consular agent at Jidda and took the opportunity to travel in Arabia, where he took a particular interest in Himyaritic, an old South Arabian language. See his contributions in the article, 'Pièces relatives aux inscriptions himyarites découvertes à Sana'â, à Kariba, à Mareb etc. par M. Arnaud', *Journal Asiatique* 6 (1845), pp. 169–237, esp. pp. 182–237. For more biographical details, see *Dictionnaire de biographie française*, vol. 14 (Paris, 1979), cols 1234–5. Renouard's positive comments about him in this letter contrast with the misguided remarks he made in his letter of 21 December 1844 (see vol. 1). Charles Forster (1789–1871) was the paternal grandfather of E. M. Forster, the novelist. For Hincks's severe criticism of Forster's views on the Himyaritic inscriptions, see Renouard's letters of 21 and 30 December 1844, and 6 January 1845; and William Palmer's letter of 11 January 1845 in vol. 1. (In Renouard's letter of 21 December 1844, he writes that 'a French M.D. who had ingratiated himself with Arabs, visited Máreb & Sabá & many other parts of Yemen about two years ago . . .'. In n. 13 to this letter I expressed the view that Renouard must be referring to Fulgence Fresnel but noted that he was not an M.D. However, Carl Phillips (CNRS, Paris) has kindly pointed out to me that Joseph Thomas Arnaud (1812–82), M.D., whose name occurs in the title of Fresnel's article above, was the explorer who collected the inscriptions.
6   'Abd al-Raḥmān b. Muḥammad Ibn Khaldūn, *Histoire des Berbères et des dynasties musulmanes de l'Afrique Septentrionale: Texte arabe*, 2 vols (Algiers, 1847–51). The translation by W. MacGuckin de Slane was published under the same title in four volumes at Algiers in 1852–6.
7   Henry Rawlinson's brother, George Rawlinson, historian and Church of England clergyman, was preparing *The History of Herodotus*, 4 vols (London, 1858–60). Henry Rawlinson and the Egyptologist John Gardner Wilkinson wrote supplementary essays for the volume. See Renouard's letter of 15 October 1851 below.

8   The 'Median' part, that is the Elamite text, was published by Edwin Norris, 'Memoir on the Scythic Version of the Behistun Inscription', *Journal of the Royal Asiatic Society* 15 (1853/1855), pp. 1–213. The long-awaited Babylonian text finally appeared in January 1852: Rawlinson, *Memoir on the Babylonian and Assyrian Inscriptions* (London, 1851) = *Journal of the Royal Asiatic Society* 14 (1851) [entire issue].

FROM EMMANUEL DE ROUGÉ
9 OCTOBER 1851

Paris 9 octobre

Monsieur,

Je profite de l'heureuse occasion que j'ai eu de faire connaissance avec le révérend Charles Graves pour vous envoyer un exemplaire de mon mémoire sur l'inscription d'Elithiyie.[1] Je crains que vous n'ayez pas reçu quelques travaux antérieurs que j'ai eu l'honneur de vous expédier par la voie de la librairie. Vous verrez que j'ai lu avec tout le soin qu'il méritait votre remarquable travail sur l'alphabet.[2] Je regrette bien de ne pas connaître quelques-uns de vos travaux et surtout ce que vous avez lu au congrès de Winchester; il m'a été impossible de trouver cela en France.[3] Je vous serais infiniment obligé si vous pouviez me le communiquer.

J'ai l'honneur M. le Docteur, de saisir cette occasion pour vous témoigner la haute estime que m'ont inspirée vos travaux littéraires, et vous assurer de mes considérations les plus distinguées.

Vte. Emmanuel de Rougé

Rue du Bac 120 Paris.

MS: GIO/H 154

1   *Mémoire sur l'inscription du tombeau d'Ahmes, chef des nautoniers* (Paris, 1851). Charles Graves (1812–99) was a member of the Royal Irish Academy and Professor of Mathematics at Trinity College, Dublin from 1843. He became Bishop of Limerick in 1866.
2   'An Attempt to Ascertain the Number, Names, and Powers, of the Letters of the Hieroglyphic, or Ancient Egyptian Alphabet; Grounded on the Establishment of a New Principle in the Use of Phonetic Characters', *Transactions of the Royal Irish Academy* 21 (1847), polite literature, pp. 132–232 + 3 plates. The separately printed copies and the volume appeared in 1847.
3   'On Certain Egyptian Papyri in the British Museum', *Transactions of the British Archaeological Association, at its Second Annual Congress, held at Winchester, August 1845* (1846), pp. 246–63 + 1 plate. The article contains an account of seven manuscripts in Hieratic, published by the British Museum. See *Select Papyri in the Hieratic Character from the Collections of the British Museum*, 3 parts (London, 1841–4).

FROM GEORGE CECIL RENOUARD
15 OCTOBER 1851

Swanscombe, Dartford
Wed. 15th. Octr. 1851

My dear Sir,

I never was more agreeably surprised than on learning from your letter of the 13th. (Monday last) which reached me by this evening's post, that the Zeitschrift had arrived safely at Killyleagh and I hope you will not put off more pressing matters on acc^t of it, as I do not wish for it immediately: but had it been unhappily lost in consequence of its insufficient clothing,[1] it would have been necessary to take immediate steps to replace [it], as after a short period such periodicals are not to be procured.

I am at present busy with D^r. Barth's papers from Bornú which he reached by the route of Ákedez (perhaps a corruption of Moḳaddes).[2] He is a very painstaking man but wants *tact* & judgement. His itineraries, when calculated, will be very useful: but his style is rambling, diffuse & his English very indifferent with a strong spice of intentional elegance & picturesqueness (if such a word can be allowed).

He has sent copious vocabularies of the Bornú & Ákedez (pron. Agedez or Agedess) languages & was much astonished on learning that the latter is identical with the Kissúr spoken at Tumbuktú. Of the truth of that information I possess an irrefragable proof, as I have a short vocabulary collected from a native of the place & another Negro whom I saw at Smyrna in 1813.[3]

I saw enough of M^r. Forster's last production in the Literary Gazette (where it is much lauded) to feel little inclination to read, much less to purchase it.[4] I am glad you approve of what *Beer* did. His early death is much to be deplored, as his performances gave great promise.[5] He has however had an able follower in D^r. Fried. Tuch, Prof^r. of Divinity at Leipzig, who has given a Versuch einer Erklärung von einen und zwanzig sinaitische Inschriften in the Zeitschrift der Deutschen morgenländischen Gesellschaft (III. S. 129–215).[6] I have only just looked at parts of it, but saw that he treads carefully in Beer's steps.

The report respecting R.'s projected ed. of Herodotus may be all good for nothing. He has, it seems, a relation in some college at Oxford who is said to be engaged with him in the work.[7] I hope to see him tomorrow, or next day, & shall ask about it. I heard many complaints at Paris of his holding back his Assyrian texts so long, but I feel assured something will soon appear as he must leave England for Baghdád in at least a fortnight's time, & I think I saw several proofs full of Median type on his table six months ago.

. . .[8]

Th. 16th. Oct^r.

You will be glad to hear that a new Egyptian room has been opened in the Louvre and M. Lenormant has commenced his lectures.[9] In a short time I hope to send you some hieroglyphic news.

If you have Porter's Travels in Georgia &c., look at pl. in vol. II, p. 123.[10] Löwenstern finds in it a chronological series of kings of Babylon, Jugaeus, Mardoc, Arcianus & Belibus (Yugva, Morotakh, Arsjan & Beplikh). I should like to hear what you think of it. Rawlinson, I find, left London for Paris & Conple in his way to Baghdád, about 10 days ago. So that he will soon be at work again.

Farewell, my dear Sir,
& believe me
your faithful & much obliged
G. C. Renouard

MS: GIO/H 462

---

1   Renouard wrote 'cloathing' in error.

2   Renouard appears to have edited the article 'Progress of the African Mission, Consisting of Messrs Richardson, Barth, Overweg to Central Africa', *Journal of the Royal Geographical Society of London* 21 (1851), pp. 130–221. Note in that article the spellings Agádéz and Timbuktu. Renouard writes Tumbuktú (see French Tombouctou). Heinrich Barth (1821–65), German geographer and one of the great explorers of Africa, travelled through west Africa from 1850 to 1855 on an expedition sponsored by the British government. His colleagues, the explorer James Richardson, and the geologist and astronomer Adolf Overweg, died during the journey. His *Travels and Discoveries in North and Central Africa*, 5 vols (London, 1857–8) is of major importance.

3   From 1810 to 1814 Renouard was chaplain to the factory at Smyrna.

4   Renouard had seen a review of C. Forster, *The One Primeval Language Traced experimentally through Ancient Inscriptions, in Alphabetic Characters of Lost Powers, from the Four Continents; Including the Voice of Israel from the Rocks of Sinai* (London, 1851), in *Literary Gazette*, no. 1790 (10 May 1851), pp. 323–4. He was obviously aware of Hincks's forthcoming trenchant review, 'The One Primeval Language', in *Dublin University Magazine* 39 (Feb. 1852), pp. 226–34.

5   Eduard Friedrich Ferdinand Beer, orientalist and professor at Leipzig, was born on 15 June 1805 and died on 5 April 1841. He deciphered the Nabataean script. See his *Inscriptiones veteres litteris et lingua hucusque incognitis ad Montem Sinai magno numero servatae* (Studia asiatica 3; Leipzig, 1840). The decipherment of Nabataean is discussed by J. F. Healey, 'The Decipherment of Alphabetic Scripts', in Cathcart (ed.), *The Edward Hincks Bicentenary Lectures*, pp. 86–91.

6   F. Tuch, 'Ein und zwanzig sinaitische Inschriften', *Zeitschrift der deutschen morgenländischen Gesellschaft* 3 (1849), pp. 129–215.

7   The relation was his brother George Rawlinson. See Renouard's letter of 9 October 1851, n. 7, above.

8   I have omitted a short paragraph in which Renouard has copied rather badly some lines of cuneiform from an article by F. de Saulcy.

9   Charles Lenormant (1802–59) was Professor of Egyptian Archaeology at the Collège de France from 1848.

10   R. Ker Porter, *Travels in Georgia, Persia, Armenia, Ancient Babylon, &c. &c. during the Years 1817, 1818, 1819, and 1820*, vol. 2 (London, 1822).

FROM GEORGE CECIL RENOUARD
15 NOVEMBER 1851

Swanscombe, Dartford, Kent
Sat. 15 Novr. 1851

My dear Sir,

I am happy to inform you of the safe arrival of the two Zeitschrifts this evening, & am very much obliged to you for returning them so soon. I had indeed scarcely looked at them, which will account to you for my having failed to notice the mention of your 'Remarks on the Enchorial language of Egypt'.[1] It had not escaped my eye, though it did not occur to my recollection when I dispatched my last packet of the Zeitschrift. That was sent off in a great hurry when my head was full of Tuch and Sinai. You will observe that the paper to which you refer (III. p. 262) is by D^r. Brugsch himself. Though printed 1849, it was probably written in 1847. The writer had not then seen your more important papers on his subjects. At least so I conjecture. In the earlier publications of the D. M. G. there were reviews of Brugsch's works by Seyffarth, who is never noticed in the paper to which you refer. He spoke civilly but not very favourably of Brugsch, whom he naturally blamed as misled by his veneration for Champollion, whose system S. was still labouring to supplant. I believe he has found little favour with his countrymen & seems to have withdrawn from the D. M. G. of which he was at first the Librarian. Under colour of being merely the mouthpiece of his friend Spohn, he seems to have substituted a new system of his own, not quite different from that of Champollion, but professedly opposed to almost every portion of it.[2] Requiescat in pace![3]

I suppose you already know that 'John Wilson of Bombay' is a missionary of the Scotch Church & was in 1843, President of the B. Branch of the R. Asiat. Soc. – the Rev^d. J. W. D. D.[4] If you have not seen his 'Pársí Religion', you will, when you have leisure, read it with satisfaction as it [is] spoken of by Spiegel (a very competent judge) as one of the very best works on the subject.[5] It is controversial & gives the Zend texts in the original language. His 'Lands of the Bible', I know only by name.[6] I believe I once met him in the Library at the India House just after his return from India two years ago. He then said that in Bombay they considered the ancient Arabic inscriptions as Christian. In that point they were certainly wrong. I suppose I told you that I passed some very agreeable hours with Fresnel (of Himyarí celebrity) while I was at Paris.

There are some good remarks on the Turin Papyrus (if by that name you mean Champollion's Rituel funéraire & Lepsius's Todtenbuch) in, I think, some late numbers of the Revue Archéologique (VI, 525, 660; VII, 559) by M. de Rougé who, if I remember rightly, has fallen into your views without, I fear, knowing your writings.[7]

I hoped to have received a copy of Rawlinson's forthcoming volume this evening, as I believe it w^{d}. be laid before the meeting of the R.A.S. this afternoon.[8] I do not wonder at the disappointment of yourself & the French & German orientalists in consequence of his withholding his treasures, but surely as he was the first not only to transcribe, but to interpret them, it is reasonable that he should endeavour to prevent the anticipation of his results. The impatience to which he has thus given birth, will cause those results to be scrutinised with no ordinary severity.

I should be highly delighted were I able to attend the Meeting at Belfast next year;[9] but my journies to Ireland & France, have told me in language I would willingly misinterpret, that I am too old for traveling. My health, though not perfect, is such as scarcely interferes with my pursuits; & it has this year been on the whole better than it was for some years past. So much so that when no unusual exertion is required I am apt to forget how many years have passed over the head of

my dear Sir,
your faithful septuagenarian[10]
G. C. Renouard

P.S. Were it not now late, I could have added some things which will perhaps interest you: such as poor Cooley's Map of the terra incognita in southern & almost central Africa which will, I trust, appear in the course of 8 or 10 weeks.[11] I say poor Cooley, for he has almost entirely lost his hearing, & a few days ago had a very narrow escape of being run over in crossing a street. He says that tho' aware of his danger, his limbs w^{d}. not move fast enough to allow him to get out of the way! Can we be surprised should paralysis seize him?[12]

MS: GIO/H 464

1   'The Enchorial Language of Egypt', *Dublin University Review* 1/3 (July 1833) (offprint, 14 pp.). It is reprinted in Cathcart (ed.), *The Edward Hincks Bicentenary Lectures*, pp. 216–27. Heinrich Brugsch refers to this article in his paper 'Die demotische Schrift der alten Ägypter und ihre Monumente', *Zeitschrift der deutschen morgenländischen Gesellschaft* 3 (1849), pp. 262–72, see pp. 263–4.

2   Friedrich August Wilhelm Spohn (1792–1824), who became Professor of Oriental Languages at Leipzig, claimed to have discovered how to decipher ancient Egyptian hieroglyphs. Gustav Seyffarth edited his works *De lingua et literis veterum Aegyptiorum*, 2 pts (Leipzig, 1825–31) and *Brevis defensio hieroglyphices inventae a F. A. G. Spohn et G. Seyffarth* (Leipzig, 1827). Seyffarth was convinced that he himself had deciphered the hieroglyphs before Champollion and made rather fantastic claims about the language, but very few scholars took him seriously. One of Hincks's later correspondents, Peter le Page Renouf, wrote an article exposing the errors in Seyffarth's views. See 'Seyffarth and Uhleman on Egyptian Hieroglyphics', *The Atlantis* 2/3 (Jan. 1859), pp. 74–97 (= *The Life-Work of Sir Peter le Page Renouf*, vol. 1, ed. G. Maspero and W. H. Rylands [Paris,

1902], pp. 1–31). Seyffarth replied angrily in his article 'Champollion and Renouf', *Transactions of the Academy of Science of St. Louis* 1 (1860), pp. 539–69, which brought a rejoinder from Renouf, 'Dr Seyffarth and the *Atlantis* on Egyptology', *The Atlantis* 3/6 ([Jan.], 1862), pp. 306–37 (= *The Life-Work of Sir Peter le Page Renouf*, vol. 1, pp. 33–80). See *The Letters of Peter le Page Renouf (1822–1897)*, vol. 3, *Dublin (1854–1864)*, ed. K. J. Cathcart (Dublin, 2003), pp. 90–1, 113, 128.

3   Seyffarth was still alive, so Renouard must be referring to Spohn.

4   John Wilson (1804–75), Scottish missionary and orientalist, spent most of his life in India. Compelled by ill-health, he returned to Scotland and was there from 1843 to 1847. Renouard mentions him in his letter of 21 December 1844. See *ODNB*.

5   J. Wilson, *The Pársi Religion, as Contained in the Zand-Avastá, and Propounded and Defended by the Zoroastrians of India and Persia, Unfolded, Refuted, and Contrasted with Christianity* (Bombay, 1843).

6   J. Wilson, *The Lands of the Bible Visited and Described in an Extensive Journey*, 2 vols (Edinburgh, 1847).

7   See E. de Rougé on R. Lepsius, *Das Todtenbuch der Ägypter nach dem hieroglyphischen Papyrus in Turin* (Leipzig, 1842) in *Revue archéologique* 6/2 (1850), pp. 525–39, 660–8; and a letter to the editor on Champollion-Figeac's views in *Revue archéologique* 7/2 (1851), pp. 559–66.

8   *Memoir on the Babylonian and Assyrian Inscriptions.*

9   In September 1852 a meeting of the British Association for the Advancement of Science was held at Belfast.

10   Renouard was born on 7 September 1780.

11   W. D. Cooley, *Inner Africa Laid Open, in an Attempt to Trace the Chief Lines of Communication across that Continent South of the Equator* (London, 1852).

12   In his letter of 23 October 1849, Renouard made a similar remark: 'You will be sorry to hear that poor Cooley is almost totally deaf. I cannot but apprehend rammollissement du cerveau: he however says it is no such thing.' Cooley died in 1883, outliving Renouard by many years.

FROM GEORGE CECIL RENOUARD

21 NOVEMBER 1851

Swanscombe, Fr. 21 Novr. 1851

My dear Sir,

The work by Col. Rawlinson to which I alluded, is the continuation of his copies & versions of the *Titulus Bagistanensis*, if, I am not greatly in error. I understand it to be the Assyrian (Median?) & perhaps Babylonian text which nobody but himself, has seen, & which is, of course the great source of all our speculations on the other texts in that language & character.

It will form a thin 8vo vol. & be, I conclude, a continuation of the series already begun & sold separately, though forming a part of the Journal of the Asiatic Society. M^r. Norris who has carried it through the press for him, can give you a complete acc^t. of it. I suppose it is not yet out, as M^r. Neale, clerk to the Asiat. Soc. promised to send me a copy by the post as soon as it was delivered, & he then expected to see it on the table at the Soc^{ty}.'s meeting on Saturday last.[1]

You do not say where subscriptions to Sir G. Wilkinson's work are received. I shall inquire in London and get my name added to the list; though the little I have seen & heard of Sir G. W. had not inspired me with a favourable notion of his ability. I do not know where he now is. You, however, seem to think well of the work which is to me a very sufficient recommendation.

I am glad to find you speak so favorably of Brugsch. A slight inspection of one of his last works lowered him considerably in my scale: and I have no hesitation in saying that he would do much better not to imitate his French friends & too many of his own countrymen, in commencing a new work before he has finished his former ones. I have the first vol. of an excellent ed. of Theophrastus's Hist. Plant.[2] & two of a capital ed: of Strabo[3] which have never been continued. In those cases, I believe, thro' no fault of the editors: but De Saulcy has left three or four or more works in such a state that his books cannot be bound; & I suppose he will attend to nothing now, but his Journey on the Dead Sea.[4] I am not sorry that he quarrelled with the Minister & was not sent with Fresnel to Baghdád.

Champ.'s Dict. is merely a vocabulary arranged under different heads, it is not therefore easy to find out particular words, and I have not succeeded in finding *sensen*; but I may have overlooked it.[5] His brother who is but a poor creature should have added an alphabetical index. Every allowance must be made for the difficulties under which he laboured in consequence of Ch.'s premature death & Salvolini's dishonesty.[6]

Cooley is recovered from his fall & his map is in progress. I heartily pity him, but he is not entirely without fault. He has often been his own enemy.

I do not feel much inclined to send any papers to the Brit. Assoc. but I hope you will meet our Sec. (D[r]. Norton Shaw) there, who is a very clever West Indian Name Surgeon & a great Linguist in the Northern Tongues.[7] The post hour requires my closing this letter with my unfeigned tho' usual assurances of the regard of,

my dear Sir,
your faithful
G. C. Renouard

MS: GIO/H 465

1   The Babylonian text was about to be published in *Memoir on the Babylonian and Assyrian Inscriptions*. The Elamite version was published later by Norris: 'Memoir on the Scythic Version of the Behistun Inscription', *Journal of the Royal Asiatic Society* 15 (1853/1855), pp. 1–213.

2   *Theophrasti Eresii opera quae supersunt omnia*, vol. 1, *Historia plantarum*, ed. F. Wimmer (Bratislava, 1842).

3   Perhaps Renouard had vols 1 and 2 of *Strabonis Geographica*, ed. G. Kramer (Berlin, 1844, 1847), and thought vol. 3 was not going to be published. It appeared in 1852.

4   See Renouard's letter of 18 September, n. 10.

5   J. F. Champollion, *Dictionnaire égyptien en écriture hiéroglyphique* (Paris, 1841).

6   François Salvolini (1809–38) became notorious when it was discovered that he had stolen some of Champollion's important manuscripts and published discoveries as his own.

7   Norton Shaw (d. 1868) was secretary of the Royal Geographical Society from 1849 to 1863 and published *Catalogue of the Library of the Royal Geographical Society Corrected to May 1851* (London, 1852).

FROM CHARLES MACDOUALL[1]
12 DECEMBER 1851

5 University Square Belfast
December 12/51[2]

Rev^d. & dear Sir,

The Greek Chair in the University of Edinburgh being vacant, & more desirable in some respects – especially to a native of the city & an alumnus of the college – than that in Queen's College Belfast, I have become a candidate, in the hope that, if I were again appointed a Professor there, the Test, to escape which I resigned the Oriental Chair in /48, might not be again enforced against me. As it is of great importance that I should carry across written attestations to the value which has been attached to my prelections *here,* & that these should be from men preeminently qualified to make a statement on such a subject – among whom you are conspicuous – I would willingly hope that sufficient information may have been received by you in your intercourse with society to warrant your making such a statement as might be available; and, in case it has, if, besides a few testimonials from parties in or about town whose opportunities of observing have been closer, I were favoured with one from you, I should feel deeply obliged. That you might be enabled to refer at the same time to my intellectual calibre & critical skill, I should have used the freedom to transmit one or two of my published contributions to Greek – especially Scriptural – criticism & philology, had any copies been here. As it is, I can only offer for your acceptance a Lecture the origin of which is mentioned in the preliminary notice. In case you should read it through – or as far as page 37 – I ought to state, in explanation of the absence of your own name, that, when it was written, I had only heard of your paper (in the Royal Irish Academy's Transactions) on cuneiform characters as also hieroglyphics, but had seen no other than the one in the Journal of the Royal Asiatic Society, so that they were quite a novelty to me, & their perusal quite a treat, when I obtained access to them a few days before you delivered a lecture, upon the Khorsabad Inscriptions, in the Museum.[3] I trust that you persevere, notwithstanding the want of adequate encouragement, in your arduous studies for which so very few possess the requisite talents & acquirements. The last notice of yours which I have seen in print was a letter in the *Athenaeum,* about the middle of September, on various interesting points. With deep respect believe me to remain

Rev^d. & dear Sir
Yours very truly
Charles MacDouall

MS: GIO/H 263

1   Charles MacDouall (1813–83) was born and educated in Edinburgh. He was elected Professor of
Hebrew at Edinburgh University in 1847, but the Edinburgh Presbytery blocked his appointment. He was
Professor of Latin at Queen's College, Belfast for a year in 1849–50 and became Professor of Greek in
1850, a post he held until his retirement in 1878.

2   The date looks like 1852 but this is surely a mistake. Warren R. Dawson has 1853 in his Catalogue of
the Hincks Papers at the Griffith Institute. This is also incorrect. MacDouall refers to a letter from
Hincks to *The Athenaeum* 'about the middle of September'. Almost certainly this is the letter written on
8 September 1851 that was published in *The Athenaeum*, no. 1246 (13 Sept. 1851), p. 977. Furthermore, the
Chair of Greek at Edinburgh was filled in 1852 with the appointment of John Suart Blackie (1809–95).

3   MacDouall may be referring to Hincks's lecture on 'The Recently Discovered Assyrian Inscriptions,
and the Mode of Deciphering Them', delivered on 30 January 1850 to the Belfast Natural History and
Philosophical Society. A lengthy summary was published in *Literary Gazette*, no. 1725 (9 Feb. 1850),
pp. 110–11. See Samuel Sharpe's letter of 28 February 1850, esp. n. 9.

TO THE EDITOR OF THE ATHENAEUM
22 DECEMBER 1851

Killyleagh C° Down
22^d December 1851

[Sir,]

The following identification will, I dare say, interest many of your readers. The king who is represented in the second line of the sculptures on the obelisk is no other than Jehu, king of Israel.[1] He is called *Ya.u.a* the son of *Kh'u.um.r'i.i*; that is *yēhû'* the son of *'omrî*, or according to the English version, *Jehu* the son of *Omri*.[2] The name of his supposed father is precisely that which appears in the cuneatic name of Samaria, *Bit-Khumri*, as identified by Col. Rawlinson. It is true that Jehu was neither the son nor the grandson of Omri; nor is it probable that he was connected with his family at all; but the king of Assyria could not know this. He found him on the throne where Omri had sat; and this was a sufficient reason for his calling him his son. As a corroboration of this identification, I observe that Hazael, the king of Syria, the known contemporary of Jehu, is repeatedly mentioned on the obelisk and in the bull inscription of the same king. He waged war with him in his eighteenth and twenty-first years. Col. Rawlinson calls this king *Khazakan*; but the four characters which compose the name are according to my syllabary *Khâ.já* (or *dzá*).*a'h.il*, the last being here the ideograph for 'God'.[3] This name would be in Hebrew $h^a z\bar{a}'\bar{e}l$, which is clearly[4] the Biblical name of the king. From this identification, it follows that the date of the obelisk is, according to the chronology in the margin of our Bibles, about 875 B.C., leaving an interval of less than 150 years between it and the accession of Sargon, the Khorsabad king.[5]

I am, &c.
Edw. Hincks.

Source: The Athenaeum, no. 1251 (27 Dec. 1851), pp. 1384–5

1   Hincks announces that he has identified the name of Jehu, King of Israel, on the Black Obelisk of Shalmaneser III. His diary entry of 21 December 1851 says: 'Thought of an identification of one of the obelisk captives – with Jehu, king of Israel, and satisfying myself on the point wrote a letter to the Athenaeum announcing it'. (Read 'captives' for Davidson's meaningless 'Cophetus'! See *Edward Hincks*, p. 167). The discovery of this four-sided black limestone obelisk by Layard in 1846 during his excavations at Kalḫu is regarded by Old Testament scholars as one of the most exciting archaeological finds of the nineteenth century because of the mention of 'Jehu son of Omri'. Hincks published a translation of the text on the obelisk in 1853; see 'The Nimrûd Obelisk', *Dublin University Magazine* 42 (Oct. 1853), pp. 420–6. For a discussion of Hincks's translation, see Cathcart, 'The Age of Decipherment: the Old Testament and the Ancient Near East in the Nineteenth Century', in J. A. Emerton (ed.), *Congress Volume: Cambridge 1995* (Leiden, 1997), pp. 81–95, esp. pp. 88–90.

2   Rawlinson thought Akkadian *IA-ú-a mar Ḫu-um-ri-i* was 'Yahua, son of Hubiri, a prince of whom there was no mention in the annals, and of whose native country therefore I am ignorant'. See 'On the Inscriptions of Assyria and Babylonia', *Journal of the Royal Asiatic Society* 12 (1850), p. 447.

3   Heather D. Baker (ed.), *The Prosopography of the Neo-Assyrian Empire*, vol. 2, part 1, Ḫ–K (Helsinki, 2000), p. 467, gives the spelling *ḫa-za-a'*-DINGIR and reads DINGIR as the absolute for -*il*, for the Black Obelisk.

4   The text has 'nearly' but surely it should be 'clearly'.

5   Shalmaneser III reigned from 858 B.C. to 824 B.C. Sargon's accession date is 722.

TO LORD CLARENDON
23 DECEMBER 1851

Killyleagh C° Down
23ᵈ. Decʳ. 1851

My Lord,

I have been strongly recommended by some of my friends to write to your Excellency concerning the vacant Provostship & I have yielded to the arguments laid before me.[1] My friends think that I ought to inform your Excellency of the views with which I would undertake the office, as well as of the grounds on which my appointment would be generally approved by all liberal persons unconnected with the present fellows. I make this exception because of course the fellows & their friends would *prefer* an appointment that would give them steps.

I have first to observe that I do not seek the office (as might perhaps be represented) under the impression that it would afford me leisure for literary pursuits. I am fully aware that its duties would occupy almost the entire of my time, so that *personally* I could do next to nothing in the way of archaeological investigation. At present, however, I am equally incapacitated for carrying on these pursuits; and my appointment to the Provostship would place me in a position to induce younger men to carry them on in my stead, to whom I could render important assistance by my unpublished papers and orally. These however are minor considerations, to which I only refer with a view to prevent misconception.

In seeking the office of Provost, I have in view the devoting myself to the discharge of its important duties, which I confidently expect that I should be able to do with credit to myself & with advantage to the College & to the Country. It is my decided conviction, & it has long been so, that the College requires reform; & that, though greatly improved of late years, and though in advance of the English universities in those points where it most deficient, it does not yet meet the just requirement of the age. The public expects that the Univ^y. Commission will recommend some important changes; and the infusion of new blood, if I may use the expression, into the flowering beds of the College at this particular crisis would be regarded with satisfaction by members. While most, if not all, of those who are high up in the present list of fellows are prejudiced against changes which public opinion requires, there is no liberal measure against which *I* am committed; and there are many of which every one that knows me knows that I have long been an advocate. If my views on the education question were not well known to be thoroughly liberal, I should not have been twice elected, as I have been, a vice president of the Royal Belfast Institution.

The right of the Crown to select a provost, without being confined to the actual fellows, is undoubted; and in the case of its being exercised in favour of an ex-fellow & a Doctor in Divinity, no objection could fairly be made. When D^r. Elrington was appointed Provost in 1811 from a college living, the fellows complained for a time; but public opinion did not go with them.[2] In time English colleges where the nomination to the headship is with the Crown, it is by no means a matter of course that a fellow should be appointed.

As respects the popularity of the appointment even in college, I would observe that College men are apt to lay much stress on literary reputation, & that there are not many[3] of the present fellows better known in the literary world than myself. I obtained my fellowship too under peculiarly creditable circumstances – at an early age – on the first occasion not only when I *was* a candidate but when by the college statutes I could have been a candidate & over three men who were afterwards fellows & who were all distinguished – D^r. Robinson & the late Messrs. Phelan & Harte.[4]

I should add that, if appointed to this office, it would give me pleasure to take a part in the business of the National Board, as one of the Commissioners, in case it were thought desirable that the Provost should do so, as it was in the case of my late friend, Dr Sadleir.[5]

I fear I have exceeded the proper bounds of a letter; and yet I have perhaps omitted to notice some points on which I ought to have spoken. If, however, your Excellency wishes for further information or explanation on any point I shall be happy to afford it either by letter or in a personal interview at any time after the present week that you may do me the honor to appoint.

[Edw. Hincks]

MS: PRONI D/1558/1/1/66 (Hincks's copy)

1   This is the only mention of Hincks's interest in becoming Provost of Trinity College, Dublin. Davidson, *Edward Hincks*, makes no reference to Hincks's application for the post.
2   Thomas Elrington (1760–1835) was appointed Provost of Trinity College, Dublin by the Duke of Richmond in November 1811. He became Bishop of Limerick in 1820 and Bishop of Leighlin and Ferns in 1822. See *ODNB*.
3   At first Hincks wrote 'very few' but crossed the words out.
4   Thomas Romney Robinson (1793–1882), Irish astronomer and physicist. See *ODNB*. William Phelan (1789–1830) matriculated at Trinity College, Dublin on 4 June 1806, B.A. 1810, M.A. 1814, fellow 1817, B.D. 1821; he became Rector of Ardtrea after Hincks's move to Killyleagh in 1826. Henry Hickman Harte (1790–1848), mathematician, matriculated at Trinity College, Dublin on 7 July 1806, B.A. 1811, fellow 1819, M.A. 1823; Rector of Cappagh, Co. Tyrone, 1831. He published English translations of works by the French scholars Pierre Simon Laplace and Siméon Denis Poisson. See *ODNB*.
5   Franc (formerly Francis) Sadleir (1775–1851) was Provost of Trinity College, Dublin from 1837 till his death on 14 December 1851. Like Hincks he was an advocate of Catholic emancipation. See *ODNB*.

≈≈≈

TO WILLIAM JOHN CAMPBELL ALLEN[1]
DECEMBER 1851

My dear Mr Allen,[2]

I called on Mr Tennent but was sorry to find he was not well, & in bed.[3] I wished a few words with you. I leave the copy of what I wrote to Lord C.[4] which I wish you to read & perhaps you may show it to him if he will take the trouble to look at it. It will show how matters stand. What I alluded to as change which I had advocated & *was known* to have advocate[d] are mainly these two: the partial opening of the corporation to those who are not of the E. C.[5] & the making the peculiar course of divinity students more distinct than it is, from the course of instruction in Arts, exempting those who are not divinity students (but scholars) from attending on it & also moving a great obstacle which at present lies in the way of graduates of another university being able to attend it. I don't think Mr Tennent could write with advantage to the L. L.[6] at present, but if this application of mine fail, perhaps he would forward & endorse a letter which I would in that case write to him on my prospects.

One thing, however, if he thinks he could do it with propriety, he might do with great benefit at present: speak in my favour to the Bishop.[7] I think he has been deceived (unintentionally I hope) by persons who have got his ear. The supposed neglected state of the parish of Killyleagh is altogether a delusion. I doubt if there be half a dozen parishes in the diocese in a more satisfactory state. Of his inference, Lord Dufferin knows nothing of his own knowledge – persons have deceived *him* & through him the bishop.[8] Mr Ward too is a violent (almost fanatical) party man & would go any length to obtain a curate who would propagate his peculiar views.[9]

I have in my parish fanatical Evangelicals & fanatical Puseyites & I think I deserve great credit for having kept the peace between them in the trying times we have gone through & with all but these two extreme parties, I believe I am very popular – the people are *quite satisfied* though they are *told* that they ought not to be so. The Puseyites grumble at having no 'pastoral care' i.e. no spiritual direction & auricular confession & the Evangelicals say that 'the gospel is not preached', that is, the five points of Calvinism, but the *people generally* are satisfied. Were a curate appointed then of the Mr Ward's school six months ago, I have little doubt he would have driven some of the others to popery. The Bp was offended at my expressing such a fear to him; but have there not been such secessions elsewhere as to give reason for fear. The storm (thank God!) has now in great measure passed away; & it has been weathered in Killyleagh, as well as anywhere else. Am I to be *blamed* for this?

I write in great haste having barely time to finish in time for [the] train & no time to revise. You will therefore be *cautious* how you use this, but I have written as I feel & as I believe to be the case. I am sure that if the Bishop would *investigate the matter fairly*, he would see more cause to praise than to blame

Yours very truly
Edw. Hincks

1   William John Campbell Allen was born in Belfast on 22 October 1810, the only son of Allen and Jane Campbell. He was educated at the Royal Belfast Academical Institution and later became joint secretary of the college from 1838 to 1878. He was the first registrar of Queen's College, Belfast, being appointed in 1849 but resigning after a disagreement with the president in 1852. See Hincks's letter of 28 December below.

2   Hincks wrote the first page of his letter on the reverse of a notice which reads as follows: 'Belfast, December, 1851. Mr. John Grogan begs to inform his friends that he has added to his Veterinary Establishment and Horse Repository, a Forge, in which the best Workmen shall be kept, and Shoeing done on the most approved principles. Castle Yard, Castle Buildings.'

3   Robert James Tennent was M.P. for Belfast from 1847 to 1852.

4   Clarendon.

5   E.C. = Established Church.

6   L.L. = Lord Lieutenant.

7   Robert Bent Knox (1808–93) was nominated by Lord Clarendon to the see of Down, Connor, and Dromore in March 1849. He became Archbishop of Armagh in 1886. See *ODNB*.

8   Frederick Temple Hamilton-Temple Blackwood (1826–1902), first marquess of Dufferin and Ava, succeeded his father in 1841 as fifth Baron Dufferin in the Irish peerage. In 1850 he was created Baron Clandeboye, of Clandeboye, Co. Down, in the peerage of the United Kingdom. On 23 October 1862 he married Hariot Georgina Rowan Hamilton (1843–1936), eldest child of Archibald Rowan Hamilton of Killyleagh Castle. Hincks officiated at the wedding which took place in the castle. See the report in the *Belfast News-Letter*, Issue 15418 (25 Oct. 1862).

9   Perhaps the Rev. Hon. Henry Ward (1795–1874), Rector of Killinchy, Co. Down. See J. B. Leslie and H. B. Swanzy, *Biographical Succession Lists of the Clergy of Diocese of Down* (Enniskillen, 1936), p. 40.

FROM JOHN GARDNER WILKINSON
24 DECEMBER 1851

Aldermaston
Newbury
Berks
24 Decr 1851

Dear Sir,

The Turin papyrus will be ready for the subscribers as soon as the plates have been sent to the printers which will be I trust at the beginning of next week. Will you have the kindness to give your directions to the printer Mr Richards 37 Great Queen Street Drury Lane about sending your copy.[1] I would do this for you with much pleasure but you probably know some one in London to whose care it may be desirable to consign it. Mr Richards receives the subscriptions. I am happy in having this opportunity of thanking you for your kind & valuable assistance which has given to the text an importance it would not otherwise have possessed.

Allow me to offer you the compliments of the Season.

Believe me
Dear Sir
Yours very truly
Gardner Wilkinson

MS: GIO/H 552

1   The printer's name was Thomas Richards.

TO THE EDITOR OF THE ATHENAEUM
29 DECEMBER 1851

Killyleagh C^o Down
29th. Dec. 1851

[Sir,]

Since I addressed you on the 22^d. inst., I have found the name of a second king of Israel in the Nimrud inscriptions published by the British Museum. In the south-western palace there is a series of slabs, brought from the centre of the mound, but of later date than the obelisk and the colossal bulls, which are of the age of Jehu. These slabs contain annals of a king, whose name does not appear. Col. Rawlinson stated confidently that he was the Khorsabad king, Sargon; but from comparing the transactions assigned to the same regnal years in this series and at Khorsabad, I felt satisfied that he laboured under a mistake. On looking over the names of certain kings who paid tribute in the eighth year of this king's reign (B.M. pl. 50, l. 10), I found a name which is decisive on the question, *Mi.na.kh'i.im.mi Sâ.mi.ri.n'â.ayi*; that is *mnḥm* of *shmrwn*, Menahem of Samaria, masoretically *Shômerôn*.[1] The final *mi* in the king's name is added as a case ending, so that the name exactly corresponds with the Hebrew. This name proves that the slabs belonged to Pul, who is mentioned in 2 Kings xv. 19, 20, as having imposed tribute upon Menahem. He was the predecessor of Sargon, and of a different family; which accounts for his slabs having been removed, and his name having (it is said) been defaced by Esarhaddon, the grandson of Sargon, who built this palace. It proves also the identity of the *Samirina* and the *Bit-Khumria** of the inscriptions, which I before considered improbable; and the consequent fact that the 27,280 men mentioned in Botta, pl. 145, l. 12, as having been carried into captivity by Sargon, were Israelites. They appear from the inscription not to have been inhabitants of Samaria itself, but of rural districts or provincial towns. This identifies the deportation spoken of with that in the reign of Pekah, recorded in 2 Kings xv. 29, and attributed to Tiglath-Pileser, who was consequently the same as Sargon, the builder of Khorsabad. I pointed out this identification in my paper on the Khorsabad inscriptions; and I think it inconsistent with Col. Rawlinson's assumption that the Khorsabad king was the Shalmaneser of Scripture. The latter I take it to be the son of Sargon, an elder brother of Sennacherib, as I mentioned in the paper referred to. I must also dissent from Col. Rawlinson's opinion that the deportation of the Israelites was in the *first* year of Sargon. The inscription where it is mentioned does not give the chronology of the events which it records, and other inscriptions seem to me to show that it must have occurred at a more advanced period of his reign.

I am, &c.
Edw. Hincks

* The *â* at the end of these names, as also at the end of *Sar.ghi.nâ*, and probably of *Yâ.u.â*, is inflectional.

Source: The Athenaeum, no. 1262 (3 Jan. 1852), p. 26

1   See H. Tadmor, *The Inscriptions of Tiglath-Pileser III King of Assyria* (Jerusalem, 1994), p. 12, n. 11, where Hincks's discovery is mentioned. Tadmor's index lists the occurrences of the name Menahem in the inscriptions.

1852

Athenaeum Club
Jany. 8, 1852

My dear Sir,

I have read with great interest your recent communication to the Athenaeum and I much regret, after what has more than once taken place with your discoveries, that they should not be more generally known.[1] I have with me a large mass of very important materials collected during my last journey. I have already made out several important facts but my time is so much occupied with putting together the various particulars of the last expedition for publication that I have really not time to devote to the analysis of the inscriptions or to obtaining the necessary ground-work for that analysis. It strikes me that together we might do a great deal and I am anxious to know whether you are contemplating a journey to London that I might have the pleasure of consulting you on several points. Or should you not, could I by any means spend a day or two in your neighbourhood? All I require is this. I will show you all the inscriptions I have, and will assign to you all the discoveries that you may make. I am very desirous that these discoveries should appear in my own work. As the material was obtained by me I think it but fair that I should have the advantage of bringing these discoveries first before the public, but at the same time they *shall be given in your name*, & as your discoveries. It would, I think, be far preferable, at the same time, to publish them altogether and in this form. They would be more generally circulated & cause greater interest. As the materials are in my hands & not accessible to any one, we need not fear being forestalled. Had not Col^l. Rawlinson *by an accident* obtained possession of my papers he would not have been able to publish his last discovery relating to Hezekiah.[2] I think it a pity that discoveries of such interest & importance should be as it were, lost in the columns of the Athenaeum. All, therefore, that I would ask you is not to communicate any of the results of an inspection of my papers, but to allow me to announce them in my own work, not as mine, but as *your* discoveries. If you would kindly drop me a line to M^r. Murray's (50 Albemarle Street) to let me know whether you approve of my suggestion I should feel much obliged.[3] I could, I dare say, manage to come over to Ireland between this time and the middle of Feb^y.

Believe me,
Yours faithfully,
A. H. Layard

MS: GIO/H 214

1   See Hincks's letter of 29 December 1851. It was published in *The Athenaeum*, no. 1262 (3 Jan. 1852), p. 26.
2   *The Athenaeum*, no. 1243 (23 Aug. 1851), pp. 902–3.
3   This was the address of the publishing house of John Murray. See *ODNB*.

TO AUSTEN HENRY LAYARD
10 JANUARY 1852

Killyleagh C° Down
10th. Jany. 1852

My dear Sir,

I this morning received your letter of the 8th. It will give me great pleasure to accede to the arrangements you propose. I am quite aware that the Athenaeum is not a good medium of communicating discoveries, but in my circumstances it was the only one open to me. Your name will commend circulation to whatever you publish & I shall be most happy to have any discoveries which I make in your papers communicated to the public through your work, which would be the proper channel for them. I promise you that I will not publish them previously in any paper.

As to our meeting, which seems necessary, it would be quite impossible for me to go to London at present; but you speak of being able to visit Ireland before the middle of February. The best way would be to come to Killyleagh, where I would be happy to see you at the Rectory. M^rs. Hincks is not at home at present & there are other objections to your coming *immediately*; but the first fortnight in February would I suppose suit you. If not, the last week in January might answer. Any time in fact after the 25th. would suit me, but I would be glad you would let me know the precise time when you would favour me with your visit a little before hand.

I must caution you (assuming you will come) against coming from *Newry here* across the country, which the maps might lead you to think desirable & which some have attempted. In winter it is in every respect the worst possible route. You should go to *Belfast* (either by Liverpool or Fleetwood from where there are boats, or by Holyhead & Dublin, where return tickets, allowing a fortnight for the journey & allowing it to be *broken* would be available) from which there is a railway to *Comber* within 12 English miles of this – a coach meeting one of the trains. Although this is a circuitous route, it is far the best in every respect. I mention this as you will not find the information in Bradshaw, which however will tell you all you want to know of the different routes to *Belfast*.[1]

Believe me
My dear Sir
Yours very faithfully
Edw. Hincks

MS: BL Add. 38980, ff. 207–8

1   Bradshaw's Railway Guide was a timetable of all trains running in Britain. It was first issued at Manchester in 1839 by George Bradshaw (1801–53), printer and engraver.

FROM AUSTEN HENRY LAYARD
15 JANUARY 1852

Orton Longueville
N[r]. Peterboro'
Jan[y]. 15 1852

My dear Sir,

I am much obliged to you for your kind note. It would suit me better to visit Ireland a little later than at the present moment. I will write to you again when I can settle my plans and give you full notice. But I must request that you will not put yourself to the least inconvenience on my account, nor would I intrude upon you so far as to accept your invitation to the Rectory. I have no doubt that I can find a room in an inn of some kind and all I want to ask of you is a little of your valuable time.

I am much indebted to you for the directions you give me for reaching Killyleagh. I have some friends at Belfast and it would suit me well to pass thro' that place.

Believe me
Yours very faithfully
A. H. Layard

MS: GIO/H 215

꩜

FROM AUSTEN HENRY LAYARD
5 FEBRUARY 1852

Athenaeum Club
London
Feby. 5, 1852

My dear Sir,

I have already informed you that my projected trip to Ireland was suddenly stopped by a summons to London. I regret to say that there is no chance of my being able to pay you a visit from which I expected so much pleasure and advantage as I am obliged to leave England immediately for the Embassy at Paris to which I have just been attached.[1] I must endeavour to copy out and send you such inscriptions as I believe to be of immediate consequence and if you could kindly give these your consideration & let me know what they contain according to your opinion, I shall be greatly obliged to you. I will send you all the information in my power connected with them. I send this from Paris and as soon as I have time I will not fail to communicate with you. I much regret that these unavoidable circumstances have prevented me having the pleasure of meeting you, & I trust that my not coming has not placed you to any inconvenience.

Believe me,
My dear Sir,
Yours very faithfully
A. H. Layard

MS: GIO/H 216

---

1   Lord Cowley (Henry Richard Charles Wellesley, 1804–84), who had been appointed ambassador to Paris, offered Layard the post of Secretary, but the latter preferred the post of Under-Secretary of State at the Foreign Office offered to him by Lord Granville, an offer which surprised many since Layard had no previous experience. On 20 February 1852, however, the government was defeated and Layard had to resign his post. See Layard's letter of 11 February 1852 to a friend in *Sir A. Henry Layard, G.C.B., D.C.L.: Autobiography and Letters*, vol. 2 (London, 1903), p. 239.

❧

TO AUSTEN HENRY LAYARD
9 FEBRUARY 1852

Killyleagh C° Down
9th. Feb[y]. 1852

My dear Sir,

I was sorry on my own account to hear of your having to give up your visit to this place; though I suppose the appointment is to your advantage, and that I should congratulate you on it. Your expected arrival did not cause any inconvenience. I had written to the Dublin post-office to say that the time would suit & that I should expect you at the Rectory.

I shall be ready to look over the inscriptions as soon as they arrive; more so now than six or seven weeks hence, when the approach of Easter will increase my parochial labours. I think it would be well if you sent an instalment *soon*, in order that if there be any difficulty as to the copying of the characters, we may come to an understanding about it before more are copied. You perhaps proceed on a different system from what I should do. Accurately to copy every wedge would be tedious & it is hard to say what is the best mode of abbreviating. I use the numbers of Fisher's list for most of those which occur in it, & for those which do not, *one* value (if any be known) which I would, in transcribing an inscription, adhere to in all instances even tho' the value of the character changed.[1] Some characters in Fisher's list might be represented by their values when these are perfectly distinct from all others, so that no two characters should be represented alike unless they were clearly calligraphic variations . . .[2] You have, however, your own way of transcribing & I am curious to know what it is. If in any point illegible, I would like to make my marks on it before you copy *much*.

Believe me
My dear Sir
Yours very faithfully
Edw. Hincks

MS: BL Add. 38980, ff. 228–9

1   *A Collection of All the Characters Simple and Compound with their Modifications . . . by Thos. Fisher.* For full reference, see Renouard's letter of 7 March 1850, n. 4.
2   I have omitted the sample signs which Hincks provided by way of illustration.

FROM GEORGE CECIL RENOUARD
10 FEBRUARY 1852

> Swanscombe Dartford Kent
> Tuesd. 10 Feb^ry. 1852

My dear Sir,

I am much pleased with the *thrashing* you have given to M^r. F. but you are quite right in supposing him proof against conviction.[1]

As the post hour is almost come & you will be anxious to hear the fate of your packet, I write this sh^rt. note to thank you for it & inclose some *blue-heads*, as it is a shame that you sh^d. have sixpence postage to pay for this very acceptable bk you sent to

> my dear Sir
> your much obliged & faithful
> G. C. Renouard

P.S. I shall have several things to add in a longer letter soon.

MS: GIO/H 466

1 'The One Primeval Language': a review of C. Forster, *The One Primeval Language Traced experimentally through Ancient Inscriptions* . . . (London, 1851), in *Dublin University Magazine* 39 (Feb. 1852), pp. 226–34. Hincks begins his review as follows: 'One of the most curiously constituted minds that has ever existed must be that of the Rev. Charles Forster. It seems to be absolutely proof against argument; as to everything, at least, which relates to ancient languages. When a philological crotchet has once taken possession of it, we question if there be any reasoning, however conclusive it may appear to others, sufficiently powerful to dislodge it.' For criticism of Forster's earlier work, see Renouard's letter of 9 October 1851, n. 5.

FROM GEORGE CECIL RENOUARD
13 FEBRUARY 1852

> Swanscombe, Dartford, Kent
> Fr. 13, Feb^ry. 1852

My dear Sir,

When I was at Paris in May last, they had just received M^r. Forster's 'One Primeval' book, &, as you may suppose, knew how to estimate it. Having had a taste of the author's speculations on the Ḥiṣni Ghoráb inscriptions, I had no curiosity to see this new publication of his 'fantasies' & was more than satisfied with reports in the Lit. Gazette, Bell's Weekly Mess^r. &c. &c. I have not, indeed, found time or inclination to read the sober parts of his 'Geogr. of Arabia', which are, I am told, well done.[1] Your castigation was needed, for his eloquence, good intentions & unhesitating confidence in the correctness of his theory, are so taking with those who are incapable of weighing the evidence on which his conclusions rest, that all who have 'more zeal than knowledge' are likely to be carried away. By his Mohammedanism unveiled, I was for a time, almost persuaded to be a Forsterite myself.[2] In his private character he is a truly amicable &, I believe, humble Christian & was well deserving of the esteem & friendship of his patron B^p. Jebb;[3] but a powerful imagination has been too much for him, & ill regulated zeal has blinded him & brought him into 'the Quag of Absurdity'.

Not having read Rödiger's Tract (unless it be the same as his paper in Lassen's Zeitschrift) I do not know exactly the state of the *désagrément* between him & his Master, Gesenius, but that G. thought he had reason to complain you will see by a letter from him, which I enclose, & R. appears not to have acted very ingenuously in the affair.[4]

I may remark, en passant, that you say 'the well-known Ethiopic characters ... in their *Geez*, or vowelless forms'. Where do you find Geez thus defined? I always understood it to be the native name of the Aethiopic alphabet, which, with that extraordinary man Prof^r. Murray (Bruce's Trav. ii, 348, n. 8vo ed.), I take to have been derived from the Greek through the Koptic.[5] The vowels appear to have been originally written above or below, abbreviated as in Syriac, & subsequently appended to the letters. I am surprised you take no notice of *Fresnel's* copies & *reading* of 56 Ḥimyarí inscriptions, found in the Sabaean Territory besides those at San'á & Ḥiṣn Ghoráb, published in 1845 (Journ. Asiat. vi, p.169–237).[6] F. observes that Ḥ. Ghor. & Ṣan'. are the only inscriptions which bear a date: the latter 573, the former 640 (A.D. 495 & 562). He supposes these to be of Jewish, most of the others of pagan origin, & the date to mark the introduction of Judaism into Yaman 700 years before Moḥ^d. about 78 B.C. (Pococke, Spec. Hist. Arab. p. 60[7]).

There are several of the *imaginary* alphabets of Ibn Waḥshīyah (published by M. de Hammer in 1806[8]) which are more or less borrowed from the Ḥimyarí (Musnad) but his musnad has scarcely a trace of it, & still less in the 23 pages of similar inventions in the chapter on Alphabets in the Iṣlaḥu-l ghalaṭát printed at Coñple in 1806.[9]

With regard to Forster's hallucinations respecting the scratchings on Wádí-l-mokatteb, you know what I should say from our former correspondence:[10] but I have to thank you for telling me that the author of the 'Lands of the Bible' is the D^r. Wilson whose book on the Pársí faith I have & esteem.[11] I was indebted for a knowledge of it to a young & able Bavarian, Prof^r. *Spiegel* of Erlangen whose

Pársí Grammatik contains veritably multum in parvo, & for the history of the Persian language is an invaluable work.[12] He has in a note pointed out a false view maintained by D^r. Wilson of whose work he speaks with great praise.

I fear your letter of the 24th Novr. 1851 was never answered; it was mislaid but not forgotten. It has been found very opportunely, as on Tuesday next, I hope to be in London & will then subscribe to the 'Turin Book of Kings'.[13] When I read Lepsius's Dissertation 'Über den ersten Ägyptischen' &c, your notice of it had entirely escaped my recollection, & I am well pleased to find you confirm my view of it.[14] If L. had seen your paper, he ought to have noticed it. I fear he hurries his publications onwards too fast & makes 'more haste than good speed'. I do not think he w^d. intentionally make dishonest omissions.

Your next sentence is remarkable 'You will see that Rawlinson will before long, be found to adopt my views as to the cuneatic characters representing syllables': and of that very act you justly complain in your letter which I received on the last day of January. I suspect from a passage in your letter of Novr. that I had probably then mentioned to you Brugsch's 'Transmigrations'.[15] I could not find *sinsin* in Champollion's Dict., though I dare say it is somewhere there. The book is greatly in want of an index; but his brother seems to have thought it his imperative duty to leave it as imperfect as he found it. Salvolini, whose dishonesty & hypocrisy (truly Italian) would hardly have been suspected, had his papers returned, unexamined, to Italy, very probably destroyed some parts of it. I have now got Champollion Figeac's report of the transactions which led to the recovery of his brother's missing MSS. & lamentable is the picture it presents of a man who had done quite enough to gain 'an honest fame', persevering while death stared him in the face, in bewailing the loss of papers which he had secreted & publishing some of the best of their contents as his own discoveries. Ch. Figeac writes with quite as much temper & fairness as could have been expected, but the evidence against Salvolini is irresistible. Without Champollion's 'Notice sur les MSS. autographes' it is impossible to judge what part of Salv.'s decyphering is his own & what is not.[16]

You surprise me by saying that you expect Layard at Killyleagh. I thought he was returned to Baghdad, where, I suppose, Rawlinson, Fresnel & Oppert now are.[17] I am behind hand with my correspondence with them, and I fear R. w^d. not give them a very cordial reception; as he was very angry with Botta for communicating the Babylonian inscription over *Bardiya* to *de Saulcy* & much displeased at Oppert for presuming to comment on his Persian text & version before he had seen R.'s second edition. R. will gain nothing by such exclusiveness.[18]

The Dublin Univ. Mag. contains an amusing article on a poet who gave myself & others a long chaw for years.[19] The Edinb. Rev. of Darwin's last work shewed very conclusively that his versification & probably the germ of 'the Loves of the Plants' were caught from an anonymous poem called 'Universal Beauty'.[20]

Nobody could find it or had heard of it & one of my friends suggested that it was a hoax. At last (years afterwards) I found it in the bookseller & ascertained its existence. Watt afterwards told me it was by the author of the Fool of Quality a book much read in the infant days of

> my dear Sir
> Yours faithfully
> G. C. Renouard

Not being able to find Gesenius's letter now; I must send it some other time.

MS: GIO/H 467

1   C. Forster, *The Historical Geography of Arabia; or, The Patriarchal Evidences of Revealed Religion: a Memoir, with Illustrative Maps; and an Appendix, Containing Translations, with an Alphabet and Glossary of the Himyaritic Inscriptions recently Discovered in Hadramaut*, 2 vols (London, 1844). See Renouard's letter of 9 October 1851, n. 5. Renouard's 'Ḥiṣni Ghoráb' is Ḥuṣn al-Ghurāb on the coast of Yemen.

2   C. Forster, *Mahometism Unveiled: an Inquiry in which that Arch-Heresy, its Diffusion and Continuance, are Examined on a New Principle, Tending to Confirm the Evidence, and Aid the Propagation of the Christian Faith* (London, 1829).

3   Forster was a protégé of Bishop John Jebb of Limerick, who wrote a favourable review of his book. See 'Forster on Arabia', in *Quarterly Review* 74 (Oct. 1844), pp. 325–58.

4   Renouard had already discussed the work of Wilhelm Gesenius and Emil Rödiger with Hincks in his letter of 21 December 1844 (vol. I, pp. 99–103). Note the following publications by the two great Semitists: W. Gesenius, 'Himjaritische Sprache und Schrift, und Entzifferung der letzteren', *Allgemeine Literatur-Zeitung 1841*, nos 123–6 (July 1841), cols 369–99 + 'Nachtrag'; *Ergänzungsblätter* 64 (July 1841), cols 511–12; E. Rödiger, 'Notiz über die himjaritische Schrift nebst doppeltem Alphabet derselben', *Zeitschrift für die Kunde des Morgenlandes* I (1837), pp. 332–40; *Versuch über die himjaritischen Schriftmonumente* (Halle, 1841); 'Exkurs über die von Lieut. Wellsted bekannt gemachten himjaritischen Inschriften', in *J. R. Wellsted's Reisen in Arabien*, ed. and tr. E. Rödiger, vol. 2 (Halle, 1842), pp. 352–411.

5   Alexander Murray (1775–1813), Scottish linguist, prepared the second and third editions of the work of the traveller James Bruce (1730–94), *Travels to Discover the Source of the Nile in the Years 1768–1773*, 8 vols (2nd edn; Edinburgh, 1804–5; 3rd edn; Edinburgh, 1813). Renouard's reference is to the 2nd edn, vol. 2, pp. 348–50, n. †. In the 3rd edn, pp. 339–41, n. *, there is a long footnote by Murray on Ethiopic/Ge'ez. Today we are confident that the Ge'ez (Ethiopic) script is derived from the South Arabian Sabaean/Minaean script. It is used to write Amharic. See Getachew Haile, 'Ethiopic Writing', in P. T. Daniels and W. Bright (eds), *The World's Writing Systems* (New York, 1996), pp. 569–76.

6   F. Fresnel, 'Pièces relatives aux inscriptions himyarites découvertes à Sana'â, à Kariba, à Mareb etc. par M. Arnaud', *Journal Asiatique* 6 (1845), pp. 169–237, esp. pp. 182–237.

7   E. Pococke, *Specimen historiae Arabum, sive, Gregorii Abul Farajii Malatiensis de origine moribus Arabum succincta narratio* (Oxford, 1650).

8   Aḥmad b. 'Alī Ibn Waḥshīyah, *Ancient Alphabets and Hieroglyphic Characters Explained*, tr. J. Hammer-Purgstall (London, 1806).

9   Meḥmed Hafīd, *al-Durar al-muntakhabāt al-manthūrah fī iṣlāḥ al-ghalaṭāt al-mashhūrah* (Istanbul, 1221/1806–7).

10   C. Forster, *The One Primeval Language Traced experimentally through Ancient Inscriptions, in Alphabetic Characters of Lost Powers, from the Four Continents; including the Voice of Israel from the Rocks of Sinai* (London, 1851). See Renouard's letter of 15 October 1851, n. 4.

11   J. Wilson, *The Lands of the Bible Visited and Described in an Extensive Journey. . .* (Edinburgh, 1847); idem, *The Pársi Religion, as Contained in the Zand-Avastá, and Propounded and Defended by the Zoroastrians of India and Persia, Unfolded, Refuted, and Contrasted with Christianity* (Bombay, 1843).

12   F. Spiegel, *Grammatik der Pârsisprache nebst Sprachproben* (Leipzig, 1851).

13   Wilkinson, *The Fragments of the Hieratic Papyrus at Turin: Containing the Names of Egyptian Kings, with the Hieratic Inscription at the Back.*

14   R. Lepsius, 'Über den ersten Ägyptischen Götterkreis und seine geschichtlich-mythologische Entstehung', *Abhandlungen der Königlichen Akademie der Wissenschaften zu Berlin* 1851 (1852), historisch-philologische Klasse, pp. 157–214 + 4 plates. On pp. 196–202 the author discusses the erasure of names, including that on monuments at Karnak. He fails to mention Hincks's paper 'On the Defacement of Divine and Royal Names on Egyptian Monuments', *Transactions of the Royal Irish Academy* 21 (1846), polite literature, pp. 105–13.

15   H. Brugsch (ed.), *Saï an Sinsin, sive, Liber metempsychosis veterum Aegyptiorum e duabus papyris funebribus hieraticis signis exaratis* (Berlin, 1851).

16   J. J. Champollion-Figeac, *Notice sur les manuscrits autographes de Champollion le jeune, perdus en l'année 1832, et retrouvés en 1840* (Paris, 1842). Renouard had already mentioned François Salvolini's notoriety in his letter of 21 November 1851.

17   Oppert left Marseille on 9 October 1851 on the ship *Hellespont*. He arrived in Mosul on 1 March 1852 and Baghdad on 27 May 1852.

18   See Rawlinson, *Memoir on the Babylonian and Assyrian Inscriptions*, preliminary note: 'It is the more important, indeed, that I should thus assert my claim to consideration for amended readings, as a series of papers are being now published by Mons. Oppert in the *Journal Asiatique*, on the Persian Behistun Inscriptions, which take cognizance alone of the original translation and meagre notes appended to my Analysis of the Persian text; and which systematically ignore the many corrections, and the diffuse etymological illustrations contained in the Vocabulary subsequently published. This is, I think, to say the least of it, uncandid; and as I should be sorry to see the present papers subjected to similar scrutiny, I have thought it necessary formally, at the outset, to protest against such a system of criticism.' See n. 2 to Renouard's letter of 18 September 1851 for details of Oppert's articles that upset Rawlinson.

19   'Our Portrait Gallery. No. LXVII. Henry Brooke', *Dublin University Magazine* 39 (Feb. 1852), pp. 200–14 + 1 etching. Henry Brooke (*c.*1703–83), Irish writer and playwright, was born in Rantavan, Co. Cavan. Renouard is referring here to his poem, 'Universal Beauty', written in 1728, and his novel, *The Fool of Quality*, 5 vols (London, 1766–70). See *ODNB*.

20   Erasmus Darwin (1732–1802), physician and natural philosopher, had a keen interest in botany and translated the writings of the Swedish naturalist Carl Linnaeus. He published anonymously a lengthy poetic work, *The Loves of Plants* in 1789. It was reprinted in 1791 as part 2 of *The Botanic Garden* (part 1 was another poetic work, *The Economy of Vegetation*). Renouard is referring to Thomas Thomson's review article on Anna Seward, *Memoirs of the Life of Dr Darwin, chiefly during His Residence at Lichfield* (London, 1804), in *Edinburgh Review* 4 (Apr. 1804), pp. 230–41, esp. pp. 238–41.

FROM THOMAS ROMNEY ROBINSON
28 FEBRUARY 1852

Observatory
Feb. 28. 52

My Dear Hincks,

I only returned last night; so you see I have lost no time in replying to yours. I do not wonder that you feel much hurt at being thus plucked to dress out others. Your opponent however has three great advantages over you. He is a trained diplomatic; a scotch-man, and a London lion; he therefore may do things with impunity which the public would not suffer in a country parson.[1] But Magna est Veritas. However truth is sometimes very unpalatable and must be administered with a little attention to the fancies of the patient. For this reason I have pencilled some modifications of your text (which I hope you may be able to decipher); in which without weakening the force of your statements I have a little softened their causticity.[2] It is an old dodge that when people cannot refute a charge, they endeavour to get it out of sight by raising a dust of personality; and while the world runs after the interloper, the original game is passed by: this is not improbable in the present case, and therefore, I think, if you look at the alterations which I have suggested in this point of view, you will agree with me that they leave your argument quite as strong, and less likely to be excepted against than as the passages at present stand.

I am very glad to find that you will have this paper ready so soon. I see from the Athenaeum that Grotefend is again in the field, but too little of his work is stated there to give me an idea of his method.[3]

Ever yours
T. R. Robinson

MS: GIO/H 482

1   Robinson is describing Henry Creswicke Rawlinson, but the latter was not a 'scotch-man'.
2   Hincks was preparing his very important paper 'On the Assyrio-Babylonian Phonetic Characters', which was published in *Transactions of the Royal Irish Academy* 22 (1852), polite literature, pp. 293–370. Robinson was president of the Royal Irish Academy and read Hincks's paper.
3   At a meeting of the Syro-Egyptian Society on 16 February 1852, a paper by Grotefend was read 'On the Builders of the Palaces at Khorsabad and Kouyunjik'. This was a translation (by Renouard) of 'Die Erbauer der Paläste in Khorsabad und Kujjundshik', *Abhandlungen der königlichen Gesellschaft der Wissenschaften zu Göttingen* 4 (1850), historisch-philologische Klasse, pp. 201–6. See *The Athenaeum*, no. 1269 (21 Feb. 1852), p. 230.

FROM EMMANUEL DE ROUGÉ
22 MARCH 1852

Paris 22 mars 1852

Je suis fâché, Monsieur, d'avoir si mal su votre adresse; désormais, si vous me permettez de continuer cette correspondance, je vous adresserai mes lettres à Killyleagh directement. Veuillez me dire, pour les ouvrages, s'il suffit de vous les adresser à Dublin à l'académie, ou bien si vous préférez qu'ils soient confiés aux soins de quelque correspondant. Si vous avez, de votre côté, l'extrême obligeance de m'envoyer vos brochures;[1] je vous prierai de me les faire tenir par l'entreprise de la maison Longman et Brown, à Londres. Il faudrait que vous eussiez l'obligeance de mettre une première enveloppe à mon nom et une seconde adressée à MM. Stassin et Xavier, *rue du Coq*, 9 à Paris.[2] C'est à peu près la seule voie sûre et prompte pour faire parvenir en France les ouvrages nouveaux que l'on veut envoyer à quelqu'un. Je désire infiniment, Monsieur, avoir la collection complète des vôtres; je suppose que l'on vend le volume du congrès de Winchester[3] et je vais donner ordre qu'on me l'envoie ainsi que la publication de S. G. Wilkinson que vous avez la bonté de m'indiquer dans votre lettre.[4]

Vous pouvez, Monsieur, m'écrire en anglais, quoique je n'ai pas assez l'usage de cette langue pour la parler ou l'écrire correctement, je la lis sans difficulté. L'embarras où l'on se trouve pour avoir les différents travaux épars dans les publications des sociétés savantes, fait que l'on ne rend pas toujours une justice suffisante aux travaux de ses devanciers. C'est ainsi que ma réfutation de M. Bunsen, insérée dans les annales de philosophie chrétienne, est restée à peu près inconnue à l'étranger.[5] M. Lepsius, à qui je l'avais envoyée par l'entremise de M. Letronne, ne s'est pas fait faute de l'employer sans la citer.

Je vous remercie de vos observations sur quelques passages de mon mémoire; je laisse exprès écouler un certain temps avant de publier la seconde partie, afin que la critique puisse s'exercer suffisamment sur l'ensemble et sur les détails. Je vous avoue que je tiens encore pour l'opinion de Champollion quant aux signes 𓍹𓏏𓏤𓏤 ; je pense qu'il a remontré juste (quant au sens) en le traduisant *sa sainteté* dans la locution 𓊹𓋴 qui désigne le roi et quelque fois un dieu dans les textes sacrés. Cet usage et l'autre groupe 𓊹𓍛 (ta sainteté?) que j'ai employés dans les textes funéraires en parlant au défunt glorifié ne me semblent pas convenir au mot serviteur. Les deux passages des papyrus que vous m'indiquez peuvent, ce me semble, se traduire convenablement par le mot *prêtre*. En tous cas, la lecture reste un mystère jusqu'ici, c'est fâcheux pour un mot si important.

Vous avez vu que je diffère aussi d'avis avec vous, Monsieur, quant à la nature du groupe 𓊪𓏏 ; je persiste à penser que ⌢ est la particule ⲕⲉ et que 𓍢 représente le sujet dans les cas que j'ai cités et non pas un régime de verbe réciproque ou passif. En effet le verbe, dans ce paradigme, a souvent, comme dans le paradigme simple, son régime direct, ce qui nous assure que nous avons bien affaire à une voie active. Les exemples sont nombreux; permettez moi de vous indiquer seulement, dans le rituel funéraire ch. 125 l. 1, 2, le passage suivant qui répète deux fois organe[?] de construction.

𓊖𓏏𓏛𓇋𓄿𓏏𓊖𓏏𓏛𓇋𓄿𓏏𓈖 &c

Les deux verbes sont également rendus par le futur de la voie active dans le pp. démot. de la bibliothèque de Paris, qui contient le texte correspondant et que M. Brugsch a publié. Les deux régimes directs 𓈖 exigent en effet cette traduction.

J'espère faire paraître prochainement la traduction complète de ce chapitre si important dans une chrestomathie égyptienne que je prépare; les difficultés d'impression qui m'ont arrêté jusqu'ici ne sont pas encore entièrement levées, mais notre charactère égyptien de l'imprimerie nationale se complète tous les jours, et j'espère pouvoir bientôt entreprendre cette publication. Je vais enfin mettre au jour tout ce que j'ai pu retrouver dans les papiers de Champollion des notes sur le rituel; c'est un hommage et une justice qui doit être rendue à notre premier maître, et qui ne sera pas sans profit pour la science.

J'ai l'honneur M. de vous assurer de ma considération la plus distinguée.
Vᵗᵉ. Emmanuel de Rougé
conservateur au Musée du Louvre

MS: GIO|H 155

1   Some years ago I bought from a second-hand bookseller in California one of the actual offprints which Hincks sent to Emmanuel de Rougé. Hincks wrote 'With the compliments of the author' on the fly-leaf.
2   Stassin et Xavier, rue du Coq, St Honoré 9, was a publishing house in Paris.
3   Hincks, 'On Certain Egyptian Papyri in the British Museum', *Transactions of the British Archaeological Association, at its Second Annual Congress, Held at Winchester, August 1845* (1846), pp. 246–63 + 1 plate. De Rougé had already asked Hincks for a copy of this paper in his letter of 9 October 1851.
4   Wilkinson, *The Fragments of the Hieratic Papyrus at Turin: Containing the Names of Egyptian Kings, with the Hieratic Inscription at the Back.*
5   De Rougé's article was in six parts: 'Examen de l'ouvrage de M. le Chevalier de Bunsen intitulé La place de l'Égypte dans l'histoire de l'humanité', *Annales de philosophie chrétienne*, 3rd ser., 13 (1846), pp. 432–58; 14 (1846), pp. 355–77; 15 (1847), pp. 44–65, 165–93, 405–38; 16 (1847), pp. 7–30.

TO AUSTEN HENRY LAYARD
3 APRIL 1852

Killyleagh Cᵒ Down
3ʳᵈ. April 1852

My dear Sir,
    I had the pleasure to receive your letter this morning & hope this will catch you in London. I wrote to you at the British Embassy Paris on the 9ᵗʰ. February. I took it for granted that on the change of your plans taking place it would be sent to you in London; & presumed that pressure of business was the cause of your not replying to it. I could have managed to go to London last month & should probably be able to do so shortly after your return. I think it likely I could go for what some call 'the Parson's week' & others 'the parson's fortnight'[1] – leaving

home on Monday morning the 19th. & returning on Saturday the 1st. May. In order to accomplish this, however, I should have notice, as it is not always possible to obtain a supply.

I am sorry to hear that Col. Rawlinson has obtained copies of the inscriptions & will thus be likely to anticipate your publication. I hope, however, you will lose no time in pressing it forward.

I feel much obliged to you for your kind intentions towards me. If I had a little conversation with you, perhaps something might be suggested which would assist me. The great matter would be to enable me to keep a permanent curate. There are often many days together that I am unable to look at any inscriptions, and what I know passes out of my head. If I could give my whole time except on Sundays to those matters, I think I would be able to serve the cause of literature greatly.

Hoping that I will be able to arrange matters so as to see you soon, I will not now enter into particulars as to what might be done for me, so as to liberate me from week day parochial labours. There is no doubt that Lord Derby could accomplish what I seek, & that in more ways than one; but he will have many claims on him; & as I am not of his party, he will be more likely to attend to them than to me.[2]

Believe me
My dear Sir
Yours very truly
Edw. Hincks

MS: BL Add. 38981, ff. 1–2

1   The parson's week was 'a holiday period of about thirteen days and including only one Sunday, humorously regarded as the longest holiday available to a parson who was excused one Sunday's duties' (*OED*). The usage is obsolete.
2   Lord Derby formed a minority government in February 1852 following the collapse of Lord John Russell's government. In December 1852 his government collapsed. He is regarded as the father of the modern Conservative Party in Britain.

TO AUSTEN HENRY LAYARD
10 APRIL 1852

Killyleagh C° Down
10th. April 1852

My dear Sir,

I had your letter of the 7th. in course. I feel greatly obliged for the interest you have taken in my behalf. *Hoping* to see you on the 21st., or 22nd. but seeing much uncertainty as to the whether I shall do so or no,[1] I have thought it well to send you a few copies of papers which were signed in my behalf two years ago.

They were presented to Lord John Russell with a letter requesting that they might be attended to, signed by Mr. Thornely and Mr. Tennent on the one side of the house & on the other by Mr. Hamilton (now Secy of the treasury) Mr. Napier (now Rt. Honble. A. G. for Ireland) Sir R. Inglis & Col. Chatterton.[2]

I was led to hope that I might have had some promotion in the Irish church. But in fact no good *government living* fell vacant during these two years. The deanery (about which you may recollect that I wrote to you & which was worth about two hundred a year) turned out not to be tenable with a living.[3] There was nothing else vacant at the disposal of government except such great prizes as could only be given for *clerical* merit & as would give no leisure for literary pursuits from their arduous duties.

At my time of life (near 60) I would be most anxious to train up others to pursue what I have begun. I believe there is a great deal, *which would die with me*, known to myself alone, some of it in papers which in their present state would be unintelligible to others. A professorship, or something of the sort, instituted for me, would I think, be the best way of doing what I want. But were I in more independent circumstances & enabled to keep a curate, either by a civil list pension or by a good government living in *England*, I would contrive to give such instruction as I think should be given, provided only I could meet persons willing to receive it.

I may add that my clerical income, which was above £800 a year when I began life, is now only £554 *net*; & in three years time will be reduced to about £370 or £380; while the incumbrences for my life are £182, which will somewhat increase as I grow older; so that if I live three years more I shall be one of the poorest men in the country; and yet still bound to live in the same large house with which I started, and to keep it in perfect repair. Under these circumstances, no reasonable person can expect that I should keep a curate, or incur other expenses needful to the proper prosecution of my studies, *without some aid*. When I could afford it, before my income was reduced, I spent a good deal in pursuit of knowledge, and in a way that has never brought me any pecuniary return. I would have continued to do so if I could; but the reduction in my income has rendered it impossible for me to continue it.

Believe me
My dear Sir
Yours very truly
Edw. Hincks

P.S. If the interval 20–30 April would not suit you, I could be in town with equal convenience 4–14 May.

MS: BL Add. 38981, ff. 13–15

1   Hincks's diary for 29 April 1852 says: 'To Layard, at whose house I remained about 4 hours. He speaks of Disraeli as a person who would take an interest in my favour'.
2   See Thomas Romney Robinson's letter dated June 1850 and especially Hincks's diary entry in n. 3.
3   See Hincks's letter of 22 August 1851 and reference to the Deanery of Armagh.

TO THE EDITOR OF THE ATHENAEUM
APRIL 1852

Killyleagh C° Down
April 1852

[Sir,]
   In reply to Hibernicus I must begin with pointing out an error that he has committed.[1] He seems not to be aware that the inscriptions of Darius are *trilinguar*. The characters of the first and the third kind of inscriptions are altogether different. The inscriptions of the *first* kind are Indo-European, and to these the Assyrian bear no resemblance in either character or language. Their resemblance is to the inscriptions of the *third* kind, which are Semitic. A reference to the Report of the British Association for 1850 (p. 140 of the 'Transactions of the Sections') will show him the nature of the Assyrio-Babylonian cuneatic writing.[2] The views expressed on that occasion have been adopted by Col. Rawlinson in his recent 'Memoir';[3] though altogether different from what he had advanced in his 'Commentary' published a few months before the meeting of the British Association in 1850.

I am, &c.
Edw. Hincks

Source: The Athenaeum, no. 1278 (24 Apr. 1852), p. 466

1   A letter signed 'Hibernicus' appeared in *The Athenaeum*, no. 1274 (27 Mar. 1852), p. 363. The writer wished 'to ascertain, the characters being identical, how the Assyrian inscriptions differ from the Persian ones...' and asked, 'Did the Assyrians employ an Indo-Germanic or a Semitic?'
2   'On the Language and Mode of Writing of the Ancient Assyrians', *Report of the Twentieth Meeting of the British Association for the Advancement of Science; Held at Edinburgh in July and August 1850* (1851), p. 140 + 1 plate.
3   *Memoir on the Babylonian and Assyrian Inscriptions*, pp. 1–16 ('The Alphabet').

TO HENRY ELLIS[1]

4 MAY 1852

Killyleagh C° Down

4th. May 1852

My dear Sir Henry,

I have to request that you will, at the next meeting of the Trustees of the British Museum, direct their attention to the importance of ascertaining the exact weights of two Babylonian Half-talents which I was shown in the collection.[2] There is no means of weighing them with sufficient accuracy within the Museum & they cannot be sent out of it without special permission. The exact weight of the Babylonian Talent (which was probably the same as the Jewish) would, it appears to me, be an important datum; and the mean of the two weights, supposing them to be nearly the same, would give it with great accuracy.[3] There are also some small weights in the Museum, which were I have no doubt, some fraction of the Talent. And the weighing of them & of the large ones would not only determine what this unknown fraction is, but would give values of the Talent which might be compared with the others.

I would beg leave also to suggest the copying some of the minute Babylonian inscriptions by a process that would at the same time exhibit them on a flat surface and magnify them – the first step towards deciphering them. I spoke about this to Lord Rosse when I had the honor of a conversation with him on Thursday last; and he seemed to think such a process might be carried out.

As it is probable that if any astronomical records exist among the Assyrio-Babylonian inscriptions, they are on some of those small cylindroids; and as they are *in their present state* illegible, even if the language in which they are written were thoroughly known, I cannot but think that an attempt to obtain enlarged copies of some of them would be very useful.

I remain

My dear Henry

Yours very faithfully

Edw. Hincks

MS: BM/CA/OP 47:19

1   Henry Ellis (1777–1869) was Principal Librarian at the British Museum from 1827 to 1856.

2   Hincks's diary for 29 April 1852 says: 'Royal Society and Lord Rosse – had a long conversation with him about the inscriptions, and suggested training persons to read them, weighing the half-talents, and taking photographs of the cylinders. He mentioned this to some persons skilled in these matters, who thought it feasible, and the attempt will probably be made.' William Parsons (1800–67) was third earl of Rosse.

3   The Babylonian talent, *biltum* = *c.* 30 kg. It was divided into 60 minas, and the mina was divided into 60 shekels. See conveniently R. Caplice, *Introduction to Akkadian* (3rd rev. edn; Rome, 1988), p. 95. For a full discussion of Mesopotamiam weights and measures, see M. Powell, 'Masse und Gewichte', *Reallexikon der Assyriologie*, ed. D. O. Edzard, vol. 7 (Berlin, 1987–90), pp. 457–517 (note *biltum* on p. 508).

TO AUSTEN HENRY LAYARD
8 MAY 1852

Killyleagh C° Down
8th. May 1852

My dear Sir,
    I have just received your letter of the 6th. and need scarcely tell you that I feel greatly obliged by the interest you take in my affairs.
    You ask 'whether there be any definite position, either in the church or connected with the universities or in any other way that could be asked for'. My friends have applied for a pension on the civil list. This is, I believe, the only way of assisting me that is open to the Government, consistently with their established rule of confining church preferment in England to those who have been born on the favoured side of the channel. Of course, I should infinitely prefer an augmentation of my clerical income to any pension; and as it would involve a change of residence, it would remove a great disadvantage under which I must labour while confined to this remote spot; but, though of pure English descent, I have had the misfortune to be *born* in Ireland; and am consequently not to be provided for in England. So, at least, Lord John Russell laid down as a principle.
    As to promotion in the Irish branch of the church, it is out of the question. Should any fall vacant, it ought to be, and *must* be given to some of these opponents of the National System of Education who have been so long excluded as such from promotion. In Ireland I am known as one of the oldest and steadiest supporters of the National System;[1] and for *me* to be promoted in Ireland by the present Government would be regarded as monstrous. In England, if the objection as to my *birth* were overlooked, that which I have just mentioned would not lie. I should *there* be only known as a clergyman unconnected with any party in the church, of *via media safe* views, and promoted for my literary merits. The Education controversy in England is totally different from that in Ireland, and I should not interfere in it. In this case, what I should wish would be a benefice with cure of souls, where by keeping a curate I should be at leisure for literary

pursuits on every day but Sunday. There are many livings in the gift of the Premier falling vacant every year; and if I was appointed to one I dare say the College would allow the Government to appoint my successor *here*.

I dare say too that I should be able to exchange Killyleagh for some small living in England, so that if I had a sinecure, tenable with a living offered to me, it would place me in a position to devote myself to those archaeological matters which (it seems to be admitted on all hands) it is of great consequence to investigate and I am, from my previous studies, peculiarly qualified to investigate successfully.

I saw in a late newspaper that the sinecure rectory of Gedney was vacant and in the gift of the government.[2] If its value be what the newspaper stated, it would, if conferred on me, enable me to do all that I could in any case do; as I should have sufficient income and be unfettered in the way of residence. This, however, is too good to be hoped for. Probably what I have just mentioned is already disposed of; but other preferments are constantly falling vacant.

I think I could have an abstract of the contents of the inscription of Sennacherib ready in a very short time after I received it. How would you think of publishing it so as to anticipate Col. R.? I am satisfied that the long inscription which I saw at your rooms is what he had seen & what he gave the contents of last year.

I have marked that this should be forwarded to you. Perhaps you may be able to send the *first sheet* where it may be of use. This, however, depends on your own judgment as to its being judiciously written, & as to the person to whom you send it being likely to advance my cause. I doubt Sir E. Tennent having influence.[3] M^r. D'Israeli of course would have great influence & I think you said you knew *him*. Perhaps before this reaches you, you would have seen my brother & talked the matter over with him.[4] I fear there is no chance whatever of Gedney or indeed of anything but the pension, which tho' a present relief would only be so for 2½ years when my income *here* will be cut down more than I could expect to receive in the way of a pension.

Believe me
My dear Sir
Yours very truly
Edw. Hincks

MS: BL Add. 38981, ff. 36–38

1   See D. H. Akenson, *The Irish Education Experiment: The National System of Education in the Nineteenth Century* (Toronto and London, 1970).

2   Gedney in Lincolnshire is famed for its beautiful church of St Mary Magdalen.

3   James Emerson Tennent (1804–69) was born in Belfast. He attended Trinity College, Dublin and was called to the bar at Lincoln's Inn. In 1831 he married Laetitia Tennent (her cousin was Robert James Tennent, M.P. for Belfast) and in 1832 he assumed the name and arms of his father-in-law in Co. Fermanagh. He was

M.P. for Belfast from 1832 to 1845 and for Lisburn from 1852. He served as colonial secretary in Ceylon from 1845 to 1850, and secretary to the Board of Trade from 1852 to 1867. Charles Dickens dedicated *Our Mutual Friend* to him and attended his funeral.

4   Francis Hincks (1807–85) was co-leader of Canada with Augustin-Norbert Morin from 1851 to 1854. See W. G. Ormsby, 'Sir Francis Hincks', in J. M. S. Careless (ed.), *The Pre-Confederation Premiers: Ontario Government Leaders, 1841–1867* (Toronto, 1980), pp. 148–96; 'Sir Francis Hincks', in *Dictionary of Canadian Biography*, vol. 11 (Toronto, 1982); also F. Hincks, *Reminiscences of his Public Life* (Montreal, 1884).

TO AUSTEN HENRY LAYARD
8 MAY 1852

Killyleagh
8th. May 1852
7 pm

My dear Sir,

In reference to what I wrote to you this morning respecting the sinecure rectory of Gedney, it has occurred to me. Might it not be made the endowment of a professorship established for me in Cambridge (which it is near) or in King's College London which might be a preferable site? If that arrangement could not be effected, I would if it were given me accept such a professorship and do the duty of it at an almost nominal salary; but I think the endowing a professorship permanently would ensure that some one or more persons would apply themselves to the study of the ancient languages in which these inscriptions are written, with a view to succeed me.

Though ecclesiastically a sinecure, I w^{d} if it were given to me, work as hard as I could both in studying the inscriptions & in training others to do so; thus there would be no opprobrium connected with the receipt of the income; and I should hope that my appointment would be considered creditable to the Ministers who made it.

Believe me
My dear Sir
Yours very faithfully
Edw. Hincks

MS: BL Add. 38981, f. 40

TO SAMUEL BIRCH
8 MAY 1852

Killyleagh C° Down
8th. May 1852

My dear Sir,

I have to thank you for your paper on the gold mines which I received this morning & have looked over.[1] What a strange mistake Lepsius made about the plan of the mines! I believe, however, that there *are* among the Turin papyri plans of royal tombs at Biban el Malach; & they ought to be examined & published. Seyffarth's statements on the subject are so positive that I cannot doubt that there is something in them. He gives the lengths & breadths of the different chambers, which he says compared in measurement to those of known tombs. One (I think) is that of Rameses III. The whole statement has every appearance of truth & it is quite evident that what Lepsius published is not the MS (or one of the MSS) referred to by Seyffarth; although I can easily conceive that Lepsius enquired at Turin for the plan of the tomb that Seyffarth had seen & that they gave him *as such* the plan of the mines.

I spoke to Sir H. Ellis about the weights & he advised me to write him an official letter which I did. I am curious to know the result; & what ratio the small weights bear to the large one.

Believe me
My dear Sir
Yours very truly
Edw. Hincks

MS: BM/ANE/Corr. 2706

---

1   S. Birch, 'Upon an Historical Tablet of Rameses II, 19th Dynasty, Relating to the Gold Mines of Aethiopia', *Archaeologia* 34 (1852), pp. 357–91.

FROM HENRY ELLIS
17 MAY 1852

British Museum
17th. May 1852

My dear Sir,

I had no opportunity till Saturday last the 15th. inst. of laying your letter of 4th. inst. before our Trustees. They are fully aware of the importance of ascertaining the exact weights of the Babylonian half-talents as well as of the fraction of the talent to which you drew their attention and M^r. Hawkins has been directed to see that they are sent to the Mint for weighing.[1]

Upon the other part of your letter, suggesting the copying some of the minute Babylonish inscriptions by a process that could at the same time exhibit them on a flat surface and magnify them, Lord Rosse himself made a communication to the Trustees on Saturday last.

I have the honour to be, Sir,
Your most obedient servant,
Henry Ellis
Principal Librarian

Source: Davidson, *Edward Hincks*, p. 1722

1 Edward Hawkins (1780–1867) was keeper in the Department of Antiquities at the British Museum.
2 There is a copy of this letter in the British Museum's central archives: BM/CA/LB 41: 154–5.

TO AUSTEN HENRY LAYARD
22 MAY 1852

Killyleagh C° Down
22^d. May 1852

My dear Sir,

I received your letter of the 17th. in course. I now write a few lines, merely to say that, my brother having left London, *Sir Ja^s. Emerson Tennent*, who has expressed himself most warmly in my favour, is now the person to be consulted with.[1] I have written to him by this post, pointing out the impossibility of an arrangement which had been suggested & stating what I thought might be done, as in my letter to you. If you could see him, & the Chancellor of the Exchequer, and if you could succeed in interesting the latter in my behalf, something might perhaps be done.

The Trustees of the British Museum have given directions for the two 'half-talents' to be weighed at the Mint. I wish the same could be done with the other articles, on which you told me that the weights were specified.

Believe me
My dear Sir
Yours very truly
Edw. Hincks

I have a new theory about the Khorsabad king. I make him the Αρκιανος of the Canon, son of Ιουγαιος & brother of Μαρδοκεμπαδος – all the names being corrupted.[2] I find that he 'conquered Babylon in his 12th. year, after the Chaldean king had occupied it 12 years'. And he speaks of his father 'as having reigned over Assyria & Chaldea'. The father, it w^d seem, divided his dominions between them; Babylon being at first in the Chaldean portion (where the inscriptions on the reverses of the slabs were executed) & afterwards in the Assyrian.

I cannot, however, agree with R. that this Chaldean King is the Merodach Baladan of Scripture.[3] If he be, we must give up the extract from Alexander Polyhistor, preserved by Eusebius, as *wholly erroneous*; & I have been in the habit of regarding it as an authentic record. I am very anxious for the Sennacherib inscription.

MS: BL Add. 38981, ff. 46–7

1   See Hincks's two letters of 8 May 1852 to Layard.
2   See Hincks's Table of the kings of Judah, Israel, Assyria, Babylon and Egypt in his article 'On the Khorsabad Inscriptions', *Transactions of the Royal Irish Academy* 22 (1850), polite literature, p. 55.
3   Rawlinson, *Memoir of the Babylonian and Assyrian Inscriptions*, p. 12.

TO HENRY ELLIS
I JUNE 1852

Killyleagh C° Down
1ˢᵗ. June 1852

Dear Sir Henry,

My brother, Mʳ. Hincks of Canada, communicated to me, before he left England, a letter from the Earl of Derby, in which his Lordship mentions his having referred to the Trustees of the British Museum a request of mine to be employed as a decypherer of Assyrian inscriptions in the Museum, and that you had informed him, that, if I should be resident in London, they would be prepared to take into consideration the terms on which my services, which they thought valuable, might be obtained.

The fact is that no such request was made by me.[1] The application made on my behalf was for a pension on the civil list; and if the Assyrian inscriptions were alluded to, it could only have been with a view to shew the desirableness of my being relieved from pecuniary embarrassment (caused by the various acts of the legislature affecting Irish tithes) and enabled by keeping a curate to devote more time to them, and under more favourable circumstances, than I can do at present.

I deem it necessary to explain this matter, as there would have been an obvious impropriety in my making such a proposal without my having previously ascertained if my diocesan would sanction it.

On receiving this intimation that it might be considered desirable to engage my services at the Museum, I applied to the Bishop of Down for a licence for non-residence till the end of December 1853; in the hope that, if they should be found so valuable as I trusted that they would be, some arrangement might be effected during that interval, by which I could become a permanent resident in London.

The Bishop expressed himself as 'desirous to aid the Trustees of the British Museum in their endeavour to secure my valuable services in connexion with the antiquarian department'; and offered to give me a licence for the time mentioned; but has stipulated that I should in that case allow *him* to nominate the curate who should have charge of the parish in my absence. He says, he requires this in all cases of non-residence on licence.

I believe this claim to be unusual. Indeed, I never heard of it being made by any other Bishop; but the Bishop of Down will not abandon it; and it is a claim to which I cannot conscientiously submit.[2] Holding the cure of souls in the parish, I cannot allow a curate to take my place, who would give instructions to my parishioners of a different character from what I have myself given, and what I believe to be in accordance with the Articles and Liturgy of the Church.

My application to the Bishop is thus *virtually refused*; and, of course, I cannot enter into any such engagement as Lord Derby appeared to contemplate.

I think it right, however, to mention, that I could engage to reside *two months* in the year in London, during which I would place my whole time at the disposal of the Trustees. In that case, it would be necessary that I should engage a curate assistant for the entire year; but I could surely do *something* for the Museum, in one way or another, *while here*, in the remaining ten months. *For these* I should, however, expect no compensation beyond the curate's salary; which would, I suppose, be £80 a year. What addition should be made to this for my services in London, my expenses while there, and my travelling backwards and forwards, I would leave altogether to the Trustees. For the first year, indeed, I should be satisfied with my expenses out of pocket.

Believe me to remain,
Dear Sir Henry,
Yours very faithfully
Edw. Hincks

MS: BM/CA/OP 48

1   Hincks seems to have been very annoyed by his brother Francis's representations to Lord Derby.
2   Robert Bent Knox was the Bishop of Down, Connor, and Dromore.

TO JAMES EMERSON TENNENT[1]
1 JUNE 1852

Killyleagh C° Down
1st. June 1852

My dear Sir James,
I received your very friendly letter on the 29th. and feel greatly obliged to you for the trouble you have taken in my behalf.

Meanwhile, I have had some correspondence with the Bishop, the result of which is a letter to the Trustees of the British Museum, with a proposal. I send it to you that you may see how matters stood, & if the negotiation should have taken such a turn as that you consider it unnecessary or undesirable to send it, you have my authority to hold it back.

I cannot help flattering myself that if I were engaged for a time, my services would be found so valuable, that the Trustees would be unwilling to lose them; & as the worst of all the acts affecting my parish (Lord Derby's own of 32) will not take effect in it till 1854, I have felt myself at liberty to offer for the first year what I could not do normally.

Believe me
Yours very faithfully
Edw. Hincks

MS: BM/CA/OP 48

1   See Hincks's letter of 8 May 1852 to Layard.

⮜⮞

EDWARD BOOTLE-WILBRAHAM[1] TO AUSTEN HENRY LAYARD
9 JUNE 1852

Downing St. June 9/52

Sir,

I have been directed by the Earl of Derby to acknowledge the receipt of your letter of the 7th inst: respecting the appointment of the Rev'd D^r. Hincks to the Office of Decypherer of the Assyrian Inscriptions at the British Museum; and to acquaint you that Lord Derby has been already in communication with the Authorities of that Institution on the subject – which correspondence I inclose, by his desire. By it you will perceive that D^r. Hincks' brother explicitly stipulated for him that he should retain his living in Ireland, to which Lord Derby doubted the Bishop assenting.

The Living to which D^r. Hincks alludes (Gedney) in his letter to Sir E. Tennent, is suppressed as a sinecure; and Lord Derby has no means of finding him another near London. I return to you the letter from D^r. Hincks to Sir E. Tennent; and beg that you will have the goodness to return to me the correspondence which I have herewith enclosed.

I have the Honor to be, Sir
Your obedient Servant
E. B. Wilbraham

June 9/52

Sir,[2]

I beg to return you the correspondence relative to D^r. Hincks. I trust you will have the goodness to carry my best thanks to the Earl of Derby for his kind

attention to my note, & I express my regret that circumstances prevent the acceptance of D[r]. Hincks' services.

BL Add. 38981, f. 56

1   Col. the Hon. Edward Bootle-Wilbraham was Lord Derby's private secretary. He was the son of
Edward Bootle Wilbraham (1771–1853), 1st Baron Skelmersdale; his sister, Emma Caroline (1805–76), was
Lord Derby's wife.
2   This note was written by Layard.

TO THE EDITOR OF THE TIMES
12 JUNE 1852

Killyleagh [C° Down]
12th. June [1852]

Sir,

You will, I hope, allow me to lay before your readers a few remarks on the
explanations offered by Lord Derby and Sir John Pakington on last Monday
night in reference to M[r] Hincks' letter of the 31st. ult.[1]

It was assumed, both by Lord Derby and by Sir John Pakington, that the letter
contained a charge of discourtesy against the latter; and some pains were taken to
show that for such a charge there was no foundation. The fact, however, is that
such a charge was never made by M[r] Hincks. There is not a syllable in the letter
which you published (and the publication of which apart from the correspondence
of which it was a portion, is, I must say, to be regretted) which can, according to
any fair interpretation of it, be supposed to convey it.

The charge in the letter is want of confidence, not of courtesy or consideration. It is remarkable that Lord Derby, in vindicating his colleague from the
latter charge – which is an imaginary one – has acknowledged the facts on which
the former, which was already made, rested. 'Sir John Pakington', he says, 'had
corresponded with the deputation upon the same terms as any Secretary of State
would have received an accredited Minister from a foreign State'. That is just
what M[r] Hincks complained of. M[r] Chandler and he were treated as 'Ministers
from foreign States', not as 'sworn confidential advisers of the Crown'. He
complains that communications 'were made to the Colonial-office, on the
subject of this railway, hostile to the views of the Governments and Legislatures
of the three provinces of Canada, Nova Scotia, and New Brunswick, supported
as those views are by the Queen's able representatives in those provinces'; and

that no communication of those papers had been made to them, who might possibly have been able to expose the fallacy of the statements contained in them, and the anti-British motives of the persons who made them.

Whether the Colonial-office was right or wrong in this withholding of confidence is a question on which I do not feel myself qualified or called upon to enter. Neither will I say anything on the general merits of the proposed railway, or on the justice of Mr Hincks' complaint, that, after seven weeks' delay, he was led to anticipate a further delay for an indefinite period by a conversation in the House of Commons on the evening before he wrote. Mr Hincks himself may perhaps vindicate the correctness of his views on some or all of these points; but with them I have nothing to do.

I confine myself to the one subject – the charge of discourtesy which he is supposed to have made against the Colonial Secretary, but which he never did make. I have reason to think that he would greatly regret this extraordinary mistake which has arisen, and I have therefore thought it right to request you to publish these few sentences in his vindication.

I am, &c.,
Edw. Hincks

Source: The Times, Issue 21143 (16 June 1852), p. 6

1   See Francis Hincks's letter of 1 May 1852 to Sir John S. Pakington, M.P., Principal Secretary of State for the Colonies, published in *The Times*, Issue 21129 (31 May 1852), p. 6. Francis Hincks was co-leader of Canada with Augustin-Norbert Morin. John Somerset Pakington (1799–1880) joined Lord Derby's government in February 1852. According to Paul Chilcott (*ODNB*), he was 'regarded by his colleagues as prone to muddled thinking'.

FROM HENRY ELLIS
15 JUNE 1852

[British Museum]
15th. June 1852

Sir,
    I have to acknowledge the receipt of your letter dated 1st. inst. which has been laid before the Trustees of the British Museum, and I am directed by the Trustees to inform you with reference to your letter, that the whole subject connected with the means of decyphering the Assyrian inscriptions is under the consideration of the Trustees. As suggested in your letter of 4th. May the Trustees have had the

Assyrian duckformed weights accurately weighed, and I am directed to send you the inclosed result, as reported by M^r. Hawkins.

I have the honour to be, Sir,
Your most obedient servant,
Henry Ellis
Principal Librarian

[Enclosure: Table of Weights (copy)]

Department of Antiquities
1st. June 1852

M^r. Hawkins reports that the Assyrian duck formed weights have been weighed at the mint with the following results.

lb	oz	dwts	gr
40	4	4	4
39	1	1	6
	6	2	3
	5	14	13
	4	2	3
		13	17.36
		4	23.36
		4	19.65
		3	15.65
		3	8.65
		1	15.625
		1	9.85
		1	6.75

Edwd. Hawkins

MSS: BM/CA/LB 42:24–5
BL Add. 39077, f. 63r

TO AUSTEN HENRY LAYARD
16 JUNE 1852

Killyleagh C° Down
16th. June 1852

My dear Sir,

I was this morning favoured with your letter of the 14th. I feel much obliged by all the trouble you have taken. There are just two things I think you ought to know; though I don't know that they need influence you as to any future step. D^r. Robinson (President of the Royal Irish Academy &c.[1]) has written a very strong letter in my favour to M^r. Napier,[2] who (on the 8th.) promised to speak to Lord Derby & also to Lord Eglinton about me.[3] The latter I should think could do nothing. If the letter from D^r. R. be transmitted to Lord Derby (as I suppose it would be) it might have weight. No one ought to have greater influence in such matters than D^r. Robinson, himself a man of *first rate* talents, and recognised by every body as such.

The other point I wished to mention was that any statement made by my brother in his letter to Lord Derby was *unauthorised by me*. He wrote *without consulting me* & I was greatly surprised to hear of his having done so. He shewed me Lord Derby's reply, but had no copy of his own letter. He seemed to think that Lord D. must have misunderstood him.[4]

So far from my considering it a *sine qua non* that I should hold my Irish living & a situation in the British Museum contemporaneously, I feel that such an arrangement could only exist, either *as a temporary one*, or under the very great disadvantage of my being only two months out of twelve at the Museum. In the remaining ten I could do something with copies of inscriptions & in the way of editing or attending to the press; but (comparatively speaking) very little.

While I hold the cure of souls, I consider it necessary that it should be *adequately provided for under my superintendence*. I would not resign it *permanently* to any curate, with whom I could not share the Sunday labours, and take counsel as to those of the week days. Nor would I resign it even temporarily to any curate who was not selected by myself.

If this can be accomplished by the exercise of Gov^t. or other patronage in my favour (for a simple exchange of Killyleagh for a benefice in or near London would scarcely bring me what would pay the curate) I am ready to give my services in any way that can be pointed out (either in connexion with the Museum, with a professorship, or both combined) or having a *moderate income secured to me for life*. A person at my age could not, of course, be expected to give up a *benefice* for a *salary*, to be only paid during health & capability to work. The *ordinary* pension for superannuation would not be applicable to a man of 60.

I was not aware that sinecures were *necessarily* abolished, though as a general rule they should be. I saw *very lately* an appointment by the Chancellor to a sinecure rectory near Canterbury which a M^r. Ireland had held.[5] I should have thought that, tho' it might not be desirable to keep up such things in ordinary cases, *my* appointment would have satisfied the public. I am, however, quite ignorant of the law on the subject.

I am anxious to examine the inscription which you speak of sending. I suppose the next Athenaeum will contain an abstract of Rawlinson's new views. Did I tell you that I was wrong in identifying the king of Lord Aberdeen's stone with that of the hexagonal piece in the museum published [in] plates 20–29? The latter must be Esarhaddon's. The name on the former agrees as to its last part; but the name of the God differs. *Shamish.akh.adna* I conjecture, in place of *Asshur-akh-adna*. The initial element in the name on the hexagonal prism is wanting; but it must certainly be supplied as *Asshur*. It is possible that the Σαοσδυχινος of the Canon may be a corruption of *Shamashakhadna* or *-adin* in the unemphatic form. I have established a point of some interest with respect to the way of calculating the regnal years of the different kings.

> Believe me
> My dear Sir
> Yours very faithfully
> Edw. Hincks

P.S. Perhaps you could let me see what my brother wrote to Lord Derby. I have as yet heard nothing as to the fate of the letter which I wrote to Sir H. Ellis & which (as I understand from Sir E. Tennent) Sir R. Inglis took away, intending to bring it to the Museum.

If there has been any serious mistake in my brother's letter, would you recommend *my* writing to correct & disavow it?

MS: BL Add. 38981, ff. 61–3

---

1   Thomas Romney Robinson (1793–1882), Irish astronomer and physicist, who has been mentioned several times in these letters, was president of the Royal Irish Academy from 1851 to 1856.
2   Joseph Napier was attorney-general for Ireland in 1852.
3   Archibald William Montgomerie (1812–61), 13th Earl of Eglinton, was a staunch Tory and in 1852 became lord lieutenant of Ireland under Lord Derby.
4   See Hincks's letter of 1 June to Sir Henry Ellis at the British Museum.
5   Edmond Stanley Ireland was rector of Bicknor, Kent. See *The Clergy List for 1851* (London, 1851), p. 140. He died on 18 June 1851. His replacement was J. A. P. Linskill. See *Gentlemen's Magazine* (Apr. 1852), p. 398.

$\mathbf{\approx}$

TO AUSTEN HENRY LAYARD
21 JUNE 1852

Killyleagh C° Down
21st. June 1852

My dear Sir,

I received the first part of the inscription on the 18th. & have been busy with it. The 17 first lines are for the most part contained in the inscription on Bellino's cylinder, of which I have a good lithograph, published by Grotefend, as well as the copy in the B. M. collection.[1] When we come to the 3^d. year which contains the Syrian campaign (the most interesting part of all) the cylinder fails us & the inscription you sent me is in a sad state of mutilation. Nevertheless I have made out a great part of it. But what do you propose that I shall do? Rawlinson has sent over a translation; & they will be publishing it immediately. When will your new work be out? I could let you have a translation of this inscription *very soon*; & it might be well to have the two out together, or at least independently of one another that the public might judge how far they agreed & thence of the soundness of our methods. I hope to have the remainder of the inscription on Wednesday. There is so much uncertainty in a copy of a copy that I don't know how far queries can be answered with any satisfaction.[2]

As to my own affair, I believe my only chance *for the present* is with the trustees of the Museum. I had a letter from Sir H. Ellis some days ago to say that the matter was still under consideration. Lord Derby, it is evident, will do nothing. He has too many claims on him.

Believe me
My dear Sir
Yours very faithfully
Edw. Hincks

MS: BL Add. 38981, ff. 66–7

---

1   'Bemerkungen zur Inschrift eines Thongefässes mit ninivitischer Keilschrift', *Abhandlungen der königlichen Gesellschaft der Wissenschaften zu Göttingen* 4 (1850), historisch-philologische Klasse, pp. 175–93.
2   I have omitted some lines which contain queries about unclear signs in Layard's copy.

FROM RICHARD LEPSIUS
25 JUNE 1852

Berlin le 25 juin 1852

Cher Monsieur,

J'ai envoyé à M. Bunsen un exemplaire de mes 'Lettres écrites d'Egypte, d'Éthiopie et de la Péninsule du Mont Sinai'[1] en le priant de le faire parvenir entre vos mains. Veuillez l'accepter comme un faible signe de ma haute estime et de l'intérêt avec lequel j'ai toujours suivi [vos études] aussi judicieuses que savantes. Je crois vous avoir déjà dit mes remerciemens pour les différens ouvrages que vous avez bien voulu m'adresser de temps en temps. Si nos opinions diffèrent souvent sur des questions qui nous intéressent tous les deux également, j'espère que les points où nous sommes d'accord, s'augmenteront en même temps et que la vérité sort bien rarement à la conviction de tous sans avoir passé par la critique des adversaires.

Dans les annotations[2] de mes 'Lettres' j'ai repris une autre fois la question du Mont Sinai et je m'imagine que je l'ai à peu près vidé[e]. Je serais bien aise de connaître votre opinion sur cette question, si vous trouvez le temps d'en prendre connaissance.

J'ai lu il y a quelques mois dans notre académie une dissertation sur la 12me. Dynastie, et je lirai en peu de semaines une autre sur les Ptolémées selon les monumens hiéroglyphiques. Je ne manquerai pas de vous les faire parvenir lorsqu'elles seront imprimées. Un mémoire sur le 1er. ordre des dieux égyptiens vous sera parvenu, j'espère, dans son temps.

Veuillez me croire
Votre très dévoué
R. Lepsius

8 nouvelles livraisons, 80 planches, de notre grand recueil de monumens égyptiens paraissent à présent. L'ancien règne est maintenant complet, et le 1er. volume de la 3me. division, laquelle comprend le reste des 30 dynasties, se termine avec Aménophis III.[3]

MS: GIO/H 256

---

1   Briefe aus Ägypten, Äthiopien und der Halbinsel des Sinai: geschrieben in den Jahren 1842–1845, während der auf Befehl sr. Majestät des Königs Friedrich Wilhelm IV von Preussen ausgeführten wissenschaftlichen Expedition (Berlin, 1852); the English translation appeared under the title Discoveries in Egypt, Ethiopia and the Peninsula of Sinai, in the years 1842–1845 during the Mission Sent out by His Majesty, Frederick William IV of Prussia, ed., with notes, by K. R. H. Mackenzie (London, 1852).
2   Lepsius wrote 'adnotations' which must be wrong.

3   In this postscript Lepsius is referring to the early parts of *Denkmäler aus Ägypten und Äthiopien*, 6 pts in 12 vols (Berlin, 1849–59).

※

TO AUSTEN HENRY LAYARD
28 JUNE 1852

28ᵗʰ. June 1852

My dear Sir,

I have had a good deal to do, so that I could not translate more of the inscription than the above; and as I am likely to be engaged all this week I think it best to send it to you as it stands.

Yours very truly
Edw. Hincks

[Enclosure:]

Annals of Sennacherib 1ˢᵗ. year (lines 3–9)[1]

At the commencement of my campaigns, when Marduk-Baladan, king of Qarduniyas[a], with the forces of Susa as his helpers, was in the *district*[b] of —, I gained a victory over him. *He retreated to save his life.* I obtained possession of chariots, wagons, horses and *mares*[c], which were abandoned. I penetrated to his palace beside Babylon. I opened his treasury. I made a spoil of gold and silver, vessels of gold and silver, precious stones, and the male and female attendants of his palace. I subdued his principal towns, the fortresses of Chaldea, and the small towns attached to them.[d] I carried away their spoils. Afterwards, I subdued the Syrians on the banks of the Tigris and the Euphrates[e]. I carried away their spoils.

In the latter part of my campaign, I received a large tribute from the chief of the city Khararat. *I received the submission* of the inhabitants of Khirim, who were *rebels*.[f] *I did not take their lives.* I appointed the city to be *a holy one*. I settled on one ox, ten sheep, ten *goats* and twenty *lambs*, as *its consecration-offering to* the Gods of Assyria.

N.B. Expressions of which the meaning is not fully established are underlined.

a   Qarduniyas was the capital of Chaldea; and seems to have been situated near the modern Basra.[2]

b   or *neighbourhood*; the town mentioned was in Babylonia; and occurs among the places in that region subdued by Sargon on all the Khorsabad bulls. I have not been able to identify it.

c   Bellino's cylinder adds 'asses, camels and riding-horses with equipments for war'.

d   Bellino's cylinder says that he took 79 principal towns and 820 small ones.

e   Bellino's cylinder mentions seventeen tribes before the Aramaeans or Syrians, which may have been a generic term. Some of these tribes are, however, clearly Arabs; as the Hagarenes *(Khagaranu)* and the Nabatheans *(Nabatu)*. Their spoil amounted to 208,000 men and women, 7,200 horses and mares, 11,063 asses, 5,230 camels, 120,100 oxen and 800,600 sheep.

f   This word is applied in Bellino's cylinder to the nations next mentioned, as well as the people of Khirim, and in both cases it is added 'who for a long time had not been obedient to my authority', or 'to the kings my fathers'. I cannot conjecture where this city was; but I observe that, contrary to the usual custom when cities were appointed to be holy ones, neither its name nor its inhabitants were changed. Khirim may mean 'holy'; and its having been already a holy place may account for the peculiar favour shown to it.

3rd. year latter part (lines 27–32)

Hezekiah *(Khazaqi'a'u)* of Judah *(Ya'uday)* not being obedient to my authority, I went to and subdued 36 of his towns, that were principal fortresses, and small towns attached to them which they did not number. I carried away their spoil. I shut himself up within Jerusalem *(Ursalimma)* his capital city. I severed from his country the fortresses, and *the remainder of* his towns which I had spoiled. I gave them to the Kings of Ascalon, Ekron and Gaza, *making* his country *small.* I added to the former tribute imposed on their countries a new tribute the form of which I settled. Hezekiah *returned to his obedience to me.* I caused to be brought to Nineveh the . . . . . . and the . . . . . . 11,000 soldiers, who *were collected together* in Jerusalem his capital city, together with 30 talents of gold, 800 talents of silver . . . . . . the treasures of his palace [his sons] and his daughters with the people of his palace, *both male and female* . . . (The conclusion is illegible; and several words, which appear to signify classes of persons, occur in this passage, that I have not met with elsewhere.)

MS: BL Add. 38981, ff. 70–1

1   See Layard, *Discoveries in the Ruins of Nineveh and Babylon* (London, 1853), pp. 139–44.
2   Kar-Dunias was the Kassite administrative capital. It was near the later Baghdad, not Basra.

FROM AUSTEN HENRY LAYARD
29 JUNE 1852

9 Little Ryder St
June 29/52

My dear Sir,

I have been so much occupied with my election return at Aylesbury that I have been quite unable to send an earlier answer to your note of the 21st with the queries on the inscription.[1] I am afraid I can only answer one of your questions on reference to my copy of the inscription. At line 25 the character is most distinctly copied 𒂊 and it would seem that this is incorrect. The stone has been so much mutilated that I have no idea of the characters which filled up the broken spaces in lines 22–28. In the 28 line, at the end of the mutilation, 𒉿 is given as *part* of a mutilated character which may [be] broken 𒈾. 𒂊 is plain; the two following characters are merely conjectural. Rawlinson writes that he has had to restore the inscription almost entirely. There are paper casts of it in the Museum. Immediately after the election I am going to finish my book which I hope to bring out as soon as the publishing season in the autumn commences. I beg you will do as you think best for your own reputation of prior discovery with the material I send you. It would, of course, be of great interest to me if I could include any translation you were good eno' to send me in my work. But I should be very sorry to risk any priority of discovery to which you might have claim. I shall return to trouble you with several short inscriptions, rather epigraphs as I go on, which I hope you will kindly examine for me. You have, of course, seen Rawlinson's last paper which is now published in a separate form by the Asiatic Society.

I have communicated again with Sir E. Tennent concerning your affairs & he has again promised to do his best. I wish I had it in my power to send you some definite information. Politics seem to engross everyone.

Yours faithfully,
A. H. Layard

MS: GIO/H 217

---

1   Layard was elected Liberal M.P. for Aylesbury on 7 July. See his letter to a friend dated 9 July 1852 and the letter he received from Lord Granville dated 17 July 1852 in *Sir A. Henry Layard, G.C.B., D.C.L.: Autobiography and Letters*, vol. 2 (London, 1903), pp. 240–1.

High Cliffe – Christchurch
July 15/52

My dear Sir,

I have been so occupied with the struggles & excitement of a contested election (which has fortunately ended in a triumph as far as I am concerned) that I have neglected acknowledging the translation of the first part of the inscription which you were good eno' to send me. I locked it up in London before going to Aylesbury and not having returned since I have been unable to make use of it. I shall be in town on Thursday again. Pray let me know what you would wish me to do with it. Whether to publish it at once or to keep it for my work, which I hope will be out in the autumn. I shall be most happy to do what you think most advantageous to yourself with it. I am now going to devote myself to my MS, & hope to have it completely finished before Parliament meets. I shall take the liberty of troubling you as I go on with various epigraphs & short inscriptions & requesting you to have the great kindness to give me your opinion of them. You have, of course, seen Rawlinson's paper.

Believe me,
Yours very faithfully,
A. H. Layard

I wish I could send you a copy of the big inscription from Nimroud.

MS: GIO/H 218

Killyleagh C° Down
17th. July 1852

My dear Sir,

I received your letter of the 13th. yesterday.[1] Allow me to congratulate you on your success at Aylesbury. It gave me great pleasure to see it announced in the papers. I have not been able to look at the inscriptions since I sent you the

translation. Rawlinson's book has not reached me yet. I did not see the advertisement of it in time to order it by parcel which my bookseller got at the end of last
month. He will have another by the middle of this – that is, leaving London then;
& I expect it on Monday. I don't think it would be of any service to me to publish
the translation separately from your work. Perhaps, if you wish it, I may finish the
annals before you are ready for them. I wish much, however, that I could have
some conversation with you as to the *latter* part of the inscription, which contains a
description of the Kouyunjik palace. I have little doubt that with the help of your
local knowledge I could make out the names of the different kinds of stone and the
different objects made of them and of wood. I have recognised in the Babylonian
inscription the *gilt-poles* for the *awnings* at the top of the palace, and (though with
less certainty) the bridges or corredors[2] and the couches. The lions & bulls must
certainly be mentioned in the Kouyunjik inscription. I am now satisfied that
the reign of Sargon (who was *not* Shalmaneser) began in the 6th. Hezekiah or
721 B.C. – that of Sennacherib in the 23rd. Hezekiah 704 B.C. & that the expedition against Judea was not in the 14th. year according to the present copies of the
2d Book of Kings but in the 25th. The 14th. was the date of Hezekiah's illness & of
the embassy of Merodach Baladan, which immediately followed it; but these
preceded the expedition of Sennacherib. 'In those days' is an indefinite expression
meaning no more than 'in the course of this reign'.

Could you not manage to come over to the British Association meeting on
the 1st. September?[3] If you do, I wish you would spend a few days *here* either
before or after it; and we can talk the matter over. When I see Rawlinson's new
work I shall better know how far he agrees with me & where his system is
vulnerable. Perhaps, too, I may be led to correct my own views in some points.

I have heard nothing further of my proposal to the British Museum than that
it was 'under consideration'. I don't know whether I am to expect any further
answer. Perhaps the election bustle has prevented there being any decision; but I
rather fear the decision has been adverse to me, & that I am to *infer this* from my
having no answer.

I should fear there might be jealousy of my being called in, on the part of
those now employed in the Museum, though they must be sensible that they
want more hands. On reading an advertisement for a London clergyman who
desired to exchange to the country, it occurred to me that if I could be sure of a
situation in the Museum I might make such an exchange. I fear the clergyman
advertising in this instance would not be satisfied with Killyleagh; but I have
written a few lines, stating its value & circumstances, so that if he thinks it worth
his notice, he can write to me. I take the liberty of inclosing this to you and would
thank you to commit it to the Postoffice or to the flames, according as you think,
or do not think, that a desirable engagement with the Museum could be formed
if I had a London parish. The part of London is not stated in the advertisement;

117

& indeed I think it by no means likely that the negotiations would be successful. As I could not make such a sacrifice of income as it would involve, without something additional to look forward to, it would not be using the advertizer well to send him the letter, unless there was a reasonable prospect that I could have it in my power to carry on the negotiation.

I will willingly do what I can in respect to any short epigraphs or other inscriptions that you send me.

I think you would like the B.A. meeting & shall be disappointed if I do not then see you.

Believe me
My dear Sir
Yours very faithfully
Edw. Hincks

MS: BL Add. 38981 ff. 92–4

1   Hincks seems to have misread the date of Layard's letter which is dated 15 July 1852.
2   It is surprising to find this older spelling being used by Hincks.
3   The British Association for the Advancement of Science was held in Belfast.

FROM JAMES RANKIN[1]
22 JULY 1852

Greenholm 22^d. July 1852

Rev^d. Sir,

In the University of Glasgow the Cleland Gold Medal of next session is proposed for the best Essay on 'The Bearing of Recent Discoveries at Nineveh on the Evidences of Scripture History'.[2] For some time past I have been engaged on the above collecting materials from the Bible, from the vols. of Botta, Layard, Bonomi,[3] Fletcher,[4] Eadie[5] and from various papers in the Athenaeum. So far as the sculptures are concerned I am satisfied; but I believe that the inscriptions though not the most showy are the most valuable part of the late discoveries, as says Rawlinson in 'The Inscr: of Ass: & Bab:' note 2 page [4]70.[6] All the inscriptions throw light on Assyrian Hist:, but I find such assertions & denials of coincidence concerning the same passages of inscr: said to bear upon that part of Ass: Hist: touched on by Scripture, that I am afraid to employ them at all as confirmations of Holy Writ. Can you tell me whether beyond the identification of the names Jehu, Beth-Omri, Hazael, Jerusalem, Judah &c any names & events of Biblical interest have been generally agreed on or, though not so, have such proofs of coincidence as may be relied on: and, if additional identifications have been made, where accounts of them are to be found? I have not been able as yet to get at your work upon the 'Khorsabad Inscriptions' and so know nothing of its contents but what I may guess from the title.[7] I got lately 'Rawlinson's Commentary' but since 1850 his opinion of the Obelisk date has been changed & many of his explanations of the legend upon it are thus superseded. Being not yet acquainted with German, any thing by Prof: Grotefend or others in that language will be unavailable to me, but of English or French books on the subject I shall be most glad to hear.

Considering the prominent part which, from various sources, I see you have taken in the decipherment of the cuneatic, I have ventured to write this note. Being an entire stranger I feel considerable hesitation in troubling you at all in the matter. But should you deem this request intrusive and impertinent, I dont know if you will have done so unreasonably: and in excuse I have only to plead youth & the interest I have taken in the subject. With sincere wishes for your continued success in the restoration of Assyrian history.

I am, Rev^d. Sir,
Your most obed^t. servant,
James Rankin

P.S. Should you favour me with any information as to the above, please address to James Rankin A. M. Greenholm by Uddingstone Lanarkshire Scotland.

MS: GIO/H 404

1   James Rankin (1831–1902), A.M., D.D., was minister of Muthill, a small Scottish village, three miles from Crieff in Perthshire. He was a fellow of the Royal Asiatic Society.

2   In 1840 James Cleland, LL.D. (Glasgow 1846) founded a medal to be awarded in alternate years to a student of divinity and a student of physics for the best essay on a prescribed subject.

3   J. Bonomi, *Nineveh and Its Palaces. The Discoveries of Botta and Layard Applied to the Elucidation of Holy Writ* (London, 1852).

4   J. P. Fletcher, *Narrative of a Two Years' Residence at Nineveh, and Travels in Mesopotamia, Assyria, and Syria*, 2 vols (2nd edn; London, 1850); *Notes from Nineveh, and Travels in Mesopotamia, Assyria, and Syria* (London, 1850).

5   J. Eadie, *A Biblical Cyclopaedia, or, Dictionary of Eastern Antiquties* (3rd edn; Glasgow, 1851).

6   'On the Inscriptions of Assyria and Babylonia', *Journal of the Royal Asiatic Society* 12 (1850), p. 470, n.

7   'On the Khorsabad Inscriptions', *Transactions of the Royal Irish Academy* 22 (1850), polite literature, pp. 3–72.

FROM AUSTEN HENRY LAYARD
23 JULY 1852

Canford Manor[1]
Wimborne
July 23/52

My dear Sir,

I am greatly obliged to you for your letter of the 17th. July. What you told me of the inscriptions interests me exceedingly, especially that which relates to the description of the Palaces. I have so much to do in order to get my MS sufficiently advanced before the meeting of Parliament to publish towards the end of the autumn that I almost despair of being able to run over to Ireland. I must wait & see how I get on. I should have more pleasure in spending two or three quiet days with you than in being at Belfast to meet the Association.

I should indeed be very glad of any translations you can send me of the inscriptions. I am anxious to place as perfect a copy as I can make by comparing three together of the Sennacherib Inscriptions at *Ninive* in your hands & shall set about it as soon as I have a leisure hour.

It would be very important for me to learn how far the upper part of the Palace was constructed of wood. A restoration I have lately made is founded upon the presumption that a large portion of the upper part of the building was of this material. I have thought it as such to put the letter you inclosed to me in the post. At any rate there can be no harm in communicating with the advertiser. I am undoubtedly inclined to think that the trustees of the B. M. would gladly avail themselves of your valuable services, particularly as they have now ordered the last batch of inscriptions to be prepared for the press. I will lose no opportunity of pushing the matter, but I am heartily disgusted with the Trustees.[2]

Believe me
Yours faithfully
A. H. Layard

MS: GIO/H 220

1   Canford Manor, today a school, is situated in East Dorset near the market town of Wimborne Minster. It traces its origin to Saxon times and grew to prominence in the Middle Ages. The Norman church has survived but the main building is from the nineteenth century. It was the home of Sir John and Lady Charlotte Guest from 1846.
2   Layard was annoyed with the Trustees on several accounts. Some months earlier they sought explanations about his use and ownership of a private collection of Assyrian antiquities. See J. M. Russell, *From Nineveh to New York: The Strange Story of the Assyrian Reliefs in the Metropolitan Museum and the Hidden Masterpiece at Canford School* (New Haven, 1997), pp. 79–80; G. Waterfield, *Layard of Nineveh* (London, 1963), p. 232.

꩜

FROM AUSTEN HENRY LAYARD
23 JULY 1852

Canford Manor
Wimborne
July 23/52

My dear Sir,

I venture to trouble you as I proceed with my MS. and I now send you a few epigraphs and a couple of inscriptions from bricks. I am particularly desirous of having your opinion & any version you would be kind eno' to give me, of the short inscriptions on the basreliefs representing the building of the Palace. Perhaps you would kindly return me these inscriptions when you have quite done with them. You are, of course, quite at liberty to keep copies.

Believe me,
Yours faithfully,
A. H. Layard

MS: GIO/H 219

꩜

TO AUSTEN HENRY LAYARD
26 JULY 1852

Killyleagh C° Down
26th. July 1852

My dear Sir,

I received your two letters yesterday & today. I should be very glad [if] you would come over here either at the Association or *after it*, which would perhaps be the best. Before it, I shall not have so much leisure & besides Mrs. Hincks will have a succession of lady visitors.

The small inscriptions you sent me are *most important*. They determine in the first instance *most decidedly* the word for 'bulls' & [. . .]¹ that for 'lions', as I had settled in my own mind that two words stood for these two & only doubted which was which. The mound puzzles me. I had conjectured that the thing mentioned in that inscription as being made was 'the wall'. If you are sure it is 'the *mound*' the determination is of great importance as the word occurs in different forms very commonly.

What was the mound made of? Rawlinson has translated it 'cut stones' which applied to a mound is absurd. Is it 'sun dried bricks'? The word for 'baked bricks' I know, but that for the sun-dried I had not ascertained previously. If so, however, it must be translated '*small* sun-dried bricks' & such bricks of a larger kind are mentioned elsewhere. Have you met them? Again what were the bulls made of? Marble? Are there not two materials used for them? If I recollect right, two are mentioned on the bulls. Images are spoken of in the bull inscriptions as being along with the bulls. What do they represent? & of what are they made? The word which occurs in the inscription on the entrance to the inclosure I take to signify 'castle' or 'fort'.[2]

It is of great importance to settle the signification of as many objects as possible in order to have a complete translation. I see by the Athenaeum that Rawlinson has rectified some of his errors, & that he has found the annals of Tiglath Pileser which is of great importance. I have *demonstrated* that the first year of Sargon began in Feb. 721 B.C. & that of Sennacherib in Feb. 703 B.C. I fear Rawlinson will correct his errors on this point before I can have mine out in any other channel than the Athenaeum, & *that* channel I do not like.[3] He has failed to recognize the name *Belibus* which occurs in the Canon, & in the inscription on Bellino's cylinder, as that of the person whom Sennacherib made king of Chaldea in place of Merodach Baladan in his first year. He reads the name *Bel adon* in place of *Bil. ebu*.

I find I have no note paper in the house: so I am forced to use this.

Believe me
Yours very faithfully
Edw. Hincks

You have placed over two inscriptions 'fragment on similar subject'. Surely what is *here* represented is not a *bull*. It seems a *wooden* object brought from Lebanon.

MS: BL Add.38981, ff. 97–8

---

1 The word is illegible.
2 Hincks sent Layard the following notes on the same matter, but we do not know in which letter they were enclosed.

The *pilu pesu* of which the (mound?) was made can scarcely be *bricks*.

We have distinctly in the bull inscription 'bulls of (marble?) with bulls and (lions?) of (rubble stone?)'. Again 'bulls and images ... of (marble?) which were made of single stones'.

The bulls were thus made of two substances one in solid blocks, the other put together. Does this agree with what you have observed? Is the solid material marble? or what?

The lions seem to have been made in 3 materials; the two that the bulls were made of & a third.

I shall be anxious for your opinion on the above points.

See MS: BL Add.38981, f. 3. *pīlu pēṣu* is 'white limestone'.

3   Hincks wrote a letter to the editor of *The Athenaeum* on 26 July but the latter declined to publish it. See my note (n. 4) to Hincks's letter of 2 August 1852 to Layard. As a result Hincks avoided publishing in *The Athenaeum* until October 1856.

FROM JAMES RANKIN
28 JULY 1852

Greenholm by Uddingstone Lanarksh:
28th July 1852

Revd. Sir,

I have just received your letter of the 24th. The information which it contains will, I think, be very serviceable to me. For the prompt & satisfactory manner in which you have been kind enough to reply to me I now tender you my warmest thanks. The embarrassment I felt in discovering the titles of works on the subject, rather than any very sanguine hope of receiving an answer made me at first think of applying to you at all. I have experienced ere now the readiness of scholars in communicating information, but never an instance so creditable as the present, where I am utterly unknown to you. But that which in the world of life is brought about by formal & circuitous introductions, is, in the world of letters, at once accomplished by an intuitive & secret sympathy. It would appear that not only 'omnes artes quae ad humanitatem pertinent, habere quoddam commune vinculum, et quasi cognatione quâdam inter se contineri' but that the same bond that unites the studies themselves unites also those who cultivate them.[1]

With sincere respect,
I am Sir,
Yours gratefully,
James Rankin

MS: GIO/H 405

1   See Cicero, *Pro Archia*, 2: Etenim omnes artes, quae ad humanitatem pertinent, habent quoddam commune vinculum, et quasi cognatione quadam inter se continentur, 'Indeed, all the arts which have any bearing upon culture, have some common bond by which they are united to one another' (tr. N. H. Watts, Loeb edition).

FROM AUSTEN HENRY LAYARD
JULY 1852[1]

[My dear Sir,]

...[2]

I am so much interested in what you tell me that I have made up my mind to run over to Ireland in September after the meeting of the association. I will then bring all my inscriptions & my drawings & I have no doubt that we shall make some curious discoveries. Rawlinson is, I am convinced, wrong on many points. I shall send you inscriptions as I go on. I am anxious to get my book ready to print or at least to commence printing before coming to you & it will then be out early in the autumn, with such additions as you will kindly allow me to make. I am anxious that you should have the Bellino inscription & the long historical inscription from Nimroud. Pray do not let anyone know of these discoveries. Rawlinson has not the material and cannot forestall you & I should have great satisfaction in announcing them.

Believe me,
Yours faithfully,
A. H. Layard

MS: GIO/H 223

1   The first part of this letter is missing so the date is uncertain, but Hincks's letter of 2 August 1852 seems to be the reply.

2   In the first page of the second sheet of the letter (the first sheet is missing), there is a sketch of a tent or pavilion accompanied by the following words: 'I should much like to have your opinion of it. Behind the king is a tent, felt hut or pavilion. I do not exactly know which – it is evidently supported by cords & in this shape.' He asks Hincks to translate the epigraph over the tent. See Layard, *Discoveries in the Ruins of Nineveh and Babylon*, p. 151, n. *: 'the *tent* (?) (the word seems to read *sarata*) of Sennacherib, king of Assyria.' See J. M. Russell, *The Writing on the Wall: Studies in the Architectural Context of Late Assyrian Palace Inscriptions* (Winona Lake IND, 1999), p. 288. The Akkadian word for 'tent' is *zāratu*.

FROM AUSTEN HENRY LAYARD
1 AUGUST 1852

Canford Manor
Wimborne
Augt. 1/52

My dear Sir,

I send you the inscription on the breast of the small figure of a king or priest found in a small temple near the North West Palace of Nimroud.[1] I am leaving Canford for a few days. Should you have anything pressing to communicate please direct to No. 9 Little Ryder Street, otherwise a letter sent here will be kept for me. Lady Charlotte Guest, with whom I am staying, begs me to say that should a journey eastward suit you in the autumn and if you would spend a few days with them in Dorsetshire they should be most happy to see you, and you could have my inscriptions at your command and at the leisure you liked. We have just been building a Nineveh Porch here and have several fine sculptures from Nimroud, amongst them a colossal lion and bull.[2]

By the way what is your opinion as to the site of Lachish? Pray let me know when the meeting of the association is expected to be on.

Believe me,
Yours faithfully,
A. H. Layard

MS: GIO/H 221

---

1   The inscription is on a statue of Ashurnasirpal II which is in the British Museum.
2   For a detailed account of Layard's gift of the Assyrian sculptures to Lady Charlotte Guest and the building of the 'Nineveh Porch' at Canford Manor, see J. M. Russell, *From Nineveh to New York: The Strange Story of the Assyrian Reliefs in the Metropolitan Museum and the Hidden Masterpiece at Canford School* (New Haven, 1997). Workmen began to erect the lion and bull, which had been cut in pieces to facilitate transportation, on Friday 23 July.

⚜

TO AUSTEN HENRY LAYARD

2 AUGUST 1852

Killyleagh, 2ⁿᵈ. Augᵗ. 1852

My dear Sir,

I have received your letter of the 29ᵗʰ. I am glad you are coming over here & that we shall be able to talk the matter over respecting the inscriptions.

I send you a few memoranda – perhaps I may as well defer writing any more till we meet. I am satisfied, however, that your Lachish is the real one & that R. is mistaken.[1] His Lachish is *Aliqqu* out of which it is impossible to make Heb. *lākîsh*. But *Lakitsu/lākîs* ought not to be objected to – the *u* is a case ending (the other place would be simply *Aliq/ʾālîq*) & must be struck off. I *at first* was unwilling to admit that *ts s* could represent the Hebrew *sh*; & for this reason I objected to R's identification of *tsamirina* with Samaria *shōmerôn*;* but after this was corroborated by my discovery of the name of Menahem with the title 'King of Tsamirina', this objection fell to the ground;[2] & we have again *Werutstsalimmo* for *yᵉrûshālēm*. The *ts* is then no objection to the reading. Whereas Rawlinson's word has no *s* atall & has a *q* for *k*; & though the *k* series is often used for the *q* series, the latter is never used for the former.

You will observe that I equate *ts* to *s*; I have *proved* that the final character in the name of the city is both *tsu* and *sû*; *ṣ* is *ch* or *tsh*. I think you may *confidently maintain* that the name of the besieged city is Lachish & that it is the city referred to in the Bible.[3]

I have sent a note to the Athenaeum with the outline of my chronological system.[4] I now think that Shalmaneser (of whom no monument has yet been found) besieged Samaria in his last year 722 & that Sargon took it in his *second* 720; that Merodach Baladan's embassy to Hezekiah was in 712, but Sennacherib's expedition not till 701, the 14ᵗʰ. year of Hezekiah (the date of the former event) having been established in the text of 2 Kings 18 for the 25ᵗʰ. year the true date of the expedition.

I think I shall exhibit a map of the countries mentioned in the inscriptions at the Geographical Section of the British Association & read a short paper illustrative of it. My chief points of difference from R. are that *I* make the Kharkhar of the Khorsabad inscriptions (which he imagines to be Van) to be at the extreme south of Assyria, somewhere about Helwan, & that Illipi[5] (which he makes to be Azerbijan) I make to be Luristan. On these points I have most positive evidence. See p. 20.[6] Marukarta is a *false* reading. Yatnan is Asia Minor.[7] It is where the Tyrians fled to *by sea*, as represented at Kouyunjik. His notion of its being Rhinocolura is quite erroneous.[8] He has strangely mistaken the name Tyre for Abiri;[9] the text has clearly 'he fled from Tyre to Yatnan'. And the line of seacoast is described as from Yatnan to the borders of Egypt, which would be absurd if these two places coincided. This shows that R. is not *infallible*; though he & his friends at the R.A.S. sometimes write as if he was. I will leave the Lachish question to *you*, not touching on it in my paper.

I fear the election will enable the present people to keep in power – for some time at least. A [. . .] party has been returned in Ireland which will oppose them & would oppose equally the Whigs if they succeeded them. The only mode of conciliating them would be by the repeal of the Ecclesiastical Tithes Bill (the most absurd of all measures in my humble opinion) & I fear there is no chance of the parliament now elected doing that. To follow it up, which they would be more likely to do, will lead to fresh mischief. The prospects of Ireland in particular, for which two sets of fanatics are fighting & of a portion of which each set seems to have possession, is particularly gloomy. I heartily wish I was out of it, though I am probably placed in the very best spot in it; but of this I fear there is but little chance.

Believe me
My dear Sir
Yours very faithfully
Edw. Hincks

* I never objected to his reading of Bit-'Omri (as he says I did in his note p. 9). On the contrary all the characters in this name had been previously correctly valued by me.

MS: BL Add. 38981, ff. 99–101

1   Rawlinson, *Outline of the History of Assyria, as Collected from the Inscriptions Discovered by A. H. Layard in the Ruins of Nineveh* (London, 1852), pp. 23–4; 26. (See *Journal of the Royal Asiatic Society* 14.)

2   For Hincks's identification of the name of Menahem, see his letter of 29 December 1851 to *The Athenaeum*. On the sibilants in the Neo-Assyrian and Hebrew words for Samaria, see A. R. Millard, 'Assyrian Royal Names in Biblical Hebrew', *Journal of Semitic Studies* 21 (1976), p. 4.

3   Hincks's identification of the name of Lachish is mentioned by C. Uehlinger, 'Clio in a World of Pictures – Another Look at the Lachish Reliefs from Sennacherib's Southwest Palace at Nineveh', in L. L. Grabbe (ed.), *'Like a Bird in a Cage': The Invasion of Sennacherib in 701 B.C.* (Edinburgh, 2004), pp. 221–307; see p. 222, n. 4.

4   Hincks's letter of 26 July was not published. In November 1852 Hincks wrote in one of his articles: 'I sent a statement of what I considered an important and interesting discovery to *The Athenaeum*, the editor of which acknowledged the receipt of my letter on the 31st July, but did not publish it. I had, however, an opportunity, of which I availed myself, of making it public at the meeting of the British Association on the 2nd September.' See 'On the Assyrio-Babylonian Phonetic Characters', *Transactions of the Royal Irish Academy* 22 (1852), polite literature, p. 369, n.*. Note also the remarks in his letter of 19 October 1852.

5   Elsewhere Hincks spells the name Yelappi. See 'On Assyrio-Babylonian Phonetic Characters', p. 350. Layard, *Discoveries in the Ruins of Nineveh and Babylon*, p. 141 has Illibi.

6   This paragraph deals with the remarks by Rawlinson, *Outline of the History of Assyria*, pp. 20–1.

7   Hincks wrote, then crossed out: 'or Cyprus'. This was unfortunate because Yadnana is indeed Cyprus. By the time his paper was ready for delivery at the September meeting of the British Association for the Advancement of Science in Belfast, he had changed his mind further and read 'Yavan' for 'Yatnan'. He was correct in recognising that Akk. *yawan* and *yawnaya* meant 'Ionia' and 'Ionians' respectively, but this was no

basis for abandoning the reading Yadnan (Hincks's Yatnan). In his paper 'On the Assyrio-Babylonian Phonetic Characters', p. 351, in opposition to Rawlinson's Rhinocolura, he wrote: 'The place intended is not Rhinocolura, or any country bordering on Egypt, but the isles of Greece, or at any rate Cyprus.' See his letter of 3 September 1852 to Layard and the bibliography in n. 4 to that letter.

8   *Outline of the History of Assyria*, p. 22.

9   Rawlinson's version in *Outline of the History of Assyria*, p. 21 is: 'On my approach from *Abiri* he fled to *Yetnan*, which was on the sea coast'. His 'Tyre, Sidon and *Yabna* (Jabneh of Scripture)' (pp. 26–7) was plainly wrong as Hincks pointed out.

⁓⳩⳩⁓

TO AUSTEN HENRY LAYARD
6 AUGUST 1852

Killyleagh C° Down
6th. August 1852

My dear Sir,

I received your letter of the 1st. in course. I feel much obliged to Lady Charlotte Guest for her invitation; but having no curate I am a prisoner here, were there no other obstacle to my going so far from home. The meeting of the British Association will terminate on the 8th. Sept^r. & I suppose all will be quiet by the 12th. The inscription on the breast of the king is much the same as in the standard inscription. It refers to the visit to Lebanon which is also recorded on the altar. What I took to be determinatives before the word bulls are not so; but mean I think more 'figures' or 'likenesses'. The word occurs detached from 'bulls'!

The word 'statues' or 'images' occurs in the great bull inscription, but not in any of the epigraphs. I have met with mention of 'bulls, fashioned (or carved) of mountain-stone', in the Khorsabad inscriptions. They are in connection with the word for 'gates'. We will discuss the passage when we meet. I have made some corrections in what I sent you, though not of much moment. I would be glad you would not print any thing of mine without letting me see the proof, as I am always gaining ground more or less rapidly. I was vexed at Capt Smyth publishing in his Aedes Hartwellianae translations of mine, which, though very good for the time they were written (which was not mentioned), I could have corrected in many points.[1]

Lachish is to the South or S.S.E. of Jerusalem and at no great distance, within the limits of the tribe of Judah. Jerome says it is 7 miles South of Eleutheropolis.[2] I suppose Robinson's map will give the site, & also that his work will mention whether the *precise* site is determinable, or can only be approximated to. I have not his work within reach.[3]

Yours very truly
Edw. Hincks

MS: BL Add. 38981, ff. 102–3

1   W. H. Smith, *Aedes Hartwellianae, or Notices of the Manor and Mansion of Hartwell* (London, 1851), p. 140.
2   Eusebius, *Onomasticon*, 120.20, says that Lachish was a village on the seventh mile along the road from Eleutheropolis to Gaza. On this basis William Foxwell Albright suggested that ancient Lachish was modern Tell el-Duweir, not Tell el-Hesi, as proposed by Claude Reignier Conder. Tell el-Duweir is surely right. See G. I. Davies, 'Tell ed-Duweir = Ancient Lachish: a Response to G. W. Ahlström', *Palestine Exploration Quarterly* 114 (1982), pp. 25–8; 'Tell ed-Duweir: not Libnah but Lachish', *Palestine Exploration Quarterly* 117 (1985), pp. 92–6.
3   The map is in E. Robinson, *Biblical Researches in Palestine, Mount Sinai and Arabia Petraea: a Journal of Travels in the Year 1838* (London, 1841).

FROM AUSTEN HENRY LAYARD
7 AUGUST 1852

9 Little Ryder St
Augt. 7/52

My dear Sir,

Your letter of the 2nd. Augt. only reached me yesterday, as I have been into Wales. I hasten to thank you for it. The translations you are good eno' to send me are of the highest interest. I have no doubt that when I can show you other materials and explain to you many things connected with the buildings we shall make even more important discoveries. I am very glad to find that you are of opinion that this is the real Lachish. That list of towns in Joshua,[1] given apparently with reference to their relative positions, must surely be of use in determining some of the names in the Kuyunjik inscriptions. Pray do not trouble yourself to write again, except once and to say when the meeting of the Association will be on and when you will be at Killyleagh. I shall lose no time in joining you there as I am very anxious to get my work as far advanced as possible & to publish in November or December. I shall continue to send you such small inscriptions & epigraphs as occur to me as I write my MS.

I am afraid the recent proceedings in Ireland on the part of the Roman Catholic priesthood has raised such a feeling in England against the Irish that it will be very difficult for those who are friends of religious liberty to be of any use to them.[2] The prospect is indeed gloomy, and the present Ministry most unfortunately have done all in their power to pander to the lax intolerant passions of a large section of the English public.

Yours faithfully,
A. H. Layard

MS: GIO/H 222

1   Joshua 12: 7–24.
2   Layard shared the feelings of his cousin Lady Charlotte Guest who was particularly outspoken against the Catholic Church in Ireland.

≈≈≈

FROM JAMES WHATMAN BOSANQUET
13 AUGUST 1852

Claysmore Enfield
13th. Augt. 1852

My Dear Sir,

Your letter of the 10th., received yesterday, has interested me more than I can express. You say that you have 'identified the Belibus, and *three* years after the Apronadius (as correctly written) of the Canon, in the Assyrian annals, thus settling the chronology with absolute certainty'. But you do not say, whether your discovery confirms or disproves the current system. If the names of these kings are found in the annals of Sennacherib, then is the received system of chronology correct? If in the reign of Sargina (Shalmanezer), there is the system for which I contend, and which is founded upon the dates of Demetrius, & Polyhistor, confirmed beyond dispute; and the authority of those two historians immediately rises in the scale of authoritative history. If the dates of Demetrius for the invasion of Sennacherib, and for the carrying away of the ten tribes, are proved to be correct, we must receive with extreme deference his date for the destruction of Jerusalem. If Polyhistor proves to be correct in his date for the 1st. of Sennacherib, we may feel satisfied he wrote from authentic sources, and we shall look with interest for the identification of Sardinapalus with Nabopolassar, which rests upon his authority. Pray have the kindness, therefore, to let me know, in the annals of which king you discover the names of Belibus & Apronadius, with a reference also to the place in the inscriptions where they are found.

You need not conclude that the Editor of the Athenaeum will not insert your letter.[1] Mine on the eclipse of Thales was in his hands, and in type, in May last.[2] I trust that I shall see your letter in the Athenaeum of to-morrow. I think, however, that you have been most unfairly treated, in the endeavour which has been made to rob you of several discoveries, in which you had undoubtedly precedence of Rawlinson, however much we are indebted to him for most valuable discoveries.

With regard to the eclipse of Thales, you ask me, 'how could an error in the latitude of the moon in the year B.C. 310, affect the course of the shadow in the eclipses of B.C. 610, and 585?'.[3] You will find that all three eclipses happened at the same mode, viz. north ascending. They are, therefore, all affected in the same direction, by any alternation of the position of the moon. If the latitude of the moon was more north, than the tables make it, in the year B.C. 310, it was still more north at the dates of the two other eclipses. You will observe in a letter of mine to the Athenaeum of 25 Octr. last, that M^r. Airy has discovered a difference amounting to a minute of arc in a century in the true position of the moon's node.[4] And he considers that in the time of Thales the accumulation of error must have been to the extent of 27 degrees, throwing the shadow north or south to the extent of 200 miles. I had a letter from M^r. Airy dated 5th. June on the subject of the eclipse of Thales, in which he says, 'I will forthwith commence calculations'. I have no doubt therefore we shall soon know the truth, from the most accurate source.

Believe me
My dear Sir
Yours very truly
J. W. Bosanquet

MS: GIO/H 56

1   See Hincks's letter of 2 August 1852 to Layard, n. 4.
2   'The Eclipse of Thales Identified with that of the 28th May, B.C. 585, by means of the Eclipse of Agathocles', *The Athenaeum*, no. 1293 (7 Aug. 1852), pp. 846–7.
3   Hincks and Bosanquet corresponded on the subject of Thales's eclipse for some years. For a modern discussion, see A. A. Mosshammer, 'Thales' Eclipse', *Journal of the American Philological Association* 111 (1981), pp. 145–55.
4   'Assyrian Antiquities': A Letter to the Editor, *The Athenaeum*, no. 1252 (25 Oct. 1851), pp. 1119–20. George Biddell Airy (1801–92) was astronomer royal from 1835 to 1881. See *ODNB*.

TO AUSTEN HENRY LAYARD
14 AUGUST 1852

Killyleagh 14th. Augt. 1852

My dear Sir,
    I am not sure of your address but hope that this will reach you and that you will favour me with an early answer.
    I have determined on confining my remarks in my British Association paper to the quarter from Media to the Persian gulf & I am now anxious to make what I say of this quarter as complete as possible. Col. Chesney, who knows it well, will be in the chair.[1] I can obtain no information as to some points which I require to know, namely how far up the Shat el Arab the tide runs, its *breadth* at Basra & below it & whether its banks are sufficiently dry for towns to have been built on

them. Perhaps these questions are complicated by another; how far can we suppose that this is the *ancient* channel of the River?

The view which I am going to maintain is that 𒀭 𒄑 𒈠 𒇬 𒅆 is not the *Persian Gulf* (as Rawlinson supposes & as he builds so much on) but the *Shat el Arab* – or its ancient representative, '*bartu marratu* the bitter (or salt) *river*'.[2] Chaldea only reached to *this*, not to the gulf. Sargon speaks of having pushed his conquests to *the salt river*; but Sennacherib in the bull inscription (though not on Bellino's cylinder) speaks of reigning as far as 'the lower sea of the rising sun', the true description of the Persian Gulf, the upper being the Caspian. To this he pushed his conquests in his 6th. year, the last of which the bull gives his annals, having taken *two* cities of the name of Nagit on opposite sides of the Salt River. Rawlinson has overlooked the fact of there being *two* cities, like North & South Shields. So named & makes one which he imagines to be beyond the Persian gulf, & even thought of making an Indian voyage out of a simple crossing the river! See p. 26.[3]

I wish to be in possession of the facts above referred to, not only with a view to be able to answer questions as to detail but to find if possible the sites of the two Nagits both of which belonged to Susa. They commanded the mouth of the river; but how far off is a question. Basra is some 60 miles. I should think the Nagits were much lower down; but Qarduniyas the capital of Chaldea *above* Basra.

Believe me
Yours very truly
Edw. Hincks

MS: BL Add. 38981, ff. 104–5

1   Francis Rawdon Chesney (1789–1872), army officer and explorer, was born in Annalong, Co. Down. From 1829 to 1832 he travelled in the Near East. For four months in 1831 he sailed by raft down the Euphrates and made a survey.

2   Layard reports Hincks's conjecture. See *Discoveries in the Ruins of Nineveh and Babylon*, p. 441; also p. 145. 'Bartu' is now read *nartu*.

3   Rawlinson, *Outline of the History of Assyria*, pp. 26–7.

~∽

FROM AUSTEN HENRY LAYARD
17 AUGUST 1852

High Cliff Christchurch.
Augt. 17. 1852

My dear Sir,

I received your note this morning and send you an answer at once hoping that it will reach you in time.

The tide of the Shat al Arab extends up both rivers: on the Tigris to the Tomb of Ezra; I do not know exactly up to what point on the Euphrates. Chesney's map will tell you exactly the distance of this tomb from the Persian Gulf. I should say about 150 or 160 miles, i.e. about 30 miles from Korna, the place of junction of the two rivers. The water is salt far down Basrah when the tide is coming in. I cannot tell you the exact breadth of the Shat at Basrah. I should say nearly a mile. Basrah, you know, stands on a canal at some distance from the river.

The mouth of the river has undoubtedly undergone numerous changes. It would I think be impossible now to trace all the old outlet. Within the last fifty years, as you will see by reference to my map & Chesney's maps & the maps published in earlier works, the river has undergone a very considerable change in this part of its course. I am inclined to believe that the Chaldean lake mentioned by the ancient geographers may have been an immense marsh formed by the . . .[1] rivers such as the Tigris & Euphrates somewhere about the modern district of Haweiza.[2] You will find some notes on this part of the country in my memoir published in the Journal of the Geographical Society which may interest you.[3] I have no doubt that cities may have stood on the banks of the Shat el Arab and did do so. There are numerous traces of ruins, most of them are laid down, I fancy, on Chesney's maps. I do not think the river has undergone many changes in its channel near Basrah for many hundred years. There was probably always a city near the mouth of the river. I should say somewhere near the junction of the Karoon, whose waters are celebrated for their purity. I think in my memoir I have given a list of the mounds occurring about this spot.

I am obliged to be at Aylesbury on the 16th. of Septr. and on the 17th. I shall start for Ireland if it will be convenient for you to receive me at that time. I think you recommended me to go to Dublin & thence to Belfast. May I trouble you to drop me a line to say whether you will be at Killyleagh the 18th. or 19th.? Address to 9 Little Ryder Street.

Yours faithfully,
A. H. Layard

MS: GIO/H 224

1   The word is illegible.
2   The writing is unclear but the place name seems to be Haweiza (see Arabic Ḥawīza; Ḥuwayza).
3   'A Description of the Province of Khúzistán', *Journal of the Royal Geographical Society of London* 16 (1846), pp. 1–105.

FROM JAMES WHATMAN BOSANQUET
21 AUGUST 1852

Claysmore
21st. Augt. 1852

My dear Sir,

Many thanks to you for your last letter, kindly informing me of your views of Assyrian Chronology. I am just setting off for Paris, but I will not lose a moment in letting you know, as it is a point bearing so closely on your inquiries, that M^r. Hind of the Observatory in the Regent's Park has just sent me a note to say that he has calculated the path of the eclipse of B.C. 585, and finds it to be as nearly as possible that which I have suggested, and that the eclipse of B.C. 610 cannot stand for that of Thales.[1]

I hardly venture to question any of your deductions from the contents of the Assyrian inscriptions; but I think you will see the necessity now of lowering the reign of Nebuchadnezzar, and if so, of Hezekiah. It strikes me that you have left a strong position for a weaker. If Sargon is really Arcianus, the reign of Hezekiah must be lowered 11 years. But allow me to suggest that Sargon's reign at Babylon, which you say is proved by the inscriptions may have been in the time of the first interregnum, and what Polyhistor records must fall in the second. Rabshakah I suspect came against Jerusalem in the 7th. of Sennacherib. This point can be decided by Colonel Taylor's cylinder. Pray let me hear again from you on the subject.

& Believe me
Yours very truly
J. W. Bosanquet

MS: GIO/H 57

1   John Russell Hind (1823–95), astronomer. See *ODNB*.

᷾᷾

High Cliff, Christchurch
Augt. 21, 1852

My dear Sir,

I send you some epigraphs from a very interesting series of bas reliefs – pray do not let them go further than yourself. They can remain with you until we meet. As I go on I think I shall find many illustrations of the contents of the inscriptions in the drawings from the bas reliefs – amongst others I fancy I have got the Shat al-Arab and the two towns of Nagit one on either side of the river. I shall be very glad to examine my collection with you. I feel sure we shall find many curious things.

Believe me,
Yours faithfully
A. H. Layard

MS: GIO/H 225

᷾᷾

Killyleagh 21st. Augt. 1852

My dear Sir,

I received your letter this morning & am obliged to you for the information in it. I will reconsider the whole matter with the light it affords me on Monday. Saturday is always a busy day with me & Sunday a *dies non* as to cuneatics. Monday the 20th. Sep^t. has been fixed for the Bishop's visitation. I should therefore prefer your not coming till *that evening*. There is a *tolerable* omnibus in connexion with the C^o Down train that leaves Belfast at 4.15.

You may come to Belfast either by Fleetwood, or by Holyhead & Dublin. You can have through tickets from Euston Square by *either* route. The former is much cheaper for the single journey; but by taking a *return ticket* the latter w^d come as cheap. The advantages are a shorter voyage & that you can fix your time better. The Holyhead conveyance goes 3 times each day – the Fleetwood only 4 times (I believe) in the week.

I hope you will bring the drawings as well as the inscriptions. I am anxious to see the *action* in the Lachish sculptures.[1] The king is sitting on the throne of judgment – *within* Lachish – is it not? But what are the people doing?[2]

I have fully identified the word for gypsum & am nearly sure of the shell lime-stone. Do you know where this was quarried?

Yours very truly
Edw. Hincks

MS: BL Add. 38981, ff. 111–12

1   The bibliography on the Lachish sculptures is extensive; see C. Uehlinger, 'Clio in a World of Pictures – Another Look at the Lachish Reliefs from Sennacherib's Southwest Palace at Nineveh', in L. L. Grabbe (ed.), *'Like a Bird in a Cage': The Invasion of Sennacherib in 701 B.C.* (Edinburgh, 2004), pp. 221–307.
2   Hincks translated the epitaph of Sennacherib sitting in judgement for Layard, *Discoveries in the Ruins of Nineveh and Babylon*, p. 152. See Russell, *The Writing on the Wall*, pp. 137–8, 287–8; Uehlinger, 'Clio in a World of Pictures', pp. 286–8.

TO AUSTEN HENRY LAYARD
3 SEPTEMBER 1852

[Belfast]
Sept^r

My dear Sir,

I write a hurried line from the British Association just to say that D^r. Robinson, to whom I gave your address (Romney Robinson, Astronomer at Armagh & President of the R.I. Academy), will write to you about me. He has got a promise from Ld Rosse to do what he can for me, & almost a promise that something *shall* be done. Ld Rosse told D^r. R. to write him a letter that he might have to shew to justify a grant from the civil contingencies & to make it as strong as possible. I believe he is well disposed to do so; & he wishes something from you that he can *quote*. It would be well if you would mention along with what you say of yourself, what you told me that you had heard of me in Paris. D^r. Lloyd of Dublin has also had a letter speaking very highly of me from a person in Berlin & has promised to look for it & send it to D^r. Robinson.[1] I had a paper on the language of Assyria on Thursday[2] & one today on the site of *Ancient Mines* which latter will be repeated in the Belfast papers of which I will send you one.[3] I maintain that what Rawlinson calls Yatnan & supposes to be Rhinocolura was Yavan, the Grecian islands, & that the Yavnai whom he makes to be the people of Jabneh were the Greeks of the islands.[4] I exhibited your print of the Tyrians crossing the sea as the fulfilment of the prophecy in Isaiah 23.12.

We had some discussion as to the five nations who paid tribute on the obelisk. Col. Chesney appears satisfied that the 5th were Persians.

> In great haste
> Yours very truly
> Edw. Hincks

MS: BL Add. 38981, ff. 200–1

1   Humphrey Lloyd (1800–81) was Professor of Natural and Experimental Philosophy at Trinity College, Dublin from 1831 to 1843 and president of the Royal Irish Academy from 1846 to 1851. He became provost of Trinity in 1867.

2   'On the Ethnological Bearing of the Recent Discoveries in Connexion with the Assyrian Inscriptions', *Report of the Twenty-Second Meeting of the British Association for the Advancement of Science; Held at Belfast in September 1852* (1853), pp. 85–7.

3   'On Certain Ancient Mines', *Report of the Twenty-Second Meeting of the British Association for the Advancement of Science; Held at Belfast in September 1852* (1853), pp. 110–12. There is a draft of this paper among Hincks's 'Miscellaneous Papers' at the Griffith Institute, Oxford: MS GIO/H 572. The paper appeared in the *Belfast Mercury* (Sept. 1852).

4   Layard mentioned Hincks's identification of Yavan in his *Discoveries in the Ruins of Nineveh and Babylon*, p. 142. For reliable, detailed studies of the Akkadian *yawan* and *yawnaya/yawanaya*, see J. A. Brinkman, 'The Akkadian Words for "Ionia" and "Ionian"', in *Daidalikon: Studies in Honor of Raymond V. Schroeder, S.J.* (Wauconda, Ill., 1989), pp. 53–71; R. Rollinger, 'Zur Bezeichnung von "Griechen" in Keilschrifttexten', *Revue d'Assyriologie* 91 (1997), pp. 167–72 and especially Rollinger's article 'The Ancient Greeks and the Impact of the Ancient Near East: Textual Evidence and Historical Perspective (*ca.* 750–650 B.C.)', in R. M. Whiting (ed.), *Melammu Symposia II* (Helsinki, 2001), pp. 223–64. On Greeks in Palestine, see A. C. Hagedorn, '"Who Would Invite a Stranger from Abroad?": The Presence of Greeks in Palestine in Old Testament Times', in R. P. Gordon and J. C. de Moor (eds), *The Old Testament in Its World* (Leiden, 2005), pp. 68–93.

FROM AUSTEN HENRY LAYARD
8 SEPTEMBER 1852

Canford
Sept[r]. 8. 1852

My dear Sir,

I have only just received your note, with one from D[r]. Robinson. I lose no time in replying to them both. I do most sincerely trust that his endeavours will be successful. I need scarcely say that I should be most happy to do anything in my power to aid, and that as soon as Parliament meets and people are again in London I will endeavour to interest them in the matter.

I will endeavour to be at Killyleagh on the 20th. or 21st. as you suggest. I should be very sorry to place you to the least inconvenience and if there be an hotel or any inn that I could stay at for a few days I should feel obliged by your hiring me a room. I will bring all my drawings & inscriptions with me and I hope to do a good deal towards finalising my MS whilst with you. I am very curious to see your papers to which I have only seen illusions in the public prints.

As I have much to do in Town I may be detained for a day or so more than I expect – pray, therefore, do not make any preparations for me.

Yours faithfully
A. H. Layard

MS: GIO/H 226

FROM GEORGE CECIL RENOUARD
22 SEPTEMBER 1852

Swanscombe, Dartford
W. 22 Septr. 1852

My dear Sir,

I ought to have thanked you two or three days ago for your letter of 17th. inst^t. which reached me on the 19th. & the Belfast Mercury[1] by 7 Septr. 52, my birthday,[2] which came that or the next day.

It was fortunate that I had got the Morgenländische Zeitschrift together, here below, & knew where to look for the paper in question, for locomotion has been so troublesome to me of late, that I might have been obliged to send messengers to seek it who would have had no small difficulty in understanding what book I wanted. I hope that testimony which is clear & decisive will have its due weight. You have, doubtless, collected similar evidence from other writers respecting your labours on the cuneatic inscriptions. I have read with great satisfaction your report to the B.A. at Belfast & hope you will not be put to great inconvenience by my detaining the paper, till I have copied that report. I saw no notice of it in the Lit. Gazette, & the Athenaeum I do not now see.[3]

Having lately had occasion to look into the Cheval. Bunsen's attempt to fix in some degree the Egyptian Chronol. I sought in vain for two valuable tracts of yours, which D^r. Lee lent me some time ago, on that subject, containing very important strictures on Lepsius' arrangement of the Turin MSS. May I beg the favour of the titles of those tracts (extracts from periodicals), & whether they are procurable in London?[4] I am anxious to see your labours better known in Germany, & hope you are in correspondence with the 'Deutsche Morgenländische Gesellschaft', whose quarterly Zeitschrift does not deteriorate.

Much has been done respecting the *Zend*, especially by Spiegel, whose writings seems to me nearly the most candid, judicious & *exhausting* which have yet issued from the German school. He has printed a corrected text of a considerable part of the Avesta, as you will see in the Zeitschrift, which you probably have, or at least can get from Belfast:[5] & *Westergaard* one of the great authorities for the Zend, has printed the commencement of an improved ed. with an English version commentary & notes, at Copenhagen.[6]

It strikes me that in proportion as he finds it needful to tread in your steps, R. seems to forget whose steps they are, and if a good parallel of the progress made by him & yourself in fixing (or endeavouring to fix) the value of the cuneatic letters & ascertaining the expression of the language used [were set out[7]], it w^d. appear more clearly to whom we are most indebted. I suppose you have seen Grotefend's last tracts on 'an inscription on an earthen vase', & a plan of the palace of Khorsábád.[8] I think he will assent to most of your present results.

You will be glad to hear that I hope I am mending, though still far from recovered.

Believe, my dear Sir,
Very gratefully & faithfully
Geo. Cecil Renouard

MS: GIO/H 468

1   The *Belfast Mercury* with Hincks's paper on 'On Certain Ancient Mines'. See Hincks's letter of 3 September 1852, n. 3.

2   Renouard was born on 7 September 1780.

3   Hincks presented three papers at the September meeting of the British Association for the Advancement of Science in Belfast: 'On the Ethnological Bearing of the Recent Discoveries in Connexion with the Assyrian Inscriptions'; 'On the Forms of the Personal Pronouns of the Two First Persons in the Indian, European, Syro-Arabic, and Egyptian Languages'; and 'On Certain Ancient Mines'. See the abstracts of them in *Report of the Twenty-Second Meeting of the British Association for the Advancement of Science; Held at Belfast in September 1852* (1853), pp. 85–7, 88, 110–12. Abstracts of the first and second papers were also published in *Journal of the Ethnological Society of London* 3 (1854), pp. 210–14, 218–20.

4   'On the Portion of the Turin Book of Kings which Corresponds to the Sixth Dynasty of Manetho'; 'On the Portion of the Turin Book of Kings which Follows that Corresponding to the Twelfth Dynasty of Manetho', *Transactions of the Royal Society of Literature*, 2nd ser., 3 (1850), pp. 128–38, 139–50.

5   *Avesta: die heiligen Schriften der Parsen*, tr. F. Spiegel, vol. 1 (Leipzig, 1852). Volumes 2 and 3 were published in 1859 and 1863.

6   *Zendavesta, or the Religious Books of the Zoroastrians*, tr. with commentary by N. L. Westergaard, vol. 1 (Copenhagen, 1852–4).

7   I have inserted these words, for it is clear that Renouard has accidentally omitted part of his sentence.

8   Perhaps Renouard is referring to Grotefend's paper, 'Die Erbauer der Paläste in Khorsabad und Kujjundshik', *Abhandlungen der königlichen Gesellschaft der Wissenschaften zu Göttingen* 4 (1850), historisch-

philologische Klasse, pp. 201–6. Renouard's translation of it, 'On the Builders of the Palaces at Khorsabad and Kouyunjik', was read at a meeting of the Syro-Egyptian Society on 16 February. It followed Grotefend's article 'Bemerkungen zur Inschrift eines Thongefässes mit ninivitischer Keilschrift', *Abhandlungen der königlichen Gesellschaft der Wissenschaften zu Göttingen* 4 (1850), historisch-philologische Klasse, pp. 175–93.

FROM AUSTEN HENRY LAYARD
23 SEPTEMBER 1852

Ampthill
Septr 23/52

My dear Sir,

I scarcely know how to excuse myself to you for showing so great a want of punctuality but I am unfortunately again detained for a day or two and must go to Aylesbury again before leaving England on business of importance. So that I fear it will be Friday or Saturday before I can start and you must not expect me until Thursday. I hope to start by the earliest tram on Saturday morning or on Friday night. They are about the registry for the Boro' of Aylesbury and I am obliged to be near for the moment. I most sincerely trust that I have not placed you to inconvenience.

Believe me,
Yours faithfully
A. H. Layard

May I trouble you to keep my letters that may be forwarded to you for me.

MS: GIO/H 227

AUSTEN HENRY LAYARD TO BENJAMIN AUSTEN[1]
3 OCTOBER 1852

Killyleagh, Co. Down
Oct. 3, 1852

[My dear Sir,]

I write you a line to announce my safe arrival here and temporary domestication with Dr. Hincks. I should have done so before but have been emersed day and night in cuneiforms. My departure from England was delayed much longer than I had anticipated.[2] There was a most agreeable party assembled at the Parke's and I offered, it must be confessed, very little resistance to the persuasions of the company to remain on – my visit lasting nine days instead of three![3]

I left England yesterday (Saturday) week, and reached Belfast Sunday evening having spent a few dull hours in Dublin. At Belfast I had a short interview with the lovely recluses, of whom you heard me speak, and then came on here by, what is called, a mail outside car – a very primitive locomotive. The Doctor received me very hospitably and the neighbourhood has been gay with festivities in my honor.

The small town stands on Lough Strangford, a very picturesque sheet of water in the midst of hills. The Lord of the village is one Capt. Hamilton who has lately rebuilt an old castle in the medieval French style and has carried it out well inside and out.[4] This building overhanging the village gives importance to the place and adds considerably to its appearance which would be commonplace eno' but for its aristocratic protector. On the Lough about four miles from this Lord de Ros (Lady Cowley's brother) has a pretty place,[5] and in the neighbourhood I have found a cousin of Lady Canning's, married to an Irish landholder, so that I am amongst friends. Everyone is most kind and hospitable, well deserving the reputation on this score that their countrymen have earned. I am obliged daily to decline invitations. The place swarms with young ladies, who at our dinners here have a proportion of three to one of the ruder sex, speaking a rich brogue and having very free and somewhat universally affectionate manners, which render their society peculiarly agreeable.

The Doctor is an original and spends his whole time in his study. We work hard and have already made good progress. He is wonderfully acute and logical, and has already made greater progress than I anticipated in decyphering. He has grown up daughters, and a wife who is absent. I hope to get through my work in about ten days more and return to England.[6]

[Believe me
Yours faithfully
A. H. Layard]

MS: BL Add.38948, ff. 4–5

1   Benjamin Austen (died 1861) was Layard's uncle. See Waterfield, *Layard of Nineveh, passim*. There is a plate opposite p. 22 with portraits of Sara and Benjamin Austen by the Irish artist Daniel Maclise.

2   Layard left England on 25 September 1852 and spent a few days in Belfast.

3   Layard's short letter to Hincks, dated 23 September, was written at Ampthill. Ampthill Park was the residence of James Parke (1782–1868), Baron Wensleydale. If Layard spent the best part of nine days with the Parke family, then it makes a mockery of the excuses he gave to Hincks for his continued delay in travelling to Ireland.

4   Captain Archibald Rowan Hamilton (1818–60). His wife was Catherine Anne, *née* Caldwell (1820–1919).

5   William FitzGerald de Ros, 23rd Baron de Ros (1797–1874), was the son of Lord Henry FitzGerald and Charlotte Boyle, Baroness de Ros. His sister, the Hon. Olivia Cecilia (1807–85), married Henry

Richard Charles Wellesley, first Earl Cowley (1804–84). The principal family residence in Ireland was at Old Court, Strangford, Co. Down. The Baron is said by some to be the originator of the Oxford and Cambridge boat race.

6   Layard left Belfast on 13 October.

TO JOHN COUCH ADAMS[1]
16 OCTOBER 1852

Killyleagh C° Down
16th. Octr. 1852

My dear M^r. Adams,

I wrote the paper which accompanies this shortly after our meeting in Belfast.[2] I put it aside not knowing whereabouts you would be found. I have just met with it & think it best to send it at once to Cambridge where you will no doubt now be.[3]

Yours very truly
Edw. Hincks

[Enclosure:]

The Eclipse of 28th. May 585 was that foretold by Thales; and Herodotus erroneously supposed it to be that which terminated the Median war with the Lydians, and partially altered his history with a view to make it consistent with this; substituting Alyaltes for Sadyaltes his father as the king in whose reign it occurred. He is, however, inconsistent with himself, making it to have happened under Cyaxares (who reigned from 636 to 596) and before the capture of Nineveh by the Medes & Babylonians, which was most certainly in 626. I therefore propose for examination the eclipse of 19th. September 626 (528 lunations before the eclipse of Thales) as probably that which terminated the war. It was *total* on the earth, not annular; and if the tables in general use would make the track of the centre of the shadow to pass within any reasonable distance of Cappadocia (the probable site of the battle) the tables may be safely corrected so as to bring it there.

As for the eclipse of Thales, the tables should not be corrected so as to make it total in Cappadocia, but 'on the banks of the Hellespont', the locality indicated by Theon; nor is there any proof that it was visible to the east of this.

That Thales predicted this eclipse is certain; but two things appear certain also; one, that he did not predict it on correct principles of calculation, which he could not apply; and the other that he did not predict it by the Chaldean Sasor, as no similar eclipse occurred in 603. I conceive that he was *accidentally* right; but that an astronomical fact is assumed in his prediction which may perhaps be applied to correct the tables.

Thales seems to have known the general principles on which eclipses depend; but to have had false notions as to the relative magnitudes of the sun & moon. He probably thought that there was a greater analogy than there is between the eclipses of the sun & the moon; that solar eclipses were visible over much greater areas than they are; & that when total in one planet they should be so in others. Probably his idea was that when the moon was in the node, or within a semidiameter, at the time of conjunction, the eclipse would be total; and that the rareness of total eclipses of the sun arose from the rareness of this conjunction at the node. I take it, therefore, that the prediction of Thales was grounded on his having calculated that the conjunction of the 28th. May 585 would happen when the sun & moon were *at* the node; which he erroneously supposed to be both necessary & sufficient for a total eclipse taking place. We may assume then that on the 28th. May 585 (taking into account *mean motions* & *secular equations*, but not *periodic equations*) the ☉ & ☾ were at their conjunction *very close* to ☊.

A calculation as to this point would be interesting for the condit of Thales; & it might indicate the direction at least in which error existed in the tables. The constants of the acceleration of the moon are perhaps not ascertained with the accuracy of which they are susceptible; & the ratio of the absolute acceleration to the acceleration relative to the node is still more doubtful.

It is more important, however, to calculate this eclipse of Thales with all the equations taken into account, as they would affect the observation. The statement of Theon that it was total on the banks of the Hellespont is very definite, & it should not be neutralized by any fancies respecting the Medo-Lydian war, which are assuredly grounded on a mistake of Herodotus. If the calculated eclipse would not be total on the Hellespont, an error in the tables is certain.

It would be well worthwhile also to calculate the eclipse of Sept 19. 628; rudely in the first instance, but if it should turn out that it would be central near Asia Minor, then with accuracy. There would be another eclipse Sept 30. 629 which might also be tried. Indeed, I think it is the more likely of the two to answer the conditions, provided that the cone of the shadow would reach the earth. I have no tables whatever to refer to; & therefore can do no more than suggest matters to others, who may be able to go through the calculations required.

MS: St JCLC/Adams 9/22/1

1   John Couch Adams (1819–92), astronomer and mathematician, was president of the Royal Astronomical Society from 1851 to 1853. He was a fellow of St John's College, Cambridge, 1843–52, and a fellow of Pembroke College, Cambridge from 1853.

2   The entry for 3 September in Hincks's diary says: 'Saw Adams, and spoke to him about the Eclipse of Thales, and that of the Lydian War, which I maintain to be different.'

3   Hincks wrote the paper in mid-September. His diary says on 16 September: 'All day at calculations connected with the Eclipse of Thales', and on 17 September: 'At Eclipse of Lydo-Median War – finished paper on the eclipse, to be substituted for the letter to Adams, which I had written, and now burned.'

⨂

FROM AUSTEN HENRY LAYARD
16 OCTOBER 1852

Athenaeum Club
Octr. 16/52

My dear D^r. Hincks,

I reached London safely on Thursday night, having left Belfast, as I had intended, on Wednesday by the Fleetwood Boat.[1] Yesterday I went to the Museum and regret to find that there is no meeting of the Trustees before the middle of next month, nor any means whatever of having anything done in the meanwhile. I fear, therefore, that we must again wait. As soon as Parliament meets I shall be able to speak to several of the Trustees and will then push the matter on as much as possible.

I looked over a few of the tablets in the Museum and I am inclined to think that there are other fragments with the tables of characters & their equivalents. I see a letter from Rawlinson in which he states that the French had found two perfect cylinders at Khorsabad of Sargon, which he believed would give the entire Samaritan campaign, & that he (Rawlinson) had found annals of Esarhaddon (he does not say where) giving an account of that monarch's conquest of Egypt & the Aethiopians. He also says that he has found data which make the Assyrian annals correspond exactly with Ptolemy's tables & that he can fix positively the accession of Sennacherib to 702 B.C.[2] Nothing has been done in Babylonia.

Pray give my kindest regards to your daughters &

Believe me
Yours very faithfully,
A. H. Layard

MS: GIO/H 228

1   Layard left Killyleagh on 11 October and Belfast on 13 October. Hincks wrote in his diary: 'He has got a great deal from me, for which I hope he will give me proper credit. I have trusted entirely to his honour.' Layard fulfilled the promise he made in his letter of 8 January 1852. In *Discoveries in the Ruins of Nineveh and Babylon*, he fully acknowledged his indebtedness to Hincks: 'I must here remind the reader that any new discoveries in the cuneiform inscriptions referred to in the text are to be attributed to Dr Hincks' (p. 139).

2   Sennacherib's accession date is 705 B.C., so Hincks's 703 B.C. and Rawlinson's 702 B.C. were close.

TO AUSTEN HENRY LAYARD
19 OCTOBER 1852

Killyleagh C° Down
19th. Octr. 1852

My dear Sir,

I received your letter yesterday morning. I am sorry that there is so little prospect of my having the power to examine the Museum's treasures where I have no doubt I should find a rich harvest. I see by today's paper that Parliament is to recess much earlier than was expected. Even so, however, arrangements could not be made so as to give me much more than a week before Christmas. I believe I have mentioned to you that *this year* I would go without a greater remuneration than would defray my expenses in London & payment to curate here.

With respect to editing, as it might activate jealousy, you might state, if the matter was spoken of, that I would not expect to be the *nominal* editor. M^r. Hawkins' name as that of the head of the department would appear as in the volumes already issued.[1]

I am hurrying forward my Irish Academy paper; but printing is slow here – 5 sheets, as I estimate, are yet to be printed, of which three are not in type; the others are in process of revision.[2] I hope you have not mentioned the progress of this paper at the Asiatic Society. I would like to have it out before Rawlinson's next publication.

I have been so much hurried since you left that I have not had time to go over the different copies of Sennacherib's annals. The recovery of Sargon's is a great point; & I think the French would be more likely to hasten the publication than Rawlinson, who would keep it to himself. I have no doubt the conquest of Samaria in the second year will be recorded in them.

I see Rawlinson has adopted my views as to the accession of Sennacherib. He differs *one* year, owing to his taking a different view from what I do of the arrangement of Ptolemy's canon. It is wise if my letter of the 26th. July to the Editor of Athenaeum, which he *burned*, was not shewn to some body who communicated its contents to Rawlinson.[3]

I send you some notes of Sharman Crawford's on his bill, which M^r. Martin thought would interest you.[4]

Believe me
Yours very faithfully
Edw. Hincks

In the BM series pl. 19, Esarhaddon calls himself 'king of Egypt, conqueror of Ethiopia', *sar Miṣir kamish Milukhe.*[5]

MS: BL Add. 38981, ff. 185–6

1   Edward Hawkins, keeper in the Department of Antiquities at the British Museum.
2   'On the Assyrio-Babylonian Phonetic Characters', *Transactions of the Royal Irish Academy* 22 (1852), polite literature, pp. 293–370.
3   See Hincks's letter of 2 August 1852, esp. n. 4.
4   William Sharman Crawford (1781–1861), politician and landlord in Co. Down, was M.P. for Dundalk in 1835 and for Rochdale from 1841 to 1852. He favoured Catholic emancipation and was a strong supporter of tenants' rights in Ulster. He failed to be elected in his own Co. Down in 1852 and William Shee (1804–68), an Irish politician and judge, reintroduced the Tenant Right Bill on 25 November 1852. See *ODNB*.
5   The phrase is *šar* ^*mat*^*muṣur kāmû šar* ^*mat*^*meluḫ*, 'King of Egypt, who defeated the king of Meluhha (= Kush).' It is found on the reverse side of a colossus in the Southwest Palace at Nimrud (Kalḫu). See Russell, *The Writing on the Wall*, pp. 291–2.

FROM HENRY TATTAM[1]
4 NOVEMBER 1852

Stanford Rivers,
Romford, Essex,
Novr. 4th. 1852

My dear Sir,

Years have passed away since I had the pleasure of hearing from you, and since I last addressed you.[2] But I venture to hope you will pardon the liberty I am taking in addressing you on the present occasion.

I requested our mutual friend D[r]. Lee, when coming to Belfast, to ask you if you had seen M[r]. Forster's work on the Sinaitic Inscriptions, and if so, to further ask you what is your opinion respecting his translation of them?[3] But he either forgot the message, or had not the opportunity of putting the question to you. I should here have let the subject, had not Sir Harry Verney,[4] who knew of my message by D[r]. Lee, and who has taken a deep interest in the subject, which has been increased by the Chevalier Bunsen, who has been visiting him, having given him his views on the subject, wished to know the result. May I venture to ask you whether you have seen M[r]. Forster's publications, and what you think of them? They have not come in my way, and I do not like to purchase them, without some opinion on the subject.

D[r]. Lee spoke in the warmest terms of commendation of your very interesting paper read before the meeting in Belfast, which he said was published, and has promised me I should see it, but I have not yet had that pleasure.[5]

My health has not been good the last three years, but it is now much better. I however finished the publication of the Major Prophets in Coptic and Latin,[6] and am now preparing for a second edition of my Lexicon, which I hope, if I am spared, to render as perfect as possible.[7]

I remain, My dear Sir,
yours very truly,
Henry Tattam

MS: GIO/H 537

1   Henry Tattam (1789–1868), Coptic scholar, was archdeacon of Bedford from 1845 to 1866, but he was non-resident most of the time, for in 1849 he was presented by the crown with the living of Stanford Rivers in Essex.

2   There are letters from Tattam for the years 1833 and 1841. See vol. 1.

3   Tattam was not aware obviously of Hincks's review article, 'The One Primeval Language', a review of C. Forster, *The One Primeval Language Traced Experimentally through Ancient Inscriptions* . . . (London, 1851), in *Dublin University Magazine* 39 (Feb. 1852), pp. 226–34. See Renouard's letter of 10 February 1852.

4   Sir Harry Verney (formerly Calvert) (1801–94), politician.

5   See Hincks's letter of 3 September 1852 to Layard.

6   *Prophetae Maiores in dialecto linguae Aegyptiacae Memphitica seu Coptica*, 2 vols (Oxford, 1852).

7   A second edition was not published.

FROM AUSTEN HENRY LAYARD
5 NOVEMBER 1852

Athenaeum
Nov[r]. 5/52

My dear Sir,
    I must apologise for having left your letter so long unanswered. I had scarcely arrived in London when I was summoned to Wales and have been there ever since, much occupied with various matters.[1] It was only yesterday that I returned to town to take my seat in Parliament which I have done today. I will now lose no time in seeing Sir R. Inglis & others who may be useful in promoting your views, but I fear that it will be utterly impossible to do anything before Christmas. No one of the Trustees can take it upon himself to make such an arrangement as you propose, nor, am I assured by Sir Henry Ellis, could it be done without a meeting of a considerable number & not a mere business meeting. I cannot understand this nor indeed can I understand many other of their arrangements. You may depend upon it that I will not let the occasion pass if I can be of the slightest use & I will write again on the subject the moment I have anything to communicate.

As far as I can hear you need have no fear of being anticipated in anything by Rawlinson. I have not mentioned your forthcoming work. I shall be glad to see it out as I think you ought to have the credit so justly due to you for your discoveries. I am now printing & I will send you such proof sheets relating to matters you were good eno' to help me in as soon as I receive them.

I shall be very glad of any account of the edifices from the inscriptions or any other information you can kindly send me. I do not think anything of Rawlinson's will be published before Christmas. I have taken every opportunity of attributing to their right source many of the discoveries of which you had been deprived & shall continue to do so.

Your daughters will also have thought that I have soon forgotten my promises to them, but such is not the case. I have promised them autographs they asked for and hope to send them one day. I have not yet had time to make copies of the cylinders in the B.M.

With kind regards to all the members of your family,

I am, yours sincerely,
A. H. Layard

MS: GIO/H 229

1   After his return from Ireland, Layard spent some time with Lady Charlotte Guest and her family. Sir John Guest had become seriously ill in early September and on the doctor's advice he was moved to his house in Dowlais, Glamorgan. He died of kidney failure on 26 November.

FROM AUSTEN HENRY LAYARD
8 NOVEMBER 1852

9 Little Ryder Street
Novr. 8/52

My dear D^r. Hincks,

I send you a few proof sheets of my forthcoming book which particularly refer to you. I should be much obliged if you would kindly cast your eye over them and see whether I have stated anything that you do not approve or is not consistent with the fact. As I am printing as fast as I can to have [it] by Christmas will you kindly return me the sheets as soon as you conveniently can? I will keep back the printing until I hear from you. If you have done anything more with the Bavian inscription or with the account of the building of the palaces or the bulls I should indeed be glad of any information.

I saw Sir Robert Inglis on Saturday & it was agreed that I should address a strong letter to the Trustees next Saturday (their first day of meeting) on the subject of your offer.[1] I shall do this & do trust with some results.

I hope M[rs]. Hincks & your daughters are well; with kind regards to them,

I am
Yours very truly
A. H. Layard

Will you kindly thank M[r]. Martin, with my compliments, for the notes on Crawford's Tenant Right Bill he sent me. Am I to return them?

MS: GIO/H 230

---

1   Sir Robert Henry Inglis (1786–1855), M.P. for Oxford University, was one of Hincks's supporters. See Thomas Romney Robinson's letter dated June 1850.

TO AUSTEN HENRY LAYARD
10 NOVEMBER 1852

Killyleagh
10th. Nov[r]. 1852

My dear M[r]. Layard,
    I received your packet this morning. I feel much obliged to you for the very handsome manner you have spoken of me. I must endeavour to deserve it. I will send you tomorrow the sheet p. 145 &c; today I can only send the two leaves. I have suggested some corrections. I wish to have them such as not to disturb the page. A note added in 126 will make up for what must be omitted in the present note. Suppose you make the latter:
Compare the Hebrew *khoresh*, a thick wood. Or perhaps quarries; compare &c.[1]
    If it were left to me, however, I would dismiss this last suggestion altogether & you would then have some little room at the top of the next page for bringing in the name Kauser as a possible or probable derivation of Khasri – the last vowel of which is not radical. You might also omit the two lines above the 3[d] fragment.

In great haste.
Yours very truly
Edw. Hincks

Last line in p. 125, transports *thither* (?). The doubt only attaches to the last word.[2]
p. 126, l. 12
which, as the gods* willed, were found in the land of Balad for the *walls* &c.
*A peculiar deity is mentioned who probably presided over the earth, but his name is as yet unknown; it is here denoted by a monogram.
l.16
the people of the *forests* (?) 'kharshane'
The note in this page must be *shortened* & *modified*. Kharshane is applied to places not persons.
p. 127, l. 5 through the Kharri (or Khasri).
Please modify the next sentence so as to make the reading Kauser your own. It is the same river mentioned on the 3ᵈ fragment l. 30 after 'brought up'; you may insert 'on sledges (?) <boats(?)> I caused to mount': (the verb used is elsewhere applied to embarking in ships – & to riding on horses).[3]
l. 23 taken up (to the top of the mound) from the Tigris – no *doubt* should be expressed; the words in the parentheses are explanatory.

MS: BL Add. 38981, ff. 147–8

---

1   See Layard, *Discoveries in the Ruins of Nineveh and Babylon*, p. 117, n. ‡: 'Compare the Hebrew ḥōresh, khersh, a thick wood, or, perhaps ḥārāsh, a stonecutter, or a workman in stone or wood.'
2   These notes are found in Layard, *Discoveries in the Ruins of Nineveh and Babylon*, pp. 117–18.
3   See Layard, *Discoveries in the Ruins of Nineveh and Babylon*, p. 118: 'Sennacherib, king of Assyria . . . (some object, the nature not ascertained) of wood, which from the Tigris I caused to be brought up (through?) the Kharri, or Khasri, on sledges (or boats), I caused to be carried (or to mount).'

AUSTEN HENRY LAYARD TO HENRY ELLIS
II NOVEMBER 1852

9 Little Ryder Street
Novʳ. 11 1852

Sir,
    The Trustees, I am informed, signified some time ago their wish that the inscriptions brought from Assyria last year should be prepared for the press and published. I much regret that important occupations have hitherto prevented me fulfilling my promise to put them into such form as would allow of immediate printing. I now venture to suggest to the Trustees that it would be of very great advantage if the services of the Revᵈ. Dʳ. Hincks could be secured not only for the superintendence of the publication of the inscriptions copied by me, but for

copying the inscribed tablets now in the British Museum. During a recent residence with D^r. Hincks I have been able to appreciate that extraordinary sagacity and learning which have enabled him to make the most important discoveries that have hitherto been made in the investigation of the cuneiform characters. To show the highly interesting & valuable nature of the inscribed nature of the tablets in the Museum I need only mention that having by chance copies of the fragments with me in Ireland D^r. Hincks discovered them to be a table of the alphabetic values of the cuneiform characters, and a calendar. It can scarcely be doubted that an examination of the large collection in the British Museum will lead to the most important results.

D^r. Hincks has offered to come to London at once and to commence the superintendence of the publication of the inscriptions upon the payment due of his expenses, and I am authorised on his behalf to make that offer for the remainder of this year. The Trustees might then be able to judge how far his services might be considered useful.

May I be allowed to suggest that instructions should be now given to M^r. Hormuzd Rassam to move the interesting bas reliefs that want of means compelled me to leave behind?[1]

I am Sir
Yours very obediently
A. H. Layard

MS: BM/CA/OP 48:24

1 Hormuzd Rassam (1826–1910) was born in Mosul of Chaldaean Christian parents. He assisted Layard in his excavations in Mesopotamia, and afterwards he undertook explorations for the British Museum.

TO AUSTEN HENRY LAYARD
17 NOVEMBER 1852

Killyleagh
17th. Novr. 1852

My dear M^r. Layard,

I send you on the other side a paragraph to be introduced in p. 228 & have made some other corrections on the page itself. I hope the printers will be able to make it out.

Is not the Khauser *itself* the canal dug by Sennacherib, & wholly (or almost &c) an artificial stream? I think there is no *reservoir* mentioned but what is called the Khusur or Ussur (one inscription has it one way & another the other). It might be better therefore to alter what you have said where I have put a pencil query.

Yours very truly
Edw. Hincks

To be introduced at this mark in p. 227.[1]
*After mentioning some canals which he had made in the south of Assyria, he speaks of the army which defended the workpeople being attacked by the king of Elam and the king of Babylon with many kings of the hills and the plains who were their allies. He defeated them in the neighbourhood of Khalul (site unde-termined). Many of the great people of the king of Elam & the son of the king of Karduniyas were either killed or taken prisoners, while the kings themselves fled to their respective countries. Sennacherib then mentions his advance to Babylon, his conquest and plunder of it, and concludes with saying

MS: BL Add. 38981, ff. 160–1

1   See Layard, *Discoveries in the Ruins of Nineveh and Babylon*, p. 212.

⁓⃑≍⃑⁓

HENRY ELLIS TO AUSTEN HENRY LAYARD
17 NOVEMBER 1852

Dear Sir,
     The Trustees had under their consideration at their last meeting, your letter of 11th. inst. in which you suggest the advantage of engaging the services of D[r]. Hincks for the superintendence of the publication of Assyrian inscriptions, and for copying the inscribed tablets now in the Museum; and I am directed by the Trustees to acquaint you, that they are prepared to consider the proposal for employing D[r]. Hincks upon some arrangement to be definitely fixed, and request that you will be so good and to communicate with D[r]. Hincks, and acquaint the Trustees with the result.

Believe me, Dear Sir,
Yours very faithfully
Henry Ellis

MS: BL Add. 38981, f. 162

FROM AUSTEN HENRY LAYARD
17 NOVEMBER 1852

Wednesday 17th Nov^r./52
9 Little Ryder St

My dear D^r. Hincks,

I have only this evening received an answer to my communication to the Trustees of the B.M. in these words: 'The Trustees are prepared to consider the proposal for employing D^r. Hincks upon some arrangement to be definitely fixed, and request that you will be good eno' to communicate with D^r. Hincks, & acquaint the Trustees with the result.'

Will you, therefore, consider the subject & write such a letter as I can place before the Trustees. I am afraid it would not be worth your while to come up before Christmas but if upon the strength of the letter you would like to come at once & work at the inscriptions trusting to having your expenses paid I think it might be managed but I have *no authority* whatever for saying so. If important results were shown to the Trustees I scarcely think they would hesitate. However you are the best judge.

Yours very faithfully,
A. H. Layard

MS: GIO/H 231

TO AUSTEN HENRY LAYARD
20 NOVEMBER 1852

Killyleagh
20th. Novr. 1852

My Dear M^r. Layard,

I received your letter this morning, containing an indication that 'the Trustees of the British Museum were prepared to consider a proposal for employing me upon some arrangement to be definitively fixed'.

The basis of any such arrangement must be that I should be employed in London for *a portion of the year*, and at Killyleagh for *the remainder*. In London I would of course do far more than here; but *here* I must be for nine months in the year, and it strikes me that I can do a great deal here in connexion with the publication of any inscriptions in the Museum that may be considered desirable. They would have to be *prepared for the press* in many instances, as well as *superintended through it*. Some inscriptions, for instance, are in characters resembling no manuscript; and an attempt to *imitate* them by *types* must prove a failure. The only way to publish them would be either to lithograph a facsimile or to substitute for the characters analogous to [. . .][1] the corresponding characters analogous to printing, which are used in other inscriptions. I mention this as a specimen of what might be done. Having rubbings or photographs of the inscriptions here, I might copy them out for the press.

It appears to me that if such an arrangement were to be adopted, I ought at the commencement of it to go to London for a short time to look over the collection, mark what should be copied, and what would be most important for publication; but that the main part of my residence in London had better be when the days were long. It would favour such an arrangement as this, that any short time I could spend in London in 1852 would not interfere with my being three months absent in 1853.

What I would then propose is that I should be engaged for a year from 1st. December 1852 to spend 13 weeks in London, of which at least 10 should be at such time in 1853 as the Trustees thought best – the remainder in December.

Being quite ignorant of the salaries given at the Museum, I would not like to propose any sum for the period of my being in London; but I should be satisfied with £75 for the *remainder* of the year; as, though I could do something while here, I do not suppose that it would be very much. It would, however, be necessary that I should engage a curate for the *entire year*.

I beg you will communicate this to the Trustees. Should the arrangement be entered into, it would be highly desirable that I should leave this so as to be in London on the 1st. Decr. as I could not leave it later than the 22nd.

Believe me to remain
Yours very faithfully
Edw. Hincks

MS: BM/CA/OP 48

1   One word is illegible.

FROM HENRY TATTAM
20 NOVEMBER 1852

Stanford Rivers
Romford,
Novr. 20th. 1852

My dear Sir,

I beg you to accept my best thanks for your very acceptable, and very satisfactory letter, which contained all that I could wish. It is a great pity that persons who are so vile should be suffered to put their productions before the world.[1]

I have for some time thought of asking you to put forth a publication on the discoveries of Nineveh, in connection with its history, particularly its sacred history, as being the only person who is capable of doing justice to the subject; and such work is absolutely required by the reading public, and I have no doubt would fully remunerate you in a pecuniary point of view. But I have just heard that a M^r. Blackburn, a Dissenter, has put forth a 1^s 6^d book on the subject, which takes and sells well; and he has published a second or third edition.[2] I have ordered the book, to see what it is, but I cannot think that his publication is the thing I wish to see. He cannot have done justice to the subject.

I do not know whether you ever come to England. If you do, I shall be most happy to see you here, and we can at all times insure you a bed. I was in a very bad state of health when I came here, and up to the present year, but I am thankful to say I am now tolerably well.

I am writing with an infamous pen, pray excuse it.

I remain, My dear Sir,
yours very truly,
Henry Tattam

MS: GIO/H 538

1   When Hincks replied to Tattam's letter of 4 November 1852, he must have dealt with the works of Charles Forster. However, Tattam's use of the word 'vile' was entirely unjustified.
2   J. Blackburn, *Nineveh: its Rise and Ruin; as Illustrated by Ancient Scriptures and Modern Discoveries* (rev. and enl. edn; London, 1852).

TO AUSTEN HENRY LAYARD
29 NOVEMBER 1852

Killyleagh C° Down
29th. Novr. 1852

My dear M^r. Layard,

I received yesterday a letter from D^r. Robinson which I wish you to *see*, that you may know how matters stand with respect to the application to the Gov^t. through Lord Naas[1] & M^r. Napier,[2] which originated at the meeting of the British Association.

What is stated in the second page as the substance of the answer to the application may, I think, be extracted for *future use*, should an opportunity present itself. *At present*, I don't think it would be advisable to take any step in the matter.

The pretense about sanctioning non-residence is clearly not the real reason of refusal. The non-residence proposed would not amount to three months in the year, *which the law allows.* That the Government have no horror of a clergyman's non-residence within this limit is evident from the fact, that at the time they gave this answer they were appointing (& with very general approbation) a clergyman to the Bishopric of Meath, who was notoriously non-resident on his benefice for a much longer time than I should &c; though within the limits recognized by law.[3]

What the *real* objection is I cannot of course know. It *may be* what D^r. R. suggests; but I should think the intrigues of Col. Rawlinson's friends, who would I am sure be glad to extinguish me, a more likely cause.

You will return the letter to me when read. It ought not to go beyond yourself. You of course know that D^r. Robinson is President of the Royal Irish Academy & was President of the British Association in 1849. He is perhaps the very first man in Ireland in respect to science & literature, & has always been a Tory in his politics, so that the refusal of the Gov^t. could not have arisen from disregard to *him*.

From the Museum I have heard nothing. I could not leave this till this day week at soonest which would give me but a fortnight now in London. This indeed would be of little consequence as I could very well give a week more in the Spring. I fear, however, that I may infer from the delay that my proposal is rejected; & indeed I think it most likely that nothing would be done without the sanction of government, which it seems is not to be had.

I have transmitted a proof of my last sheet to M^r. Bosanquet, with directions to forward it to you. It contains my chronological views & a statement respecting the number of characters valued – classified as mine & Rawlinson's. You can keep it & can perhaps find a place for this statement in your work, which I see that Murray announces for December.

Believe me
Yours very faithfully
Edw. Hincks

MS: BL Add. 38981, ff. 170–1

1   Richard Southwell Bourke, sixth earl of Mayo (1822–72) assumed the courtesy title Lord Naas when his father succeeded to the earldom of Mayo. In 1852 he was appointed chief secretary for Ireland in Lord Derby's administration. See *ODNB*.
2   Joseph Napier was the attorney-general for Ireland.
3   Joseph Henderson Singer (1786–1866), an evangelical Church of Ireland clergyman, was appointed Bishop of Meath in 1852. He was rector of Raymochy in the diocese of Raphoe in 1850 and archdeacon in 1851. But he was a fellow of Trinity College, Dublin and became Regius Professor of Divinity in 1850.

FROM HENRY ELLIS
30 NOVEMBER 1852

[British Museum]
30ᵗʰ. November 1852

Sir,
    At a meeting of the Trustees of the British Museum on Saturday last, I was directed to write to acquaint you that the Trustees will be glad to confer personally with you in the event of your being able to attend them in London either in the early part of December next or in February next, when you can examine the Assyrian remains in their possession. The Trustees will then be enabled to form an opinion as to the course which it would be desirable to pursue with respect to the proposal which you have submitted to them, they being willing under any circumstances to defray the expense of your journey and of your residence in London for four weeks.

I have the honor to be, Sir,
Your most obedient servant,
Henry Ellis
Principal Librarian

MS: BM/CA/LB 43:6 (copy)

꿍

FROM AUSTEN HENRY LAYARD
1 DECEMBER 1852

H. of C.[1] Dec[r]. 1/52

My dear D[r]. Hincks,

I return you D[r]. Robinson's note. I very greatly regret that the Government have shown such a very illiberal & narrow-minded feeling, if, as D[r]. R. suggests, the motive of their refusal is the position your brother holds in Canada.[2] As we all know the excuse they offer is absurd eno'. I have sent your letter to the Trustees of the B.M. but have not heard from them since. I will not lose a moment in letting you know when I do. I have good hope that something may be managed in that quarter. I was talking to Lord Dufferin about you last night and he hinted at a proposal which appeared to me not quite unworthy of consideration.[3] I will see him again and you may depend upon my doing all in my power to obtain what I can but consider a just acknowledgement of the services you have rendered to literature. I do not think Rawlinson's friends have any influence that they could bring to bear against you – even should they desire to stand in the way of your advancement which I doubt.

I should be very glad of your paper and will certainly make use of it.[4] I hope M[r]. Bosanquet will let me have it soon. I have printed half my book.

Kind regards to your daughters & M[rs]. Hincks.

Yours very faithfully,
A. H. Layard

MS: GIO/H 232

1   H. of C. = House of Commons
2   Francis Hincks was co-premier (with Augustin-Norbert Morin) of Canada. See Hincks's letter of 8 May 1852 to Layard, especially n. 4.
3   Frederick Temple Hamilton-Temple Blackwood (1826–1902), first marquess of Dufferin and Ava. See Hincks's letter of December 1851 to Campbell Allen.
4   'On the Assyrio-Babylonian Phonetic Characters', *Transactions of the Royal Irish Academy* 22 (1852), polite literature, pp. 293–370.

༭༒

TO AUSTEN HENRY LAYARD
4 DECEMBER 1852

Killyleagh C° Down
4th. Decr. 1852

My dear M^r. Layard,

I received your letter yesterday. I am glad you think Rawlinson's friends had no hand in the negative given by Lord Derby. I cannot think he was influenced by any hostility to my brother; but it is possible that my brother's letter (injudiciously written I fear, but which I never saw) may have led him to think that I contemplated having *permanent* employment in London along with this living, which *I never did contemplate. One* refusal may probably be considered a bar to all further applications for the same object.

The Museum Trustees have preferred communicating directly with myself to making you their medium. I had a letter from Sir H. Ellis on the 2nd. which I enclose for your perusal, as you may not have heard from them. The last paragraph is not quite satisfactory as it says nothing of the expense of a temporary curate whom I would have to engage while away, & which would cost me 8 or 10 pounds. Of this, however, I have taken no notice in my reply, in which I have stated that I could not *now* be more than 10 days in London before Christmas & that I therefore prefer waiting till February – before which I would communicate with him as to the precise time.

I have satisfied myself that the Khussur mentioned in the Bavian inscription must be the River Khaussur[1] – & that the *kharru* or *khasru* mentioned in the epigraphs as what the stones were drawn up or from was an artificial conduit between this & the Tigris running I presume to the N. of Kouyunjik. I am pretty sure that Sennacherib speaks of gardens and of tanks or fountains (*beerat*) which were fed by water conveyed *from* the Khussur, *through* this *kharru*.[2] As to the palace itself I have made little progress in translating what is said of it. There are several important words, on which the meaning depends, which I do not yet understand. A comparison with other inscriptions would probably clear the matter up; but I have not *data* here & indeed my time is a good deal occupied just at present with parochial matters.

Believe me
Yours very faithfully
Edw. Hincks

MS: BL Add. 38981, ff. 178–9

1   The river Khosr is a tributary of the Tigris.
2   Akkadian *ḫarru* = 'water channel', 'moat'; *ḫarāru* = 'to hollow out'. See *The Assyrian Dictionary*, vol. Ḫ (Chicago, 1956), pp. 114–15 sub *ḫarru* A.

FROM AUSTEN HENRY LAYARD
7 DECEMBER 1852

9 Little Ryder S^t
Decr. 7/52

My dear D^r. Hincks,

I am glad to find that we are getting nearer to something definite. It is probable that the Trustees misunderstood my last communication, in which I urged them to write to you without delay, and fancied that I had declined being a medium. However, it is perhaps as well that they should be in direct correspondence with you, and I hope you will find it possible, after Christmas, to have a personal interview with them & point out what is to be done. They do seem to be inclined to help and I will not neglect any opportunity of supporting you. I return you their letter.

Believe me,
Yours very truly
A. H. Layard

MS: GIO/H 233

FROM JAMES WHATMAN BOSANQUET
8 DECEMBER 1852

Claysmore
8. Decr. 1852

My dear Sir

I thank you much for the proof sheet of part of your forthcoming paper on Assyrian Antiquities, and have forwarded it to M^r. Layard, as you wished.[1] I am very anxious to see the remaining portion of your paper which you kindly promise to send me. You appear to have made an important discovery, in fixing the position of the reign of Sargon, but I still doubt whether the consequences you draw from it are necessary. Sargon conquers Merodach Baladan, *son of Yakin*. The presumption is, at first sight, therefore, that this Merodach is not the *son of Baladan* mentioned in Scripture. When also we find that the 14th. year of Hezekiah, in this view, does not fall within the reign of Senacherib, the presumption becomes still stronger. There is also a little awkwardness in arranging the details of the conquest of Samaria, and the reign of Esarhaddon in Babylon, supposing his father to have reigned as many as 10 years.

These difficulties, however, all disappear, if, leaving Sargon & Mardoc-Empadus as you place them, you take Mesesse-Merodac to be the *son of Baladan* of Scripture. *The third year of Senacherib*, who conquers this king in his own first year, will then fall in the very year which Demetrius, 200 years before Christ, fixed as the true date of the Senacherib's attack on Jerusalem, in the 14th. year of Hezekiah, viz. Feb. B.C. 688. Giving also about 16 years to the reign of Senacherib, and 8 to Esarhaddon, the death of this latter king will fall in the very year of the death of Asaradinas of the Canon. This view agrees well with the date of the eclipse of Thales also.

Believe me
Yours very truly
J. W. Bosanquet

MS: GIO/H 61

1   See Hincks, 'On the Assyrio-Babylonian Phonetic Characters', *Transactions of the Royal Irish Academy* 22 (1852), polite literature, pp. 293–370, esp. pp. 364–70 on the matters discussed in this letter.

FROM JOHN COUCH ADAMS
10 DECEMBER 1852

S^t. John's Coll.
Decr. 10 1852

My dear Sir,
    I am much obliged to you for your letter respecting the Eclipse of Thales, about which I feel much interest.[1] The Astronomer Royal has been making some very elaborate calculations lately on the subject & in conversation with him a short time since, I mentioned your idea that the Eclipse of Thales was not that described by Herodotus, & also stated that you had met with a passage in Theon which asserts that the eclipse of Thales was total at the Hellespont. He was not aware of the eclipse being alluded to by Theon, & much wished to have a reference to the passage. I shall therefore be much indebted to you, if you will kindly communicate this reference. I

am loth to admit the mistake of Herodotus. Perhaps the two accounts are not incompatible. Is anything very definite known about the locality of the battle?

Yours very truly
J. C. Adams

MS: GIO/H1

1   See Hincks's letter of 16 October 1852.

TO AUSTEN HENRY LAYARD
11 DECEMBER 1852

Killyleagh
11th. Dec[r]. 1852

My dear M[r]. Layard,

I send you back the proof – I have made a good many changes. I have also put some things on a separate sheet which you can attend to or not as you think best. You will I expect have a letter from Dublin giving you the page of my paper on the Khorsabad inscriptions where I give the name of Sennacherib.[1] I don't think I read it in the Van paper but I have not a copy of that at hand.[2]

I hope to send you a proof of the last sheet of my paper containing the chronological part in a day or two.[3] I have referred to it in the proof. I think my paper will be out certainly this month. There is now only one sheet containing cuneatic characters & one of ordinary printing remaining: & both of these are in a forward state. In the hope that something may be done on Saturday, as the result of your letters, I have been making enquiries as to the possibility of obtaining assistance in my parish between this & Christmas. I will not commit myself, however, till I hear. I am now working at the Khorsabad bull inscription – for a comparison of which with the Kouyunjik bulls & with the great inscription at the India House I hope to clear up the architecture – but there are many important words of which I do not yet know the meaning.

Yours very truly
Edw. Hincks

MS: BL Add. 38981, ff. 180–1

1   'On the Khorsabad Inscriptions', *Transactions of the Royal Irish Academy* 22 (1850), polite literature, pp. 34–5.
2   'On the Inscriptions at Van', *Journal of the Royal Asiatic Society* 9 (1848), pp. 387–449.
3   'On the Assyrio-Babylonian Phonetic Characters', *Transactions of the Royal Irish Academy* 22 (1852), polite literature, pp. 293–70. The 'chronological part' is on pp. 364–70.

～∂❀∂～

FROM EDWARD CLIBBORN
17 DECEMBER 1852

Royal Irish Academy
19, Dawson S^t. Dublin
17 Dec. 1852

Dear Sir,

In reply to yours rec^d. today [I] have to say in the first place that I am very ill with influenza & am only half alive to business &c. &c.[1] I am ready to take charge of any copies of your paper & do with them as you may desire.

The part of the Trans. is not only closed with your paper but the vol. also.[2] In the course of next week I calculate on its being widely diffused amongst our members and also amongst the public institutions. Though there are great numbers of persons who take a great interest in the progress you are making yet there are few who have set themselves to work so as to be able to understand or feel the importance of your labours. Here the inscriptions which your alphabets enable people to read are *not* to be found so your essays are like keys the locks of which are missing. I will ask at Hodges & Smith's & try & find out what non-residents may be anxious to get the papers. I think the Roman Catholic people at Maynooth[3] might like a copy & the Jesuits at Clongowes.[4] If there were any people in either institution likely to take the thing up, I think I could find them out thro' M^r. Curry or some other parts in Dublin.[5] There is a very Irish Review published in Dublin by one Kelly in Grafton S^t; its Editor M^r. Gilbert is a very fine clever young man most willing to put his fingers into John Bull's pockets & make him shell out other peoples goods.[6] I wish we had a popular sketch of what you have done in the arrow head writing showing how without the advantages of Major Rawlinson you had done so & so – the case for this success *must* be made eminently *national*. I think Gilbert would do great justice to the subject if he had the materials to work upon, & it would do good in many ways. I hope to see him soon, when I am able to be out & will tell him what a grand grievance we may make out of this business & hear his opinion on the matter.

I have not seen the work of M^r. Kitto's you refer to.[7] There is some mistake about its title. I have seen two but neither of them was the right one.

I am tired & must conclude.
Yours &c.
E. Clibborn

MS: GIO/H 117

1   Edward Clibborn was assistant secretary and assistant librarian at the Royal Irish Academy for nearly forty years.

2   'On the Assyrio-Babylonian Phonetic Characters', *Transactions of the Royal Irish Academy* 22 (1852), polite literature, pp. 293–370. This paper was read to the Academy on 24 May 1852. A postscript is dated 5 November. It was published separately as *A List of Assyrio-Babylonian Characters and their Phonetic Values* (Dublin, 1852).

3   Maynooth College was founded in 1795 as a seminary for the education of Catholic priests.

4   Clongowes Wood College was founded by the Society of Jesus (Jesuits) in 1814.

5   Eugene O'Curry (1794–1862), a distinguished Irish scholar, became Professor of Archaeology and Irish History at the Catholic University in Dublin in 1854.

6   John Thomas Gilbert (1829–98), historian and antiquary, was among the founders of the *Irish Quarterly Review* in 1851. In volumes 2 and 3 of this review, from March 1852 to September 1853, he published eight articles on 'The Streets of Dublin', which later formed the basis of his *History of the City of Dublin*, 3 vols (Dublin, 1854–9). He was a member of the Irish Archaeological and Celtic Society and of the Royal Irish Academy. His widow, Rosa Mulholland, Lady Gilbert, wrote his biography: *Life of Sir John T. Gilbert* (London, 1905). See further Mary Clark, Yvonne Desmond and Nodlaig P. Hardiman (eds), *Sir John T. Gilbert 1829–1898: Historian, Archivist and Librarian. Papers and Letters Delivered during the Centenary Year, 1998* (Dublin, 1999); and *ODNB*.

7   John Kitto (1804–54), writer and missionary, edited the *Journal of Sacred Literature* from 1845 to 1853.

FROM AUSTEN HENRY LAYARD
21 DECEMBER 1852

9 Little Ryder S^t
Decr. 21/52

My dear D^r. Hincks,

Would you kindly give me a slight notion of the contents of the great Nebuchadnezzar tablets in the India House? I merely wish to know what they are supposed to comprise as I indirectly allude to them in my account of Babylon.[1]

There is no news yet as to the probable form of the new ministry. Lord John is very averse to joining a Peelite government & there is on the whole much disinclination on the part of the Liberal party to merge with that which was originally a conservative section.[2]

I hope that your family are well. Pray give them my very kind regards. I am not getting on as quickly as I might with my book. It is the printers' fault.

Have you done anything about the architecture?

Yours very truly,
A. H. Layard

MS: GIO/H 234

1   See Layard, *Discoveries in the Ruins of Nineveh and Babylon*, pp. 529–30.
2   The government of Lord Derby fell in December 1852.

TO AUSTEN HENRY LAYARD
23 DECEMBER 1852

Killyleagh C° Down
23rd. Decr. 1852

My dear M^r. Layard,

I received your letter this morning. I lose no time in forwarding to you a sketch of what is on the stone at the India House. Do you know any thing of this Bit Shaqqathu? Rawlinson professed ignorance of its site when he published his commentary. He then read the name 'Bit Digla'.[1] He told me at Edinburgh that he had found the name on bricks (I think) at either Niffer or Warka.[2] I see in the very commencement of his outline that he supposes it to be Senkereh[3] – what he reads So or Sikkaru is *most certainly* Shaqqathu. All therefore that he says here as to its supposed identification with Sankarah falls to the ground. In looking for this (which I did since I began this note) I stumbled on the very unfair statement he makes respecting me in p. 9 of this Outline (or is it Norris? or the Asiatic Society that is to be considered as making the statement?). He says I 'impugned the readings of Samaria & *Beth Khumri*'.[4] On the contrary I admitted the *latter* from the *first*, & only objected to the former that it began with *s* while the Hebrew began with *sh* – (the very same objection which there is in the case of Lachish). I gave this up when I found the name of 'Munakhimmi Samirinay', & finding that the objection does not lie in this case, nor in that of Sargon – (*sh* in Assyrian, *s* in Hebrew) I cannot admit it to apply in the case of Lachish – more especially as Rawlinson's Lachish is not *lākîs* but *'allaqqish* while yours is *lākîs* differing *only* in the consonants.

By the time this reaches you – the crisis will probably be over; and I hope it will terminate to your satisfaction. If the coalition falls to the ground through the Peelites requiring too much it will be a sad misfortune, as Ld Derby will be strengthened.

You will receive this I suppose on Christmas morning of which I wish you many happy returns.

Believe me
Yours very truly
Edw. Hincks

[Enclosure:][5]

The inscriptions of Nebuchadnezzar are partly in a peculiar uncial character and partly in a cursive character similar to that like the Achaemenian inscriptions of the third kind. The former is found on the bricks in all parts of Babylon, and on the great inscription at the India House, which is on a rectangular prism having ten columns. The principal inscriptions in the cursive character are:

1st. the fragment in the British Museum of which an engraving is published in Sir R. K. Porter's Travels in Georgia &c Pl. 78 (Vol. II p. 395). I pointed out in the Literary Gazette of 25th. July 1846 that this contained transcripts of two portions of the Inscription at the India House.

2nd. an inscription in two columns on a clay cylinder published by Mr. Rich in his Ruins of Babylon and

3d. an inscription in three columns on a cylinder published by Director Grotefend in the Göttingen Transactions.[6] They are all much to the same purpose. The Great Inscription begins with the name and titles of Nebuchadnezzar the Great (whose reign began according to Ptolemy's Canon in 604 B.C.). He is called 'Nabukudurruchur king of Babylon, son of Nabubaluchur king of Babylon'. From the omission of his grandfather's name we may safely infer that he was not a king. The subsequent part of the inscription contains no notice of any foreign conquests, but speaks of the building of various temples and palaces in addition to the walls of Babylon and Borsippa. Mention is also made of works at Bit-Shaqqathu and Bit-Zida*;[7] but whether these were distinct cities may be doubted. The ornaments of some of the palaces & temples appear to have been very rich. If the inscription could be perfectly translated, a good deal would be known about the architecture; but the precise meaning of some important words is yet unascertained. The walls were built of burned bricks & bitumen lined with gypsum and other materials. Some appear to have been wainscotted. Over these walls was woodwork, and at the top an awning sustained by poles. Some of the woodwork is said to have been gilt, other parts silvered; and Lebanon is mentioned as the place from which part of it at least was brought.

Marduk appears in these inscriptions as the principal deity, holding that place which Ashur held in the Assyrian inscriptions. He is called 'the great lord', 'lord of lords', 'Senior of the Gods' &c. Nabu seems to hold the second place. The great inscription contains a sort of thanksgiving to Marduk for what he has done already for the king & a prayer for his blessing on him & his house.

*shqt signifies in Hebrew 'rest, peace'. Bit-Shaqqathu should mean 'the peaceful house'. The meaning of Zida is less certain; Bit-zida may be 'the house of pride', or 'the house of addition, or enlargement'. These two places are mentioned together in a fragment of the Annals of Pul. B.M. 34.6. The former is also twice mentioned on the very curious stone belonging to Lord Aberdeen, the characters of which appear to be unique. It is there called 'the palace of the gods' & is mentioned in connexion with Babylon; while on Grotefend's cylinder 'the gods of Bit Shaqqathu and the gods of Babylon' are mentioned together, as distinct from one another. I have heard that this name was found on bricks at Warka or Niffer and that this has been supposed to be its side. It may be so; but on the other hand these names may be generic terms, applicable to various places.

MS: BL Add. 38981, ff. 190–3

1   Rawlinson, 'On the Inscriptions of Assyria and Babylonia', *Journal of the Royal Asiatic Society* 12 (1850), p. 477, n. 1. Both are probably readings of Esagila, which was an important temple complex in Babylon, dedicated to the god Marduk.

2   See C. B. F. Walker, *Cuneiform Brick Inscriptions in the British Museum, the Ashmolean Museum, Oxford, the City of Birmingham Museums and Art Gallery, the City of Bristol Museum and Art Gallery* (London, 1981), esp. nos 101 and 102.

3   *Outline of the History of Assyria*, p. 3.

4   Ibid., p. 9.

5   Most of this enclosure is reproduced in Layard, *Discoveries in the Ruins of Nineveh and Babylon*, pp. 529–30.

6   'Bemerkungen zur Inschrift eines Thongefässes mit babylonischer Keilschrift', *Abhandlungen der königlichen Gesellschaft der Wissenschaften zu Göttingen* 4 (1850), historisch-philologische Klasse, pp. 3–18 + 2 plates.

7   The name is now read Ezida, which was the name of the temple of Nabû, the god of writing and wisdom. The oldest and most important temple was at Borsippa (Birs Nimrud).

FROM JAMES RANKIN
27 DECEMBER 1852

Greenholm by Uddingstone Lanarkshire
27th Dec: 1852

Dear Sir,

I beg to acknowledge receipt of your letter of the 22[d]. The information which it contained will be of great use. I am particularly thankful for your correction of my mistake in reference to the obelisk. The fact which you mention respecting Rawlinson's attempt to wrest from you the honour of the discovery of the name of Jehu and the consequent settlement of the date of the obelisk is a most unprincipled affair as far as the Col: is concerned. His pretended discovery of the name of Ithbaal was a well laid scheme to support his claim. When I first read his letter in the Athenaeum it appeared to be just what was desirable, to be a beautiful confirmation of Scripture history. But according to a common saying some things are too sweet to be wholesome and it would seem to be so in this instance. Rawlinson's studying cuneatic from the text of 1 Kings XVI, 31 is a rich idea.

I should like very much to see your two papers in the last vol: of the Trans. of the R.I.Ac.[1] The University Library here is at present closed for the Christmas holidays and will not be opened till 5th Jan: Besides, I am not sure if the Trans: are to be found there. So that if you were to send to me copies of your papers I would be sure to have them in time to peruse them before finishing my essay which is due for 10th. Jan: If you should favour me with these papers be sure to let me know the whole expenses which you incur thereby both for the papers themselves and for postage so that I may transmit the same to you. On this condition only do I wish to receive them. Sorry for having troubled you so frequently and grateful for the kindness and freedom with which you have answered my letter.

I remain,
Dear Sir,
Yours ever,
James Rankin

MS: GIO/H406

1   'On the Khorsabad Inscriptions', *Transactions of the Royal Irish Academy* 22 (1852), polite literature, pp. 3–72; 'On the Assyrio-Babylonian Phonetic Characters', pp. 293–370.

FROM JOHN LEE
29 DECEMBER 1852

College, Doctors Commons
December 29, 1852

My dear Sir,

I take leave to acquaint you that I have received a parcel of papers from D^r. Grotefend of Hannover, one portion of which is for you, and another for the Royal Irish Academy – and I have packed them up together, and sent them to a Messr^s Barthes & Lowell of Great Marlborough Street, who will I hope, be able to forward them to Dublin in about a fortnight.[1] I have also written to the Librarian of the Academy at Dublin, to give him notice, and to request him to take care of your poster for you.

M^{rs}. Lee and I passed two days recently with the Reverend G. C. Renouard, at Swanscombe in Kent and we left him in his usual cheerful spirits, but weak in body – and I learn that he has since our departure, been able to resume his duties at his church on a Sunday. We hope that he will be restored to his usual health in the Spring.

We trust that you and your family may be blessed with health and enabled to commence the new year in the possession of every real worldly comfort with cheerfulness; and we regret that we did not pay our respects to you and your family on leaving Belfast for Armagh and Dublin.

> I remain
> My dear Sir
> Yours respectfully
> J. Lee

M^{rs}. Lee and I helped M^r. Renouard to arrange the four first parts of the plates of D^r. Lepsius' great work on Egypt – and we hope to revisit him in the Spring, and to hear his description of them.[2] We propose to go on the 30th. to Hartwell.

MS: GIO/H 253

1   Barthès & Lowell, foreign booksellers, 14 Great Marlborough Street, London.
2   R. Lepsius, *Denkmäler aus Ägypten und Äthiopien*, 6 pts in 12 vols (Berlin, 1849–59).

1853

❧❧❧

9 Little Ryder St

Jany. 1 1853

My dear D^r. Hincks,

I am greatly obliged to you for your last letter. I will do as you suggest & make use of your notes. I now send you in their simplest state some remarks on them. Do not mind the grammar or composition – all that I have corrected. Should anything strike you as wrong pray let me know as early as convenient as we are now printing off as fast as possible. I presume there is nothing which I have stated that requires great attention.

After all the Government have given us nothing.[1] I am somewhat disappointed, of course, but there have been so many Peelite chickens to feed & so many persons to conciliate that one who clearly supported the Ministry from principle would not have much chance of office.

Pray wish all your family the compliments of the season from me. I trust they are well.

Yours faithfully,

A. H. Layard

MS: GIO/H 235

1   See Waterfield, *Layard of Nineveh*, pp. 233–4.

～☆☆

TO AUSTEN HENRY LAYARD
3 JANUARY 1853

Killyleagh C° Down
3ᵈ. January 1853

My dear Mʳ. Layard,

Your letter of the 1ˢᵗ. which I received this morning was a great disappoint-ment to us all. We had understood that you were First Secretary with Mʳ. Lowe at the India Board. I think they would have acted more wisely in giving you an office than in making some of the appointments announced. I doubt if the ministry will keep together.[1]

With respect to the Wan kings I have made some notes on the proof sheet. Rawlinson had a letter about them in a late number of the Athenaeum (report of the meeting at the Asiatic Society).[2] He makes Argishtish a contemporary of Sargon, but he also identifies the first Melidduris (or whatever the name is) with a king mentioned on the obelisk & his father Lutibri with the king of Armenia mentioned in the annals of the obelisk king's father. Of this last identification I can say nothing except by *inference*; the other I am satisfied is a mistake (it is a false reading) & *consequently this*. It is however possible that there were two Lutibris as there were two – discoveries.

It is now within 5 weeks of my setting off for London. I hope I shall find no difficulty through not being able to procure a supply. I presume your book will be out before that.

We all wish you a happy new year.

I am
Yours very faithfully
Edw. Hincks

MS: BL Add. 38980, ff. 201–2

1   Robert Lowe, Viscount Sherbrooke (1811–92), M.P. for Kidderminster, was joint secretary to the Board of Control in the Liberal–Peelite coalition government of Lord Aberdeen. He played an important part in the introduction of the India Act of 1853.
2   Report of a meeting of the Royal Asiatic Society on 4 December 1852, *The Athenaeum*, no. 1311 (11 Dec. 1852), p. 1362. In a letter to the society dated 4 September 1852, Rawlinson proposed that six kings mentioned in the Van inscriptions were contemporary with Assyrian kings.

FROM AUSTEN HENRY LAYARD
3 JANUARY 1853

9 Little Ryder Street
Jany. 3/53

My dear D^r. Hincks,

Would you kindly give me a note upon the inscriptions on the bricks – that last brought to England and the prism. They state, I believe, that they are half a talent Babylonian – or rather 30 mina.[1]

Yours faithfully
A. H. Layard

MS: GIO/H 236

1   See Layard, *Discoveries in the Ruins of Nineveh and Babylon*, pp. 600–1, especially Hincks's note on p. 601. The original manuscript of the note is in the British Library, BL Add.39077, f. 64: 'The inscriptions on the large ducks state them to be weights of "30 mana" that is half a talent. The actual weight is something more than 480 oz. troy; which would give for the *mana* 16 oz. with a small fraction over. The Attic mana has been computed to be 14 oz. with a small fraction. It would, consequently, be to the Babylonian as 7 to 8. According to Herodotus (3.89) the Eubaean talent was to the Babylonian as 6 to 7. If this statement be correct, the Eubaean would be to the Attic as 48 to 49. Smaller ducks weigh different multiples of a unit, the relation of which to the mana is yet unascertained; as I am ignorant of the weight of the only duck on which the inscription has been published.'

FROM EDWIN NORRIS
8 JANUARY 1853

Foreign Office
8th Jany. 1853

My dear Sir,

I have just received your two publications for which I beg you to accept my sincere thanks.[1] I have no doubt that we are getting to the real understanding of the Assyrian monuments, though I sometimes doubt our *reading* them with much accuracy. I am especially interested in the sounds, as I have at this moment in the printer's hands a paper on the so-called Median inscriptions, with the alphabet of which no one knows better than you do how much there is in common with the Assyrian.[2] I think I have penetrated into the structure of the language, and I am able to inform you (or at least I have received information which induces one to believe) that very many inscriptions are found in Persis and Elymais, in dialects of the same language, but in the Assyrian character. I shall have the pleasure of forwarding you my paper, so soon as printed, with the copy of the Behistun Inscription in the so-called Median language; it will be some time printing, because the work is troublesome but it is all in hand. You may be interested to learn that I have received an inscription of Artaxerxes Mnemon in the same language and character from Susa, with very different orthography; it is equally ungrammatical with the Persian inscriptions of Artaxerxes Ochus.

I am, my dear Sir
Yours very faithfully
Edwin Norris

MS: GIO/H309

1   One of the publications must have been Hincks's paper 'On the Assyrio-Babylonian Phonetic Characters', *Transactions of the Royal Irish Academy* 22 (1852), polite literature, pp. 293–370.

2   E. Norris, 'Memoir on the Sythic Version of the Behistun Inscription', *Journal of the Royal Asiatic Society* 15 (1853/1855), pp. 1–213, 431–3. He had read Hincks's limited study of Elamite in his article 'On the First and Second Kinds of Persepolitan Writing', *Transactions of the Royal Irish Academy* 21 (1846), polite literature, pp. 114–31. The next substantial study of Elamite was published many years later by A. H. Sayce, 'The Languages of the Cuneiform Inscriptions of Elam and Media', *Transactions of the Society of Biblical Archaeology* 3 (1874), pp. 465–85.

FROM JAMES RANKIN
10 JANUARY 1853

Greenholm by Uddingstone Lanarksh:
10th. Jan: 1853

Rev.d Sir,

I received your two papers on the inscriptions and the accompanying letter of the 5th. on the 8th. The matter of these papers I have found extremely useful for my purpose. I enclose stamps for the sum you named and I beg most sincerely to thank you for the readiness with which you have uniformly satisfied my enquiries.

The subject is a new one and especially to me, but from what I have already seen of it, I feel it to have attractions beyond those of mere novelty and I can never but continue to take a deep interest in the progress of the discoveries. I have been especially [taken] with the affinity that from your papers and elsewhere I see to exist between the language of the inscriptions and the Hebrew; and my means of recognizing these will probably be extended when I shall have advanced more in Arabic. If I was master of the cuneatic syllabification so far as that has been ascertained I might perhaps hope to be able to follow the translations.

I remain,
Dear Sir,
Yours ever,
James Rankin

P.S. You need not be at the trouble of acknowledging receipt of this. I shall just take it for granted that it has reached you.

J. R.

MS: GIO/H 407

FROM GEORGE CECIL RENOUARD
11 JANUARY 1853

Swanscombe, Dartford
Tuesd^y, Jan^{ry}. 1853

My dear Sir,

I am very sorry that any impediments should have made me keep you in suspence as to the safety of the packet which came to me as a most acceptable New Year's gift, three days ago, and will supply me with matter for serious study as soon as my present engagement gives me leisure.[1]

I am much pleased to find you have abandoned M^r. Ellis's system of orthography which from the first appeared to me a most ill judged attempt at perpetuating error in defiance of all principles of etymology & analogy;[2] comparable only to the system of the Masoretes who attempted to perpetuate a local & transient pronunciation in all its least minutiae & saddled the Hebrew alphabet, one of the simplest & most invariable existing with a vocalisation as complex & arbitrary as it is difficult to reduce it within any general rules.[3] The Arabic, no doubt, shews what was the original system of the Semitic languages, but even there the *Spitzfindigkeit* of the Eastern grammarians has been at work.

I felt, from the beginning, more reliance on your reading & interpretation of the Babylonian & Assyrian (for such I apprehend the 2^d. Persepolitan should be called[4]) than the result of Col. Rawlinson's researches. But the great uncertainty of the value of many characters made me hesitate: your progress has now been such that I feel assured you are in the right path.

I have lately got a new tract by Grotefend; but have scarcely opened it. His last on an inscription on an earthenware vase, which inscription he translated through out, left no favourable impression on my mind. It might [be] compared to some versions of the inscriptions at Palmyra, by a Hebrew teacher in the West of England, printed several years ago – which differ so completely in style & import from those which[5] have been correctly read, that one sees immediately that the translator must have made a bad guess & drawn largely upon his imagination.

I had long perceived that the Col. was coming round to your readings & wish he had acknowledged his debt to you more candidly.

I am sorry you have not given the Chaldee or Syriac equivalents rather than the Hebrew: all who can read the common Hebrew character must see the similarity between the Ch. or Syr. & Hebrew: & can we believe that the Chaldee of Daniel was not the Babylonian language. The addition of the masoretic points also seems to make the resemblance less apparent.

But as the post hour approaches I must only add that the last sheet of the Arabic Pentateuch,[6] the Index to the last 10 Vols of the Geographical Journal[7] & some MSS. of a particular friend required for the press were the impediments which prevented me from sooner wishing you & your family, a happy New Year & begging you to accept the sincere thanks of,

> my dear Sir,
> Your much obliged & very faithful
> G. C. Renouard

MS: GIO/H 469

1   The packet certainly contained Hincks's paper 'On the Assyrio-Babylonian Phonetic Characters', *Transactions of the Royal Irish Academy* 22 (1852), polite literature, pp. 293–370.
2   In his paper 'On the Khorsabad Inscriptions', Hincks announced that he was adopting the phonetic alphabet of Pitman and Ellis. See A. J. Ellis, *The Essentials of Phonetics* (London, 1848), which was published by Fred Pitman. Hincks thought this book ought to be in the hands of every student of languages. See 'On the Khorsabad Inscriptions', *Transactions of the Royal Irish Academy* 22 (1850), polite literature, p. 7. Renouard expressed his strong opposition to Ellis's method in his letter of 23 November 1849 (vol. 1, pp. 310–11). Hincks's systems of transliteration often cause difficulties for those reading his publications.
3   Negative views of the Masoretic vocalisation of the Hebrew Bible were not uncommon.
4   Today we call the language of the 'second Persepolitan' inscriptions Elamite. Norris and Rawlinson called it Scythic, and Renouard Assyrian.
5   Renouard wrote: 'which those which'. I have corrected to 'from those which'.
6   'In 1848 the S.P.C.K. undertook to publish a new Arabic version [of the Bible]. It was made by Faris Al-Shidyak... under the supervision of a committee... which included S. Lee and Thomas Jarrett... This Psalter was the first portion published.' See the opening pages of the *Psalter in Arabic* (London, 1850); also the *New Testament in Arabic* (London, 1851). A translation of the Pentateuch into Arabic was prepared by

Fares al-Shidyak and Samuel Lee and was published in London between 1851 and 1857, but it was not used. In the same period an Arabic version of the Bible was being prepared by Eli Smith and Cornelius V. A. Van Dyck. See I. H. Hall, 'The Arabic Bible of Drs. Eli Smith and Cornelius V. A. Van Dyck', *Journal of the American Oriental Society* 11 (1882), pp. 276–86.

7   *General Index to the Second Ten Volumes of the Journal of the Royal Geographical Society*, compiled by G. S. Brent (London, 1853).

꙳꙳꙳

FROM JAMES WHATMAN BOSANQUET
12 JANUARY 1853

Claysmore Enfield
12th. Jany. 1853

My dear Sir,

Many thanks to you for your valuable present, containing your most recent observations on the Assyrian inscriptions.[1] I wish I was capable of fully appreciating your philological criticisms. It is only on the chronological portion that I venture to make any observations.[2] I feel quite confident that I shall be able to satisfy you, that your former view of the chronology was nearer the truth, than that which you have recently adopted. Also that it was the view entertained by all the writers of the third century before the Xtian aera including Berosus, Megasthenes, Menander, & Manetho, that it is consistent with the ascertained date of the Eclipse of Thales, and also with Scripture.

Pray therefore do not close your eyes to any further evidence from the Assyrian inscriptions which may tend towards confirming your previous views. Probably the Museum now contains materials from which the question between us might be finally settled. How is it that those valuable materials are left unheeded, while so much money is spent in binding and decorations?

Believe me
My dear Sir
Yours truly obliged
J. W. Bosanquet

MS: GIO/H 62

1   'On the Assyrio-Babylonian Phonetic Characters', *Transactions of the Royal Irish Academy* 22 (1852), polite literature, pp. 293–370.
2   Bosanquet is referring to the Appendix in Hincks's paper, pp. 364–70.

~~~

TO HENRY ELLIS
13 JANUARY 1853

Killyleagh C° Down
13th. January 1853

Sir,

Referring to your letter of the 30th. November, and to my reply of the 3rd. December, I have to state that I find it difficult to obtain a clergyman to take charge of my parish for a month on any terms; and that it would probably cost me not less than £12.

As you have not alluded to this item of expense in your letter, I am unwilling to incur it without the sanction of the Trustees. I think it right to mention at the same time that I have been offered a supply for either the 20th. or the 27th. February; and that I could thus meet the Trustees on either the 19th. or the 26th; reaching London on the Tuesday morning preceding, & remaining till the following Friday afternoon – 5 o'clock the train would leave.

As the entire expense of such a visit to London would not exceed the £12, which, on the other plan, I should have to pay for clerical assistance; and as in the event of an engagement with me being entered into, by which I should keep a permanent curate, I could make out my time in London at any period the Trustees might think best; while in the other case, it would make but little difference how long I remained; I have thought it right to suggest this arrangement.

I would be obliged to you to let me know as soon as possible whatever it be approved of; and if so which of the days that I mention would be best. If I must be the month in London, it will be desirable that I should take immediate steps to procure assistance, in order that I may leave this on the 7th. February.

I have the honor to be, Sir
Your most obedient servant
Edw. Hincks

MS: BM/CA/OP 49

FROM HENRY ELLIS
17 JANUARY 1853

[British Museum]
Jan. 17ᵗʰ. 1853

Dear Sir,

The Trustees of the British Museum have altogether objected to being party to any arrangement which might interfere with the discharge of your clerical duties as incumbent of a parish in Ireland, and in suggesting your visit for a short period to London, did so under an impression that you would be able to pay such a visit without inconvenience to yourself or the parish. As the Trustees will not hold a meeting for some days to come I cannot undertake to say how far they may be disposed to alter their view, but my strong impression is that they will not be disposed to vary the terms laid down in my letter of the 30ᵗʰ. Nov. 1852, though I feel equally confident that if a postponement of your visit to London was more suited to your convenience they would readily acquiesce in it.

I remain, Dear Sir,
Your faithful servᵗ.
Henry Ellis

MS: BM/CA/LB 43: 68

FROM AUSTEN HENRY LAYARD
19 JANUARY 1853

9 Little Ryder St
Janʸ. 19/53

My dear Dʳ. Hincks,

As you are continually referred to & quoted in the enclosed I think it as well to send it to you in case you should have any remarks or suggestions to make.[1] I have not received the tables from the printer; you shall have them the moment I do. I have been very much delayed. Pray return me the enclosed as soon as you have looked thro' it.

I hope all your family are well. Pray give them my kind remembrances & regards. I am very anxious for your visit to London. Pray do not hesitate to let me know if I can do anything to make you comfortable or to help you.

Do not mind the style or grammar of the enclosed. As it was very hastily written I have to admit it is full of errors. I will correct them all in going thru'.

Yours sincerely
A. H. Layard.

Many thanks for your paper on the cuneiform writing, which I received some days ago. I have not yet had time to go thru' it. I shall add your discussion on the date of Sargon to the enclosed.[2]

MS: GIO/H 237

1   Layard must have sent proofs of at least the first part of ch. 26 of his *Discoveries in the Ruins of Nineveh and Babylon*. Hincks is mentioned frequently on pp. 611–22.
2   'On the Assyrio-Babylonian Phonetic Characters', *Transactions of the Royal Irish Academy* 22 (1852), polite literature, pp. 293–370. The discussion of Sargon is on pp. 366–70.

FROM FANNY CORBAUX[1]
22 JANUARY 1853

22, Westbourne Place
Eaton Square, London
Jan.[y] 22.[d] 1853

Miss Fanny Corbaux presents her compliments to D.[r] Hincks, and by the same post as this note has forwarded a copy of a series of papers 'on the Rephaim and their connexion with Egyptian history' which she has lately contributed to the Journal of Sacred Literature, and of which she begs the honour of Dr Hincks' acceptance.[2]

Should they by any mischance not arrive safely, Miss Corbaux will be happy to replace them.[3]

MS: GIO/H 147

1   Marie Françoise Catherine Doetter Corbaux, known as Fanny Corbaux (1812–83), painter and biblical critic, was the daughter of Francis Corbaux, an Englishman who lived abroad for a long time. She earned her living from painting in oil and watercolours and exhibited at many academies and institutions. As a writer on biblical subjects, she published in *The Athenaeum* and the *Journal of Sacred Literature*. She also wrote the introduction to D. I. Heath, *The Exodus Papyri, with a Historical and Chronological Introduction by Miss F. Corbaux* (London, 1855). See *ODNB*.

2   'The Rephaim, and their Connexion with Egyptian History', *Journal of Sacred Literature* 1/1 (Oct. 1851), pp. 151–72; 1/2 (Jan. 1852), pp. 363–94; 2/3 (Apr. 1852), pp. 55–91 + 2 plates; 2/4 (July 1852), pp. 303–40; 3/5 (Oct. 1852), pp. 87–116; 3/6 (Jan. 1853), pp. 279–307.
3   The letter is not signed.

FROM AUSTEN HENRY LAYARD
22 JANUARY 1853

9 Little Ryder St
Jany. 22/53

My dear D^r. Hincks,

I send you the three tables of the names of kings, countries & gods.[1] Would you kindly cast your eye over them and make any suggestion or amendment that you may think fit? I am very anxious to have them back *as soon as possible* as my work must appear immediately – and we are only now waiting for these tables.

On receipt of your letter yesterday I went to the British Museum to see if I could learn anything about you, but have not succeeded in doing so. I think the Trustees should undoubtedly bear every expense connected with your visit to England and it would be most unfair were you made to suffer any loss. I do not despair of doing something. I think, not owing to any merit of my own but principally to those who have helped me, my work will excite considerable attention on account of the connection of the inscriptions & sculptures with Biblical history. It will have a considerable sale, & between ourselves Murray has already disposed of nearly the whole 8000 copies he is printing. When the public will be as they will I hope then, more alive to the great value & importance of your discoveries, the way will be easier. I am very sanguine that something will be managed in the end. If you are coming to London this year, as a usual habit, I should say accept the proposal of the Trustees in so far as to be up here in February, to meet them and to break ground at the Museum. When they have once understood the importance of retaining you, there will, I hope, be little difficulty in managing the rest. I can scarcely bear the idea of your not coming up, for I do think it is important that no time should be lost. If you come to town, ascertain whether the Trustees will meet on the Saturday you are here, which you can easily do by writing to Sir Henry Ellis.

I will not fail to speak to those members of the Government who may be of use to you. At present it is difficult to get at them, but the moment Parliament meets I shall have an opportunity.

I return you the letters. Pray give my kind regards to M^{rs}. Hincks & your family.

Believe me
Yours sincerely
A. H. Layard

MS: GIO/H 238

1   See Layard, *Discoveries in the Ruins of Nineveh and Babylon*, pp. 623–9.

TO AUSTEN HENRY LAYARD
22 JANUARY 1853

Killyleagh C° Down
22^d. Jany. 1853

My dear M^r. Layard,

I send you back the proofs.[1] There are two passages which I have marked as requiring modification. That about Samaria & Bit Khumri (which as it stands is erroneous) & that about Esarhaddon, which as it now stands would lead a person to think that the son of Sennacherib whom he made king of Babylon was the same who succeeded him in Assyria – that is M^r. Bosanquet's view & was once mine – probably it was that of Eusebius & *perhaps* that of Polyhistor whom he cites; but I think it equally likely that Eusebius altered Polyhistor's text, supposing that he thereby corrected it.[2] The two names are totally distinct Ashur-nadin the Babylonian king & Ashur-akh-adin or adina the successor at Nineveh.[3]

I presume the lists which I shall be glad to look over will finish the volume.
You will have read my letter with Sir H. Ellis before this.

Yours very truly
Edw. Hincks

MS: BL Add. 38981, f. 210

1   See Layard's letter of 19 January 1853.
2   See Layard, *Discoveries in the Ruins of Nineveh and Babylon*, pp. 613, 620–2.
3   The names of these sons of Sennacherib are now read Aššur-nādin-šumi and Aššur-aḫ-iddina (Esarhaddon).

TO AUSTEN HENRY LAYARD
24 JANUARY 1853

Killyleagh C° Down
24th. Jany. 1853

My dear Mr. Layard,

I had your letter this morning & return you the proofs. I have made some additions to the two last tables which I have endeavoured to make so distinct that the printer cannot [be] mistaken. I propose going to Belfast on Wednesday & will there try to make arrangements for my going to London; but I really doubt the *possibility* of it.

I congratulate you on the extended sale your work is about to have.

Believe me
Yours very faithfully
Edw. Hincks

My 'womenkind' claim kind remembrances

MS: BL Add. 38981, f. 212

FROM HENRY ELLIS
25 JANUARY 1853

[British Museum]
Jan. 25

Dear Sir,

I beg to acquaint you that your letter from Killyleagh of 13 January was laid before a meeting of the Trustees on Saturday last together with a copy of my answer of which the Trustees entirely approved.

I remain, Dear Sir,
Yours sincerely
Henry Ellis

MS: BM/CA/LB 43: 78–9

TO HENRY ELLIS
27 JANUARY 1853

Killyleagh C° Down
27th. January 1853

Dear Sir Henry,

I have received your letters of the 17th. (in which you tell me that 'the Trustees of the British Museum have altogether objected to being party to any arrangement which might interfere with the discharge of my clerical duties as incumbent of a parish in Ireland') and of the 25th, in which you mention that the Trustees had approved of your answer.

I must not presume to ask the Trustees to alter their determination; but I am not sure that I rightly understand it. At all events, as I am satisfied that there is most important information, historical and philological (and probably astronomical also) to be derived from the Assyrian antiquities in the Museum; and as I feel very confident that I should be able to elicit this information to a very great extent, I am desirous to explain how it happens that I am unable to do so.

I am ready to entertain any reasonable proposal for either of two things; either that I should resign my parish and become a resident in London or its immediate vicinity; or that I should permanently keep a curate-assistant, spending as much time in London as the law will allow me to be absent from my parish, and applying myself to the study of the Assyrian language and inscriptions to a greater extent than I have yet done, or than I can possibly do while I have the sole charge of this parish.

With the small income, however, which I have to maintain my family and to meet the pecuniary engagements which I have contracted, I cannot engage a permanent curate-assistant – much less could I resign my parish – *on an uncertainty*, without an engagement being *previously* made with the Trustees. If, however, I rightly understand the passage of your letter which I have quoted, the Trustees will not be a party to any such *previous* engagement.

With respect to a temporary engagement for a month, I would first observe that I have not been four weeks together absent from my parish for, I suppose, twenty years. My absences have very rarely exceeded a single Sunday with portions of the preceding and following weeks; and from the constant attention which I am in the habit of paying to my parishioners, they would feel the inconvenience of being left for four successive weeks, with only a clergyman to visit them on Sundays, much more than would be felt in many other parishes. *This*, in short, is an arrangement which I could not conscientiously make. If I leave home for a month, I must have a clergyman to reside in the parish during my absence. Now, I have hitherto failed to hear of any one, who was not absolutely objectionable, that would take the duty for so a short time. I have made further enquiries, and do not as yet absolutely despair of succeeding, though I am by no means confident of doing so. If, however, I could engage such a temporary curate at all, I should have to pay him a pretty large sum; and, on the principle that you have laid down, this would not be reimbursed to me by the Trustees.

But supposing this difficulty to be overcome, I should not think it right to enter into a *temporary engagement*, unless with a reasonable prospect of its leading to such a permanent one as I have before described. I mean to say, that if, on trial, the Trustees find that my services are valuable, they will engage them permanently in one of the ways that I have mentioned. Now if, as I understand you to say, the Trustees have insuperable objections to both these ways, I of course cannot go with any such expectation; and accordingly, I must decline going at all.

It would be too much to expect that a body of laymen should take the same views of a clergyman's duty as he takes himself; but it is on *his own* views of duty that each clergyman must act. I do not feel myself so bound to my parish as that I cannot resign the charge of it altogether; or that I cannot transfer a large portion of the duties of it to an assistant in whom I can confide. But I feel so bound to it, that, while I retain the cure of souls in it, its duties shall be properly performed. I think change of system in itself an evil; especially a double change, backwards and forwards; and accordingly, if I am likely to continue with the sole charge of this parish, I do not think that it would be for the welfare of parishioners that, even for a month, a system different to what they are accustomed to should be introduced among them.

I remain,
Dear Sir Henry,
Yours very truly
Edw. Hincks

MS: BM/CA/OP 49

⁓

FROM FANNY CORBAUX
29 JANUARY 1853

22, Westbourne Place
Eaton Square
Jany 29th. 1853

Dear Sir,

Accept my best thanks for your valuable paper on Assyrian writing, which in its matter cannot fail to prove of the highest interest both to the comparative philologist & to the biblical student.[1]

Accidental errors in the Hebrew equivalents of numerals, such as the one in the reign of Hezekiah which your Assyrian dates enable you to correct are, I apprehend, of much more frequent occurrence than the means of rectifying them & even sometimes of detecting them. A similar error (by the change of the *plural* for the *singular* form) appears in the stated age of Hezekieh at his accession, 25 years, which would make him only 11 years younger than his father & 42 years older than his eldest son. The error thus betrays itself, but it was still remaining doubtful whether Ahab's reign was *under*-stated, or Hezekieh's age *over*-stated, by 10 years. The dates of Ptolemy's canon & of the Assyrian and Babylonian kings referable to it as they appear by your researches, forbid any displacement of Hezekieh's *reign*, & therefore would seem to justify the correction of his *age* to 15 at his accession instead of 25. This is but one of a host of similar problems that might be addressed.

I remain, Dear Sir
Yours very sincerely
Fanny Corbaux

MS: GIO/H 148

---

1   'On the Assyrio-Babylonian Phonetic Characters', *Transactions of the Royal Irish* Academy 22 (1852), polite literature, pp. 293–370. Corbaux was particularly interested in the matter contained in the Postscript to the Appendix, pp. 366–70.

TO HENRY ELLIS
3 FEBRUARY 1853

Killyleagh C° Down
3rd. February 1853

Dear Sir Henry,

Not having heard from you this week, I infer that my letter of the 27th. has not yet been laid before the Trustees, and that this will be in time to accompany it.

Having further considered the matter, I would undertake to go *on trial* to the Museum for 30 days in the course of the spring, provided that I be at liberty to *divide the time*, so as not to be two Sundays in succession away from my parish. That is to say, I would go for three periods of ten days, the first of which would extend from the 15th. inst. to the 25th., the others would be after Easter. As the trains by which I should travel would be due in London at 11 am on the 15th. & would not leave till 5 pm on the 25th. I should have 10 *clear* days. I have estimated that the expense of *each* period, including travelling, residence in London & Sunday duty here, would be £11.

This would give the same amount of time in London as what was proposed in your letter of 30th. Novr; only divided in such a way as not to prejudice my parishioners, in the event of a permanent engagement with me not being concluded; and I therefore hope that the Trustees will consider this modification of what they proposed as meeting their views.

Believe me, Dear Sir Henry
Yours very truly
Edw. Hincks

MS: BM/CA/OP 49

TO AUSTEN HENRY LAYARD
3 FEBRUARY 1853

Killyleagh C° Down
3rd. February 1853

My dear Mr. Layard,

I send you copies of a paper which I had printed in a Belfast newspaper. I presume you take some interest in the subject treated of, but don't know whether your views would agree with mine. I have never seen my plan proposed; but it may have been so. Much has been written in the London papers that I have not seen. I sent a copy to the Chancellor of the Exchequer but he will hardly oblige to read it[1] – also to Mr. Hume.[2]

On Saturday I suppose the Trustees of the Museum will decide whether I go to London or not. I wrote last week to say that if they had made up their minds against engaging my services, wholly or in part, for a permanence, I would rather not disturb my parochial arrangements by going for a month.

I have since suggested that I could go for 3 periods of 10 days each, so as to be only a Sunday at a time absent, & at *intervals*. This would give them the time they originally proposed & I should hope they would not object to it. And yet, if those who are now employed could have their own way, I think they would rather that the work was not done atall than that it was done by an *interloper* like me. I fear this feeling exists; but my saying so should be between ourselves. A great deal depends on who attends the meeting of the Trustees on Saturday.

Believe me
Yours very truly
Edw. Hincks

I mentioned as the *first* of my three periods from the 15th. inst. to the 25th.

MS: BL Add. 38981, ff. 218–19

1   William Ewart Gladstone was chancellor of the exchequer from 1852 to 1855.
2   Joseph Hume (1777–1855), Scottish radical and politician, espoused views similar to Hincks's.

FROM HENRY ELLIS
3 FEBRUARY 1853

[British Museum]
3rd. February 1853

Dear Sir,

Your letter of 27th. January came duly to hand, and I should have acknowledged its receipt earlier if I had been able to make any further communication to you. I have now to acquaint you that the Trustees will meet on the 12th. inst. when your letter shall be laid before their Board.

Yours very faithfully,
Henry Ellis

MS: BM/CA/LB 43: 83

FROM AUSTEN HENRY LAYARD
8 FEBRUARY 1853

Orton Longueville
Feb^r. 8/53

My dear D^r. Hincks,

I have been travelling about England for the last ten days and that must be my excuse for not sending you an earlier reply to your letters & returning you my best thanks for the revision of the proofs. I have thankfully received all your emendations. You will have perceived that in the title of the kings &c most of the errors were printers' errors. I had not time to go over them before sending you the proofs. We are now only waiting for the performance of our contract with America to have the book out. We are bound to give the Americans a clear print right on which condition they take 2000 copies, our first edition of 8000 is pretty nigh sold off. I have had a great deal of trouble in correcting the sheets as in it several errors have slipt in.

I am looking forward with much interest to your visit to London, and I sincerely trust that it will be managed. I have not seen anything of the Trustees lately. Parliament meets on Thursday & I return to town for it. I shall then have an opportunity of seeing many people. Pray do not hesitate to let me know if I can be of any use to you. Should you on your visit be in need of a bed I dare say I could find one in my lodgings.

I am much obliged to you for your remarks on the income tax. Some alteration must be made in the apportionment, but I think the Liberal side of the House is willing to let the matter stand on this reprise to a certain extent, altho' some slight change must be made to satisfy a large class of people. No two men seem to agree upon the capitalisation question, which certainly appears to me to be surrounded with difficulties. In some respects your suggestion appears to be as good as any other that I have heard of tho' you would have the same difficulty in determining the limits to which it would have to apply. Kind regards to M[rs]. Hincks and your young ladies.

Yours very truly,
A. H. Layard

MS: GIO/H 239

❧❧

FROM SWINTON BOULT[1]
14 FEBRUARY 1853

14th. February 1853

My dear D[r]. Hincks,

I am rather surprised that you should stigmatize as 'almost the most objectionable of all', a plan of dealing with the Property & Income Tax, in such a way, as to tax in due proportion, the Capital, Profits, and Wages of the country. You may doubt if the plan will effect that object but you must allow I think that if the object be attained the plan cannot be objectionable as proposed. The proposal rests upon certain principles – if they are sound, the object must, of necessity, be gained; if they are not sound, I admit at once that the proposal is a failure and will work injustice. The principles contended for, then, are three:

1st. That Rent is the combined product of Capital and Labour.

2nd. That there is no capital and no productive labour that does not contribute to produce Rent.

3rd. That a tax on Rent will adjust itself so as to be borne in due proportion by the different elements that produce it – in other words, that it will be an equal tax on Capital Profits and Wages.

Which now of these three Principles is fallacious? and wherein does the fallacy consist? – in using the term rental, I mean to designate precisely that on which existing assessments are made, or which Parliament may hereafter enact shall so be considered – observe the Tenant is to pay. Your plan appears to me to retain all the moral evil of the existing system, which I regard, I confess, with more aversion than the pecuniary injustice, and all the un-English inquisition by which the amount of an individual's contribution may be ascertained. To this latter ingredient, you propose indeed to add this further item. Did you or did you not insure your life &c., and if you did, what did you pay? With great submission I think there would be quite as much reason in inquiring how much was paid for Bread, and how much for Butcher's meat. I do not agree with the actuaries – it appears to me quite as unjust to calculate an Income into Capital, by a process based upon observations of the happening of a given number of uncertain events, as to tax uncertain Incomes themselves as highly as those that are not precarious. A & B being of the same age and in receipt of the same Income, the actuaries would tax both alike. A dies in three months. B lives 30 years – on their own principle as it seems to me injustice is done. A has paid too much, B much too little. The Annuity case is as I put it – we compound for our Tax paying on our annual dividend in addition to our investments and so we do not pay specifically on the Annuities – but we do it practically notwithstanding.

I am glad that on one point we are agreed, and remain,
Very truly yours,
Swinton Boult

Source: Davidson, *Edward Hincks,* pp. 39–40

1   Swinton Boult (1806–76), a well-known insurance company manager, was the son of Francis Boult, a shipowner, who was a brother of Hincks's mother, *née* Anne Boult. Therefore Swinton Boult and Hincks were first cousins. See *ODNB.*

FROM HENRY ELLIS
14 FEBRUARY 1853

[British Museum]
14 Feby. 1853
Dear Sir,
    Your letters of 27th. January, and 3rd. and 5th. February were laid before our Trustees on Saturday last; when I was directed to acquaint you that they acquiesce in the terms proposed in your letter of 3rd. February.

Yours faithfully
Henry Ellis

MS: BM/CA/LB 43: 89

꙳꙳

TO HENRY ELLIS
17 FEBRUARY 1853

Killyleagh C° Down
17th. Feby. 1853

Dear Sir Henry,

I write to say that I hope to be at the Museum about noon on Tuesday next, the 22^d. inst. & to beg that every thing may be arranged for my setting to work when I arrive.[1] I will enquire for letter of instructions at the table where umbrellas &c are left.

I remain,
Yours very truly
Edw. Hincks

MS: BM/CA/OP 49

1   Hincks was at the British Museum from 22 February to 4 March.

꙳꙳

FROM HENRY ELLIS
22 FEBRUARY 1853

[British Museum]
22 Feby. 1853

Dear Sir,

Mr. Hawkins, who is at the head of our Department of Antiquities, will explain the accomodation, and afford the facilities for your researches allowed by our rules.

You are aware that at present ten to four o'clock are the Museum hours, next month they will be ten to five.

Believe me, dear Sir
Yours very faithfully
Henry Ellis

MS: BM/CA/LB 43: 96–7

THOMAS ROMNEY ROBINSON[1] TO
BISHOP CHARLES JAMES BLOMFIELD[2]
22 FEBRUARY 1853

Observatory
Armagh
Feb. 22, 53

My Lord,

I feel that I am taking a very great liberty in introducing to Your Lordship my friend Edward Hincks D.D., nor could I do it were I not emboldened by Your Lordship's high position in the Church; & still higher in Literature. He has for many years distinguished himself among the most successful in unraveling Egyptian lore, and has I think no equal in that new field which has been opened in the ruins of Assyria. This opinion I hold in common with some of the most renowned continental scholars. He is visiting London for a few days to inspect some marbles which have recently arrived, and, if possible, to arrange about settling himself in that city; as he finds it difficult to manage his decyphering without the power of frequent access to the original inscriptions. His pecuniary means are limited; and I tried in vain last year, though aided by the influence of the late Secretary and Attorney General for Ireland,[3] to obtain for him some small pension which might enable him to keep a curate, and defray the expenses of occasional visits to London.

His present plan is to exchange his parish (in the County of Down) for one within reach of the Museum; and if he effect this, I anticipate the best results in this strange research. He is admirably fitted for it; distinguished as a decipherer when a boy; a first rate orientalist, and a good mathematician.

And let me add that it seems peculiarly important that the highest places in this pursuit should be occupied [by] Christians. Even now there are signs that Infidelity will lay hold, if it can, on this Assyrian history, and deal with it as the French did with the Egyptian Zodiacs; and though truth will prevail at last it is better to prevent its being assailed at all.

This I hope will plead my excuse; and if I succeed in directing Your Lordship's attention to this good and talented man, I shall esteem myself most fortunate.

I have the honour to be
Your Lordship's Obed Servant
T. R. Robinson
President of the R. I. Academy

MS: YU/BL

1   Thomas Romney Robinson was president of the Royal Irish Academy from 1851 to 1856.
2   Charles James Blomfield (1786–1857) was bishop of London from 1828. His academic attainments are outlined in the *ODNB*.
3   In 1852 Lord Naas (Richard Southwell Bourke, sixth earl of Mayo) was appointed chief secretary for Ireland in Lord Derby's administration, and Joseph Napier became attorney-general. See Hincks's letters of 16 June 1852 and 29 November 1852 to Layard.

FROM HENRY ELLIS
4 MARCH 1853

[British Museum]
4th. March 1853

Sir Henry Ellis presents his compliments to the Rev[d]. D[r]. Hincks and requests that before his return to Ireland he would be so good as to leave with M[r]. Hawkins at the Museum, all copies of inscriptions which have been entrusted to him by M[r]. Layard.[1]

Rev[d]. D[r]. Hincks, 38 Great Russell Street

MS: BM/CA/LB 43: 107

1   Hincks wrote in his diary for 4 March: 'Called on Sir H. Ellis, who was friendly, but gave me no hope … a sad disappointment.'

TO THE PRESIDENT OF THE ROYAL IRISH ACADEMY
7 MARCH 1853

Killyleagh C° Down
7th. March 1853

My dear Sir,
    I have discovered at my recent visit to the British Museum:

1. Two fragments of syllabariums of a similar nature to that which I described in a note to my recent paper (*Transactions*, vol. xxii, P.L. p. 342).[1] One of these is in excellent preservation.

2. I obtained the complete list of the monograms, in their proper order, representing the twelve Assyrian months of thirty days and the Epagomenae. These monograms I would read provisionally by the Egyptian names of the corresponding months.[2]

3. I determined the points represented by each of the four names, which Colonel Rawlinson recognised on the Khorsabad Bulls, as representing the four cardinal points. I am not aware that the point denoted by any one name had been previously determined, any more than the place in the calendar of any one month. I have succeeded in determining them all, upon evidence that precludes further controversy.

4. I have determined the division of the manah, which was used by the Assyrians. It was sexagesimal. The manah, itself the sixtieth of the talent, contained sixty shekels. The shekel, the double shekel, and the quadruple shekel, were all represented by monograms.

I remain, my dear Sir,
Yours very truly,
Edw. Hincks

Source: Proceedings of the Royal Irish Academy 5 (1853), pp. 403–53

1   In October 1852, Hincks reported the following: 'Among some inscriptions from pieces of *terra cotta* in the British Museum, which Mr Layard recently showed me, was one which I recognized as an Assyrian syllabarium. Unfortunately, it is but a fragment; but enough remains to show its nature. It contains parts of four columns, each of which is divided by ruled lines into three series. That in the middle contains the characters to be valued; that on the left contains the values; and that on the right contains the plural form, or the value which the character would have if the plural sign were added. This syllabary, which will probably be speedily published by the authorities at the Museum, establishes a number of points on which doubts may yet linger in some minds. First, it proves that the characters are syllabic; secondly, that many values belong to the same character . . .'. See 'On the Assyrio-Babylonian Phonetic Characters', *Transactions of the Royal Irish Academy* 22 (1852), polite literature, p. 342, n.*; see also p. 335, n. *. On the importance of Hincks's discovery of this lexical list which today we refer to as 'Syllabary A' (the British Museum number is K. 62), see Daniels, 'Edward Hincks's Decipherment of Mesopotamian Cuneiform', in Cathcart (ed.), *The Edward Hincks Bicentenary Lectures* (Dublin, 1994), p. 48; 'Methods of Decipherment', in P. T. Daniels and W. Bright (eds), *The World's Writing Systems* (Oxford, 1996), p. 147. The text has been edited by R. T. Hallock, 'Syllabary A', in *Materialien zum sumerischen Lexikon*, vol. 3 (Rome, 1955), pp. 3–45.

2   The month names were usually written in Sumerian, preceded by the determinative ITI, ITU = Akkadian *warḫum*, 'month'. The names were often abbreviated, only the first sign being written, as in the list given by Hincks.

3   'List of four discoveries by E. Hincks on a visit to the British Museum, including a list of "Assyrian Months" in cuneiform "monograms", and recognition of "four cardinal points"': A Letter, *Proceedings of the Royal Irish Academy* 5 (1853), pp. 403–5. The communication was read at a meeting of the Royal Irish Academy on 16 March. I have omitted the lists of months and weights as they can be seen in the published version of the letter.

TO SAMUEL BIRCH
7 MARCH 1853

Killyleagh C° Down
7th. March 1853

My dear Sir,

I send you these copies of the Catalogue that you asked for.[1] In one of them you will find a paper which I drew this morning. I could only make two copies of it. The other I have sent to the Royal Irish Academy.[2]

I remain
Yours very truly
Edw. Hincks

MS: BM/ANE/Corr. 2710

1   *Catalogue of the Egyptian Manuscripts in the Library of Trinity College, Dublin* (Dublin, 1843).
2   See the previous letter.

FROM JAMES WHATMAN BOSANQUET
19 MARCH 1853

Claysmore Enfield
19th. Mar: 1853

My dear Sir,

I write you a few lines, as it will be interesting to you to know that Colonel Taylor's Cylinder has been found. I saw it on Friday last in the office of Alderman Finnis,[1] where it had been packed up with other curiosities, and left in the Alderman's charge by M[rs]. Taylor, and the Colonel's Executor, Captain Lynch now in India.[2] It is in beautiful preservation; but the Alderman does not feel inclined to allow any one to copy it, without the authority of the Executor. I have written to M[rs]. Taylor who is abroad, to sanction its being sent to the Asiatic Society for inspection, but I have not yet received an answer. May we expect any thing further from you on the subject of the successors of Esarhaddon at Nineveh? I was rather disappointed to find that Layard gave us nothing new on this point, though his book is deeply interesting. Is Assur-akh-bal still the reading of the name of the son of Esarhaddon?[3] and are we still to conclude that the *son*, not the *brother* of this king, as the Greeks tell us, succeeded him?

Believe me
My dear Sir
Yours very truly
J. W. Bosanquet

MS: GIO/H 63

1   Thomas Quested Finnis (1801–83), son of Robert Finnis and Elizabeth Quested, was an alderman in the city of London from 1848 to 1883 and mayor from 1856 to 1857.

2   Mrs Rosa Taylor, widow of Colonel Robert Taylor, was an Armenian, the daughter of Hovhannes Mosco, a merchant in Shiraz. She eloped with Taylor when he was a 20-year-old ensign in Bushehr, Persia, and she was only 12 years old. Henry Blosse Lynch (1807–73), explorer and soldier in Mesopotamia and India, was the son of Major Henry Blosse Lynch of Partry House, Ballinrobe, Co. Mayo and Elizabeth, daughter of Robert Finnis of Hythe, Kent. Therefore Lynch was Thomas Quested Finnis's nephew. Furthermore, he married Colonel Robert Taylor's daughter, Caroline Ann, which may explain why he was appointed Taylor's executor. For some details of his career in India, see *ODNB*.

3   The name is read *Aššur-bāni-apli* = Ashurbanipal.

FROM JAMES WHATMAN BOSANQUET
24 MARCH 1853

73, Lombard Street
24th. Mar: 1853

My dear Sir,

Many thanks to you for your answer to my inquiries. With regard to Col[l]. Taylor's cylinder, I have a letter from his widow, saying that it shall be exhibited on the return of her coexecutor Capt[n]. Lynch, who will arrive in England in the course of the year.[1] *I hope you will.* I could get you a sight of it, no doubt, from Alderman Finnis, if you should think it worth while to come into the city for the purpose.[2]

I hope you will give me the opportunity of seeing you when next in London, either at the Museum, or here, or, which I should much prefer, at my house out of town, where I could give you a bed, and bring you up in the morning in good time for the Museum. I would also read you, if inclined to listen, what I have written to prove, that your date for Sennacherib is too high.

Have you seen a little work by Joannes von Gumpach in German? – just put into my hands within this few days.[3] I see he places Sennacherib as I do, but as he takes for granted that Oltmanns' calculation of the time of Thales' eclipse was correct, he is out in all his lower dates.[4]

Believe me
My dear Sir
Yours very truly
J. W. Bosanquet

MS: GIO/H 64

1   In 1851–3 Henry Blosse Lynch commanded a squadron of vessels of the Indian navy.
2   When Hincks was in London during April to work on the cuneiform inscriptions at the British Museum, he went with Bosanquet to see Taylor's prism at Alderman Thomas Quested Finnis's office.
3   J. von Gumpach (ed.), *Hülfsbuch der rechnenden Chronologie: oder, Largeteau's abgekurzte Sonnen- und Montafeln, zum Handbrauch für Astronomen, Chronologen, Geschichtsforscher und Andere* (Heidelberg, 1853).
4   Jabbo Oltmanns (1783–1833).

TO HENRY ELLIS
6 APRIL 1853

Killyleagh C° Down
6th. April 1853

Dear Sir Henry,

I was unable to leave home last Monday, as I found it impossible to obtain a supply for the 10th. I have, however, obtained one for the 17th; and hope to be in London on Tuesday morning next.[1]

I should be very glad if the Trustees would point out what they wish that I should direct my attention to. I do not expect that much more is to be learned from the *terra cottas*, which are mostly fragments. The copies of the great historical inscriptions taken by M[r]. Layard seem to me of much greater importance. I have scarcely looked at these; as, though they were at my lodgings for a few days, I was unwilling to try my eyes by examining them by candlelight while I expected permission to take them with me to Ireland.

I should also esteem it a great favour if the Trustees would come to a decision as to engaging my services permanently. In the state of suspense in which I have been kept for so many months, I have done scarcely anything in the way of prosecuting my studies; having in the first place scarcely any time, and in the second place having no heart to pursue what I may be unable to go on with for above a few weeks. I think it necessary to observe, in consequence of an observation which you made to me when I was last in town, that I am not looking to the Trustees for any *reward* for my *past* exertions, but simply for the *means of carrying them on*. The duties of my parish are fully within the power of a single clergyman to perform; but if I have the whole of them to perform, I cannot have above a few hours in the week for study; while if I were enabled to keep a curate, I should have the greatest part of six days.

So many persons have represented to me the importance of my devoting more of my time than *I ever have done*, or that I *can do* under existing circumstances, to the Assyrian language and monuments, that if it would only involve the loss of my personal comforts I should not hesitate to engage a curate. I *cannot*, however, take this step, because it would disable me from fulfilling the pecuniary engagements which I contracted when the income of my parish was double what the legislation of the last twenty one years has made it to be.

What I ask, then, of the Trustees is merely employment for two months, or thereabouts, in London, at such a rate as would defray the expenses of that visit and the salary of a curate *for the entire year*. Suppose £120. For this I would not only work for the Museum while in London; but if I could do any thing while *here*, either in the way of assisting in the editing of inscriptions or in compiling an Assyrian lexicon, I would do it.

If this cannot be done, the sooner I am aware of the fact the better; and in that case I hope the Trustees will dispense with my paying a third visit of ten days next month, as I should on my return home wish to destroy my papers and to dismiss the subject from my thoughts. On the other hand, should they be disposed to engage me on these terms, it must be obvious that I should lose no time in looking out for a curate.

I have the honour to remain
Dear Sir Henry
Yours faithfully
Edw. Hincks

I shall hope for an answer to the first part of the letter – at the Porter's table in the Museum when I go there on Tuesday.

MS: BM/CA/OP 49

1   Hincks worked at the British Museum from 12 to 22 April. His diary for 13 April says: 'At inscription of Esarhaddon's cylinder. Compared my copy of what has been printed with the original, and found a vast number of mistakes. Found notice of an Arab queen. Wrote a paper for Royal Society of Literature on Arabian queens, refuting Rawlinson's views respecting the site of Sheba'. See his article 'On Certain Ancient Arab Queens', *Transactions of the Royal Society of Literature*, 2nd ser., 5/2 (1856), pp. 162–4 (read 13 April 1853); and my discussion of this paper in 'The Age of Decipherment: The Old Testament and the Ancient Near East in the Nineteenth Century', in J. A. Emerton (ed.), *Congress Volume: Cambridge 1995* (Leiden, 1997), pp. 90–1.

～⁊⁊⁊～

FROM HENRY ELLIS
9 APRIL 1853

[British Museum]
9th. April 1853

Dear Sir,

The Standing Committee of the Trustees of the British Museum met this day and I laid before them your letter of the 6th. inst.

In reply to your request that the Trustees would express their wish as to the points to which your attention ought to be directed, I am directed to acquaint you that as you have studied Assyrian inscriptions you must necessarily be the best judge as to the course which will most contribute to elucidate their meaning, and the Trustees therefore must leave the whole of that matter to your own discretion.

With respect to the other part of your letter the Trustees forbear to make any observation until they have had an opportunity of personal communication with you.

Believe me, Dear Sir,
Yours very faithfully,
Henry Ellis

MS: BM/CA/LB 43: 161–2

～⁊⁊⁊～

FROM HENRY ELLIS
18 APRIL 1853

[British Museum]
18th. April 1853

Dear Sir,

I am directed by the Trustees of the British Museum to acquaint you, that after the conference which they had with you on Saturday last, they resolved that the sum of £120 should be assigned to you for the duties specified in your letter to me of the 6th. inst. – the year therein mentioned commencing on the 1st. May next.[1] The Trustees request that you will commence with the Nimroud inscription thence passing to those which treat of Sennacherib: the result to be a transcript of the inscriptions, so far as it is phonetic, into English character, and a translation into the English language.

I have the honor to remain, Dear Sir
Your faithful serv^t.
Henry Ellis, Principal Librarian

MS: BM/CA/LB 44: 6–7

1   Hincks's diary for 16 April says: 'Explained my views and had to give some account of deciphering. They engaged me at £120 for the year commencing 1st May – two months to be spent in London, and to apply myself to the inscriptions for the year. It will just pay what I am out of pocket.'

TO HENRY ELLIS
4 MAY 1853

Killyleagh C° Down
4th. May 1853

Dear Sir Henry,

I send two specimens, a transcript and a translation, made from a portion of the inscription on Bellino's cylinder; and I would be obliged to you if you would ascertain from the Trustees which of the two best meets their wishes and let me know; as also whether I should make the changes suggested in the last paragraphs of what I have written.

I can have no objection to the specimens that I send being shewn to any Hebrew or Arabic scholar, who may be desirous of seeing a portion of what Sennacherib says that he did. I think, however, it would not be fair to me that they should be at present submitted to any rival decipherer.

Of course, I could not object to Col. Rawlinson, or any one else, being applied to for a *transcript* and *trans.* taken from the *same passage*; which would be a good mode of testing how far he and I agree and how far we differ; but I think he should not have a copy of mine till he has given in his own.

I hope to be in London early in next month; but as Col. Taylor's cylinder is not likely to be accessible till long after that, I hope the Trustees will not object to my dividing my time in London so that I can complete the two months later in the year when the cylinder shall have been procured.

I remain,
Dear Sir Henry,
Yours very faithfully
Edw. Hincks

[Enclosure:]

Bellino's Cylinder lines 32 (latter half) 33 and 34[1]

R.

bit.I.bar.r'u.u, na.g'u.u a.na gi.mir.ti.su, ul.tu ki.rib mati.su
Bit-Barru,   nagu   ana gimirti-su,   ultu kirib mati-su
Bit-Barru, the district through its circuit, from (connection with) his country
            (i.e. the entire district)   (an expletive)

ab.<u>dil</u>.ma; eli wichir (R.Ashur.T) we.rad.di. V.il.<u>bil.ch'a</u>.as a.na
ab<u>dil</u>ma;   eli wichir (   ) weraddi.   Il<u>bilch</u>as   ana
I severed; to the territory of Assyria annexing (it). Il<u>bilch</u>as for

ir sarru.ti u dan.n'a.at na.g'e.e shu.a.tu aj.bat.ma shum.su
ir sarruti u dannat   nage   shuatu ajbatma   shum-su
the city of government and place of strength that district I took. Its former

makh.r'a.a we.<u>sak</u>.kir.ma,   V.qar.I (Tsan) (akh'e) ir.ba at.ta.bi ni.bit.tsu
makhra   we<u>sak</u>kirma,   Qar-Tsanakhirba   attabi   nibit-su
náme   abolishing,   Qar-Tsanakhirba   I named its name
          (making strange?) (i.e. The city of Sennacherib)

I.na t'a.ay'a.ar.ti.ya, sha R.ma.d'a.aya ru.qu.ti,   sha ï.na (sar. pl.) ne (abu.
Ina tayartiya,   sha   madaya   ruquti,   sha ina (   ) ne (
In   my return,   from   distant   Medians, of whom among the kings my

pl.) ya m'a.am.man la yesh.me.u ji.kir (mate) sunu, man.da.ta.su
)-ya mamman   la yeshmau   jikir matesunu,   mandata-su
fathers no one had heard mention of their countries, their tributes

ka.bit.ta av.khar; a.na ni.ri be.lu.ti.ya we.sak.ni.tsu.nu.ti
kabitta   avkhar; ana niri belutiya   wasaknit-sunuti.
of great amount I received; to the yoke of my government making them to
submit.

In the upper of the two transcripts sent, the value of each character when known
is expressed separately, except in the case of compound ideographs where the
value of the whole group is included in a parenthesis. (*Tsan*), for instance, expresses
the joint value of three characters; (*akhe*) of two; &c. Determinative signs are

represented by capital letters conventionally selected. It is thus that I think inscriptions should be edited, when not published in facsimile; care being taken that the same English letters shall always represent the same cuneatic characters; so that by help of an index to the English characters, arranged in their alphabetical order, the original cuneatic characters can be with certainty restored. Moveable cuneatic types would then be required in the index only; with the exception of those comparatively rare cases where the value of a character was dubious or unknown. Such characters have lines under their transcriptions in the specimens I sent.

In the lower of the transcripts, the words composed of the characters are proposed to be represented, and not the characters separately. For many pupuses this is preferable. It would always, however, be easy to reduce a transcript of the former kind to one of the latter. In both these transcripts I have used *s* and *sh* to correspond to the Hebrew *shin*; *ts* to correspond to *samek*; *ch* to *sade*, and *j* to *zayin*. I have satisfied myself that these were the powers of the consonants in question in Assyrian; and I think it highly probable that they were the true values of the Hebrew letters. As, however, some of these are materially different from the values commonly assigned to them, it may perhaps be thought better to express *shin* by *sh* always, *samek* by *s* and *zayin* by *z*. I am willing to make this change if the Trustees desire it; and I am also ready to meet their wishes in taking either the upper or the lower of the transcripts for a model.[2]

Edw. Hincks Killyleagh Rectory 4th. May 1853

MS: BM/CA/OP 49

1   For the modern edition of the Akkadian text, see R. Borger, *Babylonisch-Assyrische Lesestücke*, vol. 1 (Rome, 1979), p. 72 ('Der zweite Feldzug', 11: 25–36). In the text and translation of this passage which Hincks submitted to the Trustees of the British Museum in May 1854, there are a few differences. These include *apludma* (for *abdilma*), *uraddi* (for *weraddi*), *Ilimchash* (for *Ilbilchas*), *unakkirma* (for *wesakkirma*), *ishmu* (for *yeshmau*). See MS BL Add.22097 ('Readings of Inscriptions on the Nineveh Marbles by Dr Hincks'). 'The Inscriptions of Sennacherib', received on 6 May 1854, are contained in ff. 1–27 (Hincks numbered the pages 1–53). The passage mentioned above is in ff. 5v–6r.
2   Hincks decided to retain his unusual and confusing method of transliteration and to use the lower of the two 'transcripts'. See his letter of 8 September 1851 to *The Athenaeum*.

FROM HENRY ELLIS
II MAY 1853

[British Museum]
11 May 1853

My dear Sir,

Your letter of the 4th. inst. was laid on Saturday last before our Board together with the specimens of transcripts, and translation, from a portion of the inscription on Bellino's cylinder.

The Trustees expressed their general approbation of your exertions, and directed me to acquaint you that they accede to the proposal in your letter to come at such times as may most facilitate your access to Col. Taylor's cylinder.

I remain,
My dear Sir
Yours faithfully
Henry Ellis
Principal Librarian

MS: BM/CA/LB 44: 29–30

FROM ROBERT SHIPBOY MCADAM[1]
23 MAY 1853

18 College Square
Belfast
23^d May 1853

Dear Sir,

You may have seen a notice some time since of a discovery stated to have been made of ancient MSS. at Mount Athos by a person named *Simonides*, a Greek.[2] He has come to England & has addressed the enclosed letter to Mr Masson of this town, which I send for your perusal.[3] He asserts that he has succeeded in discovering a new key to the Egyptian hieroglyphics & gives, as a sample of his method, the interpretation of an inscription in the British Museum. This I also inclose you to examine. If you thought the matter worth publishing we might make it an article in the July number of our Journal of Archaeology & give a lithograph in illustration.[4] Perhaps in this case you would yourself add an introduction & notes. However all depends on your opinion of the performance & of the man's pretensions. Be good enough to look over the whole & drop me a line when returning them.

Mr Masson is just commencing a new periodical publication in Greek, of which I send you the prospectus to look at.[5] It is of course meant for circulation chiefly among the Greeks themselves.

I am, Dear Sir,
Yours truly
Robt McAdam

P.S. We do not intend to confine our Journal to Irish subjects merely, but hope hereafter to include philology & various kindred topics.

MS: GIO/H 258

1   Robert Shipboy McAdam (1808–95), antiquarian and Irish Gaelic scholar.
2   Constantine Simonides (1820–67) was an accomplished forger. He resided in monasteries on Mount Athos between 1839 and 1841 and again in 1852. He visited England between 1853 and 1855, selling some of his forged manuscripts to unsuspecting purchasers. In 1862 he claimed to have written the Codex Sinaiticus.
3   Edward Masson was Professor of Ecclesiastical Greek in the General Assembly's College, Belfast. He resided in Greece from 1824 to 1845 and was a professor at Athens.
4   Simonides's paper was not published in *Ulster Journal of Archaeology*.
5   Edward Masson edited *The Philhellenic Banner*. Numbers appeared from June to November 1853, and in January 1854. In 1862 he edited another magazine, *The Anglo-Hellenic Witness*.

FROM JAMES WHATMAN BOSANQUET
25 MAY 1853

Claysmore Enfield

25th. May 1853

My dear Sir,
    I attended the annual meeting of the Royal Asiatic Society last Saturday, when a report from Colonel Rawlinson was read.[1] His messenger it appears had been robbed soon after leaving Baghdad, so that the chief matter was not forthcoming. You will be interested however in hearing that he has found what he calls an almanach running over 12 years, with a list of kings if I understood the report rightly. Also what he has termed a dictionary & grammar of the language and astronomical observations.

With regard to the Insurance Co. of which you spoke to me when I had the pleasure of seeing you, I have made some inquiry, and I hear that it was anxious to dispose of its business some little time ago, which looks any thing but satisfactory. I hope you are proceeding with your labours.

Believe me
Yours sincerely
J. W. Bosanquet

MS: GIO/H 65

1   Annual Report for May 1853, *Journal of the Royal Asiatic Society* 15 (1853/55), p. xvi.

TO SAMUEL BIRCH
30 MAY 1853

Killyleagh C° Down
30th. May 1853

My dear M[r]. Birch,
   You asked some time ago for additional copies of the list of Assyrian months.[1] I received a few copies of the paper that I send (at the close of which the list is given) yesterday morning. I find that no copies containing the list *alone* were struck off; which has been a disappointment to me. A copy has it seems been already sent to Sir H. Ellis & that with what I now send & with the MS. list which I sent on the 7th. March will, I hope, suffice.

Believe me
Yours very truly
Edw. Hincks

I hope to be in London in about a fortnight. I fear I cannot leave this till the 10th; but will go sooner if I can.

MS: BM/ANE/Corr. 2708

1   'List of four discoveries by E. Hincks on a visit to the British Museum, including list of "Assyrian Months" in cuneiform "monograms", and recognition of "four cardinal points"': A Letter, *Proceedings of the Royal Irish Academy* 5 (1853), pp. 403–5.

FROM EDWIN NORRIS
1 JUNE 1853

R. Asiatic Society
1st. June, 1853

My dear Sir,

Many thanks for your paper.[1] Your months coincide with those sent me by Rawlinson, except that his first ⊢⊟⊦ is your eighth, and his eighth . . .[2] I have myself been inclined to make ⊧⊟ (the ninth, which corresponds with Atriyatiya as you know) = Athyr; but as the old Egyptian year was vague, it will scarcely help us for the season. Perhaps the Jewish Adar may do more for us; and as we have a Markazana in Persian, there is another Jewish similarity. I take the name Markazana from the Median, of which I hope to send you a copy in a week or two; it has been an enormous time in the printers hands, but the last sheet is now in type.[3] My hypothesis is that the language is Ugrian or Magyar.[4]

I am very much pleased to see that your determinations in almost all cases coincide with those of Rawlinson. I expect a large booty with his next communication. His dispatch, which contained a full detail of his most recent discoveries intended for our annual report, was plundered by the Arabs and everything destroyed. This last contained the barest résumé very hurriedly dashed off. I hope he will be in London this year to dig in our Museum. He laments that he has in Baghdad only the débris. I heartily wish you were at our Museum, regularly installed; without some such measure, the immense mass of valuable material there will always be buried in those wretched cupboards.

I am, my dear Sir
Yours very truly
Edwin Norris

P.S. Are you at all interested in African languages? I have just finished a sketch of a Bornu grammar, of which I will send you a copy with the Median, if you have any curiosity to look at the language.[5] It is curious as being unlike any other negro tongue, and oddly enough, its structure, and even some of the words, are not unlike those of the Median Inscriptions. Of course you will not suspect me of imagining any connection: the oddness consists in my working at them at the same time.

MS: GIO/H310

1   See the preceding letter, n. 1.
2   The sign is illegible.

3   Norris, 'Memoir on the Scythic Version of the Behistun Inscription', *Journal of the Royal Asiatic Society* 15 (1853/1855), pp. 1–213. Separate copies of the work were circulated in 1853.
4   Hungarian (Magyar) belongs to the Finno-Ugric group of languages.
5   See E. Norris, *Grammar of the Bornu or Kanuri Language; with Dialogues, Translations, and Vocabulary* (London, 1853).

~~≈≈~~

TO AUSTEN HENRY LAYARD[1]
14 JUNE 1853

67 Great Russell St[2]
Bloomsbury
14th. June 1853

My dear M[r]. Layard,

A M[r]. Pote, whom you have perhaps heard of, tells me he is about to publish some cheap popular work containing translations of inscriptions.[3] I gave him a few specimens from the Nimroud obelisk. He urges me very strongly to give him readings of Sennacherib's, & says he intends taking some extracts that you have published. The North British have done the same.[4] I told him they were not so good as I could *now* produce; & he wants me to *correct* them for him, so as to be brought up to our *present* knowledge. I told him I would not do this, or be a party to his republishing them atall, without your consent; & I accordingly write to know what you wish done. His book will contain translations by Rawlinson (whose friends have given him licence to publish as much as he pleases), by *me* & I fear by himself also. His object is to catch the million, in which I think he will fail. *My* object is to put on record before any thing more of Rawlinson's appears, some specimens of my improved translation. I have made very great progress in the language during the last few months; & am hard at work on a complete translation of the Nimrud fragment inscription of that old *miscreant*, who records atrocities to which I believe there is no parallel to be found in the annals of cruelty; & this as if it was something greatly to his credit.[5] I reserve all this; but the specimens from the obelisk will be a curious contrast to Rawlinson's translation of the same; which is mainly conjectural & in many instances the reverse of the real meaning.[6]

I saw Col. Taylor's cylinder which seems in perfect preservation. I wonder when it will be accessible for copying. The 8 years' annals are separated by cross lines from each other & from the preliminary & final matter. It was a present to the Governor of Carchemish; & it seems Col. T. found it on the Euphrates at or near the site of that city. The cylinder in the B.M. that was in fragments has been restored to a great extent. I have not seen it since I came last; M[r]. Vaux being in

Paris.[7] The terracottas have all been removed from their former cases; so that I have not been able to do much at them. I hope, however, to look over them next week. I suspect that the cylinder in the Museum is the work of Shalmaneser, the predecessor of Sargon; but till it was put together I could not make much of it out. It also needed cleaning.

They have had a good day for trying the capabilities of the soldiers at the camp. I fear many will suffer from it. I trust this Turkish affair will be satisfactorily settled.[8]

Believe me
Yours very truly
Edw. Hincks

I am 6 hours in the Museum every day.

MS: BL Add. 38981, ff. 365–6

1   Layard was in Constantinople for a month from the beginning of April to assist the ambassador Stratford Canning (Lord Stratford de Redcliffe). This was a bad move on Layard's part. He did not get on with the ambassador and wrote: 'to be of any use one must be at the head of the Embassy, not at the beck and call of a man who suspects everything and everyone and only has one end – his own selfish views.' See Waterfield, *Layard of Nineveh*, p. 236.

2   Hincks travelled to London on 9 June and worked at the British Museum for a month. His diary for 13 June says: 'Up at 6, working at my cuneatic inscriptions. At Museum from 10 to 4. In the evening at inscriptions till bed at 9¾.' He returned to Killyleagh on 9 July, having visited the Exhibition in Dublin on the way. See J. Sproule (ed.), *The Irish Industrial Exhibition of 1853* (Dublin, 1854). According to his diary for 4 July, Hincks was in the Trinity College Library: 'at Lepsius' inscription all the morning.' So he was not at the meeting of the Royal Society of Literature in London on 6 July when his paper 'On an Ancient Cylinder in the British Museum' was read. It was published later in *Transactions of the Royal Society of Literature*, 2nd ser., 5 (1856), pp. 165–8.

3   R. G. Pote, *Nineveh: A Review of Its Ancient History and Modern Explorers* (London, [1854]). See the review by J. Bruce in *The Athenaeum*, no. 1395 (13 May 1854), p. 589.

4   [D. Brewster], Review article on A. H. Layard, *Discoveries in the Ruins of Nineveh and Babylon* (London, 1853), *North British Review* 19/37 (May 1853), pp. 255–96. On pp. 267–8 Brewster gives Hincks's (and Rawlinson's) translation of the passage mentioning Hezekiah of Judah and the siege of Jerusalem in the Annals of Sennacherib. They are taken from Layard, *Discoveries in the Ruins of Nineveh and Babylon*, pp. 143–4. See also Hincks's version in his letter of 28 June 1852 to Layard.

5   Hincks published his version in 'The Nimrûd Obelisk', *Dublin University Magazine* 42 (Oct. 1853), pp. 420–6.

6   See Rawlinson's translation of the Black Obelisk in his article 'On the Inscriptions of Assyria and Babylonia', *Journal of the Royal Asiatic Society* 12 (1850), pp. 430–48.

7   William Sandys Wright Vaux (1818–85) was an assistant in the department of antiquities at the British Museum.

8   In June 1853 an Anglo-French naval force entered the Dardanelles.

～ゝ～

TO AUSTEN HENRY LAYARD
6 AUGUST 1853

Killyleagh C° Down
6th. August 1853

My dear Mr. Layard,

I have desired a Belfast Morning to be sent to you which will I presume reach you by the same post as this. It contains a letter of mine on the Education Question, on which all the London Newspapers seem to have been writing – in complete ignorance as to the facts of the point at issue.

I have been busy at the Assyrian & have ascertained some very curious points as to the grammar of the language which admits much greater variety than the Hebrew or even the Arabic.[1] I have completely settled (to my own satisfaction) the names of the old Nimrud king & what answer to Merodach-Baladan, Sennacherib & Esarhaddon. The old king was Sardanapalus – at least the name is the same. I have found the *exact reading* of these four names & their *significance*; & till these two are *both* ascertained, I shall not feel confident of *any name*.[2] Ishmi-Dakan, 'Dagon has heard', the oldest name yet found gave me the clue to the four others.[3] The received vocalization of all of them is incorrect.

When you were here you told me something about an Ararat on the Tigris or at any rate south of the celebrated mountain of that name. I find in the inscription not only *Urardhi* (Ararat of Armenia) but *Arardhi*, which represents the name better & which seems to lie beside the Upper Tigris. It is a mountain.

Believe me
Yours very truly
Edw. Hincks

MS: BL Add. 38982, f. 57

1   The entry in Hincks diary for 9 August says: 'Arabic grammar all the evening, it differs immensely from Assyrian.'
2   As he told Layard in his letter of 14 June 1853, he was hard at work on a new translation of the Black Obelisk of Shalmaneser III, which he published some months later: 'The Nimrûd Obelisk', *Dublin University Magazine* 42 (Oct. 1853), pp. 420–6. The following extract (p. 421) illustrates exactly what he is saying about the names of Assyrian kings in this letter. I have given today's readings and translations in square brackets.

The name of the king's father is Assur-yuchura-bal, i.e. 'Assur has formed a son' [*Aššur-nāṣir-apli:* Ashurnasirpal, 'Ashur guards the inheriting son']. A name very similar to this was Assur-iddana-bal, 'Assur has given a son'; out of which the Greeks made Sardanapalus. Both names seem to have been borne by the son of Assur-akh-iddan [*Aššur-aḫ-iddina*, Esarhaddon]. The name last mentioned signifies, 'Assur has given a brother', as the name of his father, Sin-akhi-irib [*Sin-aḫḫē-erība*], signifies

'Sin has multiplied brethren' ['Sin has replaced brothers'; root R'B not RB']. We may infer that the first of these three names was given to the eldest son of his father, the next to a second son, and the last to one who had at least two elder brethren. Analagous to the first of these is the name of the Chaldean king, Marduk-bal-iddan, 'Marduk has given a son' [*Marduk-bal-iddina*, 'Marduk gave the inheriting son']. He is called, in the present text of Isaiah xxxix, 1, the son of Baladan, an unmeaning, and therefore impossible name. We may, with a high degree of probability, restore Bil-bal-iddan, 'Bil has given a son'. A similar name, Nabû-bal-iddan, 'Nabu has given a son' [*Nabû-apla-iddina*, 'Nabu gave the inheriting son'], was borne by the king of Chaldea contemporary with the father of the obelisk king.

A. R. Millard, 'Baladan, the Father of Merodach-Baladan', *Tyndale Bulletin* 21 (1971), pp. 125–6, suggests that the name *bal'ᵃdān* in Isaiah 39: 1 is *Bel-iddin(a)*, comparing an Aramaic name *bl'dn*.

3   The name is Ishme-Dagan.

FROM AUSTEN HENRY LAYARD
9 AUGUST 1853

9 Little Ryder St
Aug^r. 9/53

My dear D^r. Hincks,

Very many thanks for the Belfast paper containing your letter which I have read with great interest. I am starting on Monday or Tuesday next for Italy, & shall probably remain abroad during the Autumn months.[1] I hope to see you on my return, to find that you have made many important discoveries in the meanwhile. M^r. Rassam writes that they are still finding sculptures at Kuyunjik. The new obelisk seems to be a highly curious monument. Rawlinson thinks it belongs to the first Tiglath Pileser.[2] I wish your position at the Museum were more satisfactory. I have been speaking today about it to Lord Rosse who is very well inclined.

The traditional mountain on which the ark rested, the Jebel Judi, is near the Tigris where it enters into the plains at Jezireh, but I have never heard it called by the name of Ararat, nor can I swear that there is any other mountain which bears that name except the well known one in Armenia.

I hope M^{rs}. Hincks & your daughters are well. Pray give them my kind regards. Should you have occasion to write to me, please direct letters care of M^r. Murray in Albemarle Street, after next Sunday. The Assyrian Society promises much; they have already got much above £1000.[3]

Yours sincerely,
A. H. Layard

MS: GIO/H240

1   Layard remained in London and on 16 August he made his first major speech in parliament. It dealt with Russia's aggression. See Waterfield, *Layard of Nineveh*, pp. 237–49.
2   In February 1853 during the excavations at Kalaḥ Shergat (ancient Ashur), Hormuzd Rassam found a well-preserved clay prism. It contained the annals of the Assyrian King Tiglath-Pileser I.
3   On the 'Assyrian Exploration Fund', see M. Trolle Larsen, *The Conquest of Assyria*, p. 322.

⟨ornament⟩

FROM JOHN BOWRING[1]
19 SEPTEMBER 1853

[Dear Sir,]

I am writing a book with a view to prepare the public mind for the introduction of the Decimal System into our coinage and accountancy.[2] Perhaps you may remember that it was on a motion of mine that the florin was coined,[3] and now we have a unanimous report from the committee recommending the immediate adoption of decimal currency.[4] I want to trace the *history* of the decimal currency. I have got from Sir Gardner Wilkinson what he knows as to ancient Egypt, and from Dr. Wilson such light as he can give regarding the Hindoos.[5] But you have been exploring the Assyrian and perhaps other fields. Can you give me any light as to these regions and peoples or refer me to any sources where my imperfect knowledge may be brightened into information?

[John Bowring]

Source: Davidson, *Edward Hincks*, p. 40

1   Sir John Bowring (1792–1872), political economist, politician and linguist. He was active in the Unitarian Church and knew Hincks's brother William.
2   See J. Bowring, *The Decimal System in Numbers, Coins, and Accounts: especially with Reference to the Decimalisation of the Currency and Accountancy of the United Kingdom* (London, 1854).
3   Bowring had long favoured the introduction of decimal coinage. In April 1847 he proposed in the House of Commons the introduction of the florin (= 2 shillings = one tenth of a pound).
4   In 1853 a Select Committee of the House of Commons recommended decimalisation of the British currency.
5   I have not found any reference to a 'Dr Wilson' in Bowring's book. Perhaps he is referring to John Wilson (1804–75), Scottish missionary and orientalist, who spent most of his life in India. If he means Horace Hayman Wilson (1786–1860), who was Boden Professor of Sanskrit at Oxford, librarian at East India House, and director of the Royal Asiatic Society, then one would expect the title 'Professor Wilson'.

TO JOHN BOWRING
21 SEPTEMBER 1853

Killyleagh, Down
21ˢᵗ. Septʳ. 1853

My dear Sir,

I have just received your letter of the 19ᵗʰ. & lose no time in replying to it. The Assyrians were partial to the sexagesimal system, as opposed to the decimal, of which they were probably the inventors. They had a noun, denoting 'a sixty' analogous to our score & dozen & in expressing 360 would say 3 hundreds & a sixty. Their talent *tikun* from the root 'to weigh' contained 60 manahs *mana* from the root 'to count'; & this again contained 60 of what we may call provisionally shekels but of which the Assyrian *name* is unknown, the *monogram* only having been as yet found.[1] You will see at the end of Layard's last work a table of the Assyrian weights that have been preserved.[2] It appears from this that they had two scales slightly differing, the royal & the popular.

In the University Magazine for next month, there will be a translation of the Annals on the Nimrud Obelisk; the first translation from the Assyrian, of more than a few lines, which has any pretensions to accuracy.[3] That published some years ago by the Asiatic Society as Col. Rawlinson's was *conjectural*.[4] Since I sent this, I have been enabled to give the true translation of four or five clauses that are given as doubtful in the Magazine. I am now advancing in the knowledge of the language with great rapidity. Every day I ascertain the meaning of several new words. I am compiling a glossary or lexicon with the words digested under their roots; & when I have entered in it all that I already know, I shall have considerably above 1000 entries. I hope to be able to publish this next year. It throws vast light on all the cognate languages – (Semitic, as usually called).[5]

With respect to the decimal system, I am very desirous that it should succeed; & there is one point which I should like very much to press, if I could do it with any weight. Let the present penny stamp (for postage & receipts) stand as 5 mills; & simultaneously with the change & as a compensation for it let there be a liberal reduction of the ocean postage to 10 if not 5 mills. What the Govᵗ. would gain by the change from 1/240 to 1/200 – one fifth – would amply compensate for what they would lose by the reduction in ocean postage.

Believe me my dear Sir
Yours very truly
Edw. Hincks

MS: Bod. Lib. Eng. lett. b. 32, ff. 187–8

1   See Hincks's letter of 4 May 1852 to Henry Ellis and especially his letter of 7 March 1853 to the Royal
Irish Academy.
2   Layard, *Discoveries in the Ruins of Nineveh and Babylon*, p. 601.
3   'The Nimrûd Obelisk', *Dublin University Magazine* 42 (Oct. 1853), pp. 420–6.
4   'On the Inscriptions of Assyria and Babylonia', *Journal of the Royal Asiatic Society* 12 (1850), pp. 430–48.
5   This project seems not to have advanced very far.

TO JOHN BOWRING
21 SEPTEMBER 1853[1]

The Assyrians were partial to the sexagesimal system as opposed to the
decimal, of which they were probably the inventors. They had a noun denoting 'a
sixty'[a] analagous to our score or dozen; and, in expressing 360, would say three
hundred and a sixty. Their talent (*tikun*), from the root 'to weigh', contained 60
manah (*mana*), from the root 'to count', and this again contained 60 of what we
may call provisionally shekels, but of which the Assyrian name is unknown, the
monogram only having been yet found. With reference to the measures of the
Assyrians, I believe they invented the sexagesimal divisions still in use in this
country; but it would seem that their measurement of terrestrial lengths was
decimal. The oldest palace recorded to have been built (on Michaux's stone) is
said to have been in length three lengths, and in breadth one length and fifty – Ψ.
The palace was probably, then, as long as it was broad, which would give 100 of
the smaller measures, equal to that called 'a length'. I take the monument to be of
the date 1200 B.C. In confirmation of this, we have in the Khorsabad bulls the
cubication, 6 lengths and 50 cubits. What I here suppose to be a cubit, must be
nearly of that value; but some colossal figures are said to have been nine of this
measure in length. Now we know that the Assyrians had a cubit (in the inscription
of Nebuchadnezzar, at the India House, 680 *ammat* are mentioned), and it is
extremely unlikely that they had another measure so near this in magnitude, as
that a statue should contain nine of it – a statue spoken of as 'the pride of Khamana',
and evidently of colossal size. I regard it, then, as certain, that the smaller measures
were cubits, and 'the length', of course, 100 cubits. It is natural to suppose (and
yet not certain) that the series was continued in decimal progression. In Egypt,
however, though the square measures in the time of the Ptolemies were certainly
the arura[2] of 10,000 square cubits, the square measure of 100 square cubits, and
the cubit itself; and though we may presume the existence of measures of length
equal to the sides of the two latter, we know that the royal cubit contained

7 hands, or 28 digits; and it is possible, I may say probable, that the Assyrians had a similar division of their cubit.

a   The ancient Bohemians had *Kopa*, and the Danes to this day use *skok*, for sixty.[3]

Source: Bowring, *The Decimal System*, pp. 10–11

1   This piece was possibly an enclosure in the preceding letter from Hincks; but it may have been part of another letter. Bowring states that it he took it from the MS of a letter from Hincks.
2   The *arura* was an Egyptian measure.
3   Danish *skok* is not used today.

TO THE EDITOR OF NOTES AND QUERIES
OCTOBER 1853

Killyleagh, C° Down

Sir,

     Many questions are proposed by G. W., to which it is extremely improbable that any but a conjectural answer can ever be given.[1] That tin was in common use 2800 years ago, is certain. Probably evidence may be obtained, if it have not been so already, of its use at a still earlier period; but it is unlikely that we shall ever know who first brought it from Cornwall to Asia, and used it to harden copper. It is, however, a matter of interest to trace the mention of this metal in the ancient inscriptions, Egyptian and Assyrian, which have of late years been so successfully interpreted. Mistakes have been made from time to time, which subsequent researches have rectified. It was thought for a long time that a substance, mentioned in the hieroglyphical inscriptions very frequently, and in one instance said to have been procured from Babylon, was *tin*. This has now been ascertained to be a mistake. M[r]. Birch has proved that it was *lapis lazuli*, and that what was brought from Babylon was an artificial blue-stone in imitation of the genuine one. I am not aware whether the true hieroglyphic term for *tin* has been discovered. Mention was again supposed to have been made of *tin* in the annals of Sargon. A tribute paid to him in his seventh year by Pirhu (Pharaoh, as Col. Rawlinson rightly identifies the name; not Pihor, Bocchoris, as I at one time supposed), king of Egypt, Tsamtsai, queen of Arabia, and Idhu, ruler of the Isabeans, was supposed to have contained tin as well as gold, horses, and camels.[2] This, however, was in itself an improbable supposition. It is much more likely that incense and spices should have been yielded by the countries named than tin. It reads *anna*; and I supposed it, till very lately, to mean 'rings'. I find, however, that it signifies a metal, and that a different word has the signification 'rings'. When Assur-yuchura-bal, the founder of the north-western palace at Nimrûd, conquered the people who lived on the banks of the Orontes from the confines of Hamath to the sea, he obtained from them twenty talents of silver, half a talent of gold, one hundred talents of *anna* (tin), one hundred talents of iron, &c. His successor received from the same people all these metals, and also copper.[3]

It is already highly probable, and further discoveries may soon convert this probability to certainty, that the people just referred to (whom I incline strongly to identify with the *Shirutana* of the Egyptian inscriptions) were the merchants of the world before Tyre was called into existence; their port being what the Greeks called Seleucia, when they attempted to revive its ancient greatness. It is probably to them that the discovery of Britain is to be attributed; and it was probably from them that it received its name.

In G. W.'s communication, a derivation of the name from *barat-anac*, 'the land of tin', is suggested.[4] He does not say by whom, but he seems to disclaim it as his own. I do not recollect to have met with it before; but it appears to me, even as it stands, a far more plausible one than *bruit-tan*, 'the land of tin': the former term being supposed to be Celtic for *tin*, and the latter a termination with the sense of *land*: or than *brit-daoine*, 'the painted (or separated) people'.

I am, however, disposed to think that the name is not of Phoenician origin, but was given by their northern neighbours, whom I have mentioned as their predecessors in commerce. These were evidently of kindred origin, and spoke a language of the same class; and I think it all but certain, that in the Assyrian name for tin (*anna*) we have the name given it by this people, from whom the Assyrians obtained it. 'The land of tin' would be in their language *barat* (or probably *barit*) *anna*, from which the transition to Britannia presents no difficulty. I assume here that *b-r-t* here, without expressed vowels, is a Phoenician term for 'land of'. I assume it on the authority of the person, whoever he may be, that first gave the derivation that G. W. quotes. I have no Phoenician authority within reach: but I can readily believe the statement, knowing that *banit* would be the Assyrian word used in such a compound, and *n, r,* and *l* are perpetually interchanged in the Semitic languages, and notoriously so in this root. *Ummi banitiya*, 'of the mother who produced me', is pure Assyrian; and so would *banit-anna*, 'the producer of tin', be; all names of lands being feminine in Assyrian.

It would be curious if the true derivation of the world-renowned name of Britain should be ascertained for the first time through an Assyrian medium.

Edw. Hincks

Source: Notes and Queries 8/206 (8 Oct. 1853), pp. 344–5

1 'G. W.', 'Early Use of Tin', *Notes and Queries* 8/204 (24 Sept. 1853), p. 291.

2 In the Annals of Sargon (seventh year) we read: 'From Pir'u, the king of Musru, Samsi, the queen of Arabia, It'amra, the Sabaean – these are the kings of the seashore, and from the desert – I received as their presents, gold in the form of dust, precious stones, ivory, ebony-seeds, all kinds of aromatic substances, horses (and) camels.' See Pritchard (ed.), *Ancient Near Eastern Texts*, p. 286.

3 The Akkadian word for 'tin' is *annaku*. It is regarded as a 'culture word' by Mankowski, *Akkadian Loanwords in Biblical Hebrew*, pp. 35–6.

4 See S. Bochart, *Geographiae sacrae pars altera Chanaan seu de coloniis et sermone Phoenicum* (Cadomi [Caen], 1646), p. 720: 'Porro Bretanica mihi quidem nihil videtur esse aliud quàm *brt-'nk* Barat-anac, id est *ager*, seu terra *stanni & plumbi*.' This statement is in chapter 39 (pp. 719–26) on 'Phoenices in Britanniâ & Hiberniâ & Cassiteridibus insulis'.

FROM JOHN BOWRING

3 OCTOBER 1853

[3rd October 1853]

[Dear Sir,]

One of the great impediments to the popularity of the Decimal System is the fear that people will have to pay 5 mils for the 1d. postage stamps.

Rowland Hill tells me he *now* thinks they shall get *as much revenue* from 4 mils as from 5.[1] And according to all experience in such matters the one fifth which you anticipate would be added to the receipts by the augmentation of the charge would be practically found to be the cause rather of loss than gain. I am of opinion that when we get 250 stamps for a £stg instead of 240 we shall *use* the *additional* ten.

The ocean postage stands upon grounds of its own, and will, I trust, be carried in due time. I have long thought that *communication* is not a fit subject for taxation and I am by no means sure that it would not be *profitable* to a community that the Government should provide for the gratuitous circulation of correspondence in same way that it provides for the administration of justice.

[John Bowring]

Source: Davidson, *Edward Hincks*, p. 41

1 Rowland Hill (1795–1879) is usually connected with postal reform and credited with the invention of the postage stamp.

FROM HENRY ELLIS
11 OCTOBER 1853

British Museum
October 11th. 1853

My dear Sir,

I have the pleasure to acquaint you that I had the opportunity of laying your letter from Killyleagh of September 27th. before a committee of our Trustees on Saturday last, by whom I am directed to report to you their minute of 16th. April, and to state clearly that at May 1854 the agreement ceases.

The following are the words of the minute:

'Resolved, that the sum of £120 be assigned to the Rev^d. D^r. Hincks for the duties specified in his letter of the 6th. inst., the year therein mentioned commencing on the 1st. of May next. D^r. Hincks to commence with the Nimroud inscriptions, thence passing to those which treat of Sennacherib, the result to be a transcript of the inscriptions, so far as they are phonetic, into the English character, and a translation of them into the English language.'

The Trustees desired me to say they beg you will take your own time within the space mentioned for what you have undertaken. The Trustees have further directed me to advance the sum of thirty pounds, being the pay for three months, the rest to be paid upon the execution of the business you have in hand.

The following is the passage of your letter of 6th. April 1853 referred to in the Trustees minute of April 16th:

'What I ask the Trustees is merely employment for two months, or there-abouts in London at such a rate as would defray expenses of that visit and the salary of a curate for the entire year: suppose £120. For this I would not only work for the Museum while in London, but, if I could do any thing while here, either in the way of assisting in the editing of the inscriptions, or in compiling an Assyrian Lexicon I would do it.'

If you will have the goodness to tell me through what channel you wish the thirty pounds to be paid, I will lose no time in attending to your directions.

Very faithfully yours
Henry Ellis

MS: BM/CA/LB 44: 175–6

━━━

TO HENRY ELLIS
15 OCTOBER 1853

Killyleagh C° Down
15th. Octr. 1853

Dear Sir Henry,

I received your letter of the 11th. in course. I send a receipt for the £30 you mention; and would thank you to send me a cheque for the amount which I can get cashed at my banker's here.[1]

I hope to be at the Museum in March.

Believe me to remain
Yours very faithfully
Edw. Hincks

MS: BM/CA/OP

---

1   A draft upon the Bank of England for £30 was sent to Hincks on 17 October 1853. See *MS: BM/CA/LB 45: 6.*

━━━

TO THE EDITOR OF THE DOWNPATRICK RECORDER
14 DECEMBER 1853

[Killyleagh C° Down
14th. Dec. 1853]

Dear Sir,

Unwilling though I may be to drag the affairs of my parish before the public, especially when, as in the present instance, they redound so little to the credit of some of my parishioners, the editorial article in your paper of last week, which you have done me the honour of devoting to me, leaves me no alternative, as it is evident you have been mis-informed in many important particulars on the 'Anti-Hood Aggression' of which I do you the justice to believe, had you known 'the whole truth and nothing but the truth', you would have written in a very different spirit. Although I deny the right of any private individual to call me to account for an act so strictly personal as the wearing or non-wearing of my academic hood, I am ready, as a matter of course, to give my reasons for adopting the use of it so many years after I had been entitled to wear it.

Circumstances having rendered it probable that I should have to spend some part of this year in London, and it being not improbable that I might be called upon to officiate in one of the Churches there, it naturally suggested itself to me to provide myself with what is there universally worn; having got the hood, it did not seem to me that there was any reason against wearing it also in my own Church, many of my brother ministers in the diocese constantly doing the like; and it never having occurred to me that the Protestant spirit of any of my parishioners *could* take alarm at what every person who would give himself the trouble to inquire would find to be a mere badge of scholastic rank, not of any party in the Church. Had I foreseen the result, as I have over and over again said, I should have forborne 'to put a stumbling-block in the way of a weak brother' – but, as the affair at present stands, I consider that, having done no wrong, it would lower my character in the eyes of all right-minded persons to yield to an ignorant agitation, resulting in an act of violence contrary both to the laws of God and man.[1]

The root of this agitation moreover I believe to have lain not *within* but *without* the Church. In the course of the summer, the Presbyterians of Killyleagh saw fit to get up a course of controversial lectures, which, professing to be directed against 'Popery and Prelacy', dealt even worse with the latter than the former – our church, clergy, prayer-book, sacraments, in short all that is or ought to be dear to churchmen as such, were held up to ridicule and contempt. To these exhibitions, from which even moderate Presbyterians were known to absent themselves, some professing churchmen were inconsistent enough to resort; and thus 'with itching ears heaping to themselves teachers', they learned but too well their lesson; and forgetting that those who ridiculed the hood as 'a rag of Popery', would speak in no more respectful terms of the surplice, they have given a just cause of triumph to the enemies of the Church, by fanning the flames which *they* had kindled.

You have, however, been misinformed as to the *extent* of the agitation; it has happily been confined to a very few, while from many of my parishioners I have experienced a sympathy which has been very consoling. That it should exist still is a matter of deep regret to me; and I earnestly hope and pray that, in the near approach of that holy season which speaks to every Christian of peace and good-will, we may all 'have grace to lay aside all hatred and prejudice, and whatsoever is contrary to godly union and concord', and that, 'seriously laying to heart the great danger we are in by reason of our unhappy divisions', we may endeavour henceforth to 'dwell in love', and 'have peace one with another'.

With many apologies for this lengthened communication,

I am, dear sir
Faithfully yours
Edw. Hincks, Rector of Killyleagh

P.S. My curate, Mr Butterworth, did not wear *his* hood until I had procured mine. To charge him with *vanity* for using it seems singularly unreasonable. When every clergyman had *some* degree, what vanity would there be in wearing the hood of the *lowest* degree.[2]

Source: *Belfast News-Letter,* Issue 11975 (19 Dec. 1853)

1   In the *Belfast News-Letter,* Issue 11968 (2 Dec. 1853), an article entitled 'Alleged Puseyism in Killyleagh' appeared with extracts from, and comments on, an account of the alleged protests made against Hincks and his curate by some of his parishioners. I give the first paragraph of the article here.

We observe in the League Utensil of yesterday a would-be witty paragraph, introduced on the *quasi* authority of a Killyleagh correspondent, who subscribes himself 'a Puritan', in which a stupid and ignorant attack is made upon the Rector of Killyleagh and his curate, under the heading of 'An Irish Method of Suppressing Puseyism'. The materials of this paragraph may, perhaps, have been furnished by a *bona fide* correspondent, but it has evidently been *doctored* by the editorial process, as its pedantic verbosity fully testifies. The story is to this effect: that the curate of the parish, an English gentleman, 'after the Puseyite fashion of his own country', thought fit, after his induction, to ornament his back, during Divine service, with 'a goatskin, ecclesiastically styled a hood'. Immediately the rector himself appeared in church with a red hood – that the congregation, 'fancying they beheld the mystic Babylon in the scarlet of their pastor, resolving to have no fellowship with the works of darkness, protested against what they considered Popish innovation, and left the church' – and, finally, that, 'on Sabbath week, one or more of them entered the vestry, and left the "mystic Babylon" of the rector's hood a spectacle of shreds and patches'.

2   Albert Nelson Butterworth (1826–86) was educated at St John's College, Cambridge, where he took his B.A. degree in 1850. See Leslie and Swanzy, *Biographical Succession Lists of the Clergy of Diocese of Down,* p. 146.

FROM JAMES WHATMAN BOSANQUET
21 DECEMBER 1853

Claysmore Enfield
21st. Decr. 1853

My dear Sir,

I am extremely sorry to hear that you are compelled to relinquish your labours in the Assyrian field, which will be a real loss to the public.[1] There are so few who have the time or talent for such studies, that no one who has made such progress as you can be spared. I only wish that I had time and ability enough to take up vigorously your pursuit. If you really intend to bring your Assyrian books and papers to sale, pray let me have a catalogue of them. I have a friend who is making rapid progress in Egyptian antiquities, and who I think might possibly turn his head to Assyrian. Perhaps there may be some that might be useful to me.

I hope I shall have the pleasure of seeing you when you come to London. I trust you are getting on with the annals of Assurachbal. By the bye may not

Assarac-bal the third, be the Saracus of Polyhistor & Abzdianas? Suppose him to have been an infant at the death of Esarhaddon, and that his throne was usurped by Adramelec, & Axerdis, and also by Nabopolassar, or Sardanapalus, may not the Scythians have supported the expelled though lawful heir, and have sustained him on the throne till the final destruction of Nineveh? Nothing is more common in these days, and the East never changes, than for the foreign interfering power to set up any pretender to the throne, and to make use of him as a vassal prince. I am now investigating the authority for our assumed date B.C. 536 for the first of Cyrus in Babylon. I think there is no solid ground for this date.

Believe me
My dear Sir
Yours very truly
J. W. Bosanquet

MS: GIO/H 67

1   The cause was the reduction in Hincks's clerical income. See Hincks's letter of 23 January 1854 to Layard.

TO THE EDITOR OF THE ATHENAEUM
DECEMBER 1853

ASSYRIAN LANGUAGE. REV. DR. E. HINCKS would dispose of a number of BOOKS and MSS connected with the ASSYRIAN LANGUAGE, and would also give *viva voce* instruction therein, to a Gentleman who may be willing to devote himself to this important study, and who, from his age, antecedents, and present position, may appear to him likely to succeed in it. Apply to him at the Rectory, Killyleagh, Co. Down, before the 21st January.[1]

Source: The Athenaeum, no. 1365 (24 Dec. 1853), p. 1533

1   This advertisement was also published in *Notes and Queries* 8/217 (24 Dec. 1853), p. 634, and 8/218 (31 Dec. 1853), p. 658. In *Notes and Queries* 8/218 (31 Dec. 1853), p. 656, the following comment appeared:

> Our readers, we have no doubt, shared the regret with which we read the advertisement in our columns last week from the Rev. Dr Hincks, who, from the want of encouragement, and in the face of peculiarly adverse circumstances, is compelled to withdraw from the field of Assyrian discovery; and who is advertising for some competent person who will work out what he has in progress. Although Assyrian literature may at present be discouraged by the Church and neglected by the Universities, there can be little doubt that it must ere long assume a very different position: and we therefore trust that some means may yet be taken to prevent Dr Hincks' withdrawal from a field of study in which he has been so successful.

1854

FROM 'J. R.' TO THE EDITOR OF THE ATHENAEUM
JANUARY 1854

[Sir,]

I have just seen a paragraph in a provincial paper, to the effect that Dr. Edward Hincks has relinquished the study of the cuneiform inscriptions, in consequence of the discouragement of the Government and the jealousy of certain parties connected with Oriental discovery. I have no personal acquaintance with Dr. Hincks, but I am acquainted with the value of his Egyptian and Assyrian researches; and I must say, that I could not read his paragraph without strong feeling. The decipherment of the cuneiform inscriptions is a matter of high religious as well as historical importance, and the public are interested in having no unnecessary obstacles thrown in the way of their speedy translation; besides this, if the report alluded to be true, the great principle of literary equality has been infringed at the expense of an individual. Can any of your readers state the facts of the case?

'J. R.'

Source: The Athenaeum, no. 1368 (14 Jan. 1854), p. 56

TO THE EDITOR OF THE ATHENAEUM
16 JANUARY 1854

[Killyleagh, C° Down]
[16ᵗʰ. January 1854]

[Sir,]

The letter of your correspondent 'J. R.' requires some notice. Various paragraphs on the subject to which he alludes have appeared in London and provincial papers; and it may suffice to say that there was none of them the writer of which was perfectly acquainted with 'the facts of the case'. It is not necessary nor desirable that these should be published. Your correspondent, however, may be

glad to know that D^r. E. Hincks does not *now* contemplate that speedy abandon-
ment of his cuneatic investigations which he did when his advertisement appeared
in your paper of the 24th. ult.;[1] and that the necessity which he believed to exist for
the announcement then made was not caused by Col. Rawlinson or his friends.

[I am, my dear Sir
Yours faithfully
Edw. Hincks]

Source: The Athenaeum, no. 1369 (21 Jan. 1854), p. 89

1   A rather curious exchange took place between Baron Bunsen, the Prussian ambassador to the Court of
St James, and the editor of *The Athenaeum*. In a letter dated 23 January 1854 Bunsen wrote: 'Sir, I have only
learned this moment that you have been induced by an advertisement of Messrs. Sotheby to suppose that the
library of a Foreign Minister which is advertised for sale is mine. Will you oblige me by inserting my assurance
that I know nothing of that library or its sale, and that I never intended to sell my literary or any other
collections. I have the honour to be, &c. Bunsen.' The editor replied that he was not aware that any statement
to such effect had appeared in his paper. See *The Athenaeum*, no. 1370 (28 Jan. 1854), p. 120. Was Hincks's
advertisement in *The Athenaeum*, no. 1365 (24 Dec. 1853), p. 1533, mistakenly thought to be Bunsen's?

TO AUSTEN HENRY LAYARD
23 JANUARY 1854

Killyleagh C° Down
23^d. Jany. 1854

My dear M^r. Layard,
    I had your letter yesterday morning. I am glad you are returned & am obliged by
the interest you express about me. You would probably see by the Athenaeum of
Saturday that I have relinquished *for the present* my design of disposing of my books
& quitting the field of Assyrian discovery.[1] The cause which seemed to compel me
to take the step in question has been *averted for the present*. It will not, indeed, *now*
act till the spring or summer of 1855; though, as a prudent man I should begin to act
in reference to it some months sooner. I am, however, now vigourously at work &
will as far as possible banish all thoughts of the future. I know I have some friends
who will exert themselves for me on both sides of the water; & I hope something
will be done for me of a *permanent nature*, before the period I have referred to
arrives, supposing that that period should find me still living. The cause of my
difficulties – a cause, which unless permanently removed, will renew its pressure
from time to time – is the very heavy reduction of my clerical income. This is not
the mere result of the repeal of the corn laws, nor is it any thing at all analagous to

what affects the English clergy. It arises from an Irish measure, passed many years since but only coming into question now, by which the parish of which I have the misfortune to be rector is affected to an *extent* that not one parish in 600 is, & with a degree of *injustice* that not one in 50 is. Taking into account the two together the *injustice* & *the extent*, I believe my case to be literally *unique*. What I have lost by hostile legislation is *fully* 300 a year; & taking into account the insurances that I effected for the benefit of my family – some at rates progressively increasing (for it *used to be thought* that a clergyman might as he grew older look to have his income *increased*; but at all events *depreciation* would only affect his successor) & also the annuities on which I raised money for buildings & furnishing my glebe house, I should be reduced to *indigence*. In fact I could not support my family & discharge my pecuniary engagements. Under the pressure of pecuniary embarrassments I find it to be impossible to work at the inscriptions or at anything that requires active mental exertion. *For the present*, this embarrassment is removed; & I trust that something *permanent* will be done before it recurs.

As to this, however, I am making dependence on the exertions of my friends; & as you have kindly offered to do what you can & have asked me to suggest a way, I do so.

There are three ways in which I could be relieved. The first is by an amendment of the act, by which I suffer so severely. I drew up a statement on this subject & submitted it to Sir John Young & others.[2] Sir John says it cannot be done & I would not press it, under the circumstances; though I should like this better than the *second* way.

The 2^d. way is by giving me a pension from the civil list. This could not be done till after the 20th. June; but that would be time enough. I believe this is the most feasible plan; & with ruin staring me in the face – tho' at a distance – I ought not to object to it. Since I would rather anything else was done for me. The public are accustomed to criticize pensions – to give one to a clergyman would excite criticism of an unpleasant nature; it would be asked why was he not provided for by advancing him in his profession? A stigma would then be cast on him? Besides the consideration must be expressed; & if the pension were of such magnitude, as would enable me to carry on my literary pursuits with advantage (& less than 200 a year w^d not do *this*, though 600 might keep my family from being broken up) something would have to be stated at which Col. R's friends might take outrage. My own wishes are, therefore, very strong in favour of the 3^d. way – by a promotion in my profession. The pension would be a *pis-aller*; not to be *refused*, but certainly not a very desirable thing, if anything else could be got. When I say promotion, I don't mean to any *dignity*. On anything of this kind I set no value. In Ireland, it would be almost a matter of necessity; for the Gov^t. has retained no good living in Ireland without a deanery attached to it; but I should infinitely prefer a plain English living within a reasonable distance of London to an Irish deanery. Now such livings exist in considerable number, in the gift of the crown. They are every now & then falling vacant; and it would not be a *very great* matter to give such a one to me, the presentation to Killyleagh being given in return for it. The College have promised me that they will assent to this arrangement, *whatever be the age of the presenter*. A living so desirably situated as Killyleagh & with such a good house & £400 a year, clear of all deductions, attached to it, would not be a bad thing to give away. Many would be glad of it, though from my antecedents it will not suffice to keep me from indigence.

If there were any chance whatever of such an arrangement taking place, it would be desirable that I should become known as a clergyman during my visits to London. Could you suggest or conceive any way of obtaining for me employment on Sundays? *Remunerative* employment may not be to hand; nor would this be an object of much importance; but I should like a place where I could officiate on Sundays. I am tired of wandering about; & indeed I feel uncomfortable at being idle on Sundays. I propose being in London by the second (& perhaps the first) Sunday in March & to continue there about a month. I must be home the Sunday before Easter. My views being *via media* or broad church, I sh^d not like to have anything to do with the *extreme factions* – nor would they with me.

There is a publication which seems to have a good circulation 'the Journal of Sacred Literature'. I have made arrangements for publishing in the next number what, I hope, will attract attention to Assyrian studies: 'Illustrations of Hebrew Roots from the Assyrian Language'. I maintain that, independently of the documents which may exist in the language, *the language itself* throws light on the Hebrew – probably to at least as great a degree as the Arabic.[3]

My idea would be to try for the *living* up to June; if *that* should be unobtainable (of which an opinion can by that time be formed on good grounds), a pension can then be pressed for. I need scarcely say that with an English country living I could do far more than I can here; from my being able to run up at all times to London in case of there being anything of peculiar interest. A curate for the weekday duties & to assist on Sundays is a matter of course.

By the way, I saw in the paper last week the death of M^{rs}. Taylor widow of a Col. Taylor of the H.E.I.C.S. Is this the possessor of the Cylinder? The importance of obtaining access to this is immense.[4]

Believe me
Yours very faithfully
Edw. Hincks

MS: BL Add. 38982, ff. 124–7

1   See Hincks's letter of 16 January 1854 to *The Athenaeum*.
2   Sir John Young (1807–76) was M.P. for Cavan, 1831–55, and chief secretary for Ireland under Lord Aberdeen from 1852 to 1855.

3   Among the Hincks papers at the Griffith Institute there is a fragment of a manuscript headed 'Hebrew Roots Compared with Assyrian' (GIO/H 558), which contains some interesting notes by Hincks. See Cathcart, 'Some Nineteenth- and Twentieth-Century Views on Comparative Semitic Lexicography', in M. Kropp and A. Wagner (eds), *'Schnittpunkt' Ugarit* (Frankfurt, 1999), pp. 1–8.

4   The obituary which Hincks read was that of Joanna Taylor, who died at Walton-on-Thames, on 11 January 1854, aged 88, relict of Colonel John Taylor. Mrs Rosa Taylor, widow of Colonel Robert Taylor, possessor of the prism, did not die until 1876.

⇜⇝

TO THE TRUSTEES OF THE BRITISH MUSEUM
31 MARCH 1854

31st. March 1854

My Lords and Gentlemen,

By the first of May next, I trust I shall be able to lay before you a transcription and translation of a large portion of the historical inscriptions of Assur-yuchura-bal, the builder of the northern palace of Nimrûd,[1] and of Sin-akhi-irib, the Sennacherib of the Bible.[2] This I propose to do in manuscript; but I have thought it right, before I leave town, to present to each of you a report on the cylinders and portions of cylinders in the Museum, and on some terra-cotta tablets which I have examined.

The oldest cylinder is that of Tiglath Pileser I, who must have reigned above 1100 years before Christ.[3] In his reign Nineveh was taken by the Babylonians, and certain images of gods were carried to Babylon, which were brought back by Sennacherib when he took Babylon in the first year of his reign, or 703 B.C. The interval between these events is stated in the Bavian inscriptions to have been 418 years.

This king not only mentions the names of four of his direct ancestors, whom he calls kings of Assyria, but speaks of two personages who lived at a remote period, Shamsi-Bin, and Ishmi-Dakan his father,[4] to whom he gives the title *patātsi Asura*, which I rendered, in some measure conjecturally, 'priest (or champion) of Assur'.[5] The writer of the cylinder speaks of this personage as 'his father', or remote ancestor, and speaks of his having built a temple on the mound of Kalah Shergât, which in the course of 641 years had fallen into decay, and was taken down by the writer's great-grandfather. It lay in ruins for sixty years, when the writer rebuilt it. He says also that he found tablets of his father Shamsi-Bin, and put them up in the new temple along with his own. It appears from these dates, that Shamsi-Bin lived in the 19th century B.C. Considering the extreme importance of these facts, I announced them to the Royal Society of Literature in the summer of last year, and published them at greater length in the *Dublin University Magazine* for October.[6]

226

It has been alleged, however, that I have 'rather disfigured than illustrated history' by this publication, as I have failed to perceive, what has been lately announced with great confidence, that it is to the Chaldean and not the Assyrian empire that this antiquity belongs. I certainly have failed to see this supposed fact; and it appears to me that it is a passing fancy, which its author will very speedily find himself compelled to abandon. That the personages who bore the above title ruled in Babylonia, as well as in Assyria, I of course do not question, after the discovery of Ishmi-Dakan's bricks at Um Qeyr;[7] but I regard him as Assyrian, not Chaldean. At any rate he was considered, by the old Assyrian dynasty to which Tiglath Pileser I belonged, as one of their ancestors, and not a foreign lord. The evidence for this latter opinion is in my mind of no weight whatever.[a]

Among some fragments of cylinders found among the tablets, I observed one which belongs to a second cylinder of this king. It contains apparently the same text; but the lines are differently divided, and in one place several words are omitted. The characters on this fragment are far larger than those of the cylinders first found, and are much more deeply cut. I have called both of these pillars cylinders. Strictly speaking, however, they are octagonal prisms.

The so-called cylinder next to them in point of antiquity, which is in the shape of a barrel, is the work of Sennacherib; and contains the annals of the two first years of his reign. Of these I have prepared a transcription and translation.[8] I will here make an observation as to its date, which occurs in the last line of the inscription, erroneously made the first in the copy published by Grotefend.

The Assyrians had two modes of describing their years.[b] They sometimes used the years of their kings' reigns, and sometimes they described them by the name of the superintendent, as he may be called, an officer who was annually appointed, but whose duties are as yet unknown. On the Nimrûd obelisk both forms are used. In general the king dates by his regnal years; but in compliment to his favourite general, Dikut-Assur, in place of saying, 'in my fourth year', he says, 'Under the superintendence of Dikut-Assur'.[9] It may be observed that on the bulls, executed many years before the obelisk, and before Dikut-Assur had distinguished himself as a general, we have the ordinary form, 'in my fourth year'. The builder of the north-west palace never dates by his regnal years. Later monarchs seem to do so always in their historical inscriptions; but I believe their cylinders are always dated by the superintendents. The present cylinder is said to have been written 'when Nabu-liah, governor of Irbahil (Arbela), was superindent'.[10]

In the hasty glance which alone I was permitted to take at Col. Taylor's cylinder, I observed that it ended with the words 'governor of Carchemish'. It occurred to me at the time, that the cylinder was a present to this governor, and that the place where it was found might perhaps determine the site of Carchemish; which was most certainly not at or near Circesium, but on the opposite bank of the Euphrates, and much higher up. I am now, however, decidedly of opinion that the inscription ended with a date, 'under the superintendence of so-and-so, governor of Carchemish'.

There is a cylinder, or rather a hexagonal prism, of which only a few lines at the top and bottom of each column are injured, which contains an account of the conquests of Esarhaddon in the early part of his reign; not arranged, however, in the form of annals. He mentions, among other things, that he made a certain woman queen of Udumi, or Edom.[11]

The lower part of another hexagonal prism, of the same reign, and containing for the most part the same text, but somewhat enlarged, is also in the Museum. It contains a list of twelve kings of the sea-coast, and ten kings of 'Yatnan (or Yavan) in the middle of the sea', which is wanting on the other cylinder. The first two names are Bahlu, king of Tyre, and *Minatsi sar Yahuda*, Manasseh, king of Judah. Other names of places on the coast that I have made out are, Udumi (Edom),ᶜ Gaza, Ascalon, Ekron, Gubal, and Azdod.[12] The characters on this cylinder are in many places completely filled up with a hard substance which cannot be removed by a brush, and which renders them illegible. In consequence of this, I have been unable to satisfy myself as to a single name of king or kingdom among the insular ones. It is possible that some chemical means of cleaning the cylinder may be found.

Two fragments of a cylinder which I suspect to belong to this reign may be here mentioned. One of them has preserved a very interesting fact. It is dated 'under the superintendence of Nabu-bil-akhi'-su (*i.e.* Nabu is lord of his brethren) governor of Tsamirina (*i.e.* Samaria)'.[13] It appears that at the time when this inscription was made Samaria had an Assyrian governor, of such rank as to entitle him to hold an office which could not from its nature have been conferred on very many.

Of the son of Esarhaddon, whom we may provisionally call Assur-yuchura-bal II, there are fragments of at least three cylinders containing apparently the same text. I have found a passage on three fragments, and another on two. The inscription seems to be in the form of annals, extending to the sixth year at least. It contains an account of the war against the kings of Susa, who were successively slain and beheaded. The transport of the head of Teumman, the first king who was killed, is represented on the bas-relief in the new gallery; and it is stated in an epigraph whose head it was. The cylinder states that the king put the head on his Temple-Bar, 'on the top of a great gate in the centre of Nineveh'. The pulling-out of the tongues and peeling-off the skins of some blasphemers, which is represented in another bas-relief, and described in the epigraph, occurs in another fragment of the cylinder. It appears that these blasphemers were of the country of Gambul, which lay, I believe, between Luristan and the Tigris; and that they were executed at Arbela.[14] A grandson of Merodach Baladan, and various members of the royal family of Susa, including Ummanibi who submitted, and was allowed to reign, are mentioned in these fragments. It is much to be desired that more of them should be sought for; as the cylinders, if completed, would furnish us with most interesting information.

The remaining cylinders in the Museum are of the reign of Nebuchadnezzar; and are all barrel-shaped, like that of Sennacherib. The largest of them contains an inscription, which has not yet been published. The inscription on the second has been published by Rich; and also by Grotefend, from another cylinder containing the same text. There are several small cylinders containing a short inscription. It appears to be the same in all; and from a comparison of them an exact copy might probably be obtained; but these cylinders are all in very bad preservation. There is also in the Museum the fragment of a cylinder, the inscription of which has been published by Ker Porter. The whole of its contents is comprehended in the great inscription at the India House.

All the inscriptions of Nebuchadnezzar contain descriptions of the buildings, which he made or completed at Babylon or other places; and prayers to his gods, especially to Marduk. There is nothing in any of them of an historical character; and I cannot imagine where Dr Grotefend could have supposed that the extraordinary statement existed, which he published last summer, respecting the sacrifice of his son.

Along with the Babylonian cylinders is a collection of clay tablets brought from Babylonia, which are chiefly interesting from being dated in known years. The dates extend from the reign of Esarhaddon to that of Artaxerxes, or from 676 to 454 B.C.

They contain the month, the day of the month, the year of the king's reign, and the king's name and titles. Esarhaddon is called 'king of Assyria', the other kings before Cyrus, 'king of Babylon'. The Persian kings are styled 'king of Babylon, king of the provinces', or simply 'king of the provinces'.[15]

The following dates may be depended on, as those of fifteen of these tablets:

1st month, 25th day, 4th year of Esarhaddon (Assur-akh-iddin, i.e. Assur has given a brother).[d]

11th month, 27th day, 7th year of Nabopolassar (Nabû-bin-yuchur, i.e. Nabo has formed a son). This and the following tablets are from Warka.[e]

1st month, 5th day, 20th year of do. or Nebuchadnezzar (Nabû-kudurri-yuchur, i.e. Nabo has formed troops). Name partly defaced.

11th month, 6th day, 9th year of Nabonidus (Nabû-nahid, i.e. Nabo is glorious).

1st	"	17th	"	3rd	"	Cyrus
12th	"	28th	"	3rd	"	do
5th	"	9th	"	4th	"	do
8th	"	23rd	"	5th	"	do
6th	"	11th	"	3rd	"	Darius
1st	"	1st	"	7th	"	do
9th	"	17th	"	17th	"	do
3rd	"	11th	"	3rd	"	Artaxerxes
2nd	"	1st	"	10th	"	do

There are four other tablets, on which the kings' names are illegible; I suspect from the appearance of some fragments of them that remain, that one was Nabonidus and another Artaxerxes.

These tablets appear to me to contain deeds of sale or acknowledgments of money lent. They abound in proper names; some of which are those of the parties to the contract, some those of witnesses, and the last is that of the writer of the deed, who was an official person. The name of the father of each person is added to his own. Many of them are impressed with seals; over and under which are written, 'the seal of such a person', who is always one who is named in the body of the contract. An Assyrian contract, which has four seals, united together, is of an unknown age, being dated by the annual superintendent.

Besides these, there is in the Museum a large collection of Assyrian terra-cotta tablets – not less, I should suppose, than a thousand. The contents of these are very various, and many of them are, it cannot be doubted, of great importance. I have examined between thirty and forty. Some of these contain prayers to different gods, or directions for their worship. Others are almanacs, having a separate line for each day in each month, after which is a short remark, and sometimes a second. It is possible that some of these remarks may be astro-nomical; but I rather think that they are astrological, pointing out what it would be lucky to do or to forbear to do on those days; as in the Egyptian almanac on one of the Sallier papyri.

A few tablets appear to be astronomical, containing the ideograph for 'star' over and over again. It would be desirable that all these were collected together, in order that by comparing them with one another their meaning may be elicited.

When I first visited the Museum in the course of last spring, I saw a tablet, which appeared to me of great importance. Combining my recollection of its contents with my present knowledge, I should say that it contained a list of the annual superintendents, with the year of the king's reign to which they belonged. Four or five kings were named on the two sides of the tablet, with the number of years of at least three complete. The position of the tablets in the presses has been since altered; and it would probably take many days to search out for this. If it be what I suppose it to be, it is of the highest importance.

The tablets, however, for which I have sought most anxiously, are those of a philological character, calculated to throw light on the Assyrian language, and thus to facilitate the reading and translation of the inscriptions. My first knowl-edge of such tablets was derived from the copy of an inscription on a tablet which M^r. Layard had made, and which he showed to me in the autumn of 1852. It contained values of the more complicated Assyrian characters expressed by more simple ones; and what had long been suspected by me, and indeed all but demonstrated from the inscriptions, though very much opposed to what might be inferred *à priori*, was here distinctly stated. The same character had in many instances two or more distinct values, which it expressed under different circumstances.

The tablet copied by M^r. Layard being only a fragment, I searched carefully for other similar ones, which might possibly have formed parts of it, as well as for itself. I could not find what M^r. Layard copied; but I found two pieces of similar tablets, one of which was in very good preservation.

I met with several other tablets, mostly fragments, on which interesting information was contained. By one of them I learned the words to which the ideographs for 'year' and 'month' corresponded. They are *sha-na-at* and *a-ra-akh*. These are the forms in the singular number in the state of construction, which was regarded as the theme; but they admitted, of course, the usual inflections. Another tablet gave some numeral forms, apparently ordinal adverbs; namely, *shul-ush-ti* 'thirdly', *ri-ba-a-ti* 'fourthly', *kha-an-ish-ti* 'fifthly', and *is-ri-ti* 'tenthly'. Another tablet equated the monogram which is often used at the end of Sennacherib, to *i-ri-ib*.

What struck me, however, as of peculiar importance among these tablets, was the confirmation given to the view of the Assyrian verb which I had been led to take by a careful study of the forms in the inscriptions. I found two large fragments of tablets illustrating this subject, to say nothing of a third, of which no use can be made unless the piece from which it was broken can be found.

One of the tablets gives different tenses of the same verb in the same conjugation. Thus, it gives *ish-ku-un*, *ish-ku-nu*, *i-sa-kan*, *i-sa-ka-nu*; which are the simple preterite, the preterperfect, the present, and the future of the verb ŠKN.^{f} It gives also the cuneatic symbols by which these several tenses were designated.

The other tablet gives the same verb in the passive, the simple active, and the intensive conjugations; and, as before, it gives the symbols by which the Assyrians denoted those several conjugations. Some of the examples given are curiously irregular. The passive is sometimes in the niphal and sometimes in the puhal form, and sometimes in what cannot be reduced to either. Thus, *yachab* is given as the passive of a verb, of which *yuchchib* is given as the simple active; and of this the intensive is given *yuraddi*, which would come regularly from *irdi*! If this irregularity had not been distinctly laid down on Assyrian authority, I suppose no one would have ventured to infer it from observation. It may well stand with such anomalous verbs as *fero*, *tuli*, *latum*, in Latin, or our own *go*, *went*.

I must now conclude, and have the honour to be,

My Lords and Gentlemen,
Your most obedient humble servant,
Edw. Hincks

a   What reliance can be placed on the statements of Berosus – coming to us, as they do, at third or fourth hand, and through writers, the good faith of one of whom at least is with reason suspected? Besides, it is a part of his evidence, which ought not to be lost sight of in estimating his credibility, that his Chaldean dynasty was preceded by a native Assyrian one of 30,000 and odd years! As to the notion that the first three characters in the above title 'represent the Arian word *patis*, lord', it surprises me not a little that Colonel Rawlinson could entertain it for a moment. Were an Assyrian to represent that word, it is incredible that he would use *tsi* to represent the last syllable, when he had characters for *is* and *ish*. For what purpose could the final vowel be added? The word being in the state of construction must terminate with a radical letter; and though *patis*, a foreign nominative, might perhaps be considered a root, *patisi* is evidently not so. I deny, however, that the second character in this title can signify *ti*, when it is not followed by a syllable beginning with *i*. Before a consonant (other than *m* or *n* which it doubled) it seems to have been sounded like the French *ten* or *tin*, which I express by *tā*. The last character also has been proved to contain a double consonant, *ts*. *Patātsi* is not *patis*, but is a participle, or *nomen agentis*, of the root *PS'* (or some other, of which the two first radicals are the same and the third some other weak letter) in one of the intensive conjugations. It is a pure Assyrian form, whatever the precise meaning may be.

   Again, *Asura*, 'of Assur', is not to be confounded with *Mat-Asura*, 'of Assyria' (the land of Assur). These are always distinguished in Tiglath Pileser's inscriptions; and the god, not the country, is here named. It is, indeed, my belief that this title, 'patātsi Assura', is identical with the [title] 'priest of Assur', borne by Assur-yuchura-bal, his father and his son; these two monograms being the representatives of the words in the text.

b   It has been stated that some early kings dated by the years of a cycle. I have seen no instance of this, and am inclined to think that the statement is erroneous. I have no doubt, however, that the Assyrians used a sixty-year cycle in their computation of time.

c   This assembling of the kings occurred before the capture of Edom mentioned above, when the king here spoken of was replaced by a queen.

d   It appears from his very name, that Esarhaddon was the second son of Sennacherib. He ought not to be confounded, as he has been, with his elder brother Assur-nadin *i.e.* Assur is bounteous), whom his father made king of Babylon in his fourth year. This was the Apronadius of Ptolemy's Canon; and he probably died at the end of the short reign which the Canon assigns him.

e   In the Babylonian inscriptions of his son, he is always called Nabu-bal-yuchur. The difference is dialectic. I can by no means agree with Colonel Rawlinson that the proper value of the character here used was *pal*. Its use as the first element in the name of Ben-hadad has far more weight with me than a Greek transcription.

f   This root is designated by the symbol **Ψ** ; and it appears that a large number of Assyrian roots had special characters to represent them; which they sometimes did alone and sometimes with the addition of a termination. Thus, the symbol here given, when followed by *an*, is to be read *atsakan*, or according to the context, by some other inflexion of this tense. When followed by *un*, it is to be read *askun, iskun, iskunu*, &c.; by *in* it is *sakin*; by *nu, saknu*. It appears from the Assyrian tablet referred to, that this character represented another root as well as ŠKN; namely, ŠRK. To denote this, it would, I presume, be joined with a syllable containing *k*. I have never met an instance of this in an Assyrian inscription; but the combinations with characters containing *n*, that I have just given, are common. The tablets give a great number of roots with their symbols, which were often composed of two characters, being compound ideographs.

Source: Report to the Trustees of the British Museum[16]

1   Ashurnasirpal II (883–859 B.C.), builder of the north-west palace at Nimrud.

2   Akkadian *Sin-aḫḫē-erība*; Hebrew *Sanḥerīb*.

3   Prism of Tiglath-Pileser I (1114–1076 B.C.).

4   Hincks has the names the wrong way round. Ishme-Dagan (1780–1741 B.C.) was son of Shamshi Adad (1813–1781 B.C.).

5   This is now read *iššakku Aššur*.

6   'The Nimrûd Obelisk', *Dublin University Magazine* 42 (Oct. 1853), pp. 420–6, esp. pp. 421–2.

7   The ancient city of Ur was at modern Tell al-Muqayyar, which is variously written as Um Mugheir, Um Qeyr or Muqeyyer in nineteenth- and twentieth-century publications.

8   Hincks's transcription and translation of 'The Inscriptions of Sennacherib' (and the annals of Shalmaneser III) were received at the British Museum on 6 May 1854. They are contained in MS BL Add.22097 ff. 1–27 (Hincks numbered the pages 1–53). The annals of the first two years of Sennacherib's reign are in ff. 2v–6r (pp. 4–11).

9   *Dikkut Aššur* is now read *Dayyān-Aššur*. Hincks's 'superintendence' is for *limmu*, 'eponym official'.

10 Nabû-liah is now read Nabû-lē'i.

11 Esarhaddon made Tabua, who had grown up in Nineveh, queen in Adummatu in Northern Arabia. See Pritchard (ed.), *Ancient Near Eastern Texts*, p. 291.

12 See the texts of the Syro-Palestinian Campaign of Esarhaddon in Pritchard (ed.), *Ancient Near Eastern Texts*, p. 291. For Hincks's views on Yadnan and Yawan, see his letters of 2 August and 3 September 1852 to Layard.

13 The name is now read Nabû-šar-aḫḫēšu.

14 The king is Ashurbanipal and the campaign was against Teumman of Elam and Dunanu of Gambulu. See the detailed examination of the epigraphs and the reliefs in Russell, *The Writing on the Wall*, pp. 154–209.

15 For the eponyms see A. R. Millard, *The Eponyms of the Assyrian Empire 910–612 B.C.* (State Archives of Assyria Studies 2; Helsinki, 1994). In Assyriology, the term eponym refers to a functionary (called *limmu*) who gave his name to his year of office.

16 *Report to the Trustees of the British Museum Respecting Certain Cylinders and Terra-Cotta Tablets, with Cuneiform Inscriptions* (London, 1854). See also *Literary Gazette*, no. 1944 (22 Apr. 1854), pp. 375–7.

~~~

TO AUSTEN HENRY LAYARD
8 APRIL 1854

<div align="right">

Rectory Killyleagh
C° Down
8th. April 1854
</div>

My dear M^r. Layard,

On my return home, I found a letter from Sheffield, from a nephew of mine who is on the Committee of the Mechanics' Institute.[1] He asked me to propose to you to give them a Lecture (in the Easter recess, or at such later period as may suit you) on your Eastern Researches. Lord Cavendish gave them one some months ago.[2] They have been making great exertions latterly to revive the Institute & place it on a better footing. This has incurred expense & Mechanics' Institute (you are aware) can barely support their ordinary expenses. Extraordinary ones require aid from without. I am not myself interested for Sheffield & venture to ask you to go there as a favour to me. If, however, you fancy giving the lecture, you have a fair yes to do so. In that case you can write to 'Rev. Thos. Hincks – Sheffield' to mention the time of your going. Should you not go you could oblige me by simply saying to him that *business* required your presence in London or elsewhere as the case may be.

Private

As to my own matter, I was greatly surprised to find on my return home that Lord A.[3] had offered a pension of £100 a year. He has been greatly pressed for others & the friend who informed me of it speaks of it as honourable, not for its amount but for the competition; & as a recognition of my service. What shall I do in the matter? My brother will be here next week & I shall consult *him*;[4] but *you* perhaps are in a better position to give advice. Would the acceptance of it be a bar to my obtaining anything else hereafter? Would it entitle them to say that I am a man who have *had my reward*; tho' with £50 less than I had before, losing £150 & being granted £100.

I should have time to hear from you before I see my brother.

Believe me
Yours very faithfully
Edw. Hincks

While acceptance is doubtful, it would be wrong to speak of it.

<div align="right">

MS: BL Add. 38982, ff. 186–7
</div>

1   Thomas Hincks (1818–99), Unitarian minister and naturalist, was born on 15 July 1818 in Exeter, son of William Hincks (Edward Hincks's brother) and Maria Ann Yandell. He was educated at the Unitarian Manchester College in York and London University where he took a B.A. In 1868 he permanently lost his voice and had to retire from the ministry. He was a very good zoologist and continued his research throughout his life. In 1872 he was elected a Fellow of the Royal Society. Among his many publications, special mention must be made of *A History of the British Hydroid Zoophytes*, 2 vols (London, 1868) and *A History of the British Marine Polyzoa*, 2 vols (London, 1880).

2   Lord George Henry Cavendish (1810–80) succeeded his brother William, Lord Cavendish of Keighley, as M.P. for North Derbyshire in 1834, when the latter became the seventh Duke of Devonshire. He lived at Ashford Hall but the Cavendish family/Dukes of Devonshire also owned Lismore Castle in County Waterford.

3   Lord Aberdeen was prime minister from December 1852 until January 1855.

4   Francis Hincks was co-premier (with Augustin-Norbert Morin) of Canada.

FROM AUSTEN HENRY LAYARD
12 APRIL 1854

Canford, Wimborne[1]
April 12/54

My dear D[r]. Hincks,

I found your letter when passing thro' town yesterday & not having time to answer it I have been compelled to put you to the inconvenience of a delay in receiving my reply. I am sorry to say that I cannot possibly go to Sheffield. I have already had an application from the Mechanics' Institute there which I am obliged to decline owing to the number of engagements & continual occupations that I have, and now I am going over to Paris where I shall probably remain until the end of the Easter recess. This completely precludes the possibility of my undertaking a lecture at Sheffield. I have innumerable applications of the same kind which I am obliged to decline.

I am truly glad to hear that Lord Aberdeen has offered you the pension of £100 a year, such as it is, but as your friend justly observes it is a recognition of your merits & services and as such is an honorable distinction. I cannot suppose that it in any way interferes with your obtaining any church or other advancement hereafter, but I cannot speak positively on the matter & have no one near to ask, but surely it cannot do so. The sum is so trifling in itself that no one can expect that a man situated as you are should renounce any prospects for it. All I can say is that none of your friends will rejoice more than I shall if you can accept it and if it proves of any assistance towards enabling you to carry out the important researches in which you are engaged. As to it being offered to you there can be but one opinion.

You will hear that Rawlinson has discovered the name of Semiramis – she was the wife of Pul according to him.[2] I hope you will continue to have your translations before the public so as to secure the honor of your own discoveries. Rawlinson will be home at the beginning of May.

I hope you found all your family well, my very kind regards to them.

Yours very faithfully,
A. H. Layard

MS: GIO/H 241

[1]   In March 1854 Layard took Lady Charlotte Guest to see the Nineveh Court which was being erected in the new Crystal Palace at Sydenham near London. In April the architect James Fergusson, who designed the Nineveh Court, accompanied Layard on a visit to Canford Manor. See Russell, *From Nineveh to New York*, pp. 117–18.

[2]   'Babylonian Discoveries', *The Athenaeum*, no. 1377 (18 Mar. 1854), pp. 341–3. He read the royal Assyrian name Adad-nirari III as Phal-lukha, based on Septuagint Φαλωχ, and equated it with Pul, the name of the Assyrian king mentioned in the Old Testament at 2 Kings 15: 19. See Hincks's letter dated May 1856 to the editor of the *Monthly Review*. Today we know that Pul was a name for Tiglath-Pileser III. Adad-nirari's mother, not his wife, was Sammu-ramat = Semiramis of Greek legend. On 'The Curious Case of Missing King Pul' in the nineteenth century, see S. W. Hollaway, 'Biblical Assyria and Other Anxieties in the British Empire', *Journal of Religion and Society* 3 (2001), pp. 1–19, esp. pp. 8–11.

TO AUSTEN HENRY LAYARD
15 APRIL 1854

Killyleagh C° Down
15th. April 1854

My dear M[r]. Layard,

I received your letter this morning. I had previously, acting on the advice of all my friends here, *accepted with thanks* the pension offered. I am glad that your advice agrees with theirs.

As you say you are going to Paris, it occurs to me to write to you to beg that you will ascertain whether they have any Assyrian inscriptions there *inedited*, & which a stranger like myself would be *allowed access to*. Michaux's stone is, I know, there, & a *cast* of Col. Taylor's cylinder.[1] Probably they have something from Khorsabad, not included in Botta's publication. If so & that no difficulty would be made in the way of giving me access to them, I would take a run over in June. I think I ought to do something for my pension *this year*, when it is in fact an accession of income beyond what I have been accustomed to; & Paris seems a mine which has hitherto been unexplored. I cannot leave this till the middle of June; so you will have ample time to make enquiries.

I hope to have my translation near ready by the 1st. but have been engaged with clerical duty since my return – shall be so till Monday coming & then expect relatives.[2] So that it will be Thursday before I can set regularly to work.

Believe me
Yours very faithfully
Edw. Hincks

MS: BL Add. 38982, f. 191

1   Michaux's Stone is basalt and ovoid in shape. It is 43 cm high and 61 cm in circumference, and there are two columns of cuneiform text in the lower part of the stone. It is named after Michaux, the traveller who discovered it near the Tigris and not far from Ctesiphon. It was brought to France in 1800 and is kept in the Louvre at Paris.
2   As he told Layard in his letter of 8 April 1854, Hincks was expecting his brother Francis to visit him, so other relatives would surely be there also.

FROM EDWIN NORRIS
15 APRIL 1854

R. Asiatic Society
15th April, 1854

My dear Sir,

I am about to print my little paper on the readings upon the Lion and Duck weights from Assyria,[1] of which you may have seen some notice in the Athenaeum. I had there come to the conclusion that the shekel was the fiftieth part of the manah. I had not seen, of course, when I read my paper, your note on the value of the shekel and the monograms expressing its multiples, printed in vol. V. of the R. Irish Academy's Proceedings, p. 405. Your values of ⊢⟨⟨⟨𝄇 and 𝍦𝄇 coincide very nearly though not quite exactly with what I have said in my paper and I am inclined to go with your estimates of 1/15 and 1/30 rather than my own of 1/16 and 1/32, but I should like to know whether your estimate is from positive testimony that the maneh contained 60 shekels, or only an inference from the actual value of the weights [as] marked in the Museum. I should be very glad to print any note from you on this matter as an appendix to my paper. I have already written a half dozen lines, merely mentioning your discovery, but I could do no more than mention it. I should like to add something more, and will do so if you will furnish me with any thing you wish to be said on the subject.

I am
My dear Sir
Yours very faithfully
Edwin Norris

MS: GIO/H 312

1   'On the Assyrian and Babylonian Weights', *Journal of the Royal Asiatic Society* 16 (1854), pp. 215–26. On p. 218, n. 3 there is a note from Hincks on what he had written previously on Babylonian Weights. See the following letter.

TO THE PRESIDENT OF THE ROYAL IRISH ACADEMY
19 APRIL 1854

Killyleagh C° Down
19th. April 1854

My dear Sir,

In my communication, printed in the Proceedings of March 16, 1853, I mentioned three subdivisions of the manah.[1] I have lately discovered a still smaller one, equivalent to about 4·3 grains. It was the thirtieth part of the shekel, or 1–1800th of the manah. The monogram which represented it was , and I propose to call it a gerah. The Assyrian name of none of these subdivisions of the manah has yet been discovered. It seems to me probable that the Assyrians kept their accounts in manahs, and in what I call shekels and gerahs – the sixtieth and eighteen hundredth parts of the manah. I infer this from a sort of memorandum which I met with on a terra cotta tablet in the British Museum. It is to this effect:

1 shekel, ... 6 gerahs[2]
10 shekels, ... 2 shekels
1 manah, ... 12 shekels

It is evident, from the remainder of the lines being identical, that the same ratio exists between the two weights in each line; and this appears, from the second line, to be the ratio of five to one. It follows that the weights in the second line are ten times those in the first; and those in the third are six times those in the second.

This requires that the manah should be equal to sixty shekels, and the shekel to thirty gerahs.

> I remain, my dear Sir,
> Yours very truly
> Edw. Hincks

Source: Proceedings of the Royal Irish Academy 6 (1854), p. 72

1   See Hincks's letter of 7 March 1853 to the President of the Royal Irish Academy. It was published in *Proceedings of the Royal Irish Academy* 5 (1853), pp. 403–5.
2   Hincks is using the Hebrew term. According to M. Powell, 'Masse und Gewichte', *Reallexikon der Assyriologie*, vol. 7 (Berlin, 1987–90), p. 512, *girû*, 'carat' was worth ½₄ of a shekel. It has no monogram and the sign that Hincks gives is not recognised. One sixth of a shekel is *suddû*.

FROM GEORGE CECIL RENOUARD
27 APRIL 1854

Swanscombe, Dartford
27th. April 1854

My dear Sir,

I was much pleased to hear of your labours at the B.M. & delighted to find, notwithstanding my demerits, that you had not forgotten me. I read your address with great eagerness & satisfaction, intended to thank [you] for it, immediately, but have been so much interrupted as to have had no time to do so & have in the mean time either hidden or lent the precious pages, so that I have not them now before me. I sincerely hope I shall live to see the larger work you speak of, & hope the Assyrian society will bring ample accessions to the stores already at your hand.

I hope yourself & all your family are enjoying good health, & shall be much disappointed, if I miss seeing you, when next on this side of the Channel.

I suppose you know that D^r. Lee has lost his wife – she was uneducated but had much natural goodness & was a thoroughly well principled woman.[1] To him she was inestimable.

> Believe me,
> My dear Sir,
> Your much obliged & faithful
> G. C. Renouard

P.S. There is a paper worth reading 'Über die zweite Art der achämenischen Keilschrift' by Prof^r. Holtzmann in the 2nd. part of the VIIIth vol. of the D. M. Gesellschaft (Leipzig, 1854).[2] He disallows the Skythische Element & so do I. But who has shewn that Skythische = Türkische?[3]

MS: GIO/H 470

---

1   John Lee's first wife, Cecilia Rutter (1782–1854), died on 1 April. Lee married Louisa Catherine Heath in 1855. According to *ODNB* her name was Louisa Catherine Wilkinson.
2   A. Holtzmann, 'Über die zweite Art der achämenischen Keilschrift', *Zeitschrift der deutschen morgenländischen Gesellschaft* 8 (1854), pp. 329–45.
3   See extracts from Rawlinson's letters read at a meeting of the Royal Asiatic Society on 5 February 1853 in *The Athenaeum*, no. 1321 (19 Feb. 1853), p. 228; no. 1338 (18 June 1853), pp. 741–2.

FROM AUSTEN HENRY LAYARD
27 APRIL 1854

H. of. C.
April 27/54

My dear D^r. Hincks,

I am much rejoiced that you accepted the pension and that it was offered to you in the manner you describe. It is honorable to both parties.

I am just returned from Paris, where your note was forwarded to me but not in time to make all the enquiries I could have wished. As far as I could learn from M. Mercier & others no new inscriptions have been received in France.[1] I saw the last drawings & plans sent by Place but there were no inscriptions.[2] The cast of M^{rs}. Taylor's cylinder is at the Louvre & that is the only inedited monumental inscription of any interest with which I am acquainted at Paris. I confess I think you would have a much wider & more promising field for research & discovery among the treasures of the British Museum, which are still almost unexplored.

I shall be very glad when your translation appears.

Believe me,
in haste,
yours very faithfully,
A. H. Layard.

MS: GIO/H 242

1   Joseph Lemercier opened the most important lithographic press at Paris in 1828.
2   Victor Place (1818–75); see M. Pilet, *Un pionnier de l'assyriologie. Victor Place* (Cahiers de la société asiatique, 16; Paris, 1962).

TO HENRY ELLIS
3 MAY 1854

Killyleagh C° Down
3rd. May 1854

Dear Sir Henry,

I am sorry that several unexpected occurrences have rendered it impossible for me to complete the translations at the time proposed.

I send you, however, now (in an open cover) a book containing the Annals of Sennacherib, so far as I have been able to translate them; and also a revised translation of the Annals on the Nimroud Obelisk, and some passages from pieces of cylinders. I hope that these last will be accepted to make up for any deficiency in the former.

In about a fortnight I hope to send you a similar book, containing the annals of the early Nimroud king.[1]

I remain,
Yours very faithfully,
Edw. Hincks

MS: BM/CA/OP

1   Hincks wrote his 'transcriptions' and translations in two hardcover exercise books. In the British Library the books are catalogued as one manuscript: MS BL Add.22097 ('Readings of Inscriptions on the Nineveh Marbles by Dr Hincks'). 'The Inscriptions of Sennacherib' (and the annals of Shalmaneser III), received on 6 May 1854, are contained in ff. 1–27 (Hincks numbered the pages 1–53). 'The Inscriptions of Assur-yuchura-bal [Asshurnasirpal II]', received on 20 May 1854, are in ff. 28–56 (Hincks's pages 1–58; note that pp. 28 and 31 are out of order: f. 41 = pp. 26–7; f. 42 = pp. 29–30; f. 43 = pp. 28, 31).

TO AUSTEN HENRY LAYARD
6 MAY 1854

Killyleagh C° Down
6th. May 1854

My dear M^r. Layard,

I received your letter a week ago. You are probably right in thinking that the unexplored treasures of the British Museum present a fairer prospect than there is at Paris. The question, however, arises – are the former accessible? I may be unreasonably suspicious; but I thought I could plainly see a wish to *keep from me* what was likely to be of interest. I have now sent a MS book to the Trustees, containing translations from the inscriptions of Sennacherib, his grandson & the obelisk king; & will follow it in a week or ten days by a translation of the annals from the N.W. palace. In these I have given not merely a translation but ample notes, containing a mass of information, which is (I flatter myself – indeed I have no doubt) of *very great value*. I have no idea that the Trustees will *publish* it. I think it very possible that they will allow the use of it privately to some of their present employees, so as to enable him to carry on the work – *I being cast off*; and it may be that the person thus benefitted may not acknowledge the source from which he derives information. Rawlinson will be in London, I suppose, forthwith. He was expected before this.[1] And I have no doubt that *he* will be allowed free access to all that the Museum possesses.[2]

I am thus *disheartened* with respect to London. Paris possesses the cast of Taylor's cylinder, the stone of Michaux & the originals of some of Botta's inscriptions; the published copies of which are by no means accurate. All these are worth examination; & I think the two first might be made interesting to the public. As to their accessibility, Lord Clarendon has promised me a letter to Lord Cowley, which would (I should think) pass me to all I wished to see.[3] It is possible too that, tho' there are no *very recent* inscriptions, there are others than what I have mentioned, not included in Botta's publications.

The drawback to my going to Paris is that I do not readily speak the language, or understand it when spoken; though I *read* it without difficulty. Is it not the case, however, that English is very generally spoken by *some* one at the Museums & other public places? or that interpreters may be had?[4]

I have stated to you the *pros* & *cons*; & should like to know your opinion, after reading what I have said.

There is another matter on which you would oblige me by giving me information. M^r. Bland had a notice on the paper for a bill on Irish Ratecharges for Tuesday last. It fell to the ground through the count out. I should like to know if he has renewed it & *for what day*.[5] It is a subject of immense interest to me – *any* change in the law w^d be beneficial to me & a *fair* change in it would be *very much so*. The Irish clergy generally, would, however, like matters to remain as they are.

The vote paper, which you will have on Monday, will I presume enable you to answer this question.

> Believe me
> Yours very faithfully
> Edw. Hincks

MS: BL Add. 38982, ff. 206–7

1   After recovering from a broken collar-bone, incurred through a fall from his horse during a hunt, Rawlinson left Baghdad in March 1855, travelled to Bombay, and after a fortnight's rest there, he made the long journey home to England, where he arrived in May. See Lesley Adkins, *Empires of the Plain: Henry Rawlinson and the Lost Languages of Babylon* (London, 2004), p. 333.

2   Hincks's words may appear as evidence of paranoia. Subsequent events will show that his suspicions were not unfounded.

3   Lord Clarendon, formerly Viceroy of Ireland, was now Foreign Secretary and Lord Cowley was British ambassador to Paris. In 1852 the latter offered Layard the post of secretary at the embassy. See Layard's letter of 5 February 1852 n. 1.

4   Hincks abandoned his plan to visit Paris. See his letter of 17 June 1854 to Layard. He never travelled outside the British Isles during his whole lifetime.

5   Loftus Henry Bland was born in August 1805 in Blandsfort, Queen's County (today Co. Laois), son of John Bland. He was educated at Trinity College, Cambridge: B.A. 1825, M.A. 1828. He was called to the Irish bar in 1829 and was M.P. for King's County (today Co. Offaly) from 1852 to 1868. He died at 33 Merrion Square, Dublin on 21 January 1872.

FROM HENRY ELLIS
19 MAY 1854

[British Museum]
19th. May 1854

My dear Sir,

Your letter, together with the book containing the Annals of Sennacherib, so far as you have been enabled to translate them, were laid before our Trustees at their meeting on Saturday last: the Trustees desire me to say they will be glad to receive the remainder when you shall have finished.

> I remain, My dear Sir,
> very faithfully yours,
> Henry Ellis

MS: BM/CA/LB 46: 2

FROM HENRY ELLIS
20 MAY 1854

[British Museum]
20th. May 1854

My dear Sir,

I wrote to you by yesterday's post acknowledging the receipt of the pink covered copy-book containing the first portion of the Annals of Sennacherib; and today I have to acknowledge the receipt of a second copy book containing the Nimrud Annals, specimens of mode of transcriptions &c.[1] As far as I at present know our Trustees will not meet again till the middle of next month in committee; but your fresh communication, with the request for payment, shall, when they meet, be laid before them.

I remain, My dear Sir,
very faithfully yours,
Henry Ellis

MS: BM/CA/LB 46: 3–4

1   On the front inside cover of the second copy-book (MS BL Add.22097), the following notes are written: 'Received at the British Museum on 20th. May 1854. A. Panizzi'; 'Deposited in the Dep^t. of MSS by order of committee 1 Augt. 1857.' Anthony Panizzi (1797–1879) was keeper of printed books at the British Museum from 1837 to 1856 and principal librarian from 1856 to 1866. See the *ODNB* where he is described as 'the outstanding figure in the history of the British Museum'.

FROM HENRY ELLIS
6 JUNE 1854

British Museum
June 6th. 1854

My dear Sir,

I inclose the form of receipt for the amount agreed by the Trustees to be paid to you upon the completion of your engagement, and have only to request that the receipt may be duly stamped & signed by you, and then forwarded to me, when the money shall be paid to your order.

Believe me, My dear Sir,
Yours faithfully,
Henry Ellis

MS: BM/CA/LB 46: 24

᾿᾿᾿

HENRY ELLIS TO HENRY CRESWICKE RAWLINSON[1]
7 JUNE 1854

British Museum
7th. June 1854

D[r]. Hincks having completed his connexion with the Museum has given as the result of his labour, two manuscripts, one a book containing the Annals of Sennacherib as far as he has been able to translate them; and also a revised translation of the annals on the Nimrud Obelisk, and some passages from pieces of cylinders: the other book containing the Nimrud annals, and at the end specimens of the mode of transcription, which D[r]. Hincks thinks preferable to all others for the editing of inscriptions: with specimens also of the two lists and the three indices which would be necessary in order to render such transcriptions available to those who desire to study them.

The Trustees intend to keep these manuscripts, and upon your return you will be able to compare them with the originals. If however you should wish to see these translations *before your return*, the Trustees will direct a copy to be made and transmitted to you.

Henry Ellis

MS: BM/CA/LB 46: 26–8

1   Extract from copy.

᾿᾿᾿

FROM HENRY ELLIS
10 JUNE 1854

10th. June 1854

My dear Sir,
You have sent me a receipt, but you have not told me to whom, how, or where I am to pay the money. Till your directions have been received on this point, I do not feel authorized to send the payment.

Yours very faithfully,
Henry Ellis

MS: BM/CA/LB 46: 34

TO AUSTEN HENRY LAYARD
17 JUNE 1854

Killyleagh C° Down
17th. June 1854

My dear Mr. Layard,

I had your letter of the 8th. ult°. & have given up the Paris expedition according to what is evidently your opinion. I shall be at liberty to leave this after the 25th. and would go to London, had I any prospect of doing any good there. I will write to Sir H. Ellis for leave to have access to the terracottas as I had previously. My *engagement* is at an end; & this time I mean to go at my own expense. I have heard nothing as to my translations. The day of meeting of the Trustees falling on that when the Crystal Palace was opened, nothing was probably done but what was strictly matter of course.[1] Probably they may not meet again till the 8th. July. I have no expectation that *they* will publish any thing; & I doubt whether the public would take so much interest in the translations as to make it worth a bookseller's while to publish them. I have sent a chronological piece to the Journal of Sacred Literature, a quarterly publication.[2] It will be out in a fortnight. I have also sent a grammatical paper to the Irish Academy which is to be read on the 26th.[3]

So much for cuneatic matters. Let me now venture to request that you will do what you can for Mr. Bland's bill. I see that it was on the paper for second reading on Wednesday & again on Thursday. Probably, however, this is only with a view to fix the day when it is to be actually discussed. I will put on another paper some hints about it. What I would feel particularly obliged by your doing is to make the following suggestion which appears a reasonable one.

In case Sir John Young[4] meets the bill by a *promise* to bring forward a government measure next session, would he object to a short bill being now passed *to suspend for a year* the revision of compositions which were registered under Mr. Goulburn's act 'as invariable whatever the price of corn might be'?[5] In a very large proportion of these compositions, the present year is that in which the revision should take place. If the bill be deferred a year, a fraud will be committed this coming autumn; and the bill of next session will be too late to prevent it. Of course, this short bill should provide for the revisions which are postponed this year being made next year, if the promised bill is not passed.

Parish A in the accompanying memorandum is unfortunately my own; Parish B you will, I dare say, hear of from Mr. Bland.

Between ourselves, I fear Sir John Young. His uncle has got a parish in which the tithes are compounded under the new act of Lord Derby's; *he* may wish matters left in statu quo; & how far Sir John may be influenced by his representations I cannot say.

The second object of the proposed bill does not (I conceive) apply to parishes compounded under M^r. Stanley's act.[6] That act was fair enough with respect to uncompounded parishes, in which it established equitable compositions; but it *shamefully violated the pledged faith of parliament* (& of Mr Goulburn *personally* in some cases) where compositions had been made under the old acts.

> Believe me
> Yours very faithfully
> Edw. Hincks

[Enclosure:]

Memorandum

M^r. Bland's bill is not one of hostility to the Irish clergy, *as may perhaps be represented*, but one for *avoiding litigation*, and *remedying monstrous injustice*.

It avoids litigation by making that revision a matter of course (as it is in England) which is now affected by a proceeding which is always regarded as one of hostility & which sets the tithe owner & the tithe payer at as much enmity with one another as the old tithe system did. It remedies the most monstrous injustice that was probably ever committed by Parliament. It *deals with it*, at least; and if it do not remedy it, *as brought in*, it may do so, *as amended in committee*.

Let one specimen suffice. The facts are *certain* & *can be proved before a committee*. There were two parishes, the incumbents and parishioners of which intended to make *precisely the same bargain*. Under M^r. Goulburn's act, they stood towards each other in *precisely the same position*; but under M^r. Stanley's (which is now in force) the one composition was made 80 per cent higher than the other! For every £100 in parish A, the rector will receive only £75; while for every £100 in parish B he will receive £135; *supposing a revision to be made this year under the existing law*. The proposed bill deals with this atrocious injustice (as it is in *both* instances; in the one case to the Clergyman, in the other to the parishioners) & may or ought to make the reduction *the same* in *each case*, & to about £93 as I estimate.

The bill having some such effect, of course *some* tithe owners will make an outcry against it. I don't know what part the members for the University will take.

MS: BL Add. 38982, ff. 255–7

1   The Crystal Palace, which housed the Great Exhibition in 1851–2, was reopened at Sydenham on 10 June 1854.
2   'Chronology of the Reigns of Sargon and Sennacherib', *Journal of Sacred Literature* 6/12 (July 1854), pp. 393–410.

3   'On the Personal Pronouns of the Assyrian and Other Languages, especially Hebrew', *Transactions of the Royal Irish Academy* 23 (1854), part 2, pp. 3–9.
4   Sir John Young, Baron Lisgar (1807–76) was M.P. for Cavan, 1831–55, and chief secretary for Ireland under Lord Aberdeen from 1852 to 1855. He became governor-general of Canada in 1869.
5   Henry Goulburn (1784–1856) was chief secretary for Ireland from 1821 to 1827. He steered the Irish Tithe Composition Bill through parliament in 1823.
6   When Edward Stanley, fourteenth earl of Derby (1799–1869), was chief secretary for Ireland from 1830 to 1834, he introduced measures intended to reform tithes in Ireland.

FROM HENRY ELLIS
27 JUNE 1854

British Museum
27th. June 1854

My dear Sir,

I have shewn your letter of 24th. June to M^r. Hawkins who, as the Head of the Department of Antiquities, regulates any extra-facilities which may be wished for by applicants.[1] He begs me to say he will be very happy to allow the access you request to the terra-cotta inscriptions for your own private study, in the same way in which you had the opportunity of consulting them when *employed* under arrangement with the Trustees. Of course you are fully aware that your engagement with the Trustees is entirely at an end.

I remain, My dear Sir,
Very faithfully yours,
Henry Ellis

MS: BM/CA/LB 46: 46

1   Edward Hawkins, keeper in the Department of Antiquities at the British Museum.

≈≈≈

TO THE EDITOR OF THE LITERARY GAZETTE
24 JULY 1854

Killyleagh C° Down
24th July 1854

Sir,

During a recent visit to London, I was enabled, through the courtesy of the gentlemen in the Antiquarian department of the British Museum, to examine many of the terra cotta tablets with cuneiform inscriptions;[1] and as I know that several persons take a deep interest in their contents, I think it will be proper to send you a short report of what I met with, which may serve as a continuation of that which you published in the 'Literary Gazette' of 22nd April.[2]

I examined about 150 tablets, which have been numbered for reference, and of which photograph copies have been made. A few of these – I did not count how many, but suppose about a tenth – are in the Babylonian character, similar to that on the contracts published by Grotefend in the first four volumes of the 'Zeitschrift für die Kunde des Morgenlandes'.[3] The remainder are in the ordinary Assyrian character, which differs from the Babylonian something more than our italic character differs from the Roman, but not as much as it differs from the German. The Babylonian tablets mostly belong to a class of which there may be from 20 to 30 in the 150 that have been photographed. At the head of each of these there is an introductory formula, containing the name of some private individual, which varies in the different tablets, who invokes blessings on 'the king my lord' from different deities, as Nabyu and Marduk, or Assur, Shamas and Marduk. The former are named on Babylonian, the latter on Assyrian tablets. Not finding on a cursory inspection of these tablets, that they contain either king's name or date, and believing them to relate to the affairs of private individuals, I put them aside as of less interest than others. I do not doubt, however, that they contain matter which would repay the person who should have leisure and opportunity to study them.

Of the remaining inscriptions almost all are more or less mutilated. This is deeply to be regretted. We may hope, however, that an examination of the remaining tablets in the Museum collection (of which I understand there are above 800) will bring out fragments which can be connected with those that have been numbered and photographed. Among these I found a fragment of an historical document, relating to the war of Assur-yuchura-bal against the Elymeans (K. 30).[4] It contains the same text, or at any rate refers to the same events, as the fragments of cylinders which I mentioned in my Report to the Trustees. Another fragment relates to the marches of Tukulti-bal-itsri (Tiglath Pileser) the Second, the commencement of whose reign, which probably lasted above forty years, was about 770 B.C., and also, it would seem, to those of some of his predecessors.[5] It contains the conclusions of about ninety consecutive lines, the beginning and ending of the inscription being altogether wanting, and the beginnings of all the lines that remain. In its present state it affords some geographical information, but if the remaining portions could be recovered it would be of immense value as an historical document.

Several tablets relate to the calendar, and from these I have ascertained what has surprised me not a little.[6] Notwithstanding the reference made by Ptolemy of the eclipses observed at Babylon to the months of a wandering year, resembling that of the Egyptians (the correctness of which reference I had supposed to be established by my observing, that in more than one instance there were thirty days assigned to consecutive Assyrian months); I have now obtained positive proof that the Assyrians used a lunar year, consisting of twelve or thirteen months, each of which contained nominally thirty days. Of course, every sixty-third or sixty-fourth day in the calendar was omitted, in the same manner as it was in the Grecian calendars of Meton and Calippus. I found in a sort of calendar that the first day of the month was called the *arakh*, or 'new moon', as well as the month itself; I found that the thirteenth month, which I had supposed to consist of the five epagomenae, had as many days as that which preceded it; and, lastly, I found a tablet (K. 90) which contained an estimate of the magnitude of the illuminated portion of the lunar disk on each of the thirty days of the month.[7] This is not very creditable to the mathematical knowledge of the Assyrians; but it is a sufficiently close approximation to leave no doubt as to what was intended. On the first day they estimated that five parts were visible out of the 240 into which they divided the disk. On the second day they doubled this, counting ten parts. In like manner they counted twenty on the third day, forty on the fourth, and eighty on the fifth. They then substituted an arithmetical for a geometrical series, adding sixteen parts each day till the fifteenth, when the whole 240 were visible. They took sixteen parts away on each of the next ten days, so that they had eighty parts, or one-third of the disk, visible on the 25th. day. The latter part of the inscription is injured, so that I cannot be very positive. I believe, however, that they halved what was visible on each of the next four days, so as to have five on the 29th. as well as on the first, the thirtieth day being altogether dark.

I have very little doubt that the Assyrian year began with the new moon which followed, or which was nearest to, the vernal equinox. This appears from the second month being that which corresponded to the Persian *Thuravahara*, a name which, as Benfey has pointed out, signifies 'the heat of spring'; while the ninth month was that which corresponded to the Persian *Atriyadiya*, a name which seems to signify 'the commencement of fire', indicating the first month of the winter. In confirmation of this, I observe that the character for this ninth month is phonetically *kan*, and that the Syriac *Kanun* is the third month from the autumnal equinox; also that *Ab* is the fifth month from the vernal equinox in Hebrew and Syriac, and that one phonetic value of the Assyrian character for the fifth month is *ab*. The dates of the commencements of the different campaigns, which are given in the Nimrûd annals, appear to me also to agree with this date of the commencement of the year better than with any other.

With respect to the position of the intercalary month, it is curious that while more than one tablet places it at the end of the year, that is, before the vernal equinox, which was its place in the Syrian and Hebrew year, it is distinctly placed in K. 160, at the end of the sixth month.[8] It is probable that a change took place in the calendar between the making of the earlier and the later chronological tablets. It is not easy, however, to say which was the earlier. The law of the intercalation, and the precise rule for omitting days in the alternate months, remain to be ascertained.

I mentioned in my Report that the Syllabarium, M^r. Layard's copy of which first indicated the nature of such a document, had not been found. I am happy to say that it has been met with since, and that it appears in this series as K. 62.[9] Another Syllabarium which I had not seen before is also numbered as K. 110.[10] Both of these are inscribed on both sides. The other Syllabarium, which I met with in the spring of 1853, is K. 144. It is inscribed on one side only. These three fragments belong to three different Syllabariums, so that it is more than ever desirable to obtain the deficient portions. A very large number of values is, however, determined from the fragments which we already possess.

There are other inscriptions of a philological or lexicographical character which possess much interest for me, but with which I will not trouble the public. Some inscriptions are in praise of different deities, and some appear to be poetical. I will only notice the conclusions of some of the inscriptions, as bearing on the royal succession after Esarhaddon. Many of the tablets, of which the conclusion is preserved, after the hymns, or whatever they may be, that they contain, present to us a formula, which consists of a sort of title, indicating the contents of the main part of the inscription;[11] some connecting words, which vary in different tablets, and which appear to me to indicate the compartment in the royal library where the tablet was to be placed; and then 'the palace of Assur-bani-bal, the great king, the powerful king, the king of the provinces, the king of Assyria'; with the addition of a number of fanciful titles, varying on different tablets, such as 'who looks for help to Assur and Nina', 'whose ears Nabyu and Urmitu have opened wide'.[12] By the multiplying of titles of this sort, and by leaving wide intervals between the words, the writer of the tablet contrived to fill up even a very large space, should such remain vacant at the close of the regular inscription. There are other tablets in which a similar large space remains after the regular inscription, but it appears for the most part blank, the following words being alone legible: 'The property of Assur-yuchura-bal,[13] the king of the provinces, the king of Assyria.' These words are incised after the tablet was burned, and, as it appears to me, on a surface which had been smoothed to receive them. Comparing what I observed in this instance with what I have observed on several well-known Egyptian monuments, I have no doubt that where this name now occurs there were formerly the name and titles of Assur-bani-bal, as they now appear on κ. 131 or κ. 155. These were inscribed before the tablet was baked. Afterwards, Assur-yuchura-bal, having made himself master of the palace, caused the name and titles of his predecessor to be scraped away, and had his own name incised on the surface so left smooth.

The name of Assur-yuchura-bal does not appear in Mʳ. Layard's list of kings of Assyria; nor am I aware that Col. Rawlinson has noticed him as distinct from his successor in any of his publications. We know that his successor (whom I formerly called Assur-akh-bal, but now Assur-yuchura-bal)* was the son of Esarhaddon, and it became a question of some interest who his predecessor was. He does not give his genealogy on any of these tablets, nor elswhere, so far as I know; but I think there can be no doubt that he was a son of Esarhaddon, and that he and Assur-yuchura-bal reigned at the same time in different portions of the empire, the former possessing Nineveh in the first instance, and being succeeded there by the latter. If it were not for the defacement of the one name, and the substitution of the other, and that in a manner which is evidently disrespectful, I should have not questioned the identity of the two kings; and some may think it unreasonable for me to do so even now, inasmuch as it is certain that both kings claimed the glory of the conquest of Elymais, which must have happened, according to my view of the matter, while Assur-bani-bal was reigning at Nineveh, and Assur-yuchura-bal in some other part of the empire. It is possible that the latter may have carried on the war in person; but it seems to me more likely, that having dethroned his brother, and wishing himself to be considered as the immediate successor of his father, he caused himself to be represented as gaining victories, which were in reality those of his brother, or of his brother's generals. I incline to think that the king who is commemorated on the very remarkable stone in the possession of Lord Aberdeen, and who there calls himself ruler of Babylon (which neither Assur-bani-bal nor Assur-yuchura-bal ever does), was a third son of Esarhaddon, and that Babylon was separated from Assyria on the death of Esarhaddon, in 667 B.C. The Saosdukhin of Ptolemy's Canon seems a possible corruption of Shamas-akh-iddan; but it is not easy to derive it from any name beginning with Assur. The order of succession, however, and the length of the different reigns between 667 B.C. and 625 B.C., when Nineveh was taken by the Medes and Babylonians, is now, and is likely to remain, very obscure. The most likely way of clearing it up would be, I think, the collection of tablets with dates in regnal years, similar to those of which I gave a list in my Report ('Lit. Gaz.' 22ⁿᵈ April, p. 375). The date of the capture of Nineveh appears to me quite certain; and of course I hold that

Herodotus committed a gross blunder, either in placing the Lydian war before the capture of Nineveh, or in identifying the eclipse which terminated that war with the eclipse which Thales foretold. As to the Scythian conquest, it must have occurred about the middle of the interval of forty-two years that I have mentioned; but no allusion to it has been met with on the monuments, nor do I think that any is to be expected. The Assyrian kings carefully recorded their successes; but as to their reverses, they were as carefully silent.

I am, &c.,
Edw. Hincks.

*The monogram which represents the second element in the name, as it is written on the sculptures in the British Museum representing the conquest of Elymais; and on these tablets is one which is used in no other proper name that I am aware of, except that of the father of Nebuchadnezzar, where it represents the last syllable *yuchur*. In the present name it ought perhaps to be read as a participle rather than as an aorist, judging from the analogy of Assur-bani-bal, of which the second element is written phonetically in K. 131.

Source: Literary Gazette, no. 1959 (5 Aug. 1854), pp. 707–8[14]

1   Hincks was in London from 5 to 21 July to continue his research at the British Museum, although he was no longer under contract with the Trustees. On 9 or 16 July he preached for Renouard at Swanscombe.
2   See Hincks's letter of 31 March 1854 to the Trustees of the British Museum above, which was published in *Literary Gazette,* no. 1944 (22 Apr. 1854), pp. 375–7 = *Report to the Trustees of the British Museum Respecting Certain Cylinders and Terra-Cotta Tablets, with Cuneiform Inscriptions* (London, 1854).
3   G. F. Grotefend, 'Urkunden in babylonischer Keilschrift', *Zeitschrift für die Kunde des Morgenlandes* I (1837), pp. 212–22; 2 (1839), pp. 177–89; 3 (1840), pp. 179–83; 4 (1842), pp. 43–57. Each article is accompanied by a lithograph.
4   See A. K. Grayson, *Assyrian Rulers of the Early First Millennium B.C.: I (1114–859 B.C.)* (Toronto, 1991), p. 233.
5   Tiglath-Pileser III (745–727 B.C.) (Assyrian: *Tukulti-apil-ešarra*).
6   On the Assyrian calendar, see J. M. Steele, *Calendars and Years: Astronomy and Time in the Ancient Near East* (Oxford, 2007).
7   Hincks discusses the tablet K. 90 in his paper 'On the Assyrian Mythology', *Transactions of the Royal Irish Academy* 22 (1855), polite literature, pp. 406–7. He returned to it in 1865; see his paper 'On the Assyrio-Babylonian Measures of Time', *Transactions of the Royal Irish Academy* 24 (1865), polite literature, p. 16. For bibliography of nineteenth-century studies of the text, see C. Bezold, *Catalogue of the Cuneiform Tablets in the Kouyunjik Collection of the British Museum,* vol. 1 (London, 1889), p. 24.
8   Hincks was fascinated by the contents of tablet K. 160. In June 1860 he published 'On Certain Babylonian Observations of the Planet Venus', *Monthly Notices of the Royal Astronomical Society,* vol. 20 (1860), pp. 319–20; see also the abstract of his paper 'On Some Recorded Observations of the Planet Venus in the Seventh Century before Christ', *Report of the Thirtieth Meeting of the British Association for the Advancement of Science; held at Oxford in June and July 1860* (1861), pp. 35–6. In 1864 he published a paper on 'Series of Observations of Disappearances and Reappearances of Venus, Recorded on Tablets in the British Museum, Marked K. 160', *Astronomische Nachrichten* 63 (1865), cols 223–4 (dated 2 Sept.1864). The

article accompanies his letter of 13 September to the editor, cols 221–3. See also 'On the Assyrio-Babylonian Measures of Time', *Transactions of the Royal Irish Academy* 24 (1865), p. 21. The fundamental edition of this tablet today is Erica Reiner and D. Pingree, *The Venus Tablet of Ammiṣaduqa* (Babylonian Planetary Omens, Part 1; Malibu, 1975). See also C. B. F. Walker, 'Notes on the Venus Tablet of Ammiṣaduqa', *Journal of Cuneiform Studies* 36 (1984), pp. 64–6.

9   On the importance of Hincks's discovery of this lexical list which today we refer to as 'Syllabary A' (the British Museum number is K. 62), see Hincks's letter of 7 March 1853 to the Royal Irish Academy, especially n. 1 and the bibliography there.

10   The British Museum tablet K. 110 has been edited by B. Landsberger, 'Das Vokabular Sᵇ'. Nach einem Manuskript von H. S. Schuster' in *Materialien zum sumerischen Lexikon*, vol. 3 (Rome, 1955), pp. 129–53. For a description of the tablet and a bibliography of nineteenth-century studies of the text after Hincks, see C. Bezold, *Catalogue of the Cuneiform Tablets in the Kouyunjik Collection of the British Museum*, vol. 1 (London, 1889), p. 28.

11   See H. Hunger, *Babylonische und assyrische Kolophone* (Alter Orient und Altes Testament 2; Kevelaer and Neukirchen-Vluyn, 1968).

12   'Urmitu' is now read Tašmetu. She was the spouse of Nabu, the god of wisdom and writing.

13   Aššur-nāṣir-apli (Ashurnasirpal).

14   This letter was also published in *Journal of Sacred Literature* 7/13 (Oct. 1854), pp. 231–4.

FROM GEORGE CECIL RENOUARD

3 AUGUST 1854

Swanscombe, Dartford

Th. 3. Aug. 1854

My dear Sir,

I am very sorry that continual interruptions have prevented me from thanking you sooner for the Zeitschrift which you returned & very particularly for 'the Chronology of Sargon & Sennacherib' which is the *first* very satisfactory fruit gathered from our Assyrian & Babylonian nurseries.[1]

To your proposed correction of the Masoretic text of 2 Kings xviii, 13, no valid objection can be made as numerals, even in words, are peculiarly liable to error, but if 24ᵗʰ. in lieu of 14ᵗʰ. could be admitted the changes would only be from *hē* to *mēm* (for I suppose the *yōd* to have been commonly omitted) which would amount to scarcely any thing & even if the place of the numerals must be changed, the alteration wᵈ. be small.[2] It does not appear that there is MSS. authority for any change: but we have no Heb. MSS. of undoubted or probable antiquity. The LXX agrees with the Masoretic text which shews the antiquity of the error.

I hope you have got a note of all your papers in the various periodicals thro' which they are dispersed. Many, if not all, of them will deserve to be reprinted & would make a goodly volume: at all events an index to them would be extremely useful and perhaps be admitted into 'Notes & Queries' or some such collection.

I sincerely hope the report of your pension is not erroneous[3] & that you can safely receive the congratulations of,

my dear Sir,
yours faithfully,
G. C. Renouard

MS: GIO/H 471

1   Hincks, 'Chronology of the Reigns of Sargon and Sennacherib', *Journal of Sacred Literature* 6/12 (July 1854), pp. 393–410.

2   In rejecting the numeral 'fourteen' in 2 Kings 18: 13, Hincks was trying to solve a crux that has bothered modern commentators since the discovery of the Assyrian texts. See the discussion in M. Cogan and H. Tadmor, *II Kings* (The Anchor Bible, 11; Garden City, N.Y., 1988), p. 228.

3   See Hincks's letter of 8 April 1854 to Layard in which he mentions that Lord Aberdeen has offered him a pension of £100.

FROM RICHARD CULL[1]
13 OCTOBER 1854

13, Tavistock Street
Bedford Square
13 Oct. 1854

My dear Sir,
The reading of your excellent paper 'On the location of the ancient Chaldeans' was deferred at Liverpool hoping we should see you, and after being on the list two days was at last left amongst others unread.[2] I brought the paper with me to town and I await your instructions on it. If you would like it to be read at our Society (Ethnological) I will lay it before the Council on Wednesday; should you however desire me to return you the paper unread I will do so at once.

I ought to have written to you on the subject ere this, but since my return from a short tour after leaving Liverpool I have been much occupied.

I am
Yours very truly
Richard Cull

MS: GIO/H 150

1   Richard Cull was a fellow and secretary of the Ethnological Society. He followed Hincks's studies with great interest. After Hincks's death, he published 'On a *t* Conjugation, Such as Exists in Assyrian, Shown to Be a Character of Early Shemitic Speech, by its Vestiges Found in the Hebrew, Phoenician, Aramaic, and Arabic Languages', *Transactions of the Society of Biblical Archaeology* 2 (1873), pp. 83–109.
2   Due to illness Hincks was unable to attend the meeting of the British Association for the Advancement of Science at Liverpool in August–September.

FROM GEORGE CECIL RENOUARD
16 OCTOBER 1854

Swanscombe
Dartford, Kent
M. 16 Octr. 1854

My dear Sir,

I am happy to hear that the books reached you safely & beg they may have a place on your shelves where they will be much better used than while on mine. The Aethiopic Grammar & Dict^y have had the singular good fortune to pass some years in Abyssinia, in the custody of M. *Antoine d'Abbadie*, to whom the marginal notes in the latter, are due.[1] You, perhaps, will think, with me that they are an improvement, rather than a deterioration to the book.

I have read most of Bunsen's three vols, without being satisfied with his system.[2] His criticisms & emendation of Strabo's text are bold, & plausible, but unsatisfactory, & such, from Kramer's gentle hint, I have no doubt, he deemed them. Bunsen, though handsomely treated by De Rougé, is rarely commended by him.[3] Not so, D^r. Hincks; with some of whose writings M. de Rougé is well acquainted & acknowledges his obligations to them. His paper on the stele of Ahmes (Amasis?) is the most complete & honest attempt at reading & translating a hieroglyphic inscription, I had yet seen.[4] With Lenormant's writings, I am little acquainted, but to judge from his pupils he must be a worthy successor of Champollion.[5] I have occasionally looked into Lepsius's 'Introduction' to his great work on the Aegyptian Chronology; & have been satisfied with what I have seen.[6] His plates I have, but they are fearfully voluminous, much too large to be manageable; & are almost a dead letter without the text.[7] Brugsch, as, I suppose, you know, has been in Egypt, & has probably collected much there:[8] but his self sufficiency & jealousy will, I fear, prevent him from doing all he might to advance our knowledge of ancient Egyptian literature. I hope soon to reperuse & study your paper on the pronominal roots.[9] It will, when duly weighed, greatly enlarge & correct the views of,

my dear Sir,
your much obliged & faithful
G. C. Renouard

The Amharic or Amharénya as, it seems, the natives call it, is, I apprehend, an *African*, not a *Semitic* tongue:[10] the language spoken in Tigré, differs little from the Geez or ancient Aethiopic.

MS: GIO/H 472

1   Antoine Thomson d'Abbadie was born in Dublin on 3 January 1810 of an Abyssinian father and an Irish mother. He visited Brazil in 1837, and from 1840 to 1848 he travelled in Abyssinia. He died on 19 March 1897 at Paris. See *Dictionnaire de biographie française*, vol. 1 (Paris, 1929–32), cols 35–41. Manfred Kropp thinks that the volumes annotated by d'Abbadie are C. W. Isenberg's *Dictionary of the Amharic Language* (London, 1841) and *Grammar of the Amharic Language* (London, 1842). See M. Kropp, 'From Manuscripts to the Computer: Ethiopic Studies in the Last 150 Years', in K. J. Cathcart (ed.), *The Edward Hincks Bicentenary Lectures*, pp. 114–37, here 115. The only alternatives are H. Ludolf's, *Grammatica Aethiopica* (Frankfurt, 1661; 2nd edn, 1702) and *Lexicon Aethiopico-Latinum* (Frankfurt, 1699); or his *Grammatica linguae Amharicae* (Frankfurt, 1698) and *Lexicon Amharico-Latinum* (Frankfurt, 1698).

2   The first three volumes of C. C. J. Bunsen, *Ägyptens Stelle in der Weltgeschichte*, 5 vols (Hamburg, 1845–57).

3   This may be a reference to Emmanuel de Rougé's long review of Bunsen's work in *Annales de philosophie chrétienne*. See his letter to Hincks of 22 March 1852, n. 5.

4   E. de Rougé, *Mémoire sur l'inscription du tombeau d'Ahmes, chef des nautoniers* (Paris, 1851).

5   Charles Lenormant (1802–59) was Professor of Egyptian Archaeology at the Collège de France from 1848.

6   R. Lepsius, *Die Chronologie der Ägypter* (Berlin, 1849).

7   The plates in Lepsius, *Denkmäler aus Ägypten und Äthiopien*, 6 pts in 12 vols (Berlin, 1849–59).

8   The King of Prussia sent Heinrich Brugsch to Egypt in 1853.

9   'On the Personal Pronouns of the Assyrian and Other Languages, especially Hebrew', *Transactions of the Royal Irish Academy* 23 (1854), part 2, pp. 3–9.

10   Renouard is mistaken in his view here.

TO SAMUEL BIRCH
18 OCTOBER 1854

Killyleagh C° Down
18th. Oct[r] 1854

My dear M[r]. Birch,
    I send you half a dozen copies of a little paper I have just had printed.[1] You will I hope accept one yourself & can give them to the fellows in the Museum who are interested in these matters, & one to the Royal Society of Literature. I have just been reading Bunsen's book & am surprised to find that he still clings to that broken reed Eratosthenes.[2]

Yours very truly
Edw. Hincks

MS: KJC 1

1   'On the Personal Pronouns of the Assyrian and Other Languages, especially Hebrew', *Transactions of the Royal Irish Academy* 23 (1854), part 2, pp. 3–9.

2   Eratosthenes (276–194 B.C.), a Greek scholar.

FROM FRIEDRICH MAX MÜLLER[1]

24 NOVEMBER 1854

Taylor Institution
Oxford
Nov. 24, 54

Reverend Sir,

I have to thank you for your interesting article on Assyrian pronouns[2] which I received yesterday from Prof. Phillips.[3] I wish I had known it before because I should have liked to insert these Assyrian pronouns in a comparative list of pronominal affixes which I lately published, and I am looking forward with great interest to your promised article on the Assyrian verb, because it is a subject of immense importance in the history of language as exhibiting the earliest growth of a Semitic dialect fixed by contemporaneous evidence, and of no less importance for the philosophy of language as disclosing one of the earliest crystallizations of human thought in its struggle after grammatical forms.[4]

I take the liberty of sending you some small publications of my own, 'On the Languages of the Seat of War', of which I hope soon to send you a second edition;[5] 'Proposals for a Missionary Alphabet',[6] & 'On the Classification of the Turanian Languages'.[7] In the last I have given my theory of the verb in the three great families of language, Turanian, Semitic & Arian, and from what I can gather from your article on the pronouns, I believe that I have been working in the same direction which you point out as the only safe one for arriving at a solution of the highest problems of language.

[end of letter missing]

MS: GIO/H 276

1   Friedrich Max Müller (1823–1900), Sanskritist and philologist, was Taylorian Professor of Modern European Languages at Oxford from 1854 to 1868, and Professor of Comparative Philology from 1868. He was a fellow of All Souls College. One of his great works was the edition of the text of the Rig-Veda: *Rig-Veda-Samhita: The Sacred Hymns of the Brâhmans together with the Commentary of Sayanachara*, 6 vols (London, 1849–74); 2nd edn in 4 vols, 1890–2. He published translations of some texts.

2   Hincks, 'On the Personal Pronouns of the Assyrian and Other Languages, especially Hebrew', *Transactions of the Royal Irish Academy* 23 (1854), part 2, pp. 3–9.

3   George Phillips (1804–92), orientalist.

4   See Hincks's important articles 'On Assyrian Verbs', *Journal of Sacred Literature* 1/2 (July 1855), pp. 381–93; 2/3 (Oct. 1855), pp. 141–62; 3/5 (Apr. 1856), pp. 152–71; 3/6 (July 1856), pp. 392–403.
5   F. Max Müller, *The Languages of the Seat of War in the East* (London, 1854); the second edition was published in 1855.
6   *Proposals for a Missionary Alphabet* (London, 1854). According to the subtitle, it was submitted to the 'alphabetical conferences held at the residence of Chevalier Bunsen in January 1854'. Another work entitled *Proposals for a Uniform Missionary Alphabet* is dated 'Oxford, Christmas, 1853'.
7   *On the Classification of the Turanian Languages* (London, 1854). This work, which has 266 pages, is in the form of a letter to 'Chevalier Bunsen', and the signed copy presented by the author to the Taylor Institute library in Oxford is dated 9 October 1854.

TO EDWIN NORRIS
29 NOVEMBER 1854

Killyleagh C° Down
29th. Nov. 1854

Dear Sir,

I observe that a communication from Colonel Rawlinson was read at the last meeting of the Royal Asiatic Society, containing what he conceived to be rectifications of statements made by me in a report and letter of mine published in the Literary Gazette.[1] I trust the Society will accept a communication from me, tending to show that these are by no means rectifications.

Of Colonel Rawlinson's two objections, the first is of little importance. He says that the true name of the eldest son of Sennacherib is not *Assur-nadin*, but *Assur-nadin-iddin*. I have met with this name in three different forms in three different Bull inscriptions copied by M[r]. Layard. In one the name is distinctly *Assur-nadin*. In the other two an addition to this is found, which I at first read *sumi*.[2] Afterwards, I found an explanation of the *whole* conclusion of the name on a tablet in the British Museum, from which I inferred that it should be pronounced *nadin*, without any addition. Unfortunately I have mislaid my notes of the inscription on this tablet; and I am therefore unable to give my reasons for thus reading it more specifically than I have done. It is a matter of but little moment.

All the other points of difference to which Colonel Rawlinson has referred in his communication may be reduced to this: A certain royal name appears on tablets in the British Museum, and on bricks found at Babylon on the river side, which Colonel Rawlinson believes to be a variant of the name of *Nabu-nahid* (or, as he calls him, *Nabu-nit*), who began to reign in 555 B.C.;[3] but which I believe to be a variant of the name of Nabopolassar, who began to reign seventy years earlier.[4] The question is, which of us is right? That it is one or other of those kings seems pretty evident; for the father of this king is mentioned, and he was not a king. He was, according to Colonel Rawlinson, *Nabu-dirba*, and filled the high office of *'rubu-emga'*. Colonel Rawlinson has adduced, in support of his theory, a statement of Berosus that Nabunit executed some considerable works at Babylon; but Berosus mentions the outer walls of the city as all that he built; whereas the bricks are from the river side. On the other hand, in the great inscription at the India House, Nebuchadnezzar distinctly mentions these works by the river side, as having been completed by himself; they having been commenced by his father, Nabopolassar, whose bricks might, therefore, be naturally expected to be found in their foundations. Besides, if M^r. Layard's copies be correct, the final character in the disputed name is interchanged with one which is interchanged with the character which ordinarily expresses the last element in the names of Nebuchadnezzar and his father – *yuchur*, as I read it.[5] On these grounds, I must retain my opinion as to the person to whom this name belongs; and of course I attach no weight to the objections brought against my other readings, that they are dependent upon, or connected with, this. The rectification which appears to me most needed is that Colonel Rawlinson should cease to attribute to Nabunahid the bricks and the buildings and the parentage of Nabopolassar.

Believe me to remain.
Yours very truly,
Edw. Hincks

Source: Journal of the Royal Asiatic Society 15 (1855), pp. 402–3[6]

1   H. C. Rawlinson, 'On the Orthography of Some of the Later Royal Names of Assyrian and Babylonian History', *Journal of the Royal Asiatic Society* 15 (1855), pp. 398–402. Read on 18 November 1854.
2   Hincks read the name correctly the first time: Assur-nadin-sumi. Note, however, that he uses *s* for *š*.
3   Nabonidus (556–539 B.C.): *Nabû-nā'id*.
4   Nabopolassar (625–605 B.C.): *Nabû-apla-uṣur*.
5   *yuchur*, i.e. *yuṣur*, is now read *uṣur*.
6   'Letter from Dr Hincks, in Reply to Colonel Rawlinson's Note on the Successor of Sennacherib', *Journal of the Royal Asiatic Society* 15 (1855), pp. 402–3. It was read before the Royal Asiatic Society on 2 December 1854. See *Literary Gazette*, no. 1979 (23 Dec. 1854), p. 1101.

FROM EDWIN NORRIS
5 DECEMBER 1854

R. Asiatic Society
5th. December 1854

My dear Sir,

Your communication was read at our Saturday's meeting, and an abstract is gone to the literary journals, which will appear on Saturday next. The notices you sometimes see in the Daily Papers, do not emanate from me. Light is beaming in upon us from the collision of opinions, which I am by no means sorry to see. I wish the Germans would also take the matter up.

I am My dear Sir
Yours sincerely
Edwin Norris

MS: GIO/H 313

FROM FRIEDRICH MAX MÜLLER
8 DECEMBER 1854

9 Park Place
Oxford
Dec. 8

My dear Sir,

Your letter arrived here while I was away from Oxford, otherwise I should have answered it before now. I am extremely sorry that my parcel should have put you to any expense. I told my servant to pay for it which he did, and though I was struck by the smallness of the postage charged, I never thought that you would have had to pay anything additional. I should be obliged to you if you would let me know how much you had to pay, for I could not think of letting you pay for my parcels or letters.

What you write about the Assyrian verb is of great interest to me.[1] I certainly expected the Assyrian to express the *temporal* distinctions of verbs in the same manner as the Semitic languages by means of predicative or subjective pronominal affixes. All the verbal forms which you mention in your letter are first persons &

therefore I do not know whether in Assyrian the persons are expressed differently in the historical tenses from what they are in the present & future. All would depend on this. If this difference does not exist, my theory on the influence of the personal affixes on the temporal meaning of verbal forms, would not be affected by the Assyrian. The Assyrian would on this point stand alone in the Semitic family, whether it had *not yet* developed this peculiar feature in the formation of verbs or whether it had lost it. If the Assyrian *has* different sets of pronominal affixes for historical & aoristical tenses, it will be necessary to account for this difference for in no language, whether Arian, Turanian or Semitic, can this difference be considered as the result of accident. I am anxiously waiting for your article on the Assyrian verb, whether it confirms my expectations or not. A theory can only be framed to account for facts which are known; if what I consider the Semitic process of distinguishing historical & aoristical tenses cannot be traced in Assyrian, the very absence of it will be of importance for fixing the right relation of Assyrian to the rest of the Semitic family.

The forms which you mention as occurring in the Van inscriptions, look certainly Arian, though I see no peculiar similarity with Armenian, at least in its modern form. In the declensions the Armenian has been defaced more than most Arian dialects

	Arm.	Van		Arm.	Van
Nom.	karg	x dis		karg'q	x dias
Gen.	kargi ⎫	Locative		kargatz ⎫	x dinan
Dat.	kargi ⎭	"		kargatz ⎭	
Acc.	z'karg	x din		z'karg's	x diâ
Abl.	i kargê	x dada		———	x diasta
Inst.	kargav			kargavq	

I do not know how far the alphabet of the Van inscriptions has been decyphered, but if the reading of these terminations is firmly established I should certainly expect the Van dialect to turn out a scion of the Arian family. Single forms and isolated coincidences in roots, however, must of course be treated as indications rather than proofs, and I feel as yet the same hesitation with regard to the so-called Scythic dialect. I see in the latter a small number of forms pointing to the Turanian system of grammar, but as yet I see no decisive proofs. I cannot judge of the correctness of Norris's alphabet. I took it for granted that his letters were right and I shall never have time to enter into this part of the inquiry. The only points of interest which I find in all these inscriptions are the grammatical forms, though I know very well that there is a mutual relation between the decyphering & the grammar of inscriptions, and that the niceties of an alphabet are never fully made out until a thorough knowledge of the grammar leads us to know what the proper power of each letter should be. I know nothing analogous to the employment of the copulative and relative enclitics which you point out in Assyrian & the Scythic or so called Scythic inscriptions. The only thing that occurs to me with regard to the copulative enclitic is that perhaps the *va* might be the usual Semitic copula, joined to the preceding word in *writing*, as we find in the Achaemenian inscriptions also, words which have no accent, joined to a preceding word with which grammatically & syntactically they have no connection whatsoever. I mean such phrases as *hushiya | kshatram | frabara |*, instead of *huwa | shiya | kshatram | frabara*, he gave him the kingdom. Or *Adamsham | kshayathîya | aham*, instead of *Adam | sham | kshayathiya aham*, I was the king of them. In these cases the enclisis is only graphic, perhaps produced by the accent; but whether something of the same kind might have taken place in Assyrian, I cannot tell without knowing more of the phonetical & grammatical character of your inscriptions than I do.

And now I come to the last point – your views of the relationship of the Arian languages. I cannot see any difference between your opinion as far as I can gather from your letter, & what I have held myself ever since I arrived at any independent views on comparative philology. If there is a difference between us, it can only be a difference of terms; but if it should be more I have no doubt that here, as everywhere else, truth will find its way to prevail in the end, however great the authorities which for a time oppose it. You remark on the absurdity of deriving *avis* from Sanskrit *vis*; I have pointed out the same thing over & over again. It would be as absurd as deriving 'otto' from 'huit'. According to my conviction Sanskrit, Persian, Greek, Latin, German & Slavonic stand to one another in exactly the same relation as French, Italian, Spanish, Wallachian. Italian was not derived from French, nor French from Italian, and there is no Latin word which had once been Greek (foreign terms excepted) nor any Greek word that had once been Latin. To treat Sanskrit as the original source of the Arian languages would be as great a mistake as deriving the Romance dialects from Provençal. Provençal is indeed the elder sister, but it has never grown, since its first establishment as an independent dialect, into either Italian or Spanish. The same applies to Sanskrit. Sanskrit has indeed many grammatical forms more primitive than any of the other Arian dialects, but on several points it is decidedly more corrupt than Latin or Greek. There must have been a language from which Sanskrit, Greek & Latin descended, as French, Spanish & Italian did from Latin, but that language is lost, and we can only arrive at a general conception of what it was by putting together what is common to all Arian dialects, Sanskrit & the rest, & dropping what is peculiar to each. Now you say, that the Lithuanian or something similar existed earlier than Sanskrit. I accept the second alternative, the something similar; I object to the name Lithuanian, because the language we call Lithuanian though certainly very primitive on some points, is so corrupt on others that it could in no way be taken as the type of the Arian family. It would be easier to explain the

Lithuanian as a secondary formation of Sanskrit, than to derive Sanskrit from Lithuanian. But both attempts I should consider as historically wrong and grammatically indefensible, while everything becomes clear and intelligible if we look upon all the ancient Arian dialects as independent nationalisations of the common Arian speech, the admission of which, as a historical language, is as necessary as the admission of a language, like Latin would be, even if we had no information of its historical existence, but saw only the different diverging dialects of French, Italian, & Spanish, all necessitating a common origin by what they have in common & by their respective peculiarities. The only point therefore in your letter against which I must protest until I see your proofs for it, is that 'Sanskrit is *derived* from the Lithuano-Slavonic family'. If this could be demonstrated, it would certainly modify considerably the present view of the history of the Arian or Indo-European family; but any new light on this subject will be welcome, and as to myself, I can only say that no prejudice of any kind shall ever prevent me from bearing testimony to what I think demonstrated, from whatever quarters it may come, and however opposed it may be to opinions which I have myself defended. Let science at least be kept free from any party-feeling & let the discovery of truth be the only object to which we devote our lives & our energies.

Believe me
Yours very faithfully
M. Müller

MS: GIO/H277

1   Hincks was preparing his paper 'On Assyrian Verbs', *Journal of Sacred Literature* 1/2 (July 1855), pp. 381–93, and probably wrote to Müller for his opinion on various points of grammar.

FROM FRIEDRICH MAX MÜLLER
25 DECEMBER 1854

9 Park Place, Oxford
Xmas day, 54

My dear Sir,
    Your Assyrian paradigm has been of great interest to me. I only wish I had been able to ask you some questions about it in person, as it is very difficult to do so by letter. You give two forms of the Assyrian Verb,

the one	the other
ipkul	*ipakal*
tapkul	*tapakal*
tapkul	*tapakal*
tapkuli	*tapakali*
apkul	*apakal*

You call the one the Simple Preterite, the other the Present. Here I cannot follow. There may be a difference of meaning in these two sets of verbal forms as used in Assyrian, but with regard to the formation process by which they are produced, I look upon both as identical. They are both formed according to the principle of the Hebrew Future or Aorist. In this I believe, we agree. Now what I desired was to see a paradigm of the Assyrian verb corresponding in form, whatever its meaning may be, with the Hebrew Preterite. The characteristic distinction of these two Semitic tenses is after all simply this – without asking any questions as to how they were originally formed – that in the Preterite the initial elements of the root remain intact throughout, while in the Future changes take place not only at the end but also at the beginning of the verbal root. The first question then is this: does the Assyrian possess two sets of verbal forms, subject to the same formal distinction? I believe you, for besides the third persons singular formed like *ipkul* you mention *lamad, tapkul, lamdat*. Here then it seems to me that the pronominal element which express the feminine of the third person is *t* in both, only used as a possessive or, as I call it, predicative suffix in *lamdat*, and as a subjective prefix in *tapkul*. I believe this *t* belongs to the demonstrative pronoun, & that the Assyrian *si*, the Hebrew *hi*, was changeable to *ti*. The interchange between *h* & *t*, in feminine terminations is frequent in Hebrew; besides there can be no doubt that the Hebrew *kathelah* = Ethiopic *kathelat*, i.e. *h* = *t*. Now it is curious that exactly the same variety of *t, s, h*, exists for the demonstrat. pronoun in the Arian languages:

Sanskrit *sâ* – *s*

Greek η – *h*

Greek τη – *t*

Your position, if I understand it right, is that in *lamdat*, the *t* is a pronominal suffix, but that in *tapkul*, the *t* was 'adopted arbitrarily to indicate the persons of verbs, and that it is not the subject which precedes its verbal predicate'. Here then we differ.

What applies to these two persons, applies to the rest. I agree with you that some of the changes which must be admitted in order to trace the personal preformatives, or the subjective prefixes, back to their original form are violent, as for instance in the first person, where a mere vowel-breathing remains in place of *aku*. Yet we know as a fact that *aku*, or *ego*, or *ah/am*, becomes *I* in English, why not *i* in Assyrian *apkul*? I should rather admit any changes, than the possibility of arbitrary additions in the formation of verbal forms. Nothing can exist in language which had not originally & in its first conception a purpose or meaning. It is impossible to discover always this original intention, but I should rather admit the most violent changes of an originally organic or organised form, than allow anything inorganic or arbitrary in language. Besides I believe to have shown that the only means which language had, to express anything as past, was the use of possessive pronominal prefixes or suffixes (in modern languages possessive verbs, like *habeo, teneo*). In the Semitic languages the possessive pronouns must be suffixed, in the Arian, prefixed. The mere predicative tense, on the contrary requires the pronouns as subjects, and hence, even a priori, I should say, that as in Arian, the subjective pronominal elements are suffixed, so in Semitic they must be prefixed. However it is impossible to reason this out on paper.

The Lithuanian certainly possesses some forms more primitive than Sanskrit, but this proves nothing as to Sanskrit being subordinate or derived from Lithuanian, as little as huit & otto, proves French derived from Italian. Where Sanskrit superadds a second corruption to one existing in Lithuanian, the first corruption will have taken place in Sanskrit as well as Lithuanian, the second is peculiar to Sanskrit. Thus in Ital. *sette* & French *sept*, the loss of the Latin *p* took place in both – the French added to this the suppression of the last syllable, but we could never say that the French *sept* (pronounced *set*) was derived from *sette*; both are derived from *septem*, their Latin type, and so are Sanskrit & Lithuanian from their common Arian type.

With regard to *râg, râga, râgan, râgya* etc, all forms existing in Sanskrit, I must agree with Bopp that *a, an, ya* are all derivative elements. Declension begins with the *s* of the nominative, as *reg/s, reg/u/s, reg/an/s* corrupted into *rego, reg/iu/m*. In *râg/an, an* is the derivative element, as in the participle *rag/ant/, ant*; or in *reg/ent/s, ent*. A step beyond brings us to *antia*, as in *fulg/entiu/s*, & so *in infinitum*. I read your article on the cuneiform inscriptions in Lassen's Zeitschrift a long time ago. I have no copy of it here but have written for one to Germany. The theory on the internal vowels was first announced, I thought, by Oppert; he published it before Rawlinson, but in '47; therefore your paper in 1846 would properly claim a priority. I wish you would publish a collection of your papers published in the Transactions of your Academy. It is impossible to get at them now without going to a Library, which is a great inconvenience.

I remain
Yours very truly,
M. Müller

MS: GIO/H 280

1855

〰〰

Claysmore

2nd. Jany. 1855

Dear Sir,

Thank you for your last letter describing the exact position of the question between you and Colonel Rawlinson with regard to Nabunit.[1] It is an extremely interesting and important question. If your view is correct, Nabopolassar must at one time have changed his name. Now this seems to have been a common practice with conquered kings at the particular period in question. Eliakim was changed to Jehoiakim by his supreme Lord Pharaoh Neco[2] & Daniel and his three companions, probably of royal blood, had their names changed as soon as they became captive, and were set in high positions.[3] Zedekiah, if I remember, had his name also changed.[4] If Nabopolassar was as I suppose the last King of Nineveh of the *Assyrian line*, conquered by the Scythians, may they not have changed his title to mark his position of vassal?[5] On the expulsion of the Scythians, which was about the time of [. . .],[6] i.e. shortly before his death, he would reassume of course his former title, and his son would of course always so designate him. In this view also, the bricks in all the buildings set up in Babylon by the father of Nebuchadnezzar during his vassalage, would bear the subordinate title *Nabunit*, or *Labynetus* as Herodotus distinctly calls him.

Are you aware that the Assyrian Fund Society have in their possession some new inscriptions lately sent over by M^{r}. Loftus?[7] One is an inscription on a brick from Warka. I saw them at Murray's a few days ago, and recommended that they should be published for your benefit as well as other students. Enclosed is the answer I have received. If you would give me the form of the disputed characters, I would compare them with the brick inscription, and let you know the result, if you like.

Believe me
Yours very truly
J. W. Bosanquet

MS: GIO/H 70

1   See Hincks's letter of 29 November 1854 to Edwin Norris, which was published in *Journal of the Royal Asiatic Society* 15 (1855), pp. 402–3. He makes the same point against Rawlinson's 'Nabunit' in his article 'On the Assyrian Mythology', *Transactions of the Royal Irish Academy* 22 (1855), polite literature, pp. 405–22, here p. 412, n. *.
2   See 2 Kings 23: 34; 2 Chronicles 36: 4.
3   See Daniel 1: 7.
4   When the Babylonian King Nebuchadnezzar II captured Jerusalem in 597 B.C., he placed Mattaniah on the throne of Judah and changed his name to Zedekiah. See 2 Kings 24: 10–17.
5   When the city of Ashur fell in 614 B.C., the king was Sin-šar-iškun. When Nineveh fell in 612 B.C., the king was Aššur-uballiṭ. Nabopolassar was the first king of the Chaldean dynasty of the Neo-Babylonian empire and he joined forces with the Median King Cyaxares.
6   Bosanquet forgot to insert the name.
7   William Kennett Loftus (*c.*1821–58), archaeologist, travelled in Mesopotamia and in 1853 he excavated at Warka for three months. In 1855 he returned to England and brought back dozens of cuneiform tablets. He published *Travels and Researches in Chaldaea and Susiana* (London, 1857). See S. Harbottle, 'W. K. Loftus: An Archaeologist from Newcastle', *Archaeologia Aeliana*, 5th ser., 1 (1958), pp. 195–217.

FROM JAMES WHATMAN BOSANQUET
18 JANUARY 1855

Claysmore
18th. Jany. 1855

Dear Sir,

I called on M^r. Murray in Albemarle Street on Tuesday last, and examined the inscriptions on the bricks I mentioned to you as having been sent over by M^r. Loftus. I did not find any of the characters contained in your letter in the inscriptions: but thinking you would like to see the inscriptions themselves, M^r. Murray has promised to have three of them copied for you. They are only of a few lines each. I mentioned your readiness to pay a reasonable sum to any one who would copy for you what you may require, and I think if you would write to M^r. John Murray, he would be able to put you in communication with someone who would undertake this for you.[1]

Have you heard that Rawlinson has excavated a corner of the Temple of Belus – Birs Nimroud – from top to bottom, finding all the seven stages complete, though much decomposed?[2] A cylinder of Nebuchadnezzar was taken out at one corner, which stated that he had just repaired this temple which had fallen to decay since it was originally built 42 cycles before his time. What is a cycle? 12, 30, or 60 years?

Believe me
Yours very truly
J. W. Bosanquet

MS: GIO/H 71

1   The publishing house of John Murray was at 50 Albemarle Street, London.

2   H. C. Rawlinson, 'On the Birs Nimrud, or the Great Temple of Borsippa', *Journal of the Royal Asiatic Society* 18 (1855/61), pp. 1–34. Read 13 January 1855; see *The Athenaeum*, no. 1421 (20 Jan. 1855), p. 84; no. 1447 (21 July 1855), p. 846.

⁓⁓

FROM JAMES WHATMAN BOSANQUET
27 JANUARY 1855

Claysmore

27th. Jan: 1855

My dear Sir,

I send you enclosed a note received yesterday, containing the characters which you are in search of. I would not attempt to copy them, lest, by imperfect copying, I might misrepresent them. Pray therefore have the goodness to return the note, with any comments which you may be inclined to make upon them. They appear to me so totally unlike the characters which you sent me in your note of the 9th, that I cannot but suppose that they represent a different king from the one represented by your characters. And yet Rawlinson supposes his characters to represent the Nabonidus father of Nabuchodnosor of the Behistun inscription, whose name is spelt, as regards the last character (which is the important one) according to one of your three ways.[1]

It is inconceivable that the same name should be spelt in six different ways. Many begin to doubt whether the names have in any case been truly deciphered. I do not fall into this view of the case. But I cannot satisfy myself with regard to Nebuchadnezzar or his father.

Believe me
My dear Sir
Yours very truly
J. W. Bosanquet

MS: GIO/H 72

1   Nabopolassar was the father of Nebuchadnezzar II.

FROM JAMES WHATMAN BOSANQUET
17 FEBRUARY 1855

Claysmore Enfield
17th. Feb: 1855

My dear Sir,

M^r. Murray has sent me the enclosed tracings from some of the inscriptions which seemed to me most likely to contain the king's name you are in search of. I have taken copies, and shall be much obliged by your letting me know what you think of them. Is not N°. 27 from a brick of Nabopolassar? If so, whose son is he? Have we got in N°. 30 the Bel-shar-assar of Rawlinson? If so, is he son or grandson of Nabonit?[1]

I hear that Rawlinson has orders from Government to return, and that he will be home in May. I was at the Asiatic Society today, but there was nothing new.

Believe me, My dear Sir
Yours very truly
J. W. Bosanquet

MS: GIO/H 73

1   Belshazzar was the son of Nabonidus (556–539 B.C.), the last king of Babylon.

FROM WILLIAM HENRY SCOTT[1]
5 MARCH 1855

Edinburgh
March 5th 1855

Sir,

In a letter which I received some time ago from my good friend Mr Sainthill of Cork,[2] he informed me that you have done me the honour to coincide in my reading of a legend on some Parthian coins. It may be perhaps agreeable to you to see and consider the reasons which have led me to this reading, and I have the honour to transmit to you a copy of the paper, of which I beg your acceptance[3]

and remain, Sir,
your obedient servant
W. H. Scott

MS: GIO/H 489

1   William Henry Scott was born in Edinburgh on 13 February 1831. He was an M. D. of Edinburgh University and contributed many articles to the *Numismatic Chronicle*. He died in Edinburgh on 4 October 1855.
2   Richard Sainthill, wine-merchant, antiquarian and numismatist. See Hincks's letter of 9 July 1851 to Sainthill.
3   'On Parthian Coins', *Numismatic Chronicle* 17 (1855), pp. 131–73.

FROM GEORGE REDFORD[1]
22 MARCH 1855

Worcester
Mar. 22, 1855

Rev. & Dear Sir,

From what I know of your reputation & attainments as an Egyptologist I am sure you are able to afford a fellow student & inquirer valuable assistance & from what I believe of your character as a gentleman & a Christian I have no doubt you will. The case I wish to consult you upon is this. A very considerable vol. has lately appeared in this country from the pens of two American infidels (D[r] Nott & G. Gliddon Esq.) entitled 'Types of Mankind', which very plausibly assails the *Chronology* of the Holy Scriptures, as well as attempts to refute the account of the human creation.[2] I am preparing something that is intended to be a refutation. The chief stress of the argument rests upon Lepsius & Bunsen who by the bye totally disagree in their Egyptian Chronology, as doubtless you know perfectly well.

My anxiety is to know your opinion of M[r] Osburn's corrections of the dynasties in his work on the Monumental Hist. of Egypt.[3] Do you consider him a safe authority on the question? May I ask also your candid opinion of M[r] R. S. Poole's papers in the Literary Gazette of 1849 since republished?[4] Can one confide in his calculations?

Any instruction or information upon the monumental chronology will be considered a great favour

My Dear & Rev[d] Sir
Yrs Ad[y] & respectf[ully]
Geo. Redford

My address is D[r] Redford New St Worcester

MS: GIO/H 408

1   George Redford (1785–1860), was Congregational minister of Angel Street Chapel, Worcester from 1826 to 1856. He was an LL.D. of the University of Glasgow and was awarded a D.D. by Amherst College, Massachusetts. See *ODNB*.

2   J. C. Nott and G. R. Gliddon, *Types of Mankind: or, Ethnological Researches, Based upon the Ancient Monuments, Paintings, Sculptures, and Crania of Races, and upon their Natural, Geographical, Philological, and Biblical History* (Philadelphia, 1854). This infamous book, which was reprinted many times, was racist and popularised the polygenist theory. Josiah Clark Nott (1804–73) was an American physician and surgeon. He is credited with the discovery that insects were responsible for the transmission of yellow fever. Nott's co-author, George Robbins Gliddon, was formerly US Consul in Egypt.

3   See W. Osburn, *The Monumental History of Egypt as Recorded on the Ruins of her Temples, Palaces, and Tombs*, 2 vols (London, 1854).

4   Reginald Stuart Poole (1832–95), English Egyptologist and numismatist, entered the service of the British Museum in 1852. As a teenager he published twelve articles in the *Literary Gazette* for 1849. They were republished in his *Horae Aegyptiacae: or, the Chronology of Ancient Egypt* (London, 1851).

FROM GEORGE CECIL RENOUARD

12 APRIL 1855

Swanscombe, Dartford
12th. April, 1855

My dear Sir,

A very cursory perusal of your two papers, which reached me yesterday evening, has given me much satisfaction: particularly the latter, which very satisfactorily removes some considerable difficulties in the only secure part of the Egyptian chronology, & points out oversights which, if left uncorrected, would not fail to mislead.[1] Miss Fanny Corbaux (an artist, I believe) has bestowed much labour on the interpretation & chronology of the wars commemorated in the sculptures at Thebes.[2] I have not had time to examine her papers carefully, but my impression is, that she has been unguarded in many of her conclusions. Whether Lepsius is retarded by insoluble problems, or the backwardness in the Prussian Gov^t. as to the supply of requisite funds, I know not: but some months have elapsed, since any part of his plates has appeared,[3] & his 'personal narrative' has never issued from the press, for his popular volume is only a lively precursor of what he can &, I hope, will give.[4]

From Brugsch, I expect more: but his 'Prospectus', only, has yet been announced.[5] In France little has been done: by Champollion's successor in 'the Egyptological Chair', almost nothing:[6] but M. de Rougé has done much & well; & from him more may be expected.[7]

Perhaps you will think me very daring when I say of Assyrian &c. interpretations, 'Adhuc sub iudice lis est'.[8] Of your own system, I have always said, it bears the strongest marks of probability, of any: but I am much tempted to question the universal application of the *syllabic* power of the symbols (for such I suppose them to be) to the extent, to which it has been carried by yourself, Rawlinson, Oppert &c. &c.

I cannot stomach the masc. adjectives ending in *iya* in the Babylonian & Assyrian (which I suspect to have been also a Semitic dialect) & I have little doubt that your former rendering of *iy(a)* by *í*, in the Persian was right. I cannot believe that the Babylonian in the time of Nebúkadneṣṣar, was much closer to the Chald. of Daniel, than has yet been shown.

The Masoretic punctuation, which was evidently the work of time & an idle attempt to perpetuate the peculiar pronunciation of an almost extinct language & perhaps a local dialect, has been a heavy drag on the Hebrew student. The Arab. *punctuation* (which is perhaps more modern) is clearly much more simple* &, therefore, analogous, to the structure of the language; but it gives a more probable result: e.g. the plural differs from the sing. & the fem. from the masc. only by the addition of a termination: for the irregular plurals, as they are called, are nothing but nouns of multitude in the singular & *fem.*, or rather *neuter* gender.

The germ of these notions I got from some lists thrown out by Sir W^m. Jones.[9] I cannot agree with you in the power you give to certain letters, *sāmek, ṣādē, zayin* &c. & I doubt exceedingly the existence of the compound sounds of *ch, j* & *zh* in very early times: *jim* is still *g* in 'gate', 'good', 'get', 'girl' & 'gullet' in Egypt, many parts of Arabia, & in the Æthiopic. Had it as now sounded been known to the Greeks, would not they have expressed it (as they do now) by τζ or τσ?

Have you looked at the inscribed (wooden) dishes in the Brit. Mus. brought from Babylonia? Their character is almost the same as the common Heb. but, I think, they are *pagan* not Jewish or Xtian works. What is their age?[10]

You have no doubt, seen Brugsch's 'Ägyptische Studien' in the last n°. of the D.M.G.'s Zeitschrift S. 193, & w^d. be pleased with it.[11]

I shall be happy to hear good acc^ts. of yourself & family & am, myself, quite as well as can be expected by a Septuagenarian who has the pleasure of calling you the friend of

Your humble servant
Geo. Cec. Renouard

P.S. This is the first summer's day we have yet had, though it still blows a gale from the W.S.W.

* The Arab. points are only the *ḥurūf al-'illa* abridged, as you no doubt have observed.[12]

MS: GIO/H 473

1   See the articles 'On the Assyrian Mythology' and 'On the Chronology of the Twenty-sixth Egyptian Dynasty, and of the Commencement of the Twenty-seventh', *Transactions of the Royal Irish Academy* 22 (1855), polite literature pp. 405–22; 423–36.

2   See Fanny Corbaux's letter of 22 January 1853.

3   R. Lepsius, *Denkmäler aus Ägypten und Äthiopien*, 6 pts in 12 vols (Berlin, 1849–59).

4   See Lepsius's letter of 25 June 1852. *Briefe aus Ägypten, Äthiopien und der Halbinsel des Sinai: geschrieben in den Jahren 1842–1845, während der auf Befehl sr. Majestät des Königs Friedrich Wilhelm IV von Preussen ausgeführten wissenschaftlichen Expedition* (Berlin, 1852); the English translation appeared under the title *Discoveries in Egypt, Ethiopia and the Peninsula of Sinai, in the years 1842–1845 during the Mission Sent out by his Majesty, Frederick William IV of Prussia*, edited, with notes, by K. R. H. Mackenzie (London, 1852).

5   The prospectus for Brugsch's *Grammaire démotique* (Berlin, 1855) was published in *Zeitschrift der deutschen morgenländischen Gesellschaft* 9 (1855), pp. 645–6; and that for his *Monumens de l'Égypte, décrits, commentés et reproduits pendant le séjour qu'il a fait dans ce pays en 1853 et 1854* (Berlin, 1857) on pp. 318–20.

6   Charles Lenormant (1802–59) was Professor of Egyptology at Collège de France from 1848.

7   Emmanuel de Rougé; see Birch's letter of 16 February 1850, n. 2.

8   Horace, *Ars poetica*, line 78: Grammatici certant et adhuc sub iudice lis est, 'Grammarians are wrangling and the jury is still out'. See the commentary in C. O. Brink, *Horace on Poetry: The 'Ars Poetica'* (Cambridge, 1971), p. 167.

9   Sir William Jones (1746–94), judge, orientalist, and English philologist, is associated with the foundation of Indo-European linguistics.

10   Incantation bowls were not made of wood. See J. B. Segal, with a contribution by Erica C. D. Hunter, *Catalogue of the Aramaic and Mandaic Incantation Bowls in the British Museum* (London, 2000). Hunter discusses the manufacture of the bowls on pp. 163–8.

11   Brugsch, 'Ägyptische Studien', *Zeitschrift der deutschen morgenländischen Gesellschaft* 9 (1855), pp. 193–213, 492–517.

12   In Arabic *ḥurūf al-'illa* are the letters of weakness, the weak vowel letters /'/, /y/ and /w/.

FROM JAMES WHATMAN BOSANQUET

7 MAY 1855

Claysmore 7th. May 1855

My dear Sir,

Many thanks to you for your paper upon the 'Chronology of the 26 Egyptian dynasty &c.',[1] which I have read with much attention, in the hope of being able to fix with some degree of certainty the date of the conquest of Egypt by Cambyses, which is a most important date.

You take as your foundation the reign of Darius: placing his first year, with the canon, in B.C. 521. You then show, from the Apis inscription, that Cambyses reigned 9 y^{rs}. in Egypt, making his first y^r. B.C. 530. But B.C. 530 is the forty first y^r. of Amasis as *you* place it, and Amasis was dead before Cambyses came into Egypt. I can understand Cambyses himself counting the years of his reign from the time of Cyrus. But I cannot understand how the Egyptian priests should do so. I think the direct evidence of the Apis inscription is, that Cambyses reigned 9 y^{rs}. in Egypt, and that the Apis born in his 5th. y^r. lived to the 4th. y^r. of Darius, *counted from the death of Cambyses*. This 4th. of Darius need not be the 4th. from his accession in Persia. Darius had a first y^r. in Persia, another first y^r. in Babylon, when 62 y^{rs}. old, and there is no reason why he may not have had a first in Egypt distinct from the rest.

Why are we to believe Herodotus in preference to Ctesias, who tells us that Cambyses reigned 18 years? Clemens Alexandrinus gives him 19 y^{rs}.; Africanus, in the passage amended by you, gives him 15. Syncellus who places the conquest of Egypt in B.C. 516, though he gives but 8 y^{rs}., favours the result I would come to. If, as Ctesias says, Cambyses reigned 18 ^{yrs}. and the Egyptian monument proves that the last 9 were in Egypt, Cambyses must have conquered Egypt in B.C. 520, from about which time Darius began to count his reign in Persia, being left governor there, we may suppose, in the absence of Cambyses, who never returned. The year B.C. 520 accords well with my view of the chronology. If we make the length of the 26th. dynasty, with Africanus, *150 years* & 6 months, and add 18 more for Tirhakah, 520 + 150 + 18 = B.C. 688 which is the year after Sennacherib's invasion of Judaea, as *I* place it. I do not think you are justified in placing Tirhakah as reigning in Egypt at the time of Sennacherib. Clearly he was then king of *Ethiopia*. Besides which Herodotus tells us that Sethos was still on the throne in Egypt.

There is thus great difficulty involved in your view, viz, that the first y^r. of Tirhakah on the throne of Egypt is B.C. 710, ten years before Sennacherib's invasion of Judaea; whereas he had not even ascended the throne of Egypt, when he came out against Sennacherib, in B.C. 700, as you place it. Either, therefore, you place Sennacherib too low, or Tirhakah too high. I think them both too high.[2] To make your views consistent it seems to me that you must only give 18 y^{rs}. to Tirhakah, and accept the 150 years of Africanus for the 26th. dynasty by lapping over the reigns of Apries and Amosis.

Believe me
My dear Sir
Yours very truly
J. W. Bosanquet

MS: GIO/H 75

---

1   'On the Chronology of the Twenty-Sixth Egyptian Dynasty, and of the Commencement of the Twenty-Seventh', *Transactions of the Royal Irish Academy* 22 (1855), polite literature, pp. 423–36.

2   Tarhaqa was King of Egypt (twenty-fifth dynasty) from 690 B.C. to 664 B.C. Sennacherib was King of Assyria from 705 B.C. to 681 B.C.

FROM SAMUEL BIRCH
21 MAY 1855

BM
21 May, 1855

My dear Dr Hincks,

I enclose you the rubbing of one side of the apex of a broken obelisk which has arrived with the recent additions at the Museum, the other sides are too much obliterated to take by this process. But perhaps I shall be able to have a wet paper impression made of them. I have distributed all the copies of your papers except one on the Egyptian XXVI dynasty.[1] Lepsius is at present here, and I have given him also a copy. But I shall want *two* more on the Assyrian mythology, one for Mr Norris the other for Colonel Rawlinson if you do not send them yourself.[2] Among others I have given Mr Fox Talbot a copy of both papers. He has made *great* progress in the Assyrian and is going to publish some extended texts with interlinear translations.[3] Lepsius will send you his last correction on the epoch of the conquest of Egypt by Cambyses. He has reverted to the old date 525 as he discovered from De Rougé that there is no date of the 4th year of Cambyses on any Egyptian monument, and he has now seen the very tablet or Sarcophagus of the Apis who died in the reign of Cambyses. He was misled as to the 4th year by an assertion of M. Brugsch. He therefore returns to the 6th year of Cambyses as marking the conquest of Egypt.

I suppose that you have seen Mr Heath's work on the 'Exodus Papyri';[4] his theories are perfectly extravagant about Moses & the brick fields, although *some* of his translations appear to me to be correct.

Believe me to remain
My dear Dr Hincks
Yours very truly
Samuel Birch

MS: GIO/H 37

1 'On the Chronology of the Twenty-sixth Egyptian Dynasty, and of the Commencement of the Twenty-seventh', *Transactions of the Royal Irish Academy* 22 (1855), polite literature, pp. 423–36.
2 'On the Assyrian Mythology', *Transactions of the Royal Irish Academy* 22 (1855), polite literature, pp. 405–22.
3 William Henry Fox Talbot (1800–77), pioneer in photography, began to publish articles on cuneiform inscriptions in 1856. See the series 'On the Assyrian Inscriptions' [I–IV], *Journal of Biblical Literature* 2/4 (Jan. 1856), pp. 414–25; 3/5 (Apr. 1856), pp. 188–94; 3/6 (July 1856), pp. 422–6; 4/7 (Oct. 1856), pp. 164–70. His *Assyrian Texts Translated, No. 1* (London, 1856) was not continued, but he published translations in later years.

4   D. I. Heath, *The Exodus Papyri, with a Historical and Chronological Introduction by Miss F. Corbaux* (London, 1855). See the entry for Dunbar Isidore Heath (1816–88) in W. R. Dawson and E. P. Uphill, *Who Was Who in Egyptology*, ed. M. L. Bierbrier (3rd edn; London, 1995).

FROM BENJAMIN GOLDING
26 JUNE 1855

Biddenham, Bedford
26 June /55

Rev. & dear Sir,

I received a few days since by post, a copy of your paper on the Chronology of ye Twenty-Sixth Egyptian Dynasty, published in ye Transactions of the Royal Irish Academy.[1] As ye envelope had ye Killyleagh post-mark, I can only suppose that it was obligingly sent by you, in connection with my communication to ye Journal of Sacred Literature.[2] If so, pray accept my best thanks.

Your identification of Amenirtas with ye Ammeris of Eusebius affords a key to what is otherwise inexplicable; & I am glad that you so thoroughly reject ye principal notion that ye Biblical Kush was in Asia.[3]

May I be pardoned for trespassing upon your time with one or two remarks. About six or seven years ago I sent two papers which were inserted in one of our monthly religious periodicals. The first was in reference to ye identification of Shishak with Shishonk;[4] and, *as a consequence*, ye identification of ye Ethiopian Zerah (Zerach. 2. Chron. XIV. 9 and XVI. 8[5]) with the Osorthon (*Osorcho*) who succeeded Shishonk. If Shishonk who reigned 21 years, ascended ye throne about ye third or fourth of Rehoboam, ye reign of Osorcho would be nearly contemporaneous with ye first fifteen years of Asa.

My second paper had reference to the *Sukkiim 2. Chron. XII. 3.[6] It appeared to me that ye Lubim were y^e *Western* Africans & that y^e *Sukkiim were possibly situated to ye south of Egypt, & ye north of Ethiopia, somewhere in Nubia. On examining ye map, I found a district named *Sukkoth*; & it appeared to me that, as ye name of Berber, & one or two other extant names had been discovered in Egyptian records more ancient than ye time of Shishonk, it was not impossible that *Sukkoth* may have been a name equally ancient. If, in ye course of your own examination of ancient Egyptian history, you have found any thing tending to elucidate these points, it would perhaps be doing good service to ye cause of Biblical truth, if you would communicate them briefly to ye Athenaeum, & more fully to ye Journal of Sacred Literature.

I read with much pleasure your paper in yᵉ Journal of Sacred Literature for last July on yᵉ Chronology of Sargon & Sennacherib;⁷ & thank you for yᵉ courtesy with which you declined noticing any mistakes in Assyrian chronology in my letter on Eusebius & Niebuhr – mistakes which were almost unavoidable from my want of access to more perfect documents.⁸

In yᵉ Quarterly Journal of Prophecy for last January in a paper from me on Darius the Mede & Darius Hystaspes, & in yᵉ Correspondence of yᵉ April Number, I have availed myself of your interesting discovery of yᵉ name of Dayukku (Δηιόκης) in yᵉ Assyrian monuments.⁹

I have never had yᵉ pleasure of reading your paper on yᵉ Egyptian Stele.¹⁰

Please excuse yᵉ liberty which has just been taken in thus trespassing upon your time; & believe me to be,

Rev. & Dear Sir,
Yʳ. faithful Servᵗ.
B. Golding

*The Septuagint translate as if yᵉ word had been written with *kaph* & not with *qōph*.

MS: GIO/H 164

1   See previous letter, n. 1.

2   [B. Golding], 'Niebuhr and Eusebius': A Letter to the Editor, *Journal of Sacred Literature* 5/10 (Jan. 1854), pp. 489–508. The letter is dated 12 August 1853, and the P.S. 5 September 1853.

3   See 'On the Chronology of the Twenty-sixth Egyptian Dynasty', pp. 431, 434.

4   Shishak = the Egyptian Shoshenq I.

5   The biblical references are confused here. Zerah the Cushite is found at 2 Chronicles 14:9 (Eng. 14: 8), but not at 2 Chronicles 16: 8.

6   *Sukkiyyîm* are listed between the *lûbîm*, 'Libyans' and *kûšîm*, 'Cushites'. The Greek Septuagint translation has Τρωγλοδύται, 'troglodytes', 'cave-dwellers', which is hardly correct. Modern scholars suggest that the Sukkiim were mercenaries in the Egyptian army.

7   Hincks, 'Chronology of the Reigns of Sargon and Sennacherib', *Journal of Sacred Literature* 6/12 (July 1854), pp. 393–410.

8   See n. 3 above.

9   'Darius the Mede, and Darius Hystaspes', *Quarterly Journal of Prophecy* 7 (1855), pp. 34–43; Letter to the Editor, pp. 196–8. In the latter Golding cites Hincks's letter of 24 July 1854 to the editor of the *Literary Gazette*. This was published in *Literary Gazette*, no. 1959 (5 Aug. 1854), pp. 707–8, but also in the *Journal of Sacred Literature* 7/13 (Oct. 1854), pp. 231–4, where Golding may have seen it.

10   'On the Egyptian Stele, or Tablet', *Transactions of the Royal Irish Academy* 19 (1842), polite literature, pp. 49–71.

FROM JULIUS OPPERT[1]
28 JUNE 1855

2 York Street
Covent Garden
London[2]
June 28, 1855

Reverend Doctor,

Since long time I wished to enter into scientific intercourse with a man whose first rate sagacity I esteemed by his writing. I take now the liberty to apply directly to your kindness in account of the claim you addressed to me in your last valuable paper on the Assyrian mythology.[3]

I am at first delighted to assure you that you were right in attributing the idea of 'Nergal' to the sign ⌐𝅘 .[4] Indeed the name on the account of which you quote me, is written ⊢⊣ ⌐𝅘 ⊨𝅘 𝆶𝆶 , and I believed to be sure when I identified it, in the month of February 1853, the Babylonian king's name Neriglissor.[5]

I am working on the Assyrian inscriptions' problems since almost two years and a half, and I think now to undertake, under the auspices of the French government, my great publication on my Babylonian discoveries.[6] I regret to say that I had no cognizance of your most esteemable papers until my returning to Europe, and now I know only your memoir on the inscriptions of 1852.[7] It is therefore very likely that in the short letters I wrote to my friends from Babylon and in the few articles I published since in the French Athenaeum and other where else, I may have expressed opinions similar to those you published before. My independence nevertheless is certain and I believe that I can only congratulate myself of my having met with you in the same opinion. It is also my custom to quote the scholars I adopt the opinions, and I am rich enough to plunder no body. I was therefore astonished to find in your paper that you speak of my 'alleged discoveries since longtime published in that country'.[8] The discovery of the Chaldean numbers, the elucidation of the Babylonian topography, the explanation of the reason and the necessity of the polyphony, the several restrictive principles I introduce in the cuneiform inscriptions and my independant interpretations of this documents have never been contested as my own property neither in France nor in Germany. I can give you the assurance that I acknowledge plainly your priority when I adopt your opinions published before mine but you will also concede my right of stating that I arrived often at the same conclusions independently from your writings.

In a lecture held at the Institute of France in the Academy of Inscriptions and Belles Lettres I have exposed the principles of my decipherings and they are somewhat different from yours.[9] I do admit the polyphony, of course, and I believe to

279

have given the first the reason of that very extravagant phenomenon; but I do not admit any syllabical homophony.[10] There is, for instance, only one syllabical expression for *sar*; that the king was represented by the sound of *sarru* in Assyrian (but not in Scythic or in Armenian or in Susian) is a coincidence, and you may not find in any word the syllable *sar* represented by the monogramme for king, except in the compound proper names, where the word itself is a constitutive element.

But perhaps I have been too long and I beg your pardon for my prolixity and for my bad English, and am, with the most respectful sentiments,

yours most obediently
Dr. Julius Oppert

MS: GIO/H 387

1   Julius Oppert (1825–1905). For biographical details, see Renouard's letter of 18 September 1851, n. 2. Hincks was in London in early July and met Oppert at the British Museum. Davidson, *Edward Hincks*, p. 191, thinks they may have worked together.
2   Oppert was examining the cuneiform materials at the British Museum. See his *Rapport adressé à Son Excellence M. le Ministre de l'Instruction publique et des Cultes, par M. Jules Oppert, chargé d'une mission scientifique en Angleterre* (Paris, 1856). He mentions Hincks on pp. 9, 32.
3   'On the Assyrian Mythology', *Transactions of the Royal Irish Academy* 22 (1855), polite literature, pp. 405–22.
4   Ibid., p. 415.
5   Neriglissar (Nergal- šarra-uṣur), king of Babylon (560–556 B.C.). See Oppert's letter to F. de Saulcy, dated 14 April 1853 at Hillah in *L'Athenaeum français* 2, no. 23 (4 June 1853), pp. 543–4.
6   *Expédition scientifique en Mésopotamie exécutée par ordre du gouvernement de 1851 à 1854 par MM. F. Fresnel, F. Thomas et J. Oppert*, 2 vols (Paris, 1859–63).
7   Hincks, 'On the Assyrio-Babylonian Phonetic Characters', *Transactions of the Royal Irish Academy* 22 (1852), polite literature, pp. 293–370.
8   In his paper 'On the Assyrian Mythology', p. 416, n.*, Hincks wrote: 'M. Oppert has announced, as an important discovery, that this character denoted that what follows was to be taken ideographically, and not phonetically. In this he is altogether mistaken. Other alleged discoveries of M. Oppert have been long since published in this country.'
9   The Address, 'Sur l'origine des inscriptions cunéiformes', was communicated to the Académie des inscriptions et belles lettres on 20 October 1854.
10   I have omitted some cuneiform signs which Oppert provided as examples.

FROM GEORGE ROBBINS GLIDDON
29 JUNE 1855

29.6.55

Dear Sir,
    Doubts as to your true address have alone retarded my acknowledgements to you for the valuable roll of 'tirages à part'[1] received, on 16th. inst., through our friend Mr. Birch. Permit me to thank you cordially for documents on which,

coming from you and emanating from your learned pen, I have long learned to set a very high value.

One of the sets I devote to my old colleague in Egypt M^r. Prisse (now at Paris, Rue d'Enfer, 88),[2] who had specially enjoined me to ask if you ever received the copy of his *Papyrus*, sent through M^r. Birch years ago.[3] I told him that, if you had, nothing but unacquaintance with his address could be the cause why he had not heard from you. I hope to make good use of the others; and, in hopes that I may be favoured with a copy of your future distributions, beg leave to add my permanent London address: care of

Messrs. Trübner & Co.

Publishers,

12, Paternoster Row.

The subjects of your vast studies intensely interest me, and it is owing solely to my wandering life in America that I have not been able to follow you except at long intervals.

With great respect, I am,

Dear Sir,

Mo. sincerely y^{rs}.

Gliddon.

MS: GIO/H 163

1   Tirages à part = 'offprints'.

2   Émile Prisse d'Avennes (1807–79), French Egyptologist. See M. Dewachter, *Un avesnois: l'égyptologue Prisse d'Avennes (1807–1879): études et documents inédits* (Avesnes-sur-Helpe [Nord], 1988).

3   The book was E. Prisse D'Avennes, *Fac-simile d'un papyrus égyptien en caractères hiératiques* (Paris, 1847). See Birch's letter of 11 January 1848 (vol. 1, p. 226) informing Hincks that he has received a present for him from Prisse d'Avennes.

FROM BENJAMIN GOLDING
3 JULY 1855

Biddenham
3. July. 55

Rev. & dear Sir,

Pray accept my best thanks for your obliging reply to my hasty letter.

I had, in writing it, overlooked ye fact of Jeroboam having fled from Solomon to the court of *Shishak*. In my former paper on ye chronology of Shishak &

Rehoboam, I had employed this fact to shew ye great probability of Shishak's having invaded Judea at ye instigation of Jeroboam.

You will see in ye present number of ye Journal of Sacred Literature a very unsatisfactory (in my judgment) communication from a writer who advocates ye views of M^r. Forster, rather than those of yourself & Col. Rawlinson.[1] I hope that ye perusal of your paper on Assyrian verbs will shake his confidence in M^r. F.'s theory.[2]

Not an unimportant result of the Assyrian investigations has been to establish ye real character of ye apocryphal book of Tobit who seems to tell us that not very much more than fifty-five days elapsed from the return of Sennacherib from Judea to his murder by his sons.[3] At ye same time ye language of Berossus, as quoted by Josephus, would perhaps lead us to think that in his days there was a popular tradition that Sennacherib did not long survive his return from Palestine.[4]

I know not if you have at all turned your attention to ye subject of Nahum's birth place, & his statement that at some time previously, Thebes, though protected by Ethiopia & Egypt, had been taken & sacked by ye Assyrians, aided by Put & Lubim.[5] I am disposed to think that Nahum's place of residence was ye Assyrian Al-Kosh. If I do not mistake, Gesenius mentions two traditions of an Al-Kosh in Palestine which neutralise each other – one by Jerome, & ye other by Pseudo-Epiphanius.[6] It is perhaps interesting that Capernaum is represented in ye Hebrew version of ye New Testament by *kepar-naḥḥûm* ye village of Nahum or ye town [of] Nahum.[7] It is not very probable that ye name of Nahum's birth-place, or residence where he prophesied, even if in ye kingdom of ye twelve tribes, should have survived ye Assyrian & Roman conquests.

I thank you for your remarks on ye Sukkiim, & believe your view to be correct. Pray excuse this second intrusion & believe me to be

Rev. & dear Sir
Y^r. faithful serv^t
B. Golding

I hope you will be able to publish before long, your investigations in the Egyptian chronology: it is a subject of great importance to Biblical students.

MS: GIO/H 165

---

1   [T. Myers], 'The Nineveh Inscriptions', *Journal of Sacred Literature* 1/2 (July 1855), pp. 365–81. The article is signed T.M., The Vicarage, Sheriff Hutton, near York. For Thomas Myers, see *The Clergy List for 1855* (London, 1855), pp. 201, (191). He supported the views of Forster as laid out in *The One Primeval Language Traced experimentally through Ancient Inscriptions, in Alphabetic Characters of Lost Powers, from the Four Continents; Including the Voice of Israel from the Rocks of Sinai* (London, 1851). Henry Fox Talbot responded to Myers's views in his article 'On the Assyrian Inscriptions [I]', *Journal of Sacred Literature* 2/4 (Jan. 1856), pp. 414–25 (a letter dated October 1855).

2   'On Assyrian Verbs', *Journal of Sacred Literature* 1/2 (July 1855), pp. 381–93.

3   See Tobit 1: 21, 'Forty days did not pass before two of his sons killed Sennacherib, and they fled to the mountains of Ararat.' Although Tobit was originally composed in Aramaic, the best text we have is the Greek 'long recension' found, for example, in Codex Sinaiticus. In the Greek 'short recension', this verse has 'fifty days', and the Old Latin has 'forty-five days'. See J. A. Fitzmyer, *Tobit* (Berlin, 2003), pp. 120–1. 'Fifty-five' is found in only a few Greek manuscripts.

4   See Josephus, *Antiquities*, x, 1–23.

5   Nahum 3: 8–9.

6   See W. Gesenius, *Thesaurus philologicus criticus linguae hebraeae et chaldaeae Veteris Testamenti*, vol. 3 (Leipzig, 1842), p. 1211; and note the discussion of Nahum's birthplace and residence in the recent works, W. Rudolph, *Micha–Nahum–Habakuk–Zephanja* (Guttersloh, 1975), p. 149; K. Spronk, *Nahum* (Kampen, 1997), pp. 31–2.

7   These words are not clearly written: they look like 'of yᵉ town Nahum', which do not make sense.

TO SAMUEL BIRCH
8 AUGUST 1855

Killyleagh
8ᵗʰ. August 1855

My dear Dʳ Birch,

I had your note of the 6ᵗʰ this morning. If the photographs be sent to Longman's under cover to Mʳ. Henry Greer (not Green) High Street Belfast, they will reach me in course of time.

I sent you two copies of my paper on the Animals[1] – also one to your care for Dʳ Oppert. I was doubtful if the latter be still in London. These are proofs. It will appear, I believe, in quarto form with cuneatic characters inserted & probably notes.

Yours very truly
Edw. Hincks

⸸ ⸗ ⸋⊙ Ra-mes-su, Sol genuit eum. Is it not so?[2] Why then write it Rameses, as every one does?

MS: BM/ANE/Corr. 2709

1   'On Certain Animals mentioned in the Assyrian Inscriptions', *Proceedings of the Royal Irish Academy* 6 (1855), pp. 251–60.

2   Rameses means 'Re has fashioned him'.

FROM SAMUEL BIRCH
AUGUST 1855

My dear Dr Hincks,

I have transmitted to Messrs Longman your copy of the photographs received from the Sec.^y Sir H. Ellis and I should feel obliged if you would drop him a line acknowledging their safe arrival. M. Oppert has continued his labours, and has made some remarkable discoveries – he has found a table of the *pronouns* affix, and several other syllabaries. As many of the *tablets have been cleaned of the calcareous incrustations since you saw them* he has been able to arrive at further results. From all that I can hear the proposed publication mentioned in the Athenaeum is a long way off, as the proposal has yet to be submitted to the government, and the work could not be commenced till next year.[1] The *Sennacherib* hexagon of Mrs Taylor has been purchased by the Museum, it is in excellent preservation, and very full of matter.[2]

As Ramses appears to have been very young when found I have regarded the ◦𓁷𓏤𓊨 Remessu as 'the nascent sun'.

Mr. Oppert every day finds pieces which he unites to others, and many of the photographs are only of portions which can now be completed. I must close this letter although there are many questions I wish to ask you, but I will defer my queries till next letter.

Believe me to remain
Yours very truly
S. Birch

MS: GIO/H 38

1   An announcement of the publication of Assyrian inscriptions by the Trustees of the British Museum was published in *The Athenaeum*, no. 1448 (28 July 1855), pp. 874–5.
2   In June 1855 Henry Rawlinson informed the Trustees of the British Museum that he had acquired the prism for £250.

FROM JOHN GEMMEL[1]
28 SEPTEMBER 1855

Fairlie, by Greenock
Sept[r]. 28th., 1855

Rev[d]. Dear Sir,

I read a paper in the Ethnological Section of the British Association, that recently met in Glasgow. The paper was on the inscriptions of two seals, as given in the 155th. page of Layard's second expedition to Nineveh.[2] I unfolded my views at some length in my paper, which I had earnestly wished both you and Colonel Rawlinson to hear. But you were not in Glasgow, and he was not present at the time. The substance is this:

The middle line of the seal, first from the left, I read as = *Kodrosr*; reading from right to left: the star with the emblem before and after it, I take to be a cursive monogram for *Nebo*. The full word I read as Nebokodrosr.[3] Although found at Nineveh, I endeavour to show that Nineveh underwent not an immediate but gradual decay; and that it is no more improbable that a seal of a king of Babylon should be found there, than the sign manual of a monarch of the United Kingdom, at Dublin or Edinburgh Castle.[4]

The second seal has evidently on it the Phoenician or Old Aramaic *yeleq*, *scarabaei species*, and the sculpture plainly represents the same creature.[5] The meaning is given by Buxtorf,[6] under the verb *lqq*: and, as the word *yeleq*, is given by Nahum (ch. 3. 15–19) in predicting the desolation of Nineveh, I could not but be struck with the seal of the *yeleq*, and its representation, and the very monarch's name, who must have witnessed its overthrow, being dug up from the spot, at the distance of twenty four centuries at least, since the fulfilment of the prophecy.

As for the representation on the lowest part of the seal, on which I read *Nebokodrosr*, I cannot say that I understand it.

If, at your leisure, you could let me know whether you agree with me, as to those two seals, thus carrying the cursive character back to the period of six hundred years B.C., you will much oblige,

Rev[d]. Sir,
Yours sincerely,
John Gemmel
Minister at Fairlie by Greenock.

MS: GIO/H 159

1   John Gemmel was one of the hundreds of ministers in the Church of Scotland who broke away to form the Free Church of Scotland in the Disruption of 1843.

2   The title of Gemmel's paper, 'On the Deciphering of Inscriptions on Two Seals, found by Mr. Layard at Koyunjik', is given in the *Report of the Twenty-Fifth Meeting of the British Association for the Advancement of Science; held at Glasgow in September 1855* (London, 1856), p. 145. There are drawings of these alleged Phoenician seals in Layard, *Discoveries in the Ruins of Nineveh and Babylon*, p. 155.

3   This seal impression has been published in recent years by N. Avigad and B. Sass, *Corpus of West Semitic Stamp Seals* (Jerusalem, 1997), no. 837 (p. 313). The inscription is read *l'tr'zr*, 'Belonging to Atarazar', and the script is Phoenician-Aramaic.

4   A royal sign manual was the autograph signature of the sovereign serving to authenticate a document.

5   The inscription on this seal is very difficult to make out. In Avigad and Sass, *Corpus of West Semitic Seals*, no. 796 (p. 298), it is read *ḥnn*, 'Hanan'. This is a long way from Gemmel's *yeleq*, 'locust'!

6   See J. Buxtorf, *Lexicon hebraicum et chaldaicum* (London, 1646), p. 378.

FROM CHARLES WILLIAM WALL
1 OCTOBER 1855

Trin: Coll: Octr. 1 – 55

Dear Hincks,

I have just received the letter at the other side of this from Sir W^m, by which you will perceive he is ready to make the calculations for you, as soon as you furnish him with the particulars – his address is the Observatory, Dunsink, Dublin.[1]

At last I have got into the press so far as to give Mr Gill materials for the first sheet, but they go on here at our printing office miserably slow and give me a vast deal of trouble during the process.[2]

Yours very truly,
Chas Wm Wall

[Enclosure: William Rowan Hamilton to Charles William Wall]

Observatory of T.C.D.
September 29th 1855

My Dr Wall,

I delayed longer than I had intended at the Glasgow Observatory, but have received your letter on my return, & I shall be delighted if I can in any way cooperate with Dr Hincks.[3]

In some haste
Most truly yours,
W. Rowan Hamilton

MS: GIO/H 543

1   William Rowan Hamilton (1805–65), distinguished mathematician, was Andrews Professor of Astronomy at Trinity College, Dublin from 1827. He was also astronomer royal of Ireland and director of the

Dunsink observatory near Dublin. He and Hincks had corresponded frequently in 1838 about a paper Hincks was preparing for publication. See the letters in vol. 1, pp. 37–48. The article was 'On the Years and Cycles Used by the Ancient Egyptians', *Transactions of the Royal Irish Academy* 18 (1838), polite literature, pp. 153–98.

2   Wall is referring to the printing of part 3 of his work *An Examination of the Ancient Orthography of the Jews: and of the Original State of the Text of the Hebrew Bible* (London and Dublin, 1856). Part 1 was published in 1835, and part 2/1–2 in 1840–1.

3   In July 1856 Hincks prepared for publication a paper which had been read at a meeting of the Royal Irish Academy on 12 November 1855 entitled: 'On a Tablet in the British Museum, Recording in Cuneatic Characters, an Astronomical Observation; with Incidental Remarks on the Assyrian Numerals, Divisions of Time, and Measures of Length'. It appeared in *Transactions of the Royal Irish Academy* 23 (1856), part 2, pp. 31–47. In a footnote dated 21 July 1856 on p. 47 he wrote: 'I had hoped that the astronomical calculations for this paper would have been made by a friend. When disappointed as to this, I had to make them myself; and, not being a practical astronomer, I neglected to allow for the effect of refraction.' It is clear from this that the 'friend' was William Rowan Hamilton who did not see Wall's letter containing Hincks's request for the calculations until his return to Dublin.

FROM JOHN GEMMEL
3 OCTOBER 1855

Fairlie, by Greenock
Octr. 3rd. 1855

Rev^d. & Dear Sir,

I have just received your favour of 29th. Sept^r., and I send you a printed report of my paper, in the Scottish Guardian Newspaper, which I think will obviate some of your objections by anticipation.

I read the word *qwdrwšr*, or *q'dr'šr*, taking the last letter as a final *resh*; or *k*. The Phoenician & old Aramean *k* & *q* are like one another.[1] Allow me to quote the following words of your friend M^r. Norris: 'I fear we may now assume that the cursive legends found here and there upon Assyrian and Babylonian relics, are either notes of explanation or translations, and not transcripts' (Journal of Royal Asiatic Society, vol. XVI. p. 226).[2]

The cursive and cuneatic characters on a monument of the reign of Sennacherib will not therefore prove that the cursive letters are of the age of Sennacherib; for, they may have been added afterwards, as an explanation. But an isolated name, such as Nebochodrosar, or Sennacherib, in such characters will prove their antiquity; for, standing alone, they bear in their very face the impress of an original. That is the deduction that I draw from the reading that I have endeavoured to substantiate.

I remain, Rev^d. & Dear Sir,
With much regard, Yours faithfully,
John Gemmel

MS: GIO/H 160

1   This is incorrect and it is difficult to see how Gemmel came to this view.
2   E. Norris, 'On the Assyrian and Babylonian Weights', *Journal of the Royal Asiatic Society* 16 (1854), p. 226.

FROM JOHN GEMMEL
15 OCTOBER 1855

Fairlie by Greenock
October 15th, 1855

Rev^d. and Dear Sir,

Your favour of the 4th. ins^t. came to hand just as I was preparing to go to our Synod in Glasgow; and other matters have occupied me until now. Most men will be disposed to bow at once to your opinion as to the philological question in hand; and none would be more willing to do so, than myself.

At the same time, allow me to say, that the question is not whether a great king would be careful to spell his name *correctly*, but how *actually* do we find it spelled? There can be no doubt that the king of whom we speak has his name spelled in two different ways, in the Hebrew scriptures, at once *nebūkadne'ṣar*, and *nebûkadre'ṣar*. Is it not possible, that it may be spelled in more than one way, on the Assyrian remains?

Your etymology, *nbyw-kwdr-ywṣr*, 'Nebo-a-host-has formed', is ingenious, and would seem to fasten down the orthography to *nbywkwdrwṣr*. But, Rawlinson gives another derivation; Ar. *qudr*, 'power', and Heb. *ṣûr*, 'a refuge'.[1] Nork, in his Lexicon, gives a third; the Persian, *choda*, 'god', and Heb. *ṣar*, Russian Czaar, Heb. *śar*, 'prince' (see under *bl-š'ṣr*, and *nbw-kdn'-ṣr*).[2] And, to my mind, a better derivation than any of those might be had from the Scythic: *Nabu-ku-tar-ru-sar*, 'Nebo-the king's-son- (is) prince'.

It must, I think, be acknowledged that there is a close connection, in certain shades of their meaning, between all the three roots *zûr*, and *śûr*, and *ṣûr*; and one of those roots, and not *yāṣar*, 'to form', is, I suspect, the origin of the syllable *zar* or *sar*, occurring in a number of proper names found in the Hebrew Scriptures. Thus:

Belša'-ṣṣar, Bel's-prince,
Šešba-ṣṣar, (*ššbyn-ṣr*) companion- of the prince,
Belṭeša-ṣṣar, Beltis's-prince,
Nebû-zar-'adān, Nebo's-prince-of the camp,
and, so, in like manner,
Nebo-ku-tar-ru-sar, Nebo- the king's son-(is) prince.

It is plain, I conceive, that the root *sar*, or *zar*, is the same in all those and similar words, and rendered as *prince* gives a distinct meaning. But, if into such words, we attempt to carry out the verb *yāṣar*, 'to form', we shall find that it will not do.

But whilst I hold ṣar to be in meaning equivalent to śar, I am *now* free to acknowledge that the last syllable of the inscription on the clay seal must be read ṣar, as the first letter may be *kaph*, not *qōph*. The *'ayin* will then be the only difference between my reading and what you urge as the true orthography. But, as the letters on the Babylonian cups are confessedly an admixture of Syriac, Palmyrine, and Phoenician characters (Layard, 2ⁿᵈ. Expedit. p. 510[3]) may it not be that the letter o on the clay seal is just the Syriac *wau*? And, if so, the latter portion of the word runs just as you would have it; *kwdrwṣr*. As for the previous part of the name, made up as I suppose of a monogram, that of course is only to be substantiated by some other instance or instances of the same sort. All that we can say at present is, the monogram, however probable, remains in the category of things not proven.

I have again to thank you for your attention; and I cannot conclude without expressing the wish that you had expressed in *cursive* characters, as nearly as possible, the name of that monarch, or referred me to the place where I might find it. For, I fear that my slender knowledge, in this new line of study, may at once have misled me, and given you some trouble.

I remain,
Revᵈ. Dear Sir,
Yours sincerely
John Gemmel

MS: GIO/H 161

1   Hincks and Rawlinson were both wide of the mark.
2   F. Nork, *Vollständiges hebräisch-chaldäisch-rabbinisches Wörterbuch über das alte Testament, die Thargumim, Midrashim, und den Talmud* (Grimma, 1842), pp. 108, 400.
3   Layard, *Discoveries in the Ruins of Nineveh and Babylon*, p. 510.

DIARY OF VISIT TO LORD DUFFERIN[1]
1–4 NOVEMBER 1855

1ˢᵗ. Novʳ. To Terrace, and fixed with Anne[2] to go to the Inauguration.[3] Dinner at 6 – all went off well – left at 11 after Lord Dufferin had returned thanks.
2ⁿᵈ. Novʳ. To Commercial Buildings at 11.4 – addresses to Lᵈ. Lᵗ. one from Belfast Institution – kept in background.
3ʳᵈ. Novʳ. By 4.20 train to Newtownards – Sir E. Tennent in carriage; he spoke to me. I thanked him for his assistance 2 years ago.[4] Lord D. very friendly – got fixed

at dinner between Major Bagot and D^r. Marshall, with both of whom I had a good deal of conversation – then large concert party – there were about 30 at dinner – probably 200 or more at concert. After it cold irregular supper. L^d. L^t. spoke to me asking questions about Killyleagh and Col. Rawlinson. Slipped off to bed at 12.

4th. Novr. Breakfast at 10 – between Lord D. and M^r. Ker. At 11 to Church. M^r. Binney preached – Evangelical[5] – stayed to sacrament, being the only one of the party who did so. Walked to Clandeboye – had a good deal of talk with M^r. Ker for a while in the study – Dinner at 7 – next to L^d. L^t. – D^r. McCosh on other side – got on well.[6] Lord D. read prayers to his whole family and visitors.

5th. Family prayers at 9 1/2 – breakfast soon after. Car came for me, but I had to wait – Mummy case – translated the inscription which was that of a priest of Memphis of a family of which I know other members.

Source: Davidson, *Edward Hincks*, p. 19

1   Lord Dufferin (fifth Baron in the Irish peerage) was Baron Clandeboye, of Clandeboye, Co. Down, in the peerage of the United Kingdom. See Hincks's letter of December 1851 to Campbell Allen, n. 8.

2   Hincks's sister Anne lived with her father at 7 Murray's Terrace, Belfast.

3   This was the inauguration of George William Frederick Howard, seventh earl of Carlisle (1802–64), who was appointed Lord Lieutenant of Ireland by the Liberal Prime Minister, Lord Palmerston in February 1855.

4   Sir James Emerson Tennent, an Ulster M.P., was particularly supportive of Hincks in 1852. See the correspondence for that year.

5   Richard Binney was born 1808 in Halifax, Nova Scotia, son of Hon. Hibbert Newton Binney and Lucy Binney, *née* Creighton. He entered Trinity College, Dublin in 1841, B.A. 1844, M.A. 1848. He was curate at Killeevan (Clogher), 1848–9 and Rural Dean of Bangor, 1849–75. He was conferred with a D.C.L. at King's College, Nova Scotia in 1857. He died at Stanley House, Holywood, Co. Down on 14 January 1876. See J. B. Leslie and H. B. Swanzy, *Biographical Succession Lists of the Clergy of Diocese of Down* (Enniskillen, 1936), p. 97.

6   James McCosh (1811–94) was Professor of Logic and Metaphysics at Queen's College, Belfast from 1852. In 1868 he became President of the College of New Jersey (now Princeton University).

FROM JOHN BAKER GREENE[1]

9 NOVEMBER 1855

Paris le 9 Novembre 1855

Monsieur,

Me trouvant en Égypte l'année dernière, j'obtins de S. A. Saïd Pacha, l'autorisation de faire des fouilles; j'entrepris quelques travaux à Thèbes, et je parvins à recueillir les textes de plusieurs inscriptions que je publie aujourd'hui.

Ces textes ayant paru présenter un assez grand intérêt et contenir des faits nouveaux, je me permets de vous adresser un exemplaire de ma publication que je vous prie de vouloir bien accepter.[2]

Veuillez agréer, Monsieur, l'expression de mon profond respect.

John B. Greene
10 Rue de la Grange Batelière

à Monsieur Hincks
Membre de la Société Royale de Littérature de Londres

MS: GIO/H 172

---

1   John Baker Greene (*c.*1830–*c.*1886), English surgeon, barrister and excavator, studied Egyptology under Emmanuel de Rougé. He excavated at Thebes in 1854. See Dawson and Uphill, *Who Was Who in Egyptology.*
2   J. B. Greene, *Fouilles exécutées à Thèbes dans l'année 1855* (Paris, 1855).

FROM SAMUEL BIRCH
19 NOVEMBER 1855

British Museum
19 Novr. 1855

My dear Dr Hincks,

In answer to an inquiry in one of your former letters about one of the series of photographs, I believe that the missing one of your series was not completed. Mr. Fenton our photographer departed to the Crimea to take photographs of Sebastopol . . .[1] &c. and the Trustees having exhausted their fund the work was temporarily abandoned. He has however returned, and told me that as time and weather permits he is completing the entire series.[2]

I do not know whether M. Oppert has told you the results of his examination of our collection but he has recognised several pieces and found entire vocabularies and conjugations. Some of the photographs are only fragments of these, and it is necessary to bear this in mind when examining them.

Believe me
My dear Dr Hincks
Yours sincerely
S. Birch

MS: GIO/H 39

1 There is an illegible word here.
2 Roger Fenton (1819–69) took up photography in 1852 and two years later became the official photographer at the British Museum. From 8 March to 26 June 1855 he was in Balaklava to make an official photographic record of the Crimean War. See *ODNB* and the bibliography there. An entry in Hincks's diary on 11 September 1855 says: 'Sebastopol taken, rejoicings in town, Church bells rung, and I let off a blue light.'

TO AUSTEN HENRY LAYARD
DECEMBER 1855

Private

My dear M^r. Layard,

On looking over again what I have written I fear I have asked what may be considered too much. The Government has, however, very little good patronage in the church to which a titular deanery is not attached. I mentioned Down as likely to be vacant, a preferment without this dignity, giving me the same income, w^d be equally welcome; & some such might fall vacant even before Down.[1]

I sh^d really be ashamed to urge my claims on an additional £600 a year pension tho' of course if offered I should not refuse it. There are so many literary men who cannot be *otherwise* assisted; however there may be an addition made to *my* income from the members of the church; & I solemnly declare I think I *deserve* it. I have been for 35 years a liberal, *conscientious* & *painstaking* clergyman of a parish, few more so, though the Bp of Down may, to serve his own selfish purposes, perhaps endeavour to discredit me.[2] He told me some years ago that he knew that Lord Clarendon intended to promote me & I heard from good authority that Lord C. actually recommended me on one occasion to Lord J. Russell. I cannot say how this may be; but I feel that I have done nothing to make me unworthy of a step in the church; & a parish of 700 a year *net* (which is the full value of Down) even tho' a titular deanery is attached to it is no great prize for the oldest, or one of the oldest, liberal clergymen in Ireland. I mean by oldest of longest standing as a liberal – of course I do not speak in respect to age.

I write this for yourself privately. Do as you please with respect to the other letter – you now know my wishes.

Believe me again
Yours very faithfully
Edw. Hincks

MS: BL Add. 38984 ff. 194–5

1   Theophilus Blakely was born in 1769, son of Robert Blakely of Dublin and Mary Blakely, *née* Cusack. He entered Trinity College, Dublin in 1784, but took his B.A. at Trinity Hall, Cambridge in 1794 and was called to the Irish Bar in 1795. He was ordained in 1804, served as Dean of Connor, 1811–24, of Achonry, 1824–39, and was Dean of Down from 1839 until his death on 1 December 1855. See Leslie and Swanzy, *Biographical Succession Lists of the Clergy of Diocese of Down*, p. 24. His successor was Thomas Woodward, who was born in Tipperary in 1814, son of Henry Woodward and Melesina Woodward, *née* Lovett. He entered Trinity College, Dublin in 1831, and was scholar 1834, B.A. 1837, M.A. 1849. He was ordained in 1841 and was vicar of Mullingar from 1851 until his appointment as Dean of Down in 1856. He died in London on 30 September 1875. See Leslie and Swanzy, *Biographical Succession Lists of the Clergy of Diocese of Down*, p. 25.

2   Robert Bent Knox (1808–93) became Bishop of Down, Connor, and Dromore in 1849. In the entry for him in the *ODNB*, uncomplimentary remarks by James Henthorn Todd are quoted. They accord with Hincks's low opinion of the bishop. In 1886 Knox was appointed Archbishop of Armagh.

1856

Confidential

British Museum
17th Jany. 1856

Dear D^r. Hincks,

As one of the Committee managing the Monthly Review I received with the greatest pleasure your letter of the 15th to the Editor of that Review.[2] I am authorized to say that we shall be highly gratified by any communication from you and will give it insertion to the utmost our space allows. We are determined to be independent & admit for discussion. The contributors are all churchmen & scholars and we have rigorously excluded all newspaper writers.

We can receive matter for an original article or correspondence until the 23rd *at latest*.

I directed a copy to be sent to you before your letter arrived. As you have thus received two copies perhaps you will kindly give one to any of your friends who might be likely to promote the objects of the Review.

Pray pardon haste & believe me,

Dear D^r. Hincks,
With much respect
Yours very sincerely
R. Stuart Poole

Any communication should be sent to me at 4 Hereford Road, Bayswater, as letters directed to the Museum are often locked up there all night.

MS: GIO/H 396

---

1 Reginald Stuart Poole (1832–95), numismatist, entered the service of the British Museum in 1852.
2 *The Monthly Review* was published in 1856 (vol. 1) and 1857 (seven issues only).

~~~~~

TO THE EDITOR OF THE MONTHLY REVIEW[1]
JANUARY 1856

Killyleagh C° Down
January

Sir,

In your opening number there is a notice of Colonel Rawlinson's Researches, on which I hope you will allow me space for a few observations.[2] Among many questionable statements which are made on the Colonel's authority, the existence of Assyrian syllabaries and the use of ideographs, single and compound, by the Assyrians are denied. On these two points I naturally feel a peculiar interest, inasmuch as the existence of syllabaries was discovered by me in 1852 (see 'Nineveh and Babylon', p. 345,[3] and 'Transactions of the Royal Irish Academy', vol. xxii, p. 342[4]); and the use of simple and compound ideographs had been pointed out and explained by me in 1849, in my paper (in the same Transactions) on the Khorsabad Inscriptions.[5] Feeling satisfied, as I do, that on both these points I was and am right, and that Colonel Rawlinson in condemning me has fallen into error himself, I naturally wish to impress this view upon others; and I send you this communication, in the hope that you will publish it, as impartial seekers after truth with whomsoever it is to be found.

By a syllabary, I understand a tablet in which characters are arranged in three vertical columns; the middle one containing in each line *one* character, simple or complex, and those on the left and right containing explanations of it by the help of other characters. The explanations give the different ways in which the character might be read in an Assyrian inscription; and considering the multiplicity of the characters, and the number of different values which some of them admitted, such an aid to the memory was an indispensable requisite in an Assyrian library. Of this description are the tablets marked K. 62, 110, and 144; and if Colonel Rawlinson supposes, as your article certainly intimates, that these are comparative alphabets of the Assyrian language, and of the Scythian, or any other, I must hold him to be grievously deceived.[6]

There are other tablets, however, which contain inscriptions in two parallel columns, in place of three; such, for example, as those marked K. 39, 45, and 52; which contain in the left-hand column a word or words, and in the right-hand column an explanation of what preceded it. These tablets were never called 'syllabaries' by *me*; I always spoke of them as philological or lexicographical. The tablets of this class which I was able to examine, were very few, and almost all of them were mutilated on one side or other, so that I certainly failed to perceive what Colonel Rawlinson discovered; that the words explained, though many of them were in common use in Assyrian writings, were not Assyrian, but of some foreign language. The passage in your article, however, in which it is stated that Col. Rawlinson 'at once perceived' the nature of these tablets, is calculated to convey a very false impression: it implies that he made the discovery on *first seeing* such tablets; whereas the fact is that he had in his possession, at Baghdad, a great number of these tablets, which he had been quoting for a long time in his various publications, as syllabaries. To come, however, to the main question, had the Assyrians syllabaries? This is a question as to a matter of fact; and must be answered by producing specimens of the entries, occurring in what I take to be such. In K. 110,[7] we have in the central column, a character which clearly represents 'a gate'; on a line with this, on the right, is the word *babu*, a well known Semitic word for *a gate*; and here I may observe, that in the right-hand columns, almost all the words are *inflected*, and may therefore be set down as Assyrian, for nouns of the so-called Scythic language have no inflexions. On the left, over against *babu*, is *ka*; now I admit that *ka* signifies *a gate* in the language of the bilingual tablets. I will call this provisionally Akkadian;[8] I object to the term 'Scythic', as I see two distinct reasons for rejecting Col. Rawlinson's Scythic hypothesis, either of which would, I think, suffice. Judging, then, from this example, it might be inferred that the central column contained the character, and that its Akkadian value was given on the left, and its Assyrian on the right.

This, however, would be a very rash inference. My view of the matter is equally consistent with the facts before us; I take the word on the right to be intended to suggest what the character would signify, as *a word*. The one form *babu* is written down, but the Assyrian student would, according to the context, read it *babu, babi, baba,* or (in construction) *bab*: what is on the left hand of the character denotes its value, as *the syllabic element of a word*. It may have acquired this value from the Akkadian language, as in the instance before us; or, as in other instances, its syllabic value may be of Assyrian origin.

As an example of the latter, I take from the same tablet, *muk* on the left, and *mukku* on the right; the character denoted the Assyrian word *mukku* (in construction, *muk*) from the root *MKK*; it probably signified the tumbling down of a building; but it also denoted *muk* as a syllable, as in *muktabli* from *QBL*.

Take now the case of characters with double values: in K. 62,[9] we have *bar* on the left, or *bâru* on the right; *mas* on the left, *mâsu* on the right; the same character being in both cases in the centre. It is quite clear that Colonel Rawlinson must admit, either that the Assyrian and the Akkadian languages had a large part of their glossaries in common, or that a large part of the tablets in question were not bilingual.

Again, we have in K. 110, the character that is so commonly used for 'great', between *gal* and *rabu*. The latter, to be varied, of course, according to gender and case, is the way in which the character is to be read as a word; its syllabic value is *gal* or *qal*, as in the word *isaqal*, 'he weighs'. That this value was derived from the Akkadian (or rather from a more ancient form of it than we have in the bilingual tablets, for in these *gula*, not *gal*, is 'great') I admit. I must remark, that I have long held it as an established fact, that the Assyrians derived many of the values of their characters, as well as their system of writing, from a people which spoke a non-Semitic language. When at the Edinburgh meeting of the British Association in 1850, I announced that the Assyrio-Babylonian characters represented *syllables*, and not *letters*, as Colonel Rawlinson and everybody else had previously supposed; I accompanied the announcement with what I considered a necessary consequence of this newly discovered fact, that this mode of writing must have originated with a non-Semitic people, as no Semitic people could have invented a system of writing so uncongenial to their language.[10] Colonel Rawlinson, who was present, and followed me, represented this to the audience as a ridiculous opinion. I at that time gave reasons for thinking it more probable that the people from whom the Assyrians learned this mode of writing was an Indo-European people, than the Egyptians. It is now pretty evident that it was neither of these, but a people who spoke a language akin to (or rather, perhaps, a more ancient form of) the Akkadian of the bilingual tablets. As to the affinities of this language, if any there be, I abstain at present from offering any *positive* opinion. That Colonel Rawlinson is on a wrong scent is, I think, quite certain. I fear I have already asked too much of your space, I will, therefore, conclude with summing up my view, which is this: though Akkadian words are occasionally found in the right-hand column of the syllabaries, it is very far from being the case that all the words in these columns are Akkadian; and when Akkadian words do occur there, it is not by way of illustrating the Akkadian language, but because the characters were used with those values in Assyrian words.

Next month, if you will permit me, I propose to show that the Assyrians, in writing their language, used *bonâ fide* ideographs, simple and compound, along with phonetic characters.

I am, &c.,
Edw. Hincks

Source: Monthly Review 1 (Feb. 1856), pp. 130–2

1   'Are There Any Assyrian Syllabaries?': A Letter to the Editor, *Monthly Review* 1 (Feb. 1856), pp. 130–2.

2   'Colonel Rawlinson's Researches', *Monthly Review* 1 (Jan. 1856), pp. 44–7. The author was, perhaps, William Sandys Wright Vaux, a friend of Rawlinson's.

3   Layard, *Discoveries in the Ruins of Nineveh and Babylon*, p. 345.

4   'On the Assyrio-Babylonian Phonetic Characters', *Transactions of the Royal Irish Academy* 22 (1852), polite literature, p. 342, n.*.

5   'On the Khorsabad Inscriptions', *Transactions of the Royal Irish Academy* 22 (1850), polite literature, pp. 3–72.

6   For Hincks's study of the tablets containing syllabaries, see his letter of 7 March 1853 to the Royal Irish Academy and that of 24 July 1854 to the editor of the *Literary Gazette*.

7   The British Museum tablet K. 110 has been edited by B. Landsberger, 'Das Vokabular S^b. Nach einem Manuskript von H. S. Schuster' in *Materialien zum sumerischen Lexikon*, vol. 3 (Rome, 1955), pp. 129–53.

8   The language which Hincks 'provisionally' called 'Akkadian' is today called Sumerian.

9   The tablet K. 62 has been edited by R. T. Hallock, 'Syllabary A', in *Materialien zum sumerischen Lexikon*, vol. 3 (Rome, 1955), pp. 3–45.

10   Hincks announced these important discoveries in his paper 'On the Language and Mode of Writing of the Ancient Assyrians'. See the *Report of the Twentieth Meeting of the British Association for the Advancement of Science; Held at Edinburgh in July and August 1850* (1851), p. 140 + 1 plate.

FROM JAMES WHATMAN BOSANQUET
21 JANUARY 1856

Claysmore Enfield
21st Jany. 1856

My dear Sir,

I have the pleasure of sending you enclosed a copy of my reply to an attack upon my chronology in the October number of the Journal of Sacred Literature.[1] The more I consider the subject the more satisfied I am that the dates about the period I have examined should be lowered considerably. Is it true that you have discovered on Assyrian monuments a reference to the 30th. y^r. of Tiglath Pileser the second?

On Saturday last I had the pleasure of suggesting that you should be enrolled as an honorary member of the Royal Asiatic Society, in consideration of your great services to the public in the investigation of the cuneiform inscriptions. Colonel Rawlinson immediately acquiesced, and proposed your name, and I hope to see you duly elected at the next meeting on the 2nd. February.[2]

Believe me
My dear Sir
Yours very truly
J. W. Bosanquet

MS: GIO/H 77

1   See J. W. Bosanquet, 'Who was Darius the Son of Ahasuerus, of the Seed of the Medes? Daniel IX, 1': A Letter to the Editor, *Journal of Sacred Literature* 2/4 (Jan. 1856), pp. 393–403 (dated 5 Nov. 1855). This was a reply to 'G. B.', 'The Dial of Ahaz and the Embassy from Merodach Baladan': A Letter to the Editor, *Journal of Sacred Literature* 2/3 (Oct. 1855), pp. 163–79, which was in turn a reply to Bosanquet, 'The Dial of Ahaz, and Scriptural Chronology': A Letter to the Editor, *Journal of Sacred Literature* 1/2 (July 1855), pp. 407–13 (dated

23 April 1855). The debate revolved around a paper Bosanquet delivered at a meeting of the Royal Asiatic Society that was later published as 'Chronology of the Reigns of Tiglath Pileser, Sargon, Shalmanezer and Sennacherib, in Connexion with the Phenomenon Seen on the Dial of Ahaz', *Journal of the Royal Asiatic Society* 15 (1855), pp. 277–96. As it happened, Bosanquet's latest paper drew Hincks into the debate.

2   See below the letter of 2 February 1856 from Richard Clarke, secretary of the Royal Asiatic Society.

TO JOACHIM MÉNANT[1]
23 JANUARY 1856

Killyleagh County Down
23rd. January 1856

My dear Sir,

I received this afternoon your kind letter, and I immediately sent you two packets of my publications including the last. I suspect that the former packet is at M. Hardel's at Caen;[2] and you will probably receive it some time. If so, you can give the duplicates to any friend that takes an interest in these matters. I sent copies of my last paper to M. Renan and D^r. Oppert.[3]

I hope to write to you fully next week & will then send you a list of my publications with their dates. If you have not a copy of my paper of 1852 in the Transactions of the Royal Irish Academy, I will send it to you.[4] You quote it, but perhaps at second hand. At present I have a great deal to do, and shall not have leisure to write for a week.

I am greatly obliged to you for the parcel of books which you sent me.

Believe me
Yours very truly
Edw. Hincks

MS: BIF 3785

1   Joachim Ménant (1820–99), French magistrate and Assyriologist.
2   Aimable-Augustin Hardel (1802–64) was a well-known printer and publisher at Caen.
3   Ernest Renan (1823–92), French philosopher and orientalist.
4   'On the Assyrio-Babylonian Phonetic Characters', *Transactions of the Royal Irish Academy* 22 (1852), polite literature, pp. 293–370.

※※

TO HERMANN BROCKHAUS[1]
24 JANUARY 1856

Killyleagh
24th. Jany. 1856

Dear Sir,

I send you for the Zeitschrift d. D.M.G. specimens of the language of the bilingual tablets of clay in the British Museum.[2] Colonel Rawlinson calls this language Scythian; and has compared it to the Mongolian and Mandschu; but the specimens of these languages which Gabelentz has given in the Zeitschrift f.d.K.d.M. do not seem to me in any respect similar to these.[3]

I have exactly followed the spelling of the Assyrian words, as given on the tablets. There is an occasional irregularity observable in respect to the doubling of consonants; which is sometimes neglected in Pihel, and introduced in Kal. The substitution of k for k, when in immediate contact with s, is in accordance with a euphonic rule.

In the sixth specimen the Akkadian text (as Col. Rawlinson has also proposed to call this) seems to be paraphrased rather than translated. I have given German versions of both. I think this passage relates to colonization. In the seventh specimen also there is a slight variation; the Akkadian inserting an objective pronoun.

The verbs admit no change for number or person. The aorist is the root, to which suffixes are added for the other tenses. The plural of nouns is formed by the addition of *ua* or *wa*; and the singular sometimes takes a after it, which seems to express not only the genitive but any oblique case. The prepositions of the Indo-European and Semitic languages are replaced by postpositions.

In this respect and in some others, the language resembles the Turanian languages;[4] but it has much which resembles the Assyrian also. *Sar gina* signifies in Akkadian 'the true king'; and the equivalent Assyrian would be *sarru kinu*. This was the name of the founder of the last Assyrian dynasty; who was clearly of Akkadian origin.

I am,
Dear Sir
Yours faithfully & respectfully
Edw. Hincks D.D.

I must beg of you to correct my German; which is, I fear, often erroneous. I must also trouble you to have such portion of the above letter as you may deem fit for publication translated into German. Doubtful readings are underlined, and should be in italic type.

To Professor D^r. H. Brockhaus.

MS: NStuUBG/80 Cod.Ms.philos. 198:146

1   Hermann Brockhaus (1806–77), Indologist and Iranist, studied oriental languages at Leipzig, Göttingen and Bonn. He spent the years 1829–35 doing research in Copenhagen, Paris, London and Oxford. From 1839 he lectured at the University of Jena and in 1841 he became professor at the University of Leipzig. See *Neue Deutsche Biographie*, vol. 2 (Berlin, 1955), pp. 626–7; E. Franco at http://www.uni.leipzig.de/campus2009/jubilaeen/ 2006/brockhaus.html

2   In May 1855 Hincks was made an honorary member of the Deutsche Morgenländische Gesellschaft. Brockhaus, the editor, must have been pleased to publish Hincks's article in the society's journal. The article accompanied a German translation of this letter. See next letter.

3   See H. C. von der Gabelentz, 'Versuch über eine alte mongolische Inschrift', *Zeitschrift für die Kunde des Morgenlandes* 2 (1839), pp. 1–21 (note the 'Nachtrag', 3 [1840], pp. 225–5); 'Mandschu-sinesische Grammatik nach dem Sô-ho-pián-lân', *Zeitschrift für die Kunde des Morgenlandes* 3 (1840), pp. 88–104. The entry for 11 January in Hincks's diary reads: 'All morning at Magyar and Mongolian, trying to understand the general principles of this class of language, with a view to compare them with that of the bilingual tablets.'

4   Turanian = Turkic. The association of Sumerian (Hincks's 'Akkadian') with Turkic was made on the understanding that both languages were agglutinative.

꠱꠱꠱

TO HERMANN BROCKHAUS[1]
24 JANUARY 1856

Killyleagh
d. 24. Jan. 1856

[Prof. Brockhaus,]
    Ich schicke Ihnen hier für die Zeitschrift der D. M. G Sprachproben von den zweisprachigen Thontafeln im Britischen Museum. Oberst Rawlinson nennt die erste dieser beiden Sprachen scythisch und vergleicht sie mit den Mongolischen und Mandschu; aber die Proben dieser Sprachen, welche v. d. Gabelentz in der Zeitschrift f. d. K. d. M. gegeben hat, scheinen mir des vorliegenden in keiner Hinsicht ähnlich zu seyn. Ich habe die einzelnen Buchstaben der assyrischen Worte genau so, wie sie auf den Tafeln stehen, wiedergegeben. Man bemerkt darin hier und da eine Unregelmässigkeit in Beziehung auf die Verdoppelung der Consonanten, die bisweilen im Piel nicht, dagegen aber im Kal stattfindet. Für ḳ, wenn es in unmittelbare Berührung mit *s* kommt, tritt nach einem euphonischen Gesetze *k* ein. In der sechsten Probe scheint der akkadische Text (wie ihn Oberst Rawlinson ebenfalls zu nennen vorgeschlagen hat) mehr paraphrasirt als übersetzt

zu sein. Diese Stelle bezieht sich, wie ich glaube, auf Anlegung einer Colonie. In der siebenten Probe findet sich gleichfalls eine, jedoch nur unbedeutende Verschiedenheit zwischen den beiden Texten, indem der akkadische ein Accusativ-Pronomen einschiebt. Die Verba bleiben durch alle Numeri und Personen unverändert. Der Aorist ist die Wurzel, an welche für die andere Tempora Flexionsendungen treten. Der Plural der Nomina bildet sich durch Anhängung von *ua* oder *wa*; der Singular nimmt bisweilen den Auslaut *a* an, der nicht bloss den Genitiv, sondern jeden beliebigen casus obliquus auszudrücken scheint. Die Präpositionen der indo-europäischen und semitischen Sprachen werden durch Postpositionen ersetzt. Hierin und in einigen andern Punkten schliesst sich die Sprache dem Turanischen an; sie hat aber auch vieles, was dem Assyrischen entspricht. Sargina bedeutet im Akkadischen 'der wahre König'; der gleichbedeutende assyrische Ausdruck würde *sarru kinu* seyn. So hiess der Stifter der letzten assyrischen Dynastie, der offenbar von akkadischer Herkunft war. Zweifelhafte Lesarten sind unterstrichen und mögen in Cursivschrift gesetzt werden.

I.[2]

In lal	in lalis	in lal'i	in lal'*ikum*
Iskul	iskulu	isaḳal	isaḳalu
Er wog	Er hat gewogen	Er wäge	Er wird wägen

2.

In nu	in nuis
Junakir od. jusanni	junakiru od. jusannu'u
Er hasste (od. vielleicht verfeindete)	Er hat gehasst
in nuri	in nuri*kum*
junakkar od. jusanna	junakaru od. jusannu'u
Er hasse	Er wird hassen

3.

In nan lalis	in nan lal'i
Iskulu-su	isaḳal-su
Er hat es gewogen	Er wäge es

4.

Namgab	namgaba-ni	namgaba-ni-ku
ipṭiru	ipṭiru-su	ana ipṭiri-su
die Befreiung	seine Befreiung	für seine Befreiung
namgaba-ni-ku	k.[a]	in lal
ana ki"	k.	iskul
für ebend	das Silber	er wog

Nam bildet das Nomen actionis; gab ist die Wurzel. Ein Bruchstück hat [Bab] gab / ipṭuru / Sie befreiten. Eine gleichbedeutende Vorsylbe ist nicht sicher; vielleicht ʻil.

5.

ʻilkudu	ʻilkuldua-ni	ʻilkulduni	in ci (tsi)
sibirtu	sibirta-su	ki"	iddin
das Lösegeld	sein Lösegeld (acc.)	ebend.	er gab

6.

namgankua	namgankua-ku	namgankua-ku
assabutu	ana assabuti	ana assabuti
die Wohnsitze	für die Wohnsitze	für die Wohnsitze od.
		die Niederlassungen

namgan bildet das Nomen passionis.[3]

tuntan tudu	Man vergleiche in ku
jusik'i (tschi)	jusisib
er führte heraus, Assy. oder	er setzte
sie wurden herausgeführt, Akkad.	[Hebr. hôšēb]

7.

idrapa-ua-ni	ban nab lalʻi
manaḫti-su	isaḳalu
seine Geschenke	sie wägen dieselben

8.

idrapa-ua-ni	ban nan cimu
seine Geschenke	sie gaben dieselben

ban findet sich vor einem Dental, und bab vor den übrigen Consonanten.

9.

Id gʻida-mu	*id* kappu-mu
Ina imni-ya	ina sumili-ya
An meiner rechten Hand	an meiner linken Hand

a) Die Lesung unbekannt.

Source: *Zeitschrift der deutschen morgenländischen Gesellschaft* 10 (1856), pp. 516–18

1   The first part is a German version of the preceding letter.
2   All but one of these texts have been edited by Benno Landsberger in *Die Serie ana ittišu* (Materialien
zum sumerischen Lexikon; Rome, 1937): 1 = I, ii 1–4 (p. 5); 2 = I, iii 58–65 (p. 11); 3 = I, ii 6–7 (p. 5); 4 = II, iv
16–19 (p. 29); 5 = II, iv 24–6 (p. 29); 6 = IV, iv 5–8 (p. 64); last entry = I, iii 71 (p. 11); 7 = IV, iv 34–5 (p. 66); 8 = I,
iii 34–5 (p. 61). Text no. 9 is not identified .
3   In the original this line is printed after text 5, but it must be a printing error.

TO THE EDITOR OF THE ILLUSTRATED LONDON NEWS
28 JANUARY 1856

[Killyleagh
28ᵗʰ. Janʸ. 1856]
[Dear Sir,]
    As you have introduced my name into your notice of the inscription over the
libation, you will perhaps admit my translation of it, which differs considerably
from Colonel Rawlinson's.[1] Those who are in doubt as to the fact of these inscriptions having been deciphered may like to compare the two versions.
    'I am Assur-bani-bal, King of the Provinces, King of Assyria, whom Assur
and Jarpanit[2] have welcomed to (or made successful in) the extended valleys.
When I had killed sixty huge lions, I raised over them a strong wooden altar for
Ishtar, who presided over archery. I poured over them a libation. I sacrificed over
them a goat.'[3]

I am, &c.
Edw. Hincks

Source: Illustrated London News 28/783 (9 Feb. 1856), p. 158

1   See 'Inscription from Nineveh', *Illustrated London News* 28/781 (26 Jan. 1856), p. 102. Rawlinson's
translation, which is given after introductory remarks by the editor, reads as follows: 'I am Assur-bani-pal,
the Supreme Monarch, the King of Assyria, who, having been excited by the inscrutable divinities Assur
and Beltis, have slain four lions. I have erected over them an altar sacred to Ishtar (Ashtareth), the goddess
of war. I have offered a holocaust over them. I sacrificed a kid (?) over them.' A facsimile of W. Boutcher's
drawing of the libation scene with the inscription was published in 'Recent Discoveries at Nineveh', *Illustrated
London News* 28/780 (19 Jan. 1856), pp. 63–4. Photographs of the scene are published in R. D. Barnett,
Sculptures from the North Palace of Assurbanipal at Nineveh (668–627 B.C.) (London, 1976), plates 56–9.
2   'Jarpanit' is Hincks's way of writing Zarpanit, consort of the god Marduk.
3   Russell, *The Writing on the Wall*, p. 202, gives the following translation: 'I, Assurbanipal, king of the
world, king of Assyria, whom Assur and Mulissu have granted exalted strength. The lions that I killed: I
held the fierce bow of Istar, lady of battle, over them, I set up an offering over them, (and) I made a libation
over them', which is taken with revisions from Pamela Gerardi, 'Epigraphs and Assyrian Palace Reliefs:
The Development of the Epigraphic Text', *Journal of Cuneiform Studies* 40 (1988), p. 28.

～～

TO AUSTEN HENRY LAYARD
29 JANUARY 1856

Killyleagh
29th. Jany. 1856

My dear M^r. Layard,

I received your letter this evening & lose no time in sending you for Signor Bardelli three of my publications – all of which I have new copies.[1] That on the pronouns would perhaps be the most satisfactory to him.[2]

I am glad to hear you have been enjoying yourself in Italy. We are all well here; but my career as a discoverer is, I believe, at an end. I am, however, shewing that I am alive, writing in the Journal of Sacred Literature[3] & Monthly Review,[4] & I have sent a translation to the Illustrated London News which after their very impertinent article last Saturday they can scarcely decline to print.[5] Col. Rawlinson will have every thing his own way at the Museum. His translations appearing simultaneously with the originals, no rival translator can enter the field with any chance of success.

Believe me
Yours very faithfully
Edw. Hincks

I could not make out what Library in Florence Signor Bardelli is connected with. There is a MS of Tacitus at Florence about which I feel much interested.[6] It is the only one which contains the early part of the Annals & I wish to know whether the mutilation in the 6th. Book has the appearance of being *intentional*.[7] I suspect that it was so; but I am not aware that the MS was ever examined with a view to that question. Of course the object of the mutilator was, according to my view of the matter, to destroy the testimony of Tacitus respecting our Lord's ministry & death. Whether this was done by an enemy or a misjudging friend must remain a question.

MS: BL Add. 38984, ff. 225–6

1   Giuseppe Bardelli (1815–85) wrote *Biografia del professore Ippolito Rosellini* (Florence, 1843) and edited *Daniel copto-memphitice* (Pisa, 1849).
2   'On the Personal Pronouns of the Assyrian and other Languages, especially Hebrew', *Transactions of the Royal Irish Academy* 23 (1854), part 2, pp. 3–9.
3   Hincks was continuing his important series of articles on Assyrian grammar which began to appear in 1855. See 'On Assyrian Verbs', *Journal of Sacred Literature* 1/2 (July 1855), pp. 381–93; 2/3 (Oct. 1855), pp. 141–62; 3/5 (Apr. 1856), pp. 152–71; 3/6 (July 1856), pp. 392–403.

4   See Hincks's letter to the editor of the *Monthly Review* above.

5   The translation was published on 9 February. See the previous letter.

6   See Layard's reply of 1 February 1856 below.

7   Hincks is referring to the famous Medicean manuscript of Tacitus' *Annales*, MS Plut.68.1 (often referred to as M 1) in the Biblioteca Medicea Laurenziana in Florence. See L. D. Reynolds, *Texts and Transmission: A Survey of the Latin Classics* (Oxford, 1983), pp. 406–11; C. Ando, 'Tacitus, Annales VI: Beginning and End', *American Journal of Philology* 118 (1997), pp. 285–303.

FROM LOVELL AUGUSTUS REEVE[1]
30 JANUARY 1856

[London; 5 Henrietta Street, Covent Garden
Jan. 30th, 1856]

[Dear Sir,]

If I understand the matter rightly there is very good material for an argument, but Mr. Harle's name is not sufficiently known in connection with the subject to guarantee its importance.[2]

I mentioned to Mr. Harle that I am preparing to introduce into the Literary Gazette a page wood engraving every week, and that if the question could be submitted in the form of a communication from *yourself*, in controversion of Col. Rawlinson's statement, I would have no objections to accompanying it with a facsimile of so much of the inscription as would fill our usual quarto page (engraved of course at my own expence) either of the natural size, or, if desirable for the sake of getting more into the same space, on a reduced scale.

[I am, dear Sir,
Your faithful Servant,
Lovell Reeve]

Source: Davidson, *Edward Hincks*, p. 194

1   Lovell Augustus Reeve (1814–65), conchologist and publisher, was the editor of the *Literary Gazette* from 1850 to 1856.

2   Charles Ebenezer Harle, with an address at Cross Street, Islington, London, began to write to Hincks in October 1855. Seventeen letters, the last written on 19 April 1856, reveal his obsession with an erroneous statement by Rawlinson to the effect that there is a reference to the madness of Nebuchadnezzar in the East India House inscription (see Hincks's letter of 9 February 1856 to Reeve below). The letters tell us nothing about Hincks but they do give the impression that Harle himself was a rather excitable person and probably a nuisance. Davidson, *Edward Hincks*, pp. 191–2, writes: 'His letters display an excitement about the subject and an animus against Rawlinson which suggest a not very well-balanced mind'. Hincks was very cautious about any proposals by Harle.

FROM AUSTEN HENRY LAYARD[1]
1 FEBRUARY 1856

9 Little Ryder Street
Feby. 1/56

My dear D^r. Hincks,

The pamphlets which you have been good eno' to send, arrived safely this morning. I will forward them at once to Sgn^r. Bardelli & will at the same time ask him the question about the Tacitus MS. Sgn^r. B. is librarian at the celebrated library of San Lorenzo, which contains, as you know of course, one of the most valuable collections of MSS in the world. I was very sorry to find you saying that you have given up the task of a discoverer in cuneiforms. Why should you? You will have the published inscriptions as soon as any one else, and will certainly bring to bear upon them as much, if not more, learning and acuteness than any other investigator. I quite feel with you as to the advantages given to others, & as you know, have always lamented it, doing my very best to serve you.

Has Lord Carlisle never communicated with you? He promised very certainly to do so.[2]

Yours sincerely
A. H. Layard

MS: GIO/H 243

1   Letters from Layard are few after this date. His interests were now in business and politics. In 1856 he became chairman of the newly established Ottoman Bank and his appointment provoked comments such as: 'We shall see him soon Nineveh-bulling and bear-ing on the Stock-Exchange.' See *Illustrated London News* 28/785 (16 Feb. 1856), p. 171.
2   George William Frederick Howard, seventh earl of Carlisle (1802–64). In February 1855 he was appointed Lord Lieutenant of Ireland by the Liberal Prime Minister, Lord Palmerston. A man with literary and theological interests, he would certainly have been acquainted in a general way with Hincks's research. A motion in favour of the disestablishment of the Church of Ireland was defeated in the House of Commons in May 1856 and Hincks almost certainly voiced his views on the subject in the months preceding the vote.

FROM RICHARD CLARKE[1]
2 FEBRUARY 1856

Royal Asiatic Society
5, Burlington Street
London, 2 Feby. 1856

Sir,

I have the honour to inform you that in consideration of the great services rendered by you to the objects on which the Society is engaged, especially in the study and development of the cuneatic inscriptions, to which such deep interest necessarily attaches itself, the Society has this day unanimously elected you an Honorary Member.

I have the honour to be, Sir,

Your most obedient,
humble servant,
R. Clarke
Hon^y. Sec^y.

P.S. The diploma of appointment will be forwarded as soon as it is duly prepared and signed.

Source: Davidson, *Edward Hincks*, pp. 195–6

1   Richard Clarke was secretary of the Royal Asatic Society from 1839, and treasurer from 1857.

FROM JAMES WHATMAN BOSANQUET
4 FEBRUARY 1856

Claysmore
4th. Feb: 1856

My dear Sir,

I trust you will excuse me for having disobeyed your positive injunction to withdraw your name from the ballot at the R. Asiatic Society.[1] I felt so satisfied that every member of the Society would be gratified at seeing your name enrolled amongst its members, and that it would have been a reflection upon the body to suppose it possible, that one who had done such eminent service in the cause of

Asiatic literature could be rejected by them, that I could not venture to interfere in the ordinary course of election on Saturday last. The result, which will be communicated to you in an official form, I hope will be gratifying to you. In the meantime pray take no notice of this intimation, as of course I am not authorized individually to announce to you, what must be considered as proceeding from the whole body, not from any particular individuals.

We had an interesting lecture from Sir Henry Rawlinson at the meeting.[2] Amongst other things he mentioned a new inscription in the reign of Phuluka, which arrived after he had left Baghdad, from which he has satisfied himself that Phuluka is the Pul of Scripture, and ceased to reign about the year B.C. 747, when his wife Semiramis set up the government at Babylon.[3] He also now admits that Menahem of Samaria who gave tribute to Pul, and also to Tiglath Pileser in his 8th. year, must be placed chronologically as I have placed him in a paper of mine, which I believe you have got 'on the Dial of Ahaz &c'.[4] But if Menahem began to reign in B.C. 747–6, I do not see how the inference which I have drawn can be avoided, *that Sennacherib began to reign in B.C. 693–2*, without falsifying the whole of chapters XV–XXVI of II Kings.[5]

You say that you have leisure for work, but cannot procure the material. You are entitled I consider to a copy of all that is distributed gratis by the Museum to savants in Assyrian language. I do not know whether any thing has been recently published. I should be most happy to afford you any little assistance in my power in procuring copies if any thing has been published. I should be very glad to examine the translations you speak of in the British Museum. The obelisk inscription I have in the Dublin Un^{ty}. Magazine.[6] Is there any other for which I may inquire?

Believe me
My dear Sir
Yours very truly
J. W. Bosanquet

MS: GIO/H 78

1   Hincks had strict rules about joining societies. In a letter dated 6 July 1844 to John Lee, he wrote: 'I this morning received your letter relative to the Syro-Egyptian Society. I am willing to join it, if I be one of the first 101; but I have made it a rule never to join societies on inferior terms than the original members.' See vol. I, p. 97.

2   See the account of Rawlinson's lecture in *The Athenaeum*, no. 1476 (9 Feb. 1856), pp. 174–5. Rawlinson claimed for himself discoveries made at Nimrud and Kouyunjik that were clearly made by others. There was a sharp exchange of letters between James Radford (on behalf of William Kennett Loftus) and Rawlinson regarding 'Assyrian Discovery' in *The Athenaeum*, no. 1478 (23 Feb. 1856), pp. 232–3. A week later there was a grovelling apology by Rawlinson when he conceded that the inscription of Pul was discovered by Loftus in February 1855. See *The Athenaeum*, no. 1479 (1 Mar. 1856), p. 265. But Rawlinson was incorrigible and he landed himself in more trouble with his long report on 'Assyrian Antiquities', *The Athenaeum*, no. 1484 (5 Apr. 1856), pp. 426–8. Within a week he had to write again to concede that certain important discoveries were made by Hormuzd Rassam. For example, the North palace of Kuyunjik was discovered by Rassam; and

Rawlinson admits that, if notice had been taken of his recommendations, the palace would not have been uncovered. See *The Athenaeum*, no. 1485 (12 Apr. 1856), p. 461. Loftus's quarrel with Rawlinson is noted by S. Harbottle, 'W. K. Loftus: An Archaeologist from Newcastle', *Archaeologia Aeliana*, 5th ser., 1 (1973), pp. 212–13. On the strained relations between Rassam and Rawlinson, see M. Trolle Larsen, *The Conquest of Assyria*, pp. 330–2.

3   Rawlinson had already announced this wrong identification in 1854: 'Babylonian Discoveries', *The Athenaeum*, no. 1377 (18 Mar. 1854), pp. 341–3. See Hincks's letter to Layard of 12 April 1854, esp. n. 2.

4   'The Dial of Ahaz, and Scriptural Chronology': A Letter to the Editor, *Journal of Sacred Literature* 1/2 (July 1855), pp. 407–13 (dated 23 April 1855).

5   There are only twenty-five chapters in 2 Kings, so Bosanquet has made an error.

6   Hincks, 'The Nimrûd Obelisk', *Dublin University Magazine* 42 (Oct. 1853), pp. 420–6. A version of this inscription was also included in the translations he prepared for the Trustees of the British Museum in 1853–4.

TO LOVELL AUGUSTUS REEVE
9 FEBRUARY 1856

Killyleagh C° Down
9ᵗʰ. February 1856

Dear Sir,

I now reply to your letter of the 30ᵗʰ. ult°.

Although the statement which Col. Rawlinson has repeatedly made as to the madness of Nebuchadnezzar being referred to in the India House inscription is utterly unfounded, I doubt much whether there would be any advantage in calling public attention to *this error, alone* or *principally*.[1] It might be noticed as an additional one if other unfounded statements were to be brought forward.

My reasons are these. A charge against the Col. must either be for *mala fides* or ignorance. The former can scarcely be substantiated as respects this passage. At any rate, it cannot be made by *me*, because I have in an article which was in the hands of the Editor of the Journal of Sacred Literature in December (& which would I hoped have appeared in the January number & will I trust in the April one) endeavoured to trace the mistakes which gave rise to this opinon to their origin.[2] Neither can a charge of ignorance be made, before the *general public* as a tribunal, grounded on this passage; as it must be admitted to be obscure; a *complete* translation of it cannot be given; although it can be so far translated as to show (in my opinion at least) that it will by no means admit the sense assigned to it. The general public could not be made to see the points of the argument; & no illustration (such as you spoke of) would be of the slightest assistance.

There is, however, another matter on which a charge of *mala fides* can be substantiated against Col. R. In the Athenaeum for 27ᵗʰ. Decʳ. 1851, I announced the discovery of the names of Jehu king of Israel & of Hazael king of Syria his contemporary on the Nimrud Obelisk.[3] By this *double* identification, I fixed on sure grounds the date of the obelisk. This Athenaeum, or a letter stating its contents would in course of post reach Constantinople on the 10th. January & Col. R. would have it at Baghdad about the 21st. On the 6th. March a letter from him was read (date not mentioned; but there was full time for it to be written subsequent to the receipt of the Athenaeum of the 27th. Dec[r].) containing an announcement that he had fixed the date of the Nimrud Obelisk by a *treble* identification of names viz. Jehu and Hazael on the Obelisk, & *Ithbaal*, father in law of Ahab, *king of the Sidonians*, in the inscription of the father of the king who made the Obelisk.[4] At that time no copy of these last inscriptions had reached Europe. They are new here; and the name of Ithbaal does not occur there atall! The Sidonians are mentioned; but their king is *not named*! By this artifice Col. R. has obtained the credit which properly belongs to me of having determined the date of this obelisk. His pretended *treble* identification superseded my real *double* one; & he has been quoted where I ought to have been so. Of course, I consider this very unfair treatment; & yet I cannot think it advisable to bring forward such a charge against him without some special reason, which certainly does not exist at the present time.

The only way in which I should like to come forward in opposition to him would be as to the translation of some cuneatic text. In the Illustrated London News of the 26th. ult° there was a copy of an inscription which occurs over the representation of a libation, with a translation by Sir H. Rawlinson which is in several respects erroneous.[5] I think the public would like to see a specimen of Assyrian writing with the reading of the characters one by one, and the translation of the words one by one; & this inscription seems well calculated for the purpose. It consists of three lines, each of which might be divided into two, & the characters reduced to such a size as that half a line would occupy the width of a page of the L. Gazette. There are *at least* two errors in the copy in the Ill. L. News; so that a reference to the original in the British Museum would be necessary. I would then divide the characters by small vertical lines underneath & the words by longer lines, placing under each character its phonetic value in Roman types, and again under each word its literal translation – Col. Rawlinson's translation as given in the Ill. L. News being added. My idea would be that the wood cut containing the inscription should be divided into six lines, which should be printed separately. I have not the Ill. L. News at hand to refer to; but the following will give an idea of what I mean.

a na ku   *(Assur) (bani) (bal)
  I   am   Assurbanibal

It would have to be explained that a star was used under determinative signs which were not sounded, and that the phonetic values of ideographs were placed between parenthesis.

If you think well of this, you might insert it any day that you had not anything of greater interest to occupy your page of illustration. In the meantime, the cut might be prepared and when a proof was sent to me, I would supply the letter parse for the lines that are to intervene.[6]

I have an article on Egyptian chronology, in which I flatter myself that I have carried back the period during which we have certain definite knowledge 110 years. I send it to you; & if you think well of inserting it I would be glad that you could send me a proof, as there are many difficult words & I do not write very distinctly. I don't wish it to appear till the new series commences.[7]

I send you also a POrder in your favour for half a year's subscription for the stamped copy of the new series.

& remain Dear Sir
Yours faithfully
Edw. Hincks

MS: KJC 2

1   Rawlinson gave his views on the madness of Nebuchadnezzar in an address to the Royal Institution in May 1855. See *Journal of Sacred Literature* 1/2 (July 1855), p. 488. For a modern scholarly view of Nebuchadnezzar's madness in the Old Testament at Daniel 4, see M. Henze, *The Madness of King Nebuchadnezzar* (Leiden, 1999), especially chapter 2, 'The Babylonian Prehistory', pp. 51–99. On pp. 92–3 Henze mentions the medical explanation of the madness as 'lycanthropy' (a form of insanity in which the patient imagines himself a beast), advocated by Edward Bouverie Pusey and James A. Montgomery. Rawlinson soon abandoned his misguided interpretation of the East India House inscription. See G. Rawlinson, *The History of Herodotus*, vol. 1 (London, 1858), p. 516, n. 4.

2   'On Assyrian Verbs', *Journal of Sacred Literature* 3/5 (Apr. 1856), pp. 152–71, here pp. 166–7.

3   See Hincks's letter of 22 December 1851 to the editor of *The Athenaeum*. It was published in *The Athenaeum*, no. 1251 (27 Dec. 1851), pp. 1384–5.

4   Rawlinson's letter was read at a meeting of the Royal Asiatic Society on 6 March 1852. The substance of it was published in *The Athenaeum*, no. 1274 (27 Mar. 1852), p. 357.

5   See 'Inscription from Nineveh', *Illustrated London News* 28/781 (26 Jan. 1856), p. 102, which contains Rawlinson's translation, and Hincks's version of the inscription in his letter of 28 January 1856 to the editor of *Illustrated London News* above.

6   Nothing came of this suggestion. The editorship of the *Literary Gazette* changed hands in 1856.

7   'On the Chronology of the Egyptian Dynasties prior to the Reign of Psammetichus', *Literary Gazette*

5   (29 Mar. 1856), pp. 111–12. It was also published in *Journal of Sacred Literature* 3/6 (July 1856), pp. 472–5.

FROM ALFRED JONES[1]
10 FEBRUARY 1856

Aske's Hospital
Feb. 10. 1856

Reverend Doctor,

May I venture to ask you the form of the name of Tsin-akhi'-irib in the cunei-form inscriptions; its derivation and meaning?

Not having the honour of your acquaintance, I sincerely beg your forgiveness in troubling you: but I am writing a Hebrew Onomasticon, of which the enclosed is the Prospectus, and not having met with the derivation of the above name any where, and knowing your valuable labours in this field of research, I have ventured to trouble you.[2]

Believe me to be,
Reverend doctor,
Yours obediently,
Alfred Jones

MS: GIO/H 206

1   Alfred Jones (died 31 December 1896) was educated at King's College, London. He was ordained priest in 1850 and served at the church of St Matthew's, Westminster until 1853. He was chaplain of Aske's Hospital, Hoxton, London, 1854–74; B.D. Lambeth 1877, and D.D. 1883. He was Vicar of Carrington, Cheshire, 1877–82. See F. Boase, *Modern English Biography* vol. 5 (Truro, 1897), p. 786 (= *Supplement*, vol. 2).
2   See A. Jones, *The Proper Names of the Old Testament Scriptures Expounded and Illustrated* (London, 1856). At the beginning of the entry on 'Sennacherib' (p. 318), the following is provided by Hincks: '"San has multiplied brethren". The name of Sennacherib consists of three elements, *tsin* (i.e. *sen* name of a god), *akhi'* (for *akhim, 'ahhîm*, brethren), the acc. pl., the final nasal being elided before the following vowel), and *irib*, which is a verb in the third sing. masc. of the aorist, or simple preterit'. Today the name is read *Sin-aḫḫē-erība*, 'Sin has supplied brothers'. Hincks also provided part of the entry on Sargon (pp. 315–16).

TO THE EDITOR OF THE JOURNAL OF SACRED LITERATURE
FEBRUARY 1856

Killyleagh C° Down

Dear Sir,

M^r. Bosanquet, in his letter which appeared in your last number, has, no doubt unintentionally, misrepresented a matter in which he introduces my name.[1] You will allow me, in the first place, to rectify this. He says (in p. 395), 'Africanus, in transcribing the dynasties of Manetho, writes, "Cambyses, in the *fifth* year of his reign (which D^r. Hincks proposes to read, *ninth* θ for ε) over Persia, reigned over Egypt for six years", together eleven years.'

What Africanus says is distinctly this, if we are to go by the only MS. copies of his text that we possess: 'Cambyses reigned five years (ἔτη ε) over his own kingdom of the Persians, and six years over Egypt.' I have proposed to read θ for ε: that is, 'nine' for 'five' – supposing that Africanus gives the whole period of his reign over Persia from his father's death to his own, as well as the portion of this period during which he reigned over Egypt. The reason why Africanus has given this double date is because Cambyses counted the years of his reign in Egypt from his father's death, and not from his conquest of the country.

Eusebius quotes Manetho as saying, 'Cambyses, in the fifth year of his reign, reigned (i.e., began to reign) over Egypt, three years.'

Bunsen, in the Appendix of Authorities, which he has given in the first volume of his *Egypt's Place*, has substituted the ἔτει of Eusebius in Africanus's text, in place of the ἔτη which appears in all previous editions, so far as I am aware, as well as in the MSS. Strangely enough, he has not apprised his readers of the change that he has made. M^r. Bosanquet, I presume, used Bunsen's edition; and he has applied to *his altered text*, the connection which I applied to the genuine text.

Having set this matter to rights, so far as it affects myself, I will add a few words respecting M^r. Bosanquet's theory. He endeavours to show that Cambyses must have reigned for a longer period than is generally admitted. It appears, however, from a contemporary record that has been recently discovered by M. Mariette, and which may be seen in the Salle d'Apis at the Louvre, s. 2274, that a certain Apis was born in the fifth year of Cambyses, that he lived eight years, and that he died in the fourth year of Darius. This is a positive proof that the Egyptian year, which was in its beginning the ninth of Cambyses, was the next year before that which was in its latter part of the first of Darius. The latter part of the former of these years, and the former part of the latter, would compose the reign of the Majean imposter who called himself Bardis.

With respect to the length of the reign of Darius, there surely ought to be no difference of opinion. The inscriptions on the Cosseir road determine it to have lasted thirty-six years. Added to this we have two eclipses recorded by Ptolemy, as having occurred in the 20th and 31st years of Darius, which identify those years beyond all dispute as the 246th and 257th years of Nabonassar. His first year would accordingly be the 227th of Nabonassar, and the first of Xerxes, his successor, the 263rd. This would begin the 23rd December, 486 years B.C.; and we know from Grecian history that Xerxes began to reign at this very time.

The intervals between the first years of the Saite kings Psammitichus I, Necho II, Psammitichus II, Apries, and Amasis, are all determined by contemporary inscriptions with *absolute certainty*. They were 54, 15, 6, and 19 years, making the first four of these reigns together equal to ninety-four years. Whether the reign of Amasis, of his son, and of Cambyses, amounted to fifty or to fifty-one years, is the only point on which the contemporary monuments allow us to doubt. I can conceive no chronological fact better authenticated than that the death of Nichao (the Pharaoh Necho of Scripture) occurred in the summer of either 597 or 596 B.C. The former date I consider the more probable; but any other is absolutely inadmissible. Now, as the death of Josiah occurred before this, and evidently a considerable time before it – probably not long after 612 or 611, when Nechao began to reign – I cannot be influenced by any of M^r. Bosanquet's arguments, plausible as some of them are, to bring down the date of the conquest of

Jerusalem to 560 B.C., and consequently that of the death of Josiah to 582 B.C. It is enough for me that this last date is at least fourteen years after the death of Nechao, by whom Josiah was slain. It is quite unnecessary to follow M^r. Bosanquet through the arguments by which he attempts to establish a system of chronology which is inconsistent with contemporary monuments; but I will make a few observations on what may be inferred from two inscriptions of Darius, which have at least as much bearing on the question as that which he quotes.

In the inscription on the tomb of Darius, at Nakshi Rustam, the N. R. of Lassen, after the usual statement that he was 'the son of Hystaspes, an Achaemenian', he proceeds: 'a Persian, the son of a Persian, an Arian, of Arian descent.' How is it possible to suppose that this was the same Darius, who was called in the book of Daniel 'the son of Ahashuerus, of the seed of the Medes?' and I may add, who was so called, by way of distinction from another Darius, the Persian?

M^r. Bosanquet quotes the inscription 'H' of Niebuhr and Lassen, as proving that at one time Darius was only king of the province of Persia. As the Behistun inscription describes the manner in which Darius became possessed of the whole kingdom, formerly held by Cambyses, and usurped by the pretender Bardis, it appears strange to bring forward a small inscription in opposition to its statements. There is, however, no real opposition. The inscription in question is one of four, two of which are in the Persian character and language, one in the Median or Elymaean, and one in the Babylonian. The contents of all four are different; in which respect this set of inscriptions differs from all others; the series of inscriptions met with everywhere else consisting of a Persian, an Elymaean, and a Babylonian, conveying the same meaning. This being the case, it seems to me exceedingly unfair to draw an inference from the absence of a statement in *one* inscription of a set, when that statement is most distinctly made in *another*. A list of the provinces subject to Darius is given in the inscription 'I' (which is parallel to 'H', and which was evidently of the same date with it), which is similar to that at Behistun, but more extensive, because it includes India, which Darius conquered after the date of the Behistun inscription. These inscriptions were first translated by Lassen in 1845. Some corrections were made by Holtzmann and myself, in 1846; and in 1847 Colonel Rawlinson published translations of them, as did also Benfey. The following translations are slightly altered from those of Benfey:

'H'

'Very powerful is Ormazda, the greatest of the gods. He Darius king; he gave to him the kingdom. Darius is king by the grace of Ormazda. Darius the king declares. This land of Persia, which Ormazda has given me, is beautiful, rich in horses, rich in men; by the help of Ormazda and of me Darius the King, it dreads no enemy. Darius the king declares. May Ormazda and the domestic gods bring me help; and may Ormazda defend this land from war, from pestilence, and from lying (i.e. false religion!). May no war nor pestilence, nor lying, come upon this land! This is what I ask of Ormazda and the domestic gods grant me this!'

Looking no further than to the above, I can by no means agree with M^r. Bosanquet that this inscription is 'in an humbler tone' than the others; that 'it affords the most distinct evidence that Darius was once merely king of the province of Persia; and that he must have been subordinate to some greater king, who was sovereign of the empire'. I proceed, however, to the inscription that accompanies it:

'I'

'I am Darius, the powerful king, the king of these many provinces, the son of Hystaspes, the Achaemenian. Darius the king declares. These are the provinces which I have acquired along with (or by the help of it?) this Persian people; which have trembled before me, and which pay me tribute: Susiana, Media, Babylon, Arabia, Assyria, Egypt, Armenia, Cappadocia, Saparda, the Ionians both of the land and of the sea; and the provinces to the East, Asagartia, Parthia, Taranga, Asia, Bactria, Sogdania, Chorasma, Sattagydia, Arachoria, India, Gundara, Sacia, and Macia. Darius the king declares. If it be thy will that we shall not fear an enemy, defend thou this Persian people; and if this Persian people be defended, then, oh! greatly to be praised Ormazda, may an existence of the utmost length be allotted to this palace.'

Whatever uncertainty may attach to the concluding passage in this inscription, and there is certainly some, the inference to be drawn from the whole is beyond a doubt that the writer was Darius, when king of Persia, and asserting the pre-eminence of Persia over Media and all other provinces; but Darius also, when his power was at its height, and when the whole empire was under his government.

More than was necessary has been said already; but I must, as I conclude, beg M^r. Bosanquet to consider how early in his reign Darius appears, from Herodotus, to have invaded Scythia, and to have had satraps in Asia Minor who had intercourse with Greeks. Is it possible that he could then have been only a dependent prince of Persia Proper?

I am , &c.,
Edw. Hincks

Source: Journal of Sacred Literature 3/5 (Apr. 1856), pp. 181–4

1   Bosanquet, 'Who Was Darius the Son of Ahasuerus, of the Seed of the Medes? Daniel IX, 1': A Letter to the Editor, *Journal of Sacred Literature* 2/4 (Jan. 1856), pp. 393–403 (dated 5 Nov. 1855).

FROM JAMES WHATMAN BOSANQUET
11 FEBRUARY 1856

Claysmore Enfield
11th. Feb: 1856

My dear Sir,

Pray do not think that I shall ever take umbrage at your contradicting any thing which I may have advanced, if you think it in error. I would much rather be proved to be in error, than allowed to retain an opinion known to be false. I am quite sure of gaining something valuable from any observations of yours, and look forward therefore with interest to what you have written in reply to me. You have of course seen Sir Henry Rawlinson's observations in the last Athenaeum upon the inscription of Pul.[1] They differ slightly from what I had represented him to say. As you and he substantially concur in the same view of the case, I may remark that your view of the chronology involves three contradictions of contemporary evidence.

1st. The capture of Samaria by *Sargon*, is assumed to be the capture spoken of in Scripture as in the reign of *Shalmanezer*.

2nd. The evidence of the inscription itself is denied, which informs us that Tiglath Pileser took tribute of Menahem in his 8th. year.

3rd. The reigns of the kings of Judah must be altered to suit the confusion thus introduced.

On the other hand I say that Nitocris was alarmed at the power of the Medes '*who had taken Nineveh*'. She reigned therefore after the eclipse of B.C. 585, say about 582. Semiramis reigned 5 generations earlier than Nitocris = 165 years, which added to B.C. 582 brings us to B.C. 747, which gives us the probable time of the death of Pul. There is no absolute proof that Pul died in that year. But assuming that he died in 747, and that Tiglath Pileser came to the throne in 746,[2] the reign of Menahem becomes fixed as running from 747 to 738, and the reign of Sennacherib becomes fixed as beginning in B.C. 693–2, for which there is so much direct evidence. My friend M^r. Heath who met M^r. Airy a few days ago, tells me that he concurs in my view of the eclipse of B.C. 689 as marking the date of the 14th. or 13th. of Hezekiah.[3]

Believe me
My dear Sir
Yours very truly
J. W. Bosanquet

MS: GIO/H79

1   *The Athenaeum*, no. 1476 (9 Feb. 1856), pp. 174–5. See Bosanquet's letter of 4 February 1856, n. 2.
2   Pul = Tiglath-Pileser III.
3   Dunbar Isidore Heath, a clergyman, held misguided views on the biblical Exodus and Egyptian papyri. See Birch's letter of 21 May 1855. George Biddell Airy was Astronomer Royal.

FROM LOVELL AUGUSTUS REEVE
12 FEBRUARY 1856

> 5 Henrietta Street
> Covent Garden, London
> Feb. 12th. 1856

Dear Sir,

I am much obliged by the favour of your communication, and will gladly accept your proposal to give a facsimile of the inscription to which you refer. It would not do to give the cuneatic *only* as it has been given so recently in the Illus. News, but to give it, as you propose, with the interpretation beneath each line would show a *reason* for its repetition and make it of much more general interest.

I enclose a copy of it, and will feel obliged if you will mark the errors, and how you would like to have it divided.[1]

I am, dear Sir,
Your faithful Servant,
Lovell Reeve

Will send proof of the other article in a few days.[2] I have to acknowledge, with thanks, the receipt of P.O. order 10/10 for six months subscription to Lit. Gazette commencing with the N°. for March 1st. I will refer the draftsman to the original slab in the Museum, but it would assist him, if you could point out the errors.

MS: GIO/H 410

1   Nothing seems to have come of this proposal.
2   Hincks, 'On the Chronology of the Egyptian Dynasties prior to the Reign of Psammetichus', *Literary Gazette* 5 (29 Mar. 1856), pp. 111–12. This article was also published in *Journal of Sacred Literature* 3/6 (July 1856), pp. 472–5.

TO AUSTEN HENRY LAYARD
18 FEBRUARY 1856

Private & Confidential

Killyleagh
18th. Feby. 1856

My dear M^r. Layard,

In your letter of the 1st. you asked if Lord Cavendish had ever communicated with me, as he promised you that he would do.

I send you two of his letters & I would be much obliged to you if you would return them – in a week or ten days – with your candid opinion as to whether you think I have any prospects from him, by which I ought to be influenced.

The one is a reply to a letter which my brother, now Governor of Barbados, wrote to him when in England last year.[1] I called on Lord Cavendish in Dublin in July last. He was very friendly & I had a good deal of talk with him. From what passed then I augured well.

In the beginning of November I was asked to Lord Dufferin's house to meet the Lord Lieutenant when on his tour to the north.[2] He then appeared very friendly. Lord Dufferin spoke to him about me; but in what manner I of course cannot tell.

In the beginning of December the Dean of Down died.[3] I wrote applying for the deanery, having been assured by persons who ought to know that it was considered necessary to apply for anything that was wished for. Lord Cavendish wrote me the other letter in reply; & the question is: Am I to consider it a refusal of *any ecclesiastical preferment*? or only that *on this occasion* another was preferred? The deanery was given to a M^r. Woodward,[4] who had Lord Monck's interest (a Lord of the Treasury)[5] & whose father never received anything from the Gov^t; though by almost universal consent he ought to have been made a bishop. M^r. W. himself also is a deserving man.

The mere fact then of being passed over on this occasion is not discouraging *in itself*; but it occurred to myself & also to a friend to whom I shewed Lord Cavendish's letters, that it was *peculiarly worded*; & was intended to intimate that he had heard reports to my disadvantage as a clergyman; & did not feel that he could conscientiously promote me. As to the origin of such reports I can only offer surmises; but it is pretty obvious that the Bishop of Down wanted the Deanery for his brother & that Lord Dufferin wanted it for his uncle. The most approved way of slandering a person is to express the greatest possible regard for him, but to declare his being unfit for whatever office may be vacant. I sincerely believe that I would have been better suited to the deanery of Down than almost anyone else; at the same time, its duties would have been *engrossing*, leaving me no time whatever for literature; & I should have had some annoyances to struggle against. I therefore cannot much regret having lost it, if it were only *it* that I have lost.

I have barely room to say, that what is *now* to be decided is, shall I go to the Lord Lieutenant's house on the 5th. March? Not to go would be to relinquish all chance. To go would involve both trouble & expense, though neither *very* great. What do you advise?

> Believe me
> Yours very faithfully
> Edw. Hincks

MS: BL Add. 38984, ff. 257–8

1   Sir Francis Hincks became Governor of Barbados and the Winward Islands in 1856.
2   Lord Carlisle was the lord lieutenant of Ireland. See Hincks's Diary of Visit to Lord Dufferin, 1–4 November 1855 above.
3   Theophilus Blakely was Dean of Down from 1839 until his death on 1 December 1855.
4   Thomas Woodward; see his biographical details in n. 1 to Hincks's letter to Layard, dated December 1855.
5   Charles Stanley Monck (1819–94) was a lord of the treasury in Lord Palmerston's government, 1855–8. Later he became governor general of Canada and in 1866 he was created a peer as Baron Monck of Ballytrammon, Co. Wexford. See *ODNB*.

᷾᷾᷾

FROM REGINALD STUART POOLE
1 MARCH 1856

British Museum
1st March 1856

My dear D[r]. Hincks,

I did not reply to your kind note at once as I was anxious to see Col. Rawlinson on the subject. He said that you had misapprehended him through our statement, and therefore I judged that in obedience to your instructions I must omit the paper, lest I should by its insertion put you in a false position.[1] I did not succeed in seeing Sir H. until just before publication. I retain your paper and if I do not hear from you, next time you write, that you wish it returned, I will destroy it. I trust that this circumstance will not prevent your sending us any communication of the kind in future for we shall always be very glad of such valuable papers. We are not at all afraid of learned matter but anxious on the contrary to gain it.

> Pray believe me
> My dear D[r]. Hincks
> Yours very sincerely
> R. Stuart Poole

If you write to me at Hereford Road please add *Bayswater* otherwise there is a little delay in the post.

MS: GIO/H 397

1   Poole was an editor of the *Monthly Review*.

FROM REGINALD STUART POOLE
4 APRIL 1856

4 Hereford Road
Bayswater
4 April 1856

My dear D^r. Hincks,

In a paper of Dr. Lepsius's (Über eine Hieroglyphische Inschrift am Tempel von Edfu¹) which has just reached me I find that he mentions a discovery of yours without giving you the credit of it, and apparently wishes it to be understood that it is partly his & partly M. Mariette's. Lepsius must or *ought* to have known that you had long before shewn that Necht-her-heb was a king of the 30th Dynasty.² One would be ready to pass over such things were not the Germans so tenacious of their own rights and so utterly regardless of ours. I wish you could do me the favour of sending me a note explaining your own priority of discovery & I could put it in the next Monthly Review. I cite on the next page all that Lepsius says on the subject.³ M^r. Bosanquet asked me to suggest to you a paper on a comparison of the language of the Book of Daniel with that of the Babylonian inscriptions, if the study be sufficiently advanced for this. Any thing from you – the more learned the better – would be a great help to the Monthly Review.

Pray believe me
My dear D^r. Hincks
Yours most sincerely
R. Stuart Poole

Vaux has a paper on the new Nineveh marbles in our next number.⁴ I am hard at work on the art. Hieroglyphics for the Encyclopedia Britannica, as to which I shall take the liberty of asking you a question or two. I must add that the Edfu inscription is of Ptolemy Alexander 1st & earlier sovereigns are mentioned as former benefactors of the temple.

MS: GIO/H 398

1   R. Lepsius, 'Über eine hieroglyphische Inschrift am Tempel von Edfu (Appollinopolis Magna) in welcher der Besitz dieses Temples an Ländereien unter der Regierung Ptolemaeus XI Alexander I verzeichnet ist', *Abhandlungen der königlichen Akademie der Wissenschaften zu Berlin* (1855), pp. 69–114.

2   See Hincks's letter of 29 November 1841 to John Lee (vol. 1, pp. 58–60).

3   I have omitted the postscript with the citations from Lepsius's work and some cartouches.

4   'New Assyrian Sculptures at the British Museum', *Monthly Review* 1 (May 1856), pp. 309–18. William Sandys Wright Vaux had a keen interest in the ancient Near East and published *Nineveh and Persepolis* (London, 1850).

FROM REGINALD STUART POOLE
9 APRIL 1856

4 Hereford Road, Bayswater
9th Apr. 1856

My dear D^r. Hincks,

Thank you very much for your kind note in reply to mine. I shall much like to review Lepsius in the June number, and shall be very much obliged to you for a review of Brugsch for the *May* number.[1] I gave your message to Birch who promised to write to you and gave me some information which added to what I before knew enables me to answer your question as to the seasons satisfactorily.

Champollion's reading of the names of the seasons appears to have been universally accepted until recently, when different ones have been proposed. M^r. Nash in the Transactions of the Syro-Egyptian Society (vol. i. p^t. ii.) reads them respectively [as] *sha* the 'rise', *hert* the 'appearance' or ploughing when the Nile is falling and agricultural operations are carried on, and *nutuat* or 'irrigation', denoting the time when the land is watered by artificial irrigation.[2] He makes the year to commence with the summer solstice. These particulars I have taken from a paper in the Journ. Sac. Lit. but will not fail to collate them with the memoir of M^r. Nash before posting this note.[3]

M^r. Birch in his 'Observations on an Egyptian Calendar, of the reign of Philip Ardaeus' (Archaeological Journal n°. 26.) remarks that the difficulties about the meaning of the names of these seasons are very great: 'They have been perceived by most inquirers. It is not possible to discuss such a question in a note, but I think that *sha* means "the rise", i.e., the season of the rise; *her*, the "coming forth" or overflow, and *aru*, "the river" or low Nile, and that all three seasons refer to the state of the Nile. The whole question will be found ably discussed, with the hypotheses full that have preceded, by M^r. Nash, Pap. of Syro-Egypt. Soc., Lond. 1850. When the Calendar was formed, the 1st Thoth, or of the "rise", must have corresponded with the solstice.'[4] Lepsius in his Chronologie der Ägypter (i. p. 134) translates the names of the seasons 'Frühlingsjahreszeit', 'Fruchtjahreszeit', 'Wasserjahreszeit'.[5]

I hope I have not at all misapprehended your meaning as to Brugsch. I did not know of his book till the day before yesterday & at once ordered it of my book-sellers but it has not yet reached me.

Pray believe me
My dear D^r. Hincks
Yours most truly
R. Stuart Poole

I am much obliged by the full extract you have given me abt. the kings of the 30th Dyn.

p.s. If you should think a plate advisable for your review I would do one myself by the anastatic group which is much cheaper & more satisfactorily accurate than wood cuts. In that case I should be glad to have the drawings (roughly sketched) at the latest before the 20th. Of course one would require not less than 5 or 6 groups to fill the page. I am convinced that if the Monthly Review is to be the organ for Egyptian & Assyrian discoveries it must be illustrated, until we have hiero-glyphic types and can get casts of the cuneiform ones.

I add a line from the Museum to say that I have been unable to get a sight of the Trans. Syro-Eg. Soc. which are not in the Library here. I will manage to see them tomorrow or the day after & write to you again Friday morning.

MS: GIO/H399

1   Review article on H. Brugsch, *Nouvelles recherches sur la division de l'année des anciens Égyptiens* (Berlin, 1856), in *Monthly Review* 1 (May 1856), pp. 281–90 + plate 1 (facing p. 322). Hincks's diary for 15 April says: 'At calculations respecting my review of Brugsch – continued my review and inquiries con-nected with it till bedtime.' He had written a short review of H. Brugsch, *Grammaire démotique* (Berlin, 1855) in *Monthly Review* 1 (Apr. 1856), p. 242.
2   D. W. Nash, 'On the Antiquity of the Egyptian Calendar', *Original Papers Read before the Syro-Egyptian Society*, vol. 1/part 2 (London, 1850), pp. 29–57.
3   F. Corbaux, 'Historical Origin of the Passover', *Journal of Sacred Literature* 7/14 (Jan. 1855), pp. 281–97. Poole took the details from p. 296.
4   S. Birch, 'Observations on an Egyptian Calendar, of the Reign of Philip Ardaeus', *Archaeological Journal* 7 (1850), pp. 111–20.
5   *Die Chronologie der Ägypter*, vol. 1 (Berlin, 1849), p. 134. I have omitted a lengthy citation from Lepsius's work and Poole's comments on the subject.

﹏﹏

FROM REGINALD STUART POOLE
18 APRIL 1856

British Museum
18th. April 1856

My dear D^r. Hincks,

I am fortunately able to lend you a copy of M^r. Nash's paper which I send by this post.[1] You will see that he confounds the *inundation* with the *rise*. The rise does not commence in Egypt at the Summer Solstice much less does the inundation. He is also wrong in making the time of the commencement of the overflow very variable. It is not so and the objection which he here urges against my system is at least as equally applicable to his own. The article mentioning this is in the viith vol. of the J.S.L. p. 296.[2]

The materials for a plate will be in good time to-morrow by either post. I have looked to the question as to Roman Chronology and will write to you in detail tomorrow.[3] The article will be in time on Wednesday but I should rather have it on Tuesday.

Sir Gardner Wilkinson has sent me some most interesting chronological and historical matter from Egypt which I mean to put in the next number. He sends his notes for the purpose.

Pray believe me
My dear D^r. Hincks
Yours very truly
R. Stuart Poole

MS: GIO/H 400

1   See the preceding letter, n. 2.
2   See the preceding letter, n. 3.
3   There is a letter from Poole dated 19 April 1856 'bearing on the length of Vespasian's reign & the consulate of Commodus & Priscus' (*MS: GIO/H 401*), which I have not included. Poole thanks Hincks for his 'careful drawings'. This is a reference to the plate which accompanied the review article on Brugsch's *Nouvelles recherches*. See the previous letter, n. 1.

꾸ᵕᔥᕦ

FROM REGINALD STUART POOLE
25 APRIL 1856

BritishMuseum
25th Apr. 1856

My dear D^r. Hincks,

Thank you very much for the remaining portion of your MS.[1] I will not fail to correct the proofs myself very carefully. I have Brugsch's book.

Your remarks about Oppert are very interesting to me and I shall gladly make use of them in illustration of Lepsius's drawings next month for the June n°.

Pray pardon unavoidable haste, and, with many thanks,

Believe me,
My dear D^r. Hincks
Yours most truly
R. Stuart Poole

MS: GIO/H 402

1   See the preceding letters.

꾸ᵕᔥᕦ

TO THE EDITOR OF THE MONTHLY REVIEW[1]
MAY 1856

Sir,

In your May number there is an article on the Nineveh Marbles, in which, having spoken of the husband of Semiramis as a king, 'about whose name, date, and lineage there can be now scarcely be more doubt than about Edward IV or Henry VIII', you introduce my name as that of a scholar, who agreed with Sir H. Rawlinson in his views as to this king.[2] You will, I hope, allow me to disclaim in your valuable journal any such agreement. I believe that Sir Henry and I are now fully agreed as to Assyrio-Babylonian chronology subsequent to 723 B.C., when Shalmaneser was king. He has adopted my views as to Sargon having been distinct from Shalmaneser, and having succeeded him during the siege of Samaria in 721 B.C.; and I have adopted his views as to the son and grandson of Esarhaddon. I am still, however, at variance with him respecting Tiglath Pileser, the predecessor of Shalmaneser, whom I believe to have occupied the throne of Assyria more than twenty years before 747 B.C., which is Sir Henry's date of his accession; and I can by no means admit that the biblical Pul, *his* predecessor, was identical with the husband of Semiramis.[3]

You say that I coincide with Sir H. Rawlinson as to the interpretation of the inscription on the statue, on which his whole theory is based. I have never seen a copy of the inscription, and therefore I am not prepared to express any such coincidence. It is not likely, however, that I should differ from Sir Henry on any material point in the inscription, except the reading of the king's name: on this I certainly am inclined to differ from him. I have never seen any *proof* that the correct reading of the name of the god, which is the first element in the royal name, was *Pul* or *Pal*. The latter part of the name is *ûkh*, i.e. 'is friendly (or in alliance)', connected with *akh*, 'a brother'. As to this, I presume there can be no question. Now, as the Septuagint translators write Φαλωχ for this king's name, it is certainly possible that the name in question may have been thus read. Even in that case, however, I should not admit that the two kings, the husband of Semiramis, and the predecessor of Tiglath Pileser, were the same person. The latter I consider to have been the restorer of the Assyrian Monarchy, after its overthrow by the Medes and Babylonians, under Arbaces and Belesis, in 816 B.C., or before it; while the former must have belonged to the old Assyrian dynasty, which was then overthrown. It is not in itself improbable, that the restorer of the monarchy should assume, if he had not previously borne, the name of the last king; but I see no proof that he did so; and I feel more inclined to read the name on the statue *Nin-ukh*, the Greek tradition being that Semiramis was the wife of Ninus. You say (p. 316) that 'the *only* Assyrian sovereigns mentioned in any of the Greek chronologists, who reigned conjointly, are Belochus and Semiramis'.[4] Where is this statement to be found? Certainly not in any of the Greek lists published by Cory. In a Latin catalogue of Assyrian kings, Cory gives 'XVIII Actosai et Semiramis femina XXIII; XIX Bilochus XXV'.[5] Here, Bilochus is the son, not the husband, of Semiramis. Besides, *Bilochus* rather represents the Assyrian name *Bil-ukh*, which appears to have been in actual use, than *Pal-ukh*.[6] It may be said that the Assyrians would confound these two names. It is, indeed, quite certain, that what was pronounced with a *b* at Babylon, was often pronounced with a *p* at Nineveh; and that the western nations generally adopted the latter pronunciation. But what is to be inferred from this? That it is exceedingly improbable that the Assyrians had two names of deities so likely to be confounded as *Bil* and *Pal* or *Phal*; that Palukh, the Biblical Pul, the predecessor of Tiglath Pileser, was probably *Bil-ukh*; and that the husband of Semiramis was *Nin-ukh*.

As Menahem paid tribute to Pul in the beginning of his reign, and to Tiglath Pileser before its close, in the seventh year of the reign of that monarch, it follows that the the first year of Tiglath Pileser was about the third year of Menahem, or the

42nd. of Uzziah, that is, according to the received chronology, 768 B.C. Forty-seven years seem little enough for the two reigns of Tiglath Pileser and Shalmaneser.

Assuming the chronology of the Books of Kings to be correct, the King of Assyria, to whom Jehu paid tribute, must have begun to reign about 900 B.C. This is quite consistent with the reign of his grandson having lasted to 843 B.C., the most probable date of the capture of Nineveh by Arbaces and Belesis; or even to 816, its latest date, according to any ancient authority. I remark also that, according to the Latin list, published by Cory from Scaliger, and according to Castor, Ninus was the name of the last king of the old Assyrian monarchy. The Greeks seem to have substituted the name of Sardanapalus, to which they assigned a peculiar meaning derived from thir own language, rendering it suitable to the luxurious and wanton character of the king. It is by no means unlikely that this Ninus would call his wife Semiramis, after the name of the wife of his distinguished ancestor, the son of Belus, the founder of the monarchy. I take these two names to have been *Bil-ukh* and *Nin-ukh*; unless indeed, they were *Bil* and *Nin* alone; those kings being mythic personages, who were identified with those two gods.

In conclusion, while Sir H. Rawlinson believes that a Chaldean monarchy existed from remote antiquity, to about 1279 B.C., when it was succeeded by an Assyrian one, and that Pul, the last king of this Assyrian dynasty, reigned till 747 B.C., when he was dethroned by his wife, Semiramis, Tiglath Pileser, and Nabonassar; a theory which he admits to be inconsistent with the numbers in the Second Book of Kings, I recognize one Assyrian monarchy extending from the earliest records, to the latter end of the ninth century B.C., when it was overthrown by a foreign invasion; and another, which commenced after an interval of some fifty years, at least, and lasted till 721 B.C. when the dynasty was changed; and from that, under the new dynasty, till 625 B.C., when it was overthrown.

With respect to the alleged non-Semitic names of the early kings, I am strongly inclined to think that it will turn out that there are *none such*. Ismi-Dagon, one of the oldest, is certainly a pure Semitic name; and so, I am persuaded, is that of the builder of Warka, which Sir Henry Rawlinson reads Sin-shada. I have a tracing of an inscription on a brick, containing this name, and it appears to me evident that the last character is not 'da', as Sir Henry supposes, but a calligraphic variant of 'makh'.[7] I read *Tsin-shamakh*, 'Tsin rejoices', which is purely Semitic. The only non-Semitic royal name with which I am acquainted is that of Sargon.

Edw. Hincks

Source: Monthly Review 1 (1856), pp. 381–3

1   'The Pul of the Bible': A Letter to the Editor, *Monthly Review* 1 (June 1856), pp. 381–3.
2   [W. S. W. Vaux], 'New Assyrian Sculptures at the British Museum', *Monthly Review* 1 (May 1856), pp. 309–18.

3 Rawlinson had made claims about Pul already in 1854. See Layard's letter of 12 April 1854. Rawlinson's date for the accession of Tiglath-Pileser III was nearly correct, however.
4 'New Assyrian Sculptures at the British Museum', *Monthly Review* I (June 1856), p. 316.
5 I. P. Cory, *Ancient Fragments* (2nd edn; London, 1832), p. 77.
6 See Layard's letter of 12 April 1854, n. 2 for Palukh as Adadnirari III.
7 The name is probably that of Sin-Kašid, an Old Babylonian ruler of Uruk (Warka). See Walker, *Cuneiform Brick Inscriptions*, nos 53 and 54.

FROM JOHN STEVELLY[1]
15 AUGUST 1856

Cheltenham, August 15th, 1856

Dear Edward,

The evening you left Cheltenham I had a long and most satisfactory conversation with Dr. Whewell on your business.[2] He authorized me to assure you that if there was anything in his manner which was offensive to you that a wish to be so was the farthest from his thoughts, and that he had a very high respect for your character and acquirements, though not having the pleasure of a personal acquaintance with you. He also informed me that he had during the course of that day gone from place to place in the hope of seeing you, to assure you that he had understood from some friends that his manner and expressions had the appearance of being rude to you; he wished to assure yourself that he had not the slightest intention to be so, and would even beg your pardon if any expression which fell from him when he addressed you hurt your feelings or was calculated to do so when fairly explained.

Under these circumstances I told him I thought you would feel much gratified in meeting and receiving the explanation from him and that I should be very happy indeed to be the medium of bringing you together.

When I went to your Hotel however you had just flown, and (like you) had only written,

Miss Hincks

44 Regent Square,

in his departure book. I take for granted however you are not Miss Hincks.

Yours faithfully,
John Stevelly.

P.S. I told the President of our Society of the satisfactory issue of my speaking to Whewell. It gratified him much.

Source: Davidson, *Edward Hincks,* p. 199

1   John Stevelly was born in Cork in 1794 or 1795, son of George Stevelly. He was educated at Trinity College, Dublin: B.A. 1817, M.A. 1827, LL.B., LL.D. 1844. From 1823 he was Professor of Natural Philosophy at the Royal Belfast Academical Institution and later at Queen's College, Belfast. He died in 1868.

2   Hincks attended the meeting of the British Association at Cheltenham where he read a paper 'On the Eclipse of the Sun Mentioned in the First Book of Herodotus'. See the abstract in *Report of the Twenty-Sixth Meeting of the British Association for the Advancement of Science; held at Cheltenham in August 1856* (London, 1857), p. 27. There were sharp exchanges between Hincks and William Whewell, the historian and philosopher of science, who became Master of Trinity College, Cambridge. Whewell was a man of great learning but he had a reputation for arrogance. Hincks wrote in his diary for 8 August: 'It seems Airy had published a paper on the subject, of which I was ignorant. A person interrupted me and when I had finished referred to Airy's paper as what I ought to have known. When I asked what his view was he refused to inform me. I found afterwards it was Whewell. Wrote to the President of Section A, explaining my ignorance of Airy's paper, and complaining of Whewell's want of courtesy.' Stevelly's mediation should have brought the incident to a close, but Bosanquet stirred the row up again. See Hincks's letter of 13 October 1856 to the editor of *The Athenaeum* in reply to Bosanquet's letter of 23 August in the same journal. The argument between Hincks and Bosanquet rumbled on during 1857. In spite of their differences they remained correspondents.

FROM FRIEDRICH MAX MÜLLER
16 AUGUST 1856

55 St. John Street, Oxford
Aug 16. 56

My dear Dr. Hincks,

I should have liked very much to attend the meeting of the British Association at Cheltenham, but I was prevented by the visit of some of my relations from Germany who are staying with me at Oxford. You seem to have had a very interesting meeting, and Assyrian and Babylonian philology seems to have been in great favour. I wish I could follow your interesting researches, but my time is so much taken up by the Veda, that I can only glance at your papers which you were so kind as to send to me.[1] The older one gets the more one sees the necessity of becoming one-sided in one's studies, and a proper division of labour is perhaps the best for a real advancement of science. Yet I see that it is impossible to ignore that new field of philological science which you and Rawlinson have opened, and I hope that some day I may be able to take up your valuable papers in good earnest, and see what new light they throw on the darkest periods in the history of the language. Accept my best thanks for your kind present, and believe me with sincere respect

Yours very truly
M. Müller

MS: GIO/H 278

1   One of Max Müller's great works was his edition of the text of the Rig-Veda: *Rig-Veda-Samhita: the Sacred Hymns of the Brâhmans together with the Commentary of Sayanachara*, 6 vols (London, 1849–74); 2nd edn, 4 vols, 1890–92.

⤳⤳⤳

FROM GEORGE CECIL RENOUARD
19 AUGUST 1856

Swanscombe, Dartford
19th. Augt. 1856

My dear Sir,

I hardly know how to apologize sufficiently for not having long since told you how much I was pleased by your remembrance of me, & by the very important work which you have begun. My first business was to procure D^r. Burgess's periodical in order to see what went before: but innumerable interruptions have stopped my progress, tho' what I have seen has given me much satisfaction.[1]

I know not whether I ever mentioned *Oppert's* name to you. Papers by him on the Persepolitan inscriptions in the Journ. Asiatique & a small tract on them, an exercise for a degree at Berlin, I believe, gave me a very favourable opinion of his powers & industry, & I congratulated poor Fresnel on having such a coadjutor in his Mission to Nineveh & Babylon.[2] I saw him with Birch, for a few minutes in the Brit. Mus. last year.

Knowing, as you do, my former opinions of the Assyrian *decyphrations*, you will not be surprised to hear that I was much pleased to find my notions confirmed by a Table given by Oppert in the Deutsche Morgenländische Zeitschrift, early this year.[3] In it, he has given the figures of a very large number of cuneatic characters definitely determined, with the initials of those who succeeded in determining them. Of the latter I was pleased to find that you had been half as successful as himself & R. about 1/3. The exact proportions I cannot add, having mislaid the paper on which I noted them. In the last number of the Journal Asiatique, there is a short but stringent review of a tract on the subject, in which he gives an analysis of the cuneatic names of *Sanḥerib*, & names Dr Hincks as the first & Sir H. R. as the next successful decypherer.[4]

I have not yet found the text of which he has given a copy in the common Hebrew characters.[5]

I lately had a visit from D^r. Barth who went from Bornú to Tombuktú & was much pleased with him. He is so young that I should not have thought it possible for him to have gone half so far.[6] I suppose you know of his meeting D^r (M^r.) Bakie on the Bénué.[7]

330

You will be sorry to hear that on the 12th. of Novr. by leaning against a glazed door which was not fastened & appeared like a window, I fell backwards down a stone stair-case & injured my right shoulder so severely that since that time I have never had a proper night's rest. I am now trying the efficacy of a galvanic chair; but in nearly a month's trial, its utility is scarcely perceptible. I am still a constant sufferer.

Till Saturday last the weather here was unusually warm & dry. On that night a heavy thunderstorm from 11 till 2 A.M. brought a cold north wind, much rain & very gloomy weather. Just now, the Sun who seemed to have forsaken us, appears inclined to show his face.

I sincerely hope yourself & family are now well & beg my kind remembrances to the ladies. Col. Chesney, a friend of mine, is now resident at no great distance from Newry.[8] I begged him to give my kind regards to you, should you meet; but I fear he is not near enough to Bally Leagh [sic][9] to be called your neighbour. I wish he may complete his book but I fear he embraced too wide a circuit.[10]

Believe me, my dear Sir,
Your much obliged & faithful
Geo: C. Renouard

P.S. This day sennight[11] the Bishop of Rochester will come to consecrate a new church lately finished at Greenhithe, but I, alas! am too much an invalid to be one of the attendants![12]

MS: GIO/H 474

1   Renouard must be referring to the important series of articles 'On Assyrian Verbs' which Hincks published in *Journal of Sacred Literature*. See Hincks's letter of 29 January 1856 to Layard, n. 3. The perodical was edited by Henry Burgess (1808–86), Church of England clergyman and scholar.

2   See Renouard's letter of 18 September 1851, esp. nn. 2–3; and 9 October 1851, n. 5. Fulgence Fresnel died on 30 November 1855 in Baghdad.

3   J. Oppert, 'Schreiben des Hrn. Dr. Julius Oppert an den Präsidenten der Hamburger Orientalisten-Versammlung und an Prof. Brockhaus', *Zeitschrift der deutschen morgenländischen Gesellschaft* 10 (1856), pp. 288–92. There are two short letters dated 2 September 1855 and 4 December 1855. The second letter is accompanied by a large plate of signs with values and a page of explanation.

4   J. Oppert, Review of J. Brandis, *Über den historischen Gewinn aus der Entzifferung der assyrischen Inschriften. Nebst einer Übersicht über die Grundzüge des assyrisch-babylonischen Keilinschriftsystems* (Berlin, 1856), in *Journal Asiatique*, 5th ser., 7 (Apr.–May 1856), pp. 438–43.

5   The Hebrew version is on p. 291 of the article in n. 3.

6   Heinrich Barth, the great German explorer, travelled through West Africa from 1850 to 1855. See Renouard's letter of 15 October 1851, n. 2. He was only thirty-four when Renouard met him in London.

7   William Balfour Baikie (occasionally written Bakie) (1825–64) was surgeon and naturalist to the Niger expedition in 1854. The Bénué is the longest tributary of the Niger.

8   Francis Rawdon Chesney, army officer and explorer, had moved to his residence 'Packelot' near Kilkeel, Co. Down in 1851. Renouard gives the impression that the move was more recent. Chesney was preparing to set out once more for the Near East. See [Louisa Chesney], *The Life of the Late General F. R. Chesney*, by his wife and daughter, ed. S. Lane-Poole (2nd edn; London, 1893), ch. 18 'The Euphrates Railway Expedition (1856)', pp. 422–6.

9   'Bally Leagh' for Killyleagh suggests that Renouard was suffering in more ways than one after his accident.

10   See Chesney, *Narrative of the Euphrates Expedition: Carried on by Order of the British Government during the Years 1835, 1836, and 1837* (London, 1868). The author had previously published *The Expedition for the Survey of the Rivers Euphrates and Tigris, carried on by Order of the British Government during the Years 1835, 1836, and 1837: Preceded by Geographical and Historical Notices*, 2 vols (London, 1850).

11   Sennight: a week from this day.

12   George Murray (1784–1860) was the Bishop of Rochester from 1827 until his death. The church of St Mary at Greenhithe was built in 1855–6 and the parish was created in 1857 from Renouard's church of SS Peter and Paul. John Fuller Russell (1813–84), ecclesiastical historian, was presented to the Rectory of Greehithe, near Swanscombe, Kent, by Sidney Sussex College, Cambridge.

FROM HENRY FOX TALBOT[1]
28 AUGUST 1856

Lacock Abbey, Chippenham
Aug. 28, 1856

Dear Sir,

I presume that I am indebted to you for a copy of your last paper in the Transactions of the Royal Irish Academy which I received by the post.[2] I have read it with much interest and instruction. Before reading it however, I took the inscription presented in your first page, and endeavoured to make a translation of it myself, according to a practice which I frequently adopt and find to be very useful. The following is the translation which I thus obtained using two letters to designate words that remained doubtful.

On the 6th day of the month

A and B
were weighed (*meshkulu*) root *mshql*
six kasbu of A
six kasbu of B

on the Reverse
May Nebo and Merodach bless
the king my master!

On then looking at your translat[n]. I had the satisfact[n]. to find that the *Reverse* agreed entirely.[3] With regard to the prior part of the inscription I regard your interpretation as highly curious, but not altogether proved as yet. I desiderate proof that *mushi* signifies *the night*, for though you refer to line 47 of Bellino's cylinder I think it is open to a question whether that passage bears the meaning you ascribe to it 'in the course of a night'.[4]

Your explanation of the *colophon* to Bellino's cylinder appears to me very successful and very curious so also the explanation of 𒀭 ⟶ as a 'unit'. I observe that in p. 39 you quote a work of Grotefend's thus, Gr. 2.6. Allow me to enquire where this inscription is to be found, as I have never met with a copy of it?

Believe me to be
Yours very truly
H. F. Talbot

P.S. I think the word for 'night' may have sounded *vushi* rather than *mushi*. And if so, I think I can offer an explanation of the word. It is neither more nor less than the Egyptian word for night, *ushi* or *oushi*, which the Assyrians have borrowed as they have 2 or 3 other undoubted Egyptian words. See Tattam's Egyp^n. Dict^y. p. 368. *Eushi* in another dialect. Also found in Arabic according to Tattam.[5]

MS: GIO/H503

1 William Henry Fox Talbot did not use the name William and disliked the use of the name Fox. He always signed his name 'H. F. Talbot'.
2 'On a Tablet in the British Museum, Recording in Cuneatic Characters, an Astronomical Observation; with Incidental Remarks on the Assyrian Numerals, Divisions of Time, and Measures of Length', *Transactions of the Royal Irish Academy* 23 (1856), part 2, pp. 31–47 (read 12 Nov. 1855).
3 Actually Hincks has 'may they draw near' (*iqrubu*).
4 Hincks was right about '*mushi*'. Akkadian *mūšu* means 'night'; *urru u mūšu*, 'day and night'. See Arabic *masā*', 'evening'; and Hebrew *'emeš*, 'last night'. Hincks returned to this British Museum tablet K. 15 in 1865. See his paper 'On the Assyrio-Babylonian Measures of Time', *Transactions of the Royal Irish Academy* 24 (1865), polite literature, pp. 13–16. There is an important note in the *Addenda et Corrigenda*.
5 Talbot published these remarks in his article 'On the Assyrian Inscriptions. No. IV', *Journal of Sacred Literature* 4/7 (Oct. 1856), pp. 164–70. See p. 170, 'On the Assyrian Term for "Night"'.

FROM SIDNEY SMITH[1]
29 AUGUST 1856

Brookeboro
29 Aug 56

My dear Dr Hincks,

A friend of mine Mr Robert Ross of Bladensburg (Rostrevor)[2] who has travelled much in the East has sent me a copy of one of his inscriptions which he copied in the Wady Mokatteb & from a tomb at Petra, and I venture to suppose that they may be of some interest to you, & accordingly send the tracing & his letter also explanatory of the same.

Yours very faithfully
Sidney Smith D.D.
Ex. F.T.C.D

[Enclosure: Robert Ross to Sidney Smith]

Carraigbhan
Rostrevor
Augt 18th. 1856

My dear Sir,
According to my promise I send you a copy of some of the inscriptions I found in the vallies on my way from Cairo to Sinai & from thence eastward to Akabah. I, II, III, IV (of IV the end is defaced) V were found in the Wadhee Mokatteb; VI & VII in the Wadhee Ledja on the way to the stone pointed out by the monks at Sinai as the Rock struck by Moses. VIII & IX in one of the vallies eastward from Sinai on the road to Akabah; X I am not quite sure where I copied it. Those on sheet B are copied from a tomb at Petra.[3] The tomb lies nearly opposite the fissure of the amphitheatre & inside of a square chamber cut into the rock on the side opposite the entrance is an arched recess with two cavities cut in the rocky floor capable of containing each a corpse within this that is deeper. In the interior is a square recess in whose floor is excavated a large cavity or grave capable of containing several corpses. In the floor of the larger entrance chamber 15 holes 7 feet long 2 feet wide & 9 or ten feet deep have been sunk; they are not placed with any regard to order, but seem to have been sunk in the rock as they were required & so as to get as many as possible in the least possible space. Those nearest the wall on the northern side are parallel to the wall & above them two small pyramidal ornaments as in page B. I. have been carved in rilievo. B. I. is engraved on two pyramidal ornaments as represented. B. II is engraved below a similar pyramidal ornament which has been left quite blank. B. III. is carved in the Rock over a tank cut for the reception of water on the way to what the Arabs call Ma Deir; not far from this inscription was an excavated chamber, but whether for the repose of the living or of the dead is uncertain. I would think for the living. Any further information w^h. I can give I shall be happy to supply.

Believe me
My dear Sir
Yours very truly
R. Ross of Bladensburg

MSS: GIO/H 497, 484a, 485, 498

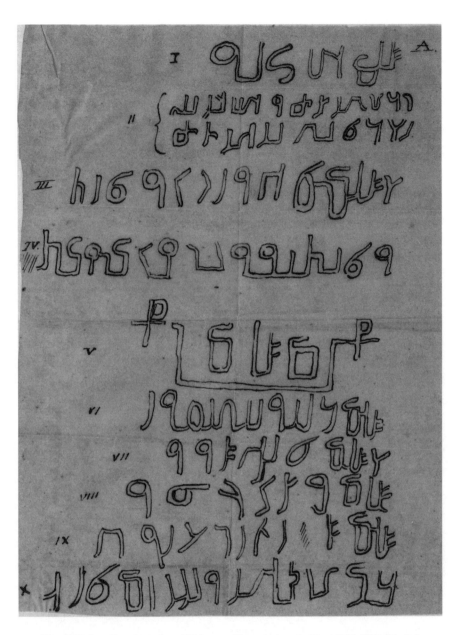

Plate I: Robert Ross's copies of Nabataean inscriptions in Sinai (MS: GIO /H 485)

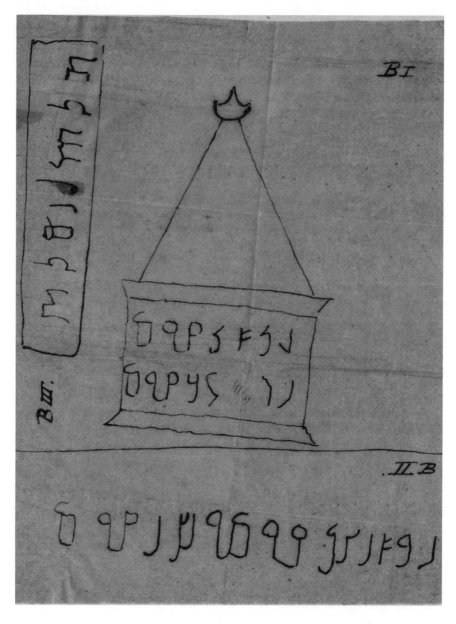

Plate II: Robert Ross's copies of Nabataean inscriptions in Petra (MS: GIO/H 498)

1　George Sidney Smith was born in 1805 in Edinburgh, son of George Smith. He was educated at Trinity College, Dublin: scholar 1823, B.A. 1825, fellow 1831, M.A. 1832, B.D. and D.D. 1840. He was Professor of Biblical Greek at Trinity College in 1838; Rector of Aghalurcher, Co. Fermanagh, 1838–67. He married Charlotte Lees and they lived at Ardushin House near Brookeborough, Co. Fermanagh. Charitie Lees Bancroft (1841–1923), the Irish-American hymn writer, was one of their daughters. From 1867 Smith was Rector of Drumragh, Co. Tyrone and Canon of St Columba's Cathedral, Derry. He died on 14 August 1875. See G. D. Burtchaell and T. U. Sadleir (eds), *Alumni Dublinenses* (Dublin, 1935), p. 758.

2　Robert Ross of Bladensburg was the son of Major-General Robert Ross (1766–1814), who commanded the expeditionary forces against the Americans in 1814 and infamously ordered his troops to burn down all the public buildings in Washington. See *ODNB*. David Ross, brother of the writer of a letter, also travelled in the Mediterranean. See D. Whitehead, 'David Ross of Bladensburg: a Nineteenth-Century Ulsterman in the Mediterranean', *Hermathena* 164 (1998), pp. 89–99; idem, 'From Smyrna to Stewartstown: a Numismatist's Epigraphic Notebook', *Proceedings of the Royal Irish Academy* 99C (1999), pp. 73–113. Robert Skeffington Ross (1847–92), a Jesuit priest, was the eldest son of David Ross and the Hon. Harriet Margaret Foster Skeffington and not the son of our Robert Ross, his brother, as thought by the writer of the lengthy obituary, 'Father Robert Skeffington Ross', *Letters and Notices* 21/107 (Apr. 1892), pp. 355–70.

3　See the plate(s) with Ross's drawings of the inscriptions. Some of the Nabataean inscriptions which Ross copied can be found in G. F. Grey, *Inscriptions from the Waady el Muketteb, or the Written Valley; copied in 1820 by the Rev. G. F. Grey* (London, 1832). Also published in *Transactions of the Royal Society of Literature* 2/1 (1834), pp. 147–8 + 14 plates. See further *Corpus inscriptionum semiticarum*, part 2 (Paris, 1881–), nos 490–3233.

TO HENRY FOX TALBOT

I SEPTEMBER 1856

Killyleagh C° Down

1ˢᵗ. September 1856

Dear Sir,

I received your letter this morning. The inscription which I quoted as Gr. 2.6 is that on a cylinder of Nebuchadnezzar (in the British Museum, I believe; but I am not quite sure). It is in three columns, and has much in common with the Great Inscription at the India House. Grotefend published a lithograph of it.[1]

I like your idea that *mushi* is *wushi* = 𒐊𒑊 *usha*. So, *tum* is 𒁹 Tum, the god of Heliopolis. By the way, Rapikh, where the battle was fought between Sargon and the king of Gaza, was not Heliopolis but Raphia, where a great battle was afterwards fought between the Syrians and Macedonio-Egyptians.[2] I pointed this out to Rawlinson many years ago, when he first published the reading 'Rabek i.e. Heliopolis'. Allow me to say that the king of Gaza who was here defeated was named Khanun i.e. Heb. ḥānûn, a name which we meet with (applied to different persons) [in] 2 Sam 10.1 & Neh. 3.13 & 30. Gesenius rightly compares it with the Punic Ḥanno. This is certainly not the same name as Canaan, Heb. kᵉna'an. The Assyrians would not have overlooked the 'ayin in this last name; & kaph is *never* confounded or interchanged with ḥēt.

That the kajab was a measure of time is obvious from comparing Botta 41.38 with 41.28. In the former place we have 'a voyage of 30 kajbu (2½ days)'; in the latter 'a voyage of 7 days' in similar connexions. This & the known meaning of *tum* leave no doubt as to what *mushi* or *wushi* signifies.

    I remain
    Dear Sir
    Yours very truly
    Edw. Hincks

MS: BL Talbot 31425

1   See G. F. Grotefend, 'Bemerkungen zur Inschrift eines Thongefässes mit babylonischer Keilschrift', *Abhandlungen der königlichen Gesellschaft der Wissenschaften zu Göttingen* 4 (1850), historisch-philologische Klasse, pp. 3–18 + 2 plates.
2   In the Annals of Sargon II (721–705 B.C.) we read: 'Hanno, king of Gaza and also Sib'e the turtan of Egypt set out from Rapihu against me to deliver a decisive battle. I defeated them'. See Pritchard (ed.), *Ancient Near Eastern Texts Relating to the Old Testament*, p. 285. In 217 B.C. Ptolemy IV of Egypt defeated Antiochus III, king of the Seleucid empire.

FROM HENRY FOX TALBOT

5 SEPTEMBER 1856

Lacock Sept. 5. 1856

Dear Sir,

    I thank you for pointing out that Rapikhu is Ραφεια, and not Heliopolis of Egypt. This clears up a very important point, and explains why the king of Gaza took a prominent part in the battle, Rapheia being I believe close to Gaza. You may very possibly also be right with respect to the name of Hanno, a decidedly Phoenician name, and therefore suitable for a king of Gaza. Is the lithograph of Grotefend to which you refer, a loose sheet or published in any work, and under what title?

    With respect to your argument, that if the Assyrians had meant to transcribe the name of Canaan they would not have overlooked the *Ain*, I think that is not so certain. In the very important name of Baal, the Chaldeans omitted the Ain and wrote simply *bl*.

    I remain, Dear Sir
    Very truly yours
    H. F. Talbot

MS: GIO/H 504

~~≈≋

TO HENRY FOX TALBOT
8 SEPTEMBER 1856

Killyleagh C° Down
8ᵗʰ. Sept 1856

Dear Sir,

I received your letter this morning. I am not quite sure of the title of Grotefend's work about which you ask; my copy being imperfect. It is (I believe) 'Bemerkungen (or Beiträgen) zur Inschrift eines Thongefässes mit babylonischer Keilschrift'[1] & is a thin quarto of much the same size as the one with a similar title (substituting ninivitischer for babylonischer) which contains Bellino's inscriptions. Both are to be had separately; as well as in the Göttingen Papers, where they were published. I think the price of each is 2/–.

As to Bel I have slightly modified my views since I wrote what appears in p. 45 of my last paper.[2] I think the *Assyrians* paid more attention to the *'ayin* than the Babylonians did, and *they* would not be likely to omit it in a proper name.

You take Ithamar for one word, the name of the king of Saba.[3] Is not, however, the true reading *Idah amara Tsabahaya*, 'Idah, Amir of Saba', *amara* representing the Arabic *amir*? If we suppose the words which I have divided to form one name (which is certainly very *possible*, & only for its length, the more putable supposition of the two) it must, I think, be read Idahamara in 5 syllables. The first consonant is more probably *d* than anything else; but might be *t*, or *ṭ*. After that we must have *'alif*, *ha* or *'ayin* before the *m*. Ithamar appears to me quite inadmissible; & I really feel jealous of the admission of scriptural names where they are not identical with those of the inscriptions. It tends to discredit the real coincidences which are so numerous.

Believe me
Dear Sir
Yours very truly
Edw. Hincks

MS: BL Talbot 31425

1   G. F. Grotefend, 'Bemerkungen zur Inschrift eines Thongefässes mit babylonischer Keilschrift', *Abhandlungen der königlichen Gesellschaft der Wissenschaften zu Göttingen* 4 (1848), pp. 3–18 + 2 plates.
2   'On a Tablet in the British Museum, Recording in Cuneatic Characters, an Astronomical Observation; with Incidental Remarks on the Assyrian Numerals, Divisions of Time, and Measures of Length', *Transactions of the Royal Irish Academy* 23 (1856), part 2, pp. 31–47.
3   H. F. Talbot, 'On the Assyrian Inscriptions. III', *Journal of Sacred Literature* 3/6 (July 1856), pp. 422–6, esp. the section on 'The Queen of Sheba, and her Visit to Solomon'. Talbot's reading of the name was more or less correct. Today we write It'amra the Sabaean.

FROM HENRY FOX TALBOT
12 SEPTEMBER 1856

London Sept 12. 1856

Dear Sir,

You appear to require that a Scriptural name should be *identical* with one found in the inscriptions before they can become fit subjects of comparison. But upon that principle we should have to exclude, in the passage under consideration, the name of Pharaoh king of Egypt, first proposed I believe by yourself, and which I admit to be highly probable, although not identical with the Assyrian form.

In fact, had the name of Pharaoh been unknown to us it would not have been suggested by the Assyrian. I must maintain the correctness, or at least high probability, of the reading Ithamar. As for the title *amir* I once thought that I had found that title used in the inscript^{ns}., but discovered my error in so rendering it, and I now think that word was unknown to the Assyrians at least that we have no example of it. Many thanks for pointing out the title of Grotefend's work. Perhaps I may be able to find a copy in the B. Museum.

Believe me
Dear Sir
Yours very truly
H. F. Talbot

MS: GIO/H 505

TO THE EDITOR OF THE ATHENAEUM
13 OCTOBER 1856

Killyleagh, C° Down
Oct. 13th. 1856

[Dear Sir,]

In the *Athenaeum* of the 23rd. August you have inserted a letter of M^{r}. Bosanquet's, in which I am attacked with some severity.[1] You will, I trust, insert this reply of mine. I have delayed it only with the view to make myself as completely master of the subject as I could before writing. M^{r}. Bosanquet says that he is surprised at my questioning a result at which M^r. Airy and M^r. Hind had arrived.[2] I am about to surprise him still further; and I shall probably surprise also some worthy members of Section A., who seem to believe in D^r. Whewell's infallibility.[3] M^r. Bosanquet's letter implies that I entertain doubts as to the eclipse of 603 B.C. being that of Herodotus. Perhaps I expressed myself doubtingly in my paper, but I entertain no doubts now. As confidently as I believe that on the evening of this present 13th. of October, 1856, the greater part of the moon will be immersed in the earth's shadow, so confidently do I believe that about 10 o'clock of the morning of the 18th. May, 603 B.C. the shadow of the moon passed over a field of battle in Turkey in Asia, where the Medes and Lydians had just commenced fighting, causing a total eclipse of the sun, which terminated the battle, as it was understood to intimate the will of Heaven that it should not take place. An eclipse at sunset, like that of the 28th. of May, 585, after the parties had been all day fighting, could never have been regarded as an intimation of Divine displeasure at their fighting. Besides, the placing this battle 585 B.C. is absolutely inconsistent with well-established chronological facts. The eclipse of 603 is the only eclipse which was visible in Western Asia, within the chronological limits which confine us, that had the two characteristics, of being total and being in the morning. I, therefore, as I said, feel the most perfect confidence that this was the eclipse intended by Herodotus; and I am confident, also, that the time is not far distant when astronomers generally will accept this fact, and will rectify their tables by making them to agree with it. I admit that, according to the tables which astronomers now use, the moon's longitude would be a considerable number of minutes greater than it must have been according to my belief: I cannot say precisely *how many* minutes, because (*as I correctly stated at Cheltenham*) no calculation of this eclipse has been published since that made by M^r. Baily, who used tables which are now superseded.[4] The circumstance which I have mentioned, however, presents no difficulty to my mind. I believe that the additions to the different arguments which are found in Damoiseau's Table IV are all much greater than they ought to be, the error being nearly proportioned to the square number of centuries between the epoch of the tables and that for which the moon's place is sought. Consequently, the tabular longitude of the moon always comes out greater than her longitude actually was. The supposition that an error like this exists in Damoiseau's tables is not to be slighted as a mere fancy of mine. M^r. Adams has published a paper in the *Philosophical Transactions* for 1853, in which he proves from theory that this is actually the case;[5] and I can put no other construction on recent statements that have been made respecting an error in his calculations that M. Plana has discovered than that he has adopted, in part at least, M^r. Adams' views. M^r. Bosanquet says that M^r. Airy has found that the correction of the tables which would be necessary to meet M. Plana's present views would 'very slightly vary' the path of the shadow in 585. *That* is a matter of little or no importance, as it is absolutely impossible that *that* could be the eclipse of Herodotus; but what is of real importance is this: How would it affect the eclipse of 603? I suspect that it would make a considerable change in the path of the shadow in *it*. I know that a correction in accordance with M^r. Adams' views would do so.

I abstain from details, which would only be interesting for mathematicians; but I feel that I have good grounds for the confidence which I have expressed in the correctness of my views. I must now say a few words on the occurrence at Cheltenham, to which M^r. Bosanquet has referred, and which is noticed in your report of the proceedings. When I stated that the eclipse of 603 had not been calculated with the improved tables, and expressed a wish that it should be so, D^r. Whewell told me that it had been calculated, and that all the particulars respecting it would be found in a paper of M^r. Airy's in the *Philosophical Transactions* for 1853.[6] He said that the path of the shadow in this eclipse was laid down by M^r. Airy on a map. I asked him to point it out to me, as determined by *him*, on the large map, on which I had just pointed it out, as it was laid down by a writer in the *Philosophical Transactions*, whose name I now forget; which path, I expressed my belief, would be found not to differ much from the true one. D^r. Whewell refused to gratify me, referring me to the volume of the *Transactions* in which M^r. Airy's paper was, which, he intimated, contained all that I had been reading to the Section. He said that he *could* point it out, but that he did not choose to do so. Some weeks after the meeting at Cheltenham, I procured the part of the *Philosophical Transactions* which contained M^r. Airy's paper; and I discovered, to my great surprise, that the entire of D^r. Whewell's statement was erroneous. The paper contained no calculations respecting the eclipse of 603, and no map exhibiting the track of the shadow in it. The only eclipse among those which have been thought to be that referred to by Herodotus, as to which M^r. Airy gave any calculations, or laid down any track of the shadow, was that of 585. It was evident that D^r. Whewell had, through some strange misapprehension, confounded the two eclipses of 603 and 585; and that he imagined I was giving a very absurd representation of the circumstances of the latter, when I was speaking of those of the former. If D^r. Whewell had complied with my request to communicate to me the information which he supposed that he possessed, he would have been led to see the mistake under which he was himself labouring. It is scarcely necessary for me to add that D^r. Whewell's suggestion that my paper was a *réchauffé* of M^r. Airy's was completely erroneous. It was grounded on the mistake that I have just pointed out. He imagined that I was arguing in favour of the same eclipse for which M^r. Airy had argued, as being that of Herodotus; the fact being that I was arguing in favour of a different eclipse, and against that of M^r. Airy. The only point on which I took the same ground with M^r. Airy was in opposing M^r. Baily's view that the eclipse of 610 B.C. was that of Herodotus; and if I had read M^r. Airy's paper I should probably have said less on this point, which I should have taken as proved.

As to everything else connected with the eclipse, I have, I believe, sufficiently expressed my total dissent from M^r. Airy's views.

I am, etc.
Edw. Hincks

Source: The Athenaeum, no. 1513 (25 Oct. 1856), pp. 1308–9

1   J. W. Bosanquet, Letter to the Editor, dated 21 August 1856, *The Athenaeum*, no. 1504 (23 Aug. 1856), p. 1053.

2   George Biddell Airy (1801–92) and John Russell Hind (1823–95), astronomers.

3   For Hincks's disagreement with William Whewell during the meeting at Cheltenham, see John Stevelly's letter of 15 August 1856.

4   See F. Baily, 'On the Solar Eclipse which is said to have been Predicted by Thales', *Philosophical Transactions of the Royal Society of London* 101 (1811), pp. 220–41.

5   J. C. Adams, 'On the Secular Variation of the Moon's Mean Motion' *Philosophical Transactions of the Royal Society of London* 143 (1853), pp. 397–406.

6   G. B. Airy, 'On the Eclipses of Agathocles, Thales, and Xerxes', *Philosophical Transactions of the Royal Society of London* 143 (1853), pp. 179–200.

FROM GEORGE CECIL RENOUARD
20 NOVEMBER 1856

Swanscombe, Dartford, Kent
20 November 1856

My dear Sir,

Three months would not have elapsed before I made my very sincere acknowledgements for your so kindly remembering me, had I not hoped to study your grammar of the Assyrian language which will hereafter be considered as, perhaps, the *most remarkable result* of Rassam's discovery of the sculptures at Nineveh, & will certainly place your name where it ought to be – at the head of the revealers of the secrets thus unexpectedly brought to light.[1] The grammar, however, has not advanced so rapidly as I expected, & a desire to verify Oppert's reading of a Van Inscription, & various other avocations have, for this long period, caused me to adjourn my answer.

As my accident last year (on the 12th. of November) is almost the first thing which meets my eye in your letter, I will begin by telling you that I am now nearly recovered from its consequences; suffer little or no pain in my shoulder; have rest at night, scarcely disturbed, & can put my arm behind my back: so that I have the greatest reason to be thankful: but a tendency to asthma (a family complaint) makes walking very troublesome, & nearly confines me to the house.[2]

You will congratulate me on being able to attend at the consecration of S[t]. Mary's at Greenhithe.[3] The weather was as the French say, magnificent; the congregation large, & the sum collected far beyond my expectations, & enough to clear off a large part of our debts. With what I have seen of the incumbent of S[t]. Mary's, I am well pleased, though I knew nothing of him before he was chosen by the committee for the building of the new church, & I suspect he is more inclined to Puseyism than I like; but I am still more opposed to the opposite extreme.

I hope you have now resumed your cuneatic labours. The Museum has, I see, published facsimiles of several papyri, that of Sallier among the rest; & Oppert is fully aware of what he owes to you. The Museum, I hope, you will again visit in summer, & I trust you will receive more materials to work upon. Your paper on the astronomical *laterculus*, I should like to see in *a very portable form* with a plate of the *tile*,[4] as it would be generally interesting; &, in a very small compass, conveys a very clear notion of the almost incredible difficulties impeding the decypherment of the Assyrian records; & a striking illustration of Pliny's (N. H. vii, 56) account of those records.[5] It is short enough to be read by those who would not look at a longer paper, & clear enough to be comprehended by any reader.

Oppert, indeed, has leisure & materials, which you have not; but I hope that your materials are now more abundant than they were when you wrote, & I cannot but think that a continuation of your labours will be fitly appreciated. You must be aware that the reliance once given to every thing said by others on this subject is no longer what it was; & by pointing out the chronological errors you allude to, you will do a most important service to all who may otherwise be misled by them.

A short & *popular* acc[t]. of what has been done in this arduous undertaking, & especially a note of what was primarily done by yourself, would probably be printed in the *Athenaeum*, & be very useful.

Perhaps O. had not seen what you had written since 1852. Your writings are unhappily so much scattered, & that in periodicals, not (till you opened man's eyes) ever heard of abroad, that it was long before one student in ten, even in England, was aware of them. Without mentioning any other name you might give the values discovered by yourself & the dates of their discovery.

As I did not know the Journal of Sacred Literature was published by D[r]. Burgess after the death of D[r]. Kitto,[6] I had never seen or heard of your communications to it, till I received your first instalment in July.[7] At that time I believe I did not know what Oppert had done. I thought favourably of his paper on the *Persian Cuneatics*, & congratulated Fresnel on having so able a coadjutor. From O.'s manner when he told me of Fresnel's death, I fear he was not a very cordial aid to him in his investigations.[8] I was never less taken by any man's manner, than O.'s; but that has nothing to do with his *powers*. His Report to the French Minister which gives his views at greater length I have not seen.[9] That & the great work on Abyssinian by *Lefebvre, Petit & Dillon* seem scarcely known in London.[10]

Believe me, my dear Sir,
Your much obliged & faithful
Geo. C. Renouard

P.S. As you have the *D. M. Zeitschrift*, you have seen Oppert's last short but instructive paper – s. 802.[11]

The 'Archives des Missions Scientifiques' is, I presume, a very modern publication.[12] It is not mentioned in my latest catalogues.

MS: GIO/H 475

1   Renouard is probably referring to Hincks's papers 'On Assyrian Verbs', *Journal of Sacred Literature* 1/2 (July 1855), pp. 381–93; 2/3 (Oct. 1855), pp. 141–62; 3/5 (Apr. 1856), pp. 152–71; 3/6 (July 1856), pp. 392–403.

2   There is an account of the accident in Renouard's letter of 19 August 1856.

3   See Renouard's letter of 19 August 1856.

4   Although Latin *laterculus* does have the meaning 'tile', 'small brick' or 'hard baked tablet' are preferable for describing the small clay tablet studied by Hincks in his paper, 'On a Tablet in the British Museum, Recording in Cuneatic Characters, an Astronomical Observation; with Incidental Remarks on the Assyrian Numerals, Divisions of Time, and Measures of Length', *Transactions of the Royal Irish Academy* 23 (1856), part 2, pp. 31–47. Read 12 November 1855.

5   Pliny, *Naturalis Historia*, VII, 56. *Laterculus* occurs in one passage: 'On the other side Epigenes, an authority of the first rank, teaches that the Babylonians had astronomical observations for 730,000 years inscribed on baked bricks (*coctilibus laterculis*)' (tr. H. Rackham, Loeb edition).

6   John Kitto (1804–54) edited the *Journal of Sacred Literature* from 1845 to 1853.

7   Renouard is referring again to the series on Assyrian verbs. See n. 1.

8   Fulgence Fresnel died in Baghdad on 30 November 1855. According to Larsen, *The Conquest of Assyria*, p. 309, he was 'described by all as a most kind and considerate man ...'.

9   *Rapport adressé à Son Excellence M. le Ministre de l'Instruction publique et des Cultes, par M. Jules Oppert, chargé d'une mission scientifiique en Angleterre* (Paris, 1856).

10   T. Lefebvre, A. Petit, L. R. Quartin-Dillon *et al.*, *Voyage en Abyssinie exécuté pendant les années 1839, 1840, 1841, 1842, 1843*, 6 vols (Paris, 1845–51).

11   Oppert, 'Schreiben des Hrn. Dr. J. Oppert an Prof. Brockhaus', *Zeitschrift der deutschen morgen-ländischen Gesellschaft* 10 (1856), pp. 802–6.

12   Publication of the *Archives des missions scientifiques* began in 1850.

FROM EDWIN NORRIS
24 NOVEMBER 1856

Royal Asiatic Society
5, New Burlington Street
London 24th. Nov^r., 1856

My dear Sir,

I should say that a paper of the kind you mention would be of much interest; I have not any doubt that it would be acceptable to the Society, and forthwith printed in the Journal.[1] Of course I speak unofficially, because no paper can be ordered for printing until it has been laid before the Council; but in this case I am persuaded that it would be only a form. Could you send such a paper soon? I should like to know, that I may communicate your intentions to such of the members as are interested in these researches.

Believe me
Yours sincerely
Edwin Norris

MS: GIO/H 314

---

1   In his letter of 9 January 1857 Norris asked Hincks about his promised paper on the Van inscriptions for *Journal of the Royal Asiatic Society*. Hincks made three attempts to write the paper. Two drafts (each nine pages long) were written in 1856–7 and a third one (five pages plus a plate) in 1861. They are preserved in the Griffith Institute, Oxford. See MS GIO/H 571.

TO THE EDITOR OF THE JOURNAL OF SACRED LITERATURE
6 DECEMBER 1856

Killyleagh C° Down
6th. Dec. 1856

Sir,

You have inserted in different numbers of the *Journal* letters of M^r. Bosanquet in defence of his chronological views, and of others in opposition to them. Enough has, I believe, been said on the objections brought against his system. They are such as have not been, and I am persuaded, cannot be removed. On these, therefore, I will not now speak. There are two points, however, on which he relies very confidently, as in favour of his system; and I admit that if either of the statements which he puts forward could be substantiated, it would render it difficult, if not impossible, to maintain the received system. It seems, therefore, of importance that these two statements should be examined; and I purpose to examine them in the present letter.

I. It is alleged by M^r. Bosanquet that Herodotus, although he states the reigns of the four Median kings to be such that their sum would be 150 years, says that they reigned in all only 128 years. From the accession of Dejoces to the conquest of Astyages by Cyrus was, according to Herodotus, *as cited by M^r. Bosanquet*, only 128 years. From this he infers that the reign of Astyages which Herodotus states to have been 35 years, was in reality only 13 years; and that for 22 years after his conquest he was allowed to retain the title of king. According to this view, Cyaxares, the father of Astyages, must have died in 573 B.C., and not in 595 B.C., as Herodotus has generally been understood as stating. On the general merits of the system of which this is a branch, I do not, as I said, intend to speak. I confine myself to the one point of denying that Herodotus said what M^r. Bosanquet attributes to him. *Beloe's* Herodotus certainly contains such a statement, but not the original Greek.

The passage occurs in the First Book, chap. 130. Beloe's translation is this; and, by the way, the first sentence strikes at the root of M^r. Bosanquet's argument from the passage, as it clearly makes the 35 years attributed to Astyages terminate at his deposition, and not at his death: 'After a reign of 35 years, Astyages was thus deposed. To his cruelty of temper the Medes owed the loss of their power, after possessing for the space of 128 years all that part of Asia which lies beyond the Halys, *deducting from this period* the short interval of the Scythian dominion.'[1] The words which I have italicized appear to me a mistranslation. I would substitute '*exclusive of*'. The Greek is παρεξ ἢ ὅσον οἱ Σκύθαι ἦρχον. The force of παρεξ is well known to be 'outside of, *extra*'. We have the derivative παρεκτος in 2 Cor. xi 28: 'Besides these things which are *without*', i.e. 'besides *extra* work', as contrasted with *daily* occupation. So in Acts xxvi 29, 'exclusive of these bonds'. I believe that no instance can be produced in which παρεξ is so used as to justify Beloe's translation of this passage.

How then, it may be asked, is the difference in the statements of Herodotus to be accounted for? According to my view, the interval from the accession of Dejoces to the dethronement of Astyages is the sum of the four reigns 53 (chap. 102) + 22 (chap. 102) + 40 (chap. 106) + 35 (chap. 130) = 150. He gives it the sum of 128 (chap. 130) + 28 (chap. 106) which would be 156. It appears to me evident that this last number is a mistake for 22, the true duration of the Scythian dominion. Whether Astyages reigned 35 years only, as stated by Herodotus, or 38, as stated by later writers, who were possibly better informed, I will not now discuss. I content myself with saying that the *latest possible* date of the death of Cyaxares is 595 B.C.

II. M^r. Bosanquet relies on an alleged astronomical proof of his theory. He says that M^r. Airy has *proved* that the eclipse which terminated the Lydian war occurred on the 28th. May, 585 B.C.; and that as Cyaxares was king of Media at the time of that eclipse, the received chronology, by which he was at least ten years dead, must be false. If M^r. Airy had really *proved* what M^r. Bosanquet says that he has proved, it would, I grant, be impossible to maintain the received chronology; but I deny the fact. M^r. Airy has *asserted* it, but he has given no *proof* of his assertion. I say this with the full knowledge of the paper on this eclipse that M^r. Airy has published in the *Philosophical Transactions*, the most inconclusive paper on a mathematical subject which I have ever perused. M^r. Airy sets out with an hypothesis, which is not only arbitrary, but in the highest degree improbable, not to say absurd; and on this hypothesis his entire argument rests. Deny the hypothesis, and the whole of what he says in support of his position comes to nought.

To make this plain, it will be necessary that I should introduce some mathematical statements; but I will take care that they be of the most simple kind.

In order to find the place of the moon in latitude and longitude at any particular time, it is necessary to know its mean longitude, and also four other elements. By variously combining these, we obtain the arguments of a number of tables, from which we take out equations or corrections, to be applied to the mean longitude, and also the latitude and its equations. The tables are so constructed that this process, though laborious, presents no real difficulty; and, the five elements being known, the moon's place as derived from them may be confidently relied on.

Before the series of tables by which the longitude, latitude, motion, &c., are computed from the five elements, there are tables given by which the five elements are to be found for any instant. That these tables give the elements with sufficient accuracy for any time within the last hundred years is not to be doubted; but whether or not they can be depended on for giving the elements at remote periods is another matter. The case stands thus. Let us consider the mean elongation of the moon, that is, the excess of her mean longitude over that of the sun. It is certain from physical astronomy that this may be expressed as follows: t denoting the time.

$$A + Bt + g \left(\tfrac{1}{2}xt^2 + \tfrac{1}{3}yt^3 + \&c. \right)$$

A and B are known from observation with very great accuracy; and the sum of the terms A x Bt can be computed from the tables with great facility, whether t be positive or negative; that is, whether the time for which the moon's place is required be before or after the epoch of the tables. Another table is given, from which the supplementary terms involving the squares and cubes of the time may be taken. As the coefficients of these are extremely small, it has been thought best to consider t in these terms as having for its unit a century. The first of these small terms for one century is considerably less than half what the term Bt is for one minute. From this it will be evident that an error might exist in the estimated value of the coefficient ($\frac{1}{2}gx$), which would be very sensible when multiplied by the square of a large number of centuries, but which might escape notice when multiplied by the square of a fraction of a century. Four seconds in the value of $\frac{1}{2}gx$ would be one second only for fifty years, but would amount to forty minutes for 2450 years. It is not then by *recent* observations that the value of this coefficient can be known. We must depend on theory, aided by *remote* observations.

Now the coefficient contains as factors two quantities, g and x. The value of g depends on the lunar theory. Laplace considered only one term of it, and made it about $\frac{1}{119}$. M. Plana pursued the investigation of its value much further, making it to consist of several terms; the first and principal of which he made $\frac{1}{127}$. M^r. Adams has published a paper in which he shews that his predecessors had overlooked quantities which ought to have been taken into account, and which would diminish the term, accordingly he made to be about $\frac{1}{153}$. M^r. Adams has not yet calculated the other terms of g; nor has he calculated the value of h and k, which will presently be mentioned. He has not yet, at any rate, published the result of his calculation.

So much for one of the factors in the co-efficient of the square of the time. As to the other there is at least equal uncertainty. The quantity which I have called x is in fact the sum of five quantities, originating in the action of the five planets on the earth's orbit, and of course proportional to the masses of these planets. But the masses of Venus and Mars are only known approximately, and that of Mercury can only be conjectured. The last named planet has been assumed to have a very great density, more than two and a half times that of the Earth, while the density of Venus is supposed to be something less than that of the Earth. It seems probable that the density of Venus should be a little increased, and that of Mercury greatly diminished. Both these changes would diminish the value of x. The mean anomaly of the moon and the argument of her latitude are expressed by time similar to that above given. We may express the former by $C + Dt + h \left(\frac{1}{2}xt^2 + \frac{1}{3}yt^3 \right)$ and the latter by $E + Ft + k \left(\frac{1}{2}xt^2 + \frac{1}{3}yt^3 \right)$.

The quantities C, D, E and F are discoverable from modern observations, and may be considered as accurately known, or at least very nearly so. M^r. Airy thinks that the Greenwich observations from which F was computed are not to be altogether depended on, and that a small error may exist in it. The quantities h and k depend on the lunat theory, consist of a number of terms, and have been computed by M. Plana. M. Damoiseau, whose lunar tables are in use, has also computed these quantities, and I believe he makes k considerably greater than M. Plana does. He differs from him less as to g and h.

I have thought it necessary to give this explanation before I proceeded to shew the fallacy – I may say, the sophistry – of M^r. Airy's paper. I will begin with stating the points in which M^r. Airy agrees with his opponents. On the 15th. of

August 310 B.C. (civil reckoning) in the morning, the fleet of Agattocles was passed by the Moon's shadow, causing a total eclipse of the Sun which made the stars visible. The fleet must have been in or about E. Long. 15°.30´ and N. Lat. (geometric) about 37°.45´. According to Damoiseau's lunar tables, the Moon's shadow would not have passed over this spot. A correction must, therefore, be applied to the Moon's place as given by the tables; and a similar correction, but of increased magnitude, must be applied to the moon's tabular place in any ancient eclipse, such as that of the Lydian war.

So far all persons would agree; but now comes the difference. The Moon's place as given by the tables may be corrected, so as to make the shadow pass over the required spot, *in an unlimited number of different ways*; and M^r. Airy has selected that particular way, though *a priori* the most improbable of all, which would suit the eclipse of 28th. May, 585; an eclipse which, in defiance of chronology, he was *predetermined* to make the eclipse of the Lydian war. The Moon's shadow might be made to pass over the fleet by giving a certain increase to the Moon's latitude, leaving her longitude unchanged; by diminishing her longitude to a certain extent, leaving her latitude unchanged; by diminishing her longitude to less extent than this and increasing her latitude; and again by diminishing her longitude to a greater extent, and diminishing her latitude also. The first of these corrections, which M^r. Airy has adopted, supposes that the secular equations (that is, the terms expressed by small letters in the above values, which involve the squares and higher power of the time) are laid down with perfect accuracy in Damoiseau's tables, and must not be called in question, but that the quantity F, on which the latitude depends, may be altered. Accordingly, he assumes an error in this quantity, which is quite inconsistent with modern observations. Surely, it is infinitely more probable to suppose that the value of F, as deduced from modern observations, is correct, and that the error exists in the secular equations, as to the amount of which there is every reason to expect material error; seeing that, so far as they depend on calculation at all, the best mathematicians of the day are disagreed about them, and that there are data used in the calculation which are admitted to be uncertain and even conjectural.

Supposing, however, that we corrected the Moon's place in the eclipse of Agattocles by means of the *secular equations*, which would require that these equations should be diminished according to M^r. Adams's view (and also, perhaps, by correcting the masses of Mercury and Venus) we should have to diminish these equations also in the eclipses of 585 B.C., 603 B.C., and 610 B.C. The dimunition would be greater than in the eclipse of Agattocles nearly in the ratio of 64 to 49. The effect of this would be to destroy all pretensions that the first named of these there would have to be the eclipse of the Lydian war. If we admitted this correction, it would not be visible at all in any part of Asia Minor where we can suppose that the battle could have been fought. *Therefore*, I may venture to say, M^r. Airy will not admit this most probable supposition. On the contrary, the effect of the correction of the secular equations, such as would suit the eclipse of Agattocles, might bring the shadow of the Moon over a possible field of battle in either 603 or 610. If the value of $g, h,$ and k were *settled* by the consent of the astronomers, we might be able to tell to which of these two the eclipse of Agattocles would point us. At present we cannot do so. I myself, however, entertain no doubt that the eclipse of 18th. May 603 was that which terminated the Lydian war. It occurred in the morning, when the two armies would be commencing their battle, and its occurrence at this time might naturally be considered as an intimation from heaven that they ought not to fight. Besides, this date best suits the chronology, as has been shewn by your correspondent J. F. in your last number.[2]

I am, &c.,
Edw. Hincks

P.S. According to the estimate given by M^r. Adams in his paper, the Moon's mean longitude at the time of the eclipse of Agattocles would be, in virtue of the correction made by him, about twelve minutes and a quarter less than the tables make it. This correction *alone* would throw the Moon's shadow about six degrees and a half of longitude to the west of its course, as given by the tables; and would thus bring its northern part over the fleet. It seems to me probable that the other corrections suggested would greatly increase this effect, bringing the central or southern part of the shadow over the fleet.

Source: Journal of Sacred Literature 4/8 (Jan. 1857), pp. 462–6

1   *Herodotus Translated from the Greek, with Notes,* tr. W. Beloe, vol. 1 (3rd edn; London, 1812), p. 178. Hincks would quickly spot the mistranslations by Beloe. On the inferior quality of the latter's work, see the remarks of Tom Winnifrith in S. Gillespie and D. Hopkins (eds), *The Oxford History of Translation in English,* vol. 3, *1660–1790* (Oxford, 2005), p. 277.
2   A Letter to the Editor [signed J. F.], *Journal of Sacred Literature* 4/7 (Oct. 1856), pp. 177–8.

～⁊ⅅ⋐ゝ

FROM HENRY FOX TALBOT
DECEMBER 1856

Lacock Abbey
Chippenham
December 1856

Dear Sir,

I beg your acceptance of a copy of the 1ˢᵗ number of my new publication on the cuneiform inscriptions.[1] I could have wished to have added a commentary explanatory of the text and supporting my translation by the necessary arguments, but finding that this would delay indefinitely the publication, I thought I should be doing more real service to the science by publishing a little at a time. I have been perusing the journals of the German scientific societies to see what they have been doing in the way of cuneiform discovery during the last 2 or 3 years. It is surprising how *very little* I can find nor is that little, very instructive. I have read Brandis's work, but though it appears the production of a man of sound judgment in many things, it is evidently premature.[2] If he had read and considered what has been published in England he would have seen that many things are quite plain and clear, which to him appear still enveloped in the greatest obscurity.

In the department of hieroglyphic literature on the contrary, there are some writers of brilliant talent such as Brugsch and Lepsius and de Rougé in France. Before publishing in the Journal of S. Lit^re. on the Assyrian verb, did you publish on the *pronouns* also?[3] If so, can you favor me with a reference to the volume where published, as I have not seen it.

Believe me
Yours very truly
H. F. Talbot

MS: GIO/H 506

1   H. F. Talbot, 'On the Assyrian Inscriptions [No. 1]', *Journal of Sacred Literature* 2/4 (Jan. 1856), pp. 414–25.
2   J. Brandis, *Über den historischen Gewinn aus der Entzifferung der assyrischen Inschriften. Nebst einer Übersicht über die Grundzüge des assyrisch-babylonischen Keilinschriftsystems* (Berlin, 1856). Previously he published *Rerum Assyriarum tempora emendata, commentatio* (Bonn, 1853).
3   Talbot may have heard about Hincks's article 'On the Personal Pronouns of the Assyrian and Other Languages, especially Hebrew', *Transactions of the Royal Irish Academy* 23 (1854), part 2, pp. 3–9.

~~~

TO HENRY FOX TALBOT
20 DECEMBER 1856

Killyleagh, C° Down
20th. Decr. 1856

Dear Sir,

I this morning received your letter & pamphlet for which I feel much obliged to you.[1] I send you a copy of my paper on the Pronouns.[2] I should wish too that you read a paper of mine on the Annals of Sargon and Sennacherib in the Journal of Sacred Literature 2nd. Ser^s. vol. 6th. a note in which (p. 408 or 16 of the separate paper) anticipates some remarks of yours in pages 27 & 31.[3] As I happen to have a separate copy of this paper, & as you may not have the last volumes of the Journal, I send it to you with the other. This paper was dated in January, tho' not published till July, 1854.

You seem not to have heard of my translations of the Annals of Assur-akh-bal (if that be his name) & of Sennacherib from Bellino's cylinder. I prepared these with great labour under an agreement with the Trustees of the British Museum; and I certainly understood when I parted with the copyright of them to them, that they would *publish* them. This was not indeed stipulated for; but had I not believed that it would be done, I would never have made the agreement with them that I did. I allowed them to dictate to me on what I was to be occupied; I was kept from studying the all-important *tablets*, on the ground that the Trustees divined that 'I should *confine myself* to the Inscriptions of Sennacherib and those of the North West Palace'; though the tablets were of the nature of a *lexicon* & would have enabled me to improve my translations. And when in the summer of 1854 I sent it, as carefully prepared as I could (and certainly *beyond all comparison* superior to any thing that had previously come out), I found that it was to be kept locked up in the private repositories of the Secretary or Principal Librarian; not being even put among the 'Additional Manuscripts', where it would be accessible to the public. By whom it was seen I know not; but I know that D^r. Oppert published much *as his own*, which was contained in my manuscripts; and I certainly think (supposing even that he did not see them) that my delivery of the MSS to a public body like the Trustees of the Museum was a *publication*, so as to entitle me to claim priority.[4]

I don't suppose that it was with a view to D^r. Oppert's benefit that my MSS were *burked*;[5] but I am very sure that there is a strong feeling in some influential quarters that *I* must be kept out of the field of discovery, whoever may occupy it. I am now laid on the shelf; and I never expect to have again the means of pursuing my discoveries.

You have, I think, hit off the meaning in one or two places where I missed it; in others I think you have not been as successful as I think I was; but in more numerous instances we agree.

On the first page you are puzzled with *ṣumbi, chumbi, ṣammᵉbê*; or as the Hebrews wrote it *ṣabbê, chabbê* 'lecticae' – Gesenius p. 1147. This I translated 'waggons'.[6]

D[r]. Oppert is the only continental scholar, as far as I am aware, who has made any progress in cuneatics. The three you mention seem to have gone far ahead of any in this country as to hieroglyphics. Birch of late seems to have been asleep. Perhaps, however, he has something great in preparation.

I remain
Dear Sir
Yours very truly
Edw. Hincks

MS: BL Talbot 31426

1   H. F. Talbot, *Assyrian Texts Translated, No. 1* (London, 1856).
2   'On the Personal Pronouns of the Assyrian and Other Languages, especially Hebrew', *Transactions of the Royal Irish Academy* 23 (1854), part 2, pp. 3–9.
3   'Chronology of the Reigns of Sargon and Sennacherib', *Journal of Sacred Literature* 6/12 (July 1854), pp. 393–410.
4   There is no evidence that Oppert was given access to Hincks's translations and notes.
5   The verb 'to burke' means 'to smother, suppress quietly' (*OED*).
6   Hincks was right to compare Akk. *ṣumbu, ṣubbu*, 'waggon' with Heb. *ṣāb*, plur. *ṣabbîm*, 'wagon'. The Hebrew word is a loanword from Akkadian, which in turn *may* be a loan from Elamite. See Mankowski, *Akkadian Loanwords in Biblical Hebrew*, pp. 130–1.

FROM WILLIAM WRIGHT[1]
20 DECEMBER 1856

19 Trinity College
Dublin
20th. Decr. 1856

Dear Sir,

I have to acknowledge the receipt of your communication, which I shall forward at once to D[r]. Böttcher of Dresden.[2]

I am very glad that you have responded to the advertisement, since, I am sorry to say, none other of our distinguished orientalists (Lane, Cureton, Rawlinson,

&c.³) has paid any attention to it. I really think that such a work as Dʳ. B's is deserving of all encouragement, as it is very desirable to know *all* that different scholars have written, and articles in Reviews often escape the notice even of the most painstaking student.

> I remain,
> Dear Sir,
> Yours very truly
> Wm Wright

MS: GIO/H 553

1   William Wright (1830–89), distinguished Semitist, was Professor of Arabic at Trinity College, Dublin from 1856 to 1861 and at Cambridge from 1870 until his death. See *ODNB*.

2   Friedrich Böttcher was born on 25 October 1801 in Dresden. He was educated at the Landesschule Meißen and Leipzig University. Afterwards he was a teacher in the gymnasium in Dresden from 1824. His best known work is *Ausführliches Lehrbuch der hebräischen Sprache*, ed. F. Mühlhau, 2 vols (Leipzig, 1866–8).

3   Edward William Lane (1801–76), Arabic scholar; William Cureton (1808–64), Syriac scholar.

FROM HENRY FOX TALBOT
24 DECEMBER 1856

Lacock Abbey
24 Dec/56

Dear Sir,

I am particularly obliged to you for copies of your papers on the Personal Pronouns, and on the Annals of Sargon and Sennacherib, neither of which I had previously seen.¹

I was not aware that you had prepared for the trustees of the British Museum a translation of Bellino's cylinder. But Mʳ. Vaux informed me one day, that you had prepared a translation of the annals of Ashurakhbal, or of some considerable part of them.² I said to him, that I could not ask to look at your MS. without your express sanction and permission, to which he replied 'Of course not', and thus, as you were in Ireland, the MS. remained unseen by me. But as nothing can exceed the kindness and courtesy of Mʳ. Vaux, I feel sure that if it depended on him to order the publication of your most valuable labours now in MS. in the hands of the Trustees, he would not hesitate to do so immediately, or to take any other step calculated to place your rights of priority in a just light.³ They are incontestable. The delivery of your MS. to a public body like the trustees of the B. Musᵐ. was a publication. I am extremely glad to learn from you that there is a *general* agreement between your version and mine as to the meaning of Bellino & the 'Annals' (you do not say whether you translated Esarhaddon likewise) and it appears to me that it may be possible to turn this circumstance to a very useful account, and thus to educe good out of evil, which is always a great satisfaction. The object at present clearly is to carry a conviction of the truth of these discoveries to the minds of the numerous candid and enlightened scholars of this country who are at present hesitating whether to admit them or not (many conceiving that *something* has been ascertained with certainty but that no reliance can be placed on the translation of whole texts). Now, nothing is so convincing an argument to the minds of most people as the agreement between two independent translators. If therefore you think fit to memorialize the Trustees to print your translation, I am sure I should be most happy to second it by pointing out the great advantage which must accrue to science from the publication of independent translations, to say nothing of the justice of the thing which must be evident. You might easily add in notes, those improvements which your increased knowledge at the present day enables you to effect in your translation. As I know several of the Trustees personally I should have great pleasure in aiding your request.

I thank you for pointing out the probable meaning of *ṣumbi* as 'waggons'.[4] In p. 17 of your pamphlet you have clearly anticipated my remarks on the identity of Sirki and Kerkesiah.[5] If my work is continued, I will point out this in N° 2.[6] In the meanwhile I look upon it as a most satisfactory agreement, with respect to a fundamental point of the geography. I see also that you have identified the city of Khazazi, which had escaped me.

I remain
Dear Sir
Yours very truly
H. F. Talbot

MS: GIO/H 507

1   'On the Personal Pronouns of the Assyrian and Other Languages, especially Hebrew', *Transactions of the Royal Irish Academy* 23 (1854), part 2, pp. 3–9; 'Chronology of the Reigns of Sargon and Sennacherib', *Journal of Sacred Literature* 6/12 (July 1854), pp. 393–410.

2   Hincks's British Museum translations.

3   Talbot was clearly not aware of the negative aspects of Vaux's character. See Vaux's letter of 1 November 1854 to Rawlinson in Larsen, *The Conquest of Assyria*, p. 336.

4   See Hincks's letter of 20 December 1856 to Talbot.

5   'Chronology of the Reigns of Sargon and Sennacherib', *Journal of Sacred Literature* 6/12 (July 1854), pp. 408–9, n. i = pp. 16–17, n. i of the offprint which Hincks had sent to Talbot.

6   Talbot published *Assyrian Texts Translated, No. 1* (London, 1856), but the work was not continued.

TO HENRY FOX TALBOT
27 DECEMBER 1856

Killyleagh, Co. Down
27th. Decr. 1856

Dear Sir,

I duly received your letter of the 24th. I gave in to the Trustees a translation of that part of Bellino's cylinder which contains the annals of Sennacherib's two first years. I *have* a translation (with gaps here & there) of the part relating to the building, & also of the cylinder of Esarhaddon; but *these* I have not in any manner published. I am rather surprised at M^r. Vaux's remark, which you mention, that 'of course, he could not shew the translation without my consent'. I do not consider myself to have any control over it, or any rights beyond what any of the public may have. I think the translations & accompanying remarks *ought to be* where *any one* may see them; & this is what I desire respecting them. As to their publication, I am not prepared to undertake it on my own account; but I think that, for the reasons you mention, it would be well that portions of the translations should be published, (& perhaps in parallel columns) along with other translations made independently of them, either by you or by Sir H. Rawlinson. The latter, I believe, has a translation in preparation of all that has been lithographed by the Trustees of the Museum; i.e. the annals from the N.W. Palace, the Octagon of Tiglath Pileser I & the Hexagon of Col. Taylor. The lithographs are finished, or the next thing to it; but I understand they are held back from the public till the translation is ready to accompany them. I think the Trustees would have better served the public interests if they had allowed copies of the lithographs to be circulated among those who were likely to be translators; so that the translations might appear together; or if they had encouraged a *joint translation*, on which all would be agreed. It is quite evident that two or three persons, acting in concert, and comparing their views, would produce a better translation than any of them would do singly.[1] Sir H. Rawlinson has abandoned (I believe, I may say, in consequence of *my* criticisms) two fanciful views that he brought forward in 1855, both at the Royal Institution & at the Glasgow meeting of the British Association; viz. 1st. the Birs Nimrûd being the tower of Babel, and its original building being 42 cycles before Nebuchadnezzar's time (2520 years, as supposed) & 2dly. there being an allusion to Nebuchadnezzar's madness in the E. I. House inscription col. 7. Is it not possible that he may bring forward something equally absurd in his translation of the documents that are lithographed? At any rate he may make many minor mistakes; but what then? The *public* will fare worse; but the *individual* will have more credit. Besides, it would be considered an ungracious[2] proceeding to criticise

357

a translation which is on the whole good; & of which all that can be said is that it is not so good as it might have been made. It is not my present intention to meddle with it. I may have a laugh in private over any mistakes that I notice; but I will not enlighten the public in respect to them. The public has given its voice in favour of Sir Henry's exclusiveness; and if it be deceived in consequence thereof, let it be so!

I remain
Dear Sir,
Yours very truly
Edw. Hincks

P.S. If any of my translation were to be published, I sh^d like to have an opportunity of correcting it (by footnotes) where I now see that it requires correction – letting the translation as given in remain in the text. But it would depend on the good feeling towards me of the person who published it, whether this should be allowed me.

MS: BL Talbot 31428

1   See the correspondence for 1857 in vol. 3 for an account of how Hincks, Oppert, Rawlinson and Talbot prepared translations of parts of the inscription of Tiglath-Pileser I.
2   Hincks wrote 'invidious' in his copy of the letter (GIO/H 508).

BIBLIOGRAPHY

Adams, J. C., 'On the Secular Variation of the Moon's Mean Motion', *Philosophical Transactions of the Royal Society of London* 143 (1853), pp. 397–406.

Adkins, Lesley, *Empires of the Plain: Henry Rawlinson and the Lost Languages of Babylon* (London, 2004).

Ainsworth, W. F., *Researches in Assyria, Babylonia and Chaldaea, Forming Part of the Labour of the Euphrates Expedition* (London, 1838).

——, *Travels and Researches in Asia Minor, Mesopotamia, Chaldea, and Armenia*, 2 vols (London, 1842).

——, *Travels in the Track of Ten Thousand Greeks* (London, 1844).

——, 'Layard's Assyrian Researches': a review article on A. H. Layard, *Nineveh and Its Remains*, 2 vols (London, 1849), in *New Monthly Magazine* 85 (1849), pp. 240–51.

——, 'The City of Nimrod', *New Monthly Magazine* 88 (1850), pp. 326–31.

——, 'Remarks on the Topography of Nineveh', *Original Papers Read before the Syro-Egyptian Society of London*, vol. 1/2 (London, 1850), pp. 15–26.

Airy, G. B., 'On the Eclipses of Agathocles, Thales, and Xerxes', *Philosophical Transactions of the Royal Society of London* 143 (1853), pp. 179–200.

Akenson, D. H., *The Irish Education Experiment: the National System of Education in the Nineteenth Century* (Toronto and London, 1970).

Ando, C., 'Tacitus, Annales VI: Beginning and End', *American Journal of Philology* 118 (1997), pp. 285–303.

[Anon.] 'Progress of the African Mission, Consisting of Messrs Richardson, Barth, Overweg to Central Africa', *Journal of the Royal Geographical Society of London* 21 (1851), pp. 130–221.

The Assyrian Dictionary (Chicago, 1956–).

Avesta: die heiligen Schriften der Parsen, tr. F. Spiegel, 3 vols (Leipzig, 1852–63).

Avigad, N. and Sass, B., *Corpus of West Semitic Stamp Seals* (Jerusalem, 1997).

Baily, F., 'On the Solar Eclipse which is said to have been predicted by Thales', *Philosophical Transactions of the Royal Society of London* 101 (1811), pp. 220–41.

Baker, Heather D. (ed.), *The Prosopography of the Neo-Assyrian Empire*, vol. 2, part 1, H–K (Helsinki, 1998).

Bardelli, G., *Biografía del professore Ippolito Rosellini* (Florence, 1843).

——, (ed.), *Daniel copto-memphitice* (Pisa, 1849).

Barnett, R. D., *Sculptures from the North Palace of Assurbanipal at Nineveh (668–627 B. C.)* (London, 1976).

Barr, J., *Comparative Philology and the Text of the Old Testament* (Oxford, 1968).

Barth, H., *Travels and Discoveries in North and Central Africa*, 5 vols (London, 1857–8).

Bartlett, W. H., *Forty Days in the Desert on the Track of the Israelites or, A Journey from Cairo, by Wady Feiran, to Mount Sinai and Petra* (London, 1848).

Beer, E. F. F., *Inscriptiones veteres litteris et lingua hucusque incognitis ad montem Sinai magno numero servatae* (Studia Asiatica 3; Leipzig, 1840).

Bezold, C., 'Julius Oppert', *Zeitschrift für Assyriologie* 19 (1905), pp. 169–73.

Birch, S., 'Letter to Mr. George R. Gliddon, on Various Archaeological Criteria for Determining the Relative Epochs of Mummies', dated 23 December 1848, in G. R. Gliddon, *Otia Aegyptiaca: Discourses on Egyptian Archaeology and Hieroglyphical Literature* (London, 1849), pp. 78–87.

——, 'Observations on an Egyptian Calendar, of the Reign of Philip Ardaeus', *Archaeological Journal* 7 (1850), pp. 111–20.

——, 'Upon an Historical Tablet of Rameses II, 19th Dynasty, Relating to the Gold Mines of Aethiopia', *Archaeologia* 34 (1852), pp. 357–91.

Blackburn, J., *Nineveh: its Rise and Ruin; as Illustrated by Ancient Scriptures and Modern Discoveries* (rev. and enl. edn; London, 1852).

Boase, F., *Modern English Biography*, 6 vols (Truro, 1892–1921).

Bochart, S., *Geographiae sacrae pars altera Chanaan seu de coloniis et sermone Phoenicum* (Cadomi [Caen], 1646).

Bonomi, J., *Nineveh and Its Palaces: The Discoveries of Botta and Layard Applied to the Elucidation of Holy Writ* (London, 1852).

Borger, R., *Babylonisch-Assyrische Lesestücke*, 2 vols (3rd edn; Rome, 2006).

Bosanquet, J. W., *Chronology of the Times of Daniel, Ezra, and Nehemiah* (London, 1848).

——, 'The Eclipse of Thales Identified with That of the 28th May, B.C. 585, by means of the Eclipse of Agathocles', *The Athenaeum*, no. 1293 (7 Aug. 1852), pp. 846–7.

——, 'Chronology of the Reigns of Tiglath Pileser, Sargon, Shalmanezer and Sennacherib, in Connexion with the Phenomenon Seen on the Dial of Ahaz', *Journal of the Royal Asiatic Society* 15 (1855), pp. 277–96.

——, 'The Dial of Ahaz, and Scriptural Chronology': A Letter to the Editor, *Journal of Sacred Literature* 1/2 (July 1855), pp. 407–13.

——, 'Who was Darius the Son of Ahasuerus, of the Seed of the Medes? Daniel IX, 1': A Letter to the Editor, *Journal of Sacred Literature* 2/4 (Jan. 1856), pp. 393–403.

Böttcher, F., *Ausführliches Lehrbuch der hebräischen Sprache*, ed. F. Mühlhau, 2 vols (Leipzig, 1866–8).

Bowring, J., *The Decimal System in Numbers, Coins, and Accounts: especially with Reference to the Decimalisation of the Currency and Accountancy of the United Kingdom* (London, 1854).

Brandis, J., *Rerum Assyriarum tempora emendata, commentatio* (Bonn, 1853).

——, *Über den historischen Gewinn aus der Entzifferung der assyrischen Inschriften. Nebst einer Übersicht über die Grundzüge des assyrisch-biblischen Keilinschriftsystems* (Berlin, 1856).

[Brewster, D.], Review article on A. H. Layard, *Discoveries in the Ruins of Nineveh and Babylon* (London, 1853), *North British Review* 19/37 (May 1853), pp. 255–96.

Brinkman, J. A., 'Merodach-Baladan II', *Studies Presented to A. Leo Oppenheim* (Chicago, 1964), pp. 6–53.

——, 'The Akkadian Words for "Ionia" and "Ionian"', in *Daidalikon: Studies in Honor of Raymond V. Schroeder, S.J.* (Wauconda, Ill., 1989), pp. 53–71.

Bruce, J., *Travels to Discover the Source of the Nile in the Years 1768–1773*, 8 vols (2nd edn; Edinburgh, 1804–5; 3rd edn; Edinburgh, 1813).

Brugsch, H., 'Die demotische Schrift der alten Ägypter und ihre Monumente', *Zeitschrift der deutschen morgenländischen Gesellschaft* 3 (1849), pp. 262–72.

——(ed.), *Saï an Sinsin, sive, Liber metempsychosis veterum Aegyptiorum e duabus papyris funebribus hieraticis signis exaratis* (Berlin, 1851).

——, 'Ägyptische Studien', *Zeitschrift der deutschen morgenländischen Gesellschaft* 9 (1855), pp. 193–213, 492–517.

——, *Grammaire démotique* (Berlin, 1855).

——, *Monumens de l'Égypte, décrits, commentés et reproduits pendant le séjour qu'il a fait dans ce pays en 1853 et 1854* (Berlin, 1857).

Buffon, G. L. L., *Histoire naturelle, générale, et particulière, avec la description du cabinet du Roi*, vol. 11 (Paris, 1754).

Bunsen, C. C. J., *Ägyptens Stelle in der Weltgeschichte*, 5 vols (Hamburg, 1845–57).

Burtchaell, G. D. and Sadleir, T. U. (eds), *Alumni Dublinenses* (Dublin, 1935).

Buxtorf, J., *Lexicon hebraicum et chaldaicum* (London, 1646).

Calmet, A., *An Historical, Critical, Geographical, Chronological, and Etymological Dictionary of the Hebrew Bible in Three Volumes*, vol. 1 (London, 1732).

——, *Calmet's Dictionary of the Bible by the Late Mr. Charles Taylor with the Fragments Incorporated* (11th edn; London, 1849).

Caplice, R., *Introduction to Akkadian* (3rd rev. edn; Rome, 1988).

Carpenter, W. B., 'On the Mutual Relations of the Vital and Physical Forces', *Philosophical Transactions of the Royal Society of London* 140 (1850), pp. 727–57.

Cathcart, K. J. (ed.), *The Edward Hincks Bicentenary Lectures* (Dublin, 1994).

——, 'The Age of Decipherment: the Old Testament and the Ancient Near East in the Nineteenth Century', in J. A. Emerton (ed.), *Congress Volume: Cambridge 1995* (Leiden, 1997), pp. 81–95.

——, 'Some Nineteenth- and Twentieth-Century Views on Comparative Semitic Lexicography', in M. Kropp and A. Wagner (eds), *'Schnittpunkt' Ugarit* (Frankfurt, 1999), pp. 1–8.

Champollion, J. F., *Dictionnaire égyptien en écriture hiéroglyphique* (Paris, 1841).

Champollion-Figeac, J. J., *Notice sur les manuscrits autographes de Champollion le jeune, perdus en l'année 1832, et retrouvés en 1840* (Paris, 1842).

——, 'De la table manuelle des rois et des dynasties d'Égypte ou papyrus royal de Turin, de ses fragments originaux de ses copies et de ses interprétations', *Revue Archéologique* 7/2 (1851), pp. 397–407, 461–72, 589–99, 653–5.

Chesney, F. R., *The Expedition for the Survey of the Rivers Euphrates and Tigris, Carried on by Order of the British Government during the Years 1835, 1836, and 1837: Preceded by Geographical and Historical Notices*, 2 vols (London, 1850).

——, *Narrative of the Euphrates Expedition: Carried on by Order of the British Government during the Years 1835, 1836, and 1837* (London, 1868).

[Chesney, Louisa], *The Life of the Late General F. R. Chesney, by his wife and daughter*, ed. S. Lane-Poole (2nd edn; London, 1893).

Clark, Mary, Desmond, Yvonne and Hardiman, Nodlaig P. (eds), *Sir John T. Gilbert 1829–1898: Historian, Archivist and Librarian. Papers and Letters Delivered during the Centenary Year, 1998* (Dublin, 1999).

Cogan, M., and Tadmor, H., *II Kings* (The Anchor Bible 11; Garden City, N.Y., 1988), pp. 259–61.

Cooley, W. D., *Inner Africa Laid Open, in an Attempt to Trace the Chief Lines of Communication across that Continent South of the Equator* (London, 1852).

Corbaux, F., 'The Rephaim, and their Connexion with Egyptian History', *Journal of Sacred Literature* 1/1 (Oct. 1851), pp. 151–72; 1/2 (Jan. 1852), pp. 363–94; 2/3 (Apr. 1852), pp. 55–91 + 2 plates; 2/4 (July 1852), pp. 303–40; 3/5 (Oct. 1852), pp. 87–116; 3/6 (Jan. 1853), pp. 279–307.

——, 'Historical Origin of the Passover', *Journal of Sacred Literature* 7/14 (Jan. 1855), pp. 281–97.

Cull, R., 'On a *t* Conjugation, Such as Exists in Assyrian, Shown to Be a Character of Early Shemitic Speech, by its Vestiges Found in the Hebrew, Phoenician, Aramaic, and Arabic Languages', *Transactions of the Society of Biblical Archaeology* 2 (1873), pp. 83–109.

Daniels, P. T., 'Edward Hincks's Decipherment of Mesopotamian Cuneiform', in K. J. Cathcart (ed.), *The Edward Hincks Bicentenary Lectures* (Dublin, 1994), pp. 30–57.

Daniels, P. T. and Bright, W. (eds), *The World's Writing Systems* (New York and Oxford, 1996).

Davidson, E. F., *Edward Hincks: A Selection from his Correspondence with a Memoir* (London, 1933).

Davidson, S., *A Treatise on Biblical Criticism*, vol. 1, *The Old Testament* (Edinburgh, 1852).

——, *The Autobiography and Diary of Samuel Davidson*, ed. Anne J. Davidson (Edinburgh, 1899).

Davies, G. I., 'Tell ed-Duweir = Ancient Lachish: a Response to G. W. Ahlström', *Palestine Exploration Quarterly* 114 (1982), pp. 25–8.

——, 'Tell ed-Duweir: not Libnah but Lachish', *Palestine Exploration Quarterly* 117 (1985), pp. 92–6.

Dawson, W. R. and Uphill, E. P., *Who Was Who in Egyptology*, ed. M. L. Bierbrier (3rd edn; London, 1995).

Dennis, G., *The Cities and Cemeteries of Etruria*, vol. 1 (London, 1848).

Dennis, J. N. C. M., *Einleitung in die Bücherkunde*, 2 vols (Vienna, 1795–6).

Desâtîr or Sacred Writings of the Ancient Persian Prophets, published by Mulla Firuz bin Kaus (Bombay, 1818).

Dictionary of Ulster Biography (Belfast, 1993).

Dictionnaire de biographie française (Paris, 1933–).

Eadie, J., *A Biblical Cyclopaedia, or, Dictionary of Eastern Antiquties* (3rd edn; Glasgow, 1851).

Eckhel, J., *Doctrina numorum veterum*, vol. 3 (Vienna, 1794).

Ellis, A. J., *The Essentials of Phonetics* (London, 1848).

——, *The Fonetic Nuz* (London, 1849).

England, T. R. (ed.), *A Short Memoir of an Antique Medal, Bearing on One Side the Representation of the Head of Christ, and on the Other a Curious Hebrew Inscription, lately Found at Friar's Walk, near the City of Cork, in Ireland; Containing Some Letters and Observations of Different Men of Learning on the Subject; with Divers Translations of the Characters thereon* (London, 1819).

Expédition scientifique en Mésopotamie exécutée par ordre du gouvernement de 1851 à 1854 par MM. F. Fresnel, F. Thomas et J. Oppert, 2 vols (Paris, 1859–63).

Fisher, T., *A Collection of All the Characters Simple and Compound with their Modifications, which Appear in the Inscription on a Stone Found among the Ruins of Ancient Babylon Sent, in the Year 1801, as a Present to Sir Hugh Inglis Bart. by Harford Jones Esqr., then the*

Honorable the East India Company's Resident in Bagdad; and, now Deposited in the Company's Library in Leadenhall Street, London. Collected Etched and Published June the 1st 1807 by Thos. Fisher. A single sheet 28 x 35.5 cm. 287.

Fleming, W. E. C., *Armagh Clergy 1880–2000* (Dublin, 2001).

Fletcher, J. P., *Narrative of a Two Years' Residence at Nineveh, and Travels in Mesopotamia, Assyria, and Syria*, 2 vols (2nd edn; London, 1850).

——, *Notes from Nineveh, and Travels in Mesopotamia, Assyria, and Syria* (London, 1850).

Forster, C., *Mahometism Unveiled: An Inquiry in which that Arch-Heresy, its Diffusion and Continuance, Are Examined on a New Principle, Tending to Confirm the Evidence, and Aid the Propagation of the Christian Faith* (London, 1829).

——, *The Historical Geography of Arabia; or, The Patriarchal Evidences of Revealed Religion: a Memoir, with Illustrative Maps; and an Appendix, Containing Translations, with an Alphabet and Glossary of the Himyaritic Inscriptions recently Discovered in Hadramaut*, 2 vols (London, 1844).

——, *The One Primeval Language Traced experimentally through Ancient Inscriptions, in Alphabetic Characters of Lost Powers, from the Four Continents; Including the Voice of Israel from the Rocks of Sinai* (London, 1851).

Fresnel, F., 'Pièces relatives aux inscriptions himyarites découvertes à Sana'â, à Kariba, à Mareb etc. par M. Arnaud', *Journal Asiatique* 6 (1845), pp. 169–237.

Gabelentz, H. C. von der, 'Versuch über eine alte mongolische Inschrift', *Zeitschrift für die Kunde des Morgenlandes* 2 (1839), pp. 1–21 ('Nachtrag', 3 [1840], pp. 225–5).

——, 'Mandschu-sinesische Grammatik nach dem Sô-ho-pián-lân', *Zeitschrift für die Kunde des Morgenlandes* 3 (1840), pp. 88–104.

Gerardi, Pamela, 'Epigraphs and Assyrian Palace Reliefs: The Development of the Epigraphic Text', *Journal of Cuneiform Studies* 40 (1988), pp. 1–35.

Gesenius, W., *Thesaurus philologicus criticus linguae hebraeae et chaldaeae Veteris Testamenti*, 3 vols (Leipzig, 1835–58).

——, *Scripturae linguaeque phoeniciae* (Leipzig, 1837).

——, 'Himjaritische Sprache und Schrift, und Entzifferung der letzteren', *Allgemeine Literatur-Zeitung 1841*, nos 123–6 (July 1841), cols 369–99 + 'Nachtrag'; *Ergänzungsblätter* 64 (July 1841), cols 511–12.

Gillespie, S. and Hopkins, D. (eds), *The Oxford History of Translation in English*, vol. 3, *1660–1790* (Oxford, 2005).

[Golding, B.], 'Niebuhr and Eusebius': A Letter to the Editor, *Journal of Sacred Literature* 5/10 (Jan. 1854), pp. 489–508.

——, 'Darius the Mede, and Darius Hystaspes', *Quarterly Journal of Prophecy* 7 (1855), pp. 34–43.

Grayson, A. K., *Assyrian Rulers of the Early First Millennium B.C.: I (1114–859 B.C.)* (Toronto, 1991).

Greene, J. B., *Fouilles exécutées à Thèbes dans l'année 1855* (Paris, 1855).

Grey, G. F., *Inscriptions from the Waady el Muketteb, or the Written Valley; Copied in 1820 by the Rev. G. F. Grey* (London, 1832). Also published in *Transactions of the Royal Society of Literature* 2/1 (1834), pp. 147–8 + 14 plates.

Grotefend, G. F., 'Urkunden in babylonischer Keilschrift', *Zeitschrift für die Kunde des Morgenlandes* 1 (1837), pp. 212–22; 2 (1839), pp. 177–89; 3 (1840), pp. 179–83; 4 (1842), pp. 43–57.

——, 'Bemerkungen zur Inschrift eines Thongefässes mit babylonischer Keilschrift', *Abhandlungen der königlichen Gesellschaft der Wissenschaften zu Göttingen* 4 (1850), historisch-philologische Klasse, pp. 3–18 + 2 plates.

——, 'Bemerkungen zur Inschrift eines Thongefässes mit ninivitischer Keilschrift', *Abhandlungen der königlichen Gesellschaft der Wissenschaften zu Göttingen* 4 (1850), historisch-philologische Klasse, pp. 175–93.

——, 'Keil-Inschriften aus der Gegend von Niniveh, nebst einem persischen Siegel', *Zeitschrift für die Kunde des Morgenlandes* 7 (1850), pp. 63–70 + 1 lithograph.

——, 'Die Erbauer der Paläste in Khorsabad und Kujjundshik', *Abhandlungen der königlichen Gesellschaft der Wissenschaften zu Göttingen* 4 (1850), historisch-philologische Klasse, pp. 201–6.

Gumpach, J. von (ed.), *Hülfsbuch der rechnenden Chronologie: oder, Largeteau's abgekurzte Sonnen- und Mondtafeln, zum Handgebrauch für Astronomen, Chronologen, Geschichtsforscher und Andere* (Heidelberg, 1853).

Hafid, Meḥmed, *al-Durar al-muntakhabāt al-manthūrah fī iṣlāḥ al-ghalaṭāt al-mashhūrah* (Istanbul, 1221/1806–7).

Hagedorn, A. C., '"Who Would Invite a Stranger from Abroad?": The Presence of Greeks in Palestine in Old Testament Times', in R. P. Gordon and J. C. de Moor (eds), *The Old Testament in Its World* (Leiden, 2005), pp. 68–93.

Hallock, R. T., 'Syllabary A', in *Materialien zum sumerischen Lexikon*, vol. 3 (Rome, 1955), pp. 3–45.

Harbottle, S., 'W. K. Loftus: An Archaeologist from Newcastle', *Archaeologia Aeliana*, 5th ser., 1 (1958), pp. 195–217.

Haupt, P., 'Contributions to the History of Assyriology, with Special Reference to the Works of Sir Henry Rawlinson', *Johns Hopkins University Circulars* 8/72 (Apr. 1889), pp. 57–62.

Healey, J. F., 'The Decipherment of Alphabetic Scripts', in K. J. Cathcart (ed.), *The Edward Hincks Bicentenary Lectures* (Dublin, 1994), pp. 75–93.

Heath, D. I., *The Exodus Papyri*, with a Historical and Chronological Introduction by Miss F. Corbaux (London, 1855).

Henze, M., *The Madness of King Nebuchadnezzar* (Leiden, 1999).

Herbst, J. G., *Historisch-Kritische Einleitung in die Heilige Schriften des Alten Testaments. 1. Allgemeine Einleitung* (Karlsruhe and Freiburg, 1840).

Hincks, E., 'The Enchorial Language of Egypt', *Dublin University Review* 1/3 (July 1833) (offprint, 14 pp.). Reprinted in K. J. Cathcart (ed.), *The Edward Hincks Bicentenary Lectures* (Dublin, 1994), pp. 216–27.

——, 'On the Years and Cycles Used by the Ancient Egyptians', *Transactions of the Royal Irish Academy* 18 (1838), polite literature, pp. 153–98.

——, 'On the Egyptian Stele, or Tablet', *Transactions of the Royal Irish Academy* 19 (1842), polite literature, pp. 49–71.

——, *Catalogue of the Egyptian Manuscripts in the Library of Trinity College, Dublin* (Dublin, 1843).

——, 'The Pretended Patriarchal Inscriptions of Arabia': a review of C. Forster, *The Historical Geography of Arabia: or the Patriarchal Evidences of Revealed Religion; a Memoir, with Illustrative Maps; and an Appendix, Containing Translations, with an Alphabet and Glossary, of the Himyaritic Inscriptions recently Discovered in Hadramaut*, 2 vols (London 1844), in *Dublin University Magazine* 24 (Dec. 1844), pp. 724–40.

——, 'On Certain Egyptian Papyri in the British Museum', *Transactions of the British Archaeological Association, at its Second Annual Congress, Held at Winchester, August 1845* (1846), pp. 246–63 + 1 plate.

——, 'On the First and Second Kinds of Persepolitan Writing', *Transactions of the Royal Irish Academy* 21 (1846), polite literature, pp. 114–31.

——, 'An Attempt to Ascertain the Number, Names, and Powers, of the Letters of the Hieroglyphic, or Ancient Egyptian Alphabet; Grounded on the Establishment of a New Principle in the Use of Phonetic Characters', *Transactions of the Royal Irish Academy* 21 (1847), polite literature, pp. 132–232 + 3 plates.

——, 'On the Three Kinds of Persepolitan Writing, and on the Babylonian Lapidary Characters', *Transactions of the Royal Irish Academy* 21 (1847), polite literature, pp. 233–48.

——, 'On the Third Persepolitan Writing, and on the Mode of Expressing Numerals in Cuneatic Characters' *Transactions of the Royal Irish Academy* 21 (1847), polite literature, pp. 249–56.

——, 'On the Inscriptions at Van', *Journal of the Royal Asiatic Society* 9 (1848), pp. 387–449.

——, 'On the Portion of the Turin Book of Kings, which Corresponds to the First Five Dynasties of Manetho', *Proceedings of the Royal Society of Literature* 1 (1848), pp. 275–7.

——, 'On the Khorsabad Inscriptions', *Transactions of the Royal Irish Academy* 22 (1850), polite literature, pp. 3–72.

——, 'On the Portion of the Turin Book of Kings which Corresponds to the Sixth Dynasty of Manetho', *Transactions of the Royal Society of Literature*, 2nd ser., 3 (1850), pp. 128–38. Dated 7 March 1846; read 12 March 1846.

——, 'On the Portion of the Turin Book of Kings which Follows that Corresponding to the Twelfth Dynasty of Manetho', *Transactions of the Royal Society of Literature*, 2nd ser., 3 (1850), pp. 139–50. Dated 5 May 1846; read 28 May 1846.

——, 'The Recently Discovered Assyrian Inscriptions, and the Mode of Deciphering Them': A lecture delivered on 30 January 1850 to the Belfast Natural History and Philosophical Society [long summary], *Literary Gazette*, no. 1725 (9 Feb. 1850), pp. 110–11.

——, *Special Meeting of the Natural History and Philosophical Society of Belfast, Held in the Music-Hall, on Wednesday, the 23rd of October, 1850, Relative to Two Mummies Transmitted from Thebes, by Sir James Emerson Tennent, and Unrolled in the Museum, on the 17th and 18th of the Above Month* (Belfast, 1850) (From *The Northern Whig*, 24 Oct. 1850), pp. 10–18.

——, 'Observations on the Turin Papyrus', in J. Gardner Wilkinson, *The Fragments of the Hieratic Papyrus at Turin: Containing the Names of Egyptian Kings, with the Hieratic Inscription at the Back* (London, 1851), pp. 47–60.

——, 'On the Language and Mode of Writing of the Ancient Assyrians', *Report of the Twentieth Meeting of the British Association for the Advancement of Science; Held at Edinburgh in July and August 1850* (1851), p. 140 + 1 plate.

——, 'On the Assyrio-Babylonian Phonetic Characters', *Transactions of the Royal Irish Academy* 22 (1852), polite literature, pp. 293–370.

——, 'The One Primeval Language': A review of C. Forster, *The One Primeval Language Traced experimentally through Ancient Inscriptions in Alphabetic Characters of Lost Powers, from the Four Continents; Including the Voice of Israel from the Rocks of Sinai* (London, 1851), in *Dublin University Magazine* 39 (Feb. 1852), pp. 226–34.

Hincks, E., 'On the Ethnological Bearing of the Recent Discoveries in Connexion with the Assyrian Inscriptions', *Report of the Twenty-Second Meeting of the British Association for the Advancement of Science; held at Belfast in September 1852* (1853), pp. 85–7.

——, 'On Certain Ancient Mines', *Report of the Twenty-Second Meeting of the British Association for the Advancement of Science; held at Belfast in September 1852* (1853), pp. 110–12.

——, 'List of four discoveries by E. Hincks on a visit to the British Museum, including a list of "Assyrian Months" in cuneiform "monograms", and recognition of "four cardinal points"', *Proceedings of the Royal Irish Academy* 5 (1853), pp. 403–5.

——, 'The Nimrûd Obelisk', *Dublin University Magazine* 42 (Oct. 1853), pp. 420–6.

——, 'On the Personal Pronouns of the Assyrian and Other Languages, especially Hebrew', *Transactions of the Royal Irish Academy* 23 (1854), part 2, pp. 3–9.

——, 'Chronology of the Reigns of Sargon and Sennacherib', *Journal of Sacred Literature* 6/12 (July 1854), pp. 393–410.

——, *Report to the Trustees of the British Museum Respecting Certain Cylinders and Terra-Cotta Tablets, with Cuneiform Inscriptions* (London, 1854).

——, 'Letter from Dr Hincks, in Reply to Colonel Rawlinson's Note on the Successor of Sennacherib', *Journal of the Royal Asiatic Society* 15 (1855), pp. 402–3.

——, 'On the Assyrian Mythology', *Transactions of the Royal Irish Academy* 22 (1855), polite literature, pp. 405–22.

——, 'On Certain Animals mentioned in the Assyrian Inscriptions', *Proceedings of the Royal Irish Academy* 6 (1855), pp. 251–60.

——, 'On the Chronology of the Twenty-sixth Egyptian Dynasty, and of the Commencement of the Twenty-seventh', *Transactions of the Royal Irish Academy* 22 (1855), polite literature, pp. 423–36.

——, 'On Assyrian Verbs', *Journal of Sacred Literature* 1/2 (July 1855), pp. 381–93; 2/3 (Oct. 1855), pp. 141–62; 3/5 (Apr. 1856), pp. 152–71; 3/6 (July 1856), pp. 392–403.

——, 'Are There Any Assyrian Syllabaries?': A Letter to the Editor, *Monthly Review* 1 (Feb. 1856), pp. 130–2.

——, 'On the Chronology of the Egyptian Dynasties prior to the Reign of Psammetichus', *Literary Gazette* 5 (29 Mar. 1856), pp. 111–12. Also published in *Journal of Sacred Literature* 3/6 (July 1856), pp. 472–5.

——, 'On a Tablet in the British Museum, Recording in Cuneatic Characters, an Astronomical Observation; with Incidental Remarks on the Assyrian Numerals, Divisions of Time, and Measures of Length', *Transactions of the Royal Irish Academy* 23 (1856), part 2, pp. 31–47. Read 12 November 1855.

——, 'On Certain Ancient Arab Queens', *Transactions of the Royal Society of Literature*, 2nd ser., 5/2 (1856), pp. 162–4. Read 13 April 1853.

——, 'On the Assyrio-Babylonian Measures of Time', *Transactions of the Royal Irish Academy* 24 (1865), polite literature, pp. 13–24.

——, Review article on H. Brugsch, *Nouvelles recherches sur la division de l'année des anciens Égyptiens* (Berlin, 1856), in *Monthly Review* 1 (May 1856), pp. 281–90 + plate 1 (facing p. 322).

——, 'Brief des Herrn Dr. Edw. Hincks an Prof. Brockhaus', *Zeitschrift der deutschen morgenländischen Gesellschaft* 10 (1856), pp. 516–18.

Hincks, F., *Reminiscences of his Public Life* (Montreal, 1884).

Hincks, T., *A History of the British Hydroid Zoophytes*, 2 vols (London, 1868).

——, *A History of the British Marine Polyzoa*, 2 vols (London, 1880).

Hollaway, S. W., 'Biblical Assyria and Other Anxieties in the British Empire', *Journal of Religion and Society* 3 (2001), pp. 1–19.

Holtzmann, A., *Beiträge zur Erklärung der persischen Keilinschriften* (Karlsruhe, 1845).

——, 'Über die zweite Art der achämenidischen Keilschrift', *Zeitschrift der deutschen morgenländischen Gesellschaft* 5 (1851), pp. 145–78.

——, 'Über die zweite Art der achämenischen Keilschrift', *Zeitschrift der deutschen morgenländischen Gesellschaft* 8 (1854), pp. 329–45.

Houghton, W. E. (ed.), *The Wellesley Index of Victorian Periodicals, 1824–1900*, 5 vols (Toronto and London, 1966–89).

Hunger, H., *Babylonische und assyrische Kolophone* (Alter Orient und Altes Testament 2; Kevelaer and Neukirchen-Vluyn, 1968).

Hupfeld, H., 'Kritische Beleuchtung einiger dunteln und misverstanden Stellen der alttestamentlichen Textgeschichte', *Theologische Studien und Kritiken* 3 (1830), pp. 247–301.

Ibn Khaldūn, 'Abd al-Raḥmān b. Muḥammad, *Histoire des Berbères et des dynasties musulmanes de l'Afrique Septentrionale: Texte arabe*, 2 vols (Algiers, 1847, 1851). Eng. tr. by W. MacGuckin de Slane under the same title in 4 vols (Algiers, 1852–6).

Ibn Waḥshīyah, Aḥmad b. 'Alī, *Ancient Alphabets and Hieroglyphic Characters Explained*, tr. J. Hammer-Purgstall (London, 1806).

Isenberg, C. W., *Dictionary of the Amharic Language* (London, 1841).

——, *Grammar of the Amharic Language* (London, 1842).

[Jebb, J.], 'Forster on Arabia': a review of C. Forster, *The Historical Geography of Arabia; or, The Patriarchal Evidences of Revealed Religion: a Memoir, with Illustrative Maps; and an Appendix, Containing Translations, with an Alphabet and Glossary, of the Himyaritic Inscriptions recently Discovered in Hadramaut*, 2 vols (London, 1844), in *Quarterly Review* 74 (Oct. 1844), pp. 325–58.

Jones, A., *The Proper Names of the Old Testament Scriptures Expounded and Illustrated* (London, 1856).

Kaiser, O., *Zwischen Reaktion und Revolution: Hermann Hupfeld (1796–1866) – ein deutsches Professorenleben* (Göttingen, 2005).

Kenrick, J., *Ancient Egypt under the Pharaohs*, 2 vols (London, 1850).

Kircher, A., *Lingua Aegyptiaca restituta* (Rome, 1643).

Kopp, U. F., *Bilder und Schriften der Vorzeit*, 2 vols (Mannheim, 1819–21).

Kropp, M., 'From Manuscripts to the Computer: Ethiopic Studies in the Last 150 Years', in K. J. Cathcart (ed.), *The Edward Hincks Bicentenary Lectures*, pp. 114–37.

Landsberger, B., *Die Serie ana ittišu* (Materialien zum sumerischen Lexikon; Rome, 1937).

——, 'Das Vokabular S^b. Nach einem Manuskript von H. S. Schuster' in *Materialien zum sumerischen Lexikon*, vol. 3 (Rome, 1955), pp. 129–53.

Larsen, M. Trolle, *The Conquest of Assyria: Excavations in an Antique Land 1840–1860* (London, 1996).

——, 'Hincks versus Rawlinson: The Decipherment of the Cuneiform System of Writing', in B. Magnusson *et al.* (eds), *Ultra terminum vagari: Scritti in onore di Carl Nylander* (Rome, 1997), pp. 339–56.

Layard, A. H., 'A Description of the Province of Khúzistán', *Journal of the Royal Geographical Society of London* 16 (1846), pp. 1–105.

Layard, A. H., *Inscriptions in the Cuneiform Character from Assyrian Monuments Discovered by A. H. Layard* (London, 1851).

——, *Discoveries in the Ruins of Nineveh and Babylon* (London, 1853).

——, *Sir A. Henry Layard, G.C.B., D.C.L.: Autobiography and Letters*, ed. W. N Bruce, 2 vols (London, 1903).

Lefebvre, T., Petit, A. and Quartin-Dillon, L. R. *et al.*, *Voyage en Abyssinie exécuté pendant les années 1839, 1840, 1841, 1842, 1843*, 6 vols (Paris, 1845–51).

Lepsius, R., *Inscriptiones umbricae et oscae* (Leipzig, 1841).

——, *Auswahl der Wichtigsten Urkunden des ägyptischen Alterthums* (Leipzig, 1842).

——, *Das Todtenbuch der Ägypter nach dem hieroglyphischen Papyrus in Turin* (Leipzig, 1842).

——, *Die Chronologie der Ägypter*, vol. 1 (Berlin, 1849).

——, *Denkmäler aus Ägypten und Äthiopien*, 6 pts in 12 vols (Berlin, 1849–59).

——, 'Über den ersten Ägyptischen Götterkreis und seine geschichtlich-mythologische Entstehung', *Abhandlungen der Königlichen Akademie der Wissenschaften zu Berlin 1851* (1852), philologisch-historische Klasse, pp. 157–214 + 4 plates.

——, *Briefe aus Ägypten, Äthiopien und der Halbinsel des Sinai: geschrieben in den Jahren 1842–1845, während der auf Befehl sr. Majestät des Königs Friedrich Wilhelm IV von Preussen ausgeführten wissenschaftlichen Expedition* (Berlin, 1852); Eng. tr. *Discoveries in Egypt, Ethiopia and the Peninsula of Sinai, in the years 1842–1845 during the Mission Sent out by his Majesty, Frederick William IV of Prussia*, ed. with notes by K. R. H. Mackenzie (London, 1852).

——, 'Über eine hieroglyphische Inschrift am Tempel von Edfu (Appollinopolis Magna) in welcher der Besitz dieses Temples an Ländereien unter der Regierung Ptolemaeus XI Alexander I verzeichnet ist', *Abhandlungen der Königlichen Akademie der Wissenschaften zu Berlin* (1855), pp. 69–114.

Leslie, J. B., *Clogher Clergy and Parishes* (Enniskillen, 1929).

Leslie, J. B. and Swanzy, H. B., *Biographical Succession Lists of the Clergy of Diocese of Down* (Enniskillen, 1936).

Loftus, W. K., *Travels and Researches in Chaldaea and Susiana* (London, 1857).

Longpérier, A. de, 'Lettre à M. Isidore Löwenstern sur les inscriptions cunéiformes de l'Assyrie', dated 20 September 1847, *Revue archéologique* 4/2 (1848), pp. 501–7.

Löwenstern, I., *Exposé des éléments constitutifs du système de la troisième écriture cunéiforme de Persépolis* (Paris, 1847).

——, 'Note sur une table généalogique des rois de Babylone dans Ker-Porter', *Revue archéologique* 6/2 (1850), pp. 417–20.

——, 'Remarques sur la deuxième écriture cunéiforme de Persépolis', *Revue archéologique* 6/2 (1850), pp. 687–728.

Ludolf, H., *Grammatica Aethiopica* (Frankfurt, 1661; 2nd edn, 1702).

——, *Grammatica linguae Amharicae* (Frankfurt, 1698).

——, *Lexicon Amharico-Latinum* (Frankfurt, 1698).

——, *Lexicon Aethiopico-Latinum* (Frankfurt, 1699).

Malék, J., 'The Original Version of the Royal Canon of Turin', *Journal of Egyptian Archaeology* 68 (1982), pp. 93–106.

Mankowski, P. V., *Akkadian Loanwords in Biblical Hebrew* (Harvard Semitic Studies 47; Winona Lake IND, 2000).

Mildenberg, L., *The Coinage of the Bar Kokhba War* (Aarau, 1984).

——, *Vestigia Leonis: Studien zur antik Numismatik Israels, Palästinas und der östlichen Mittelmeerwelt*, eds U. Hübner and E. A. Knauf (Freiburg Schweiz and Göttingen, 1998), pp. 161–241.

Millard, A. R., 'Baladan, the Father of Merodach-Baladan', *Tyndale Bulletin* 22 (1971), pp. 125–6.

——, 'Assyrian Royal Names in Biblical Hebrew', *Journal of Semitic Studies* 21 (1976), pp. 1–14.

——, *The Eponyms of the Assyrian Empire 910–612 B.C.* (State Archives of Assyria Studies 2; Helsinki, 1994).

Milman, H. H., 'Persian and Assyrian Inscriptions': a review article on H. C. Rawlinson, *The Persian Cuneiform Inscriptions at Behistun* (London, 1846–7); C. Lassen and N. L. Westergaard, *Über die Keilinschriften der ersten und zweiten Gattung* (Bonn, 1845); N. L. Westergaard, *On the Deciphering of the Second Achaemenian or Median Species of Arrow-head Writing* (Copenhagen, 1844); A. Holtzmann, *Beiträge zur Erklärung der persischen Keilinschriften* (Karlsruhe, 1845); F. Hitzig, *Die Grabinschrift des Darius in Nakshi Rustam erläutert* (Zurich, 1847); J. Mohl, *Lettres de M. Botta sur les découvertes à Khorsabad* (Paris, 1845), in *Quarterly Review* 79 (Dec. 1846–Mar. 1847), pp. 413–49.

——, Review article of A. H. Layard, *Nineveh and Its Remains* (London, 1848), in *Quarterly Review* 84 (Dec. 1848), pp. 106–53.

Mosshammer, A. A., 'Thales' Eclipse', *Journal of the American Philological Association* 111 (1981), pp. 145–55.

Müller, F. Max, *Rig-Veda-Samhita: the Sacred Hymns of the Brâhmans together with the Commentary of Sayanachara*, 6 vols (London, 1849–74; 2nd edn in 4 vols, 1890–2).

——, *The Languages of the Seat of War in the East* (London, 1854; 2nd edn 1855).

Müller, K. O., *Die Etrusker* (Breslau, 1828).

Muss-Arnoldt, W., 'The Works of Jules Oppert', *Beiträge für Assyriologie* 2 (1894), pp. 523–56.

[Myers, T.], 'The Nineveh Inscriptions', *Journal of Sacred Literature* 1/2 (July 1855), pp. 365–81.

Nash, D. W., 'On the Antiquity of the Egyptian Calendar', *Original Papers Read before the Syro-Egyptian Society of London*, vol. 1/2 (London, 1850), pp. 29–57.

Neue Deutsche Biographie (Berlin, 1952–).

Nork, F., *Vollständiges hebräisch-chaldäisch-rabbinisches Wörterbuch über das alte Testament, die Thargumim, Midrashim, und den Talmud* (Grimma, 1842).

Norris, E., *Grammar of the Bornu or Kanuri Language; with Dialogues, Translations, and Vocabulary* (London, 1853).

——, 'Memoir on the Sythic Version of the Behistun Inscription', *Journal of the Royal Asiatic Society* 15 (1853/55), pp. 1–213.

——, 'On the Assyrian and Babylonian Weights', *Journal of the Royal Asiatic Society* 16 (1854), pp. 215–26.

Oppert, J., 'Mémoire sur les inscriptions des Achéménides, conçues dans l'idiome des anciens Perses', *Journal Asiatique*, 4th ser., 17 (1851), pp. 255–96, 378–430, 534–67; 18 (1851), pp. 56–83, 322–66, 553–84; 19 (1852), pp. 140–215.

——, *Rapport adressé à Son Excellence M. le Ministre de l'Instruction publique et des Cultes, par M. Jules Oppert, chargé d'une mission scientifiique en Angleterre* (Paris, 1856).

——, 'Schreiben des Hrn. Dr. Julius Oppert an den Präsidenten der Hamburger Orientalisten – Versammlung und an Prof. Brockhaus', *Zeitschrift der deutschen morgenländischen Gesellschaft* 10 (1856), pp. 288–92.

Oppert, J., Review of J. Brandis, *Über den historischen Gewinn aus der Entzifferung der assyrischen Inschriften. Nebst einer Übersicht über die Grundzüge des assyrisch-babylonischen Keilinschriftsystems* (Berlin, 1856), in *Journal Asiatique*, 5th ser., 7 (Apr.–May 1856), pp. 438–43.

——, *Expédition scientifique en Mésopotamie exécutée par ordre du gouvernement de 1851 à 1854 par MM. F. Fresnel, F. Thomas et J. Oppert*, 2 vols (Paris, 1859–63).

Ormsby, W. G., 'Sir Francis Hincks', in J. M. S. Careless (ed.), *The Pre-Confederation Premiers: Ontario Government Leaders, 1841–1867* (Toronto, 1980), pp. 148–96.

——, 'Sir Francis Hincks', in *Dictionary of Canadian Biography*, vol. 11 (Toronto, 1982).

Osburn, W., *The Monumental History of Egypt as Recorded on the Ruins of her Temples, Palaces, and Tombs*, 2 vols (London, 1854).

'Our Portrait Gallery. No. LXVII. Henry Brooke', *Dublin University Magazine* 39 (Feb. 1852), pp. 200–14 + 1 etching.

Oxford Dictionary of National Biography (Oxford, 2004).

Pérez Bayer, F., *De numis hebraeo-samaritanis* (Valencia, 1781).

Pilet, M., *Un pionnier de l'assyriologie. Victor Place* (Cahiers de la société asiatique, 16; Paris, 1962).

Pococke, E., *Specimen historiae Arabum, sive, Gregorii Abul Farajii Malatiensis de origine moribus Arabum succincta narratio* (Oxford, 1650).

Poole, R. S., *Horae Aegyptiacae: or, the Chronology of Ancient Egypt* (London, 1851).

Pope, M., *The Story of Decipherment* (London, 1975).

Porter, R. Ker, *Travels in Georgia, Persia, Armenia, Ancient Babylon, &c. &c. during the Years 1817, 1818, 1819, and 1820*, 2 vols (London, 1821–2).

Pote, R. G., *Nineveh: a Review of Its Ancient History and Modern Explorers* (London, [1854]).

Powell, M., 'Masse und Gewichte', *Reallexikon der Assyriologie*, ed. D. O. Edzard, vol. 7 (Berlin, 1987–90), pp. 457–517.

Prisse D'Avennes, E., *Fac-simile d'un papyrus égyptien en caractères hiératiques* (Paris, 1847).

Pritchard, J. B. (ed.), *The Ancient Near East in Pictures Relating to the Old Testament* (2nd edn; Princeton, 1969).

——, *Ancient Near Eastern Texts Relating to the Old Testament* (3rd edn; Princeton, 1969).

Quatremère, E. M., *Recherches critiques et historiques sur la langue et littérature de l'Égypte* (Paris, 1808).

Radner, Karen, 'Der Gott Salmānu ("Šulmānu") und seine Beziehung zur Stadt Dūr-Katlimmu', *Die Welt des Orients* 29 (1998), pp. 33–51.

Rawlinson, G. (tr. and ed.), *The History of Herodotus*, 4 vols (London, 1858–60).

Rawlinson, H. C., 'On the Inscriptions of Assyria and Babylonia', *Journal of the Royal Asiatic Society* 12 (1850), pp. 401–83. Published separately as *A Commentary on the Cuneiform Inscriptions of Babylonia and Assyria; including Readings of the Inscription of the Nimrud Obelisk, and a Brief Notice of the Ancient Kings of Nineveh and Babylon* (London, 1850).

——, 'Memoir on the Babylonian and Assyrian Inscriptions', *Journal of the Royal Asiatic Society* 14 (1851 [publ. 1852]), pp. i–civ, 1–16.

——, 'Notes on Some Paper Casts of Cuneiform Inscriptions upon the Sculptured Rock at Behistun Exhibited to the Society of Antiquaries', *Archaeologia* 34 (1852), pp. 73–6.

——, *Outline of the History of Assyria, as Collected from the Inscriptions Discovered by A. H. Layard in the Ruins of Nineveh* (London, 1852) (= *Journal of the Royal Asiatic Society 14*).

——, 'Babylonian Discoveries', *The Athenaeum*, no. 1377 (18 Mar. 1854), pp. 341–3.

——, 'On the Birs Nimrud, or the Great Temple of Borsippa', *Journal of the Royal Asiatic Society* 18 (1861), pp. 1–34.

Reiner, Erica and Pingree, D., *The Venus Tablet of Ammiṣaduqa* (Babylonian Planetary Omens, Part 1; Malibu, 1975).

Renouf, P. le Page, 'Seyffarth and Uhleman on Egyptian Hieroglyphics', *The Atlantis* 2/3 (Jan. 1859), pp. 74–97 = G. Maspero and W. H. Rylands (eds), *The Life-Work of Sir Peter le Page Renouf*, vol. 1 (Paris, 1902), pp. 1–31.

——, 'Dr Seyffarth and the *Atlantis* on Egyptology', *The Atlantis* 3/6 ([Jan.], 1862), pp. 306–37 = G. Maspero and W. H. Rylands (eds), *The Life-Work of Sir Peter le Page Renouf*, vol. 1 (Paris, 1902), pp. 33–80.

Report of the Commissioners Appointed to Enquire into the Constitution and Government of the British Museum (London, 1850).

Reynolds, L. D., *Texts and Transmission: A Survey of the Latin Classics* (Oxford, 1983).

Robinson, E., *Biblical Researches in Palestine, Mount Sinai and Arabia Petraea: A Journal of Travels in the Year 1838* (London, 1841).

Rödiger, E., 'Notiz über die himjaritische Schrift nebst doppeltem Alphabet derselben', *Zeitschrift für die Kunde des Morgenlandes* 1 (1837), pp. 332–40.

——, *Versuch über die himjaritischen Schriftmonumente* (Halle, 1841).

——, 'Exkurs über die von Lieut. Wellsted bekannt gemachten himjaritischen Inschriften', in *J. R. Wellsted's Reisen in Arabien*, ed. and tr. E. Rödiger, vol. 2 (Halle, 1842), pp. 352–411.

Rogerson, J., *Old Testament Criticism in the Nineteenth Century: England and Germany* (London, 1984).

Rollinger, R., 'Zur Bezeichnung von "Griechen" in Keilschrifttexten', *Revue d'Assyriologie* 91 (1997), pp. 167–72.

——, 'The Ancient Greeks and the Impact of the Ancient Near East: Textual Evidence and Historical Perspective (ca. 750–650 B.C.)', in R. M. Whiting (ed.), *Melammu Symposia II* (Helsinki, 2001), pp. 223–64.

Rougé, E. de, 'Examen de l'ouvrage de M. le Chevalier de Bunsen intitulé La place de l'Égypte dans l'histoire de l'humanité', *Annales de philosophie chrétienne*, 3rd ser., 13 (1846), pp. 432–58; 14 (1846), pp. 355–77; 15 (1847), pp. 44–65, 165–93, 405–38; 16 (1847), pp. 7–30.

——, 'Inscription des rochers de Semné', *Revue archéologique* 5/1 (1848), pp. 311–14.

——, 'Lettre à M. Leemans sur une stèle égyptienne du Musée de Leyde', *Revue archéologique* 6/2 (1850), pp. 557–75.

——, *Mémoire sur l'inscription du tombeau d'Ahmes, chef des nautoniers* (Paris, 1851).

Rudolph, W., *Micha–Nahum–Habakuk–Zephanja* (Guttersloh, 1975).

Russell, J. M., *From Nineveh to New York: The Strange Story of the Assyrian Reliefs in the Metropolitan Museum and the Hidden Masterpiece at Canford School* (New Haven, 1997).

——, *The Writing on the Wall: Studies in the Architectural Context of Late Assyrian Palace Inscriptions* (Winona Lake IND, 1999).

Sainthill, R., *An Olla Podrida; or, Scraps, Numismatic, Antiquarian, and Literary*, 2 vols. (London, 1844–53).

——, 'The Use of the Samaritan Language by the Jews until the Reign of Hadrian, Deduced from the Coins of Judea'. A Letter to J. B. Bergne, *An Olla Podrida*, vol. 2 (1853), pp. 25–40.

Saulcy, F. de, *Voyage autour de la mer morte et dans les terres bibliques: exécuté de décembre 1850 à avril 1851*, 2 vols (Paris, 1853); Eng. tr.: E. *Narrative of a Journey round the Dead Sea and in the Bible Lands in 1850 and 1851*, tr. E. de Warren, 2 vols (London, 1853).

Sayce, A. H., 'The Languages of the Cuneiform Inscriptions of Elam and Media', *Transactions of the Society of Biblical Archaeology* 3 (1874), pp. 465–85.

Scott, W. H., 'On Parthian Coins', *Numismatic Chronicle* 17 (1855), pp. 131–73.

Secchi, G., 'Tesoretto di etruschi arredi funebri in oro posseduti al signor cav. Giampietro Campana', *Bullettino dell'instituto di corrispondenza archeologica*, nos 1–2 (Jan.–Feb. 1846), pp. 3–16. Also published separately as *Descrizione d'alquanti Etruschi arredi in oro posseduti dal sig. cav. Giampietro Campana e interpretazione d'una epigrafe etrusca sopra una fibula con alfabeto etrusco della tazzetta borghesiana scoperto nell'anno CDDCCCXLV* (Rome, 1846).

Segal, J. B., with a contribution by Erica C. D. Hunter, *Catalogue of the Aramaic and Mandaic Incantation Bowls in the British Museum* (London, 2000).

Seyffarth, G., *Remarks upon an Egyptian History, in Egyptian Characters, in the Royal Museum at Turin* (London, 1828).

——, 'Champollion and Renouf', *Transactions of the Academy of Science of St. Louis* 1 (1860), pp. 539–69.

Smyth, W. H., *Aedes Hartwellianae, or Notices of the Manor and Mansion of Hartwell* (London, 1851).

Solé, R. and Valbelle, Dominique, *The Rosetta Stone: The Story of the Decoding of Hieroglyphics* (London, 2002).

Spiegel, F., *Grammatik der Pârsisprache nebst Sprachproben* (Leipzig, 1851).

Spronk, K., *Nahum* (Kampen, 1997).

Steele, J. M., *Calendars and Years: Astronomy and Time in the Ancient Near East* (Oxford, 2007).

Strabonis Geographica, ed. G. Kramer, 3 vols (Berlin, 1844–52).

Strabonis rerum geographicarum libri XVII, ed. I. Casaubon (Paris, 1620).

Swinton, J., 'Observations upon Two Antient Etruscan Coins, never before Illustrated or Explained': A Letter to T. Birch, *Philosophical Transactions* 54 (1764), pp. 99–106.

Tadmor, H., *The Inscriptions of Tiglath-Pileser III King of Assyria* (Jerusalem, 1994).

Talbot, H. F. 'On the Assyrian Inscriptions'[I–IV], *Journal of Sacred Literature* 2/4 (Jan. 1856), pp. 414–25; 3/5 (Apr. 1856), pp. 188–94; 3/6 (July 1856), pp. 422–6; 4/7 (Oct. 1856), pp. 164–70.

——, *Assyrian Texts Translated, No. 1* (London, 1856).

Tattam, H., *Prophetae Maiores in dialecto linguae Aegyptiacae Memphitica seu Coptica*, 2 vols (Oxford, 1852).

Theophrasti Eresii opera quae supersunt omnia, vol. 1, *Historia plantarum*, ed. F. Wimmer (Bratislava, 1842).

[Thomson, T.], Review article on Anna Seward, *Memoirs of the Life of Dr Darwin, chiefly during His Residence at Lichfield* (London, 1804), in *Edinburgh Review* 4 (Apr. 1804), pp. 230–41.

Tuch, F., 'Ein und zwanzig sinaitische Inschriften', *Zeitschrift der deutschen morgenländischen Gesellschaft* 3 (1849), pp. 129–215.

Uehlinger, C., 'Clio in a World of Pictures – Another Look at the Lachish Reliefs from Sennacherib's Southwest Palace at Nineveh', in L. L. Grabbe (ed.), *'Like a Bird in a Cage': The Invasion of Sennacherib in 701 B.C.* (Edinburgh, 2004), pp. 221–307.

Vaux, W. S. W., *Nineveh and Persepolis* (London, 1850).

——, 'New Assyrian Sculptures at the British Museum', *Monthly Review* 1 (May 1856), pp. 309–18.

Walker, C. B. F., 'Notes on the Venus Tablet of Ammiṣaduqa', *Journal of Cuneiform Studies* 36 (1984), pp. 64–6.

Wall, C. W., *An Examination of the Ancient Orthography of the Jews: and of the Original State of the Text of the Hebrew Bible*, 3 parts (London and Dublin, 1835–56).

——, 'On the Different Kinds of Cuneiform Writing in the Triple Inscriptions of the Persians, and on the Language Transmitted through the First Kind', *Transactions of the Royal Irish Academy* 21 (1848), polite literature, pp. 257–314.

Waterfield, G., *Layard of Nineveh* (London, 1963).

Whitehead, D., 'David Ross of Bladensburg: a Nineteenth-Century Ulsterman in the Mediterranean', *Hermathena* 164 (1998), pp. 89–99.

——, 'From Smyrna to Stewartstown: a Numismatist's Epigraphic Notebook', *Proceedings of the Royal Irish Academy* 99C (1999), pp. 73–113.

Wilkinson, J. G., *The Architecture of Ancient Egypt: in which the Columns are Arranged in Orders and the Temples Classified, with Remarks on the Early Progress of Architecture, etc.*, 2 vols (London, 1850).

——, *The Fragments of the Hieratic Papyrus at Turin: Containing the Names of Egyptian Kings, with the Hieratic Inscription at the Back*, 2 vols (London, 1851).

Wilson, J., *The Pársi Religion, as Contained in the Zand-Avastá, and Propounded and Defended by the Zoroastrians of India and Persia, Unfolded, Refuted, and Contrasted with Christianity* (Bombay, 1843).

——, *The Lands of the Bible Visited and Described in an Extensive Journey*, 2 vols (Edinburgh, 1847).

Zendavesta, or the Religious Books of the Zoroastrians, tr. with commentary by N. L. Westergaard, vol. 1 (Copenhagen, 1852–4).

INDEX